John Elvey

AF843571

TABLE OF CONTENTS

Preface	V
Introductory Remarks	VII
Subject List	IX
PART 1	
Alphabetical Bibliography	1
PART 2	
Subject Classification	109
PART 3	
List of Authors	281

PREFACE

Integer Programming is one of the most important areas in optimization theory. Since its origin in the fifties, it has made such rapid progress that after only a few decades its literature is now too vast to keep track of. Therefore in 1970 I initiated the collection and classification of all available publications on integer programming scattered throughout the literature. This work was performed during the years 1970-75 in the Institut für Ökonometrie und Operations Research, Bonn, mainly by Claus Kastning, and culminated in the classified bibliography entitled "Integer Programming and Related Areas", Lecture Notes in Economics and Mathematical Systems, Vol. 128.

After the tragic death of Claus Kastning, his work was continued by Dirk Hausmann, who collected all articles published after 1975 as well as earlier publications that were missed in the first volume. Due to the ever increasing research in integer programming, it seems appropriate to publish the second volume of the bibliography already after three years. The deadline set was March of 1978, until then 3162 new publications of 2742 authors were compiled and classified by 6404 descriptor entries. Together the two volumes of the bibliography cover 7866 titles.

The form of the second volume is nearly identical to that of the first volume. However some changes were made in the subject list in order to take into account new directions of research that have developed in the meantime.

Finally I would like to thank Reinhardt Euler, who did the final work for this new volume and who is now preparing the third volume. Thanks are also due to the programmers of the institute for their extensive efforts in programming and punching.

Bonn, June 1978 Bernhard Korte

INTRODUCTORY REMARKS

This second volume supplements the first volume " Integer Programming and Related Areas " , Lecture Notes in Economics and Mathematical Systems No. 128 , of the bibliography on integer programming. It contains the publications appearing in the period from 1976 to the beginning of 1978, as well as several previous publications, which were not included in the first volume.

The following topics are covered :

- theory and methods of general integer programming ;
- combinatorial and graph theoretical optimization problems ;
- applications of integer programming .

A detailed list of the subjects included is given at the end of the introduction. Textbooks and proceedings volumes have been included only if they deal principally with integer programming.

PART 1 : This part comprises the alphabetical bibliography which lists all publications alphabetically by the first author and chronologically for each author, and includes the full bibliographical data plus a reference code, which reflects the author's name and the year of publication. The purpose of the code is to enable the user to identify publications referred to in Part 2 and Part 3.
Bibliographical data are given as follows :

a) Textbook :

ref. code author(s)
title. Where published: publisher(s), no.of pages. (year).

b) Article in a journal :

ref. code author(s)
title. Journal, vol.no., pages. (year).

c) Article in a proceedings volume :

ref. code author(s)

title. IN: Editor(s) : title of the proceedings volume, pages. Where published: publisher(s). (year).

d) Paper :

ref. code author(s)

title. classification of the paper. Where published: publisher, no. of pages. (year).

Publications in English, French or German appear under their original title. Titles of publications written in other languages have been translated into English and the original language is indicated in brackets.

PART 2 : In this part the publications have been classified according to 57 subjects; within each subject the ordering is again alphabetical by the first author.

The subject list contains 57 headings and for 73 further topics the corresponding subject headings have been indicated. The publications appear under all relevant headings; the references are restricted to name(s) of author(s), title and reference code, with which the full bibliographical data can be retrieved in Part 1.

PART 3 : The purpose of this part is to enable the user to locate publications for which only a coauthor´s name is known. It consists of an alphabetical catalogue of all the authors appearing in Part 1, and lists for each author the reference codes of all his publications.

SUBJECT LIST

ACYCLIC SUBGRAPH PROBLEM , ORDERINGS , RELATIONS

ADJACENCY ON INTEGER POLYHEDRA

aggregation	→	diophantine systems and aggregation

ALGEBRAICAL METHODS AND ALGEBRAIZATION OF PROBLEMS

algorithm (codes)	→	computer codes
allocation	→	location and allocation
anti blocking	→	duality in integer programming

APPLICATIONS OF INTEGER PROGRAMMING

approximative algorithms	→	heuristics
assembly line balancing	→	scheduling

ASSIGNMENT

backtracking methods	→	branch and bound
balanced matrices	→	unimodular, balanced, and perfect matrices
Benders´ method	→	decomposition ; mixed integer programming
bibliographies	→	surveys, bibliographies
bivalent programming	→	zero-one programming
blocking	→	duality in integer programming

BOOLEAN METHODS

BOTTLENECK PROBLEMS

BRANCH AND BOUND

branchings	→	graph theoretic optimization problems
chance constrained methods	→	stochastic integer programming

Chinese postman → routing

clustering → partitioning ; combinatorial optimization problems

COLORING PROBLEMS

COMBINATORIAL ALGORITHMS (UNLESS SPECIFIED IN PARTICULAR)

COMBINATORIAL OPTIMIZATION PROBLEMS (UNLESS SPECIFIED IN PARTICULAR)

COMPLEXITY (PROBABILISTIC ANALYSIS)

COMPLEXITY (WORST CASE ANALYSIS)

COMPUTATIONAL RESULTS

COMPUTER CODES

cones → group theoretic approach

convex integer programming → nonlinear integer programming

corner polyhedra → group theoretic approach ; integer polyhedra

COVERING

CPM → network flows

crew scheduling → scheduling ; applications of integer programming

CUTTING PLANES

cutting stock → combinatorial optimization problems

cyclic groups → group theoretic approach

DECOMPOSITION

DIOPHANTINE SYSTEMS AND AGGREGATION

DISJUNCTIVE PROGRAMMING

DUALITY IN INTEGER PROGRAMMING

DYNAMIC PROGRAMMING

edge covering → covering

enumeration methods → branch and bound

FACETS OF INTEGER POLYHEDRA

FIXED CHARGE

flow shop scheduling → scheduling

flows in networks → network flows

fractional integer programming → nonlinear integer programming

generalized networks → network flows

graph isomorphism → graph theoretic results related to integer programming

GRAPH THEORETIC OPTIMIZATION PROBLEMS (UNLESS SPECIFIED IN PARTICULAR)

GRAPH THEORETIC RESULTS RELATED TO INTEGER PROGRAMMING

GROUP THEORETIC APPROACH

Hamiltonian cycle → graph theoretic results related to integer programming ; travelling salesman

Hermite normal form → diophantine systems and aggregation ; group theoretic approach

HEURISTICS (PROBABILISTIC ANALYSIS)

HEURISTICS (WITHOUT ANALYSIS)

HEURISTICS (WORST CASE ANALYSIS)

Hitchcock Koopmans Problem → transportation

Hungarian method → assignment

implicit enumeration → branch and bound

independence systems → matroids and independence systems

integer linear programming over cones (ILPC) → group theoretic approach

INTEGER POLYHEDRA (UNLESS SPECIFIED IN PARTICULAR)

INTEGER PROGRAMMING (GENERAL)

job shop scheduling → scheduling

KNAPSACK

Lagrangean relaxations → relaxations

large scale programming → decomposition

layout problems → location and allocation ; combinatorial optimization problems

LOCATION AND ALLOCATION

longest paths → shortest paths

machine scheduling → scheduling

marriage problem → assignment

MATCHING

MATROIDS AND INDEPENDENCE SYSTEMS

maximal flow problem → network flows

MIXED INTEGER PROGRAMMING

modulo optimization problems → group theoretic approach

MPM → network flows

MULTICOMMODITY FLOW PROBLEMS

MULTIDIMENSIONAL TRANSPORTATION

multiterminal network flows → network flows

NETWORK FLOWS

node covering → covering

node packing → packing

NONLINEAR INTEGER PROGRAMMING

orderings → acyclic subgraph problem, orderings, relations

PACKING

PARAMETRIC INTEGER PROGRAMMING

PARTITIONING

penalty methods	→	branch and bound
perfect graphs	→	packing ; covering ; coloring problems
perfect matrices	→	unimodular, balanced, and perfect matrices
PERT	→	network flows
p-median problem	→	location and allocation ; combinatorial optimization problems

PROBABILISTIC ANALYSIS IN INTEGER PROGRAMMING

programming and storing techniques	→	computer codes
quadratic integer programming	→	nonlinear integer programming
reduction of inequalities	→	diophantine systems and aggregation
relations	→	acyclic subgraph problem, orderings, relations

RELAXATIONS

ROUTING

SCHEDULING

sensitivity analysis	→	parametric integer programming
sequencing	→	scheduling

SHORTEST PATHS

Smith normal form	→	diophantine systems and aggregation ; group theoretic approach

SPANNING TREE

stable sets	→	packing
Steiner problems	→	graph theoretic optimization problems ; combinatorial optimization problems

STOCHASTIC INTEGER PROGRAMMING

subgradient optimization → relaxations

submodular functions → matroids and independence systems

SURVEYS , BIBLIOGRAPHIES

TEXTBOOKS

time tabling → scheduling ; assignment

totally unimodular matrices → unimodular, balanced, and perfect matrices

TRANSPORTATION

transshipment → transportation

TRAVELLING SALESMAN

triangularization → branch and bound

UNIMODULAR , BALANCED , AND PERFECT MATRICES

vertex covering → covering

vertex packing → packing

ZERO-ONE PROGRAMMING (GENERAL)

```
*************************************************************************************************************





PPPPPPPPPPP    AAAAAAAAAA   RRRRRRRRRRR    TTTTTTTTTTTT                          11
PPPPPPPPPPPP  AAAAAAAAAAAA  RRRRRRRRRRRR   TTTTTTTTTTTT                         111
PP        PP  AA        AA  RR        RR        TT                             1111
PP        PP  AA        AA  RR        RR        TT                               11
PP        PP  AA        AA  RR        RR        TT                               11
PPPPPPPPPPPP  AAAAAAAAAAAA  RRRRRRRRRRRR        TT                               11
PPPPPPPPPPP   AAAAAAAAAAAA  RRRRRRRRRRR         TT                               11
PP            AA        AA  RR    RR            TT                               11
PP            AA        AA  RR     RR           TT                               11
PP            AA        AA  RR      RR          TT                               11
PP            AA        AA  RR       RR         TT                               11
PP            AA        AA  RR        RR        TT                               11





                          ALPHABETIC BIBLIOGRAPHY

                          ***********************
                          ***********************
```

****

AASH 1976 -1 AASHTIANI, H. A.; MAGNANTI, T. L.
IMPLEMENTING PRIMAL-DUAL NETWORK FLOW ALGORITHMS. WORKING PAPER OR 055-76. CAMBRIDGE: M. I. T. (1976).

ABAD 1976 -1 ABADIE, J.; DAYAN, H.; AKOKA, J.
QUELQUES EXPERIENCES NUMERIQUES SUR LA PROGRAMMATION NON LINEAIRE EN NOMBRES ENTIERS. REVUE FRANCAISE D'AUTOMATIQUE, D'INFORMATIQUE ET DE RECHERCHE OPERATIONNELLE 10, NO. 10, 65-70. (1976).

ABEL 1977 -1 ABELLA, R. J.; BOVA, T. E.
OPTIMAL PLANT ALLOCATION OF STOCKKEEPING UNITS. SAN FRANCISCO (51. ORSA-MEETING + TIMS, MAY 1977) 10 P. (1977).

ACKE 1977 -1 ACKERMAN, L. J.; LUSS, H.; BERKOWITZ, R. S.
APPLICATION OF SEQUENCING POLICIES TO TELEPHONE SWITCHING FACILITIES. IEEE TRANSACTIONS ON SYSTEMS, MAN, AND CYBERNETICS SMC-7, 604-609. (1977).

ACKE 1977 -2 ACKERMAN, L. J.; LUSS, H.; BERKOWITZ, R. S.
IMPLICIT ENUMERATION ALGORITHM FOR SEQUENCING POLICIES APPLIED TO TELEPHONE SWITCHING FACILITIES, AN. SAN FRANCISCO (51. ORSA-MEETING + TIMS, MAY 1977) 19 P. (1977).

ACKEA 1963 -1 ACKERMAN, S. S.
EVEN-FLOW. A SCHEDULING METHOD FOR REDUCING LATENESS IN JOB SHOPS. MANAGEMENT TECHNOLOGY 3. (1963).

ADAM 1977 -1 ADAM, N.; SURKIS, J.
COMPARISON OF CAPACITY PLANNING TECHNIQUES IN A JOB SHOP CONTROL SYSTEM, A. MANAGEMENT SCIENCE 23, 1011-1015. (1977).

ADAMA 1976 -1 ADAMSKA-TCHON, J.; JACAK, W.; TCHON, K.
FORMAL DESCRIPTION AND ALGORITHMIZATION OF A CERTAIN TRANSPORT PROBLEM (POLISH), A. POLITECHNIKA WROCLAWSKA, INSTYTUT CYBERNETYKI TECHNICZNEJ, PRACE NAUKOWE, NO. 25, SERIA STUDIA I MATERIALY, NO. 17, 181-208. (1976).

ADEL 1975 -1 ADEL'SON-VEL'SKIJ, G. M.; DINITS, E. A.; KARZANOV, A. V.
FLOW ALGORITHMS (RUSSIAN). MOSCOW: IZDAT. NAUKA, 119 P. (1975).

ADLE 1975 -1 ADLER, H. J.
BESTIMMUNG VON TEILKOMPONENTEN EINES NETZWERKES BEI GLEICHZEITIGER MINIMIERUNG DER SCHNITTSTELLEN. ANGEWANDTE INFORMATIK 17, 52-54. (1975).

ADOL 1976 -1 ADOLPHSON, D. L.
NOTE ON MINIMUM COST CONVEX FLOWS, A. NAVAL RESEARCH LOGISTICS QUARTERLY 23, 713-714. (1976).

ADOL 1977 -1 ADOLPHSON, D. L.
SINGLE MACHINE JOB SEQUENCING WITH PRECEDENCE CONSTRAINTS. SIAM JOURNAL ON COMPUTING 6, 40-54. (1977).

ADOLA 1977 -1 ADOLPHSON, D. L.; THOMAS, G. N.
LINEAR TIME ALGORITHM FOR A 2 X N TRANSPORTATION PROBLEM, A. SIAM JOURNAL ON COMPUTING 6, 481-486. (1977).

ADRA 1976 -1 ADRABINSKI, A.
ALGORITHM 45: A HEURISTIC ALGORITHM FOR THE TRAVELING-SALESMAN PROBLEM. ZASTOSOWANIA MATEMATYKI 15, 223-244. (1976).

AGAP 1976 -1 AGAPOVA, L. V.
ALLOCATION PROBLEM FOR RESOURCES (RUSSIAN). IZVESTIJA AKADEMII NAUK SSSR, TEHNICESKAJA KIBERNETIKA, NO. 4, 14-21. (1976).

AGAR 1976 -1 AGARWAL, A. K.; ARORA, S. R.
SYNTHESIS OF MULTITERMINAL COMMUNICATION NETS: FINDING ONE OR ALL SOLUTIONS. IEEE TRANSACTIONS ON CIRCUITS AND SYSTEMS CS-23, 141-146. (1976).

AGIN 1976 -1 AGIN, N. I.
SUMMARY OF A NOVEL APPROACH TO THE WAREHOUSE LOCATION PROBLEM. MIAMI (50. ORSA-MEETING + TIMS, NOVEMBER 1976) 4 P. (1976).

AGINA 1975 -1 AGIN, N. I.; CULLEN, D. E.
ALGORITHM FOR TRANSPORTATION ROUTING AND VEHICLE LOADING, AN. LOGISTICS 1, 1-20. (1975).

AGNE 1976 -1 AGNEW, C. E.
ON QUADRATIC ADAPTIVE ROUTING ALGORITHMS. COMMUNICATIONS OF THE ACM 19, 18-22. (1976).

AGRA 1976 -1 AGRAWAL, S. C.
ALTERNATE METHOD ON INTEGER SOLUTIONS TO QUADRATIC PROGRAM BY A BRANCH AND BOUND TECHNIQUE, AN. TRABAJOS DE ESTADISTICA Y DE INVESTIGACION OPERATIVA 27, 185-192. (1976).

AGRA 1976 -2 AGRAWAL, S. C.
ON INTEGER SOLUTIONS TO LINEAR FRACTIONAL FUNCTIONALS BY A BRANCH AND BOUND TECHNIQUE. ACTA CIENCIA INDICA 2, 75-78. (1976).

AGRE 1977 -1 AGRESTI, W. W.
DYNAMIC PROGRAMMING FOR COMPUTER REGISTER ALLOCATION. COMPUTERS AND OPERATIONS RESEARCH 4, 101-110. (1977).

AHO 1977 -1 AHO, A. V.; GAREY, M. R.; HWANG, F. K.
RECTILINEAR STEINER TREES: EFFICIENT SPECIAL-CASE ALGORITHMS. NETWORKS 7, 37-58. (1977).

AHO A 1974 -1 AHO, A. V.; HIRSCHBERG, D. S.; ULLMAN, J. D.
BOUNDS ON THE COMPLEXITY OF THE LONGEST COMMON SUBSEQUENCE PROBLEM. IN: 15TH ANNUAL SYMPOSIUM ON SWITCHING AND AUTOMATA THEORY, NEW ORLEANS 1974, 104-109. LONG BEACH: INSTITUTE OF ELECTRICAL AND ELECTRONICS ENGINEERS; ALSO PUBLISHED IN: JOURNAL OF THE ACM 23, 1976, 1-13. (1974).

AHO B 1974 -1 AHO, A. V.; HOPCROFT, J. E.; ULLMAN, J. D.
DESIGN AND ANALYSIS OF COMPUTER ALGORITHMS, THE. READING, LONDON U.A.: ADDISON-WESLEY, 470 P. (1974).

AHO C 1976 -1 AHO, A. V.; ULLMAN, J. D.
NODE LISTINGS FOR REDUCIBLE FLOW GRAPHS. JOURNAL OF COMPUTER AND SYSTEM SCIENCES 13, 286-299. (1976).

AHRE 1975 -1 AHRENS, J. H.; FINKE, G.
MERGING AND SORTING APPLIED TO THE ZERO-ONE KNAPSACK PROBLEM. OPERATIONS RESEARCH 23, 1099-1109. (1975).

AHREA 1972 -1 AHRENS, W.
BINAERE OPTIMIERUNG ALS HILFSMITTEL BEI DER PLANUNG REGIONALER ABWASSERBESEITIGUNGSSYSTEME. II, DIE. TECHNISCHER BERICHT NR. 8. KARLSRUHE : UNIV. KARLSRUHE, INSTITUT FUER SIEDLUNGSWASSERWIRTSCHAFT. (1972).

AHREA 1974 -1 AHRENS, W.
OPIMIERUNGSVERFAHREN ZUR LOESUNG NICHTLINEARER INVESTITIONSPROBLEME - ANGEWANDT AUF DAS PROBLEM DER TECHNISCHEN UND WIRTSCHAFTLICHEN OPTIMIERUNG BEI DER BILDUNG VON ZWECKVERBAENDEN ZUR ABWASSERBESEITIGUNG. DISSERTATION, KARLSRUHE:UNIV. KARLSRUHE, TECHNISCHE HOCHSCHULE. (1974).

AHREA 1974 -2 AHRENS, W.
SOLVING OF THE REGIONALIZATION PROBLEM BY CERTAIN BRANCH-AND-BOUND METHODS. ECE-SYMPOSIUM ON THE USE OF COMPUTER TECHNIQUES AND AUTOMATION FOR WATER RESOURCES SYSTEMS, WASHINGTON. (1974).

****

AHREA 1975 -1 AHRENS, W.
OPTIMIERUNGSVERFAHREN ZUR LOESUNG NICHTLINEARER INVESTITIONSPROBLEME. QUANTITATIVE METHODEN DER UNTERNEHMUNGSPLANUNG; 4. MEISENHEIM: HAIN, 205 P. (1975).

AHREA 1976 -1 AHRENS, W.
NUMERICAL EXPERIENCE WITH HEURISTIC RULES IN SPECIFIC BINARY OPTIMIZATION ALGORITHMS. IN: A. PREKOPA (ED.): PROGRESS IN OPERATIONS RESEARCH, 19-50, AMSTERDAM, LONDON: NORTH HOLLAND. (1976).

AIGN 1976 -1 AIGNER, M.
KOMBINATORIK II. MATROIDE UND TRANSVERSALTHEORIE. BERLIN, HEIDELBERG, NEW YORK: SPRINGER, 324 P. (1976).

AKAD 1976 -1 AKADEMIA NAUK SSSR, TSENTR. EHKON.-MAT. INST.
SOME QUESTIONS CONCERNING THE STUDY OF TRANSPORTATION FLOWS (RUSSIAN). MOSCOW: TSENTR. EHKON.-MAT. INST., 94 P. (1976).

AKER 1967 -1 AKERS, S. A. JR.
MODIFICATION OF LEE'S PATH CONNECTION ALGORITHM, A. IEEE TRANSACTIONS ON ELECTRONIC COMPUTERS EC-16, 97-98. (1967).

AKHP 1977 -1 AKHPATELOV, E. A.; CHERENIN, V. P.
OPTIMAL SEQUENCE FOR A SINGLE-DEVICE SERVICE (RUSSIAN), THE. ZURNAL VYCISLITEL'NOI MATEMATEKI I MATEMATICESKOJ FIZIKI 17, 328-338. (1977).

AKIN 1974 -1 AKINC, U.
BRANCH AND BOUND PROCEDURE FOR SOLVING WAREHOUSE LOCATION PROBLEMS WITH CAPACITY CONSTRAINTS, A. DISSERTATION. CHAPEL HILL: UNIV. OF NORTH CAROLINA, 236 P. (1974).

AKINA 1977 -1 AKINC, U.; KHUMAWALA, B. M.
EFFICIENT BRANCH AND BOUND ALGORITHM FOR THE CAPACITATED WAREHOUSE LOCATION PROBLEM, AN. MANAGEMENT SCIENCE 23, 585-594. (1977).

AKOP 1974 -1 AKOPJAN, G. C.
COVERS AND JOINS OF THE VERTICES OF A GRAPH BY EDGES (RUSSIAN). ZURNAL VYCISLITEL'NOI MATEMATIKI I MATEMATICESKOI FIZIKI 14, 461-469 + 533-534. (1974).

AKUL 1977 -1 AKULIC, F. E.
DECOMPOSITION METHOD FOR PROBLEMS OF MAXIMAL FLOW IN LARGE RANGED NETWORKS (RUSSIAN), A. VESCI AKADEMII NAVUK BSSR, SERYJA FIZIKA-MATEMATYCNYH NAVUK, NO. 4, 27-30. (1977).

ALAV 1977 -1 ALAVI, Y.; BEHZAD, M.; LESNIAK-FOSTER, L. M.; NORDHAUS, E. A.
TOTAL MATCHINGS AND TOTAL COVERINGS OF GRAPHS. JOURNAL OF GRAPH THEORY 1, 135-140. (1977).

ALBE 1977 -1 ALBERS, S.
MIXED INTEGER NONLINEAR PROGRAMMING PROCEDURE FOR SIMULTANEOUSLY LOCATING MULTIPLE PRODUCTS IN AN ATTRIBUTE SPACE, A. IN: R. HENN, P. KALL, B. KORTE, O. KRAFFT ET AL. (EDS.): OPERATIONS RESEARCH VERFAHREN XXVI, METHODS OF OPERATIONS RESEARCH, 897-898. MEISENHEIM: HAIN. (1977).

ALBEA 1976 -1 ALBERTSON, M.; BOLLOBAS, B.; TUCKER, S.
INDEPENDENCE RATIO AND MAXIMUM DEGREE OF A GRAPH, THE. IN: F. HOFFMAN, L. LESNIAK, R. MULLIN ET AL. (EDS.): PROCEEDINGS OF THE 7TH SOUTHEASTERN CONFERENCE ON COMBINATORICS, GRAPH THEORY, AND COMPUTING, BATON ROUGE 1976, 43-50. CONGRESSUS NUMERANTIUM 17, WINNIPEG: UTILITAS MATHEMATICA. (1976).

ALBEB 1974 -1 ALBERTSON, M. O.
FINDING AN INDEPENDENT SET IN A PLANAR GRAPH. IN: R. A. BARI, F. HARARY (EDS.): GRAPHS AND COMBINATORICS, PROCEEDINGS OF THE CAPITAL CONFERENCE ON GRAPH THEORY AND COMBINATORICS, WASHINGTON 1973, 173-179. LECTURE NOTES IN MATHEMATICS 406, BERLIN: SPRINGER. (1974).

ALBEB 1976 -1 ALBERTSON, M. O.
LOWER BOUND FOR THE INDEPENDENCE NUMBER OF A PLANAR GRAPH, A. JOURNAL OF COMBINATORIAL THEORY 20, SERIES B, 84-93. (1976).

ALBEC 1976 -1 ALBERTSON, M. O.; BERMAN, D. M.
ACYCLIC CHROMATIC NUMBER, THE. IN: F. HOFFMAN, L. LESNIAK, R. MULLIN ET AL. (EDS.): PROCEEDINGS OF THE 7TH SOUTHEASTERN CONFERENCE ON COMBINATORICS, GRAPH THEORY, AND COMPUTING, BATON ROUGE 1976, 51-60. CONGRESSUS NUMERANTIUM 17, WINNIPEG: UTILITAS MATHEMATICA. (1976).

ALBED 1975 -1 ALBERTSON, M. O.; HUTCHINSON, J. P.
MAXIMUM SIZE OF AN INDEPENDENT SET IN A NONPLANAR GRAPH, THE. BULLETIN OF THE AMERICAN MATHEMATICAL SOCIETY 81, 554-555. (1975).

ALCO 1975 -1 ALCOUFFE, A.; ENJALBERT, M.; MURATET, G.
METHODES DE RESOLUTION DU PROBLEME DE TRANSPORT ET DE PRODUCTION D'UNE ENTREPRISE A ETABLISSEMENTS MULTIPLES EN PRESENCE DE COUTS FIXES. REVUE FRANCAISE D'AUTOMATIQUE, D'INFORMATIQUE ET DE RECHERCHE OPERATIONNELLE 9, NO. V-3, 41-56. (1975).

ALCOA 1976 -1 ALCOUFFE, A.; MURATET, G.
OPTIMAL LOCATION OF PLANTS. MANAGEMENT SCIENCE 23, 267-274. (1976).

ALEK 1974 -1 ALEKSEYEV, O. G.
INCREASING THE EFFICIENCY OF A DYNAMIC PROGRAMMING METHOD IN OPTIMAL STAND-BY PROBLEMS (RUSSIAN). IZVESTIJA AKADEMII NAUK SSSR, TEHNICESKAJA KIBERNETIKA, NO. 1, 107-111 + 236-237; TRANSLATED AS ENGINEERING CYBERNETICS 12, 1974, 89-94. (1974).

ALEK 1977 -1 ALEKSEYEV, O. G.
CLASS OF INTEGER PROGRAMMING PROBLEMS (RUSSIAN), A. ZURNAL VYCISLITEL'NOI MATEMATIKI I MATEMATICESKOI FIZIKI 17, 1047-1052 + 1095. (1977).

ALEKA 1977 -1 ALEKSEYEV, O. G.; MERZHANOV, V. YU.
ALGORITHM FOR JOB ALLOCATION ACCORDING TO A MINIMAX CRITERION (RUSSIAN), AN. UPRAVLJAEME SISTEMY MACH., NO. 1, 36-41. (1977).

ALEKB 1976 -1 ALEKSEYEV, O. G.; VOLODOS, I. F.
COMBINED USE OF DYNAMIC PROGRAMMING AND BRANCH-AND-BOUND METHODS IN DISCRETE-PROGRAMMING PROBLEMS (RUSSIAN). AVTOMATIKA I TELEMEHANIKA, NO. 4, 92-100; TRANSLATED AS AUTOMATION AND REMOTE CONTROL 37, 557-565. (1976).

ALEKC 1964 -1 ALEKSEYEV, O. G.; YAKUSHEV, V. I.
ALGORITHM OF OPTIMAL MACHINE SCHEDULING (RUSSIAN). IZVESTIJA AKADEMII NAUK SSSR. TEHNICESKAJA KIBERNETIKA 3, 62-68. (1964).

ALEKD 1974 -1 ALEKSEYEVA, S. M.
METHODS OF SCHEDULE THEORY IN TECHNOLOGICAL CHECKING PROBLEMS (RUSSIAN). IZVESTIJA AKADEMII NAUK SSSR, TEHNICESKAJA KIBERNETIKA, NO. 6, 139-142 + 222; TRANSLATED AS ENGINEERING CYBERNETICS 12, 1974, NO. 6, 106-109. (1974).

ALEKE 1971 -1 ALEKSEYEVA, S. M.; ALEKSEYEV, O. G.
ALGORITHM FOR FINDING THE SHORTEST PATH IN A NETWORK GRAPH (TRANSLATED FROM RUSSIAN), AN. U.S.S.R. COMPUTATIONAL MATHEMATICS AND MATHEMATICAL PHYSICS 11, 336-345. (1971).

ALEM 1972 -1 ALEMANY, R.; DIEGO, A.
SEMITENSIONS IN THE SOLVING OF THE "POTENTIAL'S RESTRICTED PROBLEM". REVISTA DE LA UNION MATEMATICA ARGENTINA 26, 235-249. (1972).

ALEM 1973 -1 ALEMANY, R.; DIEGO, A.
TENSIONS ET SEMITENSIONS SUR UN GRAPHE. UNIVERSIDAD NACIONAL DE TUCUMAN, FACULTAD DE CIENCIAS EXACTAS Y TECNOLOGIA, REVISTA, SERIES A, MATEMATICAS Y FISICA TEORICA 23, 221-227. (1973).

****

ALES 1975 -1 ALESKEROV, S. A.; GRIGOROVIC, A. F.
MULTICOMMODITY FLOWS IN COMMUNICATION NETWORKS (RUSSIAN). DOKLADY AKADEMII NAUK SSSR 220, 1286-1288. (1975).

ALI 1977 -1 ALI, A. I.; HELGASON, R. V.; KENNINGTON, J. L.; LALL, H. S.
PRIMAL SIMPLEX NETWORK CODES: STATE-OF-THE-ART IMPLEMENTATION TECHNOLOGY (REVISED VERSION). TECHNICAL REPORT IEOR 76014. DALLAS: SOUTHERN METHODIST UNIV., DEPT. OF INDUSTRIAL ENG. AND OPERATIONS RESEARCH, 30 P. (1977).

ALIA 1976 -1 ALIA, G.; MAESTRINI, P.
PROCEDURE TO DETERMINE OPTIMAL PARTITIONS OF WEIGHTED HYPERGRAPHS THROUGH A NETWORK-FLOW ANALOGY, A. CALCOLO 13, 191-211. (1976).

ALIB 1971 -1 ALIBEKOV, B. I.
APPROXIMATE METHOD FOR THE SOLUTION OF A MULTIEXTREMAL ALLOCATION PROBLEM (RUSSIAN), AN. IN: PROCEEDINGS OF THE FOURTH WINTER SCHOOL ON MATHEMATICAL PROGRAMMING AND RELATED QUESTIONS (RUSSIAN), DROGOBYCH 1971, NO. 1, 121-136. MOSCOW: MOSKOV. INZ.-STROITEL. INST. (1971).

ALIB 1975 -1 ALIBEKOV, B. I.
ON AN ALLOCATION PROBLEM WITH CONSTRAINED CAPACITIES (RUSSIAN). EKONOMIKA I MATEMATICHESKIE METODY 11, 535-541. (1975).

ALLE 1976 -1 ALLEN, B.; MUNRO, I.
SELF-ORGANIZING BINARY SEARCH TREES. IN: 17TH ANNUAL SYMPOSIUM ON FOUNDATIONS OF COMPUTER SCIENCE, HOUSTON 1976, 166-172. LONG BEACH: INSTITUTE OF ELECTRICAL AND ELECTRONICS ENGINEERS. (1976).

ALLEA 1976 -1 ALLENDE, S.
TYPE OF IMPORTANT FUNCTIONS IN INTEGER PROGRAMMING (SPANISH), A. INVESTIGACION OPERACIONAL, NO. 17, 21-25. (1976).

ALSP 1976 -1 ALSPACH, B.; MASON, D. W.; PULLMAN, N. J.
PATH NUMBERS OF TOURNAMENTS. JOURNAL OF COMBINATORIAL THEORY 20, SERIES B, 222-228. (1976).

ALTH 1975 -1 ALTHERR, W.
ALGORITHM FOR ENUMERATING ALL VERTICES OF A CONVEX POLYHEDRON, AN. COMPUTING 15, 181-194. (1975).

ALTH 1976 -1 ALTHERR, W.
ECKPUNKTBESTIMMUNG KONVEXER POLYEDER. IN T. M. LIEBLING, M. ROESSLER (EDS.): KOMBINATORISCHE ENTSCHEIDUNGSPROBLEME: METHODEN UND ANWENDUNG, 184-195. LECTURE NOTES IN ECONOMICS AND MATHEMATICAL SYSTEMS 153, BERLIN: SPRINGER. (1976).

ALTHA 1976 -1 ALTHERR, W.; MAYER, J.
OPTIMALER EINSATZ EINES BETRIEBSEIGENEN STRASSENTRANSPORTSYSTEMS. IN: T. M. LIEBLING, M. ROESSLER (EDS.): KOMBINATORISCHE ENTSCHEIDUNGSPROBLEME: METHODEN UND ANWENDUNG, 70-80. LECTURE NOTES IN ECONOMICS AND MATHEMATICAL SYSTEMS 153, BERLIN: SPRINGER. (1976).

AM 1976 -1 AM, T. D.; TSUKIYAMA, S.; SHIRAKAWA, I.; OZAKI, H.
ALGORITHM FOR GENERATING ALL THE PATHS BETWEEN TWO VERTICES IN A DIGRAPH AND ITS APPLICATION, AN. TECHNOLOGICAL REPORTS OF THE OSAKA UNIV. 26, NO. 1308-1336, 411-418. (1976).

AMIN 1973 -1 AMIN, A. T.; HAKIMI, S. L.
MAXIMALLY EDGE-CONNECTED GRAPHS WITH LEAST NUMBER OF MINIMUM EDGE CUT-SETS. IN: PROCEEDINGS OF THE SEVENTH ANNUAL PRINCETON CONFERENCE ON INFORMATION SCIENCES AND SYSTEMS, 409. (1973).

ANDE 1970 -1 ANDERSON, S.
GRAPH THEORY AND FINITE COMBINATORICS. CHICAGO: MARKHAM. (1970).

ANDEA 1976 -1 ANDERSSEN, R. S.; BRENT, R. P.
COMPLEXITY OF COMPUTATIONAL PROBLEM SOLVING, THE. ST. LUCIA: UNIV. OF QUEENSLAND PRESS, 262 P. (1976). (PROCEEDINGS OF A SEMINAR HELD AT THE AUSTRALIAN NATIONAL UNIVERSITY, DECEMBER 1974)

ANDO 1976 -1 ANDON, F. I.; VARTANYAN, G. M.
PROBLEM OF PACKAGING DEMANDS WITH TIMING MARKS IN INFORMATION SYSTEMS (RUSSIAN). KIBERNETIKA, NO. 4, 20-25. (1976).

ANDR 1975 -1 ANDRAS, B.
SOLUTION ALGORITHM FOR A MINIMAL PATH PROBLEM IN A MULTITERMINAL GENERALIZED NETWORK (HUNGARIAN), A. ALKALMAZOTT MATEMATIKAI LAPOK, NO, 1-2, 109-115. (1975).

ANDRA 1976 -1 ANDRONICO, A.; RANDI, P.; SCALAS, M. R.
INTERACTIVE CODE FOR MIXED INTEGER LINEAR PROGRAMS, AN. SERIE A-35. PISA: UNIV. DI PISA, DIPT. DI RICERCA OPERATIVA E SCIENZE STATISTICHE, 19 P. (1976).

ANEJ 1977 -1 ANEJA, Y. P.; VEMUGANTI, R. R.
ROW GENERATION SCHEME FOR FINDING A MULTI-COMMODITY MINIMUM DISCONNECTING SET, A. MANAGEMENT SCIENCE 23, 652-659. (1977).

ANGL 1977 -1 ANGLUIN, D.; VALIANT, L. G.
FAST PROBABILISTIC ALGORITHMS FOR HAMILTONIAN CIRCUITS AND MATCHINGS. IN: PROCEEDINGS OF THE NINTH ANNUAL A.C.M. SYMPOSIUM ON THEORY OF COMPUTING, 30-41. (1977).

ANIS 1976 -1 ANISIMOV, V. V.
ON OPTIMUM PACKING OF TREES (RUSSIAN). KIBERNETIKA, NO. 3, 89-91. (1976).

ANN 1977 -1 ANN, T.-K.
TRAITEMENT DU PROBLEME DE VOYAGEUR DE COMMERCE PAR UNE METHODE DE PROGRAMMATION LINEAIRE. DISSERTATION. NANCY: UNIV. NANCY I, 135 P. (1977).

ANZA 1974 -1 ANZAI, Y.
ON INTEGER FRACTIONAL PROGRAMMING. JOURNAL OF THE OPERATIONS RESEARCH SOCIETY OF JAPAN 17, 49-66. (1974).

AOSH 1977 -1 AOSHIMA, K.; IRI, M.
COMMENTS ON F. HADLOCK'S PAPER: "FINDING A MAXIMUM CUT OF A PLANAR GRAPH IN POLYNOMIAL TIME". SIAM JOURNAL ON COMPUTING 6, 86-87. (1977).

APPE 1976 -1 APPEL, K.; HAKEN, W.
EVERY PLANAR MAP IS FOUR COLORABLE. BULLETIN OF THE AMERICAN MATHEMATICAL SOCIETY 82, 711-712. (1976).

ARDI 1977 -1 ARDITTI, J. C.
COMPARABILITY GRAPHS AND THE DIMENSION OF ORDERED SETS(EXTENDED ABSTRACT). IN: R. HENN, P. KALL, B. KORTE, O. KRAFFT ET AL. (EDS.): OPERATIONS RESEARCH VERFAHREN XXV, METHODS OF OPERATIONS RESEARCH, 293-295. MEISENHEIM: HAIN. (1977).

ARDIA 1975 -1 ARDITTI, J.-C.
GRAPHES DE COMPARABILITE D'ARBRES ET D'ARBORESCENCES. DISSERTATION. PARIS: UNIV. PARIS XI, 131 P. (1975).

ARDIA 1976 -1 ARDITTI, J.-C.
PROBLEMES DE GRAPHE ET LA COMPLEXITE DES ALGORITHMES, LES. IN: ANAL. APPL. INF. JOURN. COMB. INF., BORDEAUX 1975, 7-22. BORDEAUX: UNIV. BORDEAUX I. (1976).

ARET 1974 -1 ARET, L.
SOLUTION OF A CERTAIN MULTI-INDEX PROBLEM OF TRANSPORTATION TYPE (RUSSIAN), THE. EESTI NSV TEADUSTE AKADEEMIA TOIMETISED. FUEUESIKA-MATEMAATIKA 23, 155-159. (1974).

ARIS 1977 -1 ARISAWA, S.; ELMAGHRABY, S. E.
"HUB" AND "WHEEL" SCHEDULING PROBLEMS, I. THE "HUB" SCHEDULING PROBLEM: THE MYOPIC CASE, THE. TRANSPORTATION SCIENCE 11, 124-146. (1977).

****

ARIS 1977 -2 ARISAWA, S.; ELMAGHRABY, S. E.
"HUB" AND "WHEEL" SCHEDULING PROBLEMS, II. THE HUB OPERATION SCHEDULING PROBLEM (HOSP): MULTI-PERIOD AND INFINITE HORIZON, AND THE WHEEL OPERATION SCHEDULING PROBLEM (WOSP), THE. TRANSPORTATION SCIENCE 11, 147-165. (1977).

ARJO 1975 -1 ARJOMANDI, E.; CORNEIL, D. G.
PARALLEL COMPUTATIONS IN GRAPH THEORY. IN: 16TH ANNUAL SYMPOSIUM ON FOUNDATIONS OF COMPUTER SCIENCE, BERKELEY 1975, 13-18. NEW YORK: INSTITUTE OF ELECTRICAL AND ELECTRONICS ENGINEERS. (1975).

ARKH 1975 -1 ARKHIPOWA, T. T.; SEGIENKO, I. V.
METHOD OF POSSIBLE AND A METHOD OF DROP VECTOR IN THE INTEGER PROGRAMMING (RUSSIAN), A. KIBERNETIKA, NO. 5, 125-128. (1975).

ARLA 1973 -1 ARLAZAROV, V. L.; USKOV, A. V.; FARADZEV, I. A.
ALGORITHM FOR FINDING ALL SIMPLE CYCLES IN A DIRECTED GRAPH (RUSSIAN), AN. IN: A. A. FRIDMAN (ED.): STUDIES IN DISCRETE MATHEMATICS (RUSSIAN), 178-183. MOSCOW: IZDAT. NAUKA. (1973).

ARMS 1978 -1 ARMSTRONG, R. D.
NONBASIC BRANCHING AND DUAL RELAXATION IN A BRANCH-AND-BOUND MIXED INTEGER PROGRAMMING ALGORITHM. NEW ZEALAND OPERATIONAL RESEARCH 6, 59-74. (1978).

ARMSA 1975 -1 ARMSTRONG, R. D.; CHARNES, A.; PHILLIPS, F.
PAGE CUTS FOR INTEGER INTERVAL LINEAR PROGRAMMING. AUSTIN: UNIV. OF TEXAS, CENTER FOR CYBERNETIC STUDIES, 19 P. (1975). (RESEARCH REPORT CCS 199)

ARMSB 1976 -1 ARMSTRONG, R. D.; COOK, W. D.
OPTIMAL LINKING OF TASKS AND SCHEDULE DEVELOPMENT FOR MILITARY AIR TRANSPORT. INFOR 14, 13-31. (1976).

ARMSB 1977 -1 ARMSTRONG, R. D.; COOK, W. D.
COMMITTEE RANKING PROBLEM USING ORDINAL UTILITIES, A. RESEARCH REPORT CCS 298. AUSTIN: UNIV. OF TEXAS, CENTER FOR CYBERNETIC STUDIES, 30 P. (1977).

ARMSB 1977 -2 ARMSTRONG, R. D.; COOK, W. D.
STRATEGY SELECTION IN ORDINAL RANKING PROBLEMS. RESEARCH REPORT CCS 299. AUSTIN: UNIV. OF TEXAS, CENTER FOR CYBERNETIC STUDIES, 23 P. (1977).

ARMSC 1977 -1 ARMSTRONG, R. D.; COOK, W. D.; PALACIOS-GOMESZ, F. E.
APPLICATION OF QUASI-INTEGER PROGRAMMING TO A CAPITAL BUDGETING PROBLEM IN PAVEMENT MAINTENANCE, AN. RESEARCH REPORT CCS 285. AUSTIN: UNIV. OF TEXAS, CENTER FOR CYBERNETIC STUDIES, 27 P. (1977).

ARMSD 1976 -1 ARMSTRONG, R. D.; FROME, E. L.
BRANCH-AND-BOUND SOLUTION OF A RESTRICTED LEAST SQUARES PROBLEM, A. TECHNOMETRICS 18, 447-450. (1976).

ARMSE 1975 -1 ARMSTRONG, R. D.; WILLIS, C. E.
SIMULTANEOUS INVESTMENT-ALLOCATION - AN APPLICATION OF GENERALIZED BENDERS DECOMPOSITION TO WATER PLANNING. RESEARCH REPORT CCS 243. AUSTIN: UNIV. OF TEXAS, CENTER FOR CYBERNETIC STUDIES, 20 P. (1975).

ARMSE 1977 -1 ARMSTRONG, R. D.; WILLIS, C. E.
SIMULTANEOUS INVESTMENT AND ALLOCATION DECISIONS APPLIED TO WATER PLANNING. MANAGEMENT SCIENCE 23, 1080-1088. (1977).

ARNO 1974 -1 ARNOLD, L. R.; BELLMORE, M.
GENERATED CUT FOR PRIMAL INTEGER PROGRAMMING, A. OPERATIONS RESEARCH 22, 137-143. (1974).

ARNO 1974 -2 ARNOLD, L. R.; BELLMORE, M.
ITERATION SKIPPING IN PRIMAL INTEGER PROGAMMING. OPERATIONS RESEARCH 22, 129-136. (1974).

AROR 1977 -1 ARORA, S. R.
NOTE ON FRACTIONAL FIXED CHARGE PROBLEMS, A. NEW ZEALAND OPERATIONAL RESEARCH 5, 66-71. (1977).

ARORA 1977 -1 ARORA, S. R.; PURI, M. C.
ENUMERATION TECHNIQUE FOR THE SET COVERING PROBLEM WITH A LINEAR FRACTIONAL FUNCTIONAL AS ITS OBJECTIVE FUNCTION. ZEITSCHRIFT FUER ANGEWANDTE MATHEMATIK UND MECHANIK 57, 181-186. (1977).

ARORB 1976 -1 ARORA, S. R.; SWARUP, K.; PURI, M. C.
ENUMERATIVE TECHNIQUE FOR THE SET COVERING PROBLEM, AN. RICERCA OPERATIVA 6, 3, 13-22; ALSO PUBLISHED IN NEW ZEALAND OPERATINAL RESEARCH 5, 1977, 119-128. (1976).

ARORB 1977 -1 ARORA, S. R.; SWARUP, K.; PURI, M. C.
CUTTING PLANE TECHNIQUE FOR SET COVERING PROBLEM WITH LINEAR FRACTIONAL FUNCTIONAL. ZEITSCHRIFT FUER ANGEWANDTE MATHEMATIK UND MECHANIK 57, 597-602. (1977).

ARUI 1971 -1 ARUIN, L. I.
EDGES OF THE CONVEX LINEAR HULL OF THE SET OF INTEGER POINTS THAT BELONG TO THE SET OF ADMISSIBLE SOLUTIONS OF AN INTEGER LINEAR PROGRAMMING PROBLEM (RUSSIAN), THE. IN: PROCEEDINGS OF THE 4TH WINTER SCHOOL ON MATHEMATICAL PROGRAMMING AND RELATED QUESTIONS, DROGOBYCH 1971, NO. 1 (RUSSIAN), 137-148. MOSCOW: INZ.-STROITEL. INST. (1971).

ASAN 1976 -1 ASANO, T.; SHIBUI, M.; TAKANAMI, I.
GENERAL RESULTS ON TOUR LENGTHS IN MACHINES AND DIGRAPHS. SIAM JOURNAL ON COMPUTING 5, 629-645. (1976).

ASHT 1977 -1 ASHTON, W. B.; CLARK, G. M.
LOCATION AND SIZING OF URBAN FREIGHT CONSOLIDATION TERMINALS WITH MULTIPLE PLANNING PERIODS, THE. SAN FRANCISCO (51. ORSA-MEETING + TIMS, MAY 1977) 36 P. (1977).

ASSA 1975 -1 ASSAD, A.
MULTICOMMODITY NETWORK FLOWS - A SURVEY. WORKING PAPER OR 045-75. CAMBRIDGE: MASSACHUSETTS INST. OF TECHNOLOGY, OPERATIONS RESEARCH CENTER. (1975).

ASSAA 1976 -1 ASSAD, A. A.
MULTICOMMODITY NETWORK FLOWS - COMPUTATIONAL EXPERIENCE. WORKING PAPER OR 058-76. CAMBRIDGE: MASSACHUSETTS INST. OF TECHNOLOGY, OPERATIONS RESEARCH CENTER. (1976).

ASTA 1971 -1 ASTAFEVA, N. A.
ALGORITHM FOR DETERMINING THE SHORTEST ROUTES THROUGH A CONNECTING NETWORK (RUSSIAN). MEKH. AUTOM. UPR. 4, 6-8. (1971).

ASTAA 1976 -1 ASTAHOV, N. D.
GRADIENT ALGORITHMS FOR SOLVING SOME TRANSPORTATION PROBLEMS (RUSSIAN). ZURNAL VYCISLITEL'NOJ MATEMATIKI I MATEMATICESKOJ FIZIKI 16, 1146-1154. (1976).

ATTM 1973 -1 ATTMANN, I.
NUMERISCHE ERFAHRUNGEN MIT BALAS-ALGORITHMEN. STUDIENARBEIT, KARLSRUHE: UNIV. KARLSRUHE, INSTITUT FUER WIRTSCHAFTSTHEORIE UND OPERATIONS RESEARCH. (1973).

AUSI 1977 -1 AUSIELLO, G.
ON THE STRUCTURE AND PROPERTIES OF NP-COMPLETE PROBLEMS AND THEIR ASSOCIATED OPTIMIZATION PROBLEMS. IN: J. GRUSKA (ED.): MATHEMATICAL FOUNDATIONS OF COMPUTER SCIENCE 1977, PROCEEDINGS OF THE 6TH SYMPOSIUM, TATRANSKA LOMNICA 1977, 1-16. LECTURE NOTES IN COMPUTER SCIENCE 53, BERLIN: SPRINGER. (1977).

***

AUSIA 1977 -1 AUSIELLO, G.; D'ATRI, A.; GAUDIANO, M.; PROTASI, M.
CLASSES OF STRUCTURALLY ISOMORPHIC NP-OPTIMIZATION PROBLEMS. IN: J. GRUSKA (ED.): MATHEMATICAL FOUNDATIONS OF COMPUTER SCIENCE 1977, PROCEEDINGS OF THE 6TH SYMPOSIUM, TATRANSKA LOMNICA 1977, 222-230. LECTURE NOTES IN COMPUTER SCIENCE 53, BERLIN: SPRINGER. (1977).

AUSIB 1977 -1 AUSIELLO, G.; D'ATRI, A.; PROTASI, M.
ON THE STRUCTURE OF COMBINATORIAL PROBLEMS AND STRUCTURE PRESERVING REDUCTIONS. IN: A. SALOMAA, M. STEINBY (EDS.): AUTOMATA, LANGUAGES AND PROGRAMMING, 4TH COLLOQUIUM, TURKU 1977, 45-60. LECTURE NOTES IN COMPUTER SCIENCE 52, BERLIN: SPRINGER. (1977).

AUST 1976 -1 AUST, R. J.
DYNAMIC PROGRAMMING BRANCH AND BOUND ALGORITHM FOR PURE INTEGER PROGRAMMING, A. COMPUTERS AND OPERATIONS RESEARCH 3, 27-38. (1976).

AUST 1976 -2 AUST, R. J.
IMPROVEMENT ALGORITHM FOR SCHOOL TIMETABLING, AN. COMPUTER JOURNAL 19, 339-343. (1976).

AVDO 1975 -1 AVDONIN, L. N.
SOLUTION ALGORITHM OF ONE-PRODUCT ENTERPRISE ALLOCATION PROBLEM (RUSSIAN). EKONOMIKA I MATEMATICHESKIE METODY 11, 590-591. (1975).

AVRI 1977 -1 AVRIEL, M.; GUROVICH, V.
CONDENSATION ALGORITHM FOR A CLASS OF ALGEBRAIC PROGRAMS, A. OPERATIONS RESEARCH 25, 78-88. (1977).

AZAR 1977 -1 AZARYAN, L. L.; LEBEDEV, S. S.; MESTETZKY, L. M.
INTEGER TRANSPORT TYPE PROBLEMS SOLVING BASED ON GENERALIZED LAGRANGE MULTIPLIERS (RUSSIAN). EKONOMIKA I MATEMATICESKIE METODY 13, 723-731. (1977).

BABA 1977 -1 BABAT, L. G.
ON INTEGER LINEAR AGGREGATION (RUSSIAN). EKONOMIKA I MATEMATICHESKIE METODY 13, 599-601. (1977).

BABAA 1975 -1 BABAYEV, D. A.
METHOD OF HYPERSPHERES FOR THE SOLUTION OF DISCRETE PROGRAMMING PROBLEMS (RUSSIAN), THE. DOKLADY AKADEMII NAUK SSSR 221, 1009-1012. (1975).

BABAA 1976 -1 BABAYEV, D. A.
METHOD OF HYPERSPHERES FOR BOOLEAN PROGRAMMING PROBLEMS (RUSSIAN). ZURNAL VYCISLITEL'NOI MATEMATIKI I MATEMATICESKOI FIZIKI 16, 1414-1426. (1976).

BABU 1977 -1 BABU, A. J. G.; HSIA, W. S.
MORE EFFICIENT ALGORITHM FOR AN OPTIMAL TOUR OF A HEALTH CARE CONSUMER WITH MULTIPLE HEALTH CARE FACILITIES IN EACH COUNTY, A. IN: NONLINEAR SYSTEMS AND APPLICATIONS. PROCEEDINGS OF AN INTERNAT. CONFERENCE, ARLINGTON 1976, 373-380. NEW YORK: ACADEMIC PRESS. (1977).

BABUA 1976 -1 BABUSHKIN, A. I.; BASHTA, A. L.; BELOV, I. S.
SCHEDULING A THREE-MACHINE JOB-SHOP (RUSSIAN). AVTOMATIKA I TELEMEHANIKA, NO. 7, 154-158; TRANSLATED AS AUTOMATION AND REMOTE CONTROL 37, 1976, 1109-1113. (1976).

BABUA 1977 -1 BABUSHKIN, A. I.; BASHTA, A. L.; BELOV, I. S.
CONSTRUCTION OF THE SCHEDULE FOR THE COUNTER ROUTES PROBLEM (RUSSIAN). KIBERNETIKA, NO. 4, 130-135. (1977).

BABUB 1972 -1 BABUSIAUX, D.
ESTABLISHING A NETWORK OF WAREHOUSES FOR FINISHED PRODUCTS, TAKING INTO ACCOUNT SEASONAL FLUCTUATIONS IN DEMAND. REGIONAL AND URBAN ECONOMICS 1, 337-353. (1972).

BACH 1976 -1 BACHEM, A.
ALGORITHMUS ZUR MODULO-OPTIMIERUNG STRUKTURIERTER MATRIZEN, EIN. IN: J. KOHLAS, O. SEIFERT, P. STAEHLY, H.-J. ZIMMERMANN (EDS.): PROCEEDINGS IN OPERATIONS RESEARCH 5, 43-45. WUERZBURG: PHYSICA. (1976).

BACH 1976 -2 BACHEM, A.
APPROXIMATION GANZZAHLIGER POLYEDER DURCH CORNER POLYEDER. ZEITSCHRIFT FUER ANGEWANDTE MATHEMATIK UND MECHANIK 56, NO. 3, T332-T333. (1976).

BACH 1976 -3 BACHEM, A.
BEITRAEGE ZUR THEORIE DER CORNER POLYEDER. MATHEMATICAL SYSTEMS IN ECONOMICS, 29. MEISENHEIM: HAIN, 179 P. (1976).

BACH 1976 -4 BACHEM, A.
REDUCTION AND DECOMPOSITION OF MODULO OPTIMIZATION PROBLEMS. IN: W. OETTLI, K. RITTER (EDS.): OPTIMIZATION AND OPERATIONS RESEARCH, PROCEEDINGS OF A CONFERENCE HELD AT OBERWOLFACH, 1975, 1-8. LECTURE NOTES IN ECONOMICS AND MATHEMATICAL SYSTEMS 117, BERLIN, HEIDELBERG, NEW YORK: SPRINGER. (1976).

BACH 1977 -1 BACHEM, A.
FUNDAMENTALPUNKTALGORITHMEN ZUR LOESUNG GANZZAHLIGER PROGRAMMIERUNGSAUFGABEN. IN: R. HENN, P. KALL, B. KORTE, O. KRAFFT ET AL. (EDS.): OPERATIONS RESEARCH VERFAHREN XXIII, METHODS OF OPERATIONS RESEARCH, 1-17. MEISENHEIM: HAIN. (1977).

BACH 1977 -2 BACHEM, A.
FUNDAMENTALPUNKTALGORITHMEN ZUR LOESUNG GANZZAHLIGER PROGRAMMIERUNGSAUFGABEN(EXTENDED ABSTRACT). IN: R. HENN, P. KALL, B. KORTE, O. KRAFFT ET AL. (EDS.): OPERATIONS RESEARCH VERFAHREN XXV, METHODS OF OPERATIONS RESEARCH, 141-143, MEISENHEIM: HAIN. (1977).

BACH 1977 -3 BACHEM, A.
REDUCTION AND DECOMPOSITION OF INTEGER PROGRAMS OVER CONES. IN: P. L. HAMMER, E. L. JOHNSON, B. H. KORTE, G. L. NEMHAUSER (EDS.): STUDIES IN INTEGER PROGRAMMING, 1-12. ANNALS OF DISCRETE MATHEMATICS 1, AMSTERDAM: NORTH-HOLLAND. (1977).

BACH 1978 -1 BACHEM, A.
THEOREM OF MINKOWSKI FOR POLYHEDRAL MONOIDS AND AGGREGATED LINEAR DIOPHANTINE SYSTEMS, THE. IN: R. HENN, B. KORTE, W. OETTLI (EDS.): OPTIMIZATION AND OPERATIONS RESEARCH, PROCEEDINGS, BONN 1977, 1-13. LECTURE NOTES IN ECONOMICS AND MATHEMATICAL SYSTEMS 157. BERLIN, HEIDELBERG, NEW YORK: SPRINGER. (1978).

BACHA 1977 -1 BACHEM, A.; KORTE, B.
ALGORITHM FOR QUADRATIC OPTIMIZATION OVER TRANSPORTATION POLYTOPES, AN. REPORT 7759-OR. BONN: UNIV. BONN, INST. F. OEKONOMETRIE UND OPERATIONS RESEARCH, 5 P. (1977).

BACHA 1977 -2 BACHEM, A.; KORTE, B.
ESTIMATING INPUT-OUTPUT-MATRICES. REPORT 7784-OR. BONN: UNIV. BONN, INST. F. OEKONOMETRIE UND OPERATIONS RESEARCH. (1977).

BACHA 1977 -3 BACHEM, A.; KORTE, B.
QUADRATIC OPTIMIZATION OVER TRANSPORTATION POLYTOPES. REPORT 7767-OR. BONN: UNIV. BONN, INST. F. OEKONOMETRIE UND OPERATIONS RESEARCH, 22 P. (1977).

BACHA 1977 -4 BACHEM, A.; KORTE, B.
QUADRATISCHE OPTIMIERUNG UEBER TRANSPORTPOLYEDER ZUR LOESUNG VON SCHAETZPROBLEMEN VON INPUT-OUTPUT MATRIZEN. REPORT 7772-OR. BONN: UNIV. BONN, INST. F. OEKONOMETRIE UND OPERATIONS RESEARCH, 10 P. (1977).

BACHB 1977 -1 BACHEM, A.; SCHRADER, R.
MINIMAL INEQUALITIES AND DUALITY FOR MIXED INTEGER PROGRAMS. REPORT 7787-OR. BONN: UNIV. BONN, INST. F. OEKONOMETRIE UND OPERATIONS RESEARCH (1977).

****

BACHC 1977 -1 BACHEM, A.; SIMON, H.
OPTIMAL PRODUCT POSITIONING IN AN ATTRIBUTE SPACE. REPORT NO. 7777-OR. BONN: UNIV. BONN, INSTITUT FUER OEKONOMETRIE UND OPERATIONS RESEARCH, 12 P. (1977).

BAER 1976 -1 BAER, G.
UMFORMUNGSMOEGLICHKEITEN FUER DIE RESTRIKTIONEN EINES 0-1-OPTIMIERUNGSPROBLEMS. ELEKTRONISCHE INFORMATIONSVERARBEITUNG UND KYBERNETIK 12, 607-610. (1976).

BAER 1976 -2 BAER, G.
ZUR LINEARISIERUNG BELIEBIGER 0-1-OPTIMIERUNGSPROBLEME. MATHEMATISCHE OPERATIONSFORSCHUNG UND STATISTIK 7, 181-196. (1976).

BAGB 1976 -1 BAGBY, G.; THESEN, A.
NETWORK FLOW MODEL FOR ALLOCATION OF LAND USES TO SECTORS, A. MANAGEMENT SCIENCE 22, 1221-1228. (1976).

BAGG 1973 -1 BAGGA, P. C.
TWO MACHINE SCHEDULING PROBLEM WHEN PASSING ALLOWED. METRIKA 20, 36-40. (1973).

BAGGA 1967 -1 BAGGA, P. C.; CHAKRAVARTI, N. K.
OPTIMAL M-STAGE PRODUCTION SCHEDULE. CANADIAN OPERATIONS RESEARCH SOCIETY JOURNAL 6, 71-78. (1967).

BAKE 1976 -1 BAKER, K. R.
WORKFORCE ALLOCATION IN CYCLICAL SCHEDULING PROBLEMS: A SURVEY. OPERATIONAL RESEARCH QUARTERLY 27, 155-168. (1976).

BAKE 1977 -1 BAKER, K. R.
COMPUTATIONAL EXPERIENCE WITH A SEQUENCING ALGORITHM ADAPTED TO THE TARDINESS PROBLEM. AIIE TRANSACTIONS 9, 32-35. (1977).

BAKEA 1977 -1 BAKER, K. R.; BURNS, R. N.; CARTER, M.
STAFF SCHEDULING WITH DAY-OFF AND WORKSTRETCH CONSTRAINTS. G.S.B.A. WORKING PAPER NO. 178. DURHAM: DUKE UNIV., GRADUATE SCHOOL OF BUSINESS ADMINISTRATION, 23 P. (1977).

BAKEB 1977 -1 BAKER, K. R.; MAGAZINE, M.
WORKFORCE SCHEDULING WITH CYCLIC DEMANDS AND DAY-OFF CONSTRAINTS. MANAGEMENT SCIENCE 24, 161-167. (1977).

BAKEC 1975 -1 BAKER, T.; GILL, J.; SOLOVAY, R.
RELATIVIZATIONS OF THE P = ? NP QUESTION. SIAM JOURNAL ON COMPUTING 4, 431-442. (1975).

BAKO 1975 -1 BAKO, A.
SOLUTION OF A MULTITERMINAL MINIMAL PATH PROBLEM IN A GENERALIZED NETWORK (HUNGARIAN). ALKALMAZOTT MATEMATIKAI LAPOK 1, 109-115. (1975).

BAKO 1976 -1 BAKO, A.
ALL PATHS IN AN ACTIVITY NETWORK. MATHEMATISCHE OPERATIONSFORSCHUNG UND STATISTIK 7, 851-858. (1976).

BAKO 1976 -2 BAKO, A.
SOLUTION OF THE MULTITERMINAL MINIMAL PATH PROBLEM IN A NETWORK WITH GAINS. IN: A. PREKOPA (ED.): PROGRESS IN OPERATIONS RESEARCH, 63-69, AMSTERDAM, LONDON: NORTH HOLLAND. (1976).

BALA 1976 -1 BALACHANDRAN, V.
INTEGER GENERALIZED TRANSPORTATION MODEL FOR OPTIMAL JOB ASSIGNMENT IN COMPUTER NETWORKS, AN. OPERATIONS RESEARCH 24, 742-759. (1976).

BALAA 1977 -1 BALACHANDRAN, V.; DITTMAN, D.
OVERHEAD ALLOCATION VIA MATHEMATICAL PROGRAMMING. DISCUSSION PAPER NO. 299. EVANSTON: NORTHWESTERN UNIV., CENTER FOR MATHEMATICAL STUDIES IN ECONOMICS AND MANAGEMENT SCIENCE, 26 P. (1977).

BALAB 1977 -1 BALACHANDRAN, V.; JAIN, S.
MATHEMATICAL-PROGRAMMING MODEL FOR OPTIMAL SELECTION OF TOURISM PLANS AND GATEWAYS FOR THE AIRLINE INDUSTRY, A. DISCUSSION PAPER NO. 298. EVANSTON: NORTHWESTERN UNIV., CENTER FOR MATHEMATICAL STUDIES IN BUSINESS AND ECONOMICS AND MANAGEMENT SCIENCE, 25 P. (1977).

BALAC 1976 -1 BALACHANDRAN, V.; PERRY, A.
TRANSPORTATION TYPE PROBLEMS WITH QUANTITY DISCOUNTS. NAVAL RESEARCH LOGISTICS QUARTERLY 23, 195-210. (1976).

BALAD 1977 -1 BALACHANDRAN, V.; SRINIVASAN, V.; THOMPSON, G. L.
SURVEY OF THE APPLICATIONS OF THE OPERATOR THEORY OF PARAMETRIC PROGRAMMING FOR THE TRANSPORTATION AND GENERALIZED TRANSPORTATION PROBLEM, A. DISCUSSION PAPER NO. 300. EVANSTON: NORTHWESTERN UNIV., CENTER FOR MATHEMATICAL STUDIES IN ECONOMICS AND MANAGEMENT SCIENCE, 46 P. (1977).

BALAE 1975 -1 BALACHANDRAN, V.; THOMPSON, G. L.
OPERATOR THEORY OF PARAMETRIC PROGRAMMING FOR THE GENERALIZED TRANSPORTATION PROBLEM - 3 - WEIGHT OPERATORS, AN. NAVAL RESEARCH LOGISTICS QUARTERLY 22, 297-316. (1975).

BALAE 1975 -2 BALACHANDRAN, V.; THOMPSON, G. L.
OPERATOR THEORY OF PARAMETRIC PROGRAMMING FOR THE GENERALIZED TRANSPORTATION PROBLEM - 4 - GLOBAL OPERATORS, AN. NAVAL RESEARCH LOGISTICS QUARTERLY 22, 317-340. (1975).

BALAE 1975 -3 BALACHANDRAN, V.; THOMPSON, G. L.
OPERATOR THEORY OF PARAMETRIC PROGRAMMING FOR THE GENERALIZED TRANSPORTATION PROBLEM: 1. BASIC THEORY, AN. NAVAL RESEARCH LOGISTICS QUARTERLY 22, 79-100. (1975).

BALAE 1975 -4 BALACHANDRAN, V.; THOMPSON, G. L.
OPERATOR THEORY OF PARAMETRIC PROGRAMMING FOR THE GENERALIZED TRANSPORTATION PROBLEM: 2. RIM, COST AND BOUND OPERATORS, AN. NAVAL RESEARCH LOGISTICS QUARTERLY 22, 101-125. (1975).

BALAF 1977 -1 BALAGURU, P. N.; RAMAKRISHNAN, V.; NAAMAN, A. E.
COST AND TIME SCALED NETWORKS - GRAPHICAL MODEL. J. CONSTR. DIV. 103, NO. CO-2, 179-190. (1977).

BALAG 1968 -1 BALAS, E.
PROJECT SCHEDULING WITH RESOURCE CONSTRAINTS. AD679621, JUN 1968, 28 P. (1968).

BALAG 1973 -1 BALAS, E.
ON THE USE OF INTERSECTION CUTS AND OUTER POLARS IN BRANCH-AND-BOUND. TALK GIVEN AT THE MPS CONFERENCE IN STANFORD, AUGUST 1973. (1973).

BALAG 1974 -1 BALAS, E.
DISJUNCTIVE PROGRAMMING: FACETS OF THE CONVEX HULL OF FEASIBLE POINTS. MANAGEMENT SCIENCES RESEARCH REPORT 348. PITTSBURGH: CARNEGIE-MELLON UNIVERSITY. (1974).

BALAG 1974 -2 BALAS, E.
INTERSECTION CUTS FROM DISJUNCTIVE CONSTRAINTS. MANAGEMENT SCIENCES RESEARCH REPORT 330. PITTSBURGH: CARNEGIE-MELLON UNIVERSITY. (1974).

BALAG 1975 -1 BALAS, E.
BIVALENT PROGRAMMING BY IMPLICIT ENUMERATION. IN: J. BELZER, A. G. HOLZMAN, A. KENT (EDS.): ENCYCLOPEDIA OF COMPUTER SCIENCE AND TECHNOLOGY, VOL. 2, AN/FSQ TO BAL, 479-494. NEW YORK: DEKKER. (1975).

BALAG 1975 -2 BALAS, E.
DISJUNCTIVE PROGRAMMING: CUTTING PLANES FROM LOGICAL CONDITIONS. IN: O. L. MANGASARIAN, R. R. MEYER, P. M. ROBINSON (EDS.): NONLINEAR PROGRAMMING 2, 279-312. NEW YORK: ACADEMIC PRESS. (1975).

BALAG 1975 -3 BALAS, E.
LINEAR CHARACTERIZATION OF PERMUTATION VECTORS, A. MANAGEMENT SCIENCE RESEARCH REPORT 364. PITTSBURGH: CARNEGIE-MELLON UNIV., GRADUATE SCHOOL OF INDUSTRIAL ADMINISTRATION, 4 P. (1975).

BALAG 1976 -1 BALAS, E.
FACETS OF ONE-DIMENSIONAL AND MULTI-DIMENSIONAL KNAPSACK POLYTOPES. IN: PROCEEDINGS OF THE CONFERENCE ON MATHEMATICAL PROGRAMMING AND ITS APPLICATIONS, INDAM, ROME 1974 (ITALIAN), 11-34. SYMPOSIA MATHEMATICA 19, LONDON: ACADEMIC PRESS. (1976).

BALAG 1976 -2 BALAS, E.
NOTE ON DUALITY IN DISJUNCTIVE PROGRAMMING, A. MANAGEMENT SCIENCES RESEARCH REPORT NO.398. PITTSBURGH: CARNEGIE-MELLON UNIV., GRADUATE SCHOOL OF INDUSTRIAL ADMINISTRATION, 8 P. (1976).

BALAG 1977 -1 BALAS, E.
DISJUNCTIVE PROGRAMMING. VANCOUVER (CONFERENCE ON DISCRETE OPTIMIZATION, AUGUST 1977) 66 P. (1977).

BALAG 1977 -2 BALAS, E.
SET COVERING WITH CUTTING PLANES FROM CONDITIONAL BOUNDS (REVISED VERSION). MANAGEMENT SCIENCES RESEARCH REPORT NO. 399. PITTSBURGH: CARNEGIE-MELLON UNIV., GRADUATE SCHOOL OF INDUSTRIAL ADMINISTRATION, 48 P. (1977).

BALAG 1977 -3 BALAS, E.
SOME VALID INEQUALITIES FOR THE SET PARTITIONING PROBLEM. IN: P. L. HAMMER, E. L. JOHNSON, B. H. KORTE, G. L. NEMHAUSER (EDS.): STUDIES IN INTEGER PROGRAMMING, 13-48. ANNALS OF DISCRETE MATHEMATICS 1, AMSTERDAM: NORTH-HOLLAND. (1977).

BALAH 1974 -1 BALAS, E.; GERRITSEN, R.; PADBERG, M. W.
ALL-BINARY COLUMN GENERATING ALGORITHM FOR SET PARTITIONING, AN. PAPER PRESENTED AT THE 45TH JOINT NATIONAL MEETING OF ORSA AND TIMS, BOSTON 1974. (1974).

BALAI 1969 -1 BALAS, E.; JEROSLOW, R.
ON THE STRUCTURE OF THE UNIT HYPERCUBE. MANAGEMENT SCIENCE RESEARCH REPORT 198, PITTSBURGH: CARNEGIE MELLON UNIV. (1969).

BALAJ 1975 -1 BALAS, E.; PADBERG, M. W.
SET PARTITIONING: A SURVEY. IN: B. ROY (ED.): COMBINATORIAL PROGRAMMING: METHODS AND APPLICATIONS. PROCEEDINGS OF THE NATO ADVANCED STUDY INST., VERSAILLES 1974, 205-258. DORDRECHT: REIDEL. (1975).

BALAJ 1976 -1 BALAS, E.; PADBERG, M. W.
SET PARTITIONING: A SURVEY. SIAM REVIEW 18, 710-760. (1976).

BALAK 1977 -1 BALAS, E.; SAMUELSSON, H.
NODE COVERING ALGORITHM, A. NAVAL RESEARCH LOGISTICS QUARTERLY 24, 213-234. (1977).

BALAL 1975 -1 BALAS, E.; ZEMEL, E.
ALL THE FACETS OF ZERO-ONE PROGRAMMING POLYTOPES WITH POSITIVE COEFFICIENTS. PITTSBURGH: CARNEGIE-MELLON UNIV., GRADUATE SCHOOL OF INDUSTRIAL ADMINISTRATION, 21 P. (1975). (MANAGEMENT SCIENCES RESEARCH REPORT NO. 374)

BALAL 1976 -1 BALAS, E.; ZEMEL, E.
CRITICAL CUTSETS OF GRAPHS AND CANONICAL FACETS OF SET PACKING POLYTOPES. MANAGEMENT SCIENCES RESEARCH REPORT NO. 385. PITTSBURGH: CARNEGIE-MELLON UNIV., GRADUATE SCHOOL OF INDUSTRIAL ADMINISTRATION, 10 P. (1976).

BALAL 1977 -1 BALAS, E.; ZEMEL, E.
GRAPH SUBSTITUTION AND SET PACKING POLYTOPES. NETWORKS 7, 267-284. (1977).

BALAM 1975 -1 BALAS, E.; ZOLTNERS, A.
INTERSECTION CUTS FROM OUTER POLARS OF TRUNCATED CUBES. NAVAL RESEARCH LOGISTICS QUARTERLY 22, 477-496. (1975).

BALE 1970 -1 BALERIAUX, H.; JAMOULLE, E.; DOULLIEZ, P.; KELECOM, J. VAN
RECHERCHE DE LA POLITIQUE OPTIMALE D'INVESTISSEMENT D'UN RESEAU ELECTRIQUE EN EXPANSION PAR UNE METHODE DE DECISION SEQUENTIELLE. C.I.G.R.E., RAPPORT NO. 32-08. (1970).

BALI 1964 -1 BALINSKI, M. L.
ON FINDING INTEGER SOLUTIONS TO LINEAR PROGRAMS. MATEMATICA. (1964).

BALIA 1977 -1 BALINSKI, M. L.; YOUNG, H. P.
ON HUNTINGTON METHODS OF APPORTIONMENT. SIAM JOURNAL ON APPLIED MATHEMATICS 33, 607-618. (1977).

BALL 1977 -1 BALL, B. C.; WEBSTER, D. B.
OPTIMAL SCHEDULING FOR EVEN-NUMBERED TEAM ATHLETIC CONFERENCES. AIIE TRANSACTIONS 9, 161-169. (1977).

BALLA 1977 -1 BALL, M. O.; NEMHAUSER, G. L.
MATROIDS AND A RELIABILITY ANALYSIS PROBLEM. CORE DISCUSSION PAPER 7732. HEVERLEE: CENTER FOR OPERATIONS RESEARCH AND ECONOMETRICS, 35 P. (1977).

BALLB 1970 -1 BALLOU, R. H.
COMPUTATIONAL LIMITATIONS OF DYNAMIC PROGRAMMING FOR WAREHOUSE LOCATION: A COMMENT. JOURNAL OF MARKETING RESEARCH 7, 263-264. (1970).

BAND 1975 -1 BANDLER, J. W.; CHEN, J. H. K.; DALSGAARD, P.; LIU, P. C.
TOLOPT - A PROGRAM FOR OPTIMAL, CONTINUOUS OR DISCRETE, DESIGN CENTERING AND TOLERANCING, PART 1 - USER'S GUIDE, PART 2 - FORTRAN LISTING. HAMILTON: MCMASTER UNIV., PART 1: 47 P., PART 2: 27 P. (1975).

BANDA 1975 -1 BANDLER, J. W.; LIU, P. C.; CHEN, J. H. K.
WORST CASE NETWORK TOLERANCE OPTIMIZATION. IEEE TRANSACTIONS ON MICROWAVE THEORY AND TECHNIQUES 23, 630-641. (1975).

BANE 1965 -1 BANERJEE, B. P.
SINGLE FACILITY SEQUENCING WITH RANDOM EXECUTION TIMES. OPERATIONS RESEARCH 13, 358-364. (1965).

BANS 1976 -1 BANSAL, S. P.
GLOBAL LOWER BOUND IN THE FLOW-SHOP PROBLEM. OPSEARCH 13, 115-118. (1976).

BANS 1977 -1 BANSAL, S. P.
MINIMIZING THE SUM OF COMPLETION TIMES OF N JOBS OVER M MACHINES IN A FLOWSHOP - A BRANCH AND BOUND APPROACH. AIIE TRANSACTIONS 9, 306-311. (1977).

BAR 1975 -1 BAR, G.
UEBER DEN EINSATZ LOGISCHER MITTEL ZUR LINEARISIERUNG DISKRETER OPTIMIERUNGSPROBLEME. ELEKTRONISCHE INFORMATIONSVERARBEITUNG UND KYBERNETIK 11, 631-634. (1975).

BARA 1972 -1 BARANKIEWICZ, M. N.
CERTAIN GENERALIZATIONS OF THE MULTISTAGE TRANSPORTATION PROBLEM (POLISH). ZESZYTY NAUKOWE SZKOLY GLOWNEJ PLANOWANIA I STATYSTYKI, NO. 86, 105-119. (1972).

BARB 1970 -1 BARBER, G. M.
DYNAMIC LOCATIONAL ANALYSIS: THE FIRM AND ITS OPERATING POLICY. DISCUSSION PAPER 9. TORONTO: UNIV. OF TORONTO, DEPT. OF GEOGRAPHY. (1970).

BARB 1975 -1 BARBER, G. M.
MATHEMATICAL PROGRAMMING APPROACH TO A NETWORK DEVELOPMENT PROGRAM, A. ECONOMIC GEOGRAPHY 51, 128-141. (1975).

BARL 1978 -1 BARLOW, J. F.
ALGORITHM FOR THE SOLUTION OF THE POSTMAN'S PROBLEM, AN. COMPUTER JOURNAL 21, 86-89. (1978).

BARN 1977 -1 BARNES, J. W.; BRENNAN, J. J.
IMPROVED ALGORITHM FOR SCHEDULING JOBS ON IDENTICAL MACHINES, AN. AIIE TRANSACTIONS 9, 25-31. (1977).

BARR 1977 -1 BARR, R. S.; GLOVER, F.; KLINGMAN, D.
ALTERNATING BASIS ALGORITHM FOR ASSIGNMENT PROBLEMS, THE. TECHNICAL REPORT IEOR 77002. DALLAS: SOUTHERN METHODIST UNIV., DEPT. OF INDUSTRIAL ENGINEERING AND OPERATIONS RESEARCH, 26 P. (1977).

BARR 1977 -2 BARR, R. S.; GLOVER, F.; KLINGMAN, D.
ALTERNATING BASIS ALGORITHM FOR ASSIGNMENT PROBLEMS, THE. RESEARCH REPORT CCS 263. AUSTIN: UNIV. OF TEXAS, CENTER FOR CYBERNETIC STUDIES, 21 P. (1977).

BARR 1977 -3 BARR, R. S.; GLOVER, F.; KLINGMAN, D.
GENERALIZED ALTERNATING PATH ALGORITHM FOR TRANSPORTATION PROBLEMS, THE. RESEARCH REPORT CCS 282. AUSTIN: UNIV. OF TEXAS, CENTER FOR CYBERNETIC STUDIES, 19 P. (1977).

BARR 1977 -4 BARR, R. S.; GLOVER, F.; KLINGMAN, D.
NEW ALTERNATING BASIS ALGORITHM FOR SEMI-ASSIGNMENT NETWORKS, A. RESEARCH REPORT CCS 264. AUSTIN: UNIV. OF TEXAS, CENTER FOR CYBERNETIC STUDIES, 29 P. (1977).

BARRA 1975 -1 BARR, R. S.; ROSS, G. T.
LINKED LIST DATA STRUCTURE FOR A BINARY KNAPSACK ALGORITHM, A. RESEARCH REPORT CCS 232. AUSTIN: UNIV. OF TEXAS, CENTER FOR CYBERNETIC STUDIES, 26 P. (1975)

BARRB 1976 -1 BARRER, J.; CHANDRASEKARAN, R.
ADJACENT PAIRWISE INTERCHANGE (API) ALGORITHM: A CLARIFICATION OF THE LITERATURE, THE. MIAMI (50. ORSA-MEETING + TIMS, NOVEMBER 1976) 15 P. (1976).

BARRC 1977 -1 BARRY, D.; ROBINSON, D. F.
OPTIMAL BATCH SIZES FOR URGENT DISPATCH. UTILITAS MATHEMATICA 11, 347-356. (1977).

BARRD 1977 -1 BARRY, D. L.; ROBINSON, D. F.
HEURISTIC FOR SCHEDULING JOBS ON SEVERAL IDENTICAL MACHINES, A. NEW ZEALAND OPERATIONAL RESEARCH 5, 45-53. (1977).

BARS 1974 -1 BARSKIY, A. B.
METHOD OF OPTIMAL DISTRIBUTION OF OPERATIONS IN A HOMOGENEOUS COMPUTATIONAL SYSTEM (RUSSIAN), A. IZVESTIJA AKADEMII NAUK SSSR, TEHNICESKAJA KIBERNETIKA, NO. 5, 133-138; TRANSLATED AS ENGINEERING CYBERNETICS 12, 1974, NO. 5, 110-115. (1974).

BART 1975 -1 BARTHES, J. P.
BRANCHING METHODS IN COMBINATORIAL OPTIMIZATION. IN: P. RINALDI (ED.): TOPICS IN COMBINATORIAL OPTIMIZATION, 117-143, WIEN, NEW YORK: SPRINGER. (1975).

BARTA 1977 -1 BARTHOLDI, J. J. III; ORLIN, J. B.; RATLIFF, H. D.
CIRCULAR 1'S AND CYCLIC STAFFING. RESEARCH REPORT 77-11. GAINESVILLE: UNIV. OF FLORIDA, INDUSTRIAL AND SYSTEMS ENGINEERING DEPT., 24 P. (1977).

BARTB 1977 -1 BARTHOLDI, J. J. III; RATLIFF, H. D.
FIELD GUIDE TO IDENTIFYING NETWORK FLOW AND MATCHING PROBLEMS, A. RESEARCH REPORT NO. 77-12. GAINESVILLE: UNIV. OF FLORIDA, INDUSTRIAL AND SYSTEMS ENGINEERING DEPT., 33 P. (1977).

BARTB 1977 -2 BARTHOLDI, J. J. III; RATLIFF, H. D.
UNNETWORKS, WITH APPLICATIONS TO IDLE TIME SCHEDULING. RESEARCH REPORT NO. 77-4. GAINESVILLE: UNIV. OF FLORIDA, INDUSTRIAL AND SYSTEMS ENGINEERING DEPT., 17 P. (1977).

BASC 1977 -1 BASCA, O. C.
ON ALGORITHMS OF MINIMUM PATHS IN GRAPHS. ECONOMIC COMPUTATION AND ECONONOMIC CYBERNETICS STUDIES AND RESEARCH, NO. 2, 91-101. (1977).

BATT 1977 -1 BATTAGLIA, F.
LOCATION OF CENTERS IN AN EUCLIDEAN SPACE (ITALIAN), THE. RICERCA OPERATIVA 7, NO. 2, 27-38. (1977).

BAUE 1973 -1 BAUERSFELD, G.; ESSMANN, CH.; LOEHLE, H.
ALGORITHMUS ZUR FESTSTELLUNG DER ISOMORPHIE VON ENDLICHEN, ZUSAMMENHAENGENDEN GRAPHEN. COMPUTING 11, 159-168. (1973).

BAUM 1977 -1 BAUM, S.; TROTTER, L. E. JR.
INTEGER ROUNDING AND POLYHEDRAL DECOMPOSITION FOR TOTALLY UNIMODULAR SYSTEMS. IN: R. HENN, B. KORTE, W. OETTLI (EDS.): OPTIMIZATION AND OPERATIONS RESEARCH, PROCEEDINGS, BONN 1977, 25-23. LECTURE NOTES IN ECONOMICS AND MATHEMATICAL SYSTEMS 157. BERLIN, HEIDELBERG, NEW YORK: SPRINGER. (1977).

BAUMA 1975 -1 BAUMGARTNER, J. E.; DAYKIN, D. E.; GUTMAN, I.; KLEITMAN, D. J. ET AL.
PROBLEMS. IN: A. HAYNAL, R. RADO, V. T. SOS (EDS.): INFINITE AND FINITE SETS, COLLOQ. KESZTHELY 1973, VOL. 3, 1547-1555. COLLOQ. MATH. SOC. JANOS BOLYAI 10, AMSTERDAM: NORTH-HOLLAND. (1975).

BAUT 1973 -1 BAUTZ, M.; BIESS, G.
ALGORITHMUS ZUR ERMITTLUNG ALLER ELEMENTARKREISE EINES GERICHTETEN GRAPHEN, EIN. WISSENSCHAFTLICHE ZEITSCHRIFT DER TECHNISCHEN HOCHSCHULE FUER CHEMIE LEUNA-MERSEBURG 15, 146-147. (1973).

BAZA 1975 -1 BAZARAA, M. S.
COMPUTERIZED LAYOUT DESIGN: A BRANCH AND BOUND APPROACH. AIIE TRANSACTIONS 7, 432-438. (1975).

BAZAA 1977 -1 BAZARAA, M. S.; ELSHAFEI, A. N.
ON THE USE OF FICTITIOUS BOUNDS IN TREE SEARCH ALGORITHMS. MANAGEMENT SCIENCE 23, 904-908. (1977).

BAZAB 1977 -1 BAZARAA, M. S.; GOODE, J. J.
TRAVELING SALESMAN: A DUALITY APPROACH, THE. MATHEMATICAL PROGRAMMING 13, 221-237. (1977).

BAZAC 1977 -1 BAZARAA, M. S.; JARVIS, J. J.
LINEAR PROGRAMMING AND NETWORK FLOWS. NEW YORK, LONDON U.A.: WILEY, 565 P. (1977).

BEAL 1977 -1 BEALE, E. M. L.
BRANCH AND BOUND METHODS. VANCOUVER (CONFERENCE ON DISCRETE OPTIMIZATION, AUGUST 1977) 36 P. (1977).

BEAL 1977 -2 BEALE, E. M. L.
INTEGER PROGRAMMING. IN: THE STATE OF THE ART IN NUMERICAL ANALYSIS, PROCEEDINGS OF A CONFERENCE, HESLINGTON 1976, 409-448. LONDON: ACADEMIC PRESS. (1977).

BEAN 1976 -1 BEAN, D. R.
RECURSIVE EULER AND HAMILTONIAN PATHS. PROCEEDINGS OF THE AMERICAN MATHEMATICAL SOCIETY 55, 385-394. (1976).

BEANA 1970 -1 BEAN, D. W. T.
MINIMAL EDGE SETS MEETING ALL CIRCUITS OF A GRAPH. IN: R. C. MULLIN, K. B. REID, D. P. ROSELLE (EDS.): PROCEEDINGS OF THE LOUISIANA CONFERENCE ON COMBINATORICS, GRAPH THEORY, AND COMPUTING, BATON ROUGE 1970, 1-46. CONGRESSUS NUMERANTIUM 1, WINNIPEG: UTILITAS MATHEMATICA. (1970).

BEAR 1959 -1 BEARDWOOD, J.; HALTON, J. H.; HAMMERSLEY, J. M.
SHORTEST PATH THROUGH MANY POINTS, THE. PROCEEDINGS OF THE CAMBRIDGE PHILOSOPHICAL SOCIETY 55, 229-327. (1959).

****

BECK 1976 -1 BECK, L. L.; MATULA, D. W.
CLIQUE DETERMINATION IN LARGE DENSE GRAPHS. PAPER PRESENTED AT 7TH SOUTHEASTERN CONFERENCE ON COMBINATORICS, GRAPH THEORY, AND COMPUTING, BATON ROUGE 1976. (1976).

BECKA 1974 -1 BECKER, L.; YEH, W. W.-G.
OPTIMAL TIMING, SEQUENCING, AND SIZING OF MULTIPLE RESERVOIR SURFACE WATER SUPPLY FACILITIES. WATER RESOURCES RESEARCH 10, 57-62. (1974).

BECKA 1974 -2 BECKER, L.; YEH, W. W.-G.
TIMING AND SIZING OF COMPLEX WATER RESOURCE SYSTEMS. JOURNAL OF THE HYDRAULICS DIVISION, PROCEEDINGS OF THE AMERICAN SOCIETY OF CIVIL ENGINEERS 100, NO. HY10, 1457-1470. (1974).

BECKB 1956 -1 BECKMANN, M.; MAC GUIRE, C. B.; WINSTON, C. B.
STUDIES IN THE ECONOMICS OF TRANSPORTATION. NEW HAVEN: YALE UNIVERSITY PRESS. (1956).

BECKC 1976 -1 BECKMANN, M. J.; BURNS, L. D.
THEORY OF HOUSEHOLD AUTOMOBILE ALLOCATION DECISIONS, A. TRANSPORTATION SCIENCE 10, 72-84. (1976).

BEEN 1976 -1 BEENHAKKER, H. L.; NARAYANAN, V.
ALGORITHMS FOR SCHEDULING PROJECTS WITH LIMITED RESOURCES. THE ENGINEERING ECONOMIST 21, 119-140. (1976).

BEHA 1969 -1 BEHARA, D. N.
COMPUTATIONAL MATRIX ALGORITHM FOR SHORTEST ROUTE PROBLEM, A. CANADIAN MATHEMATICAL BULLETIN 12, 697. (1969).

BELE 1977 -1 BELETSKIJ, S. A.
ON ONE CLASS OF EXTREMAL COMBINATORY PROBLEM AND THEIR SOLUTION BY THE METHOD OF DICHOTOMY (RUSSIAN). KIBERNETIKA, NO. 3, 117-122. (1977).

BELK 1965 -1 BELKINA, M. V.; GROMOVA, A. P.; KYUNNAPU, E. P.; OSKALKOV, I. O.
OPTIMIZATION OF OPERATIONAL SEQUENCES. AUTOMATION AND REMOTE CONTROL 26, 2009-2024. (1965).

BELL 1973 -1 BELL, D. E.
RESOLUTION OF DUALITY GAPS IN DISCRETE OPTIMIZATION, THE. TECHNICAL REPORT 81. CAMBRIDGE: M.I.T., DEPT. OPERATIONS RESEARCH CENTER. (1973).

BELL 1977 -1 BELL, D. E.
SIMPLE ALGORITHM FOR INTEGER PROGRAMS USING GROUP CONSTRAINTS, A. OPERATIONAL RESEARCH QUARTERLY 28, 453-458. (1977).

BELLA 1975 -1 BELL, D. E.; SHAPIRO, J. F.
FINITELY CONVERGENT DUALITY THEORY FOR ZERO-ONE INTEGER PROGRAMMING, A. RESEARCH MEMORANDUM 75-33. LAXENBURG: INTERNAT. INST. FOR APPLIED SYSTEMS ANALYSIS, 27 P. (1975).

BELLA 1977 -1 BELL, D. E.; SHAPIRO, J. F.
CONVERGENT DUALITY THEORY FOR INTEGER PROGRAMMING, A. OPERATIONS RESEARCH 25, 419-434. (1977).

BELLB 1965 -1 BELL, E. J.
PRIMAL-DUAL DECOMPOSITION PROGRAMMING. DISSERTATION, REPORT ORC 65-92. BERKELEY: UNIV. OF CALIFORNIA, OPERATIONS RESEARCH CENTER. (1965).

BELLC 1965 -1 BELLMORE, M.
OPTIMAL ALLOCATION OF RESOURCES IN A COMMUNICATIONS NETWORK. DISSERTATION. BALTIMORE: JOHN HOPKINS UNIV. (1965).

BELLD 1970 -1 BELLMORE, M.; JENSEN, P.
IMPLICIT ENUMERATION SCHEME FOR PROPER CUT GENERATION, AN. TECHNOMETRICS 12, 775-788. (1970).

BELO 1977 -1 BELOUSOV, E. G.
INTRODUCTION TO CONVEX ANALYSIS AND INTEGER PROGRAMMING (RUSSIAN). MOSCOW: IZDAT MOSKOV. UNIV., 196 P. (1977).

BELOA 1974 -1 BELOV, I. S.; STOLIN, JA. N.
ALGORITHM FOR THE SINGLE-ROUTE SCHEDULING PROBLEM (RUSSIAN), AN. IN: MATHEMATICAL ECONOMICS AND FUNCTIONAL ANALYSIS (RUSSIAN), 248-257. MOSCOW: IZDAT. NAUKA. (1974).

BENI 1977 -1 BENICHOU, M.; GAUTHIER, J. M.; HENTGES, G.; RIBIERE, G.
EFFICIENT SOLUTION OF LARGE-SCALE LINEAR PROGRAMMING PROBLEMS - SOME ALGORITHMIC TECHNIQUES AND COMPUTATIONAL RESULTS, THE. MATHEMATICAL PROGRAMMING 13, 280-322. (1977).

BENIA 1976 -1 BENITO, F.; GROEFLIN, H.
OPTIMIERUNGSPROBLEME MIT NICHT POLYNOMIAL BEGRENZTEN ALGORITHMEN (NP-PROBLEME). IN: T. M. LIEBLING, M. ROESSLER (EDS.):KOMBINATORISCHE ENTSCHEIDUNGSPROBLEME: METHODEN UND ANWENDUNG, 81-121. LECTURE NOTES IN ECONOMICS AND MATHEMATICAL SYSTEMS 153, BERLIN: SPRINGER. (1976).

BENN 1976 -1 BENNETT, G. R.; BYRD, J. JR.
TRAINABLE HEURISTIC PROCEDURE FOR THE ASSEMBLY LINE BALANCING PROBLEM, A. AIIE TRANSACTIONS 8, 195-201. (1976).

BENNA 1972 -1 BENNETT, J. M.; EDWARDS, J.
EXPERIENCE WITH A METHOD FOR HANDLING LARGE STRUCTURAL LINEAR PROGRAMMING PROBLEMS WITH INTEGER RESTRICTIONS. IN: R. P. ANDERSSEN, L. P. JENNINGS, D. M. RYAN (EDS.): OPTIMIZATION, PROCEEDINGS OF A SEMINAR ON OPTIMIZATION, CANBERRA 1971, 129-151. ST. LUCIA: UNIV. OF QUEENSLAND PRESS. (1972).

BENNB 1976 -1 BENNINGER, H.; LIEBLING, T. M.
OPTIMIERUNGSPROBLEME MIT POLYNOMIAL BEGRENZTEN ALGORITHMEN. IN: T. M. LIEBLING, M. ROESSLER (EDS.): KOMBINATORISCHE ENTSCHEIDUNGSPROBLEME: METHODEN UND ANWENDUNG. 14-55. LECTURE NOTES IN ECONOMICS AND MATHEMATICAL SYSTEMS 153, BERLIN: SPRINGER. (1976).

BERE 1974 -1 BERESNEV, V. L.
IMPLICIT ENUMERATION ALGORITHM FOR AN ASSIGNMENT AND STANDARDIZATION PROBLEM (RUSSIAN). UPRAVLJAEME SISTEMY, NO 12, 24-34. (1974).

BERG 1973 -1 BERGE, C.
INTRODUCTION A LA THEORIE DES HYPERGRAPHES. MONTREAL: PRESSES DE L'UNIVERSITE DE MONTREAL, 110 P. (1973).

BERG 1973 -2 BERGE, C.
NOTES SUR LES BONNES COLORATIONS D'UN HYPERGRAPHE. CAHIERS DU CENTRE D'ETUDES DE RECHERCHE OPERATIONNELLE 15, 219-223. (1973).

BERG 1975 -1 BERGE, C.
NOMBRES DE COLORATION DE L'HYPERGRAPHE H-PARTI COMPLET. ANNALI DI MATEMATICA PURA ED APPLICATA 103, 3-9. (1975).

BERG 1975 -2 BERGE, C.
ON THE GOOD K-COLORINGS OF A HYPERGRAPH. IN: A. HAJNAL, R. RADO, V. T. SOS (EDS.): INFINITE AND FINITE SETS, COLLOQ. KESZTHELY 1973, VOL. 1, 159-163. COLLOQ. MATH. SOC. JANOS BOLYAI 10, AMSTERDAM: NORTH-HOLLAND. (1975).

BERG 1977 -1 BERGE, C.
THEORIE FRACTIONNAIRE DES GRAPHES. VANCOUVER (CONFERENCE ON DISCRETE OPTIMIZATION, AUGUST 1977) 85 P. (1977).

BERG 1978 -1 BERGE, C.
ALGORITHMS AND EXTREMAL PROBLEMS FOR EQUIPARTITE COLORINGS IN GRAPHS AND HYPERGRAPHS (ABSTRACT). IN: ALSPACH, B.;HELL, P.;MILLER, D. J.(EDS.): ALGORITHMIC ASPECTS OF COMBINATORICS. ANNALS OF DISCRETE MATHEMATICS 2, 149-150. (1978).

****

BERGA 1976 -1 BERGE, C.; HOFFMAN, A. J.
MULTICOLORATIONS DANS LES HYPERGRAPHES UNIMODULAIRES ET MATRICES DONT LES COEFFICIENTS SONT DES ENSEMBLES. IBM RESEARCH REPORT 27047. YORKTOWN HEIGHTS: IBM RESEARCH CENTER, 9 P. (1976).

BERGB 1977 -1 BERGE, C.; JOHNSON, E. L.
COLORING THE EDGES OF A HYPERGRAPH AND LINEAR PROGRAMMING TECHNIQUES. IN: P. L. HAMMER, E. L. JOHNSON, B. H. KORTE, G. L. NEMHAUSER (EDS.): STUDIES IN INTEGER PROGRAMMING, 65-78. ANNALS OF DISCRETE MATHEMATICS 1, AMSTERDAM: NORTH-HOLLAND. (1977).

BERGC 1966 -1 BERGENDAHL, G.
MODELS FOR CAPACITY PLANNING. DISSERTATION. LUND: UNIVERSITY OF LUND. (1966).

BERGC 1968 -1 BERGENDAHL, G.
CAPACITY PLANNING MODELS FOR SYSTEMS OF PLANT AND ROAD INVESTMENTS. THE SWEDISH JOURNAL OF ECONOMICS 70, 94-105. (1968).

BERGC 1969 -1 BERGENDAHL, G.
COMBINED LINEAR AND DYNAMIC PROGRAMMING MODEL FOR INTERDEPENDENT ROAD INVESTMENT PLANNING, A. TRANSPORTATION RESEARCH 3, 211-228. (1969).

BERGD 1974 -1 BERGENDORFF, H.
MODEL TO EVALUATE AND PLAN PROJECTS IN FORESTRY AND FOREST INDUSTRIES, A. WASHINGTON: INTERNATIONAL BANK FOR RECONSTRUCTION AND DEVELOPMENT, DEVELOPMENT RESEARCH CENTER. (1974).

BERGE 1975 -1 BERGMANN, D. R.
INTEGER PROGRAMMING FORMULATION FOR DERIVING MINIMUM DISPATCH INTERVALS ON A GUIDEWAY ACCOMODATING THROUGH AND LOCAL PUBLIC TRANSPORTATION SERVICES. TRANSPORTATION PLANNING AND TECHNOLOGY 3, 27-30. (1975).

BERL 1976 -1 BERLIN, G. N.; REVELLE, C. S.; ELZINGA, D. J.
DETERMINING AMBULANCE-HOSPITAL LOCATIONS FOR ON-SCENE AND HOSPITAL SERVICES. ENVIRONMENT AND PLANNING 8, SERIES A, 553-561. (1976).

BERM 1975 -1 BERMOND, J. C.
1-GRAPHES REGULIERS MINIMAUX DE GIRTH DONNE. CAHIERS DU CENTRE D'ETUDES DE RECHERCHE OPERATIONNELLE 17, 125-135. (1975).

BERM 1976 -1 BERMOND, J. C.
ON HAMILTONIAN WALKS. IN: C. ST. J. A. NASH-WILLIAMS (ED.): PROCEEDINGS OF THE 5TH BRITISH COMBINATORIAL CONFERENCE, ABERDEEN 1975, 41-52. CONGRESSUS NUMERANTIUM 15, WINNIPEG: UTILITAS MATHEMATICA. (1976).

BERMA 1974 -1 BERMOND, J.-C.
NOMBRE CHROMATIQUE TOTAL DU GRAPHE R-PARTI COMPLET. JOURNAL OF THE LONDON MATHEMATICAL SOCIETY 9, 279-285. (1974).

BERMB 1976 -1 BERMOND, J.-C.; KODRATOFF, Y.
HEURISTIQUE POUR LE CALCUL DE L'INDICE DE TRANSITIVITE D'UN TOURNOI, UNE. REVUE FRANCAISE D'AUTOMATIQUE, D'INFORMATIQUE ET DE RECHERCHE OPERATIONNELLE 10, NO. 3, 83-92. (1976).

BERN 1958 -1 BERNERS-LEE, C. M.
INDUSTRIAL APPLICATION OF COMPUTERS, AN. THE ENGINEER, 5278. (1958).

BERR 1977 -1 BERRY, W. L.
AUTHOR'S REPLY TO GRAVES' NOTE ON THE PAPER "CRITICAL RATIO SCHEDULING: AN EXPERIMENTAL ANALYSIS". MANAGEMENT SCIENCE 23, 1360-1361. (1977).

BERRA 1976 -1 BERRY, W. L.; FINLAY, R. A.
CRITICAL RATIO SCHEDULING WITH QUEUE WAITING TIME INFORMATION: AN EXPERIMENTAL ANALYSIS. AIIE TRANSACTIONS 8, 161-168. (1976).

BERRB 1975 -1 BERRY, W. L.; RAO, V.
CRITICAL RATIO SCHEDULING: AN EXPERIMENTAL ANALYSIS. MANAGEMENT SCIENCE 22, 192-201. (1975).

BERS 1976 -1 BERSTEL, J.
COMPLEXITE DE QUELQUES ALGORITHMES DE THEORIE DES GRAPHES. IN: ANAL. APPL. INF. JOURN. COMB. INF., BORDEAUX 1975, 31-39. BORDEAUX: UNIV. BORDEAUX I. (1976).

BERZ 1971 -1 BERZTISS, A. T.
DATA STRUCTURES, THEORY AND PRACTICE. NEW YORK: ACADEMIC PRESS, 442 P. (1971).

BERZ 1973 -1 BERZTISS, A. T.
BACKTRACK PROCEDURE FOR ISOMORPHISM OF DIRECTED GRAPHS, A. JOURNAL OF THE ASSOCIATION FOR COMPUTING MACHINERY 20, 365-377. (1973).

BEST 1976 -1 BESTWICK, P. F.; HASTINGS, N. A. J.
NEW BOUND FOR MACHINE SCHEDULING, A. OPERATIONAL RESEARCH QUARTERLY 27, 479-488. (1976).

BESW 1977 -1 BESWICK, C. A.
ALLOCATING SELLING EFFORT VIA DYNAMIC PROGRAMMING. MANAGEMENT SCIENCE 23, 667-678. (1977).

BEUE 1975 -1 BEUERMANN-WEHMEYER, A.
PRIMAL-DUAL-BEZIEHUNGEN IN GRAPHENALGORITHMEN. BEITRAEGE ZUR DATENVERARBEITUNG UND UNTERNEHMENSFORSCHUNG, 12. MEISENHEIM: HAIN, 241 P. (1975).

BEYE 1970 -1 BEYER, K.
EINIGE VERFAHREN ZUR TERMINPLANUNG IN NETZPLAENEN MIT BESCHRAENKTER RESOURCENVERFUEGBARKEIT. WISSENSCHAFTLICHE ZEITSCHRIFT DER HOCHSCHULE FUER VERKEHRSWESEN FRIEDRICH LIST IN DRESDEN 17, 309-319. (1970).

BHAL 1970 -1 BHALLA, H. S.
MULTI-TIME PERIOD FACILITIES LOCATION: A HEURISTIC ALGORITHM WITH APPLICATIONS TO WASTE WATER TREATMENT SYSTEMS. DISSERTATION. AMHERST: UNIV. OF MASSACHUSETTS. (1970).

BHAT 1976 -1 BHATIA, H. L.; SWARUP, K.; PURI, M. C.
TIME-COST TRADE-OFF IN A TRANSPORTATION PROBLEM. OPSEARCH 13, 129-142. (1976).

BHAU 1974 -1 BHAUMIK, G.; JENSON, P.
COMPUTATIONALLY EFFICIENT ALGORITHM FOR THE NETWORK WITH GAINS PROBLEM, A. WORKING PAPER. AUSTIN: UNIV. OF TEXAS, DEPT. OF MECHANICAL ENGINEERING. (1974).

BHAUA 1974 -1 BHAUMIK, G.; STANTON, L. W.
APPLICATION OF STOPPING RULES THROUGH MONTE CARLO SIMULATION TO THE TRAVELING SALESMAN PROBLEM, AN. FULLERTON: CALIFORNIA STATE UNIV., SCHOOL OF BUSINESS ADMINISTRATION AND ECONOMICS, 7 P. (1974).

BICH 1972 -1 BICHARA, J.
FLUESSE IN NETZWERKEN. BERLIN: TECHNISCHE UNIVERSITAET BERLIN, FACHBEREICH INFORMATIK. (1972).

BIDE 1975 -1 BIDER, I. G.; BOL'SHAKOV, I. A.
NON-MONOTONIC STRATEGIES FOR AN ALLOCATION OF LIMITED RESOURCES (RUSSIAN). IZVESTIJA AKADEMII NAUK SSSR, TEHNICESKAJA KIBERNETIKA, NO. 5, 27-37. (1975).

BIES 1976 -1 BIESS, G.
GRAPHENTHEORIE. MATHEMATIK FUER INGENIEURE, NATURWISSENSCHAFTLER, OEKONOMEN UND LANDWIRTE 21/2. LEIPZIG:TEUBNER, 80 P. (1976).

BIESA 1977 -1 BIESS, G.; JANICKE, W.; MEISSLER, H. G.
ALGORITHMUS ZUR BESTIMMUNG ALLER ELEMENTARKREISE EINES GERICHTETEN GRAPHEN, EIN. WISSENSCHAFTLICHE ZEITSCHRIFT DER TECHNISCHEN HOCHSCHULE "CARL SCHORLEMMER" LEUNA-MERSEBURG 19, 103-111. (1977).

****

BILD 1970 -1 BILDE, O.
NONLINEAR AND DISCRETE PROGRAMMING. DISSERTATION 9, IMSOR. LYNGBY: THE TECHNICAL UNIV. OF DENMARK. (1970).

BILDA 1967 -1 BILDE, O.; KRARUP, J.
COMPUTATION OF THE OPTIMAL LOCATION OF PRODUCTION DATES (DANISH). IMSOR RESEARCH REPORT. LYNGBY: THE TECHNICAL UNIV. OF DENMARK. (1967).

BILDA 1977 -1 BILDE, O.; KRARUP, J.
SHARP LOWER BOUNDS AND EFFICIENT ALGORITHMS FOR THE SIMPLE PLANT LOCATION PROBLEM. IN: P. L. HAMMER, E. L. JOHNSON, B. H. KORTE, G. L. NEMHAUSER (EDS.): STUDIES IN INTEGER PROGRAMMING, 79-98. ANNALS OF DISCRETE MATHEMATICS 1, AMSTERDAM: NORTH-HOLLAND. (1977).

BIND 1961 -1 BINDSCHEDLER, A. E.; MOORE, J. M.
OPTIMAL LOCATION OF NEW MACHINES IN EXISTING PLANT LAYOUTS. JOURNAL OF INDUSTRIAL ENGINEERING 12, 41. (1961).

BION 1973 -1 BIONDI, E.
OPTIMUM DELIVERY PROBLEMS. CONTROL AND CYBERNETICS 2, NO. 3-4, 5-12. (1973).

BIONA 1969 -1 BIONDI, E.; DIVIETI, L.; ROVEDA, C.; SCHMID, R.
OPTIMAL SOLUTION FOR A CLASS OF ASSIGNMENT AND TRANSPORTATION PROBLEMS. IN: L. A. ZADEH, L. W. NEUSTADT, A. V. BALAKRISHNAN (EDS.): COMPUTING METHODS IN OPTIMIZATION PROBLEMS - 2, 35-46. NEW YORK: ACADEMIC PRESS. (1969).

BIRD 1976 -1 BIRD, C. G.
ON COST ALLOCATION FOR A SPANNING TREE: A GAME THEORETIC APPROACH. NETWORKS 6, 335-350. (1976).

BIRDA 1976 -1 BIRD, R.
PROGRAMS AND MACHINES. AN INTRODUCTION TO THE THEORY OF COMPUTING. WILEY SERIES IN COMPUTING. LONDON, NEW YORK U. A.: WILEY, 214 P. (1976).

BITR 1977 -1 BITRAN, G. R.
LINEAR MULTIPLE OBJECTIVE PROGRAMS WITH ZERO-ONE VARIABLES. MATHEMATICAL PROGRAMMING 13, 121-139. (1977).

BIXB 1975 -1 BIXBY, R. E.
MINIMUM NUMBER OF EDGES AND VERTICES IN A GRAPH WITH EDGE CONNECTIVITY N AND M N-BONDS, THE. NETWORKS 5, 253-298. (1975).

BIXB 1977 -1 BIXBY, R. E.
KURATOWSKI'S AND WAGNER'S THEOREMS FOR MATROIDS. JOURNAL OF COMBINATORIAL THEORY 22, SERIES B, 31-53. (1977).

BIXB 1978 -1 BIXBY, R. E.
SOLUTION TO A MATROID PROBLEM OF KNUTH, THE. DISCRETE MATHEMATICS 21, 87-88. (1978).

BJOE 1977 -1 BJOERKLUND, B.; HOLMBERG, G.
VEHICLE ROUTING AND SCHEDULING FOR DELIVERING NEWSPAPERS IN THE STOCKHOLM METROPOLITAN AREA. IN: ADVANCES IN OPERATIONS RESEARCH, PROCEEDINGS OF EURO II, 2ND EUROPEAN CONGRESS IN OPERATIONS RESEARCH, STOCKHOLM 1976, 45-48. (1977).

BLAI 1976 -1 BLAIR, C.
TWO RULES FOR DEDUCING VALID INEQUALITIES FOR 0-1 PROBLEMS. SIAM JOURNAL ON APPLIED MATHEMATICS 31, 614-617. (1976).

BLAIA 1975 -1 BLAIR, C. E.
TOPICS IN INTEGER PROGRAMMING. DISSERTATION. PITTSBURGH: CARNEGIE-MELLON UNIVERSITY, 27 P. (1975).

BLAIB 1976 -1 BLAIR, C. E.; JEROSLOW, R. G.
TREELESS SEARCHES. PITTSBURGH: CARNEGIE-MELLON UNIV., 19 P. (1976).

BLAIB 1977 -1 BLAIR, C. E.; JEROSLOW, R. G.
VALUE FUNCTION OF A MIXED INTEGER PROGRAM: I, THE. DISCRETE MATHEMATICS 19, 121-138. (1977).

BLAN 1977 -1 BLAND, R. G.
COMBINATORIAL ABSTRACTION OF LINEAR PROGRAMMING, A. JOURNAL OF COMBINATORIAL THEORY 23, SERIES B, 33-57. (1977).

BLANA 1978 -1 BLAND, R. G.; VERGNAS, M. LAS
ORIENTABILITY OF MATROIDS. JOURNAL OF COMBINATORIAL THEORY 24, SERIES B, 94-123. (1978).

BLAS 1967 -1 BLASYUK, B. A.
OPTIMAL ASSIGNMENT OF PROCESSING OF ITEMS ON THREE SEQUENTIAL MACHINES (RUSSIAN). IZVESTIJA AKADEMII NAUK SSSR. TEHNICESKAJA KIBERNETIKA, 4, 79-84. (1967).

BLAZ 1976 -1 BLAZEWICZ, J.
DEADLINE SCHEDULING OF TASKS - A SURVEY. FOUNDATIONS OF CONTROL ENGINEERING 1, NO. 4, 203-216. (1976).

BLAZA 1976 -1 BLAZEWICZ, J.; CELLARY, W.
SOME COMPUTATIONAL PROBLEMS OF SCHEDULING DEPENDENT AND PREEMPTABLE TASKS TO MINIMIZE SCHEDULE LENGTH. FOUNDATIONS OF CONTROL ENGINEERING 1, NO. 2, 75-82. (1976).

BLAZB 1976 -1 BLAZEWICZ, J.; CELLARY, W.; SLOWINSKI, R.; WEGLARZ, J.
DETERMINISTIC PROBLEMS OF SCHEDULING TASKS ON PARALLEL PROCESSORS, I. SETS OF INDEPENDENT TASKS (POLISH). PODSTAWY STEROWANIA 6, 155-177. (1976).

BLAZB 1976 -2 BLAZEWICZ, J.; CELLARY, W.; SLOWINSKI, R.; WEGLARZ, J.
DETERMINISTIC PROBLEMS OF SCHEDULING TASKS ON PARALLEL PROCESSORS, II. SETS OF DEPENDENT TASKS (POLISH). PODSTAWY STEROWANIA 6, 297-319. (1976).

BLAZC 1976 -1 BLAZHKEVICH, B. I.; SEMOTYUK, V. N.
TREE SEARCH ALGORITHM IN A GRAPH (RUSSIAN). OTBOR I PEREDACA INFORMACII, NO. 48, 51-56. (1976).

BLEW 1977 -1 BLEWETT, W. J.; HU, T. C.
TREE DECOMPOSITION ALGORITHM FOR LARGE NETWORKS. NETWORKS 7, 289-296. (1977).

BLIN 1973 -1 BLIN, J. M.
FURTHER PROCEDURE FOR ORDERING AN INPUT-OUTPUT MATRIX: SOME EMPIRICAL EVIDENCE, A. ECONOMICS OF PLANNING 13, 121-130. (1973).

BLIN 1976 -1 BLIN, J. M.
LINEAR ASSIGNMENT FORMULATION OF THE MULTIATTRIBUTE DECISION PROBLEM, A. REVUE FRANCAISE D'AUTOMATIQUE, D'INFORMATIQUE, ET DE RECHERCHE OPERATIONNELLE 10, NO. 6, 21-32. (1976).

BLINA 1976 -1 BLIN, J.-M.
ASSIGNMENT MODELS IN VOTING THEORY. DISCUSSION PAPER NO. 237. EVANSTON: NORTHWESTERN UNIV., CENTER FOR MATHEMATICAL STUDIES IN ECONOMICS AND MANAGEMENT SCIENCE, 21 P. (1976).

BLUM 1967 -1 BLUM, M.
MACHINE INDEPENDENT THEORY OF THE COMPLEXITY OF RECURSIVE FUNCTIONS, A. JOURNAL OF THE ACM 14, 322-336. (1967).

BOCK 1957 -1 BOCK, F. C.
ALGORITHM FOR FINDING SHORTEST PATHS BETWEEN PAIRS OF POINTS IN A NETWORK. PHASE REPORT 1, PROJECT E 084. CHICAGO: ARMOUR RESEARCH FOUNDATION. (1957).

BOCKA 1975 -1 BOCKO, G. D.; KASIN, V. I.; MAKAROV, L. I.
ALGORITHMS FOR SUCCESSIVE ALLOCATION AND SHORTEST TRACING (RUSSIAN). VYCISLITEL'NYE SISTEMY, VYP. 64, 73-81 + 181. (1975).

BODI 1976 -1 BODIN, L. D.; ROSENFIELD, D.
ESTIMATION OF THE OPERATING COST OF MASS TRANSIT SYSTEMS. REPORT NO. WAHCUPS-TMTA-1-76. WASHINGTON: U.S. DEPT. OF TRANSPORTATION, 146 P. (1976).

BOEB 1976 -1 BOEBION, J.; PUN, L.
SERIES-PARALLEL MULTIPLE-CRITERIA MODEL FOR A SCHEDULING PROBLEM IN THE DRESS-MAKING INDUSTRY, A. IN: H. THIRIEZ, P. ZIONTS (EDS.): MULTIPLE CRITERIA DECISION MAKING, PROCEEDINGS OF A CONFERENCE, JOUY-EN-JOSAS 1975, 305-318. LECTURE NOTES IN ECONOMICS AND MATHEMATICAL SYSTEMS 130, BERLIN: SPRINGER. (1976).

BOER 1977 -1 BOERE, N, J,
AIR CANADA SAVES WITH AIRCRAFT MAINTENANCE SCHEDULING. INTERFACES 7, NO. 3, 1-13. (1977).

BOES 1976 -1 BOESCH, F.; BUTLER, D. E.
ON THE SYNTHESIS OF MINIMUM COST RELIABLE NETWORKS AND MULTIGRAPHS. MIAMI (50. ORSA-MEETING + TIMS, NOVEMBER 1976) 12 P. (1976).

BOESA 1976 -1 BOESCH, F.; HARARY, F.
LINE REMOVAL ALGORITHMS FOR GRAPHS AND THEIR DEGREE LISTS. SPECIAL ISSUE ON LARGE-SCALE NETWORKS AND SYSTEMS. IEEE TRANSACTIONS ON CIRCUITS AND SYSTEMS CAS-23, 778-782. (1976).

BOESB 1974 -1 BOESCH, F. T.; CHEN, S.; MAC HUGH, J. A. M.
ON COVERING THE POINTS OF A GRAPH WITH POINT DISJOINT PATHS. IN: R. A. BARI, F. HARARY (EDS.): GRAPHS AND COMBINATORICS, PROCEEDINGS OF THE CAPITAL CONFERENCE ON GRAPH THEORY AND COMBINATORICS, WASHINGTON 1973, 201-212. LECTURE NOTES IN MATHEMATICS 406, BERLIN: SPRINGER. (1974).

BOESC 1977 -1 BOESCH, F. T.; GIMPEL, J. F.
COVERING THE POINTS OF A DIGRAPH WITH POINT-DISJOINT PATHS AND ITS APPLICATION TO CODE OPTIMIZATION. JOURNAL OF THE ASSOCIATION FOR COMPUTING MACHINERY 24, 192-198. (1977).

BOET 1974 -1 BOETTCHER, H.; DEHNERT, G.; DOMSCHKE, W.
VERFAHREN ZUR BESTIMMUNG OPTIMALER STANDORTE FUER REGIONALE SYSTEME. DISCUSSION PAPER, KARLSRUHE: UNIV. KARLSRUHE, INSTITUT FUER WIRTSCHAFTSTHEORIE UND OPERATIONS RESEARCH. (1974).

BOGA 1976 -1 BOGARDI, J. J.; DUCKSTEIN, L.; PLATE, E. J.
SCHEDULING AND SEQUENCING THE CONSTRUCTION OF FLOOD CONTROL RESERVOIRS. MIAMI (50. ORSA-MEETING + TIMS, NOVEMBER 1976) 35 P. (1976).

BOHE 1977 -1 BOHEME, G.
ANWENDUNGSORIENTIERTE MATHEMATIK. BD. 4: AKTUELLE ANWENDUNGEN DER MATHEMATIK. BERLIN, HEIDELBERG, NEW YORK: SPRINGER, 258 P. (1977).

BOLK 1976 -1 BOLKER, E. D.
SIMPLICIAL GEOMETRY AND TRANSPORTATION POLYTOPES. TRANSACTIONS OF THE AMERICAN MATHEMATICAL SOCIETY 217, 121-142. (1976).

BOLL 1977 -1 BOLLOBAS, B.
EXTREMAL PROBLEMS IN GRAPH THEORY. JOURNAL OF GRAPH THEORY 1, 117-124. (1977).

BOLLA 1976 -1 BOLLOBAS, B.; DAYKIN, D. E.; ERDOES, P.
SETS OF INDEPENDENT EDGES OF A HYPERGRAPH. QUARTERLY JOURNAL OF MATHEMATICS 27, NO. 105, 25-32. (1976).

BOLLB 1976 -1 BOLLOBAS, B.; ELDRIDGE, S. E.
MAXIMAL MATCHINGS IN GRAPHS WITH GIVEN MINIMAL AND MAXIMAL DEGREES. MATHEMATICAL PROCEEDINGS OF THE CAMBRIDGE PHILOSOPHICAL SOCIETY 79, 221-234. (1976).

BOLLC 1975 -1 BOLLOBAS, B.; ERDOES, P.
EXTREMAL PROBLEM OF GRAPHS WITH DIAMETER 2, AN. MATHEMATICS MAGAZINE 48, 281-283. (1975).

BOLLC 1976 -1 BOLLOBAS, B.; ERDOES, P.
ALTERNATING HAMILTONIAN CYCLES. ISRAEL JOURNAL OF MATHEMATICS 23, 126-131. (1976).

BOLLC 1976 -2 BOLLOBAS, B.; ERDOES, P.
CLIQUES IN RANDOM GRAPHS. MATHEMATICAL PROCEEDINGS OF THE CAMBRIDGE PHILOSOPHICAL SOCIETY 80, 419-427. (1976).

BOLLD 1974 -1 BOLLOBAS, B.; ERDOES, P.; STRAUS, E. G.
COMPLETE SUBGRAPHS OF CHROMATIC GRAPHS AND HYPERGRAPHS. UTILITAS MATHEMATICA 6, 343-347. (1974).

BONA 1976 -1 BONA, B.; INAUDI, D.; MAURO, V.
"FUZZY" APPROACH TO A PROBLEM IN RESIDENTIAL LOCATION (ITALIAN), A. RICERCA OPERATIVA 6, NO. 4, 3-14. (1976).

BOND 1978 -1 BONDY, J. A.
EXTREMAL GRAPHS WITH PRESCRIBED COVERING NUMBER. JOURNAL OF COMBINATORIAL THEORY 24, SERIES B, 51-52. (1978).

BONDA 1976 -1 BONDY, J. A.; CHVATAL, V.
METHOD IN GRAPH THEORY, A. DISCRETE MATHEMATICS 15, 111-135. (1976).

BONN 1974 -1 BONNET, R.; ERDOES, P.
CHROMATIC INDEX OF AN INFINITE COMPLETE HYPERGRAPH: A PARTITION THEOREM, THE. IN: C. BERGE, D. RAY-CHAUDHURI (EDS.): HYPERGRAPH SEMINAR, PROCEEDINGS OF THE FIRST WORKING SEMINAR, COLUMBUS 1972, 54-60. LECTURE NOTES IN MATHEMATICS 411, BERLIN: SPRINGER. (1974).

BONNA 1976 -1 BONNEY, M. C.; GUNDRY, S. W.
SOLUTIONS TO THE CONSTRAINED FLOWSHOP SEQUENCING PROBLEM. OPERATIONAL RESEARCH QUARTERLY 27, 869-884. (1976).

BOOK 1974 -1 BOOK, R. V.
COMPARING TIME AND SPACE. IN: PROCEEDINGS OF THE EIGHTH ANNUAL PRINCETON CONFERENCE ON INFORMATION SCIENCES AND SYSTEMS, 163-167. (1974).

BOOKA 1977 -1 BOOKBINDER, J. H.; SETHI, S. P.
DYNAMIC TRANSPORTATION PROBLEM: A SURVEY, THE. WORKING PAPER 77-12. TORONTO: UNIV. OF TORONTO, FACULTY OF MANAGEMENT STUDIES, 36 P. (1977).

BOOT 1976 -1 BOOTH, K. S.
PROBLEMS POLYNOMIALLY EQUIVALENT TO GRAPH ISOMORPHISM. IN PROCEEDINGS OF THE 7TH SYMPOSIUM ON NEW DIRECTIONS AND RECENT RESULTS IN ALGORITHMS AND COMPLEXITY, PITTSBURGH: CARNEGIE-MELLON UNIV. (1976).

BOOTA 1975 -1 BOOTH, K. S.; LUEKER, G. S.
LINEAR ALGORITHMS TO RECOGNIZE INTERVAL GRAPHS AND TEST FOR CONSECUTIVE ONES PROPERTY. IN: PROCEEDINGS OF THE 7TH SYMPOSIUM ON NEW DIRECTIONS AND RECENT RESULTS IN ALGORITHMS AND COMPLEXITY, PITTSBURGH: CARNEGIE-MELLON UNIVERSITY. (1975).

BOOTA 1976 -1 BOOTH, K. S.; LUEKER, G. S.
TESTING FOR THE CONSECUTIVE ONES PROPERTY, INTERVAL GRAPHS, AND GRAPH PLANARITY USING PQ-TREE ALGORITHMS. JOURNAL OF COMPUTER AND SYSTEM SCIENCES 13, 335-379. (1976).

BORI 1975 -1 BORIES, F.
ETUDE DU NOMBRE ACHROMATIQUE DE CERTAINS GRAPHES. CAHIERS DU CENTRE D'ETUDES DE RECHERCHE OPERATIONNELLE 17, 155-171. (1975).

BORI 1975 -2 BORIES, F.
SUR QUELQUES PROBLEMES DE COLORATIONS COMPLETES DE SOMMETS ET D'ARETES DE GRAPHES ET D'HYPERGRAPHES. DISSERTATION. PARIS: UNIV. PARIS VI, 80 P. (1975).

****

BORIA 1976 -1 BORIES, F.; JOLIVET, J.-L.
ACHROMATIC NUMBER OF A HYPERGRAPH, THE. IN: C. ST. J. A. NASH-WILLIAMS (ED.): PROCEEDINGS OF THE 5TH BRITISH COMBINATORIAL CONFERENCE, ABERDEEN 1975, 91-100. CONGRESSUS NUMERANTIUM 15, WINNIPEG: UTILITAS MATHEMATICA. (1976).

BORIB 1966 -1 BORILLE, M.
REMARKS ON THE DETERMINATION OF THE OPTIMAL ALGORITHM TO FIND THE SHORTEST PATH IN A CIRCUITLESS GRAPH. IN: E. R. CAIANEIELLO (ED.): AUTOMATA THEORY. NEW YORK: ACADEMIC PRESS. (1966).

BORO 1973 -1 BORODIN, A. B.
COMPUTATIONAL COMPLEXITY: THEORY AND PRACTICE. IN: A. V. AHO (ED.): CURRENTS IN THE THEORY OF COMPUTING, 35-89. ENGLEWOOD CLIFFS: PRENTICE-HALL. (1973).

BORO 1973 -2 BORODIN, A. B.
ON THE NUMBER OF ARITHMETICS REQUIRED TO COMPUTE CERTAIN FUNCTIONS. IN: COMPLEXITY OF SEQUENTIAL AND PARALLEL NUMERICAL ALGORITHMS, 149-180. NEW YORK: ACADEMIC PRESS. (1973).

BOROA 1972 -1 BORODIN, A. B.; MUNRO, I.
NOTES ON EFFICIENT AND OPTIMAL ALGORITHMS. UNPUBLISHED MANUSCRIPT. (1972).

BOROB 1977 -1 BORODIN, O. V.; KOSTOCHKA, A. V.
ON AN UPPER BOUND OF A GRAPH'S CHROMATIC NUMBER, DEPENDING ON THE GRAPH'S DEGREE AND DENSITY. JOURNAL OF COMBINATORIAL THEORY 23, SERIES B, 247-250. (1977).

BOROC 1974 -1 BORODIN, O. V.; MEL'NIKOV, L. S.
CHROMATIC NUMBER OF A PSEUDOSURFACE (RUSSIAN), THE. DISKRETNYI ANALIZ, VYP. 24, ISSLEDOVANIJA STRUKTURY I FUNKCIONIROVANIJA UPRAVLJAJUSCIH SISTEM, 8-20 + 95. (1974).

BOROD 1976 -1 BOROVITS, I.; EIN-DOR, P.
TWO ISSUES IN THE EVALUATION OF JOB-SHOP SCHEDULES. OMEGA 4, 313-320. (1976).

BOROE 1972 -1 BOROWIECKI, M.
HAMILTONIAN CYCLES IN CONJUNCTION OF TWO GRAPHS. PRACE NAUKOWE INSTYTUTU MATEMATYKI I FIZYKI TEORETYCZNEJ POLITECHNIKI WROCLAWSKIEJ, SERIA STUDIA I MATERIALY NO. 6, ZAGADNIENIA KOMBINATORYCZNE 19-26. (1972).

BORU 1926 -1 BORUVKA, O.
ON A MINIMAL PROBLEM. PRACE MORAVSKE PRIDOVEDECKE SPDECNOSTI 3. (1926).

BOSA 1976 -1 BOSAK, J.; NESETRIL, J.
COMPLETE AND PSEUDOCOMPLETE COLOURINGS OF A GRAPH. MATHEMATICA SLOVACA 26, NO. 3, 171-184. (1976).

BOSE 1973 -1 BOSE, R. C.
CHARACTERIZATION PROBLEMS OF COMBINATORIAL GRAPH THEORY. IN: J. N. SRIVASTAVA, F. HARARY, C. R. RAO ET AL. (EDS.): A SURVEY OF COMBINATORIAL THEORY, 31-51. AMSTERDAM, NORTH-HOLLAND. (1973).

BOTT 1974 -1 BOTT, H.
ANWENDUNGSMOEGLICHKEITEN DER PSEUDO-BOOLSCHEN OPTIMIERUNG AUF PROBLEME DER FERTIGUNSWIRTSCHAFT UND IHRE RECHENTECHNISCHE REALISIERUNG. DIPLOMARBEIT, KARLSRUHE: UNIV. KARLSRUHE, INSTITUT FUER FERTIGUNGSWIRTSCHAFT UND ARBEITSWISSENSCHAFTEN. (1974).

BOUR 1969 -1 BOURGOIN, M. H.; HEURGON, E. M. J.
STUDY AND COMPARISON OF ALGORITHMS OF THE SHORTEST PATH THROUGH PLANNED EXPERIMENTS. H. J. M. LOMBAERS (ED.): PROJECT PLANNING BY NETWORK ANALYSIS, 106-118. AMSTERDAM: NORTH-HOLLAND PUBLISHING CO. (1969).

BOURA 1970 -1 BOURKOV, V. N.
RESOLUTION DE QUELQUES PROBLEMES D'OPTIMISATION. METRA 9, 129-139 + 166-167 + 173 + 179 + 185. (1970).

BOWM 1974 -1 BOWMAN, V. J.; STARR, J. H.
TWO-COMPARABLE PRIME IMPLICANTS AND CANONICAL SWITCHING FUNCTIONS. MANAGEMENT SCIENCES RESEARCH REPORT 328. PITTSBURGH: CARNEGIE-MELLON UNIVERSITY, GRADUATE SCHOOL OF INDUSTRIAL ADMINISTRATION. (1974).

BOWMA 1977 -1 BOWMAN, V. J. JR.; STARR, J. H.
PARTIAL ORDERINGS IN IMPLICIT ENUMERATION. IN: P. L. HAMMER, E. L. JOHNSON, B. H. KORTE, G. L. NEMHAUSER (EDS.): STUDIES IN INTEGER PROGRAMMING, 99-116. ANNALS OF DISCRETE MATHEMATICS 1, AMSTERDAM: NORTH-HOLLAND. (1977).

BOX 1976 -1 BOX, R. M.
PRESQUE-DUALITE EN NOMBRES ENTIERS, LA. IN: J. KOHLAS, O. SEIFERT, P. STAEHLY, H.-J. ZIMMERMANN (EDS): PROCEEDINGS IN OPERATIONS RESEARCH 5, 50-54. WUERZBURG: PHYSICA. (1976).

BOYC 1973 -1 BOYCE, D. E.; FARHI, A.; WEISCHEDEL, R.
OPTIMAL NETWORK PROBLEM: A BRANCH-AND-BOUND ALGORITHM. ENVIRONMENT AND PLANNING 5, 519-533. (1973).

BOYCA 1975 -1 BOYCE, W. M.
IMPROVED PROGRAM FOR THE FULL STEINER TREE PROBLEM, AN. BELL LABORATORIES TECHNICAL MEMORANDUM. (1975).

BOYCB 1973 -1 BOYCE, W. M.; SEERY, J. B.
STEINER 72, AN IMPROVED VERSION OF COCKAYNE AND SCHILLER'S PROGRAM STEINER FOR THE MINIMAL NETWORK PROBLEM. BELL LABORATORIES TECHNICAL MEMORANDUM. (1973).

BRAC 1975 -1 BRACE, A.; DAYKIN, D. E.
FINITE SET COVERING THEOREM, IV, A. IN: A. HAJNAL, R. RADO, V. T. SOS (EDS.): INFINITE AND FINITE SETS, COLLOQ. KESZTHELY 1973, VOL. 1, 199-203. COLLOQ. MATH. SOC. JANOS BOLYAI 10, AMSTERDAM: NORTH-HOLLAND. (1975).

BRAD 1976 -1 BRADLEY, G.; BROWN, G.; GRAVES, G.
DESIGN AND IMPLEMENTATION OF LARGE SCALE PRIMAL TRANSSHIPMENT ALGORITHMS. TECHNICAL REPORT NPS55BZBW76091. MONTEREY: NAVAL POSTGRADUATE SCHOOL. (1976).

BRADA 1971 -1 BRADLEY, G. H.
HEURISTIC SOLUTION METHODS AND TRANSFORMED INTEGER LINEAR PROGRAMMING PROBLEMS. IN: PROCEEDINGS OF THE FIFTH ANNUAL PRINCETON CONFERENCE ON INFORMATION SCIENCES AND SYSTEMS, 204-209. (1971).

BRADB 1969 -1 BRADLEY, G. H.; WAHI, P. N.
INTEGER PROGRAMMING TEST PROBLEMS. REPORT 28. NEW HAVEN: YALE UNIVERSITY. (1969).

BRADC 1965 -1 BRADLEY, S. P.
SOLUTION TECHNIQUES FOR THE TRAFFIC ASSIGNMENT PROBLEM. TECHN. REPORT ORC 65-35. BERKELEY: UNIV. OF CALIFORNIA, OPERATIONS RESEARCH CENTER. (1965).

BRADD 1977 -1 BRADLEY, S. P.; HAX, A. C.; MAGNANTI, T. L.
APPLIED MATHEMATICAL PROGRAMMING. READING, AMSTERDAM U.A.: ADDISON-WESLEY, 735 P. (1977).

BRAE 1975 -1 BRAEMERT, D.
GRAPH THEORY: A SELECTED BIBLIOGRAPHY. MATCH, NO. 1, 207-210. (1975).

BRAN 1971 -1 BRANDMAN, E. I.
CERTAIN METHODS FOR THE SOLUTION OF PROBLEMS OF PROJECTED ALLOCATION OF CAPITAL INVESTMENTS WITHIN AN INDUSTRY (RUSSIAN). IN: PROCEEDINGS OF THE FOURTH WINTER SCHOOL ON MATHEMATICAL PROGRAMMING AND RELATED QUESTIONS (RUSSIAN), DROGOBYCH 1971, NO. 1, 149-157. MOSCOW: MOSKOV. INZ.-STROITEL INST. (1971).

****

BRAT 1975 -1 BRATLEY, P.; FLORIAN, M.; ROBILLARD, P.
SCHEDULING WITH EARLIEST START AND DUE DATE CONSTRAINTS ON MULTIPLE MACHINES. NAVAL RESEARCH LOGISTICS QUARTERLY 22, 165-173. (1975).

BRAU 1974 -1 BRAUN, J.
EDMONDS' MATCHING ALGORITHMUS. REPORT NO. 7432-OR. BONN: UNIV. BONN, INSTITUT FUER OEKONOMETRIE UND OPERATIONS RESEARCH 34 P. (1974).

BREG 1972 -1 BREGMAN, L. M.
CERTAIN APPROACH TO INTEGER PROGRAMMING PROBLEMS (RUSSIAN), A. IN: OPERATIONS RESEARCH AND STATISTICAL MODELING, NO. 1 (RUSSIAN), 42-47. LENINGRAD: IZDAT. LENINGRAD UNIV. (1972).

BREGA 1974 -1 BREGMAN, T. T.
BRANCH AND BOUND METHOD FOR THE SOLUTION OF ALLOCATION PROBLEMS (RUSSIAN), A. IN: I. V. ROMANOVSKII (ED.): OPERATIONS RESEARCH AND STATISTICAL SIMULATION, NO. 2 (RUSSIAN), 81-91 + 151. LENINGRAD: IZDAT. LENINGRAD UNIV. (1974).

BREI 1976 -1 BREITENBACH, R. B.
COMPUTER ALGORITHM FOR REFINING GRID LOCATIONAL ANALYSIS, A. THE LOGISTICS AND TRANSPORTATION REVIEW 12, 154-169. (1976).

BREL 1977 -1 BRELAZ, D.
PROBLEME D'HORAIRE DE LABORATOIRE DE LANGUES, UN. REVUE BELGE DE STATISTIQUE, D'INFORMATIQUE ET DE RECHERCHE OPERATIONNELLE 17, NO. 3, 1-8. (1977).

BRELA 1977 -1 BRELAZ, D.; NICOLIER, Y.; WERRA, D. DE
COMPACTNESS AND BALANCING IN SCHEDULING. ZEITSCHRIFT FUER OPERATIONS RESEARCH 21, SERIES A 65-74. (1977).

BRES 1976 -1 BRESLAW, J. A.
LINEAR PROGRAMMING SOLUTION TO THE FACULTY ASSIGNMENT PROBLEM, A. SOCIO-ECONOMIC PLANNING SCIENCES 10, 227-230. (1976).

BRIC 1977 -1 BRICKER, D. L.
REFORMULATION OF SPECIAL ORDERED SETS FOR IMPLICIT ENUMERATION ALGORITHMS, WITH APPLICATIONS IN NONCONVEX SEPARABLE PROGRAMMING. AIIE TRANSACTIONS 9, 195-203. (1977).

BRIT 1976 -1 BRITAVSKIJ, G. M.; JUKHIMENKO, B. I.
METHOD OF BRANCH AND BOUND FOR OPTIMIZATION OF AUTOMATIC LINES STRUCTURE PARAMETERS (RUSSIAN), A. KIBERNETIKA, NO. 2, 102-104. (1976).

BROC 1976 -1 BROCKLEHURST, E. R.
GENERALIZED BRANCH AND BOUND: A METHOD FOR INTEGER LINEAR PROGRAMS. IN: K. B. HALEY (ED.): OPERATIONAL RESEARCH '75. PROCEEDINGS OF THE 7TH IFORS INTERNATIONAL CONFERENCE ON OPERATIONAL RESEARCH, TOKYO, KYOTO 1975, 141-162. AMSTERDAM: NORTH-HOLLAND. (1976).

BROW 1976 -1 BROWN, D. B.
ALLOCATION OF FEDERAL HIGHWAY SAFETY FUNDS USING DYNAMIC PROGRAMMING, THE. AIIE TRANSACTIONS 8, 461-466. (1976).

BROWA 1976 -1 BROWN, D. P.
CIRCUIT AND UNIMODULAR MATRICES. SIAM JOURNAL ON APPLIED MATHEMATICS 31, 468-473. (1976).

BROWA 1977 -1 BROWN, D. P.
COMPOUND AND UNIMODULAR MATRICES. DISCRETE MATHEMATICS 19, 1-6. (1977).

BROWB 1976 -1 BROWN, J. R.
ALGORITHM FOR R-BASES OF A DIRECTED GRAPH, AN. NETWORKS 6, 185-188. (1976).

BROWC 1973 -1 BROWN, W.
GENERALIZED PLANT LOCATION PROBLEM, A. DISSERTATION. CLEVELAND: CASE WESTERN RESERVE UNIV., DEPT. OF OPERATIONS RESEARCH. (1973).

BROWD 1976 -1 BROWNELL, W. S.; LOWERRE, J. M.
SCHEDULING OF WORK FORCES REQUIRED IN CONTINUOUS OPERATIONS UNDER ALTERNATIVE LABOR POLICIES. MANAGEMENT SCIENCE 22, 597-605. (1976).

BRUA 1975 -1 BRUALDI, R. A.
MATROIDS INDUCED BY DIRECTED GRAPHS, A SURVEY. IN: M. FIEDLER (ED.): RECENT ADVANCES IN GRAPH THEORY, PROCEEDINGS OF THE 2ND CZECHOSLOVAK SYMPOSIUM, PRAGUE 1974. PRAGUE: ACADEMIA. (1975).

BRUA 1975 -2 BRUALDI, R. A.
TRANSVERSAL THEORY AND GRAPHS. IN: D. R. FULKERSON (ED.): STUDIES IN GRAPH THEORY, PART 1, 23-88. STUDIES IN MATHEMATICS 11, WASHINGTON: MATHEMATICAL ASSOCIATION OF AMERICA. (1975).

BRUAA 1976 -1 BRUALDI, R. A.; GIBSON, P. M.
ASSIGNMENT POLYTOPE, THE. MATHEMATICAL PROGRAMMING 11, 97-101. (1976).

BRUAA 1977 -1 BRUALDI, R. A.; GIBSON, P. M.
CONVEX POLYHEDRA OF DOUBLY STOCHASTIC MATRICES III. AFFINE AND COMBINATORIAL PROPERTIES OF OMEGA N. JOURNAL OF COMBINATORIAL THEORY 22, SERIES A, 338-351. (1977).

BRUAA 1977 -2 BRUALDI, R. A.; GIBSON, P. M.
CONVEX POLYHEDRA OF DOUBLY STOCHASTIC MATRICES, 1. APPLICATIONS OF THE PERMANENT FUNCTION. JOURNAL OF COMBINATORIAL THEORY 22, SERIES A, 194-230. (1977).

BRUAA 1977 -3 BRUALDI, R. A.; GIBSON, P. M.
CONVEX POLYHEDRA OF DOUBLY STOCHASTIC MATRICES: II. GRAPH OF OMEGA N. JOURNAL OF COMBINATORIAL THEORY 22, SERIES B, 175-198. (1977).

BRUAB 1978 -1 BRUALDI, R. A.; LEE, G. M.
ON THE TRUNCATED ASSIGNMENT POLYTOPE. LINEAR ALGEBRA AND ITS APPLICATIONS 19, 33-62. (1978).

BRUC 1972 -1 BRUCKER, P.
ALL SHORTEST DISTANCES IN NETWORKS WITH A LARGE NUMBER OF STRONG COMPONENTS. PAPER PRESENTED AT THE 41ST NATIONAL ORSA-MEETING, NEW ORLEANS 1972. (1972).

BRUC 1975 -1 BRUCKER, P.
GANZZAHLIGE LINEARE PROGRAMMIERUNG MIT OEKONOMISCHEN ANWENDUNGEN. MATHEMATICAL SYSTEMS IN ECONOMICS 16, MEISENHEIM: HAIN, 130 P. (1975).

BRUC 1976 -1 BRUCKER, P.
ANMERKUNGEN ZU HEURISTISCHEN VERFAHREN. IN: H. N. DATHE, P. MERTENS, F. D. PESCHANEL ET AL. (EDS.): PROCEEDINGS IN OPERATIONS RESEARCH 6, 668-676. WUERZBURG: PHYSICA. (1976).

BRUC 1976 -2 BRUCKER, P.
KOMPLEXITAET VON SCHEDULING PROBLEMEN, DIE. IN: J. KOHLAS, O. SEIFERT, P. STAEHLY, H.-J. ZIMMERMANN (EDS): PROCEEDINGS IN OPERATIONS RESEARCH 5, 357-368. WUERZBURG: PHYSICA. (1976).

BRUC 1976 -3 BRUCKER, P.
NP-VOLLSTAENDIGE OPERATIONS RESEARCH PROBLEME. IN: H. NOLTEMEIER (ED.): GRAPHEN, ALGORITHMEN, DATENSTRUKTUREN. ERGEBNISSE DES WORKSHOPS WG 76. 2. FACHTAGUNG UEBER GRAPHENTHEORETISCHE KONZEPTE DER INFORMATIK, GOETTINGEN 1976, 135-148. MUENCHEN: HANSER. (1976).

BRUC 1977 -1 BRUCKER, P.
PROZESSOR SCHEDULING IN VERBINDUNG MIT MULTIPROGRAMMING. IN: R. HENN, P. KALL, B. KORTE, O. KRAFFT ET AL. (EDS.): OPERATIONS RESEARCH VERFAHREN XXVI, METHODS OF OPERATIONS RESEARCH, 919-928. MEISENHEIM: HAIN. (1977).

BRUC 1978 -1 BRUCKER, P.
ON THE COMPLEXITY OF CLUSTERING PROBLEMS. IN: R. HENN, B. KORTE, W. OETTLI (EDS.): OPTIMIZATION AND OPERATIONS RESEARCH, PROCEEDINGS, BONN 1977. LECTURE NOTES IN ECONOMICS AND MATHEMATICAL SYSTEMS 157. BERLIN, HEIDELBERG, NEW YORK: SPRINGER, 45-54. (1978).

BRUCA 1977 -1 BRUCKER, P.; GAREY, M. R.; JOHNSON, D. S.
SCHEDULING EQUAL-LENGTH TASKS UNDER TREE-LIKE PRECEDENCE CONSTRAINTS TO MINIMIZE MAXIMUM LATENESS. MATHEMATICS OF OPERATIONS RESEARCH 2, 275-284. (1977).

BRUCB 1975 -1 BRUCKER, P.; LENSTRA, J. K.; RINNOOY KAN, A. H. G.
COMPLEXITY OF MACHINE SCHEDULING PROBLEMS. RAPPORT NO. 15. DELFT: GRADUATE SCHOOL OF MANAGEMENT, 29 P. (1975).

BRUN 1968 -1 BRUNO, J.
MATROIDS, GRAPHS, AND RESISTANCE NETWORKS. DISSERTATION. NEW YORK: CITY COLLEGE OF THE CITY UNIV. OF NEW YORK. (1968).

BRUNA 1976 -1 BRUNO, J.; COFFMAN, E. G. JR.; JOHNSON, D. B.
ON BATCH SCHEDULING OF JOBS WITH STOCHASTIC SERVICE TIMES AND COST STRUCTURES ON A SINGLE SERVER. JOURNAL OF COMPUTER AND SYSTEM SCIENCES 12, 319-335. (1976).

BRUNB 1974 -1 BRUNO, J.; COFFMAN, E. G. JR.; SETHI, R.
ALGORITHMS FOR MINIMIZING MEAN FLOW TIME. IN: J. L. ROSENFELD (ED): INFORMATION PROCESSING 74, PROCEEDINGS OF THE IFIP CONGRESS, STOCKHOLM 1974, 504-510. AMSTERDAM: NORTH-HOLLAND. (1974).

BRUNC 1975 -1 BRUNO, J.; HOFRI, M.
ON SCHEDULING CHAINS OF JOBS ON ONE PROCESSOR WITH LIMITED PREEMPTION. SIAM JOURNAL ON COMPUTING 4, 478-490. (1975).

BRUND 1976 -1 BRUNO, J.; WEINBERG, L.
GENERALIZED NETWORKS: NETWORKS EMBEDDED ON A MATROID, PART 2. NETWORKS 6, 231-272. (1976).

BRUND 1976 -2 L BRUNO, J.; WEINBERG, L.
GENERALIZED NETWORKS: NETWORKS EMDEDDED ON A MATROID, PART I. NETWORKS 6, 53-94. (1976).

BRUNE 1976 -1 BRUNO, J. L.
SCHEDULING ALGORITHMS FOR MINIMISING THE MEAN WEIGHTED FLOW-TIME CRITERION. IN: E. G. COFFMAN JR. (ED.): COMPUTER AND JOB/SHOP SCHEDULING THEORY, CHAPTER THREE. (1976).

BRUNE 1976 -2 BRUNO, J. L.
SEQUENCING JOBS WITH STOCHASTIC TASK STRUCTURES ON A SINGLE MACHINE. JOURNAL OF THE ASSOCIATION FOR COMPUTING MACHINERY 23, 655-664. (1976).

BRUNF 1975 -1 BRUNS, M.
REIHENFOLGEPROBLEM DER NETZPLANTECHNIK, DAS. VDI-ZEITSCHRIFT 117, NO. 11, 529-533. (1975).

BRUT 1970 -1 BRUTER, C. P.
MATROIDES. NOUVEL OUTIL MATHEMATIQUE, LES. PARIS: DUNOD, 192 P. (1970).

BULA 1974 -1 BULATOV, V. P.
NEW METHOD FOR THE SOLUTION OF DISCRETE PROGRAMMING PROBLEMS (RUSSIAN), A. IN: OPTIMIZATION METHODS AND THEIR APPLICATIONS, ALL-UNION SUMMER SEMINAR, KHAKUSY 1972 (RUSSIAN), 150-154 + 182. IRKUTSK: SIBIRSK. ENERGET. INST. SIBIRSK. OTDEL. AKAD. NAUK SSSR. (1974).

BULF 1976 -1 BULFIN, R. L.; PARKER, R. G.
ON A TWO FACILITY SCHEDULING PROBLEM WITH SEQUENCE DEPENDENT PROCESSING TIME. AIIE TRANSACTIONS 8, 202-209. (1976).

BULF 1976 -2 BULFIN, R. L.; PARKER, R. G.
ON THE ASSIGNMENT AND SEQUENCING OF OPERATIONS ON A CREW-SERVED PROJECT. NAVAL RESEARCH LOGISTICS QUARTERLY 23, 261-270. (1976).

BULFA 1977 -1 BULFIN, R. L.; SHETTY, C. M.; PARKER, R. G.
COMPUTATIONAL RESULTS WITH A BRANCH-AND-BOUND ALGORITHM FOR THE GENERAL KNAPSACK PROBLEM. SAN FRANCISCO (51. ORSA-MEETING + TIMS, MAY 1977) 16 P. (1977).

BULL 1970 -1 BULLINGHAM, J. M.; MATHEWS, P. A.
ELECTRONIC SIMULATOR FOR QUICKEST ROUTES IN A ROAD NETWORK. TRAFFIC ENGINEERS AND CONTROL 12, 5, 240-243. (1970).

BUNT 1977 -1 BUNT, R. B.; HUME, J. N. P.
ADAPTIVE PROCESSOR SCHEDULING BASED ON APPROXIMATING DEMAND DISTRIBUTION. INFOR 15, 135-147. (1977).

BUR' 1976 -1 BUR'JAN, S. B.; SEROV, S. Š.; UZDEMIR, A. P.
DYNAMIC INDUSTRIAL-PLANT LOCATION PROBLEM AND A NUMERICAL METHOD FOR SOLVING IT, II (RUSSIAN). AVTOMATIKA I TELEMEHANIKA, NO. 5, 112-121; TRANSLATED AS AUTOMATION AND REMOTE CONTROL 37, 1976, 741-750. (1976).

BUR' 1976 -2 BUR'JAN, S. B.; SEROV, S. S.; UZDEMIR, A. P.
DYNAMIC INDUSTRIAL-PLANT-LOCATION PROBLEM AND A NUMERICAL METHOD FOR SOLVING IT, III (RUSSIAN). AVTOMATIKA I TELEMEHANIKA, NO. 8, 131-138; TRANSLATED AS AUTOMATION AND REMOTE CONTROL 37, 1976, 1252-1259. (1976).

BURD 1977 -1 BURDET, C.-A.; JOHNSON, E. L.
SUBADDITIVE APPROACH TO SOLVE LINEAR INTEGER PROGRAMS, A. IN: P. L. HAMMER, E. L. JOHNSON, B. H. KORTE, G. L. NEMHAUSER (EDS.): STUDIES IN INTEGER PROGRAMMING, 117-144. ANNALS OF DISCRETE MATHEMATICS 1, AMSTERDAM: NORTH-HOLLAND. (1977).

BURDA 1976 -1 BURDYUK, V. YA.; TROFIMOV, V. N.
GENERALIZATION OF RESULTS BY GILMORE-GOMORY CONCERNING THE TRAVELLING SALESMAN PROBLEM (RUSSIAN). IZVESTIJA AKADEMII NAUK SSSR, TEHNICESKAJA KIBERNETIKA, NO. 3, 16-22. (1976).

BURK 1973 -1 BURKARD, R. E.
METHOD FOR MIXED-INTEGER CONVEX PROGRAMMING, A. IN: PROCEEDINGS OF THE FOURTH CONFERENCE ON PROBABILITY THEORY. BUCHAREST: EDITURA ACADEMIEI REPUBLICII SOCIALISTE ROMANIA. (1973).

BURK 1976 -1 BURKARD, R. E.
FLUESSE IN NETZWERKEN MIT ALLGEMEINEN KOSTEN. IN: H. NOLTEMEIER (ED.): GRAPHEN, ALGORITHMEN, DATENSTRUKTUREN. ERGEBNISSE DES WORKSHOPS WG 76. 2. FACHTAGUNG UEBER GRAPHENTHEORETISCHE KONZEPTE DER INFORMATIK, GOETTINGEN 1976, 123-134. MUENCHEN: HANSER. (1976).

BURK 1976 -2 BURKARD, R. E.
GENERAL HUNGARIAN METHOD FOR THE ALGEBRAIC TRANSPORTATION PROBLEM, A. REPORT 76-6. KOELN: UNIV. KOELN, MATHEMATISCHES INSTITUT, 19 P. (1976).

BURK 1976 -3 BURKARD, R. E.
PERTURBATION METHOD FOR SOLVING QUADRATIC ASSIGNMENT AND BOTTLENECK PROBLEMS, A. REPORT 76-2. KOELN: UNIV. KOELN, MATHEMATISCHES INSTITUT, 17 P. (1976).

BURK 1977 -1 BURKARD, R. E.
TRAVELLING SALESMAN AND ASSIGNMENT PROBLEMS: A SURVEY. VANCOUVER (CONFERENCE ON DISCRETE OPTIMIZATION, AUGUST 1977) 28 P. (1977).

BURK 1977 -2 BURKARD, R. E.
ZUR BESTIMMUNG VON SCHREIBMASCHINENTASTATUREN DURCH QUADRATISCHE ZUORDNUNGSPROBLEME. IN: R. HENN, P. KALL, B. KORTE, O. KRAFFT ET AL. (EDS.): OPERATIONS RESEARCH VERFAHREN XXV, METHODS OF OPERATIONS RESEARCH, 144-146, MEISENHEIM: HAIN. (1977).

BURKA 1976 -1 BURKARD, R. E.; HAHN, W.; ZIMMERMANN, U.
ALGEBRAIC APPROACH TO ASSIGNMENT PROBLEMS, AN (REVIDIERT). REPORT 74-1. KOELN: UNIV. KOELN, MATHEMATISCHES INSTITUT, 17 P. (1976).

BURKA 1977 -1 BURKARD, R. E.; HAHN, W.; ZIMMERMANN, U.
ALGEBRAIC APPROACH TO ASSIGNMENT PROBLEMS, AN. MATHEMATICAL PROGRAMMING 12, 318-327. (1977).

BURKB 1976 -1 BURKARD, R. E.; HAMACHER, H.; ZIMMERMANN, U.
ALGEBRAIC NETWORK FLOW PROBLEM, THE. REPORT 76-7. KOELN: UNIV. KOELN, MATHEMATISCHES INSTITUT, 28 P. (1976).

****

BURKB 1977 -1 BURKARD, R. E.; HAMACHER, H.; ZIMMERMANN, U.
BEMERKUNGEN ZUM ALGEBRAISCHEN FLUSSPROBLEM, EINIGE. IN: R. HENN, P. KALL, B. KORTE, O. KRAFFT ET AL. (EDS.): OPERATIONS RESEARCH VERFAHREN XXIII, METHODS OF OPERATIONS RESEARCH, 252-254. MEISENHEIM: HAIN. (1977).

BURKB 1977 -2 BURKARD, R. E.; HAMACHER, H.; ZIMMERMANN, U.
FLUSSPROBLEME MIT ALLGEMEINEN KOSTEN. IN: L. COLLATZ, G. MEINARDUS, W. WETTERLING (EDS.): NUMERISCHE METHODEN BEI OPTIMIERUNGSAUFGABEN BAND 3, OPTIMIERUNG BEI GRAPHENTHEORETISCHEN UND GANZZAHLIGEN PROBLEMEN, 9-22. INTERNATIONAL SERIES OF NUMERICAL MATHEMATICS 36, BASEL: BIRKHAEUSER. (1977).

BURKC 1977 -1 BURKARD, R. E.; OFFERMANN, J.
ENTWURF VON SCHREIBMASCHINENTASTATUREN MITTELS QUADRATISCHER ZUORDNUNGSPROBLEME. ZEITSCHRIFT FUER OPERATIONS RESEARCH 21, SERIE B, B121-B132. (1977).

BURKD 1976 -1 BURKARD, R. E.; STRATMANN, K.-H.
NUMERICAL INVESTIGATIONS ON QUADRATIC ASSIGNMENT PROBLEMS. REPORT 76-3. KOELN: UNIV. KOELN, MATHEMATISCHES INSTITUT, 28 P. (1976).

BURKE 1977 -1 BURKARD, R. E.; ZIMMERMANN, U.
WEAKLY ADMISSIBLE TRANSFORMATIONS FOR SOLVING ALGEBRAIC ASSIGNMENT AND TRANSPORTATION PROBLEMS. REPORT 77-3. KOELN: UNIV. KOELN, MATHEMATISCHES INSTITUT, 21 P. (1977).

BURKE 1978 -1 BURKARD, R. E.; ZIMMERMANN, U.
SOLUTION OF ALGEBRAIC ASSIGNMENT AND TRANSPORTATION PROBLEMS, THE. IN: R. HENN, B. KORTE, W. OETTLI (EDS.): OPTIMIZATION AND OPERATIONS RESEARCH, PROCEEDINGS, BONN 1977. LECTURE NOTES IN ECONOMICS AND MATHEMATICAL SYSTEMS 157. BERLIN, HEIDELBERG, NEW YORK: SPRINGER, 55-65. (1978).

BURKF 1973 -1 BURKHARD, W. A.
NONRECURSIVE TREE TRAVERSAL ALGORITHM. IN: PROCEEDINGS OF THE SEVENTH ANNUAL PRINCETON CONFERENCE ON INFORMATION SCIENCES AND SYSTEMS, 403-405. (1973).

BURKF 1975 -1 BURKHARD, W. A.
NONRECURSIVE TRAVERSALS OF TREES. COMPUTER JOURNAL 18, 227-230. (1975).

BURKG 1971 -1 BURKHARD, W. A.; KROON, F. W.
TOWARD A WEAKLY INVARIANT COMPLEXITY THEORY. IN: CONFERENCE RECORD 1971, 12TH ANNUAL SYMPOSIUM ON SWITCHING AND AUTOMATA THEORY, EAST LANSING 1971, 24-32. NORTHRIDGE: INSTITUTE OF ELECTRICAL AND ELECTRONICS ENGINEERS. (1971).

BURKH 1975 -1 BURKOV, V. N.; GROPPEN, V. O.
MAXIMUM CIRCULATION AND MINIMUM SECTION ON THE PLANAR ORGRAPHS (RUSSIAN). KIBERNETIKA, NO. 6, 85-89. (1975).

BURKI 1972 -1 BURKOV, V. N.; LAZEBNIK, A. I.; KHRANOVICH, I. L.
METHOD OF BRANCHES AND BOUNDS AS A REGULAR METHOD FOR THE SOLUTION OF IRREGULAR MATHEMATICAL PROGRAMMING PROBLEMS, I (RUSSIAN). AVTOMATIKA I TELEMEHANIKA, NO. 7, 169-177. (1972).

BURKJ 1966 -1 BURKOV, V. N.; LOVETSKII, S. E.
HEURISTIC APPROACH TO THE SOLUTION OF DYNAMIC PROGRAMMING PROBLEMS IN THE ALLOCATION OF RESOURCES (RUSSIAN). AVTOMATIKA I TELEMEHANIKA 5, 82-90. (1966).

BURKJ 1968 -1 BURKOV, V. N.; LOVETSKII, S. E.
METHODS FOR THE SOLUTION OF EXTREMAL PROBLEMS OF COMBINATORIAL TYPE (REVIEW) (RUSSIAN). AVTOMATIKA I TELEMEHANIKA, NO. 11, 68-93; TRANSLATED AS AUTOMATION AND REMOTE CONTROL, NO. 11, 1968, 1785-1806. (1968).

BURN 1976 -1 BURNESS, R. C.; WHITE, J. A.
TRAVELING SALESMAN LOCATION PROBLEM, THE. TRANSPORTATION SCIENCE 10, 348-360. (1976).

BURNA 1975 -1 BURNS, F.; ROOKER, J.
SPECIAL CASE OF THE 3 X N FLOW SHOP PROBLEM, A. NAVAL RESEARCH LOGISTICS QUARTERLY 22, 811-818. (1975).

BURNA 1976 -1 BURNS, F.; ROOKER, J.
JOHNSON'S THREE-MACHINE FLOW-SHOP CONJECTURE. OPERATIONS RESEARCH 24, 578-580. (1976).

BURNB 1976 -1 BURNS, R. N.
SCHEDULING TO MINIMIZE THE WEIGHTED SUM OF COMPLETION TIMES WITH SECONDARY CRITERIA. NAVAL RESEARCH LOGISTICS QUARTERLY 23, 125-130. (1976).

BURNB 1977 -1 BURNS, R. N.
STATIONARY POINTS OF HEURISTICS IN PERMUTATION SCHEDULING. RESEARCH REPORT CORR 77/14. WATERLOO: UNIV. OF WATERLOO, DEPT. OF COMBINATORICS AND OPTIMIZATION, 15 P. (1977).

BURNC 1976 -1 BURNS, R. N.; HAFF, C. E.
COMBINATORIAL RANKING PROBLEM, A. AEQUATIONES MATHEMATICAE 14, 351-355. (1976).

BURR 1974 -1 BURR, S. A.
GENERALIZED RAMSEY THEORY FOR GRAPHS - A SURVEY. IN: R. A. BARI, F. HARARY (EDS.): GRAPHS AND COMBINATORICS, PROCEEDINGS OF THE CAPITAL CONFERENCE ON GRAPH THEORY AND COMBINATORICS, WASHINGTON 1973, 52-75. LECTURE NOTES IN MATHEMATICS 406, BERLIN: SPRINGER. (1974).

BURRA 1976 -1 BURRELL, J. E.
MULTIPLE ROUTE ASSIGNMENT: A COMPARISON OF TWO METHODS. IN: M. A. FLORIAN (ED.): TRAFFIC EQUILIBRIUM METHODS, PROCEEDINGS OF AN INTERNATIONAL SYMPOSIUM, MONTREAL 1974, 229-239. LECTURE NOTES IN ECONOMICS AND MATHEMATICAL SYSTEMS 118, BERLIN: SPRINGER. (1976).

BURS 1976 -1 BURSHTEJN, M. I.
CHROMATIC NUMBER OF PLANAR HYPERGRAPHES (RUSSIAN), THE. SAKHARTHVELOS SSR MECNIEREBATHA AKADEMII MOAMBE 84, 45-48. (1976).

BURS 1976 -2 BURSHTEJN, M. I.
ON A CLASS OF BICHROMATIC HYPERGRAPHS AND THE ESTIMATION OF THE CHROMATIC NUMBER (RUSSIAN). SOOBSCENIJA AKADEMII NAUK GRUZINSKOI SSR 81, 41-44. (1976).

BURT 1970 -1 BURT, J. M. JR.
ALL SHORTEST DISTANCES IN LARGE SERIAL NETWORKS. WORKING PAPER NO. 160. LOS ANGELES: UNIV. OF CALIFORNIA,WESTERN MANAGEMENT SCIENCE INSTITUTE. (1970).

BURTA 1977 -1 BURTIN, YU. D.
PROBABILITY OF THE CONNECTIVITY OF A RANDOM SUBGRAPH OF AN M-DIMENSIONAL CUBE (RUSSIAN). PROBLEMY PEREDACI INFORMACII 3, NO. 2, 90-95. (1977).

BURTB 1976 -1 BURTON, J. S.; GLEASON, J. M.
ALGORITHM FOR THE DYNAMIC LOT SIZE PROBLEM WITH PRICE BREAKS, AN. LUBBOCK: TEXAS TECH. UNIV., COLLEGE OF BUSINESS ADMINISTRATION, 7 P. (1976). (R 76-4)

BUSH 1976 -1 BUSHELL, G.; FERLAND, J.; GUERIN, G.; NASTANSKY, L.
ENGINE SCHEDULING PROBLEM IN A RAILWAY NETWORK, THE. INFOR 14, 121-138. (1976).

BUTA 1968 -1 BUTAS, L. F.
DIRECTIONALLY ORIENTED SHORTEST PATH ALGORITHM, A. TRANSPORTATION RESEARCH 2, NO. 3, 253-268. (1968).

BUTE 1974 -1 BUTEN, R. E.
ANALYSIS OF THE FLOW-SHOP SCHEDULING PROBLEM IN A SYSTEM WITH TWO CLASSES OF PROCESSORS. IN: PROCEEDINGS OF THE EIGHTH ANNUAL PRINCETON CONFERENCE ON INFORMATION SCIENCES AND SYSTEMS, 587-591. (1974).

BUTL 1965 -1 BUTLAR, W. H.
SHORT INTERVAL SCHEDULING; TURNS LARGE JOBS INTO SMALL TASKS. FACTORY 123, 102-104. (1965).

BUTR 1964 -1 BUTRIMENKO, A. V.
ON SEARCHING FOR THE SHORTEST PATHS ALONG A GRAPH DURING VARIATIONS IN IT. ENGINEERING CYBERNETICS 6, 50-53. (1964).

BUZA 1976 -1 BUZACOTT, J. A.
PRODUCTION CAPACITY OF JOB SHOPS WITH LIMITED STORAGE SPACE, THE. INTERNATIONAL JOURNAL OF PRODUCTION RESEARCH 14, 597-606. (1976).

BUZY 1976 -1 BUZYCKII, P. L.
CERTAIN CUT AND BRANCH FOR PRIMAL INTEGER PROGRAMMING ALGORITHMS (RUSSIAN), A. IN: MATHEMATICAL PROGRAMMING AND RELATED QUESTIONS. PROCEEDINGS OF THE SEVENTH WINTER SCHOOL, DROGOBYCH 1974, NUMERICAL METHODS (RUSSIAN), 129-132. MOSCOW: CENTRAL. EKONOM.-MAT. INST. AKAD. NAUK SSSR. (1976).

BUZY 1976 -2 BUZYCKII, P. L.
DIRECT ALGORITHM OF INTEGER PROGRAMMING FOR A PROBLEM WITH TWO-SIDED CONSTRAINTS (RUSSIAN), A. IN: STUDIES IN DISCRETE OPTIMIZATION (RUSSIAN), 88-97. MOSCOW: IZDAT. NAUKA. (1976).

BUZYA 1976 -1 BUZYCKII, P. L.; VOTJAKOV, A. A.
DIRECT ALGORITHM FOR INTEGER PROGRAMMING (RUSSIAN), A. IN: STUDIES IN DISCRETE OPTIMIZATION (RUSSIAN), 68-88. MOSCOW: IZDAT. NAUKA. (1976).

CABO 1977 -1 CABOT, A. V.; HARNETT, D. L.
INTRODUCTION TO MANAGEMENT SCIENCE, AN. READING, AMSTERDAM U.A.: ADDISON-WESLEY, 524 P. (1977).

CABOA 1976 -1 CABOT, V. A.
NONCONVEX PROGRAMMING PROBLEM SOLVED BY GENERALIZED BENDERS DECOMPOSITION, A. BLOOMINGTON: INDIANA UNIV., GRADUATE SCHOOL OF BUSINESS, 15 P. (1976).

CACC 1976 -1 CACCETTA, L.
PROBLEM IN EXTREMAL GRAPH THEORY, A. ARS COMBINATORIA 2, 33-56. (1976).

CAHI 1975 -1 CAHIT, I.; CAHIT, R.
MINIMUM SPANNING TREE. PROCEEDINGS IEEE 122, 1196. (1975).

CALY 1974 -1 CALYKOV, V.
CERTAIN HEURISTIC ALGORITHM FOR THE SOLUTION OF THE TRAVELLING SALESMAN PROBLEM (RUSSIAN), A. IN: MATHEMATICAL METHODS OF SOLUTION OF ECONOMIC PROBLEMS (RUSSIAN), 49-55, SUPPLEMENT TO EKONOMIKA I MATEMATICESKIE METODY, NO. 5. MOSCOW: IZDAT. NAUKA. (1974).

CAMB 1976 -1 CAMBINI, A.
APPLICATION OF THE METHOD OF DYNAMIC PROGRAMMING TO AN OPTIMIZATION PROBLEM OVER A FINITE ABELIAN GROUP, AN. PISA: UNIV. DI PISA, DIPT. DI RICERCA OPERATIVA E SCIENZE STATISTICHE, 26 P. (1976).

CAME 1973 -1 CAMERINI, P. M.; FRATTA, L.; MAFFIOLI, F.
HEURISTICALLY GUIDED ALGORITHM FOR THE TRAVELING SALESMAN PROBLEM, A. JOURNAL OF THE INSTITUTION OF COMPUTER SCIENCES 4, 31-35. (1973).

CAME 1974 -1 CAMERINI, P. M.; FRATTA, L.; MAFFIOLI, F.
RELAXATION METHODS IMPROVED BY MODIFIED GRADIENT TECHNIQUES. PAPER PRESENTED AT THE CONFERENCE ON OPERATIONS RESEARCH, EGER 1974. (1974).

CAMEA 1974 -1 CAMERINI, P. M.; MAFFIOLI, F.
BOUNDS FOR 3 MATROID INTERSECTION PROBLEMS. INFORMATION PROCESSING LETTERS 3, 81-83. (1974).

CAMEB 1976 -1 CAMERON, A. V.; VOILS, L. R.
HEURISTIC FOR TRANSPORTATION ROUTING AND VEHICLE LOADING UNDER UNCERTAINTY, A. MIAMI (50. ORSA-MEETING + TIMS, NOVEMBER 1976) 9 P. (1976).

CAMEC 1975 -1 CAMERON, P. J.
MINIMAL EDGE-COLOURINGS OF COMPLETE GRAPHS. JOURNAL OF THE LONDON MATHEMATICAL SOCIETY 11, 337-346. (1975).

CAMP 1970 -1 CAMPBELL, H. G.; DUDEK, R. A.; SMITH, M. L.
HEURISTIC ALGORITHM FOR THE N-JOB, M-MACHINE SEQUENCING PROBLEM. MANAGEMENT SCIENCE 16, 630-637. (1970).

CANT 1974 -1 CANTOR, D. G.; GERLA, M.
OPTIMAL ROUTING IN A PACKET-SWITCHED COMPUTER NETWORK. IEEE TRANSACTIONS ON COMPUTERS C-23, 1062-1068. (1974).

CARD 1971 -1 CARDENAS, M. A.
SYSTEMS APPROACH TO INVESTMENT PLANNING IN THE ELECTRICAL INDUSTRY, THE. DISSERTATION. LOS ANGELES: UNIV. OF CALIFORNIA. (1971).

CARL 1977 -1 CARLTON, D. W.
BIRTHS OF SINGLE ESTABLISHMENT FIRMS AND REGIONAL VARIATION IN ECONOMIC COSTS. REPORT 7729. CHICAGO: UNIV. OF CHICAGO, CENTER FOR MATHEMATICAL STUDIES IN BUSINESS AND ECONOMICS, 24 P. (1977).

CARL 1977 -2 CARLTON, D. W.
LOCATION DECISIONS OF MANUFACTURING FIRMS. REPORT 7728. CHICAGO: UNIV. OF CHICAGO, CENTER FOR MATHEMATICAL STUDIES IN BUSINESS AND ECONOMICS, 17 P. (1977).

CARL 1977 -3 CARLTON, D. W.
MODELS OF NEW BUSINESS LOCATION. REPORT 7756. CHICAGO: UNIV. OF CHICAGO, CENTER FOR MATHEMATICAL STUDIES IN BUSINESS AND ECONOMICS, 53 P. (1977).

CARL 1977 -4 CARLTON, D. W.
MODELS OF SINGLE ESTABLISHMENT BIRTHS, STANDARD INDUSTRIAL CLASSIFICATION 3079. REPORT 7730. CHICAGO: UNIV. OF CHICAGO, CENTER FOR MATHEMATICAL STUDIES IN BUSINESS AND ECONOMICS, 26 P. (1977).

CARN 1976 -1 CARNALL, C. A.; WILD, R.
LOCATION OF VARIABLE WORK STATIONS AND THE PERFORMANCE OF PRODUCTION FLOW LINES, THE. INTERNATIONAL JOURNAL OF PRODUCTION RESEARCH 14, 703-710. (1976).

CARP 1977 -1 CARPANETO, G.; TOTH, P.
EFFICIENT ALGORITHM FOR THE ASYMMETRIC TRAVELLING SALESMAN PROBLEM, AN. ATLANTA (52. ORSA-MEETING + TIMS, NOVEMBER 1977) 10 P. (1977).

CARPA 1976 -1 CARPENITO, T. A.; WHITE, J. A.
ALLOCATION OF NON-IDENTICAL MACHINES AMONG NON-IDENTICAL SERVERS, THE. INTERNATIONAL JOURNAL OF PRODUCTION RESEARCH 14, 429-436. (1976).

CARR 1971 -1 CARRE, B. A.
ELIMINATION METHOD FOR MINIMAL-COST NETWORK FLOW PROBLEMS, AN. IN: J. K. REID (ED.): LARGE SPARSE SETS OF LINEAR EQUATIONS, PROCEEDINGS OF THE OXFORD CONFERENCE 1970, 191-209. LONDON: ACADEMIC PRESS. (1971).

CARRA 1965 -1 CARROLL, D. C.
HEURISTIC SEQUENCING OF SINGLE AND MULTIPLE COMPONENT JOBS. DISSERTATION, CAMBRIDGE, MASSACHUSETTS: M. I. T. SLOAN SCHOOL MANAGEMENT, (UNPUBLISHED). (1965).

CARS 1977 -1 CARSON, J. S.; LAW, A. M.
NOTE ON SPIRA'S ALGORITHM FOR THE ALL-PAIRS SHORTEST-PATH PROBLEM, A. SIAM JOURNAL ON COMPUTING 6, 696-699. (1977).

CARU 1965 -1 CARUSO, F. R.
ASSEMBLY LINE BALANCING FOR IMPROVED PROFITS. AUTOMATION 12, 48-52. (1965).

****

CARV 1977 -1 CARVALLO, M.
SUR UN PROBLEME D'AFFECTATION. ALGORITHME DE RESOLUTION. COMPTES RENDUS HEBDOMADAIRES DES SEANCES DE L'ACADEMIE DES SCIENCES 284, SERIES A, A1543-A1545. (1977).

CAVA 1976 -1 CAVAILLE, J. B.; DELMAS, J.
ORDONNANCEMENT OPTIMAL A TRES COURT TERME D'UN ATELIER DE DECOUPE. REPORT DGRST-7470819. PARIS: CENTRE NATIONALE DE LA RECHERCHE SCIENTIFIQUE, 24 P. (1976).

CAZA 1975 -1 CAZALET, J.; MEHR, P.
AMELIORATION DE L'UTILISATION D'UNE FLOTTE AERIENNE PARTAGEE ENTRE DEUX BASES PRINCIPALES (GENERATION DE CONVOYAGES). REVUE FRANCAISE D'AUTOMATIQUE, D'INFORMATIQUE ET DE RECHERCHE OPERATIONNELLE 9, NO. V-3, 57-68. (1975).

CEDE 1974 -1 CEDERBAUM, I.
OPTIMAL BACKBOARD ORDERING THROUGH THE SHORTEST PATH ALGORITHM. IN: PROCEEDINGS OF THE 1974 IEEE INTERNATIONAL SYMPOSIUM ON CIRCUITS AND SYSTEMS, 186-190. NEW YORK: INTERNAT. INST. OF ELECTRICAL AND ELECTRONICAL ENGINEERS. (1974).

CEDE 1976 -1 CEDERBAUM, I.
EXTREMAL PROBLEM IN THE THEORY OF N-CUBE GRAPHS, AN. MATRIX AND TENSOR QUARTERLY 27, NO. 2, 71-74. (1976).

CERF 1974 -1 CERF, V. G.; COWAN, D. D.; MULLIN, R. C.; STANTON, R. G.
LOWER BOUND ON THE AVERAGE SHORTEST PATH LENGTH IN REGULAR GRAPHS, A. NETWORKS 4, 335-342. (1974).

CHAI 1975 -1 CHAIKEN, J. M.
PATROL ALLOCATION METHODOLOGY FOR POLICE DEPARTMENTS. RAND-R-1852-HUD, SANTA MONICA: RAND CORP., 69 P. (1975).

CHAIA 1976 -1 CHAIKEN, J. M.; DORMANT, P.
PATROL CAR ALLOCATION MODEL, A. RAND-P-5594, SANTA MONICA: RAND CORPORATION. (1976).

CHAL 1976 -1 CHALMET, L.
SOLVING ASSIGNMENT PROBLEMS IN WAREHOUSES (DUTCH). REVUE BELGE DE STATISTIQUE, D'INFORMATIQUE ET DE RECHERCHE OPERATIONNELLE 16, NO. 3, 1-36. (1976).

CHALA 1975 -1 CHALMET, L.; GELDERS, L.
ALGORITHM FOR GENERATING GOOD WAREHOUSE LAYOUTS, AN. PAPER PRESENTED AT THE FIRST EUROPEAN CONGRESS ON OPERATIONS RESEARCH, EURO 1, BRUSSELS 1975. (1975).

CHALB 1976 -1 CHALMET, L. G.; GELDERS, L. F.
LAGRANGEAN RELAXATIONS FOR SOLVING A WAREHOUSING MODEL. WERKDOKUMENT 76-15. MIAMI (50. ORSA-MEETING + TIMS, NOVEMBER 1976) AND HEVERLEE: KATHOLIEKE UNIV. LEUVEN, FAKULTEIT TOEGEPASTE WETENSCHAPPEN, 10 P. (1976).

CHAM 1976 -1 CHAMBERS, M. L.; DYSON, R. G.
CUTTING STOCK PROBLEM IN THE FLAT GLASS INDUSTRY - SELECTION OF STOCK SIZES, THE. OPERATIONAL RESEARCH QUARTERLY 27, 949-958. (1976).

CHAMA 1977 -1 CHAMPEAUX, D. DE; SINT, L.
IMPROVED BIDIRECTIONAL HEURISTIC SEARCH ALGORITHM, AN. JOURNAL OF THE ASSOCIATION FOR COMPUTING MACHINERY 24, 177-191. (1977).

CHAN 1976 -1 CHAN, A. W.; FRANCIS, R. L.
LEAST TOTAL DISTANCE FACILITY CONFIGURATION PROBLEM INVOLVING LATTICE POINTS, A. MANAGEMENT SCIENCE 22, 778-787. (1976).

CHAN 1976 -2 CHAN, A. W.; FRANCIS, R. L.
ROUND-TRIP LOCATION PROBLEM ON A TREE GRAPH, A. TRANSPORTATION SCIENCE 10, 35-51. (1976).

CHANA 1976 -1 CHAN, Y.
TRANSPORTATION NETWORK INVESTMENT PROBLEM, THE; A SURVEY OF TREE SEARCH SOLUTION ALGORITHMS. TECHNICAL NOTE 105. PHILADELPHIA (49. ORSA-MEETING + TIMS, MARCH/APRIL 1976) AND UNIVERSITY PARK: PENNSYLVANIA STATE UNIV., TRANSPORTATION INST., 32 P. (1976).

CHANB 1969 -1 CHAN, Y. P.
OPTIMAL TRAVEL TIME REDUCTION IN A TRANSPORT NETWORK. AN APPLICATION OF NETWORK AGGREGATION AND BRANCH AND BOUND TECHNIQUES. RESEARCH REPORT R, DEPARTMENT OF CIVIL ENGINEERING, M.I.T. (1969).

CHANC 1977 -1 CHAND, M.; AGRAWAL, S. C.
ON INTEGER SOLUTIONS TO COMPLEMENTARY PROGRAMMING PROBLEMS. OPSEARCH 14, 99-107. (1977).

CHAND 1976 -1 CHANDRA, A. K.; HIRSCHBERG, D. S.; WONG, C. K.
APPROXIMATE ALGORITHMS FOR SOME GENERALIZED KNAPSACK PROBLEMS. THEORETICAL COMPUTER SCIENCE 3, 293-304. (1976).

CHANE 1976 -1 CHANDRA, A. K.; MARKOWSKY, G.
ON THE NUMBER OF PRIME IMPLICANTS. YORKTOWN HEIGHTS: IBM RESEARCH CENTER, 7 P. (1976).

CHANF 1977 -1 CHANDRASEKARAN, R.
MINIMAL RATIO SPANNING TREES. NETWORKS 7, 335-342. (1977).

CHANG 1975 -1 CHANDRASEKARAN, R.; EMMONS, H.
SEQUENCING JOBS FOR A SINGLE WORKER TO PROCESS AND DELIVER. PAPER PRESENTED AT THE 48TH JOINT NATIONAL MEETING OF ORSA AND TIMS, LAS VEGAS 1975, 11 P. (1975).

CHANH 1975 -1 CHANDRASEKARAN, R.; MITTAL, A. K.
OPTIMAL REARRANGEMENT OF OBJECT. PAPER PRESENTED AT THE 48TH JOINT NATIONAL MEETING OF ORSA AND TIMS, LAS VEGAS 1975, 15 P. (1975).

CHANI 1977 -1 CHANDRASEKARAN, R.; RAO, S. S.
SPECIAL CASE OF THE TRANSPORTATION PROBLEM, A. OPERATIONS RESEARCH 25, 525-528. (1977).

CHANJ 1976 -1 CHANG, L.; KORSH, J. F.
CANONICAL COIN CHANGING AND GREEDY SOLUTIONS. JOURNAL OF THE ASSOCIATION FOR COMPUTING MACHINERY 23, 418-422. (1976).

CHANK 1977 -1 CHANG, R. L. P.; PAVLIDIS, T.
FUZZY DECISION TREES. ATLANTA (52. ORSA-MEETING + TIMS, NOVEMBER 1977) 11 P. (1977).

CHAR 1975 -1 CHARNES, A.; GLOVER, F.; KLINGMAN, D.
PAST, PRESENT AND FUTURE OF LARGE SCALE TRANSSHIPMENT COMPUTER CODES AND APPLICATIONS. BERLIN: INTERNATIONAL INSTITUTE OF MANAGEMENT, 26 P. (1975). (PREPRINT SERIES I/75-11)

CHARA 1975 -1 CHARNES, A.; GRANOT, D.; GRANOT, F.
ON IMPROVING BOUNDS FOR VARIABLES IN LINEAR INTEGER PROGRAMS BY SURROGATE CONSTRAINTS. INFOR 13, 260-269. (1975).

CHARA 1977 -1 CHARNES, A.; GRANOT, D.; GRANOT, F.
ON INTERSECTION CUTS IN INTERVAL INTEGER LINEAR PROGRAMMING. OPERATIONS RESEARCH 25, 352-355. (1977).

CHARB 1977 -1 CHARTRAND, G.
GRAPHS AS MATHEMATICAL MODELS. BOSTON: PRINDLE, WEBER & SCHMIDT, 294 P. (1977).

CHARC 1973 -1 CHARTRAND, G.; KAPOOR, S. F.; KRONK, H. V.
MANY FACETS OF HAMILTONIAN GRAPHS, THE. THE MATHEMATICS STUDENT 41, 327-336. (1973).

CHARD 1967 -1 CHARTRES, B. A.
ANMERKUNG ZU DEM AUFSATZ VON NICHOLSON (COMPUTER JOURNAL 9, 1966, 275). COMPUTER JOURNAL 10, 118-119. (1967).

CHAT 1975 -1 CHATY, GUY.
CHEMINEMENTS REMARQUABLES DANS LES GRAPHES: EXISTENCE, OBTENTION, CONSERVATION. DISSERTATION. PARIS (1975).

****

CHEB 1975 -1 CHEBOTAREV, A. S.
SEQUENTIAL OPTIMIZATION IN ONE SCHEDULING PROBLEM (RUSSIAN). IZVESTIJA AKADEMII NAUK SSSR. TEHNICESKAJA KIBERNETIKA, NO. 1, 32-38 + 219; TRANSLATED AS ENGINEERING CYBERNETICS 13, 1975, NO. 1, 21-27. (1975).

CHEBA 1975 -1 CHEBOTAREV, A. S.; SHIBANOV, S. E.
PROBABILISTIC PROBLEM OF SEQUENCING PROJECTS IN SEVERAL STEPS. IZVESTIJA AKADEMII NAUK SSSR, TEHNICESKAJA KIBERNETIKA, NO. 4, 44-47. (1975).

CHEI 1975 -1 CHEIN, M.
REMARQUE SUR L'ALGORITHME DE DUCAMP POUR LA RECHERCHE DE TOUS LES ORDRES TOTAUX CONTENANT UN ORDRE PARTIEL. INF. MATHEMATIQUES ET SCIENCES HUMAINES, NO. 51, 47-50. (1975).

CHEN 1975 -1 CHEN, C. C.; DAYKIN, D. E.
HAMILTONIAN CYCLES IN A LINE-COLOURED MULTIGRAPH AND A RELATED UNCOLOURED GRAPH. NANTA MATHEMATICA 8, NO. 2, 76-77. (1975).

CHENA 1977 -1 CHEN, DER-S.
ALGORITHM FOR A MANPOWER SCHEDULING MODEL, AN. SAN FRANCISCO (51. ORSA-MEETING + TIMS, MAY 1977) 23 P. (1977).

CHENB 1976 -1 CHEN, DER-S.; ZIONTS, S.
COMPARISON OF SOME ALGORITHMS FOR SOLVING THE GROUP THEORETIC INTEGER PROGRAMMING PROBLEM. OPERATIONS RESEARCH 24, 1120-1128. (1976).

CHENC 1975 -1 CHEN, I. N.
NODE-ELIMINATION METHOD FOR FINDING ALL SIMPLE CYCLES IN A DIRECTED GRAPH, A. IN: F. HOFFMAN, R. C. MULLIN, R. B. LEVOW ET AL. (EDS.): PROCEEDINGS OF THE 6TH SOUTHEASTERN CONFERENCE ON COMBINATORICS, GRAPH THEORY, AND COMPUTING, BOCA RATON 1975, 147-160. CONGRESSUS NUMERANTIUM 14, WINNIPEG: UTILITAS MATHEMATICA. (1975).

CHEND 1974 -1 CHEN, J. H. K.
DISOPT - A GENERAL PROGRAM FOR CONTINUOUS AND DISCRETE NONLINEAR PROGRAMMING PROBLEMS. DISSERTATION. HAMILTON: MCMASTER UNIV., 80 P. (1974).

CHEND 1975 -1 CHEN, J. H. K.
DISOPT - A GENERAL PROGRAM FOR CONTINUOUS AND DIŠCRETE NONLINEAR PROGRAMMING PROBLEMS. INTERNAL REPORTS IN SIMULATION, OPTIMIZATION AND CONTROL, SOC-2 9. HAMILTON: MCMASTER UNIV., FACULTY OF ENGINEERING, 79 P. (1975).

CHENE 1976 -1 CHEN, R.; CROWDER, H.; JOHNSON, E. L.
INTEGER PROGRAMMING FORMULATION OF THE INSTALLATION SCHEDULING PROBLEM, AN. IBM RESEARCH REPORT RC 6196. YORKTOWN HEIGHTS: IBM RESEARCH CENTER, 7 P. (1976).

CHENF 1977 -1 CHEN, S.; SAIGAL, R.
PRIMAL ALGORITHM FOR SOLVING A CAPACITATED NETWORK FLOW PROBLEM WITH ADDITIONAL LINEAR CONSTRAINTS, A. NETWORKS 7, 59-80. (1977).

CHENG 1969 -1 CHEN, W. K.
GENERATION OF TREES AND COTREES OF A GRAPH BY DECOMPOSITION. PROCEEDINGS OF THE INSTITUTION OF ELECTRICAL ENGINEERS. ELECTRONICS, POWER, CONTROL & SCIENCE 116, 1639-1643. (1969).

CHENH 1976 -1 CHEN, W.-K.
APPLIED GRAPH THEORY. GRAPHS AND ELECTRICAL NETWORKS (2ND REVISED ED.). APPLIED MATHEMATICS AND MECHANICS, 13. AMSTERDAM, NEW YORK, OXFORD: NORTH-HOLLAND, 542 P. (1976).

CHENI 1971 -1 CHEN, Y. E.; SHEN, V. Y.
SCHEDULING PROBLEM IN A MULTIPROCESSING ENVIRONMENT WITH TWO CLASSES OF PROCESSORS. IN: PROCEEDINGS OF THE FIFTH ANNUAL PRINCETON CONFERENCE ON INFORMATION SCIENCES AND SYSTEMS, 110-115. (1971).

CHER 1970 -1 CHERENIN, V. P.
ALLOCATION OF WAREHOUSES WITH REGARD TO DEMANDS (RUSSIAN). IN: OPERATIONS RESEARCH (MODELS, SYSTEMS, SOLUTENS) NO. 1 (RUSSIAN), 128-159. MOSCOW: VYCISL. CENTER AKAD. NAUK SSSR. (1970).

CHERA 1977 -1 CHERENKOV, A. P.
PROBLEM OF DIFFERENT TYPES OF RESOURCES ALLOCATION (RUSSIAN), A. ZURNAL VYCISLITEL'NOJ MATEMATEKI I MATEMATICESKOJ FIZIKI 17, 320-327. (1977).

CHERB 1976 -1 CHERITON, D.; TARJAN, R. E.
FINDING MINIMUM SPANNING TREES. SIAM JOURNAL ON COMPUTING 5, 724-742. (1976).

CHERC 1976 -1 CHERKASSKII, B. V.
MULTIPORT TWO-PRODUCT PROBLEMS (RUSSIAN). IN: STUDIES IN DISCRETE OPTIMIZATION (RUSSIAN), 261-290. MOSCOW: IZDAT. NAUKA. (1976).

CHERD 1976 -1 CHERNYSHEV, YU. YU.; NASEKIN, V. YA.
ON SOLUTION OF THE PROBLEM OF COVERING BY GRADIENT ALGORITHM (RUSSIAN). KIBERNETIKA, NO. 4, 85-88. (1976).

CHERE 1975 -1 CHERVAK, N. K.
FINITE ALGORITHM FOR THE TRUNCATION METHOD FOR INTEGER LINEAR PROGRAMMING PROBLEMS WITH NONINTEGER TARGET FUNCTION (RUSSIAN), A. UKRAINSKII MATEMATICESKII ZURNAL 27, 134-137 + 144. (1975).

CHERF 1972 -1 CHERVAK, YU. YU.
DETERMINATION OF THE OPTIMAL INTEGER CONVEX LINEAR COMBINATION OF POINTS. THE GENERAL DISCRETE CASE (RUSSIAN). IN: MATHEMATICAL METHODS FOR THE SOLUTION OF ECONOMIC PROBLEMS (RUSSIAN), 130-132, SUPPLEMENT TO EKONOMIKA I MATEMATICESKIE METODY. MOSCOW: IZDAT. NAUKA. (1972).

CHERF 1973 -1 CHERVAK, YU. YU.
METHOD OF LEXICOGRAPHIC SEARCH FOR THE SOLUTION OF DISCRETE LINEAR PROGRAMMING PROBLEMS (RUSSIAN), THE. EKONOMIKA I MATEMATICESKIE METODY 9, 1137-1141. (1973).

CHERF 1974 -1 CHERVAK, YU. YU.
REMARKS ON THE FIRST GOMORY ALGORITHM (RUSSIAN). EKONOMIKA I MATEMATICESKIE METODY 10, 340-342. (1974).

CHERF 1976 -1 CHERVAK, YU. YU.
KNAPSACK PROBLEMS THAT PERMIT A SIMPLE METHOD FOR FINDING THEIR SOLUTIONS (RUSSIAN). EKONOMIKA I MATEMATICESKIE METODY 12, 184-186. (1976).

CHERF 1977 -1 CHERVAK, YU. YU.
ON AN IMPROVED VARIANT OF GOMORY'S ALGORITHM (RUSSIAN). EKONOMIKA I MATEMATICHESKIE METODY 13, 391-394. (1977).

CHEU 1976 -1 CHEUNG, T.-Y.
NETWORK FLOW APPROACH TO STEINER PROBLEMS, A. TECHNICAL REPORT TR 76-03. OTTAWA: UNIV. OF OTTAWA, DEPT. OF COMPUTER SCIENCE, 22 P. (1976).

CHEU 1976 -2 CHEUNG, T.-Y.
STEPPING-STONE PARALLEL-CUT METHOD FOR INTEGER PROGRAMMING, A. TECHNICAL REPORT TR 76-01. MIAMI (50. ORSA-MEETING + TIMS, NOVEMBER 1976) AND OTTAWA: UNIV. OF OTTAWA, DEPT. OF COMPUTER SCIENCE, 23 P. (1976).

CHEV 1977 -1 CHEVALIER, A.
PROGRAMMATION DYNAMIQUE. DE LA MAITRISE DE LA CONJONCTURE A LA PLANIFICATION, LA. PARIS: DUNOD, 218 P. (1977).

CHEW 1975 -1 CHEW, K. L.; YAP. K. Y.
OPTIMAL PRODUCTION RUN LENGTHS FOR TWO-PRODUCT SCHEDULING. NANTA MATHEMATICA 8, NO. 1, 1-13. (1975).

****

CHEWA 1974 -1 CHEW, K. L.; YAP, K. Y.
NON-OPTIMAL FLOW-SHOP SCHEDULES. NANTA MATHEMATICA 7, NO. 2, 75-83. (1974).

CHIA 1975 -1 CHIANG, C. H.
MODEL FOR MULTI-PERIOD REGIONAL WASTEWATER PLANNING, A. DISSERTATION. CHAPEL HILL: UNIV. OF NORTH CAROLINA. (1975).

CHIAA 1975 -1 CHIAPPA, R. A.
CHARACTERISATION DES GRAPHES ADJOINTS HAMILTONIENS. MATHEMATICA JAPONICAE 20, 53-56. (1975).

CHIS 1977 -1 CHISMAN, J. A.
OPTIMIZING THE SHIPPING FUNCTION. INDUSTRIAL ENGINEERING 9, NO. 3, 38-41. (1977).

CHOE 1976 -1 CHOE, S. R.
ON A NEW OPTIMAL DECISION METHOD FOR SCHEDULING PROBLEMS (KOREAN). SU-HAK KWA MUL-LI 20, NO. 2, 37-45. (1976).

CHOU 1975 -1 CHOU, W.
COMPUTER COMMUNICATION NETWORKS - THE PARTS MAKE UP THE WHOLE, 1975 NATIONAL COMPUTER CONFERENCE. IN: AFIPS CONFERENCE PROCEEDINGS, 119-128. MONTVALE: AFIPS PRESS. (1975).

CHOUA 1973 -1 CHOU, W.; KERSHENBAUM, A.
UNIFIED ALGORITHM FOR DESIGNING MULTIDROP TELEPROCESSING NETWORKS, A. IN: PROCEEDINGS OF THE THIRD IEEE SYMPOSIUM ON DATA NETWORKS ANALYSIS AND DESIGN, ST. PETERSBURG 1973, 148-156. NEW YORK: INSTITUTE OF ELECTRICAL AND ELECTRONICS ENGINEERS. (1973).

CHOW 1976 -1 CHOW, T. S.
LOWER BOUNDS ON THE COMPLEXITY OF 0-1-VALUED RECURSIVE FUNCTIONS. INFORMATION AND CONTROL 31, 17-42. (1976).

CHOY 1977 -1 CHOY, D. M.; WONG, C. K.
BOUNDS FOR OPTIMAL ALPHA-BETA BINARY TREES. BIT 17, 1-15. (1977).

CHRI 1973 -1 CHRISTOFIDES, N.
OPTIMUM TRANSVERSAL OF A GRAPH, THE. OMEGA 1, 719-734. (1973).

CHRI 1974 -1 CHRISTOFIDES, N.
HAMILTONIAN CIRCUITS AND THE TRAVELLING SALESMAN PROBLEM. LONDON: IMPERIAL COLLEGE, DEPT. OF MANAGEMENT SCIENCE, 23 P. (1974).

CHRI 1975 -1 CHRISTOFIDES, N.
HAMILTONIAN CIRCUITS AND THE TRAVELLING SALESMAN PROBLEM: A SURVEY. IN: B. ROY (ED.): COMBINATORIAL PROGRAMMING: METHODS AND APPLICATIONS. PROCEEDINGS OF THE NATO ADVANCED STUDY INST., VERSAILLES 1974, 149-172. DORDRECHT: REIDEL. (1975).

CHRI 1976 -1 CHRISTOFIDES, N.
VEHICLE ROUTING PROBLEM, THE. REVUE FRANCAISE D'AUTOMATIQUE, D'INFORMATIQUE ET DE RECHERCHE OPERATIONNELLE 10, NO. 2, 55-70. (1976).

CHRI 1976 -2 CHRISTOFIDES, N.
WORST-CASE ANALYSIS OF A NEW HEURISTIC FOR THE TRAVELLING SALESMAN PROBLEM. MIAMI (50. ORSA-MEETING + TIMS, NOVEMBER 1976) 5 P. (1976).

CHRIA 1977 -1 CHRISTOFIDES, N.; WHITLOCK, C.
ALGORITHM FOR TWO-DIMENSIONAL CUTTING PROBLEMS, AN. OPERATIONS RESEARCH 25, 30-44. (1977).

CHRIB 1976 -1 CHRISTOS, H.; PAPADIMITRIOU, H.; STEIGLITZ, K.
SOME COMPLEXITY RESULTS FOR THE TRAVELING SALESMAN PROBLEM. IN: PROCEEDINGS OF THE EIGHTH ANNUAL A.C.M. SYMPOSIUM ON THEORY OF COMPUTING, 1-9. (1976).

CHU 1965 -1 CHU, Y. J.; LIU T. H.
ON THE SHORTEST ARBORESCENCE OF DIRECTED GRAPHS. SCIENTIA SINICA 14, 1396-1400. (1965).

CHUA 1965 -1 CHUANG, Y. A.; SMITH, W. G.
DYNAMIC PROGRAMMING MODEL FOR COMBINED PRODUCTION, DISTRIBUTION AND STORAGE. JOURNAL OF INDUSTRIAL ENGINEERING 17, 7-13. (1965).

CHUN 1978 -1 CHUNG, F. R. K.; GRAHAM, R. L.
STEINER TREES FOR LADDERS. IN: ALSPACH, B.;HELL, P.;MILLER, D. J.(EDS.): ALGORITHMIC ASPECTS OF COMBINATORICS. ANNALS OF DISCRETE MATHEMATICS 2, 155-160. (1978).

CHUR 1976 -1 CHURCH, R. L.; GARFINKEL, R. S.
ON LOCATING AN OBNOXIOUS FACILITY ON A NETWORK: THE MAXIMUM MEDIAN PROBLEM. PAPER PRESENTED AT THE 49TH JOINT NATIONAL MEETING OF ORSA AND TIMS, PHILADELPHIA 1976, 15 P. (1976).

CHURA 1977 -1 CHURCH, R. L.; MEADOWS, M. E.
RESULTS OF A NEW APPROACH TO SOLVING THE P-MEDIAN PROBLEM WITH MAXIMUM DISTANCE CONSTRAINTS. GEOGRAPHICAL ANALYSIS 9, 364-378. (1977).

CHURB 1976 -1 CHURCH, R. L.; REVELLE, C. S.
THEORETICAL AND COMPUTATIONAL LINKS BETWEEN THE P-MEDIAN, LOCATION SET-COVERING, AND THE MAXIMAL COVERING LOCATION PROBLEM. GEOGRAPHICAL ANALYSIS 8, 406-415. (1976).

CHVA 1975 -1 CHVATAL, V.
SOME LINEAR PROGRAMMING ASPECTS OF COMBINATORICS. IN: D. CORNEIL, E. MENDELSOHN (EDS.): PROCEEDINGS OF THE CONFERENCE ON ALGEBRAIC ASPECTS OF COMBINATORICS, TORONTO 1975, 2-30. CONGRESSUS NUMERANTIUM 13, WINNIPEG: UTILITAS MATHEMATICA. (1975).

CHVA 1976 -1 CHVATAL, V.
ON THE STRONG PERFECT GRAPH CONJECTURE. JOURNAL OF COMBINATORIAL THEORY 20, SERIES B, 139-141. (1976).

CHVA 1977 -1 CHVATAL, V.
DETERMINING THE STABILITY NUMBER OF A GRAPH. SIAM JOURNAL ON COMPUTING 6, 643-662. (1977).

CHVAA 1978 -1 CHVATAL, V.; ERDOES P.
BIASED POSITIONAL GAMES. IN: ALSPACH, B.;HELL, P.;MILLER, D. J.(EDS.): ALGORITHMIC ASPECTS OF COMBINATORICS. ANNALS OF DISCRETE MATHEMATICS 2, 221-230. (1978).

CHVAB 1978 -1 CHVATAL, V.; GAREY, M. R.; JOHNSON, D. S.
TWO RESULTS CONCERNING MULTICOLORING. IN: ALSPACH, B.;HELL, P.;MILLER, D. J.(EDS.): ALGORITHMIC ASPECTS OF COMBINATORICS. ANNALS OF DISCRETE MATHEMATICS 2, 151-154. (1978).

CHVAC 1977 -1 CHVATAL, V.; HAMMER, P. L.
AGGREGATION OF INEQUALITIES IN INTEGER PROGRAMMING. IN: P. L. HAMMER, E. L. JOHNSON, B. H. KORTE, G. L. NEMHAUSER (EDS.): STUDIES IN INTEGER PROGRAMMING, 145-162. ANNALS OF DISCRETE MATHEMATICS 1, AMSTERDAM: NORTH-HOLLAND. (1977).

CHVAD 1976 -1 CHVATAL, V.; HANSON, D.
DEGREES AND MATCHINGS. JOURNAL OF COMBINATORIAL THEORY 20, SERIES B, 128-138. (1976).

CIOB 1975 -1 CIOBANU, G.
GENERAL ALGORITHM FOR SOLVING FLOW SHOP SCHEDULING, A. ECONOMIC COMPUTATION AND ECONOMIC CYBERNETICS STUDIES AND RESEARCH, NO. 3, 207-212. (1975).

CIOBA 1977 -1 CIOBOTARU, S.
LINGUISTIC METHOD FOR DETERMINING THE HAMILTONIAN PATHS (RUMANIAN), A. STUDII SI CERCETARI MATEMATICE 29, 103-112. (1977).

CIRI 1975 -1 CIRILLO, F.
NEW METHOD FOR THE SCHEDULING OF A PROJECT (ITALIAN), A. RIC., MAT. PURE APPL. 26, 25-36. (1975).

CLAR 1976 -1 CLARK, C. R.
VEHICLE FLEET PROBLEM, A. RICERCA OPERATIVA 6, NO. 3, 49-56. (1976).

****

CLAU 1977 -1 CLAUDER, G.; JAENICKE, W.
ZUR GRAPHENTHEORETISCHEN LOESUNG EINER KLASSE VON DISKRETEN OPTIMIERUNGSPROBLEMEN. MATHEMATISCHE OPERATIONSFORSCHUNG UND STATISTIK 8, SERIES OPTIMIZATION, 323-328. (1977).

CLAUA 1975 -1 CLAUS, A.; KLEITMAN, D. J.
HEURISTIC METHODS FOR SOLVING LARGE SCALE NETWORK ROUTING PROBLEMS: THE TELPACKING PROBLEM. STUDIES IN APPLIED MATHEMATICS 54, 1, 17-29. (1975).

COCK 1975 -1 COCK, J. DE
SEARCH PROCEDURE FOR THE CONSTRUCTION OF SCHOOL TIMETABLES, A. PAPER PRESENTED AT THE FIRST EUROPEAN CONGRESS ON OPERATIONS RESEARCH, EURO 1, BRUSSELS 1975. (1975).

COCKA 1976 -1 COCKAYNE, E.; HEDETNIEMI, S.
LINEAR ALGORITHM FOR THE MAXIMUM WEIGHT OF AN INDEPENDENT SET IN A TREE, A. IN: F. HOFFMAN, L. LESNIAK, R. MULLIN ET AL. (EDS.): PROCEEDINGS OF THE 7TH SOUTHEASTERN CONFERENCE ON COMBINATORICS, GRAPH THEORY, AND COMPUTING, BATON ROUGE 1976, 217-228. CONGRESSUS NUMERANTIUM 17, WINNIPEG: UTILITAS MATHEMATICA. (1976).

COCKB 1975 -1 COCKAYNE, E. J.; HEDETNIEMI, S. T.
OPTIMAL DOMINATION IN GRAPHS. IEEE TRANSACTIONS ON CIRCUITS AND SYSTEMS CS-22, 855-857. (1975).

COCKB 1976 -1 COCKAYNE, E. J.; HEDETNIEMI, S. T.
DISJOINT INDEPENDENT DOMINATING SETS IN GRAPHS. DISCRETE MATHEMATICS 15, 213-222. (1976).

COCKB 1977 -1 COCKAYNE, E. J.; HEDETNIEMI, S. T.
TOWARDS A THEORY OF DOMINATION IN GRAPHS. NETWORKS 7, 247-262. (1977).

COCKC 1977 -1 COCKAYNE, E. J.; ROBERTS, F. D. K.
COMPUTATION OF DOMINATING PARTITIONS. INFOR 15, 94-106. (1977).

CODY 1976 -1 CODY, R. A.; COFFMAN, E. G. JR.
RECORD ALLOCATION FOR MINIMIZING EXPECTED RETRIEVAL COSTS ON DRUM-LIKE STORAGE DEVICES. JOURNAL OF THE ASSOCIATION FOR COMPUTING MACHINERY 23, 103-115. (1976).

COFF 1975 -1 COFFMAN, E. G. JR.
DETERMINISTIC SEQUENCING: COMPLEXITY AND OPTIMAL ALGORITHMS. IN: COMPUTING SYSTEMS 1975, PROCEEDINGS OF THE 4TH TEXAS CONF., AUSTIN 1975, PAPER 5B-1, 12 P. AUSTIN: UNIV. OF TEXAS. (1975).

COFF 1976 -1 COFFMAN, E. G. JR.
COMPUTER AND JOB/SHOP SCHEDULING THEORY. NEW YORK: WILEY, 299 P. (1976).

COFFA 1976 -1 COFFMAN, E. G. JR.; SETHI, R.
ALGORITHMS MINIMIZING MEAN FLOW TIME SCHEDULE-LENGTH PROPERTIES. ACTA INFORMATICA 6, 1-14. (1976).

COGI 1976 -1 COGIS, O.
FORMALISATION ET GENERALISATION D'UN PRINCIPE CLASSIQUE D'ALGORITHME DE RECHERCHE D'ISOMORPHISME ENTRE DEUX GRAPHES. IN: ANAL. APPL. INF. JOURN. COMB. INF., BORDEAUX 1975, 69-78. BORDEAUX: UNIV. BORDEAUX I. (1976).

COHE 1973 -1 COHEN, J.
SURVEY ON THE WAREHOUSE LOCATION PROBLEM. WORKING PAPER OR 022-73. CAMBRIDGE: MASS. INST. OF TECHNOLOGY, SLOAN SCHOOL OF MANAGEMENT. (1973).

COI 1975 -1 COI, S.; SERMAN, A. I.
APPROXIMATE METHOD FOR THE SOLUTION OF INTEGER PROBLEMS OF LINEAR PROGRAMMING (THE METHOD OF ADMISSIBLE COMPONENTS) (RUSSIAN), AN. IN: CODING THEORY AND INFORMATION MODELING, NO. 2 (RUSSIAN), 88-101 + 246. ALMA-ATA: IZDAT. NAUKA KAZAH. SSR. (1975).

COLL 1975 -1 COLLATZ, L.
BEZIEHUNGEN ZWISCHEN GRAPHEN, GEOMETRIE UND KOMBINATORIK, EINIGE. SER. INTERNATION. ANAL. NUMER. 29, 27-56. (1975).

COME 1966 -1 COMES, G.; BELLON, J.-L.
ETUDE D'UNE POLITIQUE SEQUENTIELLE D'INVESTISSEMENTS POUR LA RESTRUCTURATION D'UN RESEAU D'USINES. IN: D. B. HERTZ, J. MELESE (EDS.): PROCEEDINGS OF THE FOURTH INTERNATIONAL CONFERENCE ON OPERATIONAL RESEARCH, 825-833. NEW YORK: WILEY. (1966).

CONT 1976 -1 CONTI, A.; ROSSETTO, S.
LAMINATION OF SEAMLESS TUBES: A HEURISTIC METHOD FOR CHOOSING THE BARS (ITALIAN). RICERCA OPERATIVA 6, NO. 4, 37-46. (1976).

CONW 1966 -1 CONWAY, R. W.; MAXWELL, W. L.; OLDZIEY, J. W.
SEQUENCING AGAINST DUE-DATES. 4 TH PROCEEDING OF THE IFORS CONFERENCE 1966. NEW YORK: WILEY. (1966).

COOK 1974 -1 COOK, R. J.
POINT-ARBORICITY AND GIRTH. JOURNAL OF THE LONDON MATHEMATICAL SOCIETY 8, 322-324. (1974).

COOK 1975 -1 COOK, R. J.
CHROMATIC NUMBER AND GIRTH. PERIODICA MATHEMATICA HUNGARICA 6, 103-107. (1975).

COOKA 1974 -1 COOK, S. A.; RECKHOW, R. A.
ON THE LENGTHS OF PROOFS IN THE PROPOSITIONAL CALCULUS. IN: PROCEEDINGS OF THE SIXTH ANNUAL ACM SYMPOSIUM ON THEORY OF COMPUTING, 135-148. (1974).

COOP 1963 -1 COOPER, L.
LOCATION-ALLOCATION PROBLEM. OPERATIONS RESEARCH 11, 331-343. (1963).

COOP 1975 -1 COOPER, L.
EFFICIENT HEURISTIC ALGORITHM FOR THE TRANSPORTATION-LOCATION PROBLEM, AN. TECHNICAL REPORT IEOR 75017. DALLAS: SOUTHERN METHODIST UNIV., DEPT. OF INDUSTRIAL ENGINEERING, 12 P. (1975).

COOP 1975 -2 COOPER, L.
FIXED CHARGE PROBLEM, I. A NEW HEURISTIC METHOD, THE. COMPUTERS AND MATHEMATICS WITH APPLICATIONS 1, 89-95. (1975).

COOPA 1977 -1 COOPER, L.; BHAT, U. N.; LEBLANC, L. J.
INTRODUCTION TO OPERATIONS RESEARCH MODELS. PHILADELPHIA, LONDON, TORONTO: SAUNDERS, 404 P. (1977).

COOPB 1974 -1 COOPER, L.; COOPER, M. W.
ALL INTEGER LINEAR PROGRAMMING - A NEW APPROACH VIA DYNAMIC PROGRAMMING. TECHNICAL REPORT CP-74023. DALLAS: SOUTHERN METHODIST UNIV., INST. OF TECHNOLOGY, 21 P. (1974).

COOPB 1975 -1 COOPER, L.; COOPER, M. W.
NON-LINEAR INTEGER PROGRAMMING. COMPUTERS AND MATHEMATICS WITH APPLICATIONS 1, 215-222. (1975).

COOPB 1976 -1 COOPER, L.; COOPER, M. W.
EXACT ALGORITHM FOR THE CONCAVE TRANSPORTATION PROBLEM, AN. COMPUTERS AND MATHEMATICS WITH APPLICATIONS 2, 49-61. (1976).

COOPC 1977 -1 COOPER, L.; LEBLANC, L. J.
STOCHASTIC TRANSPORTATION PROBLEMS AND OTHER NETWORK RELATED CONVEX PROBLEMS. NAVAL RESEARCH LOGISTICS QUARTERLY 24, 327-338. (1977).

COOPD 1977 -1 COOPER, M. W.
IMPROVED ALGORITHM FOR NON-LINEAR INTEGER PROGRAMMING, AN. TECHNICAL REPORT IEOR 77005. DALLAS: SOUTHERN METHODIST UNIV., DEPT. OF INDUSTRIAL ENGINEERING AND OPERATIONS RESEARCH, 15 P. (1977).

CORB 1972 -1
****

CORB 1972 -1 CORBAN, A.
THREE DIMENSIONAL STOCHASTIC TRANSPORTATION PROBLEM (RUMANIAN), THE. STUDII SI CERCETARI MATEMATICE 24, 513-517. (1972).

CORB 1973 -1 CORBAN, A.
TRANSPORTATION PROBLEMS WITH FIXED COSTS (RUMANIAN). STUDII SI CERCETARI MATEMATICE 25, 637-641. (1973).

CORB 1975 -1 CORBAN, A.
MULTIDIMENSIONAL TRANSPORTATION PROBLEM, II (RUMANIAN), A. STUDII SI CERCETARI MATEMATICE 27, 413-424. (1975).

CORB 1975 -2 CORBAN, A.
TIME CRITERION IN TRANSPORTATION PROBLEMS (RUMANIAN), THE. STUDII SI CERCETARI MATEMATICE 27, 651-654. (1975).

CORN 1968 -1 CORNEIL, D. G.
GRAPH ISOMORPHISM. DISSERTATION, TECHNICAL REPORT. TORONTO: UNIV. OF TORONTO, DEPT. OF COMPUTER SCIENCE. (1968).

CORN 1970 -1 CORNEIL, D. G.
GRAPH ISOMORPHISM. TECHNICAL REPORT NO. 18, UNIV. OF TORONTO, DEPARTMENT OF COMPUTER SCIENCE. (1970).

CORN 1972 -1 CORNEIL, D. G.
ALGORITHM FOR DETERMINING THE AUTOMORPHISM PARTITIONING OF AN UNDIRECTED GRAPH, AN. BIT 12, 161-171. (1972).

CORN 1974 -1 CORNEIL, D. G.
ANALYSIS OF GRAPH THEORETICAL ALGORITHMS, THE. IN: F. HOFFMAN, R. A. KINGSLEY, R. B. LEVOW, R. C. MULLIN (EDS.): PROCEEDINGS OF THE FIFTH SOUTHEASTERN CONF. ON COMBINATORICS, GRAPH THEORY, AND COMPUTING, BOCA RATON 1974, 3-38. CONGRESSUS NUMERANTIUM 10, WINNIPEG: UTILITAS MATHEMATICA. (1974).

CORNA 1970 -1 CORNEIL, D. G.; GOTLIEB, C. C.
EFFICIENT ALGORITHM FOR GRAPH ISOMORPHISM, AN. JOURNAL OF THE ASSOCIATION FOR COMPUTING MACHINERY 20, 51-64. (1970).

CORNB 1978 -1 CORNEIL, D. G.; MATHON, R. A.
ALGORITHMIC TECHNIQUES FOR THE GENERATION AND ANALYSIS OF STRONGLY REGULAR GRAPHS AND OTHER COMBINATORIAL CONFIGURATIONS. IN: ALSPACH, B.;HELL, P.;MILLER, D. J.(EDS.): ALGORITHMIC ASPECTS OF COMBINATORICS. ANNALS OF DISCRETE MATHEMATICS 2, 1-32. (1978).

CORNC 1977 -1 CORNUEJOLS, G.; FISHER, M.; NEMHAUSER, G. L.
ON THE UNCAPACITATED LOCATION PROBLEM. IN: P. L. HAMMER, E. L. JOHNSON, B. H. KORTE, G. L. NEMHAUSER (EDS.): STUDIES IN INTEGER PROGRAMMING, 163-178. ANNALS OF DISCRETE MATHEMATICS 1, AMSTERDAM: NORTH-HOLLAND. (1977).

CORND 1976 -1 CORNUEJOLS, G.; FISHER, M. L.; NEMHAUSER, G. L.
ANALYSIS OF HEURISTICS AND RELAXATIONS FOR THE UNCAPACITATED LOCATION PROBLEM, AN. WORKING PAPER 76-02-01. PHILADELPHIA: UNIV. OF PENNSYLVANIA, DEPT. OF DECISION SCIENCES, 48 P. (1976).

CORND 1977 -1 CORNUEJOLS, G.; FISHER, M. L.; NEMHAUSER, G. L.
LOCATION OF BANK ACCOUNTS TO OPTIMIZE FLOAT: AN ANALYTIC STUDY OF EXACT AND APPROXIMATE ALGORITHMS. MANAGEMENT SCIENCE 23, 789-810. (1977).

CORNE 1978 -1 CORNUEJOLS, G.; NEMHAUSER, G. L.
TIGHT BOUNDS FOR CHRISTOFIDES' TRAVELING SALESMAN HEURISTIC. MATHEMATICAL PROGRAMMING 14, 116-121. (1978).

COT 1974 -1 COT, N.
LINEAR-TIME ORDERING PROCEDURE WITH APPLICATIONS TO VARIABLE-LENGTH ENCODING, A. IN: PROCEEDINGS OF THE EIGHTH ANNUAL PRINCETON CONFERENCE ON INFORMATION SCIENCES AND SYSTEMS, 460-463. (1974).

COTT 1974 -1 COTTLE, R.; KRARUP, J.
OPTIMIZATION METHODS FOR RESOURCE ALLOCATION. LONDON: ENGLISH UNIV. PRESS, 440 P. (1974).

COVE 1976 -1 COVERDALE, I. L.; WHARTON, F.
IMPROVED HEURISTIC PROCEDURE FOR A NONLINEAR CUTTING STOCK PROBLEM, AN. MANAGEMENT SCIENCE 23, 78-86. (1976).

CRAN 1976 -1 CRANDALL, K. C.
PROBABILISTIC TIME SCHEDULING. PROCEEDINGS OF THE AMERICAN SOCIETY OF CIVIL ENGINEERS, J. CONSTR. DIV. 102, 415-423. (1976).

CRAP 1970 -1 CRAPO, H. H.; ROTA, G. C.
ON THE FOUNDATIONS OF COMBINATORIAL THEORY: COMBINATORIAL GEOMETRIES. CAMBRIDGE: MIT PRESS. (1970).

CROU 1976 -1 CROUCHER, J. S.
OPTIMAL RESOURCE ALLOCATION AMONG AREAS OF EQUAL VALUE. NEW ZEALAND OPERATIONAL RESEARCH 4, 80-91. (1976).

CROW 1976 -1 CROWDER, H.
COMPUTATIONAL IMPROVEMENTS FOR SUBGRADIENT OPTIMIZATION. IN: PROCEEDINGS OF THE CONFERENCE ON MATHEMATICAL PROGRAMMING AND ITS APPLICATIONS, INDAM, ROME 1974 (ITALIAN), 357-372. SYMPOSIA MATHEMATICA 19, LONDON: ACADEMIC PRESS. (1976).

CRUM 1976 -1 CRUM, R.
CASH MANAGEMENT IN THE MULTINATIONAL FIRM: A CONSTRAINED GENERALIZED NETWORK APPROACH. WORKING PAPER. GAINESVILLE: UNIVERSITY OF FLORIDA. (1976).

CRUMA 1977 -1 CRUM, R. L.; KLINGMAN, D.; TAVIS, L. A.
OPERATIONAL APPROACH TO INTEGRATED WORKING CAPITAL PLANNING, AN. RESEARCH REPORT CCS 290. AUSTIN: UNIV. OF TEXAS, CENTER FOR CYBERNETIC STUDIES, 23 P. (1977).

CRUS 1975 -1 CRUSE, A. B.
PROOF OF FULKERSON'S CHARACTERIZATION OF PERMUTATION MATRICES, A. LINEAR ALGEBRA AND ITS APPLICATIONS 12, 21-28. (1975).

CUNN 1975 -1 CUNNINGHAM, W.
CHORDS AND DISJOINT PATHS IN MATROIDS. BALTIMORE: JOHN HOPKINS UNIV., DEPT. OF MATHEMATICAL SCIENCES, 55 P. (1975).

CUNNA 1974 -1 CUNNINGHAM, W. H.
COMBINATORIAL DECOMPOSITION THEORY, A. DISSERTATION. WATERLOO: UNIV. OF WATERLOO. (1974).

CUNNA 1975 -1 CUNNINGHAM, W. H.
UNBOUNDED MATROID INTERSECTION POLYHEDRON, AN. TECHNICAL REPORT 234. BALTIMORE:JOHN HOPKINS UNIV. DEPT. OF APPLIED MATHEMATICS. (1975).

CUNNA 1976 -1 CUNNINGHAM, W. H.
NETWORK SIMPLEX METHOD, A. MATHEMATICAL PROGRAMMING 11, 105-116. (1976).

CUNNB 1976 -1 CUNNINGHAM, W. H.; MARSH, A. B. III.
PRIMAL ALGORITHM FOR OPTIMUM MATCHING, A. TECHNICAL REPORT 262. BALTIMORE: JOHN HOPKINS UNIV., DEPT. OF MATHEMATICAL SCIENCES. (1976).

CVET 1976 -1 CVETKOVIC, D. M.; LACKOVIC, I. B.; SIMIC, S. K.
GRAPH EQUATIONS, GRAPH INEQUALITIES AND A FIXED POINT THEOREM. INSTITUT MATHEMATIQUE, PUBLICATIONS 20, 59-66. (1976).

****

CZER 1977 -1 CZERNIAKIEWICZ, A.
MINIMAL GRAPHS WITH CYCLIC COLOR GROUP. DISCRETE MATHEMATICS 19, 201-212. (1977).

DAEN 1977 -1 DAENINCK, G.; SMEERS, Y.
USING SHORTEST PATHS IN SOME TRANSSHIPMENT PROBLEM WITH CONCAVE COSTS. MATHEMATICAL PROGRAMMING 12, 18-25. (1977).

DAFE 1976 -1 DAFERMOS, S. C.
INTEGRATED EQUILIBRIUM FLOW MODELS FOR TRANSPORTATION PLANNING. IN: M. A. FLORIAN (ED.): TRAFFIC EQUILIBRIUM METHODS, PROCEEDINGS OF AN INTERNATIONAL SYMPOSIUM, MONTREAL 1974, 106-118. LECTURE NOTES IN ECONOMICS AND MATHEMATICAL SYSTEMS 118, BERLIN: SPRINGER. (1976).

DAGA 1977 -1 DAGANZO, C. F.; SHEFFI, Y.
ON STOCHASTIC MODELS OF TRAFFIC ASSIGNMENT. TRANSPORTATION SCIENCE 11, 253-274. (1977).

DALA 1976 -1 DALAL, Y. K.
DISTRIBUTED ALGORITHM FOR CONSTRUCTING MINIMAL SPANNING TREES IN COMPUTER-COMMUNICATION NETWORKS, A. IN: COMPUTING SYSTEMS 1976, PROCEEDINGS OF THE 5TH TEXAS CONF., AUSTIN 1976, 78-85. AUSTIN: UNIV. OF TEXAS. (1976).

DAMM 1977 -1 DAMM, E.
KUERZESTE WEG IN EINEM GITTERNETZ MIT HINDERNISSEN, DER. ANGEWANDTE INFORMATIK 19, 213-216. (1977).

DANA 1977 -1 DANARAJ, G.; KLEE, V.
CONNECTEDNESS GAME AND THE C-COMPLEXITY OF CERTAIN GRAPHS, THE. SIAM JOURNAL ON APPLIED MATHEMATICS 32, 431-442. (1977).

DANI 1977 -1 DANILENKO, S. E.; KAGAN, B. M.; SAHUNJANC, T. G.
DISCRETE ALGORITHMS IN OPTIMIZATION PROBLEMS (RUSSIAN). AVTOMATIKA I VYCISLITEL'NAJA TEHNIKA, NO. 2, 39-42 + 93. (1977).

DANN 1977 -1 DANNENBRING, D. G.
EVALUATION OF FLOW SHOP SEQUENCING HEURISTICS, AN. MANAGEMENT SCIENCE 23, 1174-1182. (1977).

DANN 1977 -2 DANNENBRING, D. G.
PROCEDURES FOR ESTIMATING OPTIMAL SOLUTION VALUES FOR LARGE COMBINATORIAL PROBLEMS. MANAGEMENT SCIENCE 23, 1273-1283. (1977).

DANT 1967 -1 DANTZIG, G. B.
ALL SHORTEST ROUTES IN A GRAPH. IN: P. ROSENSTIEHL (ED.): THEORY OF GRAPHS, 91-92. NEW YORK: GORDON AND BREACH (1967).

DANT 1975 -1 DANTZIG, G. B.
ON THE SHORTEST ROUTE THROUGH A NETWORK. IN: D. R. FULKERSON (ED.): STUDIES IN GRAPH THEORY, PART 1, 89-93. STUDIES IN MATHEMATICS 11, WASHINGTON: MATHEMATICAL ASSOCIATION OF AMERICA. (1975).

DANTA 1966 -1 DANTZIG, G. B.; BLATTNER, W. O.; RAO, M. R.
ALL SHORTEST ROUTES FROM A FIXED ORIGIN IN A GRAPH. TECHNICAL REPORT 66-2. STANFORD: STANFORD UNIV., OPERATIONS RESEARCH HOUSE. (1966).

DANTB 1975 -1 DANTZIG, G. B.; EAVES, B. C.
FOURIER-MOTZKIN ELIMINATION AND ITS DUAL WITH APPLICATION TO INTEGER PROGRAMMING. IN: B. ROY (ED.): COMBINATORIAL PROGRAMMING: METHODS AND APPLICATIONS. PROCEEDINGS OF THE NATO ADVANCED STUDY INST., VERSAILLES 1974, 93-102. DORDRECHT: REIDEL. (1975).

DANTC 1974 -1 DANTZIG, G. B.; FRIEL, J.; GOLIGHTLY, R.; HARVEY, R. P.; MAC KNIGHT, R. D.
SOLUTION OF A LARGE SCALE AIRFORCE ORDNANCE PLANNING PROBLEM BY MATHEMATICAL PROGRAMMING. PAPER PRESENTED AT THE 45TH JOINT NATIONAL MEETING OF ORSA AND TIMS, BOSTON 1974, 25 P. (1974).

DAR- 1975 -1 DAR-EL, E. M.; COTHER, R. F.
ASSEMBLY LINE SEQUENCING FOR MODEL MIX. INTERNATIONAL JOURNAL OF PRODUCTION RESEARCH 13, 463-477. (1975).

DAR-A 1977 -1 DAR-EL, E. M.; TUR, Y.
MULTI-RESOURCE PROJECT SCHEDULING ALGORITHM, A. AIIE TRANSACTIONS 9, 44-52. (1977).

DAVI 1975 -1 DAVIES, J.
ON BASES OF INDEPENDENCE STRUCTURES INTERSECTING IN A SET OF PRESCRIBED CARDINALITY. JOURNAL OF THE LONDON MATHEMATICAL SOCIETY 12, 455-458. (1975).

DAVIA 1975 -1 DAVIES, O.
SIX LOCATION STRATEGIES WITH INCREASING DEMAND LEVELS. PAPER PRESENTED AT THE 71ST ANNUAL MEETING OF THE ASSOCIATION OF AMERICAN GEOGRAPHERS, MILWAUKEE 1975. (1975).

DAVIB 1976 -1 DAVIES, O.; THOMAS, E. A. C.
FACILITY LOCATION WITH CHANGING DEMAND LEVELS. GEOGRAPHICAL ANALYSIS 8, 376-394. (1976).

DAVIC 1977 -1 DAVIS, K. R.; SIMMONS, L. F.
IMPROVING ASSEMBLY LINE EFFICIENCY: A DYNAMIC PROGRAMMING-HEURISTIC APPROACH. COMPUTERS AND OPERATIONS RESEARCH 4, 75-88. (1977).

DAVID 1977 -1 DAVIS, K. R.; WALTERS, J. E.
ADDRESSING THE N/1 SCHEDULING PROBLEM - A HEURISTIC APPROACH. COMPUTERS AND OPERATIONS RESEARCH 4, 89-100. (1977).

DAVIE 1976 -1 DAVIS, R. P.; TERRELL, M. P.
APPROXIMATION TECHNIQUE FOR PSEUDO-BOOLEAN MAXIMIZATION PROBLEMS, AN. AIIE TRANSACTIONS 8, 365-368. (1976).

DAVY 1972 -1 DAVYDOV, E. G.; SIGAL, I. KH.
APPLICATION OF THE PENALTY FUNCTION METHOD IN INTEGER PROGRAMMING PROBLEMS (RUSSIAN). IZVESTIJA AKADEMII NAUK SSSR, TEHNICESKAJA KIBERNETIKA, NO. 1, 28-31; TRANSLATED AS ENGINEERING CYBERNETICS 10, 1972, 21-24. (1972).

DAY 1975 -1 DAY, J. E.; HOTTENSTEIN, M. P.
IMPACT OF ADVANCING DUE DATES IN A PURE JOB SHOP, THE. INTERNATIONAL JOURNAL OF PRODUCTION RESEARCH 13, 603-614. (1975).

DAY A 1977 -1 DAY, W. H. E.
VALIDITY OF CLUSTERS FORMED BY GRAPH-THEORETIC CLUSTER METHODS. MATHEMATICAL BIOSCIENCES 36, 299-318. (1977).

DAYK 1976 -1 DAYKIN, D. E.
CORRIGENDUM AND ADDENDUM TO: "PSEUDO-MATCHINGS OF A BIPARTITE GRAPH" BY A. BRACE AND DAYKIN (PROCEEDINGS OF THE AMS 42, 1974, 28-32). PROCEEDINGS OF THE AMERICAN MATHEMATICAL SOCIETY 56, 380-382. (1976).

DAYK 1976 -2 DAYKIN, D. E.
GRAPHS WITH CYCLES HAVING ADJACENT LINES DIFFERENT COLORS. JOURNAL OF COMBINATORIAL THEORY 20, SERIES B, 149-152. (1976).

DEAR 1977 -1 DEAR, R. G.
EFFICIENT SEQUENCING AND SCHEDULING OF AIRCRAFT IN THE TERMINAL AREA, THE. SAN FRANCISCO (51. ORSA-MEETING + TIMS, MAY 1977) 19 P. (1977).

DEARA 1976 -1 DEARING, P. M.; FRANCIS, R. L.; LOWE, T. J.
CONVEX LOCATION PROBLEMS ON TREE NETWORKS. OPERATIONS RESEARCH 24, 628-642. (1976).

DEARB 1975 -1 DEARING, P. M.; LANGFORD, G. J.
MULTIFACILITY TOTAL COST LOCATION PROBLEM ON A TREE NETWORK, THE. TECHNICAL REPORT 209. CLEMSON: CLEMSON UNIV., DEPT. OF MATHEMATICAL SCIENCES, 21 P. (1975).

DEBA 1976 -1 DEBANNE, J. G.
GENERALIZED NETWORK AND INTEGER PROGRAMMING IN TECHNOLOGY ASSESSMENT AND PLANNING AN ENERGY SCENARIO STUDY. WORKING PAPER NO. 76-13. OTTAWA: UNIV. OF OTTAWA, FACULTY OF MANAGEMENT SCIENCES, 29 P. (1976).

DEHN 1976 -1 DEHNERT, G.
BRANCH AND BOUND METHOD FOR SOLVING FIXED CHARGE NETWORK FLOW PROBLEMS, A. IN: A. PREKOPA (ED.): PROGRESS IN OPERATIONS RESEARCH, 213-229, AMSTERDAM, LONDON: NORTH HOLLAND. (1976).

DEIN 1969 -1 DEININGER, R. A.
UEBER DIE PLANUNG VON INTERKOMMUNALEN SYSTEMEN VON KLAERANLAGEN. GWF-WASSER/ABWASSER 110, 1443-1445. (1969).

DEINA 1971 -1 DEININGER, R. A.; SU, S. Y.
REGIONAL WASTE WATER TREATMENT SYSTEMS. ASCE ANNUAL AND NATIONAL ENVIRONMENTAL ENGINEERING MEETING, ST. LOUIS, MISSOURI. (1971).

DEJO 1977 -1 DEJON, B.
BESTIMMUNG VON R KUERZESTEN WEGEN IN NETZWERKEN UNTER NEBENBEDINGUNGEN: VERFAHREN VOM HOFFMAN-PAVLEY-TYP. IN: L. COLLATZ, G. MEINARDUS, W. WETTERLING (EDS.): NUMERISCHE METHODEN BEI OPTIMIERUNGSAUFGABEN BAND 3, OPTIMIERUNG BEI GRAPHENTHEORETISCHEN UND GANZZAHLIGEN PROBLEMEN, 47-54. INTERNATIONAL SERIES OF NUMERICAL MATHEMATICS 36, BASEL: BIRKHAEUSER (1977).

DEJOA 1976 -1 DEJON, B.; GATTINGER, M.
INNER AND OUTER ALGORITHMS FOR DETERMINING R SHORTEST PATHS UNDER SIDE CONDITIONS. ERLANGEN: UNIV. ERLANGEN-NUERNBERG, 19 P. (1976).

DEJOA 1977 -1 DEJON, B.; GATTINGER, M.
BESTIMMUNG VON R-EXTREMA IN HALBGEORDNETEN MENGEN. IN: R. HENN, P. KALL, B. KORTE, O. KRAFFT ET AL. (EDS.): OPERATIONS RESEARCH VERFAHREN XXV, METHODS OF OPERATIONS RESEARCH, 296-302. MEISENHEIM: HAIN. (1977).

DEMB 1976 -1 DEMBO, R.; MULVEY, J.
ON THE ANALYSIS AND COMPARISON OF MATHEMATICAL PROGRAMMINNG ALGORITHMS AND SOFTWARE. WORKING PAPER HBS 76-19. BOSTON: HARVARD UNIV. (1976).

DEME 1968 -1 DEMENCHUK, V. M.
FLOW INTERPRETATION AND ALGORITHM FOR SOLVING THE PROBLEM OF TIME-OPTIMAL ALLOCATION OF LIMITED RESOURCES. AUTOMATION AND REMOTE CONTROL 10, 1697-1702. (1968).

DEMI 1976 -1 DEMIDENKO, V. M.
SPECIAL CASE OF THE TRAVELLING SALESMAN PROBLEM (RUSSIAN), A. VESCI AKADEMII NAVUK BSSR, SERIJA FIZIKA-MATEMATYCNYH NAVUK, NO. 5, 28-32 + 139. (1976).

DEMIA 1976 -1 DEMIN, V. K.; CHEBOTAREV, A. S.
OPTIMAL SERVICE OF A DETERMINISTIC FLOW OF DEMANDS, I. LOWER BOUND (RUSSIAN). IZVESTIJA AKADEMII NAUK SSSR, TEHNICESKAJA KIBERNETIKA, NO. 4, 93-99. (1976).

DEO 1969 -1 DEO, N.
EXTENSIVE ENGLISH LANGUAGE BIBLIOGRAPHY ON GRAPH THEORY AND ITS APPLICATIONS, AN. TECHNICAL REPORT 32-1413. PASADENA: CALIFORNIA INST. OF TECHNOLOGY, JET PROPULSION LABORATORY. (1969).

DEO A 1977 -1 DEO, N.; DAVIS, J. M.; LORD, R. E.
NEW ALGORITHM FOR DIGRAPH ISOMORPHISM, A. BIT 17, 16-30. (1977).

DERI 1975 -1 DERIGS, U.
RANGZUORDNUNGSPROBLEME. DISSERTATION. KOELN: UNIV. KOELN, MATHEMATISCHES INSTITUT, 13 P. (1975).

DERI 1977 -1 DERIGS, U.
DUALITY AND THE ALGEBRAIC MATCHING PROBLEM. AACHEN (2. SYMPOSIUM UEBER OPERATIONS RESEARCH) 12 P. (1977).

DERI 1977 -2 DERIGS, U.
RANGZUORDNUNGPROBLEME. ZEITSCHRIFT FUER OPERATIONS RESEARCH 21, SERIES A, 75-84. (1977).

DERI 1977 -3 DERIGS, U.
RANGZUORDNUNGSPROBLEME UND DAS STABLE MARRIAGE PROBLEM. IN: R. HENN, P. KALL, B. KORTE, O. KRAFFT ET AL. (EDS.): OPERATIONS RESEARCH VERFAHREN XXV, METHODS OF OPERATIONS RESEARCH, 147-155, MEISENHEIM: HAIN. (1977).

DERI 1978 -1 DERIGS, U.
ALGEBRAISCHE MATCHINGPROBLEME. DISSERTATION. KOELN: UNIV. KOELN, MATHEMATISCH-NATURWISSENSCHAFTLICHE FAKULTAET, 155 P. (1978).

DERI 1978 -2 DERIGS, U.
ON SOLVING SYMMETRIC ASSIGNMENT AND PERFECT MATCHING PROBLEMS WITH ALGEBRAIC OBJECTIVES. IN: R. HENN, B. KORTE, W. OETTLI (EDS.): OPTIMIZATION AND OPERATIONS RESEARCH, PROCEEDINGS, BONN 1977, 79-86. LECTURE NOTES IN ECONOMICS AND MATHEMATICAL SYSTEMS 157. BERLIN, HEIDELBERG, NEW YORK: SPRINGER. (1978).

DERIA 1978 -1 DERIGS, U.; ZIMMERMANN, U.
AUGMENTING PATH METHOD FOR SOLVING LINEAR BOTTLENECK ASSIGNMENT PROBLEMS, AN. COMPUTING 19, 285-296. (1978).

DERM 1975 -1 DERMAN, C.; LIEBERMAN, G. J.; ROSS, S. M.
OPTIMAL ALLOCATION OF RESOURCES IN SYSTEMS. IN: R. E. BARLOW, J. B. FUSSELL, N. D. SINGPURWALLA (EDS): RELIABILITY AND FAULT TREE ANALYSIS, CONFERENCE BERKELEY 1974, 307-324. PHILADELPHIA: SOCIETY FOR INDUSTRIAL AND APPLIED MATHEMATICS. (1975).

DERM 1975 -2 DERMAN, C.; LIEBERMAN, G. J.; ROSS, S. M.
STOCHASTIC SEQUENTIAL ALLOCATION MODEL, A. OPERATIONS RESEARCH 23, 1120-1130. (1975).

DERM 1976 -1 DERMAN, C.; LIEBERMAN, G. J.; ROSS, S. M.
OPTIMAL SYSTEM ALLOCATIONS WITH PENALTY COSTS. MANAGEMENT SCIENCE 23, 399-403. (1976).

DESC 1964 -1 DESCAMPS, R.; CHEVIGNON, P.
ALGORITHMS FOR OPTIMIZING A GENERAL CLASS OF SCHEDULING PROBLEMS WITH LIMITED RESOURCES (FRENCH). METRA 5, 603-618. (1964).

DESL 1970 -1 DESLER, J. F.; HAKIMI, S. L.
ON FINDING A MAXIMUM INTERNALLY STABLE SET OF A GRAPH. IN: PROCEEDINGS OF THE FOURTH ANNUAL PRINCETON CONFERENCE ON INFORMATION SCIENCES AND SYSTEMS, 459-462. (1970).

DETR 1976 -1 DETROYE, J.
SYNTHESE OPTIMALE D'UN RESEAU PARCOURU PAR UN MULTIFLOT. DISSERTATION. PARIS: UNIV. PARIS VI, 123 P. (1976).

DEWE 1976 -1 DEWESS, G.
KONVEXITAET UND OPTIMIERUNG IN METRISCHEN GRAPHEN. WISSENSCHAFTLICHE ZEITSCHRIFT DER TECHNISCHEN HOCHSCHULE ILMENAU 22, NO. 5, 43-59. (1976).

DEY 1976 -1 DEY, A.; MIDHA, C. K.
GENERALISED BALANCED MATRICES AND THEIR APPLICATIONS. UTILITAS MATHEMATICA 10, 139-149. (1976).

DIAL 1970 -1 DIAL, R.
PROBABILISTIC ASSIGNMENT: A MULTIPATH TRAFFIC ASSIGNMENT MODEL WHICH OBVIATES PATH ENUMERATION. DISSERTATION. SEATTLE: UNIVERSITY OF WASHINGTON. (1970).

DIAL 1971 -1 DIAL, R.
PROBABILISTIC MULTIPATH TRAFFIC ASSIGNMENT MODEL WHICH OBVIATES PATH ENUMERATION, A. TRANSPORTATION RESEARCH 5, 83-111. (1971).

DIALA 1977 -1 DIAL, R.; GLOVER, F.; KARNEY, D.; KLINGMAN, D.
COMPUTATIONAL ANALYSIS OF ALTERNATIVE ALGORITHMS AND LABELING TECHNIQUES FOR FINDING SHORTEST PATH TREES, A. RESEARCH REPORT CCS 291. AUSTIN: UNIV. OF TEXAS CENTER FOR CYBERNETIC STUDIES, 60 P. (1977).

DICK 1971 -1 DICKSON, L. E.
HISTORY OF THE THEORY OF NUMBERS. VOLUME 2 DIOPHANTINE ANALYSIS (REPRINT OF THE 1919 EDITION). NEW YORK: CHELSEA, 803 P. (1971). B

DIEH 1972 -1 DIEHR, G.
ALGORITHM FOR THE P-MEDIANS PROBLEM, AN. WORKING PAPER 191. LOS ANGELES: UNIV. OF CALIFORNIA, WESTERN MANAGEMENT SCIENCE INSTITUTE. (1972).

DIGU 1976 -1 DIGUGLIELMO, F.
ESTIMATES OF THE DUALITY GAP FOR DISCRETE OPTIMIZATION. MONTPELLIER: SEMINAIRE D'ANALYSE CONVEXE, 27 P. (1976).

DIGU 1976 -2 DIGUGLIELMO, F.
ESTIMATION DU SAUT DE DUALITE EN OPTIMISATION DISCRETE. COMPUTING REVIEWS 282, 883-886. (1976).

DIJK 1960 -1 DIJKSTRA, E. W.
SOME THEOREMS ON SPANNING SUBTREES OF A GRAPH. PROC. AMSTERDAM 22, 196-199. (1960).

DIJK 1976 -1 DIJKSTRA, E. W.
DISCIPLINE OF PROGRAMMING, A. PRENTICE-HALL SERIES IN AUTOMATIC COMPUTATION. ENGLEWOOD CLIFFS: PRENTICE HALL, 217 P. (1976).

DIKE 1964 -1 DIKE, S. H.
PROJECT SCHEDULING WITH RESOURCE CONSTRAINTS. IEEE TRANSACTION ON ENGINEERING MANAGEMENT 11, 155-157. (1964).

DINE 1972 -1 DINESCU, C.
DETERMINING THE MEAN FLOW IN A TRANSPORTATION NETWORK (RUMANIAN). ANALELE UNIVERSITATII BUCURESTI. MATEMATICA-MECANICA 21, NO. 2, 29-36. (1972).

DINE 1976 -1 DINESCU, C.
OPTIMAL ELEMENTARY PATHS IN NETWORKS WITH CIRCUITS. ANALELE UNIVERSITATII BUCURESTI, STIINTELE NATURII 25, 7-14. (1976).

DINI 1970 -1 DINITS, E. A.
ALGORITHM FOR SOLUTION OF A PROBLEM OF MAXIMUM FLOW IN A NETWORK WITH POWER ESTIMATOR. SOVIET MATHEMATICS DOKLADY, 11, 1277-1280. (1970).

DINI 1973 -1 DINITS, E. A.
ALGORITHM FOR THE DIGITWISE DECREASE OF DISCREPANCIES, AND TRANSPORT PROBLEMS (RUSSIAN), AN. IN: A. A. FRIDMAN (ED.): STUDIES IN DISCRETE MATHEMATICS (RUSSIAN), 46-57. MOSCOW: IZDAT. NAUKA. (1973).

DINI 1976 -1 DINITS, E. A.
SOLUTION OF TWO SCHEDULING PROBLEMS (RUSSIAN), THE. IN: STUDIES IN DISCRETE OPTIMIZATION (RUSSIAN), 333-348. MOSCOW: IZDAT. NAUKA. (1976).

DINIA 1977 -1 DINITS, E. A.; ZAJTSEV, M. A.
ALGORITHMS WHICH GENERATE NON-ISOMORPHIC TREES (RUSSIAN). AVOMATIKA I TELEMEHANIKA, NO. 4, 121-126. (1977).

DINK 1976 -1 DINKEL, J. J.; KLEINDORFER, G. B.; KOCHENBERGER, G. A.; WONG, S.-N.
ENVIRONMENTAL INSPECTION ROUTES AND THE CONSTRAINED TRAVELLING SYSTEM SALESMAN PROBLEM. COMPUTERS AND OPERATIONS RESEARCH 3, 269-282. (1976).

DIXO 1976 -1 DIXON, E. T.; GOODMAN, S. E.
ALGORITHM FOR THE LONGEST CYCLE PROBLEM, AN. NETWORKS 6, 139-150. (1976).

DOBK 1976 -1 DOBKIN, D.
NONLINEAR LOWER BOUND ON LINEAR SEARCH TREE PROGRAMS FOR SOLVING KNAPSACK PROBLEMS, A. JOURNAL OF COMPUTER AND SYSTEM SCIENCES 13, 69-73. (1976).

DOBKA 1975 -1 DOBKIN, D.; LIPTON, R.
ON THE COMPLEXITY OF COMPUTATIONS UNDER VARYING SETS OF PRIMITIVE OPERATIONS. IN: H. BRAKHAGE (ED.): AUTOMATA THEORY AND FORMAL LANGUAGES, 2ND GI CONFERENCE, KAISERSLAUTERN 1975, 110-117. LECTURE NOTES IN COMPUTER SCIENCE 33, BERLIN: SPRINGER. (1975).

DOBKB 1976 -1 DOBKIN, D.; LIPTON, R. J.
LOWER BOUND OF 1/(N POWER 2) ON LINEAR SEARCH PROGRAMS FOR THE KNAPSACK PROBLEM, A. IN: A. MAZURKIEWICZ (ED.): MATHEMATICAL FOUNDATIONS OF COMPUTER SCIENCE 1976, PROCEEDINGS OF THE 5TH SYMPOSIUM, GDANSK 1976, 265-269. LECTURE NOTES IN COMPUTER SCIENCE 45, BERLIN: SPRINGER. (1976).

DOER 1978 -1 DOERFLER, W.
MULTIPLE COVERS OF HYPERGRAPHS. KLAGENFURT: UNIV. FUER BILDUNGSWISSENSCHAFTEN, 17 P. (1978).

DOERA 1977 -1 DOERSCH, R. H.; PATTERSON, J. H.
SCHEDULING A PROJECT TO MAXIMIZE ITS PRESENT VALUE: A ZERO-ONE PROGRAMMING APPROACH. MANAGEMENT SCIENCE 23, 882-889. (1977).

DOLO 1976 -1 DOLODARENKO, V. A.; FEDAN, E. I.
ASSIGNMENT PROBLEM WITH A NONLINEAR STOCHASTIC OBJECTIVE FUNCTION (RUSSIAN), THE. AVTOMATIKA, NO. 3, 29-32; TRANSLATED AS SOVIET AUTOMATIC CONTROL 9, NO. 3, 24-27. (1976).

DOMS 1974 -1 DOMSCHKE, W.
GRAPHENTHEORETISCHE VERFAHREN UND IHRE ANWENDUNGEN ZUR LOESUNG VON ZUORDNUNGS- UND STANDORTPROBLEMEN. HABILITATIONSSCHRIFT, KARLSRUHE: UNIV. KARLSRUHE. (1974).

DOMS 1975 -1 DOMSCHKE, W.
VERFAHREN ZUR BESTIMMUNG OPTIMALER BETRIEBS- UND LAGERSTANDORTE. IN: E. BAHKE (ED.): MATERIALFLUSSYSTEME BD II. MATERIALFLUSSMODELLE, CHAPTER 5, MAINZ: KRAUSSKOPF. (1975).

DOMS 1976 -1 DOMSCHKE, W.
NEW ALGORITHM FOR MINIMAL COST FLOW PROBLEMS AND THEIR APPLICATIONS. IN: A. PREKOPA (ED.): PLOGRESS IN OPERATIONS RESEARCH, 269-287, AMSTERDAM, LONDON: NORTH HOLLAND. (1976).

DONE 1976 -1 DONETS, G. A.
ON THE PROPERTIES OF COLORED PLANE TRIANGULATIONS (RUSSIAN). UPRAVLJAEME SISTEMY, NO. 11, 55-60. (1976).

DORA 1966 -1 DORAN, J. E.; MICHIE, D.
EXPERIMENTS WITH THE GRAPH TRAVERSER PROGRAM. PROCEEDINGS OF THE ROYAL SOCIETY 294, 1437. (1966).

DOUL 1970 -1 DOULLIEZ, P.
POLITIQUE OPTIMALE D'INVESTISSEMENT DANS UN RESEAU DE TRANSPORT. DISSERTATION (C.O.R.E. REPRINT NO. 69, 264-284). LOUVAIN: UNIV. CATHOLIQUE DE LOUVAIN. (1970).

DOULA 1969 -1 DOULLIEZ, P.; RAO, M. R.
MAXIMUM FLOW IN A MULTITERMINAL NETWORK WITH INCREASING REQUIREMENTS AND ARCS SUBJECT TO FAILURES. C.O.R.E. DISCUSSIONS PAPER 6913. HEVERLEE: INTERNAT. CENTER FOR MANAGEMENT SCIENCES. (1969).

DOULB 1976 -1 DOULLIEZ, P. J.; RAO, M. R.
NETWORK PROBLEM WITH PARAMETRIC DEMANDS AND ARC FAILURES - A LABELING ALGORITHM. CAHIERS DU CENTRE D'ETUDES DE RECHERCHE OPERATIONNELLE 18, 419-438. (1976).

DOWL 1977 -1 DOWLING, T. A.; DENIG, W. A.
GEOMETRIES ASSOCIATED WITH PARTIALLY ORDERED SETS. IN: M. AIGNER (ED.): HIGHER COMBINATORICS, PROCEEDINGS OF THE NATO ADVANCED STUDY INSTITUTE, BERLIN 1976, 103-116. DORDRECHT: REIDEL. (1977).

DOYL 1971 -1 DOYLE, M. S.
MINIMUM PATH COVERING ALGORITHMS. IN: PROCEEDINGS OF THE MANITOBA CONFERENCE ON NUMERICAL MATHEMATICS, WINNIPEG 1971, 127-139. WINNIPEG: UNIV. OF MANITOBA, DEPT. OF COMPUTER SCIENCE. (1971).

DRAG 1976 -1 DRAGAN, I.
NETWORK FLOW MODEL FOR SOLVING A COOPERATIVE GAME, A. IN: PROCEEDINGS OF THE CONFERENCE ON MATHEMATICAL PROGRAMMING AND ITS APPLICATIONS, INDAM, ROME 1974 (ITALIAN), 215-227. SYMPOSIA MATHEMATICA 19, LONDON: ACADEMIC PRESS. (1976).

DREY 1977 -1 DREYFUS, S. E.; LAW A. M.
ART AND THEORY OF DYNAMIC PROGRAMMING, THE. MATHEMATICS IN SCIENCE AND ENGINEERING 130. NEW YORK, SAN FRANCISCO, LONDON: ACADEMIC PRESS, 284 P. (1977).

DRIS 1975 -1 DRISCOLL, W. C.
SCHEDULING PRODUCTION ON ONE MACHINE WITH CHANGEOVER TIMES. PAPER PRESENTED AT THE 48TH JOINT NATIONAL MEETING OF ORSA AND TIMS, LAS VEGAS 1975, 20 P. (1975).

DRIS 1976 -1 DRISCOLL, W. C.
BRANCH-BOUND ALGORITHM FOR SCHEDULING PRODUCTION ON ONE MACHINE WITH CHANGEOVER TIMES, A. MIAMI (50. ORSA-MEETING + TIMS, NOVEMBER 1976) 37 P. (1976).

DROE 1976 -1 DROESKE, F.; FRENNE, A. DE
ETUDE DE BORNES DU NOMBRE CHROMATIQUE D'UN GRAPHE, I. IN: ANNLS APPL. INF. JOURN. COMB. INF., BORDEAUX 1975, 113-119. BORDEAUX: UNIV. BORDEAUX I. (1976).

DRUF 1975 -1 DRUFFEL, L. E.
GRAPH RELATED ALGORITHMS: ISOMORPHISM, AUTOMORPHISM AND CONTAINMENT. TECHNICAL REPORT 75-1. VANDERBILT UNIVERSITY, SCHOOL OF ENGINEERING, SYSTEM AND INFORMATION SCIENCE. (1975).

DRUT 1975 -1 DRUTOV, V. E.
SOLUTION OF A SCHEDULING PROBLEM (RUSSIAN). IZVESTIJA AKADEMII NAUK SSSR. TEHNICESKAJA KIBERNETIKA, NO. 6, 43-46. (1975).

DUBO 1976 -1 DUBOV, YU. A.; EPIKHOVA, N. V.
ON AN ALGORITHM SEARCHING OPTIMAL SOLUTIONS IN THE PROBLEMS OF AUTOMATIC DESIGN (RUSSIAN). AVTOMATIKA I TELEMEHANIKA, NO. 8, 95-100. (1976).

DUEC 1977 -1 DUECK, W.
EINFLUSSBEREICH DISKRETER OPTIMIERUNGSMETHODEN, DER. IN: T. D. BERESNEWA, W. DUECK, K.-H. ELSTER ET AL. (EDS.): MATHEMATISCH-OEKONOMISCHE METHODEN UND MODELLE, 166-195. BERLIN: DIE WIRTSCHAFT. (1977).

DUER 1976 -1 DUERRE, K.
EFFIZIENZ VON FAERBUNGSALGORITHMEN. IN: U. PAPE (ED.): GRAPHEN-SPRACHEN UND ALGORITHMEN AUF GRAPHEN, 1. FACHTAGUNG UEBER GRAPHENTHEORETISCHE KONZEPTE DER INFORMATIK, BERLIN 1975, 163-174. MUENCHEN: HANSER. (1976).

DUER 1976 -2 DUERRE, K.
FAST REVERSIBLE CODING AND INDEXING OF COLORATIONS. COMPUTING 16, 271-279. (1976).

DUNS 1975 -1 DUNSTAN, F. D. J.
GREEDY ALGORITHMS FOR OPTIMISATION PROBLEMS. PAPER PRESENTED AT THE FIRST EUROPEAN CONGRESS ON OPERATIONS RESEARCH, EURO 1, BRUSSELS 1975. (1975).

DUNS 1976 -1 DUNSTAN, F. D. J.
SEQUENTIAL COLOURINGS OF GRAPHS. IN: C. ST. J. A. NASH-WILLIAMS (ED.): PROCEEDINGS OF THE 5TH BRITISH COMBINATORIAL CONFERENCE, ABERDEEN 1975, 151-158. CONGRESSUS NUMERANTIUM 15, WINNIPEG: UTILITAS MATHEMATICA. (1976).

DUNS 1977 -1 DUNSTAN, F. D. J.
ALGORITHM FOR SOLVING A RESOURCE ALLOCATION PROBLEM, AN. OPERATIONAL RESEARCH QUARTERLY 28, 839-852. (1977).

DURA 1973 -1 DURAND, J. A. A.
POLYHEDRAL NEOPOLARITIES. DISSERTATION. WATERLOO: UNIV. OF WATERLOO, FACULTY OF MATHEMATICS, DEPT. OF APPLIED ANALYSIS AND COMPUTER SCIENCE. (1973).

DWYE 1975 -1 DWYER, P. S.
TRANSPORTATION PROBLEMS WITH SOME X(IJ) NEGATIVE AND TRANSSHIPMENT PROBLEMS. NAVAL RESEARCH LOGISTICS QUARTERLY 22, 751-776. (1975).

DYKS 1977 -1 DYKSTRA, D. P.; RIGGS, J. L.
APPLICATION OF FACILITIES LOCATION THEORY TO THE DESIGN OF FOREST HARVESTING AREAS, AN. AIIE TRANSACTIONS 9, 270-277. (1977).

DYMA 1975 -1 DYMARSKII, JA. S.; PRUDOVSKII, B. D.; STALBO, A. K.
SEPARABLE INTEGER-VALUED PROBLEM (RUSSIAN), A. IZVESTIJA AKADEMII NAUK SSSR, TEHNICESKAJA KIBERNETIKA, NO. 5, 23-26; TRANSLATED AS ENGINEERING CYBERNETICS 13, 1975, NO. 5, 19-22. (1975).

DZIK 1974 -1 DZIKIEWICS, J.
ALGORITHM FOR FINDING THE TRANSITIVE CLOSURE OF A DIGRAPH, AN. WROCLAW: TECHNICAL UNIVERSITY OF WROCLAW, POLAND. (1974).

DZIKA 1974 -1 DZIKIEWICS, J.; SYSLO, M. M.
COMPUTIONAL EXPERIENCES WITH SOME TRANSITIVE CLOSURE ALGORITHMS. WROCLAW: TECHNICAL UNIVERSITY OF WROCLAW, POLAND. (1974).

ECKE 1977 -1 ECKER, K.
ORGANISATION VON PARALLELEN PROZESSEN. THEORIE DETERMINISTISCHER SCHEDULES. REIHE INFORMATIK 23. MANNHEIM, WIEN, ZUERICH: BIBLIOGRAPHISCHES INSTITUT, 280 P. (1977).

ECON 1977 -1 ECONOMIDES, S. C.; DUESTERHOEFT, W. C.
CONSTRUCTION SCHEDULING SYSTEM WITH RESOURCE ALLOCATION FOR ELECTRIC UTILITIES, A. IEEE TRANSACTIONS ON ENGINEERING MANAGEMENT EM-24, 2-6. (1977).

EDEL 1970 -1 EDELBERG, M.
INTEGRAL CONVEX POLYHEDRA AND AN APPROACH TO INTEGRALIZATION. DISSERTATION. CAMBRIDGE: MASSACHUSETTS INST. OF TECHNOLOGY, 170 P. (1970).

EDEN 1977 -1 EDEN, B.; PAPE, U.
PLANAR. EIN ALGORITHMUS ZUR FESTSTELLUNG DER PLANARITAET VON GRAPHEN UND ZUR ERMITTLUNG PLANARER TEILGRAPHEN. BERICHT 77-12. BERLIN: T.U. BERLIN, INSTITUT FUER ANGEWANDTE INFORMATIK, 30 P. (1977).

EDENA 1976 -1 EDEN, B. N.
HEURISTISCHES VERFAHREN ZUR DISTRIKTERMITTLUNG IN BEWERTETEN GRAPHEN UND SEINE ANWENDUNG AUF DIE BERECHNUNG VON SCHULEINZUGSBEREICHEN, EIN. DISSERTATION. BERLIN: TECHNISCHE UNIV. BERLIN, 71 P. (1976).

EDENB 1978 -1 EDEN, B. N.; PAPE, U.
HEURISTISCHES VERFAHREN ZUR DISTRIKTERMITTLUNG IN BEWERTETEN GRAPHEN, EIN. ANGEWANDTE INFORMATIK 20, 66-73. (1978).

EDMO 1967 -1 EDMONDS, J.
INTRODUCTION TO MATCHING, AN. IN: MIMEOGRAPHED LECTURE NOTES, ENGINEERING SUMMER CONFERENCE, ANN ARBOR, 1967. ANN ARBOR: UNIV. OF MICHIGAN. (1967).

EDMO 1968 -1 EDMONDS, J.
MATROID PARTITION. IN: G. B. DANTZIG, A. F. VEINOTT (EDS.): MATHEMATICS OF THE DECISION SCIENCES, PART 1, 335-345. PROVIDENCE: AMERICAN MATHEMATICAL SOCIETY. (1968).
EDMO 1975 -1 EDMONDS, J.
SOME WELL-SOLVED PROBLEMS IN COMBINATORIAL OPTIMIZATION. IN: B. ROY (ED.): COMBINATORIAL PROGRAMMING: METHODS AND APPLICATIONS. PROCEEDINGS OF THE NATO ADVANCED STUDY INST., VERSAILLES 1974, 285-302. DORDRECHT: REIDEL. (1975).
EDMO 1977 -1 EDMONDS, J.
MATROID INTERSECTION. VANCOUVER (CONFERENCE ON DISCRETE OPTIMZATION, AUGUST 1977) 20 P. (1977).
EDMOA 1977 -1 EDMONDS, J.; GILES, R.
MIN-MAX RELATION FOR SUBMODULAR FUNCTIONS ON GRAPHS, A. IN: P. L. HAMMER, E. L. JOHNSON, B. H. KORTE, G. L. NEMHAUSER (EDS.): STUDIES IN INTEGER PROGRAMMING, 185-204. ANNALS OF DISCRETE MATHEMATICS 1, AMSTERDAM: NORTH-HOLLAND. (1977).
EDMOB 1969 -1 EDMONDS, J.; JOHNSON F. L.; LOCKHART, S.
BLOSSOM I, A CODE FOR MATCHING. UNPUBLISHED REPORT, IBM RESEARCH CENTER, YORKTOWN HEIGHTS. (1969).
EGIA 1976 -1 EGIAZARJAN, E. V.
CERTAIN CLASS OF SYSTEMS OF BOOLEAN EQUATIONS (RUSSIAN), A. ZURNAL VYCISLITEL'NOI MATEMATIKI I MATEMATICESKIE FIZIKI 16, 1073-1077 + 1088. (1976).
EGOR 1977 -1 EGOROVA, N. V.; PATKIN, B. P.; SLOBODTCHIKOVA, R. I.
MODIFIED ALGORITHM OF ASSIGNMENT PROBLEM SOLUTION (RUSSIAN). EKONOMIKA I MATEMATICESKIE METODY 13, 782-783. (1977).
EILO 1975 -1 EILON S.; CHOWDHURY, I. G.; SERGHIOU, S. S.
EXPERIMENTS WITH THE SI RULE IN JOB-SHOP SCHEDULING. SIMULATION 24, 45-48. (1975).
EILOA 1975 -1 EILON, S.; CHOWDHURY, I. G.
STUDIES IN A SIMULATED JOB SHOP. PROCEEDINGS OF THE INSTITUTION OF MECHANICAL ENGINEERS 189, 417-425. (1975).
EILOA 1976 -1 EILON, S.; CHOWDHURY, I. G.
CONDITIONS FOR OPTIMAL SCHEDULING FOR THE THREE-MACHINE PROBLEM. JOURNAL OF THE INSTITUTE OF MATHEMATICS AND ITS APPLICATIONS 17, 1-4. (1976).
EILOA 1976 -2 EILON, S.; CHOWDHURY, I. G.
DUE DATES IN JOB SHOP SCHEDULING. INTERNATIONAL JOURNAL OF PRODUCTION RESEARCH 14, 223-237. (1976).
EILOA 1977 -1 EILON, S.; CHOWDHURY, I. G.
MINIMISING WAITING TIME VARIANCE IN THE SINGLE MACHINE PROBLEM. MANAGEMENT SCIENCE 23, 567-575. (1977).
EILOB 1967 -1 EILON, S.; HODGSON, M.
JOB SHOP SCHEDULING WITH DUE DATES. INTERNATIONAL JOURNAL OF PRODUCTION RESEARCH 6. (1967).
EINB 1977 -1 EINBU, J. M.
ON SHIH'S INCREMENTAL METHOD IN RESOURCE ALLOCATIONS. OPERATIONAL RESEARCH QUARTERLY 28, 459-462. (1977).
EL-S 1975 -1 EL-SHAIEB, A. M.
MIXED-INTEGER PROGRAMMING MODEL FOR ECONOMIC RESIDENTIAL ENERGY USE, A. PAPER PRESENTED AT THE 48TH JOINT NATIONAL MEETING OF ORSA AND TIMS, LAS VEGAS 1975, 15 P. (1975).
ELAM 1977 -1 ELAM, J.; GLOVER, F.; KLINGMAN, D.
STRONGLY CONVERGENT PRIMAL ALGORITHM FOR GENERALIZED NETWORKS, A. RESEARCH REPORT CCS 288. AUSTIN: UNIV. OF TEXAS, CENTER FOR CYBERNETIC STUDIES, 43 P. (1977).
ELDR 1976 -1 ELDRIDGE, S. E.; BOLLOBAS, B.
MAXIMAL MATCHINGS IN GRAPHS WITH GIVEN MAXIMAL AND MINIMAL DEGREES. IN: C. ST. J. A. NASH-WILLIAMS (ED.): PROCEEDINGS OF THE 5TH BRITISH COMBINATORIAL CONFERENCE, ABERDEEN 1975, 165-168. CONGRESSUS NUMERANTIUM 15, WINNIPEG: UTILITAS MATHEMATICA. (1976).
ELLE 1973 -1 ELLERVIK, P.
ANOTHER METHOD FOR THE SOLUTION OF COMBINATORIAL 0-1 LINEAR PROBLEMS (DANISH). PAPER PRESENTED AT THE NORDISH OPERATIONAL RESEARCH CONGRESS NOAK-73, COPENHAGEN 1973. (1973).
ELLW 1972 -1 ELLWEIN, L. B.
FLEXIBLE ENUMERATION SCHEME FOR ZERO-ONE PROGRAMMING, A. BETHESDA: NATIONAL CANCER INSTITUTE, 18 P. (1972).
ELSH 1975 -1 ELSHAFEI, A. N.
APPROACH TO LOCATIONAL ANALYSIS, AN. OPERATIONAL RESEARCH QUARTERLY 26, 167-181. (1975).
ELSH 1977 -1 ELSHAFEI, A. N.
HOSPITAL LAYOUT AS A QUADRATIC ASSIGNMENT PROBLEM. OPERATIONAL RESEARCH QUARTERLY 28, 167-180. (1977).
ELSHA 1974 -1 ELSHAFEI, A. N.; HALEY, K. B.
FACILITIES LOCATION: SOME FORMULATIONS, METHODS OF SOLUTION, APPLICATIONS, AND COMPUTATIONAL EXPERIENCE. OPERATIONS RESEARCH REPORT 90. RALEIGH: NORTH CAROLINA STATE UNIVERSITY. (1974).
ELZI 1976 -1 ELZINGA, J.; HEARN, D.; RANDOLPH, W. D.
MINIMAX MULTIFACILITY LOCATION WITH EUCLIDEAN DISTANCES. GAINESVILLE: UNIV. OF FLORIDA, 22 P. (1976).
ELZIA 1976 -1 ELZINGA, J.; VEMUGANTI, R. R.
OPTIMAL SCHEDULING OF A SALES FORCE. OPSEARCH 13, 79-92. (1976).
EMDE 1975 -1 EMDE BOAS, P. V.
PRESERVING ORDER IN A FOREST IN LESS THAN LOGARITHMIC TIME. IN: 16TH ANNUAL SYMPOSIUM ON FOUNDATIONS OF COMPUTER SCIENCE, BERKELEY 1975, 75-84. NEW YORK: INSTITUTE OF ELECTRICAL AND ELECTRONICS ENGINEERS. (1975).
EMEL 1976 -1 EMELICHEV, V. A.; KRAVTSOV, M. K.
CERTAIN PROPERTIES OF TRANSPORTATION POLYHEDRA WITH A MAXIMAL NUMBER OF VERTICES (RUSSIAN). DOKLADY AKADEMII NAUK SSSR 228, 1025-1028. (1976).
EMEL 1976 -2 EMELICHEV, V. A.; KRAVTSOV, M. K.
ON ENUMERATIVE PROBLEMS ON TRANSPORT POLYHEDRA (RUSSIAN). VESCI AKADEMII NAVUK BSSR, SERYJA FIZIKA-MATEMATYCNYH NAVUK, NO. 1, 9-13 + 138. (1976).
EMELA 1976 -1 EMELICHEV, V. A.; KRAVTSOV, M. K.; AVERBUH, N. D.
MAXIMAL NUMBER OF VERTICES OF A TRANSPORTATION POLYHEDRON (RUSSIAN), THE. DOKLADY AKADEMII NAUK BSSR 20, 509-521 + 572. (1976).
EMELB 1974 -1 EMELICHEVA, V. S.; KONONEKO, A. M.
MINIMAL NUMBER OF VERTICES AND EDGES OF THE POLYTOPE OF A PLANAR TRANSPORTATION PROBLEM (RUSSIAN), THE. DOKLADY AKADEMII NAVUK BSSR 18, 781-784 + 858-859. (1974).
EMELC 1974 -1 EMELICHEVA, V. S.; KONONENKO, A. M.
MINIMAL NUMBER OF INTEGER VERTICES OF THE POLYTOPE OF A MULTIINDEX TRANSPORTATION PROBLEM (RUSSIAN), THE. VESCI AKADEMII NAVUK BSSR. SERYJA FIZIKA-MATEMATYCNYH NAVUK, NO. 3, 124-126 + 143. (1974).
EMMO 1968 -1 EMMONS, H.
ONE MACHINE SEQUENCING TO MINIMIZE TOTAL TARDINESS. TECHNICAL REPORT NO. 52, ITHACA: CORNELL UNIVERSITY, DEPARTEMENT OF OPERATIONS RESEARCH. (1968).

****

EPIH 1975 -1 EPIHIN, V. V.; HARKEVIC, A. D.
ON THE QUESTION OF THE CONSTRUCTION OF OPTIMAL GRAPHS (RUSSIAN). IN: METHODS OF TELETRAFFIC THEORY IN INFORMATION DISTRIBUTION SYSTEMS (RUSSIAN), 146-157. MOSCOW: IZDAT. NAUKA. (1975).

EPLE 1977 -1 EPLETT, W. J. R.
RESULT ON TOURNAMENTS WITH AN APPLICATION TO COUNTING, A. BULLETIN OF THE AUSTRALIAN MATHEMATICAL SOCIETY 16, 273-278. (1977).

EPLEA 1970 -1 EPLEY, D. L.
SCHEDULING PROBLEMS IN COMPUTER SYSTEMS. P203-8, 3 RD PROCEEDINGS OF THE INTERNATIONAL CONFERENCE ON SYSTEM SCIENCE. (1970).

ERDO 1976 -1 ERDOES, P.
PROBLEMS AND RESULTS IN GRAPH THEORY AND COMBINATORIAL ANALYSIS. IN: C. ST. J. A. NASH-WILLIAMS (ED.): PROCEEDINGS OF THE 5TH BRITISH COMBINATORIAL CONFERENCE, ABERDEEN 1975, 169-192. CONGRESSUS NUMERANTIUM 15, WINNIPEG: UTILITAS MATHEMATICA. (1976).

ERDOA 1976 -1 ERDOES, P.; FAUDREE, R. J.; ROUSSEAU, C. C.; SCHELP, R. H.
GENERALIZED RAMSEY THEORY FOR MULTIPLE COLORS. JOURNAL OF COMBINATORIAL THEORY 20, SERIES B, 2TO-264. (1976).

ERDOB 1966 -1 ERDOES, P.; HAJNAL, A.
ON CHROMATIC NUMBERS OF GRAPHS AND SET-SYSTEMS. ACTA MATHEMATICA ACADEMIAE SCIENTIARUM HUNGARICAE 17, 61-99. (1966).

ERDOC 1977 -1 ERDOES, P.; MEIR, A.
ON TOTAL MATCHING NUMBERS AND TOTAL COVERING NUMBERS OF COMPLEMENTARY GRAPHS. DISCRETE MATHEMATICS 19, 229-234. (1977).

ERDOD 1959 -1 ERDOES, P.; RENYI, A.
ON RANDOM GRAPHS, I. PUBLICATIONES MATHEMATICAE 6, 290-297. (1959).

ERDOE 1974 -1 ERDOES, P.; SPENCER, J.
PROBABILISTIC METHODS IN COMBINATORICS. NEW YORK: ACADEMIC PRESS. (1974).

ERDOF 1977 -1 ERDOES, P.; WILSON, R. J.
ON THE CHROMATIC INDEX OF ALMOST ALL GRAPHS. JOURNAL OF COMBINATORIAL THEORY 23, SERIES B, 255-257. (1977).

ERLA 1966 -1 ERLANDER, S.
NONLINEAR TRANSPORTATION AND INVENTORY PROBLEM. STOCKHOLM: RESEARCH REPORT OF THE UNIVERSITY OF STOCKHOLM. (1966).

ERLE 1967 -1 ERLENKOTTER, D.
TWO PRODUCING AREAS-DYNAMIC PROGRAMMING SOLUTIONS. IN: A. P. MANNE (ED.): INVESTMENTS FOR CAPACITY EXPANSION: SIZE, LOCATION, AND TIME-PHASING, 210-227. CAMBRIDGE: M.I.T. PRESS. (1967).

ERLE 1972 -1 ERLENKOTTER, D.
CAPACITY PLANNING FOR COMPETITIVE EXPANSIONS: A REDUCED DYNAMIC PROGRAMMING APPROACH. DISCUSSION PAPER 25. LOS ANGELES: UNIV. OF CALIFORNIA, OPERATIONS RESEARCH STUDY CENTER, GRADUATE SCHOOL OF MANAGEMENT. (1972).

ERLE 1972 -2 ERLENKOTTER, D.
ECONOMIC INTEGRATION AND DYNAMIC LOCATION PLANNING. THE SWEDISH JOURNAL OF ECONOMICS 74, 8-18. (1972).

ERLE 1973 -1 ERLENKOTTER, D.
DYNAMIC MULTI-LOCATION MODELS FOR CAPACITY EXPANSION. IN: PROCEEDINGS OF THE 1973 JOINT AUTOMATIC CONTROL CONFERENCE, COLUMBUS 1973, 514-516. (1973).

ERLE 1973 -2 ERLENKOTTER, D.
SEQUENCING OF INTERDEPENDENT HYDROELECTRIC PROJECTS. WATER RESOURCES RESEARCH 9, 21-27. (1973).

ERLE 1974 -1 ERLENKOTTER, D.
DYNAMIC FACILITY LOCATION AND SIMPLE NETWORK MODELS. MANAGEMENT SCIENCE 20, 1311. (1974).

ERLE 1974 -2 ERLENKOTTER, D.
DYNAMIC PROGRAMMING APPROACH TO CAPACITY EXPANSION WITH SPECIALIZATION, A. MANAGEMENT SCIENCE 21, 360-362. (1974).

ERLE 1975 -1 ERLENKOTTER, D.
BIBLIOGRAPHY ON DYNAMIC LOCATION MODELS. DISCUSSION PAPER 55. LOS ANGELES: UNIV. OF CALIFORNIA, MANAGEMENT SCIENCE STUDY CENTER, GRADUATE SCHOOL OF MANAGEMENT, 10 P. (1975).

ERLE 1975 -2 ERLENKOTTER, D.
CAPACITY PLANNING FOR LARGE MULTILOCATION SYSTEMS: APPROXIMATE AND INCOMPLETE DYNAMIC PROGRAMMING APPROACHES. MANAGEMENT SCIENCE 22, 274-285. (1975).

ERLE 1975 -3 ERLENKOTTER, D.
COMMENT ON "OPTIMAL TIMING, SEQUENCING, AND SIZING OF MULTIPLE RESERVOIR SURFACE WATER SUPPLY FACILITIES" BY L. BECKER AND W. W.-G. YEH. WATER RESOURCES RESEARCH 11, 380-381. (1975).

ERLE 1975 -4 ERLENKOTTER, D.
COORDINATING SCALE AND SEQUENCING DECISIONS FOR WATER RESOURCES PROJECTS. WORKING PAPER 227. LOS ANGELES: UNIV. OF CALIFORNIA, WESTERN MANAGEMENT SCIENCE INSTITUTE. (1975).

ERLE 1975 -5 ERLENKOTTER, D.
OPTIMAL INVESTMENT SCHEDULING FOR PUBLIC SECTOR PROJECTS. WORKING PAPER 241. LOS ANGELES: UNIV. OF CALIFORNIA, WESTERN MANAGEMENT SCIENCE INSTITUTE. (1975).

ERLE 1976 -1 ERLENKOTTER, D.
DUAL-BASED PROCEDURE FOR UNCAPACITATED FACILITY LOCATION, A. WORKING PAPER NO. 261. LOS ANGELES: UNIV. OF CALIFORNIA, WESTERN MANAGEMENT SCIENCE INSTITUTE, 38 P. (1976).

ERLE 1976 -2 ERLENKOTTER, D.
FACILITY LOCATION WITH PRICE-SENSITIVE DEMANDS: PRIVATE, PUBLIC, AND QUASI-PUBLIC. WORKING PAPER NO. 255. LOS ANGELES: UNIV. OF CALIFORNIA, WESTERN MANAGEMENT SCIENCE INSTITUTE, 22 P. (1976).

ERLEA 1975 -1 ERLENKOTTER, D.; ROGERS, J. S.
SEQUENCING COMPETITIVE EXPANSION PROJECTS. WORKING PAPER 234. LOS ANGELES: UNIVERSITY OF CALIFORNIA, WESTERN MANAGEMENT SCIENCE INSTITUTE. (1975).

ERLEB 1975 -1 ERLENKOTTER, D.; TRIPPI, R. R.
OPTIMAL INVESTMENT SCHEDULING WITH PRICE-SENSITIVE DYNAMIC DEMAND. WORKING PAPER 235. LOS ANGELES: UNIV. OF CALIFORNIA, WESTERN MANAGEMENT SCIENCE INSTITUTE. (1975).

ERMO 1976 -1 ERMOL'EV, YU. M.; KRIVETS, T. A.; PETUKHOV, V. S.
PROBLEM OF PLANNING SEA SHIPPINGS OF EMPTY CONTAINERS (RUSSIAN), A. KIBERNETIKA, NO. 4, 145-146. (1976).

ERSC 1976 -1 ERSCHLER, J.
ANALYSE SOUS CONTRAINTES ET AIDE A LA DECISION POUR CERTAINS PROBLEMES D'ORDONNANCEMENT. DISSERTATION. TOULOUSE: UNIV. TOULOUSE, 183 P. (1976).

ESCH 1975 -1 ESCHENBACH, T.; CARLSON, R. C.
CAPACITATED MULTI-PERIOD LOCATION-ALLOCATION PROBLEM, THE. TECHNICAL REPORT SOL 75-27. STANFORD: STANFORD UNIV., DEPT. OF OPERATIONS RESEARCH, 19 P. (1975).

ESCHA 1975 -1 ESCHENBACH, T. G.
MULTI-PERIOD LOCATION-ALLOCATION PROBLEM, THE. DISSERTATION. STANFORD: STANFORD UNIVERSITY. (1975).

****

ESCHB 1974 -1 ESCHER, H. K.
SEQUENTIELLE ZUORDNUNGSPROBLEME. OPERATIONS RESEARCH VERFAHREN 19, 1-13. (1974).
ESCU 1975 -1 ESCUDERO, L. F.
CONTINUOUS, INTEGER, AND MIXED LINEAR PROGRAMMING, VOL. 1 (SPANISH). BILBAO: DEUSTO. (1975).
ESCUA 1976 -1 ESCUDERO, L. F.; VAZQUUEZ MUNIZ, A.
USE OF MIXED INTEGER PROGRAMMING FOR THE EVALUATION OF SOME ALTERNATE AIR POLLUTION ABATEMENT POLICIES, THE. IN: J. CEA (ED.): OPTIMIZATION TECHNIQUES, MODELING AND OPTIMIZATION IN THE SERVICE OF MAN, PART 1. PROCEEDINGS OF THE 7TH IFIP CONFERENCE, NICE 1975, 264-272. LECTURE NOTES IN COMPUTER SCIENCE 40, BERLIN: SPRINGER. (1976).
ESKE 1975 -1 ESKEW, J. D.; PARKER, R. G.
COMPUTATIONAL EXPERIENCE WITH A COST-BASED ALGORITHM FOR THE SHOP SCHEDULING PROBLEM. OPERATIONAL RESEARCH QUARTERLY 26, NO 1, 211-215. (1975).
ESOG 1977 -1 ESOGBUE, A. O.; MARKS, B. R.
DYNAMIC PROGRAMMING MODELS OF THE NONSERIAL CRITICAL PATH-COST PROBLEM. MANAGEMENT SCIENCE 24, 200-209. (1977).
ESWA 1976 -1 ESWARAN, K. P.; TARJAN, R. E.
AUGMENTATION PROBLEMS. SIAM JOURNAL ON COMPUTING 5, 653-665. (1976).
ETCH 1976 -1 ETCHEBERRY, J.
NEW BRANCHING AND BOUNDING STRATEGIES FOR 0-1 INTEGER LINEAR PROGRAMMING. SERIE DOCUMENTOS DE TRABAJOS 76/11. SANTIAGO: UNIV. DE CHILE, FACULTAD DE SIENCIAS FISICAS Y MATEMATICAS, 16 P. (1976).
ETCH 1976 -2 ETCHEBERRY, J.
SET REPRESENTATION PROBLEM: A NEW IMPLICIT ENUMERATION ALGORITHM, THE (REVISED VERSION). SERIE DOCUMENTOD DE TRABAJO 76/18. SANTIAGO: UNIV. DE CHILE, FACULTAD DE CIENCIAS FISICAS Y MATEMATICS, 35 P. (1976).
ETCH 1977 -1 ETCHEBERRY, J.
SET-COVERING PROBLEM: A NEW IMPLICIT ENUMERATION ALGORITHM, THE. OPERATIONS RESEARCH 25, 760-772. (1977).
EULE 1977 -1 EULER, R.
BESTIMMUNG MAXIMALER ELEMENTE IN MONOTONEN MENGENSYSTEMEN, DIE. REPORT 7761-OR. BONN: UNIV. BONN, INST. F. OEKONOMETRIE UND OPERATIONS RESEARCH (1977).
EULE 1978 -1 EULER, R.
GENERATING MAXIMUM MEMBERS OF INDEPENDENCE SYSTEMS. IN: R. HENN, B. KORTE, W. OETTLI (EDS.): OPTIMIZATION AND OPERATIONS RESEARCH, PROCEEDINGS, BONN 1977. LECTURE NOTES IN ECONOMICS AND MATHEMATICAL SYSTEMS 157. BERLIN, HEIDELBERG, NEW YORK: SPRINGER, 87-94. (1978).
EULE 1978 -2 EULER, R.
ON RANK FUNCTIONS OF GENERAL INDEPENDENCE SYSTEMS. REPORT 7893-OR. BONN: UNIV. BONN, INST. F. OEKONOMETRIE UND OPERATIONS RESEARCH, 8 P. (1978).
EVAN 1974 -1 EVANS, J. P.
USE OF A LOCATION MODEL IN A REGIONALIZATION STUDY, THE. MIMEO SERIES, CHAPEL HILL: UNIVERSITY OF NORTH CAROLINA. (1974).
EVANA 1976 -1 EVANS, J. R.
COMBINATORIAL EQUIVALENCE BETWEEN A CLASS OF MULTICOMMODITY FLOW PROBLEMS AND THE CAPACITATED TRANSPORTATION PROBLEM, A. MATHEMATICAL PROGRAMMING 10, 401-404. (1976).
EVANA 1976 -2 EVANS, J. R.
FUNDAMENTAL CIRCUITS IN COMPLETE GRAPHS AND THE TRAVELING SALESMAN PROBLEM. MIAM (50. ORSA-MEETING + TIMS, NOVEMBER 1976) 18 P. (1976).
EVANA 1976 -3 EVANS, J. R.
MAXIMUM FLOW IN PROBABILISTIC GRAPHS - THE DISCRETE CASE. NETWORKS 6, 161-184. (1976).
EVANA 1976 -4 EVANS, J. R.
TRANSPORTATION PROBLEM AND ITS EXTENSIONS: A SURVEY, THE. PAPER PRESENTED AT THE 49TH JOINT NATIONAL MEETING OF ORSA AND TIMS, PHILADELPHIA 1976, 29 P. (1976).
EVANA 1977 -1 EVANS, J. R.
SOME NETWORK FLOW MODELS AND HEURISTICS FOR MULTIPRODUCT PRODUCTION AND INVENTORY PLANNING. AIIE TRANSACTIONS 9, 75-81. (1977).
EVANA 1977 -2 EVANS, J. R.
SURVEY OF INTEGRALITY IN MULTICOMMODITY NETWORKS, A. CINCINNATI: UNIV. OF CINCINNATI, DEPT. OF QUANTITATIVE ANALYSIS, 49 P. (1977).
EVANB 1975 -1 EVANS, J. R.; JARVIS, J. J.
IMPROVED FORMULATION AND A HEURISTIC ALGORITHM FOR THE INTEGER MULTICOMMODITY TRANSPORTATION PROBLEM, AN. INDUSTRIAL AND SYSTEMS ENGINEERING REPORT SERIES J-75-14. ATLANTA: GEORGIA INST. OF TECHNOLOGY, 17 P. (1975).
EVANB 1977 -1 EVANS, J. R.; JARVIS, J. J.
OUT-OF-KILTER BASED HEURISTIC FOR THE INTEGER MULTICOMMODITY TRANSPORTATION PROBLEM, AN. COMPUTERS AND OPERATIONS RESEARCH 4, 13-20. (1977).
EVANC 1976 -1 EVANS, J. R.; MARTIN, K.
FEASIBLE FLOWS IN MULTICOMMODITY NETWORKS WITH LOWER BOUNDS. MIAMI (50. ORSA-MEETING + TIMS, NOVEMBER 1976) 16 P. (1976).
EVEN 1974 -1 EVEN, S.
MAXIMUM FLOW IN A NETWORK AND THE CONNECTIVITY OF A GRAPH. IN: PROCEEDINGS OF THE EIGHTH ANNUAL PRINCETON CONFERENCE ON INFORMATION SCIENCES AND SYSTEMS, 470-472. (1974).
EVENA 1975 -1 EVEN, S.; ITAI, A.; SHAMIR, A.
ON THE COMPLEXITY OF TIMETABLE AND MULTI-COMMODITY FLOW PROBLEMS. IN: 16TH ANNUAL SYMPOSIUM ON FOUNDATIONS OF COMPUTER SCIENCE, BERKLEY 1975, 184-193. NEW YORK: INSTITUTE OF ELECTRICAL AND ELECTRONICS ENGINEERS; ALSO PUBLISHED IN SIAM JOURNAL ON COMPUTING 5, 1976, 691-703. (1975).
EVENB 1975 -1 EVEN, S.; KARIV, O.
O(N POWER 2.5) ALGORITHM FOR MAXIMUM MATCHING IN GENERAL GRAPHS, AN. IN: 16TH ANNUAL SYMPOSIUM ON FOUNDATIONS OF COMPUTER SCIENCE, BERKELEY 1975, 100-112. NEW YORK: INSTITUTE OF ELECTRICAL AND ELECTRONICS ENGINEERS. (1975).
EVENC 1970 -1 EVEN, S.; LEMPEL, A.; PNUELI, A.
PERMUTATION GRAPHS AND TRANSITIVE GRAPHS. IN: PROCEEDINGS OF THE FOURTH ANNUAL PRINCETON CONFERENCE ON INFORMATION SCIENCES AND SYSTEMS, 463-467. (1970).
EVEND 1975 -1 EVEN, S.; TARJAN, R. E.
COMBINATORIAL PROBLEM WHICH IS COMPLETE IN POLYNOMIAL SPACE, A. IN: 7TH ANNUAL ACM SYMPOSIUM ON THEORY OF COMPUTING, ALBUQUERQUE 1975, 66-71. NEW YORK: ASSOCIATION FOR COMPUTING MACHINERY. ALSO PUBLISHED IN JOURNAL OF THE ACM 23, 1976, 710-719. (1975).
EVEND 1975 -2 EVEN, S.; TARJAN, R. E.
NETWORK FLOW AND TESTING GRAPH CONNECTIVITY. SIAM JOURNAL ON COMPUTING 4, 507-518. (1975).
EVST 1969 -1 EVSTIGNEEV, V. A.; JASAKOVA, V. P.
CERTAIN ALGORITHM ON GRAPHS (RUSSIAN), A. IN: CONTROL SYSTEMS, NO. 2 (RUSSIAN), 9-11. NOVOSIBIRSK: AKAD. NAUK SSSR SIBIRSK. OTDEL. INST. MAT. INST. KATALIZA. (1969).
EYST 1973 -1 EYSTER, J. W.; WHITE, J. A.; WIERVILLE, W. W.
ON SOLVING MULTIFACILITY LOCATION PROBLEMS USING A HYPERBOLOID APPROXIMATION PROCEDURE. AIIE TRANSACTIONS 5, 1-6. (1973).

FABI 1976 -1 FABIAN, CS.
SOME ALGORITHMS FOR NONLINEAR INTEGER PROGRAMMING. ECONOMIC COMPUTATION AND ECONOMIC CYBERNETICS STUDIES AND RESEARCH, NO. 3, 97-102. (1976).

FABIA 1977 -1 FABIAN, CS.; MOGA, S.
TWO SIMULATION PROCEDURES FOR SOLVING CONVEX INTEGER PROGRAMMING PROBLEMS. ECONOMIC COMPUTATION AND ECONOMIC CYBERNETICS STUDIES AND RESEARCH, NO. 3, 83-90. (1977).

FABR 1966 -1 FABRYCKY, W. J.; SHAMBLIN, J. E.
PROBABILITY BASED SEQUENCING ALGORITHM. INDUSTRIAL ENGINEERING 16, 308-312. (1966).

FALK 1963 -1 FALKENHAUSEN, H. VON
VERFAHREN ZUR PROGNOSE DER VERKEHRSVERTEILUNG IN EINEM GEPLANTEN STRASSENNETZ, EIN. UNTERNEHMENSFORSCHUNG 7, 75-88. (1963).

FAN 1964 -1 FAN, L. T.; WANG, G. S.
APPLICATION OF DISCRETE MAXIMUM PRINCIPLE TO TRANSPORTATION PROBLEM. JOURNAL OF MATHEMATICAL PHYSICS 43, 255-260. (1964).

FARB 1965 -1 FARBEY, B. A.; LAND, A. H.; MURCHLAND, J. D.
CASCADE ALGORITHM FOR FINDING THE MINIMUM DISTANCES ON A GRAPH, THE. REPORT LSE-TNT-19. LONDON: LONDON SCHOOL OF ECONOMICS. (1965).

FAUD 1974 -1 FAUDREE, R. J.; SCHELP, R. H.
PATH CONNECTED GRAPHS. ACTA MATHEMATICA ACADEMIAE SCIENTIARUM HUNGARICAE 25, 313-319. (1974).

FAUR 1976 -1 FAURE, R.; ROUCAIROL, C.; TOLLA, P.
CHEMINS ET FLOTS, ORDONNANCEMENTS. RECHERCHE OPERATIONNELLE APPLIQUEE, 1. PARIS: GAUTHIER-VILLARS, 233 P. (1976).

FAYA 1976 -1 FAYARD, D.; PLATEAU, G.
TECHNIQUE DE RESOLUTION DU PROBLEME DU KNAPSACK EN VARIABLES BIVALENTES. ELECTRICITE DE FRANCE, BULLETIN DE LA DIRECTION DES ETUDES ET RECHERCHES, SERIES C, NO. 1, 51-75. (1976).

FEDE 1975 -1 FEDERICO, P. J.
NUMBER OF POLYHEDRA, THE. PHILIPS RESEARCH REPORTS 30, SPECIAL NUMBER, 220-231. (1975).

FEDO 1975 -1 FEDORENKO, R. P.
ON M-MATRICES REDUCING TO SIMPLEST UNIMODULATED ONE (RUSSIAN). EKONOMIKA I MATEMATICHESKIE METODY 11, 1210-1213. (1975).

FEIC 1975 -1 FEICHT, E.; HECK, J.; PAPE, U.
ALGORITHMEN VON EDMONDS UND GABOW ZUR ERMITTLUNG DES MAXIMALEN MATCHINGS IN SCHLICHTEN UND ENDLICHEN GRAPHEN. REPORT 75-15. BERLIN: TECHNISCHE UNIV. BERLIN, FACHBEREICH 20 - KYBERNETIK, 45 P. (1975).

FEIG 1967 -1 FEIGIN, L. I.
EFFECT OF DISTURBANCES ON SCHEDULING. ENGINEERING CYBERNETICS, 11-15. (1967).

FEIG 1972 -1 FEIGIN, L. I.
GENERAL PROBLEM OF SCHEDULING THEORY WITH INCOMPLETE INFORMATION (RUSSIAN). AVTOMATIKA I TELEMEHANIKA, NO. 3, 110-116; TRANSLATED AS AUTOMATION AND REMOTE CONTROL 33, 1972, NO. 3, 445-451. (1972).

FEIN 1977 -1 FEINBERG, I.
HEURISTIC APPROACH TO WAREHOUSE LOCATION USING MULTIPLE CENTERS OF GRAVITY, A. SAN FRANCISCO (51. ORSA-MEETING + TIMS, MAY 1977) 15 P. (1977).

FENS 1967 -1 FENSKE, R. W.
SCHEDULING COMPUTER OPERATIONS. DATA PROCESSING MANAGEMENT 9, 26-28; 38-40; 36-38. (1967).

FERN 1975 -1 FERNANDEZ, E. B.; LANG, T.
COMPUTATION OF LOWER BOUNDS FOR MULTIPROCESSOR SCHEDULES. IBM JOURNAL OF RESEARCH AND DEVELOPMENT 19, 435-444. (1975).

FERN 1976 -1 FERNANDEZ, E. B.; LANG, T.
SCHEDULING AS A GRAPH TRANSFORMATION. IBM JOURNAL OF RESEARCH AND DEVELOPMENT 20, 551-559. (1976).

FERS 1976 -1 FERSTER, V. B.
DECREASING ALGORITHM FOR THE CUTTING PLANE METHOD FOR THE SOLUTION OF INTEGER PROGRAMMING PROBLEMS (RUSSIAN), A. IN: STUDIES IN DISCRETE OPTIMIZATION (RUSSIAN), 50-53. MOSCOW: IZDAT. NAUKA. (1976).

FEW 1955 -1 FEW, L.
SHORTEST PATH AND THE SHORTEST ROAD THROUGH N POINTS, THE. MATHEMATIKA 2, 141-144. (1955).

FIDU 1977 -1 FIDUCCIA, C. M.; ZALCSTEIN, Y.
ALGEBRAS HAVING LINEAR MULTIPLICATIVE COMPLEXITIES. JOURNAL OF THE ASSOCIATION FOR COMPUTING MACHINERY 24, 311-331. (1977).

FILL 1977 -1 FILL, J. A.; REINGOLD, E. M.
SOLUTIONS MANUAL. COMBINATORIAL ALGORITHMS: THEORY AND PRACTICE (BY E.M. REINGOLD, J. NIEVERGELT, N. DEO). ENGLEWOOD CLIFFS: PRENTICE-HALL, 311 P. (1977).

FILLA 1976 -1 FILLER, M. F.
ALGORITHM FOR FINDING SHORTEST PATHS (RUSSIAN), AN. IN: STUDIES IN DISCRETE OPTIMIZATION (RUSSIAN), 359-364. MOSCOW: IZDAT. NAUKA. (1976).

FINK 1975 -1 FINKE, G.
PURE FIXED COST DISTRIBUTION PROBLEM, THE. IN: PROCEEDINGS OF THE 5TH MANITOBA CONFERENCE ON NUMERICAL MATHEMATICS, WINNIPEG 1975, 305-318. CONGRESSUS NUMERANTIUM 16, WINNIPEG:UTILITAS MATHEMATICA. (1975).

FINKA 1976 -1 FINKEL'SHTEIN, YU. YU.
APPROXIMATIVE METHODS AND APPLIED PROBLEMS IN DISCRETE PROGRAMMING (RUSSIAN). MOSCOW: IZDAT. NAUKA, 264 P. (1976).

FINKA 1977 -1 FINKEL'SHTEIN, YU. YU.
E-APPROACH TO THE MULTIDIMENSIONAL KNAPSACK PROBLEM: THE POLYNOMIAL GROWTH OF THE TREE OF BRANCHING (RUSSIAN), THE. ZURNAL VYCISLITEL'NOI MATEMATIKI I MATEMATICESKOI FIZIKI 17, 1040-1041. (1977).

FIOR 1973 -1 FIORINI, S.; WILSON, R. J.
ON THE CHROMATIC INDEX OF A GRAPH, I. CAHIERS DU CENTRE D'ETUDES DE RECHERCHE OPERATIONNELLE 15, 253-262. (1973).

FIOR 1974 -1 FIORINI, S.; WILSON, R. J.
ON THE CHROMATIC INDEX OF A GRAPH, II. IN: T. P. MAC DONOUGH, V. C. MAVRON (EDS.): COMBINATORICS. PROCEEDINGS OF THE BRITISH COMBINATORIAL CONFERENCE, ABERYSTWYTH 1973, 37-51. LONDON MATHEMATICAL SOCIETY LECTURE NOTE SERIES 13, LONDON: CAMBRIDGE UNIV. PRESS. (1974).

FIOR 1976 -1 FIORINI, S.; WILSON, R. J.
EDGE-COLOURINGS OF GRAPHS - SOME APPLICATIONS. IN: C. ST. J. A. NASH-WILLIAMS (ED.): PROCEEDINGS OF THE 5TH BRITISH COMBINATORIAL CONFERENCE, ABERDEEN 1975, 193-202. CONGRESSUS NUMERANTIUM 15, WINNIPEG: UTILITAS MATHEMATICA. (1976).

FISB 1971 -1 FISBEIN, S. M.
ON THE SOLUTION OF A CERTAIN COMBINATORIAL PROBLEM (RUSSIAN). IN: N. I. ROSINA (ED.): MATHEMATICAL METHODS FOR THE STUDY AND OPTIMIZATION OF SYSTEMS, PROC. OF THE SEMINARS ON THEORY OF OPTIMAL DECISIONS, KIEV 1970 (RUSSIAN), 177-180. KIEW: AKAD. NAUK UKRAIN. SSR INST. KIBERNET. (1971).

****

FISC 1971 -1 FISCHER, M. J.; MEYER, A. R.
BOOLEAN MATRIX MULTIPLICATION AND TRANSITIVE CLOSURE. IN: CONFERENCE RECORD 1971, 12TH ANNUAL SYMPOSIUM ON SWITCHING AND AUTOMATA THEORY, EAST LANSING 1971, 129-131. NORTHRIDGE: INSTITUTE OF ELECTRICAL AND ELECTRONICS ENGINEERS. (1971).

FISH 1976 -1 FISHBURN, P. C.
ON LINEAR EXTENSION MAJORITY GRAPHS OF PARTIAL ORDERS. JOURNAL OF COMBINATORIAL THEORY 21, SERIES B, 65-70. (1976).

FISHA 1976 -1 FISHER, M. L.
DUAL ALGORITHM FOR THE ONE-MACHINE SCHEDULING PROBLEM, A. MATHEMATICAL PROGRAMMING 11, 229-251. (1976).

FISHB 1977 -1 FISHER, M. L.; NEMHAUSER, G. L.; WOLSEY, L. A.
ANALYSIS OF APPROXIMATIONS FOR MAXIMIZING SUBMODULAR FUNCTIONS. IN: R. HENN, P. KALL, B. KORTE, O. KRAFFT ET AL. (EDS.): OPERATIONS RESEARCH VERFAHREN XXV, METHODS OF OPERATIONS RESEARCH, 156-156, MEISENHEIM: HAIN. (1977).

FISHC 1973 -1 FISHER, M. L.; SHAPIRO, J. F.
CONSTRUCTIVE DUALITY FOR DISCRETE OPTIMIZATION. REPORT 7321, CHICAGO: UNIVERSITY OF CHICAGO, CENTER FOR MATHEMATICAL STUDIES IN BUSINESS AND ECONOMICS, 47 P. (1973).

FISK 1976 -1 FISK, J.
ALGORITHM FOR ASSIGNING USES TO SOURCES IN A TRANSPORTATION TABLEAU, AN. WORKING PAPER 42. ALBANY: STATE UNIVERSITY OF NEW YORK. (1976).

FISK 1976 -2 FISK, J.
SOLUTION PROCEDURE FOR A SPECIAL CLASS OF TRANSPORTATION PROBLEM CONTAINING MULTIPLE CHOICE CONSTRAINTS, A. PAPER PRESENTED AT THE 49TH JOINT NATIONAL MEETING OF ORSA AND TIMS, PHILADELPHIA 1976, 19 P. (1976).

FISK 1977 -1 FISK, J.
SOLUTION PROCEDURE FOR A SPECIAL TYPE OF CAPACITATED WAREHOUSE LOCATION PROBLEM, A. THE LOGISTICS AND TRANSPORTATION REVIEW 13, 305-320. (1977).

FISKA 1977 -1 FISK, J.; MACKEOWN, P.
DETERMINING THE MAXIMUM DEGENERATE SOLUTION FOR TRANSPORTATION PROBLEMS. SAN FRANCISCO (51. ORSA-MEETING + TIMS, MAY 1977) 15 P. (1977).

FISKB 1977 -1 FISK, J.; MACKEOWN, P. G.
APPLICATION OF INTEGER PROGRAMMING TO THE ASSIGNMENT OF STUDENTS TO SCHOOLS, AN. ATLANTA (52. ORSA-MEETING + TIMS, NOVEMBER 1977) 14 P. (1977).

FISKC 1977 -1 FISK, J. C.
INITIAL BOUNDING PROCEDURE FOR USE WITH 0-1 SINGLE KNAPSACK ALGORITHMS, AN. OPSEARCH 14, 88-98. (1977).

FISKD 1977 -1 FISK, S.
GEOMETRIC COLORING THEORY. ADVANCES IN MATHEMATICS 24, 298-340. (1977).

FITZ 1974 -1 FITZGERALD, C. H.
OPTIMAL INDEXING OF THE VERTICES OF GRAPHS. MATHEMATICS OF COMPUTATION 28, 825-831. (1974).

FLEI 1976 -1 FLEISCHNER, H.
GEMEINSAME BASIS FUER DIE THEORIE DER EULERSCHEN GRAPHEN UND DEN SATZ VON PETERSEN, EINE. MONATSHEFTE FUER MATHEMATIK 81, 267-278. (1976).

FLEI 1976 -2 FLEISCHNER, H.
IN THE SQUARE OF GRAPHS, HAMILTONICITY AND PANCYCLICITY, HAMILTONIAN CONNECTEDNESS AND PANCONNECTEDNESS ARE EQUIVALENT CONCEPTS. MONATSHEFTE FUER MATHEMATIK 82, 125-149. (1976).

FLOR 1976 -1 FLORIAN, M.; NGUYEN, S.
APPLICATION AND VALIDATION OF EQUILIBRIUM TRIP ASSIGNMENT METHODS, AN. TRANSPORTATION SCIENCE 10, 374-390. (1976).

FLORA 1970 -1 FLORIAN, M.; ROBILLARD, P.
HYPERBOLIC PROGRAMMING WITH BIVALENT VARIABLES. PUBLICATION 41, MONTREAL: UNIV. DE MONTREAL, DEPARTMENT D'INFORMATIQUE. (1970).

FOLD 1975 -1 FOLDES, S.; HAMMER, P. L.
ON A CLASS OF MATROID-PRODUCING GRAPHS. REPORT 7541-OR. BONN: UNIV. BONN, INST. F. OEKONOMETRIE UND OPERATIONS RESEARCH, 26 P. (1975).

FOLD 1976 -1 FOLDES, S.; HAMMER, P. L.
DIL WORTH OF A GRAPH, THE. REPORT 7644-OR. BONN: UNIV. BONN, INST. F. OEKONOMETRIE UND OPERATIONS RESEARCH, 16 P. (1976).

FOLD 1976 -2 FOLDES, S.; HAMMER, P. L.
SPLIT GRAPHS. REPORT 7642-OR. BONN: UNIV. BONN, INST. F. OEKONOMETRIE UND OPERATIONS RESEARCH, 6 P. (1976).

FOLD 1977 -1 FOLDES, S.; HAMMER, P. L.
SPLIT GRAPHS HAVING DILWORTH NUMBER TWO. CANADIAN JOURNAL OF MATHEMATICS 29, 666-672. (1977).

FOLD 1978 -1 FOLDES, S.; HAMMER, P. L.
DILWORTH NUMBER OF A GRAPH, THE. IN: ALSPACH, B.;HELL, P.;MILLER, D. J.(EDS.): ALGORITHMIC ASPECTS OF COMBINATORICS. ANNALS OF DISCRETE MATHEMATICS 2, 211-220. (1978).

FONG 1977 -1 FONG, A. C.; ULLMAN, J. D.
FINDING THE DEPTH OF A FLOW GRAPH. JOURNAL OF COMPUTER AND SYSTEM SCIENCES 15, 300-309. (1977).

FONGA 1974 -1 FONG, C. O.
MULTI-REGION DYNAMIC CAPACITY EXPANSION PROBLEM: EXACT AND HEURISTIC APPROACHES, THE. DISSERTATION. ROCHESTER: UNIVERSITY OF ROCHESTER. (1974).

FONGB 1975 -1 FONG, C. O.; RAO, M. R.
PARAMETRIC STUDIES IN TRANSPORTATION-TYPE PROBLEMS. NAVAL RESEARCH LOGISTICS QUARTERLY 22, 355-364. (1975).

FONGC 1974 -1 FONG, C. O.; SRINIVASAN, V.
MULTI-REGION DYNAMIC CAPACITY EXPANSION PROBLEM - I, THE. RESEARCH PAPER 227. STANFORD: STANFORD UNIVERSITY. (1974).

FONGC 1974 -2 FONG, C. O.; SRINIVASAN, V.
MULTI-REGION DYNAMIC CAPACITY EXPANSION PROBLEM - II, THE. RESEARCH PAPER 228. STANFORD: STANFORD UNIVERSITY. (1974).

FONGC 1977 -1 FONG, C. O.; SRINIVASAN, V.
DETERMINING ALL NONDEGENERATE SHADOW PRICES FOR THE TRANSPORTATION PROBLEM. TRANSPORTATION SCIENCE 11, 199-222. (1977).

FONT 1976 -1 FONTET, M.
AUTOMORPHISMES DE GRAPHES ET PLANARITE. ASTERISQUE, NO. 38-39, 73-90. (1976).

FOOT 1965 -1 FOOT, D. H. S.
SHORTEST ROUTE PROBLEM: ALGOL-PROGRAMS, AND DISCUSSION OF COMPUTATIONAL PROBLEMS IN LARGE TRANSPORT NETWORK APPLICATIONS, THE. DISCUSSION PAPER 10. BRISTOL: UNIVERSITY OF BRISTOL, DEPT. OF ECONOMICS. (1965).

FORC 1974 -1 FORCINA, J. L.
DYNAMIC LOCATION STRATEGIES FOR PUBLIC FACILITIES. DISSERTATION. BERKELEY: UNIV. OF CALIFORNIA, DEPT. OF CIVIL ENGINEERING. (1974).

****

FORG 1974 -1 FORGO, F.
SHADOW PRICES AND DECOMPOSITION FOR INTEGER PROGRAMS. REPORT DM 74-6, BUDAPEST: KARL MARX UNIV. OF ECONOMICS, 31 P. (1974).

FORG 1974 -2 FORGO, F.
SOME TRANSFORMATIONS OF INTEGER PROGRAMMING PROBLEMS (HUNGARIAN). SZIGMA. MATEMATIKAI-KOEZGAZDASAGI FOLYOIRAT 7, 271-282. (1974).

FORG 1976 -1 FORGO, F.
ON AN EXAMPLE OF DUALITY AND DECOMPOSITION METHODS IN INTEGER PROGRAMMING (HUNGARIAN). SZIGMA. MATEMATIKAI-KOZGAZDASAGI FOLYOIRAT 9, 53-68. (1976).

FORG 1976 -2 FORGO, F.
SOME TRANSFORMATIONS OF INTEGER PROGRAMMING PROBLEMS. IN: A. PREKOPA (ED.): PROGRESS IN OPERATIONS RESEARCH, 333-351, AMSTERDAM, LONDON: NORTH HOLLAND. (1976).

FOST 1976 -1 FOSTER, B. A.; RYAN, D. M.
INTEGER PROGRAMMING APPROACH TO THE VEHICLE SCHEDULING PROBLEM, AN. OPERATIONAL RESEARCH QUARTERLY 27, 367-384. (1976).

FOUL 1976 -1 FOULDS, L. R.
RELATIONSHIP BETWEEN GRAPH THEORY AND OPERATIONAL RESEARCH, THE. NEW ZEALAND MATHEMATICS MAGAZINE 13, 20-24. (1976).

FOULA 1976 -1 FOULDS, L. R.; ROBINSON, D. F.
STRATEGY FOR SOLVING THE PLANT LAYOUT PROBLEM, A. OPERATIONAL RESEARCH QUARTERLY 27, 845-856. (1976).

FOUR 1971 -1 FOURNIER, J. C.
REPRESENTATION SUR UN CORPS DES MATROIDES D'ORDRE <8 OU =8. IN: C. P. BRUTER (ED.): THEORIE DES MATROIDES, 50-61. LECTURE NOTES IN MATHEMATICS 211, BERLIN: SPRINGER. (1971).

FOUR 1973 -1 FOURNIER, J. C.
COLORATIONS DES ARETES D'UN GRAPHE. CAHIERS DU CENTRE D'ETUDES DE RECHERCHE OPERATIONNELLE 15, 311-314. (1973).

FOUR 1975 -1 FOURNIER, J. C.
SUR LES STABLES D'UN GRAPHE. CAHIERS DU CENTRE D'ETUDES DE RECHERCHE OPERATIONNELLE 17, 193-195. (1975).

FOUR 1976 -1 FOURNIER, J. C.
PROPRIETES COMBINATOIRES ET ALGEBRIQUES DES GRAPHES PLANAIRES. IN: ANAL. APPL. INF. JOURN. COMB. INF., BORDEAUX 1975, 153-179. BORDEAUX: UNIV. BORDEAUX I. (1976).

FOX 1975 -1 FOX, B.
MORE ON KTH SHORTEST PATHS. COMMUNICATIONS OF THE A.C.M. 18, 279. (1975).

FOX A 1975 -1 FOX, B. L.
PROBABILISTIC NETWORKS. PAPER PRESENTED AT THE 48TH JOINT NATIONAL MEETING OF ORSA AND TIMS, LAS VEGAS 1975, 16 P. (1975).

FOX A 1976 -1 FOX, B. L.
DATA STRUCTURES AND COMPUTER SCIENCE TECHNIQUES IN OPERATIONS RESEARCH. PUBLICATION 245, UNIV. DE MONTREAL, DEPT. D'INFORMATIQUE. (1976).

FOX B 1977 -1 FOX, B. L.; LENSTRA, J. K.; RINNOOY KAN, A. H. G.; SCHRAGE, L. E.
BRANCHING FROM THE LARGEST UPPER BOUND: FOLKLORE AND FACTS. BW 64/77. AMSTERDAM: MATHEMATISCH CENTRUM, 6 P. (1977).

FRAN 1963 -1 FRANCIS, R. L.
NOTE ON THE OPTIMUM LOCATION OF NEW MACHINES IN EXISTING PLANT LAYOUTS, A. JOURNAL OF INDUSTRIAL ENGINEERING 14, 57-58. (1963).

FRAN 1974 -1 FRANCIS, R. L.
FACILITY LAYOUT AND LOCATION. ENGLEWOOD CLIFFS: PRENTICE-HALL. (1974).

FRANA 1976 -1 FRANCIS, R. L.; LOWE, T. J.; RATLIFF, H. D.
DISTANCE CONSTRAINTS FOR TREE NETWORK MULTIFACILITY LOCATION PROBLEMS. RESEARCH REPORT NO. 76-2.GAINESVILLE: UNIV. OF FLORIDA, INDUSTRIAL AND SYSTEMS ENGINEERING DEPT., 40 P. (1976).

FRANB 1974 -1 FRANCIS, R. L.; WHITE, J. A.
FACILITY LAYOUT AND LOCATION: AN ANALYTIC APPROACH. ENGLEWOOD CLIFFS: PRENTICE HALL. (1974).

FRANC 1976 -1 FRANCON, J.
ARBRES BINAIRES DE RECHERCHE: PROPRIETES COMBINATOIRES ET APPLICATIONS. REVUE FRANCAISE D'AUTOMATIQUE, D'INFORMATIQUE, ET DE RECHERCHE OPERATIONNELLE 10, NO. 12, 35-52. (1976).

FRAND 1976 -1 FRANK, A.
SOME POLYNOMIAL ALGORITHMS FOR CERTAIN GRAPHS AND HYPERGRAPHS. IN: C. ST. J. A. NASH-WILLIAMS (ED.): PROCEEDINGS OF THE 5TH BRITISH COMBINATORIAL CONFERENCE, ABERDEEN 1975, 211-226. CONGRESSUS NUMERANTIUM 15, WINNIPEG: UTILITAS MATHEMATICA. (1976).

FRAND 1977 -1 FRANK, A.
ON A CLASS OF BALANCED HYPERGRAPHS. DISCRETE MATHEMATICS 20, 11-20. (1977).

FRANE 1969 -1 FRANK, H.
MAXIMALLY RELIABLE NODE WEIGHTED GRAPHS. IN: PROCEEDINGS OF THE THIRD ANNUAL PRINCETON CONFERENCE ON INFORMATION SCIENCES AND SYSTEMS, 1-6. (1969).

FRANF 1971 -1 FRANK, H.; CHOU, W.
HEURISTIC APPROACH TO ROUTING IN COMPUTER NETWORKS, A. IN: PROCEEDINGS OF THE FIFTH ANNUAL PRINCETON CONFERENCE ON INFORMATION SCIENCES AND SYSTEMS, 197-203. (1971).

FRANG 1970 -1 FRANK, H.; FRISCH, I.
NETWORK ANALYSIS. SCIENTIFIC AMERICAN. (1970).

FRANH 1977 -1 FRANK, O.
ESTIMATION OF GRAPH TOTALS. SCANDINAVIAN JOURNAL OF STATISTICS 4, 81-89. (1977).

FRAT 1975 -1 FRATTA, L.
BASIC ANALYTICAL TECHNIQUES AND ROUTING PROCEDURES. IN: COMPUTER COMMUNICATION NETWORKS, PROC. OF A NATO ADVANCED STUDY INST., BRIGHTON 1973, 35-62. NATO ADVANCED STUDY INST. SERIES E: APPL. SCI. 4, LEYDEN: NOORDHOFF. (1975).

FRATA 1972 -1 FRATTA, L.; MAFFIOLI, F.
ON THE SHORTEST HAMILTONIAN CHAIN OF A NETWORK. 2ND INTERNATIONAL SYMPOSIUM ON NETWORK THEORY, HERCEG NOVI. (1972).

FRATB 1975 -1 FRATTA, L.; MONTANARI, U.
VERTEX ELIMINATION ALGORITHM FOR ENUMERATING ALL SIMPLE PATHS IN A GRAPH, A. NETWORKS 5, 151-178. (1975).

FRAZ 1972 -1 FRAZER, W. D.
ANALYSIS OF COMBINATORY ALGORITHMS - A SAMPLE OF CURRENT METHODOLOGY. IN: SPRING JOINT COMPUTER CONFERENCES, 483-491. (1972).

FRED 1976 -1 FREDERICKSON, G. N.; HECHT, M. S.; KIM, C. E.
APPROXIMATION ALGORITHMS FOR SOME ROUTING PROBLEMS. IN: 17TH ANNUAL SYMPOSIUM ON FOUNDATIONS OF COMPUTER SCIENCE, HOUSTON 1976, 216-227. LONG BEACH: INSTITUTE OF ELECTRICAL AND ELECTRONICS ENGINEERS. (1976).

FREDA 1975 -1 FREDMAN, M. L.
ON THE DECISION TREE COMPLEXITY OF THE SHORTEST PATH PROBLEMS. IN: 16TH ANNUAL SYMPOSIUM ON FOUNDATIONS OF COMPUTER SCIENCE, BERKELEY 1975, 98-99. NEW YORK: INSTITUTE OF ELECTRICAL AND ELECTRONICS ENGINEERS. (1975).

FREDA 1975 -2 FREDMAN, M. L.
SOME ARITHMETICAL PACKING PROBLEMS. JOURNAL OF THE LONDON MATHEMATICAL SOCIETY 11, 389-398. (1975).

FREDA 1976 -1 FREDMAN, M. L.
NEW BOUNDS ON THE COMPLEXITY OF THE SHORTEST PATH PROBLEM. SIAM JOURNAL ON COMPUTING 5, 83-89. (1976).

FREE 1977 -1 FREELAND, J. R.; MOORE, J. H.
IMPLICATIONS OF RESOURCE DIRECTIVE ALLOCATION MODELS FOR ORGANIZATIONAL DESIGN. MANAGEMENT SCIENCE 23, 1050-1059. (1977).

FREH 1975 -1 FREHEL, J.
PROBLEME DE PARTITION SOUS CONTRAINTE, LE. IN: D. ROY (ED.): COMBINATORIAL PROGRAMMING: METHODS AND APPLICATIONS. PROCEEDINGS OF THE NATO ADVANCED STUDY INST., VERSAILLES 1974, 269-274. DORDRECHT: REIDEL. (1975).

FREH 1975 -2 FREHEL, J.
REGULARISATION DANS LES PROBLEMES COMBINATOIRES, PROBLEMES DE TOURNEES, DE PARTITION ET DE RECOUVREMENT. METRA 14, 723-734. (1975).

FREI 1973 -1 FREIMER, M.; SCHWARZ, L. B.; SRINIVASAN, V.
NOTE ON OPTIMAL SINGLE CYCLE POLICIES FOR THE ONE-WAREHOUSE N-RETAILER INVENTORY PROBLEM, A. ROCHESTER: UNIV. OF ROCHESTER, GRADUATE SCHOOL OF MANAGEMENT, 6 P. (1973). (WORKING PAPER NO. 7308)

FRID 1977 -1 FRIDMAN, A. A.
ON SOME MODERN ASPECTS IN DISCRETE OPTIMIZATION (RUSSIAN). EKONOMIKA I MATEMATICESKIE METODY 13, 1115-1131. (1977).

FRIDA 1973 -1 FRIDMAN, A. A.; LITVAK, B. G.; RAPPOPORT, A. M.
ABSOLUTELY INTEGER LINEAR PROGRAMMING PROBLEMS (RUSSIAN). IN: A. A. FRIDMAN (ED.): STUDIES IN DISCRETE MATHEMATICS (RUSSIAN), 27-40. MOSCOW: IZDAT. NAUKA. (1973).

FRIDB 1976 -1 FRIDMAN, A. A.; VOTJAKOV, A. A.
REGULAR DISCRETE PROBLEMS AND THE CUTTING PLANE METHOD (RUSSIAN). IN: STUDIES IN DISCRETE OPTIMIZATION (RUSSIAN), 19-50. MOSCOW: IZDAT. NAUKA. (1976).

FRIE 1975 -1 FRIEDMAN, M.
SCHEDULING A TRANSPORTATION NETWORK. OMEGA 3, 751-755. (1975).

FRIE 1976 -1 FRIEDMAN, M.
ON THE ANALYSIS AND SOLUTION OF CERTAIN GEOGRAPHICAL OPTIMAL COVERING PROBLEMS. COMPUTERS AND OPERATIONS RESEARCH 3, 283-294. (1976).

FRIEA 1975 -1 FRIEZE, A. M.
BOTTLENECK LINEAR PROGRAMMING. OPERATIONAL RESEARCH QUARTERLY 26, 871-874. (1975).

FRIEA 1976 -1 FRIEZE, A. M.
SHORTEST PATH ALGORITHMS FOR KNAPSACK TYPE PROBLEMS. MATHEMATICAL PROGRAMMING 11, 150-157. (1976).

FRIEA 1977 -1 FRIEZE, A. M.
ALGORITHMS FOR GENERALISATIONS OF MATROIDS. LONDON: REPORT, QUEEN MARY COLLEGE. (1977).

FRIEA 1977 -2 FRIEZE, A. M.
GENERALISATION OF THE GREEDY ALGORITHM FOR MATROIDS, A. LONDON: QUEEN MARY COLLEGE, 13 P. (1977).

FRIEA 1977 -3 FRIEZE, A. M.
MINIMUM PATHS IN DIRECTED GRAPHS. OPERATIONAL RESEARCH QUARTERLY 28, 339-346. (1977).

FRIS 1963 -1 FRISCH, I. T.
OPTIMIZATION OF COMMUNICATION NETS WITH SWITCHING. JOURNAL OF THE FRANKLIN INSTITUTE, 275, 405-430. (1963).

FRIS 1963 -2 FRISCH, I. T.
OPTIMUM ROUTES IN COMMUNICATION SYSTEMS WITH CHANNEL CAPACITIES AND CHANNEL RELIABILITIES. IEEE TRANSACTIONS ON COMMUNICATION SYSTEMS CS-11, 241-244. (1963).

FROM 1976 -1 FROMOVITZ, S.; GASS, S. I.
SCHEDULING OF SUBMARINES FOR FLEET OPERATIONS SUPPORT, THE. WORKING PAPER SERIES MS/S 76-001. COLLEGE PARK: UNIV. OF MARYLAND, COLLEGE OF BUSINESS AND MANAGEMENT, 18 P. (1976).

FRON 1975 -1 FRONC, M. B.
ADDRESSED FLOWS IN SPECIAL CASES OF PLANAR NETWORKS (CZECH). EKONOMICKO-MATEMATICKY OBZOR 11, 415-422. (1975).

FRUM 1976 -1 FRUMKIN, M. A.
ALGORITHMS FOR THE SOLUTION IN INTEGERS OF SYSTEMS OF LINEAR EQUATIONS (RUSSIAN). IN: STUDIES IN DISCRETE OPTIMIZATION (RUSSIAN), 97-127. MOSCOW: IZDAT. NAUKA. (1976).

FRUM 1976 -2 FRUMKIN, M. A.
EXPERIMENTAL TIME ALGORITHM FOR THE TRIANGULARIZATION OF AN INTEGER MATRIX (RUSSIN), AN. EKONOMIKA I MATEMATICESKIE METODY 12, 173-178. (1976).

FUJI 1966 -1 FUJII, A.; TAKAZAWA, T.
EXAMPLES OF OR METHODS IN A MASS PRODUCTION FACTORY (JAPANESE). KEIEI KAGAKU 9, 76-89. (1966).

FUJIA 1963 -1 FUJISAWA, T.
MAXIMUM FLOWS IN A LOSSY NETWORK. IN: PROCEEDINGS OF THE ALLERTON CONFERENCE ON CIRCUIT AND SYSTEM THEORY, URBANA, 383-385. (1963).

FUJIB 1974 -1 FUJISAWA, T.; ORINO, H.
EFFICIENT ALGORITHM OF FINDING A MINIMAL TRIANGULATION OF A GRAPH, AN. IN: PROCEEDINGS OF THE 1974 IEEE INTERNATIONAL SYMPOSIUM ON CIRCUITS AND SYSTEMS, 172-175. NEW YORK: INTERNAT. INST. OF ELECTRICAL AND ELECTRONICAL ENGINEERS. (1974).

FUJIC 1977 -1 FUJISHIGE, S.
ALGORITHM FOR FINDING AN OPTIMAL INDEPENDENT LINKAGE, AN. JOURNAL OF THE OPERATIONS RESEARCH SOCIETY OF JAPAN 20, 59-75. (1977).

FUJIC 1977 -2 FUJISHIGE, S.
PRIMAL APPROACH TO THE INDEPENDENT ASSIGNMENT PROBLEM, A. JOURNAL OF THE OPERATIONS RESEARCH SOCIETY OF JAPAN 20, 1-15. (1977).

FUKU 1976 -1 FUKUDA, Y.
IMPLICIT ENUMERATION ALGORITHM FOR A STORAGE ALLOCATION PROBLEM, AN. PAPER PRESENTED AT THE 49TH JOINT NATIONAL MEETING OF ORSA AND TIMS, PHILADELPHIA 1976, 37 P. (1976).

FULK 1975 -1 FULKERSON, D. R.
FLOW NETWORKS AND COMBINATORIAL OPERATIONS RESEARCH. IN: D. R. FULKERSON (ED.): STUDIES IN GRAPH THEORY, PART 1, 139-171. STUDIES IN MATHEMATICS 11, WASHINGTON: MATHEMATICAL ASSOCIATION OF AMERICA. (1975).

FULK 1975 -2 FULKERSON, D. R.
STUDIES IN GRAPH THEORY, I. STUDIES IN MATHEMATICS, 11, WASHINGTON: MATHEMATICAL ASSOCIATION OF AMERICA, 199 P. (1975).

FULK 1975 -3 FULKERSON, D. R.
STUDIES IN GRAPH THEORY, II. STUDIES IN MATHEMATICS 12,WASHINGTON: MATHEMATICAL ASSOCIATION OF AMERICA, 443 P. (1975).

FULKA 1976 -1 FULKERSON, D. R.; HARDING, G. C.
ON EDGE-DISJOINT BRANCHINGS. NETWORKS 6, 97-104. (1976).

****

FULL 1976 -1 FULLER, R. V.; JARVIS, J. J.
SELF-DUAL LINEAR PROGRAMMING MODEL FOR THE EQUILIBRIUM TRAFFIC ASSIGNMENT PROBLEM, A. REPORT SERIES J-76-6. ATLANTA: GEORGIA INST. OF TECHNOLOGY, 14 P. (1976).

FUNK 1968 -1 FUNK, M. L.; TILLMAN, F. A.
OPTIMAL CONSTRUCTION STAGING BY DYNAMIC PROGRAMMING. JOURNAL OF THE HIGHWAY DIVISION, PROCEEDINGS OF THE AMERICAN SOCIETY OF CIVIL ENGINEERS 94, NO. HW2, 255-265. (1968).

GABA 1975 -1 GABASOV, R.; KIRILLOVA, F. M.
CONSTRUCTION OF SUBOPTIMAL PLANS OF A TRANSPORTATION PROBLEM (RUSSIAN), THE. IZVESTIJA AKADEMII NAUK SSSR, TEHNICESKAJA KIBERNETIKA, NO. 6, 38-42 + 219; TRANSLATED AS ENGINEERING CYBERNETICS 13, 1975, NO. 6, 34-38. (1975).

GABB 1974 -1 GABBAY, H.
OVERVIEW OF VEHICULAR SCHEDULING PROBLEMS, AN. TECHNICAL REPORT 103. CAMBRIDGE: M.I.T., OPERATIONS RESEARCH CENTER. (1974).

GABO 1976 -1 GABOVICH, E. JA.
CONSTANT PROBLEMS OF DISCRETE PROGRAMMING ON SUBSTITUTION SETS (RUSSIAN). KIBERNETIKA, NO. 5, 128-134. (1976).

GABOA 1976 -1 GABOW, H.
DECOMPOSING SYMMETRIC EXCHANGES IN MATROID BASES. MATHEMATICAL PROGRAMMING 10, 271-276. (1976).

GABOB 1976 -1 GABOW, H. N.
EFFICIENT IMPLEMENTATION OF EDMONDS' ALGORITHM FOR MAXIMUM MATCHING ON GRAPHS, AN. JOURNAL OF THE ASSOCIATION FOR COMPUTING MACHINERY 23, 221-234. (1976).

GABOB 1976 -2 GABOW, H. N.
USING EULER PARTITIONS TO EDGE COLOR BIPARTITE MULTIGRAPHS. INTERNATIONAL JOURNAL OF COMPUTER AND INFORMATION SCIENCE 5, 4, 345-355. (1976).

GABOB 1977 -1 GABOW, H. N.
TWO ALGORITHMS FOR GENERATING WEIGHTED SPANNING TREES IN ORDER. SIAM JOURNAL ON COMPUTING 6, 139-150. (1977).

GAGN 1967 -1 GAGNON, G.
MODEL FOR FLOWING PASSENGERS OVER AIRLINE NETWORKS. TRANSPORTATION SCIENCE 1, 232-248. (1967).

GAIH 1977 -1 GAIHA, P.; GUPTA, S. K.
ADJACENT VERTICES ON A PERMUTOHEDRON. SIAM JOURNAL ON APPLIED MATHEMATICS 32, 323-327. (1977).

GAIK 1976 -1 GAIKOV, N. E.
APPROXIMATE SOLUTION OF THE GENERAL ASSIGNMENT PROBLEM AND OF THE MINIMIZATION PROBLEM OF THE SCALAR MATRIX PRODUCT (RUSSIAN), THE. VESCI AKADEMII NAVUK BSSR, SERYJA FIZIKA-MATEMATYCNYH NAVUK, NO. 2, 108-110 + 143. (1976).

GAIK 1976 -2 GAIKOV, N. E.
MINIMIZATION OF A QUADRATIC FORM WITH BOOLEAN VARIABLES ON A SYMMETRIC GROUP (RUSSIAN), THE. DOKLADY AKADEMII NAUK BSSR 20, 30-33 + 92. (1976).

GALI 1975 -1 GALIL, Z.
ON ENUMERATION PROCEDURES FOR THEOREM PROVING AND FOR INTEGER PROGRAMMING. IBM RESEARCH PAPER RC 5719. YORKTOWN HEIGHTS: IBM RESEARCH CENTER, 32 P. (1975).

GALI 1976 -1 GALIL, Z.
HIERARCHIES OF COMPLETE PROBLEMS. ACTA INFORMATICA 6, 77-88. (1976).

GALL 1977 -1 GALLAGER, R. G.
MINIMUM DELAY ROUTING ALGORITHM USING DISTRIBUTED COMPUTATION. SPECIAL ISSUE ON COMPUTER COMMUNICATIONS, A. IEEE TRANSACTIONS ON COMMUNICATIONS COM-25, 73-85. (1977).

GALLA 1975 -1 GALLAGHER, J. M.
NUCLEAR FUEL REPROCESSING EXPANSION STRATEGIES. TECHNICAL REPORT SOL-75-11. STANFORD: STANFORD UNIVERSITY, SYSTEMS OPTIMIZATION LABORATORY. (1975).

GALLB 1976 -1 GALLO, G.; HAMMER, P. L.; SIMEONE, B.
QUADRATIC KNAPSACK PROBLEMS. RESEARCH REPORT 76-43. WATERLOO: UNIV. OF WATERLOO, DEPT. OF COMBINATORICS AND OPTIMIZATION, 24 P. (1976).

GALLC 1972 -1 GALLO, G.; MARTINO. E.; SIMEONE, B.
GROUP OPTIMIZATION ALGORITHMS AND SOME NUMERICAL RESULTS VIA A BRANCH AND BOUND APPROACH. PAPER PRESENTED AT THE AICA CONFERENCE ON SIMULATION TECHNIQUES AND ALGORITHMS, MILANO 1972. (1972).

GALLD 1976 -1 GALLUS, G.
HEURISTISCHE VERFAHREN ZUR LOESUNG ALLGEMEINER GANZZAHLIGER LINEARER OPTIMIERUNGSPROBLEME (EIN UEBERBLICK). ZEITSCHRIFT FUER OPERATIONS RESEARCH 20, SERIE A, 89-104. (1976).

GALLD 1976 -2 GALLUS, G.
HEURISTISCHE VERFAHREN ZUR LOESUNG LINEARER OPTIMIERUNGSPROBLEME MIT GANZZAHLIGEN VARIABLEN. IN: J. KOHLAS, O. SEIFERT, P. STAEHLY, H.-J. ZIMMERMANN (EDS): PROCEEDINGS IN OPERATIONS RESEARCH 5, 46-47. WUERZBURG: PHYSICA. (1976).

GARB 1976 -1 GARBAI, L.; MOLNAR, L.
OPTIMIZATION OF URBAN PUBLIC UTILITY NETWORKS BY DISCRETE DYNAMIC PROGRAMMING. IN: A. PREKOPA (ED.): PROGRESS IN OPERATIONS RESEARCH, 373-390, AMSTERDAM, LONDON: NORTH HOLLAND. (1976).

GARE 1976 -1 GAREY, M. R.; GRAHAM, R. L.; JOHNSON, D. S.
SOME NP-COMPLETE GEOMETRIC PROBLEMS. IN: EIGHTH ANNUAL ACM SYMPOSIUM ON THEORY OF COMPUTING, HERSHEY 1976, 10-22. NEW YORK: ASSOCIATION FOR COMPUTING MACHINERY. (1976).

GARE 1977 -1 GAREY, M. R.; GRAHAM, R. L.; JOHNSON, D. S.
COMPLEXITY OF COMPUTING STEINER MINIMAL TREES, THE. SIAM JOURNAL ON APPLIED MATHEMATICS 32, 835-859. (1977).

GAREA 1976 -1 GAREY, M. R.; GRAHAM, R. L.; JOHNSON, D. S.; YAO, A. C.-C.
RESOURCE CONSTRAINED SCHEDULING AS GENERALIZED BIN PACKING. JOURNAL OF COMBINATORIAL THEORY 21, SERIES A, 257-298. (1976).

GAREB 1974 -1 GAREY, M. R.; JOHNSON, D. S.
COMPLEXITY RESULTS FOR MULTIPROCESSOR SCHEDULING UNDER RESOURCE CONSTRAINTS. IN: PROCEEDINGS OF THE EIGHTH ANNUAL PRINCETON CONFERENCE ON INFORMATION SCIENCES AND SYSTEMS, 168-172. (1974).

GAREB 1976 -1 GAREY, M. R.; JOHNSON, D. S.
APPROXIMATION ALGORITHMS FOR COMBINATORIAL PROBLEMS: AN ANNOTATED BIBLIOGRAPHY. IN: J. F. TRAUB (ED.): ALGORITHMS AND COMPLEXITY, NEW DIRECTIONS AND RECENT RESULTS, 41-52. NEW YORK: ACADEMIC PRESS. (1976).

GAREB 1976 -2 GAREY, M. R.; JOHNSON, D. S.
APPROXIMATION ALGORITHMS FOR COMBINATORIAL PROBLEMS: PROSPECTS AND LIMITATIONS. PAPER PRESENTED AT THE SYMPOSIUM ON ALGORITHMS AND COMPLEXITY, CARNEGIE MELLON INST., PITTSBURGH. (1976).

GAREB 1976 -3 GAREY, M. R.; JOHNSON, D. S.
COMPLEXITY OF NEAR-OPTIMAL GRAPH COLORING, THE. JOURNAL OF THE ASSOCIATION FOR COMPUTING MACHINERY 23, 43-49. (1976).

****

GAREB 1976 -4 GAREY, M. R.; JOHNSON, D. S.
SCHEDULING TASKS WITH NONUNIFORM DEADLINES ON TWO PROCESSORS. JOURNAL OF THE ASSOCIATION FOR COMPUTING MACHINERY 23, 461-467. (1976).

GAREB 1976 -5 GAREY, M. R.; JOHNSON, D. S.
SOME SIMPLIFIED NP-COMPLETE GRAPH PROBLEMS. THEORETICAL COMPUTER SCIENCE 1, 237-267. (1976).

GAREB 1976 -6 GAREY, M. R.; JOHNSON, D. S.
STRONG NP-COMPLETENESS RESULTS: MOTIVATION, EXAMPLES AND IMPLICATIONS. MIAMI (50. ORSA-MEETING + TIMS, NOVEMBER 1976), 22 P. (1976).

GAREB 1977 -1 GAREY, M. R.; JOHNSON, D. S.
RECTILINEAR STEINER TREE PROBLEM IS NP-COMPLETE, THE. SIAM JOURNAL ON APPLIED MATHEMATICS 32, 826-834. (1977).

GAREB 1977 -2 GAREY, M. R.; JOHNSON, D. S.
TWO-PROCESSOR SCHEDULING WITH START-TIMES AND DEADLINES. SIAM JOURNAL ON COMPUTING 6, 416-426. (1977).

GAREC 1976 -1 GAREY, M. R.; JOHNSON, D. S.; SETHI, R.
COMPLEXITY OF FLOWSHOP AND JOBSHOP SCHEDULING, THE. MATHEMATICS OF OPERATIONS RESEARCH 1, 117-129. (1976).

GARED 1976 -1 GAREY, M. R.; JOHNSON, D. S.; TARJAN, R. E.
PLANAR HAMILTONIAN CIRCUIT PROBLEM IS NP-COMPLETE, THE. SIAM JOURNAL ON COMPUTING 5, 704-714. (1976).

GARF 1977 -1 GARFINKEL, R. S.
MINIMIZING WALLPAPER WASTE, PART 1: A CLASS OF TRAVELING SALESMAN PROBLEMS. OPERATIONS RESEARCH 25, 741-751. (1977).

GARFA 1976 -1 GARFINKEL, R. S.; GILBERT, K. C.
BOTTLENECK TRAVELING SALESMAN PROBLEM, THE. WORKING PAPER NO. 32. MIAMI (50. ORSA-MEETING + TIMS, NOVEMBER 1976) AND KNOXVILLE: UNIV. OF TENNESSEE, COLLEGE OF BUSINESS ADMINISTRATION, 23 P. (1976).

GARFB 1973 -1 GARFINKEL, R. S.; NEEBE, A. W.; RAO, M. R.
ALGORITHM FOR THE M-MEDIAN PLANT LOCATION PROBLEM, AN. ROCHESTER: UNIV. OF ROCHESTER, GRADUATE SCHOOL OF MANAGEMENT, 30 P. (1973). (WORKING PAPER SERIES NO. 7312)

GARFB 1977 -1 GARFINKEL, R. S.; NEEBE, A. W.; RAO, M. R.
M-CENTER PROBLEM: MINIMAX FACILITY LOCATION, THE. MANAGEMENT SCIENCE 23, 1123-1142. (1977).

GARS 1977 -1 GARSIA, A. M.; WACHS, M. L.
NEW ALGORITHM FOR MINIMUM COST BINARY TREES, A. SIAM JOURNAL ON COMPUTING 6, 622-642. (1977).

GART 1976 -1 GARTNER, N.; GOLDEN, B.; WONG, R.
MODELLING AND OPTIMIZATION FOR TRANSPORTATION SYSTEMS PLANNINGS AND OPERATIONS. WINNIPEG: PROCEEDINGS OF THE INTERNATIONAL SYMPOSIUM ON LARGE ENGINEERING SYSTEMS. (1976).

GARTA 1975 -1 GARTNER, N. H.; LITTLE, J. D. C.; GABBAY, H.
OPTIMIZATION OF TRAFFIC SIGNAL SETTINGS BY MIXED-INTEGER LINEAR PROGRAMMING. PART I: THE NETWORK COORDINATION PROBLEM, PART II: THE NETWORK SYNCHRONIZATION PROBLEM. TRANSPORTATION SCIENCE 9, 321-363. (1975).

GATE 1971 -1 GATELY, D. T.
INVESTMENT PLANNING FOR THE ELECTRIC POWER INDUSTRY: A MIXED-INTEGER PROGRAMMING APPROACH, WITH APPLICATIONS TO SOUTHERN INDIA. DISSERTATION. PRINCETON: PRINCETON UNIVERSITY. (1971).

GATT 1976 -1 GATTINGER, M.
INNERE UND AEUSSERE ALGORITHMEN ZUR BESTIMMUNG R KUERZESTER WEGE UNTER NEBENBEDINGUNGEN. ERLANGEN: UNIVERSITAET ERLANGEN, INSTITUT FUER ANGEWANDTE MATHEMATIK, 3 P. (1976).

GAUL 1974 -1 GAUL, W.
R-SATISFYING MULTI-COMMODITY FLOWS AND CAPACITY-ALTERATIONS OF NETWORKS. MATHEMATICA BALKANICA 4, 219-224. (1974).

GAUL 1976 -1 GAUL, W.
ON CONSTRAINED SHORTEST-ROUTE PROBLEMS. IN: W. OETTLI, K. RITTER (EDS.): OPTIMIZATION AND OPERATIONS RESEARCH, PROCEEDINGS OF A CONFERENCE HELD AT OBERWOLFACH, 1975, 83-91. LECTURE NOTES IN ECONOMICS AND MATHEMATICAL SYSTEMS 117, BERLIN, HEIDELBERG, NEW YORK: SPRINGER. (1976).

GAUL 1976 -2 GAUL, W.
UEBER FLUSSPROBLEME IN NETZWERKEN. ZEITSCHRIFT FUER ANGEWANDTE MATHEMATIK UND MECHANIK 56, NO. 3, T340-T342. (1976).

GAUL 1976 -3 GAUL, W.
ZUR BERECHNUNG GLEICHZEITIGER FLUESSE VERSCHIEDENER BESCHAFFENHEIT. IN: J. KOHLAS, O. SEIFERT, P. STAEHLY, H.-J. ZIMMERMANN (EDS): PROCEEDINGS IN OPERATIONS RESEARCH 5, 113-118. WUERZBURG: PHYSICA. (1976).

GAUL 1977 -1 GAUL, W.
PROBLEME BEI GLEICHZEITIGEN FLUESSEN IN GRAPHEN. IN: R. HENN, P. KALL, B. KORTE, O. KRAFFT ET AL. (EDS.): OPERATIONS RESEARCH VERFAHREN XXIII, METHODS OF OPERATIONS RESEARCH, 36-48. MEISENHEIM: HAIN. (1977).

GAUL 1977 -2 GAUL, W.
UEBER WEGE MAXIMALER KAPAZITAET(EXTENDED ABSTRACT). IN: R. HENN, P. KALL, B. KORTE, O. KRAFFT ET AL. (EDS.): OPERATIONS RESEARCH VERFAHREN XXV, METHODS OF OPERATIONS RESEARCH, 303-305. MEISENHEIM: HAIN. (1977).

GAUL 1977 -3 GAUL, W.
ZUR LOESUNG VON WEGPROBLEMEN MIT ZEITABHAENGIGEN BESCHRAENKUNGEN. ZEITSCHRIFT FUER ANGEWANDTE MATHEMATIK UND MECHANIK 57, NO. 5, T323-T325. (1977).

GAUT 1977 -1 GAUTHIER, J.-M.; RIBIERE, G.
EXPERIMENTS IN MIXED-INTEGER LINEAR PROGRAMMING USING PSEUDOCOSTS. MATHEMATICAL PROGRAMMING 12, 26-47. (1977).

GAVI 1977 -1 GAVISH, B.; SCHWEITZER, P.; SHLIFER, E.
ZERO PIVOT PHENOMENON IN TRANSPORTATION AND ASSIGNMENT PROBLEMS AND ITS COMPUTATIONAL IMPLICATIONS, THE. MATHEMATICAL PROGRAMMING 12, 226-240. (1977).

GAVR 1977 -1 GAVRIL, F.
ALGORITHMS ON CLIQUE SEPARABLE GRAPHS. DISCRETE MATHEMATICS 19, 159-166. (1977).

GAZI 1975 -1 GAZIS, D. C.
TRANSPORTATION NETWORKS. NETWORKS 5, 75-79. (1975).

GELD 1976 -1 GELDERS, L.; WASSENHOVE, L. VAN
SHORT-TIME PLANNING OF THE "JOB-SHOP" TYPE FOR A PRODUCTION SITE (DUTCH). INGENIEURSBLAD, NO. 8, 246-251. (1976).

GELDA 1976 -1 GELDERS, L. F.
SOLVING AN OVERALL PRODUCTION PLANNING PROBLEM IN A SIMPLE JOB-SHOP. IN: A. PREKOPA (ED.): PROGRESS IN OPERATIONS RESEARCH, 391-402, AMSTERDAM, LONDON: NORTH HOLLAND. (1976).

GELDB 1975 -1 GELDMACHER, R. C.; LIU, C. S.
PRODUCT OF INDEPENDENT SETS OF THE VERTEX-INCIDENCE VECTORS OF A GRAPH, THE. IN: PROCEEDINGS OF THE SIXTH SOUTHEASTERN CONFERENCE ON COMBINATORICS, GRAPH THEORY AND COMPUTING. WINNIPEG: UTILITAS MATHEMATICA, 341-343. (1975).

GELL 1974 -1 GELLER, D.; KRONK, H.
FURTHER RESULTS ON THE ACHROMATIC NUMBER. FUNDAMENTA MATHEMATICAE 85, 285-290. (1974).

GELLA 1976 -1 GELLER, D. P.
R-TUPLE COLORINGS OF UNIQUELY COLORABLE GRAPHS. DISCRETE MATHEMATICS 16, 9-12. (1976).

GELLB 1974 -1 GELLER, D. P.; HILTON, A. J. W.
HOW TO COLOUR THE LINES OF A BIGRAPH. NETWORKS 4, 281-282. (1974).

GEN 1974 -1 GEN, M.; OKUNO, H.
OPTIMIZATION OF REDUNDANT ALLOCATION AND SELECTION PROBLEMS OF SYSTEMS RELIABILITY BY MEANS OF 0-1 PROGRAMMING (JAPANESE). KEIEI-KAGAKU 18, NO. 3-4, 132-147. (1974).

GEN A 1976 -1 GEN, M.; OKUNO, H.; SHINOFUJI, S.
OPTIMIZING METHOD IN SYSTEM RELIABILITY WITH FAILURE-MODES BY IMPLICIT ENUMERATION ALGORITHM, AN. JOURNAL OF THE OPERATIONS RESEARCH SOCIETY OF JAPAN 19, NO. 2, 99-116. (1976).

GEN B 1975 -1 GEN, M.; SASAKI, M.
OPTIMAL ALLOCATION OF SYSTEM AVAILABILITY. SYSTEMS-COMPUTERS-CONTROLS 6, 9-16. (1975).

GEOF 1974 -1 GEOFFRION, A. M.
GUIDED TOUR OF RECENT PRACTICAL ADVANCES IN INTEGER LINEAR PROGRAMMING, A. ACM SIGMAP NEWSLETTER 17, 22-32; ALSO PUBLISHED IN OMEGA 4, 1976, 49-58. (1974).

GEOF 1975 -1 GEOFFRION, A. M.
GUIDE TO COMPUTER-ASSISTED METHODS FOR DISTRIBUTION SYSTEMS PLANNING, A. SLOAN MANAGEMENT REVIEW 16, NO. 2, 17-41. (1975).

GEOF 1976 -1 GEOFFRION, A. M.
CUSTOMER AGGREGATION IN DISTRIBUTION MODELING. WORKING PAPER NO. 259. LOS ANGELES: UNIV. OF CALIF. WESTERN MANAGEMENT SCIENCE INSTITUTE, 17 P. (1976).

GEOF 1977 -1 GEOFFRION, A. M.
HOW CAN SPECIALIZED DISCRETE AND CONVEX OPTIMIZATION METHODS BE MARRIED. IN: P. L. HAMMER, E. L. JOHNSON, B. H. KORTE, G. L. NEMHAUSER (EDS.): STUDIES IN INTEGER PROGRAMMING, 205-220. ANNALS OF DISCRETE MATHEMATICS 1, AMSTERDAM: NORTH-HOLLAND. (1977).

GEOFA 1973 -1 GEOFFRION, A. M.; GRAVES, G. W.
MULTICOMMODITY DISTRIBUTION SYSTEM DESIGN BY BENDERS DECOMPOSITION. LOS ANGELES: UNIV. OF CALIFORNIA, WESTERN MANAGEMENT SCIENCE INSTITUTE, 49 P. (1973).

GEOFA 1976 -1 GEOFFRION, A. M.; GRAVES, G. W.
SCHEDULING PARALLEL PRODUCTION LINES WITH CHANGEOVER COSTS: PRACTICAL APPLICATION OF A QUADRATIC ASSIGNMENT/LP APPROACH. OPERATIONS RESEARCH 24, 595-610. (1976).

GEOFB 1977 -1 GEOFFRION, A. M.; GRAVES, G. W.; LEE, S.-J.
STRATEGIC DISTRIBUTION SYSTEM PLANNING: A STATUS REPORT. WORKING PAPER NO. 272. LOS ANGELES: UNIV. OF CALIFORNIA, WESTERN MANAGEMENT SCIENCE INSTITUTE, 25 P. (1977).

GEOFC 1977 -1 GEOFFRION, A. M.; MAC BRIDE, R.
LAGRANGEAN RELAXATION APPLIED TO FACILITY LOCATION PROBLEMS. WORKING PAPER NO. 263. LOS ANGELES: UNIV. OF CALIFORNIA, WESTERN MANAGEMENT SCIENCE INSTITUTE, 23 P. (1977).

GEOFD 1977 -1 GEOFFRION, A. M.; NAUSS, R.
PARAMETRIC AND POSTOPTIMALITY ANALYSIS IN INTEGER LINEAR PROGRAMMING. MANAGEMENT SCIENCE 23, 453-466. (1977).

GEOL 1975 -1 GEOLECJAN, G. G.
OPTIMAL ASSIGNMENTS AND THEIR PROPERTIES (RUSSIAN). AKADEMIJA NAUK ARMJANSKOI SSR, DOKLADY 61, NO. 2, 65-69. (1975).

GEOR 1971 -1 GEORGE, J. A.
APPLICATION OF INTEGER PROGRAMMING TO PLANT LOCATION AND INDIVISIBILITIES, AN. NEW ZEALAND ECONOMIC PAPERS 5, 107-119. (1971).

GEORA 1974 -1 GEORGOBIANI, D. A.; TABLIASHVILI, O. M.
ALLOCATION PROBLEM IN TERMS OF DYNAMIC PROGRAMMING. QUESTIONS ON MATHEMATICAL AND ENGINEERING CYBERNETICS (RUSSIAN), THE. SAKHARTHVELOS SSR MECNIEREBATHA AKADEMIA GAMOTHVLITHI CENTRIS SROMEBI 13/14, 12-16. (1974).

GERM 1971 -1 GERMEIER, JU. B.; SUHORUCENKO, E. M.
OPTIMAL ALLOCATION OF RESOURCES WHILE STRIVING TO REACH A PRESCRIBED EFFICIENCY LEVEL ON ALL THE SUBDIVISIONS (RUSSIAN). IN: A. I. BERG (ED.): CYBERNETICS - IN THE SERVICE OF COMMUNISM, VOL. 6: ECONOMIC MODELS, MATHEMATICAL METHODS OF OPERATIONS RESEARCH (RUSSIAN), 233-248. MOSCOW: ENERGIJA. (1971).

GERO 1970 -1 GEROV, G. H.
SOLUTION OF A CERTAIN SCHEDULING PROBLEM IN A SPECIAL CASE (RUSSIAN). GODISNIK NA VISSITE TEHNICESKI UCEBNI ZAVEDENIJA. MATEMATIKA 6, KN. 2, 93-100. (1970).

GERR 1976 -1 GERRITSEN, R.
EFFICIENT MANIPULATION OF BINARY ARRAYS. 76-05-04. PHILADELPHIA: UNIV. OF PENNSYLVANIA, DEPT. OF DECISION SCIENCES, 26 P. (1976).

GERT 1977 -1 GERTSBACH, I.
PROBABILISTIC IDEAS IN HEURISTIC ALGORITHMS FOR SOLVING SOME SCHEDULING PROBLEMS. NAVAL RESEARCH LOGISTICS QUARTERLY 24, 339-348. (1977).

GERTA 1977 -1 GERTSBACH, I.; GUREVICH, YU.
CONSTRUCTING AN OPTIMAL FLEET FOR A TRANSPORTATION SCHEDULE. TRANSPORTATION SCIENCE 11, 20-36. (1977).

GHIG 1976 -1 GHIGGI, C.; PULIAFITO, P. P.; ZOPPOLI, R.
COMBINATORIAL METHOD FOR HEALTH-CARE DISTRICTING, A. IN: J. CEA (ED.): OPTIMIZATION TECHNIQUES, MODELING AND OPTIMIZATION IN THE SERVICE OF MAN, PART 1. PROCEEDINGS OF THE 7TH IFIP CONFERENCE, NICE 1975, 116-130. LECTURE NOTES IN COMPUTER SCIENCE 40, BERLIN: SPRINGER. (1976).

GHIL 1973 -1 GHILARDOTTI, P.
BILP CODE: BOUNDED INTEGER LINEAR PROGRAMMING CODE USER'S INSTRUCTIONS WITH SOME NOTES AND FIRST COMPUTATIONAL RESULTS. MILANO: CENTRO RICERCA DI AUTOMATICA, 36 P. (1973). (INTERNAL REPORT G1/42-4)

GHIU 1975 -1 GHIUS, V.
SUPPLY AND DEMAND THEOREM FOR THE FLOW OF TWO PRODUCTS (RUMANIAN), A. BULETINUL INSTITUTULUI POLITEHNIC DIN IASI, SERIE NOUA 21, 75-78. (1975).

GHOS 1970 -1 GHOSH, A.; SARKAR, H.
INPUT-OUTPUT MATRIX AS A SPATIAL CONFIGURATION, AN. ECONOMICS OF PLANNING 10, 133-142. (1970).

GIAN 1976 -1 GIANNESSI, F.
DECOMPOSITION METHOD FOR THE TRAVELLING SALESMAN PROBLEM AND ITS APPLICATIONS, A. SERIE A-40. PISA: UNIV. DI PISA, DIPT. DI RICERCA OPERATIVA E SCIENZE STATISTICHE, 55 P. (1976).

GIANA 1976 -1 GIANNESSI, F.; NICCOLUCCI, F.
CONNECTIONS BETWEEN NONLINEAR AND INTEGER PROGRAMMING PROBLEMS. IN: PROCEEDINGS OF THE CONFERENCE ON MATHEMATICAL PROGRAMMING AND ITS APPLICATIONS, INDAM, ROME 1974 (ITALIAN), 161-176. SYMPOSIA MATHEMATICA 19, LONDON: ACADEMIC PRESS. (1976).

GIBB 1976 -1 GIBBONS, P.
COMPUTING TECHNIQUES FOR THE CONSTRUCTION AND ANALYSIS OF BLOCK DESIGN. DISSERTATION. UNIVERSITY OF TORONTO, DEPARTMENT OF COMPUTER SCIENCE. (1976).

GIBE 1968 -1 GIBERT, A.
METHOD FOR THE TRAFFIC ASSIGNMENT PROBLEM, A. TRANSPORT NETWORK THEORY UNIT REPORT LBS - TNT 95, LONDON: LONDON BUSINESS SCHOOL. (1968).

GIBS 1976 -1 GIBSON, P. M.
FACETS OF FACES OF TRANSPORTATION POLYTOPES. IN: F. HOFFMAN, L. LESNIAK, R. MULLIN ET AL. (EDS.): PROCEEDINGS OF THE 7TH SOUTHEASTERN CONFERENCE ON COMBINATORICS, GRAPH THEORY, AND COMPUTING, BATON ROUGE 1976, 323-334. CONGRESSUS NUMERANTIUM 17, WINNIPEG: UTILITAS MATHEMATICA. (1976).

GILB 1968 -1 GILBERT, E. N.; POLLAK H. O.
STEINER MINIMAL TREES. SIAM JOURNAL ON APPLIED MATHEMATICS 16, 1-29. (1968).

GILE 1977 -1 GILES, F. R.
OPTIMUM MATCHING FORESTS 1: SPECIAL WEIGHTS. LEXINGTON: UNIV. OF KENTUCKY, DEPT. OF MATHEMATICS, 11 P. (1977).

GILE 1977 -2 GILES, F. R.
OPTIMUM MATCHING FORESTS 2: GENERAL WEIGHTS. LEXINGTON: UNIV. OF KENTUCKY, DEPT. OF MATHEMATICS, 34 P. (1977).

GILE 1977 -3 GILES, F. R.
OPTIMUM MATCHING FORESTS 3: FACETS OF MATCHING FOREST POLYHEDRA. LEXINGTON: UNIV. OF KENTUCKY, DEPT. OF MATHEMATICS, 16 P. (1977).

GILEA 1977 -1 GILES, F. R.; PULLEYBLANK, W. R.
TOTAL DUAL INTEGRALITY AND INTEGER POLYHEDRA. CORE DISCUSSION PAPER 7731. HEVERLEE: CENTER FOR OPERATIONS RESEARCH AND ECONOMETRICS, 12 P. (1977).

GILEB 1976 -1 GILES, R.
FACETS AND OTHER FACES OF BRANCHING POLYHEDRA. CORE DISCUSSION PAPER, 7621. HEVERLEE: CENTER FOR OPERATIONS RESEARCH AND ECONOMETRICS, 24 P. (1976).

GILL 1960 -1 GILL, A.
CLASS OF TRANSPORTATION PROBLEMS. JOURNAL OF MATHEMATICAL PHYSICS, 39, 76-82. (1960).

GILLA 1969 -1 GILL, A.; TRAIGER, I. L.
COMPUTATION OF OPTIMAL PATHS IN FINITE GRAPHS. IN: L. A. ZADEH, L. W. NEUSTADT, A. V. BALAKRISHNAN (EDS.): COMPUTING METHODS IN OPTIMIZATION PROBLEMS - 2, 105-118. NEW YORK: ACADEMIC PRESS. (1969).

GILLB 1977 -1 GILL, J.
COMPUTATIONAL COMPLEXITY OF PROBABILISTIC TURING MACHINES. SIAM JOURNAL ON COMPUTING 6, 675-695. (1977).

GILLC 1975 -1 GILLES, F. R.
SUBMODULAR FUNCTIONS, GRAPHS AND INTEGER POLYHEDRA. DISSERTATION. WATERLOO: UNIV. OF WATERLOO, DEPT. OF COMBINATORICS AND OPTIMIZATION, 176 P. (1975).

GILLD 1968 -1 GILLESPIE, C. M. JR.
LOCATION ABSOLUTE 2-CENTERS OF UNDIRECTED GRAPHS. DISSERTATION. MONTEREY: NAVAL POSTGRADUATED SCHOOL. (1968).

GILLE 1975 -1 GILLETT, B. E.
COMPUTERIZED CREW SCHEDULING IN LEAD MINES. PAPER PRESENTED AT THE 48TH JOINT NATIONAL MEETING OF ORSA AND TIMS, LAS VEGAS 1975, 12 P. (1975).

GILLF 1976 -1 GILLETT, B. E.; JOHNSON, J. G.
MULTI-TERMINAL VEHICLE-DISPATCH ALGORITHM. OMEGA 4, 711-718. (1976).

GILM 1977 -1 GILMORE, P. C.
CUTTING STOCK, LINEAR PROGRAMMING, KNAPSACKING, DYNAMIC PROGRAMMING AND INTEGER PROGRAMMING SOME INTERCONNECTIONS. VANCOUVER (CONFERENCE ON DISCRETE OPTIMIZATION, AUGUST 1977) 33 P. (1977).

GILS 1970 -1 GILSINN, J.; WITZGALL, C.
COMPUTER COMPARISATION OF LABELLING ALGORITHMS FOR CALCULATING SHORTEST PATH TREES. PAPER PRESENTED AT THE 37TH NATIONAL ORSA-MEETING, WASHINGTON 1970. (1970).

GIMA 1976 -1 GIMADI, E. H.; GLEBOV, N. I.; PEREPELICA, V. A.
ALGORITHMS WITH BOUNDS FOR DISCRETE OPTIMIZATION PROBLEMS (RUSSIAN).. THIRD ALL-UNION CONFERENCE ON PROBLEMS OF THEORETICAL CYBERNETICS, NOVOSIBIRSK 1974. PROBLEMY KIBERNETIKI, NO. 31, 35-42 + 237. (1976).

GIMAA 1976 -1 GIMADI, E. K.; GLEBOV, N. I.; PEREPELITSA, V. A.
ALGORITHMS WITH ESTIMATIONS FOR THE PROBLEMS OF DISCRETE OPTIMIZATION (RUSSIAN). PROBLEMY KIBERNETIKI, NO. 31, 35-42. (1976).

GIMAB 1974 -1 GIMADI, E. K.; PEREPELITSA, V. A.
ASYMPTOTIC APPROACH TO THE SOLUTION OF THE TRAVELING SALESMAN PROBLEM (RUSSIAN). UPRAVLJAEME SISTEMY, NO. 12, 35-45. (1974).

GINO 1966 -1 GINOVKER, V. D.
METHOD OF SOLUTION OF CERTAIN PROBLEMS OF OPTIMIZATION OF ASSIGNMENT (RUSSIAN). DOKLADY AKADEMII NAUK BSSR 10, 917-919. (1966).

GINZ 1977 -1 GINZBURG, M. YA.; BURTMAN, M. I.
OPERATIONAL ALLOCATION OF RESOURCES IN LARGE TRANSPORTATION NETWORKS WITH TWO ENDPOINTS AND WITHOUT CYCLES (RUSSIAN). EKONOMIKA I MATEMATICESKIE METODY 13, 302-310. (1977).

GIRA 1976 -1 GIRAUD, G.
SUR LES PROPORTIONS RESPECTIVES DE TRIANGLES UNI, BI OU TRICOLORES DANS UN TRICOLORIAGE DES ARETES DU N-EMBLE. DISCRETE MATHEMATICS 16, 13-38. (1976).

GIUL 1976 -1 GIULIANELLI, S.; LUCERTINI. M.
DECOMPOSITION TECHNIQUE IN INTEGER LINEAR PROGRAMMING, A. IN. J. CEA (ED.): OPTIMIZATION TECHNIQUES, MODELING AND OPTIMIZATION IN THE SERVICE OF MAN, PART 2. PROCEEDINGS OF THE 7TH IFIP CONFERENCE, NICE 1975, 86-97. LECTURE NOTES IN COMPUTER SCIENCE 41, BERLIN: SPRINGER. (1976).

GLAS 1978 -1 GLASSEY, C. R.
QUADRATIC NETWORK OPTIMIZATION MODEL FOR EQUILIBRIUM SINGLE COMMODITY TRADE FLOW, A. MATHEMATICAL PROGRAMMING 14, 98-107. (1978).

GLAZ 1976 -1 GLAZEBROOK, K. D.
STOCHASTIC SCHEDULING WITH ORDER CONSTRAINTS. INTERNATIONAL JOURNAL OF SYSTEMS SCIENCES 7, 657-666. (1976).

GLAZA 1976 -1 GLAZEBROOK, K. D.; NASH, P.
ON MULTI-SERVER STOCHASTIC SCHEDULING. JOURNAL OF THE ROYAL STATISTICAL SOCIETY 38, SERIES B, 67-72. (1976).

GLEA 1975 -1 GLEASON, J. M.
SET COVERING APPROACH TO BUS STOP LOCATION, A. OMEGA 3, 505-508. (1975).

GLEB 1973 -1 GLEBOV, N. I.
ON A CLASS OF CONVEX INTEGER PROGRAMMING PROBLEMS (RUSSIAN). UPRAVLJAEME SISTEMY, NO 11, 38-42. (1973).

GLEBA 1970 -1 GLEBOV, N. I.; PEREPELITSA, V. A.
LOWER AND UPPER ESTIMATES FOR A CERTAIN PROBLEM IN SCHEDULING THEORY (RUSSIAN), THE. IN: A. A. LJAPUNOV (ED.): STUDIES IN CYBERNETICS (RUSSIAN), 11-17. MOSCOW: IZDAT. SOV. RADIO. (1970).

****

GLEBB 1963 -1 GLEBSKII, Y.; SHEVCHENKO, V. N.
CONSTRUCTION OF AN OPTIMAL WORK SCHEDULE (RUSSIAN). PROBLEMY KIBERNETIKI, 275-279. (1963).
GLEN 1975 -1 GLEN, J. J.
DYNAMIC PROGRAMMING MODEL FOR WORK SCHEDULING IN A SHIPYARD, A. OPERATIONAL RESEARCH QUARTERLY 26, 787-800. (1975).
GLIC 1977 -1 GLICKMAN, T. S.; BERGER, P. D.
COST/COMPLETION-DATE TRADEOFFS IN THE TRANSPORTATION PROBLEM. OPERATIONS RESEARCH 25, 163-167. (1977).
GLOV 1971 -1 GLOVER, F.
CONVEXITY CUTS FOR MULTIPLE-CHOICE PROBLEMS. MSRS 71-1. BOULDER: UNIV. OF COLORADO. (1971).
GLOV 1973 -1 GLOVER, F.
POLYHEDRAL ANNEXATION IN MIXED INTEGER PROGRAMMING. MSRS 73-9. BOULDER: UNIV. OF COLORADO. (1973).
GLOV 1973 -2 GLOVER, F.
POLYHEDRAL CONVEXITY CUTS AND NEGATIVE EDGE EXTENSIONS. MSRS 73-6. BOULDER: UNIV. OF COLORADO. (1973).
GLOV 1975 -1 GLOVER, F.
COMPACT LP BASES FOR A CLASS OF IP PROBLEMS. MANAGEMENT SCIENCES REPORT SERIES 75-18. BOULDER: UNIV. OF COLORADO, GRADUATE SCHOOL OF BUSINESS ADMINISTRATION. (1975).
GLOV 1975 -2 GLOVER, F.
IMPROVED LINEAR INTEGER PROGRAMMING FORMULATIONS OF NONLINEAR INTEGER PROBLEMS. MANAGEMENT SCIENCE 22, 455-460. (1975).
GLOV 1975 -3 GLOVER, F.
POLYHEDRAL ANNEXATION IN MIXED INTEGER AND COMBINATORIAL PROGRAMMING. MATHEMATICAL PROGRAMMING 9, 161-188. (1975).
GLOV 1976 -1 GLOVER, F.
REDUCING THE SIZE OF SOME IP FORMULATIONS BY SUBSTITUTION. OPERATIONAL RESEARCH QUARTERLY 27, 261-264. (1976).
GLOV 1977 -1 GLOVER, F.
COMPACT LP BASES FOR A CLASS OF IP PROBLEMS. MATHEMATICAL PROGRAMMING 12, 102-109. (1977).
GLOV 1977 -2 GLOVER, F.
HEURISTICS FOR INTEGER PROGRAMMING USING SURROGATE CONSTRAINTS. DECISION SCIENCES 8, 156-166. (1977).
GLOVA 1977 -1 GLOVER, F.; HULTZ, J.; KLINGMAN, D.
IMPROVED COMPUTER-BASED PLANNING TECHNIQUES. RESEARCH REPORT CCS 283. AUSTIN: UNIV. OF TEXAS, CENTER FOR CYBERNETIC STUDIES, 49 P. (1977).
GLOVB 1977 -1 GLOVER, F.; HULTZ, J.; KLINGMAN, D.; STUTZ, J.
NEW COMPUTER-BASED PLANNING TOOL, A. RESEARCH REPORT CCS 289. AUSTIN: UNIV. OF TEXAS, CENTER FOR CYBERNETIC STUDIES, 48 P. (1977).
GLOVC 1974 -1 GLOVER, F.; KLINGMAN, D.
IMPORTANT PRACTICAL MISCONCEPTIONS OF OPTIMIZING LARGE SCALE ASSIGNMENT AND TRANSPORTATION PROBLEMS. RESEARCH REPORT CCS 195. AUSTIN: UNIV. OF TEXAS, CENTER FOR CYBERNETIC STUDIES, 8 P. (1974).
GLOVC 1975 -1 GLOVER, F.; KLINGMAN, D.
CAPSULE VIEW OF FUTURE DEVELOPMENTS ON LARGE SCALE NETWORK AND NETWORK-RELATED PROBLEMS. AUSTIN: UNIV. OF TEXAS, CENTER FOR CYBERNETIC STUDIES, 50 P. (1975). (RESEARCH REPORT CCS 238)
GLOVC 1975 -2 GLOVER, F.; KLINGMAN, D.
FINDING MINIMUM SPANNING TREES WITH A FIXED NUMBER OF LINKS AT A NODE. IN: B. ROY (ED.): COMBINATORIAL PROGRAMMING: METHODS AND APPLICATIONS. PROC. NATO ADV. STUDY INST., VERSAILLES 1974, 191-202. DORDRECHT: REIDEL; ALSO PUBL. IN: A. PREKOPA (ED.): PROGRESS IN OPER. RES., 425-440. 1976, AMSTERDAM: NORTH HOLLAND. (1975).
GLOVC 1975 -3 GLOVER, F.; KLINGMAN, D.
IMPROVED LABELING OF L. P. BASES IN NETWORKS. RESEARCH REPORT CCS 218. AUSTIN: UNIV. OF TEXAS, CENTER FOR CYBERNETIC STUDIES, 8 P. (1975).
GLOVC 1975 -4 GLOVER, F.; KLINGMAN, D.
NETWORK APPLICATION IN INDUSTRY AND GOVERNMENT. RESEARCH REPORT CCS 247. AUSIN: UNIV. OF TEXAS, CENTER FOR CYBERNETIC STUDIES, 46 P. (1975).
GLOVC 1975 -5 GLOVER, F.; KLINGMAN, D.
NEW ADVANCES IN THE SOLUTION OF LARGE SCALE NETWORK AND NETWORK-RELATED PROBLEMS. COLLOQUIA MATHEMATICA SOCIETATIS JANOS BOLYAI 12, AMSTERDAM: NORTH HOLLAND; ALSO PUBLISHED IN: A. PREKOPA (ED.): PROGRESS IN OPERATIONS RESEARCH, 441-460, 1976, AMSTERDAM, LONDON: NORTH HOLLAND. (1975).
GLOVC 1976 -1 GLOVER, F.; KLINGMAN, D.
PRACTITIONER'S GUIDE TO THE STATE OF LARGE SCALE NETWORK AND NETWORK-RELATED PROBLEMS, A. AFIPS-CONFERENCE PROCEEDINGS, VOLUME 45, 945-950. (1976).
GLOVC 1976 -2 GLOVER, F.; KLINGMAN, D.
SOME RECENT PRACTICAL MISCONCEPTIONS ABOUT THE STATE OF THE ART OF NETWORK ALGORITHMS. RESEARCH REPORT CCS 270. AUSTIN: UNIV. OF TEXAS, CENTER FOR CYBERNETIC STUDIES, 6 P. (1976).
GLOVD 1977 -1 GLOVER, F.; KLINGMAN, D.; MACMILLAN, C.
NETFORM CONCEPT: A MORE EFFECTIVE MODEL FORM AND SOLUTION PROCEDURE FOR LARGE SCALE NONLINEAR PROBLEMS, THE. RESEARCH REPORT CCS 281. AUSTIN: UNIV. OF TEXAS, CENTER FOR CYBERNETIC STUDIES, 21 P. (1977).
GLOVE 1976 -1 GLOVER, F.; KLINGMAN, D.; NAPIER, A.
EQUIVALENCE OF GENERALIZED NETWORK AND GENERALIZED TRANSPORTATION PROBLEMS. OPSEARCH 13, 93-100. (1976).
GLOVF 1973 -1 GLOVER, F.; KLINGMAN, D.; STUTZ, J.
IMPLEMENTATION AND COMPUTATIONAL STUDY OF A GENERALIZED NETWORK CODE. PAPER PRESENTED AT THE 44TH NATIONAL MEETING OF ORSA, SAN DIEGO 1973. (1973).
GLOVG 1975 -1 GLOVER, F.; MULVEY, J.
EQUIVALENCE OF THE 0-1 INTEGER PROGRAMMING PROBLEM TO DISCRETE GENERALIZED AND DISCRETE PURE NETWORK MODELS. HBS 75-46. BOSTON: HARVARD UNIV., GRADUATE SCHOOL OF BUSINES ADMINISTRATION, 6 P. (1975).
GLOVG 1975 -2 GLOVER, F.; MULVEY, J.
EQUIVALENCE OF THE 0-1 INTEGER PROGRAMMING PROBLEM TO DISCRETE GENERALIZED AND PURE NETWORKS. MSRS 75-19. BOULDER: UNIV. OF COLORADO. (1975).
GLOVH 1975 -1 GLOVER, F.; ROSS, G. T.
IMPROVED BOUNDS FOR A CLASS OF 0-1 INTEGER PROGRAMS USING COMBINED SURROGATE-LAGRANGEAN TECHNIQUES. AUSTIN: UNIV. OF TEXAS, CENTER FOR CYBERNETIC STUDIES, 21 P. (1975). (RESEARCH REPORT CCS 241)
GLOVI 1976 -1 GLOVER, F.; SOMMER, D.
INEQUALITIES FOR MIXED INTEGER PROGRAMS WITH STRUCTURE. NAVAL RESEARCH LOGISTICS QUARTERLY 23, 603-610. (1976).
GLOVJ 1976 -1 GLOVER, F.; TANGEDAHL, L.
DYNAMIC STRATEGIES FOR BRANCH AND BOUND. OMEGA 4, 571-576. (1976).

GLOVK 1972 -1 GLOVER, F., ET AL.
NOTE ON FINDING ALL SHORTEST PATHS. REPORT A D 749713. AUSTIN: UNIV. OF TEXAS, CENTER FOR CYBERNETICS STUDIES, 17 P. (1972).

GLOVL 1965 -1 GLOVER, P. C.; NORMAN, J. T.
ASSEMBLY LINE BALANCING. ASSEMBLY ENGINEERING 1, 28-32. (1965).

GODE 1976 -1 GODEFROID, P.
LOESUNG EINES GEMISCHT-GANZZAHLIGEN MODELLS ZUR OPTIMALEN BAUFINANZIERUNG, DIE. IN: J. KOHLAS, O. SEIFERT, P. STAEHLY, H.-J. ZIMMERMANN (EDS): PROCEEDINGS IN OPERATIONS RESEARCH 5, 48-49. WUERZBURG: PHYSICA. (1976).

GOL' 1974 -1 GOL'DBERG, M. K.
REMARK ON THE CHROMATIC CLASS OF A MULTIGRAPH (RUSSIAN), A. IN: NUMERICAL MATHEMATICS AND COMPUTER TECHNOLOGY, NO. V (RUSSIAN), 128-130 + 168. KHARKOV: AKAD. NAUK UKRAIN. SSR FIZ.-TEHN. INST. NIZKIH TEMPERATUR. (1974).

GOL'A 1975 -1 GOL'DBERG, M. K.; KLIPKER, I. A.
NUMBERING OF A UNIFORM TREE THAT MINIMIZES THE SUM OF THE LENGTHS OF THE EDGES (RUSSIAN), A. GAMOQENEBITHI MATEMATIKIS INSTITUTIS SEMINARI, MOHSENEBATHA ANOTACIEBI, VYP. 10, 65-71. (1975).

GOL'B 1977 -1 GOL'DGABER, E. M.; KARNAUKH, V. G.
TRANSFORMATION OF THE OPERATIONS COMPLEX NETWORK INTO A CONJUGATE NETWORK (RUSSIAN). KIBERNETIKA, NO. 5, 121-124. (1977).

GOLD 1975 -1 GOLDEN, B.
VEHICLE ROUTING PROBLEMS: FORMULATIONS AND HEURISTIC SOLUTION TECHNIQUES. TECHNICAL REPORT 113. CAMBRIDGE: M.I.T., DEPT. OPERATIONS RESEARCH. (1975).

GOLD 1976 -1 GOLDEN, B.
LARGE-SCALE VEHICLE ROUTING AND RELATED COMBINATORIAL PROBLEMS. DISSERTATION. CAMBRIDGE: M.I.T., DEPT. OPERATIONS RESEARCH. (1976).

GOLD 1976 -2 GOLDEN, B.
SHORTEST-PATH ALGORITHMS: A COMPARISON. OPERATIONS RESEARCH 24, 1164-1168. (1976).

GOLDA 1977 -1 GOLDEN, B.; BALL, M.
SHORTEST PATHS WITH EUCLIDEAN DISTANCES: AN EXPLANATORY MODEL. ATLANTA (52. ORSA-MEETING + TIMS, NOVEMBER 1977) 25 P. (1977).

GOLDB 1975 -1 GOLDEN, B. L.
MINIMUM-COST MULTICOMMODITY NETWORK FLOW PROBLEM CONCERNING IMPORTS AND EXPORTS, A. NETWORKS 5, 331-356. (1975).

GOLDB 1976 -1 GOLDEN, B. L.
APPROACHES TO THE CUTTING STOCK PROBLEM. AIIE TRANSACTIONS 8, 265-274. (1976).

GOLDB 1976 -2 GOLDEN, B. L.
RECENT DEVELOPMENTS IN VEHICLE ROUTING. CAMRIDGE: MASS. INST. OF TECHNOLOGY, OPERATIONS RESEARCH CENTER, 25 P. (1976).

GOLDB 1977 -1 GOLDEN, B. L.
EVALUATING A SEQUENTIAL VEHICLE ROUTING ALGORITHM. AIIE TRANSACTIONS 9, 204-208. (1977).

GOLDB 1977 -2 GOLDEN, B. L.
STATISTICAL APPROACH TO THE TSP, A. NETWORKS 7, 209-226. (1977).

GOLDC 1977 -1 GOLDEN, B. L.; MAGNANTI, T. L.
DETERMINISTIC NETWORK OPTIMIZATION: A BIBLIOGRAPHY. NETWORKS 7, 149-184. (1977).

GOLDD 1977 -1 GOLDEN, B. L.; MAGNANTI, T. L.; NGUYEN, H. Q.
IMPLEMENTING VEHICLE ROUTING ALGORITHMS. NETWORKS 7, 113-148. (1977).

GOLDE 1977 -1 GOLDFARB, D.
STEEPEST EDGE SIMPLEX ALGORITHMS FOR NETWORK FLOW PROBLEMS. IBM RESEARCH REPORT RC6649. YORKTOWN HEIGHTS: IBM THOMAS J. WATSON RESEARCH CENTER, 17 P. (1977).

GOLDF 1967 -1 GOLDMAN, A. J.; NEMHAUSER, G.
TRANSPORT IMPROVEMENT PROBLEM TRANSFORMABLE TO A BEST-PATH PROBLEM, A. TRANSPORTATION SCIENCE, 1, 295-307. (1967).

GOLDG 1975 -1 GOLDNER, A. M.; HARARY, F.
NOTE ON THE SMALLEST NONHAMILTONIAN MAXIMAL PLANAR GRAPH. MALAYSIAN MATHEMATICAL SOCIETY, BULLETIN 6, NO. 1, 41-42 + ERRATA: IBID., NO. 2, 1975, 33. (1975).

GOLDH 1971 -1 GOLDSTEIN, M. C.
CONSIDERATION IN THE DESIGN OF LARGE SCALE NETWORKS. IN: PROCEEDINGS OF THE FIFTH ANNUAL PRINCETON CONFERENCE ON INFORMATION SCIENCES AND SYSTEMS, 214-218. (1971).

GOLE 1974 -1 GOLENKO, D. I.; GELEVER, K. G.; KESLER, S. S.; GOFMAN, E. F.
PROBLEMS ON OPTIMAL SCHEDULING AND OPTIMAL RESOURCE ALLOCATION FOR A TECHNOLOGICAL OPERATING PROCEDURE (RUSSIAN). IN: NUMERICAL ANALYSIS, AND STATISTICAL MODELS OF COMPLEX SYSTEMS, PART II (RUSSIAN), 3-40. MOSCOW: MOSKOV. EKONOM.-STATIST. INST. (1974).

GOLL 1969 -1 GOLLER, S.
CONTRIBUTION TO THE THEORY OF PURPOSEFUL TIMETABLE MAKING (CZECH). CESKOSLOVENSKA AKADEMIE VED. EKONOMICKO-MATEMATICKY OBZOR 5, 60-71. (1969).

GOLU 1977 -1 GOLUMBIC, M. C.
COMPARABILITY GRAPHS AND A NEW MATROID. JOURNAL OF COMBINATORIAL THEORY 22, SERIES B, 68-90. (1977).

GOLU 1977 -2 GOLUMBIC, M. C.
COMPLEXITY OF COMPARABILITY GRAPH RECOGNITION AND COLORING, THE. COMPUTING 18, 199-208. (1977).

GOMO 1975 -1 GOMORY, R. E.; HU, T. C.
MULTI-TERMINAL FLOWS IN A NETWORK. IN: D. R. FULKERSON (ED.): STUDIES IN GRAPH THEORY, PART 1, 172-199. STUDIES IN MATHEMATICS 11, WASHINGTON: MATHEMATICAL ASSOCIATION OF AMERICA. (1975).

GOND 1971 -1 GONDRAN, M.
PROGRAMMATION ENTIERE NON LINEAIRE: DUALITE DIOPHANTIENNE. ELECTRICITE DE FRANCE, BULLETIN DE LA DIRECTION DES ETUDES ET RECHERCHES, SERIE C, MATHEMATIQUES, INFORMATIQUE, NO. 1, 3 + 21-33. (1971).

GOND 1974 -1 GONDRAN, M.
ALGEBRE DES CHEMINS ET ALGORITHMES. PARIS: ELECTRICITE DE FRANCE, 13 P. (1974).

GOND 1975 -1 GONDRAN, M.
ALGEBRE DES CHEMINS ET ALGORITHMES. ELECTRICITE DE FRANCE. BULLETIN DE LA DIRECTION DES ETUDES ET RECHERCHES, SERIES C, MATHEMATIQUES, INFORMATIQUE, NO. 2, 57-64. (1975).

GOND 1975 -2 GONDRAN, M.
PATH ALGEBRA AND ALGORITHMS. IN: B. ROY (ED.): COMBINATORIAL PROGRAMMING: METHODS AND APPLICATIONS. PROCEEDINGS OF THE NATO ADVANCED STUDY INST., VERSAILLES 1974, 137-148. DORDRECHT: REIDEL. (1975).

GOND 1976 -1 GONDRAN, M.
PROBLEME DU PLUS COURT CHEMIN DANS UN GRAPHE, LE. ELECTRICITE DE FRANCE, BULLETIN DE LA DIRECTION DES ETUDES ET RECHERCHES, SERIES C, SUPPLEMENT AU NO. 2, 5-33. (1976).

GOND 1976 -2 GONDRAN, M.
PROBLEMES DE PARTITION ET DE RECOUVREMENT. APPLICATIONS ET ALGORITHMES, LES. ELECTRICITE DE FRANCE, BULLETIN DE LA DIRECTION DES ETUDES ET RECHERCHES, SERIES C, NO. 2, 59-68. (1976).

GOND 1977 -1 GONDRAN, M.
GROUP THEORETIC AND LAGRANGEAN METHODS. VANCOUVER (CONFERENCE ON DISCRETE OPTIMIZATION, AUGUST 1977) 5 P. (1977).

GONDA 1977 -1 GONDRAN, M.; DOSTATNI, M.
TRAITEMENT DES EXCLUSIVES DANS PLANNEC, LE. ELECTRICITE DE FRANCE: BULLETIN DE LA DIRECTION DES ETUDES ET RECHERCHE, SERIES C, MATHEMATIQUES, INFORMATIQUE, 69-78. (1977).

GONDB 1977 -1 GONDRAN, M.; MINOUX, M.
VALEURS PROPRES ET VECTEURS PROPRES DANS LES DIOIDES ET LEUR INTERPRETATION EN THEORIE DES GRAPHES. ELECTRICITE DE FRANCE. BULLETIN DE LA DIRECTION DES ETUDES ET RECHERCHES, SERIES C, MATHEMATIQUES, INFORMATIQUE, NO. 2, 25-41. (1977).

GONZ 1962 -1 GONZALES, R. H.
SOLUTION OF THE TRAVELING SALESMAN PROBLEM BY DYNAMIC PROGRAMMING ON THE HYPERCUBE. INTERIM TECHNICAL REPORT 18. CAMBRIDGE: MASSACHUSETTS INST. OF TECHNOLOGY, OR-CENTER. (1962).

GONZA 1975 -1 GONZALEZ-VALENZUELA, F.
SIMPLE AND CAPACITATED WAREHOUSE LOCATION PROBLEMS WITH STOCHASTIC DEMANDS. DISSERTATION. STANFORD: STANFORD UNIV., 166 P. (1975).

GONZB 1977 -1 GONZALEZ, M. J. JR.
DETERMINISTIC PROCESSOR SCHEDULING. COMPUTING SURVEYS 9, NO. 3, 173-204. (1977).

GONZC 1976 -1 GONZALEZ, M. J. JR.; SOH, J. W.
PERIODIC JOB SCHEDULING IN A DISTRIBUTED PROCESSOR SYSTEM. IEEE TRANSACTIONS ON AEROSPACE AND ELECTRONICAL SYSTEMS AES-12, 530-536. (1976).

GONZD 1977 -1 GONZALEZ, T.; IBARRA, O. H.; SAHNI, S.
BOUNDS FOR LPT SCHEDULES ON UNIFORM PROCESSORS. SIAM JOURNAL ON COMPUTING 6, 155-166. (1977).

GONZE 1976 -1 GONZALEZ, T.; SAHNI, S.
OPEN SHOP SCHEDULING TO MINIMIZE FINISH TIME. JOURNAL OF THE ASSOCIATION FOR COMPUTING MACHINERY 23, 665-679. (1976).

GONZE 1978 -1 GONZALEZ, T.; SAHNI, S.
PREEMPTIVE SCHEDULING OF UNIFORM PROCESSOR SYSTEMS. JOURNAL OF THE ASSOCIATION FOR COMPUTING MACHINERY 25, 92-101. (1978).

GOOD 1965 -1 GOODMAN, R. G.; CUMMINS, R. L.
COMPUTER DETERMINATION OF ISOMORPHISM IN LINEAR GRAPHS. EIGHTH MIDWEST SYMPOSIUM ON CIRCUIT THEORY, COLORADO STATE UNIVERSITY, FORT COLLINS. (1965).

GOODA 1973 -1 GOODMAN, S.; HEDETNIEMI, S.
ON HAMILTONIAN WALKS IN GRAPHS. IN: F. HOFFMAN, R. B. LEVOW, R. P. D. THOMAS (EDS.): PROCEEDINGS OF THE 4TH SOUTHEASTERN CONFERENCE ON COMBINATORICS, GRAPH THEORY AND COMPUTING, BOCA RATON 1973, 335-342. CONGRESSUS NUMERANTIUM 8, WINNIPEG: UTILITAS MATHEMATICA. (1973).

GOODA 1974 -1 GOODMAN, S.; HEDETNIEMI, S.
ON THE HAMILTONIAN COMPLETION PROBLEM. IN: R. A. BARI, F. HARARY (EDS.): GRAPHS AND COMBINATORICS, PROCEEDINGS OF THE CAPITAL CONFERENCE ON GRAPH THEORY AND COMBINATORICS, WASHINGTON 1973, 262-272. LECTURE NOTES IN MATHEMATICS 406, BERLIN: SPRINGER. (1974).

GOODB 1974 -1 GOODMAN, S.; HEDETNIEMI, S.; SLATER, P.
SOME RESULTS ON THE OPTIMAL AND NON-OPTIMAL HAMILTONIAN COMPLETION PROBLEM. IN: F. HOFFMAN, R. A. KINGSLEY, R. B. LEVOW, R. C. MULLIN (EDS.): PROCEEDINGS OF THE FIFTH SOUTHEASTERN CONF. ON COMBINATORICS, GRAPH THEORY, AND COMPUTING, BOCA RATON 1974, 423-448. CONGRESSUS NUMERANTIUM 10, WINNIPEG: UTILITAS MATHEMATICA. (1974).

GOODC 1976 -1 GOODMAN, S.; HEDETNIEMI, S.; TARJAN, R. E.
B-MATCHINGS IN TREES. SIAM JOURNAL ON COMPUTING 5, 104-108. (1976).

GOODD 1975 -1 GOODMAN, S. E.; HEDETNIEMI, S. T.; SLATER, P. J.
ADVANCES ON THE HAMILTONIAN COMPLETION PROBLEM. JOURNAL OF THE ASSOCIATION FOR COMPUTING MACHINERY 22, 352-360. (1975).

GORD 1976 -1 GORDON, A. JA.
ALGORITHM FOR THE SOLUTION OF THE MINIMAX SCHEDULING PROBLEM (RUSSIAN), AN. IN: STUDIES IN DISCRETE OPTIMIZATION (RUSSIAN), 327-333. MOSCOW: IZDAT. NAUKA. (1976).

GORDA 1971 -1 GORDON, B. E.
MAXIMUM MATCHING PROBLEM - A COMPARISON OF THE EDMONDS AND BALINSKI ALGORITHMS, THE. ROCHESTER: UNIV. OF ROCHESTER, GRADUATE SCHOOL OF MANAGEMENT. (1971).

GORDB 1974 -1 GORDON, J. H.
HEURISTIC METHODS IN RESOURCE ALLOCATION. IN: PROJ. MANAGE. SEVENTIES. 4TH INTERNET CONGR., PARIS 1974, 499-510. PARIS: ASSOC. FR. CYBERN. ECON. TECH. (1974).

GORS 1970 -1 GORSHKOW, A. F.; KIRPICHNIKOV, V. M.
PRACTICAL ALGORITHM FOR FINDING PATHS OF A GRAPH (RUSSIAN), A. IZVESTIJA VUZ. SSSR ELEKTROMEKH. 8, 857-860. (1970).

GOYA 1975 -1 GOYAL, S. K.
SCHEDULING A MULTI-PRODUCT SINGLE MACHINE SYSTEM. A NEW APPROACH. INTERNATIONAL JOURNAL OF PRODUCTION RESEARCH 13, 487-494. (1975).

GOYA 1975 -2 GOYAL, S. K.
SCHEDULING A SINGLE MACHINE SYSTEM: A MULTI-PRODUCT MULTI-ITEM CASE. OPERATIONAL RESEARCH QUARTERLY 26, 619-627. (1975).

GOYA 1976 -1 GOYAL, S. K.
GRAPHICAL AIDS FOR SCHEDULING A MULTI-PRODUCT SINGLE MACHINE SYSTEM. OPERATIONAL RESEARCH QUARTERLY 27, 1015-1017. (1976).

GOYA 1977 -1 GOYAL, S. K.
FLOW-SHOP SEQUENCING PROBLEM WITH NO WAIT IN PROCESS AND NON-ZERO SET-UP TIMES. NEW ZEALAND OPERATIONAL RESEARCH 5, 54-58. (1977).

GRAB 1976 -1 GRABOWSKI, J.
NEW FORMULATION AND SOLUTION OF THE SEQUENCING PROBLEM: ALGORITHM, A. ZASTOSOWANIA MATEMATYKI 15, 463-474. (1976).

GRAB 1976 -2 GRABOWSKI, J.
NEW FORMULATION AND SOLUTION OF THE SEQUENCING PROBLEM: MATHEMATICAL MODEL, A. ZASTOSOWANIA MATEMATYKI 15, 325-343. (1976).

GRAB 1976 -3 GRABOWSKI, J.
SOLVING THE TRAVELLING SALESMAN PROBLEM BY USING DISJUNCTIVE GRAPHS. IN: GRAPHS HYPERGRAPHS BLOCK SYST. SYMP. COMB. ANAL. PROC., ZIELONA GORA 1976, 75-87. ZIELONA GORA. (1976).

GRABA 1975 -1 GRABOWSKI, J.; SYSLO, M. M.
NEW SOLVABLE CASES OF M-MACHINE AND N-ELEMENT SEQUENCING PROBLEMS. ZASTOSOWANIA MATEMATYKI 14, 599-606. (1975).

GRABB 1972 -1 GRABOWSKI, W.
SOLVABILITY OF AN INTEGER LINEAR PROGRAMMING PROBLEM (POLISH), THE. ZESZYTY NAUKOWE SZKOLY GLOWNEJ PLANOWANIA I STATYSTYKI, NO. 86, 71-79. (1972).

GRABB 1975 -1 GRABOWSKI, W.
SOLUTION OF THE PLANT LOCATION PROBLEM BY THE BRANCH AND BOUND METHOD (POLISH), A. PRZEGLAD STATYSTYCNY 22, 13-29. (1975).

****

GRABB 1976 -1 GRABOWSKI, W.
LEXICOGRAPHIC AND TIME MINIMIZATION IN THE TRANSPORTATION PROBLEM. ZASTOSOWANIA MATEMATYKI 15, 191-213. (1976).

GRAH 1967 -1 GRAHAM, R. L.
SOME RESULTS ON STEINER MINIMAL TREES. BELL LABORATORIES TECHNICAL MEMORANDUM. (1967).

GRAH 1976 -1 GRAHAM, R. L.
BOUNDS ON THE PERFORMANCE OF SCHEDULING ALGORITHMS. IN: E. G. COFFMAN JR. (ED.): COMPUTER AND JOB/SHOP SCHEDULING THEORY, CHAPTER FIVE. (1976).

GRAHA 1977 -1 GRAHAM, R. L.; LAWLER, E. L.; LENSTRA, J. K.; RINNOOY KAN, A. H. G.
OPTIMIZATION AND APPROXIMATION IN DETERMINISTIC SEQUENCING AND SCHEDULING: A SURVEY. VANCOUVER (CONFERENCE ON DISCRETE OPTIMIZATION, AUGUST 1977) 52 P. (1977).

GRAHB 1974 -1 GRAHAM, R. L.; ROTHSCHILD, B. L.
SOME RECENT DEVELOPMENTS IN RAMSEY THEORY. IN: M. HALL JR., J. H. VAN LINT (EDS.): COMBINATORICS, PART 2: GRAPH THEORY; FOUNDATIONS, PARTITIONS AND COMBINATORIAL GEOMETRY, 61-76. MATHEMATICAL CENTRE TRACTS 56, AMSTERDAM: MATHEMATISCH CENTRUM. (1974).

GRAN 1974 -1 GRANDA, T. M.
HEURISTIC APPROACH TO A LARGE PERSONNEL ASSIGNMENT PROBLEM, A. ARLINGTON: U.S. ARMY RESEARCH INST. FOR BEHAVIORAL AND SOCIAL SCIENCES, 24 P. (1974).

GRANA 1976 -1 GRANOT, D.; GRANOT, F.
ON SOLVING FRACTIONAL (0,1) PROGRAMS BY IMPLICIT ENUMERATION. INFOR 14, 241-249. (1976).

GRANA 1977 -1 GRANOT, D.; GRANOT, F.
ON INTEGER AND MIXED INTEGER FRACTIONAL PROGRAMMING PROBLEMS. IN: P. L. HAMMER, E. L. JOHNSON, B. H. KORTE, G. I. NEMHAUSER (EDS.): STUDIES IN INTEGER PROGRAMMING, 221-232. ANNALS OF DISCRETE MATHEMATICS 1, AMSTERDAM: NORTH-HOLLAND. (1977).

GRANB 1970 -1 GRANOT, F.; HAMMER, P.
ON THE USE OF BOOLEAN FUNCTIONS IN 0-1 PROGRAMMING. PUB. 70, HAIFA ISRAEL:TECHNION, OPERATIONS RESEARCH STATISTIC AND ECONOMICS. (1970).

GRANC 1976 -1 GRANT, D. D.
HOW TO RUN A HOCKEY TOURNAMENT - A COMBINATORIAL SCHEDULING PROBLEM. IN: C. ST. J. A. NASH-WILLIAMS (ED.): PROCEEDINGS OF THE 5TH BRITISH COMBINATORIAL CONFERENCE, ABERDEEN 1975, 239-254. CONGRESSUS NUMERANTIUM 15, WINNIPEG: UTILITAS MATHEMATICA. (1976).

GRAP 1977 -1 GRAPA, E.; BELFORD, G. G.
SOME THEOREMS TO AID IN SOLVING THE FILE ALLOCATION PROBLEM. COMMUNICATIONS OF THE ACM 20, 878-882. (1977).

GRAT 1970 -1 GRATZER, F.
COMPUTER SOLUTION OF LARGE MULTICOMMODITY FLOW PROBLEMS. DISSERTATION. PRINCETON: PRINCETON UNIVERSITY. (1970).

GRAV 1975 -1 GRAVER, J. E.
BOOLEAN DESIGNS AND SELF-DUAL MATROIDS. LINEAR ALGEBRA AND ITS APPLICATIONS 10, 111-128. (1975).

GRAV 1975 -2 GRAVER, J. E.
ON THE FOUNDATIONS OF LINEAR AND INTEGER LINEAR PROGRAMMING I. MATHEMATICAL PROGRAMMING 9, 207-226. (1975).

GRAV 1975 -3 GRAVER, J. E.
SURVEY OF THE MAXIMUM DEPTH PROBLEM FOR INDECOMPOSABLE EXACT COVERS, A. IN: A. HAJNAL, R. RADO, V. T. SOS (EDS): INFINITE AND FINITE SETS 2, COLLOQ. KESZTHELY 1973, 731-743. COLLOQ. MATH. SOC. JANOS BOLYAI 10, AMSTERDAM: NORTH-HOLLAND. (1975).

GRAVA 1977 -1 GRAVER, J. E.; WATKINS, M. E.
COMBINATORICS WITH EMPHASIS ON THE THEORY OF GRAPHS. GRADUATES TEXTS IN MATHEMATICS 54. NEW YORK, HEIDELBERG, BERLIN: SPRINGER, 351 P. (1977).

GRAVB 1976 -1 GRAVES, R. J.; THOMAS, W. H.
ANALYTICAL APPROACH TO ACADEMIC DEPARTMENT LOCATION IN A MULTI-CAMPUS UNIVERSITY, AN. SOCIO-ECONOMIC PLANNING SCIENCES 10, 93-100. (1976).

GRAVC 1977 -1 GRAVES, S. C.
NOTE ON "CRITICAL RATIO SCHEDULING: AN EXPERIMENTAL ANALYSIS", A. MANAGEMENT SCIENCE 23, 1358-1359. (1977).

GRAY 1975 -1 GRAY, P.; HEITZMAN, W. R.
DETECTIVE ALLOCATION MODEL, A. WORKING PAPER 10-3-75. LOS ANGELES: UNIVERSITY OF CALIFORNIA, GRADUATE SCHOOL OF BUSINESS ADMINISTRATION, 10 P. (1975).

GREE 1976 -1 GREEN, C. D.
DETECTION OF MISTAKES IN THE COMPARABILITY GRAPH OF A TREE, THE. IN: C. ST. J. A. NASH-WILLIAMS (ED.): PROCEEDINGS OF THE 5TH BRITISH COMBINATORIAL CONFERENCE, ABERDEEN 1975, 255-260. CONGRESSUS NUMERANTIUM 15, WINNIPEG: UTILITAS MATHEMATICA. (1976).

GREEA 1977 -1 GREENE, C.
ACYCLIC ORIENTATIONS. IN: M. AIGNER (ED.): HIGHER COMBINATORICS, PROCEEDINGS OF THE NATO ADVANCED STUDY INSTITUTE, BERLIN 1976, 65-68. DORDRECHT: REIDEL. (1977).

GREEB 1974 -1 GREENE, C.; KLEITMAN, D. J.; MAGNANTI, T. L.
COMPLEMENTARY TREES AND INDEPENDENT MATCHINGS. STUDIES IN APPLIED MATH. 53, 57-64. (1974).

GREEC 1975 -1 GREENE, C.; MAGNANTI, T. L.
SOME ABSTRACT PIVOT ALGORITHMS. SIAM JOURNAL ON APPLIED MATHEMATICS 29, 530-539. (1975).

GRIE 1976 -1 GRIER, B.
OPTIMAL NUMBER OF ALTERNATIVES AT A CHOICE POINT WITH TRAVEL TIME CONSIDERED, THE. JOURNAL OF MATHEMATICAL PSYCHOLOGY 14, 91-97. (1976).

GRIEA 1976 -1 GRIESSHABER, K.
ERZEUGUNG VON BESATZUNGSUMLAEUFEN ZU EINEM FLUGPLAN. IN: H. N. DATHE, P. MERTENS, F. D. PESCHANEL ET AL. (EDS.): PROCEEDINGS IN OPERATIONS RESEARCH 6, 446-452. WUERZBURG: PHYSICA. (1976).

GRIG 1974 -1 GRIGOR'EV, E. G.
CERTAIN MODIFICATION OF FORD'S ALGORITHM (RUSSIAN), A. TRUDY MOSKOVSKOGO ORDENA LENINA ENERGETICESKOGO INSTITUTA VYP. 178, 94-96. (1974).

GRIGA 1974 -1 GRIGOR'EVA, L. I.
MINIMIZATION OF CARGO LOSSES IN TRANSFERS WITH ALLOWANCE FOR THE DELIVERY DATES (RUSSIAN). EKONOMIKA I MATEMATICESKIE METODY 10, 93-97. (1974).

GRIM 1976 -1 GRIMMETT, G. R.
UPPER BOUND FOR THE NUMBER OF SPANNING TREES OF A GRAPH, AN. DISCRETE MATHEMATICS 16, 323-324. (1976).

GRIMA 1975 -1 GRIMMETT, G. R.; MAC DIARMID, C. J. H.
ON COLOURING RANDOM GRAPHS. MATHEMATICAL PROCEEDINGS OF THE CAMBRIDGE PHILOSOPHICAL SOCIETY 77, 313-324. (1975).

GRIS 1972 -1 GRISHUKHIN, V. P.
ESTIMATION OF THE COMPLEXITY OF THE BALAS ALGORITHM (RUSSIAN). IN: MATHEMATICAL METHODS FOR THE SOLUTION OF ECONOMIC PROBLEMS (RUSSIAN), 93-105. SUPPL. TO EKONOMIKA I MATHEMATICESKIE METHODY. MOSCOW: IZDAT. NAUKA. (1972).

GRIS 1973 -1 GRISHUKHIN, V. P.
CERTAIN CLASS OF MATCHING ALGORITHMS (RUSSIAN), A. IN: A. A. FRIDMAN (ED.): STUDIES IN DISCRETE MATHEMATICS (RUSSIAN), 189-199. MOSCOW: IZDAT. NAUKA. (1973).

GRIS 1976 -1 GRISHUKHIN, V. P.
BRANCH AND BOUNDARY ALGORITHMS IN PROBLEMS WITH BOOLEAN VARIABLES: ESTIMATION OF THEIR EFFECTIVENESS (RUSSIAN). EKONOMIKA I MATHEMATICESKIE METODY 12, NO. 4, 757-766; TRANSLATED AS MATEKON 13, 1977, NO. 4, 50-65. (1976).

GRIS 1976 -2 GRISHUKHIN, V. P.
EFFECTIVENESS OF THE BRANCH AND BOUND METHOD IN PROBLEMS WITH BOOLEAN VARIABLES (RUSSIAN). IN: STUDIES IN DISCRETE OPTIMIZATION (RUSSIAN), 203-230. MOSCOW: IZDAT. NAUKA. (1976).

GRIS 1977 -1 GRISHUKHIN, V. P.
ON INTEGER SOLUTIONS IN THE ALLOCATION PROBLEM (RUSSIAN). EKONOMIKA I MATEMATICESKIE METODY 13, 714-722. (1977).

GROE 1976 -1 GROETSCHEL, M.
UEBER GRAPHEN MIT KREISEN, DIE GEGEBENE WEGE ENTHALTEN. IN: J. KOHLAS, O. SEIFERT, P. STAEHLY, H.-J. ZIMMERMANN (EDS): PROCEEDINGS IN OPERATIONS RESEARCH 5, 119-124. WUERZBURG: PHYSICA. (1976).

GROE 1977 -1 GROETSCHEL, M.
FACETTEN VON MATROID-POLYTOPEN. IN: R. HENN, P. KALL, B. KORTE, O. KRAFFT ET AL. (EDS.): OPERATIONS RESEARCH VERFAHREN XXV, METHODS OF OPERATIONS RESEARCH, 306-313. MEISENHEIM: HAIN. (1977).

GROE 1977 -2 GROETSCHEL, M.
GRAPHS WITH CYCLES CONTAINING GIVEN PATHS. IN: P. L. HAMMER, E. L. JOHNSON, B. H. KORTE, G. L. NEMHAUSER (EDS.): STUDIES IN INTEGER PROGRAMMING, 233-246. ANNALS OF DISCRETE MATHEMATICS 1, AMSTERDAM: NORTH-HOLLAND. (1977).

GROE 1977 -3 GROETSCHEL, M.
MONOTONE 2-MATCHING POLYTOPE ON A COMPLETE GRAPH, THE. IN: R. HENN, P. KALL, B. KORTE, O. KRAFFT ET AL. (EDS.): OPERATIONS RESEARCH VERFAHREN XXIV, METHODS OF OPERATIONS RESEARCH, 72-84. MEISENHEIM: HAIN. (1977).

GROE 1977 -4 GROETSCHEL, M.
ON THE MONOTONE SYMMETRIC TRAVELLING SALESMAN PROBLEM: HYPOHAMILTONIAN/HYPOTRACEABLE GRAPHS AND FACETS. REPORT 7788-OR. BONN: UNIV. BONN, INST. F. OEKONOMETRIE UND OPERATIONS RESEARCH (1977).

GROE 1977 -5 GROETSCHEL, M.
ON THE SYMMETRIC TRAVELLING SALESMAN PROBLEM: SOLUTION OF A 120-CITY PROBLEM. REPORT NO. 7770-OR. BONN: UNIV. BONN, INSTITUT FUER OEKONOMETRIE UND OPERATIONS RESEACH, 30 P. (1977).

GROE 1977 -6 GROETSCHEL, M.
POLYEDRISCHE CHARAKTERISIERUNGEN KOMBINATORISCHER OPTIMIERUNGSPROBLEME. MATHEMATICAL SYSTEMS IN ECONOMICS 36. MEISENHEIM: HAIN, 305 P. (1977).

GROE 1977 -7 GROETSCHEL, M.
STRONG BLOCKS AND THE OPTIMUM BRANCHING PROBLEM. REPORT 7786-OR. BONN: UNIV. BONN, INST. F. OEKONOMETRIE UND OPERATIONS RESEARCH (1977).

GROEA 1977 -1 GROETSCHEL, M.; PADBERG, M. W.
LINEARE CHARAKTERISIERUNGEN VON TRAVELLING SALESMAN PROBLEMEN. ZEITSCHRIFT FUER OPERATIONS RESEARCH 21, SERIES A, 33-64. (1977).

GROEA 1977 -2 GROETSCHEL, M.; PADBERG, M. W.
ON THE SYMMETRIC TRAVELLING SALESMAN PROBLEM 1: INEQUALITIES. REPORT NO. 7768-OR. BONN: UNIV. BONN, INSTITUT FUER OEKONOMETRIE UND OPERATIONS RESEARCH, 28 P. (1977).

GROEA 1977 -3 GROETSCHEL, M.; PADBERG, M. W.
ON THE SYMMETRIC TRAVELLING SALESMAN PROBLEM 2: LIFTING THEOREMS AND FACETS. REPORT NO. 7769-OR. BONN: UNIV. BONN, INSTITUT FUER OEKONOMETRIE UND OPERATIONS RESEARCH, 36 P. (1977).

GROEA 1978 -1 GROETSCHEL, M.; PADBERG, M. W.
ON THE SYMMETRIC TRAVELLING SALESMAN PROBLEM: THEORY AND COMPUTATION. IN:R. HENN, B. KORTE, W. OETTLI (EDS.): OPTIMIZATION AND OPERATIONS RESEARCH, PROCEEDINGS, BONN 1977, 105-115. LECTURE NOTES IN ECONOMICS AND MATHEMATICAL SYSTEMS 157. BERLIN, HEIDELBERG, NEW YORK: SPRINGER. (1978).

GROS 1975 -1 GROS, J.
POWER PLANT SITING: A PARETIAN ENVIRONMENTAL APPROACH. LAXENBURG: INTERNAT. INST. FOR APPLIED SYSTEMS ANALYSIS, 281-292 (1975). (RESEARCH MEMORANDUM 75-44)

GROSA 1960 -1 GROSS, O.
BOTTLENECK ASSIGNMENT PROBLEM: AN ALGORITHM, THE. IN: P. WOLFE (ED): PROCEEDINGS OF A RAND SYMPOSIUM ON MATHEMATICAL PROGRAMMING. RAND PUBLICATION 351, 87-88. SANTA MONICA: RAND CORP. (1960).

GRUN 1974 -1 GRUNBAUM, B.
MATCHINGS IN POLYTOPAL GRAPHS. NETWORKS 4, 175-190. (1974).

GRUNA 1975 -1 GRUNDLER, G.
ALGORITHMEN ZUR LOESUNG VERSCHIEDENER VARIANTEN DES TOURENPROBLEMS. DISSERTATION. FREIBERG: BERGAKADEMIE FREIBERG. (1975).

GRUNB 1977 -1 GRUNDLER, G.; SCHOCH, M.
TOURENPROBLEM MIT ZENTRALEM FUHRPARK BEI UNTERSCHIEDLICHEN FAHRZEUGTYPEN, DAS. MATHEMATISCHE OPERATIONSFORSCHUNG UND STATISTIK 8, SERIES OPTIMIZATION, 329-338. (1977).

GRUP 1974 -1 GRUPP, H. J. C.
NEO-CLASSICAL ANALYSIS OF THE OPTIMAL LOCATION PROBLEM OF INDUSTRY (SPANISH). REV. ECON. ESTADIST. 18, NO. 1-4, 33-53. (1974).

GUBA 1975 -1 GUBA, W.
MAXIMALER LOKALER ALGORITHMUS FUER AUFGABEN DER GANZZAHLIGEN LINEAREN OPTIMIERUNG, EIN. ELEKTRONISCHE INFORMATIONSVERARBEITUNG UND KYBERNETIK 11, 594-599. (1975).

GUGL 1976 -1 GUGLIELMO, F. D.
ESTIMATES OF THE DUALITY GAP FOR DISCRETE OPTIMIZATION. TRAVAUX SEM. ANAL. CONVEXE 6, EXP. NO. 3, 27 P. (1976).

GUHA 1977 -1 GUHA, D. K.
BRANCH AND BOUND ALGORITHM FOR SOLVING TRAVELING SALESMAN PROBLEM, A. JERSEY CITY: THE PORT AUTHORITY OF NEW YORK, 18 P. (1977).

GUIG 1974 -1 GUIGNARD, M.
PREFERRED SHADOW PRICES IN 0-1 PROGRAMMING. RESEARCH REPORT. PHILADELPHIA: UNIV. OF PENNSYLVANIA, DEPT. OF STATISTICS AND OPERATIONS RESEARCH, WHARTON SCHOOL. (1974).

GUIGA 1974 -1 GUIGNARD, M.; SPIELBERG, K.
TRIANGULAR STRUCTURE IN UNCAPACITATED PLANT LOCATION. PAPER PRESENTED AT THE 46TH JOINT NATIONAL MEETING OF ORSA AND TIMS, SAN JUAN 1974. (1974).

GUIGA 1976 -1 GUIGNARD, M.; SPIELBERG, K.
EXPERIMENTAL INTERACTIVE SYSTEM FOR INTEGER PROGRAMMING, AN. TECHNICAL REPORT NO. 17. PHILADELPHIA: UNIV. OF PENNSYLVANIA, DEPT. OF STATISTICS, 11 P. (1976).

GUIGA 1976 -2 GUIGNARD, M.; SPIELBERG, K.
INTEGER PROGRAMMING IN APL. IN: H. P. CROWDER (ED.): 1976 IBM SYMPOSIUM ON MATHEMATICS AND COMPUTATION, PROCEEDINGS, YORKTOWN HEIGTS 1976, 39-50. IBM FORM 26690, YORKTOWN HEIGHTS: IBM THOMAS J. WATSON RESEARCH CENTER. (1976).

GUIGA 1977 -1 GUIGNARD, M.; SPIELBERG, K.
ALGORITHMS FOR EXPLOITING THE STRUCTURE OF THE SIMPLE PLANT LOCATION PROBLEM. IN: P. L. HAMMER, E. L. JOHNSON, B. H. KORTE, G. L. NEMHAUSER (EDS.): STUDIES IN INTEGER PROGRAMMING, 247-272. ANNALS OF DISCRETE MATHEMATICS 1, AMSTERDAM: NORTH-HOLLAND. (1977).

GUIGA 1977 -2 GUIGNARD, M.; SPIELBERG, K.
DUAL METHOD FOR THE MIXED PLANT LOCATION PROBLEM, A. TECHNICAL REPORT NO. 20. PHILADELPHIA: UNIV. OF PENNSYLVANIA, DEPT. OF STATISTICS, 45 P. (1977).

GUIGA 1977 -3 GUIGNARD, M.; SPIELBERG, K.
PROPAGATION, PENALTY IMPROVEMENT AND USE OF LOGICAL INEQUALITIES IN INTERACTIVE (0,1) PROGRAMMING. IN: R. HENN, P. KALL, B. KORTE, O. KRAFFT ET AL. (EDS.): OPERATIONS RESEARCH VERFAHREN XXV, METHODS OF OPERATIONS RESEARCH, 157-171, MEISENHEIM: HAIN. (1977).

GUIGA 1977 -4 GUIGNARD, M.; SPIELBERG, K.
REDUCTION METHODS FOR STATE ENUMERATION INTEGER PROGRAMMING. IN: P. L. HAMMER, E. L. JOHNSON, B. H. KORTE, G. L. NEMHAUSER (EDS.): STUDIES IN INTEGER PROGRAMMING, 273-286. ANNALS OF DISCRETE MATHEMATICS 1, AMSTERDAM: NORTH-HOLLAND. (1977).

GULD 1977 -1 GULDMANN, J.-M.; SHEFER, D.
OPTIMAL PLANT LOCATION AND AIR QUALITY MANAGEMENT UNDER INDIVISIBILITIES AND ECONOMIES OF SCALE. SOCIO-ECONOMIC PLANNING SCIENCES 11, 77-94. (1977).

GUNT 1977 -1 GUNTHER, T. M.
SUBGRADIENT APPROACH TO 0,1 INTEGER PROGRAMMING. SAN FRANCISCO (51. ORSA-MEETING + TIMS, MAY 1977) 25 P. (1977).

GUPT 1969 -1 GUPTA, J. N. D.
SEQUENCING N-JOBS ON M-PARALLEL PROCESSORS. OPSEARCH 6, 295-298. (1969).

GUPT 1975 -1 GUPTA, J. N. D.
OPTIMAL SCHEDULES FOR SPECIAL STRUCTURE FLOWSHOPS. NAVAL RESEARCH LOGISTICS QUARTERLY 22, 255-270. (1975).

GUPT 1976 -1 GUPTA, J. N. D.
HEURISTIC ALGORITHM FOR THE FLOWSHOP SCHEDULING PROBLEM, A. REVUE FRANCAISE D'AUTOMATIQUE, D'INFORMATIQUE, ET DE RECHERCHE OPERATIONNELLE 10, NO. 6, 63-74. (1976).

GUPT 1976 -2 GUPTA, J. N. D.
OPTIMAL FLOWSHOP SCHEDULES WITH NO INTERMEDIATE STORAGE SPACE. NAVAL RESEARCH LOGISTICS QUARTERLY 23, 235-244. (1976).

GUPTA 1969 -1 GUPTA, R. P.
INDEPENDENCE AND COVERING NUMBERS OF LINE GRAPHS AND TOTAL GRAPHS. IN: F. HARARY (ED.): PROOF TECHNIQUES IN GRAPH THEORY, 61-62. NEW YORK: ACADEMIC PRESS. (1969).

GUTE 1976 -1 GUTENBAUM, J.; NICOLO, F.
OPTIMAL AND SUB-OPTIMAL ALLOCATION OF RESOURCES IN SOME HIERARCHICAL SYSTEMS (POLISH). ARCHIWUM AUTOMATYKI I TELEMECHANIKI 21, 483-501. (1976).

GUZI 1975 -1 GUZIK, B.
SIMULATION MODEL OF THE OPTIMAL LOCATION OF A DISTRIBUTION CENTRE (POLISH). PRZEGLAD STATYSTYCZNY 22, 409-425. (1975).

GVOZ 1974 -1 GVOZDEV, S. E.
SOME MODELS AND SOLUTION METHODS FOR THE KNAPSACK PROBLEM WITH SUPPLEMENTARY CONSTRAINTS (RUSSIAN). UPRAVLJAEME SISTEMY, NO 13, 10-25. (1974).

GVOZ 1976 -1 GVOZDEV, S. E.
ON A GENERALIZATION OF THE KNAPSACK PROBLEM (RUSSIAN). UPRAVLJAEME SISTEMY, NO. 15, 16-31. (1976).

GYUL 1975 -1 GYULUMYAN, S. M.
ON CRITICAL GRAPHS IN CONNECTION WITH COLORINGS (RUSSIAN). AKADEMIJA NAUK ARMJANSKOI SSR, DOKLADY 61, NO. 1, 21-26. (1975).

HABE 1976 -1 HABENICHT, W.
AUSGEWAEHLTE ENUMERATIONSSTRATEGIEN ZUR LOESUNG GANZZAHLIGER LINEARER OPTIMIERUNGSPROBLEME. MATHEMATICAL SYSTEMS IN ECONOMICS 21, MEISENHEIM: HAIN, 168 P. (1976).

HABE 1977 -1 HABENICHT, W.
ZUR AUSWAHL VON BASISLOESUNGEN BEI VERFAHREN DER GANZZAHLIGEN OPTIMIERUNG. IN: R. HENN, P. KALL, B. KORTE, O. KRAFFT ET AL. (EDS.): OPERATIONS RESEARCH VERFAHREN XXV, METHODS OF OPERATIONS RESEARCH, 172-185, MEISENHEIM: HAIN. (1977).

HABI 1975 -1 HABIB, M.; LAMBERT, G.
NOMBRE DE SAUTS ET INDICE DE PARTITION DANS LES GRAPHES SANS CIRCUIT. CAHIERS DU CENTRE D'ETUDES DE RECHERCHE OPERATIONNELLE 17, 197-211. (1975).

HACA 1974 -1 HACATUROV, V. R.
APPROXIMATION-COMBINATORIAL METHOD, AND SOME OF ITS APPLICATIONS (RUSSIAN), AN. ZURNAL VYCISLITEL'NOI MATEMATIKI I MATEMATICESKOI FIZIKI 14, 1464-1487 + 1629. (1974).

HACAA 1976 -1 HACATUROV, V. R.; ASTAHOV, N. D.
DYNAMICAL ALLOCATION PROBLEMS (MODELS AND METHODS OF SOLUTION) (RUSSIAN). EKONOMIKA I MATEMATICESKIE METODY 12, NO. 1, 93-109. (1976).

HADJ 1974 -1 HADJERES, S.
QELQUES METHODES RECENTES DE RESOLUTION DU PROBLEME DU VOYAGEUR DE COMMERCE. DISSERTATION. PARIS: UNIV. PARIS VI, 113 P. (1974).

HADL 1974 1 HADLOCK, F. O.
MINIMUM SPANNING FORESTS OF BOUNDED TREES. IN: F. HOFFMAN, R. A. KINGSLEY, R. B. LEVOW, R. C. MULLIN (EDS.): PROCEEDINGS OF THE FIFTH SOUTHEASTERN CONF. ON COMBINATORICS, GRAPH THEORY, AND COMPUTING, BOCA RATON 1974, 449-460. CONGRESSUS NUMERANTIUM 10, WINNIPEG: UTILITAS MATHEMATICA. (1974).

HADL 1977 -1 HADLOCK, F. O.
SHORTEST PATH ALGORITHM FOR GRID GRAPHS, A. NETWORKS 7, 323-334. (1977).

HAEN 1973 -1 HAENSLER, E.
EXPERIMENTAL HEURISTIC PROCEDURE TO OPTIMIZE A TELECOMMUNICATION NETWORK UNDER NONLINEAR COST FUNCTIONS. IN: PROCEEDINGS OF THE SEVENTH ANNUAL PRINCETON CONFERENCE ON INFORMATION SCIENCES AND SYSTEMS, 130-134. (1973).

HAES 1976 -1 HAESSLER, R. W.; HOGUE, S. L.
NOTE ON THE SINGLE-MACHINE MULTI-PRODUCT LOT SCHEDULING PROBLEM, A. MANAGEMENT SCIENCE 22, 909-912. (1976).

HAFF 1973 -1 HAFF, C. E.; BURNS, R. N.
RANKING PROBLEM IN GRAPHS, A. IN: F. HOFFMAN, R. A. KINGSLEY, R. B. LEVOW ET AL. (EDS.): PROCEEDINGS OF THE 4TH SOUTHEASTERN CONFERENCE ON COMBINATORICS, GRAPH THEORY, AND COMPUTING, BOCA RATON 1974, 461-470. CONGRESSUS NUMERANTIUM 10, WINNIPEG: UTILITAS MATHEMATICA. (1973).

HAKI 1969 -1 HAKIMI, S.
ALGORITHM FOR CONSTRUCTION OF THE LEAST VULNERABLE COMMUNICATION NETWORK, AN. IEEE TRANSACTIONS ON CIRCUIT THEORY CT-16, 229-230. (1969).

****

HAKIA 1972 -1 HAKIMI, S. L.
SHORTEST PATHS IN GRAPHS - A REVIEW. IEEE INTERNAT. SYMPOSIUM ON CIRCUIT THEORY, 368-369. (1972).

HALI 1975 -1 HALIN, R.
SOME PATH PROBLEMS IN GRAPH THEORY. ABHANDLUNGEN AUS DEM MATHEMATISCHEN SEMINAR DER UNIV. HAMBURG 44, 175-186. (1975).

HALP 1976 -1 HALPERN, J.
SEQUENTIAL COVERING PROBLEM UNDER UNCERTAINTY, THE (REVISED). WORKING PAPER 76-003. MIAMI (50. ORSA-MEETING + TIMS, NOVEMBER 1976) AND TORONTO: UNIV. OF TORONTO, DEPT. OF INDUSTRIAL ENGINEERING, 36 P. (1976).

HALP 1977 -1 HALPERN, J.
GENERALIZED DYNAMIC FLOWS PROBLEM, A. WORKING PAPER NO. WP-07-77. CALGARY: UNIV. OF CALGARY, FACULTY OF BUSINESS, 48 P. (1977).

HALP 1977 -2 HALPERN, J.
SEQUENTIAL COVERING PROBLEM UNDER UNCERTAINTY, THE. INFOR 15, 76-93. (1977).

HALP 1977 -3 HALPERN, J.
SHORTEST ROUTE WITH TIME DEPENDENT LENGTH OF EDGES AND LIMITED DELAY POSSIBILITIES IN NODES. ZEITSCHRIFT FUER OPERATIONS RESEARCH 21, 117-124. (1977).

HAMM 1969 -1 HAMMER, G.
BEMERKUNG ZUM TRIPEL-ALGORITHMUS ZUR BESTIMMUNG DER KUERZESTEN WEGE IN EINEM GRAPHEN, EINE. ZEITSCHRIFT FUER WAHRSCHEINLICHKEITSTHEORIE UND VERWANDTE GEBIETE 12, 153-154. (1969).

HAMM 1969 -2 HAMMER, G.
TRIPEL-ALGORITHMUS ZUR BESTIMMUNG KUERZESTER WEGE IN EINEM GRAPHEN. ZEITSCHRIFT FUER WAHRSCHEINLICHKEITSTHEORIE UND VERWANDTE GEBIETE 12, 153. (1969).

HAMM 1976 -1 HAMMER, G.
DUALITAET BEI KOMBINATORISCHEN SUCHVERFAHREN. ARBEITSPAPIERE ZUR MATHEMATISCHEN WIRTSCHAFTSFORSCHUNG 13. (1976).

HAMM 1977 -1 HAMMER, G.
PRIMAL-DUALES SUCHVERFAHREN FUER FAERBUNGSPROBLEME, EIN. IN: R. HENN, P. KALL, B. KORTE, O. KRAFFT ET AL. (EDS.): OPERATIONS RESEARCH VERFAHREN XXV, METHODS OF OPERATIONS RESEARCH, 186-188, MEISENHEIM: HAIN. (1977).

HAMMA 1975 -1 HAMMER, P. L.
BOOLEAN ELEMENTS IN COMBINATORIAL OPTIMIZATION: A SURVEY. IN: B. ROY (ED.): COMBINATORIAL PROGRAMMING: METHODS AND APPLICATIONS. PROCEEDINGS OF THE NATO ADVANCED STUDY INST., VERSAILLES 1974, 67-92. DORDRECHT: REIDEL. (1975).

HAMMA 1976 -1 HAMMER, P. L.
BOOLEAN PROCEDURES IN OPTIMIZATION. IN: PROCEEDINGS OF THE CONFERENCE ON MATHEMATICAL PROGRAMMING AND ITS APPLICATIONS, INDAM, ROME 1974 (ITALIAN), 103-121. SYMPOSIA MATHEMATICA 19, LONDON: ACADEMIC PRESS. (1976).

HAMMA 1977 -1 HAMMER, P. L.
BOOLEAN ELEMENTS IN COMBINATORIAL OPTIMIZATION. A SURVEY. VANCOUVER (CONFERENCE ON DISCRETE OPTIMIZATION, AUGUST 1977) 41 P. (1977).

HAMMB 1977 -1 HAMMER, P. L.; JOHNSON, E. L.; PELED, U. N.
ROLE OF MASTER POLYTOPES IN THE UNIT CUBE, THE. SIAM JOURNAL ON APPLIED MATHEMATICS 32, 711-716. (1977).

HAND 1977 -1 HANDLER, G. Y.
M-CENTER PROBLEM: A RELAXATION APPROACH, THE. WORKING PAPER 58/6. TEL-AVIV: ISRAEL INSTITUTE OF BUSINESS RESEARCH. (1977).

HANS 1971 -1 HANSEN, K. H.
SEQUENCING PROBLEMS WITH EMPHASIS ON A MODIFICATION OF HELD AND KARP'S ALGORITHM. OR: AN ALGORITHM FOR THE OPTIMAL SOLUTION OF THE SYMMETRIC TRAVELING SALESMAN PROBLEM. PART 2. (DANISH). KOPENHAGEN: UNIV. OF KOPENHAGEN, 101 P. (1971).

HANS 1971 -2 HANSEN, K. H.
SEQUENCING PROBLEMS WITH EMPHASIS ON A MODIFICATION OF HELD AND KARP'S ALGORITHM. OR: AN ALGORITHM FOR THE OPTIMAL SOLUTION OF THE SYMMETRIC TRAVELING SALESMAN PROBLEM. PART 1. (DANISH). KOPENHAGEN: UNIV. OF KOPENHAGEN, 101 P. (1971).

HANSA 1975 -1 HANSEN, P.
DEGRES ET NOMBRE DE STABILITE D'UN GRAPHE. CAHIERS DU CENTRE D'ETUDES DE RECHERCHE OPERATIONNELLE 17, 213-220; ALSO PUBLISHED IN LILLE: INSTITUT D'ECONOMIE SCIENTIFIQUE ET DE GESTION, 213-220, 1976. (1975).

HANSA 1975 -2 HANSEN, P.
FONCTIONS D'EVALUATION ET PENALITES POUR LES PROGRAMMES QUADRATIQUES EN VARIABLES 0-1. IN: B. ROY (ED.): COMBINATORIAL PROGRAMMING: METHODS AND APPLICATIONS. PROCEEDINGS OF THE NATO ADVANCED STUDY INST., VERSAILLES 1974, 361-370. DORDRECHT: REIDEL. (1975).

HANSA 1975 -3 HANSEN, P.
PROCEDURES D'EXPLORATION ET D'OPTIMISATION PAR SEPARATION ET EVALUATION, 1: CLASSIFICATION ET FORME DES TESTS, LES. LILLE: INSTITUT D'ECONOMIE SCIENTIFIQUE ET DE GESTION, 47 P. (1975).

HANSA 1975 -4 HANSEN, P.
PROCEDURES D'EXPLORATION ET D'OPTIMISATION PAR SEPARATION ET EVALUATION: A SURVEY, LES. IN: B. ROY (ED.): COMBINATORIAL PROGRAMMING: METHODS AND APPLICATIONS. PROCEEDINGS OF THE NATO ADVANCED STUDY INST., VERSAILLES 1974, 29-66. DORDRECHT: REIDEL. (1975).

HANSA 1976 -1 HANSEN, P.
CASCADE ALGORITHM FOR THE LOGICAL CLOSURE OF A SET OF BINARY RELATIONS, A. INFORMATION PROCESSING LETTERS, VOL. 5, NO. 2, 50-54. (1976).

HANSA 1977 -1 HANSEN, P.
METHODS OF NONLINEAR 0-1 PROGRAMMING. VANCOUVER (CONFERENCE ON DISCRETE OPTIMIZATION, AUGUST 1977) 25 P. (1977).

HANSA 1977 -2 HANSEN, P.
SUBDEGREES AND CHROMATIC NUMBERS OF HYPERGRAPHS. IN: P. L. HAMMER, E. L. JOHNSON, B. H. KORTE, G. L. NEMHAUSER (EDS.): STUDIES IN INTEGER PROGRAMMING, 287-292. ANNALS OF DISCRETE MATHEMATICS 1, AMSTERDAM: NORTH-HOLLAND. (1977).

HANSB 1976 -1 HANSEN, P.; DELATTRE, M.
CLUSTER ANALYSIS AND GRAPH COLORING. MONS: FACULTE UNIVERSITAIRE CATHOLIQUE DE MONS, 17 P. (1976).

HANSB 1976 -2 HANSEN, P.; DELATTRE, M.
CLUSTER ANALYSIS BY GRAPH COLORING. MONS: UNIV. DE MONS, 29 P. (1976).

HANSB 1976 -3 HANSEN, P.; DELATTRE, M.
COMPLETE-LINK CLUSTER ANALYSIS BY GRAPH COLORING. MONS: FACULTE UNIVERSITAIRE CATHOLIQUE DE MONS, 25 P. (1976).

HANSC 1976 -1 HANSEN, P.; KAUFMAN, L.
PUBLIC FACILITIES LOCATION UNDER AN INVESTMENT CONSTRAINT. IN: K. B. HALEY (ED.): OPERATIONAL RESEARCH '75. PROCEEDINGS OF THE 7TH IFORS INTERNATIONAL CONFERENCE ON OPERATIONAL RESEARCH, TOKYO, KYOTO 1975, 201-210. AMSTERDAM: NORTH-HOLLAND. (1976).

****

HANSD 1972 -1 HANSEN, P.; KAUFMANN, L.
ALGORITHM FOR THE LOCATION OF CENTRAL FACILITIES, AN. IN: MATHEMATICAL PROGRAMS FOR ACTIVITY ANALYSIS, 141-158, AMSTERDAM: NORTH-HOLLAND. (1972).

HANSE 1977 -1 HANSEN, P.; THISSE, J.-F.
MULTIPLIANT LOCATION FOR PROFIT MAXIMISATION. ENVIRONMENT AND PLANNING 9, SERIES A, 63-73. (1977).

HANSF 1977 -1 HANSON, H. H.
PROCEDURE FOR ASSIGNING WORKERS TO MACHINES WHEN EACH WORKER CAN OPERATE MORE THAN ONE MACHINE, A. SAN FRANCISCO (51. ORSA-MEETING + TIMS, MAY 1977) 35 P. (1977).

HARA 1975 -1 HARARY, F.
ON MINIMAL FEEDBACK VERTEX SETS OF A DIGRAPH. IEEE TRANSACTIONS ON CIRCUITS AND SYSTEMS CAS-22, 839-840. (1975).

HARAA 1974 -1 HARARY, F.; LEMPEL, A.
ON CLIQUE-EXTREMAL (P.Q)-GRAPHS. NETWORKS 4, 371-378. (1974).

HARAB 1974 -1 HARARY, F.; MALLION, R. B.
ELEMENTARY NECESSARY CONDITION FOR HAMILTONIAN GRAPHS INVOLVING CONNECTIVITY, AN. NANTA MATHEMATICA 7, 96-101. (1974).

HARAB 1975 -1 HARARY, F.; MALLION, R. B.
ERRATA: "AN ELEMENTARY, NECESSARY CONDITION FOR HAMILTONIAN GRAPHS, INVOLVING CONNECTIVITY" (NANTA MATHEMATICA 7, 1974, NO. 2, 96-101). NANTA MATHEMATICA 8, NO. 1, 104. (1975).

HARD 1977 -1 HARDLOCK, F. O.
SHORTEST PATH ALGORITHM FOR GRID GRAPHS, A. NETWORKS 7, 323-334. (1977).

HARDA 1975 -1 HARDOUIN, D. J.
QUELQUES RESULTATS SUR "L'INDICE DE TRANSITIVITE" DE CERTAINS TOURNOIS. MATHEMATIQUES ET SCIENCES HUMAINES, NO. 51, 35-41 + 87. (1975).

HARI 1976 -1 HARITON, JA. R.
CERTAIN METHOD FOR THE SOLUTION OF DISCRETE PROGRAMMING PROBLEMS (RUSSIAN), A. IN: MATHEMATICAL PROGRAMMING AND RELATED QUESTIONS. PROCEEDINGS OF THE SEVENTH WINTER SCHOOL, DROGOBYCH 1974, CONVEX PROGRAMMING (RUSSIAN), 159-165. MOSCOW: CENTRAL. EKONOM.-MAT. INST. AKAD. NAUK SSSR. (1976).

HARM 1976 -1 HARMATUCK, D. J.
COMPARISON OF TWO APPROACHES TO STOCK LOCATION, A. THE LOGISTICS AND TRANSPORTATION REVIEW 12, 282-284. (1976).

HARR 1976 -1 HARRIS, B.
CODE FOR THE TRANSPORTATION PROBLEM OF LINEAR PROGRAMMING, A. JOURNAL OF THE ASSOCIATION FOR COMPUTING MACHINERY 23, 155-157. (1976).

HART 1968 -1 HART, P. E.; NILLSON, N. L.; RAPHAEL, B.
FORMAL BASIS FOR THE HEURISTIC DETERMINATION OF MINIMUM COST PATHS. IEEE TRANSACTIONS ON SCC 4, 100-107. (1968).

HARTA 1976 -1 HARTLEY, R. V.
OPERATIONS RESEARCH: A MANAGERIAL EMPHASIS. PACIFIC PALISADES: GOODYEAR, 795 P. (1976).

HARTB 1976 -1 HARTMANIS, J.; BERMAN. L.
ON ISOMORPHISMS AND DENSITY OF NP AND OTHER COMPLETE SETS. IN: PROCEEDINGS OF THE EIGHTH ANNUAL A.C.M. SYMPOSIUM ON THEORY OF COMPUTING, 30-40. (1976).

HARTC 1973 -1 HARTMANN, K.
ASPEKTE DER GANZZAHLIGEN LINEAREN QUOTIENTENOPTIMIERUNG, EINIGE. WISSENSCHAFTLICHE ZEITSCHRIFT DER TECHNISCHEN HOCHSCHULE FUER CHEMIE LEUNA-MERSEBURG 15, 413-418. (1973).

HARTC 1975 -1 HARTMANN, K.
GEMISCHT GANZZAHLIGE LINEARE QUOTIENTENOPTIMIERUNG NACH DEM SCHNITTVERFAHREN VON GOMORY. MATHEMATISCHE OPERATIONSFORSCHUNG UND STATISTIK 6, 845-854. (1975).

HARTD 1976 -1 HARTNELL, B. L.
SOME PROBLEMS ON SPANNING TREES. IN: B. L. HARTNELL, H. C. WILLIAMS (EDS): PROCEEDINGS OF THE 5TH MANITOBA CONFERENCE ON NUMERICAL MATHEMATICS, WINNIPEG 1975, 375-384. CONGRESSUS NUMERANTIUM 16, WINNIPEG: UTILITAS MATHEMATICA. (1976).

HASL 1975 -1 HASLINGER, A.
BALANCED GROWTH AND RELATIVE STABILITY IN DISCRETE, NONLINEAR PERSON-FLOW MODELS. WIEN: TECHNISCHE HOCHSCHULE WIEN, INSTITUT FUER UNTERNEHMENSFORSCHUNG, 31 P. (1975).

HATC 1975 -1 HATCH, R. S.
BENCH MARKS COMPARING TRANSPORTATION CODES BASED ON PRIMAL SIMPLEX AND PRIMAL-DUAL ALGORITHMS. OPERATIONS RESEARCH 23, 1167-1172. (1975).

HAUS 1977 -1 HAUSMANN, D.
ADJACENCY ON THE BRANCHING POLYHEDRON. REPORT NO. 7756-OR. BONN: UNIV. BONN, INST. FUER OEKONOMETRIE UND OPERATIONS RESEARCH, 19 P. (1977).

HAUS 1977 -2 HAUSMANN, D.
BENACHBARTE ECKEN AUF 0-1-POLYEDERN. IN: R. HENN, P. KALL, B. KORTE, O. KRAFFT ET AL. (EDS.): OPERATIONS RESEARCH VERFAHREN XXV, METHODS OF OPERATIONS RESEARCH, 189-191, MEISENHEIM: HAIN. (1977).

HAUS 1977 -3 HAUSMANN, D.
POLYNOMIALER ALGORITHMUS FUER DIE ADJAZENZUEBERPRUEFUNG AUF DEM BRANCHING-POLYEDER, EIN. REPORT 7771-OR. BONN: UNIV. BONN, INST. F. OEKONOMETRIE UND OPERATIONS RESEARCH, 6 P. (1977).

HAUS 1978 -1 HAUSMANN, D.
ADJAZENZCHARAKTERISIERUNGEN FUER POLYEDERECKEN IN DER KOMBINATORISCHEN OPTIMIERUNG. DISSERTATION, BONN: UNIVERSITY OF BONN, INSTITUT FUER OEKONOMETRIE UND OPERATIONS RESEARCH. (1978).

HAUSA 1977 -1 HAUSMANN, D.; JENKYNS, T. A.; KORTE, B.
WORST CASE ANALYSIS OF GREEDY TYPE ALGORITHMS FOR INDEPENDENCE SYSTEMS. REPORT 7781-OR. BONN: UNIV. BONN, INST. F. OEKONOMETRIE UND OPERATIONS RESEARCH, 20P. (1977).

HAUSB 1977 -1 HAUSMANN, D.; KANNAN, R.; KORTE, B.
EXPONENTIAL LOWER BOUNDS ON A CLASS OF KNAPSACK ALGORITHMS. REPORT 7789-OR. BONN: UNIV. BONN, INST. F. OEKONOMETRIE UND OPERATIONS RESEARCH. (1977).

HAUSC 1976 -1 HAUSMANN, D.; KORTE, B.
COLOURING CRITERIA FOR ADJACENCY ON 0-1-POLYHEDRA. REPORT 7647-OR. BONN: UNIV. BONN, INST. F. OEKONOMETRIE UND OPERATIONS RESEARCH, 30 P. (1976).

HAUSC 1977 -1 HAUSMANN, D.; KORTE, B.
LOWER BOUNDS ON THE WORST-CASE COMPLEXITY OF SOME ORACLE ALGORITHMS. REPORT NO. 7757-OR. BONN: UNIV. BONN, INST. FUER OEKONOMETRIE UND OPERATIONS RESEARCH, 20 P. (1977).

HAUSC 1977 -2 HAUSMANN, D.; KORTE, B.
WORST-CASE BEHAVIOUR OF POLYNOMIAL BOUNDED ALGORITHMS FOR INDEPENDENCE SYSTEMS. REPORT NO. 7760-OR. BONN: UNIV. BONN, INST. FUER OEKONOMETRIE UND OPERATIONS RESEARCH, 4 P. (1977).

HAUSC 1978 -1 HAUSMANN, D.; KORTE, B.
ORACLE-ALGORITHMS FOR FIXED POINT PROBLEMS - AN AXIOMATIC APPROACH. IN: R. HENN, B. KORTE, W. OETTLI (EDS.): OPTIMIZATION AND OPERATIONS RESEARCH, PROCEEDINGS, BONN 1977, 137-156. LECTURE NOTES IN ECONOMICS AND MATHEMATICAL SYSTEMS 157. BERLIN, HEIDELBERG, NEW YORK: SPRINGER. (1978).

HAUSC 1978 -2 HAUSMANN, D.; KORTE, B.
WORST-CASE ANALYSIS FOR A CLASS OF COMBINATORIAL ALGORITHMS. IN: J. STOER (ED.): OPTIMISATION TECHNIQUES, PROCEEDINGS OF THE 8TH IFIP CONFERENCE ON OPTIMISATION TECHNIQUES, PART 2, 216-224., SPRINGER (1978).

HAY 1978 -1 HAY, L.
CONVEX SUBSETS OF 2 POWER N AND BOUNDED TRUTH-TABLE REDUCIBILITY. DISCRETE MATHEMATICS 21, 31-46. (1978).

HEAR 1976 -1 HEARD, E. L.
DUE-DATES AND INSTANTANEOUS LOAD IN THE ONE-MACHINE SHOP. MANAGEMENT SCIENCE 23, 444-450. (1976).

HEARA 1976 -1 HEARN, D.; LOWE, T. J.
SUBGRADIENT PROCEDURE FOR THE SOLUTION OF MINIMAL PROBLEMS, A. RESEARCH REPORT NO. 76-6. GAINESVILLE: UNIV. OF FLORIDA , DEPT. OF INDUSTRIAL AND SYSTEMS ENGINEERING, 22 P. (1976).

HECH 1977 -1 HECHLER, S. H.
ON INFINITE GRAPHS WITH A SPECIFIED NUMBER OF COLORINGS. DISCRETE MATHEMATICS 19, 241-256. (1977).

HECK 1976 -1 HECK, J.
ALGORITHMUS ZUR LOESUNG DES CHINESE-POSTMAN PROBLEMS, EIN. IN: H. NOLTEMEIER (ED.): GRAPHEN, ALGORITHMEN, DATENSTRUKTUREN. ERGEBNISSE DES WORKSHOPS WG 76. 2. FACHTAGUNG UEBER GRAPHENTHEORETISCHE KONZEPTE DER INFORMATIK, GOETTINGEN 1976, 183-216. MUENCHEN: HANSER. (1976).

HEDE 1971 -1 HEDETNIEMI, S.
CHARACTERIZATIONS AND CONSTRUCTIONS OF MINIMALLY TWO-CONNECTED GRAPHS AND MINIMALLY STRONG DIGRAPHS. IN: R. C. MULLIN, K. B. REID, D. P. ROSELLE, R. P. D. THOMAS (EDS.): PROCEEDINGS OF THE 2ND LOUISIANA CONFERENCE ON COMBINATORICS, GRAPH THEORY AND COMPUTING, BATON ROUGE 1971, 257-282. CONGRESSUS NUMERANTIUM 3, WINNIPEG: UTILITAS MATHEMATICA. (1971).

HEDEA 1963 -1 HEDETNIEMI, S.; LEVEN, B. M.
DETERMINING FASTEST ROUTES USING FIXED SCHEDULES. IN: SPRING JOINT COMPUTER CONF. PROC., BALTIMORE: SPARTAN BOOK (1963).

HEIN 1976 -1 HEIN, O.
ALGORITHMUS ZUR BERECHNUNG DER MINIMALEN KANTENBASIS IN GERICHTETEN GRAPHEN, EIN. ANGEWANDTE INFORMATIK 18, 196-198. (1976).

HEIN 1977 -1 HEIN, O.
GRAPHENTHEORIE FUER ANWENDER. MANNHEIM: BIBLIOGRAPHISCHES INSTITUT, 141 P. (1977).

HEIN 1977 -2 HEIN, O.
LOESUNGSVORSCHLAG ZUM VERALLGEMEINERTEN RUNDFAHRTPROBLEM, EIN. ANGEWANDTE INFORMATIK 19, 22-25. (1977).

HEINA 1976 -1 HEINRICH, K.
NON-IMBEDDABLE PROPER COLOURING, A. IN: R. A. CASSE, W. D. WALLIS (EDS.): COMBINATORIAL MATHEMATICS IV, PROCEEDINGS OF THE 4TH AUSTRALIAN CONFERENCE, ADELAIDE 1975, 93-115. LECTURE NOTES IN MATHEMATICS 560, BERLIN: SPRINGER. (1976).

HEIS 1977 -1 HEISING, C. D.
PLANT LOCATION AND TRANSPORTATION MODELS OF THE LIGHT-WATER REACTOR FUEL CYCLE. SAN FRANCISCO (51. ORSA-MEETING + TIMS, MAY 1977) 27 P. (1977).

HELD 1965 -1 HELD, M.; KARP, R. M.
CONSTRUCTION OF DISCRETE DYNAMIC PROGRAMMING ALGORITHMS, THE. IBM SYSTEMS JOURNAL 4, 136-147. (1965).

HELG 1977 -1 HELGASON, R. V.; KENNINGTON, J. L.
EFFICIENT PROCEDURE FOR IMPLEMENTING A DUAL SIMPLEX NETWORK FLOW ALGORITHM, AN. AIIE TRANSACTIONS 9, 63-68. (1977).

HELG 1977 -2 HELGASON, R. V.; KENNINGTON, J. L.
PRODUCT FORM REPRESENTATION OF THE INVERSE OF A MULTICOMMODITY CYCLE MATRIX, A. NETWORKS 7, 297-322. (1977).

HELL 1955 -1 HELLER, I.
ON THE TRAVELING SALESMAN'S PROBLEM. NATIONAL BUREAU OF STANDARDS (ED.): PROCEEDINGS 2ND SYMPOSIUM IN LINEAR PROGRAMMING, WASHINGTON D.C. 1955, 643-665. (1955).

HELL 1956 -1 HELLER, I.
NEIGHBOUR RELATIONS ON THE CONVEX OF CYCLIC PERMUTATIONS. PACIFIC JOURNAL OF MATHEMATICS 6, 467-477. (1956).

HELLA 1960 -1 HELLER, J.; LOGEMANN, G.
ALGORITHM FOR CONSTRUCTING FEASIBLE SCHEDULES AND COMPUTING THEIR SCHEDULE TIMES. AEC REPORT NO-9488. NEW YORK: NEW YORK UNIVERSITY, AEC COMPUTER AND APPLIED MATHEMATICS CENTER, 30 P. (1960).

HEND 1976 -1 HENDERSON, P. B.; STEIGLITZ, K.
PROBLEM IN SINGLE-MACHINE SEQUENCING WITH NONLINEAR DELAY COSTS, A. MANAGEMENT SCIENCE 22, 1342-1350. (1976).

HENDA 1976 -1 HENDERSON, W. B.; BERRY, W. L.
HEURISTIC METHODS FOR TELEPHONE OPERATOR SHIFT SCHEDULING: AN EXPERIMENTAL ANALYSIS. MANAGEMENT SCIENCE 22, 1372-1380. (1976).

HENR 1967 -1 HENRY, B. B.; JONES, C. H.
LINEAR PROGRAMMING FOR PRODUCTION ALLOCATION. JOURNAL OF INDUSTRIAL ENGINEERING 18, 403-412. (1967).

HEPP 1976 -1 HEPPES, A.; LUGOSI, G.
PRODUCTION SCHEDULING TO MINIMIZE THE SET-UP TIME OF PARALLEL MACHINES WITH ALTERNATIVE CHOICES (HUNGARIAN). ALKALMAZOTT MATEMATIKAI LAPOK 2, 83-91. (1976).

HERI 1969 -1 HERING, F.
UNTERSUCHUNGEN UEBER DIE KOMBINATORISCHE STRUKTUR VON POLYEDERN. DISSERTATION. BONN: UNIV. BONN, 71 P. (1969).

HERO 1973 -1 HERON, A. P.
PROPERTY OF THE HYPERPLANES OF A MATROID AND AN EXTENSION OF DILWORTH'S THEOREM, A. JOURNAL OF MATHEMATICAL ANALYSIS AND APPLICATIONS 42, 119-131. (1973).

HERS 1965 -1 HERSHDORFER, A. M.
OPTIMAL ROUTING OF URBAN TRAFFIC. DISSERTATION. CAMBRIDGE: M.I.T. DEPT. OF CITY AND REGIONAL PLANNING. (1965).

HERZ 1975 -1 HERZOG, U.
OPTIMAL SCHEDULING STRATEGIES FOR REAL-TIME COMPUTERS. IBM JOURNAL OF RESEARCH AND DEVELOPMENT 19, 494-504. (1975).

HESS 1977 -1 HESSELBACH, J.; WIJNANDS, J.
DYNAMISCHE PROGRAMMIERUNG - METHODIK, EDV-PROGRAMM, ANWENDUNGSBEISPIEL. BAD KREUZNACH: MAX-PLANCK-INSTITUT FUER LANDARBEIT UND LANDTECHNIK, 67 P. (1977). (ARBEITEN AUS DEM MAX-PLANCK-INSTITUT FUER LANDARBEIT UND LANDTECHNIK, HEFT A-77/1)

HEUR 1975 -1 HEURGON, E.; DELORME, J.
PROBLEMES DE PARTITIONNEMENT: EXPLORATION ARBORESCENTE OU METHODE DE TRONCATURER? REVUE FRANCAISE D'AUTOMATIQUE, D'INFORMATIQUE ET DE RECHERCHE OPERATIONNELLE 9, NO. 6, 53-65. (1975).

****

HEVE 1964 -1 HEVERAN, R. D.
RESOURCE ALLOCATION. DATA PROCESSING MANAGEMENT, 21-24. (1964).

HILL 1976 -1 HILL, A. V.; WHYBARK, D. C.
COMPARING EXACT SOLUTION PROCEDURES FOR THE MULTI-VEHICLE ROUTING PROBLEM. THE LOGISTICS AND TRANSPORTATION REVIEW 12, 145-153. (1976).

HILLA 1978 -1 HILL, E. JR.
COMPARATIVE STUDY OF VERY LARGE DATA BASES, A. LECTURE NOTES IN COMPUTER SCIENCE, 59. BERLIN, HEIDELBERG, NEW YORK: SPRINGER, 140 P. (1978).

HILT 1975 -1 HILTON, A. J. W.
COLOURING THE EDGES OF A MULTIGRAPH SO THAT EACH VERTEX HAS AT MOST J, OR AT LEAST J, EDGES OF EACH COLOUR ON IT. JOURNAL OF THE LONDON MATHEMATICAL SOCIETY 12, 123-128. (1975).

HILT 1975 -2 HILTON, A. J. W.
ON VIRZING'S UPPER BOUND FOR THE CHROMATIC INDEX OF A GRAPH. CAHIERS DU CENTRE D'ETUDES DE RECHERCHE OPERATIONNELLE 17, 225-233. (1975).

HILT 1976 -1 HILTON, A. J. W.
COVER INDEX, THE CHROMATIC INDEX AND THE MINIMUM DEGREE OF A GRAPH, THE. IN: C. ST. J. A. NASH-WILLIAMS (ED.): PROCEEDINGS OF THE 5TH BRITISH COMBINATORIAL CONFERENCE, ABERDEEN 1975, 307-318. CONGRESSUS NUMERANTIUM 15, WINNIPEG: UTILITAS MATHEMATICA. (1976).

HILT 1977 -1 HILTON, A. J. W.
DEFINITIONS OF CRITICALITY WITH RESPECT TO EDGE-COLORING. JOURNAL OF GRAPH THEORY 1, 61-68. (1977).

HILTA 1973 -1 HILTON, A. J. W.; RADO. R.; SCOTT, S. H.
(<5)-COLOR THEOREM FOR PLANAR GRAPHS, A. BULLETIN OF THE LONDON MATHEMATICAL SOCIETY 5, 369-384. (1973).

HINS 1976 -1 HINSON, J.; MULHERKAR, S.
IMPROVEMENTS TO THE CLARKE AND WRIGHT ALGORITHM AS APPLIED TO AN AIRLINE SCHEDULING PROBLEM. PAPER PRESENTED AT THE MIAMI ORSA/TIMS MEETING, MIAMI 1976. (1976).

HINT 1977 -1 HINTEREGGER, J.; TINHOFER, G.
ZERLEGUNG DER KNOTENMENGEN VON GRAPHEN ZUM NACHWEIS DER ISOMORPHIE. COMPUTING 18, 351-360. (1977).

HIRC 1970 -1 HIRCHE, J.
UEBER ZWEI MODIFIKATIONEN DES DEKOMPOSITIONSALGORITHMUS VON WILLIAMS. IN: WISSENSCHAFTLICHE ZEITSCHRIFT DER UNIV. HALLE, BD. 19, HEFT 6, P. 123-129 (1970).

HIRCA 1975 -1 HIRCHE, J.; KOEHLER, J.; STIEBITZ, V.
EFFEKTIVES LOESUNGSVERFAHREN FUER DAS HYPERBOLISCHE TRANSPORTPROBLEM, EIN. IN: BEITRAEGE ZUR ANALYSIS, P. 151-156 (1975).

HIRS 1977 -1 HIRSCHBERG, D. S.
ALGORITHMS FOR THE LONGEST COMMON SUBSEQUENCE PROBLEM. JOURNAL OF THE ASSOCIATION FOR COMPUTING MACHINERY 24, 664-675. (1977).

HIRSA 1976 -1 HIRSCHBERG, D. S.; WONG, C. K.
POLYNOMIAL-TIME ALGORITHM FOR THE KNAPSACK PROBLEM WITH TWO VARIABLES, A. JOURNAL OF THE ASSOCIATION FOR COMPUTING MACHINERY 23, 147-154. (1976).

HIRSA 1976 -2 HIRSCHBERG, D. S.; WONG, C. K.
UPPER AND LOWER BOUNDS FOR GRAPH-DIAMETER PROBLEMS WITH APPLICATION TO RECORD ALLOCATION. YORKTOWN HEIGHTS: IBM RESEARCH CENTER, 14 P. (1976).

HITC 1976 -1 HITCHINGS, G. G.; COTTAM, M.
EFFICIENT HEURISTIC PROCEDURE FOR SOLVING THE LAYOUT DESIGN PROBLEM, AN. OMEGA 4, 205-214. (1976).

HITCA 1968 -1 HITCHNER, L.
COMPARATIVE INVESTIGATION OF THE COMPUTATIONAL EFFICIENCY OF SHORTEST PATH ALGORITHMS, A. TECHNICAL REPORT ORC 68-17. BERKELEY: UNIV. OF CALIFORNIA, DEPT. OPERATIONS RESEARCH. (1968).

HODG 1974 -1 HODGSON, M. J.
STAGING INTERURBAN HIGHWAY CONSTRUCTION BY DYNAMIC PROGRAMMING. ANNALS OF REGIONAL SCIENCE 8, 123-126. (1974).

HODGA 1977 -1 HODGSON, T. J.; KILPATRICK, K. E.; LONGINI, I. M.
SCHEDULING MULTISPECIALTY CLINICS USING INTEGER PROGRAMMING. AIIE TRANSACTIONS 9, 69-74. (1977).

HOFF 1976 -1 HOFFMAN, A. J.
ON LATTICE POLYHEDRA 2: CONSTRUCTION AND EXAMPLES. IBM RESEARCH REPORT RC 6268. YORKTOWN HEIGHTS: IBM RESEARCH CENTER, 9 P. (1976).

HOFF 1976 -2 HOFFMAN, A. J.
TOTAL UNIMODULARITY AND COMBINATORIAL THEOREMS. LINEAR ALGEBRA AND ITS APPLICATIONS 13, 103-108. (1976).

HOFF 1977 -1 HOFFMAN, A. J.
ON LATTICE POLYHEDRA 3: BLOCKERS AND ANTIBLOCKERS OF LATTICE CLUTTERS. IBM RESEARCH REPORT 27503. YORKTOWN HEIGHTS: IBM RESEARCH CENTER, 18 P. (1977).

HOFF 1977 -2 HOFFMAN, A. J.
ROLE OF UNIMODULARITY IN APPLYING LINEAR INEQUALITIES TO COMBINATORIAL THEOREMS, THE. IBM RESEARCH REPORT RC6650. YORKTOWN HEIGHTS: IBM THOMAS J. WATSON RESEARCH CENTER, 19 P. (1977).

HOFFA 1978 -1 HOFFMAN, A. J.; OPPENHEIM, R.
LOCAL UNIMODULARITY IN THE MATCHING POLYTOPE. IN: ALSPACH, B.;HELL, P.;MILLER, D. J.(EDS.): ALGORITHMIC ASPECTS OF COMBINATORICS. ANNALS OF DISCRETE MATHEMATICS 2, 201-210. (1978).

HOFFB 1976 -1 HOFFMAN, A. J.; SCHWARTZ, D. E.
ON LATTICE POLYHEDRA. IBM RESEARCH RC 5979. YORKTOWN HEIGHTS: IBM RESEARCH CENTER, 8 P. (1976).

HOFFC 1958 -1 HOFFMANN, W.; PAVLEY, R.
APPLICATIONS OF DIGITAL COMPUTERS TO PROBLEMS IN THE STUDY OF VEHICULAR TRAFFIC. IN: PROC. WEST. J. COMP. CONF., 159-161. (1958).

HOLL 1977 -1 HOLLOWAY, C. A.; NELSON, R. T.
INTERMITTEN PRODUCTION SCHEDULING: PRACTICAL FORMULATIONS, RELAXATION, AND DECOMPOSITION PROCEDURES. RESEARCH PAPER NO. 405. STANFORD: STANFORD UNIV., GRADUATE SCHOOL OF BUSINESS, 37 P. (1977).

HOLM 1976 -1 HOLM, S.; KLEIN, D.
PARAMETRIC ANALYSIS FOR INTEGER PROGRAMMING PROBLEMS. PAPER PRESENTED AT THE 49TH JOINT NATIONAL MEETING OF ORSA AND TIMS, PHILADELPHIA 1976, 23 P. (1976).

HOLM 1978 -1 HOLM, S.; KLEIN, D.
DISCRETE RIGHT HAND SIDE PARAMETRIZATION FOR LINEAR INTEGER PROGRAMS. EUROPEAN JOURNAL OF OPERATIONS RESEARCH 2, 50-53. (1978).

HOLMA 1977 -1 HOLMES, J. H.; HAGGETT, P.
GRAPH THEORY INTERPRETATION OF FLOW MATRICES: A NOTE ON MAXIMIZATION PROCEDURES FOR IDENTIFYING SIGNIFICANT LINKS. GEOGRAPHICAL ANALYSIS 9, 388-399. (1977).

****

HOLMB 1976 -1 HOLMES, R. A.; PARKER, R. G.
VEHICLE SCHEDULING PROCEDURE BASED UPON SAVINGS AND A SOLUTION PERTURBATION SCHEME, A. OPERATIONAL RESEARCH QUARTERLY 27, 83-92. (1976).

HOLZ 1975 -1 HOLZMAN, A. G.
MASTER TIME SCHEDULE GENERATOR - AN ALGORITHM APPROACH. PAPER PRESENTED AT THE 48TH JOINT NATIONAL MEETING OF ORSA AND TIMS, LAS VEGAS 1975, 36 P. (1975).

HOLZA 1973 -1 HOLZMANN, C. A.; NORTON, P. G.; TOBEY, M. D.
GRAPHICAL REPRESENTATION OF MATROIDS, A. SIAM JOURNAL ON APPLIED MATHEMATICS 25, 618-627. (1973).

HONG 1974 -1 HONG, M. L.
EXPERIMENTS WITH A JOB SHOP PRODUCTION MODEL COUPLING A LINEAR PROGRAM AND A HEURISTIC SCHEDULING PROCEDURE. DISSERTATION. LOS ANGELES: UNIVERSITY OF CALIFORNIA, GRADUATE SCHOOL OF MANAGEMENT. (1974).

HONGA 1977 -1 HONG, S.
ASYMMETRIC TRAVELLING SALESMAN PROBLEM AND SPANNING ARBORESCENCES OF DIRECTED GRAPHS, THE. IN: R. HENN, P. KALL, B. KORTE, O. KRAFFT ET AL. (EDS.): OPERATIONS RESEARCH VERFAHREN XXV, METHODS OF OPERATIONS RESEARCH, 192-204. MEISENHEIM: HAIN. (1977).

HONGB 1977 -1 HONG, S.; PADBERG, M. W.
NOTE ON THE SYMMETRIC MULTIPLE TRAVELING SALESMAN PROBLEM WITH FIXED CHARGES, A. OPERATIONS RESEARCH 25, 871-873. (1977).

HOPC 1971 -1 HOPCROFT, J. E.
N LOG N ALGORITHM FOR ISOMORPHISM OF PLANAR TRIPLY CONNECTED GRAPHS, A. STANFORD COMPUTER SCIENCE REPORT STAN-CS-71-192. STANFORD. (1971).

HOPCA 1971 -1 HOPCROFT, J. E.; KARP. R. M.
(N POWER 5/2) ALGORITHM FOR MAXIMUM MATCHINGS IN BIPARTITE GRAPHS, A. IN: CONFERENCE RECORD 1971, 12TH ANNUAL SYMPOSIUM ON SWITCHING AND AUTOMATA THEORY, EAST LANSING 1971, 122-125. NORTHRIDGE: INSTITUTE OF ELECTRICAL AND ELECTRONICS ENGINEERS. (1971).

HOPCB 1971 -1 HOPCROFT, J. E.; TARJAN, R. E.
(V POWER 2) ALGORITHM FOR DETERMINING ISOMORPHISM OF PLANAR GRAPHS, A. INFORMATION PROCESSING LETTERS 1, 32-34. (1971).

HOPCB 1972 -1 HOPCROFT, J. E.; TARJAN, R. E.
ISOMORPHISM OF PLANAR GRAPHS. IN: R.E. MILLER, J.W. THATCHER (EDS.): COMPLEXITY OF COMPUTER COMPUTATIONS, PROCEEDINGS, NEW YORK 1972 ,NEW YORK: PLENUM, 131-152. (1972).

HOPCB 1973 -1 HOPCROFT, J. E.; TARJAN, R. E.
DIVIDING A GRAPH INTO TRICONNECTED COMPONENTS. SIAM JOURNAL ON COMPUTING 2, 135-158. (1973).

HOPCC 1974 -1 HOPCROFT, J. E.; WONG, J. K.
LINEAR TIME ALGORITHM FOR ISOMORPHISM OF PLANAR GRAPHS. SIXTH ANNUAL ACM SYMPOSIUM ON THEORY OF COMPUTING, SEATTLE. (1974).

HOR 1976 -1 HOR, C. G.
MAXIMUM FLOWS IN A VERTEX WEIGHTED NETWORK. NANTA MATHEMATICA 9, 10-23. (1976).

HORE 1977 -1 HOREN, J. H.
EFFECT OF SCHEDULING NETWORK TELEVISION PROGRAMS ON AUDIENCE RATINGS, THE. WORKING PAPER SERIES NO. 77-17. IOWA: UNIV. OF IOWA, COLLEGE OF BUSINESS ADMINISTRATION, 26 P. (1977).

HORI 1973 -1 HORIE, Y.; FAN, L. T.
DESTINATION TYPE TRAFFIC ASSIGNMENT WITH AIR QUALITY CONSIDERATION. TRANSPORTATION PLANNING AND TECHNOLOGY 2, 1-19. (1973).

HORN 1972 -1 HORN, R. L. VAN
OPTIMIZING TREE SEARCH SIMULATOR FOR SCHEDULING MAINTENANCE ON AIRCRAFT, AN. WORKING PAPER 72-261. BRUSSELS: EUROPEAN INSTITUTE FOR ADVANCED STUDIES IN MANAGEMENT, 20 P. (1972).

HORO 1976 -1 HOROWITZ, E.; SAHNI, S.
EXACT AND APPROXIMATE ALGORITHMS FOR SCHEDULING NONIDENTICAL PROCESSORS. JOURNAL OF THE ASSOCIATION FOR COMPUTING MACHINERY 23, 317-327. (1976).

HORS 1975 -1 HORST, R.
MAXIMALE NETZWERKFLUESSE MIT KANTENGEWINNEN. OPERATIONS RESEARCH VERFAHREN 21, 113-138. (1975).

HORS 1976 -1 HORST, R.
GRAPHENTHEORETISCHER ALGORITHMUS FUER EINIGE FLUSSPROBLEME IN NETZWERKEN MIT KANTENGEWINNEN, EIN. COMPUTING 17, 121-128. (1976).

HORV 1977 -1 HORVATH, E. C.; LAM, S.; SETHI, R.
LEVEL ALGORITHM FOR PREEMPTIVE SCHEDULING, A. JOURNAL OF THE ASSOCIATION FOR COMPUTING MACHINERY 24, 32-43. (1977).

HOUC 1976 -1 HOUCK, D. J.; VEMUGANTI, R. R.
TRAVELING SALESMAN PROBLEM AND SHORTEST N-PATHS, THE. PAPER PRESENTED AT THE 49TH JOINT NATIONAL MEETING OF ORSA AND TIMS, PHILADELPHIA 1976, 11 P. (1976).

HOUC 1977 -1 HOUCK, D. J.; VEMUGANTI, R. R.
ALGORITHM FOR THE VERTEX PACKING PROBLEM, AN. OPERATIONS RESEARCH 25, 773-787. (1977).

HOWA 1970 -1 HOWARD, D. D.
DYNAMIC LOCATION-ALLOCATION ANALYSIS. DISSERTATION. CAMBRIDGE: MASSACHUSETTS INST. OF TECHNOLOGY, DEPT. OF MECHANICAL ENGINEERING. (1970).

HROU 1975 -1 HROUDA, J.
MODIFIED BENDERS' ALGORITHM (CZECH). EKONOMICKO-MATEMATICKY OBZOR 11, 423-443. (1975).

HROU 1976 -1 HROUDA, J.
BENDERS METHOD AND PARAMETRIZATION OF THE RIGHT-HAND SIDES IN THE MIXED INTEGER LINEAR PROGRAMMING PROBLEM (CZECH). THE. APLIKACE MATEMATYKI 21, 327-364. (1976).

HROU 1977 -1 HROUDA, J.
GENERALIZATION OF THE ALL-INTEGER ALGORITHM OF GOMORY (CZECH), A. EKONOMICKO-MATEMATICKY OBZOR 13, 164-178. (1977).

HSU 1973 -1 HSU, H. T.; HONKANEN, P. A.
FAST MINIMAL STORAGE ALGORITHM FOR DETERMINING ALL THE ELEMENTARY CYCLES OF A GRAPH, A. IN: PROCEEDINGS OF THE SEVENTH ANNUAL PRINCETON CONFERENCE ON INFORMATION SCIENCES AND SYSTEMS, 419-423. (1973).

HU 1965 -1 HU, T. C.
MULTI-TERMINAL SHORTEST PATHS. PAPER ORC 65-11. LA JOLLA: UNIV. OF CALIFORNIA, OPERATIONS RESEARCH CENTRE. (1965).

HU 1966 -1 HU, T. C.
DECOMPOSITION ALGORITHM FOR SHORTEST PATHS IN A NETWORK, A. RESEARCH PAPER RC-1562. YORKTOWN HEIGHTS: IBM, THOMAS J. WATSON RESEARCH CENTER. (1966).

HU 1973 -1 HU, T. C.
OPTIMUM COMMUNICATION SPANNING TREES. MRC TECHNICAL SUMMARY REPORT 1374. MADISON: UNIV. OF WISCONSIN. (1973).

HU A 1976 -1 HU, T. C.; LENARD, M. L.
OPTIMALITY OF A HEURISTIC SOLUTION FOR A CLASS OF KNAPSACK PROBLEMS. OPERATIONS RESEARCH 24, 193-196. (1976).

****

HUAN 1977 -1 HUANG, H.-C.; TROTTER, L. E. JR.
TECHNIQUE FOR DETERMINING BLOCKING AND ANTIBLOCKING POLYHEDRAL DESCRIPTIONS, A. REPORT NO. 7763-OR. BONN: UNIV. BONN, INSTITUT FUER OEKONOMETRIE UND OPERATIONS RESEARCH, 13 P. (1977).

HUBA 1975 -1 HUBAUT, X. L.
STRONGLY REGULAR GRAPHS. DISCRETE MATHEMATICS 13, 357-381. (1975).

HUBI 1975 -1 HUBICKA, E.; SYSLO, M. M.
MINIMAL BASES OF CYCLES OF A GRAPH. IN: M. FIEDLER (ED.): RECENT ADVANCES IN GRAPH THEORY. PROCEEDINGS OF THE SECOND CZECHOSLOVAK SYMPOSIUM, PRAGUE 1974, 283-293. PRAGUE: ACADEMIA. (1975).

HULM 1975 -1 HULME, B. L.
LATTICE ALGEBRA FOR FINDING SIMPLE PATHS AND CUTS IN A GRAPH, A. IN: F. HOFFMAN, R. C. MULLIN, R. B. LEVOW ET AL. (EDS.): PROCEEDINGS OF THE 6TH SOUTHEASTERN CONFERENCE ON COMBINATORICS, GRAPH THEORY, AND COMPUTING, BOCA RATON 1975, 419-428. CONGRESSUS NUMERANTIUM 14, WINNIPEG: UTILITAS MATHEMATICA. (1975).

HULS 1977 -1 HULST, W. G. VAN; LIESHOUT, J. TH. VAN
INVESTMENT/FINANCIAL PLANNING WITH ENDOGENOUS LIFETIMES: A HEURISTIC APPROACH TO MIXED-INTEGER PROGRAMMING. REVUE FRANCAISE D'AUTOMATIQUE, D'INFORMATIQUE ET RECHERCHE OPERATIONNELLE 11, NO. 1, 85-101. (1977).

HULT 1976 -1 HULTZ, J.
ALGORITHMS AND APPLICATIONS FOR GENERALIZED NETWORKS. DISSERTATION. AUSTIN: UNIV. OF TEXAS. (1976).

HULTA 1976 -1 HULTZ, J.; KLINGMAN, D.; RUSSELL, R.
ADVANCED DUAL BASIC FEASIBLE SOLUTION FOR A CLASS OF CAPACITATED GENERALIZED NETWORKS, AN. OPERATIONS RESEARCH 24, 301-313. (1976).

HUME 1960 -1 HUME, J. N. P.
SCHEDULING PRODUCTION IN JOB SHOPS. P59-68, 2 ND PROCEEDINGS OF THE COMPUTER AND DATA PROCESSING SOCIETY CONFERENCE. (1960).

HUNG 1973 -1 HUNG, H. K.
HEURISTIC ALGORITHM FOR THE MULTI-PERIOD FACILITY LOCATION PROBLEM, A. DISSERTATION. AMHERST: UNIV. OF MASSACHUSETTS. (1973).

HUNGA 1976 -1 HUNG, M. S.; FISK, J. C.
ALGORITHM FOR 0-1 MULTIPLE KNAPSACK PROBLEMS, AN. MIAMI (50. ORSA-MEETING + TIMS, NOVEMBER 1976) 13 P. (1976).

HUNGA 1977 -1 HUNG, M. S.; FISK, J. C.
ASSORTMENT PROBLEMS AND EQUIVALENT INTEGER PROGRAMMING FORMULATIONS. ALBANY: STATE UNIV. OF NEW YORK, 11 P. (1977).

HURT 1975 -1 HURTER, A. P. JR.; SCHAEFER, M. K.; WENDELL, R. E.
SOLUTIONS OF CONSTRAINED LOCATION PROBLEMS. MANAGEMENT SCIENCE 22, 51-56. (1975).

HWAN 1976 -1 HWANG, F. K.
ON STEINER MINIMAL TREES WITH RECTILINEAR DISTANCE. SIAM JOURNAL ON APPLIED MATHEMATICS 30, 104-114. (1976).

HWANA 1969 -1 HWANG, F. K.; LIN, S.
ANALYSIS OF FORD AND JOHNSON'S SORTING ALGORITHM, AN. IN: PROCEEDINGS OF THE THIRD ANNUAL PRINCETON CONFERENCE ON INFORMATION SCIENCES AND SYSTEMS, 292-296. (1969).

HYAF 1976 -1 HYAFIL, L.; RIVEST, R. L.
CONSTRUCTING OPTIMAL BINARY DECISION TREES IS NP-COMPLETE. INFORMATION PROCESSING LETTERS 5, 15-17. (1976).

HYUN 1975 -1 HYUN, K. N.
RELIABILITY OPTIMIZATION BY 0-1 PROGRAMMING FOR A SYSTEM WITH SEVERAL FAILURE MODES. IEEE TRANSACTIONS ON RELIABILITY R-24, 206-210. (1975).

HYUNA 1974 -1 HYUN, K. N.; MITSUO, G.; OKUNO, H.
OPTIMIZATION OF THE REDUNDANT ALLOCATION IN A SYSTEM WITH SEVERAL FAILURE MODES BY ZERO-ONE PROGRAMMING. IN: PROCEEDINGS OF THE 7TH HAWAII INTERNATIONAL CONFERENCE ON SYSTEM SCIENCES, HONOLULU 1974, 189-191. NORTH HOLLYWOOD: WESTERN PERIODICALS. (1974).

IAZE 1968 -1 IAZEOLLA, G.
TIME SHARING SYSTEMS: THEORETICAL TREATMENT OF A PRIORITY SCHEDULING ALGORITHM (ITALIAN). CALCOLO 5, 251-275. (1968).

IBAR 1970 -1 IBARAKI, T.
SHORTEST PATH PROBLEMS VISITING SPECIFIED NODES (JAPANESE). ELECTRONICS AND COMMUNICATIONS IN JAPAN 53, PT. A, 639-646. (1970).

IBAR 1972 -1 IBARAKI, T.
FINITE STATE REPRESENTATIONS OF CERTAIN DISCRETE OPTIMIZATION PROBLEMS. SYSTEMS-COMPUTERS-CONTROLS 3, NO. 6, 71-78. (1972).

IBAR 1972 -2 IBARAKI, T.
REPRESENTATION THEOREMS FOR EQUIVALENT OPTIMIZATION PROBLEMS. INFORMATION AND CONTROL 21, NO. 5, 397-435. (1972).

IBAR 1973 -1 IBARAKI, T.
SOLVABLE CLASSES OF DISCRETE DYNAMIC PROGRAMMING. JOURNAL OF MATHEMATICAL ANALYSIS AND APPLICATION 43, 642-693. (1973).

IBAR 1976 -1 IBARAKI, T.
COMPUTATIONAL EFFICIENCY OF APPROXIMATE BRANCH-AND-BOUND ALGORITHMS. MATHEMATICS OF OPERATIONS RESEARCH 1, 287-298. (1976).

IBAR 1976 -2 IBARAKI, T.
INTEGER PROGRAMMING FORMULATION OF COMBINATORIAL OPTIMIZATION PROBLEMS. DISCRETE MATHEMATICS 16, 39-52. (1976).

IBAR 1976 -3 IBARAKI, T.
THEORETICAL COMPARISONS OF SEARCH STRATEGIES IN BRANCH-AND-BOUND ALGORITHMS. INTERNATIONAL JOURNAL OF COMPUTER AND INFORMATION SCIENCES 5, 315-344. (1976).

IBAR 1977 -1 IBARAKI, T.
BRANCH-AND-BOUND ENUMERATION. VANCOUVER (CONFERENCE ON DISCRETE OPTIMIZATION, AUGUST 1977) 17 P. (1977).

IBAR 1977 -2 IBARAKI, T.
POWER OF DOMINANCE RELATIONS IN BRANCH-AND-BOUND ALGORITHMS, THE. JOURNAL OF THE ASSOCIATION FOR COMPUTING MACHINERY 24, 264-279. (1977).

IBARA 1976 -1 IBARAKI, T.; ISHII, H.; MINE, H.
ASSIGNMENT PROBLEM ON A NETWORK, AN. JOURNAL OF THE OPERATIONS RESEARCH SOCIETY OF JAPAN 19, NO. 1, 70-90. (1976).

IBARB 1974 -1 IBARRA, O.; SAHNI, S.
P-COMPLETE PROBLEMS AND APPROXIMATE SOLUTIONS. TECHNICAL REPORT 74-5, UNIVERSITY OF MINNESOTA. (1974).

IBARC 1974 -1 IBARRA, O. H.; KIM, C. E.
FAST APPROXIMATION ALGORITHMS FOR THE KNAPSACK AND SUM OF SUBSET PROBLEMS. TECHNICAL REPORT 74-13. MINNEAPOLIS: UNIV. OF MINNESOTA, INST. OF TECHNOLOGY, 15 P. (1974).

IBARC 1975 -1 IBARRA, O. H.; KIM, C. E.
ON TWO-PROCESSOR SCHEDULING OF ONE- OR TWO-UNIT TIME TASKS WITH PRECEDENCE CONSTRAINTS. JOURNAL OF CYBERNETICS 5, 87-109. (1975).

IBARC 1977 -1 IBARRA, O. H.; KIM, C. E.
HEURISTIC ALGORITHMS FOR SCHEDULING INDEPENDENT TASKS ON NONIDENTICAL PROCESSORS. JOURNAL OF THE ASSOCIATION FOR COMPUTING MACHINERY 24, 280-289. (1977).

IBM. 1972 -1 IBM.
INTRODUCTION TO MODELLING USING MIXED INTEGER PROGRAMMING, AN. IBM GENERAL INFORMATION MANUAL. AMSTERDAM: IBM NEDERLAND N. V., 63 P. (1972).

IGUD 1975 -1 IGUDIN, R. V.
PROBLEMS OF SEQUENCING THEORY IN TRANSPORTATION AND ALGORITHMS FOR THEIR SOLUTION (RUSSIAN). EKONOMIKA I MATEMATICESKIE METODY 11, 491-499. (1975).

IMOM 1976 -1 IMOMNAZAROV, B.
MAXIMAL FLOW AND MINIMAL CUT PROBLEM (RUSSIAN), THE. DOKLADY AKADEMII NAUK TADZIKSKOI SSR 19, NO. 9, 11-14. (1976).

INGA 1975 -1 INGARGIOLA, G.; KORSH, J. F.
ALGORITHM FOR THE SOLUTION OF 0-1 LOADING PROBLEMS, AN. OPERATIONS RESEARCH 23, 1110-1119. (1975).

INGAA 1977 -1 INGARGIOLA, G. P.; KORSH, J. F.
GENERAL ALGORITHM FOR ONE-DIMENSIONAL KNAPSACK PROBLEMS, A. OPERATIONS RESEARCH 25, 752-759. (1977).

INGL 1976 -1 INGLETON, A. W.
NON BASE ORDERABLE MATROIDS. IN: C. ST. J. A. NASH-WILLIAMS (ED.): PROCEEDINGS OF THE 5TH BRITISH COMBINATORIAL CONFERENCE, ABERDEEN 1975, 355-360. CONGRESSUS NUMERANTIUM 15, WINNIPEG: UTILITAS MATHEMATICA. (1976).

INGL 1977 -1 INGLETON, A. W.
TRANSVERSAL MATROIDS AND RELATED STRUCTURES. IN: M. AIGNER (ED.): HIGHER COMBINATORICS, PROCEEDINGS OF THE NATO ADVANCED STUDY INSTITUTE, BERLIN 1976, 117-132. DORDRECHT: REIDEL. (1977).

INTR 1976 -1 INTRATOR, J.; LEV, B.
METHODS FOR IDENTIFICATION OF VANISHING VARIABLES IN TRANSPORTATION PROBLEMS AND ITS POSSIBLE APPLICATIONS. COMPUTERS AND OPERATIONS RESEARCH 3, 295-304. (1976).

INTRA 1977 -1 INTRATOR, J.; PAROUSH, J.
SENSITIVITY ANALYSIS OF THE CLASSICAL TRANSPORTATION PROBLEM. A COMBINATORIAL APPROACH. COMPUTERS AND OPERATIONS RESEARCH 4, 213-226. (1977).

IORD 1974 -1 IORDANSKII, M. A.
MINIMAL NUMERATIONS OF THE VERTICES OF TREES (RUSSIAN). DOKLADY AKADEMII NAUK SSSR 218, 272-275; TRANSLATED AS SOVIET MATHEMATICS. DOKLADY 218, 1974, NO. 5, 1311-1315. (1974).

IORD 1976 -1 IORDANSKII, M. A.
MINIMAL NUMBERINGS OF THE VERTICES OF TREES (RUSSIAN). PROBLEMY KIBERNETIKI, NO. 31, 109-132. (1976).

IRI 1960 -1 IRI, M.
NEW METHOD OF SOLVING TRANSPORTATION-NETWORK PROBLEMS, A. JOURNAL OF THE OPERATIONS RESEARCH SOCIETY OF JAPAN 3, 27-87. (1960).

IRI 1971 -1 IRI, M.
ON AN EXTENSION OF THE MAXIMUM-FLOW MINIMUM-CUT THEOREM TO MULTICOMMODITY FLOWS. JOURNAL OF THE OPERATIONS RESEARCH SOCIETY OF JAPAN 13, 129-135. (1971).

IRI A 1975 -1 IRI, M.; TOMIZAWA, N.
UNIFYING APPROACH TO FUNDAMENTAL PROBLEMS IN NETWORK THEORY BY MEANS OF MATROIDS (JAPANESE), A. THE TRANSACTIONS OF THE INSTITUTE OF ELECTRONICS AND COMMUNICATION ENGINEERS OF JAPAN 58-A, 33-40. (1975).

IRI A 1976 -1 IRI, M.; TOMIZAWA, N.
ALGORITHM FOR FINDING AN OPTIMAL "INDEPENDENT ASSIGNMENT", AN. JOURNAL OF THE OPERATIONS RESEARCH SOCIETY OF JAPAN 19, NO. 1, 32-57. (1976).

IRWI 1971 -1 IRWIN, N. A.; DODD, N.; CUBE, H. VON
CAPACITY RESTRAINT IN ASSIGNMENT PROGRAMS. HIGHWAY RESEARCH BOARD, BULLETIN 297,. (1971).

ISHI 1975 -1 ISHII, H.; IBARAKI, T.; MINE, H.
CERTAIN TRAVELING SALESMAN PROBLEM THAT HAS AN OBJECTIVE FUNCTION OF MINIMAX TYPE (JAPANESE), A. KEIEI-KAGAKU 19, NO. 3-4, 112-122. (1975).

ISHI 1975 -2 ISHII, H.; IBARAKI, T.; MINE, H.
TRAVELING SALESMAN PROBLEMS WITH A CAPACITY CONSTRAINT OF THE DELIVERY TRUCK. MEMOIRS OF THE FACULTY OF ENGINEERING, KYOTO UNIV. 37, 252-263. (1975).

ISHI 1976 -1 ISHII, H.; IBARAKI, T.; MINE, H.
PRIMAL CUTTING PLANE ALGORITHM FOR INTEGER FRACTIONAL PROGRAMMING PROBLEMS, A. JOURNAL OF THE OPERATIONS RESEARCH SOCIETY OF JAPAN 19, 228-244. (1976).

ISHI 1977 -1 ISHII, H.; IBARAKI, T.; MINE, H.
FRACTIONAL KNAPSACK PROBLEMS. MATHEMATICAL PROGRAMMING 13, 255-271. (1977).

ITAI 1977 -1 ITAI, A.; RODEH, M.
FINDING A MINIMUM CIRCUIT IN A GRAPH. IN: PROCEEDINGS OF THE NINTH ANNUAL A.C.M. SYMPOSIUM ON THEORY OF COMPUTING, 1-10. (1977).

ITAI 1977 -2 ITAI, A.; RODEH, M.
SOME MATCHING PROBLEMS. IN: A. SALOMAA, M. STEINBY (EDS.): AUTOMATA, LANGUAGES AND PROGRAMMING, 4TH COLLOQUIUM, TURKU 1977, 258-268. LECTURE NOTES IN COMPUTER SCIENCE 52, BERLIN: SPRINGER. (1977).

IVAN 1976 -1 IVANCHEV, M. D.
ON ONE CLASS OF THE QUADRATIC ASSIGNMENT PROBLEM (RUSSIAN). ZURNAL VYCISLITEL'NOJ MATEMATIKI I MATEMATICESKOJ FIZIKI 16, 1348-1352. (1976).

IVANA 1976 -1 IVANCHEV, M. D.; MITEV, I. G.; JANEV, N. I.
REALIZATION OF THE BRANCH AND BOUND METHOD FOR THE SOLUTION OF THE GENERAL MIXED INTEGER PROGRAMMING PROBLEM (RUSSIAN), A. ZURNAL VYCISLITEL'NOI MATEMATIKI I MATEMATICESKOI FIZIKI 16, 788-192 + 823. (1976).

IVANB 1974 -1 IVANISCEV, V. V.
DECOMPOSITION METHOD FOR THE SOLUTION OF THE PROBLEM OF THE OPTIMAL USE OF MEANS OF TRANSPORTATION ON A NETWORK (RUSSIAN), A. IN: MATHEMATICAL METHODS OF SOLUTION OF ECONOMIC PROBLEMS, NO. 5 (RUSSIAN), SUPPL. TO EKONOMIKA I MATEMATICESKIE METODY, 72-79. MOSCOW: IZDAT. NAUKA. (1974).

IVANC 1974 -1 IVANOV, N. N.
REDUCTION OF A CLASS OF INTEGER LINEAR PROGRAMMING PROBLEMS TO A COVERING PROBLEM (RUSSIAN). VESCI AKADEMII NAVUK BSSR. SERYJA FIZIKA-MATEMATYCNYH NAVUK, NO. 4, 100-101 + 142. (1974).

IVANC 1975 -1 IVANOV, N. N.
CERTAIN METHOD FOR AGGREGATION OF AN INTEGER LINEAR PROGRAMMING PROBLEM (RUSSIAN), A. VESCI ACADEMII NAVUK BSSR, SERYJA FIZIKA-MATEMATYCNYH NAVUK, NO. 2, 96-98 + 141. (1975).

JACO 1973 -1 JACOB, H.
ZUR STANDORTWAHL DER UNTERNEHMUNGEN. WIESBADEN: GABLER, 293 P. (1973).

****

JACOA 1974 -1 JACOBSEN, S. K.
ALGORITHM FOR SMALL-SCALE DYNAMIC PLANT-LOCATION PROBLEMS, AN. COPENHAGEN: THE TECHNICAL UNIVERSITY OF DENMARK. (1974).

JACOA 1978 -1 JACOBSEN, S. K.
ON THE USE OF TREE-INDEXING METHODS IN TRANSPORTATION ALGORITHMS. EUROPEAN JOURNAL OF OPERATIONS RESEARCH 2, 54-65. (1978).

JACOB 1977 -1 JACOBSON, R. E.; SADOWSKI, R. F.
SCHEDULING ALGORITHM FOR THE UNBALANCED PRODUCTION LINE, A. AIIE TRANSACTIONS 9, 36-43. (1977).

JAER 1973 -1 JAERVINEN, P.
ON A ROUTING PROBLEM IN A GRAPH WITH TWO TYPES OF COSTS ASSOCIATED WITH THE EDGES. ACTA UNIVERSITATIS TAMPERENSIS, SERIA A, 50, 55 P. (1973).

JAIN 1975 -1 JAIN, R. C.
ON PSEUDO-BOOLEAN INEQUALITIES-CONSISTENCY AND SOLUTIONS. ACTA CIÉNCIA INDICA 1, 347-352. (1975).

JAINA 1975 -1 JAIN, S. K.
SIMULATION-BASED SCHEDULING AND MANAGEMENT INFORMATION SYSTEM FOR A MACHINE SHOP, A. INTERFACES 6, NO 1, PART 2, 81-96. (1975).

JAKO 1974 -1 JAKOVLEVA, M. A.
TRANSPORTATION PROBLEM WITH BORDERING (RUSSIAN), THE. OPTIMICIJA, VYP. 15, 79-89 + 185. (1974).

JAKS 1974 -1 JAKSCH, H. J.
THEORIE LINEARER WIRTSCHAFTSMODELLE 1. MATHEMATISCHE GRUNDLAGEN. UNI-TASCHENBUECHER, 293. TUEBINGEN: MOHR, 122 P. (1974).

JAKS 1975 -1 JAKSCH, H. J.
THEORIE LINEARER WIRTSCHAFTSMODELLE 2. TRANSPORT- UND STANDORTMODELLE. UNI-TASCHENBUECHER, 491. TUEBINGEN: MOHR, 146 P. (1975).

JAME 1976 -1 JAMES, K. R.; RIHA, W.
ALGORITHM 28. ALGORITHM FOR GENERATING GRAPHS OF A GIVEN PARTITION. COMPUTING 16, 153-161. (1976).

JAND 1969 -1 JANDY, G.
PROBLEMS OF LOCATION IN OPERATIONS RESEARCH MODELS. PAPERS OF THE REGIONAL SCIENCE ASSOCIATION 22, 195-202. (1969).

JANI 1976 -1 JANICKE, W.
NEUER ALGORITHMUS ZUR BESTIMMUNG EINER BASIS VON UNABHAENGIGEN ELEMENTARKREISEN IN EINEM STARK ZUSAMMENHAENGENDEN GRAPHEN, EIN. ZASTTOSOWANIA MATEMATYKI 15, 315-326. (1976).

JARA 1976 -1 JARANCEV, JU. A.
CERTAIN PROCEDURE FOR DYNAMIC MEMORY ALLOCATION IN COMPUTERS (RUSSIAN), A. AVTOMATIKA I TELEMEHANIKA, NO. 3, 161-172; TRANSLATED AS AUTOMATION AND REMOTE CONTROL 37, 1976, 457-467. (1976).

JASI 1976 -1 JASINSKA, E.; RYSZKOWSKA, E.
WAREHOUSE LOCATION IN A TRANSPORTATION PROBLEM OF TWO STAGES (POLISH). PRZEGLAD STATYSTYCZNY 23, 163-170. (1976).

JEFF 1965 -1 JEFFERIS, R. P.; FEGLEY, K. A.
APPLICATION OF DYNAMIC PROGRAMMING TO ROUTING PROBLEMS. IEEE TRANSACTIONS ON SYSTEMS, MAN, AND CYBERNETICS, 21-26. (1965).

JENK 1974 -1 JENKYNS, T. A.
MATCHOIDS: A GENERALIZATION OF MATCHINGS AND MATROIDS. DISSERTATION. WATERLOO: UNIV. OF WATERLOO. (1974).

JENK 1976 -1 JENKYNS, T. A.
EFFICACY OF THE "GREEDY" ALGORITHM, THE. IN: F. HOFFMAN, L. LESNIAK, R. MULLIN ET AL. (EDS.): PROCEEDINGS OF THE 7TH SOUTHEASTERN CONFERENCE ON COMBINATORICS, GRAPH THEORY, AND COMPUTING, BATON ROUGE 1976, 341-350. CONGRESSUS NUMERANTIUM 17, WINNIPEG: UTILITAS MATHEMATICA. (1976).

JENK 1977 -1 JENKYNS, T. A.
GREEDY TRAVELLING SALESMAN PROBLEM, THE. ST. CATHARINS: BROCK UNIVERSITY, 23 P. (1977).

JENK 1977 -2 JENKYNS, T. A.
P-SYSTEMS AS INTERSECTIONS OF MATROIDS. ST. CATHARINS: BROCK UNIVERSITY, 4 P. (1977).

JENS 1975 -1 JENSCH, P.
DYNAMISCHE OPTIMIERUNG FUER VERKEHRSNETZE MIT ZEIT- UND ORTSABHAENGIGEN ZUSTANDSGROESSEN. ZEITSCHRIFT FUER ANGEWANDTE MATHEMATIK UNG MECHANIK 55, NO. 4, T278-T279. (1975).

JERO 1977 -1 JEROSLOW, R.
BRACKETING DISCRETE PROBLEMS BY TWO PROBLEMS OF LINEAR OPTIMIZATION. IN: R. HENN, P. KALL, B. KORTE, O. KRAFFT ET AL. (EDS.): OPERATIONS RESEARCH VERFAHREN XXV, METHODES OF OPERATIONS RESEARCH, 205-216. MEISENHEIM: HAIN. (1977).

JERO 1977 -2 JEROSLOW, R.
INTRODUCTION TO THE THEORY OF CUTTING-PLANES, AN. VANCOUVER (CONFERENCE ON DISCRETE OPTIMIZATION, AUGUST 1977) 49 P. (1977).

JEROA 1977 -1 JEROSLOW, R. G.
CUTTING-PLANE GAME AND ITS ALGORITHMS, A. CORE DISCUSSION PAPER, 7724. HEVERLEE: CENTER FOR OPERATIONS RESEARCH AND ECONOMETRICS, 38 P. (1977).

JEROA 1977 -2 JEROSLOW, R. G.
CUTTING-PLANE THEORY: DISJUNCTIVE METHODS. IN: P. L. HAMMER, E. L. JOHNSON, B. H. KORTE, G. L. NEMHAUSER (EDS.): STUDIES IN INTEGER PROGRAMMING, 293-330. ANNALS OF DISCRETE MATHEMATICS 1, AMSTERDAM: NORTH-HOLLAND. (1977).

JEROA 1977 -3 JEROSLOW, R. G.
CUTTING-PLANES FOR COMPLEMENTARITY CONSTRAINTS. CORE DISCUSSION PAPER, 7707. HEVERLEE: CENTER FOR OPERATIONS RESEARCH AND ECONOMETRICS, 16 P. (1977).

JEWE 1967 -1 JEWELL, W.
MODELS FOR THE TRAFFIC ASSIGNMENT. TRANSPORTATION RESEARCH 1, 31-46. (1967).

JOHN 1976 -1 JOHNSON, C. R.; LEIGHTON, F. T.
EFFICIENT LINEAR ALGEBRAIC ALGORITHM FOR THE DETERMINATION OF ISOMORPHISM IN PAIRS OF UNDIRECTED GRAPHS, AN. JOURNAL OF RESEARCH OF THE NATIONAL BUREAU OF STANDARDS 80B, 447-484. (1976).

JOHNA 1973 -1 JOHNSON, D. B.
ALGORITHMS FOR SHORTEST PATHS. DISSERTATION. ITHACA: CORNELL UNIVERSITY. (1973).

JOHNA 1975 -1 JOHNSON, D. B.
FINDING ALL THE ELEMENTARY CIRCUITS OF A DIRECTED GRAPH. SIAM JOURNAL ON COMPUTING 4, 77-84. (1975).

JOHNA 1977 -1 JOHNSON, D. B.
EFFICIENT ALGORITHMS FOR SHORTEST PATHS IN SPARSE NETWORKS. JOURNAL OF THE ASSOCIATION FOR COMPUTING MACHINERY 24, 1-13. (1977).

JOHNB 1962 -1 JOHNSON, D. M.; DULMAGE, A. L.; MENDELSOHN, N. S.
CONNECTIVITY AND REDUCIBILITY OF GRAPHS. CANADIAN JOURNAL OF MATHEMATICS 14, 529-539. (1962).

****

JOHNC 1974 -1 JOHNSON, D. S.
WORST CASE BEHAVIOR OF GRAPH COLORING ALGORITHMS. IN: F. HOFFMAN, R. A. KINGSLEY, R. B. LEVOW ET AL. (EDS.): PROCEEDINGS OF THE 5TH SOUTHEASTERN CONFERENCE ON COMBINATORICS, GRAPH THEORY, AND COMPUTING, BOCA RATON 1974, 513-528. CONGRESSUS NUMERANTIUM 10, WINNIPEG: UTILITAS MATHEMATICA. (1974).

JOHNC 1978 -1 JOHNSON, D. S.
GRAPH COLORING ALGORITHM: BETWEEN A ROCK AND A HARD PLACE?(ABSTRACT). IN: ALSPACH, B.;HELL, P.;MILLER, D. J.(EDS.): ALGORITHMIC ASPECTS OF COMBINATORICS. ANNALS OF DISCRETE MATHEMATICS 2, 245. (1978).

JOHND 1977 -1 JOHNSON, D. S.; LENSTRA, J. K.; RINNOOY KAN, A. H. G.
COMPLEXITY OF THE NETWORK DESIGN PROBLEM, THE. BW 70/77. AMSTERDAM: MATHEMATISCH CENTRUM, 5 P. (1977).

JOHNE 1964 -1 JOHNSON, E. L.
NETWORK FLOWS, GRAPHS, AND INTEGER PROGRAMMING. BERKELEY: UNIV. OF CALIFORNIA, OPERATIONS RESEARCH CENTER, P 1-59 (1964).

JOHNE 1975 -1 JOHNSON, E. L.
FLOWS IN NETWORKS. YORKTOWN HEIGHTS: IBM RESEARCH CENTER, 37 P. (1975).

JOHNE 1975 -2 JOHNSON, E. L.
ON CUT-SET INTEGER POLYHEDRA. CAHIERS DU CENTRE D'ETUDES DE RECHERCHE OPERATIONNELLE 17, 235-251. (1975).

JOHNE 1976 -1 JOHNSON, E. L.
FACES OF POLYHEDRA OF MIXED INTEGER PROGRAMMING PROBLEMS. IN: PROCEEDINGS OF THE CONFERENCE ON MATHEMATICAL PROGRAMMING AND ITS APPLICATIONS, INDAM, ROME 1974 (ITALIAN), 289-299. SYMPOSIA MATHEMATICA 19, LONDON: ACADEMIC PRESS. (1976).

JOHNE 1976 -2 JOHNSON, E. L.
GUIDE TO SUCCESSFULLY SOLVING INTEGER PROGRAMMING PROBLEMS, A. IN: H. P. CROWDER (ED.): 1976 IBM SYMPOSIUM ON MATHEMATICS AND COMPUTATION, PROCEEDINGS, YORKTOWN HEIGTS 1976, 71-73. IBM FORM 26690, YORKTOWN HEIGHTS: IBM THOMAS J. WATSON RESEARCH CENTER. (1976).

JOHNE 1976 -3 JOHNSON, E. L.
NOTE ON MUTUALLY EXCLUSIVE SCHEDULING AND CRITICAL PATHS, A. PAPER PRESENTED AT 7TH SOUTHEASTERN CONFERENCE ON COMBINATORICS, GRAPH THEORY, AND COMPUTING, BATON ROUGE 1976. (1976).

JOHNE 1977 -1 JOHNSON, E. L.
ON THE GROUP PROBLEM AND A SUBADDITIVE APPROACH TO INTEGER PROGRAMMING. VANCOUVER (CONFERENCE ON DISCRETE OPTIMIZATION, AUGUST 1977) 25 P. (1977).

JOHNE 1978 -1 JOHNSON, E. L.
ON THE EDGE-COLORING PROPERTY FOR THE CLOSURE OF THE COMPLETE HYPERGRAPHS. IN: ALSPACH, B.; HELL, P.;MILLER, D. J.(EDS.): ALGORITHMIC ASPECTS OF COMBINATORICS. ANNALS OF DISCRETE MATHEMATICS 2, 161-172. (1978).

JOHNF 1978 -1 JOHNSON, E. L.; NEWMAN, D.; WINSTON, K.
INEQUALITY ON BINOMIAL COEFFICIENTS, AN. IN: ALSPACH, B.;HELL, P.;MILLER, D. J.(EDS.): ALGORITHMIC ASPECTS OF COMBINATORICS. ANNALS OF DISCRETE MATHEMATICS 2, 155-160. (1978).

JOHNG 1976 -1 JOHNSON, E. L.; SPIELBERG, K.
SOLUTION OF A MULTI-COMMODITY (NO-SPLITTING) PRODUCTION-DISTRIBUTION PROBLEM. WORKING PAPER. WHITE PLAINS: IBM SCIENTIFIC MARKETING. (1976).

JOHNH 1976 -1 JOHNSON, L. F.
DETERMINING CLIQUES OF A GRAPH. IN: B. L. HARTNELL, H. C. WILLIAMS (EDS): PROCEEDINGS OF THE 5TH MANITOBA CONFERENCE ON NUMERICAL MATHEMATICS, WINNIPEG 1975, 429-437. CONGRESSUS NUMERANTIUM 16, WINNIPEG: UTILITAS MATHEMATICA. (1976).

JOHNI 1974 -1 JOHNSON, P. A.
HEURISTIC FOR INTEGRATING PLANT LOCATION AND PLANT TIMING QUESTIONS, A. PAPER PRESENTED AT THE 46TH MEETING OF ORSA AND TIMS, SAN JUAN 1974. (1974).

JOHNJ 1958 -1 JOHNSON, R. E.
USE OF OPERATIONS RESEARCH TECHNIQUES IN PLANT LOCATION PROBLEMS, THE. IN: AMERICAN MANAGEMENT ASSOCIATION (EDS.): OPERATIONS RESEARCH RECONSIDERED, 120-127. NEW YORK: AMERICAN MANAGEMENT ASSOCIATION. (1958).

JOHNK 1970 -1 JOHNSON, S. C.; KERNIGHAN, B. W.
MAKING CHANGE WITH A MINIMUM NUMBER OF COINS. BELL TELEPHONE LABORATORIES, 25 P. (1970).

JOLI 1975 -1 JOLIVET, J.-L.
PROBLEMES DE CONNEXITE ET PROBLEMES HAMILTONIENS EN THEORIE DES GRAPHES. JOINT D'UNE FAMILLE DE GRAPHES ET GENERALISATION DE LA NOTION DE GRAPHE PARFAIT. DISSERTATION. PARIS: UNIV. PARIS VI, 153 P. (1975).

JONE 1973 -1 JONES, N. D.
PRELIMINARY REPORT: REDUCIBILITY AMONG COMBINATORIAL PROBLEMS IN LOG N SPACE. IN: PROCEEDINGS OF THE SEVENTH ANNUAL PRINCETON CONFERENCE ON INFORMATION SCIENCES AND SYSTEMS, 547-551. (1973).

JUCK 1976 -1 JUCKER, J. V.; CARLSON, R. C.
SIMPLE PLANT-LOCATION PROBLEM UNDER UNCERTAINTY, THE. OPERATIONS RESEARCH 24, 1045-1055. (1976).

JUCO 1974 -1 JUCOVIC, E.; OLEJNIK, F.
ON CHROMATIC AND ACHROMATIC NUMBERS OF UNIFORM HYPERGRAPHS. CASOPIS PRO PESTOVANI MATEMATIKY 99, 123-130. (1974).

JUDI 1976 -1 JUDIN, D. B.; NEMIROVSKY, A. S.
INFORMATIONAL COMPLEXITY ESTIMATION OF MATHEMATICAL PROGRAMMING PROBLEMS (RUSSIAN). EKONOMIKA I MATEMATICHESKIE METODY 12, 128-142. (1976).

JUEN 1976 -1 JUENICKE, W.
NEUER ALGORITHMUS ZUR BESTIMMUNG EINER BASIS VON UNABHAENGIGEN ELEMENTARKREISEN IN EINEM STARK ZUSAMMENHAENGENDEN GRAPHEN, EIN. ZASTOSOWANIA MATEMATYKI 15, 315-323. (1976).

JULL 1976 -1 JULLIEN, B.
DIFFERENTES APPROCHES LOGICIELLES POUR LA RESOLUTION DES PROBLEMES COMBINATOIRES EN TEMPS REEL. DISSERTATION. TOULOUSE: UNIV. TOULOUSE, 117 P. (1976).

JUNG 1976 -1 JUNGINGER, W.
NUMERISCHE ERFAHRUNGEN BEI DER LOESUNG DES DREIDIMENSIONALEN TRANSPORTPROBLEMS. COMPUTING 17, 129-146. (1976).

JUSK 1975 -1 JUSKA, F.
NUMERICAL METHOD FOR THE CONSTRUCTION OF A NETWORK OF MINIMAL LENGTH (RUSSIAN), A. TEOR. OPTIMAL. RESENII-TRUDY SEM. PROCESSY OPTIMAL. UPRAVLENIJA VI SEKCIJA VYP. 1, 53-65. (1975).

KAAZ 1973 -1 KAAZIK, YU. YA.; PREEM, M. K.
PROBLEM OF FINDING ROUTES (RUSSIAN), A. TRUDY VYCISLITEL'NOGO CENTRA. TARTUSKII GOSUDARSTVENNYI UNIVERSITET VYP. 28, 3-18. (1973).

KACP 1975 -1 KACPRZYK, J.
APPLICATION OF THE THEORY OF FUZZY SETS TO THE OPTIMAL ASSIGNMENT OF THE STAFF (POLISH). ARCHIWUM AUTOMATYKI I TELEMECHANIKI 20, 247-260. (1975).

****

KACPA 1975 -1 KACPRZYK, J.; STANCZAK, W.
APPLICATION OF THE METHOD OF MINIMAL GROUPS TO THE SOLUTION OF THE PROBLEM OF PARTITIONING A GROUP OF FIRMS INTO SUB-GROUPS (POLISH). ARCHIWUM AUTOMATYKI I TELEMECHANIKI 20, 513-526. (1975).

KACZ 1976 -1 KACZMARSKI, J.
NOMBRE DE STABILITE ET CERTAINES PROPRIETES DE GRAPHES CRITIQUES. IN: ANAL. APPL. INF. JOURN. COMB. INF., BORDEAUX 1975, 237-243. BORDEAUX: UNIV. BORDEAUX I. (1976).

KACZ 1976 -2 KACZMARSKI, J.
STABILITY OF THE SOLUTIONS OF THE TRANSPORT PROBLEM (POLISH), THE. ZESZYTY NAUKOWE POLITECHNIKI SLASKIEJ NO. 441, MATEMATYKA-FIZYKA ZESZYT 26, 99-108. (1976).

KACZ 1976 -3 KACZMARSKI, J.
TRANSPORT PROBLEM IN CONNECTION WITH THE CONSTRUCTION OF A NETWORK (POLISH), THE. ZESZYTY NAUKOWE POLITECHNIKI SLASKIEJ NO. 441, MATEMATYKA-FIZYKA ZESZYT 26, 81-98. (1976).

KAES 1976 -1 KAESTING, H.
LEXIKOGRAPHISCHES SUCHVERFAHREN ZUR LOESUNG DES BIVALENTEN KNAPSACK-PROBLEMS, EIN. IN: J. KOHLAS, O. SEIFERT, P. STAEHLY, H.-J. ZIMMERMANN (EDS): PROCEEDINGS IN OPERATIONS RESEARCH 5, 41-42. WUERZBURG: PHYSICA. (1976).

KAFU 1977 -1 KAFURA, D. G.; SHEN, V. Y.
TASK SCHEDULING ON A MULTIPROCESSOR SYSTEM WITH INDEPENDENT MEMORIES. SIAM JOURNAL ON COMPUTING 6, 167-187. (1977).

KALD 1975 -1 KALDYBAEV, S. U.
ALGORITHM FOR THE APPROXIMATION-COMBINATORIAL METHOD IN SOLVING AN ALLOCATION PROBLEM USING BOOLEAN VARIABLES (RUSSIAN), A. IZVESTIJA AKADEMII NAUK KAZAHSKOI SSR, SERIJA FIZIKO-MATEMATICESKAJA, NO. 5, 43-49 + 93. (1975).

KALL 1976 -1 KALLINA, C.; LYNN, J.
OPTIMALITY OF THE CUBE-PER-ORDER-INDEX RULE FOR STOCK LOCATION IN A DISTRIBUTION WAREHOUSE. PAPER PRESENTED AT THE 49TH JOINT NATIONAL MEETING OF ORSA AND TIMS, PHILADELPHIA 1976, 21 P. (1976).

KALM 1970 -1 KALMANSON, K.
CLASSES OF COMBINATORIAL EXTREMA IN CERTAIN METRIC SPACES. DISSERTATION. NEW YORK: CITY UNIV. OF NEW YORK. (1970).

KALM 1972 -1 KALMANSON, K.
ANALYSIS OF EXTREME HAMILTONIAN CIRCUITS, AN. TECHNICAL REPORT. UPPER MOUNTCLAIR: MOUNTCLAIR STATE COLLEGE. (1972).

KALM 1975 -1 KALMANSON, K.
EDGECONVEX CIRCUITS AND THE TRAVELING SALESMAN PROBLEM. CANADIAN JOURNAL OF MATHEMATICS 27, 1000-1010. (1975).

KALY 1976 -1 KALYUZHNYJ, B. I.; SOLDATOV, V. S.
MODEL OF SEQUENCING DISCRETE PRODUCTION PROCESSES (RUSSIAN), A. IZVESTIJA AKADEMII NAUK SSSR, TEHNICESKAJA KIBERNETIKA, NO. 6, 21-26. (1976).

KAME 1976 -1 KAMEDA, T.
ON MAXIMALLY DISTANT SPANNING TREES OF A GRAPH. COMPUTING 17, 115-120. (1976).

KAMI 1975 -1 KAMINSKAS, V.; PUKAS, V.
OPTIMAL MEAN LOAD SCHEDULING IN A POWER PLANT (RUSSIAN). STATISTICESKIE PROBLEMY UPRAVLENIJA. TRUDY SEMINARA "PROCESSY OPTIMAL'NOGO UPRAVLENIJA" V SEKCIJA, VYP. 10, 85-94. (1975).

KANN 1978 -1 KANNAN, R.; MONMA, C. L.
ON THE COMPUTATIONAL COMPLEXITY OF INTEGER PROGRAMMING PROBLEMS. IN: R. HENN, B. KORTE, W. OETTLI (EDS.): OPTIMIZATION AND OPERATIONS RESEARCH, PROCEEDINGS, BONN 1977. LECTURE NOTES IN ECONOMICS AND MATHEMATICAL SYSTEMS 157. BERLIN, HEIDELBERG, NEW YORK: SPRINGER, 161-172. (1978).

KANT 1973 -1 KANT, E.
MULTICOMMODITY FLOW ALGORITHM, A. IN: PROCEEDINGS OF THE SEVENTH ANNUAL PRINCETON CONFERENCE ON INFORMATION SCIENCES AND SYSTEMS, 120-124. (1973).

KAPL 1977 -1 KAPLAN, E. L.
ALGORITHMS FOR SOLUTION OF TRANSPORTATION PRODUCTION PROBLEMS (RUSSIAN). KIBERNETIKA, NO. 4, 136-141. (1977).

KAPLA 1976 -1 KAPLINSKII, A. I.; OHERNYSOVA, G. D.
METHOD OF ADAPTIVE ALGORITHMS FOR SOLVING OPTIMIZATION PROBLEMS WITH BOOLEAN VARIABLES (RUSSIAN). AVTOMATIKA I TELEMEHANIKA, NO. 10, 66-77; TRANSLATED AS AUTOMATION AND REMOTE CONTROL 37, 1976, 1523-1532. (1976).

KAPP 1975 -1 KAPPAUF, C. H.; KOEHLER, G. J.
POSTMAN PROBLEM, THE. PAPER PRESENTED AT THE 48TH JOINT NATIONAL MEETING OF ORSA AND TIMS, LAS VEGAS 1975, 42 P. (1975).

KARA 1976 -1 KARAGEZOV, Z. A.
DECOMPOSITIONAL ALGORITHM FOR NON-CORRECT PROBLEM OF TIME SCHEDULE OPTIMIZATION (RUSSIAN). EKONOMIKA I MATEMATICHESKIE METODY 12, 1004-1008. (1976).

KARAA 1976 -1 KARAPETJAN, S. R.
(X-0,M)-MINIMAL CUT AND PARTITIONING OF A TREE (RUSSIAN). AKADEMIJA NAUK ARMJANSKOI SSR. DOKLADY 63, 129-135. (1976).

KARI 1976 -1 KARIV, O.
O(N POWER 2.5) ALGORITHM FOR FINDING A MAXIMUM MATCHING IN A GENERAL GRAPH, AN. REHOVOT: WEIZMANN INSTITUTE OF SCIENCE, 211 P. (1976). (DISSERTATION)

KARN 1976 -1 KARNEY, D.; KLINGMAN, D.
IMPLEMENTATION AND COMPUTATIONAL STUDY ON AN IN-CORE, OUT-OF-CORE PRIMAL NETWORK CODE. OPERATIONS RESEARCH 24, 1056-1077. (1976).

KARP 1975 -1 KARP
GRAPH ALGORITHMS. IN: EC. ETE INF., 1-58. LE BREAU. (1975).

KARPA 1975 -1 KARP, R. M.
FAST APPROXIMATE SOLUTION OF HARD COMBINATORIAL PROBLEMS, THE. IN: F. HOFFMAN, R. C. MULLIN, R. B. LEVOW ET AL. (EDS.): PROCEEDINGS OF THE 6TH SOUTHEASTERN CONFERENCE ON COMBINATORICS, GRAPH THEORY, AND COMPUTING, BOCA RATON 1975, 15-34. CONGRESSUS NUMERANTIUM 14, WINNIPEG: UTILITAS MATHEMATICA. (1975).

KARPA 1975 -2 KARP, R. M.
ON THE COMPUTATIONAL COMPLEXITY OF COMBINATORIAL PROBLEMS. NETWORKS 5, 45-68. (1975).

KARPA 1976 -1 KARP, R. M.
PROBABILISTIC ANALYSIS OF SOME COMBINATORIAL SEARCH ALGORITHMS, THE. IN: J. F. TRAUB (ED.): ALGORITHMS AND COMPLEXITY, NEW DIRECTIONS AND RECENT RESULTS, 1-20. NEW YORK: ACADEMIC PRESS. (1976).

KARPA 1977 -1 KARP, R. M.
PROBABILISTIC ANALYSIS OF PARTITIONING ALGORITHMS FOR THE TRAVELING-SALESMAN PROBLEM IN THE PLANE. MATHEMATICS OF OPERATIONS RESEARCH 2, 209-224. (1977).

KARPA 1978 -1 KARP, R. M.
PATCHING ALGORITHM FOR THE NONSYMMETRIC TRAVELING-SALESMAN PROBLEM, A. MEMORANDUM NO. UCB/ERL M78/2. BERKELEY: UNIV. OF CALIFORNIA, ELECTRONICS RESEARCH LABORATORY, 26 P. (1978).

KARW 1976 -1 KARWAN, M. H.; RARDIN, R. L.
APPLICATION OF SURROGATE CONSTRAINT DUALITY IN INTEGER LINEAR PROGRAMMING, THE. PAPER PRESENTED AT THE 49TH JOINT NATIONAL MEETING OF ORSA AND TIMS, PHILADELPHIA 1976, 20 P. (1976).

KARW 1976 -2 KARWAN, M. H.; RARDIN, R. L.
AURROGATE DUAL MULTIPLIER SEARCH PROCEDURES IN INTEGER PROGRAMMING. REPORT SERIES NO. J-76-33. ATLANTA: GEORGIA INST. OF TECHNOLOGY, DEPT. OF INDUSTRIAL AND SYSTEMS ENGINEERING, 28 P. (1976).

KARW 1977 -1 KARWAN, M. H.; RARDIN, R. L.
SEARCHABILITY OF THE COMPOSITE AND MULTIPLE SURROGATE DUAL FUNCTIONS. BUFFALO: STATE UNIV. OF NEW YORK, DEPT. OF INDUSTRIAL ENGINEERING, 12 P. (1977).

KARW 1977 -2 KARWAN, M. H.; RARDIN, R. L.
SURROGATE DUALITY IN A BRANCH-AND-BOUND PROCEDURE. INDUSTRIAL AND SYSTEMS ENGINEERING REPORT SERIES NO. J-77-12. ATLANTA: GEORGIA INST. OF TECHNOLOGY, 16 P. (1977).

KARZ 1974 -1 KARZANOV, A.
DETERMINING THE MAX FLOW IN A NETWORK. SOVIET MATHEMATICS, DOKLADY 15, 434-437. (1974).

KAS 1975 -1 KAS, P.
ON QUADRATIC MINIMAL COST FLOW PROBLEMS (HUNGARIAN). ALKALMAZOTT MATEMATIKAI LAPOK 1, 101-108. (1975).

KAS' 1975 -1 KAS'JANOV, V. N.
FINDING THE HAMMOCKS IN A DIRECTED GRAPH (RUSSIAN). DOKLADY AKADEMII NAUK SSSR 221, 1020-1022. (1975).

KASA 1975 -1 KASAI, T.; KURODA, K.
NUMBER OF HAMILTONIAN CIRCUITS CONTAINING SPECIFIED EDGES IN AN INCOMPLETE GRAPH AND ITS APPLICATIONS TO MULTIGRAPHS. BULLETIN OF UNIV. OF OSAKA PREFECTURE. SERIES A, ENGINEERING AND NATURAL SCIENCES 24, 87-100. (1975).

KASP 1977 -1 KASPSHITSKAJA, M. F.; SERGIENKO, I. V.
ON THE IDEA OF LINEARITY AND CONVEXITY IN A DISCRETE SPACE. COMBINATORIAL LINEAR PROBLEMS, I (RUSSIAN). KIBERNETIKA, NO. 5, 75-81. (1977).

KASPA 1974 -1 KASPSHITSKAJA, M. F.; SERGIENKO, I. V.; HIL'CENKO, V. I.
CERTAIN APPROACH TO THE SOLUTION OF ALLOCATION PROBLEMS (RUSSIAN), A. KIBERNETIKA, NO. 5, 51-60. (1974).

KASS 1977 -1 KASSAY, F.
APPROACH TO SENSITIVITY ANALYSIS OF THE TRANSPORTATION PROBLEM (CZECH), AN. EKONOMICKO-MATEMATICKY OBZOR 13, 213-223. (1977).

KAST 1976 -1 KASTNING, C.
INTEGER PROGRAMMING AND RELATED AREAS. A CLASSIFIED BIBLIOGRAPHY. LECTURES NOTES IN ECONOMICS AND MATHEMATICAL SYSTEMS 128, BERLIN: SPRINGER, 517 P. (1976).

KASZ 1973 -1 KASZONYI, L.
ON THE STRUCTURE OF COLORING GRAPHS. ANNALES UNIVERSITATIS SCIENTIARUM BUDAPESTINENSIS DE ROLANDO EOETVOES NOMINATAE, SECTIO MATHEMATICA 16, 25-36. (1973).

KATA 1968 -1 KATAYAMA, H.; WATANABE, H.
ALGORITHM FOR THE SHORTEST PATH PROBLEM IN A NETWORK, A. WORKING PAPER CT68-15. JAPAN: INSTITUT OF ELECTRONICS AND COMMUNICATION ENGINEERS OF JAPAN. (1968).

KATC 1977 -1 KATCHALSKI, M.
INTERSECTING ALL EDGES OF CONVEX POLYTOPES BY PLANES. DISCRETE MATHEMATICS 19, 273-280. (1977).

KATS 1977 -1 KATSEV, S. B.
SOLUTION OF A GENERALIZED PROBLEM ON SET PARTITIONING (RUSSIAN). KIBERNETIKA, NO. 5, 115-120. (1977).

KATSA 1976 -1 KATSOULAS, P. C.; SADOWSKI, R. P.
BRANCH AND BOUND ALGORITHM FOR USE IN RESOURCE ALLOCATION, A. AIIE TRANSACTIONS 8, 226-233. (1976).

KATZ 1976 -1 KATZ, I.; COOPER, L.
OPTIMAL FACILITY LOCATION FOR NORMALLY AND EXPONENTIALLY DISTRIBUTED POINTS. JOURNAL OF RESEARCH OF THE NATIONAL BUREAU OF STANDARDS 80B, SERIES B, MATHEMATICAL SCIENCES, 53-72. (1976).

KAUF 1974 -1 KAUFMAN, M. T.
ALMOST-OPTIMAL ALGORITHM FOR THE ASSEMBLY LINE SCHEDULING PROBLEM, AN. IEEE TRANSACTIONS ON COMPUTERS C-23, 1169-1174. (1974).

KAUFA 1977 -1 KAUFMANN, A.; PICHAT, E.
METHODES MATHEMATIQUES NON NUMERIQUES ET LEURS ALGORITHMES. TOME 1 ALGORITHMES DE RECHERCHE DES ELEMENTS MAXIMAUX. PARIS, NEW YORK U.A.: MASSON, 189 P. (1977).

KAUFA 1977 -2 KAUFMANN, A.; PICHAT, E.
METHODES MATHEMATIQUES NON NUMERIQUES ET LEURS ALGORITHMES. TOME 2 ALGORITHMES DE RECHERCHEDE CHEMINS ET PROBLEMES ASSOCIES. PARIS, NEW YORK U.A.: MASSON, 159 P. (1977).

KAZA 1975 -1 KAZARIN, L. S.; L'VOV, G. N.; TIMOFEEV, E. A.
HEURISTIC ALGORITHM FOR CONSTRUCTING A TIME-TABLE (RUSSIAN), A. JAROSLAVSKII UNIVERSITET, VESTNIK, VYP. 9, 47-51. (1975).

KEL' 1973 -1 KEL'MANS, A. K.
SELECTION OF THE OPTIMAL VERTEX IN A GRAPH (RUSSIAN), THE. IN: A. A. FRIDMAN (ED.): STUDIES IN DISCRETE MATHEMATICS (RUSSIAN), 151-158. MOSCOW: IZDAT. NAUKA. (1973).

KEL' 1976 -1 KEL'MANS, A. K.
NUMBER OF TREES IN A GRAPH WHICH CONTAINS A GIVEN FOREST (RUSSIAN). ACTA MATHEMATICA ACADEMIAE SCIENTIARUM HUNGARICAE 27, 89-95. (1976).

KEL'A 1976 -1 KEL'MANS, A. K.; LOMONOSOV, M. V.; POLESSKIJ, V. P.
MINIMAL COVERINGS IN MATROIDS (RUSSIAN). PROBLEMY PEREDACI INFORMACII 12, NO. 3, 94-107. (1976).

KELI 1968 -1 KELIKHOV, A. N.; BERSHTEYN, L. S.; KARELIN, P.
ISOMORPHISM OF GRAPHS AND FINITE AUTOMATA. ENGINEERING CYBERNETICS 1, 124-129. (1968).

KEND 1977 -1 KENDALL, K. E.; ZIONTS, S.
SOLVING INTEGER PROGRAMMING PROBLEMS BY AGGREGATING CONSTRAINTS. OPERATIONS RESEARCH 25, 346-351. (1977).

KENDA 1975 -1 KENDRICK, D.; STOUTJESDIJK, A.
PLANNING OF INDUSTRIAL INVESTMENT PROGRAMS: A METHODOLOGY, THE. WASHINGTON: INTERNATIONAL BANK FOR RECONSTRUCTION AND DEVELOPMENT, DEVELOPMENT RESEARCH CENTER. (1975).

KENN 1975 -1 KENNINGTON, J. L.
MULTICOMMODITY NETWORK FLOWS: A STATE-OF-THE-ART SURVEY OF LINEAR MODELS AND SOLUTION TECHNIQUES. TECHNICAL REPORTS IEOR 75014 & 75015. DALLAS: SOUTHERN METHODIST UNIV., DEPT. OF INDUSTRIAL ENGINEERING AND OPERATIONS RESEARCH. (1975).

KENN 1976 -1 KENNINGTON, J. L.
FIXED-CHARGE TRANSPORTATION PROBLEM: A COMPUTATIONAL STUDY WITH A BRANCH-AND-BOUND CODE, THE. AIIE TRANSACTIONS 8, 241-247. (1976).

****

KENN 1977 -1 KENNINGTON, J. L.
SOLVING MULTICOMMODITY TRANSPORTATION PROBLEMS USING A PRIMAL PARTITIONING SIMPLEX TECHNIQUE. NAVAL RESEARCH LOGISTICS QUARTERLY 24, 309-326. (1977).
KENN 1977 -2 KENNINGTON, J. L.
STATE-OF-THE-ART SURVEY OF LINEAR COST MULTICOMMODITY NETWORK FLOWS, A (REVISED VERSION). TECHNICAL REPORT IEOR 75014. DALLAS: SOUTHERN METHODIST UNIV., DEPT. OF INDUSTRIAL ENGINEERING AND OPERATIONS RESEARCH, 50 P. (1977).
KENNA 1977 -1 KENNINGTON, J. L.; SHALABY, M.
EFFECTIVE SUBGRADIENT PROCEDURE FOR MINIMAL COST MULTICOMMODITY FLOW PROBLEMS, AN. MANAGEMENT SCIENCE 23, 994-1004. (1977).
KERN 1973 -1 KERNIGHAN, B. W; LIN, S.
HEURISTIC SOLUTION OF A SIGNAL DESIGN OPTIMIZATION PROBLEM. IN: PROCEEDINGS OF THE SEVENTH ANNUAL PRINCETON CONFERENCE ON INFORMATION SCIENCES AND SYSTEMS, 125-129. (1973).
KERS 1976 -1 KERSHENBAUM, A.; HSIEH, W.; GOLDEN, B.
CONSTRAINED ROUTING IN LARGE SPARSE NETWORKS. CONFERENCE RECORDS OF 1976 INTERN. CONF. ON COMMUNICATIONS, PHILADELPHIA 1976. (1976).
KEYS 1975 -1 KEYS, C. D.
GRAPHS CRITICAL FOR MAXIMAL BOOKTHICKNESS. PI MU EPSILON JOURNAL 6, 79-84. (1975).
KHAC 1976 -1 KHACHATUROV, V. R.; ASTAKHOV, N. B.
DYNAMIC LOCATION PROBLEMS (MODELS AND SOLUTION METHODS) (RUSSIAN). EKONOMIKA I MATEMATICHESKIE METODY, 13; TRANSLATED AS MATEKON 13, NO. 2, 1977, 68-93. (1976).
KHAI 1976 -1 KHAIKIN, N. M.; ZAK, E. I.
CONSTRUCTION OF MINIMUM TREE PACKING IN THE PROBLEM OF OPTIMUM MEMORY DISTRIBUTION (RUSSIAN). KIBERNETIKA, NO. 3, 57-63. (1976).
KHAN 1977 -1 KHAN, S. U.; BARI, A.
PROCEDURE FOR INTEGER SOLUTION TO SOME ALLOCATION PROBLEMS, A. PURE AND APPLIED MATHEMATIKA SCIENCES 5, 25-32. (1977).
KHEN 1966 -1 KHENKIN, V. E.
PROBLEM IN SCHEDULING THEORY (2 TIMES M ORDERING PROBLEM). CYBERNETICS 2, 54-57. (1966).
KHOM 1976 -1 KHOMENKO, M. P.; VIVROT, T. M.
ON THE 0-1-MATRICES (UKRAINIAN). DOPOVIDI AKADEMII NAUK UKRAINS'KOI RSR, SERIJA A, NO. 10, 887-889. (1976).
KIAN 1976 -1 KIANFAR, F.
STRONGER INEQUALITIES FOR 0-1 INTEGER PROGRAMMING: COMPUTATIONAL REFINEMENTS. OPERATIONS RESEARCH 24, 581-585. (1976).
KIBA 1976 -1 KIBALOV, E. B.; KHUTORETZKY, A. B.
VOLUME-TIME SCHEDULE PLANNING PROBLEM OF BUILDING (RUSSIAN), A. EKONOMIKA I MATEMATICHESKIE METODY 12, 1092-1103. (1976).
KICH 1975 -1 KICHA, L. I.; LITVAK, E. I.
ON AN ENUMERATION ALGORITHM FOR THE SIMPLE CIRCUITS IN A GRAPH (RUSSIAN). ZURNAL VYCISLITEL'NOI MATEMATIKI I MATEMATICESKOI METODY 15, 1077-1079. (1975).
KIKU 1975 -1 KIKUST, P. B.
EXISTENCE OF A HAMILTONIAN CYCLE IN A REGULAR GRAPH OF DEGREE 5 (RUSSIAN), THE. IN: LATVIAN MATHEMATICAL YEARBOOK 16 (RUSSIAN), 33-38 + 271. RIGA: IZDAT. ZINATNE. (1975).
KILG 1977 -1 KILGORE, W. C.
WAREHOUSE FACILITIES ANALYSIS. ATLANTA (52. ORSA-MEETING + TIMS, NOVEMBER 1977) 26 P. (1977).
KIM 1976 -1 KIM, K. V.; KRAYNOV, G. A.; SURMENEV, E. V.
ON A PROBLEM OF TIME SCHEDULE COMPILATION (RUSSIAN). EKONOMIKA I MATEMATICHESKIE METODY 12, 767-772. (1976).
KIMB 1959 -1 KIMBALL, G. E.; HOWARD, R. A.
SEQUENTIAL DECISION PROCESSES. IN: NOTES ON OPERATIONS RESEARCH. CAMBRIDGE: MASSACHUSETTS INST. OF TECHNOLOGY, OPERATIONAL RESEARCH CENTRE. (1959).
KINA 1977 -1 KINARIWALA, B.; RAO, A. G.
FLOW SWITCHING APPROACH TO THE MAXIMUM FLOW PROBLEM: I. JOURNAL OF THE ASSOCIATION FOR COMPUTING MACHINERY 24, 630-645. (1977).
KING 1976 -1 KINGSTON, P. L.; LIPTON, S. L.; STOJKA, J. P. JR.
MIXED INTEGER PROGRAMMING MODELS FOR GENERATOR MAINTENANCE SCHEDULING. MIAMI (50. ORSA-MEETING + TIMS, NOVEMBER 1976) 39 P. (1976).
KINS 1975 -1 KINSNER, W.
ALGORITHM AND A PROGRAM FOR NETWORK ROUTING PROBLEMS, AN. INTERNAL REPORTS IN SIMULATION, OPTIMZATION AND CONTROL, SOC-7 6. HAMILTON: MCMASTER UNIV., FACULTY OF ENGINEERING, 40 P. (1975).
KIRB 1966 -1 KIRBY, R. F.
MINIMUM PATH ALGORITHM FOR A ROAD NETWORK WITH TURN PENALTIES, A. IN: PROCEEDINGS OF THE THIRD AUSTRALIAN ROAD RESEARCH BOARD CONFERENCE, PART 1, 434. (1966).
KIRBA 1968 -1 KIRBY, R. F.; POTTS, R. B.
MINIMUM ROUTE PROBLEM FOR NETWORKS WITH TURN PENALTIES AND PROHIBITIONS, THE. ADELAIDE: UNIV. OF ADELAIDE, MATHEMATICS DEPARTMENT. (1968).
KIRC 1965 -1 KIRCHGAESSNER, K.
GRAPH THEORETICAL SOLUTION OF A NONLINEAR ASSIGNMENT PROBLEM (GERMAN). UNTERNEHMENSFORSCHUNG 9, 217-229. (1965).
KIRY 1975 -1 KIRYUKIN, V. V.; POLONSKIJ, D. V.
ON THE SEARCH FOR THE HAMILTONIAN CYCLES OF A GRAPH (RUSSIAN). PROBLEMY PEREDACI INFORMACII 11, NO. 3, 70-72. (1975).
KIRZ 1973 -1 KIRZNER, V. M.; RUBLINECKII, V. I.
PROCEDURE "PROCEED TO THE NEXT" IN THE TRAVELING SALESMAN PROBLEM (RUSSIAN), THE. IN: NUMERICAL MATHEMATICS AND COMPUTER TECHNOLOGY, NO. 4 (RUSSIAN), 40-41 + 137. KHARKOV: AKAD. NAUK UKRAIN. SSR FIZ.-TEHN. INST. NIZKIH TEMPERATUR. (1973).
KIRZ 1974 -1 KIRZNER, V. M.; RUBLINECKII, V. I.
UPPER BOUND OF THE LENGTH OF THE MINIMAL ROUTE IN THE TRAVELING SALESMAN PROBLEM (RUSSIAN), THE. IN: NUMERICAL MATHEMATICS AND COMPUTER TECHNOLOGY, NO. V (RUSSIAN), 120-122 + 167. KHARKOV: AKAD. NAUK UKRAIN. SSR FIZ.-TEHN. INST. NIZKIH TEMPERATUR. (1974).
KITA 1975 -1 KITAMURA, S.
ON THE EDGE-CHROMATIC NUMBER OF A GRAPH. REPORTS OF THE FACULTY OF SCIENCE AND ENGINEERING, SAGA UNIV., MATHEMATICS, NO. 3, 17-24. (1975).
KIVI 1976 -1 KIVISTIK, L.
CUTTING-PLANE METHOD FOR INTEGER LINEAR PROGRAMMING PROBLEMS WITH ADDITIONAL ALTERNATIVE CONDITIONS (RUSSIAN), THE. EESTI NSV TEADUSTE AKADEEMIA TOIMETISED, FUEUESIKA-MATEMAATIKA 25, 245-250. (1976).
KLEE 1977 -1 KLEE, V.; LARMAN, D.
FLOYD-WARSHALL REVISITED: SHORTEST RESTRICTED PATHS (PRELIMINARY VERSION). VANCOUVER (CONFERENCE ON DISCRETE OPTIMIZATION, AUGUST 1977) 21 P. (1977).

KLEEA 1976 -1 KLEE, V.; QUAIFE, H.
MINIMUM GRAPHS OF SPECIFIED DIAMETER, CONNECTIVITY AND VALENCE, I. MATHEMATICS OF OPERATIONS RESEARCH 1, 28-31. (1976).

KLEI 1969 -1 KLEITMAN, D. J.
METHODS FOR INVESTIGATING CONNECTIVITY FOR LARGE GRAPHS. IEEE TRANSACTIONS ON CIRCUIT THEORY, CT-16, 232-233. (1969).

KLEI 1970 -1 KLEITMAN, D. J.
MINIMAL NUMBER OF MULTIPLE EDGES IN REALIZATION OF AN INCIDENCE SEQUENCE WITHOUT LOOPS. SIAM JOURNAL ON APPLIED MATHEMATICS 18, 1, 25-28. (1970).

KLIT 1976 -1 KLITZ, J. K.
TRAVELING-SALESMAN PROBLEM FORMULATION AND SOLUTION FOR A MASS STORAGE SYSTEM, A. MIAMI (50. ORSA-MEETING + TIMS, NOVEMBER 1976) 15 P. (1976).

KLJA 1976 -1 KLJAUS, P. S.
STRUCTURE OF THE OPTIMAL SOLUTIONS OF SOME CLASSES OF TRAVELING SALESMAN PROBLEMS (RUSSIAN). VESCI AKADEMII NAVUK BSSR, SERYJA FIZIKA-MATEMATYCNYH NAVUK, NO. 6, 95-98 + 143. (1976).

KLUY 1977 -1 KLUYVER, C. A. DE
EXTENSION OF ELZINGA AND VEMUGANTI'S DECOMPOSITION SCHEME TO MULTIPERIOD ADVERTISING MEDIA SCHEDULING, AN. OPSEARCH 14, 204-206. (1977).

KNAP 1976 -1 KNAPEK, S.
EINSATZ HEURISTISCHER VERFAHREN IN DER GEMISCHT-GANZZAHLIGEN INVESTITIONS- UND FINANZIERUNGSPLANUNG, DER. DISSERTATION. WIEN: UNIV. WIEN, 157 P. (1976).

KNOE 1971 -1 KNOEDEL, W.
VERFAHREN ZUR FESTSTELLUNG DER ISOMORPHIE VON ENDLICHEN ZUSAMMENHAENGENDEN GRAPHEN, EIN. COMPUTING 8, 329-334. (1971).

KNUT 1973 -1 KNUTH, D. E.
MATROID PARTITIONING. TECHN. REPORT STAN-CS-73-342, DEPT. OF COMPUTER SCIENCES, STANFORD: STANFORD UNIVERSITY. (1973).

KNUT 1976 -1 KNUTH, D. E.
MARIAGES STABLES ET LEURS RELATIONS AVEC D'AUTRES PROBLEMES COMBINATOIRES. MONTREAL: PRESSES DE L'UNIV. DE MONTREAL, 106 P. (1976).

KNUTA 1973 -1 KNUTS, L.-R.
COMBINING TWO INTEGER PROGRAMMING ALGORITHMS. ACTA ACADEMIAE ABOENSIS 33, SERIES B, MATHEMATICA ET PHYSICA, NO. 12, 8 P. (1973).

KOCH 1976 -1 KOCH, E.; PERLES, M. A.
COVERING EFFICIENCY OF TREES AND K-TREES. IN: F. HOFFMAN, L. LESNIAK, R. MULLIN ET AL. (EDS.): PROCEEDINGS OF THE 7TH SOUTHEASTERN CONFERENCE ON COMBINATORICS, GRAPH THEORY, AND COMPUTING, BATON ROUGE 1976, 391-420. CONGRESSUS NUMERANTIUM 17, WINNIPEG: UTILITAS MATHEMATICA. (1976).

KOCHA 1977 -1 KOCHURA, S. N.
ON AN OPTIMIZATION ALGORITHM OF CAPITAL INVESTMENTS DISTRIBUTION USING THE BOOLEAN VARIABLES (RUSSIAN). KIBERNETIKA, NO. 4, 88-91. (1977).

KOEH 1977 -1 KOEHLER, G. J.; KUMAR, S.
LOOK AT FINDING POINTS OF A CONVEX POLYHEDRON USING RELAXATION AND SUBGRADIENT PROCEDURES, A. ATLANTA (52. ORSA-MEETING + TIMS, NOVEMBER 1977) 22 P. (1977).

KOEHA 1977 -1 KOEHLER, J.
NUTZUNG DER DUALITAET ZUR LOESUNG EINES GEMISCHTEN LINEAREN 0-1-OPTIMALPROBLEMS. WISSENSCHAFTLICHE ZEITSCHRIFT DER MARTIN-LUTHER-UNIV. HALLE-WITTENBERG, MATHEMATISCH-NATURWISSENSCHAFTLICHE REIHE 26, 145-149. (1977).

KOH 1976 -1 KOH, K. M.
ON THE SHORTEST-PATH PROBLEM AND DIJKSTRA'S ALGORITHM. MALAYSIAN MATHEMATICAL SOCIETY, BULLETIN 7, NO. 1, 46-51. (1976).

KOH A 1974 -1 KOH, K. M.; TEH, H. H.
ON DIRECTED POSTMAN PROBLEM. NANYANG UNIV. JOURNAL PART III 8/9, 14-26. (1974).

KOHL 1975 -1 KOHLER, C.
LOESUNGSMETHODE FUER TRIADISCHE OPTIMIERUNGSAUFGABEN, EINE. WISSENSCHAFTLICHE ZEITSCHRIFT DER TECHNISCHEN HOCHSCHULE KARL-MARX-STADT 17, 755-760. (1975).

KOHLA 1975 -1 KOHLER, W.
COMPUTATIONAL EXPERIENCE WITH EFFICIENT EXACT AND APPROXIMATE ALGORITHMS FOR AN NP-COMPLETE SCHEDULING PROBLEM. TECHNICAL REPORT ECE-CS-75-13. AMHERST: UNIV. OF MASSACHUSETTS. (1975).

KOHLB 1976 -1 KOHLER, W. H.; STEIGLITZ, K.
ENUMERATIVE AND ITERATIVE COMPUTATIONAL APPROACHES. IN: E. G. COFFMAN JR. (ED.): COMPUTER AND JOB/SHOP SCHEDULING THEORY, CHAPTER SIX. (1976).

KOHO 1974 -1 KOHOV, V. A.
CERTAIN INVARIANT OF GRAPHS (RUSSIAN), A. TRUDY MOSKOVSKOGO ORDENA LENINA ENERGETICESKOGO INSTITUTA, VYP. 178, 154-160. (1974).

KOLK 1963 -1 KOLKER, R.; LISS, D.; OKADA, S.
ALGEBRAIC TECHNIQUES OF PATH FINDING AND MINIMUM PATH FINDING IN GRAPHS. TH 3421, BEDFORD: MITRE CORP., 36 P. (1963).

KOMA 1975 -1 KOMARNICKI, J.
AUTOMATISCHE GENERIERUNG ANWENDUNGSSPEZIFISCHER NETZPLAN-TECHNIK-PROGRAMMSYSTEME. ANGEWANDTE INFORMATIK 17, 70-74. (1975).

KOMAA 1974 -1 KOMAROMI, E.
FINITE PRIMAL METHOD FOR THE ASSIGNEMENT PROBLEM, A. PROBLEMS OF CONTROL AND INFORMATION THEORY 3, NO. 2, 157-166. (1974).

KONO 1976 -1 KONONENKO, A. M.; MIKUL'SKII, V. E.; TRUHANOVSKII, N. N.
SOLVABILITY CONDITIONS FOR THREE-INDEX PLANAR TRANSPORT PROBLEMS (RUSSIAN). VESCI AKADEMII NAVUK BSSR, SERYJA FIZIKA-MATEMATYCNYH NAVUK, NO. 6, 24-28 + 140. (1976).

KOPY 1977 -1 KOPYLOV, G. N.
MAXIMAL PATHS AND CYCLES IN A GRAPH (RUSSIAN). DOKLADY AKADEMII NAUK SSSR 234, 19-21. (1977).

KORD 1976 -1 KORDECKI, W.
ELEMENTS OF THE THEORY OF RANDOM GRAPHS (POLISH). POLITECHNIKI WROCLAWSKIEJ, INSTYTUTU MATEMATYKI, PRACE NAUKOWE, NO. 15, SERIA MONOGRAFIE, NO. 3, 70 P. (1976).

KORN 1969 -1 KORNILOVA, L. E.
METHOD OF MINIMAL FLOW FOR THE EXACT SOLUTION OF THE PROBLEM OF MINIMIZATION OF THE MAXIMAL LEVEL OF RESOURCE UTILIZATION IN NETWORK PLANNING (RUSSIAN), THE. IN: THEORY OF OPTIMAL SOLUTIONS, PROCEEDINGS OF A SEMINAR, KIEV 1969, NO. 2 (RUSSIAN), 75-85. KIEV: AKAD. NAUK UKRAIN SSR. (1969).

KORS 1974 -1 KORSHUNOV, A. D.
ON THE NUMBER OF PAIRS OF HAMILTONIAN CYCLES IN A COMPLETE GRAPH HAVING A FIXED NUMBER OF COMMON ARCS (RUSSIAN). UPRAVLJAEME SISTEMY, NO. 13, 40-57. (1974).

KORS 1976 -1 KORSHUNOV, A. D.
SOLUTION OF THE PROBLEM OF ERDOES-RENYI ON THE HAMILTONIAN CYCLES IN UNDIRECTED GRAPHS (RUSSIAN). DOKLADY AKADEMII NAUK SSSR 228, 529-532. (1976).

****

KORSA 1976 -1 KORSHUNOV, A. D.; NINCAK, YA.
COLORING OF HYPERGRAPHS (RUSSIAN), THE. KYBERNETIKA 12, 20-30. (1976).

KORT 1966 -1 KORTANEK, K. O.; SODARDO, D.
GENERALIZED NETWORK MODEL FOR THREE DIMENSIONAL CUTTING STOCK PROBLEMS AND NEW PRODUCT ANALYSIS. JOURNAL OF INDUSTRIAL ENGINEERING 17, 572-576. (1966).

KORTA 1977 -1 KORTE B.; MONMA, C. L.
REMARKS ON A CLASSIFICATION OF ORACLE-TYPE ALGORITHMS, SOME. REPORT 7785-OR. BONN: UNIV. BONN, INST. F. OEKONOMETRIE UND OPERATIONS RESEARCH (1977).

KORTB 1976 -1 KORTE, B.
GANZZAHLIGE PROGRAMMIERUNG (STICHWORTARTIKEL FUER: HANDBUCH DER MATHEMATISCHEN WIRTSCHAFTSWISSENSCHAFTEN, HRSG. M. J. BECKMANN). BONN: UNIV. BONN, INSTITUT FUER OEKONOMETRIE UND OPERATIONS RESEARCH, 25 P. (1976). (REPORT NO. 7646 OR)

KORTB 1976 -2 KORTE, B.
INTEGER PROGRAMMING (SURVEY PAPER). REPORT NO. 7649-OR. BONN: UNIV. BONN, INST. F. OEKONOMETRIE UND OPERATIONS RESEARCH, 21 P. (1976).

KORTB 1977 -1 KORTE, B.
ANALYSE VON HEURISTISCHEN ALGORITHMEN ZUR KOMBINATORISCHEN OPTIMIERUNG. IN: R. HENN, P. KALL, B. KORTE, O. KRAFFT ET AL. (EDS.): OPERATIONS RESEARCH VERFAHREN XXV, METHODS OF OPERATIONS RESEARCH, 217-220. MEISENHEIM: HAIN. (1977).

KORTB 1977 -2 KORTE, B.
APPROXIMATIVE ALGORITHMS FOR DISCRETE OPTIMIZATION PROBLEMS. REPORT NO. 7762-OR. VANCOUVER (CONFERENCE ON DISCRETE OPTIMIZATION, AUGUST 1977) AND BONN: UNIV. BONN, INST. FUER OEKONOMETRIE UND OPERATIONS RESEARCH, 52 P. (1977).

KORTC 1977 -1 KORTE, B.; HAUSMANN, D.
ANALYSE DER GREEDY-HEURISTIK FUER UNABHAENGIGKEITSSYSTEME, EINE. IN: R. HENN, P. KALL, B. KORTE, O. KRAFFT ET AL. (EDS.): OPERATIONS RESEARCH VERFAHREN XXIII, METHODS OF OPERATIONS RESEARCH, 132-140. MEISENHEIM: HAIN. (1977).

KORTC 1978 -1 KORTE, B.; HAUSMANN, D.
ANALYSIS OF THE GREEDY HEURISTIC FOR INDEPENDENCE SYSTEMS, AN. IN: ALSPACH, B.;HELL, P.; MILLER, D. J.(EDS.): ALGORITHMIC ASPECTS OF COMBINATORICS. ANNALS OF DISCRETE MATHEMATICS 2, 65-74. (1978).

KOST 1977 -1 KOSTOCKA, A. V.
UPPER BOUNDS FOR THE CHROMATIC NUMBER OF A GRAPH IN TERMS OF ITS DEGREE, DENSITY AND GIRTH (RUSSIAN). DOKLADY AKADEMII NAUK SSSR 235, 516-518. (1977).

KOU 1977 -1 KOU, L. T.
ALPHA-SOURCE AND ALPHA-SINK O A GRAPH. IBM RESEARCH REPORT 27504. YORKTOWN HEIGHTS: IBM RESEARCH CENTER, 9 P. (1977).

KOU 1977 -2 KOU, L. T.
POLYNOMIAL COMPLETE CONSECUTIVE INFORMATION RETRIEVAL PROBLEMS. SIAM JOURNAL ON COMPUTING 6, 67-75. (1977).

KOU A 1978 -1 KOU, L. T.; STOCKMEYER, L. J.; WONG, C. K.
COVERING EDGES BY CLIQUES WITH REGARD TO KEYWORD CONFLICTS AND INTERSECTION GRAPHS. COMMUNICATIONS OF THE ACM 21, 135-138. (1978).

KOVA 1976 -1 KOVACS, L. B.
COMPUTER PROGRAM SYSTEM FOR THE SOLUTION OF PURE LINEAR DISCRETE PROBLEMS, A. IN: A. PREKOPA (ED.): PROGRESS IN OPERATIONS RESEARCH, 573-592, AMSTERDAM, LONDON: NORTH HOLLAND. (1976).

KOVAA 1975 -1 KOVALENKO, I. N.
ON THE INVARIANCE THEOREMS FOR BOOLEAN RANDOM MATRICES (RUSSIAN). KIBERNETIKA, NO. 5, 138-152. (1975).

KOVAB 1974 -1 KOVALEV, M. M.
CERTAIN INTEGER PROGRAMMING PROBLEM WITH A CONVEX SYMMETRIC TARGET FUNCTION (RUSSIAN), A. VESTNIK BELORUSSKOGO GOSUDARSTVENNOGO UNIVERSITETA IM. V. I. LENINA. NAUCNYI ZURNAL. SERIJA I. MATEMATIKA, FIZIKA, MEHANIKA, NO. 1, 64-65 + 89. (1974).

KOVAB 1977 -1 KOVALEV, M. M.
DISCRETE OPTIMIZATION (INTEGER PROGRAMMING) (RUSSIAN). MINSK: IZDAT. BELORUSS. UNIV., 191 P. (1977).

KOVC 1976 -1 KOVCAS, L. B.; RUPP, M.
GRAPH PARTITIONING FOR MATRIX INVERSION BY INTEGER PROGRAMMING. IN: A. PREKOPA (ED.): PROGRESS IN OPERATIONS RESEARCH, 593-604, AMSTERDAM, LONDON: NORTH HOLLAND. (1976).

KOZY 1972 -1 KOZYREV, V. P.
GRAPH THEORY (RUSSIAN). IN: R. V. GAMKRELIDZE (ED.): PROBABILITY THEORY. MATHEMATICAL STATISTICS. THEORETICAL CYBERNETICS 10 (RUSSIAN), 25-74 + 144. MOSCOW: AKAD. NAUK SSSR VSESOJUZ. INST. NAUCN. I TEHN. INFORMACII. (1972).

KOZY 1976 -1 KOZYREV, V. P.
FINDING ALL THE HAMILTONIAN PATHS IN A GRAPH AND INVESTIGATION OF HAMILTONIAN CONNECTED GRAPHS (RUSSIAN). ZURNAL VYCISLITEL'NOI MATEMATIKI I MATEMATICESKOI FIZIKI 16, 767-775. (1976).

KOZY 1976 -2 KOZYREV, V. P.
HAMILTONIAN CONNECTIVITY OF THE SQUARE OF A GRAPH (RUSSIAN). PROBLEMY KIBERNETIKI, NO. 31, 225-230. (1976).

KRAR 1965 -1 KRARUP, J.
TRAVELING SALESMAN I, THE; TRAVELING SALESMAN II, THE. LINGBY: DANMARKS TEKNISKE HOEJSKOLE, INST. FOR MATEMATISK STATISTIK OG OPERATIONSANALYSE, 10 P. + 7 P. (1965).

KRAR 1967 -1 KRARUP, J.
FIXED-COST AND OTHER NETWORK FLOW PROBLEMS AS RELATED TO PLANT LOCATION AND TO THE DESIGN OF TRANSPORTATION AND COMPUTER SYSTEMS. LYNGBY: DANMARKS TEKNISKE HOEJSKOLE, 162 P. (1967).

KRAR 1967 -2 KRARUP, J.
FIXED-COST NETWORK FLOW PROBLEMS. DISSERTATION 3, IMSOR. LYNGBY: THE TECHNICAL UNIV. OF DENMARK. (1967).

KRAR 1969 -1 KRARUP, J.
BRANCH-BOUND ALGORITHM FOR A WIRING PROBLEM, A. BIT 9, 133-156. (1969).

KRAR 1969 -2 KRARUP, J.
LUBRICATION OILS: OPTIMAL EXTENSION OF PRODUCTION PLANTS AND STORAGE FACILITIES (DANISH). DIF-TRANSACTIONS, 1-69. (1969).

KRAR 1972 -1 KRARUP, J.
QUADRATIC ASSIGNMENT (IN DANISH). IN: DATA, NO. 3, 4 P. (1972).

KRAR 1973 -1 KRARUP, J.
SELECTED EXAMPLES OF COMBINATORIAL OPTIMIZATION PROBLEMS. (DANISH). COPENHAGEN (TREDIE NORDISKE OPERATIONSANALYSE KONGRES, OCTOBER 1973), 4 P. (1973).

KRAR 1975 -1 KRARUP, J.
PERIPATETIC SALESMAN AND SOME RELATED UNSOLVED PROBLEMS, THE. IN: B. ROY (ED.): COMBINATORIAL PROGRAMMING: METHODS AND APPLICATIONS, PROCEEDINGS OF THE NATO ADVANCED STUDY INST., VERSAILLES 1974, 173-178. NATO ADVANCED STUDY INST. SERIES, SERIES C. MATH. AND PHYS. SCI. 19, DORDRECHT: REIDEL. (1975).

KRARA 1967 -1 KRARUP, J.; BILDE, O.
DETERMINATION OF THE OPTIMAL LOCATION OF PRODUCTION PLANTS (DANISH). KOPENHAGEN: THE TECHNICAL UNIV. OF DENMARK. (1967).

KRARA 1977 -1 KRARUP, J.; BILDE, O.
PLANT LOCATION, SET COVERING AND ECONOMIC LOT SIZE: AN O(MN)-ALGORITHM FOR STRUCTURED PROBLEMS. IN: L. COLLATZ, G. MEINARDUS, W. WETTERLING (EDS.): NUMERISCHE METHODEN BEI OPTIMIERUNGSAUFGABEN BAND 3, OPTIMIERUNG BEI GRAPHENTHEORETISCHEN UND GANZZAHLIGEN PROBLEMEN, 155-180. INTERNAT. SERIES OF NUMERICAL MATHEMATICS 36, BASEL: BIRKHAEUSER. (1977).

KRARB 1973 -1 KRARUP, J.; HARARY, F.
CLASS OF PLANAR GRAPHS RELATED TO NONSERIAL DYNAMIC PROGRAMMING, A. SCIMA 2, 17-34. (1973).

KRARC 1972 -1 KRARUP, J.; HARARY, F.; SCHWENK, A.
GRAPHS SUPPRESSIBLE TO AN EDGE. CANADIAN MATHEMATICAL BULLETIN 15, 201-204. (1972).

KRARD 1968 -1 KRARUP, J.; PRUZAN, P. M.
DISTRIBUTION MODELS AND SOLUTION METHODS - A SURVEY (DANISH). ERHVERVSOEKONOMISK TIDSSKRIFT, 10-24. (1968).

KRARD 1977 -1 KRARUP, J.; PRUZAN, P. M.
SELECTED FAMILIES OF DISCRETE LOCATION PROBLEMS PART 1: CENTER PROBLEMS PART 2: MEDIAN PROBLEMS. RAPPORT NR. 77/7. COPENHAGEN: UNIV. OF COPENHAGEN, INSTITUTE OF DATALOGY, 73 P. (1977).

KRARD 1977 -2 KRARUP, J.; PRUZAN, P. M.
SELECTED FAMILIES OF DISCRETE LOCATION PROBLEMS PART 3: THE PLANT LOCATION FAMILY. WORKING PAPER NO. [illegible] VANCOUVER (CONFERENCE ON DISCRETE OPTIMIZATION, AUGUST 1977) AND CALGARY: UNIV. OF CALGARY, FACULTY OF BUSINESS, P. 77-123 (1977).

KRAT 1977 -1 KRATKO, M. I.; REVIN, O. M.
TURING'S MACHINES OPERATING ON THE PLANE, AND STABLE ITERATIVE SYSTEMS (RUSSIAN). KIBERNETIKA, NO. 5, 146-147. (1977).

KRAV 1976 -1 KRAVTSOV, M. K.
ENUMERATION PROBLEMS ON THE DEGENERATED TRANSPORTATION POLYHEDRA (RUSSIAN). VESCI AKADEMII NAVUK BSSR, SERYJA FIZIKA-MATEMATYCNYH NAVUK, 122-123. (1976).

KRAVA 1975 -1 KRAVTSOV, M. K.; SHERMAN, A. K.; AVERBUKH, N. D.
ALGORITHM FOR THE SOLUTION OF THE ASSIGNMENT PROBLEM (RUSSIAN), AN. VESKI AKADEMII NAVUK BSSR, SERYJA FIZIKA-MATEMATYCNYH NAVUK, NO. 6, 102-105. (1975).

KREU 1969 -1 KREUZBERGER, H.
VERFAHREN ZUR LOESUNG GANZZAHLIGER LINEARER OPTIMIERUNGSPROBLEME. DISSERTATION. DARMSTADT: TECHNISCHE HOCHSCHULE. (1969).

KROG 1976 -1 KROGDAHL, S.
COMBINATORIAL PROOF FOR LAWLER'S MATROID INTERSECTION ALGORITHM, A. STANFORD: STANFORD UNIV., 33 P. (1976).

KROG 1977 -1 KROGDAHL, S.
DEPENDENCE GRAPH FOR BASES IN MATROIDS, THE. DISCRETE MATHEMATICS 19, 47-60. (1977).

KRON 1971 -1 KRONE, M.
HEURISTIC PROGRAMMING APPLIED TO SCHEDULING MODELS. IN: PROCEEDINGS OF THE FIFTH ANNUAL PRINCETON CONFERENCE ON INFORMATION SCIENCES AND SYSTEMS, 193-196. (1971).

KRUS 1964 -1 KRUSHEVSKII, A. V.; PRIMAK, M. E.
EXTREMUM OF A LINEAR FORM ON A PERMUTATION GROUP AND THE ASSIGNMENT PROBLEM (RUSSIAN). TRUDY SEMINARA PO MATEMATICESKOI FIZIKI I NELINEINYM KOLEBANIJAM 2, 372-376. (1964).

KUBA 1974 -1 KUBALE, M.
HEURISTIC ALGORITHM FOR ESTABLISHING TIME-TABLES IN A FACULTY (POLISH). ZESZYTY NAUKOWE POLITECHNIKI GDANSK. MATEMATYKA, NO. 9, 45-60. (1974).

KUBO 1976 -1 KUBO, H.; OKINO, N.
SOLUTION FOR LARGE-SCALE TRAVELING SALESMAN PROBLEMS (JAPANESE), A. JOURNAL OF THE OPERATIONS RESEARCH SOCIETY OF JAPAN 19, 272-285. (1976).

KUCE 1977 -1 KUCERA, L.
EXPECTED BEHAVIOR OF GRAPH COLORING ALGORITHMS. IN: M. KARPINSKI (ED.): FUNDAMENTALS OF COMPUTATION THEORY, PROCEEDINGS OF THE 1977 INTERNATIONAL FCT-CONFERENCE, POZNAN-KORNIK, 447-451. LECTURE NOTES IN COMPUTER SCIENCE 56, BERLIN: SPRINGER. (1977).

KUEH 1977 -1 KUEHNER, J.; HARRINGTON, J. J.
WIRKSAME ANWENDUNG DES ALGORITHMUS FUER DIE GEMISCHT-GANZZAHLIGE PROGRAMMIERUNG MPSX-MIP VON IBM ZUR LOESUNG VON GROSSEN OEFFENTLICHEN INVESTITIONSPROBLEMEN MIT VIELEN BETEILIGTEN, TEIL II, DIE. ANGEWANDTE INFORMATIK 19, 84-87. (1977).

KUEH 1977 -2 KUEHNER, J.; HARRINGTON, J. J.
WIRKSAME ANWENDUNG DES ALGORITHMUS FUER DIE GEMISCHT-GANZZAHLIGE PROGRAMMIERUNG MPSX-MIP VON IBM ZUR LOESUNG VON GROSSEN OEFFENTLICHEN INVESTITIONSPROBLEMEN MIT VIELEN BETEILIGTEN, TEIL 1, DIE. ANGEWANDTE INFORMATIK 19, 32-38. (1977).

KUEP 1976 -1 KUEPPER, H.-U.
VERGLEICHENDE ANALYSE GANZZAHLIGER PROGRAMMIERUNGSMODELLE ZUR ABLAUFPLANUNG INDUSTRIELLER UNTERNEHMUNGEN. TUEBINGEN: UNIV. TUEBINGEN, WIRTSCHAFTSWISSENSCHAFTLICHES SEMINAR, 65 P. (1976). (ARBEITSBERICHT NR. 2/1976)

KUHN 1973 -1 KUHN, W. W.
INVERSE LINE GRAPHS AND HAMILTONIAN CIRCUITS. IN: F. HOFFMAN, R. B. LEVOW, R. P. D. THOMAS (EDS.): PROCEEDINGS OF THE 4TH SOUTHEASTERN CONFERENCE ON COMBINATORICS, GRAPH THEORY AND COMPUTING, BOCA RATON 1973, 403-411. WINNIPEG: UTILITAS MATHEMATICA. (1973).

KULI 1975 -1 KULIKOWSKI, R.
OPTIMUM ALLOCATION OF RESOURCES IN DECENTRALIZED MACROECONOMIC MODEL. BULLETIN DE L'ACADEMIE POLONAISE DES SCIENCES, SERIE DES SCIENCES TECHNIQUES 23, 1039-1045. (1975).

KUND 1974 -1 KUNDU, S.
BOUNDS ON THE NUMBER OF DISJOINT SPANNING TREES. JOURNAL OF COMBINATORIAL THEORY 17, SERIES B, 199-203. (1974).

KUND 1974 -2 KUNDU, S.
DISJOINT REPRESENTATION OF TREE REALIZABLE SEQUENCES. SIAM JOURNAL ON APPLIED MATHEMATICS 26, 103-107. (1974).

KUND 1975 -1 KUNDU, S.
MINIMUM DIAMETER STEINER TREES IN A GRAPH, THE. PAPER PRESENTED AT THE 48TH JOINT NATIONAL MEETING OF ORSA AND TIMS, LOS ANGELES 1975, 22 P. (1975).

KUND 1976 -1 KUNDU, S.
LINEAR ALGORITHM FOR THE HAMILTONIAN COMPLETION NUMBER OF A TREE, A. INFORMATION PROCESSING LETTERS 5, NO. 2, 55-57. (1976).

KUNDA 1977 -1 KUNDU, S.; MISRA, J.
LINEAR TREE PARTITIONING ALGORITHM, A. SIAM JOURNAL ON COMPUTING 6, 151-154. (1977).

KUNG 1974 -1 KUNG, H. T.
ON THE COMPUTATIONAL COMPLEXITY OF FINDING THE MAXIMA OF A SET OF VECTORS. IN: 15TH ANNUAL SYMPOSIUM ON SWITCHING AND AUTOMATA THEORY, NEW ORLEANS 1974, 117-121. LONG BEACH: INSTITUTE OF ELECTRICAL AND ELECTRONICS ENGINEERS. (1974).

****

KURC 1974 -1 KURCEVIC, K. A.; KRAVTSOV, M. K.
METHOD FOR LOWERING THE DIMENSION OF THE TRANSPORTATION PROBLEM OF LINEAR PROGRAMMING (RUSSIAN), A. IN: E. G. GOL'STEIN (ED.): MATHEMATICAL METHODS IN ECONOMIC RESEARCH (RUSSIAN), 32-38. MOSCOW: IZDAT. NAUKA. (1974).

KURE 1977 -1 KUREJCHIK, V. M.; KOROLEV, A. G.
ON ONE METHOD FOR RECOGNITION OF GRAPHS ISOMORPHISM (RUSSIAN). KIBERNETIKA, NO. 2, 82-87. (1977).

KURI 1973 -1 KURISU, T.
TWO-MACHINE SEQUENCING PROBLEM WITH EXPONENTIAL PROCESSING TIMES. TECHNOLOGY REPORTS OF THE OSAKA UNIV. 23, 1-8. (1973).

KURI 1976 -1 KURISU, T.
TWO-MACHINE SCHEDULING UNDER REQUIRED PRECEDENCE AMONG JOBS. JOURNAL OF THE OPERATIONS RESEARCH SOCIETY OF JAPAN 19, NO. 1, 1-13. (1976).

KUSJ 1975 -1 KUSJURIN, M. N.
UEBER MINIMALE UEBERDECKUNGEN AUF EINER N-ELEMENTIGEN MENGE. WISSENSCHAFTLICHE ZEITSCHRIFT DER HUMBOLDT-UNIVERSITAET BERLIN. MATHEMATISCH-NATURWISSENSCHAFTLICHE REIHE 24, 729-730. (1975).

KUZJ 1977 -1 KUZJURIN, N. N.
MINIMAL COVERINGS AND MAXIMAL PACKINGS OF (K-1)-SUBSETS BY K-SUBSETS (RUSSIAN). MATEMATICESKIE ZAMETKI 21, 565-571. (1977).

KUZN 1974 -1 KUZNETSOV, V. F.
ALLOCATION OF RESOURCES FOR A GIVEN NETWORK (RUSSIAN). IN: E. G. GOL'STEIN (ED.): MATHEMATICAL METHODS IN ECONOMIC RESEARCH (RUSSIAN), 61-69. MOSCOW: IZDAT. NAUKA. (1974).

KYAN 1975 -1 KYAN, S.; ONAGA, K.
MINIMUM-DELAY MAXIMUM-GAUGE ROUTES IN A TIME-VARYING DYNAMIC NETWORK. ELECTRONICS AND COMMUNICATIONS IN JAPAN 58, NO. 3, 10-17. (1975).

LADN 1975 -1 LADNER, R. E.
ON THE STRUCTURE OF POLYNOMIAL TIME REDUCIBILITY. JOURNAL OF THE ASSOCIATION FOR COMPUTING MACHINERY 22, 155-171. (1975).

LAGE 1976 -1 LAGEWEG, B. J.; LENSTRA, J. K.; RINNOOY KAN, A. H .G.
GENERAL BOUNDING SCHEME FOR THE PERMUTATION FLOW-SHOP, A. WORKING PAPER 227. DELFT: INTERFACULTEIT BEDRIJFSKUNDE, 20 P. (1976).

LAGEA 1976 -1 LAGEWEG, B. J.; LENSTRA, J. K.; RINNOOY KAN, A. H. G.
GENERAL BOUNDING SCHEME FOR THE PERMUTATION FLOW-SHOP, A. PUBLICATION 227. MONTREAL: UNIV. DE MONTREAL, DEPT. D'INFORMATIQUE, 20 P. (1976).

LAGEA 1976 -2 LAGEWEG, B. J.; LENSTRA, J. K.; RINNOOY KAN, A. H. G.
JOB-SHOP SCHEDULING BY IMPLICIT ENUMERATION. PUBLICATION 228. MONTREAL: UNIV. DE MONTREAL, DEPT. D'INFORMATIQUE, 23 P. (1976).

LAGEA 1976 -3 LAGEWEG, B. J.; LENSTRA, J. K.; RINNOOY KAN, A. H. G.
JOB-SHOP SCHEDULING BY IMPLICIT ENUMERATION. WORKING PAPER 76/35. DELFT: INTERFACULTEIT BEDRIJFSKUNDE, 23 P. (1976).

LAGEA 1976 -4 LAGEWEG, B. J.; LENSTRA, J. K.; RINNOOY KAN, A. H. G.
MINIMIZING MAXIMUM LATENESS ON ONE MACHINE: COMPUTATIONAL EXPERIENCE AND SOME APPLICATIONS. STATISTICA NEERLANDICA 30, 25-41. (1976).

LAI 1975 -1 LAI, A. W.
HEURISTIC SOLUTION TO THE URBAN GOODS CONSOLIDATION TERMINAL SITE SELECTION DECISION, A. PAPER PRESENTED AT THE 48TH JOINT NATIONAL MEETING OF ORSA AND TIMS, LOS ANGELES 1975, 16 P. (1975).

LAM 1977 -1 LAM, S.; SETHI, R.
WORST CASE ANALYSIS OF TWO SCHEDULING ALGORITHMS. SIAM JOURNAL ON COMPUTING 6, 518-536. (1977).

LAMB 1975 -1 LAMBE, T. A.
INITIAL BASIS FOR THE TRANSPORTATION PROBLEM. PAPER PRESENTED AT THE 48TH JOINT NATIONAL MEETING OF ORSA AND TIMS, LAS VEGAS 1975, 7 P. (1975).

LAMBA 1975 -1 LAMBERT, G.
INDICE DE C-PARTITION DES SOMMETS D'UN GRAPHE. ALGORITHMES DE RECOUVREMENT ET APPLICATION AUX FICHIERS. DISSERTATION. PARIS: UNIV. PARIS VI, 214 P. (1975).

LAMBB 1963 -1 LAMBOURN, S.
RESOURCE ALLOCATION AND MULTIPROJECT SCHEDULING (RAMPS). A NEW TOOL IN PLANNING AND CONTROL. COMPUTER JOURNAL 5, 300-304. (1963).

LAND 1977 -1 LAND, A.; POWELL, S.
COMPUTER CODES FOR PROBLEMS OF INTEGER PROGRAMMING. VANCOUVER (CONFERENCE ON DISCRETE OPTIMIZATION, AUGUST 1977) 48 P. (1977).

LANDA 1978 -1 LAND, A. H.; LAPORTE, G.: M. , P.
UNIFIED FORMULATION OF THE MACHINE SCHEDULING PROBLEM, A. EUROPEAN JOURNAL OF OPERATIONS RESEARCH 2, 32-35. (1978).

LANDB 1966 -1 LAND, A. H.; MURCHLAND, J. D.
SUBSEQUENT ASSIGNMENT OF TRAFFIC TO SHORTEST PATHS, THE. REPORT LSE-TNT-27. LONDON: LONDON SCHOOL OF ECONOMICS, TRANSPORT NETWORK THEORY UNIT. (1966).

LANG 1977 -1 LANG, T.; FERNANDEZ, E. B.
IMPROVING THE COMPUTATION OF LOWER BOUNDS FOR OPTIMAL SCHEDULES. IBM JOURNAL OF RESEARCH AND DEVELOPMENT 21, 273-280. (1977).

LANGA 1962 -1 LANGEFORS, B.
ACTIVITY NETWORK FOR PLANNING AND SCHEDULING. BIT 2, 21-34. (1962).

LANGB 1973 -1 LANGLEY, R.
CONTINUOUS AND INTEGER GENERALIZED FLOW PROBLEMS. DISSERTATION. ATLANTA: GEORGIA INST. OF TECHNOLOGY. (1973).

LAPI 1976 -1 LAPIN, V. V.
ON A SCHEDULING ALGORITHM FOR SERVICING MANY DEMANDS (RUSSIAN). KIBERNETIKA, NO. 2, 90-94. (1976).

LATE 1976 -1 LATECOERE, L.; BOEBION, J.; PUN, L.
MODEL ADAPTIVE SCHEDULING FOR SOLVING PERTURBATION PROBLEMS IN THE CLOTH INDUSTRY. INTERNATIONAL JOURNAL OF PRODUCTION RESEARCH 14, 239-250. (1976).

LAUR 1975 -1 LAURIA, D. T.
REGIONAL SEWERAGE PLANNING BY MIXED INTEGER PROGRAMMING. CHAPEL HILL: UNIV. OF NORTH CAROLINA, DEPT. OF ENVIRONMENTAL SCIENCES AND ENGINEERING. (1975).

LAURA 1974 -1 LAURIERE, J. L.
PROBLEMES D'EMPLOI DU TEMPS ET ALGORITHME DE COLORATION DES HYPERGRAPHES. COMPTES RENDUS HEBDOMADAIRES DES SEANCES DE L'ACADEMIE DES SCIENCES 278, 1159-1162. (1974).

LAURA 1976 -1 LAURIERE, J. L.
LANGAGE ET UN PROGRAMME POUR ENONCER ET RESOUDRE DES PROBLEMES COMBINATOIRES, UN. DISSERTATION. PARIS: UNIV. DE PARIS 6. (1976).

LAURB 1976 -1 LAURIERE, M.
METHODE DE RESOLUTION RAPIDE POUR LE PROBLEME DU KNAPSACK, UNE. DISSERTATION. PARIS: UNIV. PARIS VI, 97 P. (1976).

LAURB 1978 -1 LAURIERE, M.
ALGORITHM FOR THE 0/1 KNAPSACK PROBLEM, AN. MATHEMATICAL PROGRAMMING 14, 1-10. (1978).

LAVE 1976 -1 LAVENBERG, S. S.; SHEDLER, G. S.
STOCHASTIC MODELING OF PROCESSOR SCHEDULING WITH APPLICATION TO DATA BASE MANAGEMENT SYSTEMS. IBM JOURNAL OF RESEARCH AND DEVELOPMENT 20, 437-448. (1976).

LAVI 1976 -1 LAVIT, C.
CLASSES DE SOLUTIONS PARAMETRIQUES D'UNE EQUATION BOOLEENNE. REVUE ROUMAINE DE MATHEMATIQUES PURES ET APPLIQUEES 21, 1049-1052. (1976).

LAWL 1971 -1 LAWLER, E. L.
MATROID INTERSECTION ALGORITHMS. ERL MEMO M-333, BERKELEY: UNIVERSITY OF CALIFORNIA. (1971).

LAWL 1971 -2 LAWLER, E. L.
MATROIDS WITH PARITY CONDITIONS: A NEW CLASS OF COMBINATORIAL OPTIMIZATION PROBLEMS. ERL MEMO M-334, BERKELEY: UNIVERSITY OF CALIFORNIA. (1971).

LAWL 1973 -1 LAWLER, E. L.
POLYNOMIAL-BOUNDED AND (APPARANTLY) NON-POLYNOMIAL-BOUNDED MATROID COMPUTATIONS. IN: R. RUSTIN (ED.): COMBINATORIAL ALGORITHMS, COURANT COMPUTER SCIENCE SYMPOSIUM NO. 9, 1972, 49-57. NEW YORK: ALGORITHMICS PRESS. (1973).

LAWL 1975 -1 LAWLER, E. L.
ALGORITHMS, GRAPHS, AND COMPLEXITY. NETWORKS 5, 89-92. (1975).

LAWL 1975 -2 LAWLER, E. L.
COMPLEXITY OF COMBINATORIAL COMPUTATIONS. IN: P. RINALDI (ED.): TOPICS IN COMBINATORIAL OPTIMIZATION, 87-95, WIEN, NEW YORK: SPRINGER. (1975).

LAWL 1975 -3 LAWLER, E. L.
COMPUTING SHORTEST PATH IN NETWORKS. IN: P. RINALDI (ED.): TOPICS IN COMBINATORIAL OPTIMIZATION, 1-5, WIEN, NEW YORK: SPRINGER. (1975).

LAWL 1975 -4 LAWLER, E. L.
INTRODUCTION TO MATROID OPTIMIZATION, AN. IN: P. RINALDI (ED.): TOPICS IN COMBINATORIAL OPTIMIZATION, 181-186, WIEN, NEW YORK: SPRINGER. (1975).

LAWL 1975 -5 LAWLER, E. L.
OPTIMAL SEQUENCING OF JOBS SUBJECT TO SERIES PARALLEL PRECEDENCE CONSTRAINTS. AMSTERDAM: MATHEMATISCH CENTRUM, 15 P. (1975). (BW 54/75)

LAWL 1975 -6 LAWLER, E. L.
OVERVIEW OF NETWORK FLOW THEORIE. IN: P. RINALDI (ED.): TOPICS IN COMBINATORIAL OPTIMIZATION, 97-107, WIEN, NEW YORK: SPRINGER. (1975).

LAWL 1975 -7 LAWLER, E. L.
QUADRATIC ASSIGNMENT PROBLEM: A BRIEF REVIEW, THE. IN: B. ROY (ED.): COMBINATORIAL PROGRAMMING: METHODS AND APPLICATIONS. PROCEEDINGS OF THE NATO ADVANCED STUDY INST., VERSAILLES 1974, 351-360. DORDRECHT: REIDEL. (1975).

LAWL 1975 -8 LAWLER, E. L.
SOME ASPECTS OF DUALITY IN COMBINATORIAL OPTIMIZATION. IN: P. RINALDI (ED.): TOPICS IN COMBINATORIAL OPTIMIZATION, 109-116, WIEN, NEW YORK: SPRINGER. (1975).

LAWL 1976 -1 LAWLER, E. L.
GRAPHICAL ALGORITHMS AND THEIR COMPLEXITY. IN: K. APT, J. DE BAKKER (EDS.): FOUNDATIONS OF COMPUTER SCIENCE II, PART 1, 3-32. MATHEMATICAL CENTRE TRACTS 81, AMSTERDAM: MATHEMATISCH CENTRUM. (1976).

LAWL 1976 -2 LAWLER, E. L.
NOTE ON THE COMPLEXITY OF THE CHROMATIC NUMBER PROBLEM, A. INFORMATION PROCESSING LETTERS 5, 66-67. (1976).

LAWL 1977 -1 LAWLER, E. L.
'PSEUDOPOLYNOMIAL' ALGORITHM FOR SEQUENCING JOBS TO MINIMIZE TOTAL TARDINESS, A. IN: P. L. HAMMER, E. L. JOHNSON, B. H. KORTE, G. L. NEMHAUSER (EDS.): STUDIES IN INTEGER PROGRAMMING, 331-342. ANNALS OF DISCRETE MATHEMATICS 1, AMSTERDAM: NORTH-HOLLAND. (1977).

LAWL 1977 -2 LAWLER, E. L.
COMMENT ON COMPUTING THE K SHORTEST PATHS IN A GRAPH. COMMUNICATIONS OF THE ACM 20, 603-604. (1977).

LAWL 1977 -3 LAWLER, E. L.
FAST APPROXIMATION ALGORITHMS FOR KNAPSACK PROBLEMS. MEMORANDUM NO. UCB/ERL M77/45. BERKELEY: UNIV. OF CALIFORNIA, COLLEGE OF ENGINEERING, 37 P. (1977).

LAWL 1977 -4 LAWLER, E. L.
SHORTEST PATH AND NETWORK FLOW ALGORITHMS. VANCOVER (CONFERENCE ON DISCRETE OPTIMIZATION, AUGUST 1977), 22 P. (1977).

LAWL 1978 -1 LAWLER, E. L.
SEQUENCING JOBS TO MINIMIZE TOTAL WEIGHTED COMPLETION TIME SUBJECT TO PRECEDENCE CONSTRAINTS. IN: ALSPACH, B.;HELL, P.;MILLER, D. J.(EDS.): ALGORITHMIC ASPECTS OF COMBINATORICS. ANNALS OF DISCRETE MATHEMATICS 2, 75-90. (1978).

LAWLA 1976 -1 LAWLER, E. L.; LABETOULLE, J.
SCHEDULING OF PARALLEL MACHINES WITH PREEMPTIONS. RAPPORT DE RECHERCHE NO. 203. LE CHESNAY: INSTITUT DE RECHERCHE D'INFORMATIQUE ET D'AUTTOMATIQUE, 25 P. (1976).

LAWLB 1976 -1 LAWLER, E. W.
INTRODUCTION TO THE COMPLEXITY OF ALGORITHMS. IN: R. T. YEH (ED.): APPLIED COMPUTATION THEORY: ANALYSIS, DESIGN, MODELING, 61-81. ENGLEWOOD CLIFFS: PRENTICE-HALL. (1976).

LAWR 1978 -1 LAWRENCE, L.
COVERING THE VERTEX SET OF A GRAPH WITH SUBGRAPHS OF SMALLER DEGREE. DISCRETE MATHEMATICS 21, 61-68. (1978).

LAWRA 1975 -1 LAWRENCE, S. M.; LAWRENCE, K. D.
LOCATING GROUPS OF INDUSTRIES IN A REGION USING AN INPUT-OUTPUT MODEL STRUCTURE AND MULTIPLE OBJECTIVE LINEAR PROGRAMMING. PAPER PRESENTED AT THE 48TH JOINT NATIONAL MEETING OF ORSA AND TIMS, LAS VEGAS 1975, 33 P. (1975).

LAZA 1977 -1 LAZAK, D.; GERRITSEN, S. M.
TOPOLOGISCHE KANAELE BEI JOB-SHOP-SCHEDULING-VERFAHREN. ANGEWANDTE INFORMATIK 19, 258-266. (1977).

LEA 1973 -1 LEA, A. C.
LOCATION-ALLOCATION MODELS: A REVIEW. DISSERTATION. TORONTO: UNIV. OF TORONTO, DEPT. OF GEOGRAPHY. (1973).

LEBE 1974 -1 LEBEDEV, S. S.
USE OF GENERALIZED RESOLVENT MULTIPLIERS FOR THE SOLUTION OF INTEGER LINEAR PROGRAMMING PROBLEMS BY THE METHOD OF ORDERED SORTING (RUSSIAN), THE. IN: MATHEMATICAL METHODS IN ECONOMIC RESEARCH (RUSSIAN), 70-78. MOSCOW: IZDAT. NAUKA. (1974).

LEBE 1975 -1 LEBEDEV, S. S.
GENERALIZED LAGRANGE MULTIPLIERS AND ESTIMATION PROBLEMS IN INTEGER PROGRAMMING (RUSSIAN). IN: PROCEEDINGS OF THE 6TH WINTER SCHOOL ON MATHEMATICAL PROGRAMMING AND RELATED QUESTIONS (RUSSIAN), DROGOBYCH 1973, 114-157. MOSCOW: CENTRAL. EKONOM.-MAT. INST. AKAD. NAUK SSSR. (1975).

LEBEA 1976 -1 LEBEDEV, S. S.; KOVALEVSKAJA, M. I.
LAGRANGE MULTIPLIERS IN THE SIMPLEST ALLOCATION PROBLEM (RUSSIAN). IN: STUDIES IN DISCRETE OPTIMIZATION (RUSSIAN), 170-180. MOSCOW: IZDAT. NAUKA. (1976).

LEBEB 1977 -1 LEBEDEV, V. YU.
METHOD FOR THE SOLUTION OF MIXED-INTEGER MATHEMATICAL PROGRAMMING PROBLEMS (RUSSIAN), THE. ZURNAL VYCISLITEL'NOI MATEMATIKI I MATEMATICESKOI FIZIKI 17, 1189-1193. (1977).

LEBEC 1977 -1 LEBEDEVA, T. T.; ROSHCHIN, V. A.; SERGIENKO, I. V.
APPROACH TO SOLUTION OF MATHEMATICAL PROGRAMMING PROBLEMS WITH MIXED VARIABLES (RUSSIAN), AN. KIBERNETIKA, NO. 2, 53-57. (1977).

LEBED 1974 -1 LEBENSOLD, K. W.
EXTREMA OF A CLASS OF FUNCTIONS ON A FINITE SET. CANADIAN JOURNAL OF MATHEMATICS 26, 806-819. (1974).

LEBL 1975 -1 LEBLANC, L. J.
ALGORITHM FOR THE DISCRETE NETWORK DESIGN PROBLEM, AN. TRANSPORTATION SCIENCE 9, 183-199. (1975).

LEBL 1976 -1 LEBLANC, L. J.
CONJUGATE GRADIENT TECHNIQUE FOR CERTAIN QUADRATIC NETWORK PROBLEMS, THE. NAVAL RESEARCH LOGISTICS QUARTERLY 23, 597-602. (1976).

LEBL 1977 -1 LEBLANC, L. J.
HEURISTIC APPROACH OF LARGE SCALE DISCRETE STOCHASTIC TRANSPORTATION-LOCATION PROBLEMS, A. COMPUTERS AND MATHEMATICS WITH APPLICATIONS 3, 87-94. (1977).

LEBLA 1974 -1 LEBLANC, L. J.; MORLOK, L. J.
ANALYSIS AND COMPARISON OF BEHAVIORIAL ASSUMPTIONS IN TRAFFIC ASSIGNMENT, AN. INTERNATIONAL SYMPOSIUM ON TRAFFIC EQUILIBRIUM METHODS, MONTREAL. (1974).

LEBLB 1974 -1 LEBLANC, L. J.; MORLOK, L. J.; PIERSKALLA, W.
ACCURATE AND EFFICIENT APPROACH TO EQUILIBRIUM TRAFFIC ASSIGNMENT ON A CONJESTED NETWORKS, AN. TRAN. RESEARCH RECORD 491, INTERACTIVE GRAPHICS AND TRANSPORTATION SYSTEMS PLANNING, 12-33. (1974).

LECH 1976 -1 LECHEVALLIER, Y.
CLASSIFICATION AUTOMATIQUE OPTIMALE SOUS CONTRAINTE D'ORDRE TOTAL. RAPPORT DE RECHERCHE NO. 200. LE CHESNAY: INSTITUT DE RECHERCHE D'INFORMATIQUE ET D'AUTOMATIQUE, 24 P. (1976).

LECHA 1970 -1 LECHNER, R. J.
DETECTION OF PRIME IMPLICANTS BY ABSTRACT HARMONIC ANALYSIS. IN: PROCEEDINGS OF THE FOURTH ANNUAL PRINCETON CONFERENCE ON INFORMATION SCIENCES AND SYSTEMS, 56. (1970).

LEE. 1962 -1 LEE. C. Y.
NOTE ON THE N-TH SHORTEST PATH PROBLEM, A. IRE TRANSACTIONS ON ELECTRONIC COMPUTERS EC-11, 572-573. (1962).

LEE 1976 -1 LEE, D. H.
LOW COST DRAINAGE NETWORKS. NETWORKS 6, 351-372. (1976).

LEE A 1976 -1 LEE, J.; BOSE, N.; HWANG, F.
USE OF STEINER'S PROBLEM IN SUBOPTIMAL ROUTING IN RECTILINEAR METRIC. IEEE TRANSACTIONS ON CIRCUITS AND SYSTEMS CAS-23, 470-476. (1976).

LEE B 1976 -1 LEE, S. M.
INTERACTIVE INTEGER GOAL PROGRAMMING: METHODS AND APPLICATION. IN: ZIONTS, P.,(ED.): LECTURE NOTES IN ECONOMICS AND MATHEMATICAL SYSTEMS 155, BERLIN: SPRINGER, 362-383. (1976).

LEEU 1978 -1 LEEUWEN, J. VAN; EMDE BOAS, P. VAN
SOME ELEMENTARY PROOFS OF LOWER BOUNDS IN COMPLEXITY THEORY. LINEAR ALGEBRA AND ITS APPLICATIONS 19, 63-80. (1978).

LEGE 1977 -1 LEGENDRE, J.-P.; MINOUX, M.
APPLICATION DE LA NOTION DE DUALITE EN PROGRAMMATION EN NOMBRES ENTIERS: SELECTION ET AFFECTATION OPTIMALES D'UNE FLOTTE D'AVIONS, UNE. REVUE FRANCAISE D'AUTOMATIQUE, D'INFORMATIQUE ET DE RECHERCHE OPERATIONELLE 11, RECHERCHE OPERATIONELLE, 201-222. (1977).

LEIF 1967 -1 LEIFMAN, L. Y. A.
PROBLEM OF ALLOCATION OF RESOURCES IN NETWORK SCHEDULING (RUSSIAN). DOKLADY AKADEMII NAUK SSSR 175, 44-46. (1967).

LEIT 1972 -1 LEITEN, A.
OBTAINING CUTS IN THE B-ALGORITHM (RUSSIAN). TARTU RIIKLIKU UELIKOOLI TOIMETISED VIH. 305, 269-272. (1972).

LEITA 1973 -1 LEITEN, A. K.
CERTAIN MODIFICATIONS OF THE TRAVELLING SALESMAN PROBLEM (RUSSIAN). TRUDY VYCISLITEL'NOGO CENTRA. TARTUSKII GOSUDARSTVENNYI UNIVERSITET, VYP. 28, 44-58. (1973).

LEITA 1974 -1 LEITEN, A. K.
CERTAIN MATHEMATICAL MODELS FOR DISCRETE PROGRAMMING (RUSSIAN). TRUDY VYCISLITEL'NOGO CENTRA, TARTUSKII GOSUDARSTVENNYI UNIVERSITET VYP. 31, 36-50. (1974).

LEMA 1971 -1 LEMAIRE, B.
PROBLEMES DE TOURNEES AVEC CONTRAINTES MULTIPLES. DISSERTATION. PARIS, UNIV. PARIS 6. (1971).

LEMA 1975 -1 LEMAIRE, B.
DENOMBREMENT DES CYCLES HAMILTONIENS DE KN ET KN,N EMPRUNTANT OU EVITANT DES ARETES DONNEES. REVUE FRANCAISE D'AUTOMATIQUE, D'INFORMATIQUE ET DE RECHERCHE OPERATIONNELLE 9, NO. 2, 101-111. (1975).

LEMA 1976 -1 LEMAIRE, B.
FONDEMENTS, GENERALISATION ET CRITIQUE DE LA NOTION D'AFFINITE (PROBLEME DU VOYAGEUR DE COMMERCE). REVUE FRANCAISE D'AUTOMATIQUE, D'INFORMATIQUE ET DE RECHERCHE OPERATIONNELLE 10, NO. 10, 43-58. (1976).

LENI 1971 -1 LENIN, A. Y. U.
ALGORITHM FOR THE SHORTEST CONNECTION BETWEEN A GROUP OF GRAPH VERTICES (TRANSLATED FROM RUSSIAN), AN. SOVIET MATHEMATICS, DOKLADY 12, 1477-1481. (1971).

LENS 1973 -1 LENSTRA, H. W.
ACYCLIC SUBGRAPH PROBLEM, THE. WORKING PAPER BW 26/73. AMSTERDAM: MATHEMATISCH CENTRUM. (1973).

LENSA 1977 -1 LENSTRA, J. K.
SEQUENCING BY ENUMERATIVE METHODS. MATHEMATICAL CENTRE TRACTS, 69. AMSTERDAM: MATHEMATISCH CENTRUM, 198 P. (1977).

LENSB 1975 -1 LENSTRA, J. K.; RINNOOY KAN, A. H. G.
ON THE JOB-SHOP SCHEDULING PROBLEM. PAPER PRESENTED AT THE FIRST EUROPEAN CONGRESS ON OPERATIONS RESEARCH, EURO 1, BRUSSELS 1975. (1975).

LENSB 1976 -1 LENSTRA, J. K.; RINNOOY KAN, A. H. G.
COMPLEXITY OF SCHEDULING UNDER PRECEDENCE CONSTRAINTS. WORKING PAPER 76/34. DELFT: GRADUATE SCHOOL OF MANAGEMENT, 20 P. (1976).

LENSB 1976 -2 LENSTRA, J. K.; RINNOOY KAN, A. H. G.
ON GENERAL ROUTING PROBLEMS. NETWORKS 6, 273-280. (1976).

LENSB 1976 -3 LENSTRA, J. K.; RINNOOY KAN, A. H. G.
TRAVELLING ON A GRAPH (DUTCH). NIEUW TIJDSCHRIFT VOOR WISKUNDE 63, 221-229. (1976).

****

LENSB 1977 -1 LENSTRA, J. K.; RINNOOY KAN, A. H. G.
COMPUTATIONAL COMPLEXITY OF DISCRETE OPTIMIZATION PROBLEMS. VANCOUVER (CONFERENCE ON DISCRETE OPTIMIZATION, AUGUST 1977) 22 P. (1977).

LENSC 1977 -1 LENSTRA, J. K.; RINNOOY KAN, A. H. G.; BRUCKER, P.
COMPLEXITY OF MACHINE SCHEDULING PROBLEMS. IN: P. L. HAMMER, E. L. JOHNSON, B. H. KORTE, G. L. NEMHAUSER (EDS.): STUDIES IN INTEGER PROGRAMMING, 343-362. ANNALS OF DISCRETE MATHEMATICS 1, AMSTERDAM: NORTH-HOLLAND. (1977).

LEON 1975 -1 LEONT'EV, V. K.
STABILITY OF THE TRAVELING SALESMAN PROBLEM (RUSSIAN). ZURNAL VYCISLITEL'NOI MATEMATIKI I MATEMATICESKOI FIZIKI 15, 1298-1309 + 1362. (1975).

LEON 1976 -1 LEONT'EV, V. K.
STABILITY IN COMBINATORIAL PROBLEMS OF SELECTION (RUSSIAN). DOKLADY AKADEMII NAUK SSSR 228, 23-25. (1976).

LEPE 1976 -1 LEPESHINSKIJ, N. A.; MALYSHKO, V. V.
GRAPHS OF ARCS AND DETERMINATION OF THE ELEMENTARY PATHS BETWEEN THE PAIRS OF VERTICES OF A GRAPH (RUSSIAN). VESCI AKADEMII NAVUK BSSR, SERYJA FIZIKA-MATEMATYCNYH NAVUK, NO. 1, 5-8. (1976).

LERD 1975 -1 LERDA, F.; MAJORANI, E.
COMMENTS ON A NOTE BY F. RUBIN: "A NOTE ON LERDA AND MAJORANI MINIMUM CROSSING ALGORITHM". CALCOLO 12, 315-316. (1975).

LERN 1969 -1 LERNER, A. Y.; MAKAROV, I. M.
OPTIMAL ALLOCATION OF RESOURCES (RUSSIAN). PAPER PRESENTED AT THE IFAC SYMPOSIUM FOR COMPUTER CONTROL OF NATURAL RESOURCES AND PUBLIC UTILITIES 1967. HAIFA: IFAC, 23 P. (1969).

LERO 1975 -1 LEROY, P.
OPTIMISATION ET GRAPHES APPLICATIONS A L'ARCHITECTURE. CAH. BUR. UNIV. RECH. OPERAT., NO. 23, 85-100. (1975).

LESN 1977 -1 LESNIAK-FOSTER, L.
SOME RECENT RESULTS IN HAMILTONIAN GRAPHS. JOURNAL OF GRAPH THEORY 1, 27-36. (1977).

LESNA 1975 -1 LESNIAK, L.
ON LONGEST PATHS IN CONNECTED GRAPHS. FUNDAMENTA MATHEMATICAE 86, 283-286. (1975).

LESNA 1975 -2 LESNIAK, L.
ON 1-EDGE HAMILTONIAN GRAPHS. BULLETIN DE L'ACADEMIE POLONAISE DES SCIENCES 23, SERIE DES SCIENCES MATHEMATIQUES, ASTRONOMIQUES ET PHYSIQUES, 1223-1226. (1975).

LETI 1976 -1 LETICHEVSKIY, A. A.
HOW TO SPEED ITERATION OF MONOTONOUS OPERATORS (RUSSIAN). KIBERNETIKA, NO. 4, 1-7. (1976).

LEV 1976 -1 LEV, O. E.
STRUCTURAL OPTIMIZATION SOLUTION TO A BRANCH-AND-BOUND PROBLEM, A. QUARTERLY OF APPLIED MATHEMATICS 34, 365-371. (1976).

LEVI 1972 -1 LEVI, G.
NOTE ON THE DERIVATION OF MAXIMAL COMMON SUBGRAPHS OF TWO DIRECTED OR UNDIRECTED GRAPHS. CALCOLO 9, 341-354. (1972).

LEVI 1974 -1 LEVI, G.
GRAPH ISOMORPHISM: A HEURISTIC EDGE-PARTITIONING-ORIENTED ALGORITHM, A. COMPUTING 12, 291-313. (1974).

LEVIA 1963 -1 LEVIN, B. M.; HEDETNIEMI, S.
DETERMINING FASTEST ROUTES USING FIXED SCHEDULES. IN: PROCEEDINGS OF THE INTERNAT. FEDERATION FOR INFORMATION PROCESSING SOCIETIES, SPRING JOINT COMP. CONF., DETROIT. BALTIMORE: SPARTAN; ALSO PUBLISHED IN: PROCEEDINGS OF THE AMERICAN FEDERATION FOR INFORMATION PROCESSING SOCIETIES 23, 1963. (1963).

LEVN 1973 -1 LEVNER, E. V.
NETWORK APPROACH TO PROBLEMS OF SCHEDULING THEORY (RUSSIAN), A. IN: A. A. FRIDMAN (ED.): STUDIES IN DISCRETE MATHEMATICS (RUSSIAN), 135-150. MOSCOW: IZDAT. NAUKA. (1973).

LEVY 1966 -1 LEVY, J.
OPTIMAL LOCATION OF NETWORKS. DISSERTATION. CHICAGO: UNIVERSITY OF CHICAGO. (1966).

LEVYA 1973 -1 LEVY, L. S.
ORDERING ON TREES - SIMPLIFIED PROOFS. IN: PROCEEDINGS OF THE SEVENTH ANNUAL PRINCETON CONFERENCE ON INFORMATION SCIENCES AND SYSTEMS, 400-402. (1973).

LEVYB 1976 -1 LEVYI, V. D.; VOLODIN, A. A.
INTEGER PROGRAMMING PROBLEM WITH VARIABLE COEFFICIENTS (TRANSLATED FROM RUSSIAN), AN. MATEKON 12, NO. 3, 75-83. (1976).

LIBU 1974 -1 LIBURA, M.
REDUCTION OF A LINEAR INTEGER PROGRAMMING PROBLEM TO GROUP-THEORETIC FORM (POLISH). ARCHIWUM AUTOMATYKI I TELEMECHANIKI 19, 229-237. (1974).

LIEB 1970 -1 LIEBLING, T.
GRAPHENTHEORIE IN PLANUNGS- UND TOURENPROBLEMEN. LECTURE NOTES IN OPERATIONS RESEARCH AND MATHEMATICAL SYSTEMS 21. BERLIN: SPRINGER. (1970).

LIEBA 1964 -1 LIEBMAN, J. C.
BRANCH-AND-BOUND ALGORITHM FOR MINIMIZING THE COSTS OF WASTE TREATMENT, SUBJECT TO EQUITY CONSTRAINS, A. IBM SCIENTIFIC COMPUTING SYMPOSIUM ON WATER AND AIR RESOURCE MANAGEMENT, NEW YORK: WATSON RESEARCH CENTER. (1964).

LIEN 1974 -1 LIEN, Y. E.
PRELIMINARY REPORT: SOME COMPLETE PROBLEMS FOR LANGUAGES RECOGNIZABLE IN NONDETERMINISTIC LOG-SPACE. IN: PROCEEDINGS OF THE EIGHTH ANNUAL PRINCETON CONFERENCE ON INFORMATION SCIENCES AND SYSTEMS, 184-188. (1974).

LIET 1977 -1 LIETAER, B. A.
PLANNING MODEL FOR UNDERGROUND MINES. AN APPLICATION IN A DEVELOPING COUNTRY, A. OMEGA 5, 149-159. (1977).

LIHA 1975 -1 LIHACEV, V. M.
CERTAIN ESTIMATE FOR THE NUMBER OF VERTICES OF A TRANSPORTATION POLYTOPE (RUSSIAN), A. VESCI AKADEMII NAVUK BSSR, SERYJA FIZIKA-MATEMATYCNYH NAVUK, NO. 2, 99 + 142. (1975).

LIHAA 1974 -1 LIHACEV, V. M.; EMELICHEV, V. A.
ON AN UPPER BOUND OF THE NUMBER OF VERTICES OF A TRANSPORTATION POLYTOPE (RUSSIAN). VESCIAKADEMII NAVUK BSSR. SERYJA FIZIKA-MATEMATYCNYH NAVUK, NO. 3, 121-123 + 143. (1974).

LIM 1977 -1 LIM, M.-H.
PRODUCT OF MATROIDS AND ITS AUTOMORPHISM GROUP, A. JOURNAL OF COMBINATORIAL THEORY 23, SERIES B, 151-163. (1977).

LIN 1977 -1 LIN, B. W. Y.; RARDIN, R. L.
DEVELOPMENT OF A PARAMETRIC GENERATING PROCEDURE FOR INTEGER PROGRAMMING TEST PROBLEMS. JOURNAL OF THE ASSOCIATION FOR COMPUTING MACHINERY 24, 465-472. (1977).

LIN A 1969 -1 LIN, P. M.; ALDERSON, G. E.
SYMBOLIC NETWORK FUNCTIONS BY A SINGLE PATH-FINDING ALGORITHM. IN: ALLERTON CONFERENCE ON CIRCUIT AND SYSTEM THEORY, 7TH ANNUAL PROCEEDINGS. (1969).

****

LIN B 1974 -1 LIN, P. M.; LEON, B. J.
IMPROVING THE EFFICIENCY OF LABELING ALGORITHMS FOR MAXIMUM FLOWS IN NETWORKS. IN: PROCEEDINGS OF THE 1974 IEEE INTERNATIONAL SYMPOSIUM ON CIRCUITS AND SYSTEMS, 162-166. NEW YORK: INTERNAT. INST. OF ELECTRICAL AND ELECTRONICAL ENGINEERS. (1974).

LIN C 1975 -1 LIN, S.
COMPUTER SOLUTIONS OF THE TSP. BELL SYSTEMS TECHNICAL JOURNAL 44, 2245-2269. (1975).

LIN D 1971 -1 LIN, S.; KERNIGHAN, B. W.
HEURISTIC TECHNIQUE FOR SOLVING A CLASS OF COMBINATORIAL OPTIMIZATION PROBLEMS, A. IN: PROCEEDINGS OF THE FIFTH ANNUAL PRINCETON CONFERENCE ON INFORMATION SCIENCES AND SYSTEMS, 210-213. (1971).

LIND 1973 -1 LIND, C. B.; LORENTSEN, R.
APPLICATION OF A MIXED INTEGER LINEAR PROGRAMMING MODEL FOR THE DIMENSIONING OF TANK TRUCK FLEETS. PAPER PRESENTED AT THE NORDISH OPERATIONAL RESEARCH CONGRESS NOAK-73, COPENHAGEN 1973. (1973).

LINDA 1977 -1 LIND, N. C.
CONTINUOUS ALLOCATION PROBLEMS. JOURNAL OF OPTIMIZATION THEORY AND APPLICATIONS 23, 101-110. (1977).

LING 1973 -1 LING, R. F.
EXPECTED NUMBER OF COMPONENTS IN RANDOM LINEAR GRAPHS, THE. ANNALS OF PROBABILITY 1, 876-881. (1973).

LING 1975 -1 LING, R. F.
EXACT PROBABILITY DISTRIBUTION ON THE CONNECTIVITY OF RANDOM GRAPHS, AN. JOURNAL OF MATHEMATICAL PSYCHOLOGY 12, 90-98. (1975).

LINK 1976 -1 LINKE, K.-F.
BEMERKUNGEN ZUR MATHEMATISCH-OEKONOMISCHEN MODELLIERUNG UND NUMERISCHEN LOESUNG VON REIHENFOLGEAUFGABEN AUS DER INDUSTRIELLEN FERTIGUNG. WISSENSCHAFTLICHE ZEITSCHRIFT DER TECHNISCHEN HOCHSCHULE OTTO VON GUERICKE, MAGDEBURG, NO 2, 211-222. (1976).

LINS 1975 -1 LINS, S.
USE OF MATROID THEORY IN OPTIMAL MULTI-ASSIGNMENT OVER TOTALLY ORDERED SETS. THE. NOTAS E COMUNICACOES DE MATEMATICA, NO. 64. RECIFE: UNIV. DE PERNAMBUCO, DEPT. DE MATEMATICA, 8 P. (1975).

LINSA 1969 -1 LINSKIY, V. S.; KORNEV, M. D.
SCHEDULING TWO-STATE PROCESSES WITHOUT A BUFFER. ENGINEERING CYBERNETICS, 43-48. (1969).

LIPT 1976 -1 LIPTON, S. L.
SERVICE CENTER LOCATION USING MIXED INTEGER PROGRAMMING. SYRACUSE: NIAGARA MOHAWK, 52 P. (1976).

LISK 1977 -1 LISKOVEC, V. A.
GENERAL ENUMERATION SCHEME FOR LABELLED GRAPHS (RUSSIAN), A. DOKLADY AKADEMII NAUK BSSR 21, NO. 6, 496-499 + 571. (1977).

LITT 1976 -1 LITTGER, K. W.
NEW APPROACH FOR THE OPTIMUM SOLUTION OF THE M BY J PRODUCTION SCHEDULING PROBLEM, A. INTERNATIONAL JOURNAL OF PRODUCTION RESEARCH 14, 77-84. (1976).

LITTA 1976 -1 LITTLE, C. H. C.; HOLTON, D. A.
NEW CHARACTERIZATION OF PLANAR GRAPHS, A. BULLETIN OF THE AMERICAN MATHEMATICAL SOCIETY 83, 137-138. (1976).

LITV 1973 -1 LITVAK, B. G.
REDUCIBILITY OF ABSOLUTELY INTEGER PROBLEMS WITH A FIXED CONSTRAINT VECTOR (RUSSIAN), THE. IN: A. A. FRIDMAN (ED.): STUDIES IN DISCRETE MATHEMATICS (RUSSIAN), 41-45. MOSCOW: IZDAT. NAUKA. (1973).

LITVA 1973 -1 LITVAK, B. G.; NAIVEL'T, A. V.
SOLUTION OF THE MULTIDIMENSIONAL KNAPSACK PROBLEM WITH SUPPLEMENTARY CONSTRAINTS (RUSSIAN), THE. IN: A. A. FRIDMAN (ED.): STUDIES IN DISCRETE MATHEMATICS (RUSSIAN), 69-83. MOSCOW: IZDAT. NAUKA. (1973).

LITVA 1975 -1 LITVAK, B. G.; NAIVEL'T, A. V.
SOLUTION OF THE INTEGER LINEAR PROGRAMMING PROBLEM BY MEANS OF A GROUP-THEORETIC APPROACH (RUSSIAN). IN: PROCEEDINGS OF THE 6TH WINTER SCHOOL ON MATHEMATICAL PROGRAMMING AND RELATED QUESTIONS (RUSSIAN), DROGOBYCH 1973, 158-166. MOSCOW: CENTRAL. EKONOM.-MAT. INST. AKAD. NAUK SSSR. (1975).

LITVA 1976 -1 LITVAK, B. G.; NAIVEL'T, A. V.
MUTUALLY INCOMPARABLE SOLUTIONS IN THE KNAPSACK PROBLEM WITH SUPPLEMENTARY BLOCK TYPE CONSTRAINTS (RUSSIAN). IN: STUDIES IN DISCRETE OPTIMIZATION (RUSSIAN), 187-203. MOSCOW: IZDAT. NAUKA. (1976).

LITVB 1975 -1 LITVAK, E. I.
ON THE PROBABILITY OF CONNECTEDNESS OF A GRAPH (RUSSIAN). IZVESTIJA AKADEMII NAUK SSSR. TEHNICESKAJA KIBERNETIKA, NO. 5, 161-165. (1975).

LITVC 1974 -1 LITVAK, E. I.; ZAVARIN, A. N.
ON THE DETERMINATION OF THE LENGTH OF THE SHORTEST PATH IN A GRAPH (RUSSIAN). IZVESTIJA AKADEMII NAUK SSSR, TEHNICESKAJA KIBERNETICA, NO. 4, 40-42 + 220; TRANSLATED AS ENGENEERING CYBERNETICS 12, 1974, NO. 4, 29-32. (1974).

LIU 1973 -1 LIU, C. L.
COMBINATORIAL STUDY OF SOME SCHEDULING ALGORITHMS (SUMMARY), A. IN: COMPUTING SYSTEMS 1973, PROCEEDINGS OF THE 2ND TEXAS CONF., AUSTIN 1973, PAPER 40, 3 P. AUSTIN: UNIV. OF TEXAS. (1973).

LIU 1975 -1 LIU, C. L.
APPROXIMATION ALGORITHMS FOR DISCRETE OPTIMIZATION PROBLEMS (EXTENDED ABSTRACT). IN: COMPUTING SYSTEMS 1975, PROCEEDINGS OF THE 4TH TEXAS CONF., AUSTIN 1975, PAPER 4B-1, 5 P. AUSTIN: UNIV. OF TEXAS. (1975).

LIU 1976 -1 LIU, C. L.
COMBINATORIAL STUDY OF SOME SCHEDULING ALGORITHMS, A. IN: R. T. YEH (ED.): APPLIED COMPUTATION THEORY: ANALYSIS, DESIGN, MODELING, 82-105. ENGLEWOOD CLIFFS: PRENTICE-HALL. (1976).

LIU A 1976 -1 LIU, J.
LARGE TEAM EFFICIENCY FROM AN INTEGER-PROGRAMMING THEORETICAL POINT OF VIEW. MIAMI (50. ORSA-MEETING + TIMS, NOVEMBER 1976) 13 P. (1976).

LIVO 1976 -1 LIVOVSCHI, L.; SOMNEA, D.
SEARCH ALGORITHM FOR THE CIRCUITS OF A GRAPH (RUMANIAN), A. STUDII SI CERCETARI MATEMATICE 28, 189-197. (1976).

LIVS 1972 -1 LIVSIK, E. M.; RUBLINECKII, V. I.
COMPERATIVE COMPLEXITY OF CERTAIN DISCRETE OPTIMIZATION (RUSSIAN), THE. IN: NUMERICAL MATHEMATICS AND COMPUTER TECHNOLOGY, NO. 3 (RUSSIAN), 78-85. AKAD. NAUK UKRAIN. SSR FIZ.-TEHN. INST. NIZKIH KARKOW TEMPERATUR. (1972).

LIVS 1972 -2 LIVSIK, E. M.; RUBLINECKII, V. I.
OPTIMAL SPLITTING OF AN ORDERED SET INTO INTERVALS (RUSSIAN), THE. IN: NUMERICAL MATHEMATICS AND COMPUTER TECHNOLOGY, NO. 3 (RUSSIAN), 86-89. KHARKOV: AKAD. NAUK UKRAIN. SSR FIZ.-TEHN. INST. NIZKIH TEMPERATUR. (1972).

****

LODI 1970 -1 LODISH, L. M.
COMPUTATIONAL LIMITATIONS OF DYNAMIC PROGRAMMING FOR WAREHOUSE LOCATION. JOURNAL OF MARKETING RESEARCH 7, 262-263. (1970).

LOHS 1976 -1 LOHSE, D.
ALGORITHMUS 31. EIN ALGORITHMUS ZUR BESTIMMUNG VON WEGANZAHLEN IN NETZEN. COMPUTING 17, 93-96. (1976).

LONG 1977 -1 LONG, C. F.
STATE OF THE ART IN SCHEDULING PRACTICE, THE. SAN FRANCISCO (51. ORSA-MEETING + TIMS, MAY 1977) 31 P. (1977).

LORE 1972 -1 LOREA, M.
ENSEMBLES STABLES DANS LES HYPERGRAPHES. COMPTES RENDUS HEBDOMADAIRES DES SEANCES DE L'ACADEMIE DES SCIENCES 275, 163-165. (1972).

LORE 1975 -1 LOREA, M.
HYPERGRAPHES ET MATROIDES. CAHIERS DU CENTRE D'ETUDES DE RECHERCHE OPERATIONNELLE 17, 289-291. (1975).

LORE 1975 -2 LOREA, M.
MATROIDES ET LEURS APPLICATIONS A DES PROBLEMES DE RECHERCHE OPERATIONNELLE, LES. REPORT (MEMOIRE). BRUXELLES: UNIV. DE BRUXELLES. (1975).

LORE 1976 -1 LOREA, M.
STRUCTURE DE CERTAINS PROBLEMES D'AFFECTATION, LA. REVUE FRANCAISE D'AUTOMATIQUE, D'INFORMATIQUE ET DE RECHERCHE OPERATIONNELLE 10, NO. 10, 59-64. (1976).

LOS 1978 -1 LOS, M.
SIMULTANEOUS OPTIMIZATION OF LAND USE AND TRANSPORTATION: A SYNTHESIS OF THE QUADRATIC ASSIGNMENT PROBLEM AND THE OPTIMAL NETWORK PROBLEM. REGIONAL SCIENCE AND URBAN ECONOMICS 8, 21-42. (1978).

LOS A 1977 -1 LOS, M.; NGUYEN, S.
SPATIAL ALLOCATION ON A NETWORK WITH CONGESTION. PUBLICATION NO. 59. MONTREAL: UNIV. DE MONTREAL, CENTRE DE RECHERCHE SUR LES TRANSPORTS, 25 P. (1977).

LOVA 1968 -1 LOVASZ, L.
ON CHROMATIC NUMBER OF FINITE SET SYSTEMS. ACTA MATHEMATICA ACADEMIAE SCIENTIARUM HUNGARICAE 19, 59-67. (1968).

LOVA 1974 -1 LOVASZ, L.
MINIMAX THEOREMS FOR HYPERGRAPHS. IN: C. BERGE, D. RAY-CHAUDHURI (EDS.): HYPERGRAPH SEMINAR, 111-126. LECTURE NOTES IN MATHEMATICS 411, BERLIN: SPRINGER. (1974).

LOVA 1975 -1 LOVASZ, L.
ON THE RATIO OF OPTIMAL INTEGRAL AND FRACTIONAL COVERS. DISCRETE MATHEMATICS 13, 383-390. (1975).

LOVA 1975 -2 LOVASZ, L.
2-MATCHINGS AND 2-COVERS OF HYPERGRAPHS. ACTA MATHEMATICA ACADEMIAE SCIENTIARUM HUNGARICAE 26, 433-444. (1975).

LOVA 1976 -1 LOVASZ, L.
COVERS, PACKINGS AND SOME HEURISTIC ALGORITHMS. IN: C. ST. J. A. NASH-WILLIAMS (ED.): PROCEEDINGS OF THE 5TH BRITISH COMBINATORIAL CONFERENCE, ABERDEEN 1975, 417-430. CONGRESSUS NUMERANTIUM 15, WINNIPEG: UTILITAS MATHEMATICA. (1976).

LOVA 1976 -2 LOVASZ, L.
ON TWO MINIMAX THEOREMS IN GRAPHS. JOURNAL OF COMBINATORIAL THEORY 21, SERIES B, 96-103. (1976).

LOVA 1977 -1 LOVASZ, L.
CERTAIN DUALITY PRINCIPLES IN INTEGER PROGRAMMING. IN: P. L. HAMMER, E. L. JOHNSON, B. H. KORTE, G. L. NEMHAUSER (EDS.): STUDIES IN INTEGER PROGRAMMING, 363-374. ANNALS OF DISCRETE MATHEMATICS 1, AMSTERDAM: NORTH-HOLLAND. (1977).

LOVA 1977 -2 LOVASZ, L.
GRAPH THEORY AND INTEGER PROGRAMMING. VANCOUVER (CONFERENCE ON DISCRETE OPTIMIZATION, AUGUST 1977) 30 P. (1977).

LOVA 1977 -3 LOVASZ, L.
ON THE SHANNON CAPACITY OF A GRAPH. PREPRINT 27/1977. BUDAPEST: HUNGARIAN ACADEMY OF SCIENCES, MATHEMATICAL INSTITUTE, 23 P. (1977).

LOVE 1976 -1 LOVE, R. F.
ONE-DIMENSIONAL FACILITY LOCATION-ALLOCATION USING DYNAMIC PROGRAMMING. MANAGEMENT SCIENCE 22, 614-617. (1976).

LOVEA 1976 -1 LOVE, R. F.; WONG, J. Y.
ON SOLVING A ONE-DIMENSIONAL SPACE ALLOCATION PROBLEM WITH INTEGER PROGRAMMING. INFOR 14, 139-143. (1976).

LOVEA 1976 -2 LOVE, R. F.; WONG, J. Y.
SOLVING QUADRATIC ASSIGNMENT PROBLEMS WITH RECTANGULAR DISTANCES AND INTEGER PROGRAMMING. NAVAL RESEARCH LOGISTICS QUARTERLY 23, 623-628. (1976).

LUBE 1975 -1 LUBELL, D.
LOCAL MATCHING IN THE FUNCTION SPACE OF A PARTIAL ORDER. JOURNAL OF COMBINATORIAL THEORY 19, SERIES A, 154-159. (1975).

LUCE 1975 -1 LUCE, B.
DUALITY METHODS FOR MAXIMAL FLOW AND SHORTEST PATH PROBLEMS. REPORT. GENERAL TELEPHONE COMP. OF CALIFORNIA. (1975).

LUDO 1971 -1 LUDOT, J.-P.; RUBINSTEIN, M.-C.
METHODE POUR LA RECHERCHE D'UNE POLITIQUE OPTIMALE D'INVESTISSEMENT A COURT TERME D'UN RESEAU DE DISTRIBUTION. BULLETIN DES ETUDES ET RECHERCHES, SERIES B, NO. 2, 39-64. (1971).

LUED 1975 -1 LUEDTKE, H.
COMPARISON OF HEURISTIC GRAPH SCHEDULING ALGORITHMS, A. IN: PROCEEDINGS OF THE 8TH HAWAII INTERNATIONAL CONFERENCE ON SYSTEMS SCIENCE, HONOLULU 1975, 39-41. NORTH-HOLLYWOOD: WESTERN PERIODICALS. (1975).

LUEDA 1976 -1 LUEDTKE, H.; SCHINDLER, S.
COMPLEXITY OF SCHEDULING ALGORITHMS FOR THREE-PROCESSOR SYSTEMS, THE. IN: E. J. NEUHOLD (ED.): GI-6. JAHRESTAGUNG, STUTTGART 1976, 388-397. INFORMATIK-FACHBERICHTE 5, BERLIN: SPRINGER. (1976).

LUND 1973 -1 LUNDQVIST, L.
INTEGRATED LOCATION-TRANSPORTATION ANALYSIS: A DECOMPOSITION APPROACH. REGIONAL AND URBAN ECONOMICS 3, 233-262. (1973).

LUND 1975 -1 LUNDQVIST, L.
TRANSPORTATION ANALYSIS AND ACTIVITY LOCATION IN LAND-USE PLANNING - WITH APPLICATIONS TO THE STOCKHOLM REGION. IN: A. KARLQVIST ET AL. (EDS.): DYNAMIC ALLOCATION OF URBAN SPACE, 227-257. LEXINGTON: LEXINGTON BOOKS. (1975).

LUSS 1975 -1 LUSS, H.
OPTIMIZATION OF A MULTIPERIOD RESOURCE ALLOCATION MODEL. OPERATIONAL RESEARCH QUARTERLY 26, 83-90. (1975).

****

LYON 1977 -1 LYONS, J. P.; YOUNG, J. P.
DEVELOPMENT AND IMPLEMENTATION OF A STAFF ALLOCATION MODEL FOR MENTAL HEALTH FACILITIES. SAN FRANCISCO (51. ORSA-MEETING + TIMS, MAY 1977) 19 P. (1977).

LYTK 1974 -1 LYTKIN, I. P.
INVERSE SCHEDULING PROBLEM (RUSSIAN), THE. IN: MATHEMATICAL METHODS OF SOLUTION OF ECONOMIC PROBLEMS (RUSSIAN), 91-100. SUPPL. TO EKONOMIKA I MATEMATICESKIE METODY NO. 5. MOSCOW: IZDAT. NAUKA. (1974).

LYTK 1974 -2 LYTKIN, I. P.
SCHEDULING OF MEANS OF TRANSPORTATION (RUSSIAN). IN: MATHEMATICAL METHODS OF SOLUTION OF ECONOMIC PROBLEMS (RUSSIAN), 101-109. SUPPL. TO EKONOMIKA I MATEMATICESKIE METODY NO. 5. MOSCOW: IZDAT. NAUKA. (1974).

MAC 1977 -1 MAC ALLISTER, J. D.; SAKAROVITCH, M.
CLASSIFICATION DE CERTAIN MATRICES 0-1. DISCRETE MATHEMATICS 20, 133-142. (1977).

MAC A 1976 -1 MAC AVANEY, K. L.
NUMBER AND STABILITY INDICES OF CN-TREES, THE. IN: L. R. A. CASSE, W. D. WALLIS (EDS.): COMBINATORIAL MATHEMATICS IV. PROCEEDINGS OF THE 4TH AUSTRALIAN CONF., ADELAIDE 1975, 142-148. LECTURE NOTES IN MATHEMATICS 560, BERLIN: SPRINGER. (1976).

MAC B 1976 -1 MAC CALL, E. H.
ZERO-ONE MIXED INTEGER PROGRAMMING USING INTERACTIVE GRAPHICS. BUDAPEST (9TH INTERNATIONAL SYMPOSIUM ON MATHEMATICAL PROGRAMMING) 62 P. (1976).

MAC C 1975 -1 MAC CARTHY, P. J.
NUMBER OF INDEPENDENT SYSTEMS OF REPRESENTATIVES, THE. JOURNAL OF THE LONDON MATHEMATICAL SOCIETY 11, NO. 2, 134-136. (1975).

MAC D 1977 -1 MAC DANIEL, D.; DEVINE, M.
MODIFIED BENDERS' PARTITIONING ALGORITHM FOR MIXED INTEGER PROGRAMMING, A. MANAGEMENT SCIENCE 24, 312-319. (1977).

MAC E 1974 -1 MAC DIARMID, C. J. H.
PATH-PARTITION STRUCTURES OF GRAPHS AND DIGRAPHS. PROCEEDINGS OF THE LONDON MATHEMATICAL SOCIETY 29, 750-768. (1974).

MAC E 1975 -1 MAC DIARMID, C. J. H.
RADO'S THEOREM FOR POLYMATROIDS. MATHEMATICAL PROCEEDINGS OF THE CAMBRIDGE PHILOSOPHICAL SOCIETY 78, 263-281. (1975).

MAC F 1977 -1 MAC GINNIS, L. F.
SURVEY OF RESULTS FOR A CLASS OF FACILITIES LOCATION PROBLEMS, A. AIIE TRANSACTIONS 9, 11-18. (1977).

MAC G 1975 -1 MAC KEOWN, P.
VERTEX RANKING PROCEDURE FOR SOLVING THE LINEAR FIXED-CHARGE PROBLEM, A. OPERATIONS RESEARCH 23, 1182-1192. (1975).

MAC H 1975 -1 MAC KEOWN, P. G.
DETERMINING ADJACENT VERTICES ON ASSIGNMENT POLYTOPES. NEW YORK: STATE UNIV. OF NEW YORK, 12 P. (1975).

MAC I 1975 -1 MAC KEOWN, P. G.; RUBIN, D. S.
ADJACENT VERTICES ON TRANSPORTATION POLYTOPES. NAVAL RESEARCH LOGISTICS QUARTERLY 22, 365-374. (1975).

MAC J 1976 -1 MAC KINNON, R. D.
OPTIMIZATION MODELS OF TRANSPORTATION NETWORK IMPROVEMENT: REVIEW AND FUTURE PROSPECTS. LAXENBURG: INTERNAT. INST. FOR APPLIED SYSTEMS ANALYSIS, 43 P. (1976).

MAC K 1966 -1 MAC LAUGHLIN, W. A.
MULTI PATH SYSTEM TRAFFIC ASSIGNMENT ALGORITHM. REPORT RB 108. WATERLOO: UNIVERSITY OF WATERLOO, 83 P. (1966).

MAC L 1977 -1 MAC NEELY, R. A.; MALSTROM, E. M.
COMPUTER GENERATED PROCESS ROUTINGS. INDUSTRIAL ENGINEERING 9, NO. 7, 32-35. (1977).

MACH 1976 -1 MACHOL, R. E.; WIEN, M.
'HARD' ASSIGNMENT PROBLEM, A. OPERATIONS RESEARCH 24, 190-192. (1976).

MADS 1976 -1 MADSEN, K.; SCHJAER-JACOBSEN, H.
SINGULARITIES IN MINIMAX OPTIMIZATION OF NETWORKS. IEEE TRANSACTIONS ON CIRCUITS AND SYSTEMS CAS-23, 456-460. (1976).

MAFF 1975 -1 MAFFIOLI, F.
MATROIDS AND HEURISTICALLY GUIDED SEARCH IN PROBLEM SOLVING. IN: PROGRESS IN CYBERNETICS AND SYSTEMS RESEARCH, 2ND EUROPEAN MEETING, VIENNA 1974, VOL. II, 151-155. WASHINGTON: HEMISPHERE. (1975).

MAFF 1975 -2 MAFFIOLI, F.
TRAVELING SALESMAN PROBLEM AND ITS IMPLICATIONS, THE. IN: P. RINALDI (ED.): TOPICS IN COMBINATORIAL OPTIMIZATION, 69-85, WIEN, NEW YORK: SPRINGER. (1975).

MAFF 1976 -1 MAFFIOLI, F.
SUBGRADIENT OPTIMIZATION, MATROID PROBLEMS AND HEURISTIC EVALUATION. IN: J. CEA (ED.): OPTIMIZATION TECHNIQUES, MODELING AND OPTIMIZATION IN THE SERVICE OF MAN, PART 2. PROCEEDINGS OF THE 7TH IFIP CONFERENCE, NICE 1975, 389-396. LECTURE NOTES IN COMPUTER SCIENCE 41, BERLIN: SPRINGER. (1976).

MAGA 1977 -1 MAGAZINE, M. J.
SCHEDULING A PATIENT TRANSPORTATION SERVICE IN A HOSPITAL. INFOR 15, 242-254. (1977).

MAGAA 1976 -1 MAGAZINE, M. J.; NUTTLE, H. L. W.
LONG DIVISION: AN EXAMPLE OF WHEN THE GREEDY ALGORITHM SOLVES A KNAPSACK PROBLEM. AIIE TRANSACTIONS 8, 283-285. (1976).

MAHM 1976 -1 MAHMOUD, S.; RIORDON, J. S.
OPTIMAL ALLOCATION OF RESOURCES IN DISTRIBUTED INFORMATION NETWORKS. ACM TRANSACTIONS ON DATABASE SYSTEMS 1, 66-78. (1976).

MAHR 1976 -1 MAHR, B.; PERL, J.
GRUENDE, TECHNIKEN UND KONSEQUENZEN DER VERALLGEMEINERUNG VON WEGEALGORITHMEN. IN: U. PAPE (ED.): GRAPHEN-SPRACHEN UND ALGORITHMEN AUF GRAPHEN, 1. FACHTAGUNG UEBER GRAPHENTHEORETISCHE KONZEPTE DER INFORMATIK, BERLIN 1975, 131-142. MUENCHEN: HANSER. (1976).

MAHU 1976 -1 MAHUAL, D. W.
LARGEST CLIQUE IN A RANDOM GRAPH, THE. PAPER PRESENTED AT 7TH SOUTHEASTERN CONFERENCE ON COMBINATORICS, GRAPH THEORY, AND COMPUTING, BATON ROUGE 1976. (1976).

MAIE 1976 -1 MAIER, S. F.
SOLVING THE TRAFFIC ASSIGNMENT AND SUB-AREA FOCUSING PROBLEMS BY GEOGRAPHIC DECOMPOSITION. PAPER PRESENTED AT THE 49TH JOINT NATIONAL MEETING OF ORSA AND TIMS, PHILADELPHIA 1976, 24 P. (1976).

MAIL 1977 -1 MAILLARD, P.
CONTRIBUTION A L'ETUDE DES PROBLEMES DE TRANSPORT. PUBLICATIONS ECONOMETRIQUES 10, 19-41 + 87. (1977).

MAIM 1975 -1 MAIMON, A.; PETERSON, R. M.
APPLICATIONS OF AN AIRLINE FLEET ASSIGNMENT MODEL. PAPER PRESENTED AT THE 48TH JOINT NATIONAL MEETING OF ORSA AND TIMS, LAS VEGAS 1975, 34 P. (1975).

****

MAIR 1977 -1 MAIRS, T. G.; WAKEFIELD, G. W.; JOHNSON, E. L.; SPIELBERG, K.
ON A PRODUCTION ALLOCATION AND DISTRIBUTION PROBLEM. IBM RESEARCH REPORT RC6675. YORKTOWN HEIGHTS: IBM THOMAS J. WATSON RESEARCH CENTER, 17 P. (1977).

MAJO 1964 -1 MAJORANI, E.
SIMPLIFICATION OF LEE'S ALGORITHM FOR SPECIAL PROBLEMS. CALCOLO 1, 242-256. (1964).

MAKI 1966 -1 MAKINO, T.
SCHEDULING PROBLEM. JOURNAL OF THE OPERATIONS RESEARCH SOCIETY OF JAPAN 8, 32-44. (1966).

MALE 1977 -1 MALE, J. W.; LIEBMAN, J. C.; ORLOFF, C. S.
IMPROVEMENT OF ORLOFF'S GENERAL ROUTING PROBLEM, AN. NETWORKS 7, 89-92. (1977).

MALI 1974 -1 MALICKA-WASOWSKA, J.
SEQUENCING MODEL WITH CONTINUOUS MACHINE OPERATION (POLISH), A. PRZEGLAD STATYSTYCZNY 21, 211-219. (1974).

MALY 1976 -1 MALYSHKO, V. V.
SOME ALGORITHMS FOR THE ANALYSIS OF ORIENTED GRAPHS; SOLUTION OF A COVERING PROBLEM BY THE GRADIENT ALGORITHM (RUSSIAN). VESCI AKADEMII NAVUK BSSR, SERYJA FIZIKA-MATEMATYCNYH NAVUK, NO. 5, 33-36. (1976).

MANA 1973 -1 MANABE, R.; KOTANI, S.
SEARCH METHODS OF THE OPTIMUM MAXIMAL TREE IN THE ORIENTED GRAPH. KEIEI-KAGAKU 17, NO. 5, 269-278. (1973).

MAND 1976 -1 MANDERS, K.; ADLEMAN, L.
NP-COMPLETE DECISION PROBLEMS FOR QUADRATIC POLYNOMIALS. IN: PROCEEDINGS OF THE EIGHTH ANNUAL A.C.M. SYMPOSIUM ON THEORY OF COMPUTING, 23-29. (1976).

MANN 1967 -1 MANNE, A. S.
INVESTMENTS FOR CAPACITY EXPANSION: SIZE, LOCATION, AND TIME-PHASING. CAMBRIDGE: M.I.T. PRESS. (1967).

MANS 1964 -1 MANSOOR, E. M.
ASSEMBLY LINE BALANCING. IMPROVEMENT ON RANKED POSITIONAL WEIGHT TECHNIQUE. JOURNAL OF INDUSTRIAL ENGINEERING 15, 73-77. (1964).

MANT 1977 -1 MANTEL, F.; RABINOWITZ, P.
APPLICATION OF INTEGER PROGRAMMING TO THE COMPUTATION OF FULLY SYMMETRIC INTEGRATION FORMULAS IN TWO AND THREE DIMENSIONS, THE. SIAM JOURNAL ON NUMERICAL ANALYSIS 14, 391-425. (1977).

MARC 1976 -1 MARCU, D.
ALGORITHM FOR DETERMINING THE STRONGLY CONNECTED COMPONENTS OF A FINITE DIRECTED GRAPH (RUMANIAN), AN. STUDII SI CERCETARI MATEMATICE 28, 57-60. (1976).

MARCA 1976 -1 MARCUK, G. I.
ENVIRONMENT, AND PROBLEMS OF OPTIMIZING THE LOCATION OF ENTERPRISES (RUSSIAN), THE. DOKLADY AKADEMII NAUK SSSR 227, 1056-1059. (1976).

MARCB 1967 -1 MARCUS, B. T.; UNGAR, A. J.
MULTI-PERIOD TWO-STAGE MODEL FOR LOCATING ALUMINA AND ALUMINUM REDUCTION CAPACITY, A. PAPER PRESENTED AT THE 14TH INTERNAT. MEETING OF TIMS, MEXICO CITY 1967. (1967).

MARCC 1972 -1 MARCUS, M.; UNDERWOOD, E. E.
NOTE ON THE MULTIPLICATIVE PROPERTY OF THE SMITH NORMAL FORM, A. JOURNAL OF RESEARCH OF THE NATIONAL BUREAU OF STANDARDS 76-B, 205-206. (1972).

MARK 1977 -1 MARK, J. W.
GLOBAL SCHEDULING APPROACH TO CONFLICT-FREE MULTIACCESS: PART II-DELAY ANALYSIS FOR A BUS ARCHITECTURE. REPORT RC6570. YORKTOWN HEIGHTS: IBM THOMAS J. WATSON RESEARCH CENTER, 35 P. (1977).

MARKA 1975 -1 MARKLAND, R. E.; SMITH, L. D.; BECKER, J. D.; HAACK, P. J.
SITE SELECTION MODEL FOR LAND DISPOSAL OF WASTEWATER, A. PAPER PRESENTED AT THE 47TH JOINT NATIONAL MEETING OF ORSA AND TIMS, CHICAGO 1975, 29 P. (1975).

MARKB 1976 -1 MARKOSYAN, A. G.; MARKOSYAN, S. E.
ALGORITHMS FOR THE VERIFICATION OF SOME PROPERTIES OF GRAPHS (RUSSIAN). KIBERNETIKA, NO. 2, 76-81. (1976).

MARKC 1976 -1 MARKOSYAN, S. E.; KARAPETYAN, I. A.
ON THE PERFECT GRAPHS (RUSSIAN). AKADEMIJA NAUK ARMJANSKOI SSR, DOKLADY 63, NO. 5, 292-296. (1976).

MARKD 1974 -1 MARKOV, L. A.; PERVOZVANSKII, A. A.
RULE FOR "BOTTLENECKS" IN THE JOHNSON PROBLEM (RUSSIAN), A. KIBERNETIKA, NO. 5, 142-144; TRANSLATED AS CYBERNETICS 10, 1974, 907-910. (1974).

MARKE 1976 -1 MARKOWSKY, G.
STRAIGHTFORWARD TECHNIQUE FOR PRODUCING MINIMAL MULTIPLE FAULT TEST SETS FOR FANOUT-FREE COMBINATIONAL CIRCUITS, A. IBM RESEARCH REPORT RC 6222. YORKTOWN HEIGHTS: IBM RESEARCH CENTER, 13 P. (1976).

MARKF 1971 -1 MARKS. D. H.; JIRKA, G. H.
ENVIRONMENTAL SCREENING MODEL FOR THE LOCATION OF POWER GENERATING FACILITIES, AN. PAPER PRESENTED AT THE 39TH NATIONAL MEETING OF ORSA, DALLAS 1971. (1971).

MARQ 1977 -1 MARQUEZ, J.; MEDINA-MORA, O.
NETWORK SOLUTION TO A GENERAL VEHICLE SCHEDULING PROBLEM, A. EUROPEAN JOURNAL OF OPERATIONAL RESEARCH 1, 255-261. (1977).

MARS 1971 -1 MARSHALL, C.
APPLIED GRAPH THEORY. NEW YORK: WILEY. (1971).

MARSA 1972 -1 MARSTEN, R. E.
ALGORITHM FOR FINDING ALMOST ALL OF THE MEDIANS OF A NETWORK, AN. REPORT. EVANSTON: NORTHWESTERN UNIV., CENTER FOR MATHEMATICAL STUDIES IN ECONOMICS AND MANAGEMENT SCIENCES. (1972).

MARSA 1975 -1 MARSTEN, R. E.
ALGORITHM FOR LARGE SET PARTITIONING PROBLEMS, AN. IN: B. ROY (ED.): COMBINATORIAL PROGRAMMING: METHODS AND APPLICATIONS. PROCEEDINGS OF THE NATO ADVANCED STUDY INST., VERSAILLES 1974, 259-268. DORDRECHT: REIDEL. (1975).

MARSB 1977 -1 MARSTEN, R. E.; MORIN, T. L.
OPTIMAL SOLUTIONS FOUND FOR SENJU AND TOYODA'S 0/1 INTEGER PROGRAMMING PROBLEMS. MANAGEMENT SCIENCE 23, 1364-1365. (1977).

MARSB 1977 -2 MARSTEN, R. E.; MORIN, T. L.
PARAMETRIC INTEGER PROGRAMMING: THE RIGHT-HAND-SIDE CASE. IN: P. L. HAMMER, E. L. JOHNSON, B. H. KORTE, G. L. NEMHAUSER (EDS.): STUDIES IN INTEGER PROGRAMMING, 375-390. ANNALS OF DISCRETE MATHEMATICS 1, AMSTERDAM: NORTH-HOLLAND. (1977).

MARSB 1978 -1 MARSTEN, R. E.; MORIN, T. L.
HYBRID APPROACH TO DISCRETE MATHEMATICAL PROGRAMMING, A. MATHEMATICAL PROGRAMMING 14, 21-40. (1978).

MART 1976 -1 MARTELLI, A.
GAUSSIAN ELIMINATION ALGORITHM FOR THE ENUMERATION OF CUT SETS IN A GRAPH, A. JOURNAL OF THE ASSOCIATION FOR COMPUTING MACHINERY 23, 58-73. (1976).

****

MARTA 1975 -1 MARTELLI, A.; MONTANARI, U.
ON THE FOUNDATIONS OF DYNAMIC PROGRAMMING. IN: P. RINALDI (ED.): TOPICS IN COMBINATORIAL OPTIMIZATION, 145-163, WIEN, NEW YORK: SPRINGER. (1975).

MARTB 1977 -1 MARTELLO, S.
ALGORITHM FOR FINDING A MINIMAL EQUIVALENT GRAPH OF A STRONGLY CONNECTED DIGRAPH, AN. ATLANTA (52. ORSA-MEETING + TIMS, NOVEMBER 1977) 8 P. (1977).

MARTC 1976 -1 MARTELLO, S.; TOTH, P.
ENUMERATIVE ALGORITHM FOR THE 0-1 KNAPSACK PROBLEM (ITALIAN), AN. RICERCA OPERATIVA 6, NO. 4, 25-36. (1976).

MARTC 1977 -1 MARTELLO, S.; TOTH, P.
COMPUTATIONAL STUDY ON LARGE-SIZE UNIDIMENSIONAL KNAPSACK PROBLEMS, A. ATLANTA (52. ORSA-MEETING + TIMS, NOVEMBER 1977) 8 P. (1977).

MARTC 1977 -2 MARTELLO, S.; TOTH, P.
SOLUTION OF THE BOUNDED AND UNBOUNDED CHANGE-MAKING PROBLEM. SAN FRANCISCO (51. ORSA-MEETING + TIMS, MAY 1977) 7 P. (1977).

MARTC 1977 -3 MARTELLO, S.; TOTH, P.
SOLUTION OF ZERO-ONE MULTIPLE KNAPSACK PROBLEM. ATLANTA (52. ORSA-MEETING + TIMS, NOVEMBER 1977) 11 P. (1977).

MARTC 1977 -4 MARTELLO, S.; TOTH, P.
UPPER BOUND FOR THE ZERO-ONE KNAPSACK PROBLEM AND A BRANCH AND BOUND ALGORITHM, AN. EUROPEAN JOURNAL OF OPERATIONAL RESEARCH 1, 169-175. (1977).

MARTD 1976 -1 MARTIN-VEGA, L. A.; RATLIFF, H. D.
SCHEDULING RULES FOR PARALLEL PROCESSORS. RESEARCH REPORT NO. 76-9. GAINESVILLE: UNIV. OF FLORIDA, DEPT. OF INDUSTRIAL AND SYSTEMS ENGINEERING, 36 P. (1976).

MARTE 1965 -1 MARTIN, B.; MANHEIM, M.
RESEARCH PROGRAM FOR COMPARISON OF TRAFFIC ASSIGNMENT TECHNIQES, A. HIGHWAY RESEARCH RECORD 88. (1965).

MARTF 1963 -1 MARTIN, B. V.
MINIMUM PATH ALGORITHMS FOR TRANSPORTATION PLANNING. RESEARCH REPORT R63-52. CAMBRIDGE: MASSACHUSETTS INST. OF TECHNOLOGY, DEPT. OF CIVIL ENGINEERING. (1963).

MARTG 1976 -1 MARTINEC, I.
ON A GENERALIZED LOCATION PROBLEM. EKONOMICKO-MATEMATICKY OBZOR 12, 325-343. (1976).

MARTH 1976 -1 MARTINEZ Y RODRIGUEZ, J. B.; RUSSELL, S. O.
SEQUENTIAL TECHNIQUES FOR PROJECT SCREENING. CANAD. J. CIV. ENGNG. 3, NO. 1, 90-97. (1976).

MARTI 1974 -1 MARTINOVA, M. K.
MINIMAL COLORATION OF A RETICULATED GRAPH (BULGARIAN). GODISNIK NA VISSITE UCEBNI ZAVEDENIJA. PRILOZNA MATEMATIKA 10, 171-176. (1974).

MARU 1976 -1 MARUYAMA, K.; TANG, D. T.; FRATTA, L.
FLOW ASSIGNMENT ALGORITHM FOR COMPUTER COMMUNICATION NETWORK DESIGN WITH DIFFERENT CLASSES OF PACKETS. YORKTOWN HEIGHTS: IBM RESEARCH CENTER. (1976).

MASO 1977 -1 MASON, J. H.
MATROIDS AS THE STUDY OF GEOMETRICAL CONFIGURATIONS. IN: M. AIGNER (ED.): HIGHER COMBINATORICS, PROCEEDINGS OF THE NATO ADVANCED STUDY INSTITUTE, BERLIN 1976, 133-178. DORDRECHT: REIDEL. (1977).

MASS 1976 -1 MASSA, L.
SEMIECONOMIC METHOD FOR THE DETERMINATION OF OPTIMAL LOT SIZES, A. NAVAL RESEARCH LOGISTICS QUARTERLY 23, 697-702. (1976).

MATE 1976 -1 MATETI, P.; DEO, N.
ON ALGORITHMS FOR ENUMERATING ALL CIRCUITS OF A GRAPH. SIAM JOURNAL ON COMPUTING 5, 90-99. (1976).

MATI 1974 -1 MATIJASEVIC, JU. V.
TEST FOR THE COLORABILITY OF THE VERTICES OF A GRAPH, FORMULATED IN TERMS OF THE ORIENTATIONS OF THE EDGES (RUSSIAN), A. DISKRETNYI ANALIZ, VYP. 26, GRAFY I TESTY, 65-71 + 85. (1974).

MATS 1971 -1 MATSUI, H.
THEORY OF TRAFFIC ASSIGNMENT BY MAXIMIZING THE PROBABILITY, THE. BULLETIN OF NAGOYA INST. OF TECHNOLOGY 23, 385-395. (1971).

MATT 1976 -1 MATTHAEUS, F. W.
SAVINGS APPROACH TO THE MULTIPLE TERMINAL DELIVERY PROBLEM. TRANSPORTATION SCIENCE 10, 216-221. (1976).

MATTA 1977 -1 MATTHEWS, L. R.
BICIRCULAR MATROIDS. QUARTERLY JOURNAL OF MATHEMATICS 28, NO. 110, 213-228. (1977).

MATTA 1978 -1 MATTHEWS, L. R.
CLOSURE IN INDEPENDENCE SYSTEMS. REPORT 7894-OR. BONN: UNIV. BONN, INST. F. OEKONOMETRIE UND OPERATIONS RESEARCH, 7 P. (1978).

MATTA 1978 -2 MATTHEWS, L. R.
MATROIDS ON THE EDGE SETS OF DIRECTED GRAPHS. IN: R. HENN, B. KORTE, W. OETTLI (EDS.): OPTIMIZATION AND OPERATIONS RESEARCH, PROCEEDINGS, BONN 1977. LECTURE NOTES IN ECONOMICS AND MATHEMATICAL SYSTEMS 157. BERLIN, HEIDELBERG, NEW YORK: SPRINGER, 193-199. (1978).

MATU 1968 -1 MATULA, D. W.
MIN-MAX THEOREM FOR GRAPHS WITH APPLICATION TO GRAPH COLORING, A. SIAM REVIEW 10, 481-482. (1968).

MATU 1975 -1 MATULA, D. W.
UNIFORM SET COVERING LEMMA, A. PROCEEDINGS OF THE AMERICAN MATHEMATICAL SOCIETY 48, 255-261. (1975).

MATU 1977 -1 MATULA, D. W.
GRAPH THEORETIC TECHNIQUES FOR CLUSTER ANALYSIS ALGORITHMS. IN: CLASSIFICATION AND CLUSTERING. PROCEEDINGS OF THE ADV. SEM., MADISON 1976, 95-129. MATH. RES. CENTER PUBL. 37, NEW YORK: ACADEMIC PRESS. (1977).

MATU 1978 -1 MATULA, D. W.
SUBTREE ISOMORPHISM IN O(N POWER 5/2). IN: ALSPACH, B.;HELL, P.;MILLER, D. J.(EDS.): ALGORITHMIC ASPECTS OF COMBINATORICS. ANNALS OF DISCRETE MATHEMATICS 2, 91-106. (1978).

MATUA 1972 -1 MATULA, D. W.; MARBLE, G.; ISAACSON, J. D.
GRAPH COLORING ALGORITHMS. IN: R. C. READ (ED.): GRAPH THEORY AND COMPUTING, 109-122. NEW YORK: ACADEMIC PRESS. (1972).

MAUR 1976 -1 MAUREL, E.; ROUX, D.; DUPONT, D.
TECHNIQUES OPERATIONNELLES D'ORDONNANCEMENT FONDEES SUR LA METHODE PERT "POTENTIELS-TACHES". PARIS: EYROLLES, 319 P. (1976).

MAURA 1973 -1 MAURER, S. B.
MATROID BASIS GRAPHS, II. JOURNAL OF COMBINATORIAL THEORY 15, SERIES B, 121-145. (1973).

MAURB 1972 -1 MAURRAS, J. F.
FLOT OPTIMAL SUR UN GRAPHE AVEC MULTIPLICATEURS. ELECTRICITE DE FRANCE, BULLETIN DE LA DIRECTION DES ETUDES ET RECHERCHES, SERIE C, MATHEMATIQUES, INFORMATIQUES, NO. 1, 65-76 + ERRATA, IBID., 1972, NO. 2, 91. (1972).

****

MAURB 1976 -1 MAURRAS, J. F.
POLYTOPES A SOMMETS DANS (0,1)N. DISSERTATION. PARIS: UNIVERSITE PARIS VII, 82 P. (1976).

MAURB 1977 -1 MAURRAS, J. F.
EXAMPLE OF DUAL POLYTOPES IN THE UNIT HYPERCUBE, AN. IN: P. L. HAMMER, E. L. JOHNSON, B. H. KORTE, G. L. NEMHAUSER (EDS.): STUDIES IN INTEGER PROGRAMMING, 391-392. ANNALS OF DISCRETE MATHEMATICS 1, AMSTERDAM: NORTH-HOLLAND. (1977).

MAURC 1975 -1 MAURRAS, J.-F.
SOME RESULTS ON THE CONVEX HULL OF THE HAMILTONIAN CYCLES OF SYMMETRIC COMPLETE GRAPHS. IN: B. ROY (ED.): COMBINATORIAL PROGRAMMING: METHODS AND APPLICATIONS. PROCCEDINGS OF THE NATO ADVANCED STUDY INST., VERSAILLES 1974, 179-190. DORDRECHT: REIDEL. (1975).

MAWD 1974 -1 MAWDESLEY, M. J.; CULLINGFORD, G.
OPTIMAL RESOURCE SCHEDULING. IN: PROJ. MANAGE. SEVENTIES. 4TH INTERNET CONGR., PARIS 1974, 511-522. PARIS: ASSOC. FR. CYBERN. ECON. TECH. (1974).

MAXW 1964 -1 MAXWELL, W. L.
SCHEDULING OF SINGLE MACHINE SYSTEMS. A REVIEW. INTERNATIONL JOURNAL OF PRODUCTION RESEARCH 3, 177-199. (1964).

MAY 1976 -1 MAY, A. D.
OPTIMIZATION TECHNIQUES APPLIED TO IMPROVING FREEWAY OPERATIONS. ENGNG. OPTIMIZ. 2, 125-136. (1976).

MAY A 1974 -1 MAY, J. G.; STONE, C. A.; WAGNER, N. S.
SCHEDULING MODEL FOR THE INBAND SIGNALING SHOP, A. WEST. ELECTR. ENGR. 18, NO. 2, 2-14. (1974).

MAYE 1976 -1 MAYER, I.
OPTIMALER FAHRPLAN FUER EINE PRIVATBAHN. IN: LIEBLING, T. M.;ROESSLER, M.(EDS.): KOMBINATORISCHE ENTSCHEIDUNGSPROBLEME: METHODEN UND ANWENDUNG. LECTURE NOTES IN ECONOMICS AND MATHEMATICAL SYSTEMS 153, BERLIN: SPRINGER, 63-69. (1976).

MEDN 1975 -1 MEDNITSKII, V. G.
THEOREM ON TRANSPORTATION ZONES IN THE PRODUCTION-TRANSPORTATION PROBLEM (RUSSIAN), THE. EKONOMIKA I MATEMATICESKIE METODY 11, 1106-1114; TRANSLATED AS MATEKON 13, NO. 1, 1076, 79-93. (1975).

MEEU 1975 -1 MEEUSEN, W.; CUYVERS, L.
CLIQUE DETECTION IN DIRECTED GRAPHS: A NEW ALGORITHM. JOURNAL OF COMPUTATIONAL AND APPLIED MATHEMATICS 1, 185-203. (1975).

MEEU 1976 -1 MEEUSEN, W.; CUYVERS, L.
CLIQUE-DETECTION IN DIRECTED GRAPHS (DUTCH). INGENIEURSBLAD 45, NO. 3, 66-74. (1976).

MEGI 1977 -1 MEGIDDO, N.
GOOD ALGORITHM FOR LEXICOGRAPHICALLY OPTIMAL FLOWS IN MULTI-TERMINAL NETWORKS, A. BULLETIN OF THE AMERICAN MATHEMATICAL SOCIETY 83, 407-409. (1977).

MEHL 1975 -1 MEHLHORN, K.
NEARLY OPTIMAL BINARY SEARCH TREES. ACTA INFORMATICA 5, 287-296. (1975).

MEIR 1973 -1 MEIR, A.; MOON, J. W.
EXPECTED NODE-INDEPENDENCE NUMBER OF RANDOM TREES, THE. IN: PROC. KON. NED. AKAD. V. WETENSCH. 76, 335-341. (1973).

MEIR 1975 -1 MEIR, A.; MOON, J. W.
EXPECTED NODE-INDEPENDENCE NUMBER OF VARIOUS TYPES OF TREES, THE. IN: M. FIEDLER (ED.): RECENT ADVANCES IN GRAPH THEORY, PROCEEDINGS OF THE 2ND CZECHOSLOVAK SYMPOSIUM, PRAGUE 1974, 351-363. PRAGUE: ACADEMIA. (1975).

MEIR 1975 -2 MEIR, A.; MOON, J. W.
RELATIONS BETWEEN PACKING AND COVERING NUMBERS OF A TREE. PACIFIC JOURNAL OF MATHEMATICS 61, 225-233. (1975).

MEIR 1976 -1 MEIR, A.; MOON, J. W.
CLIMBING CERTAIN TYPES OF ROOTED TREES, I. IN: C. ST. J. A. NASH-WILLIAMS, J. SHEEHAN (EDS.): PROCEEDINGS OF THE 5TH CONFERENCE, ABERDEEN 1975, 461-469. CONGRESSUS NUMERANTIUM 15, WINNIPEG: UTILITAS MATHEMATICA. (1976).

MEIR 1977 -1 MEIR, A.; MOON, J. W.
PACKING AND COVERING CONSTANTS FOR CERTAIN FAMILIES OF TREES, I. JOURNAL OF GRAPH THEORY 1, 157-174. (1977).

MEL' 1977 -1 MEL'NIKOV, L. S.; STEINBERG, R.
ONE COUNTEREXAMPLE FOR TWO CONJECTURES ON THREE COLORING. DISCRETE MATHEMATICS 20, 203-206. (1977).

MEL'A 1975 -1 MEL'NIKOV, O. I.
ON THE BELLMAN-JOHNSON PROBLEM (RUSSIAN). VESCI AKADEMII NAVUK BSSR, SERYJA FIZIKA-MATEMATYCNYH NAVUK, NO. 4, 43-47 + 141. (1975).

MEL'A 1976 -1 MEL'NIKOV, O. I.
PROPERTIES OF OPTIMAL SCHEDULINGS OF THE BELLMAN-JOHNSON PROBLEM FOR TWO MACHINES (RUSSIAN). VESCI AKADEMII NAVUK BSSR, SERIJA FIZIKA-MATEMATYCNYH NAVUK, NO. 3, 19-25 + 140. (1976).

MEL'A 1976 -2 MEL'NIKOV, O. I.
STABILITY OF THE OPTIMAL SCHEDULING OF THE BELLMAN-JOHNSON PROBLEM (RUSSIAN). VESCI AKADEMII NAVUK BSSR, SERYJA FIZIKA-MATEMATYCNYH NAVUK, NO. 6, 99-101 + 144. (1976).

MELI 1972 -1 MELIKHOV, A. N.; BERSHTEYN, L. S.; KUREYCHIK, V. M.; LISYAK, V. V.
PARTITION OF A GRAPH INTO PLANE SUBGRAPHS (RUSSIAN). IZVESTIJA AKADEMII NAUK SSSR, TEHNICESKAJA KIBERNETIKA, NO. 6, 166-170; TRANSLATED AS ENGINEERING CYBERNETICS 10, 1972, 1087-1090. (1972).

MELZ 1961 -1 MELZAK, Z. A.
ON THE PROBLEM OF STEINER. CANADIAN MATHEMATICAL BULLETIN 4, 143-148. (1961).

MENG 1974 -1 MENG, D.
MATCHINGS AND COVERINGS FOR GRAPHS. DISSERTATION. EAST LANSING: MICHIGAN STATE UNIVERSITY. (1974).

MENS 1975 -1 MENSCH, G.
PERSONALEINSATZPLANUNG (EINSCHL. ZUORDNUNGSMODELLE). BERLIN: INTERNATIONAL INSTITUTE OF MANAGEMENT, 13 P. (1975). (PREPRINT SERIES I/75-22)

MERC 1974 -1 MERCHANT, D. K.; NEMHAUSER, G. L.
MODEL AND AN ALGORITHM FOR THE DYNAMIC TRAFFIC ASSIGNMENT PROBLEM, A. IN PROCEEDINGS OF THE INTERNATIONAL SYMPOSIUM ON TRAFFIC EQUILIBRIUM METHODS, MONTREAL. (1974).

MERCA 1976 -1 MERCHANT, D. K.; RAO, M. R.
MAJORITY DECISIONS AND TRANSITIVITY: SOME SPECIAL CASES. MANAGEMENT SCIENCE 23, 125-130. (1976).

MERE 1975 -1 MEREKIN, YU. V.
ON A PROBLEM OF OPTIMAL ALLOCATION OF THE PRODUCTION (RUSSIAN). VYCISLITEL'NYE SISTEMY, NO. 64, 3-10. (1975).

MERL 1971 -1 MERLINO, E. L. A.
MULTI-NATIONAL CAPACITY EXPANSION MODEL FOR PRIMARY ALUMINUM PLANTS. DISSERTATION. LOS ANGELES: UNIVERSITY OF CALIFORNIA. (1971).

****

MEVE 1977 -1 MEVERT, P.; SUHL, U.
IMPLICIT ENUMERATION WITH GENERALIZED UPPER BOUNDS. IN: P. L. HAMMER, E. L. JOHNSON, B. H. KORTE, G. L. NEMHAUSER (EDS.): STUDIES IN INTEGER PROGRAMMING, 393-402. ANNALS OF DISCRETE MATHEMATICS 1, AMSTERDAM: NORTH-HOLLAND. (1977).

MEYE 1973 -1 MEYER, C. F.
LONG-RANGE SELECTION AND TIMING ANALYSIS SYSTEM FOR FACILITY LOCATION: IMPLEMENTATION, A. MANAGEMENT SCIENCE 20, 261-273. (1973).

MEYEA 1972 -1 MEYER, J. C.
ENSEMBLES STABLES MAXIMAUX DANS LES HYPERGRAPHES. COMPT. REND. ACAD. SCI. A 274, 144-147. PARIS. (1972).

MEYEA 1978 -1 MEYER, J. C.
NOMBRE CHROMATIQUE TOTAL D'UN HYPERGRAPHE. JOURNAL OF COMBINATORIAL THEORY 24, SERIES B, 44-50. (1978).

MEYEB 1975 -1 MEYER, J.-C.
ON THE GOOD COLORINGS, II. CASE OF THE COMPLETE H-PARTITE HYPERGRAPHS. IN: A. HAJNAL, R. RADO, V. T. SOS (EDS.): INFINITE AND FINITE SETS, COLLOQ. KESZTHELY 1973, VOL. 3, 1099-1113. COLLOQ. MATH. SOC. JANOS BOLYAI 10, AMSTERDAM: NORTH-HOLLAND. (1975).

MEYEB 1976 -1 MEYER, J.-C.
PROBLEMES DE COLORATIONS DANS LES HYPERGRAPHES. DISSERTATION. PARIS: UNIV. PARIS XI, 110 P. (1976).

MEYEC 1976 -1 MEYER, R. R.
MIXED INTEGER MINIMIZATION MODELS FOR PIECEWISE-LINEAR FUNCTIONS OF A SINGLE VARIABLE. DISCRETE MATHEMATICS 16, 163-172. (1976).

MEYEC 1977 -1 MEYER, R. R.
CLASS OF NONLINEAR INTEGER PROGRAMS SOLVABLE BY A SINGLE LINEAR PROGRAM, A. SIAM JOURNAL ON CONTROL AND OPTIMIZATION 15, 935-946. (1977).

MEYED 1977 -1 MEYER, R. R.; FLEISHER, J. M.
STRONG DUALITY FOR A SPECIAL CLASS OF INTEGER PROGRAMS. JOURNAL OF OPTIMIZATION THEORY AND APPLICATIONS 22, 25-30. (1977).

MEYEE 1974 -1 MEYER, W.
ON ORDINARY POINTS IN ARRANGEMENTS. ISRAEL JOURNAL OF MATHEMATICS 17, 124-135. (1974).

MEYL 1972 -1 MEYL, H.
SHORTEST ROUTE PROBLEM WITH ADDITIONAL PROBLEMS. HEBEZEUGE FOERDERMITTEL 12, NO. 2, 41-45. (1972).

MICH 1977 -1 MICHAELI, I.; POLLATSCHEK, M. A.
ON SOME NONLINEAR KNAPSACK PROBLEMS. IN: P. L. HAMMER, E. L. JOHNSON, B. H. KORTE, G. L. NEMHAUSER (EDS.): STUDIES IN INTEGER PROGRAMMING, 403-414. ANNALS OF DISCRETE MATHEMATICS 1, AMSTERDAM: NORTH-HOLLAND. (1977).

MICHA 1968 -1 MICHIE, D.; FLEMING, J. G.; OLDFIELD, J. V.
COMPARISON OF HEURISTIC, INTERACTIVE, AND UNAIDED METHODS OF SOLVING A SHORTEST-ROUTE PROBLEM, A. IN: MACHINE INTELLIGENCE 3. 245-255. EDINBURGH: OLIVER AND BOYD. (1968).

MICK 1975 -1 MICKLE, M. H.
METHOD OF OPTIMUM ROUTE REGENERATION, A. KYBERNETES 4, 197-203. (1975).

MICKA 1969 -1 MICKLE, M. H.; VOGT, W. G.
DISCRETE CONTROL AS A TRANSPORTATION PROBLEM. IN: PROCEEDINGS OF THE THIRD ANNUAL PRINCETON CONFERENCE ON INFORMATION SCIENCES AND SYSTEMS, 102-106. (1969).

MICKA 1975 -1 MICKLE, M. H.; VOGT, W. G.
ON THE DYNAMICS OF A CLASS OF OPTIMUM ROUTING PROBLEMS. JOURNAL OF THE FRANKLIN INSTITUTE 300, 7-23. (1975).

MICKB 1977 -1 MICKLE, M. H.; WANG, J. C.; HOELZEMAN, R. G.
ON THE STORAGE AND REGENERATION OF OPTIMUM ROUTES. INTERNATIONAL JOURNAL OF SYSTEMS SCIENCE 8, 739-752. (1977).

MIHA 1976 -1 MIHALYFFY, L.
NULL-SPACE METHOD FOR SOLVING THE ASSIGNMENT PROBLEM, A. IN: A. PREKOPA (ED.): PROGRESS IN OPERATIONS RESEARCH, 625-665, AMSTERDAM, LONDON: NORTH HOLLAND. (1976).

MIHO 1965 -1 MIHOC, G.; BALAS, E.
PROBLEM OF OPTIMAL SCHEDULING (RUMANIAN). STUDII SI CERCETARI MATEMATICE 17, 477-484. (1965).

MIKH 1975 -1 MIKHAJLOVA, E. D.; BALITSKAYA, E. E.
DETERMINATION OF LIMITED DEGREES OF DIRECTED GRAPHS (RUSSIAN), THE. OTBOR I PEREDACA INFORMACII, NO 44, 66-75. (1975).

MIKHA 1966 -1 MIKHALEVICH, V. S.; SHKURBA, V. V.
SEQUENTIAL SCHEMES OF OPTIMIZATION IN PROBLEMS OF SCHEDULING THE COMPLETION OF TASKS (RUSSIAN). KIBERNETIKA 2, 34-40. (1966).

MIKO 1976 -1 MIKOLA, M.
EXTREMAL CONNECTIVITY OF THE STRICTLY WEAK DIGRAPH, THE. MATHEMATICA SLOVAKA 26, 115-118. (1976).

MILI 1976 -1 MILIOTIS, P.
INTEGER PROGRAMMING APPROACHES TO THE TRAVELLING SALESMAN PROBLEM. MATHEMATICAL PROGRAMMING 10, 367-378. (1976).

MILL 1976 -1 MILLER, D. M.
ALGORITHM FOR DETERMINING THE CHROMATIC NUMBER OF A GRAPH, AN. IN: B. L. HARTNELL, H. C. WILLIAMS (EDS): PROCEEDINGS OF THE 5TH MANITOBA CONFERENCE ON NUMERICAL MATHEMATICS, WINNIPEG 1975, 533-548. CONGRESSUS NUMERANTIUM 16, WINNIPEG: UTILITAS MATHEMATICA. (1976).

MILLA 1976 -1 MILLER, D. M.; FYFFE, D. E.
ALLOCATING BUILDING INSPECTION MANPOWER FOR FIRE PREVENTION. MANAGEMENT SCIENCE 22, 1310-1319. (1976).

MILLB 1977 -1 MILLER, G. L.
GRAPH ISOMORPHISM, GENERAL REMARKS. IN: PROCEEDINGS OF THE NINTH ANNUAL A.C.M. SYMPOSIUM ON THEORY OF COMPUTING, 143-150. (1977).

MILLC 1976 -1 MILLER, H. E.
PERSONNEL SCHEDULING AND WORK ASSIGNMENT PROBLEM, A. RESEARCH REPORT 37-76-P2. TROY: RENSSELAER POLYTECHNIC INSTITUTE, 12 P. (1976).

MILLC 1976 -2 MILLER, H. E.
PERSONNEL SCHEDULING IN PUBLIC SYSTEMS: A SURVEY. SOCIO-ECONOMIC PLANNING SCIENCES 10, 241-250. (1976).

MILLD 1976 -1 MILLER, H. E.; PIERSKALLA, W. P.; RATH, G. J.
NURSE SCHEDULING USING MATHEMATICAL PROGRAMMING. OPERATIONS RESEARCH 24, 857-870. (1976).

MILLE 1973 -1 MILLS. W. H.
COVERING PROBLEMS. IN: F. HOFFMAN, R. B. LEVOW, R. P. D. THOMAS (EDS.): PROCEEDINGS OF THE FOURTH SOUTHEASTERN CONFERENCE ON COMBINATORICS, GRAPH THEORY, AND COMPUTING, BOCA RATON, FL., 1973, 23-52. CONGRESSUS NUMERANTIUM 8, WINNIPEG: UTILITAS MATHEMATICA. (1973).

MILN 1975 -1 MILNER, E. C.
TRANSVERSAL THEORY. IN: PROCEEDINGS OF THE INTERNATIONAL CONGRESS OF MATHEMATICIANS, VANCOUVER 1974, VOL. 1, 155-169. MONTREAL: CANADIAN MATHEMATICAL CONGRESS. (1975).

****

MILNA 1976 -1 MILNER, E. C.; WELSH, D. J. A.
ON THE COMPUTATIONAL COMPLEXITY OF GRAPH THEORETICAL PROPERTIES. IN: C. ST. J. A. NASH-WILLIAMS (ED.): PROCEEDINGS OF THE 5TH BRITISH COMBINATORIAL CONFERENCE, ABERDEEN 1975, 471-488. CONGRESSUS NUMERANTIUM 15, WINNIPEG: UTILITAS MATHEMATICA. (1976).

MING 1976 -1 MING-I, Y.
ON THE N JOB, M MACHINE SEQUENCING PROBLEM OF FLOW-SHOP. IN: K. B. HALEY (ED.): OPERATIONAL RESEARCH '75. PROCEEDINGS OF THE 7TH IFORS INTERNATIONAL CONFERENCE ON OPERATIONAL RESEARCH, TOKYO, KYOTO 1975, 179-200. AMSTERDAM: NORTH-HOLLAND. (1976).

MINI 1976 -1 MINIEKA, E.
FINDING THE CIRCUITS OF A MATROID. JOURNAL OF RESEARCH OF THE NATIONAL BUREAU OF STANDARDS 80B, 337-342. (1976).

MINIA 1975 -1 MINIEKA, F.
MAXIMAL NETWORK FLOWS, MATROIDS AND MATCHINGS. JOURNAL OF RESEARCH OF THE NATIONAL BUREAU OF STANDARDS 79B, 59-62. (1975).

MINIB 1975 -1 MININA, T. R.; PEREKREST, V. T.
APPROXIMATION TO THE SOLUTION OF THE TRAVELING-SALESMAN PROBLEM BY C-CYCLES. AVTOMATIKA I TELEMEHANIKA, NO. 10, 79-89; TRANSLATED AS AUTOMATION AND REMOTE CONTROL 36, 1975, 1642-1651. (1975).

MINO 1975 -1 MINOLI, D.
COMBINATORIAL GRAPH COMPLEXITY. ATTI DELLA ACCADEMIA NAZIONALE DEI LINCEI. RENDICONTI, CLASSE DI SCIENZE FISICHE, MATEMATICHE E NATURALI 59, 651-661. (1975).

MINOA 1975 -1 MINOUX, M.
FLOT COURT-CHEMIN AVEC CONTRAINTES: ALGORITHMES ET APPLICATIONS. ANNALES DES TELECOMMUNICATIONS 30, NO. 11-12, 383-394. (1975).

MINOA 1975 -2 MINOUX, M.
RESOLUTION DES PROBLEMES DE MULTIFLOTS EN NOMBRES ENTIERS DANS LES GRANDS RESEAUX. REVUE FRANCAISE D'AUTOMATIQUE, D'INFORMATIQUE ET DE RECHERCHE OPERATIONNELLE 9, NO. V-3, 21-40. (1975).

MINOA 1976 -1 MINOUX, M.
CARACTERISATION DES COUPLAGES DE POIDS MAXIMAL ET DE CARDINALITE FIXEE DANS LES GRAPHES PONDERES FINIS, UNE. COMPTES RENDUS HEBDOMADAIRES DES SEANCES DE L'ACADEMIE DES SCIENCES 282, SERIES A, AI, A353-A354. (1976).

MINOA 1976 -2 MINOUX, M.
FLOTS EQUILIBRES ET FLOTS AVEC SECURITE. ELECTRICITE DE FRANCE, BULLETIN DE LA DIRECTION DES ETUDES ET RECHERCHES, SERIES C, NO. 1, 5-16. (1976).

MINOA 1976 -3 MINOUX, M.
MULTIFLOTS DE COUT MINIMAL AVEC FONCTIONS DE COUT CONCAVES. ANNALES DES TELECOMMUNICATIONS 31, 77-92. (1976).

MINOA 1976 -4 MINOUX, M.
STRUCTURES ALGEBRIQUES GENERALISEES DES PROBLEMES DE CHEMINEMENT DANS LES GRAPHES FINIS, ALGORITHMES ET APPLICATIONS. COMPTES RENDUS HEBDOMADAIRES DES SEANCES DE L'ACADEMIE DES SCIENCES 282, SERIES A, NO. 1, AI, A17-A18. (1976).

MINOA 1976 -5 MINOUX, M.
STRUCTURES ALGEBRIQUES GENERALISEES DES PROBLEMES DE CHEMINEMENT DANS LES GRAPHES. REVUE FRANCAISE D'AUTOMATIQUE, D'INFORMATIQUE, ET DE RECHERCHE OPERATIONNELLE 10, NO. 6, 33-62. (1976).

MINOA 1977 -1 MINOUX, M.
ALGORITHMES GLOUTONS ET ALGORITHMES GLOUTONS ACCELERES POUR LA RESOLUTION DES GRANDS PROBLEMES COMBINATOIRES. ELECTRICITE DE FRANCE: BULLETIN DE LA DIRECTION DES ETUDES ET RECHERCHE, SERIES C, MATHEMATIQUES, INFORMATIQUE, 41-58. (1977).

MINT 1975 -1 MINTY, G. J.
ON THE AXIOMATIC FOUNDATIONS OF THE THEORIES OF DIRECTED LINEAR GRAPHS, ELECTRICAL NETWORKS AND NETWORK-PROGRAMMING. IN: D. R. FULKERSON (ED.): STUDIES IN GRAPH THEORY, PART 2, 246-300. STUDIES IN MATHEMATICS 12, WASHINGTON: MATHEMATICAL ASSOCIATION OF AMERICA. (1975).

MIRC 1974 -1 MIRCHANDANI, P. B.
LOCATION OF FACILITIES ON A STOCHASTIC NETWORK. OR 039-74. CAMBRIDGE: MASSACHUSETTS INST. OF TECHNOLOGY, OPERATIONS RESEARCH CENTER, 15 P. (1974).

MIRC 1976 -1 MIRCHANDANI, P. B.
SHORTEST DISTANCE AND RELIABILITY OF PROBABILISTIC NETWORKS. COMPUTERS AND OPERATIONS RESEARCH 3, 347-356. (1976).

MIRS 1971 -1 MIRSKY, L.
TRANSVERSAL THEORY. NEW YORK: ACADEMIC PRESS. (1971).

MITC 1975 -1 MITCHELL, S.; HEDETNIEMI, S.; GOODMAN, S.
SOME LINEAR ALGORITHMS ON TREES. IN: F. HOFFMAN, R. C. MULLIN, R. B. LEVOW ET AL. (EDS.): PROCEEDINGS OF THE 6TH SOUTHEASTERN CONFERENCE ON COMBINATORICS, GRAPH THEORY, AND COMPUTING, BOCA RATON 1975, 467-484. CONGRESSUS NUMERANTIUM 14, WINNIPEG: UTILITAS MATHEMATICA. (1975).

MITCA 1976 -1 MITCHEN, J.
ON VARIOUS ALGORITHMS FOR ESTIMATING THE CHROMATIC NUMBER OF A GRAPH. COMPUTER JOURNAL 19, 182-183. (1976).

MITE 1977 -1 MITEV, J. G.
APPLICATION OF THE BRANCH AND BOUND METHOD TO SOME DISCRETE LINEAR PROGRAMMING PROBLEMS (RUSSIAN). ZURNAL VYCISLITEL'NOI MATEMATEKI I MATEMATICESKOJ FIZIKI 17, 518-521. (1977).

MITE 1977 -2 MITEV, J. G.
ON SOME KINDS OF PENALTIES IN DISCRETE PROGRAMMING (RUSSIAN). ZURNAL VYCISLITEL'NOI MATEMATIKI I MATEMATICESKOI FIZIKI 17, 1597-1598. (1977).

MITE 1977 -3 MITEV, J. G.
SOME ALGORITHMS FOR SOLVING INTEGER NON-LINEAR MATHEMATICAL PROGRAMMING PROBLEMS (RUSSIAN). ZURNAL VYCISLITEL'NOI MATEMATIKI I MATEMATICESKOI FIZIKI 17, 1042-1046. (1977).

MITT 1977 -1 MITTAL, B. S.; BAGGA, P. C.
PRIORITY PROBLEM IN SEQUENCING WITH STOCHASTIC SERVICE TIMES, A. OPSEARCH 14, 19-28. (1977).

MITTA 1967 -1 MITTAL, M. L.
OPTIMAL SEQUENCING OF MULTISTAGE FLOW SHOP OPERATION. CALCUTTA BRANCH OPERATIONS RESEARCH SOCIETY OF INDIA, 121-130. (1967).

MITTB 1973 -1 MITTEN, L. G.; WARBURTON, A. R.
IMPLICIT ENUMERATION PROCEDURES. WORKING PAPER NO. 251. VANCOUVER: UNIV. OF BRITISH COLUMBIA, FACULTY OF COMMERCE AND BUSINESS ADMINISTRATION, 40 P. (1973).

MIYA 1977 -1 MIYAKAWA, M.; YUBA, T.; SUGITO, Y.; HOSHI, M.
OPTIMUM SEQUENCE TREES. SIAM JOURNAL ON COMPUTING 6, 201-234. (1977).

MJEL 1976 -1 MJELDE, K. M.
EVALUATION AND INCREMENTAL DETERMINATION OF ALMOST OPTIMAL ALLOCATIONS OF RESOURCES. OPERATIONAL RESEARCH QUARTERLY 27, 581-588. (1976).

MOJE 1976 -1 MOJENA, R.; VOLLMANN, T. E.; OKAMOTO, Y.
ON PREDICTING COMPUTATIONAL TIME OF A BRANCH AND BOUND ALGORITHM FOR THE ASSIGNMENT OF FACILITIES. DECISION SCIENCES 7, 856-867. (1976).

MOLE 1976 -1 MOLE, R. H.; JAMESON, S. R.
SEQUENTIAL ROUTE-BUILDING ALGORITHM EMPLOYING A GENERALISED SAVINGS CRITERION, A. OPERATIONAL RESEARCH QUARTERLY 27, 503-512. (1976).

MOLN 1976 -1 MOLNAR, Z.; TRCKA, V.
SOLVING ZERO-ONE HYPERBOLIC PROGRAMMING PROBLEM BY MEANS OF BRANCH AND BOUND METHOD (CZECH). EKONOMICKO-MATEMATICKY OBZOR 12, 428-437. (1976).

MOLZ 1976 -1 MOLZBERGER, P.
HEURISTISCHER ALGORITHMUS ZUR WEGOPTIMIERUNG BEI RECHTWINKELIGEN, ACHSENPARALLELEN HINDERNISSEN, EIN. IN: H. N. DATHE, P. MERTENS, F. D. PESCHANEL ET AL. (EDS.): PROCEEDINGS IN OPERATIONS RESEARCH 6, 280-290. WUERZBURG: PHYSICA. (1976).

MONA 1976 -1 MONAHAN, J. P.
PRODUCTIONS SCHEDULING OF PERISHABLE PRODUCTS. INTERNATIONAL JOURNAL OF PRODUCT RESEARCH 14, 689-697. (1976).

MONM 1977 -1 MONMA, C. L.
OPTIMAL MXN FLOW SHOP SEQUENCING WITH PRECEDENCE CONSTRAINTS AND LAG TIMES. REPORT NO. 7778-OR. BONN: UNIV. BONN, INSTITUT FUER OEKONOMETRIE UND OPERATIONS RESEARCH, 24 P. (1977).

MONT 1976 -1 MONTICONE, L. C.
SOLUTION TO AN OPTIMAL ROUTING PROBLEM IN A LARGE SCALE NETWORK VIA A MINIMUM PERMUTATION SET. IN: NINTH ANNUAL ASILOMAR CONFERENCE ON CIRCUITS, SYSTEMS, AND COMPUTERS, PACIFIC GROVE 1975, 77-82. NORTH HOLLYWOOD: WESTERN PERIODICALS. (1976).

MOOR 1976 -1 MOORE, M. H.
ON THE FASTEST ROUTE FOR CONVOY-TYPE TRAFFIC IN FLOWRATE-CONSTRAINED NETWORKS. TRANSPORTATION SCIENCE 10, 113-124. (1976).

MOORA 1973 -1 MOORE, S. A.; PEARSON, J. M.; LINDSAY, M. A.
TRANSPORTATION PROBLEMS. COMPUTER APPL. 1, NO. 3-4, 118-135. (1973).

MOORB 1976 -1 MOORE, T. G.
SOLVING LARGE 0-1 PROBLEMS. MIAMI (50. ORSA-MEETING + TIMS, NOVEMBER 1976) 22 P. (1976).

MORE 1977 -1 MOREFIELD, C. L.
APPLICATION OF 0-1 INTEGER PROGRAMMING TO MULTITARGET TRACKING PROBLEMS. IEEE TRANSACTIONS ON AUTOMATIC CONTROL AC-22, 302-311. (1977).

MORI 1975 -1 MORIN, T. , L.; MARSTEN, R. E.
HYBRID DYNAMIC PROGRAMMING/BRANCH-AND-BOUND APPROACH TO A CLASS OF SEQUENCING PROBLEMS, A. LAS VEGAS (48. ORSA-MEETING + TIMS, NOVEMBER 1975) 20 P. (1975).

MORIA 1976 -1 MORIN, T. L.; MARSTEN, R. E.
BRANCH-AND-BOUND STRATEGIES FOR DYNAMIC PROGRAMMING. OPERATIONS RESEARCH 24, 611-627. (1976).

MORO 1974 -1 MOROZENSKII, L. YU.
ASYMPTOTIC BEHAVIOUR OF THE LENGTH OF THE PATH OF A TRAVELING SALEMAN WHEN THE CITIES ARE RANDOMLY DISTRIBUTED (RUSSIAN), THE. TEORIJA VEROJATNOSTEI I EE PRIMENENIJA 19, 828-831. (1974).

MOROA 1976 -1 MOROZOV, S. A.
PROBLEM OF FINDING THE SHORTEST TREE OF CONNECTION WITH CONSTRAINTS (RUSSIAN), THE. UPRAVLJAEMYE SISTEMY, NO. 15, 40-47. (1976).

MORP 1971 -1 MORPURGO, R.
HEURISTIC METHOD FOR THE VERIFICATION OF THE ISOMORPHISM OF TWO SIMPLE AND NON ORIENTED GRAPHS, A. CALCOLO 8, 1-31. (1971).

MORR 1977 -1 MORRILL, R. L.; SYMONS, J.
EFFICIENCY AND EQUITY ASPECTS OF OPTIMUM LOCATION. GEOGRAPHICAL ANALYSIS 9, 215-225. (1977).

MORRA 1972 -1 MORRIS, J. G.
FACILITIES LOCATION AND LOCATION-ALLOCATION USING RECTANGULAR DISTANCES. MADISON: UNIV. OF WISCONSIN, GRADUATE SCHOOL OF BUSINESS, 118 P. (1972). (OPERATIONS RESEARCH TECHNICAL REPORT NO. 2)

MORRB 1976 -1 MORROW, C.; GOODMAN, S.
EFFICIENT ALGORITHM FOR FINDING A LONGEST CYCLE IN A TOURNAMENT, AN. IN: F. HOFFMAN, L. LESNIAK, R. MULLIN ET AL. (EDS.): PROCEEDINGS OF THE 7TH SOUTHEASTERN CONFERENCE ON COMBINATORICS, GRAPH THEORY, AND COMPUTING, BATON ROUGE 1976, 453-462. CONGRESSUS NUMERANTIUM 17, WINNIPEG: UTILITAS MATHEMATICA. (1976).

MORT 1977 -1 MORTON, T. E.; DHARAN, B. G.
ALGORISTICS FOR SEQUENCING WITH PRECEDENCE CONSTRAINTS. WORKING PAPER 45-76-77. PITTSBURGH: CARNEGIE-MELLON UNIV., GRADUATE SCHOOL OF INDUSTRIAL ADMINISTRATION, 24 P. (1977).

MOSH 1963 -1 MOSHER, W. W. JR.
CAPACITY-RESTRAINT ALGORITHM FOR ASSIGNING FLOW TO A TRANSPORT NETWORK, A. HIGHWAY RESEARCH RECORD 6, 41-70. (1963).

MOSHA 1963 -1 MOSHMAN, J.; JOHNSON, J.; LARSEN, M.
RAMPS-TECHNIQUE FOR RESOURCE ALLOCATION AND MULTIPROJECT SCHEDULING. P17-27, PROCEEDINGS OF SPRING JOINT COMPUTER CONFERENCE. (1963).

MOSL 1976 -1 MOSLER, K. C.
OPTIMALE TRANSPORTNETZE. ZUR BESTIMMUNG IHRES KOSTENGUENSTIGSTEN STANDORTS BEI GEGEBENER NACHFRAGE. LECTURE NOTES IN ECONOMICS AND MATHEMATICAL SYSTEMS 125, BERLIN: SPRINGER, 148 P. (1976).

MUCH 1976 -1 MUCHNICK, S. S.
COMPUTATIONAL COMPLEXITY OF MULTIPLE RECURSIVE SCHEMATA. SIAM JOURNAL ON COMPUTING 5, 427-451. (1976).

MUCK 1977 -1 MUCKSTADT, J. A.; KOENIG, S. A.
APPLICATION OF LAGRANGIAN RELAXATION TO SCHEDULING IN POWER-GENERATION SYSTEMS, AN. OPERATIONS RESEARCH 25, 387-403. (1977).

MUEL 1975 -1 MUELLER-MERBACH, H.
MODELLING TECHNIQUES AND HEURISTICS FOR COMBINATORIAL PROBLEMS. IN: B. ROY (ED.): COMBINATORIAL PROGRAMMING: METHODS AND APPLICATIONS, 3-27. DORDRECHT: REIDEL. (1975).

MUEL 1975 -2 MUELLER-MERBACH, H.
ROLE OF PUZZLES IN TEACHING COMBINATORIAL PROGRAMMING, THE. IN: B. ROY (ED.): COMBINATORIAL PROGRAMMING: METHODS AND APPLICATIONS. PROCEEDINGS OF THE NATO ADVANCED STUDY INST., VERSAILLES 1974, 379-386. DORDRECHT: REIDEL. (1975).

MUEL 1976 -1 MUELLER-MERBACH, H.
ANSAETZE ZU ENTWURFSMETHODOLOGIEN FUER ALGORITHMEN DER KOMBINATORISCHEN OPTIMIERUNG. IN: H. N. DATHE, P. MERTENS, F. D. PESCHANEL ET AL. (EDS.): PROCEEDINGS IN OPERATIONS RESEARCH 6, 655-667. WUERZBURG: PHYSICA. (1976).

MUEL 1976 -2 MUELLER-MERBACH, H.
HEURISTIC PROCEDURES FOR SOLVING COMBINATORIAL OPTIMIZATION PROBLEMS IN TRANSPORTATION. TRANSPORTATION RESEARCH 10, 377-378. (1976).

MUEL 1976 -3 MUELLER-MERBACH, H.
MORPHOLOGIE HEURISTISCHER VERFAHREN. ZEITSCHRIFT FUER OPERATIONS RESEARCH 20, SERIE A, 69-88. (1976).

****

MUELA 1973 -1 MUELLER, H. CH.
UNTERSUCHUNG DER VERWENDBARKEIT DES ALGORITHMUS VON GOMORY FUER REIN GANZZAHLIGE PROBLEME BEIM DIENSTPLANPROBLEM. DISSERTATION. STUTTGART: UNIV. STUTTGART. (1973).

MULL 1974 -1 MULLABAEV, A. A.
ON AN OPTIMIZATION PROBLEM CONCERNING AN ARRANGEMENT OF TOOTHED WHEELS (RUSSIAN). IN: MAT. METHODY EHKON. ISSLED., 79-86. MOSCOW: IZDAT. NAUKA. (1974).

MULLA 1978 -1 MULLIN, R. C.; STANTON, R. G.
CHARACTERIZATION OF PSEUDO-AFFINE DESIGNS AND THEIR RELATIONS TO A PROBLEM OF CORDES, A. IN: ALSPACH, B.;HELL, P.;MILLER, D. J.(EDS.): ALGORITHMIC ASPECTS OF COMBINATORICS. ANNALS OF DISCRETE MATHEMATICS 2, 231-238. (1978).

MULV 1978 -1 MULVEY, J.
NETWORK RELAXATIONS FOR SET COVERING, SET PARTITIONING AND OTHER INTEGER PROGRAMMING PROBLEMS. RESEARCH REPORT. AUSTIN: UNIV. OF TEXAS, CENTER FOR CYBERNETIC STUDIES. (1978).

MULVA 1975 -1 MULVEY, J. M.
NETWORK RELAXATION APPROACH FOR THE SET PARTITIONING AND SET COVERING MODELS, A. BOSTON: HARVARD UNIVERSITY, GRADUATE SCHOOL OF BUSINESS ADMINISTRATION, 19 P. (1975). (HBS 75-37)

MULVA 1975 -2 MULVEY, J. M.
SPECIAL STRUCTURES IN NETWORK NODELS AND ASSOCIATED APPLICATIONS. DISSERTATION. LOS ANGELES: UNIV. OF CALIFORNIA, WESTERN MANAGEMENT SCIENCE INSTITUTE, 195 P. (1975).

MULVA 1975 -3 MULVEY, J. M.
TESTING OF LARGE-SCALE NETWORK OPTIMIZATION PROGRAM. WORKKING PAPER HBS 75-38. BOSTON: HARVARD UNIV., GRADUATE SCHOOL OF BUSINESS ADMINISTRATION, 54 P. (1975).

MUNC 1977 -1 MUNCH-ANDERSEN, B.; ZAHLE, T. U.
SCHEDULING ACCORDING TO JOB PRIORITY WITH PREVENTION OF DEADLOCK AND PERMANENT BLOCKING. ACTA INFORMATICA 8, 153-176. (1977).

MUNR 1971 -1 MUNRO, J. I.
SOME RESULTS IN THE STUDY OF ALGORITHMS. DISSERTATION, TECHNICAL REPORT. TORONTO: UNIV. OF TORONTO, DEPT. OF COMPUTER SCIENCE. (1971).

MURA 1974 -1 MURAV'EV, V. I.; ROMANOVSKII, I. V.
USE OF THE VARIABLE BASIS METHOD FOR THE SOLUTION OF THE CONTINUOUS ANALOGUE OF JOHNSON'S PROBLEM (RUSSIAN), A. IN: I. V. ROMANOVSKII (ED.): OPERATIONS RESEARCH AND STATISTICAL SIMULATION, NO. 2 (RUSSIAN), 27-37 + 149. LENINGRAD: IZDAT. LENINGRAD UNIV. (1974).

MURC 1973 -1 MURCHLAND, J.
CALCULATING THE PROBABILITY THAT A GRAPH IS DISCONNECTED. LONDON: PLANNING AND TRANSPORT RESEARCH AND COMPUTATION CO. REPORT JDM-182. (1973).

MURCA 1973 -1 MURCHLAND, J.; SHIER, D.
CALCULATING THE PROBABILITY THAT AN UNDIRECTED GRAPH IS DISCONNECTED. LONDON: PLANNING AND TRANSPORT RESEARCH AND COMPUTATION CO. REPORT JDM-186. (1973).

MURCB 1964 -1 MURCHLAND, J. D.
CIRCULATING FLOWS IN MULTI-COMMODITY NETWORKS. REPORT LSE-TNT-10. LONDON: LONDON SCHOOL OF ECONOMICS, 4 P. (1964).

MURCB 1965 -1 MURCHLAND, J. D.
PRIM'S RULE FOR CONSTRUCTING A MINIMAL SPANNING TREE. REPORT LSE-TNT-35, LONDON: LONDON SCHOOL OF ECONOMICS. (1965).

MURCB 1966 -1 MURCHLAND, J. D.
ALGOL PROCEDURE FOR ALL SHORTEST PATHS IN A SYMMETRIC NETWORK, AN. REPORT LBS-TNT-35. LONDON: LONDON BUSINESS SCHOOL, 5 P. (1966).

MURCB 1966 -2 MURCHLAND, J. D.
EXTENSION OF THE CASCADE ALGORITHM TO LARGE GRAPHS, THE. REPORT LSE-TNT-20, LONDON: LONDON SCHOOL OF ECONOMICS. (1966).

MURCB 1966 -3 MURCHLAND, J. D.
INDUCTIVE MATRIX METHOD FOR FINDING ALL SHORTEST PATHS IN A DIRECTED GRAPH, AN. REPORT LBS-TNT-25. LONDON: LONDON BUSINESS SCHOOL, 9 P. (1966).

MURCB 1967 -1 MURCHLAND, J. D.
EFFECT OF INCREASING OR DECREASING THE LENGTH OF A SINGLE ARC ON ALL SHORTEST DISTANCES IN A GRAPH, THE. REPORT LBS-TNT-26. LONDON: LONDON BUSINESS SCHOOL, 5 P. (1967).

MURCB 1967 -2 MURCHLAND, J. D.
FALSE PROPOSITION CONCERNING THE SECOND SHORTEST ELEMENTARY PATH IN A GRAPH, A. REPORT LBS-TNT-57. LONDON: LONDON BUSINESS SCHOOL, 2 P. (1967).

MURCB 1967 -3 MURCHLAND, J. D.
NUMBER OF OPERATIONS IN FIXED MATRIX METHODS FOR FINDING ALL SHORTEST DISTANCES IN A GRAPH, THE. REPORT LBS-TNT-59. LONDON: LONDON BUSINESS SCHOOL, 6 P. (1967).

MURCB 1967 -4 MURCHLAND, J. D.
TWO REMARKS ON CONGESTED ASSIGNMENT. REPORT LBS-TNT-53. LONDON: LONDON BUSINESS SCHOOL, 6 P. (1967).

MURCB 1968 -1 MURCHLAND, J. D.
ALGORITHM FOR FINDING A BASIS OF CIRCUITS IN A DIRECTED GRAPH. REPORT LBS-TNT-69, LONDON: LONDON SCHOOL OF BUSINESS STUDIES. (1968).

MURCB 1968 -2 MURCHLAND, J. D.
ALGORITHMS FOR DETERMINING ALL CONNECTIVITIES IN DIRECTED AND UNDIRECTED GRAPHS. REPORT LBS-TNT-78. LONDON: LONDON BUSINESS SCHOOL. (1968).

MURCB 1968 -3 MURCHLAND, J. D.
EASY CASE WHEN TESTING A DIRECTED GRAPH FOR THE PRESENCE OF A CIRCUIT OF NEGATIVE LENGHT, AN. REPORT LBS-TNT-6.2, LONDON: LONDON SCHOOL OF BUSINESS STUDIES. (1968).

MURCB 1969 -1 MURCHLAND, J. D.
FIXED MATRIX METHOD FOR ALL SHORTEST DISTANCES IN A DIRECTED GRAPH AND FOR THE INVERSE PROBLEM, A. REPORT LBS-TNT-91. LONDON: LONDON BUSINESS SCHOOL. (1969).

MURCB 1970 -1 MURCHLAND, J. D.
EXPLANATION OF DIAL'S SHORTEST DISTANCE ALGORITHM, AN. REPORT JDM-137. INSTITUT FUER ANGEWANDTE REAKTORPHYSIK. (1970).

MURCB 1971 -1 MURCHLAND, J. D.
UPPER BOUND FOR THE AVERAGE CANDIDATE CREATION FACTOR IN ARC TYPE SHORTEST DISTANCE CALCULATIONS, AN. REPORT JDM-150. INSTITUT FUER ANGEWANDTE REAKTORPHYSIK. (1971).

MURD 1976 -1 MURDASOV, A. B.
SIMPLE ALGORITHM FOR FINDING THE TREE CENTRE (RUSSIAN), A. KIBERNETIKA, NO. 1, 142-143. (1976).

MURP 1975 -1 MURPHY, F. H.; STOHR, E. A.
MATHEMATICAL PROGRAMMING APPROACH TO THE SCHEDULING OF CHECK SORTING OPERATIONS, A. EVANSTON: NORTHWESTERN UNIV., CENTER FOR MATHEMATICAL STUDIES IN ECONOMICS AND MANAGEMENT SCIENCE, 23 P. (1975). (DISCUSSION PAPER NO. 164)

MURP 1976 -1 MURPHY, F. H.; STOHR, E. A.
DYNAMIC PROGRAMMING ALGORITHM FOR CHECK SORTING, A (REVISED VERSION). DISCUSSION PAPER NO. 158. EVANSTON: NORTHWESTERN UNIV., CENTER FOR MATHEMATICAL STUDIES AND MANAGEMENT SCIENCE, 24 P. (1976).

****

MURT 1976 -1 MURTHY, M. S.
BULK TRANSPORTATION PROBLEM, A. OPSEARCH 13, 143-155. (1976).

MURTA 1967 -1 MURTY, K.
SYMMETRIC ASSIGNMENT PROBLEM, THE. TECH. REPORT ORC 67-12, OPERATIONS RESEARCH CENTER. BERKELEY: UNIV. OF CALIFORNIA. (1967).

MURTB 1976 -1 MURTY, K. G.
LINEAR AND COMBINATORIAL PROGRAMMING. NEW YORK, LONDON U.A.: WILEY, 567 P. (1976).

MURTC 1971 -1 MURTY, U. S. R.
ON THE NUMBER OF BASES OF A MATROID. IN: R. C. MULLIN, K. B. REID, D. P. ROSELLE, R. P. D. THOMAS (EDS.): PROCEEDINGS OF THE 2ND LOUISIANA CONFERENCE ON COMBINATORICS, GRAPH THEORY AND COMPUTING, BATON ROUGE 1971, 387-410. CONGRESSUS NUMERANTIUM 3, WINNIPEG: UTILITAS MATHEMATICA. (1971).

MUTH 1963 -1 MUTH, J. F.; THOMPSON, G. L.
INDUSTRIAL SCHEDULING. ENGLEWOOD CLIFFS: PRENTICE HALL. (1963).

MYER 1967 -1 MYERS, B. R.
MAX-CUT MIN-FLOW IN A CLASS OF DIRECTED WEIGHTED NETS. IN: PROCEEDINGS OF THE FIRST ANNUAL PRINCETON CONFERENCE ON INFORMATION SCIENCES AND SYSTEMS, 199-200. (1967).

MYER 1974 -1 MYERS, B. R.
ON SPANNING TREES, WEIGHTED COMPOSITIONS, FIBONACCI NUMBERS AND RESISTOR NETWORKS. IN: PROCEEDINGS OF THE EIGHTH ANNUAL PRINCETON CONFERENCE ON INFORMATION SCIENCES AND SYSTEMS, 532-535. (1974).

NABE 1965 -1 NABESHIMA, I.
COMPUTATIONAL SOLUTION TO THE M-MACHINE SCHEDULING PROBLEM. JOURNAL OF THE OPERATIONS RESEARCH SOCIETY OF JAPAN 7, 93-103. (1965).

NABE 1967 -1 NABESHIMA, I.
GENERAL SOLUTIONS OF SOME SEQUENCING PROBLEMS (JAPANESE). KEIEI KAGAKU 10, 234-248. (1967).

NABE 1972 -1 NABESHIMA, I.
ALGORITHMS FOR MULTIPROJECT SCHEDULING WITH RESOURCE CONSTRAINTS AND RELATED PARALLEL SCHEDULING. REPORTS OF THE UNIV. OF ELECTRO-COMMUNICATIONS 23, 29-49. (1972).

NABE 1973 -1 NABESHIMA, I.
BACKTRACK PROGRAMMING ALGORITHM AND RELIABLE HEURISTIC PROGRAMS FOR GENERAL SCHEDULING PROBLEM, A. REPORTS OF THE UNIVERSITY OF ELECTRO-COMMUNICATIONS 24, 23-36. (1973).

NABEA 1975 -1 NABESHIMA, I.; MARUYAMA, S.
COMPUTATIONAL EXPERIENCE ON MULTIPROJECT SCHEDULING WITH RESOURCE CONSTRAINTS. REPORTS OF THE UNIV. OF ELECTRO-COMMUNICATIONS 26, 51-64. (1975).

NAKA 1977 -1 NAKAMURA, T.; YAMASAKI, M.
MINIMAL CUT PROBLEMS ON AN INFINITE NETWORK. HIROSHIMA MATHEMATICAL JOURNAL 7, 597-603. (1977).

NAKAA 1975 -1 NAKANO, H.; NAKANISHI, Y.
STRUCTURE OF SOLUTION SPACE IN DISCRETE OPTIMIZATION PROBLEM AND ITS APPLICATION TO MINIMUM FEEDBACK EDGE SET PROBLEM. TECHNOLOGICAL REPORTS OF THE OSAKA UNIV. 25, NO. 1254-1275, 287-299. (1975).

NAND 1974 -1 NANDA, R.; WEINGARTEN, A.
PRACTICABLE SOLUTION FOR THE QUADRATIC ASSIGNMENT PROBLEM, A. IN: AMERICAN INST. OF INDUSTRIAL ENGINEERS (ED.): PROCEEDINGS AIIE 1974 SPRING ANNUAL CONFERENCE, 114-123. NORCROSS: AMERICAN INST. OF INDUSTRIAL ENGINEERS. (1974).

NARA 1977 -1 NARASIMHAN, S. L.
EFFECTIVENESS OF PRIORITY RULES FOR SCHEDULING MULTISTAGE CONTINUOUS FLOW SHOPS. SAN FRANCISCO (51. ORSA-MEETING + TIMS, MAY 1977) (1977).

NARU 1976 -1 NARULA, S. C.; OGBU, U. I.
ASSIGNMENT PROBLEM WITH SIMILAR RESOURCES. PAPER PRESENTED AT THE 49TH JOINT NATIONAL MEETING OF ORSA AND TIMS, PHILADELPHIA 1976, 8 P. (1976).

NARUA 1977 -1 NARULA, S. C.; OGBU, U. I.; SAMUELSSON, H. M.
ALGORITHM FOR THE P-MEDIAN PROBLEM, AN. OPERATIONS RESEARCH 25, 709-712. (1977).

NASH 1961 -1 NASH-WILLIAMS, C. ST. J. A.
EDGE-DISJOINT SPANNING TREES OF FINITE GRAPHS. JOURNAL OF THE LONDON MATHEMATICAL SOCIETY 36, 221-230. (1961).

NASH 1961 -2 NASH-WILLIAMS, C. ST. J. A.
EDGE-DISJOINT SPANNING TREES OF FINITE GRAPHS. JOURNAL OF THE LONDON MATHEMATICAL SOCIETY 36, 445-450. (1961).

NASH 1964 -1 NASH-WILLIAMS, C. ST. J. A.
DECOMPOSITION OF FINITE GRAPHS INTO FORESTS. JOURNAL OF THE LONDON MATHEMATICAL SOCIETY 39, 12 P. (1964).

NASH 1967 -1 NASH-WILLIAMS, C. ST. J. A.
APPLICATION OF MATROIDS TO GRAPH THEORY, AN. IN: THEORY OF GRAPHS. NEW YORK: GORDON AND BREACH. (1967).

NASH 1975 -1 NASH-WILLIAMS, C. ST. J. A.
HAMILTONIAN CIRCUITS. IN: STUDIES IN GRAPH THEORY, PART II, 301-360. STUDIES IN MATHEMATICS 12, WASHINGTON: MATHEMATICAL ASSOCIATION OF AMERICA. (1975).

NASHA 1977 -1 NASH, P.; GITTINS, J. C.
HAMILTONIAN APPROACH TO OPTIMAL STOCHASTIC RESOURCE ALLOCATION, A. ADVANCES IN APPLIED PROBABILITY 9, 55-68. (1977).

NAST 1975 -1 NASTANSKY, L.; GROSSE, H. W.
COMPUTATIONAL EXPERIMENTS WITH LINEAR PROGRAMMING ORIENTED IMPLICIT ENUMERATION ALGORITHMS FOR THE SOLUTION OF LINEAR (0,1) - PROGRAMS. ARBEITSPAPIER 1-1975. PADERBORN: GESAMTHOCHSCHULE PADERBORN, FACHBEREICH WIRTSCHAFTSWISSENSCHAFT, 34 P. (1975).

NAUS 1975 -1 NAUSS, R. M.
EFFICIENT ALGORITHM FOR THE CAPACITATED FACILITY LOCATION PROBLEM, AN. PAPER PRESENTED AT THE 48TH JOINT NATIONAL MEETING OF ORSA AND TIMS, LAS VEGAS 1975, 22 P. (1975).

NAUS 1976 -1 NAUSS, R. M.
ALGORITHM FOR THE 0-1 INTEGER LINEAR PROGRAMMING PROBLEM, AN. ST. LOUIS: UNIV. OF MISSOURI-ST. LOUIS, SCHOOL OF BUSINESS ADMINISTRATION, 11 P. (1976).

NAUS 1976 -2 NAUSS, R. M.
EFFICIENT ALGORITHM FOR THE 0-1 KNAPSACK PROBLEM, AN. MANAGEMENT SCIENCE 23, 27-31. (1976).

NAUS 1976 -3 NAUSS, R. M.
FIXED CHARGE TRANSPORTATION PROBLEM WITH ADDITIONAL CONSTRAINTS, THE. PHILADELPHIA (49. ORSA-MEETING + TIMS, MARCH/APRIL 1976) 10 P. (1976).

NEEB 1974 -1 NEEBE, A. W.
M-CENTER AND M-MEDIAN FACILITY LOCATION PROBLEMS, THE. DISSERTATION. ROCHESTER: UNIV. OF ROCHESTER, 163 P. (1974).

NEEBA 1975 -1 NEEBE, A. W.; RAO, M. R.
SUBGRADIENT APPROACH TO THE M-MEDIAN PROBLEM, A. TECHNICAL REPORT 75-12. CHAPEL HILL: UNIVERSITY OF NORTH CAROLINA, 15 P. (1975).

NEID 1967 -1 NEIDHART, F.
METHODS FOR SCHEDULING AND SEQUENCING PROBLEMS (GERMAN). INDUSTRIEORGAN 36, 134-156. (1967).

NELS 1977 -1 NELSON, R. T.; HOLLOWAY, C. A.; WONG, R. M.-L.
CENTRALIZED SCHEDULING AND PRIORITY IMPLEMENTATION HEURISTICS FOR A DYNAMIC JOB SHOP MODEL. AIIE TRANSACTIONS 9, 95-102. (1977).

NEMH 1976 -1 NEMHAUSER, G. L.; WOLSEY, L. A.
BEST ALGORITHMS FOR APPROXIMATING THE MAXIMUM OF A SUBMODULAR SET FUNCTION. CORE DISCUSSION PAPER, 7636. HEVERLEE: CENTER FOR OPERATIONS RESEARCH AND ECONOMETRICS, 32 P. (1976).

NEMHA 1976 -1 NEMHAUSER, G. L.; WOLSEY, L. A.; FISHER, M. L.
ANALYSIS OF APPROXIMATIONS FOR MAXIMIZING SUBMODULAR SET FUNCTIONS, AN. CORE DISCUSSION PAPER, 7618. HEVERLEE: CENTER FOR OPERATIONS RESEARCH & ECONOMETRICS, 57 P. (1976).

NEPO 1973 -1 NEPOMIASTCHY, P.
NON-COMBINATORIAL APPROACH TO THE SCHEDULING PROBLEM, A. BRUSSELS: EUROPEAN INST. FOR ADVANCED STUDIES IN MANAGEMENT, 18 P. (1973). (WORKING PAPER 73-46)

NESE 1976 -1 NESETRIL, J.; ROEDL, V.
RAMSEY PROPERTY FOR GRAPHS WITH FORBIDDEN COMPLETE SUBGRAPHS, THE. JOURNAL OF COMBINATORIAL THEORY 20, SERIES B, 243-249. (1976).

NEUM 1975 -1 NEUMANN, K.
OPERATIONS RESEARCH C. DYNAMISCHE OPTIMIERUNG. KARLSRUHE: UNIV. KARLSRUHE, INST. FUER WIRTSCHAFTSTHEORIE UND OPERATIONS RESEARCH, 261S. (1975).

NEWM 1972 -1 NEWMAN, M.
INTEGRAL MATRICES. NEW YORK: ACADEMIC PRESS. (1972).

NEWM 1974 -1 NEWMAN, M.
SMITH NORMAL FORM OF A PARTITIONED MATRIX, THE. JOURNAL OF RESEARCH OF THE NATIONAL BUREAU OF STANDARDS 78 B, [illegible] (1974).

NEY 1972 -1 NEY, P.
RESEARCH ON THE TRANSPORTATION PROBLEM, I (RUMANIAN). REVISTA DE ANALIZA NUMERICA SI TEORIA APROXIMATIEI 1, 199-209. (1972).

NEY 1973 -1 NEY, P.
RESEARCH ON THE TRANSPORTATION PROBLEM, II (RUMANIAN). REVISTA DE ANALIZA NUMERICA SI TEORIA APROXIMATIEI 2, 201-209. (1973).

NGUY 1975 -1 NGUYEN, H.
MULTI-DEPOT VEHICLE ROUTING PROBLEMS. DISSERTATION. CAMBRIDGE: M.I.T., DEPT. SLOAN SCHOOL OF MANAGEMENT. (1975).

NICH 1968 -1 NICHOLSON, D. A. J.; PULLEN, R. D.
PERMUTATION PROCEDURE FOR JOB SHOP SCHEDULING. COMPUTER JOURNAL 11, 48-56. (1968).

NICO 1975 -1 NICOGOSYAN, Z. G.
MINIMAL DECOMPOSITION OF GRAPHS IN SPANNING FORESTS (RUSSIAN). ACADEMIJA NAUK ARMJANSKOI SSR, DOKLADY 61, NO. 1, 15-20. (1975).

NICOA 1976 -1 NICOLAS, J.-L.
ALGORITHMES D'OPTIMISATION EN NOMBRES ENTIERS. IN: JOURNEES ALGORITHMIQUES, PARIS 1975, 169-182. ASTERISQUE NO. 38-39, PARIS: SOC. MATH. FRANCE. (1976).

NICOB 1976 -1 NICOLETTI, B.; NATALI, L.; SCALAS, M. R.
MANPOWER STAFFING AT AN AIRPORT: A GENERALIZED SET COVERING PROBLEM. PISA: UNIV. DI PISA, DIPT. DI RICERCA OPERATIVA E SCIENCZE STATISTICHE, 40 P. (1976).

NIEM 1974 -1 NIEMINEN, J.
LINEAR PSEUDO-BOOLEAN VIEWPOINT ON MATCHING AND OTHER CENTRAL CONCEPTS IN GRAPH THEORY, A. ZASTOSOWANIA MATEMATYKI 14, 365-369. (1974).

NIEM 1974 -2 NIEMINEN, J.
TWO BOUNDS FOR THE DOMINATION NUMBER OF A GRAPH. JOURNAL OF THE INSTITUTE OF MATHEMATICS AND ITS APPLICATIONS 14, 183-187. (1974).

NIEM 1975 -1 NIEMINEN, J.
CHARACTERIZATION OF VERTEX CUTS OF AN UNDIRECTED GRAPH, A. BULLETIN OF THE TECHNICAL UNIVERSITY OF ISTANBUL 28, NO 2, 35-39. (1975).

NIEM 1975 -2 NIEMINEN, J.
SOME OBSERVATIONS ON COVERINGS OF GRAPHS. GLASNIK MATEMATICKI 10, SERIJA III, 3-8. (1975).

NIEM 1975 -3 NIEMINEN, J.
SOME OBSERVATIONS ON THE DETERMINATION OF AN UPPER BOUND FOR THE CLIQUE NUMBER OF A GRAPH. INFORMATION PROCESSING LETTERS 4, 16-17. (1975).

NIEM 1976 -1 NIEMINEN, J.
REALIZING THE DISTANCE MATRIX OF A GRAPH. ELEKTRONISCHE INFORMATIONSVERARBEITUNG UND KYBERNETIK 12, 29-31. (1976).

NIGM 1975 -1 NIGMATULLIN, R. G.
COMPLEXITY OF THE APPROXIMATE SOLUTION OF COMBINATORIAL PROBLEMS. DOKLADY AKADEMII NAUK SSR 224, 289-292. (1975).

NIJE 1975 -1 NIJENHUIS, A.; WILF, H.
COMBINATORIAL ALGORITHMS. NEW YORK: ACADEMIC PRESS. (1975).

NINC 1974 -1 NINCAK, J.
CHROMATIC NUMBERS OF CERTAIN GRAPHS (RUSSIAN), THE. KOMBINATORNYI ANALIZ, VYP. 3, 83-89. (1974).

NJAR 1973 -1 NJARIPJA, H. K.
DIRECT ALGORITHM FOR THE SOLUTION OF INTEGER CONVEX PROGRAMMING PROBLEMS (RUSSIAN), A. TRUDY VYCISLITEL'NOGO CENTRA. TARTUSKII GOSUDARSTVENNYI UNIVERSITET, VYP. 28, 19-43. (1973).

NOLT 1975 -1 NOLTEMEIER, H.
ALGORITHM FOR THE DETERMINATION OF LONGEST DISTANCES IN A GRAPH, AN. MATHEMATICAL PROGRAMMING 9, 350-357. (1975).

NOLT 1976 -1 NOLTEMEIER, H.
REDUKTION VON PRAEZEDENZSTRUKTUREN. ZEITSCHRIFT FUER OPERATIONS RESEARCH 20, 151-159. (1976).

NOMU 1974 -1 NOMURA, S.; UI, T.; KURIYAMA, S.
STUDY OF A SEQUENCING PROBLEM THAT TAKES INTO ACCOUNT THE SWITCH-OVER TIME (JAPANESE), A. MEMOIRS OF THE OSAKA INST. OF TECHNOLOGY 19, SERIES A, SCIENCE AND TECHNOLOGY, 31-41. (1974).

NOON 1977 -1 NOONAN, F.; GIGLIO, R. J.
PLANNING ELECTRIC POWER GENERATION: A NONLINEAR MIXED INTEGER MODEL EMPLOYING BENDERS DECOMPOSITION. MANAGEMENT SCIENCE 23, 946-956. (1977).

NOOR 1975 -1 NOORVASH, S.
COVERING THE VERTICES OF A GRAPH BY VERTEX-DISJOINT PATHS. PACIFIC JOURNAL OF MATHEMATICS 58, 159-168. (1975).

NORB 1977 -1 NORBACK, J. P.; LOVE, R. F.
GEOMETRIC APPROACHES TO SOLVING THE TRAVELING SALESMAN PROBLEM. MANAGEMENT SCIENCE 23, 1208-1223. (1977).

NORD 1969 -1 NORDBECK, S.; RYSTEDT, B.
COMPUTER CARTOGRAPHY SHORTEST ROUTE PROBLEMS. IN: LUND STUDIES IN GEOGRAPHY, SER. C NR. 9. (1969).

NORM 1975 -1 NORMAN, J. M.
ELEMENTARY DYNAMIC PROGRAMMING. LONDON: ARNOLD, 92 P. (1975).

****

NORT 1976 -1 NORTHWESTERN UNIVERSITY; TRANSPORTATION CENTER LIBRARY
DOCTORAL DISSERTATION ON TRANSPORTATION (6. ISSUE, 1974-1975). EVANSTON: NORTHWESTERN UNIV., 33 P. (1976).

NOUR 1968 -1 NOUR, M. E.
NEW LINEAR PROGRAMMING TECHNIQUES FOR THE SHORTEST ROUTE PROBLEM. DISSERTATION, UNIV. MICROFILMS 69-4096. PITTSBURGH: UNIV. OF PITTSBURGH. (1968).

NOURA 1970 -1 NOUR, M. E.; RATHBONE, D. E.; GRANSBORG, B. S. M.
SHORTEST ROUTE MATRIX TECHNIQUE. IN: HAWAII INTERNATIONAL CONFERENCE ON SYSTEM SCIENCES, THIRD PROCEEDINGS, 407-410. (1970).

NOVA 1974 -1 NOVANSKY, R.; PLATEK, M.
CHROMATISCHE BELEGUNGEN UND CHROMATISCHE ZERLEGUNGEN ENDLICHER GRAPHEN (ALGORITHMEN). INFORMATION PROCESSING MACHINES 17, 193-220. (1974).

NOVI 1978 -1 NOVITZKY, N. A.
ANALYSIS OF FACTORS INFLUENCE ON INDUSTRY ALLOCATION EFFICIENCY (RUSSIAN). EKONOMIKA I MATEMATICESKIE METODY 14, 174-178. (1978).

NUGE 1964 -1 NUGENT, C. E.
SAMPLING APPROACHES TO THE SOLUTION OF THE N-BY-M STATIC SEQUENCING PROBLEM. DISSERTATION, ITHACA: CORNELL UNIVERSITY. (1964).

NUNE 1978 -1 NUNEN, J. A. E. E. VAN; WESSELS, J.
MULTI-ITEM LOT SIZE DETERMINATION AND SCHEDULING UNDER CAPACITY CONSTRAINTS. EUROPEAN JOURNAL OF OPERATIONS RESEARCH 2, 36-41. (1978).

NUNN 1976 -1 NUNNER, O.
ANWENDUNG EINES TRANSPORTMODELLS BEI DER EINBERUFUNG VON WEHRPFLICHTIGEN. IN: H. N. DATHE, P. MERTENS, F. D. PESCHANEL ET AL. (EDS.): PROCEEDINGS IN OPERATIONS RESEARCH 6, 441-445. WUERZBURG: PHYSICA. (1976).

NUNNA 1977 -1 NUNNIKHOVEN, T. S.; EMMONS, H.
SCHEDULING ON PARALLEL MACHINES TO MINIMIZE TWO CRITERIA RELATED TO JOB TARDINESS. AIIE TRANSACTIONS 9, 288-296. (1977).

NYKO 1964 -1 NYKOWSKI, I.; ZURKOWSKI, J.
PRODUCTION ALLOCATION PROBLEMS WHICH CAN BE REDUCED TO TRANSPORTATION PROBLEMS (POLISH). PRZEGLAD STATYSTYCZNY 11, 183-193. (1964).

O'BR 1977 -1 O'BRIAN, R. C.
UPPER BOUND ON THE PATH NUMBER OF A DIGRAPH, AN. JOURNAL OF COMBINATORIAL THEORY 22, SERIES B, 168-174. (1977).

O'BRA 1975 -1 O'BRIEN, J. J.
VPM SCHEDULING FOR HIGH-RISE BUILDINGS. PROCEEDINGS OF THE AMERICAN SOCIETY OF CIVIL ENGINEERS, J. CONSTR. DIV. 101, 895-905. (1975).

O'NE 1973 -1 O'NEIL, P. E.; O'NEIL, E. J.
FAST EXPECTED TIME ALGORITHM FOR BOOLEAN MATRIX MULTIPLICATION AND TRANSITIVE CLOSURE. IN: R. RUSTIN (ED.): COMBINATORIAL ALGORITHMS, 59-68. ALGORITHMICS PRESS. (1973).

OCHI 1975 -1 OCHI, T.; MINE, H.
SCHEDULING ALGORITHM OF CONTAINER LOADING AND UNLOADING MACHINES (JAPANESE). KEIEI KAGAKU 19, 227-239. (1975).

ODON 1974 -1 ODONI, A. R.
LOCATION OF FACILITIES ON A NETWORK: A SURVEY OF RESULTS. TECHN. REPORT 3-74, CAMBRIDGE: MASS. INST. OF TECHNOLOGY, OPERATIONS RESEARCH CENTER. (1974).

OFFI 1975 -1 OFFIR, J. D.
TRAVELLING SALESMAN WITH A KNAPSACK(A CALL-GIRL'S HEURISTIC), A. PAPER PRESENTED AT THE 48TH JOINT NATIONAL MEETING OF ORSA AND TIMS,LAS VEGAS 1975, 9 P. (1975).

OHTS 1976 -1 OHTSUKI, T.
FAST ALGORITHM FOR FINDING AN OPTIMAL ORDERING FOR VERTEX ELIMINATION ON A GRAPH, A. SIAM JOURNAL ON COMPUTING 5, 133-145. (1976).

OHTSA 1976 -1 OHTSUKI, T.; CHEUNG, L. K.; FUJISAWA, T.
MINIMAL TRIANGULATION OF A GRAPH AND OPTIMAL PIVOTING ORDER IN A SPARSE MATRIX. JOURNAL OF MATHEMATICAL ANALYSIS AND APPLICATIONS 54, 622-633. (1976).

OICE 1972 -1 OICER, V. I.
MAXIMAL FLOW IN A NETWORK WITH CAPACITIES (RUSSIAN). KOMBINATORNYI ANALIZ, VYP. 2, 68-72. (1972).

OLAH 1975 -1 OLAH, GY.
ENUMERATION OF SPANNING TREES CONTAINING GIVEN SUBTREES. ACTA MATHEMATICA ACADEMIAE SCIENTIARUM HUNGARICAE 26, 205-207. (1975).

OLAR 1973 -1 OLARU, E.
UEBER PERFEKTE UND KRITISCH IMPERFEKTE GRAPHEN. ANALELE STIINTIFICE ALE UNIVERSITATII "AL. I. CUZA" DIN IASI, SERIA NOUA, SECTIUNEA I A, MATEMATICA 19, 477-486. (1973).

OLSO 1976 -1 OLSON, E. N.; WENDELL, R. E.
ANALYSIS OF WAREHOUSE LOCATION IN THE BEEF INDUSTRY, AN. MIAMI (50. ORSA-MEETING + TIMS, NOVEMBER 1976) 6 P. (1976).

ONOK 1976 -1 ONOKERHORAYE, A. G.
CONCEPTUAL FRAMEWORK FOR THE LOCATION OF PUBLIC FACILITIES IN THE URBAN AREAS OF DEVELOPING COUNTRIES: THE NIGERIAN CASE, A. SOCIO-ECONOMIC PLANNING SCIENCES 10, 237-240. (1976).

OREL 1977 -1 OREL, E. N.
FINDING A SHORTEST PATH IN A GRAPH USING THE MINORANT OF THE BELLMAN FUNCTION (RUSSIAN). AVTOMATIKA I TELEMEHANIKA, NO. 2, 88-91; TRANSLATED AS AUTOMATION AND REMOTE CONTROL 38, 1977, 235-237. (1977).

ORLI 1977 -1 ORLIN, J.
MINIMAL INTEGRAL SEPARATOR OF A THRESHOLD GRAPH, THE. IN: P. L. HAMMER, E. L. JOHNSON, B. H. KORTE, G. L. NEMHAUSER (EDS.): STUDIES IN INTEGER PROGRAMMING, 415-420. ANNALS OF DISCRETE MATHEMATICS 1, AMSTERDAM: NORTH-HOLLAND. (1977).

ORLO 1976 -1 ORLOFF, C.; CAPRERA, D.
REDUCTION AND SOLUTION OF LARGE SCALE VEHICLE ROUTING PROBLEMS. TRANSPORTATION SCIENCE 10, 361-373. (1976).

ORLOA 1976 -1 ORLOFF, C. S.
ON GENERAL ROUTING PROBLEMS: COMMENTS. NETWORKS 6, 281-284. (1976).

ORLOA 1976 -2 ORLOFF, C. S.
ROUTE CONSTRAINED FLEET SCHEDULING. TRANSPORTATION SCIENCE 10, 149-168. (1976).

ORLOA 1977 -1 ORLOFF, C. S.
THEORETICAL MODEL OF NET ACCESSIBILITY IN PUBLIC FACILITY LOCATION, A. GEOGRAPHICAL ANALYSIS 9, 244-256. (1977).

ORLOB 1975 -1 ORLOVA, G. I.; DORFMAN, G. YA.
OPTIMAL PARTITIONING OF THE SET OF VERTICES OF A GRAPH BY REDUCTION TO A PROBLEM OF MAXIMAL CUT (RUSSIAN). IZVESTIJA AKADEMII NAUK SSSR, TEHNICESKAJA KIBERNETIKA, NO. 3, 59-65; TRANSLATED AS ENGINEERING CYBERNETICS 13, 1975, NO. 3, 44-50. (1975).

ORLOC 1975 -1 ORLOVA, G. I.; DORFMAN, YA. G.
REDUCING THE PROBLEMS OF OPTIMALLY PARTITIONING THE SET OF VERTICES OF A GRAPH TO A MAXIMAL CUT PROBLEM (RUSSIAN). IZVESTIJA AKADEMII NAUK SSSR, TEHNICESKAJA KIBERNETIKA, NO. 3, 59-65. (1975).

ORON 1976 -1 ORON, G.; KARMELI, D.; WALKER, W. R.
APPROXIMATION SOLUTION OF NON-LINEAR MIXED-INTEGER OPTIMIZATION, AN. PAPER PRESENTED AT THE 49TH JOINT NATIONAL MEETING OF ORSA AND TIMS, PHILADELPHIA 1976, 18 P. (1976).

ORTH 1973 -1 ORTH, E.
DEKOMPOSITIONSMETHODEN - EIN HILFSMITTEL BEI DER PLANUNG REGIONALER ABWASSERBESEITIGUNGSSYSTEME. GWF-WASSER/ABWASSER 114, 576-578. (1973).

ORTHA 1972 -1 ORTH, H.
BINAERE OPTIMIERUNG ALS HILFSMITTEL BEI DER PLANUNG REGIONALER ABWASSERBESEITIGUNSSYSTEME. I, DIE. TECHNISCHER BERICHT NR. 7, KARLSRUHE: UNIV. KARLSRUHE, INSTITUT FUER SIEDLUNGSWASSERWIRTSCHAFT. (1972).

OSLE 1974 -1 OSLEEB, J. P.
LOCATION THEORY FOR THE UNIFORM-PRICE MANUFACTURER, A. DISSERTATION. BUFFALO: STATE UNIV. OF NEW YORK, 276 P. (1974).

OSLEA 1977 -1 OSLEEB, J. P.; CROMLEY, R. G.
LOCATION-PRODUCTION-ALLOCATION PROBLEM WITH NONLINEAR PRODUCTION COSTS, THE. GEOGRAPHICAL ANALYSIS 9, 142-159. (1977).

OSTE 1976 -1 OSTERMANN, R.
QUELQUES EXPERIENCES DE CONFECTION AUTOMATIQUE D'HORAIRES. IN: J. KOHLAS, U. SEIFERT, P. STAEHLY, H. J. ZIMMERMANN (EDS.): PROCEEDINGS IN OPERATIONS RESEARCH 5, 108-112. WUERZBURG: PHYSIKA. (1976).

OSYC 1976 -1 OSYCZKA, A.
ALGORITHM OF OPTIMIZATION FOR A SPECIAL CLASS OF NETWORKS, AN. COMPUTING 16, 77-98. (1976).

OTTA 1976 -1 OTTAVIANI, M.
CUTS ON THE FEASIBLE REGION OF A LINEAR PROGRAMMING PROBLEM WITH INTEGER VARIABLES (ITALIAN). RICERCA OPERATIVA 6, NO. 1, 3-16. (1976).

OUBI 1976 -1 OUBINA, L.
LOCALISATION D'UNE PROPRIETE DU GRAPHE DES BASES D'UNE MATROIDE BINAIRE. UNIVERSIDAD NACIONAL DE LA PLATA. PUBLICACIONES DE LA FACULTAD DE CIENCIAS EXACT., NOTAS MAT., NO. 32, 1-7. (1976).

OUDH 1977 -1 OUDHEUSDEN, D. L. VAN
SHORTEST DISTANCES IN A MODIFIED TRANSPORTATION NETWORK. GEOGRAPHICAL ANALYSIS 9, 257-265. (1977).

OVCH 1976 -1 OVCHINNIKOV, V. G.
ON A PROBLEM OF INTEGER LINEAR PROGRAMMING (RUSSIAN). KIBERNETIKA, NO. 1, 131-135. (1976).

OVER 1975 -1 OVERBECK-LARISCH, M.
HAMILTONSCHE LINIEN IN DIGRAPHEN. SER. INTERNATION. ANAL. NUMER. 29, 109-116. (1975).

OVER 1976 -1 OVERBECK-LARISCH, M.
HAMILTONIAN PATHS IN ORIENTED GRAPHS. JOURNAL OF COMBINATORIAL THEORY 21, SERIES B, 76-80. (1976).

OZAW 1973 -1 OZAWA, T.; HIRAO, S.; HATTORI, Y.
SOME PROPERTIES OF MULTICOLORED-BRANCH GRAPHS. MEMOIRS OF THE FACULTY OF ENGINEERING, KYOTO UNIV. 35, 381-392. (1973).

PACA 1974 -1 PACAULT, J. F.
COMPUTING THE WEAK COMPONENTS OF A DIRECTED GRAPH. SIAM JOURNAL OF COMPUTING 3, 56-61. (1974).

PADB 1971 -1 PADBERG, M.
SIMPLE ZERO-ONE PROBLEMS: SET COVERING, MATCHING AND COVERING IN GRAPHS. MANAGEMENT SCIENCE RESEARCH REPORT 235. PITTSBURGH: CARNEGIE-MELLON UNIV. (1971).

PADBA 1973 -1 PADBERG, M. W.
NOTE ON THE CONVERGENCE OF INTEGER PROGRAMMING ALGORITHMS USING INTERSECTION CUTS, A. PREPRINT SERIES I/73-46. BERLIN: INTERNATIONAL INSTITUTE OF MANAGEMENT, 12 P. (1973).

PADBA 1975 -1 PADBERG, M. W.
CHARACTERIZATIONS OF TOTALLY UNIMODULAR, BALANCED AND PERFECT MATRICES. IN: B. ROY (ED.): COMBINATORIAL PROGRAMMING: METHODS AND APPLICATIONS. PROCEEDINGS OF THE NATO ADVANCED STUDY INST., VERSAILLES 1974, 275-284. DORDRECHT: REIDEL. (1975).

PADBA 1976 -1 PADBERG, M. W.
ON THE COMPLEXITY OF SET PACKING POLYHEDRA (REVISED). GBA WORKING PAPER NO.75-105. NEW YORK: NEW YORK UNIV., GRADUATE SCHOOL OF BUSINESS ADMINISTRATION, 19 P. (1976).

PADBA 1977 -1 PADBERG, M. W.
COVERING, PACKING AND KNAPSACK PROBLEMS. VANCOUVER (CONFERENCE ON DISCRETE OPTIMIZATION, AUGUST 1977), 40 P.; ALSO PUBLISHED IN IBM RESEARCH REPORT RC6658. 40 P. 1977, YORKTOWN HEIGHTS: IBM THOMAS J. WATSON RESEARCH CENTER. (1977).

PADBA 1977 -2 PADBERG, M. W.
ON THE COMPLEXITY OF SET PACKING POLYHEDRA. IN: P. L. HAMMER, E. L. JOHNSON, B. H. KORTE, G. L. NEMHAUSER (EDS.): STUDIES IN INTEGER PROGRAMMING, 421-434. ANNALS OF DISCRETE MATHEMATICS 1, AMSTERDAM: NORTH-HOLLAND. (1977).

PADBB 1977 -1 PADBERG, M. W.; HONG, S.
ON THE SYMMETRIC TRAVELLING SALESMAN PROBLEM: A COMPUTATIONAL STUDY. REPORT 7782-OR. BONN: UNIV. BONN, INST. F. OEKONOMETRIE UND OPERATIONS RESEARCH, 49 P.; ALSO PUBLISHED IN IBM RESEARCH REPORT 6720. 45 P. 1977, YORKTOWN HEIGHTS: IBM THOMAS J. WATSON RESEARCH CENTER. (1977).

PALA 1971 -1 PALASTI, I.
ON THE THRESHOLD DISTRIBUTION FUNCTION OF CYCLES IN A DIRECTED RANDOM GRAPH. STUDIA SCIENTIARUM MATEMATICARUM HUNGARICA 6, 67-73. (1971).

PALM 1974 -1 PALM, K.-P.
BEMERKUNG ZUM GEMISCHT-GANZZAHLIGEN SPIELBERG-MODELL, EINE. WISSENSCHAFTLICHE ZEITSCHRIFT DER TECHNISCHEN HOCHSCHULE FUER CHEMIE LEUNA-MERSEBURG 16, 72-74. (1974).

PANA 1976 -1 PANAGIOTAKOPOULOS, D.
MULTI-COMMODITY MULTI-TRANSFORMED NETWORK FLOWS WITH AN APPLICATION TO RESIDUALS MANAGEMENT. MANAGEMENT SCIENCE 22, 874-882. (1976).

PANA 1977 -1 PANAGIOTAKOPOULOS, D.
COST-TIME MODEL FOR LARGE CPM-PROJECT NETWORKS. J. CONSTR. DIV. 103, NO. CO-2, 201-211. (1977).

PANC 1976 -1 PANCHENKO, A. A.
SOME ALGORITHMS FOR SOLUTION OF QUOTIENTS OF LINEAR PROGRAMMING WITH BOOLEAN VARIABLES (RUSSIAN). KIBERNETIKA, NO. 2, 117-119. (1976).

PAND 1962 -1 PANDIT, S. N. N.
MINADDITION AND AN ALGORITHM TO FIND THE MOST RELIABLE PATHS IN A NETWORK. IRE TRANSACTIONS ON CIRCUIT THEORY CT-9, 190-191. (1962).

PANDA 1975 -1 PANDIT, S. N. N.; SUBRAHMANYAM, Y. V.
ENUMERATION OF ALL OPTIMAL JOB SEQUENCES. OPSEARCH 12, 35-39. (1975).

PANW 1977 -1 PANWALKAR, S. S.; ISKANDER, W.
SURVEY OF SCHEDULING RULES, A. OPERATIONS RESEARCH 25, 45-61. (1977).

PANWA 1975 -1 PANWALKAR, S. S.; KHAN, A. W.
IMPROVED BRANCH AND BOUND PROCEDURE FOR N X M FLOW SHOP PROBLEMS, AN. NAVAL RESEARCH LOGISTICS QUARTERLY 22, 787-790. (1975).

PANWA 1976 -1 PANWALKAR, S. S.; KHAN, A. W.
ORDERED FLOW SHOP PROBLEM WITH IDENTICAL DUE DATES AND TOTAL TARDINESS CRITERION, AN. MIAMI (50. ORSA-MEETING + TIMS, NOVEMBER 1976) 10 P. (1976).

PANWA 1976 -2 PANWALKAR, S. S.; KHAN, A. W.
ORDERED FLOW-SHOP SEQUENCING PROBLEM WITH MEAN COMPLETION TIME CRITERION, AN. INTERNATIONAL JOURNAL OF PRODUCTION RESEARCH 14, 631-636. (1976).

PANWA 1977 -1 PANWALKAR, S. S.; KHAN, A. W.
CONVEX PROPERTY OF AN ORDERED FLOW SHOP SEQUENCING PROBLEM, A. NAVAL RESEARCH LOGISTICS QUARTERLY 24, 159-162. (1977).

PAPA 1975 -1 PAPADIMITRIOU, C. H.
ON THE COMPLEXITY OF THE LOCAL STRUCTURE OF CERTAIN CONVEX POLYTOPES. WORKING PAPER CSTR 197. PRINCETON: PRINCETON UNIV., DEPT. OF ELECTRICAL ENGINEERING. (1975).

PAPA 1976 -1 PAPADIMITRIOU, C. H.
NP-COMPLETENESS OF THE BANDWIDTH MINIMIZATION PROBLEM, THE. COMPUTING 16, 263-270. (1976).

PAPA 1976 -2 PAPADIMITRIOU, C. H.
ON THE COMPLEXITY OF EDGE TRAVERSING. JOURNAL OF THE ASSOCIATION FOR COMPUTING MACHINERY 23, 544-554. (1976).

PAPA 1977 -1 PAPADIMITRIOU, C. H.
EUCLIDEAN TRAVELING SALESMAN PROBLEM IS NP-COMPLETE, THE. THEORETICAL COMPUTER SCIENCE 4, 237-244. (1977).

PAPAA 1976 -1 PAPADIMITRIOU, C. H.; STEIGLITZ, K.
SOME COMPLEXITY RESULTS FOR THE TRAVELING SALESMAN PROBLEM. IN: EIGHTH ANNUAL ACM SYMPOSIUM ON THEORY OF COMPUTING, HERSHEY 1976, 1-9. NEW YORK: ASSOCIATION FOR COMPUTING MACHINERY. (1976).

PAPAA 1977 -1 PAPADIMITRIOU, C. H.; STEIGLITZ, K.
ON THE COMPLEXITY OF LOCAL SEARCH FOR THE TRAVELING SALESMAN PROBLEM. SIAM JOURNAL ON COMPUTING 6, 76-83. (1977).

PAPE 1969 -1 PAPE, U.
KUERZESTE WEGE IN ASYMMETRISCHEN NETZWERKEN ZWISCHEN EINEM FESTEN KNOTEN UND BELIEBIGEN KNOTEN. ELEKTRONISCHE DATENVERARBEITUNG 11, 105-115. (1969).

PAPE 1969 -2 PAPE, U.
VERAENDERUNGEN IN GERICHTETEN NETZWERKEN UND KORREKTUR KUERZESTER WEGLAENGEN ZWISCHEN EINEM FESTEN KNOTEN UND ANDEREN KNOTEN. ELEKTRONISCHE DATENVERARBEITUNG 11, 205-208. (1969).

PAPE 1976 -1 PAPE, U.
DATENSTRUKTUREN FUER MENGEN IN ALGORITHMEN AUF GRAPHEN. IN: H. NOLTEMEIER (ED.): GRAPHEN, ALGORITHMEN, DATENSTRUKTUREN. ERGEBNISSE DES WORKSHOPS WG 76. 2. FACHTAGUNG UEBER GRAPHENTHEORETISCHE KONZEPTE DER INFORMATIK, GOETTINGEN 1976, 99-122. MUENCHEN: HANSER. (1976).

PAPE 1976 -2 PAPE, U.
GRAPHEN-SPRACHEN UND ALGORITHMEN AUF GRAPHEN. 1. FACHTAGUNG UEBER GRAPHENTHEORETISCHE KONZEPTE DER INFORMATIK, BERLIN, JULI 1975. APPLIED COMPUTER SCIENCE, 1. MUENCHEN, WIEN: HANSER, 236 P. (1976).

PAPE 1976 -3 PAPE, U.
MODEL OF A HIGH LEVEL DESIGN LANGUAGE FOR GRAPH ALGORITHMS, A (2ND EDITION). BERICHT NR. 75-29. BERLIN: TECHNISCHE UNIV. BERLIN, FACHBEREICH 20 - KYBERNETIK, 15 P. (1976).

PAPE 1976 -4 PAPE, U.
ZUR WIRTSCHAFTLICHKEIT VON KORREKTURVERFAHREN BEI NETZWERKVERAENDERUNGEN. BERICHT NR. 76-12. BERLIN: TECHNISCHE UNIV. BERLIN, FACHBEREICH 20 - KYBERNETIK, 26 P. (1976).

PAPE 1977 -1 PAPE, U.
BIBLIOGRAPHIE ZU "KUERZESTE WEGLAENGEN UND WEGE IN GRAPHEN", EINE (STAND: ENDE 1976). BERICHT 77-07. BERLIN: T.U. BERLIN, 13 P. (1977).

PAPEA 1976 -1 PAPERNOV, B. A.
REALIZABILITY OF MULTIPRODUCT FLOWS (RUSSIAN). IN: STUDIES IN DISCRETE OPTIMIZATION (RUSSIAN), 230-261. MOSCOW: IZDAT. NAUKA. (1976).

PAPI 1975 -1 PAPINEAU, R. L.; FRANCIS, R. L.; BARTHOLDI, J.
MINIMAX FACILITY LAYOUT PROBLEM INVOLVING DISTANCES BETWEEN AND WITHIN FACILITIES, A. AIIE TRANSACTIONS 7, 345-355. (1975).

PARI 1963 -1 PARIKH, S.; FRISCH, I.
FINDING THE MOST RELIABLE ROUTES IN COMMUNICATION SYSTEM. IEEE TRANSACTIONS ON COMMUNICATIONS AND SYSTEMS CS-11, 402-407. (1963).

PARK 1977 -1 PARKER, R. G.; DEANE, R. H.; HOLMES, R. A.
ON THE USE OF A VEHICLE ROUTING ALGORITHM FOR THE PARALLEL PROCESSOR PROBLEM WITH SEQUENCE DEPENDENT CHANGEOVER COSTS. AIIE TRANSACTIONS 9, 155-160. (1977).

PARKA 1977 -1 PARKER, R. G.; GRAVES, R. J.; SHERALI, H.
APPLICATION OF A GRAPH-PARTITIONING ALGORITHM TO THE SCHEDULING OF CURRICULUM REQUIREMENTS, THE. SOCIO-ECONOMIC PLANNING SCIENCES 11, 95-100. (1977).

PARR 1968 -1 PARRIS, R.
CODING PROBLEM FOR GRAPHS, THE. DISSERTATION. UNIVERSITY OF THE WEST INDIES. (1968).

PARRA 1967 -1 PARRIS, R.; READ, R. C.
GRAPH ISOMORPHISM AND THE CODING OF GRAPHS. SCIENTIFIC REPORT UWI/003. JAMAICA: UNIV. OF THE WEST INDIES. (1967).

PARRA 1968 -1 PARRIS, R.; READ, R. C.
CODING PROCEDURE FOR GRAPHS, A. SCIENTIFIC REPORT UWI/CC10. JAMAICA: UNIV. OF THE WEST INDIES. (1968).

PARS 1969 -1 PARSONS, J. A.
BRANCH AND BOUND ALGORITHMS: THE TRAVELING SALESMAN PROBLEM. JOURNAL OF SYSTEM MANAGEMENT 10, 38-40. (1969).

PARS 1970 -1 PARSONS, J. A.
SEQUENCING TASKS AND OPERATIONS. JOURNAL OF SYSTEM MANAGEMENT 21, 38-39. (1970).

PARS 1970 -2 PARSONS, J. A.
SEQUENCING TASKS WITH PROXIMITY CONSTRAINTS. JOUNAL OF SYSTEMS MANAGEMENT 21, 40-41. (1970).

PART 1976 -1 PARTHASARATHY, K. R.; CHOUDUM, S. A.
EDGE CLIQUE COVER NUMBER OF PRODUCTS OF GRAPHS, THE. JOURNAL OF MATHEMATICAL AND PHYSICAL SCIENCES 10, 255-261. (1976).

PASS 1975 -1 PASSY, U.; SILMAN, L.
ONE-AT-A-TIME DYNAMIC PROGRAMMING OVER POLYTOPE. AUTOMATICA 11, 499-508. (1975).

PATE 1976 -1 PATEL, A. M.; WALD, C. G. DE; COTES, L. C.
PRE-PARTITIONING AS A MEANS OF ENHANCING PAIRWISE INTERCHANGE SOLUTIONS TO QUADRATIC ASSIGNMENT PROBLEMS. COMPUTERS AND OPERATIONS RESEARCH 3, 49-56. (1976).

PATR 1974 -1 PATRINOS, A. N.; HAKIMI, S. L.
ORDERING PROBLEMS IN TREES AND THEIR APPLICATIONS TO TREE-STRUCTURED OPERATIONS OR FILES. IN: PROCEEDINGS OF THE EIGHTH ANNUAL PRINCETON CONFERENCE ON INFORMATION SCIENCES AND SYSTEMS, 456-459. (1974).

PATT 1976 -1 PATTERSON, J. H.
PROJECT SCHEDULING: THE EFFECTS OF PROBLEM STRUCTURE ON HEURISTIC PERFORMANCE. NAVAL RESEARCH LOGISTICS QUARTERLY 23, 95-124. (1976).

PATTA 1976 -1 PATTERSON, J. H.; ROTH, G. W.
SCHEDULING A PROJECT UNDER MULTIPLE RESOURCE CONSTRAINTS: A ZERO-ONE PROGRAMMING APPROACH. AIIE TRANSACTIONS 8, 449-455. (1976).

PAUC 1976 -1 PAUCK, R.
MODELL ZUR BESTIMMUNG VON PERSONALEINSATZPLAENEN BEI LUFTVERKEHRSGESELLSCHAFTEN (CREW SCHEDULING), EIN. IN: H. N. DATHE, P. MERTENS, F. D. PESCHANEL ET AL. (EDS.): PROCEEDINGS IN OPERATIONS RESEARCH 6, 453-464. WUERZBURG: PHYSICA. (1976).

PAUL 1977 -1 PAUL, W. J.
2.5N-LOWER BOUND ON THE COMBINATORIAL COMPLEXITY OF BOOLEAN FUNCTIONS, A. SIAM JOURNAL ON COMPUTING 6, 427-443. (1977).

PAULA 1976 -1 PAULAUSKAS, M. A.; RASHKILIS, A. V.; SHYARKSHNIS, G. A.
PROBLEM CONCERNING THE OPTIMAL ALLOCATION OF DEFICIENT RESOURCES FOR A DISCRETE CONSUMPTION, 2. ALGORITHMS FOR THE APPROXIMATE SOLUTION OF PROBLEMS WITH A GREAT NUMBER OF UNKNOWNS (RUSSIAN). LIETUVOS TSR MOKSLU AKADEMIJA DARB, SERIES B, NO. 3, 127-135. (1976).

PAULA 1976 -2 PAULAUSKAS, M. A.; RASHKILIS, A. V.; SHYARKSHNIS, G. A.
PROBLEM CONCERNING THE OPTIMAL ALLOCATION OF DEFICIENT RESOURCES FOR A DISCRETE CONSUMPTION, 3. HEURISTIC ALGORITHMS FOR THE SOLUTION OF THE PROBLEM (RUSSIAN). LIETUVOS TSR MOKSLU AKADEMIJA DARB, SERIES B, NO. 4, 115-123. (1976).

PAULB 1975 -1 PAULAUSKAS, M. A.; RASHKILIS, A. V.; STANIONIS, K.; SHYARKSHNIS, G. A.
PROBLEM OF THE OPTIMAL ALLOCATION OF DEFICIT RESOURCES FOR DISCRETE CONSUMPTION, 1. FORMULATION OF THE PROBLEM AND MATHEMATICAL METHODS OF SOLUTION (RUSSIAN), THE. LIETUVOS TSR MOKSLU AKADEMIJA. DARBA, SERIES B, NO. 5, 145-152. (1975).

PAYA 1977 -1 PAYAN, C.
SUR QUELQUES PROBLEMES DE COUVERTURE ET DE COUPLAGE EN COMBINATOIRE. DISSERTATION. GRENOBLE: UNIV. GRENOBLE, 259 P. (1977).

PEAR 1974 -1 PEARMAN, A.
HEURISTIC APPROACHES TO ROAD NETWORK OPTIMIZATION. ENGINEERING OPTIMIZATION 1, 1, 37-49. (1974).

PEKI 1975 -1 PEKICH, T. J.; ENSCORE, E. E.
EQUIPPING AN EMERGENCY VEHICLE TO MAXIMIZE PATIENT CARE THROUGH THE USE OF INTEGER PROGRAMMING. PAPER PRESENTED AT THE 48TH JOINT NATIONAL MEETING OF ORSA AND TIMS, LAS VEGAS 1975, 27 P. (1975).

PELE 1976 -1 PELED, U. N.
MATROIDAL GRAPHS. REPORT NO. 7648-OR. BONN: UNIV. BONN, INST. F. OEKONOMETRIE UND OPERATIONS RESEARCH, 46 P. (1976).

PELE 1977 -1 PELED, U. N.
PROPERTIES OF FACETS OF BINARY POLYTOPES. IN: P. L. HAMMER, E. L. JOHNSON, B. H. KORTE, G. L. NEMHAUSER (EDS.): STUDIES IN INTEGER PROGRAMMING, 435-456. ANNALS OF DISCRETE MATHEMATICS 1, AMSTERDAM: NORTH-HOLLAND. (1977).

PENA 1976 -1 PENAUD, J. G.
ALGORITHMES DE PLANARITE. IN: ANAL. APPL. INF. JOURN. COMB. INF., BORDEAUX 1975, 279-303. BORDEAUX: UNIV. BORDEAUX I. (1976).

PENT 1976 -1 PENTICO, D. W.
ASSORTMENT PROBLEM WITH NONLINEAR COST FUNCTIONS, THE. OPERATIONS RESEARCH 24, 1129-1142. (1976).

PERC 1975 -1 PERCOV, E. E.; BOGDANOVA, O. V.
ALGORITHM FOR THE MINIMIZATION OF A PSEUDO-BOOLEAN QUADRATIC FUNCTION (RUSSIAN), AN. TRUDY MOSKOVSKOGO ORDENA LENINA ENERGETICESKOGO INSTITUTA VYP. 216, 15-36. (1975).

PERE 1966 -1 PEREPELITSA, V. A.
PROBLEM IN SCHEDULE THEORY. CYBERNETICS 2, 60-62. (1966).

PEREA 1976 -1 PEREPELITSA, V. A.; SEVAST'YANOV, S. V.
ON A SEQUENCING PROBLEM IN A NETWORK (RUSSIAN). UPRAVLJAEME SISTEMY, NO. 15, 48-67. (1976).

PERK 1965 -1 PERKO, A.
SOME COMPUTATIONAL NOTES ON THE SHORTEST ROUTE PROBLEM. COMPUTER JOURNAL 8, 19-20. (1965).

PERL 1976 -1 PERL, J.
ZUR ALGORITHMISCHEN BEHANDLUNG SPEZIELLER NP-VOLLSTAENDIGER PROBLEME AUF GRAPHEN. IN: H. NOLTEMEIER (ED.): GRAPHEN, ALGORITHMEN, DATENSTRUKTUREN. ERGEBNISSE DES WORKSHOPS WG 76. 2. FACHTAGUNG UEBER GRAPHENTHEORETISCHE KONZEPTE DER INFORMATIK, GOETTINGEN 1976, 149-168. MUENCHEN: HANSER. (1976).

PERLA 1975 -1 PERLMAN, B. L.
BRANCH-AND-BOUND METHODS APPLIED TO REAL-TIME VEHICLE DISPATCHING. PAPER PRESENTED AT THE 48TH JOINT NATIONAL MEETING OF ORSA AND TIMS, LAS VEGAS 1975, 29 P. (1975).

PETE 1975 -1 PETER, K.
FLOWS IN NETWORKS WITH QUADRATIC ECONOMIC FUNCTIONS (HUNGARIAN). ALKALMAZOTT MATEMATIKAI LAPOK, NO. 1-2, 101-108. (1975).

PETEA 1975 -1 PETERSEN, E. R.
PRIMAL-DUAL TRAFFIC ASSIGNMENT ALGORITHM, A. MANAGEMENT SCIENCE 22, 87-95. (1975).

PETEB 1977 -1 PETERSEN, J. P.
MANUAL PROCEDURE FOR DESIGNING MAILMAN-ROUTES. COMPUTERS AND OPERATIONS RESEARCH 4, 295-302. (1977).

PETR 1976 -1 PETRICH, F.; WAGNER, K.
ALGORITHMUS FUER DAS LINEARE 0-1 OPTIMIERUNGSPROBLEM, EIN. MATHEMATISCHE OPERATIONSFORSCHUNG UND STATISTIK 7, 351-360. (1976).

PETRA 1975 -1 PETROV, I. P.; TIMOFEEV, A. V.
FINITELY CONVERGENT RECURRENCE ALGORITHMS FOR THE SOLUTION OF INTEGER INEQUALITIES IN THE PRESENCE OF CONSTRAINTS (RUSSIAN). ZURNAL VYCISLITEL'NOI MATEMATIKI I MATEMATICESKOI FIZIKI 15, 1582-1588 + 1626. (1975).

PETRB 1976 -1 PETROV, S. V.
ON A PROBLEM IN GRAPH THEORY (RUSSIAN). AVTOMATIKA I TELEMEHANIKA, NO. 5, 151-157. (1976).

PEVE 1969 -1 PEVEPELICHE, V. A.; GIMADI, E. X.
ON THE PROBLEM OF FINDING THE MINIMAL HAMILTONIAN CICUIT ON A GRAPH WITH WEIGHTED ARCS (RUSSIAN). DISKRET ANALYZ. 15, 57-65. (1969).

PFAF 1976 -1 PFAFFENBERGER, R. C.; WALKER, D. A.
MATHEMATICAL PROGRAMMING FOR ECONOMICS AND BUSINESS. AMES: IOWA STATE UNIVERSITY PRESS, 462 P. (1976).

PHIL 1976 -1 PHILLIPS, L. W.; UNGER, P. S.
MATHEMATICAL PROGRAMMING SOLUTION OF A HOIST SCHEDULING PROGRAM. AIIE TRANSACTIONS 8, 219-225. (1976).

****

PHUO 1976 -1 PHUONG, T. H.
SOLUTION OF INTEGER PROGRAMS WITH A QUADRATIC OBJECTIVE FUNCTION. JOURNAL OF THE ASSOCIATION FOR COMPUTING MACHINERY 23, 468-474. (1976).

PICA 1975 -1 PICARD, C.-F.
ARBRE MINIMAL D'UNE SOMME DE GRAPHES. CAHIERS DU CENTRE D'ETUDES DE RECHERCHE OPERATIONNELLE 17, 351-364. (1975).

PICAA 1975 -1 PICARD, J. C.; RATLIFF, H. D.
MINIMUM CUTS AND RELATED PROBLEMS. NETWORKS 5, 357-370. (1975).

PICAB 1976 -1 PICARD, J.-C.
MAXIMAL CLOSURE OF A GRAPH AND APPLICATIONS TO COMBINATORIAL PROBLEMS. MANAGEMENT SCIENCE 22, 1268-1272. (1976).

PICAC 1976 -1 PICARD, J.-C.; QUEYRANNE, M.
TIME-DEPENDENT TRAVELING SALESMAN PROBLEM AND APPLICATION TO THE TARDINESS PROBLEM IN ONE-MACHINE SCHEDULING, THE. MIAMI (50. ORSA-MEETING + TIMS, NOVEMBER 1976) 39 P. (1976).

PICAC 1977 -1 PICARD, J.-C.; QUEYRANNE, M.
NETWORKS, GRAPHS AND SOME NONLINEAR 0-1 PROGRAMMING PROBLEMS. RAPPORT TECHNIQUE EP 77-R-32. MONTREAL: ECOLE POLYTECHNIQUE DE MONTREAL, DEPT. DE GENIE INDUSTRIEL, 35 P. (1977).

PICAC 1977 -2 PICARD, J.-C.; QUEYRANNE, M.
ON THE INTEGER-VALUED VARIABLES IN THE LINEAR VERTEX PACKING PROBLEM. MATHEMATICAL PROGRAMMING 12, 97-101. (1977).

PICAD 1975 -1 PICARD, J.-C.; RATLIFF, H. D.
CUT APPROACH TO THE RECTILINEAR DISTANCE FACILITY LOCATION PROBLEM, A. RESEARCH REPORT NO. 75-7. GAINESVILLE: UNIV. OF FLORIDA, INDUSTRIAL AND SYSTEMS ENGINEERING DEPT., 24 P. (1975).

PICH 1976 -1 PICHOCKA, M.
CERTAIN OPTIMIZATION PROBLEM CONNECTED WITH GRAPH THEORY (POLISH), A. ZESZYTY NAUKOWE POLITECHNIKI SLASKIEJ NO. 441, MATEMATYKA-FIZYKA ZESZYT 26, 305-307. (1976).

PIEH 1975 -1 PIEHLER, J.
GLOBALE BETRACHTUNGEN IN DER GANZZAHLIGEN OPTIMIERUNG. MATHEMATISCHE OPERATIONSFORSCHUNG UND STATISTIK 6, 839-844. (1975).

PIEH 1976 -1 PIEHLER, J.
UEBER SCHNITTVERSCHAERFUNGEN IN DER GEMISCHT-GANZZAHLIGEN LINEAREN OPTIMIERUNG. MATHEMATISCHE OPERATIONSFORSCHUNG UND STATISTIK 7, 333-342. (1976).

PIEH 1977 -1 PIEHLER, J.
GANZZAHLIGE OPTIMIERUNG MIT KONGRUENZNEBENBEDINGUNGEN. MATHEMATISCHE OPERATIONSFORSCHUNG UND STATISTIK 8, SERIES OPTIMIZATION, 147-150. (1977).

PIER 1975 -1 PIERCE, A. R.
BIBLIOGRAPHY ON ALGORITHMS FOR SHORTEST PATH, SHORTEST SPANNING TREE, AND RELATED CIRCUIT ROUTING PROBLEMS (1956-1974). NETWORKS 5, 129-150. (1975).

PILI 1976 -1 PILIPOSJAN, T. E.
CERTAIN ASSIGNMENT PROBLEM (RUSSIAN), A. AKADEMIJA NAUK ARMJANSKOI SSR, DOKLADY 62, NO. 5, 265-268. (1976).

PINO 1976 -1 PINOCHET, M. R.
CENTRALIZATION AND DECENTRALIZATION OF DECISION-MAKING IN TRANSPORTATION NETWORKS. ACM SIGMAP NEWSLETTER, NO. 20, 54-68. (1976).

PINT 1975 -1 PINTO, J. V.
BACKGROUND AND DEVELOPMENT OF LOCATION THEORY. DISSERTATION. NORMAN: UNIV. OF OKLAHOMA, 344 P. (1975).

PIPE 1975 -1 PIPER, C.
COMPUTATIONAL STUDIES IN OPTIMIZING AND POSTOPTIMIZING LINEAR PROGRAMS IN ZERO-ONE VARIABLES. DISSERTATION. PITTSBURGH: CARNEGIE-MELLON UNIVERSITY. (1975).

PIPEA 1975 -1 PIPER, C. J.; ZOLTNERS, A. A.
IMPLICIT ENUMERATION BASED ALGORITHMS FOR POSTOPTIMIZING ZERO-ONE PROGRAMS. NAVAL RESEARCH LOGISTICS QUARTERLY 22, 791-810. (1975).

PIPEA 1976 -1 PIPER, C. J.; ZOLTNERS, A. A.
SOME EASY POSTOPTIMALITY ANALYSIS FOR ZERO-ONE PROGRAMMING. MANAGEMENT SCIENCE 22, 759-765. (1976).

PIPP 1976 -1 PIPPENGER, N.; VALIANT, L. G.
SHIFTING GRAPHS AND THEIR APPLICATIONS. JOURNAL OF THE ASSOCIATION FOR COMPUTING MACHINERY 23, 423-432. (1976).

PIRO 1975 -1 PIROSKA, T.
APPLICATION OF THE DECOMPOSITION PRINCIPLE OF DANTZIG AND WOLFE TO THE SOLUTION OF A GENERALIZED TRANSPORTATION PROBLEM (HUNGARIAN), AN. ALKALMAZOTT MATEMATIKAI LAPOK, NO. 1-2, 81-90. (1975).

PLAN 1975 -1 PLANE, D. R.; MONARCHI, D. E.; HENDRICK, T. E.
HIERARCHICAL APPROACH FOR SOME LOCATION PROBLEMS, A. BOULDER: UNIVERSITY OF COLORADO, GRADUATE SCHOOL OF BUSINESS ADMINISTRATION, 12 P. (1975). (MANAGEMENT SCIENCE REPORT SERIES NO. 75-25)

PLAX 1976 -1 PLAXTON, A.
LINEAR PROGRAMMING MODEL TO OPTIMALLY SCHEDULE CHECK PROCESSING ACTIVITIES, A. PAPER PRESENTED AT THE 49TH JOINT NATIONAL MEETING OF ORSA AND TIMS, PHILADELPHIA 1976, 31 P. (1976).

PLES 1976 -1 PLESNIK, JAN.
MINIMUM BLOCK CONTAINING A GIVEN GRAPH. ARCHIVES OF MATHEMATICS 27, 6, 668-672. (1976).

PLESA 1976 -1 PLESTCHINSKY, A. S.
E-APPROACH TO A NETWORK MODEL OPTIMIZATION (RUSSIAN). EKONOMIKA I MATEMATICHESKIE METODY 12, 1212-1215. (1976).

PLOT 1974 -1 PLOTNIKOV, V. N.; GALKIN, A. V.
METHOD FOR THE SOLUTION OF CERTAIN EXTREMAL PROBLEMS OF COMBINATORIAL TYPE (RUSSIAN), A. IN: CYBERNETICS AND COMPUTER TECHNOLOGY, NO. 23: COMPLEX CONTROL SYSTEMS (RUSSIAN), 81-84 + 157. KIEV: NAUKOVA DUMKA. (1974).

PLOTA 1974 -1 PLOTNIKOV, V. N.; ZVEREV, V. YU.
METHODS OF FAST DISTRIBUTION OF ALGORITHMS IN COMPUTATIONAL SYSTEMS (RUSSIAN). IZVESTIJA AKADEMII NAUK SSSR, TEHNICESKAJA KIBERNETIKA, NO. 3, 136-143; TRANSLATED AS ENGINEERING CYBERNETICS 12, 1974, NO. 3, 99-106. (1974).

PLUK 1976 -1 PLUKENE, D. B.
APPLICATION OF SEMI-BOOLEAN METHODS TO THE SOLUTION OF 0-1 INTEGER LINEAR PROGRAMMING PROBLEMS (RUSSIAN). TEOR. KOMECH. AVTOMATOV PRILOZH., LATIV. SSR, NO. 6, 51-61. (1976).

POCH 1976 -1 POCHTMAN, YU. M.; RAFES, R. YU.
MODIFICATION OF THE DYNAMIC PROGRAMMING ALGORITHM FOR PROBLEMS CONCERNING THE ALLOCATION OF RESOURCES (RUSSIAN). IZVESTIJA AKADEMII NAUK SSSR, TEHNICESKAJA KIBERNETIKA, NO. 5, 20-23. (1976).

POHL 1969 -1 POHL, I.
BI-DIRECTIONAL AND HEURISTIC SEARCH IN PATH PROBLEMS. SLAC REPORT 104, STANFORD : STANFORD LINEAR ACCELERATOR CENTER. (1969).

****

POHL 1970 -1 POHL, I.
HEURISTIC SEARCH VIEWED AS PATH FINDING IN A GRAPH. ARTIFICIAL INTELLIGENCE 1, 193-304. (1970).

POHL 1973 -1 POHL, I.
AVOIDANCE OF (RELATIVE) CATASTROPHE HEURISTIC COMPETENCE GENUINE DYNAMIC WEIGHTING AND COMPUTATIONAL ISSUES IN HEURISTIC PROBLEM SOLVING, THE. 3RD INTERNATIONAL JOINT CONFERENCE ON ARTIFICIAL INTELLIGENCE, STANFORD UNIVERSITY. (1973).

POLE 1976 -1 POLESSKIJ, V. P.
COVERING A FINITE GRAPH WITH A MINIMAL NUMBER OF FORESTS (RUSSIAN). PROBLEMY PEREDACI INFORMACII 12, NO 2, 76-82. (1976).

POLJ 1974 -1 POLJAK, S.
NOTE ON STABLE SETS AND COLORINGS OF GRAPHS, A. COMMENTATIONES MATHEMATICAE UNIVERSITATIS CAROLINAE 15, 307-309. (1974).

POLJA 1974 -1 POLJAKOV, A. S.
SHORTENING THE ENUMERATION IN THE SOLUTION OF A CERTAIN CLASS OF PROBLEMS IN GRAPH THEORY (RUSSIAN). VYCISLITEL'NAJA TEHNIKA V MASINOSTROENII, NO. 1, 9-16. (1974).

POLJA 1975 -1 POLJAKOV, A. S.
SEARCH ALGORITHMS FOR MAXIMAL COMPLETE SUBGRAPHS (RUSSIAN). AVTOMATIKA I VYCISLITEL'NAJA TEHNIKA, NO. 6, 1-5. (1975).

POLL 1977 -1 POLLACK, M.
SOME ELEMENTS OF THE AIRLINE FLEET PLANNING PROBLEM. TRANSPORTATION RESEARCH 11, 301-310. (1977).

POLLA 1976 -1 POLLATSCHECK, M. A.
PERSONNEL ASSIGNMENT BY MULTIOBJEKTIVE PROGRAMMING. ZEITSCHRIFT FUER OPERATIONS RESEARCH 20, 161-170. (1976).

POLY 1976 -1 POLYMERIS, A.
OPTIMIERUNG UND AUFTEILUNG DER KOSTEN REGIONALER ABWASSERVERBAENDE. IN: LIEBLING, T. M.; ROESSLER, M.(EDS.): KOMBINATORISCHE ENTSCHEIDUNGSPROBLEME: METHODEN UND ANWENDUNG. LECTURE NOTES IN ECONOMICS AND MATHEMATICAL SYSTEMS 153, BERLIN: SPRINGER, 151-169. (1976).

POPE 1976 -1 POPESCU, D.
CLASSES OF MAXIMAL NON HAMILTONIAN GRAPHS (RUMANIAN). STUDII SI CERCETARI MATEMATICE 28, 321-341. (1976).

POPEA 1976 -1 POPESCU, D. C.
ALGORITHM FOR SOLVING THE TRANSPORTATION TRIDIMENSIONAL PROBLEM, AN. ECONOMIC COMPUTATION AND ECONOMIC CYBERNETICS STUDIES AND RESEARCH, NO. 2, 41-52. (1976).

POPEB 1975 -1 POPESCU, E.
HAMILTONIAN PATHS IN A GRAPH: APPLICATIONS TO THE SOLUTION OF CERTAIN PROBLEMS OF PRODUCTION (RUMANIAN). GAZETA MATEMATICA PUBLICATIE LUNARA PENTRU TINERET 80, NO. 6, 202-206. (1975).

POPEC 1976 -1 POPESCU, M.-F.
QUELQUES RESULTATS DE REDUCIBILITE DANS LA PROGRAMMATION EN ENTIERS. REVUE ROUMAINE DE MATHEMATIQUES PURES ET APPLIQUEES 21, 751-757. (1976).

POPO 1972 -1 POPOVICIU, N.
DYNAMIC PROGRAMMING: ONE-SIZE ALLOCATION PROBLEM WITH A SUPPLEMENTARY CONSTRAINT. BULLETIN MATHEMATIQUE DE LA SOCIETE DES SCIENCES MATHEMATIQUES DE LA REPUBLIQUE SOCIALISTE DE ROUMAINE 16, NO. 4, 461-466. (1972).

POPO 1973 -1 POPOVICIU, N.
ALLOCATION PROBLEM, WITH BIVALENT ZERO-ONE VARIABLES DEFINED BY MEANS OF THE ELEMENTS OF A TRANSPORTATION PROBLEM (RUMANIAN), AN. ANALELE UNIVERSITATII BUCURESTI. MATEMATICA-MECANICA 22, NO. 1, 89-94. (1973).

PORT 1976 -1 PORTUGAL, V. M.
ASYMPTOTIC BEHAVIOUR OF SOLUTIONS FOR PROBLEMS OF THE THEORY OF SCHEDULES BY DIFFERENT CRITERIA (RUSSIAN). KIBERNETIKA, NO. 2, 132-134. (1976).

PORT 1976 -2 PORTUGAL, V. M.
ASYMPTOTIC BEHAVIOUR OF THE SCHEDULE THEORY PROBLEMS BY DIFFERENT CRITERIA (RUSSIAN). KIBERNETIKA, NO. 5, 97-99. (1976).

POTT 1972 -1 POTTS, R.; OLIVER, R.
FLOWS IN TRANSPORTATION NETWORKS. NEW YORK: ACADEMIC PRESS. (1972).

POUL 1975 -1 POULTON, M. C.; KANAFANI, A.
APPLICATION OF LOCATION MODELS TO OFF-AIRPORT TERMINALS, THE. TRANSPORTATION SCIENCE 9, 224-247. (1975).

POWE 1977 -1 POWELL, J.; GLEASON, J.; BURTON, J.
POWER GENERATOR SCHEDULING BY DYNAMIC PROGRAMMING. EUROPEAN JOURNAL OF OPERATIONAL RESEARCH 1, 154-160. (1977).

PRAT 1973 -1 PRATT, V. R.
LEAST UPPER BOUND ON CONVENTIONAL SHELLSORT, A. IN: R. RUSTIN (ED.): COMBINATORIAL ALGORITHMS, 31-38. NEW YORK: ALGORITHMICS PRESS. (1973).

PRES 1975 -1 PRESENT, M.
PROBLEMES DE PARTITIONNEMENT COMMENT UTILISER DES METHODES DE REDUCTION? PARIS: ASSOCIATION FRANCAISE POUR LA CYBERNETIQUE ECONOMIQUE ET TECHNIQUE, 21 P. (1975).

PRESA 1975 -1 PRESIA, H.-J.
UEBER FAERBUNGEN PLANARER GRAPHEN, I. MATHEMATISCHE NACHRICHTEN 65, 223-234. (1975).

PRESA 1975 -2 PRESIA, H.-J.
UEBER FAERBUNGEN PLANARER GRAPHEN, II. MATHEMATISCHE NACHRICHTEN 67, 181-198. (1975).

PRIT 1970 -1 PRITSKER, A. A. B.; GHARF, P. M.
LOCATING NEW FACILITIES WITH RESPECT TO EXISTING FACILITIES. AIIE TRANSACTIONS 2, 290-297. (1970).

PRIV 1974 -1 PRIVALOV, V. V.; TETERIN, G. P.
PROBLEM OF CHARGING A FORGE PRESS WITH 0-1-VARIABLES (RUSSIAN). IN: MAT. METHODY EHKON, ISSLED., 86-90. MOSCOW: IZDAT. NAUKA. (1974).

PROS 1974 -1 PROSKUROWSKI, A.
SEARCH FOR A UNIQUE INCIDENCE MATRIX OF A GRAPH. BIT 14, 209-226. (1974).

PUHO 1976 -1 PUHOV, G. E.; SARASIDZE, G. K.
SIMULATION OF COMBINATORIAL PROBLEMS OF DISCRETE MATHEMATICS (RUSSIAN). DOKLADY AKADEMII NAUK SSSR 230, 1297-1299. (1976).

PULL 1973 -1 PULLEYBLANK, W. R.
MULTI OPTIMIZATION IN MATCHING PROBLEMS. IN: R. P. D. THOMAS, H. C. WILLIAMS (EDS.): PROCEEDINGS OF THE SECOND MANITOBA CONFERENCE ON NUMERICAL MATHEMATICS, UNIV. OF MANITOBA, WINNIPEG 1972, 317-326. CONGRESSUS NUMERANTIUM 7, WINNIPEG: UTILITAS MATHEMATICA. (1973).

PULL 1977 -1 PULLEYBLANK, W. R.
DUAL INTEGRALITY IN B-MATCHING PROBLEMS. CORE DISCUSSION PAPER, 7717. HEVERLEE: CENTER FOR OPERATIONS RESEARCH AND ECONOMETRICS, 41 P. (1977).

PULL 1977 -2 PULLEYBLANK, W. R.
TWO-BICRITICAL GRAPHS AND NODE COVERS. CORE DISCUSSION PAPER, 7733. HEVERLEE: CENTER FOR OPERATIONS RESEARCH AND ECONOMETRICS, 23 P. (1977).

PULLA 1974 -1 PULLEYBLANK, W. R.; EDMONDS, J.
FACETS OF 1-MATCHING POLYHEDRA. IN: C. BERGE, D. RAY-CHAUDHURI (EDS.): HYPERGRAPH SEMINAR, PROCEEDINGS OF THE FIRST WORKING SEMINAR ON HYPERGRAPHS, OHIO STATE UNIV., COLUMBUS 1972, 214-242. LECTURE NOTES IN MATHEMATICS 411, BERLIN: SPRINGER. (1974).

PUN 1975 -1 PUN, L.; BOEBION, J.; LATECOERE, L.; ALMAR, A.; PHILIPE, J.; DEROUAULT, Y.
ORDONNANCEMENT ADAPTIF DANS LES ENTREPRISES INDUSTRIELLES DE TAILLE MOYENNE. REPORT DGRST-7371301. PARIS: CENTRE NATIONALE DE LA RECHERCHE SCIENTIFIQUE, 53 P. (1975).

PURC 1975 -1 PURCHECK, G. F. K.
LINEAR-PROGRAMMING METHOD FOR THE COMBINATORIAL GROUPING OF AN INCOMPLETE POWER SET, A. JOURNAL OF CYBERNETICS 5, NO. 4, 51-76. (1975).

PURI 1975 -1 PURI, M. C.; SWARUP, K.
STRONG-CUT CUTTING PLANE PROCEDURE FOR EXTREME POINT MATHEMATICAL PROGRAMMING PROBLEMS. INDIAN JOURNAL OF PURE AND APPLIED MATHEMATICS 6, 227-236. (1975).

RABI 1959 -1 RABIN, M. O.
SPEED OF COMPUTATION AND CLASSIFICATION OF RECURSIVE SETS. IN: THIRD CONVENTION SCI. SOC., ISRAEL, 1-2. (1959).

RABI 1960 -1 RABIN, M. O.
DEGREE OF DIFFICULTY OF COMPUTING A FUNCTION AND A PARTIAL ORDERING OF RECURSIVE SETS. TECHNICAL REPORT 1. JERUSALEM: O. N. R. (1960).

RABI 1976 -1 RABIN, M. O.
PROBABILISTIC ALGORITHMS. IN: J. F. TRAUB (ED.): ALGORITHMS AND COMPLEXITY, NEW DIRECTIONS AND RECENT RESULTS, 21-40. NEW YORK: ACADEMIC PRESS. (1976).

RABI 1977 -1 RABIN, M. O.
COMPLEXITY OF COMPUTATIONS. COMMUNICATIONS OF THE ACM 20, 625-633. (1977).

RADE 1975 -1 RADENSKI, A. A.
SEARCH ALGORITHM FOR ALL ISOMORPHISMS OF TWO GRAPHS, A (RUSSIAN). SERDICA 1, 167-177. (1975).

RAGH 1976 -1 RAGHAVACHARI, M.
CONSTRUCTIVE METHOD TO RECOGNIZE THE TOTAL UNIMODULARITY OF A MATRIX, A. ZEITSCHRIFT FUER OPERATIONS RESEARCH 20, SERIE A, 59-62. (1976).

RAJA 1977 -1 RAJARAMAN, M. K.
PARALLEL SEQUENCING ALGORITHM FOR MINIMIZING TOTAL COST, A. NAVAL RESEARCH LOGISTICS QUARTERLY 24, 473-482. (1977).

RAJAA 1977 -1 RAJAT, K. D.
ON GENERATION OF TEST DATA AND MINIMAL COVER OF DIRECTED GRAPHS. IN: GILCHRIST, B.(ED.): IFIP CONGRESS SERIES 7. INFORMATION PROCESSING 77. PROCEEDINGS OF THE IFIP CONGRESS 77 TORONTO. AMSTERDAM: NORTH HOLLAND, 13-16. (1977).

RAMA 1977 -1 RAMAKRISHNAN, C. S.
NOTE ON THE TIME MINIMISING TRANSPORTATION PROBLEM, A. OPSEARCH 14, 207-209. (1977).

RAMAA 1976 -1 RAMAKRISHNAN, K. G.
SOLUTION OF CAPACITATED PLANT LOCATION PROBLEMS USING BENDERS' DECOMPOSITION. CS-76-033. PULLMAN: WASHINGTON STATE UNIV., COMPUTER SCIENCE DEPT., 32 P. (1976).

RAMAB 1975 -1 RAMALINGAM, P.
SCHEDULING JOBS WITH DUE DATES AND COST FUNCTIONS ON MULTI-STAGE PROCESSORS. PAPER PRESENTED AT THE 48TH JOINT NATIONAL MEETING OF ORSA AND TIMS, LAS VEGAS 1975, 30 P. (1975).

RAMAC 1971 -1 RAMAMOORTHY, C. V.; CHANDY, K. M.; GONZALES, M. J.
TASK SCHEDULING ALGORITHMS IN A MULTIPROCESSOR ENVIRONMENT. IN: PROCEEDINGS OF THE FIFTH ANNUAL PRINCETON CONFERENCE ON INFORMATION SCIENCES AND SYSTEMS, 105-109. (1971).

RAMAD 1976 -1 RAMAMOORTHY, C. V.; FOX, T. F.; LI, H. F.
SCHEDULING PARALLEL PROCESSABLE TASKS FOR A UNIPROCESSOR. IEEE TRANSACTIONS ON COMPUTERS C-25, 485-495. (1976).

RAMAE 1975 -1 RAMASWAMY, K. H.
OPTIMAL ROUTE FOR PROCESSING A JOB LOT IN A JOB SHOP. PAPER PRESENTED AT THE 48TH JOINT NATIONAL MEETING OF ORSA AND TIMS, LAS VEGAS 1975, 13 P. (1975).

RAMAF 1976 -1 RAMASWAMY, K. V.
SCHEDULING OF N JOBS ON M DISTINCT PARALLEL PROCESSORS WITH PROCESS DEPENDENT TIMES AND COSTS. MIAMI (50. ORSA-MEETING + TIMS, NOVEMBER 1976) 13 P. (1976).

RAND 1976 -1 RANDOLPH, P.
SHORTEST PATHS THROUGH MULTIPLE CRITERIA NETWORKS. MIAMI (50. ORSA-MEETING + TIMS, NOVEMBER 1976) 20 P. (1976).

RANDA 1977 -1 RANDOLPH, P.; RINGEISEN, R.
SHORTEST PATHS THROUGH MULTIPLE CRITERIA NETWORKS. FORT WAYNE: PURDUE UNIVERSITY, DEPT. OF MATHEMATICS, 20 P. (1977).

RANDB 1976 -1 RANDOLPH, P. H.
JOB SHOP SCHEDULING: A CASE STUDY. OMEGA 4, 463-477. (1976).

RAO 1973 -1 RAO, M. R.
ON THE DIRECT SEARCH APPROACH TO THE RECTILINEAR FACILITIES LOCATION PROBLEM. AIIE TRANSACTIONS 5, 256-264. (1973).

RAO 1975 -1 RAO, M. R.
MULTI-REGION, MULTI-PERIOD CAPACITY EXPANSION MODELS: A HEURISTIC AND A SPECIAL CASE. WORKING PAPER 7528. ROCHESTER: UNIV. OF ROCHESTER, GRADUATE SCHOOL OF MANAGEMENT. (1975).

RAO 1976 -1 RAO, M. R.
ADJACENCY OF THE TRAVELING SALESMAN TOURS AND 0-1 VERTICES. SIAM JOURNAL ON APPLIED MATHEMATICS 30, 191-198. (1976).

RAO A 1974 -1 RAO, R.; RUTENBERG, D. P.
EFFICIENT COMPUTATION OF CONTINUOUS-TIME PLANT LOCATION. PAPER PRESENTED AT THE 45TH JOINT NATIONAL MEETING OF ORSA AND TIMS, BOSTON 1974. (1974).

RAPA 1959 -1 RAPAPORT, H.; ABRAMSON, P.
ANALOG COMPUTER FOR FINDING AN OPTIMUM ROUTE THROUGH A COMMUNICATIONS NETWORK, AN. IRE TRANSACTIONS CS-7, 37-42. (1959).

RAPP 1976 -1 RAPPOPORT, A. M.
TRANSPORT PROBLEMS WITH SUPPLEMENTARY CONSTRAINTS (RUSSIAN). IN: MATHEMATICAL PROGRAMMING AND RELATED QUESTIONS. PROCEEDINGS OF THE SEVENTH WINTER SCHOOL, DROGOBYCH 1974, CONVEX PROGRAMMING (RUSSIAN), 68-81. MOSCOW: CENTRAL. EKONOM.-MAT. INST. AKAD. NAUK SSSR. (1976).

RARD 1977 -1 RARDIN, R. L.; LIN, B.
WHAT MAKES INTEGER PROGRAMMING PROBLEMS HARD TO SOLVE? INDUSTRIAL AND SYSTEMS ENGINEERING REPORT SERIES J-77-13. ATLANTA: GEORGIA INST. OF TECHNOLOGY, SCHOOL OF INDUSTRIAL AND SYSTEMS ENGINEERING, 21 P. (1977).

RARDA 1976 -1 RARDIN, R. L.; UNGER, V. E.
SOLVING FIXED CHARGE NETWORK PROBLEMS WITH GROUP THEORY-BASED PENALTIES. NAVAL RESEARCH LOGISTICS QUARTERLY 23, 67-84. (1976).

RARDA 1976 -2 RARDIN, R. L.; UNGER, V. E.
SOME COMPUTATIONALLY RELEVANT GROUP THEORETIC STRUCTURES OF FIXED CHARGE PROBLEMS. MATHEMATICAL PROGRAMMING 10, 379-400 (1976).

RARDA 1976 -3 RARDIN, R. L.; UNGER, V. E.
SURROGATE CONSTRAINTS AND THE STRENGTH OF BOUNDS DERIVED FROM 0-1 BENDERS' PARTITIONING PROCEDURES. OPERATIONS RESEARCH 24, 1169-1175. (1976).

RATH 1975 -1 RATHMANN, D.
ON THE ORDER OF THE GROUP OCCURING IN THE ASYMPTOTIC INTEGER LINEAR PROGRAMMING APPROACH. DISCUSSION PAPER. GOETTINGEN: UNIV. GOETTINGEN, LEHRSTUHL FUER MATHEMATISCHE VERFAHRENSFORSCHUNG UND DATENVERARBEITUNG, 7 P. (1975).

RATH 1975 -2 RATHMANN, D.
SUCHVERFAHREN ZUM ASYMPTOTISCHEN ANSATZ FUER GANZZAHLIGE PROGRAMME. DISCUSSION PAPER 7501. GOETTINGEN: UNIV. GOETTINGEN, LEHRSTUHL FUER MATHEMATISCHE VERFAHRENSFORSCHUNG UND DATENVERARBEITUNG, 34 P. (1975).

RATH 1977 -1 RATHMANN, D.
UMFORMULIERUNG BEIM ASYMPTOTISCHEN ANSATZ FUER GANZZAHLIGE PROGRAMME. IN: R. HENN, P. KALL, B. KORTE, O. KRAFFT ET AL. (EDS.): OPERATIONS RESEARCH VERFAHREN XXV, METHODS OF OPERATIONS RESEARCH, 221-232. MEISENHEIM: HAIN. (1977).

RATH 1977 -2 RATHMANN, D.
ZUR LOESUNG DISKRETER OPTIMIERUNGSPROBLEME MIT GRUPPENTHEORETISCHEN ANSAETZEN. DISSERTATION. GOETTINGEN: UNIV. GOETTINGEN, FAKULTAET FUER WIRTSCHAFTS- UND SOZIALWISSENSCHAFTEN, 242 P. (1977).

RAVI 1977 -1 RAVINDRAN, A.; RAMASWAMI, V.
ON THE BOTTLENECK ASSIGNMENT PROBLEM. JOURNAL OF OPTIMIZATION THEORY AND APPLICATIONS 21, 451-458. (1977).

RAWA 1975 -1 RAWASWAMY, K. V.; DUDEK, R. A.
HEURISTIC ALGORITHM FOR SINGLE FACILITY SEQUENCING WITH PENALTY COST, A. PAPER PRESENTED AT THE 48TH JOINT NATIONAL MEETING OF ORSA AND TIMS, LAS VEGAS 1975, 11 P. (1975).

RAYB 1966 -1 RAYBOV, G. G.
ALGORITHM FOR THE SOLUTION OF A LABYRINTH ON A DISCRETE FIELD AND ITS APPLICATIONS (TRANSLATED FROM RUSSIAN), AN. SOVIET PHYSICS, DOKLADY 11, 110-111. (1966).

RAZH 1965 -1 RAZHUMIKHIN, B. S.
PROBLEM OF OPTIMAL RESOURCE ALLOCATION. AUTOMATION AND REMOTE CONTROL, 1216-1232. (1965).

RAZH 1967 -1 RAZHUMIKHIN, B. S.
PROBLEMS OF THE OPTIMAL ALLOCATION OF RESOURCES (RUSSIAN). AVTOMATIKA I TELEMEHANIKA, 62-74. (1967).

RAZH 1969 -1 RAZHUMIKHIN, B. S.
APPLICATION OF THE METHOD OF DYNAMIC PROGRAMMING TO CERTAIN PROBLEMS OF NETWORK OPTIMIZATION. I + II (RUSSIAN), AN. IN: THEORY OF OPTIMAL SOLUTIONS, PROC. SEMINAR, KIEV 1968, NO. 5 (RUSSIAN), 3-27. KIEV: AKAD. NAUK UKRAIN. SSR. (1969).

RAZHA 1967 -1 RAZHUMIKHIN, B. S.; STAVROKOVA, N. E.
PROBLEM OF DISTRIBUTING RESOURCES TO MINIMIZE TIME OF EXECUTION OF JOBS COMPLEX. AUTOMATION AND REMOTE CONTROL 9, 1345-1365. (1967).

READ 1978 -1 READ, R. C.
EVERY ONE A WINNER -OR- HOW TO AVOID ISOMORPHISM SEARCH WHEN CATALOGUING COMBINATORIAL CONFIGURATIONS. IN: ALSPACH, B.;HELL, P.;MILLER, D. J.(EDS.): ALGORITHMIC ASPECTS OF COMBINATORICS. ANNALS OF DISCRETE MATHEMATICS 2, 107-120. (1978).

READA 1977 -1 READ, R. C.; CORNEIL, D. G.
GRAPH ISOMORPHISM DISEASE, THE. JOURNAL OF GRAPH THEORY 1, 339-363. (1977).

READB 1975 -1 READ, R. C.; TARJAN, R. E.
BOUNDS ON BACKTRACK ALGORITHMS FOR LISTING CYCLES, PATHS, AND SPANNING TREES. NETWORKS 5, 237-252. (1975).

RECS 1976 -1 RECSKI, A.
MATROIDS AND INDEPENDENT STATE VARIABLES. BUDAPEST: RESEARCH INSTITUTE FOR TELECOMMUNICATION, 8 P. (1976).

RECS 1976 -2 RECSKI, A.
MATROIDS AND N-PORTS. BUDAPEST (9. INTERNAT. SYMPOSIUM ON MATH. PROGRAMMING) 9 P. (1976).

RECS 1976 -3 RECSKI, A.
ON RANDOM PARTITIONS. DISCRETE MATHEMATICS 16, 173-178. (1976).

REDD 1976 -1 REDDI, S. S.
INTERDEPENDENCE OF INDUSTRIAL ENGINEERING AND COMPUTING: SELECTION OF SCHEDULING CRITERIA IN COMPUTER SYSTEMS, THE. AIIE TRANSACTIONS 8, 314-319. (1976).

REDD 1976 -2 REDDI, S. S.
SEQUENCING WITH FINITE INTERMEDIATE STORAGE. MANAGEMENT SCIENCE 23, 216-217. (1976).

REEV 1976 -1 REEVES, G. R.
GOAL PROGRAMMING FORMULATIONS OF INTEGER PROGRAMMING PROBLEMS. MIAMI (50. ORSA-MEETING + TIMS, NOVEMBER 1976) 7 P. (1976).

REIN 1977 -1 REINGOLD, E. M.; NIEVERGELT, J.; DEO, N.
COMBINATORIAL ALGORITHMS: THEORY AND PRACTICE. ENGLEWOOD CLIFFS: PRENTICE-HALL, 432 P. (1977).

REKL 1977 -1 REKLAJTIS, V. K.; TAMULINAS, B. V.
AUTOMATIC CONSTRUCTION OF SEQUENCING RULES IN THE PROBLEM OF JOB SCHEDULING FOLLOWING A CALENDAR (RUSSIAN). AVTOMATIKA I TELEMEHANIKA, NO. 4, 182-186. (1977).

REYN 1977 -1 REYNER, W. S.
ANALYSIS OF A GOOD ALGORITHM FOR THE SUBTREE PROBLEM, AN. SIAM JOURNAL ON COMPUTING 6, 730-732. (1977).

RIBB 1972 -1 RIBBECK, K. F.
ROUTENSUCHEN IN STADTSTRASSENNETZEN. SCHRIFTENREIHE STRASSENBAU UND STRASSENVERKEHRSTECHNIK 134, BONN: BUNDESMINISTERIUM FUER VERKEHR. (1972).

RIBE 1971 -1 RIBEILL, G.; TABOURIER, Y.
CHEMINS EXTREMAUX DANS UN GRAPHE ENTRE TOUS COUPLES DE SOMMETS: QUELQUES ALGORITHMES ET LEUR MISE EN OEUVRE.COMPARAISONS SUR LES PERFORMANCES. DIRECTION SCIENTIFIQUE METRA, NOTE DE TRAVAIL NO. 148. (1971).

RICH 1976 -1 RICHARDSON, R.
OPTIMIZATION APPROACH TO ROUTING AIRCRAFT, AN. TRANSPORTATION SCIENCE 10, 52-71. (1976).

RICHA 1976 -1 RICHMOND, L. B.
ON HAMILTONIAN POLYGONS. JOURNAL OF COMBINATORIAL THEORY 21, SERIES B, 81-87. (1976).

RICHB 1976 -1 RICHTER, D.
ALGORITHMUS ZUR LOESUNG GEMISCHT GANZZAHLIGER KONVEXER OPTIMIERUNGSPROBLEME, EIN. MATHEMATISCHE OPERATIONSFORSCHUNG UND STATISTIK 7, 197-208. (1976).

RICHB 1976 -2 RICHTER, D.
CONTRIBUTION TO AN ALGORITHM FOR SOLVING MIXED INTEGER COVEX PROGRAMMING PROBLEMS, A. IN: A. PREKOPA (ED.): PROGRESS IN OPERATIONS RESEARCH, 771-777, AMSTERDAM, LONDON: NORTH HOLLAND. (1976).

RICHB 1976 -3 RICHTER, D.
ZUR BEHANDLUNG GANZZAHLIGER OPTIMIERUNGSAUFGABEN OHNE GANZZAHLIGKEITSBEDINGUNG AN DIE ZIELFUNKTION. MATHEMATISCHE OPERATIONSFORSCHUNG UND STATISTIK 7, 343-350. (1976).

RICHC 1976 -1 RICHTER, K.
RESSOURCENAUFSPALTUNG FUER DISKRETE OPTIMIERUNGSAUFGABEN, TEIL 1. MATHEMATISCHE OPERATIONSFORSCHUNG UND STATISTIK 7, 831-842. (1976).

RICHC 1977 -1 RICHTER, K.
RESSOURCENAUFSPALTUNG FUER DISKRETE OPTIMIERUNGSAUFGABEN, TEIL II. MATHEMATISCHE OPERATIONSFORSCHUNG UND STATISTIK 8, SERIES OPTIMIZATION, 3-16. (1977).

RIDE 1975 -1 RIDER, K. L.
PARAMETRIC MODEL FOR THE ALLOCATION OF FIRE COMPANIES: EXECUTIVE SUMMARY, A. RAND-R-1646-1-HUD, SANTA MONICA: RAND CORP, 26 P. (1975).

RIDE 1976 -1 RIDER, K. L.
PARAMETRIC MODEL FOR THE ALLOCATION OF FIRE COMPANIES IN NEW YORK CITY, A. MANAGEMENT SCIENCE 23, 146-158. (1976).

RIDL 1965 -1 RIDLEY, T. M.
INVESTMENT IN A NETWORK TO REDUCE LENGTH OF A SHORTEST ROUTE. IN PROCEEDINGS OF THE THIRD INTERNATIONAL SYMPOSIUM ON THE THEORY OF TRAFFIC FLOW. ELSEVIER, NEW YORK, 235-263. (1965).

RIHA 1976 -1 RIHA, W.; JAMES, K. R.
ALGORITHM 29. EFFICIENT ALGORITHMS FOR DOUBLY AND MULTIPLY RESTRICTED PARTITIONS. COMPUTING 14, 163-168. (1976).

RILE 1970 -1 RILEY, D.
LONGEST PATH IN A NETWORK, THE. JOURNAL OF MATHEMATICAL ANALYSIS AND APPLICATIONS 30, 636-637. (1970).

RINA 1975 -1 RINALDI, S.
OPTIMAL CONSTRAINED CYCLES IN GRAPHS. IN: P. RINALDI (ED.): TOPICS IN COMBINATORIAL OPTIMIZATION, 47-67, WIEN, NEW YORK: SPRINGER. (1975).

RINA 1975 -2 RINALDI, S.
OPTIMAL UNCONSTRAINED CYCLES IN GRAPHS. IN: P. RINALDI (ED.): TOPICS IN COMBINATORIAL OPTIMIZATION, 7-45, WIEN, NEW YORK: SPRINGER. (1975).

RINE 1977 -1 RINE, D. C.
OPTIMIZING TWO TYPES OF DISCRETE FUNCTIONS, SUBJECT TO LINEAR RESTRICTIONS. JOURNAL OF COMBINATORIAL THEORY 22, SERIES A, 1-16. (1977).

RINN 1976 -1 RINNOOY KAN, A. H. G.
ON MITTEN'S AXIOMS FOR BRANCH-AND-BOUND. OPERATIONS RESEARCH 24, 1176-1178. (1976).

RINNA 1975 -1 RINNOOY KAN, A. H. G.; LAGEWEG, B. J.; LENSTRA, J. K.
MINIMIZING TOTAL COSTS IN ONE-MACHINE SCHEDULING. IN: B. ROY (ED.): COMBINATORIAL PROGRAMMING: METHODS AND APPLICATIONS. PROCEEDINGS OF THE NATO ADVANCED STUDY INST., VERSAILLES 1974, 343-350. DORDRECHT: REIDEL. (1975).

RIVE 1976 -1 RIVEST, R. L.; VUILLEMIN, J.
ON RECOGNIZING GRAPH PROPERTIES FROM ADJACENCY MATRICES. THEORETICAL COMPUTER SCIENCES 3, 371-384. (1976).

RIVI 1975 -1 RIVIERE, M.
COVERING OF VERTICES OF A UNIFORM HYPERGRAPH BY EDGES AND STABLE SETS. IN: A. HAYNAL, R. RADO, V. T. SOS (EDS.): INFINITE AND FINITE SETS, COLLOQ. KESZTHELY 1973, VOL. 3, 1181-1197. COLLOQ. MATH. SOC. JANOS BOLYAI 10, AMSTERDAM: NORTH-HOLLAND. (1975).

ROBB 1975 -1 ROBBINS, J. A.; SHAMBLIN, J. E.; TURNER, W. C.; BYRD, D. F.
DEVELOPMENT OF AND COMPUTATIONAL EXPERIENCE WITH A COMBINATION TOUR CONSTRUCTION - TOUR IMPROVEMENT ALGORITHM FOR VEHICLE ROUTING PROBLEMS. PAPER PRESENTED AT THE 48TH JOINT NATIONAL MEETING OF ORSA AND TIMS, LAS VEGAS 1975, 22 P. (1975).

ROBE 1976 -1 ROBERS, P.; COOPER, L.
STUDY OF THE FIXED CHARGE TRANSPORTATION PROBLEM, A. COMPUTERS AND MATHEMATICS WITH APPLICATIONS 2, 125-135. (1976).

ROBEA 1976 -1 ROBERTS, J.
SUFFICIENT CONDITION FOR K-PATH HAMILTONIAN DIGRAPHS, A. BULLETIN OF THE AMERICAN MATHEMATICAL SOCIETY 82, 63-65. (1976).

ROBEB 1972 -1 ROBERTS, K. J.
DISCUSSION OF 'SELECTING SOLID WASTE DISPOSAL FACILITIES' BY HELMS AND CLARK. JOURNAL OF THE SANITARY ENGINEERING DIVISION, PROCEEDINGS OF THE AMERICAN SOCIETY OF CIVIL ENGINEERS 98, NO. SA2, 456-459. (1972).

ROBEC 1964 -1 ROBERTS, R.; FUNK, M.
TOWARD OPTIMUM METHODS OF LINK ADDITION IN TRANSPORTATION NETWORKS. DEPARTMENT OF CIVIL ENGINEERING., M. I. T. (1964).

ROBI 1972 -1 ROBINSON, D. F.
OPTIMIZATION OF FINITE SETS. IN: J. WALLIS, W. D. WALLIS (EDS.): PROCEEDINGS OF THE FIRST AUSTRALIAN CONFERENCE ON COMBINATORIAL MATHEMATICS, UNIV. OF NEWCASTLE 1972, 39-52. NEWCASTLE: UNIV. OF NEWCASTLE RESEARCH ASSOCIATES. (1972).

ROBIA 1977 -1 ROBINSON, D. R.
COMPUTATIONAL REFINEMENT FOR DISCRETE-VALUED DYNAMIC PROGRAMS WITH CONVEX FUNCTIONS, A. MANAGEMENT SCIENCE 23, 1251-1255. (1977).

ROCH 1976 -1 ROCHAT, J. C.
ON AN OPTIMIZATION PROBLEM IN QUEUEING. LAUSANNE: ECOLE POLYTECHNIQUE FEDERALE DE LAUSANNE, DEPT. DE MATHEMATIQUES, 12 P. (1976).

RODG 1977 -1 RODGERS, J. R.; MAC CARTHY, O.
SERVICE FACILITY LOCATION PROBLEM: THE LOCATION OF FARM ADVISORY OFFICERS IN N. Z., A. NEW ZEALAND OPERATIONAL RESEARCH 5, 10-34. (1977).

ROED 1976 -1 ROEDDER, W.
LEXIKOGRAPHISCHER SUCHALGORITHMUS ZUR GANZZAHLIGEN PROGRAMMIERUNG: LEXS, EIN. ZEITSCHRIFT FUER OPERATIONS RESEARCH 20, 209-218. (1976).

ROEDA 1977 -1 ROEDL, V.
ON THE CHROMATIC NUMBER OF SUBGRAPHS OF A GIVEN GRAPH. PROCEEDINGS OF THE AMERICAN MATHEMATICAL SOCIETY 64, 370-371. (1977).

ROES 1976 -1 ROESSLER, M.
CLUSTER-ANALYSE. IN: LIEBLING, T. M.;ROESSLER, M.(EDS.): KOMBINATORISCHE ENTSCHEIDUNGSPROBLEME: METHODEN UND ANWENDUNG. LECTURE NOTES IN ECONOMICS AND MATHEMATICAL SYSTEMS 153, BERLIN: SPRINGER, 196-206. (1976).

ROGE 1971 -1 ROGER, J.
UPDATING A MINIMUM SPANNING TREE. JOURNAL OF THE ROYAL STATISTICAL SOCIETY. SERIES C. APPLIED STATISTICS 20, 204-206. (1971).

ROOD 1977 -1 ROODMAN, G. M.; SCHWARZ, L. B.
EXTENSIONS OF THE MULTI-PERIOD FACILITY PHASE-OUT MODEL: NEW PROCEDURES AND APPLICATION TO A PHASE-IN/PHASE-OUT PROBLEM. AIIE TRANSACTIONS 9, 103-107. (1977).

ROOS 1975 -1 ROOSE, J. R.; RIEMERSMA, P.
HEURISTIC OPTIMISATION OF A LOCATION-ALLOCATION PROBLEM IN A TRANSPORT NETWORK WITH MODAL CHOICE. PAPER PRESENTED AT THE FIRST EUROPEAN CONGRESS ON OPERATIONS RESEARCH, EURO 1, BRUSSELS 1975. (1975).

ROSE 1976 -1 ROSE, D. J.; TARJAN, R. E.; LUEKER, G. S.
ALGORITHMIC ASPECTS OF VERTEX ELIMINATION ON GRAPHS. SIAM JOURNAL ON COMPUTING 5, 266-283. (1976).

ROSEA 1971 -1 ROSENFELD, M.
ON CERTAIN POLYTOPES ASSOCIATED WITH GRAPHS. IN: PROCEEDINGS OF SOUTHEASTERN CONFERENCE ON COMBINATORICS, GRAPH THEORY AND COMPUTING. (1971).

ROSEB 1974 -1 ROSENKRANTZ, D. J.; STEARNS, R. E.; LEWIS, P. M.
APPROXIMATE ALGORITHMS FOR THE TRAVELING SALESPERSON PROBLEM. IN: 15TH ANNUAL SYMPOSIUM ON SWITCHING AND AUTOMATA THEORY, NEW ORLEANS 1974, 33-42. LONG BEACH: INSTITUTE OF ELECTRICAL AND ELECTRONICS ENGINEERS. (1974).

ROSEB 1977 -1 ROSENKRANTZ, D. J.; STEARNS, R. E.; LEWIS, P. M.
ANALYSIS OF SEVERAL HEURISTICS FOR THE TRAVELING SALESMAN PROBLEM, AN. SIAM JOURNAL ON COMPUTING 6, 563-581. (1977).

ROSEC 1975 -1 ROSENTHAL, A.
BIT-PUSHING SHORTEST DISTANCE ALGORITHM, A. NETWORKS 5, 301-306. (1975).

ROSEC 1977 -1 ROSENTHAL, A.
NONSERIAL DYNAMIC PROGRAMMING IS OPTIMAL. IN: PROCEEDINGS OF THE NINTH ANNUAL A.C.M. SYMPOSIUM ON THEORY OF COMPUTING, 98-105. (1977).

ROSED 1977 -1 ROSENTHAL, A.; GOLDNER, A.
SMALLEST AUGMENTATIONS TO BICONNECT A GRAPH. SIAM JOURNAL ON COMPUTING 6, 55-66. (1977).

ROSH 1977 -1 ROSHCHIN, V. A.; SERGIENKO, I. V.
ON A CONVERGENT ALGORITHM FOR SOLVING THE MATHEMATICAL PROGRAMMING PROBLEMS WITH MIXED VARIABLES (RUSSIAN). KIBERNETIKA, NO. [illegible] (1977).

ROSHA 1977 -1 ROSHIN, G. B.
AN ALGORITHM FOR FINDING A MINIMAL SPANNING TREE IN A GROWING GRAPH (RUSSIAN). ZURNAL VYCISLITEL'NOI MATEMATIKI I MATEMATICESKOI FIZIKI 17, 1607-1610. (1977).

ROSS 1977 -1 ROSS, G. T.; SOLAND, R. M.
MODELING FACILITY LOCATION PROBLEMS AS GENERALIZED ASSIGNMENT PROBLEMS. MANAGEMENT SCIENCE 24, 345-357. (1977).

ROSSA 1976 -1 ROSS, G. T.; SOLAND, R. M.; ZOLTNERS, A. A.
NOTE ON THE BOUNDED INTERVAL GENERALIZED ASSIGNMENT PROBLEM, A. RESEARCH REPORT CCS 253. AUSTIN: UNIV. OF TEXAS, CENTER FOR CYBERNETIC STUDIES, 12 P. (1976).

ROSSB 1974 -1 ROSS, G. T. JR.
ALGORITHMIC AND COMPUTATIONAL RESULTS FOR CONSTRAINED TRANSPORTATION, NETWORK AND ASSIGNMENT MODELS. DISSERTATION. AUSTIN: UNIV. OF TEXAS, 114 P. (1974).

ROSSC 1975 -1 ROSSMAN, L. A.
RESOLUTION OF DUALITY GAPS WITH DYNAMIC PROGRAMMING. PAPER PRESENTED AT THE 48TH JOINT NATIONAL MEETING OF ORSA AND TIMS, LAS VEGAS 1975, 28 P. (1975).

ROTH 1968 -1 ROTHFARB, B.
COMBINATORIC METHODS FOR MULTICOMMODITY FLOWS. DISSERTATION. BERKELEY: UNIV. OF CALIFORNIA. (1968).

ROTHA 1969 -1 ROTHFARB, R.; FRISCH, I. T.
ON THE THREE-COMMODITY FLOW PROBLEM. SIAM JOURNAL ON APPLIED MATHEMATICS 17, 46-58. (1969).

ROTHB 1966 -1 ROTHKOPF, M. H.
SCHEDULING WITH RANDOM SERVICE TIMES. MANAGEMENT SCIENCE 12, 707-713. (1966).

ROTHB 1975 -1 ROTHKOPF, M. H.
NOTE ON ALLOCATING JOBS TO TWO MACHINES, A. NAVAL RESEARCH LOGISTICS QUARTERLY 22, 829-830. (1975).

ROUX 1975 -1 ROUX, M.
NOTES SUR L'ARBRE DE LONGUEUR MINIMA. REVUE DE STATISTIQUE APPLIQUEE 23, NO. 2, 29-35. (1975).

ROWE 1960 -1 ROWE, A. J.
TOWARD A THEORY OF SCHEDULING. JOURNAL OF INDUSTRIAL ENGINEERING 11, 125-136. (1960).

ROWI 1976 -1 ROWICKI, A.
ON OPTIMAL SCHEDULING FOR TWO-PROCESSOR SYSTEMS. BULLETIN DE L'ACADEMIE POLONAISE DES SCIENCES. SERIE DES SCIENCES MATHEMATIQUES, ASTRONOMIQUES ET PHYSIQUES 24, 287-293. (1976).

ROY 1966 -1 ROY, B.
CHEMINS DE LONGUEUR EXTREMALES. GESTION,NO.SPECIAL RECHERCHE OPERATIONELLE,322-335. (1966).

ROY 1966 -2 ROY, B.
CONTRIBUTION TO THE THEORY OF GRAPHS IN THE STUDY OF SCHEDULING PROBLEMS (FRENCH). 2 ND PROCEEDINGS OF THE INTERNATIONAL CONFERENCE ON OPERATIONS RESEARCH. (1966).

ROY 1967 -1 ROY, B.
NOMBRE CHROMATIQUE ET PLUS LONGS CHEMINS D'UN GRAPHE. REVUE FRANCAISE D'AUTOMATIQUE INFORMATIQUE ET RECHERCHE OPERATIONELLE,NO.5, 129-132. (1967).

ROY 1975 -1 ROY, B.
CHEMINS ET CIRCUITS: ENUMERATION ET OPTIMISATION. PARIS: UNIV. PARIS - DAUPHINE, 38 P. (1975).

ROY 1975 -2 ROY, B.
CHEMINS ET CIRCUITS: ENUMERATION ET OPTIMISATION: A SURVEY. IN: B. ROY (ED.): COMBINATORIAL PROGRAMMING: METHODS AND APPLICATIONS. PROCEEDINGS OF THE NATO ADVANCED STUDY INST., VERSAILLES 1974, 105-136. DORDRECHT: REIDEL. (1975).

ROY A 1964 -1 ROY, B.; VIGGIONI, S.
SCHEDULING PROBLEMS (ITALIAN). METRA 3, 293-304. (1964).

RUBI 1972 -1 RUBIN, D. P.
NEIGHBORING VERTICES ON CONVEX POLYHEDRAL SETS. CHAPEL HILL: UNIV. OF NORTH CAROLINA, GRADUATE SCHOOL OF BUSINESS ADMINISTRATION, 16 P. (1972).

RUBIA 1977 -1 RUBIN, D. S.
VERTEX GENERATION METHODS FOR PROBLEMS WITH LOGICAL CONSTRAINTS. IN: P. L. HAMMER, E. L. JOHNSON, B. H. KORTE, G. L. NEMHAUSER (EDS.): STUDIES IN INTEGER PROGRAMMING, 457-466. ANNALS OF DISCRETE MATHEMATICS 1, AMSTERDAM: NORTH-HOLLAND. (1977).

RUBIB 1975 -1 RUBIN, F.
IMPROVED ALGORITHM FOR TESTING THE PLANARITY OF A GRAPH, AN. IEEE TRANSACTIONS ON COMPUTERS C-24, 113-121. (1975).

RUBIC 1975 -1 RUBIN, I.
MULTITERMINAL PATH FLOWS FOR PACKET-SWITCHING COMMUNICATION NETWORKS. NETWORKS 5, 205-236. (1975).

RUBID 1971 -1 RUBINSHTEIN, M. I.
ON THE SYMMETRIC TRAVELING SALESMAN PROBLEM (RUSSIAN). AVTOMATIKA I TELEMEHANIKA, NO. 9, 126-133; TRANSLATED AS AUTOMATION AND REMOTE CONTROL, 1972, NO. 9, PART 2, 1453-1460. (1971).

RUBIE 1976 -1 RUBINSTEIN, M. C.; LUDOT, J. P.
APPLICATION DE LA THEORIE DES GRAPHES AU CHOIX DES INVESTISSEMENTS A COURT TERME DANS UN RESEAU ELECTRIQUE SUJET A AVARIES. REVUE FRANCAISE D'AUTOMATIQUE, D'INFORMATIQUE ET DE RECHERCHE OPERATIONNELLE 10, NO. 10, 27-42. (1976).

****

RUBIF 1974 -1 RUBINSTEIN, M.-C.
APPLICATION DE LA THEORIE DES GRAPHES AU CHOIX DES INVESTISSEMENTS DANS UN RESEAU D'ENERGIE ELECTRIQUE SUJET A AVARIES. THESE DU 3E CYCLE. PARIS: UNIV. PARIS 6. (1974).

RUBL 1973 -1 RUBLINECKII, V. I.
ESTIMATES OF THE ACCURACY OF PROCEDURES IN THE TRAVELING SALESMAN PROBLEM (RUSSIAN). IN: NUMERICAL MATHEMATICAS AND COMPUTER TECHNOLOGY, NO. 4 (RUSSIAN), 18-23. KHARKOV: AKAD. NAUK UKRAIN. SSR FIZ.-TEHN. INST. NIZKIH TEMPERATUR. (1973).

RUBLA 1973 -1 RUBLINECKII, V. I.; TOMCUK, E. JA.
APPROXIMATE PROCEDURE FOR THE SOLUTION OF THE TRAVELING SALESMAN PROBLEM (RUSSIAN), AN. IN: NUMERICAL MATHEMATICS AND COMPUTER TECHNOLOGY, NO. 4 (RUSSIAN), 14-17 + 136. KHARKOV: AKAD. NAUK UKRAIN. SSR FIZ.-TEHN. INST. NIZKIH TEMPERATUR. (1973).

RUIT 1968 -1 RUITER, E. R.
TRAFFIC ASSIGNMENT RESEARCH, FINAL REPORT. REPORT R 68-59. CAMBRIDGE: MASSACHUSETTS INST. OF TECHNOLOGY, DEPT. OF CIVIL ENGINEERING, TRANSPORTATION SYSTEM DIVISION. (1968).

RUSC 1977 -1 RUSCHITZKA, M.; FABRY, R. S.
UNIFYING APPROACH TO SCHEDULING, A. COMMUNICATIONS OF THE ACM 20, 469-476. (1977).

RUSCA 1976 -1 RUSCITTI, A.
CANONICAL REDUCTION AND CONSTRUCTION METHOD RELATIVE TO MAXIMAL PLANAR GRAPHS (ITALIAN), A. CALCOLO 13, 157-171. (1976).

RUSK 1977 -1 RUSKEY, F.; HU, T. C.
GENERATING BINARY TREES LEXICOGRAPHICALLY. SIAM JOURNAL ON COMPUTING 6, 745-758. (1977).

RUSN 1976 -1 RUSNAK, Z.
ATTEMPT TO SOLVE THE CLASSICAL LOCATION PROBLEM IN THE PLANE (POLISH), AN. PRZEGLAD STATYSTYCZNY 23, 531-538. (1976).

RUSS 1975 -1 RUSSEL, R.
EFFECTIVE HEURISTIC FOR THE M-TOUR TRAVELING SALESMAN PROBLEM WITH SOME SIDE CONDITIONS, AN. PAPER PRESENTED AT THE 1975 ORSA/TIMS MEETING, LAS VEGAS 1975. (1975).

RUSSA 1976 -1 RUSSELL, A. H.
GEOMETRICAL SOLUTION TO A TREE NETWORK PROBLEM, A. OPERATIONAL RESEARCH QUARTERLY 27, 351-366. (1976).

RUSSB 1977 -1 RUSSELL, R.; IGO, W.
ASSIGNMENT ROUTING PROBLEM, AN. TULSA: UNIV. OF TULSA, 19 P. (1977).

RUSSC 1977 -1 RUSSELL, R. A.
EFFECTIVE HEURISTIC FOR THE M-TOUR TRAVELING SALESMAN PROBLEM WITH SOME SIDE CONDITIONS, AN. OPERATIONS RESEARCH 25, 517-524. (1977).

RUSSD 1965 -1 RUSSO, F. J.
HEURISTIC APPROACH TO ALTERNATE ROUTING IN A JOB SHOP. DISSERTATION. CAMBRIDGE, MASSACHUSETTS: MIT. (1965).

RUYG 1976 -1 RUYGROK, J. T.
FESTLEGUNG DER REIHENFOLGE DES LOESCHENS VON SEESCHIFFEN MIT HILFE DER GANZZAHLIGEN PROGRAMMIERUNG. IN: H. N. DATHE, P. MERTENS, F. D. PESCHANEL ET AL. (EDS.): PROCEEDINGS IN OPERATIONS RESEARCH 6, 53-60. WUERZBURG: PHYSICA. (1976).

RYAN 1976 -1 RYAN, D. M.; FOSTER, B. A.; CHESHIRE, I. M.
COMPLEXITY OF REAL-WORLD SCHEDULING PROBLEMS, THE. IN: R. P. ANDERSSEN, R. P. BRENT (EDS.): THE COMPLEXITY OF COMPUTATIONAL PROBLEM SOLVING, 248-262. ST. LUCIA: UNIV. OF QUEENSLAND PRESS. (1976).

RYND 1977 -1 RYNDIN, B. N.
MODIFIED FIRST ALGORITHM OF GOMORY (RUSSIAN), A. VESTNIK LENINGRADSKOGO UNIVERSITETA, NO 1, MATEMATIKA, MEHANIKA, ASTRONOMIJA, VYP. 1, 160-163 + 168. (1977).

RYND 1977 -2 RYNDIN, B. N.
MULTIPLICATIVE VERSION OF GOMORY'S FIRST ALGORITHM AND COMBINATIONS OF IT WITH THE METHOD OF BRANCHING (RUSSIAN), A. VESTNIK LENINGRADSKOGO UNIVERSITETA, NO. 7. MATEMATIKA, MEHANIKA, ASTRONOMIJA, VYP. 2, 85-89. (1977).

RYZH 1974 -1 RYZHKOV, A. P.
ORGANIZATION OF COMPUTER COMPUTATION OF INFORMATIONALLY CONNECTED PROGRAMS (RUSSIAN). IZVESTIJA AKADEMII NAUK SSSR, TEHNICESKAJA KIBERNETIKA, NO. 1, 179-184 + 238; TRANSLATED AS ENGINEERING CYBERNETICS 12, 1974, 145-150. (1974).

RYZH 1975 -1 RYZHKOV, A. P.
ALGORITHM FOR PARTITIONING A GRAPH INTO MINIMALLY CONNECTED SUBGRAPHS (RUSSIAN), AN. IZVESTIJA AKADEMII NAUK SSSR, TEHNICESKAJA KIBERNETIKA, NO. 6, 122-128 + 222; TRANSLATED AS ENGINEERING CYBERNETICS 13, 1975, NO. 6, 96-102. (1975).

RYZH 1975 -2 RYZHKOV, A. P.
GRAPH DIVISION INTO THE MINIMAL NUMBER OF COMPLETE SUBGRAPHS (RUSSIAN). KIBERNETIKA, NO. 6, 90-95. (1975).

SAAR 1977 -1 SAARI, D. G.
APPORTIONMENT METHODS AND THE HOUSE OF REPRESENTATIVES. DISCUSSION PAPER NO. 266. EVANSTON: NORTHWESTERN UNIV., CENTER FOR MATHEMATICAL STUDIES IN ECONOMICS AND MANAGEMENT SCIENCE, 18 P. (1977).

SAAT 1977 -1 SAATCIOGLU, O.
MATHEMATICAL PROGRAMMING MODELS FOR AIRPORT SITE SELECTION. ATLANTA (52. ORSA-MEETING + TIMS, NOVEMBER 1977) 40 P. (1977).

SAATA 1975 -1 SAATY, T. L.; ALEXANDER, J. M.
OPTIMIZATION AND THE GEOMETRY OF NUMBERS: PACKING AND COVERING. SIAM REVIEW 17, 475-519. (1975).

SABI 1977 -1 SABIDUSSI, G.
TRACING GRAPHS WITHOUT BACKTRACKING. IN: R. HENN, P. KALL, B. KORTE, O. KRAFFT ET AL. (EDS.): OPERATIONS RESEARCH VERFAHREN XXV, METHODS OF OPERATIONS RESEARCH, 314-332. MEISENHEIM: HAIN. (1977).

SADY 1976 -1 SADYKOV, I. H.; KOKOTOV, V. M.
CERTAIN PROBLEM FOR COMPILING AN OPTIMAL SCHEDULE OF FREIGHT SHIPMENTS (RUSSIAN), A. EKONOMIKA I MATEMATICESKIE METODY 12, 170-173. (1976).

SAFA 1970 -1 SAFAEVA, K.
CERTAIN DESIGN PROBLEM (RUSSIAN), A. TASKENTSKII INSTITUT NARODNOGO HOZJAISTVA, NAUCNYE ZAPISKI, MATEMATIKA V PRILOZENIJAH, VYP. 34, 19-32. (1970).

SAGU 1976 -1 SAGUEZ, C.
INTEGER PROGRAMMING APPLIED TO OPTIMAL CONTROL. IN: A. PREKOPA (ED.): PROGRESS IN OPERATIONS RESEARCH, 779-800, AMSTERDAM, LONDON: NORTH HOLLAND. (1976).

SAHB 1975 -1 SAHBAZJAN, K. V.; TUSKINA, T. A.
SURVEY OF METHODS FOR COMPILING SCHEDULES FOR MULTIPROCESSOR SYSTEMS. COMPUTATIONAL METHODS OF LINEAR ALGEBRA. PARALLEL COMPUTATIONS (RUSSIAN). ZAPISKI NAUCNYH SEMINAROV LENINGRADSKOGO OTDELENIJA MATEMATICESKOGO INSTITUTA IM. V. A. STEKLOVA AKADEMII NAUK SSSR 54, 229-258 + 265. (1975).

SAHBA 1974 -1 SAHBAZJAN, K. V.; TUSKINA, T. A.; LEBEDINSKAJA, N. B.
ALGORITHM FOR COMPILING AN OPTIMAL SCHEDULE FOR A PARALLEL-SEQUENTIAL NETWORK (RUSSIAN), AN. ZAPISKI NAUCNYH SEMINAROV LENINGRADSKOGO OTDELENIJA MATEMATICESKOGO INSTITUTA IM. V. A. STEKLOVA AKADEMII NAUK SSSR 48, 200-204 + 213-214. (1974).

SAHN 1974 -1 SAHNI, S.
SOME RELATED PROBLEMS FROM NETWORK FLOWS, GAME THEORY, SOLUTIONS. IN PROCEEDINGS OF THE 15TH ANNUAL IEEE SYMPOSIUM ON SWITCHING AND AUTOMATA THEORY, 28-32. (1974).

SAHN 1975 -1 SAHNI, S.
APPROXIMATE ALGORITHMS FOR THE 0/1 KNAPSACK PROBLEM. JOURNAL OF THE ASSOCIATION FOR COMPUTING MACHINERY 22, 115-124. (1975).

SAHN 1977 -1 SAHNI, S.
GENERAL TECHNIQUES FOR COMBINATORIAL APPROXIMATION. OPERATIONS RESEARCH 25, 920-936. (1977).

SAHNA 1974 -1 SAHNI, S.; GONZALES, T.
P-COMPLETE PROBLEMS AND APPROXIMATE SOLUTIONS. IN: 15TH ANNUAL SYMPOSIUM ON SWITCHING AND AUTOMATA THEORY, NEW ORLEANS 1974, 28-32. LONG BEACH: INSTITUTE OF ELECTRICAL AND ELECTRONICS ENGINEERS. (1974).

SAHNB 1976 -1 SAHNI, S.; GONZALEZ, T.
P-COMPLETE APPROXIMATION PROBLEMS. JOURNAL OF THE ASSOCIATION FOR COMPUTING MACHINERY 23, 555-565. (1976).

SAHNC 1976 -1 SAHNI, S. K.
ALGORITHMS FOR SCHEDULING INDEPENDENT TASKS. JOURNAL OF THE ASSOCIATION FOR COMPUTING MACHINERY 23, 116-127. (1976).

SAIG 1969 -1 SAIGAL, R.
MULTICOMMODITY FLOWS IN DIRECTED NETWORKS. TECHNICAL REPORT ORC 69-14. BERKELEY: UNIV. OF CALIFORNIA, OPERATIONS RESEARCH CENTER. (1969).

SAIP 1975 -1 SAIPE, A. L.
SOLVING A (0,1) HYPERBOLIC PROGRAM BY BRANCH AND BOUND. NAVAL RESEARCH LOGISTICS QUARTERLY 22, 497-516. (1975).

SAKA 1976 -1 SAKAGUCHI, M.
SEQUENTIAL ALLOCATION PROBLEM FOR RANDOMLY APPEARING TARGETS, A. MATHEMATICA JAPONICAE 21, 89-103. (1976).

SAKAA 1976 -1 SAKAKI, T.; NAKASHIMA, K.; HATTORI, Y.
ALGORITHMS FOR FINDING IN THE LUMP BOTH BOUNDS OF THE CHROMATIC NUMBER OF A GRAPH. COMPUTER JOURNAL 19, 329-332. (1976).

SAKAB 1966 -1 SAKAROVITCH, M.
K SHORTEST ROUTES AND THE K SHORTEST CHAINS IN A GRAPH, THE. REPORT 66-32. BERKELEY: UNIV. OF CALIFORNIA, OPERATIONS RESEARCH CENTER, COLLEGE OF ENGINEERING. (1966).

SAKAB 1968 -1 SAKAROVITCH, M.
K SHORTEST CHAINS IN A GRAPH, THE. TRANSPORTATION RESEARCH 2, 1-11. (1968).

SAKAB 1976 -1 SAKAROVITCH, M.
QUASI BALANCED MATRICES - AN ADDENDUM. MATHEMATICAL PROGRAMMING 10, 405-407. (1976).

SAKAC 1970 -1 SAKATA, M.; NOGUCHI, S.; OIZUMI, J.
OPTIMAL SCHEDULING IN SINGLE PROCESSOR SYSTEMS. P708-11, 3 RD PROCEEDINGS OF THE INTERNATIONAL CONFERENCE ON SYSTEM SCIENCE. (1970).

SALE 1976 -1 SALERNO, G.
"BRANCH AND BACK-TRACK" ALGORITHM FOR THE DESIGN OF AN OPTIMAL TRANSPORTATION NETWORK (ITALIAN), A. RICERCA OPERATIVA 6, NO. 3, 23-48. (1976).

SALI 1970 -1 SALIHOV, Z. M.; PAIZIEV, E.; BURKOV, V. N.; KANCEL'SON, M. B.
AGGREGATED ALGORITHM FOR THE SOLUTION OF THE TRANSPORT PROBLEM (RUSSIAN), AN. VOPROSY KIBERNETIKI, VYP. 40, 23-29. (1970).

SALK 1975 -1 SALKIN, H. M.
INTEGER PROGRAMMING. READING, LONDON U.A.: ADDISON-WESLEY, 537 P. (1975).

SALKA 1975 -1 SALKIN, H. M.; KLUYVER, C. A. DE
KNAPSACK PROBLEM: A SURVEY, THE. NAVAL RESEARCH LOGISTICS QUARTERLY 22, 127-144. (1975).

SALKB 1972 -1 SALKIN, H. M.; SPIELBERG, K.
ADAPTIVE ALGORITHM FOR BINARY PROGRAMMING, AN. REPORT 320-3-13. NEW YORK: IBM SCIENTIFIC CENTER. (1972).

SAMP 1976 -1 SAMPATHKUMAR, E.; BHAVE, V. N.
PARTITION GRAPHS AND COLORING NUMBERS OF A GRAPH. DISCRETE MATHEMATICS 16, 57-60. (1976).

SANA 1968 -1 SANADA, H.; LINGST, S. T.; TEZUKA, Y.; KASAHARA, Y.
OPTIMAL SCHEDULING USING SLACK TIME ARRAY (JAPANESE). KEIEI KAGAKU 11, 188-204. (1968).

SANAA 1967 -1 SANADA, H.; NAKAGAMI, K.; TEZUKA, Y.; KASAHARA, Y.
SCHEDULING PROBLEM OF OPTIMAL CYCLIC PRODUCTION (JAPANESE). KEIEI KAGAKU 11, 37-46. (1967).

SANC 1976 -1 SANCHO, N. G. F.
OPTIMAL SOLUTION FOR SCHEDULING OF PRINTING AND BINDING, AN. INFOR 14, 164-168. (1976).

SAND 1975 -1 SANDEWALL, E. J.
HEURISTISCHE SUCHE: KONZEPTE UND METHODEN. IN: N. V. FINDLER (ED.): KUENSTLICHE INTELLIGENZ UND ES HEURISTISCHES PROGRAMMIEREN, 83-104. WIEN: SPRINGER. (1975).

SANDA 1975 -1 SANDI, C.
SOLUTION OF THE MACHINE LOADING PROBLEM WITH BINARY VARIABLES. IN: B. ROY (ED.): COMBINATORIAL PROGRAMMING: METHODS AND APPLICATIONS. PROCEEDINGS OF THE NATO ADVANCED STUDY INST., VERSAILLES 1974, 371-378. DORDRECHT: REIDEL. (1975).

SANK 1975 -1 SANKIBAEV, B. N.
INTEGER DISTRIBUTION PROBLEM (RUSSIAN), AN. IN: PROCEEDINGS OF THE 6TH WINTER SCHOOL ON MATHEMATICAL PROGRAMMING AND RELATED QUESTIONS (RUSSIAN), DROGOBYCH 1973, 259-289. MOSCOW: CENTRAL. EKONOM.-MAT. INST. AKAD. NAUK SSSR. (1975).

SANKA 1973 -1 SANKIBAEV, B. N.; SERMAN, A. I.
ALGORITHM FOR THE SOLUTION OF A PURELY INTEGER LINEAR PROGRAMMING PROBLEM (RUSSIAN), AN. IZVESTIJA AKADEMII NAUK KAZAHSKOI SSR. SERIJA FIZIKO-MATEMATICESKAJA, NO. 5, 87-90 + 96. (1973).

SANKB 1975 -1 SANKOFF, D.; ROUSSEAU, P.
LOCATING THE VERTICES OF A STEINER TREE IN AN ARBITRARY SPACE. MATHEMATICAL PROGRAMMING 9, 240-246. (1975).

SARV 1976 -1 SARVANOV, V. I.
MINIMIZATION OF A LINEAR FORM ON THE SET OF CYCLES IN N TERMS (RUSSIAN). VESCI AKADEMII NAVUK BSSR, SERYJA FIZIKA-MATEMATYCNYH NAVUK, NO. 4, 17-21. (1976).

SARVA 1976 -1 SARVANOV, V.I.
MEAN VALUE OF THE FUNCTIONAL OF THE ASSIGNMENT PROBLEM (RUSSIAN), THE. VESCI AKADEMII NAVUK BSSR. SERYJA FIZIKA-MATEMATYCNYH NAVUK 2, 111-114. (1976).

SAUC 1971 -1 SAUCIER, G.
ALGORITHME EFFICACE RECHERCHANT L'ISOMORPHISME DE 2 GRAPHES, UN. REVUE FRANCAISE D'AUTOMATIQUE, INFORMATIQUE ET RECHERCHE OPERATIONELLE, NO. R-3, 39-51. (1971).

SAVA 1976 -1 SAVAGE, J. E.
COMPLEXITY OF COMPUTING, THE. NEW YORK, LONDON U.A.: WILEY, 391 P. (1976).

****

SAVAA 1973 -1 SAVAGE, S. L.
SOLUTION OF DISCRETE LINEAR OPTIMIZATION PROBLEMS BY NEIGHBORHOOD SEARCH TECHNIQUES, THE. DISSERTATION. NEW HAVEN: YALE UNIV., DEPT. OF COMPUTER SCIENCE, 109 P. (1973).

SAVAA 1973 -2 SAVAGE, S. L.
STATISTICAL INDICATORS OF OPTIMALITY. IN: 14TH ANNUAL SYMPOSIUM ON SWITCHING AND AUTOMATA THEORY, IOWA CITY 1973, 85-91. NORTHRIDGE: INSTITUTE OF ELECTRICAL AND ELECTRONICS ENGINEERS. (1973).

SAVAA 1976 -1 SAVAGE, S. L.
PROBABILISTIC APPROACH TO HEURISTIC DESIGN, A. MIAMI (50. ORSA-MEETING + TIMS, NOVEMBER 1976) 10 P. (1976).

SAVAA 1976 -2 SAVAGE, S. L.
SOME THEORETICAL IMPLICATIONS OF LOCAL OPTIMIZATION. MATHEMATICAL PROGRAMMING 10, 354-366. (1976).

SAVAA 1978 -1 SAVAGE, S. L.
PROBABILISTIC APPROACH TO THE DESIGN OF HEURISTICS FOR DECISION PROBLEMS, A. REPORT 7807. CHICAGO: UNIV. OF CHICAGO, CENTER FOR MATHEMATICAL STUDIES IN BUSINESS AND ECONOMICS, 13 P. (1978).

SAVAB 1976 -1 SAVAGE, S. L.; WEINER, P.; BAGCHI, A.
NEIGHBORHOOD SEARCH ALGORITHMS FOR GUARANTEEING OPTIMAL TRAVELING SALESMAN TOURS MUST BE INEFFICIENT. JOURNAL OF COMPUTER AND SYSTEM SCIENCES 12, 25-35. (1976).

SAVI 1974 -1 SAVITCH, W. J.
NONDETERMINISTIC LOG N SPACE. IN: PROCEEDINGS OF THE EIGHTH ANNUAL PRINCETON CONFERENCE ON INFORMATION SCIENCES AND SYSTEMS, 21-23. (1974).

SAYA 1976 -1 SAYAMA, H.; FAN, L. T.; FAN, L. S.
ON LAGRANGE MULTIPLIERS AND CONSTRAINTS, I. LAGRANGIAN APPROACH. INTERNATIONAL JOURNAL OF SYSTEMS SCIENCE 7, 1283-1298. (1976).

SAYA 1976 -2 SAYAMA, H.; FAN, L. T.; FAN, L. S.
ON LAGRANGE MULTIPLIERS AND CONSTRAINTS, II. AUGMENTED LANGRANGIAN APPROACH. INTERNATIONAL JOURNAL OF SYSTEMS SCIENCE 7, 1299-1313. (1976).

SCAR 1977 -1 SCARBROUGH, J. E.
MIXED-INTEGER PROGRAMMING MODEL FOR THE SCHEDULING OF AIRCRAFT BY TIME OF DAY WITHIN A NETWORK OF AIRPORTS, A. SAN FRANCISCO (51. ORSA-MEETING + TIMS, MAY 1977) 55 P. (1977).

SCHA 1976 -1 SCHAAR, G.
ON THE MAXIMUM NUMBER OF EDGE-DISJOINT PATHS WITH ENDPOINTS BELONGING TO A GIVEN SET OF VERTICES IN A FINITE GRAPH. IN: GRAPHS HYPERGRAPHS BLOCK SYST. SYMP. COMB. ANAL. PROC., ZIELONA GORA 1976, 239-246. ZIELONA GORA. (1976).

SCHAA 1971 -1 SCHAERF, S.; SHAHAM, J.
PUBLIC TRANSPORTATION SIMULATION ALGORITHM, A. IN: ISRAEL CONFERENCE ON OPERATIONS RESEARCH, THIRD PROCEEDINGS, GORDON 1969, 439-447. (1971).

SCHAB 1972 -1 SCHALLEHN, W.; SEBASTIAN, H.-J.
REIHENFOLGEPROBLEM IN DER NETZPLANTECHNIK, EIN. ELEKTRONISCHE INFORMATIONSVERARBEITUNG UND KYBERNETIK 8, 3-29. (1972).

SCHAC 1974 -1 SCHARGE, L.
OPTIMAL SCHEDULING RULES FOR QUADRATIC WAITING COSTS. IN: PROCEEDINGS OF THE SEVENTH ANNUAL PRINCETON CONFERENCE ON INFORMATION SCIENCES AND SYSTEMS, 120-124. (1974).

SCHI 1970 -1 SCHINZINGER, R.
ON THE REDUCTION OF THE CONTINUOUS VARIABLE COMPONENT IN MIXED INTEGER LINEAR PROGRAMMING. IRVINE: UNIV. OF CALIFORNIA, SCHOOL OF ENGINEERING, 41 P. (1970).

SCHM 1976 -1 SCHMIDT, B. K.; MATTHEISS, T. H.
ENUMERATING THE SET OF INEQUALITIES DEFINING THE CONVEX HULL OF A GIVEN SET OF POINTS IN E N. MIAMI (50. ORSA-MEETING + TIMS, NOVEMBER 1976) 8 P. (1976).

SCHMA 1976 -1 SCHMIDT, D. C.; DRUFFEL, L. E.
FAST BACKTRACKING ALGORITHM TO TEST DIRECTED GRAPHS FOR ISOMORPHISM USING DISTANCE MATRICES, A. JOURNAL OF THE ASSOCIATION FOR COMPUTING MACHINERY 23, 433-445. (1976).

SCHMB 1975 -1 SCHMITZ, W.
UEBER LAENGSTE WEGE UND KREISE IN GRAPHEN. RENDICONTI DEL SEMINARIO MATEMATICO DELL'UNIVERSITA DI PADOVA 53, 97-103. (1975).

SCHN 1976 -1 SCHNORR, C. P.
ALGORITHM FOR TRANSITIVE CLOSURE WITH LINEAR EXPECTED TIME, AN. FRANKFURT: UNIV. FRANKFURT, FACHBEREICH MATHEMATIK, 8 P. (1976).

SCHN 1976 -2 SCHNORR, C. P.
NETWORK COMPLEXITY AND THE TURING MACHINE COMPLEXITY OF FINITE FUNCTIONS, THE. ACTA INFORMATICA 7, 95-107. (1976).

SCHN 1978 -1 SCHNORR, C. P.
MULTITERMINAL NETWORK FLOW AND CONNECTIVITY IN UNSYMMETRICAL NETWORKS. IN: R. HENN, B. KORTE, W. OETTLI (EDS.): OPTIMIZATION AND OPERATIONS RESEARCH, PROCEEDINGS, BONN 1977. LECTURE NOTES IN ECONOMICS AND MATHEMATICAL SYSTEMS 157. BERLIN, HEIDELBERG, NEW YORK: SPRINGER, 241-254. (1978).

SCHN 1978 -2 SCHNORR, C. P.
SATISFIABILITY IS QUASILINEAR COMPLETE IN NQL. JOURNAL OF THE ASSOCIATION FOR COMPUTING MACHINERY 25, 136-145. (1978).

SCHO 1976 -1 SCHOCH, M.
ERWEITERUNGSPRINZIP UND SEINE ANWENDUNG ZUR ENTWICKLUNG VON ALGORITHMEN FUER DIE LOESUNG KOMBINATORISCHER OPTIMIERUNGSAUFGABEN, DAS. MUENCHEN, WIEN: OLDENBOURG, 300 P. (1976).

SCHOA 1970 -1 SCHOEMAN, M. E. F.; BHAUMIK, G.
CLASS SCHEDULING MODEL. BULLETIN OF THE OPERATIONS RESEARCH SOCIETY OF AMERICA 18, 38 P. (1970).

SCHOB 1975 -1 SCHOENE, W.
ALGORITHMEN ZUR BESTIMMUNG INTERN UND EXTERN STABILER MENGEN UND DER CHROMATISCHEN ZAHL EINES ENDLICHEN GRAPHEN. WISSENSCHAFTLICHE ZEITSCHRIFT DER TECHNISCHEN HOCHSCHULE KARL-MARX-STADT 17, 269-280. (1975).

SCHOC 1976 -1 SCHOENLEIN, A.
BEZUEGLICH KERNSPEICHER- UND RECHENZEITBEDARF EFFIZIENTES BRANCH-UND-BOUND-VERFAHREN ZUR LOESUNG DES TRAVELLING-SALESMAN-PROBLEMS, EIN. ANGEWANDTE INFORMATIK 18, 100-108. (1976).

SCHOC 1977 -1 SCHOENLEIN, A.
UEBER DIE ANWENDUNG DER METHODE DES BRANCH AND BOUND AUF DAS SUMMEN-ZUORDNUNGSPROBLEM. ANGEWANDTE INFORMATIK 19, 547-552. (1977).

SCHOD 1975 -1 SCHONHEIM, J.; BIALOSTOCKI, A.
PACKING AND COVERING OF THE COMPLETE GRAPH WITH 4-CYCLES. CANADIAN MATHEMATICAL BULLETIN 18, 703-708. (1975).

SCHR 1976 -1 SCHRIJVER, L.
LINKING SYSTEMS, MATROIDS AND BIPARTITE GRAPHS. IN: C. ST. J. A. NASH-WILLIAMS (ED.): PROCEEDINGS OF THE 5TH BRITISH COMBINATORIAL CONFERENCE, ABERDEEN 1975, 541-544. CONGRESSUS NUMERANTIUM 15, WINNIPEG: UTILITAS MATHEMATICA. (1976).

****

SCHU 1968 -1 SCHULTZ, G. P.
MANAGERIAL DECISION-MAKING IN LOCAL GOVERNMENT: FACILITY PLANNING FOR SOLID WASTE COLLECTION. DISSERTATION. ITHACA: CORNELL UNIVERSITY. (1968).

SCHU 1969 -1 SCHULTZ, G. P.
FACILITY PLANNING FOR A PUBLIC SERVICE SYSTEM: DOMESTIC SOLID WASTE COLLECTION. JOURNAL OF REGIONAL SCIENCE 9, 291-307. (1969).

SCHUA 1977 -1 SCHURAWLJEW, J. I.; FINKEL'SHTEIN, YU. YU.
AUFGABENTYPEN UND LOESUNGSMETHODEN IN DER DISKRETEN OPTIMIERUNG. IN: T. D. BERESNEWA, W. DUECK, K.-H. ELSTER ET AL. (EDS.): MATHEMATISCH-OEKONOMISCHE METHODEN UND MODELLE, 196-232. BERLIN: DIE WIRTSCHAFT. (1977).

SCHW 1977 -1 SCHWARTZ, M.
COMPUTER-COMMUNICATION NETWORK. DESIGN AND ANALYSIS. ENGLEWOOD CLIFFS: PRENTICE-HALL, 372 P. (1977).

SCIO 1964 -1 SCIONS, H. I.
COMPACT REPRESENTATION OF A ROOTED TREE AND THE TRANSPORTATION PROBLEM, THE. IN: INTERNATIONAL SYMPOSIUM ON MATHEMATICAL PROGRAMMING, LONDON 1964. (1964).

SCLE 1977 -1 SCLEE, W.
MATRIXALGORITHMEN BEI STOCHASTISCHER BEWERTUNG. IN: R. HENN, P. KALL, B. KORTE, O. KRAFFT ET AL. (EDS.): OPERATIONS RESEARCH VERFAHREN XXV, METHODS OF OPERATIONS RESEARCH, 333-340. MEISENHEIM: HAIN. (1977).

SCOT 1970 -1 SCOTT, A. J.
ELEMENTARY THEORETICAL DYNAMICS OF A COMPETITIVE LOCATIONAL SYSTEM, THE. DISCUSSSION PAPER 6. TORONTO: UNIV. OF TORONTO, DEPT. OF GEOGRAPHY. (1970).

SCOT 1971 -1 SCOTT, A. J.
COMBINATORIAL PROGRAMMING, SPATIAL ANALYSIS, AND PLANNING. LONDON: METHUEN. (1971).

SCOT 1971 -2 SCOTT, A. J.
DYNAMIC LOCATION-ALLOCATION SYSTEMS: SOME BASIC PLANNING STRATEGIES. ENVIRONMENT AND PLANNING 3, 73-83. (1971).

SCOT 1971 -3 SCOTT, A. J.
OPTIMAL DECISION PROCESSES FOR A CLASS OF DYNAMIC LOCATIONAL PROBLEMS. PAPERS OF THE REGIONAL SCIENCE ASSOCIATION 26, 25-36. (1971).

SCOT 1975 -1 SCOTT, A. J.
DISCRETE DYNAMIC LOCATIONAL SYSTEMS: NOTES ON A STRUCTURAL FRAMEWORK. IN: A. KARLQVIST ET AL. (EDS.): DYNAMIC ALLOCATION OF URBAN SPACE, 121-158. LEXINGTON: LEXINGTON BOOKS. (1975).

SCOTA 1975 -1 SCOTT, S. H.
MULTIPLE NODE COLORINGS OF FINITE GRAPHS. DISSERTATION. DEPT. OF MATHEMATICS. READING: UNIV. OF READING. (1975).

SCOTB 1976 -1 SCOTT, T. B.
GRAPH COLOURING WITH PREASSIGNMENT AND UNAVAILABILITY CONSTRAINTS. ARS COMBINATORIA 2, 25-32. (1976).

SCRI 1975 -1 SCRIABIN, M.; VERGIN, R. C.
COMPARISON OF COMPUTER ALGORITHMS AND VISUAL BASED METHODS FOR PLANT LAYOUT. MANAGEMENT SCIENCE 22, 172-181. (1975).

SCWA 1968 -1 SCWARTZ, J. G.
NETWORK FLOW AND SYNTHESIS MODELS FOR TRANSPORTATION PLANNING: A SURVEY. TECHNICAL REPORT R68-49, DEPARTMENT OF CIVIL ENGINEERING, M. I. T. (1968).

SEDL 1975 -1 SEDLACEK, J.
PERFECT AND QUASIPERFECT GRAPHS (CZECH). CASOPIS PRO PESTOVANI MATEMATIKY 100, 135-141. (1975).

SEEL 1974 -1 SEELAENDER, J.
METHODE ZUR KOPPLUNG DUALER SCHNITTEBENENVERFAHREN IN DER GANZZAHLIGEN LINEAREN OPTIMIERUNG, EINE. DISSERTATION. LEUNA-MERSEBURG: TECHNISCHE HOCHSCHULE LEUNA-MERSEBURG. (1974).

SEEL 1976 -1 SEELAENDER, J.
BEMERKUNG ZUM DRITTEN ALGORITHMUS VON GOMORY ZUR LOESUNG REIN-GANZZAHLIGER LINEARER OPTIMIERUNGSPROBLEME, EINE. WISSENSCHAFTLICHE ZEITSCHRFFT DER TECHNISCHEN HOCHSCHULE LEUNA-MERSEBURG 18, 134-140. (1976).

SEEL 1977 -1 SEELAENDER, J.
UEBER EINE METHODE ZUR KOPPLUNG DUALER SCHNITTEBENENVERFAHREN IN DER GANZZAHLIGEN LINEAREN OPTIMIERUNG. MATHEMATISCHE OPERATIONSFORSCHUNG UND STATISTIK 8, SERIES OPTIMIZATION, 307-316. (1977).

SEID 1974 -1 SEIDVASSER, M. A.
ASSIGNMENTS OF GRAPH VERTICES IN LATTICES (RUSSIAN). DISKRETNYI ANALIZ. VYP. 24. ISSLEDOVANIJA STRUKTURY I FUNKCIONIROVANIJA UPRAVLJAJUSCIH SISTEM, 84-94 + 97. (1974).

SEID 1974 -2 SEIDVASSER, M. A.
LENGTH AND BREADTH OF ASSIGNMENTS OF GRAPHS IN LATTICES (RUSSIAN), THE. PROBLEMY KIBERNETIKI, NO. 29, 63-102 + 245 + ERRATA: IBID., 1975, NO. 30, 342. (1974).

SEIF 1974 -1 SEIFERAS, J. I.
NOTE ON REDUCTIONS TO DIRECTED HC, A. IN: PROCEEDINGS OF THE EIGHTH ANNUAL PRINCETON CONFERENCE ON INFORMATION SCIENCES AND SYSTEMS, 24-28. (1974).

SEME 1974 -1 SEMENOV, V. M.; HAIKIN, N. M.; MOGILA, V. G.
CERTAIN METHOD OF SOLVING A VOLUME SCHEDULING PROBLEM (RUSSIAN), A. IN: MATHEMATICAL METHODS OF SOLUTION OF ECONOMIC PROBLEMS (RUSSIAN), 80-91. SUPPL. TO EKONOMIKA I MATEMATICESKIE METODY NO. 5. MOSCOW: IZDAT. NAUKA. (1974).

SERC 1970 -1 SERCK-HANSSEN, J.
OPTIMAL PATTERNS OF LOCATION. AMSTERDAM: NORTH-HOLLAND. (1970).

SERD 1975 -1 SERDYUKOV, A. I.
ON A COVERING PROBLEM (RUSSIAN). UPRAVLJAEME SISTEMY, NO. 14, 52-58. (1975).

SERD 1976 -1 SERDYUKOV, A. I.
MUTUAL REDUCIBILITY OF SOME EXTREMAL PROBLEMS IN GRAPH THEORY (RUSSIAN). UPRAVLJAEME SISTEMY, NO. 15, 68-73. (1976).

SERG 1975 -1 SERGIENKO, I. V.
APPLICATION OF THE METHOD OF THE VECTOR OF DESCENT TO THE SOLUTION OF INTEGER PROGRAMMING PROBLEMS (RUSSIAN). UPRAVLJAEME SISTEMY MACH., NO. 3, 92-98. (1975).

SERO 1976 -1 SEROV, S. S.; UZDEMIR, A. P.
DYNAMIC INDUSTRIAL-PLANT-LOCATION PROBLEM AND NUMERICAL METHOD FOR SOLVING IT, I (RUSSIAN). AVTOMATIKA I TELEMEHANIKA, NO. 4, 120-127; TRANSLATED AS AUTOMATION AND REMOTE CONTROL 37, 1976, 583-589. (1976).

SETH 1976 -1 SETHI, R.
SCHEDULING GRAPHS ON TWO PROCESSORS. SIAM JOURNAL ON COMPUTING 5, 73-82. (1976).

SEVA 1975 -1 SEVAST'YANOV, S. V.
ASYMPTOTIC APPROACH TO SOME PROBLEMS SEQUENCING THEORY (RUSSIAN). UPRAVLJAEME SISTEMY, NO 14, 40-51. (1975).

SEYM 1975 -1 SEYMOUR, P. D.
MATROIDS, HYPERGRAPHS AND THE MAX-FLOW MIN-CUT THEOREM. DISSERTATION. OXFORD: OXFORD UNIV. (1975).

****

SEYM 1976 -1 SEYMOUR, P. D.
MAX-FLOW MIN-CUT PROPERTY IN MATROIDS, THE. IN: C. ST. J. A. NASH-WILLIAMS (ED.): PROCEEDINGS OF THE 5TH BRITISH COMBINATORIAL CONFERENCE, ABERDEEN 1975, 545-550. CONGRESSUS NUMERANTIUM 15, WINNIPEG: UTILITAS MATHEMATICA. (1976).

SEYM 1977 -1 SEYMOUR, P. D.
MATROIDS WITH THE MAX-FLOW MIN-CUT PROPERTY, THE. JOURNAL OF COMBINATORIAL THEORY 23, SERIES B, 189-222. (1977).

SHAH 1974 -1 SHAH, Y. J.; DAVIDA, G. T.; MAC CARTHY, M. K.
OPTIMUM FEATURES AND GRAPH ISOMORPHISM. IEEE TRANSACTIONS ON SYSTEMS, MAN, AND CYBERNETICS 4, 313-319. (1974).

SHAK 1977 -1 SHAKHBASJAN, K. V.; TUSHKINA, T. A.
ON STATISTICAL EFFICIENCY OF SOLUTION ALGORITHMS FOR A PROBLEM OF VERTEX DIVISION (RUSSIAN). KIBERNETIKA, NO. 5, 38-43. (1977).

SHAM 1972 -1 SHAMMA, M. M.; WILLIAMS, T. A.
CONSTRUCTING SAMPLE NETWORKS FOR ANALYZING AND COMPARING SHORTEST ROUTE ALGORITHMS. PAPER PRESENTED AT THE 42ND NATIONAL ORSA-MEETING, ATLANTIC CITY 1972. (1972).

SHAMA 1975 -1 SHAMOS, M. I.; HOEY, D.
CLOSEST-POINT PROBLEMS. IN: 16TH ANNUAL SYMPOSIUM ON FOUNDATIONS OF COMPUTER SCIENCE, BERKELEY 1975, 151-162. LONG BEACH: INSTITUTE OF ELECTRICAL AND ELECTRONICS ENGINEERS. (1975).

SHAP 1975 -1 SHAPIRO, A. D.
SOME ASYMPTOTIC PROPERTIES OF A SCHEDULE THEORY PROBLEM (RUSSIAN). EKONOMIKA I MATEMATICHESKIE METODY 11, 1214-1216. (1975).

SHAPA 1966 -1 SHAPIRO, D. M.
ALGORITHMS FOR THE SOLUTION OF THE OPTIMAL COST AND BOTTLENECK TRAVELING SALESMAN PROBLEMS. DISSERTATION. ST. LOUIS: WASHINGTON UNIVERSITY. (1966).

SHAPB 1974 -1 SHAPIRO, J. F.
NEW DUALITY THEORY FOR MIXED INTEGER PROGRAMMING, A. WORKING PAPER OR 033-74. CAMBRIDGE: MASSACHUSETTS INST. OF TECHNOLOGY, OPERATIONS RESEARCH CENTER. (1974).

SHAPB 1976 -1 SHAPIRO, J. F.
MULTIPLE CRITERIA PUBLIC INVESTMENT DECISION MAKING BY MIXED INTEGER PROGRAMMING. IN: H. THIRIEZ, P. ZIONTS (EDS.): MULTIPLE CRITERIA DECISION MAKING, 170-182. BERLIN: SPRINGER. (1976).

SHAPB 1976 -2 SHAPIRO, J. F.
NEW CONSTRUCTIVE DUALITY THEORY FOR MIXED INTEGER PROGRAMMING, A. IN: PROCEEDINGS OF THE CONFERENCE ON MATHEMATICAL PROGRAMMING AND ITS APPLICATIONS, INDAM, ROME 1974 (ITALIAN), 61-71. SYMPOSIA MATHEMATICA 19, LONDON: ACADEMIC PRESS. (1976).

SHAPB 1977 -1 SHAPIRO, J. F.
NOTE ON THE PRIMAL-DUAL AND OUT-OF-KILTER ALGORITHMS FOR NETWORK OPTIMIZATION PROBLEMS, A. NETWORKS 7, 81-88. (1977).

SHAPB 1977 -2 SHAPIRO, J. F.
SENSITIVITY ANALYSIS IN INTEGER PROGRAMMING. IN: P. L. HAMMER, E. L. JOHNSON, B. H. KORTE, G. L. NEMHAUSER (EDS.): STUDIES IN INTEGER PROGRAMMING, 467-478. ANNALS OF DISCRETE MATHEMATICS 1, AMSTERDAM: NORTH-HOLLAND. (1977).

SHAPB 1977 -3 SHAPIRO, J. F.
SURVEY OF LAGRANGEAN TECHNIQUES FOR DISCRETE OPTIMIZATION. VANCOUVER (CONFERENCE ON DISCRETE OPTIMIZATION, AUGUST 1977) 41 P. (1977).

SHAPC 1976 -1 SHAPIRO, R. D.
SCHEDULING COUPLED TASKS. WORKING PAPER HBS 76-10. BOSTON: HARVARD UNIV., GRADUATE SCHOOL OF BUSINESS ADMINISTRATION, 30 P. (1976).

SHAPD 1977 -1 SHAPIRO, S. D.
PERFORMANCE OF HEURISTIC BIN PACKING ALGORITHMS WITH SEGMENTS OF RANDOM LENGTH. INFORMATION AND CONTROL 35, 146-158. (1977).

SHAR 1975 -1 SHARIF, M. N.; AGARWAL, R. L.
ALGORITHM FOR INTEGER GOAL PROGRAMMING, AN. PAPER PRESENTED AT THE 48TH JOINT NATIONAL MEETING OF ORSA AND TIMS, LAS VEGAS 1975, 21 P. (1975).

SHARA 1975 -1 SHARIQ, S. Z.; CHACHRA, V.
CUTTING PLANE ALGORITHM FOR LEAST DISTANCE PROGRAMMING, A. PAPER PRESENTED AT THE 48TH JOINT NATIONAL MEETING OF ORSA AND TIMS, LAS VEGAS 1975, 30 P. (1975).

SHARB 1976 -1 SHARMA, J. K.; AGRAWAL, S. C.
FIXED-CHARGE TRANSPORTATION PROBLEM, A. ACTA CIENCIA INDICA 2, 278-280. (1976).

SHARB 1976 -2 SHARMA, J. K.; AGRAWAL, S. C.
TIME-COST TRADE-OFF IN A TRANSPORTATION PROBLEM. ACTA CIENCIA INDICA 2, 152-160. (1976).

SHARC 1978 -1 SHARMA, J. K.; SWARUP, K.
TIME MINIMIZATION IN TRANSPORTATION PROBLEMS. NEW ZEALAND OPERATIONAL RESEARCH 6, 75-88. (1978).

SHARD 1977 -1 SHARMA, S. R.; SWARUP, K.
TRANSPORTATION FRACTIONAL PROGRAMMING WITH RESPECT TO TIME. RICERCA OPERATIVA 8, NO. 3, 49-58. (1977).

SHARE 1975 -1 SHARP, G. P.; CHU, J.
COMBINATORIAL APPROACH TO SCHEDULING PUBLIC TRANSIT OPERATORS, A. PAPER PRESENTED AT THE 47TH JOINT NATIONAL MEETING OF ORSA AND TIMS, CHICAGO 1975, 24 P. (1975).

SHARF 1972 -1 SHARPE, R.; BROTCHIE, J. F.
URBAN SYSTEMS STUDY, AN. ROYAL AUSTRALIAN PLANNING INSTITUTE JOURNAL 10, NO. 3, 105-118. (1972).

SHARG 1975 -1 SHARPE, R.; BROTCHIE, J. F.; AHERN, P. A.
EVALUATION OF ALTERNATIVE GROWTH PATTERNS FOR MELBOURNE. IN: A. KARLQVIST ET AL. (EDS.): DYNAMIC ALLOCATION OF URBAN SPACE, 259-285. LEXINGTON: LEXINGTON BOOKS. (1975).

SHAW 1975 -1 SHAW, G. C.
QUALITY CONTROL MEASURES FOR CANADIAN AIRLINE SCHEDULES. PAPER PRESENTED AT THE 48TH JOINT NATIONAL MEETING OF ORSA AND TIMS, LAS VEGAS 1975, 24 P. (1975).

SHAWA 1974 -1 SHAW, J. F. B.
OPTIMIZATION OF TRANSPORTATION ROUTES, APPLIED TO A FLORIDA KEYS SEVEN MILE BRIDGE. DISSERTATION. CORAL GABLES: UNIV. OF MIAMI. (1974).

SHEE 1975 -1 SHEEHAN, J.
EXTREMAL PROBLEM IN FINITE GRAPH THEORY, AN. IN: A. HAYNAL, R. RADO, V. T. SOS (EDS.): INFINITE AND FINITE SETS, COLLOQ. KESZTHELY 1973, VOL. 3, 1235-1239. COLLOQ. MATH. SOC. JANOS BOLYAI 10, AMSTERDAM: NORTH-HOLLAND. (1975).

SHEL 1975 -1 SHELNESS, N. H.; STEPHENS, P. D.; WHITFIELD, H.
EDINBURG MULTI-ACCESS SYSTEM SCHEDULING AND ALLOCATION PROCEDURES IN THE RESIDENT SUPERVISOR, THE. REVUE FRANCAISE D'AUTOMATIQUE, D'INFORMATIQUE ET DE RECHERCHE OPERATIONNELLE 9, NO. B-3, 29-46. (1975).

SHEP 1974 -1 SHEPPARD, E. S.
CONCEPTUAL FRAMEWORK FOR DYNAMIC LOCATION-ALLOCATION ANALYSIS, A. ENVIRONMENT AND PLANNING 6, SERIES A, 547-564. (1974).

SHEPA 1975 -1 SHEPSLE, K. A.
CONGRESSIONAL COMMITTEE ASSIGNMENTS: AN OPTIMIZATION MODEL WITH INSTITUTIONAL CONSTRAINTS. IN: PUBLIC CHOICE, VOL. 22, 55-78. (1975).

SHEPA 1975 -2 SHEPSLE, K. A.
OPTIMIZATION MODEL OF CONGRESSIONAL COMMITTEE ASSIGNMENTS, AN. PAPER PRESENTED AT THE 48TH JOINT NATIONAL MEETING OF ORSA AND TIMS, LAS VEGAS 1975, 65 P. (1975).

SHER 1977 -1 SHERALI, A. D.; SHETTY, C. M.
RECTILINEAR DISTANCE LOCATION-ALLOCATION PROBLEM, THE. AIIE TRANSACTIONS 9, 136-143. (1977).

SHET 1975 -1 SHETTY, C. M.
BILINEAR PROGRAMMING WITH APPLICATIONS TO LOCATION-ALLOCATION PROBLEMS. INDUSTRIAL AND SYSTEMS ENGINEERING REPORT SERIES J-75-20. ATLANTA: GEORGIA INSTITUTE OF TECHNOLOGY, 28 P. (1975).

SHEV 1975 -1 SHEVCHENKO, V. N.
SOLUTION OF AN ELEMENTARY PROBLEM IN LINEAR INTEGER PROGRAMMING (RUSSIAN). UPRAVLJAEME SISTEMY, NO 14, 69-73. (1975).

SHEV 1976 -1 SHEVCHENKO, V. N.
DISCRETE ANALOGUE OF THE FARKASCH THEOREM AND AGGREGATION PROBLEM OF THE SYSTEM OF LINEAR INTEGER-VALUED EQUATIONS (RUSSIAN), A. KIBERNETIKA, NO. 2, 99-101. (1976).

SHEVA 1973 -1 SHEVCHENKO, V. N.; REMIZOVA, O. L.
CONSTRUCTION OF REGULAR CUT-OFFS IN LINEAR INTEGER PROGRAMMING (RUSSIAN), THE. GOR'KOVSKII GOSUDARSTVENNYI UNIVERSITET. UCENYE ZAPISKI, VYP. 166 TEORIJA KOLEBANII, PRIKL. MAT. I KIBERNET., 199-206. (1973).

SHIE 1974 -1 SHIER, D. R.
MAXIMIZING THE NUMBER OF SPANNING TREES IN A GRAPH WITH N NODES AND M EDGES. JOURNAL OF RESEARCH OF THE NATIONAL BUREAU OF STANDARDS 78B, 193-196. (1974).

SHIE 1976 -1 SHIER, D. R.
ALGORITHMS FOR FINDING THE K SHORTEST PATHS IN A NETWORK. PAPER PRESENTED AT THE 49TH JOINT NATIONAL MEETING OF ORSA AND TIMS, PHILADELPHIA 1976, 33 P. (1976).

SHIE 1976 -2 SHIER, D. R.
INVERTING SPARSE MATRICES BY TREE PARTITIONING. JOURNAL OF RESEARCH OF THE NATIONAL BUREAU OF STANDARDS 80B, 245-258. (1976).

SHIE 1976 -3 SHIER, D. R.
ITERATIVE METHODS FOR DETERMINING THE K SHORTEST PATHS IN A NETWORK. NETWORKS 6, 205-230. (1976).

SHIE 1977 -1 SHIER, D. R.
MIN-MAX THEOREM FOR NETWORK LOCATION PROBLEMS ON A TREE, A. SAN FRANCISCO (51. ORSA-MEETING + TIMS, MAY 1977) 17 P. (1977).

SHIE 1977 -2 SHIER, D. R.
MIN-MAX THEOREM FOR P-CENTER PROBLEMS ON A TREE, A. TRANSPORTATION SCIENCE 11, 243-252. (1977).

SHIH 1977 -1 SHIH, W.
BRANCH AND BOUND PROCEDURE FOR A CLASS OF DISCRETE RESOURCE ALLOCATION PROBLEMS WITH SEVERAL CONSTRAINTS, A. OPERATIONAL RESEARCH QUARTERLY 28, 439-452. (1977).

SHIM 1951 -1 SHIMBEL, A.
APPLICATION OF MATRIX ALGEBRA TO COMMUNICATION NETS. BULLETIN OF MATHEMATICAL BIOPHYSICS 13, 165-178. (1951).

SHIM 1953 -1 SHIMBEL, A.
STRUCTURAL PARAMETERS OF COMMUNICATION NETWORKS. BULLETIN OF MATHEMATICAL BIOPHYSICS 15. (1953).

SHIR 1976 -1 SHIRAI, Y.; IBARAKI, T.; HASEGAWA, T.
MINIMAL REPRESENTATION OF COMBINATORIAL OPTIMIZATION PROBLEMS BY FINITE-STATE POSITIVELY MONOTONE MODELS. SYSTEMS, COMPUTERS, CONTROLS 7, 70-79. (1976).

SHIRA 1969 -1 SHIREY, R. W.
IMPLEMENTATION AND ANALYSIS OF EFFICIENT GRAPH PLANARITY TESTING ALGORITHMS. DISSERTATION. MADISON: UNIV. OF WISCONSIN. (1969).

SHKU 1965 -1 SHKURBA, V. V.
COMPUTATIONAL SCHEMES OF SOLUTION OF PROBLEMES IN SCHEDULING THEORY (RUSSIAN). KIBERNETIKA, 72-76. (1965).

SHME 1975 -1 SHMELEV, V. V.
PENALTY FUNCTIONS IN LINEAR INTEGER PROGRAMMING (RUSSIAN). AVTOMATIKA I TELEMEHANIKA, NO. 9, 203-206; TRANSLATED AS AUTOMATION AND REMOTE CONTROL 36, 1975, 1566-1570. (1975).

SHOS 1974 -1 SHOSTAK, R. E.
REFUTATION GRAPHS AND RESOLUTION THEOREM PROVING. IN: PROCEEDINGS OF THE EIGHTH ANNUAL PRINCETON CONFERENCE ON INFORMATION SCIENCES AND SYSTEMS, 447-451. (1974).

SHOW 1977 -1 SHOWALTER, M. J.; KRAJEWSKI, L. J.; RITZMAN, L. P.
MANPOWER ALLOCATION IN U.S. POSTAL FACILITIES: A HEURISTIC APPROACH. COMPUTERS AND OPERATIONS RESEARCH 4, 257-270. (1977).

SHUO 1977 -1 SHUO-YEN, R. L.
EXTREMAL PROBLEM AMONG SUBSETS OF A SET, AN. JOURNAL OF COMBINATORIAL THEORY 23, SERIES A, 341-343. (1977).

SIBI 1976 -1 SIBIRSKIJ, V. K.
GENERALIZED PARAMETRIC STEINER PROBLEMS IN GRAPHS (RUSSIAN), THE. IZVESTIJA AKADEMIA NAUK MOLDAVSKOI SSR, FIZIKO TEKH. MAT. NAUK, NO. 3, 22-26. (1976).

SICK 1977 -1 SICKLE, L. VAN; CHANDY, K. M.
COMPUTATIONAL COMPLEXITY OF NETWORK DESIGN ALGORITHMS. IN: GILCHRIST, B.(ED.): IFIP CONGRESS SERIES 7. INFORMATION PROCESSING 77. PROCEEDINGS OF THE IFIP CONGRESS 77 TORONTO. AMSTERDAM: NORTH HOLLAND, 235-240. (1977).

SIDN 1977 -1 SIDNEY, J. B.
OPTIMAL SINGLE-MACHINE SCHEDULING WITH EARLINESS AND TARDINESS PENALTIES. OPERATIONS RESEARCH 25, 62-69. (1977).

SIEL 1976 -1 SIELKEN, R. M. JR.
SEQUENCING WITH SETUP COSTS BY ZERO-ONE MIXED INTEGER LINEAR PROGRAMMING. AIIE TRANSACTIONS 8, 369-371. (1976).

SIMO 1974 -1 SIMOES-PEREIRA, J. M. S.
MATROIDS, GRAPHS AND TOPOLOGY. IN: F. HOFFMAN, R. A. KINGSLEY, R. B. LEVOW, R. C. MULLIN (EDS.): PROCEEDINGS OF THE FIFTH SOUTHEASTERN CONF. ON COMBINATORICS, GRAPH THEORY, AND COMPUTING, BOCA RATON 1974, 145-155. CONGRESSUS NUMERANTIUM 10, WINNIPEG: UTILITAS MATHEMATICA. (1974).

SING 1968 -1 SINGER, S.
MULTI-CENTERS AND MULTI-MEDIANS OF A GRAPH WITH A APPLICATION TO OPTIMAL WAREHOUSE LOCATION. UNPUBLISHED PAPER, DARLEN: DUNLAP. (1968).

****

SINGA 1975 -1 SINGMASTER D.
ENUMERATING UNLABELLED HAMILTONIAN CIRCUITS. IN: NUMERISCHE METHODEN BEI GRAPHENTHEORETISCHEN UND KOMBINATORISCHEN PROBLEMEN, MATHEMATISCHES FORSCHUNGSINSTITUT OBERWOLFACH 1974, 117-130, INTERNATIONAL SERIES OF NUMERICAL MATHEMATICS 29, BASEL: BIRKHAEUSER. (1975).

SIRI 1968 -1 SIRIK, A. F.
ON THE ALGORITHM OF M. HELD AND R. M. KARP FOR THE SOLUTION OF THE TRAVELLING SALESMAN PROBLEM (RUSSIAN). IN: N. P. KUPCOV (ED.): COMPUTING METHODS AND PROGRAMMING FOR THE ELECTRONIC COMPUTERS "URAL-2", "URAL-4", NO. 2 (RUSSIAN), 32-35. SARATOV: IZDAT. SARATOV. UNIV. (1968).

SIRO 1971 -1 SIROVICH, F.
ISOMORPHISM BETWEEN GRAPHS: AN EFFICIENT ALGORITHM TO FIND ALL ISOMORPHISMS. CALCOLO 8, 301-337. (1971).

SITR 1974 -1 SITRUK, G.
CHEMINS EXTREMAUX ET K-EXTREMAUX APPLICATION AUX PROBLEMES DE FLOTS. DISSERTATION, PARIS: UNIV. OF PARIS. (1974).

SKAF 1966 -1 SKAFF, M. S.
APPROXIMATE SOLUTION OF A WEIGHTED SCHEDULING PROBLEM WITH MULTIPLE CONSTRAINTS. P257-62, 21 ST PROCEEDINGS OF THE ACM NATIONAL CONFERENCE. (1966).

SKOR 1975 -1 SKOROBOGATOV, V. A.
ANALYSIS OF THE CONNECTIVITY OF NON-ORIENTED GRAPHS (RUSSIAN). VYCHISLIT. LIST., NO. 64, 11-25. (1975).

SKUP 1976 -1 SKUPIEN, Z.
HAMILTONIAN SHORTAGE, PATH PARTITIONS OF VERTICES, AND MATCHINGS IN A GRAPH. COLLOQUIUM MATHEMATICUM 36, 305-318. (1976).

SLAT 1976 -1 SLATE, L.; SPIELBERG, K.
EXTENDED CONTROL LANGUAGE OF MPSX/370 (AND POSSIBLE APPLICATIONS), THE. PHILADELPHIA (49. ORSA-MEETING + TIMS, MARCH/APRIL 1976) 39 P. (1976).

SLATA 1975 -1 SLATER, P. J.
LEAVES OF TREES. PROCEEDINGS OF THE SIXTH SOUTHEASTERN CONFERENCE ON COMBINATORICS, GRAPH THEORY, AND COMPUTING, FLORIDA ATLANTIC UNIV., BOCA RATON, 549-559. CONGRESSUS NUMERANTIUM NO. XIV, WINNIPEG: UTILITAS MATH. (1975).

SLATA 1975 -2 SLATER, P. J.
MAXIMIN FACILITY LOCATION. JOURNAL OF RESEARCH OF THE NATIONAL BUREAU OF STANDARDS 79, SERIES B, 107-115. (1975).

SLATA 1976 -1 SLATER, P. J.
CENTRAL VERTICES IN A GRAPH. IN: F. HOFFMAN, L. LESNIAK, R. MULLIN ET AL. (EDS.): PROCEEDINGS OF THE 7TH SOUTHEASTERN CONFERENCE ON COMBINATORICS, GRAPH THEORY, AND COMPUTING, BATON ROUGE 1976, 487-498. CONGRESSUS NUMERANTIUM 17, WINNIPEG: UTILITAS MATHEMATICA. (1976).

SLATA 1976 -2 SLATER, P. J.
R-DOMINATION IN GRAPHS. JOURNAL OF THE ASSOCIATION FOR COMPUTING MACHINERY 23, 446 450. (1976).

SLATB 1976 -1 SLATER, P. J.; GOODMAN, S. E.; HEDETNIEMI, S. T.
ON THE OPTIONAL HAMILTONIAN COMPLETION PROBLEM. NETWORKS 6, 35-52. (1976).

SLOM 1975 -1 SLOMINSKI, L.
QUADRATIC ASSIGNMENT PROBLEM AS A LINEAR ZERO-ONE PROGRAMMING PROBLEM (POLISH), THE. ARCHIWUM AUTOMATYKI I TELEMECHANIKI 20, 89-105. (1975).

SMEL 1975 -1 SMELEV, V. V.
PENALTY FUNCTIONS IN LINEAR INTEGER PROGRAMMING (RUSSIAN). AVTOMATIKA I TELEMEHANIKA, NO. 9, 203-206; TRANSLATED AS AUTOMATION AND REMOTE CONTROL 36, 1975, NO. 9, 1566-1570. (1975).

SMIT 1974 -1 SMITH, D. H.
HAMILTONIAN CIRCUITS ON THE N-CUBE. CANADIAN MATHEMATICAL BULLETIN 17, 759-761. (1974).

SMITA 1976 -1 SMITH, E. M.; OLSON, R. E.
MIXED INTEGER LINEAR PROGRAMMING MODEL APPLIED TO PETROLEUM PRODUCTS BLENDING. MIAMI (50. ORSA-MEETING + TIMS, NOVEMBER 1976) 20 P. (1976).

SMITB 1977 -1 SMITH, J. F.; QUINN, M. J.
HAMILTONIAN CIRCUITS: A HIERARCHY OF EXAMPLES. MATHEMATICS MAGAZINE 50, 35-36. (1977).

SMITC 1976 -1 SMITH, K. R.; OVER, A. M. JR.; HANSEN, M. F.; GOLLADAY, F. L.; DAVENPORT, E. J.
ANALYTIC FRAMEWORK AND MEASUREMENT STRATEGY FOR INVESTIGATING OPTIMAL STAFFING IN MEDICAL PRACTICE. OPERATIONS RESEARCH 24, 815-841. (1976).

SMITD 1975 -1 SMITH, L. D.
SCHEDULING NURSING PERSONNEL: APPLICATION OF LIST PROCESSING AND PROBLEM ORIENTED DATA STRUCTURES. PAPER PRESENTED AT THE 47TH JOINT NATIONAL MEETING OF ORSA AND TIMS, CHICAGO 1975, 24 P. (1975).

SMITD 1976 -1 SMITH, L. D.
APPLICATION OF AN INTERACTIVE ALGORITHM TO DEVELOP CYCLICAL ROTATIONAL SCHEDULES FOR NURSING PERSONNEL, THE. INFOR 14, 53-70. (1976).

SMITE 1977 -1 SMITH, L. D.; WIGGINS, A.
COMPUTER-BASED NURSE SCHEDULING SYSTEM, A. COMPUTERS AND OPERATIONS RESEARCH 4, 195-212. (1977).

SMITF 1975 -1 SMITH, M. L.; PANWALKAR, S. S.; DUDEK, R. A.; ULGEN, O.
FLOWSHOP SEQUENCING WITH NONPERMUTATION SCHEDULES - A HEURISTIC APPROACH. PAPER PRESENTED AT THE 48TH JOINT NATIONAL MEETING OF ORSA AND TIMS, LAS VEGAS 1975, 10 P. (1975).

SMITG 1975 -1 SMITH, R. S.; GIGLIO, R. J.
TWO PHASE EMPLOYEE SCHEDULING ALGORITHM FOR REDUCING LABOR COSTS IN FOOD SERVICE OPERATIONS, A. PAPER PRESENTED AT THE 48TH JOINT NATIONAL MEETING OF ORSA AND TIMS, LAS VEGAS 1975, 17 P. (1975).

SMITH 1975 -1 SMITH, T. H. C.
LIFO IMPLICIT ENUMERATION ALGORITHM FOR THE ASYMMETRIC TRAVELING SALESMAN PROBLEM USING A ONE-ARBORESCENCE RELAXATION, A. MANAGEMENT SCIENCE RESEARCH REPORT 380. PITTSBURGH: CARNEGIE-MELLON UNIV. (1975).

SMITI 1977 -1 SMITH, T. H. C.; SRINIVASAN, V.; THOMPSON, G. L.
COMPUTATIONAL PERFORMANCE OF THREE SUBTOUR ELIMINATION ALGORITHMS FOR SOLVING ASYMMETRIC TRAVELING SALESMAN PROBLEMS. IN: P. L. HAMMER, E. L. JOHNSON, B. H. KORTE, G. L. NEMHAUSER (EDS.): STUDIES IN INTEGER PROGRAMMING, 495-506. ANNALS OF DISCRETE MATHEMATICS 1, AMSTERDAM: NORTH-HOLLAND. (1977).

SMITJ 1977 -1 SMITH, T. H. C.; THOMPSON, G. L.
LIFO IMPLICIT ENUMERATION SEARCH ALGORITHM FOR THE SYMMETRIC TRAVELING SALESMAN PROBLEM USING HELD AND KARP'S 1-TREE RELAXATION, A. IN: P. L. HAMMER, E. L. JOHNSON, B. H. KORTE, G. L. NEMHAUSER (EDS.): STUDIES IN INTEGER PROGRAMMING, 479-494. ANNALS OF DISCRETE MATHEMATICS 1, AMSTERDAM: NORTH-HOLLAND. (1977).

SMOL 1966 -1 SMOLYAR, L. I.
MATHEMATICAL MODELS OF THE SCHEDULING OF SERIAL PRODUCTION (RUSSIAN). EKONOMIKA I MATEMATICESKIE METODY 2, 699-710. (1966).

****

SMYT 1974 -1 SMYTH, W. F.; BENZI, W. M. L.
ALGORITHM FOR FINDING THE DIAMETER OF A GRAPH, AN. IN: JACK L. ROSENFELD (ED.): IFIP CONGRESS 74, VOL. 3: MATHEMATICAL ASPECTS OF INFORMATION PROCESSING, 500-503. AMSTERDAM, LONDON: NORTH-HOLLAND; NEW YORK: AMERICAN ELSEVIER. (1974).

SNIC 1978 -1 SNICKARS, F.
CONVEXITY AND DUALITY PROPERTIES OF A QUADRÁTIC INTRAREGIONAL LOCATION MODEL. REGIONAL SCIENCE AND URBAN ECONOMICS 8, 5-20. (1978).

SOLO 1976 -1 SOLOVAY, R. M.
ON SETS COOK-REDUCIBLE TO SPARSE SETS. SIAM JOURNAL ON COMPUTING 5, 646-652. (1976).

SOMM 1972 -1 SOMMER, D. C.
COMPUTATIONAL EXPERIENCE WITH THE OPHELIE MIXED INTEGER CODE. MINNEAPOLIS: CONTROL DATA, 9 P. (1972).

SONN 1974 -1 SONNEMANN, E.; KRAFFT, O.
INDEPENDENCE NUMBERS OF PRODUCT GRAPHS. JOURNAL OF COMBINATORIAL THEORY 17, SERIES B, 133-142. (1974).

SOUK 1975 -1 SOUKUP, J.
ON MINIMUM COST NETWORKS WITH NONLINEAR COSTS. SIAM JOURNAL ON APPLIED MATHEMATICS 29, 571-581. (1975).

SOVE 1976 -1 SOVETOV, B. YA.; TISHKIN, A. I.
DOMINANT DISTRIBUTIONS IN CERTAIN PROBLEMS OF SCHEDULING THEORY (RUSSIAN). IN: AUTOMATION OF PRODUCTION, NO. 2 (RUSSIAN), 110-114. LENINGRAD: IZDAT. LENINGRAD UNIV. (1976).

SOVE 1976 -2 SOVETOV, B. YA.; TISHKIN, A. I.
ON THE SOLUTION OF A CLASS OF PROBLEMS IN SEQUENCING THEORY (RUSSIAN). IZVESTIJA VYSSIH UCEBNYH ZAVEDENII, PRIBOROSTR. 19, NO. 12, 56-60. (1976).

SPAE 1977 -1 SPAETH, H.
NUMERISCHE ERFAHRUNGEN ZU HEURISTISCHEN LOESUNGSVERFAHREN BEIM VARIANZKRITERIUM IN DER CLUSTER-ANALYSE. ANGEWANDTE INFORMATIK 19, 67-72. (1977).

SPAN 1976 -1 SPANEL, J.
DYNAMIC PROGRAMMING FOR SOLVING ONE MULTISTATE DECISION-MAKING PROBLEM WITH INTEGER VARIABLES (CZECH). EKONOMICKO-MATEMATICKY OBZOR 12, 176-187. (1976).

SPAT 1975 -1 SPATHIS-DIMITROPOULOU, V.
ETUDE ALGEBRIQUE D'UN ARBRE. UN PROBLEME DE CIRCULATION, UNE. DISSERTATION. PARIS: UNIV. PARIS VI, 49 P. (1975).

SPEN 1976 -1 SPENCER, J.
MAXIMAL ASYMMETRY OF GRAPHS. ACTA MATHEMATICA ACADEMIAE SCIENTIARUM HUNGARICAE 27, 47-53. (1976).

SPER 1975 -1 SPERANZA, F.
COLORINGS OF A SUPERIOR SORT OF GRAPHS (ITALIAN). BOLLETINO DELLA UNIONE MATEMATICA ITALIANA 12, NO. 3, SUPPL., 53-62. (1975).

SPIE 1972 -1 SPIELBERG, K.
MINIMAL PREFERRED VARIABLE METHODS IN ZERO-ONE PROGRAMMING. REPORT 320-3013, IBM PHILADELPHIA, SCIENTIFIC CENTER. (1972).

SPIE 1973 -1 SPIELBERG, K.
MINIMAL PREFERRED INEQUALITIES, PENALTIES AND STRUCTURE IN 0-1 PROGRAMMING. PAPER PRESENTED AT THE 8TH INTERNAT. SYMPOSIUM ON MATHEMATICAL PROGRAMMING, STANFORD 1973, 17 P. (1973).

SPIE 1973 -2 SPIELBERG, K.
PREFERRED VARIABLE BRANCH-AND-BOUND METHOD. REPORT 320-3024. IBM PHILADELPHIA, SCIENTIFIC CENTER. (1973).

SPIE 1974 -1 SPIELBERG, K.
MINIMAL PREFERRED VARIABLE METHODS IN ZERO-ONE PROGRAMMING. WORKING PAPER. WHITE PLAINS: IBM SCIENTIFIC MARKETING. (1974).

SPIE 1977 -1 SPIELBERG, K.
ENUMERATIVE METHODS IN INTEGER PROGRAMMING. VANCOUVER (CONFERENCE ON DISCRETE OPTIMIZATION, AUGUST 1977) 50 P. (1977).

SPIR 1976 -1 SPIRN, J. R.
MULTI-QUEUE SCHEDULING OF TWO TASKS. ACTA INFORMATICA 7, 217-226. (1976).

SPRA 1967 -1 SPRADLIN, B. C.; PIERCE, D. A.
PRODUCTION SCHEDULING UNDER LEARNING EFFECT BY DYNAMIC PROGRAMMING. JOURNAL OF INDUSTRIAL ENGINEERING 18, 219-222. (1967).

SRIN 1976 -1 SRINIVASAN, V.
DECOMPOSITION OF A MULTI-PERIOD MEDIA SCHEDULING MODEL IN TERMS OF SINGLE PERIOD EQUIVALENTS. MANAGEMENT SCIENCE 23, 349-360. (1976).

SRINA 1976 -1 SRINIVASAN, V.; THOMPSON, G. L.
ALGORITHMS FOR MINIMIZING TOTAL COST, BOTTLENECK TIME AND BOTTLENECK SHIPMENT IN TRANSPORTATION PROBLEMS. NAVAL RESEARCH LOGISTICS QUARTERLY 23, 567-596. (1976).

SRINA 1977 -1 SRINIVASAN, V.; THOMPSON, G. L.
COST OPERATOR ALGORITHMS FOR THE TRANSPORTATION PROBLEM. MATHEMATICAL PROGRAMMING 12, 372-391. (1977).

STAC 1976 -1 STACEY, K. C.; MAC AVANEY, K. L.; SIMS, J.
STABILITY INDEX OF THE PRODUCT OF A PATH AND A TREE, THE. IN: L. R. A. CASSE, W. D. WALLIS (EDS.): COMBINATORIAL MATHEMATICS IV. PROCEEDINGS OF THE 4TH AUSTRALIAN CONF., ADELAIDE 1975, 220-231. LECTURE NOTES IN MATHEMATICS 560, BERLIN: SPRINGER. (1976).

STAD 1978 -1 STADEL, M.
REMARK ON THE TIME COMPLEXITY OF THE SUBTREE PROBLEM, A. COMPUTING 19, 297-302. (1978).

STAH 1976 -1 STAHL, S.
N-TUPLE COLORINGS AND ASSOCIATED GRAPHS. JOURNAL OF COMBINATORIAL THEORY 20, SERIES B, 185-203. (1976).

STAHA 1976 -1 STAHL, W.
SENSITIVITAETSANALYSE MEHRSTUFIGER STANDORTPROBLEME. MATHEMATICAL SYSTEMS IN ECONOMICS, 30. MEISENHEIM: HAIN, 174 P. (1976).

STAI 1977 -1 STAINTON, R. S.
CUTTING STOCK PROBLEM FOR THE STOCKHOLDER OF STEEL REINFORCEMENT BARS, THE. OPERATIONAL RESEARCH QUARTERLY 28, 139-150. (1977).

STAN 1974 -1 STANCU-MINASIAN, I. M.
THREE-DIMENSIONAL TRANSPORTATION PROBLEM WITH A SPECIAL STRUCTURED OBJECTIVE FUNCTION, A. BULLETIN MATHEMATIQUE DE LA SOCIETE DES SCIENCES MATHEMATIQUES DE LA REPUBLIQUE SOCIALISTE DE ROUMANIE 18, 385-397. (1974).

STANA 1976 -1 STANFORD, R. E.
ANALYTICAL SOLUTION OF A DYNAMIC TRANSACTION FLOW PROBLEM. MATHEMATICAL PROGRAMMING 10, 214-229. (1976).

STANB 1969 -1 STANKARD, M. F.; GUPTA, S. K.
NOTE ON BOMBERGER'S APPROACH TO LOT SIZE SCHEDULING: HEURISTIC PROPOSED. MANAGEMENT SCIENCE 15, THEORY SERIES, 449-452. (1969).

STAR 1972 -1 STAROBINEC, S. M.
ALGORITHM FOR FINDING THE GREATEST INTERNALLY STABLE SET OF A GRAPH (RUSSIAN), AN. IZVESTIJA AKADEMII NAUK SSSR. TEHNICESKAJA KIBERNETIKA, NO. 5, 135-140. (1972).

STAR 1975 -1 STAROBINEC, S. M.
LARGEST INTERNALLY STABLE SET OF A GRAPH (RUSSIAN), THE. KIBERNETIKA, NO. 2, 61-63. (1975).

STARA 1975 -1 STARR, J. H.
ZERO-ONE PROGRAMMING: A PARTIAL ORDERING AND ITS USE IN THE OPTIMIZATION OF NUMERICAL AND NON-NUMERICAL FUNCTIONS. DISSERTATION. PITTSBURGH: CARNEGIE-MELLON UNIVERSITY. (1975).

STAU 1974 -1 STAUDER, F.
SIMULATION MODELLING FOR ANALYSIS OF OPTIMUM SEQUENCING IN THE SHOP-FLOOR. IN: G. BRUCKMANN, F. FERSCHL, L. SCHMETTERER (EDS.): COMPSTAT 1974, PROCEEDINGS IN COMPUTATIONAL STATISTICS, VIENNA 1974, 393-403. WIEN: PHYSICA. (1974).

STEF 1969 -1 STEFFEN, M. R.; DENT, W.
CONSTRUCTION OF REDUNDANT NETWORKS WITH MINIMUM DISTANCE. IN: PROCEEDINGS OF THE THIRD ANNUAL PRINCETON CONFERENCE ON INFORMATION SCIENCES AND SYSTEMS, 28. (1969).

STEFA 1976 -1 STEFFENS, K.
ERGEBNISSE AUS DER TRANSVERSALENTHEORIE, I + II. JOURNAL OF COMBINATORIAL THEORY 20, SERIES A, 178-229. (1976).

STEFA 1977 -1 STEFFENS, K.
MATCHINGS IN COUNTABLE GRAPHS. CANADIAN JOURNAL OF MATHEMATICS 29, 165-168. (1977).

STEI 1969 -1 STEIGLITZ, K.; WEINER, P.; KLEITMAN, D. J.
DESIGN OF MINIMUM COST SURVIVABLE NETWORKS. IEEE TRANSACTIONS ON CIRCUIT THEORY, CT-16, 455-460. (1969).

STEIA 1976 -1 STEINBERG, D. I.
ON SOLVING LARGE-SCALE FIXED CHARGE PROBLEMS. MIAMI (50. ORSA-MEETING + TIMS, NOVEMBER 1976) 32 P. (1976).

STEIA 1977 -1 STEINBERG, D. I.
DESIGNING A HEURISTIC FOR THE FIXED-CHARGE TRANSPORTATION PROBLEM. SIUE PAPERS IN MATHEMATICS AND THE MATHEMATICAL SCIENCES, NO. 4 6. EDWARDSVILLE: SOUTHERN ILLINOIS UNIV., DEPT. OF MATHEMATICAL STUDIES, 26 P. (1977).

STEIB 1976 -1 STEINER, J.
ZUR KONSTRUKTION VON SCHNITTEN FUER GEMISCHT-GANZZAHLIGE LO-PROBLEME. MATHEMATISCHE OPERATIONSFORSCHUNG UND STATISTIK 7, 819-830. (1976).

STER 1974 -1 STERBOUL, F.
ON THE CHROMATIC NUMBER OF THE DIRECT PRODUCT OF HYPERGRAPHS. IN: C. BERGE, D. RAY-CHAUDHURI (EDS.): HYPERGRAPH SEMINAR, PROCEEDINGS OF THE FIRST WORKING SEMINAR, COLUMBUS 1972, 165-174. LECTURE NOTES IN MATHEMATICS 411, BERLIN: SPRINGER. (1974).

STER 1974 -2 STERBOUL, F.
SUR UNE CONJECTURE DE V. CHVATAL. IN: C. BERGE, D. RAY-CHAUDHURI (EDS.): HYPERGRAPH SEMINAR, PROCEEDINGS OF THE FIRST WORKING SEMINAR, COLUMBUS 1972, 152-164. LECTURE NOTES IN MATHEMATICS 411, BERLIN: SPRINGER. (1974).

STER 1975 1 STERBOUL, F.
NEW COMBINATORIAL PARAMETER, A. IN: A. HAJNAL, R. RADO, V. T. SOS (EDS): INFINITE AND FINITE SETS, COLLOQ. KESZTHELY 1973, VOL. 3, 1387-1404. COLLOQ. MATH. SOC. JANOS BOLYAI 10, AMSTERDAM: NORTH-HOLLAND. (1975).

STER 1976 -1 STERBOUL, F.
SOME EXTREMAL COLORING PROBLEMS: A SURVEY. IN: C. ST. J. A. NASH-WILLIAMS, J. SHEEHAN (EDS.): PROCEEDINGS OF THE 5TH BRITISH COMBINATORIAL CONFERENCE, ABERDEEN 1975, 579-584. CONGRESSUS NUMERANTIUM 15, WINNIPEG: UTILITAS MATHEMATICA. (1976).

STER 1977 -1 STERBOUL, F.
EXTREMAL PROBLEM IN HYPERGRAPH COLORING, AN. JOURNAL OF COMBINATORIAL THEORY 22, SERIES B, 159-164. (1977).

STERA 1977 -1 STERN, H. I.; PODRIGUEZ, E. P.; UTTER, M. L.
CYCLICAL JOB SEQUENCING ON MULTIPLE SETS OF IDENTICAL MACHINES. NAVAL RESEARCH LOGISTICS QUARTERLY 24, 137-152. (1977).

STEW 1976 -1 STEWART, W. R. JR.
DELIVERY TRUCK ROUTING PROBLEM WITH STOCHASTIC DEMANDS, THE. MS/S 76-005. COLLEGE PARK: UNIV. OF MARYLAND, COLLEGE OF BUSINESS AND MANAGEMENT, 16 P. (1976).

STIE 1976 -1 STIERLIN, C.
ZUR STANDORTBESTIMMUNG VON KRANKENWAGENSYSTEMEN. IN: J. KOHLAS, O. SEIFERT, P. STAEHLY, H.-J. ZIMMERMANN (EDS): PROCEEDINGS IN OPERATIONS RESEARCH 5, 410-411. WUERZBURG: PHYSICA. (1976).

STOC 1973 -1 STOCKMEYER, L.
PLANAR 3-COLORABILITY IS POLYNOMIAL COMPLETE. ACM SIGACT NEWS 5, NO. 3, 19-25. (1973).

STOCA 1968 -1 STOCKTON, F.
LINEARIZATION AND STANDARDIZATION OF GRAPHS. TECHNICAL PROGRESS REPORT NO. 2-68, SHELL DEVELOPMENT CO. (1968).

STOI 1977 -1 STOICA, M.
SPLITORIAL CONCEPTS USED IN SCHEDULING. ECONOMIC COMPUTATION AND ECONOMIC CYBERNETICS STUDIES AND RESEARCH, NO. 3, 91-105. (1977).

STOJ 1976 -1 STOJAN, YU. G.; GIL, N. I.
METHODS AND ALGORITHMS FOR THE PLACING OF PLANE GEOMETRIC OBJECTS (RUSSIAN). KIEV: IZDAT. NAUKOVA DUMKA, 247 P. (1976).

STOL 1977 -1 STOLZENBURG, I.
OPTIMIERUNGSVERFAHREN FUER REIHENFOLGEPROBLEME IN DER CONTAINER-SCHIFFAHRT. BERLIN: TECHNISCHE UNIVERSITAET BERLIN, FACHBEREICH INFORMATIK. (1977).

STRA 1967 -1 STRADLIN, B. C.; PIERCE, D. A.
PRODUCTION SCHEDULING UNDER A LEARNING EFFECT BY DYNAMIC PROGRAMMING. JOURNAL OF INDUSTRIAL ENGINEERING 18, 219-222. (1967).

STRAA 1976 -1 STRATMANN, K.-H.
NUMERISCHE UNTERSUCHUNGEN UEBER QUADRATISCHE ZUORDNUNGSPROBLEME. DISSERTATION. KOELN: UNIVERSITAET KOELN, MATHEMATISCHES INSTITUT. (1976).

STRAB 1976 -1 STRATMANN, K.-H.; BURKARD, R. E.
NUMERISCHE UNTERSUCHUNGEN UEBER QUADRATISCHE ZUORDNUNGSPROBLEME. REPORT 76-3. KOELN: UNIV. KOELN, MATHEMATISCHES INSTITUT, 17 P. (1976).

STRE 1974 -1 STREET, A. P.
EULERIAN WASHING MACHINES. IN: D. A. HOLTON (ED.): COMBINATORIAL MATHEMATICS, PROC. OF THE SECOND AUSTRALIAN CONF., MELBOURNE 1973, 105-18, LECTURE NOTES IN MATHEMATICS 403, BERLIN: SLRINGER. (1974).

STRE 1976 -1 STREET, A. P.
EMBEDDING PROPER COLOURINGS. IN: R. A. CASSE, W. D. WALLIS (EDS.): COMBINATORIAL MATHEMATICS IV, PROCEEDINGS OF THE 4TH AUSTRALIAN CONFERENCE, ADELAIDE 1975, 240-245. LECTURE NOTES IN MATHEMATICS 560, BERLIN: SPRINGER. (1976).

STRI 1970 -1 STRICKER, R.
PUBLIC SECTOR VEHICLE ROUTING: THE CHINESE POSTMAN PROBLEM. DISSERTATION. CAMBRIDGE: MASS. INST. OF TECHNOLOGY. DEPT. OF ELECTRONICAL ENGINEERING. (1970).

STRO 1977 -1 STROBEL, H.; GENSER, R.; ETSCHMAIER, M.
OPTIMIZATION APPLIED TO TRANSPORTATION SYSTEMS. REPORT CP-77-007. LAXENBURG: INTERNAT. INST. FOR APPLIED SYSTEMS ANALYSIS. (1977).

STROA 1976 -1 STROEKER, R. J.
ON THE DIOPHANTINE EQUATION XXX - DYY = 1. REPRINT SERIES NO. 195. ROTTERDAM: ERASMUS UNIV. ROTTERDAM, ECONOMETRIC INSTITUTE, P. 231-254 (1976).

STUH 1966 -1 STUHR, R.
ANWENDUNGEN UND LOESUNGSVERFAHREN DER GANZZAHLIGEN LINEAREN PROGRAMMIERUNG. DORTMUND: WINTERFELD STRASSE 26, 2 X KLINGELN, 298 P. (1966).

STYC 1974 -1 STYCZEN, K.
TRANSPORTATION SYSTEM IN THE RAW MATERIAL PREPARATION DEPARTMENT OF A CEMENT PLANT: MATHEMATICAL MODEL AND CONTROL ALGORITHM (POLISH), A. PODSTAWY STEROWANIA 4, 315-333. (1974).

SUHL 1975 -1 SUHL, U.
ENTWICKLUNG VON ALGORITHMEN FUER GANZZAHLIGE OPTIMIERUNGSMODELLE. DISSERTATION. BERLIN: FREIE UNIV. BERLIN. (1975).

SUHL 1976 -1 SUHL, U.
ALGORITHMUS ZUR LOESUNG EINER SPEZIELLEN KLASSE GANZZAHLIGER OPTIMIERUNGSMODELLE, EIN. IN: J. KOHLAS, O. SEIFERT, P. STAEHLY, H.-J. ZIMMERMANN (EDS): PROCEEDINGS IN OPERATIONS RESEARCH 5, 28-37. WUERZBURG: PHYSICA. (1976).

SUHL 1976 -2 SUHL, U.
HEURISTISCHE LOESUNGSVERFAHREN ZUR LOESUNG VON (0-1)-PROGRAMMEN MIT NICHT-NEGATIVEN KOEFFIZIENTEN. IN: H. N. DATHE, P. MERTENS, F. D. PESCHANEL ET AL. (EDS.): PROCEEDINGS IN OPERATIONS RESEARCH 6, 31-41. WUERZBURG: PHYSICA. (1976).

SUL' 1974 -1 SUL'GA, YU. N.
OPTIMIZATION OF THE MAINTENANCE PROCESS IN NETWORK SYSTEMS (RUSSIAN). IN: MATHEMATICAL METHODS IN ECONOMIC RESEARCH (RUSSIAN), 90-99. MOSCOW: IZDAT. NAUKA. (1974).

SUMA 1976 -1 SUMAROKOV, S. V.
SOLUTION METHOD FOR A NETWORK PROBLEM WITH SEVERAL EXTREMA (RUSSIAN). EKONOMIKA I MATEMATICESKIE METODY 12, 1016-1018. (1976).

SUMM 1976 -1 SUMMER, D. P.
MINIMAL LINE GRAPHS. GLASGOW MATHEMATICAL JOURNAL 17, 12-16. (1976).

SUPR 1975 -1 SUPRUNENKO, D. A.
ON THE TRAVELING SALESMAN PROBLEM (RUSSIAN). KIBERNETIKA, NO. 5, 121-124. (1975).

SUSS 1965 -1 SUSSENGUTH, E. H.
GRAPH THEORETIC ALGORITHM FOR MATCHING CHEMICAL STRUCTURES, A. THE JOURNAL OF CHEMICAL DOCUMENTATION 5, 36-43. (1965).

SUST 1975 -1 SUSTAKASVILI, I. I.
ALGORITHMIZATION OF THE SELECTION OF OPTIMAL VARIANTS IN STOCHASTIC NETWORK GRAPHS THAT INCLUDE CONNECTING AND DISCONNECTING PATHS (RUSSIAN). SAKHARTHVELOS SSR MECNIEREBATHA AKADEMII MOAMBE 79, 569-572. (1975).

SVAR 1975 -1 SVARTIN, S. M.
CERTAIN ALGORITHM FOR THE PLANNING OF MINIMAL TIME TRANSFERS (RUSSIAN), A. ZURNAL VYCISLITEL'NOI MATEMATIKI I MATEMATICESKOI FIZIKI 15, 321-329 + 538. (1975).

SWAI 1971 -1 SWAIN, R. W.
DECOMPOSITION ALGORITHM FOR A CLASS OF FACILITY LOCATION PROBLEM, A. DISSERTATION. ITHACA: CORNELL UNIVERSITY. (1971).

SWAI 1971 -2 SWAIN, R. W.
PARAMETRIC DECOMPOSITION APPROACH FOR THE SOLUTION OF UNCAPACITATED LOCATION PROBLEMS. COLUMBUS: OHIO STATE UNIV., DEPT. OF INDUSTRIAL ENGINEERING. (1971).

SWAI 1976 -1 SWAIN, R. W.
CENTRAL FACILITIES LOCATION WITH MULTIPLE PLANNING PERIODS. PAPER PRESENTED AT THE 49TH JOINT NATIONAL MEETING OF ORSA AND TIMS, PHILADELPHIA 1976, 15 P. (1976).

SWAN 1974 -1 SWANEK, J. J.
LINEAR INTEGER PROGRAMMING. MATHEMATICS RESEARCH CENTER REPORT 74-1, NRL REPORT 7797. WASHINGTON: NAVAL RESEARCH LABORATORY, 44 P. (1974).

SWANA 1976 -1 SWANSON, L. W.
N-DIMENSIONAL SCHEDULING. EVANSTON: NORTHWESTERN UNIV. GRADUATE SCHOOL OF MANAGEMENT, 7 P. (1976).

SWEE 1977 -1 SWEENEY, D. J.; MURPHY, R. A.
DECOMPOSITION PRINCIPLE FOR INTEGER PROGRAMS, A (REVISED VERSION). CINCINNATI: UNIV. OF CINCINNATI, DEPT. OF QUANTITATIVE ANALYSIS, 16 P. (1977).

SWEEA 1976 -1 SWEENEY, D. J.; TATHAM, R. L.
IMPROVED LONG-RUN MODEL FOR MULTIPLE WAREHOUSE LOCATION, AN. MANAGEMENT SCIENCE 22, 748-758. (1976).

SWEN 1967 -1 SWEN, G.
DEVELOPMENT OF THE SPANNING TREE CONCEPT IN GRAPH THEORY WITH APPLICATIONS TO THE SHORTEST ROUTE PROBLEM. DISSERTATION. PITTSBURGH: UNIV. OF PITTSBURGH. (1967).

SYSL 1972 -1 SYSLO, M.
ALGORITHMS 17-18: ALGORITHMS FOR SHORTEST PATH PROBLEMS. POLSKA AKADEMIA NAUK 13. (1972).

SYSL 1973 -1 SYSLO, M.
ALGORITHMS 20-22: FUNDAMENTAL SET OF CYCLES OF A GRAPH. POLSKA AKADEMIA NAUK 13, 399-409. (1973).

SYSL 1974 -1 SYSLO, M.
ALGORITHM 36: TRANSITIVE CLOSURE OF A GRAPH. POLSKA AKADEMIA NAUK 14, 477-480. (1974).

SYSLA 1975 -1 SYSLO, M. M.
APPLIED GRAPH THEORY II. EXTREMAL PATH PROBLEMS IN GRAPHS AND NETS (POLISH). ARCHIWUM AUTOMATYKI I TELEMECHANIKI 20, 287-300. (1975).

SYSLA 1976 -1 SYSLO, M. M.
GENERALIZATIONS OF THE STANDARD TRAVELLING SALESMAN PROBLEM. WROCLAW: UNIV. OF WROCLAW, INST. OF INFORMATICS, 11 P. (1976).

SZWA 1974 -1 SZWARC, W.
MIXED INTEGER LINEAR PROGRAMMING PROBLEM WHEN THE INTEGER VARIABLES ARE ZERO OR ONE, THE. MATHEMATICA 16, 183-187. (1974).

SZWA 1977 -1 SZWARC, W.
OPTIMAL TWO-MACHINE ORDERINGS IN THE 3 X N FLOW SHOP PROBLEM. OPERATIONS RESEARCH 25, 70-77. (1977).

SZWA 1977 -2 SZWARC, W.
SPECIAL CASES OF THE FLOW-SHOP PROBLEM. NAVAL RESEARCH LOGISTICS QUARTERLY 24, 483-492. (1977).

SZWAA 1977 -1 SZWARC, W.; HUTCHINSON, G. K.
JOHNSON'S APPROXIMATE METHOD FOR THE 3XN JOB SHOP PROBLEM. NAVAL RESEARCH LOGISTICS QUARTERLY 24, 153-158. (1977).

SZWAB 1976 -1 SZWARCFITER, J. L.; LAUER, P. E.
SEARCH STRATEGY FOR THE ELEMENTARY CYCLES OF A DIRECTED GRAPH, A. BIT 16, 192-204. (1976).

TABO 1969 -1 TABOURIER, Y.
SCHEDULING WITH PURELY DISJUNCTIVE CONSTRAINTS (FRENCH). REVUE FRANCAISE D'AUTOMATIQUE, INFORMATIQUE ET RECHERCHE OPERATIONELLE 3, 51-66. (1969).

TABO 1973 -1 TABOURIER, Y.
DIVERS ASPECTS DU PROBLEME DES PLUS COURTS CHEMINS PRINCIPAUX ALGORITHMES. DIRECTION SCIENTIFIQUE, GROUPE DE SPECIALITE MATHEMATIQUES, SYNTHESE ET FORMATION NO.77. (1973).

TAHA 1975 -1 TAHA, H. A.
INTEGER PROGRAMMING. NEW YORK: ACADEMIC PRESS, 380 P. (1975).

TAHA 1975 -2 TAHA, H. A.
NOTE ON THE RELATIONSHIP BETWEEN SOME SPECIALIZED BINARY ALGORITHMS AND THE ALGORITHM OF LAND AND DOIG, A. NAVAL RESEARCH LOGISTICS QUARTERLY 22, 609-614. (1975).

TAIN 1975 -1 TAINITER, M.
STATISTICAL THEORY OF CONNECTIVITY. I. BASIC DEFINITIONS AND PROPERTIES. DISCRETE MATHEMATICS 13, 391-398. (1975).

TAKA 1966 -1 TAKAHASHI, I.
TREE ALGORITHM FOR SOLVING RESOURCE ALLOCATION PROBLEMS. JOURNAL OF THE OPERATIONS RESEARCH SOCIETY OF JAPAN 8, 172-192. (1966).

TAKAA 1974 -1 TAKAHASHI, Y.
SEQUENCING MODEL WITH AN APPLICATION TO SPEED CLASS SEQUENCING IN AIR TRAFFIC CONTROL, A. JOURNAL OF THE OPERATIONS RESEARCH SOCIETY OF JAPAN 17, 1-28. (1974).

TALW 1967 -1 TALWAR, P. P.
SEQUENCING PROBLEM WITH UNCERTAIN JOB TIMES. JOURNAL OF THE OPERATIONS RESEARCH SOCIETY OF JAPAN 9, 93-97. (1967).

TAMI 1976 -1 TAMIR, A.
ON TOTALLY UNIMODULAR MATRICES. NETWORKS 6, 373-382. (1976).

TANA 1975 -1 TANAEV, V. S.; SHKURBA, V. V.
INTRODUCTION INTO SEQUENCING THEORY (RUSSIAN). MOSCOW: IZDAT. NAUKA, 256 P. (1975).

TAPI 1971 -1 TAPIERO, C. S.
TRANSPORTATION-LOCATION-ALLOCATION PROBLEMS OVER TIME. JOURNAL OF REGIONAL SCIENCE 11, 377-384. (1971).

TARJ 1971 -1 TARJAN, R. E.
DEPTH-FIRST SEARCH AND LINEAR GRAPH ALGORITHMS. IN: CONFERENCE RECORD 1971, 12TH ANNUAL SYMPOSIUM ON SWITCHING AND AUTOMATA THEORY, EAST LANSING 1971, 114-121. NORTHRIDGE: INSTITUTE OF ELECTRICAL AND ELECTRONICS ENGINEERS. (1971).

TARJ 1972 -1 TARJAN, R. E.
FINDING A MAXIMUM CLIQUE. ITHACA: CORNELL UNIV., DEPT. OF COMPUTER SCIENCE, 13 P. (1972).

TARJ 1973 -1 TARJAN, R. E.
ENUMERATION OF THE ELEMENTARY CIRCUITS OF A DIRECTED GRAPH. SIAM JOURNAL ON COMPUTING 2, 211-216. (1973).

TARJ 1973 -2 TARJAN, R. E.
FINDING DOMINATORS IN DIRECTED GRAPHS. IN: PROCEEDINGS OF THE SEVENTH ANNUAL PRINCETON CONFERENCE ON INFORMATION SCIENCES AND SYSTEMS, 414-418; ALSO PUBLISHED IN SIAM JOURNAL ON COMPUTING 3, 1974, 62-89. (1973).

TARJ 1974 -1 TARJAN, R. E.
GOOD ALGORITHM FOR EDGE-DISJOINT BRANCHING, A. INFORMATION PROCESSING LETTERS 3, 51-53. (1974).

TARJ 1975 -1 TARJAN, R. E.
FINDING EDGE-DISJOINT SPANNING TREES. IN: PROCEEDINGS OF THE EIGHTH HAWAII INTERNATIONAL CONFERENCE ON SYSTEM SCIENCES, UNIV. OF HAWAII, HONOLULU 1975, 251-252. NORTH HOLLYWOOD: WESTERN PERIODICALS. (1975).

TARJ 1975 -2 TARJAN, R. E.
FINDING MINIMUM SPANNING TREES. MEMORANDUM NO. ERL-M501. BERKELEY: UNIV. OF CALIFORNIA, COLLEGE OF ENGINEERING, 26 P. (1975).

TARJ 1976 -1 TARJAN, R. E.
EDGE-DISJOINT SPANNING TREES AND DEPTH-FIRST SEARCH. ACTA INFORMATICA 6, 171-186. (1976).

TARJ 1976 -2 TARJAN, R. E.
ITERATIVE ALGORITHMS FOR GLOBAL FLOW ANALYSIS. IN: J. F. TRAUB (ED.): ALGORITHMS AND COMPLEXITY, NEW DIRECTIONS AND RECENT RESULTS, 71-102. NEW YORK: ACADEMIC PRESS. (1976).

TARJ 1977 -1 TARJAN, R. E.
COMPLEXITY OF COMBINATORIAL ALGORITHMS. STAN-CS-77-609. STANFORD: STANFORD UNIV., COMPUTER SCIENCE DEPT., 86 P. (1977).

TARJ 1977 -2 TARJAN, R. E.
FINDING OPTIMUM BRANCHINGS. NETWORKS 7, 25-36. (1977).

TARJ 1978 -1 TARJAN, R. E.
COMPLEXITY OF MONOTONE NETWORKS FOR COMPUTING CONJUNCTIONS. IN: ALSPACH, B.;HELL, P.;MILLER, D. J.(EDS.): ALGORITHMIC ASPECTS OF COMBINATORICS. ANNALS OF DISCRETE MATHEMATICS 2, 121-134. (1978).

TARJA 1977 -1 TARJAN, R. E.; TROJANOWSKI, A. E.
FINDING A MAXIMUM INDEPENDENT SET. SIAM JOURNAL ON COMPUTING 6, 537-546. (1977).

TART 1976 -1 TARTAR, J.
OPTIMIZATION STRATEGIES FOR COMPUTER RESOURCE ALLOCATION. IN: PROCEEDINGS OF THE 5TH MANITOBA CONFERENCE ON NUMERICAL MATHEMATICS, WINNIPEG 1975, 109-136. CONGRESSUS NUMERANTIUM 16, WINNIPEG: UTILITAS MATHEMATICA. (1976).

TAYL 1975 -1 TAYLOR, P. E.
CELLOPHANE STOCK CUTTING PROBLEM AND A SEQUENTIAL DECISION MODEL, THE. PAPER PRESENTED AT THE 48TH JOINT NATIONAL MEETING OF ORSA AND TIMS, LAS VEGAS 1975, 14 P. (1975).

TCHE 1977 -1 TCHERKASSKY, B. V.
PROBLEM SOLUTION OF MULTIPRODUCT FLOWS IN A NETWORK (RUSSIAN). EKONOMIKA I MATEMATICHESKIE METODY 13, 143-151. (1977).

TEH 1976 -1 TEH, H. H.; SHEE, S. C.
ALGEBRAIC THEORY OF GRAPHS. SINGAPORE: LEE KONG CHIAN INSTITUTE OF MATHEMATICS AND COMPUTER SCIENCE, 167 P. (1976).

TEHR 1975 -1 TEHRANI, A.
ALGORITHME DE COLORATION, UN. CAHIERS DU CENTRE D'ETUDES DE RECHERCHE OPERATIONNELLE 17, 395-398. (1975).

TEPL 1975 -1 TEPLICKII, M. G.
BRANCH-AND-BOUND METHOD FOR SOLVING CERTAIN DISCRETE OPTIMAL CONTROL PROBLEMS (RUSSIAN), A. IZVESTIJA AKADEMII NAUK SSSR, TEHNICESKAJA KIBERNETIKA, NO. 5, 38-47; TRANSLATED AS ENGINEERING CYBERNETICS 13, NO. 5, 33-41. (1975).

TERN 1975 -1 TERNO, J.; GIERTH, G.; WAGNER, H.
BESTIMMUNG DES KUERZESTEN WEGES MITTELS DYNAMISCHER OPTIMIERUNG, DIE. WISSENSCHAFTLICHE ZEITSCHRIFT DER TECHNISCHEN UNIV. DRESDEN 24, 1375-1380. (1975).

TERZ 1974 -1 TERZI, D. G.
ON A PARTICULARITY OF SOME ALGORITHMS IN LINEAR INTEGER PROGRAMMING (RUSSIAN). UPRAVLJAEME SISTEMY, NO 12, 68-72. (1974).

THES 1975 -1 THESEN, A.
RECURSIVE BRANCH AND BOUND ALGORITHM FOR THE MULTIDIMENSIONAL KNAPSACK PROBLEM, A. NAVAL RESEARCH LOGISTICS QUARTERLY 22, 341-354. (1975).

THES 1976 -1 THESEN, A.
HEURISTIC SCHEDULING OF ACTIVITIES UNDER RESOURCE AND PRECEDENCE RESTRICTIONS. MANAGEMENT SCIENCE 23, 412-422. (1976).

THIR 1966 -1 THIRIET, L.; GAUSSENS, J.
METHODE DE CHOIX DES AFFECTATIONS ET DES INVESTISSEMENTS. IN: D. B. HERTZ, J. MELESE (EDS.): PROCEEDINGS OF THE FOURTH INTERNATIONAL CONFERENCE ON OPERATIONAL RESEARCH, 527-537. NEW YORK: WILEY. (1966).

THIS 1976 -1 THISSE, J.
MODELES MATHEMATIQUES D'OTIMISATION DE LA LOCALISATION DES ENTREPRISES. REPRINT. MONS: FACULTE UNIVERSITAIRE CATHOLIQUE DE MONS, 57-90. (1976).

THOM 1976 -1 THOMAS, M. E.
SURVEY OF THE STATE OF THE ART IN DYNAMIC PROGRAMMING, A. AIIE TRANSACTIONS 8, 59-69. (1976).

THOMA 1976 -1 THOMAS, R. E.
ON OPTIMUM PATH PROBLEMS. NETWORKS 6, 287-306. (1976).

THOMB 1976 -1 THOMPSON, G. L.
SOLUTIONS OF LARGE STRUCTURED TRAVELLING SALESMAN PROBLEMS (COPIES OF TRANSPARANCIES). MIAMI (50. ORSA-MEETING + TIMS, NOVEMBER 1976) 8 P. (1976).

THOMC 1976 -1 THOMSON, P. F.
ENUMERATION OF NETWORKS. INTERNATIONAL JOURNAL FOR NUMERICAL METHODS IN ENGINEERING 10, 230-235 (1976).

THOR 1966 -1 THORELLI, L. E.
ALGORITHM FOR COMPUTING ALL PATHS IN A GRAPH, AN. BIT 6, 347-349. (1966).

THUE 1976 -1 THUEMMLER, S.
KUERZESTE WEGE IN GEMISCHTEN NETZEN. ZEITSCHRIFT FUER OPERATIONS RESEARCH 20, SERIE B, B63-B76. (1976).

THUM 1975 -1 THUMER, R.
VERFAHREN ZUR AUSWAHL VON TEILNETZEN BEI DER BESTIMMUNG KUERZESTER WEGE IN VERKEHRSNETZEN, EIN. ZEITSCHRIFT FUER OPERATIONS RESEARCH 19, 149-153. (1975).

TIEN 1977 -1 TIEN, B. N.; HU, T. C.
ERROR BOUNDS AND THE APPLICABILITY OF THE GREEDY SOLUTION TO THE COIN-CHANGING PROBLEM. OPERATIONS RESEARCH 25, 404-418. (1977).

TIER 1974 -1 TIERNAN, J. C.
DUAL MULTIPLE-RESOURCE ASSIGNMENT ALGORITHM, A. IN: PROCEEDINGS OF THE 7TH HAWAII INTERNATIONAL CONFERENCE ON SYSTEM SCIENCES, HONOLULU 1974, 118-120. NORTH HOLLYWOOD: WESTERN PERIODICALS. (1974).

TIFF 1974 -1 TIFFIN, J. E.; FOLIE, G. M.
RECENT DEVELOPMENTS IN TRANSPORTATION ALGORITHMS. IN: J. M. BENNETT, M. A. NORSA, P. SUMMERSBEE (EDS.): PROCEEDINGS OF THE SIXTH AUSTRALIAN COMPUTER CONFERENCE, VOL. 2, SYDNEY 1974, 381-393. KINGSTON: AUSTRALIAN COMPUTER SOCIETY. (1974).

TILA 1976 -1 TILANUS, C. B.; GERHARDT, C.
APPLICATION OF CUTTING STOCK IN THE STEEL INDUSTRY, AN. IN: K. B. HALEY (ED.): OPERATIONAL RESEARCH '75. PROCEEDINGS OF THE 7TH IFORS INTERNATIONAL CONFERENCE ON OPERATIONAL RESEARCH, TOKYO, KYOTO 1975, 669-676. AMSTERDAM: NORTH-HOLLAND. (1976).

TILQ 1975 -1 TILQUIN, C.
ON SCHEDULING WITH EARLIEST STARTS AND DUE DATES ON A GROUP OF IDENTICAL MACHINES. NAVAL RESEARCH LOGISTICS QUARTERLY 22, 777-786. (1975).

TIMO 1975 -1 TIMOFEEV, E. A.
SOME ALGORITHMS FOR THE SOLUTION OF THE TRAVELLING SALESMAN PROBLEM ON THE PLANE WITH A LARGE NUMBER OF CITIES (RUSSIAN). JAROSLAVSKII UNIVERSITET, VESTNIK, VYP. 9, 115-118. (1975).

TIMO 1977 -1 TIMOFEEV, E. A.
CONSTRUCTION OF AN EDGE-WISE M-CONNECTED SKELETAL SUBGRAPH WITH MINIMAL LENGTH OF THE MAXIMAL EDGE (RUSSIAN), THE. DOKLADY AKADEMII NAUK SSSR 233, 1053-1055. (1977).

TIND 1977 -1 TIND, J.
BLOCKING AND ANTIBLOCKING POLYHEDRA. VANCOUVER (CONFERENCE ON DISCRETE OPTIMIZATION, AUGUST 1977) 20 P. (1977).

TIND 1977 -2 TIND, J.
ON ANTIBLOCKING SETS AND POLYHEDRA. IN: P. L. HAMMER, E. L. JOHNSON, B. H. KORTE, G. L. NEMHAUSER (EDS.): STUDIES IN INTEGER PROGRAMMING, 507-516. ANNALS OF DISCRETE MATHEMATICS 1, AMSTERDAM: NORTH-HOLLAND. (1977).

TINH 1976 -1 TINHOFER, G.
METHODEN DER ANGEWANDTEN GRAPHENTHEORIE. WIEN, NEW YORK: SPRINGER, 236 P. (1976).

TINH 1976 -2 TINHOFER, G.
ZUM ALGORITHMISCHEN NACHWEIS DER ISOMORPHIE VON ENDLICHEN GRAPHEN. IN: H. NOLTEMEIER (ED.): GRAPHEN, ALGORITHMEN, DATENSTRUKTUREN. ERGEBNISSE DES WORKSHOPS WG 76. 2. FACHTAGUNG UEBER GRAPHENTHEORETISCHE KONZEPTE DER INFORMATIK, GOETTINGEN 1976, 169-182. MUENCHEN: HANSER. (1976).

TOBI 1975 -1 TOBIN, R. L.
MINIMAL COMPLETE MATCHINGS AND NEGATIVE CYCLES. NETWORKS 5, 371-388. (1975).

TOBI 1977 -1 TOBIN, R. L.
EXTENSION OF DIAL'S ALGORITHM UTILIZING A MODEL OF TRIPMAKERS' PRECEPTIONS, AN. TRANSPORTATION RESEARCH 11, 337-342. (1977).

TODD 1976 -1 TODD, M. J.
COMBINATORIAL GENERALIZATION OF POLYTOPES, A. JOURNAL OF COMBINATORIAL THEORY 20, SERIES B, 229-242. (1976).

TODD 1977 -1 TODD, M. J.
NUMBER OF NECESSARY CONSTRAINTS IN AN INTEGER PROGRAM: A NEW PROOF OF SCARF'S THEOREM, THE. TECHNICAL REPORT NO. 355. ITHACA: CORNELL UNIV., SCHOOL OF OPERATIONS RESEARCH AND INDUSTRIAL ENGINEERING, 8 P. (1977).

TOFT 1975 -1 TOFT, B.
ON COLOUR-CRITICAL HYPERGRAPHS. IN: A. HAJNAL, R. RADO, V. T. SOS (EDS.): INFINITE AND FINITE SETS, COLLOQ. KESZTHELY 1973, VOL. 3, 1445-1457. COLLOQ. MATH. SOC. JANOS BOLYAI 10, AMSTERDAM: NORTH-HOLLAND. (1975).

TOLB 1975 -1 TOLBIN, R. L.
MINIMAL COMPLETE MATCHINGS AND NEGATIVE CYCLES, NETWORKS 5, 371-387. (1975).

TOLC 1976 -1 TOLCHAN, A. YA.
NUMERICAL PROCESSES OF GROWTH OF GRAPHS (RUSSIAN). PROBLEMY PEREDACI INFORMACII 12, 95-103. (1976).

TOME 1966 -1 TOMESCU, I.
METHOD FOR DETERMINING THE SHORTEST PATH BETWEEN TWO NODES OF A FINITE GRAPH (RUMANIAN), A. ANALELE UNIVERSITATII BUCURESTI, STIINTELE NATURII, MAT.-MEC. 15, 91-104. (1966).

TOME 1968 -1 TOMESCU, I.
SUR LE PROBLEME DU COLORIAGE DES GRAPHES GENERALISES. COMPTES RENDUS HEBDOMADAIRES DES SEANCES DE L'ACADEMIE DES SCIENCES 267, 250-252. (1968).

TOME 1975 -1 TOMESCU, I.
ALGORITHME POUR L'OBTENTION D'UNE CHAINE HAMILTONIENNE EN PARTANT DE L'ARBRE MINIMAL D'UN GRAPHE, UN. REVUE FRANCAISE D'AUTOMATIQUE, D'INFORMATIQUE ET DE RECHERCHE OPERATIONNELLE 9, NO. V-3, 5-12. (1975).

TOME 1975 -2 TOMESCU, I.
INTRODUCTION TO COMBINATORICS (TRANSLATED FROM RUMANIAN). LONDON, WELLINGBOROUGH: COLLET'S, 249 P. (1975).

TOME 1975 -3 TOMESCU, I.
PROBLEMES EXTREMAUX CONCERNANT LE NOMBRE DES COLORATIONS DES SOMMETS D'UN GRAPHE FINI. IN: B. ROY (ED.): COMBINATORIAL PROGRAMMING: METHODS AND APPLICATIONS. PROCEEDINGS OF THE NATO ADVANCED STUDY INST., VERSAILLES 1974, 327-336. DORDRECHT: REIDEL. (1975).

TOME 1976 -1 TOMESCU, I.
NOMBRE MAXIMAL DE COLORATIONS D'UN GRAPHE HAMILTONIEN, LE. DISCRETE MATHEMATICS 16, 353-360. (1976).

TOMI 1973 -1 TOMIZAWA, N.
ON A SPECIALIZATION SEQUENCE FROM GENERAL MATROIDS TO LADDER GRAPHS WITH SPECIAL EMPHASIS ON THE CHARACTERIZATION OF LADDER MATROIDS. RAAG RESEARCH NOTES NO. 191, 22 P. (1973).

TOML 1966 -1 TOMLIN, J. A.
LINEAR PROGRAMMING MODEL FOR THE ASSIGNMENT OF TRAFFIC, A. IN PROCEEDINGS OF THE THIRD CONFERENCE OF THE AUSTRALIAN ROAD RESEARCH BOARD, 3, PART 1, 263-271. (1966).

TONE 1976 -1 TONE, K.
PARAMETRIZATION OF BENDERS DECOMPOSITION, A. KEIO ENGINEERING REPORTS 29, NO. 5, 53-60. (1976).

TORR 1976 -1 TORRIERO, A.
DECOMPOSITION METHODS FOR THE TRAVELLING SALESMAN PROBLEM IN LARGE DIMENSIONS (ITALIAN). RICERCA OPERATIVA 6, NO. 1, 49-62. (1976).

TOTH 1977 -1 TOTH, P.
DYNAMIC PROGRAMMING ALGORITHM FOR THE ZERO-ONE KNAPSACK PROBLEM, A. SAN FRANCISCO (51. ORSA-MEETING + TIMS, MAY 1977) 10 P. (1977).

TOWN 1977 -1 TOWNSEND, W.
BRANCH-AND-BOUND METHOD FOR SEQUENCING PROBLEMS WITH LINEAR AND EXPONENTIAL PENALTY FUNCTIONS, A. OPERATIONAL RESEARCH QUARTERLY 28, 191-200. (1977).

TOWN 1977 -2 TOWNSEND, W.
MINIMISING THE MAXIMUM PENALTY IN THE TWO-MACHINE FLOW SHOP. MANAGEMENT SCIENCE 24, 230-234. (1977).

TOWS 1977 -1 TOWSNEND, W.
SEQUENCING N JOBS ON M MACHINES TO MINIMISE MAXIMUM TARDINESS: A BRANCH-AND-BOUND SOLUTION. MANAGEMENT SCIENCE 23, 1016-1018. (1977).

TRAN 1976 -1 TRAN DINH AM; TSUKIYAMA, S.; SHIRAKAWA, I.; OZAKI, H.
ALGORITHM FOR GENERATING ALL THE DIRECTED PATHS AND ITS APPLICATION, AN. INFORMATION PROCESSING IN JAPAN 16, 31-35. (1976).

TRIF 1973 -1 TRIFON, R.; LIVNAT, A.
SPATIAL ALLOCATION OF SCHOOLS OVER TIME IN CITIES, THE. REGIONAL AND URBAN ECONOMICS 2, 387-400. (1973).

TRIV 1976 -1 TRIVEDI, V. M.; WARNER, D. M.
BRANCH AND BOUND ALGORITHM FOR OPTIMUM ALLOCATION OF FLOAT NURSES, A. MANAGEMENT SCIENCE 22, 972-981. (1976).

TROT 1973 -1 TROTTER, L. E. JR.
USER'S INSTRUCTIONS FOR THE INTEGER PROGRAMMING CODE ENUMER8. MRC TECHNICAL SUMMARY REPORT 1347. MADISON: UNIV. OF WISCONSIN-MADISON, MATHEMATICS RESEARCH CENTER, 36 P. (1973).

TROT 1977 -1 TROTTER, L. E. JR.
ON THE GENERALITY OF MULTI-TERMINAL FLOW THEORY. IN: P. L. HAMMER, E. L. JOHNSON, B. H. KORTE, G. L. NEMHAUSER (EDS.): STUDIES IN INTEGER PROGRAMMING, 517-526. ANNALS OF DISCRETE MATHEMATICS 1, AMSTERDAM: NORTH-HOLLAND. (1977).

TROTA 1977 -1 TROTTER, L. E. JR.; MONMA, C. L.
ON A THEOREM OF FULKERSON. REPORT NO. 7775-OR. BONN: UNIV. BONN, INSTITUT FUER OEKONOMETRIE UND OPERATIONS RESEARCH, 5 P. (1977).

TROTB 1976 -1 TROTTER, W. T. JR.; BOGART, K. P.
MAXIMAL DIMENSIONAL PARTIALLY ORDERED SETS III: A CHARACTERIZATION OF HIRAGUCHI'S INEQUALITY FOR INTERVAL DIMENSION. DISCRETE MATHEMATICS 15, 389-400. (1976).

TROTB 1976 -2 TROTTER, W. T. JR.; BOGART, K. P.
ON THE COMPLEXITY OF POSETS. DISCRETE MATHEMATICS 16, 71-82. (1976).

TROTC 1976 -1 TROTTER, W. T. JR.; MOORE, J. I.; SUMNER, D. P.
DIMENSION OF A COMPARABILITY GRAPH, THE. PROCEEDINGS OF THE AMERICAN MATHEMATICAL SOCIETY 60, 35-38. (1976).

TROTD 1976 -1 TROTTER, W. T. JR.; MOORE, J. I. JR.
CHARACTERIZATION PROBLEMS FOR GRAPHS, PARTIALLY ORDERED SETS, LATTICES, AND FAMILIES OF SETS. DISCRETE MATHEMATICS 16, 361-382. (1976).

TRUB 1976 -1 TRUBIN, V. A.
EFFECTIVE ALGORITHM FOR THE SOLUTION OF THE ALLOCATION PROBLEM ON A NETWORK IN THE SHAPE OF A TREE (RUSSIAN), AN. DOKLADY AKADEMII NAUK SSSR 231, 547-550. (1976).

TRUB 1977 -1 TRUBIN, V. A.
UNIVERSALITY OF ONE CLASS OF QUADRATIC INTEGER PROBLEMS (RUSSIAN). KIBERNETIKA, NO. 2, 147. (1977).

TRUE 1976 -1 TRUEMPER, K.
ALGEBRAIC CHARACTERIZATIONS OF TOTALLY UNIMODULAR MATRICES: A FRAMEWORK. MIAMI (50. ORSA-MEETING + TIMS, NOVEMBER 1976) 21 P. (1976).

TRUE 1976 -2 TRUEMPER, K.
ALGEBRAIC CHARACTERIZATIONS OF UNIMODULAR MATRICES. MIAMI (50. ORSA-MEETING + TIMS, NOVEMBER 1976) 12 P. (1976).

TRUE 1976 -3 TRUEMPER, K.
EFFICIENT SCALING PROCEDURE FOR GAIN NETWORKS, AN. NETWORKS 6, 151-160. (1976).

TRUE 1977 -1 TRUEMPER, K.
ON MAX FLOWS WITH GAINS AND PURE MIN-COST FLOWS. SIAM JOURNAL ON APPLIED MATHEMATICS 32, 450-456. (1977).

****

VODO 1976 -1 VODOLAZOV, V. S.; MUCNIK, M. I.
CONDITIONS FOR THE SOLVABILITY OF TRANSPORT TYPE PROBLEMS WITH LIMITED CUT CAPACITIES (RUSSIAN). IN: MATHEMATICAL PROGRAMMING AND RELATED QUESTIONS. PROCEEDINGS OF THE SEVENTH WINTER SCHOOL, DROGOBYCH 1974, NUMERICAL METHODS (RUSSIAN), 158-175. MOSCOW: CENTRAL. EKONOM.-MAT. INST. AKAD. NAUK SSSR. (1976).

VOLO 1976 -1 VOLODARSKIJ, Y. A. M.; GABOVICH, E. YA.; ZAKHARIN, A. YA.
SOLVABLE CASE IN DISCRETE PROGRAMMING (RUSSIAN), A. IZVESTIJA AKADEMII NAUK SSSR, TEHNICESKAJA KIBERNETIKA, NO. 1, 34-44. (1976).

VOLOA 1976 -1 VOLOVIK, L. N.; SKACHEK, B. YA.
ON MODIFICATION OF ALGORITHM FOR TRAVELLING SALESMAN PROBLEM (RUSSIAN). ZURNAL VYCISLITEL'NOJ MATEMATIKI I MATEMATICESKOJ FIZIKI 16, 1318-1327. (1976).

VOLP 1976 -1 VOLPENTESTA, A.
FINITE GROUPS AND INTEGER LINEAR PROGRAMS. PISA: UNIV. DI PISA, DIPT. DI RICERCA OPERATIVA E SCIENZE STATISTICHE, 35 P. (1976).

VOLP 1977 -1 VOLPENTESTA, A.
UNIMODULAR SETS AND EXTREMUM PROBLEMS WITH INTEGER VARIABLES (ITALIAN). BOLLETINO DELLA UNIONE MATEMATICA ITALIANA 14, 165-173. (1977).

VOLY 1974 -1 VOLYNSKII, E. I.; VJATKIN, A. K.; KRASIN, L. A.
ASSIGNMENT PROBLEM FOR A PRODUCTION LINE WITH OPERATIVE STORAGE POINTS (RUSSIAN), THE. EKONOMIKA I MATEMATICESKIE METODY 10, 159-167. (1974).

VOTJ 1973 -1 VOTJAKOV, A. A.
EQUIVALENT TRANSFORMATIONS OF A LINEAR INTEGER PROBLEM (RUSSIAN). IN: A. A. FRIDMAN (ED.): STUDIES IN DISCRETE MATHEMATICS (RUSSIAN), 9-21. MOSCOW: IZDAT. NAUKA. (1973).

VU 1976 -1 VU, V. T.; TURNER, W. C.
SYSTEMS DESIGN FOR RURAL REFUSE COLLECTION. AIIE TRANSACTIONS 8, 84-95. (1976).

VYAT 1975 -1 VYATKIN, A. K.; KRASIN, L. A.
PROBLEM OF OPTIMIZING A HIERARCHICAL ORGANIZATION STRUCTURE (RUSSIAN), A. UPRAVLJAEME SISTEMY MACH., NO. 3, 74-77. (1975).

WAGN 1976 -1 WAGNER, R. A.
SHORTEST PATH ALGORITHM FOR EDGE-SPARSE GRAPHS, A. JOURNAL OF THE ASSOCIATION FOR COMPUTING MACHINERY 23, 50-57. (1976).

WALD 1976 -1 WALDVOGEL, J.
DATENSTRUKTUREN. IN: LIEBLING, T. M.;ROESSLER, M.(EDS.): KOMBINATORISCHE ENTSCHEIDUNGSPROBLEME: METHODEN UND ANWENDUNG. LECTURE NOTES IN ECONOMICS AND MATHEMATICAL SYSTEMS 153, BERLIN: SPRINGER, 170-183. (1976).

WALK 1976 -1 WALKER, R. C.
GRAPH-COLOURING THEOREM, A. MATHEMATICAL GAZETTE 60, NO 411, 54-57. (1976).

WALKA 1976 -1 WALKER, W. E.
HEURISTIC ADJACENT EXTREME POINT ALGORITHM FOR THE FIXED CHARGE PROBLEM, A. MANAGEMENT SCIENCE 22, 587-596. (1976).

WALL 1977 -1 WALLACH, Y.
SCHEDULING ALGORITHMS FOR CONCURRENT EXECUTION. COMPUTER JOURNAL 20, 132-136. (1977).

WALLA 1976 -1 WALLER, D. A.
GENERAL SOLUTION TO THE SPANNING TREE ENUMERATION PROBLEM IN ARBITRARY MULTIGRAPH JOINS. IEEE TRANSACTIONS ON CIRCUITS AND SYSTEMS CAS-23, 467-469. (1976).

WALS 1974 -1 WALSH, A. M.
ON FINDING A K-CLIQUE IN A RANDOM GRAPH. IN: PROCEEDINGS OF THE EIGHTH ANNUAL PRINCETON CONFERENCE ON INFORMATION SCIENCES AND SYSTEMS, 29-31. (1974).

WALSA 1976 -1 WALSH, T. S.
COMPLEXITE DES ALGORITHMES DE FLOT. IN: ANAL. APPL. INF. JOURN. COMB. INF., BORDEAUX 1975, 353-362. BORDEAUX: UNIV. BORDEAUX I. (1976).

WALT 1976 -1 WALTERS, J. R.
EXTENSION TO THE ASSORTMENT PROBLEM IN WHICH COMPONENTS ARE GROUPED TO FORM ASSEMBLIES, AN. OPERATIONAL RESEARCH QUARTERLY 27, 169-176. (1976).

WALU 1975 -1 WALUKIEWICZ, S.
EQUIVALENT CONSTRAINT USING AN OBJECTIVE FUNCTION, AN. BULLETIN DE L'ACADEMIE POLONAISE DES SCIENCES, SERIE DES SCIENCES TECHNIQUES 23, 445-451. (1975).

WALU 1975 -2 WALUKIEWICZ, S.
SIZE REDUCTION OF A BINARY KNAPSACK PROBLEM, THE. BULLETIN DE L'ACADEMIE POLONAISE DES SCIENCES, SERIE DES SCIENCES TECHNIQUES 23, 453-458. (1975).

WANG 1976 -1 WANG, D. L.
CONSTRUCTION OF A MAXIMALLY EDGE-CONNECTED GRAPH WITH PRESCRIBED DEGREES. STUDIES IN APPLIED MATHEMATICS 55, 87-92. (1976).

WANGA 1973 -1 WANG, D.-L.; KLEITMAN, D. J.
CONSTRUCTING MAXIMALLY CONNECTED GRAPHS HAVING A GIVEN DEGREE SEQUENCE. IN: PROCEEDINGS OF THE SEVENTH ANNUAL PRINCETON CONFERENCE ON INFORMATION SCIENCES AND SYSTEMS, 410-413. (1973).

WANGB 1974 -1 WANG, J. C.; MICKLE, M. H.; HOELZEMAN, R. G.
SET COVERING APPLICATION IN MINIMUM ROUTE GENERATION, A. KYBERNETES 3, NO. 4, 207-215. (1974).

WARR 1977 -1 WARREN, H. S. JR.
THREE SIMPLE NODE LIST ALGORITHMS. IBM RESEARCH REPORT 27411. YORKTOWN HEIGHTS: IBM RESEARCH CENTER, 11 P. (1977).

WASO 1976 -1 WASOWSKI, J.
SELECTION METHOD FOR A CERTAIN PROBLEM OF ORDERING PRODUCTS BY 3 MACHINES (RUSSIAN), A. DEMONSTRATIO MATHEMATICA 9, 609-613. (1976).

WASS 1975 -1 WASSON, W. D.; MAC ISAAC, R.
NEW SPANNING TREE ALGORITHM, A. IN: B. L. HARTNELL, H. C. WILLIAMS (EDS.): PROCEEDINGS OF THE 4TH MANITOBA CONFERENCE ON NUMERICAL MATHEMATICS, WINNIPEG 1974, 375-385. CONGRESSUS NUMERANTIUM 12, WINNIPEG: UTILITAS MATHEMATICA. (1975).

WATE 1976 -1 WATERS, W. G. II; TSE, A.
BIBLIOGRAPHY OF THE MAJOR TRANSPORTATION JOURNALS 1965-1974, A. THE LOGISTICS AND TRANSPORTATION REVIEW 12, NO. 2 (SPECIAL ISSUE), 1-153. (1976).

WATT 1963 -1 WATTLEWORTH, J. A.; SHULDINER, P. W.
ANALYTICAL METHODS IN TRANSPORTATION; LEFT-TURN PENALTIES IN TRAFFIC ASSIGNMENT MODELS. JOURNAL OF THE ENGINEERING AND MECHANICAL DIVISION OF THE AMERICAN SOCIETY OF CIVIL ENGINEERS 89, 97. (1963).

WDOW 1973 -1 WDOWIAK, J.
SOLUTION OF THE FIXED CHARGE PROBLEM (POLISH), THE. PRZEGLAD STATYSTYCZNY 20, 231-252. (1973).

WEEK 1976 -1 WEEKS, J. K.; FRYER, J. S.
SIMULATION STUDY OF OPERATING POLICIES IN A HYPOTHETICAL DUAL-CONSTRAINED JOB SHOP, A. MANAGEMENT SCIENCE 22, 1362-1371. (1976).

WEGE 1974 -1 WEGENER, M.
ROUTENSUCHALGORITHMEN IN POLIS. REPORT. FRANKFURT: BATELLE-INSTITUT. (1974).

****

WEGL 1975 -1 WEGLARZ, J.
CERTAIN NETWORK PROGRAMMING PROBLEM (POLISH), A. PRZEGLAD STATYSTYCZNY 22, 549-555. (1975).
WEGL 1976 -1 WEGLARZ, J.
TIME-OPTIMAL CONTROL OF RESOURCE ALLOCATION IN A COMPLEX OF OPERATIONS FRAMEWORK. IEEE TRANSACTIONS ON SYSTEMS, MAN, AND CYBERNETICS 6, 783-788. (1976).
WEGLA 1976 -1 WEGLARZ, J.; SLOWINSKI, R.
COMPUTATIONAL ASPECTS OF A CERTAIN CLASS OF TIME-OPTIMAL RESOURCE ALLOCATION PROBLEMS IN A COMPLEX OF OPERATIONS. FOUNDATIONS OF CONTROL ENGINEERING 1, 123-133. (1976).
WEIG 1973 -1 WEIGAND, M. M.
UEBER EIN VERFAHREN ZUR BESTIMMUNG VON K-KUERZESTEN WEGEN IN EINEM GRAPHEN. DISSERTATION. STUTTGART: UNIV. STUTTGART, INST. FUER INFORMATIK, 44 P. (1973).
WEIG 1976 -1 WEIGAND, M. M.
ALGORITHMUS 27. EIN NEUER ALGORITHMUS ZUR BESTIMMUNG VON K-KUERZESTEN WEGEN IN EINEM GRAPHEN. COMPUTING 16, 139-152. (1976).
WEIN 1977 -1 WEINBERG, F.
NECESSARY AND SUFFICIENT CONDITION FOR THE AGGREGATION OF LINEAR DIOPHANTINE EQUATIONS, A. ZEITSCHRIFT FUER ANGEWANDTE MATHEMATIK UND PHYSIK 28, 679-696. (1977).
WEINA 1960 -1 WEINBERG, L.
SIMPLE AND EFFICIENT ALGORITHM FOR DETERMINING ISOMORPHISM OF PLANAR TRIPLE-CONNECTED GRAPHS. IEEE TRANSACTIONS ON COMMUNICATIONS TECHNOL. CT-13, 142-148. (1960).
WEINA 1965 -1 WEINBERG, L.
PLANE REPRESENTATIONS AND CODES FOR PLANAR GRAPHS. IN: PROCEEDINGS OF THE THIRD ALLERTON CONFERENCE ON CIRCUIT AND SYSTEM THEORY, UNIVERSITY OF ILLINOIS, 733-744. (1965).
WEINA 1975 -1 WEINBERG, L.
DUALITY IN NETWORKS: ROSES, BOUQUETS, AND CUT-SETS; TREES, FORESTS AND POLYGONS. NETWORKS 5, 179-201. (1975).
WEINA 1977 -1 WEINBERG, L.
MATROIDS, GENERALIZED NETWORKS, AND ELECTRIC NETWORK SYNTHESIS. JOURNAL OF COMBINATORIAL THEORY 23, SERIES B, 106-126. (1977).
WEINB 1976 -1 WEINBERGER, D. B.
NETWORK FLOWS, MINIMUM COVERINGS, AND THE FOUR-COLOR CONJECTURE. OPERATIONS RESEARCH 24, 272-290. (1976).
WEINB 1976 -2 WEINBERGER, D. B.
TRANSVERSAL MATROID INTERSECTIONS AND RELATED PACKINGS. MATHEMATICAL PROGRAMMING 11, 164-176. (1976).
WEINC 1975 -1 WEINER, P.
HEURISTICS. NETWORKS 5, 101-104. (1975).
WEIND 1974 -1 WEINSTEIN, J.
LARGE MATCHINGS IN GRAPHS. CANADIAN JOURNAL OF MATHEMATICS 26, 1498-1508. (1974).
WEINE 1975 -1 WEINSTEIN, J. M.
RECONSTRUCTING COLORED GRAPHS. PACIFIC JOURNAL OF MATHEMATICS 57, 307-314. (1975).
WEINF 1976 -1 WEINTRAUB, A.
OPTIMAL FLOWS IN NETWORKS AND GAMES: THE MULTICOMMODITY FLOW PROBLEM IN INTEGERS. SERIE DOCUMENTOS DE TRABAJO 76/21/C. SANTIAGO: UNIV. DE CHILE, FACULTAD DE CIENCISAS FISICAS Y MATEMATICAS, 35 P. (1976).
WELL 1976 -1 WELLS, D. W.
POWER SYSTEM SCHEDULING USING INTEGER PROGRAMMING. IN: L. C. W. DIXON (ED.): OPTIMIZATION IN ACTION, PROCEEDINGS OF THE CONFERENCE ON OPTIMIZATION IN ACTION, BRISTOL 1975, 467-477.: LONDON: ACADEMIC PRESS. (1976).
WELLA 1971 -1 WELLS, M.
ELEMENTS OF COMBINATORIAL COMPUTING. NEW YORK: PERGAMON PRESS. (1971).
WELS 1970 -1 WELSH, D. J. A.
ON MATROID THEOREMS OF EDMONDS AND RADO. JOURNAL OF THE LONDON MATHEMATICAL SOCIETY 45, 251-256. (1970).
WELS 1976 -1 WELSH, D. J. A.
MATROID THEORY. L.M.S. MONOGRAPHS, 8. LONDON, NEW YORK, SAN FRANCISCO: ACADEMIC PRESS, 433 P. (1976).
WEND 1977 -1 WENDELL, R. E.; HURTER, A. P. JR.; LOWE, T. J.
EFFICIENT POINTS IN LOCATION PROBLEMS. AIIE TRANSACTIONS 9, 238-246. (1977).
WENDA 1976 -1 WENDLER, K.
ERWEITERTES TRANSPORTPROBLEM MIT PARAMETERN ALS KAPAZITAETSBESCHRAENKUNGEN, EIN. MATHEMATISCHE OPERATIONSFORSCHUNG UND STATISTIK 7, 859-870. (1976).
WERR 1974 -1 WERRA, D. DE
PARTITIONS OF GRAPHS INTO COVERINGS AND HYPERGRAPHS INTO TRANSVERSALS. IN: COMMENTARII MATHEMATICI HELVETICI, VOL. 49, FASC. 2, 174-178. (1974).
WERR 1975 -1 WERRA, D. DE
FEW REMARKS ON CHROMATIC SCHEDULING, A. IN: B. ROY (ED.): COMBINATORIAL PROGRAMMING: METHODS AND APPLICATIONS. PROCEEDINGS OF THE NATO ADVANCED STUDY INST., VERSAILLES 1974, 337-342. DORDRECHT: REIDEL. (1975).
WERR 1975 -2 WERRA, D. DE
GRAPHS AND SCHEDULING. PAPER PRESENTED AT THE FIRST EUROPEAN CONGRESS ON OPERATIONS RESEARCH, EURO 1, BRUSSELS 1975. (1975).
WERR 1975 -3 WERRA, D. DE
HOW TO COLOR A GRAPH. IN: B. ROY (ED.): COMBINATORIAL PROGRAMMING: METHODS AND APPLICATIONS. PROCEEDINGS OF THE NATO ADVANCED STUDY INST., VERSAILLES 1974, 305-326. DORDRECHT: REIDEL. (1975).
WERR 1975 -4 WERRA, D. DE
ON A PARTICULAR CONFERENCE SCHEDULING PROBLEM. INFOR 13, 308-315. (1975).
WERR 1975 -5 WERRA, D. DE
ON GOOD AND EQUITABLE EDGE COLORINGS. CAHIERS DU CENTRE D'ETUDES DE RECHERCHE OPERATIONNELLE 17, 417-426. (1975).
WERR 1976 -1 WERRA, D. DE
COLOR-FEASIBLE SEQUENCES OF ALMOST BIPARTITE GRAPHS. O.R. REPORT 19. LAUSANNE: ECOLE POLYTECHNIQUE FEDERALE DE LAUSANNE, DEPT. DE MATHEMATIQUES, 9 P. (1976).
WERR 1976 -2 WERRA, D. DE
EVERYTHING YOU ALWAYS WANTED TO KNOW ABOUT S.EX. IN: J. KOHLAS, O. SEIFERT, P. STAEHLY, H.-J. ZIMMERMANN (EDS): PROCEEDINGS IN OPERATIONS RESEARCH 5, 104-107. WUERZBURG: PHYSICA. (1976).
WERR 1976 -3 WERRA, D. DE
MULTIGRAPHS WITH QUASIWEAK ODD CYCLES. O.R. REPORT 24. LAUSANNE: ECOLE POLYTECHNIQUE FEDERALE DE LAUSANNE, DEPT. DE MATHEMATIQUES, 15 P. (1976).
WERR 1976 -4 WERRA, D. DE
ON A MULTI-PERIOD ASSIGNMENT PROBLEM. O.R. REPORT 21. LAUSANNE: ECOLE POLYTECHNIQUE FEDERALE DE LAUSANNE, DEPT. DE MATHEMATIQUES, 20 P. (1976).

****

WERR 1976 -5 WERRA, D. DE
REMARKS ON THE COLOR-FEASIBLE SEQUENCES OF A MULTIGRAPH. O.R. REPORT 20. LAUSANNE: ECOLE POLYTECHNIQUE FEDERALE DE LAUSANNE, DEPT. DE MATHEMATIQUES, 16 P. (1976).

WERR 1976 -6 WERRA, D. DE
SOME REMARKS ON GOOD COLORATIONS. JOURNAL OF COMBINATORIAL THEORY 21, SERIES B, 57-64. (1976).

WERR 1977 -1 WERRA, D. DE
SOME COLORING TECHNIQUES. IN: P. L. HAMMER, E. L. JOHNSON, B. H. KORTE, G. L. NEMHAUSER (EDS.): STUDIES IN INTEGER PROGRAMMING, 179-184. ANNALS OF DISCRETE MATHEMATICS 1, AMSTERDAM: NORTH-HOLLAND. (1977).

WERR 1977 -2 WERRA, D. DE
UNIQUENESS OF COLORINGS. ARCHIV DER MATHEMATIK 28, 330-336. (1977).

WESO 1974 -1 WESOLOWSKY, G. O.
REJOINDER. MANAGEMENT SCIENCE 20, 1312. (1974).

WESO 1977 -1 WESOLOWSKY, G. O.
PROBABILISTIC WEIGHTS IN THE ONE-DIMENSIONAL FACILITY LOCATION PROBLEM. MANAGEMENT SCIENCE 24, 224-229. (1977).

WESOA 1971 -1 WESOLOWSKY, G. O.; LOVE, R. F.
OPTIMAL LOCATION OF NEW FACILITIES USING RECTANGULAR DISTANCES, THE. OPERATIONS RESEARCH 19, 124-129. (1971).

WESOB 1975 -1 WESOLOWSKY, G. O.; TRUSCOTT, W. G.
MULTIPERIOD LOCATION-ALLOCATION PROBLEM WITH RELOCATION OF FACILITIES, THE. MANAGEMENT SCIENCE 22, 57-65. (1975).

WESOB 1976 -1 WESOLOWSKY, G. O.; TRUSCOTT, W. G.
DYNAMIC LOCATION OF MULTIPLE FACILITIES FOR A CLASS OF CONTINUOUS PROBLEMS. AIIE TRANSACTION 8, 78-83. (1976).

WESS 1972 -1 WESSELS, H.
VERGLEICH VERSCHIEDENER ANSAETZE ZUR TRIANGULATION VON INPUT-OUTPUT-MATRIZEN. IN: H. H. ASSMUSS, P. BRUCKER, J. FROHN ET AL. (EDS.): ANWENDUNGEN STATISTISCHER UND MATHEMATISCHER METHODEN AUF SOZIALWISSENSCHAFTLICHE PROBLEME, 133-148. WUERZBURG: PHYSICA. (1972).

WEST 1977 -1 WESTERBERG, C.-H.; BJORKLUND, B.; HULTMAN, E.
APPLICATION OF MIXED INTEGER PROGRAMMING IN A SWEDISH STEEL MILL, AN. INTERFACES 7, NO. 2, 39-43. (1977).

WHIT 1977 -1 WHITAKER, R. A.
THREE ALGORITHMS FOR CALCULATING SOME OR ALL OF THE SHORTEST PATHS IN A SPARSE NETWORK. GEOGRAPHICAL ANALYSIS 9, 266-276. (1977).

WHITA 1977 -1 WHITE, C. H.; WILSON, R. C.
SEQUENCE DEPENDENT SET-UP TIMES AND JOB SEQUENCING. INTERNATIONAL JOURNAL OF PRODUCTION RESEARCH 15, 191-202. (1977).

WHITB 1977 -1 WHITE, D. E.; WILLIAMSON, S. G.
RECURSIVE MATCHING ALGORITHMS AND LINEAR ORDERS ON THE SUBSET LATTICE. JOURNAL OF COMBINATORIAL THEORY 23, SERIES A, 117-127. (1977).

WHITC 1975 -1 WHITE, D. J.
NOTE ON FACULTY TIMETABLING, A. OPERATIONAL RESEARCH QUARTERLY 26, 875-878. (1975).

WHITD 1967 -1 WHITE, L. J.
MINIMUM PARTITIONS AND THE SHANNON SWITCHING GAME. UNPUBLISHED TECH. REPORT, ANN ARBOR: UNIV. OF MICHIGAN, SYSTEMS ENGINEERING LAB. (1967).

WHITD 1967 -2 WHITE, L. J.
PARAMETRIC STUDY OF MATCHINGS AND COVERINGS IN WEIGHTED GRAPHS, A. DISSERTATION. ANN ARBOR: UNIV. OF MICHIGAN. (1967).

WHITE 1966 -1 WHITE, M.
PROJECT SCHEDULING WITH LIMITED RESOURCES. EXPERIMENT IN HEURISTIC PROGRAMMING. P193-9,3 RD PROCEEDINGS OF THE AUSTRALIAN COMPUTER CONFERENCE. (1966).

WHITF 1969 -1 WHITE, W. W.
TRAVELING SALESMAN ALGORITHM. DISSERTATION. NAVAL POSTGRADUATE SCHOOL, 62 P. (1969).

WHITG 1973 -1 WHITEHEAD, E. G. JR.
COMBINATORIAL ALGORITHMS. NEW YORK: COURANT INST. OF MATHEMATICAL SCIENCES, NEW YORK UNIV., 104 P. (1973).

WILF 1978 -1 WILF, H. S.
UNIFIED SETTING FOR SELECTION ALGORITHMS (II), A. IN: ALSPACH, B.;HELL, P.;MILLER, D. J.(EDS.): ALGORITHMIC ASPECTS OF COMBINATORICS. ANNALS OF DISCRETE MATHEMATICS 2, 135-148. (1978).

WILL 1974 -1 WILLE, D.
KOMBINATORISCHE UNTERSUCHUNGEN AN (0,1)-MATRIZEN, EINIGE. ARCHIV DER MATHEMATIK 25, 216-218. (1974).

WILLA 1975 -1 WILLIAMS, H. P.
FORMULATION OF MATHEMATICAL PROGRAMMING MODELS, THE. OMEGA 3, 551-556. (1975).

WILLA 1976 -1 WILLIAMS, H. P.
FOURIER-MOTZKIN ELIMINATION EXTENSION TO INTEGER PROGRAMMING PROBLEMS. JOURNAL OF COMBINATORIAL THEORY 21, SERIES A, 118-123. (1976).

WILLA 1977 -1 WILLIAMS, H. P.
LOGICAL PROBLEMS AND INTEGER PROGRAMMING. BULLETIN OF THE INSTITUTE OF MATHEMATICS AND ITS APPLICATIONS 13, 18-20. (1977).

WILLB 1975 -1 WILLIAMS, T.
IMPROVED SHORTEST PATH ALGORITHM FOR NETWORKS IN WHICH THE MAXIMUM OUTWARD DEGREE IS LESS THAN N/2, AN. PAPER PRESENTED AT THE 1975 ORSA/TIMS MEETING, LAS VEGAS 1975. (1975).

WILLC 1972 -1 WILLIAMS, T. A.; WHITE, G. P.
COMPARING SHORTEST ROUTE ALGORITHMS FOR DIRECTED GRAPHS. PAPER PRESENTED AT THE 42ND NATIONAL ORSA-MEETING, ATLANTIC CITY 1972. (1972).

WILLC 1972 -2 WILLIAMS, T. A.; WHITE, G. P.
NOTE ON YEN'S ALGORITHM FOR FINDING THE LENGTH OF ALL COMPARING SHORTEST ROUTE ALGORITHMS FOR DIRECTED GRAPHS, A. PAPER PRESENTED AT THE 42ND NATIONAL ORSA-MEETING, ATLANTIC CITY 1972. (1972).

WILLD 1976 -1 WILLIAMSON, S. G.
RANKING ALGORITHMS FOR LISTS OF PARTITIONS. SIAM JOURNAL ON COMPUTING 5, 602-617. (1976).

WILLE 1976 -1 WILLIS, R. J.; HASTINGS, N. A. J.
PROJECT SCHEDULING WITH RESOURCE CONSTRAINTS USING BRANCH AND BOUND METHODS. OPERATIONAL RESEARCH QUARTERLY 27, 341-350. (1976).

WILS 1973 -1 WILSON, A. G.
FURTHER DEVELOPMENTS OF ENTROPY MAXIMISING TRANSPORT MODELS. TRANSPORTATION PLANNING AND TECHNIQUES 1, 183-193. (1973).

WILSA 1975 -1 WILSON, D.
MEAN COST APPROXIMATION FOR TRANSPORTATION PROBLEMS WITH STOCHASTIC DEMAND, A. NAVAL RESEARCH LOGISTICS QUARTERLY 22, 181-187. (1975).

****

WILSB 1974 -1 WILSON, D.; SMITH, R. L.
LOCATION OF STOCHASTIC SERVICES, THE. DISSERTATION. TEMPE: ARIZONA STATE UNIV., DEPT. OF INDUSTRIAL ENGINEERING, COLLEGE OF ENGINEERING AND APPLIED SCIENCES. (1974).

WILSC 1973 -1 WILSON, R. J.
INTRODUCTION TO MATROID THEORY, AN. THE AMERICAN MATHEMATICAL MONTHLY 80, 500-525. (1973).

WIN 1975 -1 WIN, S.
EXISTENZ VON GERUESTEN MIT VORGESCHRIEBENEM MAXIMALGRAD IN GRAPHEN. ABHANDLUNGEN AUS DEM MATHEMATISCHEN SEMINAR DER UNIV. HAMBURG 43, 263-267. (1975).

WING 1961 -1 WING, O.
ALGORITHMS TO FIND THE MOST RELIABLE PATH IN A NETWORK. IRE TRANSACTIONS ON CIRCUIT THEORY CT-8, 78-79. (1961).

WINGA 1965 -1 WING, O.; GOLDMAN, A. J.
MOST PROFITABLE ROUTINE BEFORE MAINTENANCE. BULLETIN OF ORSA, B82. (1965).

WITT 1978 -1 WITTEN, P.; ZIMMERMANN, H.-G.
STOCHASTISCHE TRANSPORTPROBLEME. ZEITSCHRIFT FUER OPERATIONS RESEARCH 22, SERIE A: THEORIE, 55-68. (1978).

WOEH 1973 -1 WOEHNER, W.
GUETERZUGBILDUNG ALS KOMBINATORISCHE OPTIMIERUNGSAUFGABE. WISSENSCHAFTLICHE ZEITSCHRIFT DER HOCHSCHULE FUER VERKEHRSWESEN "FRIEDRICH LIST" IN DRESDEN. DIE ANWENDUNG MATHEMATISCHER METHODEN IM TRANSPORT- UND NACHRICHTENWESEN 20, 831-840. (1973).

WOLS 1975 -1 WOLSEY, L. A.
VALID INEQUALITIES AND SUPERADDITIVITY FOR 0-1 INTEGER PROGRAMS. CORE DISCUSSION PAPER, 7523. HEVERLEE: CENTER FOR OPERATIONS RESEARCH & ECONOMETRICS, 28 P. (1975).

WOLS 1976 -1 WOLSEY, L. A.
FACETS AND STRONG VALID INEQUALITIES FOR INTEGER PROGRAMS. OPERATIONS RESEARCH 24, 367-372. (1976).

WOLS 1976 -2 WOLSEY, L. A.
FURTHER FACET GENERATING PROCEDURES FOR VERTEX PACKING POLYTOPES. MATHEMATICAL PROGRAMMING 11, 158-163. (1976).

WOLS 1976 -3 WOLSEY, L. A.
NUCLEOLUS AND KERNEL FOR SIMPLE GAMES OR SPECIAL VALID INEQUALITIES FOR 0-1 LINEAR INTEGER PROGRAMS, THE. INTERNATIONAL JOURNAL OF GAME THEORY 5, 227-238. (1976).

WONG 1976 -1 WONG, C. K.
SOME DISCRETE OPTIMIZATION PROBLEMS WITH SECONDARY STORAGE APPLICATIONS. IBM RESEARCH REPORT RC 6228. YORKTOWN HEIGHTS: IBM RESEARCH CENTER, 38 P. (1976).

WONGA 1976 -1 WONG, C. K.; YAO, A. C.-C.
COMBINATORIAL OPTIMIZATION PROBLEM RELATED TO DATA SET ALLOCATION, A. REVUE FRANCAISE D'AUTOMATIQUE, D'INFORMATIQUE ET DE RECHERCHE OPERATIONNELLE 10, SUPPLEMENT AU NO. 5, 83-95. (1976).

WONGB 1971 -1 WONG, C. K.; YUE, P. C.
UPPER-BOUNDS FOR VARIOUS TYPES OF PATH LENGTHS OF BINARY TREES. IN: PROCEEDINGS OF THE FIFTH ANNUAL PRINCETON CONFERENCE ON INFORMATION SCIENCES AND SYSTEMS, 368-372. (1971).

WONGC 1976 -1 WONG, R.
SURVEY OF SOME NETWORK DESIGN PROBLEMS, A. WORKING PAPER OR-053-76, OPERATIONS RESEARCH CENTER, M.I.T. (1976).

WONGD 1976 -1 WONGSEELASHOTE, A.
ALGEBRA FOR DETERMINING ALL PATH-VALUES IN A NETWORK WITH APPLICATION TO K-SHORTEST-PATHS PROBLEMS, AN. NETWORKS 6, 307-334. (1976).

WOOD 1974 -1 WOODALL, D. R.
EXCHANGE THEOREM FOR BASES OF MATROIDS, AN. JOURNAL OF COMBINATORIAL THEORY 16, SERIES B, 227-228. (1974).

WOOD 1975 -1 WOODALL, D. R.
INDUCTION OF MATROIDS BY GRAPHS, THE. JOURNAL OF THE LONDON MATHEMATICAL SOCIETY 10, 27-35. (1975).

WOOD 1975 -2 WOODALL, D. R.
MAXIMAL CIRCUITS OF GRAPHS II. STUDIA SCIENTIARUM MATHEMATICARUM HUNGARICA 10, NO. 1-2. 103-109. (1975).

WOOD 1976 -1 WOODALL, D. R.
MAXIMAL CIRCUITS OF GRAPHS, I. ACTA MATHEMATICA ACADEMIA SCIENTIARUM HUNGARICAE 28, 77-80. (1976).

WOOD 1977 -1 WOODALL, D. R.
NOTE ON RANK FUNCTIONS AND INTEGER PROGRAMMING, A. DISCRETE MATHEMATICS 17, 215-218. (1977).

WRIG 1975 -1 WRIGHT, E. M.
PROBABILITY OF CONNECTEDNESS OF A LARGE UNLABELLED GRAPH, THE. JOURNAL OF THE LONDON MATHEMATICAL SOCIETY 11, 13-16. (1975).

WRIG 1976 -1 WRIGHT, E. M.
PROPORTION OF UNLABELLED GRAPHS WHICH ARE HAMILTONIAN, THE. BULLETIN OF THE LONDON MATHEMATICAL SOCIETY 8, NO. 24, 241-244. (1976).

YAGE 1969 -1 YAGED, B. J.
ROUTING INFORMATION IN MINIMUM COST NETWORKS. IN: PROCEEDINGS OF THE THIRD ANNUAL PRINCETON CONFERENCE ON INFORMATION SCIENCES AND SYSTEMS, 13-15. (1969).

YAGI 1975 -1 YAGIZ, OE.
HEURISTICALLY-AIDED SINGLE BRANCH SEARCH PROCEDURES FOR PLANT LOCATION PROBLEMS. MIDDLE EAST TECHNICAL UNIV., STUDIES IN DEVELOPMENT, NO. 9, 95-132. (1975).

YAKI 1966 -1 YAKIMOV, R. M.
FUNCTIONAL EQUATION DESCRIBING THE OPTIMAL SEQUENCING OF PROCESSING OF ITEMS (RUSSIAN). SBORNIK TRUDOV VSESOJUZNOGO ZAOCNOGO POLITEHNICESKOGO INSTITUTA, 21, 62-71. (1966).

YALO 1976 -1 YALOVIJ, V. V.
ON SOME PROBLEMS IN THE ALLOCATION OF RESOURCES (UKRAINIAN). DOPOVIDI AKADEMII NAUK UKRAINS'KOI RSR, SERIJA A, NO. 11, 982-983. (1976).

YAMA 1977 -1 YAMAMOTO, M.
APPROXIMATE SOLUTION OF MACHINE SCHEDULING PROBLEMS BY DECOMPOSITION METHOD, AN. INTERNATIONAL JOURNAL OF PRODUCTION RESEARCH 15, 599-608. (1977).

YAO 1975 -1 YAO, A. C.
O (|E|LOG LOG|V|) ALGORITHM FOR FINDING MINIMUM SPANNING TREES, AN. INFORMATION PROCESSING LETTERS 4, 21-23. (1975).

YAO A 1977 -1 YAO, A. C.; AVIS, D. M.; RIVEST, R. L.
OMEGA (N POWER 2 LOG N) LOWER BOUND TO THE SHORTEST PATHS PROBLEM, AN. IN: PROCEEDINGS OF THE NINTH ANNUAL A.C.M. SYMPOSIUM ON THEORY OF COMPUTING, 11-17. (1977).

YAO B 1976 -1 YAO, A. C.; YAO, F. F.
COMPLEXITY OF SEARCHING AN ORDERED RANDOM TABLE, THE. IN: 17TH ANNUAL SYMPOSIUM ON FOUNDATIONS OF COMPUTER SCIENCE, HOUSTON 1976, 173-177. LONG BEACH: INSTITUTE OF ELECTRICAL AND ELECTRONICS ENGINEERS. (1976).

****

YEH 1976 -1 YEH, R. T.
APPLIED COMPUTATION THEORY: ANALYSIS, DESIGN, MODELING. PRENTICE-HALL SERIES IN AUTOMATIC COMPUTATION. ENGLEWOOD CLIFFS: PRENTICE-HALL, 624 P. (1976).

YELO 1976 -1 YELOWITZ, L.
EFFICIENT ALGORITHM FOR CONSTRUCTING HIERARCHICAL GRAPHS, AN. IEEE TRANSACTIONS ON SYSTEMS, MAN, AND CYBERNETICS 6, 327-329. (1976).

YEN 1969 -1 YEN, J. Y.
SOME ALGORITHMS FOR FINDING THE SHORTEST ROUTES THROUGH THE GENERAL NETWORKS. IN: L. A. ZADEH, L. W. NEUSTADT, A. V. BALAKRISHNAN (EDS.): COMPUTING METHODS IN OPTIMIZATION PROBLEMS - 2, 377-388. NEW YORK: ACADEMIC PRESS. (1969).

YEN 1971 -1 YEN, J. Y.
ON THE EFFICIENCIES OF ALGORITHMS FOR DETECTING NEGATIVE LOOPS IN NETWORKS. SANTA CLARA BUSINESS REVIEW, 52-58. (1971).

YEN 1972 -1 YEN, J. Y.
FINDING THE OPTIMAL CYCLES IN MULTIPLY WEIGHTED NETWORKS. SANTA CLARA BUSINESS REVIEW, 61-69. (1972).

YEN 1973 -1 YEN, J. Y.
FINDING REACHABILITY AND CONNECTIVITY OF DIRECTED LINEAR GRAPHS. SANTA CLARA BUSINESS REVIEW, 61-67. (1973).

YEN 1975 -1 YEN, J. Y.
SHORTEST PATH NETWORK PROBLEMS. MATHEMATICAL SYSTEMS IN ECONOMICS 18, MEISENHEIM: HAIN, 169 P. (1975).

YORM 1975 -1 YORMARK, J. S.
ACCELERATING GREENBERG'S METHOD FOR THE COMPUTATION OF KNAPSACK FUNCTIONS. JOURNAL OF MATHEMATICAL ANALYSIS AND APPLICATIONS 49, 629-637. (1975).

YOUN 1970 -1 YOUNG, G. K.; MOSELY, J. C.; EVENSON, D. E.
TIME SEQUENCING OF ELEMENT CONSTRUCTION IN A MULTI-RESERVOIR SYSTEM. WATER RESOURCES BULLETIN 6, 528-541. (1970).

YOUNA 1976 -1 YOUNG, H. P.
ON PERMUTATIONS AND PERMUTATION POLYTOPES. LAXENBURG: INTERNAT. INST. FOR APPLIED SYSTEMS ANALYSIS, 19 P. (1976).

YOUNB 1977 -1 YOUNG, P.
OPTIMIZATION AMONG PROVABLY EQUIVALENT PROGRAMS. JOURNAL OF THE ASSOCIATION FOR COMPUTING MACHINERY 24, 693-700. (1977).

YOUNC 1975 -1 YOUNG, R. D.
ECLECTIC PRIMAL ALGORITHM: CUTTING-PLANE METHOD THAT ACCOMODATES HYBRID SUBPROBLEM SOLUTION TECHNIQUES, THE. MATHEMATICAL PROGRAMMING 9, 294-312. (1975).

YUAN 1975 -1 YUAN, J. S. C.; HOREN, J. H.; WAGNER, H. M.
OPTIMAL MULTI-PRODUCT PRODUCTION SCHEDULING AND EMPLOYMENT SMOOTHING WITH DETERMINISTIC DEMANDS. MANAGEMENT SCIENCE 21, 1250-1262. (1975).

YUDI 1976 -1 YUDIN, D. B.; NEMIROVSKIJ, A. S.
ESTIMATION OF THE INFORMATION COMPLEXITY OF MATHEMATICAL PROGRAMMING PROBLEMS. EKONOMIKA I MATEMATICESKIE METODY 12, 128-142. (1976).

YUDIA 1974 -1 YUDIN, D. B.; TSOY, E. V.
INTEGER-VALUED STOCHASTIC PROGRAMMING (RUSSIAN). IZVESTIJA AKADEMII NAUK SSSR, TEHNICESKAJA KIBERNETIKA, NO. 1, 3-11 + 234; TRANSLATED AS ENGINEERING CYBERNETICS 12, 1974, 1-8. (1974).

YUVA 1976 -1 YUVAL, G.
ALGORITHM FOR FINDING ALL SHORTEST PATHS USING (N POWER 2.81) INFINITE PRECISION MULTIPLICATIONS, AN. INFORMATION PROCESSING LETTERS 4, NO. 6, 155-156. (1976).

YUVA 1976 -2 YUVAL, G.
FINDING NEAREST NEIGHBOURS. INFORMATION PROCESSING LETTERS 5, NO. 3, 63-65. (1976).

ZABL 1976 -1 ZABLUDOWSKI, A.; CHMARA, K.
METHOD FOR FINDING A PATH IN A GRAPH (POLISH), A. ZESZYTY NAUKOWE POLITECHNIKI GDANSK, ELEKTRON., NO. 38, 53-66. (1976).

ZAHN 1973 -1 ZAHN, C. T.
ALTERNATING EULER PATHS FOR PACKINGS AND COVERS. AMERICAN MATHEMATICAL MONTHLY 80, 395-403. (1973).

ZAIT 1975 -1 ZAITSEVA, Z. N.; SHTEIN, M. E.
PLOTTING OF OPTIMAL HAMILTONIAN CYCLES IN THE MULTIVERTEX GRAPHS (RUSSIAN). KIBERNETIKA, NO. 6, 143-144. (1975).

ZAIT 1976 -1 ZAITSEVA, Z. N.; SHTEIN, M. E.
HAMILTONIAN CYCLES IN SOLVING LAYING OUT PROBLEM (RUSSIAN). KIBERNETIKA, NO. 3, 70-73. (1976).

ZAK 1976 -1 ZAK, YU. A.
GENERALIZED TRAVELLING-SALESMAN PROBLEM AND ITS PRACTICAL ASPECTS (RUSSIAN), THE. AVTOMATIKA I TELEMEHANIKA, NO. 10, 120-132. (1976).

ZAK 1977 -1 ZAK, YU. A.
METHOD OF DIRECTED SCANNING IN PROBLEMS OF INTEGRAL LINEAR PROGRAMMING WITH BOOLEAN VARIABLES (RUSSIAN). KIBERNETIKA, NO. 4, 70-79. (1977).

ZALO 1975 -1 ZALOOM, V.; VATZ, D.
NOTE ON THE OPTIMAL SCHEDULING OF TWO PARALLEL PROCESSORS, A. NAVAL RESEARCH LOGISTICS QUARTERLY 22, 823-828. (1975).

ZALOA 1977 -1 ZALOOM, V. A.; SMYER, W. N.
SCHEDULING TO MINIMIZE MAKESPAN REQUIRED TO PROCESS INDEPENDENT JOBS ON PARALLEL NON-EQUIVALENT SERVERS. SAN FRANCISCO (51. ORSA-MEETING + TIMS, MAY 1977) 17 P. (1977).

ZAMB 1974 -1 ZAMBICKII, D. K.
PROBLEM OF FINDING THE OPTIMAL VERTEX IN A GRAPH (RUSSIAN), THE. PRIKLADNAJA MATEMATIKA I PROGRAMMIROVANIE VYP. 12, 61-64 + 136-137. (1974).

ZAMF 1973 -1 ZAMFIRESCU, T.
CHARACTERIZATION OF HAMILTONIAN GRAPHS, A. ATTI DELLA ACCADEMIA DELLE SCIENZE DELL'ISTITUTO DI BOLOGNA. CLASSE DI SCIENZE FISICHE. RENDICONTI 1, SERIES 13, 39-40. (1973).

ZAMF 1975 -1 ZAMFIRESCU, T.
HISTOIRE ET L'ETAT PRESENT DES BORNES CONNUES POUR P(KJ), C(KJ), P(J)- ET C(KJ)-, L'. CAHIERS DU CENTRE D'ETUDES DE RECHERCHE OPERATIONNELLE 17, 427-439. (1975).

ZAMF 1976 -1 ZAMFIRESCU, T.
ON LONGEST PATHS AND CIRCUITS IN GRAPHS. MATHEMATICA SCANDINAVICA 32, NO. 8, 211-239. (1976).

ZANC 1974 -1 ZANCEVA, F. N.
APPROXIMATION-COMBINATORIAL METHOD FOR THE CASE OF SEVERAL APPROXIMATING PROBLEMS (RUSSIAN), AN. IZVESTIJA AKADEMII NAUK KAZAHSKOI SSR. SERIJA FIZIKO-MATEMATICESKAJA, NO. 5, 77-79 + 94. (1974).

ZANG 1966 -1 ZANGWILL, W.
SHORTEST ROUTE PROBLEM UNDER EITHER CONCAVE OR CONVEX SHIPPING COSTS, THE. PAPER PRESENTED AT THE 29TH NATIONAL ORSA-MEETING, SANTA MONICA 1966. (1966).

****

ZAVI 1975 -1 ZAVILEVICH, M. L.
DIRECT GRID PROBLEM FOR A HIERARCHICAL GRID, THE. IZVESTIJA AKADEMII NAUK SSSR, TEHNICESKAJA KIBERNETIKA, NO. 2, 32-35; TRANSLATED AS ENGINEERING CYBERNETICS 13, 1975, NO. 2, 24-27. (1975).

ZAYC 1975 -1 ZAYCHENKO, YU. P.; PECHURIN, M. K.; TOMASHEVSKIY, V. M.
GENERALIZED ASSIGNMENT PROBLEM AND ITS SOLUTION ALGORITHM (UKRAINIAN), THE. AVTOMATIKA, NO. 2, 45-48; TRANSLATED AS SOVIET AUTOMATIC CONTROL 8, NO. 2, 37-40. (1975).

ZEHN 1965 -1 ZEHNDER, C. A.
COMPUTATION OF TIME TABLES FOR SCHOOLS AND TRANSPORTATION. ZUERICH: INDUSTRIE ORGAN, 88 P. (1965).

ZEVI 1976 -1 ZEVIN, A. A.
ON ARTICLE UNIFICATION PROBLEM (RUSSIAN). EKONOMIKA I MATEMATICHESKIE METODY 12, 930-940. (1976).

ZHER 1976 -1 ZHERNAK, A. N.
GROUP OF ALGORITHMS FOR THE SOLUTION OF QUASILINEAR INTEGER PROGRAMMING PROBLEMS (RUSSIAN), A. ZURNAL VYCISLITEL'NOI MATEMATIKI I MATEMATICESKOI FIZIKI 16, 1353-1359 + 1375-1376. (1976).

ZHER 1976 -2 ZHERNAK, A. N.
NEW METHOD FOR THE SOLUTION OF INTEGER PROGRAMMING PROBLEMS (RUSSIAN), A. IZVESTIJA AKADEMII NAUK SSSR, TEHNICESKAJA KIBERNETIKA, NO. 4, 51-56. (1976).

ZHUR 1977 -1 ZHURAVLEV, JU. I.
CORRECT ALGEBRAS OVER THE SETS OF INCORRECT (HEURISTIC) ALGORITHMS, II (RUSSIAN). KIBERNETIKA, NO. 6, 21-27. (1977).

ZHURA 1968 -1 ZHURZENKO, S. L.
METHOD OF SOLVING CLASS OF PROBLEMS IN SCHEDULING THEORY. ENGINEERING CYBERNETICS, 55-64. (1968).

ZIMM 1975 -1 ZIMMERMANN, H.-J.; POLLATSCHEK, M. A.
PROBABILITY DISTRIBUTION FUNCTION OF THE OPTIMUM OF A 0-1 LINEAR PROGRAM WITH RANDOMLY DISTRIBUTED COEFFICIENTS OF THE OBJECTIVE FUNCTION AND THE RIGHT-HAND SIDE, THE. OPERATIONS RESEARCH 23, 137-149. (1975).

ZIMMA 1976 -1 ZIMMERMANN, U.
BOOLE'SCHE OPTIMIERUNGSPROBLEME MIT SEPARABLER ZIELFUNKTION UND MATROIDALEN RESTRIKTIONEN. DISSERTATION. KOELN: UNIV. KOELN, MATHEMATISCH-NATURWISSENSCHAFTLICHE FAKULTAET, 174 P. (1976).

ZIMMA 1977 -1 ZIMMERMANN, U.
DUALITY AND THE ALGEBRAIC MATROID INTERSECTION PROBLEM. AACHEN (2. SYMPOSIUM UEBER OPERATIONS RESEARCH, 1977) 12 P. (1977).

ZIMMA 1977 -2 ZIMMERMANN, U.
SOME PARTIAL ORDERS RELATED TO BOOLEAN OPTIMIZATION AND THE GREEDY-ALGORITHM. IN: P. L. HAMMER, E. L. JOHNSON, B. H. KORTE, G. L. NEMHAUSER (EDS.): STUDIES IN INTEGER PROGRAMMING, 539-550. ANNALS OF DISCRETE MATHEMATICS 1, AMSTERDAM: NORTH-HOLLAND. (1977).

ZIMMA 1977 -3 ZIMMERMANN, U.
THRESHOLD METHODS FOR BOOLEAN OPTIMIZATION PROBLEMS WITH SEPARABLE OBJECTIVES. REPORT 77-12. KOELN: UNIV. KOELN, MATHEMATISCHES INSTITUT, 10 P. (1977).

ZIMMB 1967 -1 ZIMMERMANN, W.; PFAFFENZELLER, D.
SIMULTANE MENGEN- UND ABLAUFPLANUNG INDUSTRIELLER FERTIGUNGSPROGRAMME. ABLAUF UND PLANUNGSFORSCHUNG 8, 259-268. (1967).

ZION 1977 -1 ZIONTS, S.
INTEGER LINEAR PROGRAMMING WITH MULTIPLE OBJECTIVES. IN: P. L. HAMMER, E. L. JOHNSON, B. H. KORTE, G. L. NEMHAUSER (EDS.): STUDIES IN INTEGER PROGRAMMING, 551-562. ANNALS OF DISCRETE MATHEMATICS 1, AMSTERDAM: NORTH-HOLLAND. (1977).

ZUSU 1974 -1 ZUSUPBAEV, A.
PRODUCTION ALLOCATION PROBLEM WITH A CONVEX SEPARABLE FUNCTIONAL (RUSSIAN), THE. IZVESTIJA AKADEMII NAUK KIRGIZSKOI SSR, NO. 6, 14-20. (1974).

ZUSU 1977 -1 ZUSUPBAEV, A.
SOLUTION OF THE SEASONAL PRODUCTION ALLOCATION PROBLEM BY THE METHOD OF SUCCESSIVE CALCULATIONS (RUSSIAN). IZVESTIJA AKADEMII NAUK KIRGIZSKOI SSR, NO. 2, 19-27. (1977).

ZVJA 1974 -1 ZVJAGINA, R. A.
SOLUTION OF A DYNAMICAL TRANSPORTATION PROBLEM WITH INCREASED PRODUCTION (RUSSIAN), THE. OPTIMIZACIJA 15, 50-57 + 184-185. (1974).

```
***********************************************************************************************************************************
```

```
PPPPPPPPPPP     AAAAAAAAAA    RRRRRRRRRRR    TTTTTTTTTTTT                        2222222222
PPPPPPPPPPPP   AAAAAAAAAAAA   RRRRRRRRRRRR   TTTTTTTTTTTT                       222222222222
PP        PP   AA        AA   RR        RR        TT                            22        22
PP        PP   AA        AA   RR        RR        TT                                      22
PP        PP   AA        AA   RR        RR        TT                                      22
PPPPPPPPPPPP   AAAAAAAAAAAA   RRRRRRRRRRRR        TT                                    22
PPPPPPPPPPP    AAAAAAAAAAAA   RRRRRRRRRRR         TT                                  22
PP             AA        AA   RR    RR            TT                                22
PP             AA        AA   RR     RR           TT                              22
PP             AA        AA   RR      RR          TT                             22
PP             AA        AA   RR       RR         TT                            222222222222
PP             AA        AA   RR        RR        TT                            222222222222
```

SUBJECT CLASSIFICATION

```
**********************
**********************
```

****

ALSP 1976 -1 ALSPACH, B.; MASON, D. W.; PULLMAN, N. J.
PATH NUMBERS OF TOURNAMENTS.

ARDI 1977 -1 ARDITTI, J. C.
COMPARABILITY GRAPHS AND THE DIMENSION OF ORDERED SETS(EXTENDED ABSTRACT).

ARDIA 1975 -1 ARDITTI, J.-C.
GRAPHES DE COMPARABILITE D'ARBRES ET D'ARBORESCENCES.

ARMSB 1977 -1 ARMSTRONG, R. D.; COOK, W. D.
COMMITTEE RANKING PROBLEM USING ORDINAL UTILITIES, A.

ARMSB 1977 -2 ARMSTRONG, R. D.; COOK, W. D.
STRATEGY SELECTION IN ORDINAL RANKING PROBLEMS.

BALAG 1975 -3 BALAS, E.
LINEAR CHARACTERIZATION OF PERMUTATION VECTORS, A.

BALIA 1977 -1 BALINSKI, M. L.; YOUNG, H. P.
ON HUNTINGTON METHODS OF APPORTIONMENT.

BERMB 1976 -1 BERMOND, J.-C.; KODRATOFF, Y.
HEURISTIQUE POUR LE CALCUL DE L'INDICE DE TRANSITIVITE D'UN TOURNOI, UNE.

BLIN 1973 -1 BLIN, J. M.
FURTHER PROCEDURE FOR ORDERING AN INPUT-OUTPUT MATRIX: SOME EMPIRICAL EVIDENCE, A.

BOWMA 1977 -1 BOWMAN, V. J. JR.; STARR, J. H.
PARTIAL ORDERINGS IN IMPLICIT ENUMERATION.

CHEI 1975 -1 CHEIN, M.
REMARQUE SUR L'ALGORITHME DE DUCAMP POUR LA RECHERCHE DE TOUS LES ORDRES TOTAUX CONTENANT UN ORDRE PARTIEL, UNE.

DOWL 1977 -1 DOWLING, T. A.; DENIG, W. A.
GEOMETRIES ASSOCIATED WITH PARTIALLY ORDERED SETS.

DRUT 1975 -1 DRUTOV, V. E.
SOLUTION OF A SCHEDULING PROBLEM (RUSSIAN).

EPLE 1977 -1 EPLETT, W. J. R.
RESULT ON TOURNAMENTS WITH AN APPLICATION TO COUNTING, A.

FISH 1976 -1 FISHBURN, P. C.
ON LINEAR EXTENSION MAJORITY GRAPHS OF PARTIAL ORDERS.

FONG 1977 -1 FONG, A. C.; ULLMAN, J. D.
FINDING THE DEPTH OF A FLOW GRAPH.

GAIH 1977 -1 GAIHA, P.; GUPTA, S. K.
ADJACENT VERTICES ON A PERMUTOHEDRON.

GAREA 1976 -1 GAREY, M. R.; GRAHAM, R. L.; JOHNSON, D. S.; YAO, A. C.-C.
RESOURCE CONSTRAINED SCHEDULING AS GENERALIZED BIN PACKING.

GHOS 1970 -1 GHOSH, A.; SARKAR, H.
INPUT-OUTPUT MATRIX AS A SPATIAL CONFIGURATION, AN.

GOLU 1977 -1 GOLUMBIC, M. C.
COMPARABILITY GRAPHS AND A NEW MATROID.

GRANC 1976 -1 GRANT, D. D.
HOW TO RUN A HOCKEY TOURNAMENT - A COMBINATORIAL SCHEDULING PROBLEM.

GREE 1976 -1 GREEN, C. D.
DETECTION OF MISTAKES IN THE COMPARABILITY GRAPH OF A TREE, THE.

GREEA 1977 -1 GREENE, C.
ACYCLIC ORIENTATIONS.

HARA 1975 -1 HARARY, F.
ON MINIMAL FEEDBACK VERTEX SETS OF A DIGRAPH.

HARDA 1975 -1 HARDOUIN, D. J.
QUELQUES RESULTATS SUR "L'INDICE DE TRANSITIVITE" DE CERTAINS TOURNOIS.

HAUS 1978 -1 HAUSMANN, D.
ADJAZENZCHARAKTERISIERUNGEN FUER POLYEDERECKEN IN DER KOMBINATORISCHEN OPTIMIERUNG.

HAUSA 1977 -1 HAUSMANN, D.; JENKYNS, T. A.; KORTE, B.
WORST CASE ANALYSIS OF GREEDY TYPE ALGORITHMS FOR INDEPENDENCE SYSTEMS.

HAUSC 1976 -1 HAUSMANN, D.; KORTE, B.
COLOURING CRITERIA FOR ADJACENCY ON 0-1-POLYHEDRA.

HWANA 1969 -1 HWANG, F. K.; LIN, S.
ANALYSIS OF FORD AND JOHNSON'S SORTING ALGORITHM, AN.

JENK 1977 -2 JENKYNS, T. A.
P-SYSTEMS AS INTERSECTIONS OF MATROIDS.

KORTC 1977 -1 KORTE, B.; HAUSMANN, D.
ANALYSE DER GREEDY-HEURISTIK FUER UNABHAENGIGKEITSSYSTEME, EINE.

LENS 1973 -1 LENSTRA, H. W.
ACYCLIC SUBGRAPH PROBLEM, THE.

LEVYA 1973 -1 LEVY, L. S.
ORDERING ON TREES - SIMPLIFIED PROOFS.

LUBE 1975 -1 LUBELL, D.
LOCAL MATCHING IN THE FUNCTION SPACE OF A PARTIAL ORDER.

MARTB 1977 -1 MARTELLO, S.
ALGORITHM FOR FINDING A MINIMAL EQUIVALENT GRAPH OF A STRONGLY CONNECTED DIGRAPH, AN.

MERCA 1976 -1 MERCHANT, D. K.; RAO, M. R.
MAJORITY DECISIONS AND TRANSITIVITY: SOME SPECIAL CASES.

NOLT 1976 -1 NOLTEMEIER, H.
REDUKTION VON PRAEZEDENZSTRUKTUREN.

PATR 1974 -1 PATRINOS, A. N.; HAKIMI, S. L.
ORDERING PROBLEMS IN TREES AND THEIR APPLICATIONS TO TREE-STRUCTURED OPERATIONS OR FILES.

SCHN 1976 -1 SCHNORR, C. P.
ALGORITHM FOR TRANSITIVE CLOSURE WITH LINEAR EXPECTED TIME, AN.

STARA 1975 -1 STARR, J. H.
ZERO-ONE PROGRAMMING: A PARTIAL ORDERING AND ITS USE IN THE OPTIMIZATION OF NUMERICAL AND NON-NUMERICAL FUNCTIONS.

TROTB 1976 -1 TROTTER, W. T. JR.; BOGART, K. P.
MAXIMAL DIMENSIONAL PARTIALLY ORDERED SETS III: A CHARACTERIZATION OF HIRAGUCHI'S INEQUALITY FOR INTERVAL DIMENSION.

TROTB 1976 -2 TROTTER, W. T. JR.; BOGART, K. P.
ON THE COMPLEXITY OF POSETS.

TROTC 1976 -1 TROTTER, W. T. JR.; MOORE, J. I.; SUMNER, D. P.
DIMENSION OF A COMPARABILITY GRAPH, THE.

TROTD 1976 -1 TROTTER, W. T. JR.; MOORE, J. I. JR.
CHARACTERIZATION PROBLEMS FOR GRAPHS, PARTIALLY ORDERED SETS, LATTICES, AND FAMILIES OF SETS.

WESS 1972 -1 WESSELS, H.
VERGLEICH VERSCHIEDENER ANSAETZE ZUR TRIANGULATION VON INPUT-OUTPUT-MATRIZEN.

ACYCLIC SUBGRAPH PROBLEM, ORDERINGS, RELATIONS
****

WHITB 1977 -1 WHITE, D. E.; WILLIAMSON, S. G.
RECURSIVE MATCHING ALGORITHMS AND LINEAR ORDERS ON THE SUBSET LATTICE.
WILLD 1976 -1 WILLIAMSON, S. G.
RANKING ALGORITHMS FOR LISTS OF PARTITIONS.
YELO 1976 -1 YELOWITZ, L.
EFFICIENT ALGORITHM FOR CONSTRUCTING HIERARCHICAL GRAPHS, AN.
YOUNA 1976 -1 YOUNG, H. P.
ON PERMUTATIONS AND PERMUTATION POLYTOPES.
ZIMMA 1977 -2 ZIMMERMANN, U.
SOME PARTIAL ORDERS RELATED TO BOOLEAN OPTIMIZATION AND THE GREEDY-ALGORITHM.

****

ARDI 1977 -1 ARDITTI, J. C.
COMPARABILITY GRAPHS AND THE DIMENSION OF ORDERED SETS(EXTENDED ABSTRACT).
ARUI 1971 -1 ARUIN, L. I.
EDGES OF THE CONVEX LINEAR HULL OF THE SET OF INTEGER POINTS THAT BELONG TO THE SET OF ADMISSIBLE SOLUTIONS OF AN INTEGER LINEAR PROGRAMMING PROBLEM (RUSSIAN), THE.
BALAJ 1975 -1 BALAS, E.; PADBERG, M. W.
SET PARTITIONING: A SURVEY.
BALAJ 1976 -1 BALAS, E.; PADBERG, M. W.
SET PARTITIONING: A SURVEY.
BRUAA 1976 -1 BRUALDI, R. A.; GIBSON, P. M.
ASSIGNMENT POLYTOPE, THE.
BRUAA 1977 -3 BRUALDI, R. A.; GIBSON, P. M.
CONVEX POLYHEDRA OF DOUBLY STOCHASTIC MATRICES: II. GRAPH OF OMEGA N.
DAEN 1977 -1 DAENINCK, G.; SMEERS, Y.
USING SHORTEST PATHS IN SOME TRANSSHIPMENT PROBLEM WITH CONCAVE COSTS.
GAIH 1977 -1 GAIHA, P.; GUPTA, S. K.
ADJACENT VERTICES ON A PERMUTOHEDRON.
HAUS 1977 -1 HAUSMANN, D.
ADJACENCY ON THE BRANCHING POLYHEDRON.
HAUS 1978 -1 HAUSMANN, D.
ADJAZENZCHARAKTERISIERUNGEN FUER POLYEDERECKEN IN DER KOMBINATORISCHEN OPTIMIERUNG.
HAUS 1977 -2 HAUSMANN, D.
BENACHBARTE ECKEN AUF 0-1-POLYEDERN.
HAUS 1977 -3 HAUSMANN, D.
POLYNOMIALER ALGORITHMUS FUER DIE ADJAZENZUEBERPRUEFUNG AUF DEM BRANCHING-POLYEDER, EIN.
HAUSC 1976 -1 HAUSMANN, D.; KORTE, B.
COLOURING CRITERIA FOR ADJACENCY ON 0-1-POLYHEDRA.
HELL 1955 -1 HELLER, I.
ON THE TRAVELING SALESMAN'S PROBLEM.
HELL 1956 -1 HELLER, I.
NEIGHBOUR RELATIONS ON THE CONVEX OF CYCLIC PERMUTATIONS.
MAC H 1975 -1 MAC KEOWN, P. G.
DETERMINING ADJACENT VERTICES ON ASSIGNMENT POLYTOPES.
MAC I 1975 -1 MAC KEOWN, P. G.; RUBIN, D. S.
ADJACENT VERTICES ON TRANSPORTATION POLYTOPES.
PAPA 1975 -1 PAPADIMITRIOU, C. H.
ON THE COMPLEXITY OF THE LOCAL STRUCTURE OF CERTAIN CONVEX POLYTOPES.
RAO 1976 -1 RAO, M. R.
ADJACENCY OF THE TRAVELING SALESMAN TOURS AND 0-1 VERTICES.
ROSEA 1971 -1 ROSENFELD, M.
ON CERTAIN POLYTOPES ASSOCIATED WITH GRAPHS.
RUBI 1972 -1 RUBIN, D. P.
NEIGHBORING VERTICES ON CONVEX POLYHEDRAL SETS.
SAVAA 1973 -1 SAVAGE, S. L.
SOLUTION OF DISCRETE LINEAR OPTIMIZATION PROBLEMS BY NEIGHBORHOOD SEARCH TECHNIQUES, THE.
SAVAA 1973 -2 SAVAGE, S. L.
STATISTICAL INDICATORS OF OPTIMALITY.
UELK 1975 -1 UELKUECUE, A.
EFFICIENCY MEASURES OF VERTEX FOLLOWING ALGORITHMS ON POLYHEDRAL SETS.
WALKA 1976 -1 WALKER, W. E.
HEURISTIC ADJACENT EXTREME POINT ALGORITHM FOR THE FIXED CHARGE PROBLEM, A.
YOUNA 1976 -1 YOUNG, H. P.
ON PERMUTATIONS AND PERMUTATION POLYTOPES.

ALGEBRAICAL METHODS AND ALGEBRAIZATION OF PROBLEMS
****

BACH 1978 -1 BACHEM, A.
THEOREM OF MINKOWSKI FOR POLYHEDRAL MONOIDS AND AGGREGATED LINEAR DIOPHANTINE SYSTEMS, THE.

BACH 1976 -1 BACHEM, A.
ALGORITHMUS ZUR MODULO-OPTIMIERUNG STRUKTURIERTER MATRIZEN, EIN.

BACH 1976 -4 BACHEM, A.
REDUCTION AND DECOMPOSITION OF MODULO OPTIMIZATION PROBLEMS.

BLIN 1976 -1 BLIN, J. M.
LINEAR ASSIGNMENT FORMULATION OF THE MULTIATTRIBUTE DECISION PROBLEM, A.

BURD 1977 -1 BURDET, C.-A.; JOHNSON, E. L.
SUBADDITIVE APPROACH TO SOLVE LINEAR INTEGER PROGRAMS, A.

BURK 1976 -1 BURKARD, R. E.
FLUESSE IN NETZWERKEN MIT ALLGEMEINEN KOSTEN.

BURK 1976 -2 BURKARD, R. E.
GENERAL HUNGARIAN METHOD FOR THE ALGEBRAIC TRANSPORTATION PROBLEM, A.

BURKA 1976 -1 BURKARD, R. E.; HAHN, W.; ZIMMERMANN, U.
ALGEBRAIC APPROACH TO ASSIGNMENT PROBLEMS, AN (REVIDIERT).

BURKA 1977 -1 BURKARD, R. E.; HAHN, W.; ZIMMERMANN, U.
ALGEBRAIC APPROACH TO ASSIGNMENT PROBLEMS, AN.

BURKB 1976 -1 BURKARD, R. E.; HAMACHER, H.; ZIMMERMANN, U.
ALGEBRAIC NETWORK FLOW PROBLEM, THE.

BURKB 1977 -1 BURKARD, R. E.; HAMACHER, H.; ZIMMERMANN, U.
BEMERKUNGEN ZUM ALGEBRAISCHEN FLUSSPROBLEM, EINIGE.

BURKB 1977 -2 BURKARD, R. E.; HAMACHER, H.; ZIMMERMANN, U.
FLUSSPROBLEME MIT ALLGEMEINEN KOSTEN.

BURKE 1978 -1 BURKARD, R. E.; ZIMMERMANN, U.
SOLUTION OF ALGEBRAIC ASSIGNMENT AND TRANSPORTATION PROBLEMS, THE.

BURKE 1977 -1 BURKARD, R. E.; ZIMMERMANN, U.
WEAKLY ADMISSIBLE TRANSFORMATIONS FOR SOLVING ALGEBRAIC ASSIGNMENT AND TRANSPORTATION PROBLEMS.

DERI 1977 -1 DERIGS, U.
DUALITY AND THE ALGEBRAIC MATCHING PROBLEM.

DERI 1978 -1 DERIGS, U.
ALGEBRAISCHE MATCHINGPROBLEME.

GOND 1974 -1 GONDRAN, M.
ALGEBRE DES CHEMINS ET ALGORITHMES.

GOND 1975 -2 GONDRAN, M.
PATH ALGEBRA AND ALGORITHMS.

GONDB 1977 -1 GONDRAN, M.; MINOUX, M.
VALEURS PROPRES ET VECTEURS PROPRES DANS LES DIOIDES ET LEUR INTERPRETATION EN THEORIE DES GRAPHES.

HULM 1975 -1 HULME, B. L.
LATTICE ALGEBRA FOR FINDING SIMPLE PATHS AND CUTS IN A GRAPH, A.

KACPA 1975 -1 KACPRZYK, J.; STANCZAK, W.
APPLICATION OF THE METHOD OF MINIMAL GROUPS TO THE SOLUTION OF THE PROBLEM OF PARTITIONING A GROUP OF FIRMS INTO SUB-GROUPS (POLISH).

MINOA 1976 -4 MINOUX, M.
STRUCTURES ALGEBRIQUES GENERALISEES DES PROBLEMES DE CHEMINEMENT DANS LES GRAPHES FINIS, ALGORITHMES ET APPLICATIONS.

NEMHA 1976 -1 NEMHAUSER, G. L.; WOLSEY, L. A.; FISHER, M. L.
ANALYSIS OF APPROXIMATIONS FOR MAXIMIZING SUBMODULAR SET FUNCTIONS, AN.

RATH 1977 -2 RATHMANN, D.
ZUR LOESUNG DISKRETER OPTIMIERUNGSPROBLEME MIT GRUPPENTHEORETISCHEN ANSAETZEN.

RINE 1977 -1 RINE, D. C.
OPTIMIZING TWO TYPES OF DISCRETE FUNCTIONS, SUBJECT TO LINEAR RESTRICTIONS.

ROY 1975 -2 ROY, B.
CHEMINS ET CIRCUITS: ENUMERATION ET OPTIMISATION: A SURVEY.

SABI 1977 -1 SABIDUSSI, G.
TRACING GRAPHS WITHOUT BACKTRACKING.

SHIM 1951 -1 SHIMBEL, A.
APPLICATION OF MATRIX ALGEBRA TO COMMUNICATION NETS.

SPAT 1975 -1 SPATHIS-DIMITROPOULOU, V.
ETUDE ALGEBRIQUE D'UN ARBRE. UN PROBLEME DE CIRCULATION, UNE.

TINH 1976 -1 TINHOFER, G.
METHODEN DER ANGEWANDTEN GRAPHENTHEORIE.

ZHUR 1977 -1 ZHURAVLEV, JU. I.
CORRECT ALGEBRAS OVER THE SETS OF INCORRECT (HEURISTIC) ALGORITHMS, II (RUSSIAN).

ZIMMA 1976 -1 ZIMMERMANN, U.
BOOLE'SCHE OPTIMIERUNGSPROBLEME MIT SEPARABLER ZIELFUNKTION UND MATROIDALEN RESTRIKTIONEN.

ZIMMA 1977 -1 ZIMMERMANN, U.
DUALITY AND THE ALGEBRAIC MATROID INTERSECTION PROBLEM.

ZIMMA 1977 -3 ZIMMERMANN, U.
THRESHOLD METHODS FOR BOOLEAN OPTIMIZATION PROBLEMS WITH SEPARABLE OBJECTIVES.

****

ABEL 1977 -1 ABELLA, R. J.; BOVA, T. E.
OPTIMAL PLANT ALLOCATION OF STOCKKEEPING UNITS.
ACKE 1977 -1 ACKERMAN, L. J.; LUSS, H.; BERKOWITZ, R. S.
APPLICATION OF SEQUENCING POLICIES TO TELEPHONE SWITCHING FACILITIES.
ACKE 1977 -2 ACKERMAN, L. J.; LUSS, H.; BERKOWITZ, R. S.
IMPLICIT ENUMERATION ALGORITHM FOR SEQUENCING POLICIES APPLIED TO TELEPHONE SWITCHING FACILITIES, AN.
AGIN 1976 -1 AGIN, N. I.
SUMMARY OF A NOVEL APPROACH TO THE WAREHOUSE LOCATION PROBLEM.
AGRE 1977 -1 AGRESTI, W. W.
DYNAMIC PROGRAMMING FOR COMPUTER REGISTER ALLOCATION.
AHREA 1976 -1 AHRENS, W.
NUMERICAL EXPERIENCE WITH HEURISTIC RULES IN SPECIFIC BINARY OPTIMIZATION ALGORITHMS.
AHREA 1975 -1 AHRENS, W.
OPTIMIERUNGSVERFAHREN ZUR LOESUNG NICHTLINEARER INVESTITIONSPROBLEME.
AHREA 1974 -1 AHRENS, W.
OPIMIERUNGSVERFAHREN ZUR LOESUNG NICHTLINEARER INVESTITIONSPROBLEME - ANGEWANDT AUF DAS PROBLEM DER TECHNISCHEN UND WIRTSCHAFTLICHEN OPTIMIERUNG BEI DER BILDUNG VON ZWECKVERBAENDEN ZUR ABWASSERBESEITIGUNG.
AHREA 1972 -1 AHRENS, W.
BINAERE OPTIMIERUNG ALS HILFSMITTEL BEI DER PLANUNG REGIONALER ABWASSERBESEITIGUNGSSYSTEME. II, DIE.
AHREA 1974 -2 AHRENS, W.
SOLVING OF THE REGIONALIZATION PROBLEM BY CERTAIN BRANCH-AND-BOUND METHODS.
ALBE 1977 -1 ALBERS, S.
MIXED INTEGER NONLINEAR PROGRAMMING PROCEDURE FOR SIMULTANEOUSLY LOCATING MULTIPLE PRODUCTS IN AN ATTRIBUTE SPACE, A.
ALCO 1975 -1 ALCOUFFE, A.; ENJALBERT, M.; MURATET, G.
METHODES DE RESOLUTION DU PROBLEME DE TRANSPORT ET DE PRODUCTION D'UNE ENTREPRISE A ETABLISSEMENTS MULTIPLES EN PRESENCE DE COUTS FIXES.
ALTHA 1976 -1 ALTHERR, W.; MAYER, J.
OPTIMALER EINSATZ EINES BETRIEBSEIGENEN STRASSENTRANSPORTSYSTEMS.
AM 1976 -1 AM, T. D.; TSUKIYAMA, S.; SHIRAKAWA, I.; OZAKI, H.
ALGORITHM FOR GENERATING ALL THE PATHS BETWEEN TWO VERTICES IN A DIGRAPH AND ITS APPLICATION, AN.
ANDO 1976 -1 ANDON, F. I.; VARTANYAN, G. M.
PROBLEM OF PACKAGING DEMANDS WITH TIMING MARKS IN INFORMATION SYSTEMS (RUSSIAN).
ANEJ 1977 -1 ANEJA, Y. P.; VEMUGANTI, R. R.
ROW GENERATION SCHEME FOR FINDING A MULTI-COMMODITY MINIMUM DISCONNECTING SET, A.
ANIS 1976 -1 ANISIMOV, V. V.
ON OPTIMUM PACKING OF TREES (RUSSIAN).
ARMSB 1977 -1 ARMSTRONG, R. D.; COOK, W. D.
COMMITTEE RANKING PROBLEM USING ORDINAL UTILITIES, A.
ARMSB 1976 -1 ARMSTRONG, R. D.; COOK, W. D.
OPTIMAL LINKING OF TASKS AND SCHEDULE DEVELOPMENT FOR MILITARY AIR TRANSPORT.
ARMSB 1977 -2 ARMSTRONG, R. D.; COOK, W. D.
STRATEGY SELECTION IN ORDINAL RANKING PROBLEMS.
ARMSC 1977 -1 ARMSTRONG, R. D.; COOK, W. D.; PALACIOS-GOMESZ, F. E.
APPLICATION OF QUASI-INTEGER PROGRAMMING TO A CAPITAL BUDGETING PROBLEM IN PAVEMENT MAINTENANCE, AN.
ARMSE 1977 -1 ARMSTRONG, R. D.; WILLIS, C. E.
SIMULTANEOUS INVESTMENT AND ALLOCATION DECISIONS APPLIED TO WATER PLANNING.
ARMSE 1975 -1 ARMSTRONG, R. D.; WILLIS, C. E.
SIMULTANEOUS INVESTMENT-ALLOCATION - AN APPLICATION OF GENERALIZED BENDERS DECOMPOSITION TO WATER PLANNING.
ASHT 1977 -1 ASHTON, W. B.; CLARK, G. M.
LOCATION AND SIZING OF URBAN FREIGHT CONSOLIDATION TERMINALS WITH MULTIPLE PLANNING PERIODS, THE.
AUST 1976 -2 AUST, R. J.
IMPROVEMENT ALGORITHM FOR SCHOOL TIMETABLING, AN.
BABU 1977 -1 BABU, A. J. G.; HSIA, W. S.
MORE EFFICIENT ALGORITHM FOR AN OPTIMAL TOUR OF A HEALTH CARE CONSUMER WITH MULTIPLE HEALTH CARE FACILITIES IN EACH COUNTY, A.
BABUB 1972 -1 BABUSIAUX, D.
ESTABLISHING A NETWORK OF WAREHOUSES FOR FINISHED PRODUCTS, TAKING INTO ACCOUNT SEASONAL FLUCTUATIONS IN DEMAND.
BACHA 1977 -1 BACHEM, A.; KORTE, B.
ALGORITHM FOR QUADRATIC OPTIMIZATION OVER TRANSPORTATION POLYTOPES, AN.
BACHA 1977 -2 BACHEM, A.; KORTE, B.
ESTIMATING INPUT-OUTPUT-MATRICES.
BACHA 1977 -3 BACHEM, A.; KORTE, B.
QUADRATIC OPTIMIZATION OVER TRANSPORTATION POLYTOPES.
BACHA 1977 -4 BACHEM, A.; KORTE, B.
QUADRATISCHE OPTIMIERUNG UEBER TRANSPORTPOLYEDER ZUR LOESUNG VON SCHAETZPROBLEMEN VON INPUT-OUTPUT MATRIZEN.
BACHC 1977 -1 BACHEM, A.; SIMON, H.
OPTIMAL PRODUCT POSITIONING IN AN ATTRIBUTE SPACE.
BAKE 1976 -1 BAKER, K. R.
WORKFORCE ALLOCATION IN CYCLICAL SCHEDULING PROBLEMS: A SURVEY.
BAKEA 1977 -1 BAKER, K. R.; BURNS, R. N.; CARTER, M.
STAFF SCHEDULING WITH DAY-OFF AND WORKSTRETCH CONSTRAINTS.
BAKEB 1977 -1 BAKER, K. R.; MAGAZINE, M.
WORKFORCE SCHEDULING WITH CYCLIC DEMANDS AND DAY-OFF CONSTRAINTS.
BALAB 1977 -1 BALACHANDRAN, V.; JAIN, S.
MATHEMATICAL-PROGRAMMING MODEL FOR OPTIMAL SELECTION OF TOURISM PLANS AND GATEWAYS FOR THE AIRLINE INDUSTRY, A.
BALAC 1976 -1 BALACHANDRAN, V.; PERRY, A.
TRANSPORTATION TYPE PROBLEMS WITH QUANTITY DISCOUNTS.
BALAF 1977 -1 BALAGURU, P. N.; RAMAKRISHNAN, V.; NAAMAN, A. E.
COST AND TIME SCALED NETWORKS - GRAPHICAL MODEL.
BALAJ 1975 -1 BALAS, E.; PADBERG, M. W.
SET PARTITIONING: A SURVEY.
BALE 1970 -1 BALERIAUX, H.; JAMOULLE, E.; DOULLIEZ, P.; KELECOM, J. VAN
RECHERCHE DE LA POLITIQUE OPTIMALE D'INVESTISSEMENT D'UN RESEAU ELECTRIQUE EN EXPANSION PAR UNE METHODE DE DECISION SEQUENTIELLE.

APPLICATION OF INTEGER PROGRAMMING
****

BALIA 1977 -1 BALINSKI, M. L.; YOUNG, H. P.
ON HUNTINGTON METHODS OF APPORTIONMENT.

BALL 1977 -1 BALL, B. C.; WEBSTER, D. B.
OPTIMAL SCHEDULING FOR EVEN-NUMBERED TEAM ATHLETIC CONFERENCES.

BALLA 1977 -1 BALL, M. O.; NEMHAUSER, G. L.
MATROIDS AND A RELIABILITY ANALYSIS PROBLEM.

BAND 1975 -1 BANDLER, J. W.; CHEN, J. H. K.; DALSGAARD, P.; LIU, P. C.
TOLOPT - A PROGRAM FOR OPTIMAL, CONTINUOUS OR DISCRETE, DESIGN CENTERING AND TOLERANCING, PART 1 - USER'S GUIDE, PART 2 - FORTRAN LISTING.

BANDA 1975 -1 BANDLER, J. W.; LIU, P. C.; CHEN, J. H. K.
WORST CASE NETWORK TOLERANCE OPTIMIZATION.

BARB 1970 -1 BARBER, G. M.
DYNAMIC LOCATIONAL ANALYSIS: THE FIRM AND ITS OPERATING POLICY.

BARB 1975 -1 BARBER, G. M.
MATHEMATICAL PROGRAMMING APPROACH TO A NETWORK DEVELOPMENT PROGRAM, A.

BARRC 1977 -1 BARRY, D.; ROBINSON, D. F.
OPTIMAL BATCH SIZES FOR URGENT DISPATCH.

BARS 1974 -1 BARSKIY, A. B.
METHOD OF OPTIMAL DISTRIBUTION OF OPERATIONS IN A HOMOGENEOUS COMPUTATIONAL SYSTEM (RUSSIAN), A.

BATT 1977 -1 BATTAGLIA, F.
LOCATION OF CENTERS IN AN EUCLIDEAN SPACE (ITALIAN), THE.

BAZA 1975 -1 BAZARAA, M. S.
COMPUTERIZED LAYOUT DESIGN: A BRANCH AND BOUND APPROACH.

BECKA 1974 -1 BECKER, L.; YEH, W. W.-G.
OPTIMAL TIMING, SEQUENCING, AND SIZING OF MULTIPLE RESERVOIR SURFACE WATER SUPPLY FACILITIES.

BECKA 1974 -2 BECKER, L.; YEH, W. W.-G.
TIMING AND SIZING OF COMPLEX WATER RESOURCE SYSTEMS.

BECKB 1956 -1 BECKMANN, M.; MAC GUIRE, C. B.; WINSTON, C. B.
STUDIES IN THE ECONOMICS OF TRANSPORTATION.

BECKC 1976 -1 BECKMANN, M. J.; BURNS, L. D.
THEORY OF HOUSEHOLD AUTOMOBILE ALLOCATION DECISIONS, A.

BELLC 1965 -1 BELLMORE, M.
OPTIMAL ALLOCATION OF RESOURCES IN A COMMUNICATIONS NETWORK.

BERE 1974 -1 BERESNEV, V. L.
IMPLICIT ENUMERATION ALGORITHM FOR AN ASSIGNMENT AND STANDARDIZATION PROBLEM (RUSSIAN).

BERGC 1968 -1 BERGENDAHL, G.
CAPACITY PLANNING MODELS FOR SYSTEMS OF PLANT AND ROAD INVESTMENTS.

BERGC 1969 -1 BERGENDAHL, G.
COMBINED LINEAR AND DYNAMIC PROGRAMMING MODEL FOR INTERDEPENDENT ROAD INVESTMENT PLANNING, A.

BERGD 1974 -1 BERGENDORFF, H.
MODEL TO EVALUATE AND PLAN PROJECTS IN FORESTRY AND FOREST INDUSTRIES, A.

BERGE 1975 -1 BERGMANN, D. R.
INTEGER PROGRAMMING FORMULATION FOR DERIVING MINIMUM DISPATCH INTERVALS ON A GUIDEWAY ACCOMODATING THROUGH AND LOCAL PUBLIC TRANSPORTATION SERVICES.

BERL 1976 -1 BERLIN, G. N.; REVELLE, C. S.; ELZINGA, D. J.
DETERMINING AMBULANCE-HOSPITAL LOCATIONS FOR ON-SCENE AND HOSPITAL SERVICES.

BERN 1958 -1 BERNERS-LEE, C. M.
INDUSTRIAL APPLICATION OF COMPUTERS, AN.

BERRB 1975 -1 BERRY, W. L.; RAO, V.
CRITICAL RATIO SCHEDULING: AN EXPERIMENTAL ANALYSIS.

BESW 1977 -1 BESWICK, C. A.
ALLOCATING SELLING EFFORT VIA DYNAMIC PROGRAMMING.

BHAL 1970 -1 BHALLA, H. S.
MULTI-TIME PERIOD FACILITIES LOCATION: A HEURISTIC ALGORITHM WITH APPLICATIONS TO WASTE WATER TREATMENT SYSTEMS.

BIDE 1975 -1 BIDER, I. G.; BOL'SHAKOV, I. A.
NON-MONOTONIC STRATEGIES FOR AN ALLOCATION OF LIMITED RESOURCES (RUSSIAN).

BIND 1961 -1 BINDSCHEDLER, A. E.; MOORE, J. M.
OPTIMAL LOCATION OF NEW MACHINES IN EXISTING PLANT LAYOUTS.

BION 1973 -1 BIONDI, E.
OPTIMUM DELIVERY PROBLEMS.

BJOE 1977 -1 BJOERKLUND, B.; HOLMBERG, G.
VEHICLE ROUTING AND SCHEDULING FOR DELIVERING NEWSPAPERS IN THE STOCKHOLM METROPOLITAN AREA.

BLIN 1973 -1 BLIN, J. M.
FURTHER PROCEDURE FOR ORDERING AN INPUT-OUTPUT MATRIX: SOME EMPIRICAL EVIDENCE, A.

BLINA 1976 -1 BLIN, J.-M.
ASSIGNMENT MODELS IN VOTING THEORY.

BODI 1976 -1 BODIN, L. D.; ROSENFIELD, D.
ESTIMATION OF THE OPERATING COST OF MASS TRANSIT SYSTEMS.

BOEB 1976 -1 BOEBION, J.; PUN, L.
SERIES-PARALLEL MULTIPLE-CRITERIA MODEL FOR A SCHEDULING PROBLEM IN THE DRESS-MAKING INDUSTRY, A.

BOER 1977 -1 BOERE, N, J,
AIR CANADA SAVES WITH AIRCRAFT MAINTENANCE SCHEDULING.

BOESC 1977 -1 BOESCH, F. T.; GIMPEL, J. F.
COVERING THE POINTS OF A DIGRAPH WITH POINT-DISJOINT PATHS AND ITS APPLICATION TO CODE OPTIMIZATION.

BOET 1974 -1 BOETTCHER, H.; DEHNERT, G.; DOMSCHKE, W.
VERFAHREN ZUR BESTIMMUNG OPTIMALER STANDORTE FUER REGIONALE SYSTEME.

BOGA 1976 -1 BOGARDI, J. J.; DUCKSTEIN, L.; PLATE, E. J.
SCHEDULING AND SEQUENCING THE CONSTRUCTION OF FLOOD CONTROL RESERVOIRS.

BOTT 1974 -1 BOTT, H.
ANWENDUNGSMOEGLICHKEITEN DER PSEUDO-BOOLSCHEN OPTIMIERUNG AUF PROBLEME DER FERTIGUNSWIRTSCHAFT UND IHRE RECHENTECHNISCHE REALISIERUNG.

BOYC 1973 -1 BOYCE, D. E.; FARHI, A.; WEISCHEDEL, R.
OPTIMAL NETWORK PROBLEM: A BRANCH-AND-BOUND ALGORITHM.

BRADC 1965 -1 BRADLEY, S. P.
SOLUTION TECHNIQUES FOR THE TRAFFIC ASSIGNMENT PROBLEM.

BRAN 1971 -1 BRANDMAN, E. I.
CERTAIN METHODS FOR THE SOLUTION OF PROBLEMS OF PROJECTED ALLOCATION OF CAPITAL INVESTMENTS WITHIN AN INDUSTRY (RUSSIAN).

BREL 1977 -1 BRELAZ, D.
PROBLEME D'HORAIRE DE LABORATOIRE DE LANGUES, UN.
BRELA 1977 -1 BRELAZ, D.; NICOLIER, Y.; WERRA, D. DE
COMPACTNESS AND BALANCING IN SCHEDULING.
BRES 1976 -1 BRESLAW, J. A.
LINEAR PROGRAMMING SOLUTION TO THE FACULTY ASSIGNMENT PROBLEM, A.
BRIC 1977 -1 BRICKER, D. L.
REFORMULATION OF SPECIAL ORDERED SETS FOR IMPLICIT ENUMERATION ALGORITHMS, WITH APPLICATIONS IN NONCONVEX SEPARABLE PROGRAMMING.
BRIT 1976 -1 BRITAVSKIJ, G. M.; JUKHIMENKO, B. I.
METHOD OF BRANCH AND BOUND FOR OPTIMIZATION OF AUTOMATIC LINES STRUCTURE PARAMETERS (RUSSIAN), A.
BROW 1976 -1 BROWN, D. B.
ALLOCATION OF FEDERAL HIGHWAY SAFETY FUNDS USING DYNAMIC PROGRAMMING, THE.
BROWD 1976 -1 BROWNELL, W. S.; LOWERRE, J. M.
SCHEDULING OF WORK FORCES REQUIRED IN CONTINUOUS OPERATIONS UNDER ALTERNATIVE LABOR POLICIES.
BRUC 1977 -1 BRUCKER, P.
PROZESSOR SCHEDULING IN VERBINDUNG MIT MULTIPROGRAMMING.
BRUC 1975 -1 BRUCKER, P.
GANZZAHLIGE LINEARE PROGRAMMIERUNG MIT OEKONOMISCHEN ANWENDUNGEN.
BRUNF 1975 -1 BRUNS, M.
REIHENFOLGEPROBLEM DER NETZPLANTECHNIK, DAS.
BULL 1970 -1 BULLINGHAM, J. M.; MATHEWS, P. A.
ELECTRONIC SIMULATOR FOR QUICKEST ROUTES IN A ROAD NETWORK.
BUR' 1976 -2 BUR'JAN, S. B.; SEROV, S. S.; UZDEMIR, A. P.
DYNAMIC INDUSTRIAL-PLANT-LOCATION PROBLEM AND A NUMERICAL METHOD FOR SOLVING IT, III (RUSSIAN).
BURK 1977 -2 BURKARD, R. E.
ZUR BESTIMMUNG VON SCHREIBMASCHINENTASTATUREN DURCH QUADRATISCHE ZUORDNUNGSPROBLEME.
BURKC 1977 -1 BURKARD, R. E.; OFFERMANN, J.
ENTWURF VON SCHREIBMASCHINENTASTATUREN MITTELS QUADRATISCHER ZUORDNUNGSPROBLEME.
BURN 1976 -1 BURNESS, R. C.; WHITE, J. A.
TRAVELING SALESMAN LOCATION PROBLEM, THE.
BURRA 1976 -1 BURRELL, J. E.
MULTIPLE ROUTE ASSIGNMENT: A COMPARISON OF TWO METHODS.
BUSH 1976 -1 BUSHELL, G.; FERLAND, J.; GUERIN, G.; NASTANSKY, L.
ENGINE SCHEDULING PROBLEM IN A RAILWAY NETWORK, THE.
CABOA 1976 -1 CABOT, V. A.
NONCONVEX PROGRAMMING PROBLEM SOLVED BY GENERALIZED BENDERS DECOMPOSITION, A.
CAMEB 1976 -1 CAMERON, A. V.; VOILS, L. R.
HEURISTIC FOR TRANSPORTATION ROUTING AND VEHICLE LOADING UNDER UNCERTAINTY, A.
CANT 1974 -1 CANTOR, D. G.; GERLA, M.
OPTIMAL ROUTING IN A PACKET-SWITCHED COMPUTER NETWORK.
CARD 1971 -1 CARDENAS, M. A.
SYSTEMS APPROACH TO INVESTMENT PLANNING IN THE ELECTRICAL INDUSTRY, THE.
CARL 1977 -1 CARLTON, D. W.
BIRTHS OF SINGLE ESTABLISHMENT FIRMS AND REGIONAL VARIATION IN ECONOMIC COSTS.
CARL 1977 -2 CARLTON, D. W.
LOCATION DECISIONS OF MANUFACTURING FIRMS.
CARL 1977 -3 CARLTON, D. W.
MODELS OF NEW BUSINESS LOCATION.
CARL 1977 -4 CARLTON, D. W.
MODELS OF SINGLE ESTABLISHMENT BIRTHS, STANDARD INDUSTRIAL CLASSIFICATION 3079.
CARN 1976 -1 CARNALL, C. A.; WILD, R.
LOCATION OF VARIABLE WORK STATIONS AND THE PERFORMANCE OF PRODUCTION FLOW LINES, THE.
CARU 1965 -1 CARUSO, F. R.
ASSEMBLY LINE BALANCING FOR IMPROVED PROFITS.
CAVA 1976 -1 CAVAILLE, J. B.; DELMAS, J.
ORDONNANCEMENT OPTIMAL A TRES COURT TERME D'UN ATELIER DE DECOUPE.
CAZA 1975 -1 CAZALET, J.; MEHR, P.
AMELIORATION DE L'UTILISATION D'UNE FLOTTE AERIENNE PARTAGEE ENTRE DEUX BASES PRINCIPALES (GENERATION DE CONVOYAGES).
CEDE 1974 -1 CEDERBAUM, I.
OPTIMAL BACKBOARD ORDERING THROUGH THE SHORTEST PATH ALGORITHM.
CHAI 1975 -1 CHAIKEN, J. M.
PATROL ALLOCATION METHODOLOGY FOR POLICE DEPARTMENTS.
CHAIA 1976 -1 CHAIKEN, J. M.; DORMANT, P.
PATROL CAR ALLOCATION MODEL, A.
CHAL 1976 -1 CHALMET, L.
SOLVING ASSIGNMENT PROBLEMS IN WAREHOUSES (DUTCH).
CHALA 1975 -1 CHALMET, L.; GELDERS, L.
ALGORITHM FOR GENERATING GOOD WAREHOUSE LAYOUTS, AN.
CHAM 1976 -1 CHAMBERS, M. L.; DYSON, R. G.
CUTTING STOCK PROBLEM IN THE FLAT GLASS INDUSTRY - SELECTION OF STOCK SIZES, THE.
CHAMA 1977 -1 CHAMPEAUX, D. DE; SINT, L.
IMPROVED BIDIRECTIONAL HEURISTIC SEARCH ALGORITHM, AN.
CHAN 1976 -2 CHAN, A. W.; FRANCIS, R. L.
ROUND-TRIP LOCATION PROBLEM ON A TREE GRAPH, A.
CHANB 1969 -1 CHAN, Y. P.
OPTIMAL TRAVEL TIME REDUCTION IN A TRANSPORT NETWORK. AN APPLICATION OF NETWORK AGGREGATION AND BRANCH AND BOUND TECHNIQUES.
CHEB 1975 -1 CHEBOTAREV, A. S.
SEQUENTIAL OPTIMIZATION IN ONE SCHEDULING PROBLEM (RUSSIAN).
CHEBA 1975 -1 CHEBOTAREV, A. S.; SHIBANOV, S. E.
PROBABILISTIC PROBLEM OF SEQUENCING PROJECTS IN SEVERAL STEPS.
CHENA 1977 -1 CHEN, DER-S.
ALGORITHM FOR A MANPOWER SCHEDULING MODEL, AN.
CHENE 1976 -1 CHEN, R.; CROWDER, H.; JOHNSON, E. L.
INTEGER PROGRAMMING FORMULATION OF THE INSTALLATION SCHEDULING PROBLEM, AN.
CHER 1970 -1 CHERENIN, V. P.
ALLOCATION OF WAREHOUSES WITH REGARD TO DEMANDS (RUSSIAN).
CHIA 1975 -1 CHIANG, C. H.
MODEL FOR MULTI-PERIOD REGIONAL WASTEWATER PLANNING, A.

****

CHIS 1977 -1 CHISMAN, J. A.
OPTIMIZING THE SHIPPING FUNCTION.

CHOU 1975 -1 CHOU, W.
COMPUTER COMMUNICATION NETWORKS - THE PARTS MAKE UP THE WHOLE, 1975 NATIONAL COMPUTER CONFERENCE.

CHOUA 1973 -1 CHOU, W.; KERSHENBAUM, A.
UNIFIED ALGORITHM FOR DESIGNING MULTIDROP TELEPROCESSING NETWORKS, A.

CHOY 1977 -1 CHOY, D. M.; WONG, C. K.
BOUNDS FOR OPTIMAL ALPHA-BETA BINARY TREES.

CLAR 1976 -1 CLARK, C. R.
VEHICLE FLEET PROBLEM, A.

CLAU 1977 -1 CLAUDER, G.; JAENICKE, W.
ZUR GRAPHENTHEORETISCHEN LOESUNG EINER KLASSE VON DISKRETEN OPTIMIERUNGSPROBLEMEN.

COCK 1975 -1 COCK, J. DE
SEARCH PROCEDURE FOR THE CONSTRUCTION OF SCHOOL TIMETABLES, A.

COCKC 1977 -1 COCKAYNE, E. J.; ROBERTS, F. D. K.
COMPUTATION OF DOMINATING PARTITIONS.

CODY 1976 -1 CODY, R. A.; COFFMAN, E. G. JR.
RECORD ALLOCATION FOR MINIMIZING EXPECTED RETRIEVAL COSTS ON DRUM-LIKE STORAGE DEVICES.

COME 1966 -1 COMES, G.; BELLON, J.-L.
ETUDE D'UNE POLITIQUE SEQUENTIELLE D'INVESTISSEMENTS POUR LA RESTRUCTURATION D'UN RESEAU D'USINES.

CONT 1971 -1 CONTI, A.; RUSSETTO, S.
LAMINATION OF SEAMLESS TUBES: A HEURISTIC METHOD FOR CHOOSING THE BARS (ITALIAN).

CORND 1977 -1 CORNUEJOLS, G.; FISHER, M. L.; NEMHAUSER, G. L.
LOCATION OF BANK ACCOUNTS TO OPTIMIZE FLOAT: AN ANALYTIC STUDY OF EXACT AND APPROXIMATE ALGORITHMS.

COVE 1976 -1 COVERDALE, I. L.; WHARTON, F.
IMPROVED HEURISTIC PROCEDURE FOR A NONLINEAR CUTTING STOCK PROBLEM, AN.

CROU 1976 -1 CROUCHER, J. S.
OPTIMAL RESOURCE ALLOCATION AMONG AREAS OF EQUAL VALUE.

CRUM 1976 -1 CRUM, R.
CASH MANAGEMENT IN THE MULTINATIONAL FIRM: A CONSTRAINED GENERALIZED NETWORK APPROACH.

CRUMA 1977 -1 CRUM, R. L.; KLINGMAN, D.; TAVIS, L. A.
OPERATIONAL APPROACH TO INTEGRATED WORKING CAPITAL PLANNING, AN.

DAEN 1977 -1 DAENINCK, G.; SMEERS, Y.
USING SHORTEST PATHS IN SOME TRANSSHIPMENT PROBLEM WITH CONCAVE COSTS.

DAFE 1976 -1 DAFERMOS, S. C.
INTEGRATED EQUILIBRIUM FLOW MODELS FOR TRANSPORTATION PLANNING.

DALA 1976 -1 DALAL, Y. K.
DISTRIBUTED ALGORITHM FOR CONSTRUCTING MINIMAL SPANNING TREES IN COMPUTER-COMMUNICATION NETWORKS, A.

DANTC 1974 -1 DANTZIG, G. B.; FRIEL, J.; GOLIGHTLY, R.; HARVEY, R. P.; MAC KNIGHT, R. D.
SOLUTION OF A LARGE SCALE AIRFORCE ORDNANCE PLANNING PROBLEM BY MATHEMATICAL PROGRAMMING.

DAVIA 1975 -1 DAVIES, O.
SIX LOCATION STRATEGIES WITH INCREASING DEMAND LEVELS.

DAVIC 1977 -1 DAVIS, K. R.; SIMMONS, L. F.
IMPROVING ASSEMBLY LINE EFFICIENCY: A DYNAMIC PROGRAMMING-HEURISTIC APPROACH.

DEAR 1977 -1 DEAR, R. G.
EFFICIENT SEQUENCING AND SCHEDULING OF AIRCRAFT IN THE TERMINAL AREA, THE.

DEBA 1976 -1 DEBANNE, J. G.
GENERALIZED NETWORK AND INTEGER PROGRAMMING IN TECHNOLOGY ASSESSMENT AND PLANNING AN ENERGY SCENARIO STUDY.

DEIN 1969 -1 DEININGER, R. A.
UEBER DIE PLANUNG VON INTERKOMMUNALEN SYSTEMEN VON KLAERANLAGEN.

DEINA 1971 -1 DEININGER, R. A.; SU, S. Y.
REGIONAL WASTE WATER TREATMENT SYSTEMS.

DEMIA 1976 -1 DEMIN, V. K.; CHEBOTAREV, A. S.
OPTIMAL SERVICE OF A DETERMINISTIC FLOW OF DEMANDS, I. LOWER BOUND (RUSSIAN).

DERI 1975 -1 DERIGS, U.
RANGZUORDNUNGSPROBLEME.

DERI 1977 -2 DERIGS, U.
RANGZUORDNUNGPROBLEME.

DERI 1977 -3 DERIGS, U.
RANGZUORDNUNGSPROBLEME UND DAS STABLE MARRIAGE PROBLEM.

DETR 1976 -1 DETROYE, J.
SYNTHESE OPTIMALE D'UN RESEAU PARCOURU PAR UN MULTIFLOT.

DINK 1976 -1 DINKEL, J. J.; KLEINDORFER, G. B.; KOCHENBERGER, G. A.; WONG, S.-N.
ENVIRONMENTAL INSPECTION ROUTES AND THE CONSTRAINED TRAVELLING SYSTEM SALESMAN PROBLEM.

DOERA 1977 -1 DOERSCH, R. H.; PATTERSON, J. H.
SCHEDULING A PROJECT TO MAXIMIZE ITS PRESENT VALUE: A ZERO-ONE PROGRAMMING APPROACH.

DOMS 1975 -1 DOMSCHKE, W.
VERFAHREN ZUR BESTIMMUNG OPTIMALER BETRIEBS- UND LAGERSTANDORTE.

DRIS 1975 -1 DRISCOLL, W. C.
SCHEDULING PRODUCTION ON ONE MACHINE WITH CHANGEOVER TIMES.

DUBO 1976 -1 DUBOV, YU. A.; EPIKHOVA, N. V.
ON AN ALGORITHM SEARCHING OPTIMAL SOLUTIONS IN THE PROBLEMS OF AUTOMATIC DESIGN (RUSSIAN).

DYKS 1977 -1 DYKSTRA, D. P.; RIGGS, J. L.
APPLICATION OF FACILITIES LOCATION THEORY TO THE DESIGN OF FOREST HARVESTING AREAS, AN.

ECON 1977 -1 ECONOMIDES, S. C.; DUESTERHOEFT, W. C.
CONSTRUCTION SCHEDULING SYSTEM WITH RESOURCE ALLOCATION FOR ELECTRIC UTILITIES, A.

EDENA 1976 -1 EDEN, B. N.
HEURISTISCHES VERFAHREN ZUR DISTRIKTERMITTLUNG IN BEWERTETEN GRAPHEN UND SEINE ANWENDUNG AUF DIE BERECHNUNG VON SCHULEINZUGSBEREICHEN, EIN.

EL-S 1975 -1 EL-SHAIEB, A. M.
MIXED-INTEGER PROGRAMMING MODEL FOR ECONOMIC RESIDENTIAL ENERGY USE, A.

ELSH 1977 -1 ELSHAFEI, A. N.
HOSPITAL LAYOUT AS A QUADRATIC ASSIGNMENT PROBLEM.

ELSHA 1974 -1 ELSHAFEI, A. N.; HALEY, K. B.
FACILITIES LOCATION: SOME FORMULATIONS, METHODS OF SOLUTION, APPLICATIONS, AND COMPUTATIONAL EXPERIENCE.

ELZIA 1976 -1 ELZINGA, J.; VEMUGANTI, R. R.
OPTIMAL SCHEDULING OF A SALES FORCE.

EPLE 1977 -1 EPLETT, W. J. R.
RESULT ON TOURNAMENTS WITH AN APPLICATION TO COUNTING, A.

EPLEA 1970 -1 EPLEY, D. L.
SCHEDULING PROBLEMS IN COMPUTER SYSTEMS.
ERLE 1976 -1 ERLENKOTTER, D.
DUAL-BASED PROCEDURE FOR UNCAPACITATED FACILITY LOCATION, A.
ERLE 1976 -2 ERLENKOTTER, D.
FACILITY LOCATION WITH PRICE-SENSITIVE DEMANDS: PRIVATE, PUBLIC, AND QUASI-PUBLIC.
ERLE 1972 -2 ERLENKOTTER, D.
ECONOMIC INTEGRATION AND DYNAMIC LOCATION PLANNING.
ERLE 1973 -2 ERLENKOTTER, D.
SEQUENCING OF INTERDEPENDENT HYDROELECTRIC PROJECTS.
ERLE 1975 -2 ERLENKOTTER, D.
CAPACITY PLANNING FOR LARGE MULTILOCATION SYSTEMS: APPROXIMATE AND INCOMPLETE DYNAMIC PROGRAMMING APPROACHES.
ERLE 1975 -3 ERLENKOTTER, D.
COMMENT ON "OPTIMAL TIMING, SEQUENCING, AND SIZING OF MULTIPLE RESERVOIR SURFACE WATER SUPPLY FACILITIES" BY L. BECKER AND W. W.-G. YEH.
ERLE 1975 -4 ERLENKOTTER, D.
COORDINATING SCALE AND SEQUENCING DECISIONS FOR WATER RESOURCES PROJECTS.
ERLE 1975 -5 ERLENKOTTER, D.
OPTIMAL INVESTMENT SCHEDULING FOR PUBLIC SECTOR PROJECTS.
ERLEA 1975 -1 ERLENKOTTER, D.; ROGERS, J. S.
SEQUENCING COMPETITIVE EXPANSION PROJECTS.
ERLEB 1975 -1 ERLENKOTTER, D.; TRIPPI, R. R.
OPTIMAL INVESTMENT SCHEDULING WITH PRICE-SENSITIVE DYNAMIC DEMAND.
ERMO 1976 -1 ERMOL'EV, YU. M.; KRIVETS, T. A.; PETUKHOV, V. S.
PROBLEM OF PLANNING SEA SHIPPINGS OF EMPTY CONTAINERS (RUSSIAN), A.
ESCUA 1976 -1 ESCUDERO, L. F.; VAZQUUEZ MUNIZ, A.
USE OF MIXED INTEGER PROGRAMMING FOR THE EVALUATION OF SOME ALTERNATE AIR POLLUTION ABATEMENT POLICIES, THE.
ESOG 1977 -1 ESOGBUE, A. O.; MARKS, B. R.
DYNAMIC PROGRAMMING MODELS OF THE NONSERIAL CRITICAL PATH-COST PROBLEM.
FALK 1963 -1 FALKENHAUSEN, H. VON
VERFAHREN ZUR PROGNOSE DER VERKEHRSVERTEILUNG IN EINEM GEPLANTEN STRASSENNETZ, EIN.
FENS 1967 -1 FENSKE, R. W.
SCHEDULING COMPUTER OPERATIONS.
FINKA 1976 -1 FINKEL'SHTEIN, YU. YU.
APPROXIMATIVE METHODS AND APPLIED PROBLEMS IN DISCRETE PROGRAMMING (RUSSIAN).
FISK 1976 -1 FISK, J.
ALGORITHM FOR ASSIGNING USES TO SOURCES IN A TRANSPORTATION TABLEAU, AN.
FISK 1977 -1 FISK, J.
SOLUTION PROCEDURE FOR A SPECIAL TYPE OF CAPACITATED WAREHOUSE LOCATION PROBLEM, A.
FISKB 1977 -1 FISK, J.; MACKEOWN, P. G.
APPLICATION OF INTEGER PROGRAMMING TO THE ASSIGNMENT OF STUDENTS TO SCHOOLS, AN.
FLOR 1976 -1 FLORIAN, M.; NGUYEN, S.
APPLICATION AND VALIDATION OF EQUILIBRIUM TRIP ASSIGNMENT METHODS, AN.
FONGA 1974 -1 FONG, C. O.
MULTI-REGION DYNAMIC CAPACITY EXPANSION PROBLEM: EXACT AND HEURISTIC APPROACHES, THE.
FONGB 1975 -1 FONG, C. O.; RAO, M. R.
PARAMETRIC STUDIES IN TRANSPORTATION-TYPE PROBLEMS.
FONGC 1974 -1 FONG, C. O.; SRINIVASAN, V.
MULTI-REGION DYNAMIC CAPACITY EXPANSION PROBLEM - I, THE.
FONGC 1974 -2 FONG, C. O.; SRINIVASAN, V.
MULTI-REGION DYNAMIC CAPACITY EXPANSION PROBLEM - II, THE.
FORC 1974 -1 FORCINA, J. L.
DYNAMIC LOCATION STRATEGIES FOR PUBLIC FACILITIES.
FOST 1976 -1 FOSTER, B. A.; RYAN, D. M.
INTEGER PROGRAMMING APPROACH TO THE VEHICLE SCHEDULING PROBLEM, AN.
FOULA 1976 -1 FOULDS, L. R.; ROBINSON, D. F.
STRATEGY FOR SOLVING THE PLANT LAYOUT PROBLEM, A.
FRANB 1974 -1 FRANCIS, R. L.; WHITE, J. A.
FACILITY LAYOUT AND LOCATION: AN ANALYTIC APPROACH.
FREDA 1975 -2 FREDMAN, M. L.
SOME ARITHMETICAL PACKING PROBLEMS.
FRIE 1976 -1 FRIEDMAN, M.
ON THE ANALYSIS AND SOLUTION OF CERTAIN GEOGRAPHICAL OPTIMAL COVERING PROBLEMS.
FRIE 1975 -1 FRIEDMAN, M.
SCHEDULING A TRANSPORTATION NETWORK.
FRIS 1963 -2 FRISCH, I. T.
OPTIMUM ROUTES IN COMMUNICATION SYSTEMS WITH CHANNEL CAPACITIES AND CHANNEL RELIABILITIES.
FROM 1976 -1 FROMOVITZ, S.; GASS, S. I.
SCHEDULING OF SUBMARINES FOR FLEET OPERATIONS SUPPORT, THE.
FUJI 1966 -1 FUJII, A.; TAKAZAWA, T.
EXAMPLES OF OR METHODS IN A MASS PRODUCTION FACTORY (JAPANESE).
FULL 1976 -1 FULLER, R. V.; JARVIS, J. J.
SELF-DUAL LINEAR PROGRAMMING MODEL FOR THE EQUILIBRIUM TRAFFIC ASSIGNMENT PROBLEM, A.
FUNK 1968 -1 FUNK, M. L.; TILLMAN, F. A.
OPTIMAL CONSTRUCTION STAGING BY DYNAMIC PROGRAMMING.
GAGN 1967 -1 GAGNON, G.
MODEL FOR FLOWING PASSENGERS OVER AIRLINE NETWORKS.
GALLA 1975 -1 GALLAGHER, J. M.
NUCLEAR FUEL REPROCESSING EXPANSION STRATEGIES.
GALLB 1976 -1 GALLO, G.; HAMMER, P. L.; SIMEONE, B.
QUADRATIC KNAPSACK PROBLEMS.
GARF 1977 -1 GARFINKEL, R. S.
MINIMIZING WALLPAPER WASTE, PART 1: A CLASS OF TRAVELING SALESMAN PROBLEMS.
GART 1976 -1 GARTNER, N.; GOLDEN, B.; WONG, R.
MODELLING AND OPTIMIZATION FOR TRANSPORTATION SYSTEMS PLANNINGS AND OPERATIONS.
GARTA 1975 -1 GARTNER, N. H.; LITTLE, J. D. C.; GABBAY, H.
OPTIMIZATION OF TRAFFIC SIGNAL SETTINGS BY MIXED-INTEGER LINEAR PROGRAMMING. PART I: THE NETWORK COORDINATION PROBLEM, PART II: THE NETWORK SYNCHRONIZATION PROBLEM.
GATE 1971 -1 GATELY, D. T.
INVESTMENT PLANNING FOR THE ELECTRIC POWER INDUSTRY: A MIXED-INTEGER PROGRAMMING APPROACH, WITH APPLICATIONS TO SOUTHERN INDIA.
GELD 1976 -1 GELDERS, L.; WASSENHOVE, L. VAN
SHORT-TIME PLANNING OF THE "JOB-SHOP" TYPE FOR A PRODUCTION SITE (DUTCH).

APPLICATION OF INTEGER PROGRAMMING

****

GEN 1974 -1 GEN, M.; OKUNO, H.
OPTIMIZATION OF REDUNDANT ALLOCATION AND SELECTION PROBLEMS OF SYSTEMS RELIABILITY BY MEANS OF 0-1 PROGRAMMING (JAPANESE).

GEN A 1976 -1 GEN, M.; OKUNO, H.; SHINOFUJI, S.
OPTIMIZING METHOD IN SYSTEM RELIABILITY WITH FAILURE-MODES BY IMPLICIT ENUMERATION ALGORITHM, AN.

GEN B 1975 -1 GEN, M.; SASAKI, M.
OPTIMAL ALLOCATION OF SYSTEM AVAILABILITY.

GEOF 1977 -1 GEOFFRION, A. M.
HOW CAN SPECIALIZED DISCRETE AND CONVEX OPTIMIZATION METHODS BE MARRIED.

GEOF 1975 -1 GEOFFRION, A. M.
GUIDE TO COMPUTER-ASSISTED METHODS FOR DISTRIBUTION SYSTEMS PLANNING, A.

GEOFB 1977 -1 GEOFFRION, A. M.; GRAVES, G. W.; LEE, S.-J.
STRATEGIC DISTRIBUTION SYSTEM PLANNING: A STATUS REPORT.

GEOR 1971 -1 GEORGE, J. A.
APPLICATION OF INTEGER PROGRAMMING TO PLANT LOCATION AND INDIVISIBILITIES, AN.

GERTA 1977 -1 GERTSBACH, I.; GUREVICH, YU.
CONSTRUCTING AN OPTIMAL FLEET FOR A TRANSPORTATION SCHEDULE.

GHIG 1976 -1 GHIGGI, C.; PULIAFITO, P. P.; ZOPPOLI, R.
COMBINATORIAL METHOD FOR HEALTH-CARE DISTRICTING, A.

GHOS 1970 -1 GHOSH, A.; SARKAR, H.
INPUT-OUTPUT MATRIX AS A SPATIAL CONFIGURATION, AN.

GIBE 1980 -1 GIBERT, A.
METHOD FOR THE TRAFFIC ASSIGNMENT PROBLEM, A.

GILLE 1975 -1 GILLETT, B. E.
COMPUTERIZED CREW SCHEDULING IN LEAD MINES.

GLEA 1975 -1 GLEASON, J. M.
SET COVERING APPROACH TO BUS STOP LOCATION, A.

GLEN 1975 -1 GLEN, J. J.
DYNAMIC PROGRAMMING MODEL FOR WORK SCHEDULING IN A SHIPYARD, A.

GLOVA 1977 -1 GLOVER, F.; HULTZ, J.; KLINGMAN, D.
IMPROVED COMPUTER-BASED PLANNING TECHNIQUES.

GLOVC 1975 -4 GLOVER, F.; KLINGMAN, D.
NETWORK APPLICATION IN INDUSTRY AND GOVERNMENT.

GLOVC 1975 -5 GLOVER, F.; KLINGMAN, D.
NEW ADVANCES IN THE SOLUTION OF LARGE SCALE NETWORK AND NETWORK-RELATED PROBLEMS.

GLOVD 1977 -1 GLOVER, F.; KLINGMAN, D.; MACMILLAN, C.
NETFORM CONCEPT: A MORE EFFECTIVE MODEL FORM AND SOLUTION PROCEDURE FOR LARGE SCALE NONLINEAR PROBLEMS, THE.

GODE 1976 -1 GODEFROID, P.
LOESUNG EINES GEMISCHT-GANZZAHLIGEN MODELLS ZUR OPTIMALEN BAUFINANZIERUNG, DIE.

GOLDB 1976 -2 GOLDEN, B. L.
RECENT DEVELOPMENTS IN VEHICLE ROUTING.

GOLDF 1967 -1 GOLDMAN, A. J.; NEMHAUSER, G.
TRANSPORT IMPROVEMENT PROBLEM TRANSFORMABLE TO A BEST-PATH PROBLEM, A.

GOLE 1974 -1 GOLENKO, D. I.; GELEVER, K. G.; KESLER, S. S.; GOFMAN, E. F.
PROBLEMS ON OPTIMAL SCHEDULING AND OPTIMAL RESOURCE ALLOCATION FOR A TECHNOLOGICAL OPERATING PROCEDURE (RUSSIAN).

GOND 1976 -2 GONDRAN, M.
PROBLEMES DE PARTITION ET DE RECOUVREMENT. APPLICATIONS ET ALGORITHMES, LES.

GONZC 1976 -1 GONZALEZ, M. J. JR.; SOH, J. W.
PERIODIC JOB SCHEDULING IN A DISTRIBUTED PROCESSOR SYSTEM.

GRAVB 1976 -1 GRAVES, R. J.; THOMAS, W. H.
ANALYTICAL APPROACH TO ACADEMIC DEPARTMENT LOCATION IN A MULTI-CAMPUS UNIVERSITY, AN.

GRIE 1976 -1 GRIER, B.
OPTIMAL NUMBER OF ALTERNATIVES AT A CHOICE POINT WITH TRAVEL TIME CONSIDERED, THE.

GRIEA 1976 -1 GRIESSHABER, R.
ERZEUGUNG VON BESATZUNGSUMLAEUFEN ZU EINEM FLUGPLAN.

GROS 1975 -1 GROS, J.
POWER PLANT SITING: A PARETIAN ENVIRONMENTAL APPROACH.

GRUNB 1977 -1 GRUNDLER, G.; SCHOCH, M.
TOURENPROBLEM MIT ZENTRALEM FUHRPARK BEI UNTERSCHIEDLICHEN FAHRZEUGTYPEN, DAS.

GRUP 1974 -1 GRUPE, H. J. C.
NEO-CLASSICAL ANALYSIS OF THE OPTIMAL LOCATION PROBLEM OF INDUSTRY (SPANISH).

GULD 1977 -1 GULDMANN, J.-M.; SHEFER, D.
OPTIMAL PLANT LOCATION AND AIR QUALITY MANAGEMENT UNDER INDIVISIBILITIES AND ECONOMIES OF SCALE.

GUPT 1976 -2 GUPTA, J. N. D.
OPTIMAL FLOWSHOP SCHEDULES WITH NO INTERMEDIATE STORAGE SPACE.

GUTE 1976 -1 GUTENBAUM, J.; NICOLO, F.
OPTIMAL AND SUB-OPTIMAL ALLOCATION OF RESOURCES IN SOME HIERARCHICAL SYSTEMS (POLISH).

HAEN 1973 -1 HAENSLER, E.
EXPERIMENTAL HEURISTIC PROCEDURE TO OPTIMIZE A TELECOMMUNICATION NETWORK UNDER NONLINEAR COST FUNCTIONS.

HAES 1976 -1 HAESSLER, R. W.; HOGUE, S. L.
NOTE ON THE SINGLE-MACHINE MULTI-PRODUCT LOT SCHEDULING PROBLEM, A.

HAKI 1969 -1 HAKIMI, S.
ALGORITHM FOR CONSTRUCTION OF THE LEAST VULNERABLE COMMUNICATION NETWORK, AN.

HALP 1976 -1 HALPERN, J.
SEQUENTIAL COVERING PROBLEM UNDER UNCERTAINTY, THE (REVISED).

HALP 1977 -2 HALPERN, J.
SEQUENTIAL COVERING PROBLEM UNDER UNCERTAINTY, THE.

HANSD 1972 -1 HANSEN, P.; KAUFMANN, L.
ALGORITHM FOR THE LOCATION OF CENTRAL FACILITIES, AN.

HANSE 1977 -1 HANSEN, P.; THISSE, J.-F.
MULTIPLIANT LOCATION FOR PROFIT MAXIMISATION.

HANSF 1977 -1 HANSON, H. H.
PROCEDURE FOR ASSIGNING WORKERS TO MACHINES WHEN EACH WORKER CAN OPERATE MORE THAN ONE MACHINE, A.

HEIN 1976 -1 HEIN, O.
ALGORITHMUS ZUR BERECHNUNG DER MINIMALEN KANTENBASIS IN GERICHTETEN GRAPHEN, EIN.

HENDA 1976 -1 HENDERSON, W. B.; BERRY, W. L.
HEURISTIC METHODS FOR TELEPHONE OPERATOR SHIFT SCHEDULING: AN EXPERIMENTAL ANALYSIS.

****

HEPP 1976 -1 HEPPES, A.; LUGOSI, G.
PRODUCTION SCHEDULING TO MINIMIZE THE SET-UP TIME OF PARALLEL MACHINES WITH ALTERNATIVE CHOICES (HUNGARIAN).
HERS 1965 -1 HERSHDORFER,A. M.
OPTIMAL ROUTING OF URBAN TRAFFIC.
HERZ 1975 -1 HERZOG, U.
OPTIMAL SCHEDULING STRATEGIES FOR REAL-TIME COMPUTERS.
HESS 1977 -1 HESSELBACH, J.; WIJNANDS, J.
DYNAMISCHE PROGRAMMIERUNG METHODIK, EDV-PROGRAMM, ANWENDUNGSBEISPIEL.
HINS 1976 -1 HINSON, J.; MULHERKAR, S.
IMPROVEMENTS TO THE CLARKE AND WRIGHT ALGORITHM AS APPLIED TO AN AIRLINE SCHEDULING PROBLEM.
HODG 1974 -1 HODGSON, M. J.
STAGING INTERURBAN HIGHWAY CONSTRUCTION BY DYNAMIC PROGRAMMING.
HODGA 1977 -1 HODGSON, T. J.; KILPATRICK, K. E.; LONGINI, I. M.
SCHEDULING MULTISPECIALTY CLINICS USING INTEGER PROGRAMMING.
HOFFC 1958 -1 HOFFMANN, W.; PAVLEY, R.
APPLICATIONS OF DIGITAL COMPUTERS TO PROBLEMS IN THE STUDY OF VEHICULAR TRAFFIC.
HOLL 1977 -1 HOLLOWAY, C. A.; NELSON, R. T.
INTERMITTEN PRODUCTION SCHEDULING: PRACTICAL FORMULATIONS, RELAXATION, AND DECOMPOSITION PROCEDURES.
HOLMA 1977 -1 HOLMES, J. H.; HAGGETT, P.
GRAPH THEORY INTERPRETATION OF FLOW MATRICES: A NOTE ON MAXIMIZATION PROCEDURES FOR IDENTIFYING SIGNIFICANT LINKS.
HONG 1974 -1 HONG, M. L.
EXPERIMENTS WITH A JOB SHOP PRODUCTION MODEL COUPLING A LINEAR PROGRAM AND A HEURISTIC SCHEDULING PROCEDURE.
HORE 1977 -1 HOREN, J. H.
EFFECT OF SCHEDULING NETWORK TELEVISION PROGRAMS ON AUDIENCE RATINGS, THE.
HORI 1973 -1 HORIE, Y.; FAN, L. T.
DESTINATION TYPE TRAFFIC ASSIGNMENT WITH AIR QUALITY CONSIDERATION.
HORN 1972 -1 HORN, R. L. VAN
OPTIMIZING TREE SEARCH SIMULATOR FOR SCHEDULING MAINTENANCE ON AIRCRAFT, AN.
HU 1973 -1 HU, T. C.
OPTIMUM COMMUNICATION SPANNING TREES.
HULS 1977 -1 HULST, W. G. VAN; LIESHOUT, J. TH. VAN
INVESTMENT/FINANCIAL PLANNING WITH ENDOGENOUS LIFETIMES: A HEURISTIC APPROACH TO MIXED-INTEGER PROGRAMMING.
HUNG 1973 -1 HUNG, H. K.
HEURISTIC ALGORITHM FOR THE MULTI-PERIOD FACILITY LOCATION PROBLEM, A.
HUNGA 1977 -1 HUNG, M. S.; FISK, J. C.
ASSORTMENT PROBLEMS AND EQUIVALENT INTEGER PROGRAMMING FORMULATIONS.
HYUN 1975 -1 HYUN, K. N.
RELIABILITY OPTIMIZATION BY 0-1 PROGRAMMING FOR A SYSTEM WITH SEVERAL FAILURE MODES.
HYUNA 1974 -1 HYUN, K. N.; MITSUO, G.; OKUNO, H.
OPTIMIZATION OF THE REDUNDANT ALLOCATION IN A SYSTEM WITH SEVERAL FAILURE MODES BY ZERO-ONE PROGRAMMING.
IBM. 1972 -1 IBM.
INTRODUCTION TO MODELLING USING MIXED INTEGER PROGRAMMING, AN.
IGUD 1975 -1 IGUDIN, R. V.
PROBLEMS OF SEQUENCING THEORY IN TRANSPORTATION AND ALGORITHMS FOR THEIR SOLUTION (RUSSIAN).
ISHI 1975 -2 ISHII, H.; IBARAKI, T.; MINE, H.
TRAVELING SALESMAN PROBLEMS WITH A CAPACITY CONSTRAINT OF THE DELIVERY TRUCK.
JACO 1973 -1 JACOB, H.
ZUR STANDORTWAHL DER UNTERNEHMUNGEN.
JAINA 1975 -1 JAIN, S. K.
SIMULATION-BASED SCHEDULING AND MANAGEMENT INFORMATION SYSTEM FOR A MACHINE SHOP, A.
JAND 1969 -1 JANDY, G.
PROBLEMS OF LOCATION IN OPERATIONS RESEARCH MODELS.
JARA 1976 -1 JARANCEV, JU. A.
CERTAIN PROCEDURE FOR DYNAMIC MEMORY ALLOCATION IN COMPUTERS (RUSSIAN), A.
JASI 1976 -1 JASINSKA, E.; RYSZKOWSKA, E.
WAREHOUSE LOCATION IN A TRANSPORTATION PROBLEM OF TWO STAGES (POLISH).
JENS 1975 -1 JENSCH, P.
DYNAMISCHE OPTIMIERUNG FUER VERKEHRSNETZE MIT ZEIT- UND ORTSABHAENGIGEN ZUSTANDSGROESSEN.
JOHNE 1976 -2 JOHNSON, E. L.
GUIDE TO SUCCESSFULLY SOLVING INTEGER PROGRAMMING PROBLEMS, A.
JOHNG 1976 -1 JOHNSON, E. L.; SPIELBERG, K.
SOLUTION OF A MULTI-COMMODITY (NO-SPLITTING) PRODUCTION-DISTRIBUTION PROBLEM.
JOHNI 1974 -1 JOHNSON, P. A.
HEURISTIC FOR INTEGRATING PLANT LOCATION AND PLANT TIMING QUESTIONS, A.
JOHNJ 1958 -1 JOHNSON, R. E.
USE OF OPERATIONS RESEARCH TECHNIQUES IN PLANT LOCATION PROBLEMS, THE.
KACP 1975 -1 KACPRZYK, J.
APPLICATION OF THE THEORY OF FUZZY SETS TO THE OPTIMAL ASSIGNMENT OF THE STAFF (POLISH).
KACPA 1975 -1 KACPRZYK, J.; STANCZAK, W.
APPLICATION OF THE METHOD OF MINIMAL GROUPS TO THE SOLUTION OF THE PROBLEM OF PARTITIONING A GROUP OF FIRMS INTO SUB-GROUPS (POLISH).
KALY 1976 -1 KALYUZHNYJ, B. I.; SOLDATOV, V. S.
MODEL OF SEQUENCING DISCRETE PRODUCTION PROCESSES (RUSSIAN), A.
KASPA 1974 -1 KASPSHITSKAJA, M. F.; SERGIENKO, I. V.; HIL'CENKO, V. I.
CERTAIN APPROACH TO THE SOLUTION OF ALLOCATION PROBLEMS (RUSSIAN), A.
KAZA 1975 -1 KAZARIN, L. S.; L'VOV, G. N.; TIMOFEEV, E. A.
HEURISTIC ALGORITHM FOR CONSTRUCTING A TIME-TABLE (RUSSIAN), A.
KENDA 1975 -1 KENDRICK, D.; STOUTJESDIJK, A.
PLANNING OF INDUSTRIAL INVESTMENT PROGRAMS: A METHODOLOGY, THE.
KHAI 1976 -1 KHAIKIN, N. M.; ZAK, E. I.
CONSTRUCTION OF MINIMUM TREE PACKING IN THE PROBLEM OF OPTIMUM MEMORY DISTRIBUTION (RUSSIAN).
KIBA 1976 -1 KIBALOV, E. B.; KHUTORETZKY, A. B.
VOLUME-TIME SCHEDULE PLANNING PROBLEM OF BUILDING (RUSSIAN), A.
KILG 1977 -1 KILGORE, W. C.
WAREHOUSE FACILITIES ANALYSIS.
KIM 1976 -1 KIM, K. V.; KRAYNOV, G. A.; SURMENEV, E. V.
ON A PROBLEM OF TIME SCHEDULE COMPILATION (RUSSIAN).

KING 1976 -1 KINGSTON, P. L.; LIPTON, S. L.; STOJKA, J. P. JR.
MIXED INTEGER PROGRAMMING MODELS FOR GENERATOR MAINTENANCE SCHEDULING.

KIRB 1966 -1 KIRBY, R. F.
MINIMUM PATH ALGORITHM FOR A ROAD NETWORK WITH TURN PENALTIES, A.

KLEEA 1976 -1 KLEE, V.; QUAIFE, H.
MINIMUM GRAPHS OF SPECIFIED DIAMETER, CONNECTIVITY AND VALENCE, I.

KLIT 1976 -1 KLITZ, J. K.
TRAVELING-SALESMAN PROBLEM FORMULATION AND SOLUTION FOR A MASS STORAGE SYSTEM, A.

KNAP 1976 -1 KNAPEK, S.
EINSATZ HEURISTISCHER VERFAHREN IN DER GEMISCHT-GANZZAHLIGEN INVESTITIONS- UND FINANZIERUNGSPLANUNG, DER.

KOCHA 1977 -1 KOCHURA, S. N.
ON AN OPTIMIZATION ALGORITHM OF CAPITAL INVESTMENTS DISTRIBUTION USING THE BOOLEAN VARIABLES (RUSSIAN).

KORTB 1976 -2 KORTE, B.
INTEGER PROGRAMMING (SURVEY PAPER).

KOVC 1976 -1 KOVCAS, L. B.; RUPP, M.
GRAPH PARTITIONING FOR MATRIX INVERSION BY INTEGER PROGRAMMING.

KRAR 1969 -1 KRARUP, J.
BRANCH-BOUND ALGORITHM FOR A WIRING PROBLEM, A.

KRAR 1967 -1 KRARUP, J.
FIXED-COST AND OTHER NETWORK FLOW PROBLEMS AS RELATED TO PLANT LOCATION AND TO THE DESIGN OF TRANSPORTATION AND COMPUTER SYSTEMS.

KRAR 1969 -2 KRARUP, J.
LUBRICATION OILS: OPTIMAL EXTENSION OF PRODUCTION PLANTS AND STORAGE FACILITIES (DANISH).

KRARD 1968 -1 KRARUP, J.; PRUZAN, P. M.
DISTRIBUTION MODELS AND SOLUTION METHODS - A SURVEY (DANISH).

KUBA 1974 -1 KUBALE, M.
HEURISTIC ALGORITHM FOR ESTABLISHING TIME-TABLES IN A FACULTY (POLISH).

KUEH 1977 -1 KUEHNER, J.; HARRINGTON, J. J.
WIRKSAME ANWENDUNG DES ALGORITHMUS FUER DIE GEMISCHT-GANZZAHLIGE PROGRAMMIERUNG MPSX-MIP VON IBM ZUR LOESUNG VON GROSSEN OEFFENTLICHEN INVESTITIONSPROBLEMEN MIT VIELEN BETEILIGTEN, TEIL II, DIE.

KUEH 1977 -2 KUEHNER, J.; HARRINGTON, J. J.
WIRKSAME ANWENDUNG DES ALGORITHMUS FUER DIE GEMISCHT-GANZZAHLIGE PROGRAMMIERUNG MPSX-MIP VON IBM ZUR LOESUNG VON GROSSEN OEFFENTLICHEN INVESTITIONSPROBLEMEN MIT VIELEN BETEILIGTEN, TEIL 1, DIE.

KUEP 1976 -1 KUEPPER, H.-U.
VERGLEICHENDE ANALYSE GANZZAHLIGER PROGRAMMIERUNGSMODELLE ZUR ABLAUFPLANUNG INDUSTRIELLER UNTERNEHMUNGEN.

KULI 1975 -1 KULIKOWSKI, R.
OPTIMUM ALLOCATION OF RESOURCES IN DECENTRALIZED MACROECONOMIC MODEL.

KYAN 1975 -1 KYAN, S.; ONAGA, K.
MINIMUM-DELAY MAXIMUM-GAUGE ROUTES IN A TIME-VARYING DYNAMIC NETWORK.

LAGEA 1976 -4 LAGEWEG, B. J.; LENSTRA, J. K.; RINNOOY KAN, A. H. G.
MINIMIZING MAXIMUM LATENESS ON ONE MACHINE: COMPUTATIONAL EXPERIENCE AND SOME APPLICATIONS.

LAI 1975 -1 LAI, A. W.
HEURISTIC SOLUTION TO THE URBAN GOODS CONSOLIDATION TERMINAL SITE SELECTION DECISION, A.

LANDB 1966 -1 LAND, A. H.; MURCHLAND, J. D.
SUBSEQUENT ASSIGNMENT OF TRAFFIC TO SHORTEST PATHS, THE.

LATE 1976 -1 LATECOERE, L.; BOEBION, J.; PUN, L.
MODEL ADAPTIVE SCHEDULING FOR SOLVING PERTURBATION PROBLEMS IN THE CLOTH INDUSTRY.

LAUR 1975 -1 LAURIA, D. T.
REGIONAL SEWERAGE PLANNING BY MIXED INTEGER PROGRAMMING.

LAURA 1974 -1 LAURIERE, J. L.
PROBLEMES D'EMPLOI DU TEMPS ET ALGORITHME DE COLORATION DES HYPERGRAPHES.

LAVE 1976 -1 LAVENBERG, S. S.; SHEDLER, G. S.
STOCHASTIC MODELING OF PROCESSOR SCHEDULING WITH APPLICATION TO DATA BASE MANAGEMENT SYSTEMS.

LAWRA 1975 -1 LAWRENCE, S. M.; LAWRENCE, K. D.
LOCATING GROUPS OF INDUSTRIES IN A REGION USING AN INPUT-OUTPUT MODEL STRUCTURE AND MULTIPLE OBJECTIVE LINEAR PROGRAMMING.

LAZA 1977 -1 LAZAK, D.; GERRITSEN, S. M.
TOPOLOGISCHE KANAELE BEI JOB-SHOP-SCHEDULING-VERFAHREN.

LEBL 1975 -1 LEBLANC, L. J.
ALGORITHM FOR THE DISCRETE NETWORK DESIGN PROBLEM, AN.

LEBLA 1974 -1 LEBLANC, L. J.; MORLOK, L. J.
ANALYSIS AND COMPARISON OF BEHAVIORIAL ASSUMPTIONS IN TRAFFIC ASSIGNMENT, AN.

LEBLB 1974 -1 LEBLANC, L. J.; MORLOK, L. J.; PIERSKALLA, W.
ACCURATE AND EFFICIENT APPROACH TO EQUILIBRIUM TRAFFIC ASSIGNMENT ON A CONJESTED NETWORKS, AN.

LEE 1976 -1 LEE, D. H.
LOW COST DRAINAGE NETWORKS.

LEE B 1976 -1 LEE, S. M.
INTERACTIVE INTEGER GOAL PROGRAMMING: METHODS AND APPLICATION.

LEGE 1977 -1 LEGENDRE, J.-P.; MINOUX, M.
APPLICATION DE LA NOTION DE DUALITE EN PROGRAMMATION EN NOMBRES ENTIERS: SELECTION ET AFFECTATION OPTIMALES D'UNE FLOTTE D'AVIONS, UNE.

LENS 1973 -1 LENSTRA, H. W.
ACYCLIC SUBGRAPH PROBLEM, THE.

LENSA 1977 -1 LENSTRA, J. K.
SEQUENCING BY ENUMERATIVE METHODS.

LERO 1975 -1 LEROY, P.
OPTIMISATION ET GRAPHES APPLICATIONS A L'ARCHITECTURE.

LEVY 1966 -1 LEVY, J.
OPTIMAL LOCATION OF NETWORKS.

LEVYB 1976 -1 LEVYI, V. D.; VOLODIN, A. A.
INTEGER PROGRAMMING PROBLEM WITH VARIABLE COEFFICIENTS (TRANSLATED FROM RUSSIAN), AN.

LIEBA 1964 -1 LIEBMAN, J. C.
BRANCH-AND-BOUND ALGORITHM FOR MINIMIZING THE COSTS OF WASTE TREATMENT, SUBJECT TO EQUITY CONSTRAINS, A.

LIET 1977 -1 LIETAER, B. A.
PLANNING MODEL FOR UNDERGROUND MINES. AN APPLICATION IN A DEVELOPING COUNTRY, A.

****

LIND 1973 -1 LIND, C. B.; LORENTSEN, R.
APPLICATION OF A MIXED INTEGER LINEAR PROGRAMMING MODEL FOR THE DIMENSIONING OF TANK TRUCK FLEETS.

LINK 1976 -1 LINKE, K.-F.
BEMERKUNGEN ZUR MATHEMATISCH-OEKONOMISCHEN MODELLIERUNG UND NUMERISCHEN LOESUNG VON REIHENFOLGEAUFGABEN AUS DER INDUSTRIELLEN FERTIGUNG.

LIPT 1976 -1 LIPTON, S. L.
SERVICE CENTER LOCATION USING MIXED INTEGER PROGRAMMING.

LIU A 1976 -1 LIU, J.
LARGE TEAM EFFICIENCY FROM AN INTEGER-PROGRAMMING THEORETICAL POINT OF VIEW.

LONG 1977 -1 LONG, C. F.
STATE OF THE ART IN SCHEDULING PRACTICE, THE.

LOS A 1977 -1 LOS, M.; NGUYEN, S.
SPATIAL ALLOCATION ON A NETWORK WITH CONGESTION.

LOVEA 1976 -1 LOVE, R. F.; WONG, J. Y.
ON SOLVING A ONE-DIMENSIONAL SPACE ALLOCATION PROBLEM WITH INTEGER PROGRAMMING.

LUDO 1971 -1 LUDOT, J.-P.; RUBINSTEIN, M.-C.
METHODE POUR LA RECHERCHE D'UNE POLITIQUE OPTIMALE D'INVESTISSEMENT A COURT TERME D'UN RESEAU DE DISTRIBUTION.

LUND 1975 -1 LUNDQVIST, L.
TRANSPORTATION ANALYSIS AND ACTIVITY LOCATION IN LAND-USE PLANNING - WITH APPLICATIONS TO THE STOCKHOLM REGION.

LYON 1977 -1 LYONS, J. P.; YOUNG, J. P.
DEVELOPMENT AND IMPLEMENTATION OF A STAFF ALLOCATION MODEL FOR MENTAL HEALTH FACILITIES.

LYTK 1974 -1 LYTKIN, I. P.
INVERSE SCHEDULING PROBLEM (RUSSIAN), THE.

LYTK 1974 -2 LYTKIN, I. P.
SCHEDULING OF MEANS OF TRANSPORTATION (RUSSIAN).

MAC K 1966 -1 MAC LAUGHLIN, W. A.
MULTI PATH SYSTEM TRAFFIC ASSIGNMENT ALGORITHM.

MAC L 1977 -1 MAC NEELY, R. A.; MALSTROM, E. M.
COMPUTER GENERATED PROCESS ROUTINGS.

MAGA 1977 -1 MAGAZINE, M. J.
SCHEDULING A PATIENT TRANSPORTATION SERVICE IN A HOSPITAL.

MAHM 1976 -1 MAHMOUD, S.; RIORDON, J. S.
OPTIMAL ALLOCATION OF RESOURCES IN DISTRIBUTED INFORMATION NETWORKS.

MAIE 1976 -1 MAIER, S. F.
SOLVING THE TRAFFIC ASSIGNMENT AND SUB-AREA FOCUSING PROBLEMS BY GEOGRAPHIC DECOMPOSITION.

MAIM 1975 -1 MAIMON, A.; PETERSON, R. M.
APPLICATIONS OF AN AIRLINE FLEET ASSIGNMENT MODEL.

MAIR 1977 -1 MAIRS, T. G.; WAKEFIELD, G. W.; JOHNSON, E. L.; SPIELBERG, K.
ON A PRODUCTION ALLOCATION AND DISTRIBUTION PROBLEM.

MANN 1967 -1 MANNE, A. S.
INVESTMENTS FOR CAPACITY EXPANSION: SIZE, LOCATION, AND TIME-PHASING.

MANT 1977 -1 MANTEL, F.; RABINOWITZ, P.
APPLICATION OF INTEGER PROGRAMMING TO THE COMPUTATION OF FULLY SYMMETRIC INTEGRATION FORMULAS IN TWO AND THREE DIMENSIONS, THE.

MARCB 1967 -1 MARCUS, B. T.; UNGAR, A. J.
MULTI-PERIOD TWO-STAGE MODEL FOR LOCATING ALUMINA AND ALUMINUM REDUCTION CAPACITY, A.

MARKA 1975 -1 MARKLAND, R. E.; SMITH, L. D.; BECKER, J. D.; HAACK, P. J.
SITE SELECTION MODEL FOR LAND DISPOSAL OF WASTEWATER, A.

MARKE 1976 -1 MARKOWSKY, G.
STRAIGHTFORWARD TECHNIQUE FOR PRODUCING MINIMAL MULTIPLE FAULT TEST SETS FOR FANOUT-FREE COMBINATIONAL CIRCUITS, A.

MARKF 1971 -1 MARKS. D. H.; JIRKA, G. H.
ENVIRONMENTAL SCREENING MODEL FOR THE LOCATION OF POWER GENERATING FACILITIES, AN.

MARTE 1965 -1 MARTIN, B.; MANHEIM, M.
RESEARCH PROGRAM FOR COMPARISON OF TRAFFIC ASSIGNMENT TECHNIQES, A.

MARTF 1963 -1 MARTIN, B. V.
MINIMUM PATH ALGORITHMS FOR TRANSPORTATION PLANNING.

MARTH 1976 -1 MARTINEZ Y RODRIGUEZ, J. B.; RUSSELL, S. O.
SEQUENTIAL TECHNIQUES FOR PROJECT SCREENING.

MARU 1976 -1 MARUYAMA, K.; TANG, D. T.; FRATTA, L.
FLOW ASSIGNMENT ALGORITHM FOR COMPUTER COMMUNICATION NETWORK DESIGN WITH DIFFERENT CLASSES OF PACKETS.

MATS 1971 -1 MATSUI, H.
THEORY OF TRAFFIC ASSIGNMENT BY MAXIMIZING THE PROBABILITY, THE.

MAY 1976 -1 MAY, A. D.
OPTIMIZATION TECHNIQUES APPLIED TO IMPROVING FREEWAY OPERATIONS.

MAY A 1974 -1 MAY, J. G.; STONE, C. A.; WAGNER, N. S.
SCHEDULING MODEL FOR THE INBAND SIGNALING SHOP, A.

MAYE 1976 -1 MAYER, J.
OPTIMALER FAHRPLAN FUER EINE PRIVATBAHN.

MENS 1975 -1 MENSCH, G.
PERSONALEINSATZPLANUNG (EINSCHL. ZUORDNUNGSMODELLE).

MERC 1974 -1 MERCHANT, D. K.; NEMHAUSER, G. L.
MODEL AND AN ALGORITHM FOR THE DYNAMIC TRAFFIC ASSIGNMENT PROBLEM, A.

MERCA 1976 -1 MERCHANT, D. K.; RAO, M. R.
MAJORITY DECISIONS AND TRANSITIVITY: SOME SPECIAL CASES.

MERE 1975 -1 MEREKIN, YU. V.
ON A PROBLEM OF OPTIMAL ALLOCATION OF THE PRODUCTION (RUSSIAN).

MERL 1971 -1 MERLINO, E. L. A.
MULTI-NATIONAL CAPACITY EXPANSION MODEL FOR PRIMARY ALUMINUM PLANTS.

MEYE 1973 -1 MEYER, C. F.
LONG-RANGE SELECTION AND TIMING ANALYSIS SYSTEM FOR FACILITY LOCATION: IMPLEMENTATION, A.

MEYL 1972 -1 MEYL, H.
SHORTEST ROUTE PROBLEM WITH ADDITIONAL PROBLEMS.

MICK 1975 -1 MICKLE, M. H.
METHOD OF OPTIMUM ROUTE REGENERATION, A.

MILLA 1976 -1 MILLER, D. M.; FYFFE, D. E.
ALLOCATING BUILDING INSPECTION MANPOWER FOR FIRE PREVENTION.

MILLC 1976 -1 MILLER, H. E.
PERSONNEL SCHEDULING AND WORK ASSIGNMENT PROBLEM, A.

MILLC 1976 -2 MILLER, H. E.
PERSONNEL SCHEDULING IN PUBLIC SYSTEMS: A SURVEY.

****

MILLD 1976 -1 MILLER, H. E.; PIERSKALLA, W. P.; RATH, G. J.
NURSE SCHEDULING USING MATHEMATICAL PROGRAMMING.

MINOA 1975 -1 MINOUX, M.
PLUS COURT-CHEMIN AVEC CONTRAINTES: ALGORITHMES ET APPLICATIONS.

MIYA 1977 -1 MIYAKAWA, M.; YUBA, T.; SUGITO, Y.; HOSHI, M.
OPTIMUM SEQUENCE TREES.

MORE 1977 -1 MOREFIELD, C. L.
APPLICATION OF 0-1 INTEGER PROGRAMMING TO MULTITARGET TRACKING PROBLEMS.

MORR 1977 -1 MORRILL, R. L.; SYMONS, J.
EFFICIENCY AND EQUITY ASPECTS OF OPTIMUM LOCATION.

MOSH 1963 -1 MOSHER, W. W. JR.
CAPACITY-RESTRAINT ALGORITHM FOR ASSIGNING FLOW TO A TRANSPORT NETWORK, A.

MOSL 1976 -1 MOSLER, K. C.
OPTIMALE TRANSPORTNETZE. ZUR BESTIMMUNG IHRES KOSTENGUENSTIGSTEN STANDORTS BEI GEGEBENER NACHFRAGE.

MUCK 1977 -1 MUCKSTADT, J. A.; KOENIG, S. A.
APPLICATION OF LAGRANGIAN RELAXATION TO SCHEDULING IN POWER-GENERATION SYSTEMS, AN.

MUEL 1975 -1 MUELLER-MERBACH, H.
MODELLING TECHNIQUES AND HEURISTICS FOR COMBINATORIAL PROBLEMS.

MUELA 1973 -1 MUELLER, H. CH.
UNTERSUCHUNG DER VERWENDBARKEIT DES ALGORITHMUS VON GOMORY FUER REIN GANZZAHLIGE PROBLEME BEIM DIENSTPLANPROBLEM.

MULL 197[illegible] -1 MULLABAEV, A. A.
ON AN OPTIMIZATION PROBLEM CONCERNING AN ARRANGEMENT OF TOOTHED WHEELS (RUSSIAN).

MULVA 1975 -2 MULVEY, J. M.
SPECIAL STRUCTURES IN NETWORK NODELS AND ASSOCIATED APPLICATIONS.

MURP 1975 -1 MURPHY, F. H.; STOHR, E. A.
MATHEMATICAL PROGRAMMING APPROACH TO THE SCHEDULING OF CHECK SORTING OPERATIONS, A.

NGUY 1975 -1 NGUYEN, H.
MULTI-DEPOT VEHICLE ROUTING PROBLEMS.

NICOB 1976 -1 NICOLETTI, B.; NATALI, L.; SCALAS, M. R.
MANPOWER STAFFING AT AN AIRPORT: A GENERALIZED SET COVERING PROBLEM.

NOON 1977 -1 NOONAN, F.; GIGLIO, R. J.
PLANNING ELECTRIC POWER GENERATION: A NONLINEAR MIXED INTEGER MODEL EMPLOYING BENDERS DECOMPOSITION.

NORD 1969 -1 NORDBECK, S.; RYSTEDT, B.
COMPUTER CARTOGRAPHY SHORTEST ROUTE PROBLEMS.

NORT 1976 -1 NORTHWESTERN UNIVERSITY; TRANSPORTATION CENTER LIBRARY
DOCTORAL DISSERTATION ON TRANSPORTATION (6. ISSUE, 1974-1975).

NOVI 1978 -1 NOVITZKY, N. A.
ANALYSIS OF FACTORS INFLUENCE ON INDUSTRY ALLOCATION EFFICIENCY (RUSSIAN).

NUNN 1976 -1 NUNNER, O.
ANWENDUNG EINES TRANSPORTMODELLS BEI DER EINBERUFUNG VON WEHRPFLICHTIGEN.

O'BRA 1975 -1 O'BRIEN, J. J.
VPM SCHEDULING FOR HIGH-RISE BUILDINGS.

OHTS 1976 -1 OHTSUKI, T.
FAST ALGORITHM FOR FINDING AN OPTIMAL ORDERING FOR VERTEX ELIMINATION ON A GRAPH, A.

OLSO 1976 -1 OLSON, E. N.; WENDELL, R. E.
ANALYSIS OF WAREHOUSE LOCATION IN THE BEEF INDUSTRY, AN.

ONOK 1976 -1 ONOKERHORAYE, A. G.
CONCEPTUAL FRAMEWORK FOR THE LOCATION OF PUBLIC FACILITIES IN THE URBAN AREAS OF DEVELOPING COUNTRIES: THE NIGERIAN CASE, A.

ORLO 1976 -1 ORLOFF, C.; CAPRERA, D.
REDUCTION AND SOLUTION OF LARGE SCALE VEHICLE ROUTING PROBLEMS.

ORLOA 1977 -1 ORLOFF, C. S.
THEORETICAL MODEL OF NET ACCESSIBILITY IN PUBLIC FACILITY LOCATION, A.

ORLOA 1976 -2 ORLOFF, C. S.
ROUTE CONSTRAINED FLEET SCHEDULING.

ORTH 1973 -1 ORTH, E.
DEKOMPOSITIONSMETHODEN - EIN HILFSMITTEL BEI DER PLANUNG REGIONALER ABWASSERBESEITIGUNGSSYSTEME.

ORTHA 1972 -1 ORTH, H.
BINAERE OPTIMIERUNG ALS HILFSMITTEL BEI DER PLANUNG REGIONALER ABWASSERBESEITIGUNSSYSTEME. I, DIE.

OSLE 1974 -1 OSLEEB, J. P.
LOCATION THEORY FOR THE UNIFORM-PRICE MANUFACTURER, A.

OSLEA 1977 -1 OSLEEB, J. P.; CROMLEY, R. G.
LOCATION-PRODUCTION-ALLOCATION PROBLEM WITH NONLINEAR PRODUCTION COSTS, THE.

OSTE 1976 -1 OSTERMANN, R.
QUELQUES EXPERIENCES DE CONFECTION AUTOMATIQUE D'HORAIRES.

OUDH 1977 -1 OUDHEUSDEN, D. L. VAN
SHORTEST DISTANCES IN A MODIFIED TRANSPORTATION NETWORK.

PANA 1977 -1 PANAGIOTAKOPOULOS, D.
COST-TIME MODEL FOR LARGE CPM-PROJECT NETWORKS.

PANA 1976 -1 PANAGIOTAKOPOULOS, D.
MULTI-COMMODITY MULTI-TRANSFORMED NETWORK FLOWS WITH AN APPLICATION TO RESIDUALS MANAGEMENT.

PARI 1963 -1 PARIKH, S.; FRISCH, I.
FINDING THE MOST RELIABLE ROUTES IN COMMUNICATION SYSTEM.

PARKA 1977 -1 PARKER, R. G.; GRAVES, R. J.; SHERALI, H.
APPLICATION OF A GRAPH-PARTITIONING ALGORITHM TO THE SCHEDULING OF CURRICULUM REQUIREMENTS, THE.

PATE 1976 -1 PATEL, A. M.; WALD, C. G. DE; COTES, L. C.
PRE-PARTITIONING AS A MEANS OF ENHANCING PAIRWISE INTERCHANGE SOLUTIONS TO QUADRATIC ASSIGNMENT PROBLEMS.

PATTA 1976 -1 PATTERSON, J. H.; ROTH, G. W.
SCHEDULING A PROJECT UNDER MULTIPLE RESOURCE CONSTRAINTS: A ZERO-ONE PROGRAMMING APPROACH.

PAUC 1976 -1 PAUCK, R.
MODELL ZUR BESTIMMUNG VON PERSONALEINSATZPLAENEN BEI LUFTVERKEHRSGESELLSCHAFTEN (CREW SCHEDULING), EIN.

PAULA 1976 -1 PAULAUSKAS, M. A.; RASHKILIS, A. V.; SHYARKSHNIS, G. A.
PROBLEM CONCERNING THE OPTIMAL ALLOCATION OF DEFICIENT RESOURCES FOR A DISCRETE CONSUMPTION, 2. ALGORITHMS FOR THE APPROXIMATE SOLUTION OF PROBLEMS WITH A GREAT NUMBER OF UNKNOWNS (RUSSIAN).

****

PAULA 1976 -2 PAULAUSKAS, M. A.; RASHKILIS, A. V.; SHYARKSHNIS, G. A.
PROBLEM CONCERNING THE OPTIMAL ALLOCATION OF DEFICIENT RESOURCES FOR A DISCRETE CONSUMPTION, 3. HEURISTIC ALGORITHMS FOR THE SOLUTION OF THE PROBLEM (RUSSIAN).

PEAR 1974 -1 PEARMAN, A.
HEURISTIC APPROACHES TO ROAD NETWORK OPTIMIZATION.

PEKI 1975 -1 PEKICH, T. J.; ENSCORE, E. E.
EQUIPPING AN EMERGENCY VEHICLE TO MAXIMIZE PATIENT CARE THROUGH THE USE OF INTEGER PROGRAMMING.

PEREA 1976 -1 PEREPELITSA, V. A.; SEVAST'YANOV, S. V.
ON A SEQUENCING PROBLEM IN A NETWORK (RUSSIAN).

PERLA 1975 -1 PERLMAN, B. L.
BRANCH-AND-BOUND METHODS APPLIED TO REAL-TIME VEHICLE DISPATCHING.

PETEB 1977 -1 PETERSEN, J. P.
MANUAL PROCEDURE FOR DESIGNING MAILMAN-ROUTES.

PETRB 1976 -1 PETROV, S. V.
ON A PROBLEM IN GRAPH THEORY (RUSSIAN).

PLAX 1976 -1 PLAXTON, A.
LINEAR PROGRAMMING MODEL TO OPTIMALLY SCHEDULE CHECK PROCESSING ACTIVITIES, A.

PLOTA 1974 -1 PLOTNIKOV, V. N.; ZVEREV, V. YU.
METHODS OF FAST DISTRIBUTION OF ALGORITHMS IN COMPUTATIONAL SYSTEMS (RUSSIAN).

POCH 1976 -1 POCHTMAN, YU. M.; RAFES, R. YU.
MODIFICATION OF THE DYNAMIC PROGRAMMING ALGORITHM FOR PROBLEMS CONCERNING THE ALLOCATION OF RESOURCES (RUSSIAN).

POLL 1977 -1 POLLACK, M.
SOME ELEMENTS OF THE AIRLINE FLEET PLANNING PROBLEM.

POLY 1976 -1 POLYMERIS, A.
OPTIMIERUNG UND AUFTEILUNG DER KOSTEN REGIONALER ABWASSERVERBAENDE.

POPO 1973 -1 POPOVICIU, N.
ALLOCATION PROBLEM, WITH BIVALENT ZERO-ONE VARIABLES DEFINED BY MEANS OF THE ELEMENTS OF A TRANSPORTATION PROBLEM (RUMANIAN), AN.

POUL 1975 -1 POULTON, M. C.; KANAFANI, A.
APPLICATION OF LOCATION MODELS TO OFF-AIRPORT TERMINALS, THE.

POWE 1977 -1 POWELL, J.; GLEASON, J.; BURTON, J.
POWER GENERATOR SCHEDULING BY DYNAMIC PROGRAMMING.

PRAT 1973 -1 PRATT, V. R.
LEAST UPPER BOUND ON CONVENTIONAL SHELLSORT, A.

PRIV 1974 -1 PRIVALOV, V. V.; TETERIN, G. P.
PROBLEM OF CHARGING A FORGE PRESS WITH 0-1-VARIABLES (RUSSIAN).

PUN 1975 -1 PUN, L.; BOEBION, J.; LATECOERE, L.; ALMAR, A.; PHILIPE, J.; DEROUAULT, Y.
ORDONNANCEMENT ADAPTIF DANS LES ENTREPRISES INDUSTRIELLES DE TAILLE MOYENNE.

RANDD 1976 -1 RANDOLPH, P. H.
JOB SHOP SCHEDULING: A CASE STUDY.

RAO 1975 -1 RAO, M. R.
MULTI-REGION, MULTI-PERIOD CAPACITY EXPANSION MODELS: A HEURISTIC AND A SPECIAL CASE.

RAPA 1959 -1 RAPAPORT, H.; ABRAMSON, P.
ANALOG COMPUTER FOR FINDING AN OPTIMUM ROUTE THROUGH A COMMUNICATIONS NETWORK, AN.

RATH 1977 -2 RATHMANN, D.
ZUR LOESUNG DISKRETER OPTIMIERUNGSPROBLEME MIT GRUPPENTHEORETISCHEN ANSAETZEN.

RAYB 1966 -1 RAYBOV, G. G.
ALGORITHM FOR THE SOLUTION OF A LABYRINTH ON A DISCRETE FIELD AND ITS APPLICATIONS (TRANSLATED FROM RUSSIAN), AN.

RECS 1976 -1 RECSKI, A.
MATROIDS AND INDEPENDENT STATE VARIABLES.

RECS 1976 -2 RECSKI, A.
MATROIDS AND N-PORTS.

REDD 1976 -1 REDDI, S. S.
INTERDEPENDENCE OF INDUSTRIAL ENGINEERING AND COMPUTING: SELECTION OF SCHEDULING CRITERIA IN COMPUTER SYSTEMS, THE.

RIBB 1972 -1 RIBBECK, K. F.
ROUTENSUCHEN IN STADTSTRASSENNETZEN.

RICH 1976 -1 RICHARDSON, R.
OPTIMIZATION APPROACH TO ROUTING AIRCRAFT, AN.

RIDE 1976 -1 RIDER, K. L.
PARAMETRIC MODEL FOR THE ALLOCATION OF FIRE COMPANIES IN NEW YORK CITY, A.

RIDE 1975 -1 RIDER, K. L.
PARAMETRIC MODEL FOR THE ALLOCATION OF FIRE COMPANIES: EXECUTIVE SUMMARY, A.

RIDL 1965 -1 RIDLEY, T. M.
INVESTMENT IN A NETWORK TO REDUCE LENGTH OF A SHORTEST ROUTE.

ROBEB 1972 -1 ROBERTS, K. J.
DISCUSSION OF 'SELECTING SOLID WASTE DISPOSAL FACILITIES' BY HELMS AND CLARK.

ROCH 1976 -1 ROCHAT, J. C.
ON AN OPTIMIZATION PROBLEM IN QUEUEING.

RODG 1977 -1 RODGERS, J. R.; MAC CARTHY, O.
SERVICE FACILITY LOCATION PROBLEM: THE LOCATION OF FARM ADVISORY OFFICERS IN N. Z., A.

ROSE 1976 -1 ROSE, D. J.; TARJAN, R. E.; LUEKER, G. S.
ALGORITHMIC ASPECTS OF VERTEX ELIMINATION ON GRAPHS.

ROSS 1977 -1 ROSS, G. T.; SOLAND, R. M.
MODELING FACILITY LOCATION PROBLEMS AS GENERALIZED ASSIGNMENT PROBLEMS.

RUBIC 1975 -1 RUBIN, I.
MULTITERMINAL PATH FLOWS FOR PACKET-SWITCHING COMMUNICATION NETWORKS.

RUBIE 1976 -1 RUBINSTEIN, M. C.; LUDOT, J. P.
APPLICATION DE LA THEORIE DES GRAPHES AU CHOIX DES INVESTISSEMENTS A COURT TERME DANS UN RESEAU ELECTRIQUE SUJET A AVARIES.

RUBIF 1974 -1 RUBINSTEIN, M.-C.
APPLICATION DE LA THEORIE DES GRAPHES AU CHOIX DES INVESTISSEMENTS DANS UN RESEAU D'ENERGIE ELECTRIQUE SUJET A AVARIES.

RUIT 1968 -1 RUITER, E. R.
TRAFFIC ASSIGNMENT RESEARCH, FINAL REPORT.

RUSSA 1976 -1 RUSSELL, A. H.
GEOMETRICAL SOLUTION TO A TREE NETWORK PROBLEM, A.

RUSSB 1977 -1 RUSSELL, R.; IGO, W.
ASSIGNMENT ROUTING PROBLEM, AN.

RUYG 1976 -1 RUYGROK, J. T.
FESTLEGUNG DER REIHENFOLGE DES LOESCHENS VON SEESCHIFFEN MIT HILFE DER GANZZAHLIGEN PROGRAMMIERUNG.

****

RYAN 1976 -1 RYAN, D. M.; FOSTER, B. A.; CHESHIRE, I. M.
COMPLEXITY OF REAL-WORLD SCHEDULING PROBLEMS, THE.

RYZH 1974 -1 RYZHKOV, A. P.
ORGANIZATION OF COMPUTER COMPUTATION OF INFORMATIONALLY CONNECTED PROGRAMS (RUSSIAN).

SAAR 1977 -1 SAARI, D. G.
APPORTIONMENT METHODS AND THE HOUSE OF REPRESENTATIVES.

SAAT 1977 -1 SAATCIOGLU, O.
MATHEMATICAL PROGRAMMING MODELS FOR AIRPORT SITE SELECTION.

SADY 1976 -1 SADYKOV, I. H.; KOKOTOV, V. M.
CERTAIN PROBLEM FOR COMPILING AN OPTIMAL SCHEDULE OF FREIGHT SHIPMENTS (RUSSIAN), A.

SAFA 1970 -1 SAFAEVA, K.
CERTAIN DESIGN PROBLEM (RUSSIAN), A.

SAGU 1976 -1 SAGUEZ, C.
INTEGER PROGRAMMING APPLIED TO OPTIMAL CONTROL.

SAHBA 1974 -1 SAHBAZJAN, K. V.; TUSKINA, T. A.; LEBEDINSKAJA, N. B.
ALGORITHM FOR COMPILING AN OPTIMAL SCHEDULE FOR A PARALLEL-SEQUENTIAL NETWORK (RUSSIAN), AN.

SAKA 1976 -1 SAKAGUCHI, M.
SEQUENTIAL ALLOCATION PROBLEM FOR RANDOMLY APPEARING TARGETS, A.

SALE 1976 -1 SALERNO, G.
"BRANCH AND BACK-TRACK" ALGORITHM FOR THE DESIGN OF AN OPTIMAL TRANSPORTATION NETWORK (ITALIAN), A.

SANC 1976 -1 SANCHO, N. G. F.
OPTIMAL SOLUTION FOR SCHEDULING OF PRINTING AND BINDING, AN.

SANK 1975 -1 SANKIBAEV, B. N.
INTEGER DISTRIBUTION PROBLEM (RUSSIAN), AN.

SCAR 1977 -1 SCARBROUGH, J. E.
MIXED-INTEGER PROGRAMMING MODEL FOR THE SCHEDULING OF AIRCRAFT BY TIME OF DAY WITHIN A NETWORK OF AIRPORTS, A.

SCHAA 1971 -1 SCHAERF, S.; SHAHAM, J.
PUBLIC TRANSPORTATION SIMULATION ALGORITHM, A.

SCHAB 1972 -1 SCHALLEHN, W.; SEBASTIAN, H.-J.
REIHENFOLGEPROBLEM IN DER NETZPLANTECHNIK, EIN.

SCHOA 1970 -1 SCHOEMAN, M. E. F.; BHAUMIK, G.
CLASS SCHEDULING MODEL.

SCHU 1968 -1 SCHULTZ, G. P.
MANAGERIAL DECISION-MAKING IN LOCAL GOVERNMENT: FACILITY PLANNING FOR SOLID WASTE COLLECTION.

SCHU 1969 -1 SCHULTZ, G. P.
FACILITY PLANNING FOR A PUBLIC SERVICE SYSTEM: DOMESTIC SOLID WASTE COLLECTION.

SCHUA 1977 -1 SCHURAWLJEW, J. I.; FINKEL'SHTEIN, YU. YU.
AUFGABENTYPEN UND LOESUNGSMETHODEN IN DER DISKRETEN OPTIMIERUNG.

SCOT 1971 -1 SCOTT, A. J.
COMBINATORIAL PROGRAMMING, SPATIAL ANALYSIS, AND PLANNING.

SCWA 1968 -1 SCWARTZ, J. G.
NETWORK FLOW AND SYNTHESIS MODELS FOR TRANSPORTATION PLANNING: A SURVEY.

SEME 1974 -1 SEMËNOV, V. M.; HAIKIN, N. M.; MOGILA, V. G.
CERTAIN METHOD OF SOLVING A VOLUME SCHEDULING PROBLEM (RUSSIAN), A.

SERG 1975 -1 SERGIENKO, I. V.
APPLICATION OF THE METHOD OF THE VECTOR OF DESCENT TO THE SOLUTION OF INTEGER PROGRAMMING PROBLEMS (RUSSIAN).

SHAPB 1976 -1 SHAPIRO, J. F.
MULTIPLE CRITERIA PUBLIC INVESTMENT DECISION MAKING BY MIXED INTEGER PROGRAMMING.

SHARE 1975 -1 SHARP, G. P.; CHU, J.
COMBINATORIAL APPROACH TO SCHEDULING PUBLIC TRANSIT OPERATORS, A.

SHARF 1972 -1 SHARPE, R.; BROTCHIE, J. F.
URBAN SYSTEMS STUDY, AN.

SHARG 1975 -1 SHARPE, R.; BROTCHIE, J. F.; AHERN, P. A.
EVALUATION OF ALTERNATIVE GROWTH PATTERNS FOR MELBOURNE.

SHAW 1975 -1 SHAW, G. C.
QUALITY CONTROL MEASURES FOR CANADIAN AIRLINE SCHEDULES.

SHAWA 1974 -1 SHAW, J. F. B.
OPTIMIZATION OF TRANSPORTATION ROUTES, APPLIED TO A FLORIDA KEYS SEVEN MILE BRIDGE.

SHEL 1975 -1 SHELNESS, N. H.; STEPHENS, P. D.; WHITFIELD, H.
EDINBURG MULTI-ACCESS SYSTEM SCHEDULING AND ALLOCATION PROCEDURES IN THE RESIDENT SUPERVISOR, THE.

SHEPA 1975 -1 SHEPSLE, K. A.
CONGRESSIONAL COMMITTEE ASSIGNMENTS: AN OPTIMIZATION MODEL WITH INSTITUTIONAL CONSTRAINTS.

SHEPA 1975 -2 SHEPSLE, K. A.
OPTIMIZATION MODEL OF CONGRESSIONAL COMMITTEE ASSIGNMENTS, AN.

SHIE 1976 -2 SHIER, D. R.
INVERTING SPARSE MATRICES BY TREE PARTITIONING.

SHIH 1977 -1 SHIH, W.
BRANCH AND BOUND PROCEDURE FOR A CLASS OF DISCRETE RESOURCE ALLOCATION PROBLEMS WITH SEVERAL CONSTRAINTS, A.

SHIM 1951 -1 SHIMBEL, A.
APPLICATION OF MATRIX ALGEBRA TO COMMUNICATION NETS.

SHIM 1953 -1 SHIMBEL, A.
STRUCTURAL PARAMETERS OF COMMUNICATION NETWORKS.

SHOW 1977 -1 SHOWALTER, M. J.; KRAJEWSKI, L. J.; RITZMAN, L. P.
MANPOWER ALLOCATION IN U.S. POSTAL FACILITIES: A HEURISTIC APPROACH.

SMITA 1976 -1 SMITH, E. M.; OLSON, R. E.
MIXED INTEGER LINEAR PROGRAMMING MODEL APPLIED TO PETROLEUM PRODUCTS BLENDING.

SMITC 1976 -1 SMITH, K. R.; OVER, A. M. JR.; HANSEN, M. F.; GOLLADAY, F. L.; DAVENPORT, E. J.
ANALYTIC FRAMEWORK AND MEASUREMENT STRATEGY FOR INVESTIGATING OPTIMAL STAFFING IN MEDICAL PRACTICE.

SMITD 1976 -1 SMITH, L. D.
APPLICATION OF AN INTERACTIVE ALGORITHM TO DEVELOP CYCLICAL ROTATIONAL SCHEDULES FOR NURSING PERSONNEL, THE.

SMITD 1975 -1 SMITH, L. D.
SCHEDULING NURSING PERSONNEL: APPLICATION OF LIST PROCESSING AND PROBLEM ORIENTED DATA STRUCTURES.

SMITE 1977 -1 SMITH, L. D.; WIGGINS, A.
COMPUTER-BASED NURSE SCHEDULING SYSTEM, A.

****

SMITG 1975 -1 SMITH, R. S.; GIGLIO, R. J.
TWO PHASE EMPLOYEE SCHEDULING ALGORITHM FOR REDUCING LABOR COSTS IN FOOD SERVICE OPERATIONS, A.
SNIC 1978 -1 SNICKARS, F.
CONVEXITY AND DUALITY PROPERTIES OF A QUADRATIC INTRAREGIONAL LOCATION MODEL.
SPAN 1976 -1 SPANEL, J.
DYNAMIC PROGRAMMING FOR SOLVING ONE MULTISTATE DECISION-MAKING PROBLEM WITH INTEGER VARIABLES (CZECH).
STAHA 1976 -1 STAHL, W.
SENSITIVITAETSANALYSE MEHRSTUFIGER STANDORTPROBLEME.
STAI 1977 -1 STAINTON, R. S.
CUTTING STOCK PROBLEM FOR THE STOCKHOLDER OF STEEL REINFORCEMENT BARS, THE.
STAU 1974 -1 STAUDER, E.
SIMULATION MODELLING FOR ANALYSIS OF OPTIMUM SEQUENCING IN THE SHOP-FLOOR.
STIE 1976 -1 STIERLIN, C.
ZUR STANDORTBESTIMMUNG VON KRANKENWAGENSYSTEMEN.
STOL 1977 -1 STOLZENBURG, I.
OPTIMIERUNGSVERFAHREN FUER REIHENFOLGEPROBLEME IN DER CONTAINER-SCHIFFAHRT.
STRAA 1976 -1 STRATMANN, K.-H.
NUMERISCHE UNTERSUCHUNGEN UEBER QUADRATISCHE ZUORDNUNGSPROBLEME.
STRI 1970 -1 STRICKER, R.
PUBLIC SECTOR VEHICLE ROUTING: THE CHINESE POSTMAN PROBLEM.
STYC 1974 -1 STYCZEN, K.
TRANSPORTATION SYSTEM IN THE RAW MATERIAL PREPARATION DEPARTMENT OF A CEMENT PLANT: MATHEMATICAL MODEL AND CONTROL ALGORITHM (POLISH), A.
SUSS 1965 -1 SUSSENGUTH, E. H.
GRAPH THEORETIC ALGORITHM FOR MATCHING CHEMICAL STRUCTURES, A.
SWAI 1976 -1 SWAIN, R. W.
CENTRAL FACILITIES LOCATION WITH MULTIPLE PLANNING PERIODS.
SWEE 1977 -1 SWEENEY, D. J.; MURPHY, R. A.
DECOMPOSITION PRINCIPLE FOR INTEGER PROGRAMS, A (REVISED VERSION).
SWEEA 1976 -1 SWEENEY, D. J.; TATHAM, R. L.
IMPROVED LONG-RUN MODEL FOR MULTIPLE WAREHOUSE LOCATION, AN.
TAKAA 1974 -1 TAKAHASHI, Y.
SEQUENCING MODEL WITH AN APPLICATION TO SPEED CLASS SEQUENCING IN AIR TRAFFIC CONTROL, A.
TARJ 1976 -2 TARJAN, R. E.
ITERATIVE ALGORITHMS FOR GLOBAL FLOW ANALYSIS.
TART 1976 -1 TARTAR, J.
OPTIMIZATION STRATEGIES FOR COMPUTER RESOURCE ALLOCATION.
TAYL 1975 -1 TAYLOR, P. E.
CELLOPHANE STOCK CUTTING PROBLEM AND A SEQUENTIAL DECISION MODEL, THE.
THIR 1966 -1 THIRIET, L.; GAUSSENS, J.
METHODE DE CHOIX DES AFFECTATIONS ET DES INVESTISSEMENTS.
THUM 1975 -1 THUMER, R.
VERFAHREN ZUR AUSWAHL VON TEILNETZEN BEI DER BESTIMMUNG KUERZESTER WEGE IN VERKEHRSNETZEN, EIN.
TIFF 1974 -1 TIFFIN, J. E.; FOLIE, G. M.
RECENT DEVELOPMENTS IN TRANSPORTATION ALGORITHMS.
TILA 1976 -1 TILANUS, C. B.; GERHARDT, C.
APPLICATION OF CUTTING STOCK IN THE STEEL INDUSTRY, AN.
TIND 1977 -2 TIND, J.
ON ANTIBLOCKING SETS AND POLYHEDRA.
TOBI 1977 -1 TOBIN, R. L.
EXTENSION OF DIAL'S ALGORITHM UTILIZING A MODEL OF TRIPMAKERS' PRECEPTIONS, AN.
TOML 1966 -1 TOMLIN, J. A.
LINEAR PROGRAMMING MODEL FOR THE ASSIGNMENT OF TRAFFIC, A.
TRAN 1976 -1 TRAN DINH AM; TSUKIYAMA, S.; SHIRAKAWA, I.; OZAKI, H.
ALGORITHM FOR GENERATING ALL THE DIRECTED PATHS AND ITS APPLICATION, AN.
TRIF 1973 -1 TRIFON, R.; LIVNAT, A.
SPATIAL ALLOCATION OF SCHOOLS OVER TIME IN CITIES, THE.
TRIV 1976 -1 TRIVEDI, V. M.; WARNER, D. M.
BRANCH AND BOUND ALGORITHM FOR OPTIMUM ALLOCATION OF FLOAT NURSES, A.
TRYG 1976 -1 TRYGAR, T. A.; WRIGHT, G.
APPLICATION OF DECOMPOSITION TECHNIQUES TO THE CALCULATION OF A MINIMUM DISTANCE MATRIX FOR A LARGE SCALE STREET NETWORK MODEL, THE.
TUER 1977 -1 TUERK, P.
OPTIMIERUNG VON VERSORGUNGSNETZEN.
TURC 1974 -1 TURCAT, C.
ISOMORPHISME, IMMERSION ET RECOUVREMENT DE GRAPHES.
TURK 1977 -1 TURK, P.
OPTIMIERUNG VON VERSORGUNGSNETZEN.
TURNA 1975 -1 TURNER, W. C.; HOUGLAND, E. S.
OPTIMAL ROUTING OF SOLID WASTE COLLECTION VEHICLES, THE.
TURNB 1976 -1 TURNER, W. C.; TERRELL, M. P.
VEHICLE ROUTING ACTIVITY AT OKLAHOMA STATE UNIVERSITY.
ULLM 1976 -1 ULLMAN, J. D.
COMPLEXITY OF CODE GENERATION, THE.
ULRI 1975 -1 ULRICH, L. Y.
HEURISTIC ALGORITHMS FOR SOLVING A LARGE SCALE MULTICOMMODITY FLOW PROBLEM ON A NETWORK WITH A STEP FUNCTION COST.
UNIT 1964 -1 UNITED STATES DEPT. OF COMMERCE BUREAU OF PUBLIC ROADS.
TRAFFIC ASSIGNMENT MANUAL.
VASE 1975 -1 VASERSHTEJN, L. N.; SHERMAN, E. B.
FORMING VARIANTS IN THE DISCRETE PROBLEMS OF OPTIMAL PLANNING BY BRANCHING (RUSSIAN).
VERGA 1977 -1 VERGIN, R. C.; SCRIABIN, M.
MAINTENANCE SCHEDULING FOR MULTICOMPONENT EQUIPMENT.
VINC 1975 -1 VINCENTELLI, S. A.
OPTIMAL BORDERED TRIANGULAR FORM OF A SPARSE MATRIX, A.
VINO 1975 -1 VINOGRADSKAJA, S. V.
DETERMINATION OF THE OPTIMAL NUMBER OF SPARE PARTS FOR AUTOMATED MECHANISMS USED IN LARGE-SCALE INDUSTRIAL PLANTS (RUSSIAN).
VISH 1976 -1 VISHNEVSKIJ, L. D.
ON A SCHEDULING PROBLEM FOLLOWING A CALENDAR (RUSSIAN).
VOLY 1974 -1 VOLYNSKII, E. I.; VJATKIN, A. K.; KRASIN, L. A.
ASSIGNMENT PROBLEM FOR A PRODUCTION LINE WITH OPERATIVE STORAGE POINTS (RUSSIAN), THE.

****

VU 1976 -1 VU, V. T.; TURNER, W. C.
SYSTEMS DESIGN FOR RURAL REFUSE COLLECTION.

VYAT 1975 -1 VYATKIN, A. K.; KRASIN, L. A.
PROBLEM OF OPTIMIZING A HIERARCHICAL ORGANIZATION STRUCTURE (RUSSIAN), A.

WALT 1976 -1 WALTERS, J. R.
EXTENSION TO THE ASSORTMENT PROBLEM IN WHICH COMPONENTS ARE GROUPED TO FORM ASSEMBLIES, AN.

WANGB 1974 -1 WANG, J. C.; MICKLE, M. H.; HOELZEMAN, R. G.
SET COVERING APPLICATION IN MINIMUM ROUTE GENERATION, A.

WATT 1963 -1 WATTLEWORTH, J. A.; SHULDINER, P. W.
ANALYTICAL METHODS IN TRANSPORTATION; LEFT-TURN PENALTIES IN TRAFFIC ASSIGNMENT MODELS.

WDOW 1973 -1 WDOWIAK, J.
SOLUTION OF THE FIXED CHARGE PROBLEM (POLISH), THE.

WEGL 1976 -1 WEGLARZ, J.
TIME-OPTIMAL CONTROL OF RESOURCE ALLOCATION IN A COMPLEX OF OPERATIONS FRAMEWORK.

WELL 1976 -1 WELLS, D. W.
POWER SYSTEM SCHEDULING USING INTEGER PROGRAMMING.

WERR 1976 -2 WERRA, D. DE
EVERYTHING YOU ALWAYS WANTED TO KNOW ABOUT S.EX.

WERR 1975 -4 WERRA, D. DE
ON A PARTICULAR CONFERENCE SCHEDULING PROBLEM.

WESS 1972 -1 WESSELS, H.
VERGLEICH VERSCHIEDENER ANSAETZE ZUR TRIANGULATION VON INPUT-OUTPUT-MATRIZEN.

WEST [illegible] WESTERBERG, C.-H.; BJORKLUND, B.; HULTMAN, E.
APPLICATION OF MIXED INTEGER PROGRAMMING IN A SWEDISH STEEL MILL, AN.

WHITC 1975 -1 WHITE, D. J.
NOTE ON FACULTY TIMETABLING, A.

WILLE 1976 -1 WILLIS, R. J.; HASTINGS, N. A. J.
PROJECT SCHEDULING WITH RESOURCE CONSTRAINTS USING BRANCH AND BOUND METHODS.

WILS 1973 -1 WILSON, A. G.
FURTHER DEVELOPMENTS OF ENTROPY MAXIMISING TRANSPORT MODELS.

WILSB 1974 -1 WILSON, D.; SMITH, R. L.
LOCATION OF STOCHASTIC SERVICES, THE.

WING 1961 -1 WING, O.
ALGORITHMS TO FIND THE MOST RELIABLE PATH IN A NETWORK.

WINGA 1965 -1 WING, O.; GOLDMAN, A. J.
MOST PROFITABLE ROUTINE BEFORE MAINTENANCE.

WOEH 1973 -1 WOEHNER, W.
GUETERZUGBILDUNG ALS KOMBINATORISCHE OPTIMIERUNGSAUFGABE.

WONG 1976 -1 WONG, C. K.
SOME DISCRETE OPTIMIZATION PROBLEMS WITH SECONDARY STORAGE APPLICATIONS.

WONGA 1976 -1 WONG, C. K.; YAO, A. C.-C.
COMBINATORIAL OPTIMIZATION PROBLEM RELATED TO DATA SET ALLOCATION, A.

YALO 1976 -1 YALOVIJ, V. V.
ON SOME PROBLEMS IN THE ALLOCATION OF RESOURCES (UKRAINIAN).

YOUN 1970 -1 YOUNG, G. K.; MOSELY, J. C.; EVENSON, D. E.
TIME SEQUENCING OF ELEMENT CONSTRUCTION IN A MULTI-RESERVOIR SYSTEM.

ZAIT 1976 -1 ZAITSEVA, Z. N.; SHTEIN, M. E.
HAMILTONIAN CYCLES IN SOLVING LAYING OUT PROBLEM (RUSSIAN).

ZANG 1966 -1 ZANGWILL, W.
SHORTEST ROUTE PROBLEM UNDER EITHER CONCAVE OR CONVEX SHIPPING COSTS, THE.

ZAYC 1975 -1 ZAYCHENKO, YU. P.; PECHURIN, M. K.; TOMASHEVSKIY, V. M.
GENERALIZED ASSIGNMENT PROBLEM AND ITS SOLUTION ALGORITHM (UKRAINIAN), THE.

ZEVI 1976 -1 C ZEVIN, A. A.
ON ARTICLE UNIFICATION PROBLEM (RUSSIAN).

ZUSU 1977 -1 ZUSUPBAEV, A.
SOLUTION OF THE SEASONAL PRODUCTION ALLOCATION PROBLEM BY THE METHOD OF SUCCESSIVE CALCULATIONS (RUSSIAN).

****

ABEL 1977 -1 ABELLA, R. J.; BOVA, T. E.
OPTIMAL PLANT ALLOCATION OF STOCKKEEPING UNITS.

AUST 1976 -2 AUST, R. J.
IMPROVEMENT ALGORITHM FOR SCHOOL TIMETABLING, AN.

BALA 1976 -1 BALACHANDRAN, V.
INTEGER GENERALIZED TRANSPORTATION MODEL FOR OPTIMAL JOB ASSIGNMENT IN COMPUTER NETWORKS, AN.

BAND 1975 -1 BANDLER, J. W.; CHEN, J. H. K.; DALSGAARD, P.; LIU, P. C.
TOLOPT - A PROGRAM FOR OPTIMAL, CONTINUOUS OR DISCRETE, DESIGN CENTERING AND TOLERANCING, PART 1 - USER'S GUIDE, PART 2 - FORTRAN LISTING.

BARR 1977 -1 BARR, R. S.; GLOVER, F.; KLINGMAN, D.
ALTERNATING BASIS ALGORITHM FOR ASSIGNMENT PROBLEMS, THE.

BARR 1977 -2 BARR, R. S.; GLOVER, F.; KLINGMAN, D.
ALTERNATING BASIS ALGORITHM FOR ASSIGNMENT PROBLEMS, THE.

BARR 1977 -4 BARR, R. S.; GLOVER, F.; KLINGMAN, D.
NEW ALTERNATING BASIS ALGORITHM FOR SEMI-ASSIGNMENT NETWORKS, A.

BAZAA 1977 -1 BAZARAA, M. S.; ELSHAFEI, A. N.
ON THE USE OF FICTITIOUS BOUNDS IN TREE SEARCH ALGORITHMS.

BELLC 1965 -1 BELLMORE, M.
OPTIMAL ALLOCATION OF RESOURCES IN A COMMUNICATIONS NETWORK.

BIONA 1969 -1 BIONDI, E.; DIVIETI, L.; ROVEDA, C.; SCHMID, R.
OPTIMAL SOLUTION FOR A CLASS OF ASSIGNMENT AND TRANSPORTATION PROBLEMS.

BLAS 1967 -1 BLASYUK, B. A.
OPTIMAL ASSIGNMENT OF PROCESSING OF ITEMS ON THREE SEQUENTIAL MACHINES (RUSSIAN).

BLIN 1973 -1 BLIN, J. M.
FURTHER PROCEDURE FOR ORDERING AN INPUT-OUTPUT MATRIX: SOME EMPIRICAL EVIDENCE, A.

BLINA 1976 -1 BLIN, J.-M.
ASSIGNMENT MODELS IN VOTING THEORY.

BRADC 1965 -1 BRADLEY, S. P.
SOLUTION TECHNIQUES FOR THE TRAFFIC ASSIGNMENT PROBLEM.

BRES 1976 -1 BRESLAW, J. A.
LINEAR PROGRAMMING SOLUTION TO THE FACULTY ASSIGNMENT PROBLEM, A.

BRUAA 1976 -1 BRUALDI, R. A.; GIBSON, P. M.
ASSIGNMENT POLYTOPE, THE.

BRUAA 1977 -3 BRUALDI, R. A.; GIBSON, P. M.
CONVEX POLYHEDRA OF DOUBLY STOCHASTIC MATRICES: II. GRAPH OF OMEGA N.

BRUAB 1978 -1 BRUALDI, R. A.; LEE, G. M.
ON THE TRUNCATED ASSIGNMENT POLYTOPE.

BULF 1976 -2 BULFIN, R. L.; PARKER, R. G.
ON THE ASSIGNMENT AND SEQUENCING OF OPERATIONS ON A CREW-SERVED PROJECT.

BURDA 1976 -1 BURDYUK, V. YA.; TROFIMOV, V. N.
GENERALIZATION OF RESULTS BY GILMORE-GOMORY CONCERNING THE TRAVELLING SALESMAN PROBLEM (RUSSIAN).

BURK 1977 -1 BURKARD, R. E.
TRAVELLING SALESMAN AND ASSIGNMENT PROBLEMS: A SURVEY.

BURK 1977 -2 BURKARD, R. E.
ZUR BESTIMMUNG VON SCHREIBMASCHINENTASTATUREN DURCH QUADRATISCHE ZUORDNUNGSPROBLEME.

BURK 1976 -2 BURKARD, R. E.
GENERAL HUNGARIAN METHOD FOR THE ALGEBRAIC TRANSPORTATION PROBLEM, A.

BURK 1976 -3 BURKARD, R. E.
PERTURBATION METHOD FOR SOLVING QUADRATIC ASSIGNMENT AND BOTTLENECK PROBLEMS, A.

BURKA 1977 -1 BURKARD, R. E.; HAHN, W.; ZIMMERMANN, U.
ALGEBRAIC APPROACH TO ASSIGNMENT PROBLEMS, AN.

BURKA 1976 -1 BURKARD, R. E.; HAHN, W.; ZIMMERMANN, U.
ALGEBRAIC APPROACH TO ASSIGNMENT PROBLEMS, AN (REVIDIERT).

BURKC 1977 -1 BURKARD, R. E.; OFFERMANN, J.
ENTWURF VON SCHREIBMASCHINENTASTATUREN MITTELS QUADRATISCHER ZUORDNUNGSPROBLEME.

BURKD 1976 -1 BURKARD, R. E.; STRATMANN, K.-H.
NUMERICAL INVESTIGATIONS ON QUADRATIC ASSIGNMENT PROBLEMS.

BURKE 1977 -1 BURKARD, R. E.; ZIMMERMANN, U.
WEAKLY ADMISSIBLE TRANSFORMATIONS FOR SOLVING ALGEBRAIC ASSIGNMENT AND TRANSPORTATION PROBLEMS.

BURKE 1978 -1 BURKARD, R. E.; ZIMMERMANN, U.
SOLUTION OF ALGEBRAIC ASSIGNMENT AND TRANSPORTATION PROBLEMS, THE.

BURRA 1976 -1 BURRELL, J. E.
MULTIPLE ROUTE ASSIGNMENT: A COMPARISON OF TWO METHODS.

CARP 1977 -1 CARPANETO, G.; TOTH, P.
EFFICIENT ALGORITHM FOR THE ASYMMETRIC TRAVELLING SALESMAN PROBLEM, AN.

CARV 1977 -1 CARVALLO, M.
SUR UN PROBLEME D'AFFECTATION. ALGORITHME DE RESOLUTION.

CHAL 1976 -1 CHALMET, L.
SOLVING ASSIGNMENT PROBLEMS IN WAREHOUSES (DUTCH).

COCK 1975 -1 COCK, J. DE
SEARCH PROCEDURE FOR THE CONSTRUCTION OF SCHOOL TIMETABLES, A.

DAGA 1977 -1 DAGANZO, C. F.; SHEFFI, Y.
ON STOCHASTIC MODELS OF TRAFFIC ASSIGNMENT.

DERI 1975 -1 DERIGS, U.
RANGZUORDNUNGSPROBLEME.

DERI 1977 -2 DERIGS, U.
RANGZUORDNUNGPROBLEME.

DERI 1978 -2 DERIGS, U.
ON SOLVING SYMMETRIC ASSIGNMENT AND PERFECT MATCHING PROBLEMS WITH ALGEBRAIC OBJECTIVES.

DERI 1977 -3 DERIGS, U.
RANGZUORDNUNGSPROBLEME UND DAS STABLE MARRIAGE PROBLEM.

DERIA 1978 -1 DERIGS, U.; ZIMMERMANN, U.
AUGMENTING PATH METHOD FOR SOLVING LINEAR BOTTLENECK ASSIGNMENT PROBLEMS, AN.

DIAL 1971 -1 DIAL, R.
PROBABILISTIC MULTIPATH TRAFFIC ASSIGNMENT MODEL WHICH OBVIATES PATH ENUMERATION, A.

DIAL 1970 -1 DIAL, R.
PROBABILISTIC ASSIGNMENT: A MULTIPATH TRAFFIC ASSIGNMENT MODEL WHICH OBVIATES PATH ENUMERATION.

DOLO 1976 -1 DOLODARENKO, V. A.; FEDAN, E. I.
ASSIGNMENT PROBLEM WITH A NONLINEAR STOCHASTIC OBJECTIVE FUNCTION (RUSSIAN), THE.

****

DOMS 1974 -1 DOMSCHKE, W.
GRAPHENTHEORETISCHE VERFAHREN UND IHRE ANWENDUNGEN ZUR LOESUNG VON ZUORDNUNGS- UND STANDORTPROBLEMEN.
EGOR 1977 -1 EGOROVA, N. V.; PATKIN, B. P.; SLOBODTCHIKOVA, R. I.
MODIFIED ALGORITHM OF ASSIGNMENT PROBLEM SOLUTION (RUSSIAN).
ELSH 1977 -1 ELSHAFEI, A. N.
HOSPITAL LAYOUT AS A QUADRATIC ASSIGNMENT PROBLEM.
ELZIA 1976 -1 ELZINGA, J.; VEMUGANTI, R. R.
OPTIMAL SCHEDULING OF A SALES FORCE.
ESCHB 1974 -1 ESCHER, H. K.
SEQUENTIELLE ZUORDNUNGSPROBLEME.
EVENA 1975 -1 EVEN, S.; ITAI, A.; SHAMIR, A.
ON THE COMPLEXITY OF TIMETABLE AND MULTI-COMMODITY FLOW PROBLEMS.
FAUR 1976 -1 FAURE, R.; ROUCAIROL, C.; TOLLA, P.
CHEMINS ET FLOTS, ORDONNANCEMENTS.
FISK 1976 -1 FISK, J.
ALGORITHM FOR ASSIGNING USES TO SOURCES IN A TRANSPORTATION TABLEAU, AN.
FISKB 1977 -1 FISK, J.; MACKEOWN, P. G.
APPLICATION OF INTEGER PROGRAMMING TO THE ASSIGNMENT OF STUDENTS TO SCHOOLS, AN.
FLOR 1976 -1 FLORIAN, M.; NGUYEN, S.
APPLICATION AND VALIDATION OF EQUILIBRIUM TRIP ASSIGNMENT METHODS, AN.
FUJIC 1977 -2 FUJISHIGE, S.
PRIMAL APPROACH TO THE INDEPENDENT ASSIGNMENT PROBLEM, A.
FULL 1976 -1 FULLER, R. V.; JARVIS, J. J.
SELF-DUAL LINEAR PROGRAMMING MODEL FOR THE EQUILIBRIUM TRAFFIC ASSIGNMENT PROBLEM, A.
GAIK 1976 -1 GAIKOV, N. E.
APPROXIMATE SOLUTION OF THE GENERAL ASSIGNMENT PROBLEM AND OF THE MINIMIZATION PROBLEM OF THE SCALAR MATRIX PRODUCT (RUSSIAN), THE.
GAVI 1977 -1 GAVISH, B.; SCHWEITZER, P.; SHLIFER, E.
ZERO PIVOT PHENOMENON IN TRANSPORTATION AND ASSIGNMENT PROBLEMS AND ITS COMPUTATIONAL IMPLICATIONS, THE.
GEOF 1977 -1 GEOFFRION, A. M.
HOW CAN SPECIALIZED DISCRETE AND CONVEX OPTIMIZATION METHODS BE MARRIED.
GEOFA 1976 -1 GEOFFRION, A. M.; GRAVES, G. W.
SCHEDULING PARALLEL PRODUCTION LINES WITH CHANGEOVER COSTS: PRACTICAL APPLICATION OF A QUADRATIC ASSIGNMENT/LP APPROACH.
GEOL 1975 -1 GEOLECJAN, G. G.
OPTIMAL ASSIGNMENTS AND THEIR PROPERTIES (RUSSIAN).
GHOS 1970 -1 GHOSH, A.; SARKAR, H.
INPUT-OUTPUT MATRIX AS A SPATIAL CONFIGURATION, AN.
GIBE 1968 -1 GIBERT, A.
METHOD FOR THE TRAFFIC ASSIGNMENT PROBLEM, A.
GINO 1966 -1 GINOVKER, V. D.
METHOD OF SOLUTION OF CERTAIN PROBLEMS OF OPTIMIZATION OF ASSIGNMENT (RUSSIAN).
GLOVC 1974 -1 GLOVER, F.; KLINGMAN, D.
IMPORTANT PRACTICAL MISCONCEPTIONS OF OPTIMIZING LARGE SCALE ASSIGNMENT AND TRANSPORTATION PROBLEMS.
GLOVC 1976 -1 GLOVER, F.; KLINGMAN, D.
PRACTITIONER'S GUIDE TO THE STATE OF LARGE SCALE NETWORK AND NETWORK-RELATED PROBLEMS, A.
GLOVC 1976 -2 GLOVER, F.; KLINGMAN, D.
SOME RECENT PRACTICAL MISCONCEPTIONS ABOUT THE STATE OF THE ART OF NETWORK ALGORITHMS.
GRAN 1974 -1 GRANDA, T. M.
HEURISTIC APPROACH TO A LARGE PERSONNEL ASSIGNMENT PROBLEM, A.
GROSA 1960 -1 GROSS, O.
BOTTLENECK ASSIGNMENT PROBLEM: AN ALGORITHM, THE.
HANSF 1977 -1 HANSON, H. H.
PROCEDURE FOR ASSIGNING WORKERS TO MACHINES WHEN EACH WORKER CAN OPERATE MORE THAN ONE MACHINE, A.
HATC 1975 -1 HATCH, R. S.
BENCH MARKS COMPARING TRANSPORTATION CODES BASED ON PRIMAL SIMPLEX AND PRIMAL-DUAL ATGORITHMS.
HAUS 1978 -1 HAUSMANN, D.
ADJAZENZCHARAKTERISIERUNGEN FUER POLYEDERECKEN IN DER KOMBINATORISCHEN OPTIMIERUNG.
HAUSC 1976 -1 HAUSMANN, D.; KORTE, B.
COLOURING CRITERIA FOR ADJACENCY ON 0-1-POLYHEDRA.
HELL 1956 -1 HELLER, I.
NEIGHBOUR RELATIONS ON THE CONVEX OF CYCLIC PERMUTATIONS.
HELL 1955 -1 HELLER, I.
ON THE TRAVELING SALESMAN'S PROBLEM.
HORI 1973 -1 HORIE, Y.; FAN, L. T.
DESTINATION TYPE TRAFFIC ASSIGNMENT WITH AIR QUALITY CONSIDERATION.
IBARA 1976 -1 IBARAKI, T.; ISHII, H.; MINE, H.
ASSIGNMENT PROBLEM ON A NETWORK, AN.
IRI A 1976 -1 IRI, M.; TOMIZAWA, N.
ALGORITHM FOR FINDING AN OPTIMAL "INDEPENDENT ASSIGNMENT", AN.
IRWI 1971 -1 IRWIN, N. A.; DODD, N.; CUBE, H. VON
CAPACITY RESTRAINT IN ASSIGNMENT PROGRAMS.
IVAN 1976 -1 IVANCHEV, M. D.
ON ONE CLASS OF THE QUADRATIC ASSIGNMENT PROBLEM (RUSSIAN).
JACOB 1977 -1 JACOBSON, R. E.; SADOWSKI, R. F.
SCHEDULING ALGORITHM FOR THE UNBALANCED PRODUCTION LINE, A.
JEWE 1967 -1 JEWELL, W.
MODELS FOR THE TRAFFIC ASSIGNMENT.
KACP 1975 -1 KACPRZYK, J.
APPLICATION OF THE THEORY OF FUZZY SETS TO THE OPTIMAL ASSIGNMENT OF THE STAFF (POLISH).
KARPA 1978 -1 KARP, R. M.
PATCHING ALGORITHM FOR THE NONSYMMETRIC TRAVELING-SALESMAN PROBLEM, A.
KIRC 1965 -1 KIRCHGAESSNER, K.
GRAPH THEORETICAL SOLUTION OF A NONLINEAR ASSIGNMENT PROBLEM (GERMAN).
KNUT 1976 -1 KNUTH, D. E.
MARIAGES STABLES ET LEURS RELATIONS AVEC D'AUTRES PROBLEMES COMBINATOIRES.
KOMAA 1974 -1 KOMAROMI, E.
FINITE PRIMAL METHOD FOR THE ASSIGNEMENT PROBLEM, A.
KRAR 1972 -1 KRARUP, J.
QUADRATIC ASSIGNMENT (IN DANISH).

****

KRAVA 1975 -1 KRAVTSOV, M. K.; SHERMAN, A. K.; AVERBUKH, N. D.
ALGORITHM FOR THE SOLUTION OF THE ASSIGNMENT PROBLEM (RUSSIAN), AN.

KRUS 1964 -1 KRUSHEVSKII, A. V.; PRIMAK, M. E.
EXTREMUM OF A LINEAR FORM ON A PERMUTATION GROUP AND THE ASSIGNMENT PROBLEM (RUSSIAN).

KUBA 1974 -1 KUBALE, M.
HEURISTIC ALGORITHM FOR ESTABLISHING TIME-TABLES IN A FACULTY (POLISH).

LAWL 1975 -7 LAWLER, E. L.
QUADRATIC ASSIGNMENT PROBLEM: A BRIEF REVIEW, THE.

LEBLA 1974 -1 LEBLANC, L. J.; MORLOK, L. J.
ANALYSIS AND COMPARISON OF BEHAVIORIAL ASSUMPTIONS IN TRAFFIC ASSIGNMENT, AN.

LEBLB 1974 -1 LEBLANC, L. J.; MORLOK, L. J.; PIERSKALLA, W.
ACCURATE AND EFFICIENT APPROACH TO EQUILIBRIUM TRAFFIC ASSIGNMENT ON A CONJESTED NETWORKS, AN.

LENSA 1977 -1 LENSTRA, J. K.
SEQUENCING BY ENUMERATIVE METHODS.

LEON 1976 -1 LEONT'EV, V. K.
STABILITY IN COMBINATORIAL PROBLEMS OF SELECTION (RUSSIAN).

LINS 1975 -1 LINS, S.
USE OF MATROID THEORY IN OPTIMAL MULTI-ASSIGNMENT OVER TOTALLY ORDERED SETS. THE.

LORE 1976 -1 LOREA, M.
STRUCTURE DE CERTAINS PROBLEMES D'AFFECTATION, LA.

LOS 1978 -1 LOS, M.
SIMULTANEOUS OPTIMIZATION OF LAND USE AND TRANSPORTATION: A SYNTHESIS OF THE QUADRATIC ASSIGNMENT PROBLEM AND THE OPTIMAL NETWORK PROBLEM.

LOVEA 1976 -2 LOVE, R. F.; WONG, J. Y.
SOLVING QUADRATIC ASSIGNMENT PROBLEMS WITH RECTANGULAR DISTANCES AND INTEGER PROGRAMMING.

MAC H 1975 -1 MAC KEOWN, P. G.
DETERMINING ADJACENT VERTICES ON ASSIGNMENT POLYTOPES.

MACH 1976 -1 MACHOL, R. E.; WIEN, M.
'HARD' ASSIGNMENT PROBLEM, A.

MAIE 1976 -1 MAIER, S. F.
SOLVING THE TRAFFIC ASSIGNMENT AND SUB-AREA FOCUSING PROBLEMS BY GEOGRAPHIC DECOMPOSITION.

MAIM 1975 -1 MAIMON, A.; PETERSON, R. M.
APPLICATIONS OF AN AIRLINE FLEET ASSIGNMENT MODEL.

MARTE 1965 -1 MARTIN, B.; MANHEIM, M.
RESEARCH PROGRAM FOR COMPARISON OF TRAFFIC ASSIGNMENT TECHNIQES, A.

MARU 1976 -1 MARUYAMA, K.; TANG, D. T.; FRATTA, L.
FLOW ASSIGNMENT ALGORITHM FOR COMPUTER COMMUNICATION NETWORK DESIGN WITH DIFFERENT CLASSES OF PACKETS.

MATS 1971 -1 MATSUI, H.
THEORY OF TRAFFIC ASSIGNMENT BY MAXIMIZING THE PROBABILITY, THE.

MENS 1975 -1 MENSCH, C.
PERSONALEINSATZPLANUNG (EINSCHL. ZUORDNUNGSMODELLE).

MERC 1974 -1 MERCHANT, D. K.; NEMHAUSER, G. L.
MODEL AND AN ALGORITHM FOR THE DYNAMIC TRAFFIC ASSIGNMENT PROBLEM, A.

MERCA 1976 -1 MERCHANT, D. K.; RAO, M. R.
MAJORITY DECISIONS AND TRANSITIVITY: SOME SPECIAL CASES.

MEYEC 1977 -1 MEYER, R. R.
CLASS OF NONLINEAR INTEGER PROGRAMS SOLVABLE BY A SINGLE LINEAR PROGRAM, A.

MIHA 1976 -1 MIHALYFFY, L.
NULL-SPACE METHOD FOR SOLVING THE ASSIGNMENT PROBLEM, A.

MILLC 1976 -1 MILLER, H. E.
PERSONNEL SCHEDULING AND WORK ASSIGNMENT PROBLEM, A.

MILLC 1976 -2 MILLER, H. E.
PERSONNEL SCHEDULING IN PUBLIC SYSTEMS: A SURVEY.

MOJE 1976 -1 MOJENA, R.; VOLLMANN, T. E.; OKAMOTO, Y.
ON PREDICTING COMPUTATIONAL TIME OF A BRANCH AND BOUND ALGORITHM FOR THE ASSIGNMENT OF FACILITIES.

MULVA 1975 -2 MULVEY, J. M.
SPECIAL STRUCTURES IN NETWORK NODELS AND ASSOCIATED APPLICATIONS.

MURCB 1967 -4 MURCHLAND, J. D.
TWO REMARKS ON CONGESTED ASSIGNMENT.

MURTA 1967 -1 MURTY, K.
SYMMETRIC ASSIGNMENT PROBLEM, THE.

NAND 1974 -1 NANDA, R.; WEINGARTEN, A.
PRACTICABLE SOLUTION FOR THE QUADRATIC ASSIGNMENT PROBLEM, A.

NARU 1976 -1 NARULA, S. C.; OGBU, U. I.
ASSIGNMENT PROBLEM WITH SIMILAR RESOURCES.

NUNN 1976 -1 NUNNER, O.
ANWENDUNG EINES TRANSPORTMODELLS BEI DER EINBERUFUNG VON WEHRPFLICHTIGEN.

OSTE 1976 -1 OSTERMANN, R.
QUELQUES EXPERIENCES DE CONFECTION AUTOMATIQUE D'HORAIRES.

PATE 1976 -1 PATEL, A. M.; WALD, C. G. DE; COTES, L. C.
PRE-PARTITIONING AS A MEANS OF ENHANCING PAIRWISE INTERCHANGE SOLUTIONS TO QUADRATIC ASSIGNMENT PROBLEMS.

PETEA 1975 -1 PETERSEN, E. R.
PRIMAL-DUAL TRAFFIC ASSIGNMENT ALGORITHM, A.

PILI 1976 -1 PILIPOSJAN, T. E.
CERTAIN ASSIGNMENT PROBLEM (RUSSIAN), A.

PLOTA 1974 -1 PLOTNIKOV, V. N.; ZVEREV, V. YU.
METHODS OF FAST DISTRIBUTION OF ALGORITHMS IN COMPUTATIONAL SYSTEMS (RUSSIAN).

POLLA 1976 -1 POLLATSCHECK, M. A.
PERSONNEL ASSIGNMENT BY MULTIOBJEKTIVE PROGRAMMING.

RAJAA 1977 -1 RAJAT, K. D.
ON GENERATION OF TEST DATA AND MINIMAL COVER OF DIRECTED GRAPHS.

RAO 1976 -1 RAO, M. R.
ADJACENCY OF THE TRAVELING SALESMAN TOURS AND 0-1 VERTICES.

RAVI 1977 -1 RAVINDRAN, A.; RAMASWAMI, V.
ON THE BOTTLENECK ASSIGNMENT PROBLEM.

ROSS 1977 -1 ROSS, G. T.; SOLAND, R. M.
MODELING FACILITY LOCATION PROBLEMS AS GENERALIZED ASSIGNMENT PROBLEMS.

ROSSA 1976 -1 ROSS, G. T.; SOLAND, R. M.; ZOLTNERS, A. A.
NOTE ON THE BOUNDED INTERVAL GENERALIZED ASSIGNMENT PROBLEM, A.

****

ROSSB 1974 -1 ROSS, G. T. JR.
ALGORITHMIC AND COMPUTATIONAL RESULTS FOR CONSTRAINED TRANSPORTATION, NETWORK AND ASSIGNMENT MODELS.

RUIT 1968 -1 RUITER, E. R.
TRAFFIC ASSIGNMENT RESEARCH, FINAL REPORT.

RUSSB 1977 -1 RUSSELL, R.; IGO, W.
ASSIGNMENT ROUTING PROBLEM, AN.

SAHNB 1976 -1 SAHNI, S.; GONZALEZ, T.
P-COMPLETE APPROXIMATION PROBLEMS.

SANDA 1975 -1 SANDI, C.
SOLUTION OF THE MACHINE LOADING PROBLEM WITH BINARY VARIABLES.

SARVA 1976 -1 SARVANOV, V.I.
MEAN VALUE OF THE FUNCTIONAL OF THE ASSIGNMENT PROBLEM (RUSSIAN), THE.

SCHO 1976 -1 SCHOCH, M.
ERWEITERUNGSPRINZIP UND SEINE ANWENDUNG ZUR ENTWICKLUNG VON ALGORITHMEN FUER DIE LOESUNG KOMBINATORISCHER OPTIMIERUNGSAUFGABEN, DAS.

SCHOC 1977 -1 SCHOENLEIN, A.
UEBER DIE ANWENDUNG DER METHODE DES BRANCH AND BOUND AUF DAS SUMMEN-ZUORDNUNGSPROBLEM.

SHEPA 1975 -1 SHEPSLE, K. A.
CONGRESSIONAL COMMITTEE ASSIGNMENTS: AN OPTIMIZATION MODEL WITH INSTITUTIONAL CONSTRAINTS.

SHEPA 1975 -2 SHEPSLE, K. A.
OPTIMIZATION MODEL OF CONGRESSIONAL COMMITTEE ASSIGNMENTS, AN.

SLOM 1975 -1 SLOMINSKI, L.
QUADRATIC ASSIGNMENT PROBLEM AS A LINEAR ZERO-ONE PROGRAMMING PROBLEM (POLISH), THE.

STRAA 1976 -1 STRATMANN, K.-H.
NUMERISCHE UNTERSUCHUNGEN UEBER QUADRATISCHE ZUORDNUNGSPROBLEME.

STRAB 1976 -1 STRATMANN, K.-H.; BURKARD, R. E.
NUMERISCHE UNTERSUCHUNGEN UEBER QUADRATISCHE ZUORDNUNGSPROBLEME.

SWANA 1976 -1 SWANSON, L. W.
N-DIMENSIONAL SCHEDULING.

TIER 1974 -1 TIERNAN, J. C.
DUAL MULTIPLE-RESOURCE ASSIGNMENT ALGORITHM, A.

TOBI 1977 -1 TOBIN, R. L.
EXTENSION OF DIAL'S ALGORITHM UTILIZING A MODEL OF TRIPMAKERS' PRECEPTIONS, AN.

TOML 1966 -1 TOMLIN, J. A.
LINEAR PROGRAMMING MODEL FOR THE ASSIGNMENT OF TRAFFIC, A.

VISH 1976 -1 VISHNEVSKIJ, L. D.
ON A SCHEDULING PROBLEM FOLLOWING A CALENDAR (RUSSIAN).

VOLY 1974 -1 VOLYNSKII, E. I.; VJATKIN, A. K.; KRASIN, L. A.
ASSIGNMENT PROBLEM FOR A PRODUCTION LINE WITH OPERATIVE STORAGE POINTS (RUSSIAN), THE.

WERR 1976 -4 WERRA, D. DE
ON A MULTI-PERIOD ASSIGNMENT PROBLEM.

WHITC 1975 -1 WHITE, D. J.
NOTE ON FACULTY TIMETABLING, A.

ZAYC 1975 -1 ZAYCHENKO, YU. P.; PECHURIN, M. K.; TOMASHEVSKIY, V. M.
GENERALIZED ASSIGNMENT PROBLEM AND ITS SOLUTION ALGORITHM (UKRAINIAN), THE.

BABAA 1976 -1 BABAYEV, D. A.
METHOD OF HYPERSPHERES FOR BOOLEAN PROGRAMMING PROBLEMS (RUSSIAN).

BAER 1976 -2 BAER, G.
ZUR LINEARISIERUNG BELIEBIGER O-1-OPTIMIERUNGSPROBLEME.

BAR 1975 -1 BAR, G.
UEBER DEN EINSATZ LOGISCHER MITTEL ZUR LINEARISIERUNG DISKRETER OPTIMIERUNGSPROBLEME.

BART 1975 -1 BARTHES, J. P.
BRANCHING METHODS IN COMBINATORIAL OPTIMIZATION.

BOTT 1974 -1 BOTT, H.
ANWENDUNGSMOEGLICHKEITEN DER PSEUDO-BOOLSCHEN OPTIMIERUNG AUF PROBLEME DER FERTIGUNSWIRTSCHAFT UND IHRE RECHENTECHNISCHE REALISIERUNG.

BOWM 1974 -1 BOWMAN, V. J.; STARR, J. H.
TWO-COMPARABLE PRIME IMPLICANTS AND CANONICAL SWITCHING FUNCTIONS.

CHANE 1976 -1 CHANDRA, A. K.; MARKOWSKY, G.
ON THE NUMBER OF PRIME IMPLICANTS.

GAIK 1976 -2 GAIKOV, N. E.
MINIMIZATION OF A QUADRATIC FORM WITH BOOLEAN VARIABLES ON A SYMMETRIC GROUP (RUSSIAN), THE.

GRANB 1970 -1 GRANOT, F.; HAMMER, P.
ON THE USE OF BOOLEAN FUNCTIONS IN 0-1 PROGRAMMING.

GRIS 1976 -1 GRISHUKHIN, V. P.
BRANCH AND BOUNDARY ALGORITHMS IN PROBLEMS WITH BOOLEAN VARIABLES: ESTIMATION OF THEIR EFFECTIVENESS (RUSSIAN).

GRIS 1976 -2 GRISHUKHIN, V. P.
EFFECTIVENESS OF THE BRANCH AND BOUND METHOD IN PROBLEMS WITH BOOLEAN VARIABLES (RUSSIAN).

HAMMA 1976 -1 HAMMER, P. L.
BOOLEAN PROCEDURES IN OPTIMIZATION.

HAMMA 1977 -1 HAMMER, P. L.
BOOLEAN ELEMENTS IN COMBINATORIAL OPTIMIZATION. A SURVEY.

HAMMA 1975 -1 HAMMER, P. L.
BOOLEAN ELEMENTS IN COMBINATORIAL OPTIMIZATION: A SURVEY.

HANSA 1976 -1 HANSEN, P.
CASCADE ALGORITHM FOR THE LOGICAL CLOSURE OF A SET OF BINARY RELATIONS, A.

HAUS 1978 -1 HAUSMANN, D.
ADJAZENZCHARAKTERISIERUNGEN FUER POLYEDERECKEN IN DER KOMBINATORISCHEN OPTIMIERUNG.

HAY 1978 -1 HAY, L.
CONVEX SUBSETS OF 2 POWER N AND BOUNDED TRUTH-TABLE REDUCIBILITY.

JAIN 1975 -1 JAIN, R. C.
ON PSEUDO-BOOLEAN INEQUALITIES-CONSISTENCY AND SOLUTIONS.

KAPLA 1976 -1 KAPLINSKII, A. I.; OHERNYSOVA, G. D.
METHOD OF ADAPTIVE ALGORITHMS FOR SOLVING OPTIMIZATION PROBLEMS WITH BOOLEAN VARIABLES (RUSSIAN).

LAVI 1976 -1 LAVIT, C.
CLASSES DE SOLUTIONS PARAMETRIQUES D'UNE EQUATION BOOLEENNE.

LECHA 1970 -1 LECHNER, R. J.
DETECTION OF PRIME IMPLICANTS BY ABSTRACT HARMONIC ANALYSIS.

MARC 1976 -1 MARCU, D.
ALGORITHM FOR DETERMINING THE STRONGLY CONNECTED COMPONENTS OF A FINITE DIRECTED GRAPH (RUMANIAN), AN.

PERC 1975 -1 PERCOV, E. E.; BOGDANOVA, O. V.
ALGORITHM FOR THE MINIMIZATION OF A PSEUDO-BOOLEAN QUADRATIC FUNCTION (RUSSIAN), AN.

TARJ 1978 -1 TARJAN, R. E.
COMPLEXITY OF MONOTONE NETWORKS FOR COMPUTING CONJUNCTIONS.

UEDA 1977 -1 UEDA, T.
FIXING GROUPS FOR THE 2-ASUMMABLE BOOLEAN FUNCTIONS, THE.

VIZV 1975 -1 VIZVARI, B.
ENUMERATIVE ALGORITHMS IN POLYNOMIAL 0-1 PROGRAMMING (HUNGARIAN).

WHITC 1975 -1 WHITE, D. J.
NOTE ON FACULTY TIMETABLING, A.

ZAK 1977 -1 ZAK, YU. A.
METHOD OF DIRECTED SCANNING IN PROBLEMS OF INTEGRAL LINEAR PROGRAMMING WITH BOOLEAN VARIABLES (RUSSIAN).

ZHER 1976 -2 ZHERNAK, A. N.
NEW METHOD FOR THE SOLUTION OF INTEGER PROGRAMMING PROBLEMS (RUSSIAN), A.

ZIMMA 1976 -1 ZIMMERMANN, U.
BOOLE'SCHE OPTIMIERUNGSPROBLEME MIT SEPARABLER ZIELFUNKTION UND MATROIDALEN RESTRIKTIONEN.

ZIMMA 1977 -2 ZIMMERMANN, U.
SOME PARTIAL ORDERS RELATED TO BOOLEAN OPTIMIZATION AND THE GREEDY-ALGORITHM.

ZIMMA 1977 -3 ZIMMERMANN, U.
THRESHOLD METHODS FOR BOOLEAN OPTIMIZATION PROBLEMS WITH SEPARABLE OBJECTIVES.

BOTTLENECK PROBLEMS
****

BURK 1976 -3 BURKARD, R. E.
PERTURBATION METHOD FOR SOLVING QUADRATIC ASSIGNMENT AND BOTTLENECK PROBLEMS, A.

BURKB 1976 -1 BURKARD, R. E.; HAMACHER, H.; ZIMMERMANN, U.
ALGEBRAIC NETWORK FLOW PROBLEM, THE.

BURKE 1978 -1 BURKARD, R. E.; ZIMMERMANN, U.
SOLUTION OF ALGEBRAIC ASSIGNMENT AND TRANSPORTATION PROBLEMS, THE.

BURKE 1977 -1 BURKARD, R. E.; ZIMMERMANN, U.
WEAKLY ADMISSIBLE TRANSFORMATIONS FOR SOLVING ALGEBRAIC ASSIGNMENT AND TRANSPORTATION PROBLEMS.

DERI 1977 -1 DERIGS, U.
DUALITY AND THE ALGEBRAIC MATCHING PROBLEM.

DERIA 1978 -1 DERIGS, U.; ZIMMERMANN, U.
AUGMENTING PATH METHOD FOR SOLVING LINEAR BOTTLENECK ASSIGNMENT PROBLEMS, AN.

FRANA 1976 -1 FRANCIS, R. L.; LOWE, T. J.; RATLIFF, H. D.
DISTANCE CONSTRAINTS FOR TREE NETWORK MULTIFACILITY LOCATION PROBLEMS.

FRIEA 1975 -1 FRIEZE, A. M.
BOTTLENECK LINEAR PROGRAMMING.

GARFA 1976 -1 GARFINKEL, R. S.; GILBERT, K. C.
BOTTLENECK TRAVELING SALESMAN PROBLEM, THE.

GROSA 1960 -1 GROSS, O.
BOTTLENECK ASSIGNMENT PROBLEM: AN ALGORITHM, THE.

HEARA 1976 -1 HEARN, D.; LOWE, T. J.
SUBGRADIENT PROCEDURE FOR THE SOLUTION OF MINIMAL PROBLEMS, A.

ISHI 1975 -1 ISHII, H.; IBARAKI, T.; MINE, H.
CERTAIN TRAVELING SALESMAN PROBLEM THAT HAS AN OBJECTIVE FUNCTION OF MINIMAX TYPE (JAPANESE), A.

KATS 1977 -1 KATSEV, S. B.
SOLUTION OF A GENERALIZED PROBLEM ON SET PARTITIONING (RUSSIAN).

MARKD 1974 -1 MARKOV, L. A.; PERVOZVANSKII, A. A.
RULE FOR "BOTTLENECKS" IN THE JOHNSON PROBLEM (RUSSIAN), A.

MERCA 1976 -1 MERCHANT, D. K.; RAO, M. R.
MAJORITY DECISIONS AND TRANSITIVITY: SOME SPECIAL CASES.

RAND 1976 -1 RANDOLPH, P.
SHORTEST PATHS THROUGH MULTIPLE CRITERIA NETWORKS.

RANDA 1977 -1 RANDOLPH, P.; RINGEISEN, R.
SHORTEST PATHS THROUGH MULTIPLE CRITERIA NETWORKS.

RAVI 1977 -1 RAVINDRAN, A.; RAMASWAMI, V.
ON THE BOTTLENECK ASSIGNMENT PROBLEM.

SHAPA 1966 -1 SHAPIRO, D. M.
ALGORITHMS FOR THE SOLUTION OF THE OPTIMAL COST AND BOTTLENECK TRAVELING SALESMAN PROBLEMS.

SRINA 1976 -1 SRINIVASAN, V.; THOMPSON, G. L.
ALGORITHMS FOR MINIMIZING TOTAL COST, BOTTLENECK TIME AND BOTTLENECK SHIPMENT IN TRANSPORTATION PROBLEMS.

TIMO 1977 -1 TIMOFEEV, E. A.
CONSTRUCTION OF AN EDGE-WISE M-CONNECTED SKELETAL SUBGRAPH WITH MINIMAL LENGTH OF THE MAXIMAL EDGE (RUSSIAN), THE.

ZIMMA 1977 -1 ZIMMERMANN, U.
DUALITY AND THE ALGEBRAIC MATROID INTERSECTION PROBLEM.

ZIMMA 1977 -3 ZIMMERMANN, U.
THRESHOLD METHODS FOR BOOLEAN OPTIMIZATION PROBLEMS WITH SEPARABLE OBJECTIVES.

****

ACKE 1977 -2 ACKERMAN, L. J.; LUSS, H.; BERKOWITZ, R. S.
IMPLICIT ENUMERATION ALGORITHM FOR SEQUENCING POLICIES APPLIED TO TELEPHONE SWITCHING FACILITIES, AN.

AGRA 1976 -1 AGRAWAL, S. C.
ALTERNATE METHOD ON INTEGER SOLUTIONS TO QUADRATIC PROGRAM BY A BRANCH AND BOUND TECHNIQUE, AN.

AGRA 1976 -2 AGRAWAL, S. C.
ON INTEGER SOLUTIONS TO LINEAR FRACTIONAL FUNCTIONALS BY A BRANCH AND BOUND TECHNIQUE.

AHRE 1975 -1 AHRENS, J. H.; FINKE, G.
MERGING AND SORTING APPLIED TO THE ZERO-ONE KNAPSACK PROBLEM.

AHREA 1976 -1 AHRENS, W.
NUMERICAL EXPERIENCE WITH HEURISTIC RULES IN SPECIFIC BINARY OPTIMIZATION ALGORITHMS.

AHREA 1974 -2 AHRENS, W.
SOLVING OF THE REGIONALIZATION PROBLEM BY CERTAIN BRANCH-AND-BOUND METHODS.

AKIN 1974 -1 AKINC, U.
BRANCH AND BOUND PROCEDURE FOR SOLVING WAREHOUSE LOCATION PROBLEMS WITH CAPACITY CONSTRAINTS, A.

AKINA 1977 -1 AKINC, U.; KHUMAWALA, B. M.
EFFICIENT BRANCH AND BOUND ALGORITHM FOR THE CAPACITATED WAREHOUSE LOCATION PROBLEM, AN.

ALEKA 1977 -1 ALEKSEYEV, O. G.; MERZHANOV, V. YU.
ALGORITHM FOR JOB ALLOCATION ACCORDING TO A MINIMAX CRITERION (RUSSIAN), AN.

ALEKB 1976 -1 ALEKSEYEV, O. G.; VOLODOS, I. F.
COMBINED USE OF DYNAMIC PROGRAMMING AND BRANCH-AND-BOUND METHODS IN DISCRETE-PROGRAMMING PROBLEMS (RUSSIAN).

ALEKD 1974 -1 ALEKSEYEVA, S. M.
METHODS OF SCHEDULE THEORY IN TECHNOLOGICAL CHECKING PROBLEMS (RUSSIAN).

ALLE 1976 -1 ALLEN, B.; MUNRO, I.
SELF-ORGANIZING BINARY SEARCH TREES.

ARMS 1978 -1 ARMSTRONG, R. D.
NONBASIC BRANCHING AND DUAL RELAXATION IN A BRANCH-AND-BOUND MIXED INTEGER PROGRAMMING ALGORITHM.

ARMSB 1977 -1 ARMSTRONG, R. D.; COOK, W. D.
COMMITTEE RANKING PROBLEM USING ORDINAL UTILITIES, A.

ARMSD 1976 -1 ARMSTRONG, R. D.; FROME, E. L.
BRANCH-AND-BOUND SOLUTION OF A RESTRICTED LEAST SQUARES PROBLEM, A.

ARORB 1976 -1 ARORA, S. R.; SWARUP, K.; PURI, M. C.
ENUMERATIVE TECHNIQUE FOR THE SET COVERING PROBLEM, AN.

ASHT 1977 -1 ASHTON, W. B.; CLARK, G. M.
LOCATION AND SIZING OF URBAN FREIGHT CONSOLIDATION TERMINALS WITH MULTIPLE PLANNING PERIODS, THE.

ATTM 1973 -1 ATTMANN, I.
NUMERISCHE ERFAHRUNGEN MIT BALAS-ALGORITHMEN.

AUST 1976 -1 AUST, R. J.
DYNAMIC PROGRAMMING BRANCH AND BOUND ALGORITHM FOR PURE INTEGER PROGRAMMING, A.

BALAG 1973 -1 BALAS, E.
ON THE USE OF INTERSECTION CUTS AND OUTER POLARS IN BRANCH-AND-BOUND.

BALAG 1977 -1 BALAS, E.
DISJUNCTIVE PROGRAMMING.

BALAG 1975 -1 BALAS, E.
BIVALENT PROGRAMMING BY IMPLICIT ENUMERATION.

BALAG 1977 -3 BALAS, E.
SOME VALID INEQUALITIES FOR THE SET PARTITIONING PROBLEM.

BALAJ 1975 -1 BALAS, E.; PADBERG, M. W.
SET PARTITIONING: A SURVEY.

BALAJ 1976 -1 BALAS, E.; PADBERG, M. W.
SET PARTITIONING: A SURVEY.

BANS 1977 -1 BANSAL, S. P.
MINIMIZING THE SUM OF COMPLETION TIMES OF N JOBS OVER M MACHINES IN A FLOWSHOP - A BRANCH AND BOUND APPROACH.

BARS 1974 -1 BARSKIY, A. B.
METHOD OF OPTIMAL DISTRIBUTION OF OPERATIONS IN A HOMOGENEOUS COMPUTATIONAL SYSTEM (RUSSIAN), A.

BART 1975 -1 BARTHES, J. P.
BRANCHING METHODS IN COMBINATORIAL OPTIMIZATION.

BAZA 1975 -1 BAZARAA, M. S.
COMPUTERIZED LAYOUT DESIGN: A BRANCH AND BOUND APPROACH.

BAZAA 1977 -1 BAZARAA, M. S.; ELSHAFEI, A. N.
ON THE USE OF FICTITIOUS BOUNDS IN TREE SEARCH ALGORITHMS.

BAZAB 1977 -1 BAZARAA, M. S.; GOODE, J. J.
TRAVELING SALESMAN: A DUALITY APPROACH, THE.

BEAL 1977 -1 BEALE, E. M. L.
BRANCH AND BOUND METHODS.

BECK 1976 -1 BECK, L. L.; MATULA, D. W.
CLIQUE DETERMINATION IN LARGE DENSE GRAPHS.

BELLD 1970 -1 BELLMORE, M.; JENSEN, P.
IMPLICIT ENUMERATION SCHEME FOR PROPER CUT GENERATION, AN.

BENI 1977 -1 BENICHOU, M.; GAUTHIER, J. M.; HENTGES, G.; RIBIERE, G.
EFFICIENT SOLUTION OF LARGE-SCALE LINEAR PROGRAMMING PROBLEMS - SOME ALGORITHMIC TECHNIQUES AND COMPUTATIONAL RESULTS, THE.

BENNA 1972 -1 BENNETT, J. M.; EDWARDS, J.
EXPERIENCE WITH A METHOD FOR HANDLING LARGE STRUCTURAL LINEAR PROGRAMMING PROBLEMS WITH INTEGER RESTRICTIONS.

BERE 1974 -1 BERESNEV, V. L.
IMPLICIT ENUMERATION ALGORITHM FOR AN ASSIGNMENT AND STANDARDIZATION PROBLEM (RUSSIAN).

BEST 1976 -1 BESTWICK, P. F.; HASTINGS, N. A. J.
NEW BOUND FOR MACHINE SCHEDULING, A.

BILDA 1977 -1 BILDE, O.; KRARUP, J.
SHARP LOWER BOUNDS AND EFFICIENT ALGORITHMS FOR THE SIMPLE PLANT LOCATION PROBLEM.

BLAIB 1976 -1 BLAIR, C. E.; JEROSLOW, R. G.
TREELESS SEARCHES.

BOGA 1976 -1 BOGARDI, J. J.; DUCKSTEIN, L.; PLATE, E. J.
SCHEDULING AND SEQUENCING THE CONSTRUCTION OF FLOOD CONTROL RESERVOIRS.

BOLLD 1974 -1 BOLLOBAS, B.; ERDOES, P.; STRAUS, E. G.
COMPLETE SUBGRAPHS OF CHROMATIC GRAPHS AND HYPERGRAPHS.

BRANCH AND BOUND

****

BOWMA 1977 -1 BOWMAN, V. J. JR.; STARR, J. H.
PARTIAL ORDERINGS IN IMPLICIT ENUMERATION.

BOYC 1973 -1 BOYCE, D. E.; FARHI, A.; WEISCHEDEL, R.
OPTIMAL NETWORK PROBLEM: A BRANCH-AND-BOUND ALGORITHM.

BRAT 1975 -1 BRATLEY, P.; FLORIAN, M.; ROBILLARD, P.
SCHEDULING WITH EARLIEST START AND DUE DATE CONSTRAINTS ON MULTIPLE MACHINES.

BREGA 1974 -1 BREGMAN, T. T.
BRANCH AND BOUND METHOD FOR THE SOLUTION OF ALLOCATION PROBLEMS (RUSSIAN), A.

BRIC 1977 -1 BRICKER, D. L.
REFORMULATION OF SPECIAL ORDERED SETS FOR IMPLICIT ENUMERATION ALGORITHMS, WITH APPLICATIONS IN NONCONVEX SEPARABLE PROGRAMMING.

BRIT 1976 -1 BRITAVSKIJ, G. M.; JUKHIMENKO, B. I.
METHOD OF BRANCH AND BOUND FOR OPTIMIZATION OF AUTOMATIC LINES STRUCTURE PARAMETERS (RUSSIAN), A.

BROC 1976 -1 BROCKLEHURST, E. R.
GENERALIZED BRANCH AND BOUND: A METHOD FOR INTEGER LINEAR PROGRAMS.

BULF 1976 -2 BULFIN, R. L.; PARKER, R. G.
ON THE ASSIGNMENT AND SEQUENCING OF OPERATIONS ON A CREW-SERVED PROJECT.

BULFA 1977 -1 BULFIN, R. L.; SHETTY, C. M.; PARKER, R. G.
COMPUTATIONAL RESULTS WITH A BRANCH-AND BOUND ALGORITHM FOR THE GENERAL KNAPSACK PROBLEM.

BURD 1977 -1 BURDET, C.-A.; JOHNSON, E. L.
SUBADDITIVE APPROACH TO SOLVE LINEAR INTEGER PROGRAMS, A.

BURK 1972 -1 BURKOV, V. N.; LAZEBNIK, A. I.; KHRANOVICH, I. L.
METHOD OF BRANCHES AND BOUNDS AS A REGULAR METHOD FOR THE SOLUTION OF IRREGULAR MATHEMATICAL PROGRAMMING PROBLEMS, I (RUSSIAN).

BUZY 1976 -1 BUZYCKII, P. L.
CERTAIN CUT AND BRANCH FOR PRIMAL INTEGER PROGRAMMING ALGORITHMS (RUSSIAN), A.

CAMEA 1974 -1 CAMERINI, P. M.; MAFFIOLI, F.
BOUNDS FOR 3-MATROID INTERSECTION PROBLEMS.

CARP 1977 -1 CARPANETO, G.; TOTH, P.
EFFICIENT ALGORITHM FOR THE ASYMMETRIC TRAVELLING SALESMAN PROBLEM, AN.

CHAN 1976 -1 CHAN, A. W.; FRANCIS, R. L.
LEAST TOTAL DISTANCE FACILITY CONFIGURATION PROBLEM INVOLVING LATTICE POINTS, A.

CHANB 1969 -1 CHAN, Y. P.
OPTIMAL TRAVEL TIME REDUCTION IN A TRANSPORT NETWORK. AN APPLICATION OF NETWORK AGGREGATION AND BRANCH AND BOUND TECHNIQUES.

CHANK 1977 -1 CHANG, R. L. P.; PAVLIDIS, T.
FUZZY DECISION TREES.

CHERF 1973 -1 CHERVAK, YU. YU.
METHOD OF LEXICOGRAPHIC SEARCH FOR THE SOLUTION OF DISCRETE LINEAR PROGRAMMING PROBLEMS (RUSSIAN), THE.

CHRIA 1977 -1 CHRISTOFIDES, N.; WHITLOCK, C.
ALGORITHM FOR TWO-DIMENSIONAL CUTTING PROBLEMS, AN.

COCK 1975 -1 COCK, J. DE
SEARCH PROCEDURE FOR THE CONSTRUCTION OF SCHOOL TIMETABLES, A.

COCKC 1977 -1 COCKAYNE, E. J.; ROBERTS, F. D. K.
COMPUTATION OF DOMINATING PARTITIONS.

COOK 1975 -1 COOK, R. J.
CHROMATIC NUMBER AND GIRTH.

DEHN 1976 -1 DEHNERT, G.
BRANCH AND BOUND METHOD FOR SOLVING FIXED CHARGE NETWORK FLOW PROBLEMS, A.

DIXO 1976 -1 DIXON, E. T.; GOODMAN, S. E.
ALGORITHM FOR THE LONGEST CYCLE PROBLEM, AN.

DRIS 1976 -1 DRISCOLL, W. C.
BRANCH-BOUND ALGORITHM FOR SCHEDULING PRODUCTION ON ONE MACHINE WITH CHANGEOVER TIMES, A.

DUNS 1975 -1 DUNSTAN, F. D. J.
GREEDY ALGORITHMS FOR OPTIMISATION PROBLEMS.

ELLW 1972 -1 ELLWEIN, L. B.
FLEXIBLE ENUMERATION SCHEME FOR ZERO-ONE PROGRAMMING, A.

ERLE 1976 -1 ERLENKOTTER, D.
DUAL-BASED PROCEDURE FOR UNCAPACITATED FACILITY LOCATION, A.

ESCUA 1976 -1 ESCUDERO, L. F.; VAZQUEZ MUNIZ, A.
USE OF MIXED INTEGER PROGRAMMING FOR THE EVALUATION OF SOME ALTERNATE AIR POLLUTION ABATEMENT POLICIES, THE.

ETCH 1976 -1 ETCHEBERRY, J.
NEW BRANCHING AND BOUNDING STRATEGIES FOR 0-1 INTEGER LINEAR PROGRAMMING.

ETCH 1977 -1 ETCHEBERRY, J.
SET-COVERING PROBLEM: A NEW IMPLICIT ENUMERATION ALGORITHM, THE.

ETCH 1976 -2 ETCHEBERRY, J.
SET REPRESENTATION PROBLEM: A NEW IMPLICIT ENUMERATION ALGORITHM, THE (REVISED VERSION).

FISHA 1976 -1 FISHER, M. L.
DUAL ALGORITHM FOR THE ONE-MACHINE SCHEDULING PROBLEM, A.

FISK 1977 -1 FISK, J.
SOLUTION PROCEDURE FOR A SPECIAL TYPE OF CAPACITATED WAREHOUSE LOCATION PROBLEM, A.

FISKA 1977 -1 FISK, J.; MACKEOWN, P.
DETERMINING THE MAXIMUM DEGENERATE SOLUTION FOR TRANSPORTATION PROBLEMS.

FISKC 1977 -1 FISK, J. C.
INITIAL BOUNDING PROCEDURE FOR USE WITH 0-1 SINGLE KNAPSACK ALGORITHMS, AN.

FOULA 1976 -1 FOULDS, L. R.; ROBINSON, D. F.
STRATEGY FOR SOLVING THE PLANT LAYOUT PROBLEM, A.

FOX B 1977 -1 FOX, B. L.; LENSTRA, J. K.; RINNOOY KAN, A. H. G.; SCHRAGE, L. E.
BRANCHING FROM THE LARGEST UPPER BOUND: FOLKLORE AND FACTS.

FRANC 1976 -1 FRANCON, J.
ARBRES BINAIRES DE RECHERCHE: PROPRIETES COMBINATOIRES ET APPLICATIONS.

FUKU 1976 -1 FUKUDA, Y.
IMPLICIT ENUMERATION ALGORITHM FOR A STORAGE ALLOCATION PROBLEM, AN.

GALI 1975 -1 GALIL, Z.
ON ENUMERATION PROCEDURES FOR THEOREM PROVING AND FOR INTEGER PROGRAMMING.

GALLB 1976 -1 GALLO, G.; HAMMER, P. L.; SIMEONE, B.
QUADRATIC KNAPSACK PROBLEMS.

GALLC 1972 -1 GALLO, G.; MARTINO. E.; SIMEONE, B.
GROUP OPTIMIZATION ALGORITHMS AND SOME NUMERICAL RESULTS VIA A BRANCH AND BOUND APPROACH.

GARFB 1977 -1 GARFINKEL, R. S.; NEEBE, A. W.; RAO, M. R.
M-CENTER PROBLEM: MINIMAX FACILITY LOCATION, THE.

GARTA 1975 -1 GARTNER, N. H.; LITTLE, J. D. C.; GABBAY, H.
OPTIMIZATION OF TRAFFIC SIGNAL SETTINGS BY MIXED-INTEGER LINEAR PROGRAMMING. PART I: THE NETWORK COORDINATION PROBLEM, PART II: THE NETWORK SYNCHRONIZATION PROBLEM.

GAUT 1977 -1 GAUTHIER, J.-M.; RIBIERE, G.
EXPERIMENTS IN MIXED-INTEGER LINEAR PROGRAMMING USING PSEUDOCOSTS.

GEN A 1976 -1 GEN, M.; OKUNO, H.; SHINOFUJI, S.
OPTIMIZING METHOD IN SYSTEM RELIABILITY WITH FAILURE-MODES BY IMPLICIT ENUMERATION ALGORITHM, AN.

GEN B 1975 -1 GEN, M.; SASAKI, M.
OPTIMAL ALLOCATION OF SYSTEM AVAILABILITY.

GEOF 1977 -1 GEOFFRION, A. M.
HOW CAN SPECIALIZED DISCRETE AND CONVEX OPTIMIZATION METHODS BE MARRIED.

GHIG 1976 -1 GHIGGI, C.; PULIAFITO, P. P.; ZOPPOLI, R.
COMBINATORIAL METHOD FOR HEALTH-CARE DISTRICTING, A.

GLOVH 1975 -1 GLOVER, F.; ROSS, G. T.
IMPROVED BOUNDS FOR A CLASS OF 0-1 INTEGER PROGRAMS USING COMBINED SURROGATE-LAGRANGEAN TECHNIQUES.

GLOVI 1976 -1 GLOVER, F.; SOMMER, D.
INEQUALITIES FOR MIXED INTEGER PROGRAMS WITH STRUCTURE.

GLOVJ 1976 -1 GLOVER, F.; TANGEDAHL, L.
DYNAMIC STRATEGIES FOR BRANCH AND BOUND.

GRAB 1976 -3 GRABOWSKI, J.
SOLVING THE TRAVELLING SALESMAN PROBLEM BY USING DISJUNCTIVE GRAPHS.

GRABB 1975 -1 GRABOWSKI, W.
SOLUTION OF THE PLANT LOCATION PROBLEM BY THE BRANCH AND BOUND METHOD (POLISH), A.

GRANA 1976 -1 GRANOT, D.; GRANOT, F.
ON SOLVING FRACTIONAL (0,1) PROGRAMS BY IMPLICIT ENUMERATION.

GRIS 1972 -1 GRISHUKHIN, V. P.
ESTIMATION OF THE COMPLEXITY OF THE BALAS ALGORITHM (RUSSIAN).

GRIS 1976 -1 GRISHUKHIN, V. P.
BRANCH AND BOUNDARY ALGORITHMS IN PROBLEMS WITH BOOLEAN VARIABLES: ESTIMATION OF THEIR EFFECTIVENESS (RUSSIAN).

GRIS 1976 -2 GRISHUKHIN, V. P.
EFFECTIVENESS OF THE BRANCH AND BOUND METHOD IN PROBLEMS WITH BOOLEAN VARIABLES (RUSSIAN).

GUHA 1977 -1 GUHA, D. K.
BRANCH AND BOUND ALGORITHM FOR SOLVING TRAVELING SALESMAN PROBLEM, A.

GUIGA 1977 -2 GUIGNARD, M.; SPIELBERG, K.
DUAL METHOD FOR THE MIXED PLANT LOCATION PROBLEM, A.

GUIGA 1977 -3 GUIGNARD, M.; SPIELBERG, K.
PROPAGATION, PENALTY IMPROVEMENT AND USE OF LOGICAL INEQUALITIES IN INTERACTIVE (0,1) PROGRAMMING.

GUIGA 1977 -4 GUIGNARD, M.; SPIELBERG, K.
REDUCTION METHODS FOR STATE ENUMERATION INTEGER PROGRAMMING.

HABE 1976 -1 HABENICHT, W.
AUSGEWAEHLTE ENUMERATIONSSTRATEGIEN ZUR LOESUNG GANZZAHLIGER LINEARER OPTIMIERUNGSPROBLEME.

HANSA 1975 -2 HANSEN, P.
FONCTIONS D'EVALUATION ET PENALITES POUR LES PROGRAMMES QUADRATIQUES EN VARIABLES 0-1.

HANSA 1975 -3 HANSEN, P.
PROCEDURES D'EXPLORATION ET D'OPTIMISATION PAR SEPARATION ET EVALUATION, 1: CLASSIFICATION ET FORME DES TESTS, LES.

HANSA 1975 -4 HANSEN, P.
PROCEDURES D'EXPLORATION ET D'OPTIMISATION PAR SEPARATION ET EVALUATION: A SURVEY, LES.

HEARA 1976 -1 HEARN, D.; LOWE, T. J.
SUBGRADIENT PROCEDURE FOR THE SOLUTION OF MINIMAL PROBLEMS, A.

HEND 1976 -1 HENDERSON, P. B.; STEIGLITZ, K.
PROBLEM IN SINGLE-MACHINE SEQUENCING WITH NONLINEAR DELAY COSTS, A.

HILL 1976 -1 HILL, A. V.; WHYBARK, D. C.
COMPARING EXACT SOLUTION PROCEDURES FOR THE MULTI-VEHICLE ROUTING PROBLEM.

HONGA 1977 -1 HONG, S.
ASYMMETRIC TRAVELLING SALESMAN PROBLEM AND SPANNING ARBORESCENCES OF DIRECTED GRAPHS, THE.

HUNGA 1976 -1 HUNG, M. S.; FISK, J. C.
ALGORITHM FOR 0-1 MULTIPLE KNAPSACK PROBLEMS, AN.

HYUN 1975 -1 HYUN, K. N.
RELIABILITY OPTIMIZATION BY 0-1 PROGRAMMING FOR A SYSTEM WITH SEVERAL FAILURE MODES.

IBAR 1977 -1 IBARAKI, T.
BRANCH-AND-BOUND ENUMERATION.

IBAR 1976 -1 IBARAKI, T.
COMPUTATIONAL EFFICIENCY OF APPROXIMATE BRANCH-AND-BOUND ALGORITHMS.

IBAR 1977 -2 IBARAKI, T.
POWER OF DOMINANCE RELATIONS IN BRANCH-AND-BOUND ALGORITHMS, THE.

IBAR 1976 -3 IBARAKI, T.
THEORETICAL COMPARISONS OF SEARCH STRATEGIES IN BRANCH-AND-BOUND ALGORITHMS.

INGAA 1977 -1 INGARGIOLA, G. P.; KORSH, J. F.
GENERAL ALGORITHM FOR ONE-DIMENSIONAL KNAPSACK PROBLEMS, A.

IVANA 1976 -1 IVANCHEV, M. D.; MITEV, I. G.; JANEV, N. I.
REALIZATION OF THE BRANCH AND BOUND METHOD FOR THE SOLUTION OF THE GENERAL MIXED INTEGER PROGRAMMING PROBLEM (RUSSIAN), A.

JAER 1973 -1 JAERVINEN, P.
ON A ROUTING PROBLEM IN A GRAPH WITH TWO TYPES OF COSTS ASSOCIATED WITH THE EDGES.

KARW 1976 -1 KARWAN, M. H.; RARDIN, R. L.
APPLICATION OF SURROGATE CONSTRAINT DUALITY IN INTEGER LINEAR PROGRAMMING, THE.

KARW 1976 -2 KARWAN, M. H.; RARDIN, R. L.
AURROGATE DUAL MULTIPLIER SEARCH PROCEDURES IN INTEGER PROGRAMMING.

KARW 1977 -2 KARWAN, M. H.; RARDIN, R. L.
SURROGATE DUALITY IN A BRANCH-AND-BOUND PROCEDURE.

KATSA 1976 -1 KATSOULAS, P. C.; SADOWSKI, R. P.
BRANCH AND BOUND ALGORITHM FOR USE IN RESOURCE ALLOCATION, A.

KENN 1976 -1 KENNINGTON, J. L.
FIXED-CHARGE TRANSPORTATION PROBLEM: A COMPUTATIONAL STUDY WITH A BRANCH-AND-BOUND CODE, THE.

KOHLB 1976 -1 KOHLER, W. H.; STEIGLITZ, K.
ENUMERATIVE AND ITERATIVE COMPUTATIONAL APPROACHES.

KORTB 1976 -2 KORTE, B.
INTEGER PROGRAMMING (SURVEY PAPER).

****

KOVA 1976 -1 KOVACS, L. B.
COMPUTER PROGRAM SYSTEM FOR THE SOLUTION OF PURE LINEAR DISCRETE PROBLEMS, A.

KOVC 1976 -1 KOVCAS, L. B.; RUPP, M.
GRAPH PARTITIONING FOR MATRIX INVERSION BY INTEGER PROGRAMMING.

KRAR 1969 -1 KRARUP, J.
BRANCH-BOUND ALGORITHM FOR A WIRING PROBLEM, A.

LAGE 1976 -1 LAGEWEG, B. J.; LENSTRA, J. K.; RINNOOY KAN, A. H .G.
GENERAL BOUNDING SCHEME FOR THE PERMUTATION FLOW-SHOP, A.

LAGEA 1976 -1 LAGEWEG, B. J.; LENSTRA, J. K.; RINNOOY KAN, A. H. G.
GENERAL BOUNDING SCHEME FOR THE PERMUTATION FLOW-SHOP, A.

LAGEA 1976 -2 LAGEWEG, B. J.; LENSTRA, J. K.; RINNOOY KAN, A. H. G.
JOB-SHOP SCHEDULING BY IMPLICIT ENUMERATION.

LAGEA 1976 -3 LAGEWEG, B. J.; LENSTRA, J. K.; RINNOOY KAN, A. H. G.
JOB-SHOP SCHEDULING BY IMPLICIT ENUMERATION.

LAND 1977 -1 LAND, A.; POWELL, S.
COMPUTER CODES FOR PROBLEMS OF INTEGER PROGRAMMING.

LANDA 1978 -1 LAND, A. H.; LAPORTE, G.: M. , P.
UNIFIED FORMULATION OF THE MACHINE SCHEDULING PROBLEM, A.

LAURB 1978 -1 LAURIERE, M.
ALGORITHM FOR THE 0/1 KNAPSACK PROBLEM, AN.

LAURB 1976 -1 LAURIERE, M.
METHODE DE RESOLUTION RAPIDE POUR LE PROBLEME DU KNAPSACK, UNE.

LEBL 1979 -1 LEBLANC, L. J.
ALGORITHM FOR THE DISCRETE NETWORK DESIGN PROBLEM, AN.

LENS 1973 -1 LENSTRA, H. W.
ACYCLIC SUBGRAPH PROBLEM, THE.

LENSA 1977 -1 LENSTRA, J. K.
SEQUENCING BY ENUMERATIVE METHODS.

LENSB 1975 -1 LENSTRA, J. K.; RINNOOY KAN, A. H. G.
ON THE JOB-SHOP SCHEDULING PROBLEM.

LEV 1976 -1 LEV, O. E.
STRUCTURAL OPTIMIZATION SOLUTION TO A BRANCH-AND-BOUND PROBLEM, A.

LIEBA 1964 -1 LIEBMAN, J. C.
BRANCH-AND-BOUND ALGORITHM FOR MINIMIZING THE COSTS OF WASTE TREATMENT, SUBJECT TO EQUITY CONSTRAINS, A.

LISK 1977 -1 LISKOVEC, V. A.
GENERAL ENUMERATION SCHEME FOR LABELLED GRAPHS (RUSSIAN), A.

LITVA 1973 -1 LITVAK, B. G.; NAIVEL'T, A. V.
SOLUTION OF THE MULTIDIMENSIONAL KNAPSACK PROBLEM WITH SUPPLEMENTARY CONSTRAINTS (RUSSIAN), THE.

LOS 1978 -1 LOS, M.
SIMULTANEOUS OPTIMIZATION OF LAND USE AND TRANSPORTATION: A SYNTHESIS OF THE QUADRATIC ASSIGNMENT PROBLEM AND THE OPTIMAL NETWORK PROBLEM.

MAC F 1977 -1 MAC GINNIS, L. F.
SURVEY OF RESULTS FOR A CLASS OF FACILITIES LOCATION PROBLEMS, A.

MAFF 1976 -1 MAFFIOLI, F.
SUBGRADIENT OPTIMIZATION, MATROID PROBLEMS AND HEURISTIC EVALUATION.

MARSA 1975 -1 MARSTEN, R. E.
ALGORITHM FOR LARGE SET PARTITIONING PROBLEMS, AN.

MARSB 1978 -1 MARSTEN, R. E.; MORIN, T. L.
HYBRID APPROACH TO DISCRETE MATHEMATICAL PROGRAMMING, A.

MARSB 1977 -2 MARSTEN, R. E.; MORIN, T. L.
PARAMETRIC INTEGER PROGRAMMING: THE RIGHT-HAND-SIDE CASE.

MARTB 1977 -1 MARTELLO, S.
ALGORITHM FOR FINDING A MINIMAL EQUIVALENT GRAPH OF A STRONGLY CONNECTED DIGRAPH, AN.

MARTC 1976 -1 MARTELLO, S.; TOTH, P.
ENUMERATIVE ALGORITHM FOR THE O-1 KNAPSACK PROBLEM (ITALIAN), AN.

MARTC 1977 -2 MARTELLO, S.; TOTH, P.
SOLUTION OF THE BOUNDED AND UNBOUNDED CHANGE-MAKING PROBLEM.

MARTC 1977 -3 MARTELLO, S.; TOTH, P.
SOLUTION OF ZERO-ONE MULTIPLE KNAPSACK PROBLEM.

MARTC 1977 -4 MARTELLO, S.; TOTH, P.
UPPER BOUND FOR THE ZERO-ONE KNAPSACK PROBLEM AND A BRANCH AND BOUND ALGORITHM, AN.

MATI 1974 -1 MATIJASEVIC, JU. V.
TEST FOR THE COLORABILITY OF THE VERTICES OF A GRAPH, FORMULATED IN TERMS OF THE ORIENTATIONS OF THE EDGES (RUSSIAN), A.

MEIR 1976 -1 MEIR, A.; MOON, J. W.
CLIMBING CERTAIN TYPES OF ROOTED TREES, I.

MEVE 1977 -1 MEVERT, P.; SUHL, U.
IMPLICIT ENUMERATION WITH GENERALIZED UPPER BOUNDS.

MITE 1977 -1 MITEV, J. G.
APPLICATION OF THE BRANCH AND BOUND METHOD TO SOME DISCRETE LINEAR PROGRAMMING PROBLEMS (RUSSIAN).

MITE 1977 -2 MITEV, J. G.
ON SOME KINDS OF PENALTIES IN DISCRETE PROGRAMMING (RUSSIAN).

MITTB 1973 -1 MITTEN, L. G.; WARBURTON, A. R.
IMPLICIT ENUMERATION PROCEDURES.

MOJE 1976 -1 MOJENA, R.; VOLLMANN, T. E.; OKAMOTO, Y.
ON PREDICTING COMPUTATIONAL TIME OF A BRANCH AND BOUND ALGORITHM FOR THE ASSIGNMENT OF FACILITIES.

MOLN 1976 -1 MOLNAR, Z.; TRCKA, V.
SOLVING ZERO-ONE HYPERBOLIC PROGRAMMING PROBLEM BY MEANS OF BRANCH AND BOUND METHOD (CZECH).

MOORB 1976 -1 MOORE, T. G.
SOLVING LARGE 0-1 PROBLEMS.

MORI 1975 -1 MORIN, T. , L.; MARSTEN, R. E.
HYBRID DYNAMIC PROGRAMMING/BRANCH-AND-BOUND APPROACH TO A CLASS OF SEQUENCING PROBLEMS, A.

MORIA 1976 -1 MORIN, T. L.; MARSTEN, R. E.
BRANCH-AND-BOUND STRATEGIES FOR DYNAMIC PROGRAMMING.

MORT 1977 -1 MORTON, T. E.; DHARAN, B. G.
ALGORISTICS FOR SEQUENCING WITH PRECEDENCE CONSTRAINTS.

MUCK 1977 -1 MUCKSTADT, J. A.; KOENIG, S. A.
APPLICATION OF LAGRANGIAN RELAXATION TO SCHEDULING IN POWER-GENERATION SYSTEMS, AN.

MURP 1976 -1 MURPHY, F. H.; STOHR, E. A.
DYNAMIC PROGRAMMING ALGORITHM FOR CHECK SORTING, A (REVISED VERSION).

****

MURTB 1976 -1 MURTY, K. G.
LINEAR AND COMBINATORIAL PROGRAMMING.
NABE 1973 -1 NABESHIMA, I.
BACKTRACK PROGRAMMING ALGORITHM AND RELIABLE HEURISTIC PROGRAMS FOR GENERAL SCHEDULING PROBLEM, A.
NABE 1972 -1 NABESHIMA, I.
ALGORITHMS FOR MULTIPROJECT SCHEDULING WITH RESOURCE CONSTRAINTS AND RELATED PARALLEL SCHEDULING.
NABEA 1975 -1 NABESHIMA, I.; MARUYAMA, S.
COMPUTATIONAL EXPERIENCE ON MULTIPROJECT SCHEDULING WITH RESOURCE CONSTRAINTS.
NARUA 1977 -1 NARULA, S. C.; OGBU, U. I.; SAMUELSSON, H. M.
ALGORITHM FOR THE P-MEDIAN PROBLEM, AN.
NAST 1975 -1 NASTANSKY, L.; GROSSE, H. W.
COMPUTATIONAL EXPERIMENTS WITH LINEAR PROGRAMMING ORIENTED IMPLICIT ENUMERATION ALGORITHMS FOR THE SOLUTION OF LINEAR (0,1) - PROGRAMS.
NAUS 1976 -1 NAUSS, R. M.
ALGORITHM FOR THE 0-1 INTEGER LINEAR PROGRAMMING PROBLEM, AN.
NAUS 1976 -2 NAUSS, R. M.
EFFICIENT ALGORITHM FOR THE 0-1 KNAPSACK PROBLEM, AN.
NEEBA 1975 -1 NEEBE, A. W.; RAO, M. R.
SUBGRADIENT APPROACH TO THE M-MEDIAN PROBLEM, A.
NUNNA 1977 -1 NUNNIKHOVEN, T. S.; EMMONS, H.
SCHEDULING ON PARALLEL MACHINES TO MINIMIZE TWO CRITERIA RELATED TO JOB TARDINESS.
OTTA 1976 -1 OTTAVIANI, M.
CUTS ON THE FEASIBLE REGION OF A LINEAR PROGRAMMING PROBLEM WITH INTEGER VARIABLES (ITALIAN).
PADBA 1977 -1 PADBERG, M. W.
COVERING, PACKING AND KNAPSACK PROBLEMS.
PANWA 1975 -1 PANWALKAR, S. S.; KHAN, A. W.
IMPROVED BRANCH AND BOUND PROCEDURE FOR N X M FLOW SHOP PROBLEMS, AN.
PARS 1969 -1 PARSONS, J. A.
BRANCH AND BOUND ALGORITHMS: THE TRAVELING SALESMAN PROBLEM.
PATTA 1976 -1 PATTERSON, J. H.; ROTH, G. W.
SCHEDULING A PROJECT UNDER MULTIPLE RESOURCE CONSTRAINTS: A ZERO-ONE PROGRAMMING APPROACH.
PERLA 1975 -1 PERLMAN, B. L.
BRANCH-AND-BOUND METHODS APPLIED TO REAL-TIME VEHICLE DISPATCHING.
PHUO 1976 -1 PHUONG, T. H.
SOLUTION OF INTEGER PROGRAMS WITH A QUADRATIC OBJECTIVE FUNCTION.
PICAC 1976 -1 PICARD, J.-C.; QUEYRANNE, M.
TIME-DEPENDENT TRAVELING SALESMAN PROBLEM AND APPLICATION TO THE TARDINESS PROBLEM IN ONE-MACHINE SCHEDULING, THE.
PIPEA 1975 -1 PIPER, C. J.; ZOLTNERS, A. A.
IMPLICIT ENUMERATION BASED ALGORITHMS FOR POSTOPTIMIZING ZERO-ONE PROGRAMS.
PLOT 1974 -1 PLOTNIKOV, V. N.; GALKIN, A. V.
METHOD FOR THE SOLUTION OF CERTAIN EXTREMAL PROBLEMS OF COMBINATORIAL TYPE (RUSSIAN), A.
PRES 1975 -1 PRESENT, M.
PROBLEMES DE PARTITIONNEMENT COMMENT UTILISER DES METHODES DE REDUCTION?
RARD 1977 -1 RARDIN, R. L.; LIN, B.
WHAT MAKES INTEGER PROGRAMMING PROBLEMS HARD TO SOLVE?
RARDA 1976 -2 RARDIN, R. L.; UNGER, V. E.
SOME COMPUTATIONALLY RELEVANT GROUP THEORETIC STRUCTURES OF FIXED CHARGE PROBLEMS.
REIN 1977 -1 REINGOLD, E. M.; NIEVERGELT, J.; DEO, N.
COMBINATORIAL ALGORITHMS: THEORY AND PRACTICE.
RINN 1976 -1 RINNOOY KAN, A. H. G.
ON MITTEN'S AXIOMS FOR BRANCH-AND-BOUND.
ROED 1976 -1 ROEDDER, W.
LEXIKOGRAPHISCHER SUCHALGORITHMUS ZUR GANZZAHLIGEN PROGRAMMIERUNG: LEXS, EIN.
ROOD 1977 -1 ROODMAN, G. M.; SCHWARZ, L. B.
EXTENSIONS OF THE MULTI-PERIOD FACILITY PHASE-OUT MODEL: NEW PROCEDURES AND APPLICATION TO A PHASE-IN/PHASE-OUT PROBLEM.
ROSSA 1976 -1 ROSS, G. T.; SOLAND, R. M.; ZOLTNERS, A. A.
NOTE ON THE BOUNDED INTERVAL GENERALIZED ASSIGNMENT PROBLEM, A.
ROY 1975 -2 ROY, B.
CHEMINS ET CIRCUITS: ENUMERATION ET OPTIMISATION: A SURVEY.
RYND 1977 -2 RYNDIN, B. N.
MULTIPLICATIVE VERSION OF GOMORY'S FIRST ALGORITHM AND COMBINATIONS OF IT WITH THE METHOD OF BRANCHING (RUSSIAN), A.
RYZH 1975 -2 RYZHKOV, A. P.
GRAPH DIVISION INTO THE MINIMAL NUMBER OF COMPLETE SUBGRAPHS (RUSSIAN).
SAIP 1975 -1 SAIPE, A. L.
SOLVING A (0,1) HYPERBOLIC PROGRAM BY BRANCH AND BOUND.
SALE 1976 -1 SALERNO, G.
"BRANCH AND BACK-TRACK" ALGORITHM FOR THE DESIGN OF AN OPTIMAL TRANSPORTATION NETWORK (ITALIAN), A.
SALK 1975 -1 SALKIN, H. M.
INTEGER PROGRAMMING.
SALKB 1972 -1 SALKIN, H. M.; SPIELBERG, K.
ADAPTIVE ALGORITHM FOR BINARY PROGRAMMING, AN.
SCHAB 1972 -1 SCHALLEHN, W.; SEBASTIAN, H.-J.
REIHENFOLGEPROBLEM IN DER NETZPLANTECHNIK, EIN.
SCHO 1976 -1 SCHOCH, M.
ERWEITERUNGSPRINZIP UND SEINE ANWENDUNG ZUR ENTWICKLUNG VON ALGORITHMEN FUER DIE LOESUNG KOMBINATORISCHER OPTIMIERUNGSAUFGABEN, DAS.
SCHOC 1977 -1 SCHOENLEIN, A.
UEBER DIE ANWENDUNG DER METHODE DES BRANCH AND BOUND AUF DAS SUMMEN-ZUORDNUNGSPROBLEM.
SCHOC 1976 -1 SCHOENLEIN, A.
BEZUEGLICH KERNSPEICHER- UND RECHENZEITBEDARF EFFIZIENTES BRANCH-UND-BOUND-VERFAHREN ZUR LOESUNG DES TRAVELLING-SALESMAN-PROBLEMS, EIN.
SHAPB 1977 -3 SHAPIRO, J. F.
SURVEY OF LAGRANGEAN TECHNIQUES FOR DISCRETE OPTIMIZATION.
SHIH 1977 -1 SHIH, W.
BRANCH AND BOUND PROCEDURE FOR A CLASS OF DISCRETE RESOURCE ALLOCATION PROBLEMS WITH SEVERAL CONSTRAINTS, A.

****

SHIR 1976 -1 SHIRAI, Y.; IBARAKI, T.; HASEGAWA, T.
MINIMAL REPRESENTATION OF COMBINATORIAL OPTIMIZATION PROBLEMS BY FINITE-STATE POSITIVELY MONOTONE MODELS.

SMEL 1975 -1 SMELEV, V. V.
PENALTY FUNCTIONS IN LINEAR INTEGER PROGRAMMING (RUSSIAN).

SMITH 1975 -1 SMITH, T. H. C.
LIFO IMPLICIT ENUMERATION ALGORITHM FOR THE ASYMMETRIC TRAVELING SALESMAN PROBLEM USING A ONE-ARBORESCENCE RELAXATION, A.

SMITI 1977 -1 SMITH, T. H. C.; SRINIVASAN, V.; THOMPSON, G. L.
COMPUTATIONAL PERFORMANCE OF THREE SUBTOUR ELIMINATION ALGORITHMS FOR SOLVING ASYMMETRIC TRAVELING SALESMAN PROBLEMS.

SMITJ 1977 -1 SMITH, T. H. C.; THOMPSON, G. L.
LIFO IMPLICIT ENUMERATION SEARCH ALGORITHM FOR THE SYMMETRIC TRAVELING SALESMAN PROBLEM USING HELD AND KARP'S 1-TREE RELAXATION, A.

SOVE 1976 -2 SOVETOV, B. YA.; TISHKIN, A. I.
ON THE SOLUTION OF A CLASS OF PROBLEMS IN SEQUENCING THEORY (RUSSIAN).

SPIE 1972 -1 SPIELBERG, K.
MINIMAL PREFERRED VARIABLE METHODS IN ZERO-ONE PROGRAMMING.

SPIE 1973 -1 SPIELBERG, K.
MINIMAL PREFERRED INEQUALITIES, PENALTIES AND STRUCTURE IN 0-1 PROGRAMMING.

SPIE 1974 -1 SPIELBERG, K.
MINIMAL PREFERRED VARIABLE METHODS IN ZERO ONE PROGRAMMING.

SPIE 1977 -1 SPIELBERG, K.
ENUMERATIVE METHODS IN INTEGER PROGRAMMING.

SPIE 1973 -2 SPIELBERG, K.
PREFERRED VARIABLE BRANCH-AND-BOUND METHOD.

SUHL 1975 -1 SUHL, U.
ENTWICKLUNG VON ALGORITHMEN FUER GANZZAHLIGE OPTIMIERUNGSMODELLE.

SUHL 1976 -1 SUHL, U.
ALGORITHMUS ZUR LOESUNG EINER SPEZIELLEN KLASSE GANZZAHLIGER OPTIMIERUNGSMODELLE, EIN.

TAHA 1975 -1 TAHA, H. A.
INTEGER PROGRAMMING.

TEPL 1975 -1 TEPLICKII, M. G.
BRANCH-AND-BOUND METHOD FOR SOLVING CERTAIN DISCRETE OPTIMAL CONTROL PROBLEMS (RUSSIAN), A.

THES 1975 -1 THESEN, A.
RECURSIVE BRANCH AND BOUND ALGORITHM FOR THE MULTIDIMENSIONAL KNAPSACK PROBLEM, A.

TOWN 1977 -1 TOWNSEND, W.
BRANCH-AND-BOUND METHOD FOR SEQUENCING PROBLEMS WITH LINEAR AND EXPONENTIAL PENALTY FUNCTIONS, A.

TOWS 1977 -1 TOWSNEND, W.
SEQUENCING N JOBS ON M MACHINES TO MINIMISE MAXIMUM TARDINESS: A BRANCH-AND-BOUND SOLUTION.

TRIV 1976 -1 TRIVEDI, V. M.; WARNER, D. M.
BRANCH AND BOUND ALGORITHM FOR OPTIMUM ALLOCATION OF FLOAT NURSES, A.

TROT 1973 -1 TROTTER, L. E. JR.
USER'S INSTRUCTIONS FOR THE INTEGER PROGRAMMING CODE ENUMER8.

TRYP 1977 -1 TRYPIA, M.
NON-BINARY TREE SEARCH METHOD FOR THE 0-1 LINEAR PROGRAMMING PROBLEM, A.

VYAT 1975 -1 VYATKIN, A. K.; KRASIN, L. A.
PROBLEM OF OPTIMIZING A HIERARCHICAL ORGANIZATION STRUCTURE (RUSSIAN), A.

WDOW 1973 -1 WDOWIAK, J.
SOLUTION OF THE FIXED CHARGE PROBLEM (POLISH), THE.

WILLE 1976 -1 WILLIS, R. J.; HASTINGS, N. A. J.
PROJECT SCHEDULING WITH RESOURCE CONSTRAINTS USING BRANCH AND BOUND METHODS.

YAGI 1975 -1 YAGIZ, OE.
HEURISTICALLY-AIDED SINGLE BRANCH SEARCH PROCEDURES FOR PLANT LOCATION PROBLEMS.

YAO B 1976 -1 YAO, A. C.; YAO, F. F.
COMPLEXITY OF SEARCHING AN ORDERED RANDOM TABLE, THE.

ZAK 1976 -1 ZAK, YU. A.
GENERALIZED TRAVELLING-SALESMAN PROBLEM AND ITS PRACTICAL ASPECTS (RUSSIAN), THE.

ZION 1977 -1 ZIONTS, S.
INTEGER LINEAR PROGRAMMING WITH MULTIPLE OBJECTIVES.

****

ALBEC 1976 -1 ALBERTSON, M. O.; BERMAN, D. M.
ACYCLIC CHROMATIC NUMBER, THE.

APPE 1976 -1 APPEL, K.; HAKEN, W.
EVERY PLANAR MAP IS FOUR COLORABLE.

BERG 1975 -1 BERGE, C.
NOMBRES DE COLORATION DE L'HYPERGRAPHE H-PARTI COMPLET.

BERG 1973 -1 BERGE, C.
INTRODUCTION A LA THEORIE DES HYPERGRAPHES.

BERG 1978 -1 BERGE, C.
ALGORITHMS AND EXTREMAL PROBLEMS FOR EQUIPARTITE COLORINGS IN GRAPHS AND HYPERGRAPHS (ABSTRACT).

BERG 1973 -2 BERGE, C.
NOTES SUR LES BONNES COLORATIONS D'UN HYPERGRAPHE.

BERG 1975 -2 BERGE, C.
ON THE GOOD K-COLORINGS OF A HYPERGRAPH.

BERGA 1976 -1 BERGE, C.; HOFFMAN, A. J.
MULTICOLORATIONS DANS LES HYPERGRAPHES UNIMODULAIRES ET MATRICES DONT LES COEFFICIENTS SONT DES ENSEMBLES.

BERGB 1977 -1 BERGE, C.; JOHNSON, E. L.
COLORING THE EDGES OF A HYPERGRAPH AND LINEAR PROGRAMMING TECHNIQUES.

BERMA 1974 -1 BERMOND, J.-C.
NOMBRE CHROMATIQUE TOTAL DU GRAPHE R-PARTI COMPLET.

BOLL 1977 -1 BOLLOBAS, B.
EXTREMAL PROBLEMS IN GRAPH THEORY.

BONN 1974 -1 BONNET, R.; ERDOES, P.
CHROMATIC INDEX OF AN INFINITE COMPLETE HYPERGRAPH: A PARTITION THEOREM, THE.

BORI 1975 -1 BORIES, F.
ETUDE DU NOMBRE ACHROMATIQUE DE CERTAINS GRAPHES.

BORI 1975 -2 BORIES, F.
SUR QUELQUES PROBLEMES DE COLORATIONS COMPLETES DE SOMMETS ET D'ARETES DE GRAPHES ET D'HYPERGRAPHES.

BORIA 1976 -1 BORIES, F.; JOLIVET, J.-L.
ACHROMATIC NUMBER OF A HYPERGRAPH, THE.

BOROB 1977 -1 BORODIN, O. V.; KOSTOCHKA, A. V.
ON AN UPPER BOUND OF A GRAPH'S CHROMATIC NUMBER, DEPENDING ON THE GRAPH'S DEGREE AND DENSITY.

BOROC 1974 -1 BORODIN, O. V.; MEL'NIKOV, L. S.
CHROMATIC NUMBER OF A PSEUDOSURFACE (RUSSIAN), THE.

BOSA 1976 -1 BOSAK, J.; NESETRIL, J.
COMPLETE AND PSEUDOCOMPLETE COLOURINGS OF A GRAPH.

BRELA 1977 -1 BRELAZ, D.; NICOLIER, Y.; WERRA, D. DE
COMPACTNESS AND BALANCING IN SCHEDULING.

BURS 1976 -1 BURSHTEJN, M. I.
CHROMATIC NUMBER OF PLANAR HYPERGRAPHES (RUSSIAN), THE.

BURS 1976 -2 BURSHTEJN, M. I.
ON A CLASS OF BICHROMATIC HYPERGRAPHS AND THE ESTIMATION OF THE CHROMATIC NUMBER (RUSSIAN).

CAMEC 1975 -1 CAMERON, P. J.
MINIMAL EDGE-COLOURINGS OF COMPLETE GRAPHS.

CHEN 1975 -1 CHEN, C. C.; DAYKIN, D. E.
HAMILTONIAN CYCLES IN A LINE-COLOURED MULTIGRAPH AND A RELATED UNCOLOURED GRAPH.

CHVAB 1978 -1 CHVATAL, V.; GAREY, M. R.; JOHNSON, D. S.
TWO RESULTS CONCERNING MULTICOLORING.

CZER 1977 -1 CZERNIAKIEWICZ, A.
MINIMAL GRAPHS WITH CYCLIC COLOR GROUP.

DAYK 1976 -2 DAYKIN, D. E.
GRAPHS WITH CYCLES HAVING ADJACENT LINES DIFFERENT COLORS.

DROE 1976 -1 DROESKE, F.; FRENNE, A. DE
ETUDE DE BORNES DU NOMBRE CHROMATIQUE D'UN GRAPHE, I.

DUER 1976 -1 DUERRE, K.
EFFIZIENZ VON FAERBUNGSALGORITHMEN.

DUER 1976 -2 DUERRE, K.
FAST REVERSIBLE CODING AND INDEXING OF COLORATIONS.

DUNS 1976 -1 DUNSTAN, F. D. J.
SEQUENTIAL COLOURINGS OF GRAPHS.

ERDOA 1976 -1 ERDOES, P.; FAUDREE, R. J.; ROUSSEAU, C. C.; SCHELP, R. H.
GENERALIZED RAMSEY THEORY FOR MULTIPLE COLORS.

ERDOB 1966 -1 ERDOES, P.; HAJNAL, A.
ON CHROMATIC NUMBERS OF GRAPHS AND SET-SYSTEMS.

ERDOF 1977 -1 ERDOES, P.; WILSON, R. J.
ON THE CHROMATIC INDEX OF ALMOST ALL GRAPHS.

FIOR 1974 -1 FIORINI, S.; WILSON, R. J.
ON THE CHROMATIC INDEX OF A GRAPH, II.

FIOR 1976 -1 FIORINI, S.; WILSON, R. J.
EDGE-COLOURINGS OF GRAPHS - SOME APPLICATIONS.

FIOR 1973 -1 FIORINI, S.; WILSON, R. J.
ON THE CHROMATIC INDEX OF A GRAPH, I.

FISKD 1977 -1 FISK, S.
GEOMETRIC COLORING THEORY.

FOUR 1973 -1 FOURNIER, J. C.
COLORATIONS DES ARETES D'UN GRAPHE.

FULK 1975 -3 FULKERSON, D. R.
STUDIES IN GRAPH THEORY, II.

GABOB 1976 -2 GABOW, H. N.
USING EULER PARTITIONS TO EDGE COLOR BIPARTITE MULTIGRAPHS.

GAREB 1976 -3 GAREY, M. R.; JOHNSON, D. S.
COMPLEXITY OF NEAR-OPTIMAL GRAPH COLORING, THE.

GAREB 1976 -5 GAREY, M. R.; JOHNSON, D. S.
SOME SIMPLIFIED NP-COMPLETE GRAPH PROBLEMS.

GAREB 1976 -6 GAREY, M. R.; JOHNSON, D. S.
STRONG NP-COMPLETENESS RESULTS: MOTIVATION, EXAMPLES AND IMPLICATIONS.

GAVR 1977 -1 GAVRIL, F.
ALGORITHMS ON CLIQUE SEPARABLE GRAPHS.

GELL 1974 -1 GELLER, D.; KRONK, H.
FURTHER RESULTS ON THE ACHROMATIC NUMBER.

****

GELLA 1976 -1 GELLER, D. P.
R-TUPLE COLORINGS OF UNIQUELY COLORABLE GRAPHS.
GELLB 1974 -1 GELLER, D. P.; HILTON, A. J. W.
HOW TO COLOUR THE LINES OF A BIGRAPH.
GIRA 1976 -1 GIRAUD, G.
SUR LES PROPORTIONS RESPECTIVES DE TRIANGLES UNI, BI OU TRICOLORES DANS UN TRICOLORIAGE DES ARETES DU N-EMBLE.
GOL' 1974 -1 GOL'DBERG, M. K.
REMARK ON THE CHROMATIC CLASS OF A MULTIGRAPH (RUSSIAN), A.
GOLU 1977 -2 GOLUMBIC, M. C.
COMPLEXITY OF COMPARABILITY GRAPH RECOGNITION AND COLORING, THE.
GRAVA 1977 -1 GRAVER, J. E.; WATKINS, M. E.
COMBINATORICS WITH EMPHASIS ON THE THEORY OF GRAPHS.
GRIMA 1975 -1 GRIMMETT, G. R.; MAC DIARMID, C. J. H.
ON COLOURING RANDOM GRAPHS.
GYUL 1975 -1 GYULUMYAN, S. M.
ON CRITICAL GRAPHS IN CONNECTION WITH COLORINGS (RUSSIAN).
HAMM 1977 -1 HAMMER, G.
PRIMAL-DUALES SUCHVERFAHREN FUER FAERBUNGSPROBLEME, EIN.
HANSA 1977 -2 HANSEN, P.
SUBDEGREES AND CHROMATIC NUMBERS OF HYPERGRAPHS.
HANSB 1976 -1 HANSEN, P.; DELATTRE, M.
CLUSTER ANALYSIS AND GRAPH COLORING.
HANSB 1976 -2 HANSEN, P.; DELATTRE, M.
CLUSTER ANALYSIS BY GRAPH COLORING.
HANSB 1976 -3 HANSEN, P.; DELATTRE, M.
COMPLETE-LINK CLUSTER ANALYSIS BY GRAPH COLORING.
HECH 1977 -1 HECHLER, S. H.
ON INFINITE GRAPHS WITH A SPECIFIED NUMBER OF COLORINGS.
HEINA 1976 -1 HEINRICH, K.
NON-IMBEDDABLE PROPER COLOURING, A.
HILT 1976 -1 HILTON, A. J. W.
COVER INDEX, THE CHROMATIC INDEX AND THE MINIMUM DEGREE OF A GRAPH, THE.
HILT 1977 -1 HILTON, A. J. W.
DEFINITIONS OF CRITICALITY WITH RESPECT TO EDGE-COLORING.
HILT 1975 -1 HILTON, A. J. W.
COLOURING THE EDGES OF A MULTIGRAPH SO THAT EACH VERTEX HAS AT MOST J, OR AT LEAST J, EDGES OF EACH COLOUR ON IT.
HILT 1975 -2 HILTON, A. J. W.
ON VIRZING'S UPPER BOUND FOR THE CHROMATIC INDEX OF A GRAPH.
HILTA 1973 -1 HILTON, A. J. W.; RADO. R.; SCOTT, S. H.
(<5)-COLOR THEOREM FOR PLANAR GRAPHS, A.
JOHNC 1978 -1 JOHNSON, D. S.
GRAPH COLORING ALGORITHM: BETWEEN A ROCK AND A HARD PLACE?(ABSTRACT).
JOHNC 1974 -1 JOHNSON, D. S.
WORST CASE BEHAVIOR OF GRAPH COLORING ALGORITHMS.
JOHNE 1978 -1 JOHNSON, E. L.
ON THE EDGE-COLORING PROPERTY FOR THE CLOSURE OF THE COMPLETE HYPERGRAPHS.
JOHNF 1978 -1 JOHNSON, E. L.; NEWMAN, D.; WINSTON, K.
INEQUALITY ON BINOMIAL COEFFICIENTS, AN.
JUCO 1974 -1 JUCOVIC, E.; OLEJNIK, F.
ON CHROMATIC AND ACHROMATIC NUMBERS OF UNIFORM HYPERGRAPHS.
KARPA 1975 -1 KARP, R. M.
FAST APPROXIMATE SOLUTION OF HARD COMBINATORIAL PROBLEMS, THE.
KASZ 1973 -1 KASZONYI, L.
ON THE STRUCTURE OF COLORING GRAPHS.
KEYS 1975 -1 KEYS, C. D.
GRAPHS CRITICAL FOR MAXIMAL BOOKTHICKNESS.
KITA 1975 -1 KITAMURA, S.
ON THE EDGE-CHROMATIC NUMBER OF A GRAPH.
KORSA 1976 -1 KORSHUNOV, A. D.; NINCAK, YA.
COLORING OF HYPERGRAPHS (RUSSIAN), THE.
KOST 1977 -1 KOSTOCKA, A. V.
UPPER BOUNDS FOR THE CHROMATIC NUMBER OF A GRAPH IN TERMS OF ITS DEGREE, DENSITY AND GIRTH (RUSSIAN).
KOZY 1972 -1 KOZYREV, V. P.
GRAPH THEORY (RUSSIAN).
KUCE 1977 -1 KUCERA, L.
EXPECTED BEHAVIOR OF GRAPH COLORING ALGORITHMS.
LAURA 1974 -1 LAURIERE, J. L.
PROBLEMES D'EMPLOI DU TEMPS ET ALGORITHME DE COLORATION DES HYPERGRAPHES.
LAWL 1976 -2 LAWLER, E. L.
NOTE ON THE COMPLEXITY OF THE CHROMATIC NUMBER PROBLEM, A.
LAWR 1978 -1 LAWRENCE, L.
COVERING THE VERTEX SET OF A GRAPH WITH SUBGRAPHS OF SMALLER DEGREE.
LOVA 1968 -1 LOVASZ, L.
ON CHROMATIC NUMBER OF FINITE SET SYSTEMS.
MAC 1977 -1 MAC ALLISTER, J. D.; SAKAROVITCH, M.
CLASSIFICATION DE CERTAIN MATRICES 0-1.
MARTI 1974 -1 MARTINOVA, M. K.
MINIMAL COLORATION OF A RETICULATED GRAPH (BULGARIAN).
MATU 1968 -1 MATULA, D. W.
MIN-MAX THEOREM FOR GRAPHS WITH APPLICATION TO GRAPH COLORING, A.
MATU 1975 -1 MATULA, D. W.
UNIFORM SET COVERING LEMMA, A.
MATUA 1972 -1 MATULA, D. W.; MARBLE, G.; ISAACSON, J. D.
GRAPH COLORING ALGORITHMS.
MEL' 1977 -1 MEL'NIKOV, L. S.; STEINBERG, R.
ONE COUNTEREXAMPLE FOR TWO CONJECTURES ON THREE COLORING.
MEYEA 1978 -1 MEYER, J. C.
NOMBRE CHROMATIQUE TOTAL D'UN HYPERGRAPHE.
MEYEB 1976 -1 MEYER, J.-C.
PROBLEMES DE COLORATIONS DANS LES HYPERGRAPHES.
MEYEB 1975 -1 MEYER, J.-C.
ON THE GOOD COLORINGS, II. CASE OF THE COMPLETE H-PARTITE HYPERGRAPHS.

****

MILL 1976 -1 MILLER, D. M.
ALGORITHM FOR DETERMINING THE CHROMATIC NUMBER OF A GRAPH, AN.
MITCA 1976 -1 MITCHEN, J.
ON VARIOUS ALGORITHMS FOR ESTIMATING THE CHROMATIC NUMBER OF A GRAPH.
NINC 1974 -1 NINCAK, J.
CHROMATIC NUMBERS OF CERTAIN GRAPHS (RUSSIAN), THE.
NOVA 1974 -1 NOVANSKY, R.; PLATEK, M.
CHROMATISCHE BELEGUNGEN UND CHROMATISCHE ZERLEGUNGEN ENDLICHER GRAPHEN (ALGORITHMEN).
OZAW 1973 -1 OZAWA, T.; HIRAO, S.; HATTORI, Y.
SOME PROPERTIES OF MULTICOLORED-BRANCH GRAPHS.
PETRB 1976 -1 PETROV, S. V.
ON A PROBLEM IN GRAPH THEORY (RUSSIAN).
PIEH 1976 -1 PIEHLER, J.
UEBER SCHNITTVERSCHAERFUNGEN IN DER GEMISCHT-GANZZAHLIGEN LINEAREN OPTIMIERUNG.
POLJ 1974 -1 POLJAK, S.
NOTE ON STABLE SETS AND COLORINGS OF GRAPHS, A.
PRESA 1975 -1 PRESIA, H.-J.
UEBER FAERBUNGEN PLANARER GRAPHEN, I.
PRESA 1975 -2 PRESIA, H.-J.
UEBER FAERBUNGEN PLANARER GRAPHEN, II.
ROEDA 1977 -1 ROEDL, V.
ON THE CHROMATIC NUMBER OF SUBGRAPHS OF A GIVEN GRAPH.
ROY 1967 -1 ROY, B.
NOMBRE CHROMATIQUE ET PLUS LONGS CHEMINS D'UN GRAPHE.
RYZH 1975 -2 RYZHKOV, A. P.
GRAPH DIVISION INTO THE MINIMAL NUMBER OF COMPLETE SUBGRAPHS (RUSSIAN).
SAKAA 1976 -1 SAKAKI, T.; NAKASHIMA, K.; HATTORI, Y.
ALGORITHMS FOR FINDING IN THE LUMP BOTH BOUNDS OF THE CHROMATIC NUMBER OF A GRAPH.
SAMP 1976 -1 SAMPATHKUMAR, E.; BHAVE, V. N.
PARTITION GRAPHS AND COLORING NUMBERS OF A GRAPH.
SCOTA 1975 -1 SCOTT, S. H.
MULTIPLE NODE COLORINGS OF FINITE GRAPHS.
SCOTB 1976 -1 SCOTT, T. B.
GRAPH COLOURING WITH PREASSIGNMENT AND UNAVAILABILITY CONSTRAINTS.
SEDL 1975 -1 SEDLACEK, J.
PERFECT AND QUASIPERFECT GRAPHS (CZECH).
SPER 1975 -1 SPERANZA, F.
COLORINGS OF A SUPERIOR SORT OF GRAPHS (ITALIAN).
STAH 1976 -1 STAHL, S.
N-TUPLE COLORINGS AND ASSOCIATED GRAPHS.
STER 1976 -1 STERBOUL, F.
SOME EXTREMAL COLORING PROBLEMS: A SURVEY.
STER 1977 -1 STERBOUL, F.
EXTREMAL PROBLEM IN HYPERGRAPH COLORING, AN.
STER 1974 -1 STERBOUL, F.
ON THE CHROMATIC NUMBER OF THE DIRECT PRODUCT OF HYPERGRAPHS.
STOC 1973 -1 STOCKMEYER, L.
PLANAR 3-COLORABILITY IS POLYNOMIAL COMPLETE.
STRE 1976 -1 STREET, A. P.
EMBEDDING PROPER COLOURINGS.
TEHR 1975 -1 TEHRANI, A.
ALGORITHME DE COLORATION, UN.
TOFT 1975 -1 TOFT, B.
ON COLOUR-CRITICAL HYPERGRAPHS.
TOME 1976 -1 TOMESCU, I.
NOMBRE MAXIMAL DE COLORATIONS D'UN GRAPHE HAMILTONIEN, LE.
TOME 1968 -1 TOMESCU, I.
SUR LE PROBLEME DU COLORIAGE DES GRAPHES GENERALISES.
TOME 1975 -2 TOMESCU, I.
INTRODUCTION TO COMBINATORICS (TRANSLATED FROM RUMANIAN).
TOME 1975 -3 TOMESCU, I.
PROBLEMES EXTREMAUX CONCERNANT LE NOMBRE DES COLORATIONS DES SOMMETS D'UN GRAPHE FINI.
WALK 1976 -1 WALKER, R. C.
GRAPH-COLOURING THEOREM, A.
WEINB 1976 -1 WEINBERGER, D. B.
NETWORK FLOWS, MINIMUM COVERINGS, AND THE FOUR-COLOR CONJECTURE.
WEINE 1975 -1 WEINSTEIN, J. M.
RECONSTRUCTING COLORED GRAPHS.
WERR 1976 -1 WERRA, D. DE
COLOR-FEASIBLE SEQUENCES OF ALMOST BIPARTITE GRAPHS.
WERR 1977 -1 WERRA, D. DE
SOME COLORING TECHNIQUES.
WERR 1975 -1 WERRA, D. DE
FEW REMARKS ON CHROMATIC SCHEDULING, A.
WERR 1976 -2 WERRA, D. DE
EVERYTHING YOU ALWAYS WANTED TO KNOW ABOUT S.EX.
WERR 1977 -2 WERRA, D. DE
UNIQUENESS OF COLORINGS.
WERR 1975 -3 WERRA, D. DE
HOW TO COLOR A GRAPH.
WERR 1976 -3 WERRA, D. DE
MULTIGRAPHS WITH QUASIWEAK ODD CYCLES.
WERR 1975 -5 WERRA, D. DE
ON GOOD AND EQUITABLE EDGE COLORINGS.
WERR 1976 -5 WERRA, D. DE
REMARKS ON THE COLOR-FEASIBLE SEQUENCES OF A MULTIGRAPH.
WERR 1976 -6 WERRA, D. DE
SOME REMARKS ON GOOD COLORATIONS.

COMBINATORIAL ALGORITHMS (UNLESS SPECIFIED IN PARTICULAR)
****

AGAR 1976 -1 AGARWAL, A. K.; ARORA, S. R.
SYNTHESIS OF MULTITERMINAL COMMUNICATION NETS: FINDING ONE OR ALL SOLUTIONS.

AHO B 1974 -1 AHO, A. V.; HOPCROFT, J. E.; ULLMAN, J. D.
DESIGN AND ANALYSIS OF COMPUTER ALGORITHMS, THE.

ALTH 1975 -1 ALTHERR, W.
ALGORITHM FOR ENUMERATING ALL VERTICES OF A CONVEX POLYHEDRON, AN.

AM 1976 -1 AM, T. D.; TSUKIYAMA, S.; SHIRAKAWA, I.; OZAKI, H.
ALGORITHM FOR GENERATING ALL THE PATHS BETWEEN TWO VERTICES IN A DIGRAPH AND ITS APPLICATION, AN.

ARJO 1975 -1 ARJOMANDI, E.; CORNEIL, D. G.
PARALLEL COMPUTATIONS IN GRAPH THEORY.

ARLA 1973 -1 ARLAZAROV, V. L.; USKOV, A. V.; FARADZEV, I. A.
ALGORITHM FOR FINDING ALL SIMPLE CYCLES IN A DIRECTED GRAPH (RUSSIAN), AN.

BAKO 1976 -1 BAKO, A.
ALL PATHS IN AN ACTIVITY NETWORK.

BALAK 1977 -1 BALAS, E.; SAMUELSSON, H.
NODE COVERING ALGORITHM, A.

BAUE 1973 -1 BAUERSFELD, G.; ESSMANN, CH.; LOEHLE, H.
ALGORITHMUS ZUR FESTSTELLUNG DER ISOMORPHIE VON ENDLICHEN, ZUSAMMENHAENGENDEN GRAPHEN.

BAUT 1973 -1 BAUTZ, M.; BIESS, G.
ALGORITHMUS ZUR ERMITTLUNG ALLER ELEMENTARKREISE EINES GERICHTETEN GRAPHEN, EIN.

BERS 1976 -1 BERSTEL, J.
COMPLEXITE DE QUELQUES ALGORITHMES DE THEORIE DES GRAPHES.

BERZ 1973 -1 BERZTISS, A. T.
BACKTRACK PROCEDURE FOR ISOMORPHISM OF DIRECTED GRAPHS, A.

BIESA 1977 -1 BIESS, G.; JANICKE, W.; MEISSLER, H. G.
ALGORITHMUS ZUR BESTIMMUNG ALLER ELEMENTARKREISE EINES GERICHTETEN GRAPHEN, EIN.

BLAZC 1976 -1 BLAZHKEVICH, B. I.; SEMOTYUK, V. N.
TREE SEARCH ALGORITHM IN A GRAPH (RUSSIAN).

BOESA 1976 -1 BOESCH, F.; HARARY, F.
LINE REMOVAL ALGORITHMS FOR GRAPHS AND THEIR DEGREE LISTS. SPECIAL ISSUE ON LARGE-SCALE NETWORKS AND SYSTEMS.

BOOTA 1975 -1 BOOTH, K. S.; LUEKER, G. S.
LINEAR ALGORITHMS TO RECOGNIZE INTERVAL GRAPHS AND TEST FOR CONSECUTIVE ONES PROPERTY.

BROWB 1976 -1 BROWN, J. R.
ALGORITHM FOR R-BASES OF A DIRECTED GRAPH, AN.

BURKF 1975 -1 BURKHARD, W. A.
NONRECURSIVE TRAVERSALS OF TREES.

BURKF 1973 -1 BURKHARD, W. A.
NONRECURSIVE TREE TRAVERSAL ALGORITHM.

CARS 1977 -1 CARSON, J. S.; LAW, A. M.
NOTE ON SPIRA'S ALGORITHM FOR THE ALL-PAIRS SHORTEST-PATH PROBLEM, A.

CHANF 1977 -1 CHANDRASEKARAN, R.
MINIMAL RATIO SPANNING TREES.

CHEI 1975 -1 CHEIN, M.
REMARQUE SUR L'ALGORITHME DE DUCAMP POUR LA RECHERCHE DE TOUS LES ORDRES TOTAUX CONTENANT UN ORDRE PARTIEL, UNE.

CHENC 1975 -1 CHEN, I. N.
NODE-ELIMINATION METHOD FOR FINDING ALL SIMPLE CYCLES IN A DIRECTED GRAPH, A.

CHENG 1969 -1 CHEN, W. K.
GENERATION OF TREES AND COTREES OF A GRAPH BY DECOMPOSITION.

CIOBA 1977 -1 CIOBOTARU, S.
LINGUISTIC METHOD FOR DETERMINING THE HAMILTONIAN PATHS (RUMANIAN), A.

COGI 1976 -1 COGIS, O.
FORMALISATION ET GENERALISATION D'UN PRINCIPE CLASSIQUE D'ALGORITHME DE RECHERCHE D'ISOMORPHISME ENTRE DEUX GRAPHES.

COLL 1975 -1 COLLATZ, L.
BEZIEHUNGEN ZWISCHEN GRAPHEN, GEOMETRIE UND KOMBINATORIK, EINIGE.

CORN 1974 -1 CORNEIL, D. G.
ANALYSIS OF GRAPH THEORETICAL ALGORITHMS, THE.

CORN 1970 -1 CORNEIL, D. G.
GRAPH ISOMORPHISM.

CORN 1972 -1 CORNEIL, D. G.
ALGORITHM FOR DETERMINING THE AUTOMORPHISM PARTITIONING OF AN UNDIRECTED GRAPH, AN.

CORN 1968 -1 CORNEIL, D. G.
GRAPH ISOMORPHISM.

CORNA 1970 -1 CORNEIL, D. G.; GOTLIEB, C. C.
EFFICIENT ALGORITHM FOR GRAPH ISOMORPHISM, AN.

CORNB 1978 -1 CORNEIL, D. G.; MATHON, R. A.
ALGORITHMIC TECHNIQUES FOR THE GENERATION AND ANALYSIS OF STRONGLY REGULAR GRAPHS AND OTHER COMBINATORIAL CONFIGURATIONS.

COT 1974 -1 COT, N.
LINEAR-TIME ORDERING PROCEDURE WITH APPLICATIONS TO VARIABLE-LENGTH ENCODING, A.

DEO A 1977 -1 DEO, N.; DAVIS, J. M.; LORD, R. E.
NEW ALGORITHM FOR DIGRAPH ISOMORPHISM, A.

DIJK 1976 -1 DIJKSTRA, E. W.
DISCIPLINE OF PROGRAMMING, A.

DINIA 1977 -1 DINITS, E. A.; ZAJTSEV, M. A.
ALGORITHMS WHICH GENERATE NON-ISOMORPHIC TREES (RUSSIAN).

DRUF 1975 -1 DRUFFEL, L. E.
GRAPH RELATED ALGORITHMS: ISOMORPHISM, AUTOMORPHISM AND CONTAINMENT.

DZIK 1974 -1 DZIKIEWICS, J.
ALGORITHM FOR FINDING THE TRANSITIVE CLOSURE OF A DIGRAPH, AN.

DZIKA 1974 -1 DZIKIEWICS, J.; SYSLO, M. M.
COMPUTIONAL EXPERIENCES WITH SOME TRANSITIVE CLOSURE ALGORITHMS.

EDEN 1977 -1 EDEN, B.; PAPE, U.
PLANAR. EIN ALGORITHMUS ZUR FESTSTELLUNG DER PLANARITAET VON GRAPHEN UND ZUR ERMITTLUNG PLANARER TEILGRAPHEN.

EMDE 1975 -1 EMDE BOAS, P. V.
PRESERVING ORDER IN A FOREST IN LESS THAN LOGARITHMIC TIME.

EVEND 1975 -2 EVEN, S.; TARJAN, R. E.
NETWORK FLOW AND TESTING GRAPH CONNECTIVITY.

FILL 1977 -1 FILL, J. A.; REINGOLD, E. M.
SOLUTIONS MANUAL. COMBINATORIAL ALGORITHMS: THEORY AND PRACTICE (BY E.M. REINGOLD, J. NIEVERGELT, N. DEO).

FISC 1971 -1 FISCHER, M. J.; MEYER, A. R.
BOOLEAN MATRIX MULTIPLICATION AND TRANSITIVE CLOSURE.
FRAND 1977 -1 FRANK, A.
ON A CLASS OF BALANCED HYPERGRAPHS.
FRATB 1975 -1 FRATTA, L.; MONTANARI, U.
VERTEX ELIMINATION ALGORITHM FOR ENUMERATING ALL SIMPLE PATHS IN A GRAPH, A.
FRUM 1976 -2 FRUMKIN, M. A.
EXPERIMENTAL TIME ALGORITHM FOR THE TRIANGULARIZATION OF AN INTEGER MATRIX (RUSSIN), AN.
FULKA 1976 -1 FULKERSON, D. R.; HARDING, G. C.
ON EDGE-DISJOINT BRANCHINGS.
GAREB 1976 -2 GAREY, M. R.; JOHNSON, D. S.
APPROXIMATION ALGORITHMS FOR COMBINATORIAL PROBLEMS: PROSPECTS AND LIMITATIONS.
GARS 1977 -1 GARSIA, A. M.; WACHS, M. L.
NEW ALGORITHM FOR MINIMUM COST BINARY TREES, A.
GOLU 1977 -2 GOLUMBIC, M. C.
COMPLEXITY OF COMPARABILITY GRAPH RECOGNITION AND COLORING, THE.
GOOD 1965 -1 GOODMAN, R. G.; CUMMINS, R. L.
COMPUTER DETERMINATION OF ISOMORPHISM IN LINEAR GRAPHS.
GRIS 1973 -1 GRISHUKHIN, V. P.
CERTAIN CLASS OF MATCHING ALGORITHMS (RUSSIAN), A.
HEIN 1976 -1 HEIN, O.
ALGORITHMUS ZUR BERECHNUNG DER MINIMALEN KANTENBASIS IN GERICHTETEN GRAPHEN, EIN.
HINT 1977 -1 HINTEREGGER, J.; TINHOFER, G.
ZERLEGUNG DER KNOTENMENGEN VON GRAPHEN ZUM NACHWEIS DER ISOMORPHIE.
HIRS 1977 -1 HIRSCHBERG, D. S.
ALGORITHMS FOR THE LONGEST COMMON SUBSEQUENCE PROBLEM.
HOPC 1971 -1 HOPCROFT, J. E.
N LOG N ALGORITHM FOR ISOMORPHISM OF PLANAR TRIPLY CONNECTED GRAPHS, A.
HOPCB 1973 -1 HOPCROFT, J. E.; TARJAN, R. E.
DIVIDING A GRAPH INTO TRICONNECTED COMPONENTS.
HOPCB 1971 -1 HOPCROFT, J. E.; TARJAN, R. E.
(V POWER 2) ALGORITHM FOR DETERMINING ISOMORPHISM OF PLANAR GRAPHS, A.
HOPCB 1972 -1 HOPCROFT, J. E.; TARJAN, R. E.
ISOMORPHISM OF PLANAR GRAPHS.
HOPCC 1974 -1 HOPCROFT, J. E.; WONG, J. K.
LINEAR TIME ALGORITHM FOR ISOMORPHISM OF PLANAR GRAPHS.
HUBI 1975 -1 HUBICKA, E.; SYSLO, M. M.
MINIMAL BASES OF CYCLES OF A GRAPH.
JAME 1976 -1 JAMES, K. R.; RIHA, W.
ALGORITHM 28. ALGORITHM FOR GENERATING GRAPHS OF A GIVEN PARTITION.
JANI 1976 -1 JANICKE, W.
NEUER ALGORITHMUS ZUR BESTIMMUNG EINER BASIS VON UNABHAENGIGEN ELEMENTARKREISEN IN EINEM STARK ZUSAMMENHAENGENDEN GRAPHEN, EIN.
JOHN 1976 -1 JOHNSON, C. R.; LEIGHTON, F. T.
EFFICIENT LINEAR ALGEBRAIC ALGORITHM FOR THE DETERMINATION OF ISOMORPHISM IN PAIRS OF UNDIRECTED GRAPHS, AN.
JOHNA 1975 -1 JOHNSON, D. B.
FINDING ALL THE ELEMENTARY CIRCUITS OF A DIRECTED GRAPH.
JOHNH 1976 -1 JOHNSON, L. F.
DETERMINING CLIQUES OF A GRAPH.
JUEN 1976 -1 JUENICKE, W.
NEUER ALGORITHMUS ZUR BESTIMMUNG EINER BASIS VON UNABHAENGIGEN ELEMENTARKREISEN IN EINEM STARK ZUSAMMENHAENGENDEN GRAPHEN, EIN.
KAME 1976 -1 KAMEDA, T.
ON MAXIMALLY DISTANT SPANNING TREES OF A GRAPH.
KARP 1975 -1 KARP
GRAPH ALGORITHMS.
KAS' 1975 -1 KAS'JANOV, V. N.
FINDING THE HAMMOCKS IN A DIRECTED GRAPH (RUSSIAN).
KAUFA 1977 -1 KAUFMANN, A.; PICHAT, E.
METHODES MATHEMATIQUES NON NUMERIQUES ET LEURS ALGORITHMES. TOME 1 ALGORITHMES DE RECHERCHE DES ELEMENTS MAXIMAUX.
KAUFA 1977 -2 KAUFMANN, A.; PICHAT, E.
METHODES MATHEMATIQUES NON NUMERIQUES ET LEURS ALGORITHMES. TOME 2 ALGORITHMES DE RECHERCHEDE CHEMINS ET PROBLEMES ASSOCIES.
KELI 1968 -1 KELIKHOV, A. N.; BERSHTEYN, L. S.; KARELIN, P.
ISOMORPHISM OF GRAPHS AND FINITE AUTOMATA.
KHAN 1977 -1 KHAN, S. U.; BARI, A.
PROCEDURE FOR INTEGER SOLUTION TO SOME ALLOCATION PROBLEMS, A.
KHOM 1976 -1 KHOMENKO, M. P.; VIVROT, T. M.
ON THE 0-1-MATRICES (UKRAINIAN).
KICH 1975 -1 KICHA, L. I.; LITVAK, E. I.
ON AN ENUMERATION ALGORITHM FOR THE SIMPLE CIRCUITS IN A GRAPH (RUSSIAN).
KIRY 1975 -1 KIRYUKIN, V. V.; POLONSKIJ, D. V.
ON THE SEARCH FOR THE HAMILTONIAN CYCLES OF A GRAPH (RUSSIAN).
KLEI 1969 -1 KLEITMAN, D. J.
METHODS FOR INVESTIGATING CONNECTIVITY FOR LARGE GRAPHS.
KNOE 1971 -1 KNOEDEL, W.
VERFAHREN ZUR FESTSTELLUNG DER ISOMORPHIE VON ENDLICHEN ZUSAMMENHAENGENDEN GRAPHEN, EIN.
KNUT 1973 -1 KNUTH, D. E.
MATROID PARTITIONING.
KNUTA 1973 -1 KNUTS, L.-R.
COMBINING TWO INTEGER PROGRAMMING ALGORITHMS.
KOHO 1974 -1 KOHOV, V. A.
CERTAIN INVARIANT OF GRAPHS (RUSSIAN), A.
KOZY 1976 -1 KOZYREV, V. P.
FINDING ALL THE HAMILTONIAN PATHS IN A GRAPH AND INVESTIGATION OF HAMILTONIAN CONNECTED GRAPHS (RUSSIAN).
KROG 1977 -1 KROGDAHL, S.
DEPENDENCE GRAPH FOR BASES IN MATROIDS, THE.
KUHN 1973 -1 KUHN, W. W.
INVERSE LINE GRAPHS AND HAMILTONIAN CIRCUITS.
KUND 1976 -1 KUNDU, S.
LINEAR ALGORITHM FOR THE HAMILTONIAN COMPLETION NUMBER OF A TREE, A.

COMBINATORIAL ALGORITHMS (UNLESS SPECIFIED IN PARTICULAR)
****

KURE 1977 -1 KUREJCHIK, V. M.; KOROLEV, A. G.
ON ONE METHOD FOR RECOGNITION OF GRAPHS ISOMORPHISM (RUSSIAN).

KYAN 1975 -1 KYAN, S.; ONAGA, K.
MINIMUM-DELAY MAXIMUM-GAUGE ROUTES IN A TIME-VARYING DYNAMIC NETWORK.

LAWL 1976 -1 LAWLER, E. L.
GRAPHICAL ALGORITHMS AND THEIR COMPLEXITY.

LEPE 1976 -1 LEPESHINSKIJ, N. A.; MALYSHKO, V. V.
GRAPHS OF ARCS AND DETERMINATION OF THE ELEMENTARY PATHS BETWEEN THE PAIRS OF VERTICES OF A GRAPH (RUSSIAN).

LEVI 1972 -1 LEVI, G.
NOTE ON THE DERIVATION OF MAXIMAL COMMON SUBGRAPHS OF TWO DIRECTED OR UNDIRECTED GRAPHS.

LEVI 1974 -1 LEVI, G.
GRAPH ISOMORPHISM: A HEURISTIC EDGE-PARTITIONING-ORIENTED ALGORITHM, A.

LISK 1977 -1 LISKOVEC, V. A.
GENERAL ENUMERATION SCHEME FOR LABELLED GRAPHS (RUSSIAN), A.

LIVO 1976 -1 LIVOVSCHI, L.; SOMNEA, D.
SEARCH ALGORITHM FOR THE CIRCUITS OF A GRAPH (RUMANIAN), A.

LOHS 1976 -1 LOHSE, D.
ALGORITHMUS 31. EIN ALGORITHMUS ZUR BESTIMMUNG VON WEGANZAHLEN IN NETZEN.

MALY 1976 -1 MALYSHKO, V. V.
SOME ALGORITHMS FOR THE ANALYSIS OF ORIENTED GRAPHS; SOLUTION OF A COVERING PROBLEM BY THE GRADIENT ALGORITHM (RUSSIAN).

MARC 197[illegible] -1 MARCU, D.
ALGORITHM FOR DETERMINING THE STRONGLY CONNECTED COMPONENTS OF A FINITE DIRECTED GRAPH (RUMANIAN), AN.

MARKB 1976 -1 MARKOSYAN, A. G.; MARKOSYAN, S. E.
ALGORITHMS FOR THE VERIFICATION OF SOME PROPERTIES OF GRAPHS (RUSSIAN).

MATE 1976 -1 MATETI, P.; DEO, N.
ON ALGORITHMS FOR ENUMERATING ALL CIRCUITS OF A GRAPH.

MEEU 1976 -1 MEEUSEN, W.; CUYVERS, L.
CLIQUE-DETECTION IN DIRECTED GRAPHS (DUTCH).

MEEU 1975 -1 MEEUSEN, W.; CUYVERS, L.
CLIQUE DETECTION IN DIRECTED GRAPHS: A NEW ALGORITHM.

MEHL 1975 -1 MEHLHORN, K.
NEARLY OPTIMAL BINARY SEARCH TREES.

MELI 1972 -1 MELIKHOV, A. N.; BERSHTEYN, L. S.; KUREYCHIK, V. M.; LISYAK, V. V.
PARTITION OF A GRAPH INTO PLANE SUBGRAPHS (RUSSIAN).

MIKH 1975 -1 MIKHAJLOVA, E. D.; BALITSKAYA, E. E.
DETERMINATION OF LIMITED DEGREES OF DIRECTED GRAPHS (RUSSIAN), THE.

MINOA 1976 -4 MINOUX, M.
STRUCTURES ALGEBRIQUES GENERALISEES DES PROBLEMES DE CHEMINEMENT DANS LES GRAPHES FINIS, ALGORITHMES ET APPLICATIONS.

MORP 1971 -1 MORPURGO, R.
HEURISTIC METHOD FOR THE VERIFICATION OF THE ISOMORPHISM OF TWO SIMPLE AND NON ORIENTED GRAPHS, A.

MURCB 1968 -1 MURCHLAND, J. D.
ALGORITHM FOR FINDING A BASIS OF CIRCUITS IN A DIRECTED GRAPH.

MURCB 1968 -2 MURCHLAND, J. D.
ALGORITHMS FOR DETERMINING ALL CONNECTIVITIES IN DIRECTED AND UNDIRECTED GRAPHS.

MURCB 1968 -3 MURCHLAND, J. D.
EASY CASE WHEN TESTING A DIRECTED GRAPH FOR THE PRESENCE OF A CIRCUIT OF NEGATIVE LENGHT, AN.

MURD 1976 -1 MURDASOV, A. B.
SIMPLE ALGORITHM FOR FINDING THE TREE CENTRE (RUSSIAN), A.

NIEM 1976 -1 NIEMINEN, J.
REALIZING THE DISTANCE MATRIX OF A GRAPH.

NIJE 1975 -1 NIJENHUIS, A.; WILF, H.
COMBINATORIAL ALGORITHMS.

NOLT 1976 -1 NOLTEMEIER, H.
REDUKTION VON PRAEZEDENZSTRUKTUREN.

NOVA 1974 -1 NOVANSKY, R.; PLATEK, M.
CHROMATISCHE BELEGUNGEN UND CHROMATISCHE ZERLEGUNGEN ENDLICHER GRAPHEN (ALGORITHMEN).

O'NE 1973 -1 O'NEIL, P. E.; O'NEIL, E. J.
FAST EXPECTED TIME ALGORITHM FOR BOOLEAN MATRIX MULTIPLICATION AND TRANSITIVE CLOSURE.

OVER 1975 -1 OVERBECK-LARISCH, M.
HAMILTONSCHE LINIEN IN DIGRAPHEN.

PENA 1976 -1 PENAUD, J. G.
ALGORITHMES DE PLANARITE.

PETRA 1975 -1 PETROV, I. P.; TIMOFEEV, A. V.
FINITELY CONVERGENT RECURRENCE ALGORITHMS FOR THE SOLUTION OF INTEGER INEQUALITIES IN THE PRESENCE OF CONSTRAINTS (RUSSIAN).

PIPP 1976 -1 PIPPENGER, N.; VALIANT, L. G.
SHIFTING GRAPHS AND THEIR APPLICATIONS.

POLJA 1975 -1 POLJAKOV, A. S.
SEARCH ALGORITHMS FOR MAXIMAL COMPLETE SUBGRAPHS (RUSSIAN).

POPEB 1975 -1 POPESCU, E.
HAMILTONIAN PATHS IN A GRAPH: APPLICATIONS TO THE SOLUTION OF CERTAIN PROBLEMS OF PRODUCTION (RUMANIAN).

RADE 1975 -1 RADENSKI, A. A.
SEARCH ALGORITHM FOR ALL ISOMORPHISMS OF TWO GRAPHS, A (RUSSIAN).

RATH 1977 -2 RATHMANN, D.
ZUR LOESUNG DISKRETER OPTIMIERUNGSPROBLEME MIT GRUPPENTHEORETISCHEN ANSAETZEN.

READ 1978 -1 READ, R. C.
EVERY ONE A WINNER -OR- HOW TO AVOID ISOMORPHISM SEARCH WHEN CATALOGUING COMBINATORIAL CONFIGURATIONS.

READB 1975 -1 READ, R. C.; TARJAN, R. E.
BOUNDS ON BACKTRACK ALGORITHMS FOR LISTING CYCLES, PATHS, AND SPANNING TREES.

REIN 1977 -1 REINGOLD, E. M.; NIEVERGELT, J.; DEO, N.
COMBINATORIAL ALGORITHMS: THEORY AND PRACTICE.

REYN 1977 -1 REYNER, W. S.
ANALYSIS OF A GOOD ALGORITHM FOR THE SUBTREE PROBLEM, AN.

RUBIB 1975 -1 RUBIN, F.
IMPROVED ALGORITHM FOR TESTING THE PLANARITY OF A GRAPH, AN.

RUSK 1977 -1 RUSKEY, F.; HU, T. C.
GENERATING BINARY TREES LEXICOGRAPHICALLY.

RYZH 1975 -1 RYZHKOV, A. P.
ALGORITHM FOR PARTITIONING A GRAPH INTO MINIMALLY CONNECTED SUBGRAPHS (RUSSIAN), AN.

SAUC 1971 -1 SAUCIER, G.
ALGORITHME EFFICACE RECHERCHANT L'ISOMORPHISME DE 2 GRAPHES, UN.

SCHMA 1976 -1 SCHMIDT, D. C.; DRUFFEL, L. E.
FAST BACKTRACKING ALGORITHM TO TEST DIRECTED GRAPHS FOR ISOMORPHISM USING DISTANCE MATRICES, A.

SCHN 1976 -1 SCHNORR, C. P.
ALGORITHM FOR TRANSITIVE CLOSURE WITH LINEAR EXPECTED TIME, AN.

SCHOB 1975 -1 SCHOENE, W.
ALGORITHMEN ZUR BESTIMMUNG INTERN UND EXTERN STABILER MENGEN UND DER CHROMATISCHEN ZAHL EINES ENDLICHEN GRAPHEN.

SHAH 1974 -1 SHAH, Y. J.; DAVIDA, G. T.; MAC CARTHY, M. K.
OPTIMUM FEATURES AND GRAPH ISOMORPHISM.

SHIE 1976 -2 SHIER, D. R.
INVERTING SPARSE MATRICES BY TREE PARTITIONING.

SHIRA 1969 -1 SHIREY, R. W.
IMPLEMENTATION AND ANALYSIS OF EFFICIENT GRAPH PLANARITY TESTING ALGORITHMS.

SIRO 1971 -1 SIROVICH, F.
ISOMORPHISM BETWEEN GRAPHS: AN EFFICIENT ALGORITHM TO FIND ALL ISOMORPHISMS.

SKOR 1975 -1 SKOROBOGATOV, V. A.
ANALYSIS OF THE CONNECTIVITY OF NON-ORIENTED GRAPHS (RUSSIAN).

SMYT 1974 -1 SMYTH, W. F.; BENZI, W. M. L.
ALGORITHM FOR FINDING THE DIAMETER OF A GRAPH, AN.

SPAT 1975 -1 SPATHIS-DIMITROPOULOU, V.
ETUDE ALGEBRIQUE D'UN ARBRE. UN PROBLEME DE CIRCULATION, UNE.

STOCA 1968 -1 STOCKTON, F.
LINEARIZATION AND STANDARDIZATION OF GRAPHS.

SUSS 1965 -1 SUSSENGUTH, E. H.
GRAPH THEORETIC ALGORITHM FOR MATCHING CHEMICAL STRUCTURES, A.

SYSL 1974 -1 SYSLO, M.
ALGORITHM 36: TRANSITIVE CLOSURE OF A GRAPH.

SYSL 1973 -1 SYSLO, M.
ALGORITHMS 20-22: FUNDAMENTAL SET OF CYCLES OF A GRAPH.

SZWAB 1976 -1 SZWARCFITER, J. L.; LAUER, P. E.
SEARCH STRATEGY FOR THE ELEMENTARY CYCLES OF A DIRECTED GRAPH, A.

TARJ 1973 -1 TARJAN, R. E.
ENUMERATION OF THE ELEMENTARY CIRCUITS OF A DIRECTED GRAPH.

TARJ 1971 -1 TARJAN, R. E.
DEPTH-FIRST SEARCH AND LINEAR GRAPH ALGORITHMS.

TARJ 1974 -1 TARJAN, R. E.
GOOD ALGORITHM FOR EDGE-DISJOINT BRANCHING, A.

TARJ 1976 -1 TARJAN, R. E.
EDGE-DISJOINT SPANNING TREES AND DEPTH-FIRST SEARCH.

TARJ 1975 -1 TARJAN, R. E.
FINDING EDGE-DISJOINT SPANNING TREES.

TARJ 1973 -2 TARJAN, R. E.
FINDING DOMINATORS IN DIRECTED GRAPHS.

THOR 1966 -1 THORELLI, L. E.
ALGORITHM FOR COMPUTING ALL PATHS IN A GRAPH, AN.

TINH 1976 -2 TINHOFER, G.
ZUM ALGORITHMISCHEN NACHWEIS DER ISOMORPHIE VON ENDLICHEN GRAPHEN.

TOLC 1976 -1 TOLCHAN, A. YA.
NUMERICAL PROCESSES OF GROWTH OF GRAPHS (RUSSIAN).

TRUE 1976 -3 TRUEMPER, K.
EFFICIENT SCALING PROCEDURE FOR GAIN NETWORKS, AN.

TSUK 1977 -1 TSUKIYAMA, S.; IDE, M.; ARIYOSHI, H.; SHIRAKAWA, I.
NEW ALGORITHM FOR GENERATING ALL THE MAXIMAL INDEPENDENT SETS, A.

TSUKA 1975 -1 TSUKIYAMA, S.; SHIRAKAWA, I.; OZAKI, H.
ALGORITHM FOR GENERATING THE CYCLES OF A DIGRAPH, AN.

UNGE 1964 -1 UNGER, S. H.
HEURISTIC PROGRAM FOR TESTING PAIRS OF DIRECTED LINE GRAPHS FOR ISOMORPHISM, A.

USMA 1970 -1 USMANOV, A. A.
ALGORITHM FOR AUTOMATIC RETRIEVAL OF OPTIMAL STRATEGIES IN CHESS PROGRAMS (RUSSIAN), AN.

WARR 1977 -1 WARREN, H. S. JR.
THREE SIMPLE NODE LIST ALGORITHMS.

WEINA 1960 -1 WEINBERG, L.
SIMPLE AND EFFICIENT ALGORITHM FOR DETERMINING ISOMORPHISM OF PLANAR TRIPLE-CONNECTED GRAPHS.

WEINA 1965 -1 WEINBERG, L.
PLANE REPRESENTATIONS AND CODES FOR PLANAR GRAPHS.

WEINE 1975 -1 WEINSTEIN, J. M.
RECONSTRUCTING COLORED GRAPHS.

WHITG 1973 -1 WHITEHEAD, E. G. JR.
COMBINATORIAL ALGORITHMS.

WILF 1978 -1 WILF, H. S.
UNIFIED SETTING FOR SELECTION ALGORITHMS (II), A.

WILLD 1976 -1 WILLIAMSON, S. G.
RANKING ALGORITHMS FOR LISTS OF PARTITIONS.

YELO 1976 -1 YELOWITZ, L.
EFFICIENT ALGORITHM FOR CONSTRUCTING HIERARCHICAL GRAPHS, AN.

YEN 1973 -1 YEN, J. Y.
FINDING REACHABILITY AND CONNECTIVITY OF DIRECTED LINEAR GRAPHS.

YEN 1971 -1 YEN, J. Y.
ON THE EFFICIENCIES OF ALGORITHMS FOR DETECTING NEGATIVE LOOPS IN NETWORKS.

YUVA 1976 -1 YUVAL, G.
ALGORITHM FOR FINDING ALL SHORTEST PATHS USING (N POWER 2.81) INFINITE PRECISION MULTIPLICATIONS, AN.

COMBINATORIAL OPTIMIZATION PROBLEMS (UNLESS SPECIFIED IN PARTICULAR)
****

AHO A 1974 -1 AHO, A. V.; HIRSCHBERG, D. S.; ULLMAN, J. D.
BOUNDS ON THE COMPLEXITY OF THE LONGEST COMMON SUBSEQUENCE PROBLEM.

ALEM 1973 -1 ALEMANY, R.; DIEGO, A.
TENSIONS ET SEMITENSIONS SUR UN GRAPHE.

ALEM 1972 -1 ALEMANY, R.; DIEGO, A.
SEMITENSIONS IN THE SOLVING OF THE "POTENTIAL'S RESTRICTED PROBLEM".

ALSP 1976 -1 ALSPACH, B.; MASON, D. W.; PULLMAN, N. J.
PATH NUMBERS OF TOURNAMENTS.

BALE 1970 -1 BALERIAUX, H.; JAMOULLE, E.; DOULLIEZ, P.; KELECOM, J. VAN
RECHERCHE DE LA POLITIQUE OPTIMALE D'INVESTISSEMENT D'UN RESEAU ELECTRIQUE EN EXPANSION PAR UNE METHODE DE DECISION SEQUENTIELLE.

BAND 1975 -1 BANDLER, J. W.; CHEN, J. H. K.; DALSGAARD, P.; LIU, P. C.
TOLOPT - A PROGRAM FOR OPTIMAL, CONTINUOUS OR DISCRETE, DESIGN CENTERING AND TOLERANCING, PART 1 - USER'S GUIDE, PART 2 - FORTRAN LISTING.

BANDA 1975 -1 BANDLER, J. W.; LIU, P. C.; CHEN, J. H. K.
WORST CASE NETWORK TOLERANCE OPTIMIZATION.

BELE 1977 -1 BELETSKIJ, S. A.
ON ONE CLASS OF EXTREMAL COMBINATORY PROBLEM AND THEIR SOLUTION BY THE METHOD OF DICHOTOMY (RUSSIAN).

BENNB 1976 -1 BENNINGER, H.; LIEBLING, T. M.
OPTIMIERUNGSPROBLEME MIT POLYNOMIAL BEGRENZTEN ALGORITHMEN.

BION 1973 -1 BIONDI, F.
OPTIMUM DELIVERY PROBLEMS.

BRUC 1978 -1 BRUCKER, P.
ON THE COMPLEXITY OF CLUSTERING PROBLEMS.

BURTB 1976 -1 BURTON, J. S.; GLEASON, J. M.
ALGORITHM FOR THE DYNAMIC LOT SIZE PROBLEM WITH PRICE BREAKS, AN.

CHAM 1976 -1 CHAMBERS, M. L.; DYSON, R. G.
CUTTING STOCK PROBLEM IN THE FLAT GLASS INDUSTRY - SELECTION OF STOCK SIZES, THE.

CHAN 1976 -2 CHAN, A. W.; FRANCIS, R. L.
ROUND-TRIP LOCATION PROBLEM ON A TREE GRAPH, A.

CHANJ 1976 -1 CHANG, L.; KORSH, J. F.
CANONICAL COIN CHANGING AND GREEDY SOLUTIONS.

CHERF 1972 -1 CHERVAK, YU. YU.
DETERMINATION OF THE OPTIMAL INTEGER CONVEX LINEAR COMBINATION OF POINTS. THE GENERAL DISCRETE CASE (RUSSIAN).

CHEU 1976 -1 CHEUNG, T.-Y.
NETWORK FLOW APPROACH TO STEINER PROBLEMS, A.

CHIS 1977 -1 CHISMAN, J. A.
OPTIMIZING THE SHIPPING FUNCTION.

CHRIA 1977 -1 CHRISTOFIDES, N.; WHITLOCK, C.
ALGORITHM FOR TWO-DIMENSIONAL CUTTING PROBLEMS, AN.

CLAUA 1975 -1 CLAUS, A.; KLEITMAN, D. J.
HEURISTIC METHODS FOR SOLVING LARGE SCALE NETWORK ROUTING PROBLEMS: THE TELPACKING PROBLEM.

COVE 1976 -1 COVERDALE, I. L.; WHARTON, F.
IMPROVED HEURISTIC PROCEDURE FOR A NONLINEAR CUTTING STOCK PROBLEM, AN.

DETR 1976 -1 DETROYE, J.
SYNTHESE OPTIMALE D'UN RESEAU PARCOURU PAR UN MULTIFLOT.

DIEH 1972 -1 DIEHR, G.
ALGORITHM FOR THE P-MEDIANS PROBLEM, AN.

EDENA 1976 -1 EDEN, B. N.
HEURISTISCHES VERFAHREN ZUR DISTRIKTERMITTLUNG IN BEWERTETEN GRAPHEN UND SEINE ANWENDUNG AUF DIE BERECHNUNG VON SCHULEINZUGSBEREICHEN, EIN.

ERDO 1976 -1 ERDOES, P.
PROBLEMS AND RESULTS IN GRAPH THEORY AND COMBINATORIAL ANALYSIS.

EVENA 1975 -1 EVEN, S.; ITAI, A.; SHAMIR, A.
ON THE COMPLEXITY OF TIMETABLE AND MULTI-COMMODITY FLOW PROBLEMS.

FINK 1975 -1 FINKE, G.
PURE FIXED COST DISTRIBUTION PROBLEM, THE.

FISB 1971 -1 FISBEIN, S. M.
ON THE SOLUTION OF A CERTAIN COMBINATORIAL PROBLEM (RUSSIAN).

FOULA 1976 -1 FOULDS, L. R.; ROBINSON, D. F.
STRATEGY FOR SOLVING THE PLANT LAYOUT PROBLEM, A.

FRIS 1963 -1 FRISCH, I. T.
OPTIMIZATION OF COMMUNICATION NETS WITH SWITCHING.

FUJIC 1977 -1 FUJISHIGE, S.
ALGORITHM FOR FINDING AN OPTIMAL INDEPENDENT LINKAGE, AN.

GARE 1977 -1 GAREY, M. R.; GRAHAM, R. L.; JOHNSON, D. S.
COMPLEXITY OF COMPUTING STEINER MINIMAL TREES, THE.

GART 1976 -1 GARTNER, N.; GOLDEN, B.; WONG, R.
MODELLING AND OPTIMIZATION FOR TRANSPORTATION SYSTEMS PLANNINGS AND OPERATIONS.

GILLF 1976 -1 GILLETT, B. E.; JOHNSON, J. G.
MULTI-TERMINAL VEHICLE-DISPATCH ALGORITHM.

GOL'B 1977 -1 GOL'DGABER, E. M.; KARNAUKH, V. G.
TRANSFORMATION OF THE OPERATIONS COMPLEX NETWORK INTO A CONJUGATE NETWORK (RUSSIAN).

GOLDB 1976 -1 GOLDEN, B. L.
APPROACHES TO THE CUTTING STOCK PROBLEM.

GOLDF 1967 -1 GOLDMAN, A. J.; NEMHAUSER, G.
TRANSPORT IMPROVEMENT PROBLEM TRANSFORMABLE TO A BEST-PATH PROBLEM, A.

GOODD 1975 -1 GOODMAN, S. E.; HEDETNIEMI, S. T.; SLATER, P. J.
ADVANCES ON THE HAMILTONIAN COMPLETION PROBLEM.

GRAV 1975 -3 GRAVER, J. E.
SURVEY OF THE MAXIMUM DEPTH PROBLEM FOR INDECOMPOSABLE EXACT COVERS, A.

HACA 1974 -1 HACATUROV, V. R.
APPROXIMATION-COMBINATORIAL METHOD, AND SOME OF ITS APPLICATIONS (RUSSIAN), AN.

HANSB 1976 -1 HANSEN, P.; DELATTRE, M.
CLUSTER ANALYSIS AND GRAPH COLORING.

HANSB 1976 -2 HANSEN, P.; DELATTRE, M.
CLUSTER ANALYSIS BY GRAPH COLORING.

HANSB 1976 -3 HANSEN, P.; DELATTRE, M.
COMPLETE-LINK CLUSTER ANALYSIS BY GRAPH COLORING.

HITC 1976 -1 HITCHINGS, G. G.; COTTAM, M.
EFFICIENT HEURISTIC PROCEDURE FOR SOLVING THE LAYOUT DESIGN PROBLEM, AN.

HUNGA 1976 -1 HUNG, M. S.; FISK, J. C.
ALGORITHM FOR 0-1 MULTIPLE KNAPSACK PROBLEMS, AN.

****

HUNGA 1977 -1 HUNG, M. S.; FISK, J. C.
ASSORTMENT PROBLEMS AND EQUIVALENT INTEGER PROGRAMMING FORMULATIONS.
INGA 1975 -1 INGARGIOLA, G.; KORSH, J. F.
ALGORITHM FOR THE SOLUTION OF 0-1 LOADING PROBLEMS, AN.
JUSK 1975 -1 JUSKA, F.
NUMERICAL METHOD FOR THE CONSTRUCTION OF A NETWORK OF MINIMAL LENGTH (RUSSIAN), A.
KASP 1977 -1 KASPSHITSKAJA, M. F.; SERGIENKO, I. V.
ON THE IDEA OF LINEARITY AND CONVEXITY IN A DISCRETE SPACE. COMBINATORIAL LINEAR PROBLEMS, I (RUSSIAN).
KERN 1973 -1 KERNIGHAN, B. W; LIN, S.
HEURISTIC SOLUTION OF A SIGNAL DESIGN OPTIMIZATION PROBLEM.
KEYS 1975 -1 KEYS, C. D.
GRAPHS CRITICAL FOR MAXIMAL BOOKTHICKNESS.
KLEI 1970 -1 KLEITMAN, D. J.
MINIMAL NUMBER OF MULTIPLE EDGES IN REALIZATION OF AN INCIDENCE SEQUENCE WITHOUT LOOPS.
KOHL 1975 -1 KOHLER, C.
LOESUNGSMETHODE FUER TRIADISCHE OPTIMIERUNGSAUFGABEN, EINE.
KORT 1966 -1 KORTANEK, K. O.; SODARDO, D.
GENERALIZED NETWORK MODEL FOR THREE DIMENSIONAL CUTTING STOCK PROBLEMS AND NEW PRODUCT ANALYSIS.
KOU 1977 -2 KOU, L. T.
POLYNOMIAL COMPLETE CONSECUTIVE INFORMATION RETRIEVAL PROBLEMS.
KUND 1975 -1 KUNDU, S.
MINIMUM DIAMETER STEINER TREES IN A GRAPH, THE.
LAURA 1976 -1 LAURIERE, J. L.
LANGAGE ET UN PROGRAMME POUR ENONCER ET RÈSOUDRE DES PROBLEMES COMBINATOIRES, UN.
LEBED 1974 -1 LEBENSOLD, K. W.
EXTREMA OF A CLASS OF FUNCTIONS ON A FINITE SET.
LEE A 1976 -1 LEE, J.; BOSE, N.; HWANG, F.
USE OF STEINER'S PROBLEM IN SUBOPTIMAL ROUTING IN RECTILINEAR METRIC.
LEON 1976 -1 LEONT'EV, V. K.
STABILITY IN COMBINATORIAL PROBLEMS OF SELECTION (RUSSIAN).
LERD 1975 -1 LERDA, F.; MAJORANI, E.
COMMENTS ON A NOTE BY F. RUBIN: "A NOTE ON LERDA AND MAJORANI MINIMUM CROSSING ALGORITHM".
LUDO 1971 -1 LUDOT, J.-P.; RUBINSTEIN, M.-C.
METHODE POUR LA RECHERCHE D'UNE POLITIQUE OPTIMALE D'INVESTISSEMENT A COURT TERME D'UN RESEAU DE DISTRIBUTION.
MART 1976 -1 MARTELLI, A.
GAUSSIAN ELIMINATION ALGORITHM FOR THE ENUMERATION OF CUT SETS IN A GRAPH, A.
MARTC 1977 -2 MARTELLO, S.; TOTH, P.
SOLUTION OF THE BOUNDED AND UNBOUNDED CHANGE-MAKING PROBLEM.
MASS 1976 -1 MASSA, L.
SEMIECONOMIC METHOD FOR THE DETERMINATION OF OPTIMAL LOT SIZES, A.
MATS 1971 -1 MATSUI, H.
THEORY OF TRAFFIC ASSIGNMENT BY MAXIMIZING THE PROBABILITY, THE.
MEYEE 1974 -1 MEYER, W.
ON ORDINARY POINTS IN ARRANGEMENTS.
MURP 1976 -1 MURPHY, F. H.; STOHR, E. A.
DYNAMIC PROGRAMMING ALGORITHM FOR CHECK SORTING, A (REVISED VERSION).
NAKAA 1975 -1 NAKANO, H.; NAKANISHI, Y.
STRUCTURE OF SOLUTION SPACE IN DISCRETE OPTIMIZATION PROBLEM AND ITS APPLICATION TO MINIMUM FEEDBACK EDGE SET PROBLEM.
OHTS 1976 -1 OHTSUKI, T.
FAST ALGORITHM FOR FINDING AN OPTIMAL ORDERING FOR VERTEX ELIMINATION ON A GRAPH, A.
PAPA 1976 -1 PAPADIMITRIOU, C. H.
NP-COMPLETENESS OF THE BANDWIDTH MINIMIZATION PROBLEM, THE.
PEAR 1974 -1 PEARMAN, A.
HEURISTIC APPROACHES TO ROAD NETWORK OPTIMIZATION.
PUHO 1976 -1 PUHOV, G. E.; SARASIDZE, G. K.
SIMULATION OF COMBINATORIAL PROBLEMS OF DISCRETE MATHEMATICS (RUSSIAN).
RIDL 1965 -1 RIDLEY, T. M.
INVESTMENT IN A NETWORK TO REDUCE LENGTH OF A SHORTEST ROUTE.
ROBEC 1964 -1 ROBERTS, R.; FUNK, M.
TOWARD OPTIMUM METHODS OF LINK ADDITION IN TRANSPORTATION NETWORKS.
ROBI 1972 -1 ROBINSON, D. F.
OPTIMIZATION OF FINITE SETS.
ROES 1976 -1 ROESSLER, M.
CLUSTER-ANALYSE.
ROOD 1977 -1 ROODMAN, G. M.; SCHWARZ, L. B.
EXTENSIONS OF THE MULTI-PERIOD FACILITY PHASE-OUT MODEL: NEW PROCEDURES AND APPLICATION TO A PHASE-IN/PHASE-OUT PROBLEM.
ROSE 1976 -1 ROSE, D. J.; TARJAN, R. E.; LUEKER, G. S.
ALGORITHMIC ASPECTS OF VERTEX ELIMINATION ON GRAPHS.
RUBIE 1976 -1 RUBINSTEIN, M. C.; LUDOT, J. P.
APPLICATION DE LA THEORIE DES GRAPHES AU CHOIX DES INVESTISSEMENTS A COURT TERME DANS UN RESEAU ELECTRIQUE SUJET A AVARIES.
RUBIF 1974 -1 RUBINSTEIN, M.-C.
APPLICATION DE LA THEORIE DES GRAPHES AU CHOIX DES INVESTISSEMENTS DANS UN RESEAU D'ENERGIE ELECTRIQUE SUJET A AVARIES.
RUSSA 1976 -1 RUSSELL, A. H.
GEOMETRICAL SOLUTION TO A TREE NETWORK PROBLEM, A.
SAHNB 1976 -1 SAHNI, S.; GONZALEZ, T.
P-COMPLETE APPROXIMATION PROBLEMS.
SANDA 1975 -1 SANDI, C.
SOLUTION OF THE MACHINE LOADING PROBLEM WITH BINARY VARIABLES.
SANKB 1975 -1 SANKOFF, D.; ROUSSEAU, P.
LOCATING THE VERTICES OF A STEINER TREE IN AN ARBITRARY SPACE.
SARV 1976 -1 SARVANOV, V. I.
MINIMIZATION OF A LINEAR FORM ON THE SET OF CYCLES IN N TERMS (RUSSIAN).
SEID 1974 -2 SEIDVASSER, M. A.
LENGTH AND BREADTH OF ASSIGNMENTS OF GRAPHS IN LATTICES (RUSSIAN), THE.
SHAMA 1975 -1 SHAMOS, M. I.; HOEY, D.
CLOSEST-POINT PROBLEMS.

COMBINATORIAL OPTIMIZATION PROBLEMS (UNLESS SPECIFIED IN PARTICULAR)
****

SHIR 1976 -1 SHIRAI, Y.; IBARAKI, T.; HASEGAWA, T.
MINIMAL REPRESENTATION OF COMBINATORIAL OPTIMIZATION PROBLEMS BY FINITE-STATE POSITIVELY MONOTONE MODELS.

SICK 1977 -1 SICKLE, L. VAN; CHANDY, K. M.
COMPUTATIONAL COMPLEXITY OF NETWORK DESIGN ALGORITHMS.

STAI 1977 -1 STAINTON, R. S.
CUTTING STOCK PROBLEM FOR THE STOCKHOLDER OF STEEL REINFORCEMENT BARS, THE.

STEI 1969 -1 STEIGLITZ, K.; WEINER, P.; KLEITMAN, D. J.
DESIGN OF MINIMUM COST SURVIVABLE NETWORKS.

STER 1975 -1 STERBOUL, F.
NEW COMBINATORIAL PARAMETER, A.

SUMA 1976 -1 SUMAROKOV, S. V.
SOLUTION METHOD FOR A NETWORK PROBLEM WITH SEVERAL EXTREMA (RUSSIAN).

TAYL 1975 -1 TAYLOR, P. E.
CELLOPHANE STOCK CUTTING PROBLEM AND A SEQUENTIAL DECISION MODEL, THE.

THOMC 1976 -1 THOMSON, P. F.
ENUMERATION OF NETWORKS.

TILA 1976 -1 TILANUS, C. B.; GERHARDT, C.
APPLICATION OF CUTTING STOCK IN THE STEEL INDUSTRY, AN.

ULLMA 1976 -1 ULLMANN, J. R.
ALGORITHM FOR SUBGRAPH ISOMORPHISM, AN.

VINC 1975 -1 VINCENTELLI, [illegible]
OPTIMAL BORDERED TRIANGULAR FORM OF A SPARSE MATRIX, A.

WALT 1976 -1 WALTERS, J. R.
EXTENSION TO THE ASSORTMENT PROBLEM IN WHICH COMPONENTS ARE GROUPED TO FORM ASSEMBLIES, AN.

WILS 1973 -1 WILSON, A. G.
FURTHER DEVELOPMENTS OF ENTROPY MAXIMISING TRANSPORT MODELS.

WONGA 1976 -1 WONG, C. K.; YAO, A. C.-C.
COMBINATORIAL OPTIMIZATION PROBLEM RELATED TO DATA SET ALLOCATION, A.

WONGC 1976 -1 WONG, R.
SURVEY OF SOME NETWORK DESIGN PROBLEMS, A.

YUVA 1976 -2 YUVAL, G.
FINDING NEAREST NEIGHBOURS.

ZAVI 1975 -1 ZAVILEVICH, M. L.
DIRECT GRID PROBLEM FOR A HIERARCHICAL GRID, THE.

****

ALLE 1976 -1 ALLEN, B.; MUNRO, I.
SELF-ORGANIZING BINARY SEARCH TREES.

ANGL 1977 -1 ANGLUIN, D.; VALIANT, L. G.
FAST PROBABILISTIC ALGORITHMS FOR HAMILTONIAN CIRCUITS AND MATCHINGS.

FREDA 1975 -1 FREDMAN, M. L.
ON THE DECISION TREE COMPLEXITY OF THE SHORTEST PATH PROBLEMS.

GARFA 1976 -1 GARFINKEL, R. S.; GILBERT, K. C.
BOTTLENECK TRAVELING SALESMAN PROBLEM, THE.

GILLB 1977 -1 GILL, J.
COMPUTATIONAL COMPLEXITY OF PROBABILISTIC TURING MACHINES.

GIMAA 1976 -1 GIMADI, E. K.; GLEBOV, N. I.; PEREPELITSA, V. A.
ALGORITHMS WITH ESTIMATIONS FOR THE PROBLEMS OF DISCRETE OPTIMIZATION (RUSSIAN).

KRAVA 1975 -1 KRAVTSOV, M. K.; SHERMAN, A. K.; AVERBUKH, N. D.
ALGORITHM FOR THE SOLUTION OF THE ASSIGNMENT PROBLEM (RUSSIAN), AN.

MOJE 1976 -1 MOJENA, R.; VOLLMANN, T. E.; OKAMOTO, Y.
ON PREDICTING COMPUTATIONAL TIME OF A BRANCH AND BOUND ALGORITHM FOR THE ASSIGNMENT OF FACILITIES.

O'NE 1973 -1 O'NEIL, P. E.; O'NEIL, E. J.
FAST EXPECTED TIME ALGORITHM FOR BOOLEAN MATRIX MULTIPLICATION AND TRANSITIVE CLOSURE.

SCHN 1976 -1 SCHNORR, C. P.
ALGORITHM FOR TRANSITIVE CLOSURE WITH LINEAR EXPECTED TIME, AN.

TARJ 1977 -2 TARJAN, R. E.
FINDING OPTIMUM BRANCHINGS.

UELK 1975 -1 UELKUECUE, A.
EFFICIENCY MEASURES OF VERTEX FOLLOWING ALGORITHMS ON POLYHEDRAL SETS.

YAO B 1976 -1 YAO, A. C.; YAO, F. F.
COMPLEXITY OF SEARCHING AN ORDERED RANDOM TABLE, THE.

COMPLEXITY (WORST CASE ANALYSIS)
****

ADEL 1975 -1 ADEL'SON-VEL'SKIJ, G. M.; DINITS, E. A.; KARZANOV, A. V.
FLOW ALGORITHMS (RUSSIAN).
ADOL 1977 -1 ADOLPHSON, D. L.
SINGLE MACHINE JOB SEQUENCING WITH PRECEDENCE CONSTRAINTS.
AHO 1977 -1 AHO, A. V.; GAREY, M. R.; HWANG, F. K.
RECTILINEAR STEINER TREES: EFFICIENT SPECIAL-CASE ALGORITHMS.
AHO A 1974 -1 AHO, A. V.; HIRSCHBERG, D. S.; ULLMAN, J. D.
BOUNDS ON THE COMPLEXITY OF THE LONGEST COMMON SUBSEQUENCE PROBLEM.
AHO B 1974 -1 AHO, A. V.; HOPCROFT, J. E.; ULLMAN, J. D.
DESIGN AND ANALYSIS OF COMPUTER ALGORITHMS, THE.
AHO C 1976 -1 AHO, A. V.; ULLMAN, J. D.
NODE LISTINGS FOR REDUCIBLE FLOW GRAPHS.
ALLEA 1976 -1 ALLENDE, S.
TYPE OF IMPORTANT FUNCTIONS IN INTEGER PROGRAMMING (SPANISH), A.
ALTH 1975 -1 ALTHERR, W.
ALGORITHM FOR ENUMERATING ALL VERTICES OF A CONVEX POLYHEDRON, AN.
ANDEA 1976 -1 ANDERSSEN, R. S.; BRENT, R. P.
COMPLEXITY OF COMPUTATIONAL PROBLEM SOLVING, THE.
AOSH 1977 -1 AOSHIMA, K.; IRI, M.
COMMENTS ON F. HADLOCK'S PAPER: "FINDING A MAXIMUM CUT OF A PLANAR GRAPH IN POLYNOMIAL TIME".
ARDIA 1976 -1 ARDITTI, J. C.
PROBLEMES DE GRAPHE ET LA COMPLEXITE DES ALGORITHMES, LES.
ARJO 1975 -1 ARJOMANDI, E.; CORNEIL, D. G.
PARALLEL COMPUTATIONS IN GRAPH THEORY.
AUSI 1977 -1 AUSIELLO, G.
ON THE STRUCTURE AND PROPERTIES OF NP-COMPLETE PROBLEMS AND THEIR ASSOCIATED OPTIMIZATION PROBLEMS.
AUSIA 1977 -1 AUSIELLO, G.; D'ATRI, A.; GAUDIANO, M.; PROTASI, M.
CLASSES OF STRUCTURALLY ISOMORPHIC NP-OPTIMIZATION PROBLEMS.
AUSIB 1977 -1 AUSIELLO, G.; D'ATRI, A.; PROTASI, M.
ON THE STRUCTURE OF COMBINATORIAL PROBLEMS AND STRUCTURE PRESERVING REDUCTIONS.
BACH 1976 -3 BACHEM, A.
BEITRAEGE ZUR THEORIE DER CORNER POLYEDER.
BAKEC 1975 -1 BAKER, T.; GILL, J.; SOLOVAY, R.
RELATIVIZATIONS OF THE P = ? NP QUESTION.
BENIA 1976 -1 BENITO, F.; GROEFLIN, H.
OPTIMIERUNGSPROBLEME MIT NICHT POLYNOMIAL BEGRENZTEN ALGORITHMEN (NP-PROBLEME).
BENNB 1976 -1 BENNINGER, H.; LIEBLING, T. M.
OPTIMIERUNGSPROBLEME MIT POLYNOMIAL BEGRENZTEN ALGORITHMEN.
BERS 1976 -1 BERSTEL, J.
COMPLEXITE DE QUELQUES ALGORITHMES DE THEORIE DES GRAPHES.
BIRDA 1976 -1 BIRD, R.
PROGRAMS AND MACHINES. AN INTRODUCTION TO THE THEORY OF COMPUTING.
BLUM 1967 -1 BLUM, M.
MACHINE INDEPENDENT THEORY OF THE COMPLEXITY OF RECURSIVE FUNCTIONS, A.
BOESC 1977 -1 BOESCH, F. T.; GIMPEL, J. F.
COVERING THE POINTS OF A DIGRAPH WITH POINT-DISJOINT PATHS AND ITS APPLICATION TO CODE OPTIMIZATION.
BOOK 1974 -1 BOOK, R. V.
COMPARING TIME AND SPACE.
BOOT 1976 -1 BOOTH, K. S.
PROBLEMS POLYNOMIALLY EQUIVALENT TO GRAPH ISOMORPHISM.
BOOTA 1976 -1 BOOTH, K. S.; LUEKER, G. S.
TESTING FOR THE CONSECUTIVE ONES PROPERTY, INTERVAL GRAPHS, AND GRAPH PLANARITY USING PQ-TREE ALGORITHMS.
BORO 1973 -1 BORODIN, A. B.
COMPUTATIONAL COMPLEXITY: THEORY AND PRACTICE.
BORO 1973 -2 BORODIN, A. B.
ON THE NUMBER OF ARITHMETICS REQUIRED TO COMPUTE CERTAIN FUNCTIONS.
BOROA 1972 -1 BORODIN, A. B.; MUNRO, I.
NOTES ON EFFICIENT AND OPTIMAL ALGORITHMS.
BRUC 1976 -1 BRUCKER, P.
ANMERKUNGEN ZU HEURISTISCHEN VERFAHREN.
BRUC 1978 -1 BRUCKER, P.
ON THE COMPLEXITY OF CLUSTERING PROBLEMS.
BRUC 1976 -2 BRUCKER, P.
KOMPLEXITAET VON SCHEDULING PROBLEMEN, DIE.
BRUC 1976 -3 BRUCKER, P.
NP-VOLLSTAENDIGE OPERATIONS RESEARCH PROBLEME.
BRUCB 1975 -1 BRUCKER, P.; LENSTRA, J. K.; RINNOOY KAN, A. H. G.
COMPLEXITY OF MACHINE SCHEDULING PROBLEMS.
BRUNB 1974 -1 BRUNO, J.; COFFMAN, E. G. JR.; SETHI, R.
ALGORITHMS FOR MINIMIZING MEAN FLOW TIME.
BURKE 1977 -1 BURKARD, R. E.; ZIMMERMANN, U.
WEAKLY ADMISSIBLE TRANSFORMATIONS FOR SOLVING ALGEBRAIC ASSIGNMENT AND TRANSPORTATION PROBLEMS.
BURKG 1971 -1 BURKHARD, W. A.; KROON, F. W.
TOWARD A WEAKLY INVARIANT COMPLEXITY THEORY.
BURNC 1976 -1 BURNS, R. N.; HAFF, C. E.
COMBINATORIAL RANKING PROBLEM, A.
CHAMA 1977 -1 CHAMPEAUX, D. DE; SINT, L.
IMPROVED BIDIRECTIONAL HEURISTIC SEARCH ALGORITHM, AN.
CHANF 1977 -1 CHANDRASEKARAN, R.
MINIMAL RATIO SPANNING TREES.
CHERB 1976 -1 CHERITON, D.; TARJAN, R. E.
FINDING MINIMUM SPANNING TREES.
CHOW 1976 -1 CHOW, T. S.
LOWER BOUNDS ON THE COMPLEXITY OF 0-1-VALUED RECURSIVE FUNCTIONS.
CHRI 1976 -2 CHRISTOFIDES, N.
WORST-CASE ANALYSIS OF A NEW HEURISTIC FOR THE TRAVELLING SALESMAN PROBLEM.
CHRIB 1976 -1 CHRISTOS, H.; PAPADIMITRIOU, H.; STEIGLITZ, K.
SOME COMPLEXITY RESULTS FOR THE TRAVELING SALESMAN PROBLEM.
CHVAC 1977 -1 CHVATAL, V.; HAMMER, P. L.
AGGREGATION OF INEQUALITIES IN INTEGER PROGRAMMING.

****

CLAU 1977 -1 CLAUDER, G.; JAENICKE, W.
ZUR GRAPHENTHEORETISCHEN LOESUNG EINER KLASSE VON DISKRETEN OPTIMIERUNGSPROBLEMEN.
COCKA 1976 -1 COCKAYNE, E.; HEDETNIEMI, S.
LINEAR ALGORITHM FOR THE MAXIMUM WEIGHT OF AN INDEPENDENT SET IN A TREE, A.
CODY 1976 -1 CODY, R. A.; COFFMAN, E. G. JR.
RECORD ALLOCATION FOR MINIMIZING EXPECTED RETRIEVAL COSTS ON DRUM-LIKE STORAGE DEVICES.
COFF 1975 -1 COFFMAN, E. G. JR.
DETERMINISTIC SEQUENCING: COMPLEXITY AND OPTIMAL ALGORITHMS.
COFF 1976 -1 COFFMAN, E. G. JR.
COMPUTER AND JOB/SHOP SCHEDULING THEORY.
COFFA 1976 -1 COFFMAN, E. G. JR.; SETHI, R.
ALGORITHMS MINIMIZING MEAN FLOW TIME SCHEDULE-LENGTH PROPERTIES.
COOKA 1974 -1 COOK, S. A.; RECKHOW, R. A.
ON THE LENGTHS OF PROOFS IN THE PROPOSITIONAL CALCULUS.
CORNC 1977 -1 CORNUEJOLS, G.; FISHER, M.; NEMHAUSER, G. L.
ON THE UNCAPACITATED LOCATION PROBLEM.
DANA 1977 -1 DANARAJ, G.; KLEE, V.
CONNECTEDNESS GAME AND THE C-COMPLEXITY OF CERTAIN GRAPHS, THE.
DINI 1973 -1 DINITS, E. A.
ALGORITHM FOR THE DIGITWISE DECREASE OF DISCREPANCIES, AND TRANSPORT PROBLEMS (RUSSIAN), AN.
DOBK 1976 -1 DOBKIN, D.
NONLINEAR LOWER BOUND ON LINEAR SEARCH TREE PROGRAMS FOR SOLVING KNAPSACK PROBLEMS, A.
DOBKA 1975 -1 DOBKIN, D.; LIPTON, R.
ON THE COMPLEXITY OF COMPUTATIONS UNDER VARYING SETS OF PRIMITIVE OPERATIONS.
DOBKB 1976 -1 DOBKIN, D.; LIPTON, R. J.
LOWER BOUND OF 1/(N POWER 2) ON LINEAR SEARCH PROGRAMS FOR THE KNAPSACK PROBLEM, A.
ECKE 1977 -1 ECKER, K.
ORGANISATION VON PARALLELEN PROZESSEN. THEORIE DETERMINISTISCHER SCHEDULES.
ESWA 1976 -1 ESWARAN, K. P.; TARJAN, R. E.
AUGMENTATION PROBLEMS.
EVENA 1975 -1 EVEN, S.; ITAI, A.; SHAMIR, A.
ON THE COMPLEXITY OF TIMETABLE AND MULTI-COMMODITY FLOW PROBLEMS.
EVENB 1975 -1 EVEN, S.; KARIV, O.
O(N POWER 2.5) ALGORITHM FOR MAXIMUM MATCHING IN GENERAL GRAPHS, AN.
EVEND 1975 -1 EVEN, S.; TARJAN, R. E.
COMBINATORIAL PROBLEM WHICH IS COMPLETE IN POLYNOMIAL SPACE, A.
EVEND 1975 -2 EVEN, S.; TARJAN, R. E.
NETWORK FLOW AND TESTING GRAPH CONNECTIVITY.
FIDU 1977 -1 FIDUCCIA, C. M.; ZALCSTEIN, Y.
ALGEBRAS HAVING LINEAR MULTIPLICATIVE COMPLEXITIES.
FINKA 1977 -1 FINKEL'SHTEIN, YU. YU.
E-APPROACH TO THE MULTIDIMENSIONAL KNAPSACK PROBLEM: THE POLYNOMIAL GROWTH OF THE TREE OF BRANCHING (RUSSIAN), THE.
FONG 1977 -1 FONG, A. C.; ULLMAN, J. D.
FINDING THE DEPTH OF A FLOW GRAPH.
FONT 1976 -1 FONTET, M.
AUTOMORPHISMES DE GRAPHES ET PLANARITE.
FRAND 1976 -1 FRANK, A.
SOME POLYNOMIAL ALGORITHMS FOR CERTAIN GRAPHS AND HYPERGRAPHS.
FRATB 1975 -1 FRATTA, L.; MONTANARI, U.
VERTEX ELIMINATION ALGORITHM FOR ENUMERATING ALL SIMPLE PATHS IN A GRAPH, A.
FRAZ 1972 -1 FRAZER, W. D.
ANALYSIS OF COMBINATORY ALGORITHMS - A SAMPLE OF CURRENT METHODOLOGY.
FRED 1976 -1 FREDERICKSON, G. N.; HECHT, M. S.; KIM, C. E.
APPROXIMATION ALGORITHMS FOR SOME ROUTING PROBLEMS.
FREDA 1976 -1 FREDMAN, M. L.
NEW BOUNDS ON THE COMPLEXITY OF THE SHORTEST PATH PROBLEM.
FREDA 1975 -1 FREDMAN, M. L.
ON THE DECISION TREE COMPLEXITY OF THE SHORTEST PATH PROBLEMS.
FRUM 1976 -2 FRUMKIN, M. A.
EXPERIMENTAL TIME ALGORITHM FOR THE TRIANGULARIZATION OF AN INTEGER MATRIX (RUSSIN), AN.
GABOB 1977 -1 GABOW, H. N.
TWO ALGORITHMS FOR GENERATING WEIGHTED SPANNING TREES IN ORDER.
GABOB 1976 -1 GABOW, H. N.
EFFICIENT IMPLEMENTATION OF EDMONDS' ALGORITHM FOR MAXIMUM MATCHING ON GRAPHS, AN.
GABOB 1976 -2 GABOW, H. N.
USING EULER PARTITIONS TO EDGE COLOR BIPARTITE MULTIGRAPHS.
GALI 1976 -1 GALIL, Z.
HIERARCHIES OF COMPLETE PROBLEMS.
GALI 1975 -1 GALIL, Z.
ON ENUMERATION PROCEDURES FOR THEOREM PROVING AND FOR INTEGER PROGRAMMING.
GARE 1977 -1 GAREY, M. R.; GRAHAM, R. L.; JOHNSON, D. S.
COMPLEXITY OF COMPUTING STEINER MINIMAL TREES, THE.
GARE 1976 -1 GAREY, M. R.; GRAHAM, R. L.; JOHNSON, D. S.
SOME NP-COMPLETE GEOMETRIC PROBLEMS.
GAREA 1976 -1 GAREY, M. R.; GRAHAM, R. L.; JOHNSON, D. S.; YAO, A. C.-C.
RESOURCE CONSTRAINED SCHEDULING AS GENERALIZED BIN PACKING.
GAREB 1977 -1 GAREY, M. R.; JOHNSON, D. S.
RECTILINEAR STEINER TREE PROBLEM IS NP-COMPLETE, THE.
GAREB 1974 -1 GAREY, M. R.; JOHNSON, D. S.
COMPLEXITY RESULTS FOR MULTIPROCESSOR SCHEDULING UNDER RESOURCE CONSTRAINTS.
GAREB 1977 -2 GAREY, M. R.; JOHNSON, D. S.
TWO-PROCESSOR SCHEDULING WITH START-TIMES AND DEADLINES.
GAREB 1976 -3 GAREY, M. R.; JOHNSON, D. S.
COMPLEXITY OF NEAR-OPTIMAL GRAPH COLORING, THE.
GAREB 1976 -4 GAREY, M. R.; JOHNSON, D. S.
SCHEDULING TASKS WITH NONUNIFORM DEADLINES ON TWO PROCESSORS.
GAREB 1976 -5 GAREY, M. R.; JOHNSON, D. S.
SOME SIMPLIFIED NP-COMPLETE GRAPH PROBLEMS.
GAREB 1976 -6 GAREY, M. R.; JOHNSON, D. S.
STRONG NP-COMPLETENESS RESULTS: MOTIVATION, EXAMPLES AND IMPLICATIONS.
GAREC 1976 -1 GAREY, M. R.; JOHNSON, D. S.; SETHI, R.
COMPLEXITY OF FLOWSHOP AND JOBSHOP SCHEDULING, THE.
GARED 1976 -1 GAREY, M. R.; JOHNSON, D. S.; TARJAN, R. E.
PLANAR HAMILTONIAN CIRCUIT PROBLEM IS NP-COMPLETE, THE.

COMPLEXITY (WORST CASE ANALYSIS)

****

GAVR 1977 -1 GAVRIL, F.
ALGORITHMS ON CLIQUE SEPARABLE GRAPHS.

GIMA 1976 -1 GIMADI, E. H.; GLEBOV, N. I.; PEREPELICA, V. A.
ALGORITHMS WITH BOUNDS FOR DISCRETE OPTIMIZATION PROBLEMS (RUSSIAN).. THIRD ALL-UNION CONFERENCE ON PROBLEMS OF THEORETICAL CYBERNETICS, NOVOSIBIRSK 1974.

GIMAA 1976 -1 GIMADI, E. K.; GLEBOV, N. I.; PEREPELITSA, V. A.
ALGORITHMS WITH ESTIMATIONS FOR THE PROBLEMS OF DISCRETE OPTIMIZATION (RUSSIAN).

GOLU 1977 -2 GOLUMBIC, M. C.
COMPLEXITY OF COMPARABILITY GRAPH RECOGNITION AND COLORING, THE.

GONZE 1976 -1 GONZALEZ, T.; SAHNI, S.
OPEN SHOP SCHEDULING TO MINIMIZE FINISH TIME.

GOODA 1974 -1 GOODMAN, S.; HEDETNIEMI, S.
ON THE HAMILTONIAN COMPLETION PROBLEM.

GOODC 1976 -1 GOODMAN, S.; HEDETNIEMI, S.; TARJAN, R. E.
B-MATCHINGS IN TREES.

GOODD 1975 -1 GOODMAN, S. E.; HEDETNIEMI, S. T.; SLATER, P. J.
ADVANCES ON THE HAMILTONIAN COMPLETION PROBLEM.

GRAHA 1977 -1 GRAHAM, R. L.; LAWLER, E. L.; LENSTRA, J. K.; RINNOOY KAN, A. H. G.
OPTIMIZATION AND APPROXIMATION IN DETERMINISTIC SEQUENCING AND SCHEDULING: A SURVEY.

GRIS 1972 -1 GRISHUKHIN, V. P.
ESTIMATION OF THE COMPLEXITY OF THE BALAS ALGORITHM (RUSSIAN).

GVOZ 1976 -1 GVOZDEV, S. E.
ON A GENERALIZATION OF THE KNAPSACK PROBLEM (RUSSIAN).

GVOZ 1974 -1 GVOZDEV, S. E.
SOME MODELS AND SOLUTION METHODS FOR THE KNAPSACK PROBLEM WITH SUPPLEMENTARY CONSTRAINTS (RUSSIAN).

HADL 1977 -1 HADLOCK, F. O.
SHORTEST PATH ALGORITHM FOR GRID GRAPHS, A.

HANSB 1976 -1 HANSEN, P.; DELATTRE, M.
CLUSTER ANALYSIS AND GRAPH COLORING.

HANSB 1976 -2 HANSEN, P.; DELATTRE, M.
CLUSTER ANALYSIS BY GRAPH COLORING.

HANSB 1976 -3 HANSEN, P.; DELATTRE, M.
COMPLETE-LINK CLUSTER ANALYSIS BY GRAPH COLORING.

HARD 1977 -1 HARDLOCK, F. O.
SHORTEST PATH ALGORITHM FOR GRID GRAPHS, A.

HARTB 1976 -1 HARTMANIS, J.; BERMAN. L.
ON ISOMORPHISMS AND DENSITY OF NP AND OTHER COMPLETE SETS.

HAUS 1977 -3 HAUSMANN, D.
POLYNOMIALER ALGORITHMUS FUER DIE ADJAZENZUEBERPRUEFUNG AUF DEM BRANCHING-POLYEDER, EIN.

HAUSB 1977 -1 HAUSMANN, D.; KANNAN, R.; KORTE, B.
EXPONENTIAL LOWER BOUNDS ON A CLASS OF KNAPSACK ALGORITHMS.

HAUSC 1978 -1 HAUSMANN, D.; KORTE, B.
ORACLE-ALGORITHMS FOR FIXED POINT PROBLEMS - AN AXIOMATIC APPROACH.

HAUSC 1977 -1 HAUSMANN, D.; KORTE, B.
LOWER BOUNDS ON THE WORST-CASE COMPLEXITY OF SOME ORACLE ALGORITHMS.

HAUSC 1977 -2 HAUSMANN, D.; KORTE, B.
WORST-CASE BEHAVIOUR OF POLYNOMIAL BOUNDED ALGORITHMS FOR INDEPENDENCE SYSTEMS.

HAUSC 1978 -2 HAUSMANN, D.; KORTE, B.
WORST-CASE ANALYSIS FOR A CLASS OF COMBINATORIAL ALGORITHMS.

HIRSA 1976 -1 HIRSCHBERG, D. S.; WONG, C. K.
POLYNOMIAL-TIME ALGORITHM FOR THE KNAPSACK PROBLEM WITH TWO VARIABLES, A.

HOPC 1971 -1 HOPCROFT, J. E.
N LOG N ALGORITHM FOR ISOMORPHISM OF PLANAR TRIPLY CONNECTED GRAPHS, A.

HOPCA 1971 -1 HOPCROFT, J. E.; KARP. R. M.
(N POWER 5/2) ALGORITHM FOR MAXIMUM MATCHINGS IN BIPARTITE GRAPHS, A.

HOPCB 1973 -1 HOPCROFT, J. E.; TARJAN, R. E.
DIVIDING A GRAPH INTO TRICONNECTED COMPONENTS.

HOPCB 1971 -1 HOPCROFT, J. E.; TARJAN, R. E.
(V POWER 2) ALGORITHM FOR DETERMINING ISOMORPHISM OF PLANAR GRAPHS, A.

HOPCB 1972 -1 HOPCROFT, J. E.; TARJAN, R. E.
ISOMORPHISM OF PLANAR GRAPHS.

HOPCC 1974 -1 HOPCROFT, J. E.; WONG, J. K.
LINEAR TIME ALGORITHM FOR ISOMORPHISM OF PLANAR GRAPHS.

HORO 1976 -1 HOROWITZ, E.; SAHNI, S.
EXACT AND APPROXIMATE ALGORITHMS FOR SCHEDULING NONIDENTICAL PROCESSORS.

HORV 1977 -1 HORVATH, E. C.; LAM, S.; SETHI, R.
LEVEL ALGORITHM FOR PREEMPTIVE SCHEDULING, A.

HYAF 1976 -1 HYAFIL, L.; RIVEST, R. L.
CONSTRUCTING OPTIMAL BINARY DECISION TREES IS NP-COMPLETE.

IBAR 1976 -1 IBARAKI, T.
COMPUTATIONAL EFFICIENCY OF APPROXIMATE BRANCH-AND-BOUND ALGORITHMS.

IBAR 1977 -2 IBARAKI, T.
POWER OF DOMINANCE RELATIONS IN BRANCH-AND-BOUND ALGORITHMS, THE.

IBARB 1974 -1 IBARRA, O.; SAHNI, S.
P-COMPLETE PROBLEMS AND APPROXIMATE SOLUTIONS.

IBARC 1977 -1 IBARRA, O. H.; KIM, C. E.
HEURISTIC ALGORITHMS FOR SCHEDULING INDEPENDENT TASKS ON NONIDENTICAL PROCESSORS.

IBARC 1974 -1 IBARRA, O. H.; KIM, C. E.
FAST APPROXIMATION ALGORITHMS FOR THE KNAPSACK AND SUM OF SUBSET PROBLEMS.

JERO 1977 -1 JEROSLOW, R.
BRACKETING DISCRETE PROBLEMS BY TWO PROBLEMS OF LINEAR OPTIMIZATION.

JOHNA 1975 -1 JOHNSON, D. B.
FINDING ALL THE ELEMENTARY CIRCUITS OF A DIRECTED GRAPH.

JOHNA 1977 -1 JOHNSON, D. B.
EFFICIENT ALGORITHMS FOR SHORTEST PATHS IN SPARSE NETWORKS.

JOHNC 1974 -1 JOHNSON, D. S.
WORST CASE BEHAVIOR OF GRAPH COLORING ALGORITHMS.

JOHND 1977 -1 JOHNSON, D. S.; LENSTRA, J. K.; RINNOOY KAN, A. H. G.
COMPLEXITY OF THE NETWORK DESIGN PROBLEM, THE.

JONE 1973 -1 JONES, N. D.
PRELIMINARY REPORT: REDUCIBILITY AMONG COMBINATORIAL PROBLEMS IN LOG N SPACE.

JUDI 1976 -1 JUDIN, D. B.; NEMIROVSKY, A. S.
INFORMATIONAL COMPLEXITY ESTIMATION OF MATHEMATICAL PROGRAMMING PROBLEMS (RUSSIAN).

****

KANN 1978 -1 KANNAN, R.; MONMA, C. L.
ON THE COMPUTATIONAL COMPLEXITY OF INTEGER PROGRAMMING PROBLEMS.
KARI 1976 -1 KARIV, O.
O(N POWER 2.5) ALGORITHM FOR FINDING A MAXIMUM MATCHING IN A GENERAL GRAPH, AN.
KARPA 1975 -1 KARP, R. M.
FAST APPROXIMATE SOLUTION OF HARD COMBINATORIAL PROBLEMS, THE.
KARPA 1975 -2 KARP, R. M.
ON THE COMPUTATIONAL COMPLEXITY OF COMBINATORIAL PROBLEMS.
KELI 1968 -1 KELIKHOV, A. N.; BERSHTEYN, L. S.; KARELIN, P.
ISOMORPHISM OF GRAPHS AND FINITE AUTOMATA.
KORTA 1977 -1 KORTE B.; MONMA, C. L.
REMARKS ON A CLASSIFICATION OF ORACLE-TYPE ALGORITHMS, SOME.
KORTB 1977 -2 KORTE, B.
APPROXIMATIVE ALGORITHMS FOR DISCRETE OPTIMIZATION PROBLEMS.
KORTB 1976 -2 KORTE, B.
INTEGER PROGRAMMING (SURVEY PAPER).
KOU 1977 -2 KOU, L. T.
POLYNOMIAL COMPLETE CONSECUTIVE INFORMATION RETRIEVAL PROBLEMS.
KRARA 1977 -1 KRARUP, J.; BILDE, O.
PLANT LOCATION, SET COVERING AND ECONOMIC LOT SIZE: AN O(MN)-ALGORITHM FOR STRUCTURED PROBLEMS.
KRAT 1977 -1 KRAIKO, M. I.; REVIN, O. M.
TURING'S MACHINES OPERATING ON THE PLANE, AND STABLE ITERATIVE SYSTEMS (RUSSIAN).
KRAVA 1975 -1 KRAVTSOV, M. K.; SHERMAN, A. K.; AVERBUKH, N. D.
ALGORITHM FOR THE SOLUTION OF THE ASSIGNMENT PROBLEM (RUSSIAN), AN.
KROG 1976 -1 KROGDAHL, S.
COMBINATORIAL PROOF FOR LAWLER'S MATROID INTERSECTION ALGORITHM, A.
KUNG 1974 -1 KUNG, H. T.
ON THE COMPUTATIONAL COMPLEXITY OF FINDING THE MAXIMA OF A SET OF VECTORS.
LADN 1975 -1 LADNER, R. E.
ON THE STRUCTURE OF POLYNOMIAL TIME REDUCIBILITY.
LAWL 1978 -1 LAWLER, E. L.
SEQUENCING JOBS TO MINIMIZE TOTAL WEIGHTED COMPLETION TIME SUBJECT TO PRECEDENCE CONSTRAINTS.
LAWL 1977 -1 LAWLER, E. L.
'PSEUDOPOLYNOMIAL' ALGORITHM FOR SEQUENCING JOBS TO MINIMIZE TOTAL TARDINESS, A.
LAWL 1975 -1 LAWLER, E. L.
ALGORITHMS, GRAPHS, AND COMPLEXITY.
LAWL 1976 -1 LAWLER, E. L.
GRAPHICAL ALGORITHMS AND THEIR COMPLEXITY.
LAWL 1973 -1 LAWLER, E. L.
POLYNOMIAL-BOUNDED AND (APPARANTLY) NON-POLYNOMIAL-BOUNDED MATROID COMPUTATIONS.
LAWL 1976 -2 LAWLER, E. L.
NOTE ON THE COMPLEXITY OF THE CHROMATIC NUMBER PROBLEM, A.
LAWL 1975 -2 LAWLER, E. L.
COMPLEXITY OF COMBINATORIAL COMPUTATIONS.
LAWL 1975 -3 LAWLER, E. L.
COMPUTING SHORTEST PATH IN NETWORKS.
LAWLB 1976 -1 LAWLER, E. W.
INTRODUCTION TO THE COMPLEXITY OF ALGORITHMS.
LEEU 1978 -1 LEEUWEN, J. VAN; EMDE BOAS, P. VAN
SOME ELEMENTARY PROOFS OF LOWER BOUNDS IN COMPLEXITY THEORY.
LENSB 1977 -1 LENSTRA, J. K.; RINNOOY KAN, A. H. G.
COMPUTATIONAL COMPLEXITY OF DISCRETE OPTIMIZATION PROBLEMS.
LENSB 1976 -1 LENSTRA, J. K.; RINNOOY KAN, A. H. G.
COMPLEXITY OF SCHEDULING UNDER PRECEDENCE CONSTRAINTS.
LENSB 1976 -2 LENSTRA, J. K.; RINNOOY KAN, A. H. G.
ON GENERAL ROUTING PROBLEMS.
LENSC 1977 -1 LENSTRA, J. K.; RINNOOY KAN, A. H. G.; BRUCKER, P.
COMPLEXITY OF MACHINE SCHEDULING PROBLEMS.
LETI 1976 -1 LETICHEVSKIY, A. A.
HOW TO SPEED ITERATION OF MONOTONOUS OPERATORS (RUSSIAN).
LIEN 1974 -1 LIEN, Y. E.
PRELIMINARY REPORT: SOME COMPLETE PROBLEMS FOR LANGUAGES RECOGNIZABLE IN NONDETERMINISTIC LOG-SPACE.
LIVS 1972 -1 LIVSIK, E. M.; RUBLINECKII, V. I.
COMPERATIVE COMPLEXITY OF CERTAIN DISCRETE OPTIMIZATION (RUSSIAN), THE.
LIVS 1972 -2 LIVSIK, E. M.; RUBLINECKII, V. I.
OPTIMAL SPLITTING OF AN ORDERED SET INTO INTERVALS (RUSSIAN), THE.
LOVA 1976 -1 LOVASZ, L.
COVERS, PACKINGS AND SOME HEURISTIC ALGORITHMS.
MACH 1978 -1 MACHOL, R. E.; WIEN, M.
'HARD' ASSIGNMENT PROBLEM, A.
MAHR 1976 -1 MAHR, B.; PERL, J.
GRUENDE, TECHNIKEN UND KONSEQUENZEN DER VERALLGEMEINERUNG VON WEGEALGORITHMEN.
MAND 1976 -1 MANDERS, K.; ADLEMAN, L.
NP-COMPLETE DECISION PROBLEMS FOR QUADRATIC POLYNOMIALS.
MART 1976 -1 MARTELLI, A.
GAUSSIAN ELIMINATION ALGORITHM FOR THE ENUMERATION OF CUT SETS IN A GRAPH, A.
MATU 1978 -1 MATULA, D. W.
SUBTREE ISOMORPHISM IN O(N POWER 5/2).
MILNA 1976 -1 MILNER, E. C.; WELSH, D. J. A.
ON THE COMPUTATIONAL COMPLEXITY OF GRAPH THEORETICAL PROPERTIES.
MINO 1975 -1 MINOLI, D.
COMBINATORIAL GRAPH COMPLEXITY.
MITC 1975 -1 MITCHELL, S.; HEDETNIEMI, S.; GOODMAN, S.
SOME LINEAR ALGORITHMS ON TREES.
MIYA 1977 -1 MIYAKAWA, M.; YUBA, T.; SUGITO, Y.; HOSHI, M.
OPTIMUM SEQUENCE TREES.
MORRB 1976 -1 MORROW, C.; GOODMAN, S.
EFFICIENT ALGORITHM FOR FINDING A LONGEST CYCLE IN A TOURNAMENT, AN.
MUCH 1976 -1 MUCHNICK, S. S.
COMPUTATIONAL COMPLEXITY OF MULTIPLE RECURSIVE SCHEMATA.
MUNR 1971 -1 MUNRO, J. I.
SOME RESULTS IN THE STUDY OF ALGORITHMS.

NAST 1975 -1 NASTANSKY, L.; GROSSE, H. W.
COMPUTATIONAL EXPERIMENTS WITH LINEAR PROGRAMMING ORIENTED IMPLICIT ENUMERATION ALGORITHMS FOR THE SOLUTION OF LINEAR (0,1) - PROGRAMS.

NEMH 1976 -1 NEMHAUSER, G. L.; WOLSEY, L. A.
BEST ALGORITHMS FOR APPROXIMATING THE MAXIMUM OF A SUBMODULAR SET FUNCTION.

NIGM 1975 -1 NIGMATULLIN, R. G.
COMPLEXITY OF THE APPROXIMATE SOLUTION OF COMBINATORIAL PROBLEMS.

NOLT 1975 -1 NOLTEMEIER, H.
ALGORITHM FOR THE DETERMINATION OF LONGEST DISTANCES IN A GRAPH, AN.

NOLT 1976 -1 NOLTEMEIER, H.
REDUKTION VON PRAEZEDENZSTRUKTUREN.

OHTS 1976 -1 OHTSUKI, T.
FAST ALGORITHM FOR FINDING AN OPTIMAL ORDERING FOR VERTEX ELIMINATION ON A GRAPH, A.

ORLOA 1976 -1 ORLOFF, C. S.
ON GENERAL ROUTING PROBLEMS: COMMENTS.

ORLOA 1976 -2 ORLOFF, C. S.
ROUTE CONSTRAINED FLEET SCHEDULING.

PADBA 1976 -1 PADBERG, M. W.
ON THE COMPLEXITY OF SET PACKING POLYHEDRA (REVISED).

PAPA 1977 -1 PAPADIMITRIOU, C. H.
EUCLIDEAN TRAVELING SALESMAN PROBLEM IS NP-COMPLETE, THE.

PAPA 1976 -1 PAPADIMITRIOU, C. H.
NP-COMPLETENESS OF THE BANDWIDTH MINIMIZATION PROBLEM, THE.

PAPA 1975 -1 PAPADIMITRIOU, C. H.
ON THE COMPLEXITY OF THE LOCAL STRUCTURE OF CERTAIN CONVEX POLYTOPES.

PAPA 1976 -2 PAPADIMITRIOU, C. H.
ON THE COMPLEXITY OF EDGE TRAVERSING.

PAPAA 1976 -1 PAPADIMITRIOU, C. H.; STEIGLITZ, K.
SOME COMPLEXITY RESULTS FOR THE TRAVELING SALESMAN PROBLEM.

PAPAA 1977 -1 PAPADIMITRIOU, C. H.; STEIGLITZ, K.
ON THE COMPLEXITY OF LOCAL SEARCH FOR THE TRAVELING SALESMAN PROBLEM.

PAUL 1977 -1 PAUL, W. J.
2.5N-LOWER BOUND ON THE COMBINATORIAL COMPLEXITY OF BOOLEAN FUNCTIONS, A.

PENA 1976 -1 PENAUD, J. G.
ALGORITHMES DE PLANARITE.

PIPP 1976 -1 PIPPENGER, N.; VALIANT, L. G.
SHIFTING GRAPHS AND THEIR APPLICATIONS.

RABI 1959 -1 RABIN, M. O.
SPEED OF COMPUTATION AND CLASSIFICATION OF RECURSIVE SETS.

RABI 1977 -1 RABIN, M. O.
COMPLEXITY OF COMPUTATIONS.

RABI 1960 -1 RABIN, M. O.
DEGREE OF DIFFICULTY OF COMPUTING A FUNCTION AND A PARTIAL ORDERING OF RECURSIVE SETS.

RADE 1975 -1 RADENSKI, A. A.
SEARCH ALGORITHM FOR ALL ISOMORPHISMS OF TWO GRAPHS, A (RUSSIAN).

READA 1977 -1 READ, R. C.; CORNEIL, D. G.
GRAPH ISOMORPHISM DISEASE, THE.

REDD 1976 -2 REDDI, S. S.
SEQUENCING WITH FINITE INTERMEDIATE STORAGE.

REIN 1977 -1 REINGOLD, E. M.; NIEVERGELT, J.; DEO, N.
COMBINATORIAL ALGORITHMS: THEORY AND PRACTICE.

RIVE 1976 -1 RIVEST, R. L.; VUILLEMIN, J.
ON RECOGNIZING GRAPH PROPERTIES FROM ADJACENCY MATRICES.

ROSE 1976 -1 ROSE, D. J.; TARJAN, R. E.; LUEKER, G. S.
ALGORITHMIC ASPECTS OF VERTEX ELIMINATION ON GRAPHS.

ROSED 1977 -1 ROSENTHAL, A.; GOLDNER, A.
SMALLEST AUGMENTATIONS TO BICONNECT A GRAPH.

RYAN 1976 -1 RYAN, D. M.; FOSTER, B. A.; CHESHIRE, I. M.
COMPLEXITY OF REAL-WORLD SCHEDULING PROBLEMS, THE.

SAHN 1975 -1 SAHNI, S.
APPROXIMATE ALGORITHMS FOR THE 0/1 KNAPSACK PROBLEM.

SAHN 1977 -1 SAHNI, S.
GENERAL TECHNIQUES FOR COMBINATORIAL APPROXIMATION.

SAHNA 1974 -1 L SAHNI, S.; GONZALES, T.
P-COMPLETE PROBLEMS AND APPROXIMATE SOLUTIONS.

SAHNB 1976 -1 SAHNI, S.; GONZALEZ, T.
P-COMPLETE APPROXIMATION PROBLEMS.

SAHNC 1976 -1 SAHNI, S. K.
ALGORITHMS FOR SCHEDULING INDEPENDENT TASKS.

SAVA 1976 -1 SAVAGE, J. E.
COMPLEXITY OF COMPUTING, THE.

SAVAB 1976 -1 SAVAGE, S. L.; WEINER, P.; BAGCHI, A.
NEIGHBORHOOD SEARCH ALGORITHMS FOR GUARANTEEING OPTIMAL TRAVELING SALESMAN TOURS MUST BE INEFFICIENT.

SAVI 1974 -1 SAVITCH, W. J.
NONDETERMINISTIC LOG N SPACE.

SCHMA 1976 -1 SCHMIDT, D. C.; DRUFFEL, L. E.
FAST BACKTRACKING ALGORITHM TO TEST DIRECTED GRAPHS FOR ISOMORPHISM USING DISTANCE MATRICES, A.

SCHN 1976 -1 SCHNORR, C. P.
ALGORITHM FOR TRANSITIVE CLOSURE WITH LINEAR EXPECTED TIME, AN.

SCHN 1978 -2 SCHNORR, C. P.
SATISFIABILITY IS QUASILINEAR COMPLETE IN NQL.

SCHN 1976 -2 SCHNORR, C. P.
NETWORK COMPLEXITY AND THE TURING MACHINE COMPLEXITY OF FINITE FUNCTIONS, THE.

SEIF 1974 -1 SEIFERAS, J. I.
NOTE ON REDUCTIONS TO DIRECTED HC, A.

SETH 1976 -1 SETHI, R.
SCHEDULING GRAPHS ON TWO PROCESSORS.

SHOS 1974 -1 SHOSTAK, R. E.
REFUTATION GRAPHS AND RESOLUTION THEOREM PROVING.

SICK 1977 -1 SICKLE, L. VAN; CHANDY, K. M.
COMPUTATIONAL COMPLEXITY OF NETWORK DESIGN ALGORITHMS.

SKOR 1975 -1 SKOROBOGATOV, V. A.
ANALYSIS OF THE CONNECTIVITY OF NON-ORIENTED GRAPHS (RUSSIAN).

****

SLATA 1976 -2 SLATER, P. J.
R-DOMINATION IN GRAPHS.
SOLO 1976 -1 SOLOVAY, R. M.
ON SETS COOK-REDUCIBLE TO SPARSE SETS.
STAD 1978 -1 STADEL, M.
REMARK ON THE TIME COMPLEXITY OF THE SUBTREE PROBLEM, A.
STOC 1973 -1 STOCKMEYER, L.
PLANAR 3-COLORABILITY IS POLYNOMIAL COMPLETE.
SZWAB 1976 -1 SZWARCFITER, J. L.; LAUER, P. E.
SEARCH STRATEGY FOR THE ELEMENTARY CYCLES OF A DIRECTED GRAPH, A.
TARJ 1977 -1 TARJAN, R. E.
COMPLEXITY OF COMBINATORIAL ALGORITHMS.
TARJ 1971 -1 TARJAN, R. E.
DEPTH-FIRST SEARCH AND LINEAR GRAPH ALGORITHMS.
TARJ 1974 -1 TARJAN, R. E.
GOOD ALGORITHM FOR EDGE-DISJOINT BRANCHING, A.
TARJ 1978 -1 TARJAN, R. E.
COMPLEXITY OF MONOTONE NETWORKS FOR COMPUTING CONJUNCTIONS.
TARJ 1973 -1 TARJAN, R. E.
ENUMERATION OF THE ELEMENTARY CIRCUITS OF A DIRECTED GRAPH.
TARJ 1975 -1 TARJAN, R. E.
FINDING EDGE-DISJOINT SPANNING TREES.
TARJ 1972 -1 TARJAN, R. E.
FINDING A MAXIMUM CLIQUE.
TARJ 1976 -1 TARJAN, R. E.
EDGE-DISJOINT SPANNING TREES AND DEPTH-FIRST SEARCH.
TARJ 1976 -2 TARJAN, R. E.
ITERATIVE ALGORITHMS FOR GLOBAL FLOW ANALYSIS.
TARJ 1973 -2 TARJAN, R. E.
FINDING DOMINATORS IN DIRECTED GRAPHS.
TARJ 1977 -2 TARJAN, R. E.
FINDING OPTIMUM BRANCHINGS.
TARJ 1975 -2 TARJAN, R. E.
FINDING MINIMUM SPANNING TREES.
TARJA 1977 -1 TARJAN, R. E.; TROJANOWSKI, A. E.
FINDING A MAXIMUM INDEPENDENT SET.
ULLM 1976 -1 ULLMAN, J. D.
COMPLEXITY OF CODE GENERATION, THE.
ULLM 1976 -2 ULLMAN, J. D.
COMPLEXITY OF SEQUENCING PROBLEMS.
ULLMA 1976 -1 ULLMANN, J. R.
ALGORITHM FOR SUBGRAPH ISOMORPHISM, AN.
VAIR 1976 -1 VAIRAVAN, K.; MILLO, R. A. DE
ON THE COMPUTATIONAL COMPLEXITY OF A GENERALIZED SCHEDULING PROBLEM.
VALI 1976 -1 VALIANT, L. G.
GRAPH-THEORETIC PROPERTIES IN COMPUTATIONAL COMPLEXITY.
VANE 1975 -1 VANEMDEN, M. H.
ANALYSIS OF COMPLEXITY, AN (2ND PRINTING).
VANL 1976 -1 VANLEEUWEN, J.; VANEMDEBOAS, P.
SOME ELEMENTARY PROOFS OF LOWER-BOUNDS IN COMPLEXITY THEORY.
VISI 1977 -1 VISING, V. G.
ON COMPLEXITY OF THE PROBLEM OF A TRAVELLING SALESMAN (RUSSIAN).
WAGN 1976 -1 WAGNER, R. A.
SHORTEST PATH ALGORITHM FOR EDGE-SPARSE GRAPHS, A.
WALSA 1976 -1 WALSH, T. S.
COMPLEXITE DES ALGORITHMES DE FLOT.
WEGLA 1976 -1 WEGLARZ, J.; SLOWINSKI, R.
COMPUTATIONAL ASPECTS OF A CERTAIN CLASS OF TIME-OPTIMAL RESOURCE ALLOCATION PROBLEMS IN A COMPLEX OF OPERATIONS.
WEIG 1976 -1 WEIGAND, M. M.
ALGORITHMUS 27. EIN NEUER ALGORITHMUS ZUR BESTIMMUNG VON K-KUERZESTEN WEGEN IN EINEM GRAPHEN.
YAO 1975 -1 YAO, A. C.
O (|E|LOG LOG|V|) ALGORITHM FOR FINDING MINIMUM SPANNING TREES, AN.
YAO A 1977 -1 YAO, A. C.; AVIS, D. M.; RIVEST, R. L.
OMEGA (N POWER 2 LOG N) LOWER BOUND TO THE SHORTEST PATHS PROBLEM, AN.
YEH 1976 -1 YEH, R. T.
APPLIED COMPUTATION THEORY: ANALYSIS, DESIGN, MODELING.
YEN 1969 -1 YEN, J. Y.
SOME ALGORITHMS FOR FINDING THE SHORTEST ROUTES THROUGH THE GENERAL NETWORKS.
YOUNB 1977 -1 YOUNG, P.
OPTIMIZATION AMONG PROVABLY EQUIVALENT PROGRAMS.
YUDI 1976 -1 YUDIN, D. B.; NEMIROVSKIJ, A. S.
ESTIMATION OF THE INFORMATION COMPLEXITY OF MATHEMATICAL PROGRAMMING PROBLEMS.
YUVA 1976 -1 YUVAL, G.
ALGORITHM FOR FINDING ALL SHORTEST PATHS USING (N POWER 2.81) INFINITE PRECISION MULTIPLICATIONS, AN.
ZHUR 1977 -1 ZHURAVLEV, JU. I.
CORRECT ALGEBRAS OVER THE SETS OF INCORRECT (HEURISTIC) ALGORITHMS, II (RUSSIAN).

COMPUTATIONAL RESULTS

ABAD 1976 -1 ABADIE, J.; DAYAN, H.; AKOKA, J.
QUELQUES EXPERIENCES NUMERIQUES SUR LA PROGRAMMATION NON LINEAIRE EN NOMBRES ENTIERS.

ADAM 1977 -1 ADAM, N.; SURKIS, J.
COMPARISON OF CAPACITY PLANNING TECHNIQUES IN A JOB SHOP CONTROL SYSTEM, A.

AHRE 1975 -1 AHRENS, J. H.; FINKE, G.
MERGING AND SORTING APPLIED TO THE ZERO-ONE KNAPSACK PROBLEM.

AHREA 1976 -1 AHRENS, W.
NUMERICAL EXPERIENCE WITH HEURISTIC RULES IN SPECIFIC BINARY OPTIMIZATION ALGORITHMS.

AKINA 1977 -1 AKINC, U.; KHUMAWALA, B. M.
EFFICIENT BRANCH AND BOUND ALGORITHM FOR THE CAPACITATED WAREHOUSE LOCATION PROBLEM, AN.

ALBE 1977 -1 ALBERS, S.
MIXED INTEGER NONLINEAR PROGRAMMING PROCEDURE FOR SIMULTANEOUSLY LOCATING MULTIPLE PRODUCTS IN AN ATTRIBUTE SPACE, A.

ANEJ 1977 -1 ANEJA, Y. P.; VEMUGANTI, R. R.
ROW GENERATION SCHEME FOR FINDING A MULTI-COMMODITY MINIMUM DISCONNECTING SET, A.

ASHT 1977 -1 ASHTON, W. B.; CLARK, G. M.
LOCATION AND SIZING OF URBAN FREIGHT CONSOLIDATION TERMINALS WITH MULTIPLE PLANNING PERIODS, THE.

ASSAA 1976 -1 ASSAD, A. A.
MULTICOMMODITY NETWORK FLOWS - COMPUTATIONAL EXPERIENCE.

ATTM 1973 -1 ATTMANN, I.
NUMERISCHE ERFAHRUNGEN MIT BALAS-ALGORITHMEN.

AUST 1976 -1 AUST, R. J.
DYNAMIC PROGRAMMING BRANCH AND BOUND ALGORITHM FOR PURE INTEGER PROGRAMMING, A.

BAKE 1977 -1 BAKER, K. R.
COMPUTATIONAL EXPERIENCE WITH A SEQUENCING ALGORITHM ADAPTED TO THE TARDINESS PROBLEM.

BALAG 1975 -1 BALAS, E.
BIVALENT PROGRAMMING BY IMPLICIT ENUMERATION.

BANS 1977 -1 BANSAL, S. P.
MINIMIZING THE SUM OF COMPLETION TIMES OF N JOBS OVER M MACHINES IN A FLOWSHOP - A BRANCH AND BOUND APPROACH.

BARN 1977 -1 BARNES, J. W.; BRENNAN, J. J.
IMPROVED ALGORITHM FOR SCHEDULING JOBS ON IDENTICAL MACHINES, AN.

BARR 1977 -2 BARR, R. S.; GLOVER, F.; KLINGMAN, D.
ALTERNATING BASIS ALGORITHM FOR ASSIGNMENT PROBLEMS, THE.

BARR 1977 -4 BARR, R. S.; GLOVER, F.; KLINGMAN, D.
NEW ALTERNATING BASIS ALGORITHM FOR SEMI-ASSIGNMENT NETWORKS, A.

BAZAB 1977 -1 BAZARAA, M. S.; GOODE, J. J.
TRAVELING SALESMAN: A DUALITY APPROACH, THE.

BECK 1976 -1 BECK, L. L.; MATULA, D. W.
CLIQUE DETERMINATION IN LARGE DENSE GRAPHS.

BEEN 1976 -1 BEENHAKKER, H. L.; NARAYANAN, V.
ALGORITHMS FOR SCHEDULING PROJECTS WITH LIMITED RESOURCES.

BENI 1977 -1 BENICHOU, M.; GAUTHIER, J. M.; HENTGES, G.; RIBIERE, G.
EFFICIENT SOLUTION OF LARGE-SCALE LINEAR PROGRAMMING PROBLEMS - SOME ALGORITHMIC TECHNIQUES AND COMPUTATIONAL RESULTS, THE.

BENN 1976 -1 BENNETT, G. R.; BYRD, J. JR.
TRAINABLE HEURISTIC PROCEDURE FOR THE ASSEMBLY LINE BALANCING PROBLEM, A.

BENNA 1972 -1 BENNETT, J. M.; EDWARDS, J.
EXPERIENCE WITH A METHOD FOR HANDLING LARGE STRUCTURAL LINEAR PROGRAMMING PROBLEMS WITH INTEGER RESTRICTIONS.

BERRA 1976 -1 BERRY, W. L.; FINLAY, R. A.
CRITICAL RATIO SCHEDULING WITH QUEUE WAITING TIME INFORMATION: AN EXPERIMENTAL ANALYSIS.

BERRB 1975 -1 BERRY, W. L.; RAO, V.
CRITICAL RATIO SCHEDULING: AN EXPERIMENTAL ANALYSIS.

BEST 1976 -1 BESTWICK, P. F.; HASTINGS, N. A. J.
NEW BOUND FOR MACHINE SCHEDULING, A.

BHAU 1974 -1 BHAUMIK, G.; JENSON, P.
COMPUTATIONALLY EFFICIENT ALGORITHM FOR THE NETWORK WITH GAINS PROBLEM, A.

BILDA 1977 -1 BILDE, O.; KRARUP, J.
SHARP LOWER BOUNDS AND EFFICIENT ALGORITHMS FOR THE SIMPLE PLANT LOCATION PROBLEM.

BODI 1976 -1 BODIN, L. D.; ROSENFIELD, D.
ESTIMATION OF THE OPERATING COST OF MASS TRANSIT SYSTEMS.

BOUR 1969 -1 BOURGOIN, M. H.; HEURGON, E. M. J.
STUDY AND COMPARISON OF ALGORITHMS OF THE SHORTEST PATH THROUGH PLANNED EXPERIMENTS.

BRADB 1969 -1 BRADLEY, G. H.; WAHI, P. N.
INTEGER PROGRAMMING TEST PROBLEMS.

BULF 1976 -2 BULFIN, R. L.; PARKER, R. G.
ON THE ASSIGNMENT AND SEQUENCING OF OPERATIONS ON A CREW-SERVED PROJECT.

BULFA 1977 -1 BULFIN, R. L.; SHETTY, C. M.; PARKER, R. G.
COMPUTATIONAL RESULTS WITH A BRANCH-AND-BOUND ALGORITHM FOR THE GENERAL KNAPSACK PROBLEM.

BURK 1976 -3 BURKARD, R. E.
PERTURBATION METHOD FOR SOLVING QUADRATIC ASSIGNMENT AND BOTTLENECK PROBLEMS, A.

BURKD 1976 -1 BURKARD, R. E.; STRATMANN, K.-H.
NUMERICAL INVESTIGATIONS ON QUADRATIC ASSIGNMENT PROBLEMS.

BUSH 1976 -1 BUSHELL, G.; FERLAND, J.; GUERIN, G.; NASTANSKY, L.
ENGINE SCHEDULING PROBLEM IN A RAILWAY NETWORK, THE.

CARP 1977 -1 CARPANETO, G.; TOTH, P.
EFFICIENT ALGORITHM FOR THE ASYMMETRIC TRAVELLING SALESMAN PROBLEM, AN.

CARPA 1976 -1 CARPENITO, T. A.; WHITE, J. A.
ALLOCATION OF NON-IDENTICAL MACHINES AMONG NON-IDENTICAL SERVERS, THE.

CARS 1977 -1 CARSON, J. S.; LAW, A. M.
NOTE ON SPIRA'S ALGORITHM FOR THE ALL-PAIRS SHORTEST-PATH PROBLEM, A.

CHAL 1976 -1 CHALMET, L.
SOLVING ASSIGNMENT PROBLEMS IN WAREHOUSES (DUTCH).

CHALA 1975 -1 CHALMET, L.; GELDERS, L.
ALGORITHM FOR GENERATING GOOD WAREHOUSE LAYOUTS, AN.

CHAR 1975 -1 CHARNES, A.; GLOVER, F.; KLINGMAN, D.
PAST, PRESENT AND FUTURE OF LARGE SCALE TRANSSHIPMENT COMPUTER CODES AND APPLICATIONS.

CHENE 1976 -1 CHEN, R.; CROWDER, H.; JOHNSON, E. L.
INTEGER PROGRAMMING FORMULATION OF THE INSTALLATION SCHEDULING PROBLEM, AN.

CHRI 1975 -1 CHRISTOFIDES, N.
HAMILTONIAN CIRCUITS AND THE TRAVELLING SALESMAN PROBLEM: A SURVEY.

CHRIA 1977 -1 CHRISTOFIDES, N.; WHITLOCK, C.
ALGORITHM FOR TWO-DIMENSIONAL CUTTING PROBLEMS, AN.

****

CHURA 1977 -1 CHURCH, R. L.; MEADOWS, M. E.
RESULTS OF A NEW APPROACH TO SOLVING THE P-MEDIAN PROBLEM WITH MAXIMUM DISTANCE CONSTRAINTS.
CHURB 1976 -1 CHURCH, R. L.; REVELLE, C. S.
THEORETICAL AND COMPUTATIONAL LINKS BETWEEN THE P-MEDIAN, LOCATION SET-COVERING, AND THE MAXIMAL COVERING LOCATION PROBLEM.
CLAU 1977 -1 CLAUDER, G.; JAENICKE, W.
ZUR GRAPHENTHEORETISCHEN LOESUNG EINER KLASSE VON DISKRETEN OPTIMIERUNGSPROBLEMEN.
COCKC 1977 -1 COCKAYNE, E. J.; ROBERTS, F. D. K.
COMPUTATION OF DOMINATING PARTITIONS.
COOP 1975 -2 COOPER, L.
FIXED CHARGE PROBLEM, I. A NEW HEURISTIC METHOD, THE.
COOPB 1975 -1 COOPER, L.; COOPER, M. W.
NON-LINEAR INTEGER PROGRAMMING.
COOPC 1977 -1 COOPER, L.; LEBLANC, L. J.
STOCHASTIC TRANSPORTATION PROBLEMS AND OTHER NETWORK RELATED CONVEX PROBLEMS.
CORND 1977 -1 CORNUEJOLS, G.; FISHER, M. L.; NEMHAUSER, G. L.
LOCATION OF BANK ACCOUNTS TO OPTIMIZE FLOAT: AN ANALYTIC STUDY OF EXACT AND APPROXIMATE ALGORITHMS.
COVE 1976 -1 COVERDALE, I. L.; WHARTON, F.
IMPROVED HEURISTIC PROCEDURE FOR A NONLINEAR CUTTING STOCK PROBLEM, AN.
DAEN 1977 -1 DAENINCK, G.; SMEERS, Y.
USING SHORTEST PATHS IN SOME TRANSSHIPMENT PROBLEM WITH CONCAVE COSTS.
DANN 1977 -1 DANNENBRING, D. G.
EVALUATION OF FLOW SHOP SEQUENCING HEURISTICS, AN.
DANN 1977 -2 DANNENBRING, D. G.
PROCEDURES FOR ESTIMATING OPTIMAL SOLUTION VALUES FOR LARGE COMBINATORIAL PROBLEMS.
DAVID 1977 -1 DAVIS, K. R.; WALTERS, J. E.
ADDRESSING THE N/1 SCHEDULING PROBLEM - A HEURISTIC APPROACH.
DAVIE 1976 -1 DAVIS, R. P.; TERRELL, M. P.
APPROXIMATION TECHNIQUE FOR PSEUDO-BOOLEAN MAXIMIZATION PROBLEMS, AN.
DAVY 1972 -1 DAVYDOV, E. G.; SIGAL, I. KH.
APPLICATION OF THE PENALTY FUNCTION METHOD IN INTEGER PROGRAMMING PROBLEMS (RUSSIAN).
DERI 1977 -2 DERIGS, U.
RANGZUORDNUNGPROBLEME.
DERIA 1978 -1 DERIGS, U.; ZIMMERMANN, U.
AUGMENTING PATH METHOD FOR SOLVING LINEAR BOTTLENECK ASSIGNMENT PROBLEMS, AN.
DIALA 1977 -1 DIAL, R.; GLOVER, F.; KARNEY, D.; KLINGMAN, D.
COMPUTATIONAL ANALYSIS OF ALTERNATIVE ALGORITHMS AND LABELING TECHNIQUES FOR FINDING SHORTEST PATH TREES, A.
DINK 1976 -1 DINKEL, J. J.; KLEINDORFER, G. B.; KOCHENBERGER, G. A.; WONG, S.-N.
ENVIRONMENTAL INSPECTION ROUTES AND THE CONSTRAINED TRAVELLING SYSTEM SALESMAN PROBLEM.
DIXO 1976 -1 DIXON, E. T.; GOODMAN, S. E.
ALGORITHM FOR THE LONGEST CYCLE PROBLEM, AN.
DOMS 1976 -1 DOMSCHKE, W.
NEW ALGORITHM FOR MINIMAL COST FLOW PROBLEMS AND THEIR APPLICATIONS.
DORA 1966 -1 DORAN, J. E.; MICHIE, D.
EXPERIMENTS WITH THE GRAPH TRAVERSER PROGRAM.
DRIS 1976 -1 DRISCOLL, W. C.
BRANCH-BOUND ALGORITHM FOR SCHEDULING PRODUCTION ON ONE MACHINE WITH CHANGEOVER TIMES, A.
DUNS 1977 -1 DUNSTAN, F. D. J.
ALGORITHM FOR SOLVING A RESOURCE ALLOCATION PROBLEM, AN.
DUNS 1975 -1 DUNSTAN, F. D. J.
GREEDY ALGORITHMS FOR OPTIMISATION PROBLEMS.
DYKS 1977 -1 DYKSTRA, D. P.; RIGGS, J. L.
APPLICATION OF FACILITIES LOCATION THEORY TO THE DESIGN OF FOREST HARVESTING AREAS, AN.
DZIKA 1974 -1 DZIKIEWICS, J.; SYSLO, M. M.
COMPUTIONAL EXPERIENCES WITH SOME TRANSITIVE CLOSURE ALGORITHMS.
EILO 1975 -1 EILON S.; CHOWDHURY, I. G.; SERGHIOU, S. S.
EXPERIMENTS WITH THE SI RULE IN JOB-SHOP SCHEDULING.
EILOA 1977 -1 EILON, S.; CHOWDHURY, I. G.
MINIMISING WAITING TIME VARIANCE IN THE SINGLE MACHINE PROBLEM.
ELSH 1977 -1 ELSHAFEI, A. N.
HOSPITAL LAYOUT AS A QUADRATIC ASSIGNMENT PROBLEM.
ELSHA 1974 -1 ELSHAFEI, A. N.; HALEY, K. B.
FACILITIES LOCATION: SOME FORMULATIONS, METHODS OF SOLUTION, APPLICATIONS, AND COMPUTATIONAL EXPERIENCE.
ELZIA 1976 -1 ELZINGA, J.; VEMUGANTI, R. R.
OPTIMAL SCHEDULING OF A SALES FORCE.
ERLE 1976 -1 ERLENKOTTER, D.
DUAL-BASED PROCEDURE FOR UNCAPACITATED FACILITY LOCATION, A.
ESKE 1975 -1 ESKEW, J. D.; PARKER, R. G.
COMPUTATIONAL EXPERIENCE WITH A COST-BASED ALGORITHM FOR THE SHOP SCHEDULING PROBLEM.
ETCH 1977 -1 ETCHEBERRY, J.
SET-COVERING PROBLEM: A NEW IMPLICIT ENUMERATION ALGORITHM, THE.
ETCH 1976 -2 ETCHEBERRY, J.
SET REPRESENTATION PROBLEM: A NEW IMPLICIT ENUMERATION ALGORITHM, THE (REVISED VERSION).
EVANB 1977 -1 EVANS, J. R.; JARVIS, J. J.
OUT-OF-KILTER BASED HEURISTIC FOR THE INTEGER MULTICOMMODITY TRANSPORTATION PROBLEM, AN.
FABIA 1977 -1 FABIAN, CS.; MOGA, S.
TWO SIMULATION PROCEDURES FOR SOLVING CONVEX INTEGER PROGRAMMING PROBLEMS.
FISHA 1976 -1 FISHER, M. L.
DUAL ALGORITHM FOR THE ONE-MACHINE SCHEDULING PROBLEM, A.
FISK 1977 -1 FISK, J.
SOLUTION PROCEDURE FOR A SPECIAL TYPE OF CAPACITATED WAREHOUSE LOCATION PROBLEM, A.
FISKA 1977 -1 FISK, J.; MACKEOWN, P.
DETERMINING THE MAXIMUM DEGENERATE SOLUTION FOR TRANSPORTATION PROBLEMS.
FISKC 1977 -1 FISK, J. C.
INITIAL BOUNDING PROCEDURE FOR USE WITH 0-1 SINGLE KNAPSACK ALGORITHMS, AN.
FOST 1976 -1 FOSTER, B. A.; RYAN, D. M.
INTEGER PROGRAMMING APPROACH TO THE VEHICLE SCHEDULING PROBLEM, AN.
FRATB 1975 -1 FRATTA, L.; MONTANARI, U.
VERTEX ELIMINATION ALGORITHM FOR ENUMERATING ALL SIMPLE PATHS IN A GRAPH, A.
FRIEA 1976 -1 FRIEZE, A. M.
SHORTEST PATH ALGORITHMS FOR KNAPSACK TYPE PROBLEMS.

****

GALLB 1976 -1 GALLO, G.; HAMMER, P. L.; SIMEONE, B.
QUADRATIC KNAPSACK PROBLEMS.

GALLC 1972 -1 GALLO, G.; MARTINO. E.; SIMEONE, B.
GROUP OPTIMIZATION ALGORITHMS AND SOME NUMERICAL RESULTS VIA A BRANCH AND BOUND APPROACH.

GARFB 1977 -1 GARFINKEL, R. S.; NEEBE, A. W.; RAO, M. R.
M-CENTER PROBLEM: MINIMAX FACILITY LOCATION, THE.

GAUT 1977 -1 GAUTHIER, J.-M.; RIBIERE, G.
EXPERIMENTS IN MIXED-INTEGER LINEAR PROGRAMMING USING PSEUDOCOSTS.

GAVI 1977 -1 GAVISH, B.; SCHWEITZER, P.; SHLIFER, E.
ZERO PIVOT PHENOMENON IN TRANSPORTATION AND ASSIGNMENT PROBLEMS AND ITS COMPUTATIONAL IMPLICATIONS, THE.

GEN A 1976 -1 GEN, M.; OKUNO, H.; SHINOFUJI, S.
OPTIMIZING METHOD IN SYSTEM RELIABILITY WITH FAILURE-MODES BY IMPLICIT ENUMERATION ALGORITHM, AN.

GHIL 1973 -1 GHILARDOTTI, P.
BILP CODE: BOUNDED INTEGER LINEAR PROGRAMMING CODE USER'S INSTRUCTIONS WITH SOME NOTES AND FIRST COMPUTATIONAL RESULTS.

GLEN 1975 -1 GLEN, J. J.
DYNAMIC PROGRAMMING MODEL FOR WORK SCHEDULING IN A SHIPYARD, A.

GLOV 1977 -2 GLOVER, F.
HEURISTICS FOR INTEGER PROGRAMMING USING SURROGATE CONSTRAINTS.

GLOVB 1977 -1 GLOVER, F.; HULTZ, J.; KLINGMAN, D.; STUTZ, J.
NEW COMPUTER-BASED PLANNING TOOL, A.

GLOVF 1973 -1 GLOVER, F.; KLINGMAN, D.; STUTZ, J.
IMPLEMENTATION AND COMPUTATIONAL STUDY OF A GENERALIZED NETWORK CODE.

GOLD 1976 -2 GOLDEN, B.
SHORTEST-PATH ALGORITHMS: A COMPARISON.

GOLDA 1977 -1 GOLDEN, B.; BALL, M.
SHORTEST PATHS WITH EUCLIDEAN DISTANCES: AN EXPLANATORY MODEL.

GOLDB 1976 -2 GOLDEN, B. L.
RECENT DEVELOPMENTS IN VEHICLE ROUTING.

GOLDD 1977 -1 GOLDEN, B. L.; MAGNANTI, T. L.; NGUYEN, H. Q.
IMPLEMENTING VEHICLE ROUTING ALGORITHMS.

GRAT 1970 -1 GRATZER, F.
COMPUTER SOLUTION OF LARGE MULTICOMMODITY FLOW PROBLEMS.

GROE 1977 -5 GROETSCHEL, M.
ON THE SYMMETRIC TRAVELLING SALESMAN PROBLEM: SOLUTION OF A 120-CITY PROBLEM.

GROEA 1978 -1 GROETSCHEL, M.; PADBERG, M. W.
ON THE SYMMETRIC TRAVELLING SALESMAN PROBLEM: THEORY AND COMPUTATION.

GUIGA 1977 -1 GUIGNARD, M.; SPIELBERG, K.
ALGORITHMS FOR EXPLOITING THE STRUCTURE OF THE SIMPLE PLANT LOCATION PROBLEM.

GUIGA 1976 -1 GUIGNARD, M.; SPIELBERG, K.
EXPERIMENTAL INTERACTIVE SYSTEM FOR INTEGER PROGRAMMING, AN.

GUIGA 1977 -2 GUIGNARD, M.; SPIELBERG, K.
DUAL METHOD FOR THE MIXED PLANT LOCATION PROBLEM, A.

GUIGA 1977 -3 GUIGNARD, M.; SPIELBERG, K.
PROPAGATION, PENALTY IMPROVEMENT AND USE OF LOGICAL INEQUALITIES IN INTERACTIVE (0,1) PROGRAMMING.

GUIGA 1977 -4 GUIGNARD, M.; SPIELBERG, K.
REDUCTION METHODS FOR STATE ENUMERATION INTEGER PROGRAMMING.

HALP 1977 -2 HALPERN, J.
SEQUENTIAL COVERING PROBLEM UNDER UNCERTAINTY, THE.

HANS 1971 -1 HANSEN, K. H.
SEQUENCING PROBLEMS WITH EMPHASIS ON A MODIFICATION OF HELD AND KARP'S ALGORITHM. OR: AN ALGORITHM FOR THE OPTIMAL SOLUTION OF THE SYMMETRIC TRAVELING SALESMAN PROBLEM. PART 2. (DANISH).

HANS 1971 -2 HANSEN, K. H.
SEQUENCING PROBLEMS WITH EMPHASIS ON A MODIFICATION OF HELD AND KARP'S ALGORITHM. OR: AN ALGORITHM FOR THE OPTIMAL SOLUTION OF THE SYMMETRIC TRAVELING SALESMAN PROBLEM. PART 1. (DANISH).

HANSA 1976 -1 HANSEN, P.
CASCADE ALGORITHM FOR THE LOGICAL CLOSURE OF A SET OF BINARY RELATIONS, A.

HANSA 1977 -2 HANSEN, P.
SUBDEGREES AND CHROMATIC NUMBERS OF HYPERGRAPHS.

HANSB 1976 -2 HANSEN, P.; DELATTRE, M.
CLUSTER ANALYSIS BY GRAPH COLORING.

HANSB 1976 -3 HANSEN, P.; DELATTRE, M.
COMPLETE-LINK CLUSTER ANALYSIS BY GRAPH COLORING.

HANSC 1976 -1 HANSEN, P.; KAUFMAN, L.
PUBLIC FACILITIES LOCATION UNDER AN INVESTMENT CONSTRAINT.

HARR 1976 -1 HARRIS, B.
CODE FOR THE TRANSPORTATION PROBLEM OF LINEAR PROGRAMMING, A.

HATC 1975 -1 HATCH, R. S.
BENCH MARKS COMPARING TRANSPORTATION CODES BASED ON PRIMAL SIMPLEX AND PRIMAL-DUAL ALGORITHMS.

HEAR 1976 -1 HEARD, E. L.
DUE-DATES AND INSTANTANEOUS LOAD IN THE ONE-MACHINE SHOP.

HEIS 1977 -1 HEISING, C. D.
PLANT LOCATION AND TRANSPORTATION MODELS OF THE LIGHT-WATER REACTOR FUEL CYCLE.

HILL 1976 -1 HILL, A. V.; WHYBARK, D. C.
COMPARING EXACT SOLUTION PROCEDURES FOR THE MULTI-VEHICLE ROUTING PROBLEM.

HITCA 1968 -1 HITCHNER, L.
COMPARATIVE INVESTIGATION OF THE COMPUTATIONAL EFFICIENCY OF SHORTEST PATH ALGORITHMS, A.

HOLMB 1976 -1 HOLMES, R. A.; PARKER, R. G.
VEHICLE SCHEDULING PROCEDURE BASED UPON SAVINGS AND A SOLUTION PERTURBATION SCHEME, A.

HONG 1974 -1 HONG, M. L.
EXPERIMENTS WITH A JOB SHOP PRODUCTION MODEL COUPLING A LINEAR PROGRAM AND A HEURISTIC SCHEDULING PROCEDURE.

HONGA 1977 -1 HONG, S.
ASYMMETRIC TRAVELLING SALESMAN PROBLEM AND SPANNING ARBORESCENCES OF DIRECTED GRAPHS, THE.

HUNGA 1976 -1 HUNG, M. S.; FISK, J. C.
ALGORITHM FOR 0-1 MULTIPLE KNAPSACK PROBLEMS, AN.

INGAA 1977 -1 INGARGIOLA, G. P.; KORSH, J. F.
GENERAL ALGORITHM FOR ONE-DIMENSIONAL KNAPSACK PROBLEMS, A.

IVANA 1976 -1 IVANCHEV, M. D.; MITEV, I. G.; JANEV, N. I.
REALIZATION OF THE BRANCH AND BOUND METHOD FOR THE SOLUTION OF THE GENERAL MIXED INTEGER PROGRAMMING PROBLEM (RUSSIAN), A.

JAER 1973 -1 JAERVINEN, P.
ON A ROUTING PROBLEM IN A GRAPH WITH TWO TYPES OF COSTS ASSOCIATED WITH THE EDGES.

JUNG 1976 -1 JUNGINGER, W.
NUMERISCHE ERFAHRUNGEN BEI DER LOESUNG DES DREIDIMENSIONALEN TRANSPORTPROBLEMS.

KAES 1976 -1 KAESTING, H.
LEXIKOGRAPHISCHES SUCHVERFAHREN ZUR LOESUNG DES BIVALENTEN KNAPSACK-PROBLEMS, EIN.

KALD 1975 -1 KALDYBAEV, S. U.
ALGORITHM FOR THE APPROXIMATION-COMBINATORIAL METHOD IN SOLVING AN ALLOCATION PROBLEM USING BOOLEAN VARIABLES (RUSSIAN), A.

KARN 1976 -1 KARNEY, D.; KLINGMAN, D.
IMPLEMENTATION AND COMPUTATIONAL STUDY ON AN IN-CORE, OUT-OF-CORE PRIMAL NETWORK CODE.

KARW 1977 -1 KARWAN, M. H.; RARDIN, R. L.
SEARCHABILITY OF THE COMPOSITE AND MULTIPLE SURROGATE DUAL FUNCTIONS.

KATS 1977 -1 KATSEV, S. B.
SOLUTION OF A GENERALIZED PROBLEM ON SET PARTITIONING (RUSSIAN).

KENN 1977 -1 KENNINGTON, J. L.
SOLVING MULTICOMMODITY TRANSPORTATION PROBLEMS USING A PRIMAL PARTITIONING SIMPLEX TECHNIQUE.

KENN 1976 -1 KENNINGTON, J. L.
FIXED-CHARGE TRANSPORTATION PROBLEM: A COMPUTATIONAL STUDY WITH A BRANCH AND-BOUND CODE, THE.

KINA 1977 -1 KINARIWALA, B.; RAO, A. G.
FLOW SWITCHING APPROACH TO THE MAXIMUM FLOW PROBLEM: I.

KNUTA 1973 -1 KNUTS, L.-R.
COMBINING TWO INTEGER PROGRAMMING ALGORITHMS.

KOHLA 1975 -1 KOHLER, W.
COMPUTATIONAL EXPERIENCE WITH EFFICIENT EXACT AND APPROXIMATE ALGORITHMS FOR AN NP-COMPLETE SCHEDULING PROBLEM.

KREU 1969 -1 KREUZBERGER, H.
VERFAHREN ZUR LOESUNG GANZZAHLIGER LINEARER OPTIMIERUNGSPROBLEME.

KUEP 1976 -1 KUEPPER, H.-U.
VERGLEICHENDE ANALYSE GANZZAHLIGER PROGRAMMIERUNGSMODELLE ZUR ABLAUFPLANUNG INDUSTRIELLER UNTERNEHMUNGEN.

LAGEA 1976 -1 LAGEWEG, B. J.; LENSTRA, J. K.; RINNOOY KAN, A. H. G.
GENERAL BOUNDING SCHEME FOR THE PERMUTATION FLOW-SHOP, A.

LAGEA 1976 -2 LAGEWEG, B. J.; LENSTRA, J. K.; RINNOOY KAN, A. H. G.
JOB-SHOP SCHEDULING BY IMPLICIT ENUMERATION.

LAGEA 1976 -3 LAGEWEG, B. J.; LENSTRA, J. K.; RINNOOY KAN, A. H. G.
JOB-SHOP SCHEDULING BY IMPLICIT ENUMERATION.

LAGEA 1976 -4 LAGEWEG, B. J.; LENSTRA, J. K.; RINNOOY KAN, A. H. G.
MINIMIZING MAXIMUM LATENESS ON ONE MACHINE: COMPUTATIONAL EXPERIENCE AND SOME APPLICATIONS.

LAURB 1978 -1 LAURIERE, M.
ALGORITHM FOR THE 0/1 KNAPSACK PROBLEM, AN.

LEBL 1975 -1 LEBLANC, L. J.
ALGORITHM FOR THE DISCRETE NETWORK DESIGN PROBLEM, AN.

LENSA 1977 -1 LENSTRA, J. K.
SEQUENCING BY ENUMERATIVE METHODS.

LENSB 1975 -1 LENSTRA, J. K.; RINNOOY KAN, A. H. G.
ON THE JOB-SHOP SCHEDULING PROBLEM.

LIU 1973 -1 LIU, C. L.
COMBINATORIAL STUDY OF SOME SCHEDULING ALGORITHMS (SUMMARY), A.

LODI 1970 -1 LODISH, L. M.
COMPUTATIONAL LIMITATIONS OF DYNAMIC PROGRAMMING FOR WAREHOUSE LOCATION.

LOS 1978 -1 LOS, M.
SIMULTANEOUS OPTIMIZATION OF LAND USE AND TRANSPORTATION: A SYNTHESIS OF THE QUADRATIC ASSIGNMENT PROBLEM AND THE OPTIMAL NETWORK PROBLEM.

LOVE 1976 -1 LOVE, R. F.
ONE-DIMENSIONAL FACILITY LOCATION-ALLOCATION USING DYNAMIC PROGRAMMING.

LOVEA 1976 -1 LOVE, R. F.; WONG, J. Y.
ON SOLVING A ONE-DIMENSIONAL SPACE ALLOCATION PROBLEM WITH INTEGER PROGRAMMING.

LOVEA 1976 -2 LOVE, R. F.; WONG, J. Y.
SOLVING QUADRATIC ASSIGNMENT PROBLEMS WITH RECTANGULAR DISTANCES AND INTEGER PROGRAMMING.

MAC D 1977 -1 MAC DANIEL, D.; DEVINE, M.
MODIFIED BENDERS' PARTITIONING ALGORITHM FOR MIXED INTEGER PROGRAMMING, A.

MAC G 1975 -1 MAC KEOWN, P.
VERTEX RANKING PROCEDURE FOR SOLVING THE LINEAR FIXED-CHARGE PROBLEM, A.

MAHM 1976 -1 MAHMOUD, S.; RIORDON, J. S.
OPTIMAL ALLOCATION OF RESOURCES IN DISTRIBUTED INFORMATION NETWORKS.

MARSA 1975 -1 MARSTEN, R. E.
ALGORITHM FOR LARGE SET PARTITIONING PROBLEMS, AN.

MARSB 1977 -2 MARSTEN, R. E.; MORIN, T. L.
PARAMETRIC INTEGER PROGRAMMING: THE RIGHT-HAND-SIDE CASE.

MARTB 1977 -1 MARTELLO, S.
ALGORITHM FOR FINDING A MINIMAL EQUIVALENT GRAPH OF A STRONGLY CONNECTED DIGRAPH, AN.

MARTC 1977 -1 MARTELLO, S.; TOTH, P.
COMPUTATIONAL STUDY ON LARGE-SIZE UNIDIMENSIONAL KNAPSACK PROBLEMS, A.

MARTC 1976 -1 MARTELLO, S.; TOTH, P.
ENUMERATIVE ALGORITHM FOR THE 0-1 KNAPSACK PROBLEM (ITALIAN), AN.

MARTC 1977 -2 MARTELLO, S.; TOTH, P.
SOLUTION OF THE BOUNDED AND UNBOUNDED CHANGE-MAKING PROBLEM.

MARTC 1977 -3 MARTELLO, S.; TOTH, P.
SOLUTION OF ZERO-ONE MULTIPLE KNAPSACK PROBLEM.

MARTC 1977 -4 MARTELLO, S.; TOTH, P.
UPPER BOUND FOR THE ZERO-ONE KNAPSACK PROBLEM AND A BRANCH AND BOUND ALGORITHM, AN.

MARU 1976 -1 MARUYAMA, K.; TANG, D. T.; FRATTA, L.
FLOW ASSIGNMENT ALGORITHM FOR COMPUTER COMMUNICATION NETWORK DESIGN WITH DIFFERENT CLASSES OF PACKETS.

MEHL 1975 -1 MEHLHORN, K.
NEARLY OPTIMAL BINARY SEARCH TREES.

MEVE 1977 -1 MEVERT, P.; SUHL, U.
IMPLICIT ENUMERATION WITH GENERALIZED UPPER BOUNDS.

****

MICHA 1968 -1 MICHIE, D.; FLEMING, J. G.; OLDFIELD, J. V.
COMPARISON OF HEURISTIC, INTERACTIVE, AND UNAIDED METHODS OF SOLVING A SHORTEST-ROUTE PROBLEM, A.

MING 1976 -1 MING-I, Y.
ON THE N JOB, M MACHINE SEQUENCING PROBLEM OF FLOW-SHOP.

MINIB 1975 -1 MININA, T. R.; PEREKREST, V. T.
APPROXIMATION TO THE SOLUTION OF THE TRAVELING-SALESMAN PROBLEM BY C-CYCLES.

MINOA 1975 -2 MINOUX, M.
RESOLUTION DES PROBLEMES DE MULTIFLOTS EN NOMBRES ENTIERS DANS LES GRANDS RESEAUX.

MOJE 1976 -1 MOJENA, R.; VOLLMANN, T. E.; OKAMOTO, Y.
ON PREDICTING COMPUTATIONAL TIME OF A BRANCH AND BOUND ALGORITHM FOR THE ASSIGNMENT OF FACILITIES.

MORT 1977 -1 MORTON, T. E.; DHARAN, B. G.
ALGORISTICS FOR SEQUENCING WITH PRECEDENCE CONSTRAINTS.

MUCK 1977 -1 MUCKSTADT, J. A.; KOENIG, S. A.
APPLICATION OF LAGRANGIAN RELAXATION TO SCHEDULING IN POWER-GENERATION SYSTEMS, AN.

MULVA 1975 -2 MULVEY, J. M.
SPECIAL STRUCTURES IN NETWORK NODELS AND ASSOCIATED APPLICATIONS.

MULVA 1975 -3 MULVEY, J. M.
TESTING OF LARGE-SCALE NETWORK OPTIMIZATION PROGRAM.

MURP 1976 -1 MURPHY, F. H.; STOHR, E. A.
DYNAMIC PROGRAMMING ALGORITHM FOR CHECK SORTING, A (REVISED VERSION).

NABEA 1975 -1 NABESHIMA, I.; MARUYAMA, S.
COMPUTATIONAL EXPERIENCE ON MULTIPROJECT SCHEDULING WITH RESOURCE CONSTRAINTS.

NARUA 1977 -1 NARULA, S. C.; OGBU, U. I.; SAMUELSSON, H. M.
ALGORITHM FOR THE P-MEDIAN PROBLEM, AN.

NAUS 1976 -2 NAUSS, R. M.
EFFICIENT ALGORITHM FOR THE 0-1 KNAPSACK PROBLEM, AN.

NAUS 1976 -3 NAUSS, R. M.
FIXED CHARGE TRANSPORTATION PROBLEM WITH ADDITIONAL CONSTRAINTS, THE.

NELS 1977 -1 NELSON, R. T.; HOLLOWAY, C. A.; WONG, R. M.-L.
CENTRALIZED SCHEDULING AND PRIORITY IMPLEMENTATION HEURISTICS FOR A DYNAMIC JOB SHOP MODEL.

NOON 1977 -1 NOONAN, F.; GIGLIO, R. J.
PLANNING ELECTRIC POWER GENERATION: A NONLINEAR MIXED INTEGER MODEL EMPLOYING BENDERS DECOMPOSITION.

NORB 1977 -1 NORBACK, J. P.; LOVE, R. F.
GEOMETRIC APPROACHES TO SOLVING THE TRAVELING SALESMAN PROBLEM.

NUNNA 1977 -1 NUNNIKHOVEN, T. S.; EMMONS, H.
SCHEDULING ON PARALLEL MACHINES TO MINIMIZE TWO CRITERIA RELATED TO JOB TARDINESS.

OFFI 1975 -1 OFFIR, J. D.
TRAVELLING SALESMAN WITH A KNAPSACK(A CALL-GIRL'S HEURISTIC), A.

ORLO 1976 -1 ORLOFF, C.; CAPRERA, D.
REDUCTION AND SOLUTION OF LARGE SCALE VEHICLE ROUTING PROBLEMS.

OSLEA 1977 -1 OSLEEB, J. P.; CROMLEY, R. G.
LOCATION-PRODUCTION-ALLOCATION PROBLEM WITH NONLINEAR PRODUCTION COSTS, THE.

PADBB 1977 -1 PADBERG, M. W.; HONG, S.
ON THE SYMMETRIC TRAVELLING SALESMAN PROBLEM: A COMPUTATIONAL STUDY.

PANWA 1975 -1 PANWALKAR, S. S.; KHAN, A. W.
IMPROVED BRANCH AND BOUND PROCEDURE FOR N X M FLOW SHOP PROBLEMS, AN.

PARK 1977 -1 PARKER, R. G.; DEANE, R. H.; HOLMES, R. A.
ON THE USE OF A VEHICLE ROUTING ALGORITHM FOR THE PARALLEL PROCESSOR PROBLEM WITH SEQUENCE DEPENDENT CHANGEOVER COSTS.

PARKA 1977 -1 PARKER, R. G.; GRAVES, R. J.; SHERALI, H.
APPLICATION OF A GRAPH-PARTITIONING ALGORITHM TO THE SCHEDULING OF CURRICULUM REQUIREMENTS, THE.

PARRA 1967 -1 PARRIS, R.; READ, R. C.
GRAPH ISOMORPHISM AND THE CODING OF GRAPHS.

PATE 1976 -1 PATEL, A. M.; WALD, C. G. DE; COTES, L. C.
PRE-PARTITIONING AS A MEANS OF ENHANCING PAIRWISE INTERCHANGE SOLUTIONS TO QUADRATIC ASSIGNMENT PROBLEMS.

PATTA 1976 -1 PATTERSON, J. H.; ROTH, G. W.
SCHEDULING A PROJECT UNDER MULTIPLE RESOURCE CONSTRAINTS: A ZERO-ONE PROGRAMMING APPROACH.

PERK 1965 -1 PERKO, A.
SOME COMPUTATIONAL NOTES ON THE SHORTEST ROUTE PROBLEM.

PETEA 1975 -1 PETERSEN, E. R.
PRIMAL-DUAL TRAFFIC ASSIGNMENT ALGORITHM, A.

PHIL 1976 -1 PHILLIPS, L. W.; UNGER, P. S.
MATHEMATICAL PROGRAMMING SOLUTION OF A HOIST SCHEDULING PROGRAM.

PHUO 1976 -1 PHUONG, T. H.
SOLUTION OF INTEGER PROGRAMS WITH A QUADRATIC OBJECTIVE FUNCTION.

PICAC 1977 -1 PICARD, J.-C.; QUEYRANNE, M.
NETWORKS, GRAPHS AND SOME NONLINEAR 0-1 PROGRAMMING PROBLEMS.

PICAC 1976 -1 PICARD, J.-C.; QUEYRANNE, M.
TIME-DEPENDENT TRAVELING SALESMAN PROBLEM AND APPLICATION TO THE TARDINESS PROBLEM IN ONE-MACHINE SCHEDULING, THE.

PIPE 1975 -1 PIPER, C.
COMPUTATIONAL STUDIES IN OPTIMIZING AND POSTOPTIMIZING LINEAR PROGRAMS IN ZERO-ONE VARIABLES.

PIPEA 1976 -1 PIPER, C. J.; ZOLTNERS, A. A.
SOME EASY POSTOPTIMALITY ANALYSIS FOR ZERO-ONE PROGRAMMING.

RAJA 1977 -1 RAJARAMAN, M. K.
PARALLEL SEQUENCING ALGORITHM FOR MINIMIZING TOTAL COST, A.

RAMAA 1976 -1 RAMAKRISHNAN, K. G.
SOLUTION OF CAPACITATED PLANT LOCATION PROBLEMS USING BENDERS' DECOMPOSITION.

RARDA 1976 -2 RARDIN, R. L.; UNGER, V. E.
SOME COMPUTATIONALLY RELEVANT GROUP THEORETIC STRUCTURES OF FIXED CHARGE PROBLEMS.

RATH 1977 -1 RATHMANN, D.
UMFORMULIERUNG BEIM ASYMPTOTISCHEN ANSATZ FUER GANZZAHLIGE PROGRAMME.

READB 1975 -1 READ, R. C.; TARJAN, R. E.
BOUNDS ON BACKTRACK ALGORITHMS FOR LISTING CYCLES, PATHS, AND SPANNING TREES.

REYN 1977 -1 REYNER, W. S.
ANALYSIS OF A GOOD ALGORITHM FOR THE SUBTREE PROBLEM, AN.

RIBE 1971 -1 RIBEILL, G.; TABOURIER, Y.
CHEMINS EXTREMAUX DANS UN GRAPHE ENTRE TOUS COUPLES DE SOMMETS: QUELQUES ALGORITHMES ET LEUR MISE EN OEUVRE.COMPARAISONS SUR LES PERFORMANCES.

****

RICH 1976 -1 RICHARDSON, R.
OPTIMIZATION APPROACH TO ROUTING AIRCRAFT, AN.
RICHC 1977 -1 RICHTER, K.
RESSOURCENAUFSPALTUNG FUER DISKRETE OPTIMIERUNGSAUFGABEN, TEIL II.
RINNA 1975 -1 RINNOOY KAN, A. H. G.; LAGEWEG, B. J.; LENSTRA, J. K.
MINIMIZING TOTAL COSTS IN ONE-MACHINE SCHEDULING.
ROBB 1975 -1 ROBBINS, J. A.; SHAMBLIN, J. E.; TURNER, W. C.; BYRD, D. F.
DEVELOPMENT OF AND COMPUTATIONAL EXPERIENCE WITH A COMBINATION TOUR CONSTRUCTION - TOUR IMPROVEMENT ALGORITHM FOR VEHICLE ROUTING PROBLEMS.
ROED 1976 -1 ROEDDER, W.
LEXIKOGRAPHISCHER SUCHALGORITHMUS ZUR GANZZAHLIGEN PROGRAMMIERUNG: LEXS, EIN.
ROOD 1977 -1 ROODMAN, G. M.; SCHWARZ, L. B.
EXTENSIONS OF THE MULTI-PERIOD FACILITY PHASE-OUT MODEL: NEW PROCEDURES AND APPLICATION TO A PHASE-IN/PHASE-OUT PROBLEM.
ROOS 1975 -1 ROOSE, J. R.; RIEMERSMA, P.
HEURISTIC OPTIMISATION OF A LOCATION-ALLOCATION PROBLEM IN A TRANSPORT NETWORK WITH MODAL CHOICE.
ROSEC 1975 -1 ROSENTHAL, A.
BIT-PUSHING SHORTEST DISTANCE ALGORITHM, A.
ROSS 1977 -1 ROSS, G. T.; SOLAND, R. M.
MODELING FACILITY LOCATION PROBLEMS AS GENERALIZED ASSIGNMENT PROBLEMS.
ROSSB 1974 -1 ROSS, G. T. JR.
ALGORITHMIC AND COMPUTATIONAL RESULTS FOR CONSTRAINED TRANSPORTATION, NETWORK AND ASSIGNMENT MODELS.
RUSSB 1977 -1 RUSSELL, R.; IGO, W.
ASSIGNMENT ROUTING PROBLEM, AN.
RUSSC 1977 -1 RUSSELL, R. A.
EFFECTIVE HEURISTIC FOR THE M-TOUR TRAVELING SALESMAN PROBLEM WITH SOME SIDE CONDITIONS, AN.
RYAN 1976 -1 RYAN, D. M.; FOSTER, B. A.; CHESHIRE, I. M.
COMPLEXITY OF REAL-WORLD SCHEDULING PROBLEMS, THE.
SAHN 1975 -1 SAHNI, S.
APPROXIMATE ALGORITHMS FOR THE 0/1 KNAPSACK PROBLEM.
SAHN 1977 -1 SAHNI, S.
GENERAL TECHNIQUES FOR COMBINATORIAL APPROXIMATION.
SALE 1976 -1 SALERNO, G.
"BRANCH AND BACK-TRACK" ALGORITHM FOR THE DESIGN OF AN OPTIMAL TRANSPORTATION NETWORK (ITALIAN), A.
SALKA 1975 -1 SALKIN, H. M.; KLUYVER, C. A. DE
KNAPSACK PROBLEM: A SURVEY, THE.
SANDA 1975 -1 SANDI, C.
SOLUTION OF THE MACHINE LOADING PROBLEM WITH BINARY VARIABLES.
SCHMA 1976 -1 SCHMIDT, D. C.; DRUFFEL, L. E.
FAST BACKTRACKING ALGORITHM TO TEST DIRECTED GRAPHS FOR ISOMORPHISM USING DISTANCE MATRICES, A.
SCHOC 1976 -1 SCHOENLEIN, A.
BEZUEGLICH KERNSPEICHER- UND RECHENZEITBEDARF EFFIZIENTES BRANCH-UND-BOUND-VERFAHREN ZUR LOESUNG DES TRAVELLING-SALESMAN-PROBLEMS, EIN.
SCHOC 1977 -1 SCHOENLEIN, A.
UEBER DIE ANWENDUNG DER METHODE DES BRANCH AND BOUND AUF DAS SUMMEN-ZUORDNUNGSPROBLEM.
SCRI 1975 -1 SCRIABIN, M.; VERGIN, R. C.
COMPARISON OF COMPUTER ALGORITHMS AND VISUAL BASED METHODS FOR PLANT LAYOUT.
SEEL 1977 -1 SEELAENDER, J.
UEBER EINE METHODE ZUR KOPPLUNG DUALER SCHNITTEBENENVERFAHREN IN DER GANZZAHLIGEN LINEAREN OPTIMIERUNG.
SHAK 1977 -1 SHAKHBASJAN, K. V.; TUSHKINA, T. A.
ON STATISTICAL EFFICIENCY OF SOLUTION ALGORITHMS FOR A PROBLEM OF VERTEX DIVISION (RUSSIAN).
SHAM 1972 -1 SHAMMA, M. M.; WILLIAMS, T. A.
CONSTRUCTING SAMPLE NETWORKS FOR ANALYZING AND COMPARING SHORTEST ROUTE ALGORITHMS.
SHAPA 1966 -1 SHAPIRO, D. M.
ALGORITHMS FOR THE SOLUTION OF THE OPTIMAL COST AND BOTTLENECK TRAVELING SALESMAN PROBLEMS.
SHER 1977 -1 SHERALI, A. D.; SHETTY, C. M.
RECTILINEAR DISTANCE LOCATION-ALLOCATION PROBLEM, THE.
SHIR 1976 -1 SHIRAI, Y.; IBARAKI, T.; HASEGAWA, T.
MINIMAL REPRESENTATION OF COMBINATORIAL OPTIMIZATION PROBLEMS BY FINITE-STATE POSITIVELY MONOTONE MODELS.
SHIRA 1969 -1 SHIREY, R. W.
IMPLEMENTATION AND ANALYSIS OF EFFICIENT GRAPH PLANARITY TESTING ALGORITHMS.
SHKU 1965 -1 SHKURBA, V. V.
COMPUTATIONAL SCHEMES OF SOLUTION OF PROBLEMES IN SCHEDULING THEORY (RUSSIAN).
SMITH 1975 -1 SMITH, T. H. C.
LIFO IMPLICIT ENUMERATION ALGORITHM FOR THE ASYMMETRIC TRAVELING SALESMAN PROBLEM USING A ONE-ARBORESCENCE RELAXATION, A.
SMITI 1977 -1 SMITH, T. H. C.; SRINIVASAN, V.; THOMPSON, G. L.
COMPUTATIONAL PERFORMANCE OF THREE SUBTOUR ELIMINATION ALGORITHMS FOR SOLVING ASYMMETRIC TRAVELING SALESMAN PROBLEMS.
SMITJ 1977 -1 SMITH, T. H. C.; THOMPSON, G. L.
LIFO IMPLICIT ENUMERATION SEARCH ALGORITHM FOR THE SYMMETRIC TRAVELING SALESMAN PROBLEM USING HELD AND KARP'S 1-TREE RELAXATION, A.
SOMM 1972 -1 SOMMER, D. C.
COMPUTATIONAL EXPERIENCE WITH THE OPHELIE MIXED INTEGER CODE.
SPAE 1977 -1 SPAETH, H.
NUMERISCHE ERFAHRUNGEN ZU HEURISTISCHEN LOESUNGSVERFAHREN BEIM VARIANZKRITERIUM IN DER CLUSTER-ANALYSE.
SPIE 1972 -1 SPIELBERG, K.
MINIMAL PREFERRED VARIABLE METHODS IN ZERO-ONE PROGRAMMING.
SPIE 1973 -2 SPIELBERG, K.
PREFERRED VARIABLE BRANCH-AND-BOUND METHOD.
STEIA 1976 -1 STEINBERG, D. I.
ON SOLVING LARGE-SCALE FIXED CHARGE PROBLEMS.
STEIA 1977 -1 STEINBERG, D. I.
DESIGNING A HEURISTIC FOR THE FIXED-CHARGE TRANSPORTATION PROBLEM.
STERA 1977 -1 STERN, H. I.; PODRIGUEZ, E. P.; UTTER, M. L.
CYCLICAL JOB SEQUENCING ON MULTIPLE SETS OF IDENTICAL MACHINES.
STRAA 1976 -1 STRATMANN, K.-H.
NUMERISCHE UNTERSUCHUNGEN UEBER QUADRATISCHE ZUORDNUNGSPROBLEME.

****

STRAB 1976 -1 STRATMANN, K.-H.; BURKARD, R. E.
NUMERISCHE UNTERSUCHUNGEN UEBER QUADRATISCHE ZUORDNUNGSPROBLEME.

SUHL 1976 -1 SUHL, U.
ALGORITHMUS ZUR LOESUNG EINER SPEZIELLEN KLASSE GANZZAHLIGER OPTIMIERUNGSMODELLE, EIN.

SWEE 1977 -1 SWEENEY, D. J.; MURPHY, R. A.
DECOMPOSITION PRINCIPLE FOR INTEGER PROGRAMS, A (REVISED VERSION).

SWEEA 1976 -1 SWEENEY, D. J.; TATHAM, R. L.
IMPROVED LONG-RUN MODEL FOR MULTIPLE WAREHOUSE LOCATION, AN.

THES 1975 -1 THESEN, A.
RECURSIVE BRANCH AND BOUND ALGORITHM FOR THE MULTIDIMENSIONAL KNAPSACK PROBLEM, A.

TILQ 1975 -1 TILQUIN, C.
ON SCHEDULING WITH EARLIEST STARTS AND DUE DATES ON A GROUP OF IDENTICAL MACHINES.

TORR 1976 -1 TORRIERO, A.
DECOMPOSITION METHODS FOR THE TRAVELLING SALESMAN PROBLEM IN LARGE DIMENSIONS (ITALIAN).

TOTH 1977 -1 TOTH, P.
DYNAMIC PROGRAMMING ALGORITHM FOR THE ZERO-ONE KNAPSACK PROBLEM, A.

TRIV 1976 -1 TRIVEDI, V. M.; WARNER, D. M.
BRANCH AND BOUND ALGORITHM FOR OPTIMUM ALLOCATION OF FLOAT NURSES, A.

TRYP 1977 -1 TRYPIA, M.
NON-BINARY TREE SEARCH METHOD FOR THE 0-1 LINEAR PROGRAMMING PROBLEM, A.

VARS 1974 -1 VARSAVSKII, V. I.; EPIFANOV, G. V.; MININA, T. R.
USING EQUIVALENT TRANSFORMATIONS OF THE DISTANCE MATRIX TO SIMPLIFY THE SOLUTION OF THE TRAVELLING SALESMAN PROBLEM (RUSSIAN).

WALKA 1976 -1 WALKER, W. E.
HEURISTIC ADJACENT EXTREME POINT ALGORITHM FOR THE FIXED CHARGE PROBLEM, A.

WHIT 1977 -1 WHITAKER, R. A.
THREE ALGORITHMS FOR CALCULATING SOME OR ALL OF THE SHORTEST PATHS IN A SPARSE NETWORK.

WILLC 1972 -1 WILLIAMS, T. A.; WHITE, G. P.
COMPARING SHORTEST ROUTE ALGORITHMS FOR DIRECTED GRAPHS.

WILLE 1976 -1 WILLIS, R. J.; HASTINGS, N. A. J.
PROJECT SCHEDULING WITH RESOURCE CONSTRAINTS USING BRANCH AND BOUND METHODS.

YAGI 1975 -1 YAGIZ, OE.
HEURISTICALLY-AIDED SINGLE BRANCH SEARCH PROCEDURES FOR PLANT LOCATION PROBLEMS.

YEN 1971 -1 YEN, J. Y.
ON THE EFFICIENCIES OF ALGORITHMS FOR DETECTING NEGATIVE LOOPS IN NETWORKS.

ZALOA 1977 -1 ZALOOM, V. A.; SMYER, W. N.
SCHEDULING TO MINIMIZE MAKESPAN REQUIRED TO PROCESS INDEPENDENT JOBS ON PARALLEL NON-EQUIVALENT SERVERS.

ZION 1977 -1 ZIONTS, S.
INTEGER LINEAR PROGRAMMING WITH MULTIPLE OBJECTIVES.

AASH 1976 -1 AASHTIANI, H. A.; MAGNANTI, T. L.
IMPLEMENTING PRIMAL-DUAL NETWORK FLOW ALGORITHMS.
ADRA 1976 -1 ADRABINSKI, A.
ALGORITHM 45: A HEURISTIC ALGORITHM FOR THE TRAVELING-SALESMAN PROBLEM.
AHO B 1974 -1 AHO, A. V.; HOPCROFT, J. E.; ULLMAN, J. D.
DESIGN AND ANALYSIS OF COMPUTER ALGORITHMS, THE.
ALI 1977 -1 ALI, A. I.; HELGASON, R. V.; KENNINGTON, J. L.; LALL, H. S.
PRIMAL SIMPLEX NETWORK CODES: STATE-OF-THE-ART IMPLEMENTATION TECHNOLOGY (REVISED VERSION).
ANDRA 1976 -1 ANDRONICO, A.; RANDI, P.; SCALAS, M. R.
INTERACTIVE CODE FOR MIXED INTEGER LINEAR PROGRAMS, AN.
BAND 1975 -1 BANDLER, J. W.; CHEN, J. H. K.; DALSGAARD, P.; LIU, P. C.
TOLOPT - A PROGRAM FOR OPTIMAL, CONTINUOUS OR DISCRETE, DESIGN CENTERING AND TOLERANCING, PART 1 - USER'S GUIDE, PART 2 - FORTRAN LISTING.
BANDA 1975 -1 BANDLER, J. W.; LIU, P. C.; CHEN, J. H. K.
WORST CASE NETWORK TOLERANCE OPTIMIZATION.
BARRA 1975 -1 BARR, R. S.; ROSS, G. T.
LINKED LIST DATA STRUCTURE FOR A BINARY KNAPSACK ALGORITHM, A.
BAUE 1973 -1 BAUERSFELD, G.; ESSMANN, CH.; LOEHLE, H.
ALGORITHMUS ZUR FESTSTELLUNG DER ISOMORPHIE VON ENDLICHEN, ZUSAMMENHAENGENDEN GRAPHEN.
BECK 1976 -1 BECK, L. L.; MATULA, D. W.
CLIQUE DETERMINATION IN LARGE DENSE GRAPHS.
BENI 1977 -1 BENICHOU, M.; GAUTHIER, J. M.; HENTGES, G.; RIBIERE, G.
EFFICIENT SOLUTION OF LARGE-SCALE LINEAR PROGRAMMING PROBLEMS - SOME ALGORITHMIC TECHNIQUES AND COMPUTATIONAL RESULTS, THE.
BIRDA 1976 -1 BIRD, R.
PROGRAMS AND MACHINES. AN INTRODUCTION TO THE THEORY OF COMPUTING.
BOOTA 1976 -1 BOOTH, K. S.; LUEKER, G. S.
TESTING FOR THE CONSECUTIVE ONES PROPERTY, INTERVAL GRAPHS, AND GRAPH PLANARITY USING PQ-TREE ALGORITHMS.
BOTT 1974 -1 BOTT, H.
ANWENDUNGSMOEGLICHKEITEN DER PSEUDO-BOOLSCHEN OPTIMIERUNG AUF PROBLEME DER FERTIGUNSWIRTSCHAFT UND IHRE RECHENTECHNISCHE REALISIERUNG.
BOYCA 1975 -1 BOYCE, W. M.
IMPROVED PROGRAM FOR THE FULL STEINER TREE PROBLEM, AN.
BOYCB 1973 -1 BOYCE, W. M.; SEERY, J. B.
STEINER 72, AN IMPROVED VERSION OF COCKAYNE AND SCHILLER'S PROGRAM STEINER FOR THE MINIMAL NETWORK PROBLEM.
BRAD 1976 -1 BRADLEY, G.; BROWN, G.; GRAVES, G.
DESIGN AND IMPLEMENTATION OF LARGE SCALE PRIMAL TRANSSHIPMENT ALGORITHMS.
BRAN 1971 -1 BRANDMAN, E. I.
CERTAIN METHODS FOR THE SOLUTION OF PROBLEMS OF PROJECTED ALLOCATION OF CAPITAL INVESTMENTS WITHIN AN INDUSTRY (RUSSIAN).
BREI 1976 -1 BREITENBACH, R. B.
COMPUTER ALGORITHM FOR REFINING GRID LOCATIONAL ANALYSIS, A.
CHAR 1975 -1 CHARNES, A.; GLOVER, F.; KLINGMAN, D.
PAST, PRESENT AND FUTURE OF LARGE SCALE TRANSSHIPMENT COMPUTER CODES AND APPLICATIONS.
CHEND 1975 -1 CHEN, J. H. K.
DISOPT - A GENERAL PROGRAM FOR CONTINUOUS AND DISCRETE NONLINEAR PROGRAMMING PROBLEMS.
CHEND 1974 -1 CHEN, J. H. K.
DISOPT - A GENERAL PROGRAM FOR CONTINUOUS AND DISCRETE NONLINEAR PROGRAMMING PROBLEMS.
COCKA 1976 -1 COCKAYNE, E.; HEDETNIEMI, S.
LINEAR ALGORITHM FOR THE MAXIMUM WEIGHT OF AN INDEPENDENT SET IN A TREE, A.
CORN 1972 -1 CORNEIL, D. G.
ALGORITHM FOR DETERMINING THE AUTOMORPHISM PARTITIONING OF AN UNDIRECTED GRAPH, AN.
COT 1974 -1 COT, N.
LINEAR-TIME ORDERING PROCEDURE WITH APPLICATIONS TO VARIABLE-LENGTH ENCODING, A.
DEMB 1976 -1 DEMBO, R.; MULVEY, J.
ON THE ANALYSIS AND COMPARISON OF MATHEMATICAL PROGRAMMINNG ALGORITHMS AND SOFTWARE.
DERI 1978 -1 DERIGS, U.
ALGEBRAISCHE MATCHINGPROBLEME.
DERIA 1978 -1 DERIGS, U.; ZIMMERMANN, U.
AUGMENTING PATH METHOD FOR SOLVING LINEAR BOTTLENECK ASSIGNMENT PROBLEMS, AN.
DIALA 1977 -1 DIAL, R.; GLOVER, F.; KARNEY, D.; KLINGMAN, D.
COMPUTATIONAL ANALYSIS OF ALTERNATIVE ALGORITHMS AND LABELING TECHNIQUES FOR FINDING SHORTEST PATH TREES, A.
DIJK 1976 -1 DIJKSTRA, E. W.
DISCIPLINE OF PROGRAMMING, A.
EDEN 1977 -1 EDEN, B.; PAPE, U.
PLANAR. EIN ALGORITHMUS ZUR FESTSTELLUNG DER PLANARITAET VON GRAPHEN UND ZUR ERMITTLUNG PLANARER TEILGRAPHEN.
EDENA 1976 -1 EDEN, B. N.
HEURISTISCHES VERFAHREN ZUR DISTRIKTERMITTLUNG IN BEWERTETEN GRAPHEN UND SEINE ANWENDUNG AUF DIE BERECHNUNG VON SCHULEINZUGSBEREICHEN, EIN.
EDMOB 1969 -1 EDMONDS, J.; JOHNSON F. L.; LOCKHART, S.
BLOSSOM I, A CODE FOR MATCHING.
ERLE 1976 -1 ERLENKOTTER, D.
DUAL-BASED PROCEDURE FOR UNCAPACITATED FACILITY LOCATION, A.
EVST 1969 -1 EVSTIGNEEV, V. A.; JASAKOVA, V. P.
CERTAIN ALGORITHM ON GRAPHS (RUSSIAN), A.
FILL 1977 -1 FILL, J. A.; REINGOLD, E. M.
SOLUTIONS MANUAL. COMBINATORIAL ALGORITHMS: THEORY AND PRACTICE (BY E.M. REINGOLD, J. NIEVERGELT, N. DEO).
FISK 1977 -1 FISK, J.
SOLUTION PROCEDURE FOR A SPECIAL TYPE OF CAPACITATED WAREHOUSE LOCATION PROBLEM, A.
FOOT 1965 -1 FOOT, D. H. S.
SHORTEST ROUTE PROBLEM: ALGOL-PROGRAMS, AND DISCUSSION OF COMPUTATIONAL PROBLEMS IN LARGE TRANSPORT NETWORK APPLICATIONS, THE.
FOX A 1976 -1 FOX, B. L.
DATA STRUCTURES AND COMPUTER SCIENCE TECHNIQUES IN OPERATIONS RESEARCH.
GABOB 1976 -2 GABOW, H. N.
USING EULER PARTITIONS TO EDGE COLOR BIPARTITE MULTIGRAPHS.
GERR 1976 -1 GERRITSEN, R.
EFFICIENT MANIPULATION OF BINARY ARRAYS.

****

GHIL 1973 -1 GHILARDOTTI, P.
BILP CODE: BOUNDED INTEGER LINEAR PROGRAMMING CODE USER'S INSTRUCTIONS WITH SOME NOTES AND FIRST COMPUTATIONAL RESULTS.

GIBB 1976 -1 GIBBONS, P.
COMPUTING TECHNIQUES FOR THE CONSTRUCTION AND ANALYSIS OF BLOCK DESIGN.

GLOV 1975 -1 GLOVER, F.
COMPACT LP BASES FOR A CLASS OF IP PROBLEMS.

GLOVF 1973 -1 GLOVER, F.; KLINGMAN, D.; STUTZ, J.
IMPLEMENTATION AND COMPUTATIONAL STUDY OF A GENERALIZED NETWORK CODE.

GRAT 1970 -1 GRATZER, F.
COMPUTER SOLUTION OF LARGE MULTICOMMODITY FLOW PROBLEMS.

GUIGA 1976 -1 GUIGNARD, M.; SPIELBERG, K.
EXPERIMENTAL INTERACTIVE SYSTEM FOR INTEGER PROGRAMMING, AN.

GUIGA 1976 -2 GUIGNARD, M.; SPIELBERG, K.
INTEGER PROGRAMMING IN APL.

HADJ 1974 -1 HADJERES, S.
QELQUES METHODES RECENTES DE RESOLUTION DU PROBLEME DU VOYAGEUR DE COMMERCE.

HANS 1971 -1 HANSEN, K. H.
SEQUENCING PROBLEMS WITH EMPHASIS ON A MODIFICATION OF HELD AND KARP'S ALGORITHM. OR: AN ALGORITHM FOR THE OPTIMAL SOLUTION OF THE SYMMETRIC TRAVELING SALESMAN PROBLEM. PART 2 (DANISH).

HANS 1971 -2 HANSEN, K. H.
SEQUENCING PROBLEMS WITH EMPHASIS ON A MODIFICATION OF HELD AND KARP'S ALGORITHM. OR: AN ALGORITHM FOR THE OPTIMAL SOLUTION OF THE SYMMETRIC TRAVELING SALESMAN PROBLEM. PART 1. (DANISH).

HARR 1976 -1 HARRIS, B.
CODE FOR THE TRANSPORTATION PROBLEM OF LINEAR PROGRAMMING, A.

HATC 1975 -1 HATCH, R. S.
BENCH MARKS COMPARING TRANSPORTATION CODES BASED ON PRIMAL SIMPLEX AND PRIMAL-DUAL ALGORITHMS.

HAUS 1978 -1 HAUSMANN, D.
ADJAZENZCHARAKTERISIERUNGEN FUER POLYEDERECKEN IN DER KOMBINATORISCHEN OPTIMIERUNG.

HAUS 1977 -3 HAUSMANN, D.
POLYNOMIALER ALGORITHMUS FUER DIE ADJAZENZUEBERPRUEFUNG AUF DEM BRANCHING-POLYEDER, EIN.

HELG 1977 -2 HELGASON, R. V.; KENNINGTON, J. L.
PRODUCT FORM REPRESENTATION OF THE INVERSE OF A MULTICOMMODITY CYCLE MATRIX, A.

HESS 1977 -1 HESSELBACH, J.; WIJNANDS, J.
DYNAMISCHE PROGRAMMIERUNG - METHODIK, EDV-PROGRAMM, ANWENDUNGSBEISPIEL.

HILLA 1978 -1 HILL, E. JR.
COMPARATIVE STUDY OF VERY LARGE DATA BASES, A.

HOPCB 1973 -1 HOPCROFT, J. E.; TARJAN, R. E.
DIVIDING A GRAPH INTO TRICONNECTED COMPONENTS.

JAME 1976 -1 JAMES, K. R.; RIHA, W.
ALGORITHM 28. ALGORITHM FOR GENERATING GRAPHS OF A GIVEN PARTITION.

JOHN 1976 -1 JOHNSON, C. R.; LEIGHTON, F. T.
EFFICIENT LINEAR ALGEBRAIC ALGORITHM FOR THE DETERMINATION OF ISOMORPHISM IN PAIRS OF UNDIRECTED GRAPHS, AN.

JOHNA 1975 -1 JOHNSON, D. B.
FINDING ALL THE ELEMENTARY CIRCUITS OF A DIRECTED GRAPH.

KARI 1976 -1 KARIV, O.
O(N POWER 2.5) ALGORITHM FOR FINDING A MAXIMUM MATCHING IN A GENERAL GRAPH, AN.

KARN 1976 -1 KARNEY, D.; KLINGMAN, D.
IMPLEMENTATION AND COMPUTATIONAL STUDY ON AN IN-CORE, OUT-OF-CORE PRIMAL NETWORK CODE.

KILG 1977 -1 KILGORE, W. C.
WAREHOUSE FACILITIES ANALYSIS.

KINS 1975 -1 KINSNER, W.
ALGORITHM AND A PROGRAM FOR NETWORK ROUTING PROBLEMS, AN.

KOMA 1975 -1 KOMARNICKI, J.
AUTOMATISCHE GENERIERUNG ANWENDUNGSSPEZIFISCHER NETZPLAN-TECHNIK-PROGRAMMSYSTEME.

KUEH 1977 -1 KUEHNER, J.; HARRINGTON, J. J.
WIRKSAME ANWENDUNG DES ALGORITHMUS FUER DIE GEMISCHT-GANZZAHLIGE PROGRAMMIERUNG MPSX-MIP VON IBM ZUR LOESUNG VON GROSSEN OEFFENTLICHEN INVESTITIONSPROBLEMEN MIT VIELEN BETEILIGTEN, TEIL II, DIE.

KUEH 1977 -2 KUEHNER, J.; HARRINGTON, J. J.
WIRKSAME ANWENDUNG DES ALGORITHMUS FUER DIE GEMISCHT-GANZZAHLIGE PROGRAMMIERUNG MPSX-MIP VON IBM ZUR LOESUNG VON GROSSEN OEFFENTLICHEN INVESTITIONSPROBLEMEN MIT VIELEN BETEILIGTEN, TEIL 1, DIE.

LAND 1977 -1 LAND, A.; POWELL, S.
COMPUTER CODES FOR PROBLEMS OF INTEGER PROGRAMMING.

LAURA 1976 -1 LAURIERE, J. L.
LANGAGE ET UN PROGRAMME POUR ENONCER ET RESOUDRE DES PROBLEMES COMBINATOIRES, UN.

LAWL 1975 -8 LAWLER, E. L.
SOME ASPECTS OF DUALITY IN COMBINATORIAL OPTIMIZATION.

LIN B 1974 -1 LIN, P. M.; LEON, B. J.
IMPROVING THE EFFICIENCY OF LABELING ALGORITHMS FOR MAXIMUM FLOWS IN NETWORKS.

LOHS 1976 -1 LOHSE, D.
ALGORITHMUS 31. EIN ALGORITHMUS ZUR BESTIMMUNG VON WEGANZAHLEN IN NETZEN.

MAHR 1976 -1 MAHR, B.; PERL, J.
GRUENDE, TECHNIKEN UND KONSEQUENZEN DER VERALLGEMEINERUNG VON WEGEALGORITHMEN.

MART 1976 -1 MARTELLI, A.
GAUSSIAN ELIMINATION ALGORITHM FOR THE ENUMERATION OF CUT SETS IN A GRAPH, A.

MEEU 1975 -1 MEEUSEN, W.; CUYVERS, L.
CLIQUE DETECTION IN DIRECTED GRAPHS: A NEW ALGORITHM.

MEVE 1977 -1 MEVERT, P.; SUHL, U.
IMPLICIT ENUMERATION WITH GENERALIZED UPPER BOUNDS.

MIKH 1975 -1 MIKHAJLOVA, E. D.; BALITSKAYA, E. E.
DETERMINATION OF LIMITED DEGREES OF DIRECTED GRAPHS (RUSSIAN), THE.

MITC 1975 -1 MITCHELL, S.; HEDETNIEMI, S.; GOODMAN, S.
SOME LINEAR ALGORITHMS ON TREES.

MOORB 1976 -1 MOORE, T. G.
SOLVING LARGE 0-1 PROBLEMS.

MULVA 1975 -2 MULVEY, J. M.
SPECIAL STRUCTURES IN NETWORK NODELS AND ASSOCIATED APPLICATIONS.

MULVA 1975 -3 MULVEY, J. M.
TESTING OF LARGE-SCALE NETWORK OPTIMIZATION PROGRAM.

****

MUNC 1977 -1 MUNCH-ANDERSEN, B.; ZAHLE, T. U.
SCHEDULING ACCORDING TO JOB PRIORITY WITH PREVENTION OF DEADLOCK AND PERMANENT BLOCKING.

MURCB 1966 -1 MURCHLAND, J. D.
ALGOL PROCEDURE FOR ALL SHORTEST PATHS IN A SYMMETRIC NETWORK, AN.

NIJE 1975 -1 NIJENHUIS, A.; WILF, H.
COMBINATORIAL ALGORITHMS.

OSYC 1976 -1 OSYCZKA, A.
ALGORITHM OF OPTIMIZATION FOR A SPECIAL CLASS OF NETWORKS, AN.

PAPE 1976 -1 PAPE, U.
DATENSTRUKTUREN FUER MENGEN IN ALGORITHMEN AUF GRAPHEN.

PAPE 1976 -2 PAPE, U.
GRAPHEN-SPRACHEN UND ALGORITHMEN AUF GRAPHEN. 1. FACHTAGUNG UEBER GRAPHENTHEORETISCHE KONZEPTE DER INFORMATIK, BERLIN, JULI 1975.

PARR 1968 -1 PARRIS, R.
CODING PROBLEM FOR GRAPHS, THE.

PARRA 1968 -1 PARRIS, R.; READ, R. C.
CODING PROCEDURE FOR GRAPHS, A.

PATR 1974 -1 PATRINOS, A. N.; HAKIMI, S. L.
ORDERING PROBLEMS IN TREES AND THEIR APPLICATIONS TO TREE-STRUCTURED OPERATIONS OR FILES.

PLUK 1976 -1 PLUKENE, D. B.
APPLICATION OF SEMI-BOOLEAN METHODS TO THE SOLUTION OF 0-1 INTEGER LINEAR PROGRAMMING PROBLEMS (RUSSIAN).

PROS 1974 -1 PROSKUROWSKI, A.
SEARCH FOR A UNIQUE INCIDENCE MATRIX OF A GRAPH.

REIN 1977 -1 REINGOLD, E. M.; NIEVERGELT, J.; DEO, N.
COMBINATORIAL ALGORITHMS: THEORY AND PRACTICE.

RIHA 1976 -1 RIHA, W.; JAMES, K. R.
ALGORITHM 29. EFFICIENT ALGORITHMS FOR DOUBLY AND MULTIPLY RESTRICTED PARTITIONS.

RUBIB 1975 -1 RUBIN, F.
IMPROVED ALGORITHM FOR TESTING THE PLANARITY OF A GRAPH, AN.

SCIO 1964 -1 SCIONS, H. I.
COMPACT REPRESENTATION OF A ROOTED TREE AND THE TRANSPORTATION PROBLEM, THE.

SHIRA 1969 -1 SHIREY, R. W.
IMPLEMENTATION AND ANALYSIS OF EFFICIENT GRAPH PLANARITY TESTING ALGORITHMS.

SLAT 1976 -1 SLATE, L.; SPIELBERG, K.
EXTENDED CONTROL LANGUAGE OF MPSX/370 (AND POSSIBLE APPLICATIONS), THE.

SLATA 1976 -2 SLATER, P. J.
R-DOMINATION IN GRAPHS.

SMITD 1975 -1 SMITH, L. D.
SCHEDULING NURSING PERSONNEL: APPLICATION OF LIST PROCESSING AND PROBLEM ORIENTED DATA STRUCTURES.

SMITE 1977 -1 SMITH, L. D.; WIGGINS, A.
COMPUTER-BASED NURSE SCHEDULING SYSTEM, A.

SOMM 1972 -1 SOMMER, D. C.
COMPUTATIONAL EXPERIENCE WITH THE OPHELIE MIXED INTEGER CODE.

SPAE 1977 -1 SPAETH, H.
NUMERISCHE ERFAHRUNGEN ZU HEURISTISCHEN LOESUNGSVERFAHREN BEIM VARIANZKRITERIUM IN DER CLUSTER-ANALYSE.

SPIE 1977 -1 SPIELBERG, K.
ENUMERATIVE METHODS IN INTEGER PROGRAMMING.

STOCA 1968 -1 STOCKTON, F.
LINEARIZATION AND STANDARDIZATION OF GRAPHS.

SWAN 1974 -1 SWANEK, J. J.
LINEAR INTEGER PROGRAMMING.

TARJ 1973 -1 TARJAN, R. E.
ENUMERATION OF THE ELEMENTARY CIRCUITS OF A DIRECTED GRAPH.

TARJ 1973 -2 TARJAN, R. E.
FINDING DOMINATORS IN DIRECTED GRAPHS.

TROT 1973 -1 TROTTER, L. E. JR.
USER'S INSTRUCTIONS FOR THE INTEGER PROGRAMMING CODE ENUMER8.

WAGN 1976 -1 WAGNER, R. A.
SHORTEST PATH ALGORITHM FOR EDGE-SPARSE GRAPHS, A.

WALD 1976 -1 WALDVOGEL, J.
DATENSTRUKTUREN.

WALL 1977 -1 WALLACH, Y.
SCHEDULING ALGORITHMS FOR CONCURRENT EXECUTION.

WEIG 1976 -1 WEIGAND, M. M.
ALGORITHMUS 27. EIN NEUER ALGORITHMUS ZUR BESTIMMUNG VON K-KUERZESTEN WEGEN IN EINEM GRAPHEN.

COVERING
****

AKOP 1974 -1 AKOPJAN, G. C.
COVERS AND JOINS OF THE VERTICES OF A GRAPH BY EDGES (RUSSIAN).

ALAV 1977 -1 ALAVI, Y.; BEHZAD, M.; LESNIAK-FOSTER, L. M.; NORDHAUS, E. A.
TOTAL MATCHINGS AND TOTAL COVERINGS OF GRAPHS.

ANEJ 1977 -1 ANEJA, Y. P.; VEMUGANTI, R. R.
ROW GENERATION SCHEME FOR FINDING A MULTI-COMMODITY MINIMUM DISCONNECTING SET, A.

ARORA 1977 -1 ARORA, S. R.; PURI, M. C.
ENUMERATION TECHNIQUE FOR THE SET COVERING PROBLEM WITH A LINEAR FRACTIONAL FUNCTIONAL AS ITS OBJECTIVE FUNCTION.

ARORB 1976 -1 ARORA, S. R.; SWARUP, K.; PURI, M. C.
ENUMERATIVE TECHNIQUE FOR THE SET COVERING PROBLEM, AN.

ARORB 1977 -1 ARORA, S. R.; SWARUP, K.; PURI, M. C.
CUTTING PLANE TECHNIQUE FOR SET COVERING PROBLEM WITH LINEAR FRACTIONAL FUNCTIONAL.

BALAG 1977 -2 BALAS, E.
SET COVERING WITH CUTTING PLANES FROM CONDITIONAL BOUNDS (REVISED VERSION).

BALAJ 1976 -1 BALAS, E.; PADBERG, M. W.
SET PARTITIONING: A SURVEY.

BALAJ 1975 -1 BALAS, E.; PADBERG, M. W.
SET PARTITIONING: A SURVEY.

BALAK 1977 -1 BALAS, E.; SAMUELSSON, H.
NODE COVERING ALGORITHM, A.

BAZA 1975 -1 BAZARAA, M. J.
COMPUTERIZED LAYOUT DESIGN: A BRANCH AND BOUND APPROACH.

BEANA 1970 -1 BEAN, D. W. T.
MINIMAL EDGE SETS MEETING ALL CIRCUITS OF A GRAPH.

BENIA 1976 -1 BENITO, F.; GROEFLIN, H.
OPTIMIERUNGSPROBLEME MIT NICHT POLYNOMIAL BEGRENZTEN ALGORITHMEN (NP-PROBLEME).

BILDA 1977 -1 BILDE, O.; KRARUP, J.
SHARP LOWER BOUNDS AND EFFICIENT ALGORITHMS FOR THE SIMPLE PLANT LOCATION PROBLEM.

BOESB 1974 -1 BOESCH, F. T.; CHEN, S.; MAC HUGH, J. A. M.
ON COVERING THE POINTS OF A GRAPH WITH POINT DISJOINT PATHS.

BOESC 1977 -1 BOESCH, F. T.; GIMPEL, J. F.
COVERING THE POINTS OF A DIGRAPH WITH POINT-DISJOINT PATHS AND ITS APPLICATION TO CODE OPTIMIZATION.

BOND 1978 -1 BONDY, J. A.
EXTREMAL GRAPHS WITH PRESCRIBED COVERING NUMBER.

BRAC 1975 -1 BRACE, A.; DAYKIN, D. E.
FINITE SET COVERING THEOREM, IV, A.

CHERD 1976 -1 CHERNYSHEV, YU. YU.; NASEKIN, V. YA.
ON SOLUTION OF THE PROBLEM OF COVERING BY GRADIENT ALGORITHM (RUSSIAN).

CHURB 1976 -1 CHURCH, R. L.; REVELLE, C. S.
THEORETICAL AND COMPUTATIONAL LINKS BETWEEN THE P-MEDIAN, LOCATION SET-COVERING, AND THE MAXIMAL COVERING LOCATION PROBLEM.

COCKB 1975 -1 COCKAYNE, E. J.; HEDETNIEMI, S. T.
OPTIMAL DOMINATION IN GRAPHS.

DOER 1978 -1 DOERFLER, W.
MULTIPLE COVERS OF HYPERGRAPHS.

DOYL 1971 -1 DOYLE, M. S.
MINIMUM PATH COVERING ALGORITHMS.

EDMOA 1977 -1 EDMONDS, J.; GILES, R.
MIN-MAX RELATION FOR SUBMODULAR FUNCTIONS ON GRAPHS, A.

ERDOC 1977 -1 ERDOES, P.; MEIR, A.
ON TOTAL MATCHING NUMBERS AND TOTAL COVERING NUMBERS OF COMPLEMENTARY GRAPHS.

ETCH 1977 -1 ETCHEBERRY, J.
SET-COVERING PROBLEM: A NEW IMPLICIT ENUMERATION ALGORITHM, THE.

ETCH 1976 -2 ETCHEBERRY, J.
SET REPRESENTATION PROBLEM: A NEW IMPLICIT ENUMERATION ALGORITHM, THE (REVISED VERSION).

FREH 1975 -2 FREHEL, J.
REGULARISATION DANS LES PROBLEMES COMBINATOIRES, PROBLEMES DE TOURNEES, DE PARTITION ET DE RECOUVREMENT.

FRIE 1976 -1 FRIEDMAN, M.
ON THE ANALYSIS AND SOLUTION OF CERTAIN GEOGRAPHICAL OPTIMAL COVERING PROBLEMS.

GILLC 1975 -1 GILLES, F. R.
SUBMODULAR FUNCTIONS, GRAPHS AND INTEGER POLYHEDRA.

GOND 1976 -2 GONDRAN, M.
PROBLEMES DE PARTITION ET DE RECOUVREMENT. APPLICATIONS ET ALGORITHMES, LES.

GUPTA 1969 -1 GUPTA, R. P.
INDEPENDENCE AND COVERING NUMBERS OF LINE GRAPHS AND TOTAL GRAPHS.

HALP 1976 -1 HALPERN, J.
SEQUENTIAL COVERING PROBLEM UNDER UNCERTAINTY, THE (REVISED).

HALP 1977 -2 HALPERN, J.
SEQUENTIAL COVERING PROBLEM UNDER UNCERTAINTY, THE.

HAUS 1978 -1 HAUSMANN, D.
ADJAZENZCHARAKTERISIERUNGEN FUER POLYEDERECKEN IN DER KOMBINATORISCHEN OPTIMIERUNG.

HOUC 1977 -1 HOUCK, D. J.; VEMUGANTI, R. R.
ALGORITHM FOR THE VERTEX PACKING PROBLEM, AN.

IVANC 1974 -1 IVANOV, N. N.
REDUCTION OF A CLASS OF INTEGER LINEAR PROGRAMMING PROBLEMS TO A COVERING PROBLEM (RUSSIAN).

JOHNE 1964 -1 JOHNSON, E. L.
NETWORK FLOWS, GRAPHS, AND INTEGER PROGRAMMING.

KEL'A 1976 -1 KEL'MANS, A. K.; LOMONOSOV, M. V.; POLESSKIJ, V. P.
MINIMAL COVERINGS IN MATROIDS (RUSSIAN).

KOCH 1976 -1 KOCH, E.; PERLES, M. A.
COVERING EFFICIENCY OF TREES AND K-TREES.

KOU A 1978 -1 KOU, L. T.; STOCKMEYER, L. J.; WONG, C. K.
COVERING EDGES BY CLIQUES WITH REGARD TO KEYWORD CONFLICTS AND INTERSECTION GRAPHS.

KRARA 1977 -1 KRARUP, J.; BILDE, O.
PLANT LOCATION, SET COVERING AND ECONOMIC LOT SIZE: AN O(MN)-ALGORITHM FOR STRUCTURED PROBLEMS.

KUSJ 1975 -1 KUSJURIN, M. N.
UEBER MINIMALE UEBERDECKUNGEN AUF EINER N-ELEMENTIGEN MENGE.

KUZJ 1977 -1 KUZJURIN, N. N.
MINIMAL COVERINGS AND MAXIMAL PACKINGS OF (K-1)-SUBSETS BY K-SUBSETS (RUSSIAN).

LAMBA 1975 -1 LAMBERT, G.
INDICE DE C-PARTITION DES SOMMETS D'UN GRAPHE. ALGORITHMES DE RECOUVREMENT ET APPLICATION AUX FICHIERS.

LAWR 1978 -1 LAWRENCE, L.
COVERING THE VERTEX SET OF A GRAPH WITH SUBGRAPHS OF SMALLER DEGREE.

LOVA 1975 -1 LOVASZ, L.
ON THE RATIO OF OPTIMAL INTEGRAL AND FRACTIONAL COVERS.

LOVA 1976 -1 LOVASZ, L.
COVERS, PACKINGS AND SOME HEURISTIC ALGORITHMS.

LOVA 1975 -2 LOVASZ, L.
2-MATCHINGS AND 2-COVERS OF HYPERGRAPHS.

MATU 1975 -1 MATULA, D. W.
UNIFORM SET COVERING LEMMA, A.

MEIR 1977 -1 MEIR, A.; MOON, J. W.
PACKING AND COVERING CONSTANTS FOR CERTAIN FAMILIES OF TREES, I.

MEIR 1975 -2 MEIR, A.; MOON, J. W.
RELATIONS BETWEEN PACKING AND COVERING NUMBERS OF A TREE.

MENG 1974 -1 MENG, D.
MATCHINGS AND COVERINGS FOR GRAPHS.

MERCA 1976 -1 MERCHANT, D. K.; RAO, M. R.
MAJORITY DECISIONS AND TRANSITIVITY: SOME SPECIAL CASES.

MILLE 1973 -1 MILLS. W. H.
COVERING PROBLEMS.

MITC 1975 -1 MITCHELL, S.; HEDETNIEMI, S.; GOODMAN, S.
SOME LINEAR ALGORITHMS ON TREES.

MULV 1978 -1 MULVEY, J.
NETWORK RELAXATIONS FOR SET COVERING, SET PARTITIONING AND OTHER INTEGER PROGRAMMING PROBLEMS.

MULVA 1975 -1 MULVEY, J. M.
NETWORK RELAXATION APPROACH FOR THE SET PARTITIONING AND SET COVERING MODELS, A.

NICOB 1976 -1 NICOLETTI, B.; NATALI, L.; SCALAS, M. R.
MANPOWER STAFFING AT AN AIRPORT: A GENERALIZED SET COVERING PROBLEM.

NIEM 1974 -1 NIEMINEN, J.
LINEAR PSEUDO-BOOLEAN VIEWPOINT ON MATCHING AND OTHER CENTRAL CONCEPTS IN GRAPH THEORY, A.

NIEM 1975 -2 NIEMINEN, J.
SOME OBSERVATIONS ON COVERINGS OF GRAPHS.

NOOR 1975 -1 NOORVASH, S.
COVERING THE VERTICES OF A GRAPH BY VERTEX-DISJOINT PATHS.

OLAR 1973 -1 OLARU, E.
UEBER PERFEKTE UND KRITISCH IMPERFEKTE GRAPHEN.

PADB 1971 -1 PADBERG, M.
SIMPLE ZERO-ONE PROBLEMS: SET COVERING, MATCHING AND COVERING IN GRAPHS.

PADBA 1977 -1 PADBERG, M. W.
COVERING, PACKING AND KNAPSACK PROBLEMS.

PART 1976 -1 PARTHASARATHY, K. R.; CHOUDUM, S. A.
EDGE CLIQUE COVER NUMBER OF PRODUCTS OF GRAPHS, THE.

PAYA 1977 -1 PAYAN, C.
SUR QUELQUES PROBLEMES DE COUVERTURE ET DE COUPLAGE EN COMBINATOIRE.

PERL 1976 -1 PERL, J.
ZUR ALGORITHMISCHEN BEHANDLUNG SPEZIELLER NP-VOLLSTAENDIGER PROBLEME AUF GRAPHEN.

POLE 1976 -1 POLESSKIJ, V. P.
COVERING A FINITE GRAPH WITH A MINIMAL NUMBER OF FORESTS (RUSSIAN).

PULL 1977 -2 PULLEYBLANK, W. R.
TWO-BICRITICAL GRAPHS AND NODE COVERS.

RAJAA 1977 -1 RAJAT, K. D.
ON GENERATION OF TEST DATA AND MINIMAL COVER OF DIRECTED GRAPHS.

RARDA 1976 -1 RARDIN, R. L.; UNGER, V. E.
SOLVING FIXED CHARGE NETWORK PROBLEMS WITH GROUP THEORY-BASED PENALTIES.

RIVI 1975 -1 RIVIERE, M.
COVERING OF VERTICES OF A UNIFORM HYPERGRAPH BY EDGES AND STABLE SETS.

SAAT 1977 -1 SAATCIOGLU, O.
MATHEMATICAL PROGRAMMING MODELS FOR AIRPORT SITE SELECTION.

SAATA 1975 -1 SAATY, T. L.; ALEXANDER, J. M.
OPTIMIZATION AND THE GEOMETRY OF NUMBERS: PACKING AND COVERING.

SAHNA 1974 -1 SAHNI, S.; GONZALES, T.
P-COMPLETE PROBLEMS AND APPROXIMATE SOLUTIONS.

SAHNB 1976 -1 SAHNI, S.; GONZALEZ, T.
P-COMPLETE APPROXIMATION PROBLEMS.

SALK 1975 -1 SALKIN, H. M.
INTEGER PROGRAMMING.

SCHOD 1975 -1 SCHONHEIM, J.; BIALOSTOCKI, A.
PACKING AND COVERING OF THE COMPLETE GRAPH WITH 4-CYCLES.

SEDL 1975 -1 SEDLACEK, J.
PERFECT AND QUASIPERFECT GRAPHS (CZECH).

SERD 1975 -1 SERDYUKOV, A. I.
ON A COVERING PROBLEM (RUSSIAN).

TIND 1977 -2 TIND, J.
ON ANTIBLOCKING SETS AND POLYHEDRA.

TOME 1975 -2 TOMESCU, I.
INTRODUCTION TO COMBINATORICS (TRANSLATED FROM RUMANIAN).

TURC 1974 -1 TURCAT, C.
ISOMORPHISME, IMMERSION ET RECOUVREMENT DE GRAPHES.

WANGB 1974 -1 WANG, J. C.; MICKLE, M. H.; HOELZEMAN, R. G.
SET COVERING APPLICATION IN MINIMUM ROUTE GENERATION, A.

WEINB 1976 -1 WEINBERGER, D. B.
NETWORK FLOWS, MINIMUM COVERINGS, AND THE FOUR-COLOR CONJECTURE.

WERR 1974 -1 WERRA, D. DE
PARTITIONS OF GRAPHS INTO COVERINGS AND HYPERGRAPHS INTO TRANSVERSALS.

WHITD 1967 -2 WHITE, L. J.
PARAMETRIC STUDY OF MATCHINGS AND COVERINGS IN WEIGHTED GRAPHS, A.

ZAHN 1973 -1 ZAHN, C. T.
ALTERNATING EULER PATHS FOR PACKINGS AND COVERS.

CUTTING PLANES

****

ARNO 1974 -1 ARNOLD, L. R.; BELLMORE, M.
GENERATED CUT FOR PRIMAL INTEGER PROGRAMMING, A.
ARNO 1974 -2 ARNOLD, L. R.; BELLMORE, M.
ITERATION SKIPPING IN PRIMAL INTEGER PROGAMMING.
ARORB 1977 -1 ARORA, S. R.; SWARUP, K.; PURI, M. C.
CUTTING PLANE TECHNIQUE FOR SET COVERING PROBLEM WITH LINEAR FRACTIONAL FUNCTIONAL.
AVRI 1977 -1 AVRIEL, M.; GUROVICH, V.
CONDENSATION ALGORITHM FOR A CLASS OF ALGEBRAIC PROGRAMS, A.
BALAG 1973 -1 BALAS, E.
ON THE USE OF INTERSECTION CUTS AND OUTER POLARS IN BRANCH-AND-BOUND.
BALAG 1977 -1 BALAS, E.
DISJUNCTIVE PROGRAMMING.
BALAG 1977 -2 BALAS, E.
SET COVERING WITH CUTTING PLANES FROM CONDITIONAL BOUNDS (REVISED VERSION).
BALAG 1975 -2 BALAS, E.
DISJUNCTIVE PROGRAMMING: CUTTING PLANES FROM LOGICAL CONDITIONS.
BALAG 1974 -2 BALAS, E.
INTERSECTION CUTS FROM DISJUNCTIVE CONSTRAINTS.
BALAG 1977 -3 BALAS, E.
SOME VALID INEQUALITIES FOR THE SET PARTITIONING PROBLEM.
BALAH 1974 -1 BALAS, E.; GERRITSEN, R.; PADBERG, M. W.
ALL-BINARY COLUMN GENERATING ALGORITHM FOR SET PARTITIONING, AN.
BALAJ 1975 -1 BALAS, E.; PADBERG, M. W.
SET PARTITIONING: A SURVEY.
BALAJ 1976 -1 BALAS, E.; PADBERG, M. W.
SET PARTITIONING: A SURVEY.
BALAM 1975 -1 BALAS, E.; ZOLTNERS, A.
INTERSECTION CUTS FROM OUTER POLARS OF TRUNCATED CUBES.
BELL 1977 -1 BELL, D. E.
SIMPLE ALGORITHM FOR INTEGER PROGRAMS USING GROUP CONSTRAINTS, A.
BELLD 1970 -1 BELLMORE, M.; JENSEN, P.
IMPLICIT ENUMERATION SCHEME FOR PROPER CUT GENERATION, AN.
BURD 1977 -1 BURDET, C.-A.; JOHNSON, E. L.
SUBADDITIVE APPROACH TO SOLVE LINEAR INTEGER PROGRAMS, A.
BUZY 1976 -1 BUZYCKII, P. L.
CERTAIN CUT AND BRANCH FOR PRIMAL INTEGER PROGRAMMING ALGORITHMS (RUSSIAN), A.
CHANC 1977 -1 CHAND, M.; AGRAWAL, S. C.
ON INTEGER SOLUTIONS TO COMPLEMENTARY PROGRAMMING PROBLEMS.
CHARA 1977 -1 CHARNES, A.; GRANOT, D.; GRANOT, F.
ON INTERSECTION CUTS IN INTERVAL INTEGER LINEAR PROGRAMMING.
CHERE 1975 -1 CHERVAK, N. K.
FINITE ALGORITHM FOR THE TRUNCATION METHOD FOR INTEGER LINEAR PROGRAMMING PROBLEMS WITH NONINTEGER TARGET FUNCTION (RUSSIAN), A.
CHERF 1977 -1 CHERVAK, YU. YU.
ON AN IMPROVED VARIANT OF GOMORY'S ALGORITHM (RUSSIAN).
CHERF 1974 -1 CHERVAK, YU. YU.
REMARKS ON THE FIRST GOMORY ALGORITHM (RUSSIAN).
CHEU 1976 -2 CHEUNG, T.-Y.
STEPPING-STONE PARALLEL-CUT METHOD FOR INTEGER PROGRAMMING, A.
FERS 1976 -1 FERSTER, V. B.
DECREASING ALGORITHM FOR THE CUTTING PLANE METHOD FOR THE SOLUTION OF INTEGER PROGRAMMING PROBLEMS (RUSSIAN), A.
FRIDB 1976 -1 FRIDMAN, A. A.; VOTJAKOV, A. A.
REGULAR DISCRETE PROBLEMS AND THE CUTTING PLANE METHOD (RUSSIAN).
GLOV 1975 -1 GLOVER, F.
COMPACT LP BASES FOR A CLASS OF IP PROBLEMS.
GLOV 1971 -1 GLOVER, F.
CONVEXITY CUTS FOR MULTIPLE-CHOICE PROBLEMS.
GLOV 1973 -2 GLOVER, F.
POLYHEDRAL CONVEXITY CUTS AND NEGATIVE EDGE EXTENSIONS.
GLOVI 1976 -1 GLOVER, F.; SOMMER, D.
INEQUALITIES FOR MIXED INTEGER PROGRAMS WITH STRUCTURE.
GROE 1977 -5 GROETSCHEL, M.
ON THE SYMMETRIC TRAVELLING SALESMAN PROBLEM: SOLUTION OF A 120-CITY PROBLEM.
HARTC 1975 -1 HARTMANN, K.
GEMISCHT GANZZAHLIGE LINEARE QUOTIENTENOPTIMIERUNG NACH DEM SCHNITTVERFAHREN VON GOMORY.
HROU 1977 -1 HROUDA, J.
GENERALIZATION OF THE ALL-INTEGER ALGORITHM OF GOMORY (CZECH), A.
ISHI 1976 -1 ISHII, H.; IBARAKI, T.; MINE, H.
PRIMAL CUTTING PLANE ALGORITHM FOR INTEGER FRACTIONAL PROGRAMMING PROBLEMS, A.
JERO 1977 -2 JEROSLOW, R.
INTRODUCTION TO THE THEORY OF CUTTING-PLANES, AN.
JEROA 1977 -1 JEROSLOW, R. G.
CUTTING-PLANE GAME AND ITS ALGORITHMS, A.
JEROA 1977 -2 JEROSLOW, R. G.
CUTTING-PLANE THEORY: DISJUNCTIVE METHODS.
JEROA 1977 -3 JEROSLOW, R. G.
CUTTING-PLANES FOR COMPLEMENTARITY CONSTRAINTS.
JOHNE 1964 -1 JOHNSON, E. L.
NETWORK FLOWS, GRAPHS, AND INTEGER PROGRAMMING.
KIAN 1976 -1 KIANFAR, F.
STRONGER INEQUALITIES FOR 0-1 INTEGER PROGRAMMING: COMPUTATIONAL REFINEMENTS.
KIVI 1976 -1 KIVISTIK, L.
CUTTING-PLANE METHOD FOR INTEGER LINEAR PROGRAMMING PROBLEMS WITH ADDITIONAL ALTERNATIVE CONDITIONS (RUSSIAN), THE.
KORTB 1976 -2 KORTE, B.
INTEGER PROGRAMMING (SURVEY PAPER).
LEIT 1972 -1 LEITEN, A.
OBTAINING CUTS IN THE B-ALGORITHM (RUSSIAN).
LIN 1977 -1 LIN, B. W. Y.; RARDIN, R. L.
DEVELOPMENT OF A PARAMETRIC GENERATING PROCEDURE FOR INTEGER PROGRAMMING TEST PROBLEMS.
MILI 1976 -1 MILIOTIS, P.
INTEGER PROGRAMMING APPROACHES TO THE TRAVELLING SALESMAN PROBLEM.

****

MUELA 1973 -1 MUELLER, H. CH.
UNTERSUCHUNG DER VERWENDBARKEIT DES ALGORITHMUS VON GOMORY FUER REIN GANZZAHLIGE PROBLEME BEIM DIENSTPLANPROBLEM.

MURTB 1976 -1 MURTY, K. G.
LINEAR AND COMBINATORIAL PROGRAMMING.

OTTA 1976 -1 OTTAVIANI, M.
CUTS ON THE FEASIBLE REGION OF A LINEAR PROGRAMMING PROBLEM WITH INTEGER VARIABLES (ITALIAN).

PADBA 1973 -1 PADBERG, M. W.
NOTE ON THE CONVERGENCE OF INTEGER PROGRAMMING ALGORITHMS USING INTERSECTION CUTS, A.

PADBB 1977 -1 PADBERG, M. W.; HONG, S.
ON THE SYMMETRIC TRAVELLING SALESMAN PROBLEM: A COMPUTATIONAL STUDY.

PIEH 1975 -1 PIEHLER, J.
GLOBALE BETRACHTUNGEN IN DER GANZZAHLIGEN OPTIMIERUNG.

PIEH 1977 -1 PIEHLER, J.
GANZZAHLIGE OPTIMIERUNG MIT KONGRUENZNEBENBEDINGUNGEN.

PURI 1975 -1 PURI, M. C.; SWARUP, K.
STRONG-CUT CUTTING PLANE PROCEDURE FOR EXTREME POINT MATHEMATICAL PROGRAMMING PROBLEMS.

RARD 1977 -1 RARDIN, R. L.; LIN, B.
WHAT MAKES INTEGER PROGRAMMING PROBLEMS HARD TO SOLVE?

RICHB 1976 -1 RICHTER, D.
ALGORITHMUS ZUR LOESUNG GEMISCHT GANZZAHLIGER KONVEXER OPTIMIERUNGSPROBLEME, EIN.

RYND 1977 -1 RYNDIN, B. N.
MODIFIED FIRST ALGORITHM OF GOMORY (RUSSIAN), A.

SALK 1975 -1 SALKIN, H. M.
INTEGER PROGRAMMING.

SCHO 1976 -1 SCHOCH, M.
ERWEITERUNGSPRINZIP UND SEINE ANWENDUNG ZUR ENTWICKLUNG VON ALGORITHMEN FUER DIE LOESUNG KOMBINATORISCHER OPTIMIERUNGSAUFGABEN, DAS.

SEEL 1974 -1 SEELAENDER, J.
METHODE ZUR KOPPLUNG DUALER SCHNITTEBENENVERFAHREN IN DER GANZZAHLIGEN LINEAREN OPTIMIERUNG, EINE.

SEEL 1976 -1 SEELAENDER, J.
BEMERKUNG ZUM DRITTEN ALGORITHMUS VON GOMORY ZUR LOESUNG REIN-GANZZAHLIGER LINEARER OPTIMIERUNGSPROBLEME, EINE.

SEEL 1977 -1 SEELAENDER, J.
UEBER EINE METHODE ZUR KOPPLUNG DUALER SCHNITTEBENENVERFAHREN IN DER GANZZAHLIGEN LINEAREN OPTIMIERUNG.

SHARA 1975 -1 SHARIQ, S. Z.; CHACHRA, V.
CUTTING PLANE ALGORITHM FOR LEAST DISTANCE PROGRAMMING, A.

SHER 1977 -1 SHERALI, A. D.; SHETTY, C. M.
RECTILINEAR DISTANCE LOCATION-ALLOCATION PROBLEM, THE.

SHEVA 1973 -1 SHEVCHENKO, V. N.; REMIZOVA, O. L.
CONSTRUCTION OF REGULAR CUT-OFFS IN LINEAR INTEGER PROGRAMMING (RUSSIAN), THE.

STEIB 1976 -1 STEINER, J.
ZUR KONSTRUKTION VON SCHNITTEN FUER GEMISCHT-GANZZAHLIGE LO-PROBLEME.

TAHA 1975 -1 TAHA, H. A.
INTEGER PROGRAMMING.

TODD 1977 -1 TODD, M. J.
NUMBER OF NECESSARY CONSTRAINTS IN AN INTEGER PROGRAM: A NEW PROOF OF SCARF'S THEOREM, THE.

VOLP 1976 -1 VOLPENTESTA, A.
FINITE GROUPS AND INTEGER LINEAR PROGRAMS.

WHITC 1975 -1 WHITE, D. J.
NOTE ON FACULTY TIMETABLING, A.

WOLS 1976 -1 WOLSEY, L. A.
FACETS AND STRONG VALID INEQUALITIES FOR INTEGER PROGRAMS.

WOLS 1976 -3 WOLSEY, L. A.
NUCLEOLUS AND KERNEL FOR SIMPLE GAMES OR SPECIAL VALID INEQUALITIES FOR 0-1 LINEAR INTEGER PROGRAMS, THE.

YOUNC 1975 -1 YOUNG, R. D.
ECLECTIC PRIMAL ALGORITHM: CUTTING-PLANE METHOD THAT ACCOMODATES HYBRID SUBPROBLEM SOLUTION TECHNIQUES, THE.

DECOMPOSITION

****

ADLE 1975 -1 ADLER, H. J.
BESTIMMUNG VON TEILKOMPONENTEN EINES NETZWERKES BEI GLEICHZEITIGER MINIMIERUNG DER SCHNITTSTELLEN.

AKUL 1977 -1 AKULIC, F. E.
DECOMPOSITION METHOD FOR PROBLEMS OF MAXIMAL FLOW IN LARGE RANGED NETWORKS (RUSSIAN), A.

ANZA 1974 -1 ANZAI, Y.
ON INTEGER FRACTIONAL PROGRAMMING.

ARMSE 1977 -1 ARMSTRONG, R. D.; WILLIS, C. E.
SIMULTANEOUS INVESTMENT AND ALLOCATION DECISIONS APPLIED TO WATER PLANNING.

ARMSE 1975 -1 ARMSTRONG, R. D.; WILLIS, C. E.
SIMULTANEOUS INVESTMENT-ALLOCATION - AN APPLICATION OF GENERALIZED BENDERS DECOMPOSITION TO WATER PLANNING.

BACH 1976 -1 BACHEM, A.
ALGORITHMUS ZUR MODULO-OPTIMIERUNG STRUKTURIERTER MATRIZEN, EIN.

BACH 1977 -3 BACHEM, A.
REDUCTION AND DECOMPOSITION OF INTEGER PROGRAMS OVER CONES.

BACH 1976 -4 BACHEM, A.
REDUCTION AND DECOMPOSITION OF MODULO OPTIMIZATION PROBLEMS.

BAUM 1977 -1 BAUM, S.; TROTTER, L. E. JR.
INTEGER ROUNDING AND POLYHEDRAL DECOMPOSITION FOR TOTALLY UNIMODULAR SYSTEMS.

BELLB 1965 -1 BELL, E. J.
PRIMAL-DUAL DECOMPOSITION PROGRAMMING.

BLEW 1977 -1 BLEWETT, W. J.; HU, T. C.
TREE DECOMPOSITION ALGORITHM FOR LARGE NETWORKS.

BONDA 1976 -1 BONDY, J. A.; CHVATAL, V.
METHOD IN GRAPH THEORY, A.

CABOA 1976 -1 CABOT, V. A.
NONCONVEX PROGRAMMING PROBLEM SOLVED BY GENERALIZED BENDERS DECOMPOSITION, A.

CUNNA 1974 -1 CUNNINGHAM, W. H.
COMBINATORIAL DECOMPOSITION THEORY, A.

DYMA 1975 -1 DYMARSKII, JA. S.; PRUDOVSKII, B. D.; STALBO, A. K.
SEPARABLE INTEGER-VALUED PROBLEM (RUSSIAN), A.

ELZIA 1976 -1 ELZINGA, J.; VEMUGANTI, R. R.
OPTIMAL SCHEDULING OF A SALES FORCE.

FORG 1976 -1 FORGO, F.
ON AN EXAMPLE OF DUALITY AND DECOMPOSITION METHODS IN INTEGER PROGRAMMING (HUNGARIAN).

FREE 1977 -1 FREELAND, J. R.; MOORE, J. H.
IMPLICATIONS OF RESOURCE DIRECTIVE ALLOCATION MODELS FOR ORGANIZATIONAL DESIGN.

GEOFA 1973 -1 GEOFFRION, A. M.; GRAVES, G. W.
MULTICOMMODITY DISTRIBUTION SYSTEM DESIGN BY BENDERS DECOMPOSITION.

GEOFB 1977 -1 GEOFFRION, A. M.; GRAVES, G. W.; LEE, S.-J.
STRATEGIC DISTRIBUTION SYSTEM PLANNING: A STATUS REPORT.

GIAN 1976 -1 GIANNESSI, F.
DECOMPOSITION METHOD FOR THE TRAVELLING SALESMAN PROBLEM AND ITS APPLICATIONS, A.

GIUL 1976 -1 GIULIANELLI, S.; LUCERTINI. M.
DECOMPOSITION TECHNIQUE IN INTEGER LINEAR PROGRAMMING, A.

GOLDH 1971 -1 GOLDSTEIN, M. C.
CONSIDERATION IN THE DESIGN OF LARGE SCALE NETWORKS.

GUIGA 1977 -1 GUIGNARD, M.; SPIELBERG, K.
ALGORITHMS FOR EXPLOITING THE STRUCTURE OF THE SIMPLE PLANT LOCATION PROBLEM.

HROU 1975 -1 HROUDA, J.
MODIFIED BENDERS' ALGORITHM (CZECH).

HROU 1976 -1 HROUDA, J.
BENDERS METHOD AND PARAMETRIZATION OF THE RIGHT-HAND SIDES IN THE MIXED INTEGER LINEAR PROGRAMMING PROBLEM (CZECH), THE.

HU 1966 -1 HU, T. C.
DECOMPOSITION ALGORITHM FOR SHORTEST PATHS IN A NETWORK, A.

IVANB 1974 -1 IVANISCEV, V. V.
DECOMPOSITION METHOD FOR THE SOLUTION OF THE PROBLEM OF THE OPTIMAL USE OF MEANS OF TRANSPORTATION ON A NETWORK (RUSSIAN), A.

KARA 1976 -1 KARAGEZOV, Z. A.
DECOMPOSITIONAL ALGORITHM FOR NON-CORRECT PROBLEM OF TIME SCHEDULE OPTIMIZATION (RUSSIAN).

KLUY 1977 -1 KLUYVER, C. A. DE
EXTENSION OF ELZINGA AND VEMUGANTI'S DECOMPOSITION SCHEME TO MULTIPERIOD ADVERTISING MEDIA SCHEDULING, AN.

KORTB 1976 -2 KORTE, B.
INTEGER PROGRAMMING (SURVEY PAPER).

LAWL 1978 -1 LAWLER, E. L.
SEQUENCING JOBS TO MINIMIZE TOTAL WEIGHTED COMPLETION TIME SUBJECT TO PRECEDENCE CONSTRAINTS.

LUND 1973 -1 LUNDQVIST, L.
INTEGRATED LOCATION-TRANSPORTATION ANALYSIS: A DECOMPOSITION APPROACH.

MAHM 1976 -1 MAHMOUD, S.; RIORDON, J. S.
OPTIMAL ALLOCATION OF RESOURCES IN DISTRIBUTED INFORMATION NETWORKS.

MAIE 1976 -1 MAIER, S. F.
SOLVING THE TRAFFIC ASSIGNMENT AND SUB-AREA FOCUSING PROBLEMS BY GEOGRAPHIC DECOMPOSITION.

NICOB 1976 -1 NICOLETTI, B.; NATALI, L.; SCALAS, M. R.
MANPOWER STAFFING AT AN AIRPORT: A GENERALIZED SET COVERING PROBLEM.

NOON 1977 -1 NOONAN, F.; GIGLIO, R. J.
PLANNING ELECTRIC POWER GENERATION: A NONLINEAR MIXED INTEGER MODEL EMPLOYING BENDERS DECOMPOSITION.

ORTH 1973 -1 ORTH, E.
DEKOMPOSITIONSMETHODEN - EIN HILFSMITTEL BEI DER PLANUNG REGIONALER ABWASSERBESEITIGUNGSSYSTEME.

PAPE 1976 -3 PAPE, U.
MODEL OF A HIGH LEVEL DESIGN LANGUAGE FOR GRAPH ALGORITHMS, A (2ND EDITION).

PIRO 1975 -1 PIROSKA, T.
APPLICATION OF THE DECOMPOSITION PRINCIPLE OF DANTZIG AND WOLFE TO THE SOLUTION OF A GENERALIZED TRANSPORTATION PROBLEM (HUNGARIAN), AN.

RAMAA 1976 -1 RAMAKRISHNAN, K. G.
SOLUTION OF CAPACITATED PLANT LOCATION PROBLEMS USING BENDERS' DECOMPOSITION.

RARDA 1976 -3 RARDIN, R. L.; UNGER, V. E.
SURROGATE CONSTRAINTS AND THE STRENGTH OF BOUNDS DERIVED FROM 0-1 BENDERS' PARTITIONING PROCEDURES.

RICH 1976 -1 RICHARDSON, R.
OPTIMIZATION APPROACH TO ROUTING AIRCRAFT, AN.
RICHC 1977 -1 RICHTER, K.
RESSOURCENAUFSPALTUNG FUER DISKRETE OPTIMIERUNGSAUFGABEN, TEIL II.
RICHC 1976 -1 RICHTER, K.
RESSOURCENAUFSPALTUNG FUER DISKRETE OPTIMIERUNGSAUFGABEN, TEIL 1.
SLAT 1976 -1 SLATE, L.; SPIELBERG, K.
EXTENDED CONTROL LANGUAGE OF MPSX/370 (AND POSSIBLE APPLICATIONS), THE.
SRIN 1976 -1 SRINIVASAN, V.
DECOMPOSITION OF A MULTI-PERIOD MEDIA SCHEDULING MODEL IN TERMS OF SINGLE PERIOD EQUIVALENTS.
SWAI 1971 -1 SWAIN, R. W.
DECOMPOSITION ALGORITHM FOR A CLASS OF FACILITY LOCATION PROBLEM, A.
SWAI 1971 -2 SWAIN, R. W.
PARAMETRIC DECOMPOSITION APPROACH FOR THE SOLUTION OF UNCAPACITATED LOCATION PROBLEMS.
SWEE 1977 -1 SWEENEY, D. J.; MURPHY, R. A.
DECOMPOSITION PRINCIPLE FOR INTEGER PROGRAMS, A (REVISED VERSION).
THOMB 1976 -1 THOMPSON, G. L.
SOLUTIONS OF LARGE STRUCTURED TRAVELLING SALESMAN PROBLEMS (COPIES OF TRANSPARANCIES).
TONE 1976 -1 TONE, K.
PARAMETRIZATION OF BENDERS DECOMPOSITION, A.
TORR 1976 -1 TORRIERO, A.
DECOMPOSITION METHODS FOR THE TRAVELLING SALESMAN PROBLEM IN LARGE DIMENSIONS (ITALIAN).
TRYG 1976 -1 TRYGAR, T. A.; WRIGHT, G.
APPLICATION OF DECOMPOSITION TECHNIQUES TO THE CALCULATION OF A MINIMUM DISTANCE MATRIX FOR A LARGE SCALE STREET NETWORK MODEL, THE.
TURCA 1975 -1 TURCHANYI, P.
DECOMPOSITION METHOD FOR THE SOLUTION OF A TRANSPORTATION PROBLEM WITH PIECEWISE LINEAR CONVEX COST FUNCTIONS (HUNGARIAN), A.
YAMA 1977 -1 YAMAMOTO, M.
APPROXIMATE SOLUTION OF MACHINE SCHEDULING PROBLEMS BY DECOMPOSITION METHOD, AN.
ZAYC 1975 -1 ZAYCHENKO, YU. P.; PECHURIN, M. K.; TOMASHEVSKIY, V. M.
GENERALIZED ASSIGNMENT PROBLEM AND ITS SOLUTION ALGORITHM (UKRAINIAN), THE.

DIOPHANTINE SYSTEMS AND AGGREGATION

****

BABA 1977 -1 BABAT, L. G.
ON INTEGER LINEAR AGGREGATION (RUSSIAN).

BACH 1978 -1 BACHEM, A.
THEOREM OF MINKOWSKI FOR POLYHEDRAL MONOIDS AND AGGREGATED LINEAR DIOPHANTINE SYSTEMS, THE.

BACH 1977 -3 BACHEM, A.
REDUCTION AND DECOMPOSITION OF INTEGER PROGRAMS OVER CONES.

BAER 1976 -1 BAER, G.
UMFORMUNGSMOEGLICHKEITEN FUER DIE RESTRIKTIONEN EINES 0-1-OPTIMIERUNGSPROBLEMS.

BLAI 1976 -1 BLAIR, C.
TWO RULES FOR DEDUCING VALID INEQUALITIES FOR 0-1 PROBLEMS.

CHANB 1969 -1 CHAN, Y. P.
OPTIMAL TRAVEL TIME REDUCTION IN A TRANSPORT NETWORK. AN APPLICATION OF NETWORK AGGREGATION AND BRANCH AND BOUND TECHNIQUES.

CHVAC 1977 -1 CHVATAL, V.; HAMMER, P. L.
AGGREGATION OF INEQUALITIES IN INTEGER PROGRAMMING.

DICK 1971 -1 DICKSON, L. E.
HISTORY OF THE THEORY OF NUMBERS. VOLUME 2 DIOPHANTINE ANALYSIS (REPRINT OF THE 1919 EDITION).

EGIA 1976 -1 EGIAZARJAN, E. V.
CERTAIN CLASS OF SYSTEMS OF BOOLEAN EQUATIONS (RUSSIAN), A.

FORG 1974 -1 FORGO, F.
SHADOW PRICES AND DECOMPOSITION FOR INTEGER PROGRAMS.

FRUM 1976 -1 FRUMKIN, M. A.
ALGORITHMS FOR THE SOLUTION IN INTEGERS OF SYSTEMS OF LINEAR EQUATIONS (RUSSIAN).

HAMMA 1975 -1 HAMMER, P. L.
BOOLEAN ELEMENTS IN COMBINATORIAL OPTIMIZATION: A SURVEY.

IVANC 1975 -1 IVANOV, N. N.
CERTAIN METHOD FOR AGGREGATION OF AN INTEGER LINEAR PROGRAMMING PROBLEM (RUSSIAN), A.

JAIN 1975 -1 C JAIN, R. C.
ON PSEUDO-BOOLEAN INEQUALITIES-CONSISTENCY AND SOLUTIONS.

KEND 1977 -1 KENDALL, K. E.; ZIONTS, S.
SOLVING INTEGER PROGRAMMING PROBLEMS BY AGGREGATING CONSTRAINTS.

KOVAA 1975 -1 KOVALENKO, I. N.
ON THE INVARIANCE THEOREMS FOR BOOLEAN RANDOM MATRICES (RUSSIAN).

LAVI 1976 -1 LAVIT, C.
CLASSES DE SOLUTIONS PARAMETRIQUES D'UNE EQUATION BOOLEENNE.

MAND 1976 -1 MANDERS, K.; ADLEMAN, L.
NP-COMPLETE DECISION PROBLEMS FOR QUADRATIC POLYNOMIALS.

NEWM 1972 -1 NEWMAN, M.
INTEGRAL MATRICES.

PETRA 1975 -1 PETROV, I. P.; TIMOFEEV, A. V.
FINITELY CONVERGENT RECURRENCE ALGORITHMS FOR THE SOLUTION OF INTEGER INEQUALITIES IN THE PRESENCE OF CONSTRAINTS (RUSSIAN).

SHEV 1976 -1 SHEVCHENKO, V. N.
DISCRETE ANALOGUE OF THE FARKASCH THEOREM AND AGGREGATION PROBLEM OF THE SYSTEM OF LINEAR INTEGER-VALUED EQUATIONS (RUSSIAN), A.

STROA 1976 -1 STROEKER, R. J.
ON THE DIOPHANTINE EQUATION XXX - DYY = 1.

WALU 1975 -1 WALUKIEWICZ, S.
EQUIVALENT CONSTRAINT USING AN OBJECTIVE FUNCTION, AN.

WEIN 1977 -1 WEINBERG, F.
NECESSARY AND SUFFICIENT CONDITION FOR THE AGGREGATION OF LINEAR DIOPHANTINE EQUATIONS, A.

****

BALAG 1977 -1 BALAS, E.
DISJUNCTIVE PROGRAMMING.

BALAG 1974 -1 BALAS, E.
DISJUNCTIVE PROGRAMMING: FACETS OF THE CONVEX HULL OF FEASIBLE POINTS.

BALAG 1974 -2 BALAS, E.
INTERSECTION CUTS FROM DISJUNCTIVE CONSTRAINTS.

BALAG 1975 -2 BALAS, E.
DISJUNCTIVE PROGRAMMING: CUTTING PLANES FROM LOGICAL CONDITIONS.

BALAG 1976 -2 BALAS, E.
NOTE ON DUALITY IN DISJUNCTIVE PROGRAMMING, A.

BALAG 1977 -2 BALAS, E.
SET COVERING WITH CUTTING PLANES FROM CONDITIONAL BOUNDS (REVISED VERSION).

BALAG 1977 -3 BALAS, E.
SOME VALID INEQUALITIES FOR THE SET PARTITIONING PROBLEM.

JERO 1977 -1 JEROSLOW, R.
BRACKETING DISCRETE PROBLEMS BY TWO PROBLEMS OF LINEAR OPTIMIZATION.

JEROA 1977 -2 JEROSLOW, R. G.
CUTTING-PLANE THEORY: DISJUNCTIVE METHODS.

DUALITY IN INTEGER PROGRAMMING
****

BACHB 1977 -1 BACHEM, A.; SCHRADER, R.
MINIMAL INEQUALITIES AND DUALITY FOR MIXED INTEGER PROGRAMS.

BALAG 1976 -2 BALAS, E.
NOTE ON DUALITY IN DISJUNCTIVE PROGRAMMING, A.

BALAK 1977 -1 BALAS, E.; SAMUELSSON, H.
NODE COVERING ALGORITHM, A.

BAUM 1977 -1 BAUM, S.; TROTTER, L. E. JR.
INTEGER ROUNDING AND POLYHEDRAL DECOMPOSITION FOR TOTALLY UNIMODULAR SYSTEMS.

BAZAB 1977 -1 BAZARAA, M. S.; GOODE, J. J.
TRAVELING SALESMAN: A DUALITY APPROACH, THE.

BELL 1973 -1 BELL, D. E.
RESOLUTION OF DUALITY GAPS IN DISCRETE OPTIMIZATION, THE.

BELLA 1977 -1 BELL, D. E.; SHAPIRO, J. F.
CONVERGENT DUALITY THEORY FOR INTEGER PROGRAMMING, A.

BELLA 1975 -1 BELL, D. E.; SHAPIRO, J. F.
FINITELY CONVERGENT DUALITY THEORY FOR ZERO-ONE INTEGER PROGRAMMING, A.

BEUE 1975 -1 BEUERMANN-WEHMEYER, A.
PRIMAL-DUAL-BEZIEHUNGEN IN GRAPHENALGORITHMEN.

BOX 1976 -1 BOX, R. M.
PRESQUE-DUALITE EN NOMBRES ENTIERS, LA.

CHARA 1975 -1 CHARNES, A.; GRANOT, D.; GRANOT, F.
ON IMPROVING BOUNDS FOR VARIABLES IN LINEAR INTEGER PROGRAMS BY SURROGATE CONSTRAINTS.

CHVA 1975 -1 CHVATAL, V.
SOME LINEAR PROGRAMMING ASPECTS OF COMBINATORICS.

DANTB 1975 -1 DANTZIG, G. B.; EAVES, B. C.
FOURIER-MOTZKIN ELIMINATION AND ITS DUAL WITH APPLICATION TO INTEGER PROGRAMMING.

DERI 1977 -1 DERIGS, U.
DUALITY AND THE ALGEBRAIC MATCHING PROBLEM.

DIGU 1976 -1 DIGUGLIELMO, F.
ESTIMATES OF THE DUALITY GAP FOR DISCRETE OPTIMIZATION.

DIGU 1976 -2 DIGUGLIELMO, F.
ESTIMATION DU SAUT DE DUALITE EN OPTIMISATION DISCRETE.

EDMO 1975 -1 EDMONDS, J.
SOME WELL-SOLVED PROBLEMS IN COMBINATORIAL OPTIMIZATION.

EDMOA 1977 -1 EDMONDS, J.; GILES, R.
MIN-MAX RELATION FOR SUBMODULAR FUNCTIONS ON GRAPHS, A.

FISHA 1976 -1 FISHER, M. L.
DUAL ALGORITHM FOR THE ONE-MACHINE SCHEDULING PROBLEM, A.

FISHC 1973 -1 FISHER, M. L.; SHAPIRO, J. F.
CONSTRUCTIVE DUALITY FOR DISCRETE OPTIMIZATION.

FORG 1974 -1 FORGO, F.
SHADOW PRICES AND DECOMPOSITION FOR INTEGER PROGRAMS.

FORG 1976 -1 FORGO, F.
ON AN EXAMPLE OF DUALITY AND DECOMPOSITION METHODS IN INTEGER PROGRAMMING (HUNGARIAN).

GILEA 1977 -1 GILES, F. R.; PULLEYBLANK, W. R.
TOTAL DUAL INTEGRALITY AND INTEGER POLYHEDRA.

GUGL 1976 -1 GUGLIELMO, F. D.
ESTIMATES OF THE DUALITY GAP FOR DISCRETE OPTIMIZATION.

GUIGA 1977 -2 GUIGNARD, M.; SPIELBERG, K.
DUAL METHOD FOR THE MIXED PLANT LOCATION PROBLEM, A.

HAMM 1976 -1 HAMMER, G.
DUALITAET BEI KOMBINATORISCHEN SUCHVERFAHREN.

HIRC 1970 -1 HIRCHE, J.
UEBER ZWEI MODIFIKATIONEN DES DEKOMPOSITIONSALGORITHMUS VON WILLIAMS.

HUAN 1977 -1 HUANG, H.-C.; TROTTER, L. E. JR.
TECHNIQUE FOR DETERMINING BLOCKING AND ANTIBLOCKING POLYHEDRAL DESCRIPTIONS, A.

KARW 1977 -1 KARWAN, M. H.; RARDIN, R. L.
SEARCHABILITY OF THE COMPOSITE AND MULTIPLE SURROGATE DUAL FUNCTIONS.

KARW 1976 -1 KARWAN, M. H.; RARDIN, R. L.
APPLICATION OF SURROGATE CONSTRAINT DUALITY IN INTEGER LINEAR PROGRAMMING, THE.

KARW 1976 -2 KARWAN, M. H.; RARDIN, R. L.
AURROGATE DUAL MULTIPLIER SEARCH PROCEDURES IN INTEGER PROGRAMMING.

KARW 1977 -2 KARWAN, M. H.; RARDIN, R. L.
SURROGATE DUALITY IN A BRANCH-AND-BOUND PROCEDURE.

KOEHA 1977 -1 KOEHLER, J.
NUTZUNG DER DUALITAET ZUR LOESUNG EINES GEMISCHTEN LINEAREN 0-1-OPTIMALPROBLEMS.

LEGE 1977 -1 LEGENDRE, J.-P.; MINOUX, M.
APPLICATION DE LA NOTION DE DUALITE EN PROGRAMMATION EN NOMBRES ENTIERS: SELECTION ET AFFECTATION OPTIMALES D'UNE FLOTTE D'AVIONS, UNE.

LOVA 1975 -1 LOVASZ, L.
ON THE RATIO OF OPTIMAL INTEGRAL AND FRACTIONAL COVERS.

LOVA 1977 -1 LOVASZ, L.
CERTAIN DUALITY PRINCIPLES IN INTEGER PROGRAMMING.

LOVA 1974 -1 LOVASZ, L.
MINIMAX THEOREMS FOR HYPERGRAPHS.

MAURB 1977 -1 MAURRAS, J. F.
EXAMPLE OF DUAL POLYTOPES IN THE UNIT HYPERCUBE, AN.

MEYED 1977 -1 MEYER, R. R.; FLEISHER, J. M.
STRONG DUALITY FOR A SPECIAL CLASS OF INTEGER PROGRAMS.

MOORB 1976 -1 MOORE, T. G.
SOLVING LARGE 0-1 PROBLEMS.

PERL 1976 -1 PERL, J.
ZUR ALGORITHMISCHEN BEHANDLUNG SPEZIELLER NP-VOLLSTAENDIGER PROBLEME AUF GRAPHEN.

PULL 1977 -1 PULLEYBLANK, W. R.
DUAL INTEGRALITY IN B-MATCHING PROBLEMS.

RICHB 1976 -3 RICHTER, D.
ZUR BEHANDLUNG GANZZAHLIGER OPTIMIERUNGSAUFGABEN OHNE GANZZAHLIGKEITSBEDINGUNG AN DIE ZIELFUNKTION.

ROSSC 1975 -1 ROSSMAN, L. A.
RESOLUTION OF DUALITY GAPS WITH DYNAMIC PROGRAMMING.

SHAPB 1974 -1 SHAPIRO, J. F.
NEW DUALITY THEORY FOR MIXED INTEGER PROGRAMMING, A.

SHAPB 1977 -2 SHAPIRO, J. F.
SENSITIVITY ANALYSIS IN INTEGER PROGRAMMING.

****

SHAPB 1976 -2 SHAPIRO, J. F.
NEW CONSTRUCTIVE DUALITY THEORY FOR MIXED INTEGER PROGRAMMING, A.

TIND 1977 -1 TIND, J.
BLOCKING AND ANTIBLOCKING POLYHEDRA.

TIND 1977 -2 TIND, J.
ON ANTIBLOCKING SETS AND POLYHEDRA.

TROTA 1977 -1 TROTTER, L. E. JR.; MONMA, C. L.
ON A THEOREM OF FULKERSON.

VIDY 1975 -1 VIDYASANKAR, K.; YOUNGER, D.
MINIMAX EQUALITY RELATED TO THE LONGEST DIRECTED PATH IN AN ACYCLIC GRAPH, A.

ZIMMA 1977 -1 ZIMMERMANN, U.
DUALITY AND THE ALGEBRAIC MATROID INTERSECTION PROBLEM.

ZIMMA 1977 -2 ZIMMERMANN, U.
SOME PARTIAL ORDERS RELATED TO BOOLEAN OPTIMIZATION AND THE GREEDY-ALGORITHM.

DYNAMIC PROGRAMMING

ABEL 1977 -1 ABELLA, R. J.; BOVA, T. E.
OPTIMAL PLANT ALLOCATION OF STOCKKEEPING UNITS.

AGRE 1977 -1 AGRESTI, W. W.
DYNAMIC PROGRAMMING FOR COMPUTER REGISTER ALLOCATION.

ALEK 1974 -1 ALEKSEYEV, O. G.
INCREASING THE EFFICIENCY OF A DYNAMIC PROGRAMMING METHOD IN OPTIMAL STAND-BY PROBLEMS (RUSSIAN).

ALEKB 1976 -1 ALEKSEYEV, O. G.; VOLODOS, I. F.
COMBINED USE OF DYNAMIC PROGRAMMING AND BRANCH-AND-BOUND METHODS IN DISCRETE-PROGRAMMING PROBLEMS (RUSSIAN).

AUST 1976 -1 AUST, R. J.
DYNAMIC PROGRAMMING BRANCH AND BOUND ALGORITHM FOR PURE INTEGER PROGRAMMING, A.

BAKE 1977 -1 BAKER, K. R.
COMPUTATIONAL EXPERIENCE WITH A SEQUENCING ALGORITHM ADAPTED TO THE TARDINESS PROBLEM.

BALE 1970 -1 BALERIAUX, H.; JAMOULLE, E.; DOULLIEZ, P.; KELECOM, J. VAN
RECHERCHE DE LA POLITIQUE OPTIMALE D'INVESTISSEMENT D'UN RESEAU ELECTRIQUE EN EXPANSION PAR UNE METHODE DE DECISION SEQUENTIELLE.

BALLB 1970 -1 BALLOU, R. H.
COMPUTATIONAL LIMITATIONS OF DYNAMIC PROGRAMMING FOR WAREHOUSE LOCATION: A COMMENT.

BERGC 1969 -1 BERGENDAHL, G.
COMBINED LINEAR AND DYNAMIC PROGRAMMING MODEL FOR INTERDEPENDENT ROAD INVESTMENT PLANNING, A.

BESW 1977 -1 BESWICK, C. A.
ALLOCATING SELLING EFFORT VIA DYNAMIC PROGRAMMING.

BRAN 1971 -1 BRANDMAN, E. I.
CERTAIN METHODS FOR THE SOLUTION OF PROBLEMS OF PROJECTED ALLOCATION OF CAPITAL INVESTMENTS WITHIN AN INDUSTRY (RUSSIAN).

BROW 1976 -1 BROWN, D. B.
ALLOCATION OF FEDERAL HIGHWAY SAFETY FUNDS USING DYNAMIC PROGRAMMING, THE.

BURKJ 1966 -1 BURKOV, V. N.; LOVETSKII, S. E.
HEURISTIC APPROACH TO THE SOLUTION OF DYNAMIC PROGRAMMING PROBLEMS IN THE ALLOCATION OF RESOURCES (RUSSIAN).

BURTB 1976 -1 BURTON, J. S.; GLEASON, J. M.
ALGORITHM FOR THE DYNAMIC LOT SIZE PROBLEM WITH PRICE BREAKS, AN.

CAMB 1976 -1 CAMBINI, A.
APPLICATION OF THE METHOD OF DYNAMIC PROGRAMMING TO AN OPTIMIZATION PROBLEM OVER A FINITE ABELIAN GROUP, AN.

CAMEB 1976 -1 CAMERON, A. V.; VOILS, L. R.
HEURISTIC FOR TRANSPORTATION ROUTING AND VEHICLE LOADING UNDER UNCERTAINTY, A.

CHEV 1977 -1 CHEVALIER, A.
PROGRAMMATION DYNAMIQUE. DE LA MAITRISE DE LA CONJONCTURE A LA PLANIFICATION, LA.

CHRIA 1977 -1 CHRISTOFIDES, N.; WHITLOCK, C.
ALGORITHM FOR TWO-DIMENSIONAL CUTTING PROBLEMS, AN.

CHUA 1965 -1 CHUANG, Y. A.; SMITH, W. G.
DYNAMIC PROGRAMMING MODEL FOR COMBINED PRODUCTION, DISTRIBUTION AND STORAGE.

CLAU 1977 -1 CLAUDER, G.; JAENICKE, W.
ZUR GRAPHENTHEORETISCHEN LOESUNG EINER KLASSE VON DISKRETEN OPTIMIERUNGSPROBLEMEN.

COOPB 1974 -1 COOPER, L.; COOPER, M. W.
ALL INTEGER LINEAR PROGRAMMING - A NEW APPROACH VIA DYNAMIC PROGRAMMING.

COOPB 1975 -1 COOPER, L.; COOPER, M. W.
NON-LINEAR INTEGER PROGRAMMING.

COOPD 1977 -1 COOPER, M. W.
IMPROVED ALGORITHM FOR NON-LINEAR INTEGER PROGRAMMING, AN.

DAVIC 1977 -1 DAVIS, K. R.; SIMMONS, L. F.
IMPROVING ASSEMBLY LINE EFFICIENCY: A DYNAMIC PROGRAMMING-HEURISTIC APPROACH.

DERM 1976 -1 DERMAN, C.; LIEBERMAN, G. J.; ROSS, S. M.
OPTIMAL SYSTEM ALLOCATIONS WITH PENALTY COSTS.

DREY 1977 -1 DREYFUS, S. E.; LAW A. M.
ART AND THEORY OF DYNAMIC PROGRAMMING, THE.

ERLE 1974 -1 ERLENKOTTER, D.
DYNAMIC FACILITY LOCATION AND SIMPLE NETWORK MODELS.

ERLE 1973 -1 ERLENKOTTER, D.
DYNAMIC MULTI-LOCATION MODELS FOR CAPACITY EXPANSION.

ERLE 1972 -1 ERLENKOTTER, D.
CAPACITY PLANNING FOR COMPETITIVE EXPANSIONS: A REDUCED DYNAMIC PROGRAMMING APPROACH.

ERLE 1967 -1 ERLENKOTTER, D.
TWO PRODUCING AREAS-DYNAMIC PROGRAMMING SOLUTIONS.

ERLE 1975 -2 ERLENKOTTER, D.
CAPACITY PLANNING FOR LARGE MULTILOCATION SYSTEMS: APPROXIMATE AND INCOMPLETE DYNAMIC PROGRAMMING APPROACHES.

ERLE 1974 -2 ERLENKOTTER, D.
DYNAMIC PROGRAMMING APPROACH TO CAPACITY EXPANSION WITH SPECIALIZATION, A.

ESCHB 1974 -1 ESCHER, H. K.
SEQUENTIELLE ZUORDNUNGSPROBLEME.

ESOG 1977 -1 ESOGBUE, A. O.; MARKS, B. R.
DYNAMIC PROGRAMMING MODELS OF THE NONSERIAL CRITICAL PATH-COST PROBLEM.

EVST 1969 -1 EVSTIGNEEV, V. A.; JASAKOVA, V. P.
CERTAIN ALGORITHM ON GRAPHS (RUSSIAN), A.

FUNK 1968 -1 FUNK, M. L.; TILLMAN, F. A.
OPTIMAL CONSTRUCTION STAGING BY DYNAMIC PROGRAMMING.

GARB 1976 -1 GARBAI, L.; MOLNAR, L.
OPTIMIZATION OF URBAN PUBLIC UTILITY NETWORKS BY DISCRETE DYNAMIC PROGRAMMING.

GEORA 1974 -1 GEORGOBIANI, D. A.; TABLIASHVILI, O. M.
ALLOCATION PROBLEM IN TERMS OF DYNAMIC PROGRAMMING. QUESTIONS ON MATHEMATICAL AND ENGINEERING CYBERNETICS (RUSSIAN), THE.

GILM 1977 -1 GILMORE, P. C.
CUTTING STOCK, LINEAR PROGRAMMING, KNAPSACKING, DYNAMIC PROGRAMMING AND INTEGER PROGRAMMING SOME INTERCONNECTIONS.

GLEN 1975 -1 GLEN, J. J.
DYNAMIC PROGRAMMING MODEL FOR WORK SCHEDULING IN A SHIPYARD, A.

GVOZ 1976 -1 GVOZDEV, S. E.
ON A GENERALIZATION OF THE KNAPSACK PROBLEM (RUSSIAN).

HEAR 1976 -1 HEARD, E. L.
DUE-DATES AND INSTANTANEOUS LOAD IN THE ONE-MACHINE SHOP.

****

HEIN 1977 -2 HEIN, O.
LOESUNGSVORSCHLAG ZUM VERALLGEMEINERTEN RUNDFAHRTPROBLEM, EIN.
HELD 1965 -1 HELD, M.; KARP, R. M.
CONSTRUCTION OF DISCRETE DYNAMIC PROGRAMMING ALGORITHMS, THE.
HESS 1977 -1 HESSELBACH, J.; WIJNANDS, J.
DYNAMISCHE PROGRAMMIERUNG - METHODIK, EDV-PROGRAMM, ANWENDUNGSBEISPIEL.
HODG 1974 -1 HODGSON, M. J.
STAGING INTERURBAN HIGHWAY CONSTRUCTION BY DYNAMIC PROGRAMMING.
HOWA 1970 -1 HOWARD, D. D.
DYNAMIC LOCATION-ALLOCATION ANALYSIS.
HUNGA 1977 -1 HUNG, M. S.; FISK, J. C.
ASSORTMENT PROBLEMS AND EQUIVALENT INTEGER PROGRAMMING FORMULATIONS.
IBAR 1973 -1 IBARAKI, T.
SOLVABLE CLASSES OF DISCRETE DYNAMIC PROGRAMMING.
ISHI 1975 -2 ISHII, H.; IBARAKI, T.; MINE, H.
TRAVELING SALESMAN PROBLEMS WITH A CAPACITY CONSTRAINT OF THE DELIVERY TRUCK.
JACOA 1974 -1 JACOBSEN, S. K.
ALGORITHM FOR SMALL-SCALE DYNAMIC PLANT LOCATION PROBLEMS, AN.
JARA 1976 -1 JARANCEV, JU. A.
CERTAIN PROCEDURE FOR DYNAMIC MEMORY ALLOCATION IN COMPUTERS (RUSSIAN), A.
JEFF 1965 -1 JEFFERIS, R. P.; FEGLEY, K. A.
APPLICATION OF DYNAMIC PROGRAMMING TO ROUTING PROBLEMS.
KHAC 1976 -1 KHACHATUROV, V. R.; ASTAKHOV, N. B.
DYNAMIC LOCATION PROBLEMS (MODELS AND SOLUTION METHODS) (RUSSIAN).
KRARB 1973 -1 KRARUP, J.; HARARY, F.
CLASS OF PLANAR GRAPHS RELATED TO NONSERIAL DYNAMIC PROGRAMMING, A.
LAURB 1976 -1 LAURIERE, M.
METHODE DE RESOLUTION RAPIDE POUR LE PROBLEME DU KNAPSACK, UNE.
LECH 1976 -1 LECHEVALLIER, Y.
CLASSIFICATION AUTOMATIQUE OPTIMALE SOUS CONTRAINTE D'ORDRE TOTAL.
LODI 1970 -1 LODISH, L. M.
COMPUTATIONAL LIMITATIONS OF DYNAMIC PROGRAMMING FOR WAREHOUSE LOCATION.
LOVE 1976 -1 LOVE, R. F.
ONE-DIMENSIONAL FACILITY LOCATION-ALLOCATION USING DYNAMIC PROGRAMMING.
MARSB 1978 -1 MARSTEN, R. E.; MORIN, T. L.
HYBRID APPROACH TO DISCRETE MATHEMATICAL PROGRAMMING, A.
MARTA 1975 -1 MARTELLI, A.; MONTANARI, U.
ON THE FOUNDATIONS OF DYNAMIC PROGRAMMING.
MORI 1975 -1 MORIN, T. , L.; MARSTEN, R. E.
HYBRID DYNAMIC PROGRAMMING/BRANCH-AND-BOUND APPROACH TO A CLASS OF SEQUENCING PROBLEMS, A.
MORIA 1976 -1 MORIN, T. L.; MARSTEN, R. E.
BRANCH-AND-BOUND STRATEGIES FOR DYNAMIC PROGRAMMING.
MURP 1976 -1 MURPHY, F. H.; STOHR, E. A.
DYNAMIC PROGRAMMING ALGORITHM FOR CHECK SORTING, A (REVISED VERSION).
NEUM 1975 -1 NEUMANN, K.
OPERATIONS RESEARCH C. DYNAMISCHE OPTIMIERUNG.
NORM 1975 -1 NORMAN, J. M.
ELEMENTARY DYNAMIC PROGRAMMING.
PASS 1975 -1 PASSY, U.; SILMAN, L.
ONE-AT-A-TIME DYNAMIC PROGRAMMING OVER POLYTOPE.
POCH 1976 -1 POCHTMAN, YU. M.; RAFES, R. YU.
MODIFICATION OF THE DYNAMIC PROGRAMMING ALGORITHM FOR PROBLEMS CONCERNING THE ALLOCATION OF RESOURCES (RUSSIAN).
POPO 1972 -1 POPOVICIU, N.
DYNAMIC PROGRAMMING: ONE-SIZE ALLOCATION PROBLEM WITH A SUPPLEMENTARY CONSTRAINT,
POWE 1977 -1 POWELL, J.; GLEASON, J.; BURTON, J.
POWER GENERATOR SCHEDULING BY DYNAMIC PROGRAMMING.
RAZH 1969 -1 RAZHUMIKHIN, B. S.
APPLICATION OF THE METHOD OF DYNAMIC PROGRAMMING TO CERTAIN PROBLEMS OF NETWORK OPTIMIZATION. I + II (RUSSIAN), AN.
ROBIA 1977 -1 ROBINSON, D. R.
COMPUTATIONAL REFINEMENT FOR DISCRETE-VALUED DYNAMIC PROGRAMS WITH CONVEX FUNCTIONS, A.
ROSEC 1977 -1 ROSENTHAL, A.
NONSERIAL DYNAMIC PROGRAMMING IS OPTIMAL.
ROSSC 1975 -1 ROSSMAN, L. A.
RESOLUTION OF DUALITY GAPS WITH DYNAMIC PROGRAMMING.
SPAN 1976 -1 SPANEL, J.
DYNAMIC PROGRAMMING FOR SOLVING ONE MULTISTATE DECISION-MAKING PROBLEM WITH INTEGER VARIABLES (CZECH).
SPRA 1967 -1 SPRADLIN, B. C.; PIERCE, D. A.
PRODUCTION SCHEDULING UNDER LEARNING EFFECT BY DYNAMIC PROGRAMMING.
SRIN 1976 -1 SRINIVASAN, V.
DECOMPOSITION OF A MULTI-PERIOD MEDIA SCHEDULING MODEL IN TERMS OF SINGLE PERIOD EQUIVALENTS.
STRA 1967 -1 STRADLIN, B. C.; PIERCE, D. A.
PRODUCTION SCHEDULING UNDER A LEARNING EFFECT BY DYNAMIC PROGRAMMING.
SWEEA 1976 -1 SWEENEY, D. J.; TATHAM, R. L.
IMPROVED LONG-RUN MODEL FOR MULTIPLE WAREHOUSE LOCATION, AN.
TERN 1975 -1 TERNO, J.; GIERTH, G.; WAGNER, H.
BESTIMMUNG DES KUERZESTEN WEGES MITTELS DYNAMISCHER OPTIMIERUNG, DIE.
THOM 1976 -1 THOMAS, M. E.
SURVEY OF THE STATE OF THE ART IN DYNAMIC PROGRAMMING, A.
TILA 1976 -1 TILANUS, C. B.; GERHARDT, C.
APPLICATION OF CUTTING STOCK IN THE STEEL INDUSTRY, AN.
TOTH 1977 -1 TOTH, P.
DYNAMIC PROGRAMMING ALGORITHM FOR THE ZERO-ONE KNAPSACK PROBLEM, A.
TRUS 1976 -2 TRUSCOTT, W. G.
SOME DYNAMIC EXTENSIONS OF A DISCRETE LOCATION-ALLOCATION PROBLEM.
WALT 1976 -1 WALTERS, J. R.
EXTENSION TO THE ASSORTMENT PROBLEM IN WHICH COMPONENTS ARE GROUPED TO FORM ASSEMBLIES, AN.
WESOB 1975 -1 WESOLOWSKY, G. O.; TRUSCOTT, W. G.
MULTIPERIOD LOCATION-ALLOCATION PROBLEM WITH RELOCATION OF FACILITIES, THE.
ZALOA 1977 -1 ZALOOM, V. A.; SMYER, W. N.
SCHEDULING TO MINIMIZE MAKESPAN REQUIRED TO PROCESS INDEPENDENT JOBS ON PARALLEL NON-EQUIVALENT SERVERS.

FACETS OF INTEGER POLYHEDRA
****

BALAG 1976 -1 BALAS, E.
FACETS OF ONE-DIMENSIONAL AND MULTI-DIMENSIONAL KNAPSACK POLYTOPES.

BALAG 1974 -1 BALAS, E.
DISJUNCTIVE PROGRAMMING: FACETS OF THE CONVEX HULL OF FEASIBLE POINTS.

BALAG 1975 -3 BALAS, E.
LINEAR CHARACTERIZATION OF PERMUTATION VECTORS, A.

BALAG 1977 -3 BALAS, E.
SOME VALID INEQUALITIES FOR THE SET PARTITIONING PROBLEM.

BALAJ 1976 -1 BALAS, E.; PADBERG, M. W.
SET PARTITIONING: A SURVEY.

BALAJ 1975 -1 BALAS, E.; PADBERG, M. W.
SET PARTITIONING: A SURVEY.

BALAL 1975 -1 BALAS, E.; ZEMEL, E.
ALL THE FACETS OF ZERO-ONE PROGRAMMING POLYTOPES WITH POSITIVE COEFFICIENTS.

BALAL 1976 -1 BALAS, E.; ZEMEL, E.
CRITICAL CUTSETS OF GRAPHS AND CANONICAL FACETS OF SET PACKING POLYTOPES.

BALAL 1977 -1 BALAS, E.; ZEMEL, E.
GRAPH SUBSTITUTION AND SET PACKING POLYTOPES.

BRUAA 1976 -1 BRUALDI, R. A.; GIBSON, P. M.
ASSIGNMENT POLYTOPE, THE.

BRUAA 1977 -1 BRUALDI, R. A.; GIBSON, P. M.
CONVEX POLYHEDRA OF DOUBLY STOCHASTIC MATRICES III. AFFINE AND COMBINATORIAL PROPERTIES OF OMEGA N.

BRUAA 1977 -2 BRUALDI, R. A.; GIBSON, P. M.
CONVEX POLYHEDRA OF DOUBLY STOCHASTIC MATRICES, 1. APPLICATIONS OF THE PERMANENT FUNCTION.

BRUAB 1978 -1 BRUALDI, R. A.; LEE, G. M.
ON THE TRUNCATED ASSIGNMENT POLYTOPE.

EDEL 1970 -1 EDELBERG, M.
INTEGRAL CONVEX POLYHEDRA AND AN APPROACH TO INTEGRALIZATION.

GIBS 1976 -1 GIBSON, P. M.
FACETS OF FACES OF TRANSPORTATION POLYTOPES.

GILE 1977 -3 GILES, F. R.
OPTIMUM MATCHING FORESTS 3: FACETS OF MATCHING FOREST POLYHEDRA.

GILEB 1976 -1 GILES, R.
FACETS AND OTHER FACES OF BRANCHING POLYHEDRA.

GILLC 1975 -1 GILLES, F. R.
SUBMODULAR FUNCTIONS, GRAPHS AND INTEGER POLYHEDRA.

GLOV 1973 -1 GLOVER, F.
POLYHEDRAL ANNEXATION IN MIXED INTEGER PROGRAMMING.

GLOV 1975 -3 GLOVER, F.
POLYHEDRAL ANNEXATION IN MIXED INTEGER AND COMBINATORIAL PROGRAMMING.

GROE 1977 -1 GROETSCHEL, M.
FACETTEN VON MATROID-POLYTOPEN.

GROE 1977 -3 GROETSCHEL, M.
MONOTONE 2-MATCHING POLYTOPE ON A COMPLETE GRAPH, THE.

GROE 1977 -4 GROETSCHEL, M.
ON THE MONOTONE SYMMETRIC TRAVELLING SALESMAN PROBLEM: HYPOHAMILTONIAN/HYPOTRACEABLE GRAPHS AND FACETS.

GROE 1977 -5 GROETSCHEL, M.
ON THE SYMMETRIC TRAVELLING SALESMAN PROBLEM: SOLUTION OF A 120-CITY PROBLEM.

GROE 1977 -6 GROETSCHEL, M.
POLYEDRISCHE CHARAKTERISIERUNGEN KOMBINATORISCHER OPTIMIERUNGSPROBLEME.

GROE 1977 -7 GROETSCHEL, M.
STRONG BLOCKS AND THE OPTIMUM BRANCHING PROBLEM.

GROEA 1978 -1 GROETSCHEL, M.; PADBERG, M. W.
ON THE SYMMETRIC TRAVELLING SALESMAN PROBLEM: THEORY AND COMPUTATION.

GROEA 1977 -1 GROETSCHEL, M.; PADBERG, M. W.
LINEARE CHARAKTERISIERUNGEN VON TRAVELLING SALESMAN PROBLEMEN.

GROEA 1977 -2 GROETSCHEL, M.; PADBERG, M. W.
ON THE SYMMETRIC TRAVELLING SALESMAN PROBLEM 1: INEQUALITIES.

GROEA 1977 -3 GROETSCHEL, M.; PADBERG, M. W.
ON THE SYMMETRIC TRAVELLING SALESMAN PROBLEM 2: LIFTING THEOREMS AND FACETS.

HAMMA 1975 -1 HAMMER, P. L.
BOOLEAN ELEMENTS IN COMBINATORIAL OPTIMIZATION: A SURVEY.

HAMMB 1977 -1 HAMMER, P. L.; JOHNSON, E. L.; PELED, U. N.
ROLE OF MASTER POLYTOPES IN THE UNIT CUBE, THE.

JEROA 1977 -1 JEROSLOW, R. G.
CUTTING-PLANE GAME AND ITS ALGORITHMS, A.

JOHNE 1976 -1 JOHNSON, E. L.
FACES OF POLYHEDRA OF MIXED INTEGER PROGRAMMING PROBLEMS.

MAURB 1976 -1 MAURRAS, J. F.
POLYTOPES A SOMMETS DANS (0,1)N.

MAURC 1975 -1 MAURRAS, J.-F.
SOME RESULTS ON THE CONVEX HULL OF THE HAMILTONIAN CYCLES OF SYMMETRIC COMPLETE GRAPHS.

PADBA 1977 -1 PADBERG, M. W.
COVERING, PACKING AND KNAPSACK PROBLEMS.

PADBA 1976 -1 PADBERG, M. W.
ON THE COMPLEXITY OF SET PACKING POLYHEDRA (REVISED).

PADBA 1977 -2 PADBERG, M. W.
ON THE COMPLEXITY OF SET PACKING POLYHEDRA.

PADBB 1977 -1 PADBERG, M. W.; HONG, S.
ON THE SYMMETRIC TRAVELLING SALESMAN PROBLEM: A COMPUTATIONAL STUDY.

PELE 1977 -1 PELED, U. N.
PROPERTIES OF FACETS OF BINARY POLYTOPES.

PULLA 1974 -1 PULLEYBLANK, W. R.; EDMONDS, J.
FACETS OF 1-MATCHING POLYHEDRA.

SCHM 1976 -1 SCHMIDT, B. K.; MATTHEISS, T. H.
ENUMERATING THE SET OF INEQUALITIES DEFINING THE CONVEX HULL OF A GIVEN SET OF POINTS IN E N.

WOLS 1976 -1 WOLSEY, L. A.
FACETS AND STRONG VALID INEQUALITIES FOR INTEGER PROGRAMS.

WOLS 1975 -1 WOLSEY, L. A.
VALID INEQUALITIES AND SUPERADDITIVITY FOR 0-1 INTEGER PROGRAMS.

WOLS 1976 -2 WOLSEY, L. A.
FURTHER FACET GENERATING PROCEDURES FOR VERTEX PACKING POLYTOPES.

****

ALCO 1975 -1 ALCOUFFE, A.; ENJALBERT, M.; MURATET, G.
METHODES DE RESOLUTION DU PROBLEME DE TRANSPORT ET DE PRODUCTION D'UNE ENTREPRISE A ETABLISSEMENTS MULTIPLES EN PRESENCE DE COUTS FIXES.

ALCOA 1976 -1 ALCOUFFE, A.; MURATET, G.
OPTIMAL LOCATION OF PLANTS.

AROR 1977 -1 ARORA, S. R.
NOTE ON FRACTIONAL FIXED CHARGE PROBLEMS, A.

BALAC 1976 -1 BALACHANDRAN, V.; PERRY, A.
TRANSPORTATION TYPE PROBLEMS WITH QUANTITY DISCOUNTS.

BHAT 1976 -1 BHATIA, H. L.; SWARUP, K.; PURI, M. C.
TIME-COST TRADE-OFF IN A TRANSPORTATION PROBLEM.

CHEU 1976 -1 CHEUNG, T.-Y.
NETWORK FLOW APPROACH TO STEINER PROBLEMS, A.

COOP 1975 -2 COOPER, L.
FIXED CHARGE PROBLEM, I. A NEW HEURISTIC METHOD, THE.

CORB 1973 -1 CORBAN, A.
TRANSPORTATION PROBLEMS WITH FIXED COSTS (RUMANIAN).

DEHN 1976 -1 DEHNERT, G.
BRANCH AND BOUND METHOD FOR SOLVING FIXED CHARGE NETWORK FLOW PROBLEMS, A.

ERLE 1976 -1 ERLENKOTTER, D.
DUAL-BASED PROCEDURE FOR UNCAPACITATED FACILITY LOCATION, A.

ERLE 1976 -2 ERLENKOTTER, D.
FACILITY LOCATION WITH PRICE-SENSITIVE DEMANDS: PRIVATE, PUBLIC, AND QUASI-PUBLIC.

FISKA 1977 -1 FISK, J.; MACKEOWN, P.
DETERMINING THE MAXIMUM DEGENERATE SOLUTION FOR TRANSPORTATION PROBLEMS.

FORG 1976 -2 FORGO, F.
SOME TRANSFORMATIONS OF INTEGER PROGRAMMING PROBLEMS.

HONGB 1977 -1 HONG, S.; PADBERG, M. W.
NOTE ON THE SYMMETRIC MULTIPLE TRAVELING SALESMAN PROBLEM WITH FIXED CHARGES, A.

KENN 1976 -1 KENNINGTON, J. L.
FIXED-CHARGE TRANSPORTATION PROBLEM: A COMPUTATIONAL STUDY WITH A BRANCH-AND-BOUND CODE, THE.

KRAR 1967 -1 KRARUP, J.
FIXED-COST AND OTHER NETWORK FLOW PROBLEMS AS RELATED TO PLANT LOCATION AND TO THE DESIGN OF TRANSPORTATION AND COMPUTER SYSTEMS.

KRAR 1967 -2 KRARUP, J.
FIXED-COST NETWORK FLOW PROBLEMS.

MAC G 1975 -1 MAC KEOWN, P.
VERTEX RANKING PROCEDURE FOR SOLVING THE LINEAR FIXED-CHARGE PROBLEM, A.

MINOA 1976 -3 MINOUX, M.
MULTIFLOTS DE COUT MINIMAL AVEC FONCTIONS DE COUT CONCAVES.

NAUS 1976 -3 NAUSS, R. M.
FIXED CHARGE TRANSPORTATION PROBLEM WITH ADDITIONAL CONSTRAINTS, THE.

PENT 1976 -1 PENTICO, D. W.
ASSORTMENT PROBLEM WITH NONLINEAR COST FUNCTIONS, THE.

RARDA 1976 -2 RARDIN, R. L.; UNGER, V. E.
SOME COMPUTATIONALLY RELEVANT GROUP THEORETIC STRUCTURES OF FIXED CHARGE PROBLEMS.

ROBE 1976 -1 ROBERS, P.; COOPER, L.
STUDY OF THE FIXED CHARGE TRANSPORTATION PROBLEM, A.

SCHO 1976 -1 SCHOCH, M.
ERWEITERUNGSPRINZIP UND SEINE ANWENDUNG ZUR ENTWICKLUNG VON ALGORITHMEN FUER DIE LOESUNG KOMBINATORISCHER OPTIMIERUNGSAUFGABEN, DAS.

SHARB 1976 -1 SHARMA, J. K.; AGRAWAL, S. C.
FIXED-CHARGE TRANSPORTATION PROBLEM, A.

STEIA 1977 -1 STEINBERG, D. I.
DESIGNING A HEURISTIC FOR THE FIXED-CHARGE TRANSPORTATION PROBLEM.

STEIA 1976 -1 STEINBERG, D. I.
ON SOLVING LARGE-SCALE FIXED CHARGE PROBLEMS.

STUH 1966 -1 STUHR, R.
ANWENDUNGEN UND LOESUNGSVERFAHREN DER GANZZAHLIGEN LINEAREN PROGRAMMIERUNG.

WALKA 1976 -1 WALKER, W. E.
HEURISTIC ADJACENT EXTREME POINT ALGORITHM FOR THE FIXED CHARGE PROBLEM, A.

WDOW 1973 -1 WDOWIAK, J.
SOLUTION OF THE FIXED CHARGE PROBLEM (POLISH), THE.

GRAPH THEORETIC OPTIMIZATION PROBLEMS (UNLESS SPECIFIED IN PARTICULAR)
****

AHO 1977 -1 AHO, A. V.; GAREY, M. R.; HWANG, F. K.
RECTILINEAR STEINER TREES: EFFICIENT SPECIAL-CASE ALGORITHMS.

ALAV 1977 -1 ALAVI, Y.; BEHZAD, M.; LESNIAK-FOSTER, L. M.; NORDHAUS, E. A.
TOTAL MATCHINGS AND TOTAL COVERINGS OF GRAPHS.

ALSP 1976 -1 ALSPACH, B.; MASON, D. W.; PULLMAN, N. J.
PATH NUMBERS OF TOURNAMENTS.

AMIN 1973 -1 AMIN, A. T.; HAKIMI, S. L.
MAXIMALLY EDGE-CONNECTED GRAPHS WITH LEAST NUMBER OF MINIMUM EDGE CUT-SETS.

ANEJ 1977 -1 ANEJA, Y. P.; VEMUGANTI, R. R.
ROW GENERATION SCHEME FOR FINDING A MULTI-COMMODITY MINIMUM DISCONNECTING SET, A.

AOSH 1977 -1 AOSHIMA, K.; IRI, M.
COMMENTS ON F. HADLOCK'S PAPER: "FINDING A MAXIMUM CUT OF A PLANAR GRAPH IN POLYNOMIAL TIME".

APPE 1976 -1 APPEL, K.; HAKEN, W.
EVERY PLANAR MAP IS FOUR COLORABLE.

BALAF 1977 -1 BALAGURU, P. N.; RAMAKRISHNAN, V.; NAAMAN, A. E.
COST AND TIME SCALED NETWORKS - GRAPHICAL MODEL.

BALAK 1977 -1 BALAS, E.; SAMUELSSON, H.
NODE COVERING ALGORITHM, A.

BAUMA 1975 -1 BAUMGARTNER, J. E.; DAYKIN, D. E.; GUTMAN, I.; KLEITMAN, D. J. ET AL.
PROBLEMS.

BEANA 1970 -1 BEAN, D. W. T.
MINIMAL EDGE SETS MEETING ALL CIRCUITS OF A GRAPH.

BERG 1978 -1 BERGE, C.
ALGORITHMS AND EXTREMAL PROBLEMS FOR EQUIPARTITE COLORINGS IN GRAPHS AND HYPERGRAPHS (ABSTRACT).

BERM 1975 -1 BERMOND, J. C.
1-GRAPHES REGULIERS MINIMAUX DE GIRTH DONNE.

BERZ 1971 -1 BERZTISS, A. T.
DATA STRUCTURES, THEORY AND PRACTICE.

BEUE 1975 -1 BEUERMANN-WEHMEYER, A.
PRIMAL-DUAL-BEZIEHUNGEN IN GRAPHENALGORITHMEN.

BIES 1976 -1 BIESS, G.
GRAPHENTHEORIE. MATHEMATIK FUER INGENIEURE, NATURWISSENSCHAFTLER, OEKONOMEN UND LANDWIRTE 21/2.

BIXB 1975 -1 BIXBY, R. E.
MINIMUM NUMBER OF EDGES AND VERTICES IN A GRAPH WITH EDGE CONNECTIVITY N AND M N-BONDS, THE.

BOES 1976 -1 BOESCH, F.; BUTLER, D. E.
ON THE SYNTHESIS OF MINIMUM COST RELIABLE NETWORKS AND MULTIGRAPHS.

BOESB 1974 -1 BOESCH, F. T.; CHEN, S.; MAC HUGH, J. A. M.
ON COVERING THE POINTS OF A GRAPH WITH POINT DISJOINT PATHS.

BOESC 1977 -1 BOESCH, F. T.; GIMPEL, J. F.
COVERING THE POINTS OF A DIGRAPH WITH POINT-DISJOINT PATHS AND ITS APPLICATION TO CODE OPTIMIZATION.

BOLLC 1975 -1 BOLLOBAS, B.; ERDOES, P.
EXTREMAL PROBLEM OF GRAPHS WITH DIAMETER 2, AN.

BURKH 1975 -1 BURKOV, V. N.; GROPPEN, V. O.
MAXIMUM CIRCULATION AND MINIMUM SECTION ON THE PLANAR ORGRAPHS (RUSSIAN).

BURNC 1976 -1 BURNS, R. N.; HAFF, C. E.
COMBINATORIAL RANKING PROBLEM, A.

CACC 1976 -1 CACCETTA, L.
PROBLEM IN EXTREMAL GRAPH THEORY, A.

CEDE 1976 -1 CEDERBAUM, I.
EXTREMAL PROBLEM IN THE THEORY OF N-CUBE GRAPHS, AN.

CHANF 1977 -1 CHANDRASEKARAN, R.
MINIMAL RATIO SPANNING TREES.

CHEU 1976 -1 CHEUNG, T.-Y.
NETWORK FLOW APPROACH TO STEINER PROBLEMS, A.

CHOY 1977 -1 CHOY, D. M.; WONG, C. K.
BOUNDS FOR OPTIMAL ALPHA-BETA BINARY TREES.

CHRI 1973 -1 CHRISTOFIDES, N.
OPTIMUM TRANSVERSAL OF A GRAPH, THE.

CHUN 1978 -1 CHUNG, F. R. K.; GRAHAM, R. L.
STEINER TREES FOR LADDERS.

COCKB 1975 -1 COCKAYNE, E. J.; HEDETNIEMI, S. T.
OPTIMAL DOMINATION IN GRAPHS.

COCKB 1977 -1 COCKAYNE, E. J.; HEDETNIEMI, S. T.
TOWARDS A THEORY OF DOMINATION IN GRAPHS.

COCKB 1976 -1 COCKAYNE, E. J.; HEDETNIEMI, S. T.
DISJOINT INDEPENDENT DOMINATING SETS IN GRAPHS.

COCKC 1977 -1 COCKAYNE, E. J.; ROBERTS, F. D. K.
COMPUTATION OF DOMINATING PARTITIONS.

COOK 1974 -1 COOK, R. J.
POINT-ARBORICITY AND GIRTH.

CORN 1974 -1 CORNEIL, D. G.
ANALYSIS OF GRAPH THEORETICAL ALGORITHMS, THE.

DEWE 1976 -1 DEWESS, G.
KONVEXITAET UND OPTIMIERUNG IN METRISCHEN GRAPHEN.

DIJK 1976 -1 DIJKSTRA, E. W.
DISCIPLINE OF PROGRAMMING, A.

DIXO 1976 -1 DIXON, E. T.; GOODMAN, S. E.
ALGORITHM FOR THE LONGEST CYCLE PROBLEM, AN.

DOYL 1971 -1 DOYLE, M. S.
MINIMUM PATH COVERING ALGORITHMS.

EDENA 1976 -1 EDEN, B. N.
HEURISTISCHES VERFAHREN ZUR DISTRIKTERMITTLUNG IN BEWERTETEN GRAPHEN UND SEINE ANWENDUNG AUF DIE BERECHNUNG VON SCHULEINZUGSBEREICHEN, EIN.

EDENB 1978 -1 EDEN, B. N.; PAPE, U.
HEURISTISCHES VERFAHREN ZUR DISTRIKTERMITTLUNG IN BEWERTETEN GRAPHEN, EIN.

EPIH 1975 -1 EPIHIN, V. V.; HARKEVIC, A. D.
ON THE QUESTION OF THE CONSTRUCTION OF OPTIMAL GRAPHS (RUSSIAN).

ERDO 1976 -1 ERDOES, P.
PROBLEMS AND RESULTS IN GRAPH THEORY AND COMBINATORIAL ANALYSIS.

ESWA 1976 -1 ESWARAN, K. P.; TARJAN, R. E.
AUGMENTATION PROBLEMS.

EVENC 1970 -1 EVEN, S.; LEMPEL, A.; PNUELI, A.
PERMUTATION GRAPHS AND TRANSITIVE GRAPHS.
FITZ 1974 -1 FITZGERALD, C. H.
OPTIMAL INDEXING OF THE VERTICES OF GRAPHS.
FOULA 1976 -1 FOULDS, L. R.; ROBINSON, D. F.
STRATEGY FOR SOLVING THE PLANT LAYOUT PROBLEM, A.
FRAND 1976 -1 FRANK, A.
SOME POLYNOMIAL ALGORITHMS FOR CERTAIN GRAPHS AND HYPERGRAPHS.
FRANE 1969 -1 FRANK, H.
MAXIMALLY RELIABLE NODE WEIGHTED GRAPHS.
FUJIB 1974 -1 FUJISAWA, T.; ORINO, H.
EFFICIENT ALGORITHM OF FINDING A MINIMAL TRIANGULATION OF A GRAPH, AN.
FUJIC 1977 -1 FUJISHIGE, S.
ALGORITHM FOR FINDING AN OPTIMAL INDEPENDENT LINKAGE, AN.
GARE 1977 -1 GAREY, M. R.; GRAHAM, R. L.; JOHNSON, D. S.
COMPLEXITY OF COMPUTING STEINER MINIMAL TREES, THE.
GAREB 1977 -1 GAREY, M. R.; JOHNSON, D. S.
RECTILINEAR STEINER TREE PROBLEM IS NP-COMPLETE, THE.
GARFB 1977 -1 GARFINKEL, R. S.; NEEBE, A. W.; RAO, M. R.
M-CENTER PROBLEM: MINIMAX FACILITY LOCATION, THE.
GARS 1977 -1 GARSIA, A. M.; WACHS, M. L.
NEW ALGORITHM FOR MINIMUM COST BINARY TREES, A.
CAVR 1977 -1 CAVRIL, F.
ALGORITHMS ON CLIQUE SEPARABLE GRAPHS.
GEOL 1975 -1 GEOLECJAN, G. G.
OPTIMAL ASSIGNMENTS AND THEIR PROPERTIES (RUSSIAN).
GILB 1968 -1 GILBERT, E. N.; POLLAK H. O.
STEINER MINIMAL TREES.
GILE 1977 -1 GILES, F. R.
OPTIMUM MATCHING FORESTS 1: SPECIAL WEIGHTS.
GILE 1977 -2 GILES, F. R.
OPTIMUM MATCHING FORESTS 2: GENERAL WEIGHTS.
GILE 1977 -3 GILES, F. R.
OPTIMUM MATCHING FORESTS 3: FACETS OF MATCHING FOREST POLYHEDRA.
GILEB 1976 -1 GILES, R.
FACETS AND OTHER FACES OF BRANCHING POLYHEDRA.
GOL'A 1975 -1 GOL'DBERG, M. K.; KLIPKER, I. A.
NUMBERING OF A UNIFORM TREE THAT MINIMIZES THE SUM OF THE LENGTHS OF THE EDGES (RUSSIAN), A.
GONDB 1977 -1 GONDRAN, M.; MINOUX, M.
VALEURS PROPRES ET VECTEURS PROPRES DANS LES DIOIDES ET LEUR INTERPRETATION EN THEORIE DES GRAPHES.
GOODA 1974 -1 GOODMAN, S.; HEDETNIEMI, S.
ON THE HAMILTONIAN COMPLETION PROBLEM.
GOODB 1974 -1 GOODMAN, S.; HEDETNIEMI, S.; SLATER, P.
SOME RESULTS ON THE OPTIMAL AND NON-OPTIMAL HAMILTONIAN COMPLETION PROBLEM.
GOODD 1975 -1 GOODMAN, S. E.; HEDETNIEMI, S. T.; SLATER, P. J.
ADVANCES ON THE HAMILTONIAN COMPLETION PROBLEM.
GRAH 1967 -1 GRAHAM, R. L.
SOME RESULTS ON STEINER MINIMAL TREES.
HABI 1975 -1 HABIB, M.; LAMBERT, G.
NOMBRE DE SAUTS ET INDICE DE PARTITION DANS LES GRAPHES SANS CIRCUIT.
HADL 1974 -1 HADLOCK, F. O.
MINIMUM SPANNING FORESTS OF BOUNDED TREES.
HAKI 1969 -1 HAKIMI, S.
ALGORITHM FOR CONSTRUCTION OF THE LEAST VULNERABLE COMMUNICATION NETWORK, AN.
HANSB 1976 -2 HANSEN, P.; DELATTRE, M.
CLUSTER ANALYSIS BY GRAPH COLORING.
HAUS 1977 -1 HAUSMANN, D.
ADJACENCY ON THE BRANCHING POLYHEDRON.
HAUS 1977 -3 HAUSMANN, D.
POLYNOMIALER ALGORITHMUS FUER DIE ADJAZENZUEBERPRUEFUNG AUF DEM BRANCHING-POLYEDER, EIN.
HEDE 1971 -1 HEDETNIEMI, S.
CHARACTERIZATIONS AND CONSTRUCTIONS OF MINIMALLY TWO-CONNECTED GRAPHS AND MINIMALLY STRONG DIGRAPHS.
HIRSA 1976 -2 HIRSCHBERG, D. S.; WONG, C. K.
UPPER AND LOWER BOUNDS FOR GRAPH-DIAMETER PROBLEMS WITH APPLICATION TO RECORD ALLOCATION.
HUBI 1975 -1 HUBICKA, E.; SYSLO, M. M.
MINIMAL BASES OF CYCLES OF A GRAPH.
HULM 1975 -1 HULME, B. L.
LATTICE ALGEBRA FOR FINDING SIMPLE PATHS AND CUTS IN A GRAPH, A.
HWAN 1976 -1 HWANG, F. K.
ON STEINER MINIMAL TREES WITH RECTILINEAR DISTANCE.
IORD 1974 -1 IORDANSKII, M. A.
MINIMAL NUMERATIONS OF THE VERTICES OF TREES (RUSSIAN).
IORD 1976 -1 IORDANSKII, M. A.
MINIMAL NUMBERINGS OF THE VERTICES OF TREES (RUSSIAN).
ITAI 1977 -1 ITAI, A.; RODEH, M.
FINDING A MINIMUM CIRCUIT IN A GRAPH.
JOHND 1977 -1 JOHNSON, D. S.; LENSTRA, J. K.; RINNOOY KAN, A. H. G.
COMPLEXITY OF THE NETWORK DESIGN PROBLEM, THE.
JUSK 1975 -1 JUSKA, F.
NUMERICAL METHOD FOR THE CONSTRUCTION OF A NETWORK OF MINIMAL LENGTH (RUSSIAN), A.
KAAZ 1973 -1 KAAZIK, YU. YA.; PREEM, M. K.
PROBLEM OF FINDING ROUTES (RUSSIAN), A.
KAME 1976 -1 KAMEDA, T.
ON MAXIMALLY DISTANT SPANNING TREES OF A GRAPH.
KARAA 1976 -1 KARAPETJAN, S. R.
(X-O,M)-MINIMAL CUT AND PARTITIONING OF A TREE (RUSSIAN).
KARP 1975 -1 KARP
GRAPH ALGORITHMS.
KLEEA 1976 -1 KLEE, V.; QUAIFE, H.
MINIMUM GRAPHS OF SPECIFIED DIAMETER, CONNECTIVITY AND VALENCE, I.
KOPY 1977 -1 KOPYLOV, G. N.
MAXIMAL PATHS AND CYCLES IN A GRAPH (RUSSIAN).

GRAPH THEORETIC OPTIMIZATION PROBLEMS (UNLESS SPECIFIED IN PARTICULAR)
****

KUND 1975 -1 KUNDU, S.
MINIMUM DIAMETER STEINER TREES IN A GRAPH, THE.
KUNDA 1977 -1 KUNDU, S.; MISRA, J.
LINEAR TREE PARTITIONING ALGORITHM, A.
LAWL 1976 -1 LAWLER, E. L.
GRAPHICAL ALGORITHMS AND THEIR COMPLEXITY.
LAWL 1975 -1 LAWLER, E. L.
ALGORITHMS, GRAPHS, AND COMPLEXITY.
LEE 1976 -1 LEE, D. H.
LOW COST DRAINAGE NETWORKS.
LERD 1975 -1 LERDA, F.; MAJORANI, E.
COMMENTS ON A NOTE BY F. RUBIN: "A NOTE ON LERDA AND MAJORANI MINIMUM CROSSING ALGORITHM".
LIEN 1974 -1 LIEN, Y. E.
PRELIMINARY REPORT: SOME COMPLETE PROBLEMS FOR LANGUAGES RECOGNIZABLE IN NONDETERMINISTIC LOG-SPACE.
LOVA 1977 -3 LOVASZ, L.
ON THE SHANNON CAPACITY OF A GRAPH.
LUDO 1971 -1 LUDOT, J.-P.; RUBINSTEIN, M.-C.
METHODE POUR LA RECHERCHE D'UNE POLITIQUE OPTIMALE D'INVESTISSEMENT A COURT TERME D'UN RESEAU DE DISTRIBUTION.
MARTB 1977 -1 MARTELLO, S.
ALGORITHM FOR FINDING A MINIMAL EQUIVALENT GRAPH OF A STRONGLY CONNECTED DIGRAPH, AN.
MATT 1976 -1 MATTHAEUS, F. W.
SAVINGS APPROACH TO THE MULTIPLE TERMINAL DELIVERY PROBLEM.
MATU 1977 -1 MATULA, D. W.
GRAPH THEORETIC TECHNIQUES FOR CLUSTER ANALYSIS ALGORITHMS.
MEIR 1973 -1 MEIR, A.; MOON, J. W.
EXPECTED NODE-INDEPENDENCE NUMBER OF RANDOM TREES, THE.
MEIR 1975 -2 MEIR, A.; MOON, J. W.
RELATIONS BETWEEN PACKING AND COVERING NUMBERS OF A TREE.
MELZ 1961 -1 MELZAK, Z. A.
ON THE PROBLEM OF STEINER.
MORRB 1976 -1 MORROW, C.; GOODMAN, S.
EFFICIENT ALGORITHM FOR FINDING A LONGEST CYCLE IN A TOURNAMENT, AN.
NASH 1964 -1 NASH-WILLIAMS, C. ST. J. A.
DECOMPOSITION OF FINITE GRAPHS INTO FORESTS.
NASH 1961 -2 NASH-WILLIAMS, C. ST. J. A.
EDGE-DISJOINT SPANNING TREES OF FINITE GRAPHS.
NICO 1975 -1 NICOGOSYAN, Z. G.
MINIMAL DECOMPOSITION OF GRAPHS IN SPANNING FORESTS (RUSSIAN).
NIEM 1974 -2 NIEMINEN, J.
TWO BOUNDS FOR THE DOMINATION NUMBER OF A GRAPH.
NIEM 1975 -3 NIEMINEN, J.
SOME OBSERVATIONS ON THE DETERMINATION OF AN UPPER BOUND FOR THE CLIQUE NUMBER OF A GRAPH.
NOOR 1975 -1 NOORVASH, S.
COVERING THE VERTICES OF A GRAPH BY VERTEX-DISJOINT PATHS.
O'BR 1977 -1 O'BRIAN, R. C.
UPPER BOUND ON THE PATH NUMBER OF A DIGRAPH, AN.
OHTS 1976 -1 OHTSUKI, T.
FAST ALGORITHM FOR FINDING AN OPTIMAL ORDERING FOR VERTEX ELIMINATION ON A GRAPH, A.
OHTSA 1976 -1 OHTSUKI, T.; CHEUNG, L. K.; FUJISAWA, T.
MINIMAL TRIANGULATION OF A GRAPH AND OPTIMAL PIVOTING ORDER IN A SPARSE MATRIX.
ORLOC 1975 -1 ORLOVA, G. I.; DORFMAN, YA. G.
REDUCING THE PROBLEMS OF OPTIMALLY PARTITIONING THE SET OF VERTICES OF A GRAPH TO A MAXIMAL CUT PROBLEM (RUSSIAN).
OSYC 1976 -1 OSYCZKA, A.
ALGORITHM OF OPTIMIZATION FOR A SPECIAL CLASS OF NETWORKS, AN.
PANA 1977 -1 PANAGIOTAKOPOULOS, D.
COST-TIME MODEL FOR LARGE CPM-PROJECT NETWORKS.
PARI 1963 -1 PARIKH, S.; FRISCH, I.
FINDING THE MOST RELIABLE ROUTES IN COMMUNICATION SYSTEM.
PART 1976 -1 PARTHASARATHY, K. R.; CHOUDUM, S. A.
EDGE CLIQUE COVER NUMBER OF PRODUCTS OF GRAPHS, THE.
PICA 1975 -1 PICARD, C.-F.
ARBRE MINIMAL D'UNE SOMME DE GRAPHES.
PICAB 1976 -1 PICARD, J.-C.
MAXIMAL CLOSURE OF A GRAPH AND APPLICATIONS TO COMBINATORIAL PROBLEMS.
PICH 1976 -1 PICHOCKA, M.
CERTAIN OPTIMIZATION PROBLEM CONNECTED WITH GRAPH THEORY (POLISH), A.
PLES 1976 -1 PLESNIK, JAN.
MINIMUM BLOCK CONTAINING A GIVEN GRAPH.
PLESA 1976 -1 PLESTCHINSKY, A. S.
E-APPROACH TO A NETWORK MODEL OPTIMIZATION (RUSSIAN).
POLE 1976 -1 POLESSKIJ, V. P.
COVERING A FINITE GRAPH WITH A MINIMAL NUMBER OF FORESTS (RUSSIAN).
POLJA 1974 -1 POLJAKOV, A. S.
SHORTENING THE ENUMERATION IN THE SOLUTION OF A CERTAIN CLASS OF PROBLEMS IN GRAPH THEORY (RUSSIAN).
PRAT 1973 -1 PRATT, V. R.
LEAST UPPER BOUND ON CONVENTIONAL SHELLSORT, A.
RIVI 1975 -1 RIVIERE, M.
COVERING OF VERTICES OF A UNIFORM HYPERGRAPH BY EDGES AND STABLE SETS.
ROSE 1976 -1 ROSE, D. J.; TARJAN, R. E.; LUEKER, G. S.
ALGORITHMIC ASPECTS OF VERTEX ELIMINATION ON GRAPHS.
ROSED 1977 -1 ROSENTHAL, A.; GOLDNER, A.
SMALLEST AUGMENTATIONS TO BICONNECT A GRAPH.
RUBIE 1976 -1 RUBINSTEIN, M. C.; LUDOT, J. P.
APPLICATION DE LA THEORIE DES GRAPHES AU CHOIX DES INVESTISSEMENTS A COURT TERME DANS UN RESEAU ELECTRIQUE SUJET A AVARIES.
RUBIF 1974 -1 RUBINSTEIN, M.-C.
APPLICATION DE LA THEORIE DES GRAPHES AU CHOIX DES INVESTISSEMENTS DANS UN RESEAU D'ENERGIE ELECTRIQUE SUJET A AVARIES.
RUSCA 1976 -1 RUSCITTI, A.
CANONICAL REDUCTION AND CONSTRUCTION METHOD RELATIVE TO MAXIMAL PLANAR GRAPHS (ITALIAN), A.

RYZH 1975 -1 RYZHKOV, A. P.
ALGORITHM FOR PARTITIONING A GRAPH INTO MINIMALLY CONNECTED SUBGRAPHS (RUSSIAN), AN.
RYZH 1975 -2 RYZHKOV, A. P.
GRAPH DIVISION INTO THE MINIMAL NUMBER OF COMPLETE SUBGRAPHS (RUSSIAN).
SCHA 1976 -1 SCHAAR, G.
ON THE MAXIMUM NUMBER OF EDGE-DISJOINT PATHS WITH ENDPOINTS BELONGING TO A GIVEN SET OF VERTICES IN A FINITE GRAPH.
SCHMB 1975 -1 SCHMITZ, W.
UEBER LAENGSTE WEGE UND KREISE IN GRAPHEN.
SCHOB 1975 -1 SCHOENE, W.
ALGORITHMEN ZUR BESTIMMUNG INTERN UND EXTERN STABILER MENGEN UND DER CHROMATISCHEN ZAHL EINES ENDLICHEN GRAPHEN.
SEID 1974 -1 SEIDVASSER, M. A.
ASSIGNMENTS OF GRAPH VERTICES IN LATTICES (RUSSIAN).
SEID 1974 -2 SEIDVASSER, M. A.
LENGTH AND BREADTH OF ASSIGNMENTS OF GRAPHS IN LATTICES (RUSSIAN), THE.
SERD 1976 -1 SERDYUKOV, A. I.
MUTUAL REDUCIBILITY OF SOME EXTREMAL PROBLEMS IN GRAPH THEORY (RUSSIAN).
SHEE 1975 -1 SHEEHAN, J.
EXTREMAL PROBLEM IN FINITE GRAPH THEORY, AN.
SIBI 1976 -1 SIBIRSKIJ, V. K.
GENERALIZED PARAMETRIC STEINER PROBLEMS IN GRAPHS (RUSSIAN), THE.
SINGA 1975 -1 SINGMASTER D.
ENUMERATING UNLABELLED HAMILTONIAN CIRCUITS.
SLATA 1975 -1 SLATER, P. J.
LEAVES OF TREES.
SLATA 1976 -2 SLATER, P. J.
R-DOMINATION IN GRAPHS.
SLATB 1976 -1 SLATER, P. J.; GOODMAN, S. E.; HEDETNIEMI, S. T.
ON THE OPTIONAL HAMILTONIAN COMPLETION PROBLEM.
SOUK 1975 -1 SOUKUP, J.
ON MINIMUM COST NETWORKS WITH NONLINEAR COSTS.
STUH 1966 -1 STUHR, R.
ANWENDUNGEN UND LOESUNGSVERFAHREN DER GANZZAHLIGEN LINEAREN PROGRAMMIERUNG.
SUL' 1974 -1 SUL'GA, YU. N.
OPTIMIZATION OF THE MAINTENANCE PROCESS IN NETWORK SYSTEMS (RUSSIAN).
SUST 1975 -1 SUSTAKASVILI, I. I.
ALGORITHMIZATION OF THE SELECTION OF OPTIMAL VARIANTS IN STOCHASTIC NETWORK GRAPHS THAT INCLUDE CONNECTING AND DISCONNECTING PATHS (RUSSIAN).
SYSLA 1975 -1 SYSLO, M. M.
APPLIED GRAPH THEORY II. EXTREMAL PATH PROBLEMS IN GRAPHS AND NETS (POLISH).
TARJ 1973 -2 TARJAN, R. E.
FINDING DOMINATORS IN DIRECTED GRAPHS.
TARJ 1977 -2 TARJAN, R. E.
FINDING OPTIMUM BRANCHINGS.
TIMO 1977 -1 TIMOFEEV, E. A.
CONSTRUCTION OF AN EDGE-WISE M-CONNECTED SKELETAL SUBGRAPH WITH MINIMAL LENGTH OF THE MAXIMAL EDGE (RUSSIAN), THE.
TOME 1975 -1 TOMESCU, I.
ALGORITHME POUR L'OBTENTION D'UNE CHAINE HAMILTONIENNE EN PARTANT DE L'ARBRE MINIMAL D'UN GRAPHE, UN.
TRUB 1976 -1 TRUBIN, V. A.
EFFECTIVE ALGORITHM FOR THE SOLUTION OF THE ALLOCATION PROBLEM ON A NETWORK IN THE SHAPE OF A TREE (RUSSIAN), AN.
TUER 1977 -1 TUERK, P.
OPTIMIERUNG VON VERSORGUNGSNETZEN.
TURK 1977 -1 TURK, P.
OPTIMIERUNG VON VERSORGUNGSNETZEN.
TUTT 1961 -1 TUTTE, W. T.
ON THE PROBLEM OF DECOMPOSING A GRAPH INTO CONNECTED FACTORS.
ULLMA 1976 -1 ULLMANN, J. R.
ALGORITHM FOR SUBGRAPH ISOMORPHISM, AN.
URQU 1967 -1 URQUHART, R. J.
DEGREE CONSTRAINED SUBGRAPHES OF LINEAR GRAPHES.
VELIA 1975 -1 VELIKOVIC, L. L.
SYMMETRIZATION OF GRAPHS (RUSSIAN).
VENT 1975 -1 VENTRE, A. G.
EXTREMAL PROBLEM CONCERNING GRAPHS WITH A HAMILTONIAN CYCLE (ITALIAN), AN.
WALS 1974 -1 WALSH, A. M.
ON FINDING A K-CLIQUE IN A RANDOM GRAPH.
WANG 1976 -1 WANG, D. L.
CONSTRUCTION OF A MAXIMALLY EDGE-CONNECTED GRAPH WITH PRESCRIBED DEGREES.
WANGA 1973 -1 WANG, D.-L.; KLEITMAN, D. J.
CONSTRUCTING MAXIMALLY CONNECTED GRAPHS HAVING A GIVEN DEGREE SEQUENCE.
WEGL 1975 -1 WEGLARZ, J.
CERTAIN NETWORK PROGRAMMING PROBLEM (POLISH), A.
WING 1961 -1 WING, O.
ALGORITHMS TO FIND THE MOST RELIABLE PATH IN A NETWORK.
ZAHN 1973 -1 ZAHN, C. T.
ALTERNATING EULER PATHS FOR PACKINGS AND COVERS.
ZAMB 1974 -1 ZAMBICKII, D. K.
PROBLEM OF FINDING THE OPTIMAL VERTEX IN A GRAPH (RUSSIAN), THE.
ZAMF 1976 -1 ZAMFIRESCU, T.
ON LONGEST PATHS AND CIRCUITS IN GRAPHS.
ZAVI 1975 -1 ZAVILEVICH, M. L.
DIRECT GRID PROBLEM FOR A HIERARCHICAL GRID, THE.

GRAPH THEORETIC RESULTS RELATED TO INTEGER PROGRAMMING
****

ALAV 1977 -1 ALAVI, Y.; BEHZAD, M.; LESNIAK-FOSTER, L. M.; NORDHAUS, E. A.
TOTAL MATCHINGS AND TOTAL COVERINGS OF GRAPHS.
ALBEA 1976 -1 ALBERTSON, M.; BOLLOBAS, B.; TUCKER, S.
INDEPENDENCE RATIO AND MAXIMUM DEGREE OF A GRAPH, THE.
ALBEB 1976 -1 ALBERTSON, M. O.
LOWER BOUND FOR THE INDEPENDENCE NUMBER OF A PLANAR GRAPH, A.
ALBEC 1976 -1 ALBERTSON, M. O.; BERMAN, D. M.
ACYCLIC CHROMATIC NUMBER, THE.
ANDE 1970 -1 ANDERSON, S.
GRAPH THEORY AND FINITE COMBINATORICS.
ARDIA 1975 -1 ARDITTI, J.-C.
GRAPHES DE COMPARABILITE D'ARBRES ET D'ARBORESCENCES.
BALAL 1976 -1 BALAS, E.; ZEMEL, E.
CRITICAL CUTSETS OF GRAPHS AND CANONICAL FACETS OF SET PACKING POLYTOPES.
BAUE 1973 -1 BAUERSFELD, G.; ESSMANN, CH.; LOEHLE, H.
ALGORITHMUS ZUR FESTSTELLUNG DER ISOMORPHIE VON ENDLICHEN, ZUSAMMENHAENGENDEN GRAPHEN.
BAUMA 1975 -1 BAUMGARTNER, J. E.; DAYKIN, D. E.; GUTMAN, I.; KLEITMAN, D. J. ET AL.
PROBLEMS.
BEAN 1976 -1 BEAN, D. R.
RECURSIVE EULER AND HAMILTONIAN PATHS.
BEANA 1970 -1 BEAN, D. W. T.
MINIMAL EDGE SETS MEETING ALL CIRCUITS OF A GRAPH.
BERG 1977 -1 BERGE, C.
THEORIE FRACTIONNAIRE DES GRAPHES.
BERG 1973 -1 BERGE, C.
INTRODUCTION A LA THEORIE DES HYPERGRAPHES.
BERG 1973 -2 BERGE, C.
NOTES SUR LES BONNES COLORATIONS D'UN HYPERGRAPHE.
BERGB 1977 -1 BERGE, C.; JOHNSON, E. L.
COLORING THE EDGES OF A HYPERGRAPH AND LINEAR PROGRAMMING TECHNIQUES.
BERM 1976 -1 BERMOND, J. C.
ON HAMILTONIAN WALKS.
BERM 1975 -1 BERMOND, J.'C.
1-GRAPHES REGULIERS MINIMAUX DE GIRTH DONNE.
BERZ 1973 -1 BERZTISS, A. T.
BACKTRACK PROCEDURE FOR ISOMORPHISM OF DIRECTED GRAPHS, A.
BEUE 1975 -1 BEUERMANN-WEHMEYER, A.
PRIMAL-DUAL-BEZIEHUNGEN IN GRAPHENALGORITHMEN.
BOLL 1977 -1 BOLLOBAS, B.
EXTREMAL PROBLEMS IN GRAPH THEORY.
BOLLB 1976 -1 BOLLOBAS, B.; ELDRIDGE, S. E.
MAXIMAL MATCHINGS IN GRAPHS WITH GIVEN MINIMAL AND MAXIMAL DEGREES.
BOLLC 1976 -1 BOLLOBAS, B.; ERDOES, P.
ALTERNATING HAMILTONIAN CYCLES.
BOLLD 1974 -1 BOLLOBAS, B.; ERDOES, P.; STRAUS, E. G.
COMPLETE SUBGRAPHS OF CHROMATIC GRAPHS AND HYPERGRAPHS.
BOOT 1976 -1 BOOTH, K. S.
PROBLEMS POLYNOMIALLY EQUIVALENT TO GRAPH ISOMORPHISM.
BOOTA 1975 -1 BOOTH, K. S.; LUEKER, G. S.
LINEAR ALGORITHMS TO RECOGNIZE INTERVAL GRAPHS AND TEST FOR CONSECUTIVE ONES PROPERTY.
BORI 1975 -1 BORIES, F.
ETUDE DU NOMBRE ACHROMATIQUE DE CERTAINS GRAPHES.
BOROE 1972 -1 BOROWIECKI, M.
HAMILTONIAN CYCLES IN CONJUNCTION OF TWO GRAPHS.
BOSE 1973 -1 BOSE, R. C.
CHARACTERIZATION PROBLEMS OF COMBINATORIAL GRAPH THEORY.
BRAE 1975 -1 BRAEMERT, D.
GRAPH THEORY: A SELECTED BIBLIOGRAPHY.
BRUN 1968 -1 BRUNO, J.
MATROIDS, GRAPHS, AND RESISTANCE NETWORKS.
BRUND 1976 -2 BRUNO, J.; WEINBERG, L.
GENERALIZED NETWORKS: NETWORKS EMDEDDED ON A MATROID, PART I.
BURR 1974 -1 BURR, S. A.
GENERALIZED RAMSEY THEORY FOR GRAPHS - A SURVEY.
CHARB 1977 -1 CHARTRAND, G.
GRAPHS AS MATHEMATICAL MODELS.
CHARC 1973 -1 CHARTRAND, G.; KAPOOR, S. F.; KRONK, H. V.
MANY FACETS OF HAMILTONIAN GRAPHS, THE.
CHAT 1975 -1 CHATY, GUY.
CHEMINEMENTS REMARQUABLES DANS LES GRAPHES: EXISTENCE, OBTENTION, CONSERVATION.
CHEN 1975 -1 CHEN, C. C.; DAYKIN, D. E.
HAMILTONIAN CYCLES IN A LINE-COLOURED MULTIGRAPH AND A RELATED UNCOLOURED GRAPH.
CHENH 1976 -1 CHEN, W.-K.
APPLIED GRAPH THEORY. GRAPHS AND ELECTRICAL NETWORKS (2ND REVISED ED.).
CHIAA 1975 -1 CHIAPPA, R. A.
CHARACTERISATION DES GRAPHES ADJOINTS HAMILTONIENS.
CHRI 1975 -1 CHRISTOFIDES, N.
HAMILTONIAN CIRCUITS AND THE TRAVELLING SALESMAN PROBLEM: A SURVEY.
CHRI 1974 -1 CHRISTOFIDES, N.
HAMILTONIAN CIRCUITS AND THE TRAVELLING SALESMAN PROBLEM.
CHVA 1976 -1 CHVATAL, V.
ON THE STRONG PERFECT GRAPH CONJECTURE.
CHVA 1977 -1 CHVATAL, V.
DETERMINING THE STABILITY NUMBER OF A GRAPH.
CHVAA 1978 -1 CHVATAL, V.; ERDOES P.
BIASED POSITIONAL GAMES.
COCKB 1977 -1 COCKAYNE, E. J.; HEDETNIEMI, S. T.
TOWARDS A THEORY OF DOMINATION IN GRAPHS.
COLL 1975 -1 COLLATZ, L.
BEZIEHUNGEN ZWISCHEN GRAPHEN, GEOMETRIE UND KOMBINATORIK, EINIGE.
COOK 1975 -1 COOK, R. J.
CHROMATIC NUMBER AND GIRTH.
CORN 1972 -1 CORNEIL, D. G.
ALGORITHM FOR DETERMINING THE AUTOMORPHISM PARTITIONING OF AN UNDIRECTED GRAPH, AN.

****

CORN 1970 -1 CORNEIL, D. G.
GRAPH ISOMORPHISM.

CORNA 1970 -1 CORNEIL, D. G.; GOTLIEB, C. C.
EFFICIENT ALGORITHM FOR GRAPH ISOMORPHISM, AN.

CORNB 1978 -1 CORNEIL, D. G.; MATHON, R. A.
ALGORITHMIC TECHNIQUES FOR THE GENERATION AND ANALYSIS OF STRONGLY REGULAR GRAPHS AND OTHER COMBINATORIAL CONFIGURATIONS.

CVET 1976 -1 CVETKOVIC, D. M.; LACKOVIC, I. B.; SIMIC, S. K.
GRAPH EQUATIONS, GRAPH INEQUALITIES AND A FIXED POINT THEOREM.

CZER 1977 -1 CZERNIAKIEWICZ, A.
MINIMAL GRAPHS WITH CYCLIC COLOR GROUP.

DANA 1977 -1 DANARAJ, G.; KLEE, V.
CONNECTEDNESS GAME AND THE C-COMPLEXITY OF CERTAIN GRAPHS, THE.

DAY A 1977 -1 DAY, W. H. E.
VALIDITY OF CLUSTERS FORMED BY GRAPH-THEORETIC CLUSTER METHODS.

DAYK 1976 -1 DAYKIN, D. E.
CORRIGENDUM AND ADDENDUM TO: "PSEUDO-MATCHINGS OF A BIPARTITE GRAPH" BY A. BRACE AND DAYKIN (PROCEEDINGS OF THE AMS 42, 1974, 20 32).

DAYK 1976 -2 DAYKIN, D. E.
GRAPHS WITH CYCLES HAVING ADJACENT LINES DIFFERENT COLORS.

DEO 1969 -1 DEO, N.
EXTENSIVE ENGLISH LANGUAGE BIBLIOGRAPHY ON GRAPH THEORY AND ITS APPLICATIONS, AN.

DONE 1976 -1 DONETS, G. A.
ON THE PROPERTIES OF COLORED PLANE TRIANGULATIONS (RUSSIAN).

DRUF 1975 -1 DRUFFEL, L. E.
GRAPH RELATED ALGORITHMS: ISOMORPHISM, AUTOMORPHISM AND CONTAINMENT.

EPIH 1975 -1 EPIHIN, V. V.; HARKEVIC, A. D.
ON THE QUESTION OF THE CONSTRUCTION OF OPTIMAL GRAPHS (RUSSIAN).

ERDO 1976 -1 ERDOES, P.
PROBLEMS AND RESULTS IN GRAPH THEORY AND COMBINATORIAL ANALYSIS.

ERDOA 1976 -1 ERDOES, P.; FAUDREE, R. J.; ROUSSEAU, C. C.; SCHELP, R. H.
GENERALIZED RAMSEY THEORY FOR MULTIPLE COLORS.

ERDOB 1966 -1 ERDOES, P.; HAJNAL, A.
ON CHROMATIC NUMBERS OF GRAPHS AND SET-SYSTEMS.

ERDOE 1974 -1 ERDOES, P.; SPENCER, J.
PROBABILISTIC METHODS IN COMBINATORICS.

EVANA 1976 -2 EVANS, J. R.
FUNDAMENTAL CIRCUITS IN COMPLETE GRAPHS AND THE TRAVELING SALESMAN PROBLEM.

EVEN 1974 -1 EVEN, S.
MAXIMUM FLOW IN A NETWORK AND THE CONNECTIVITY OF A GRAPH.

EVENC 1970 -1 EVEN, S.; LEMPEL, A.; PNUELI, A.
PERMUTATION GRAPHS AND TRANSITIVE GRAPHS.

FAUD 1974 -1 FAUDREE, R. J.; SCHELP, R. H.
PATH CONNECTED GRAPHS.

FITZ 1974 -1 FITZGERALD, C. H.
OPTIMAL INDEXING OF THE VERTICES OF GRAPHS.

FLEI 1976 -1 FLEISCHNER, H.
GEMEINSAME BASIS FUER DIE THEORIE DER EULERSCHEN GRAPHEN UND DEN SATZ VON PETERSEN, EINE.

FLEI 1976 -2 FLEISCHNER, H.
IN THE SQUARE OF GRAPHS, HAMILTONICITY AND PANCYCLICITY, HAMILTONIAN CONNECTEDNESS AND PANCONNECTEDNESS ARE EQUIVALENT CONCEPTS.

FOLD 1975 -1 FOLDES, S.; HAMMER, P. L.
ON A CLASS OF MATROID-PRODUCING GRAPHS.

FOLD 1976 -1 FOLDES, S.; HAMMER, P. L.
DIL WORTH OF A GRAPH, THE.

FOLD 1978 -1 FOLDES, S.; HAMMER, P. L.
DILWORTH NUMBER OF A GRAPH, THE.

FOLD 1977 -1 FOLDES, S.; HAMMER, P. L.
SPLIT GRAPHS HAVING DILWORTH NUMBER TWO.

FOLD 1976 -2 FOLDES, S.; HAMMER, P. L.
SPLIT GRAPHS.

FONG 1977 -1 FONG, A. C.; ULLMAN, J. D.
FINDING THE DEPTH OF A FLOW GRAPH.

FONT 1976 -1 FONTET, M.
AUTOMORPHISMES DE GRAPHES ET PLANARITE.

FOUL 1976 -1 FOULDS, L. R.
RELATIONSHIP BETWEEN GRAPH THEORY AND OPERATIONAL RESEARCH, THE.

FOUR 1973 -1 FOURNIER, J. C.
COLORATIONS DES ARETES D'UN GRAPHE.

FOUR 1975 -1 FOURNIER, J. C.
SUR LES STABLES D'UN GRAPHE.

FOUR 1976 -1 FOURNIER, J. C.
PROPRIETES COMBINATOIRES ET ALGEBRIQUES DES GRAPHES PLANAIRES.

FULK 1975 -2 FULKERSON, D. R.
STUDIES IN GRAPH THEORY, I.

FULK 1975 -3 FULKERSON, D. R.
STUDIES IN GRAPH THEORY, II.

GARED 1976 -1 GAREY, M. R.; JOHNSON, D. S.; TARJAN, R. E.
PLANAR HAMILTONIAN CIRCUIT PROBLEM IS NP-COMPLETE, THE.

GELDB 1975 -1 GELDMACHER, R. C.; LIU, C. S.
PRODUCT OF INDEPENDENT SETS OF THE VERTEX-INCIDENCE VECTORS OF A GRAPH, THE.

GELL 1974 -1 GELLER, D.; KRONK, H.
FURTHER RESULTS ON THE ACHROMATIC NUMBER.

GELLA 1976 -1 GELLER, D. P.
R-TUPLE COLORINGS OF UNIQUELY COLORABLE GRAPHS.

GIRA 1976 -1 GIRAUD, G.
SUR LES PROPORTIONS RESPECTIVES DE TRIANGLES UNI, BI OU TRICOLORES DANS UN TRICOLORIAGE DES ARETES DU N-EMBLE.

GOL' 1974 -1 GOL'DBERG, M. K.
REMARK ON THE CHROMATIC CLASS OF A MULTIGRAPH (RUSSIAN), A.

GOLDG 1975 -1 GOLDNER, A. M.; HARARY, F.
NOTE ON THE SMALLEST NONHAMILTONIAN MAXIMAL PLANAR GRAPH.

GOLU 1977 -1 GOLUMBIC, M. C.
COMPARABILITY GRAPHS AND A NEW MATROID.

GRAPH THEORETIC RESULTS RELATED TO INTEGER PROGRAMMING
****

GOOD 1965 -1 GOODMAN, R. G.; CUMMINS, R. L.
COMPUTER DETERMINATION OF ISOMORPHISM IN LINEAR GRAPHS.

GOODA 1973 -1 GOODMAN, S.; HEDETNIEMI, S.
ON HAMILTONIAN WALKS IN GRAPHS.

GRAHB 1974 -1 GRAHAM, R. L.; ROTHSCHILD, B. L.
SOME RECENT DEVELOPMENTS IN RAMSEY THEORY.

GRAVA 1977 -1 GRAVER, J. E.; WATKINS, M. E.
COMBINATORICS WITH EMPHASIS ON THE THEORY OF GRAPHS.

GREE 1976 -1 GREEN, C. D.
DETECTION OF MISTAKES IN THE COMPARABILITY GRAPH OF A TREE, THE.

GROE 1976 -1 GROETSCHEL, M.
UEBER GRAPHEN MIT KREISEN, DIE GEGEBENE WEGE ENTHALTEN.

GROE 1977 -2 GROETSCHEL, M.
GRAPHS WITH CYCLES CONTAINING GIVEN PATHS.

GROE 1977 -4 GROETSCHEL, M.
ON THE MONOTONE SYMMETRIC TRAVELLING SALESMAN PROBLEM: HYPOHAMILTONIAN/HYPOTRACEABLE GRAPHS AND FACETS.

GROE 1977 -7 GROETSCHEL, M.
STRONG BLOCKS AND THE OPTIMUM BRANCHING PROBLEM.

GUPTA 1969 -1 GUPTA, R. P.
INDEPENDENCE AND COVERING NUMBERS OF LINE GRAPHS AND TOTAL GRAPHS

HABI 1975 -1 HABIB, M.; LAMBERT, O.
NOMBRE DE SAUTS ET INDICE DE PARTITION DANS LES GRAPHES SANS CIRCUIT.

HALI 1975 -1 HALIN, R.
SOME PATH PROBLEMS IN GRAPH THEORY.

HANSA 1975 -1 HANSEN, P.
DEGRES ET NOMBRE DE STABILITE D'UN GRAPHE.

HARAA 1974 -1 HARARY, F.; LEMPEL, A.
ON CLIQUE-EXTREMAL (P,Q)-GRAPHS.

HARAB 1974 -1 HARARY, F.; MALLION, R. B.
ELEMENTARY NECESSARY CONDITION FOR HAMILTONIAN GRAPHS INVOLVING CONNECTIVITY, AN.

HARAB 1975 -1 HARARY, F.; MALLION, R. B.
ERRATA: "AN ELEMENTARY, NECESSARY CONDITION FOR HAMILTONIAN GRAPHS, INVOLVING CONNECTIVITY" (NANTA MATHEMATICA 7, 1974, NO. 2, 96-101).

HARTD 1976 -1 HARTNELL, B. L.
SOME PROBLEMS ON SPANNING TREES.

HAUS 1977 -2 HAUSMANN, D.
BENACHBARTE ECKEN AUF 0-1-POLYEDERN.

HAUS 1977 -3 HAUSMANN, D.
POLYNOMIALER ALGORITHMUS FUER DIE ADJAZENZUEBERPRUEFUNG AUF DEM BRANCHING-POLYEDER, EIN.

HAUSA 1977 -1 HAUSMANN, D.; JENKYNS, T. A.; KORTE, B.
WORST CASE ANALYSIS OF GREEDY TYPE ALGORITHMS FOR INDEPENDENCE SYSTEMS.

HECH 1977 -1 HECHLER, S. H.
ON INFINITE GRAPHS WITH A SPECIFIED NUMBER OF COLORINGS.

HEDE 1971 -1 HEDETNIEMI, S.
CHARACTERIZATIONS AND CONSTRUCTIONS OF MINIMALLY TWO-CONNECTED GRAPHS AND MINIMALLY STRONG DIGRAPHS.

HEIN 1976 -1 HEIN, O.
ALGORITHMUS ZUR BERECHNUNG DER MINIMALEN KANTENBASIS IN GERICHTETEN GRAPHEN, EIN.

HEIN 1977 -1 HEIN, O.
GRAPHENTHEORIE FUER ANWENDER.

HILT 1977 -1 HILTON, A. J. W.
DEFINITIONS OF CRITICALITY WITH RESPECT TO EDGE-COLORING.

HILT 1975 -2 HILTON, A. J. W.
ON VIRZING'S UPPER BOUND FOR THE CHROMATIC INDEX OF A GRAPH.

HOPC 1971 -1 HOPCROFT, J. E.
N LOG N ALGORITHM FOR ISOMORPHISM OF PLANAR TRIPLY CONNECTED GRAPHS, A.

HOPCB 1972 -1 HOPCROFT, J. E.; TARJAN, R. E.
ISOMORPHISM OF PLANAR GRAPHS.

HOPCB 1971 -1 HOPCROFT, J. E.; TARJAN, R. E.
(V POWER 2) ALGORITHM FOR DETERMINING ISOMORPHISM OF PLANAR GRAPHS, A.

HOPCC 1974 -1 HOPCROFT, J. E.; WONG, J. K.
LINEAR TIME ALGORITHM FOR ISOMORPHISM OF PLANAR GRAPHS.

HSU 1973 -1 HSU, H. T.; HONKANEN, P. A.
FAST MINIMAL STORAGE ALGORITHM FOR DETERMINING ALL THE ELEMENTARY CYCLES OF A GRAPH, A.

HUBA 1975 -1 HUBAUT, X. L.
STRONGLY REGULAR GRAPHS.

JENK 1976 -1 JENKYNS, T. A.
EFFICACY OF THE "GREEDY" ALGORITHM, THE.

JOHNB 1962 -1 JOHNSON, D. M.; DULMAGE, A. L.; MENDELSOHN, N. S.
CONNECTIVITY AND REDUCIBILITY OF GRAPHS.

JOLI 1975 -1 JOLIVET, J.-L.
PROBLEMES DE CONNEXITE ET PROBLEMES HAMILTONIENS EN THEORIE DES GRAPHES. JOINT D'UNE FAMILLE DE GRAPHES ET GENERALISATION DE LA NOTION DE GRAPHE PARFAIT.

JUCO 1974 -1 JUCOVIC, E.; OLEJNIK, F.
ON CHROMATIC AND ACHROMATIC NUMBERS OF UNIFORM HYPERGRAPHS.

KACZ 1976 -1 KACZMARSKI, J.
NOMBRE DE STABILITE ET CERTAINES PROPRIETES DE GRAPHES CRITIQUES.

KALM 1975 -1 KALMANSON, K.
EDGECONVEX CIRCUITS AND THE TRAVELING SALESMAN PROBLEM.

KASA 1975 -1 KASAI, T.; KURODA, K.
NUMBER OF HAMILTONIAN CIRCUITS CONTAINING SPECIFIED EDGES IN AN INCOMPLETE GRAPH AND ITS APPLICATIONS TO MULTIGRAPHS.

KASZ 1973 -1 KASZONYI, L.
ON THE STRUCTURE OF COLORING GRAPHS.

KEL' 1976 -1 KEL'MANS, A. K.
NUMBER OF TREES IN A GRAPH WHICH CONTAINS A GIVEN FOREST (RUSSIAN).

KELI 1968 -1 KELIKHOV, A. N.; BERSHTEYN, L. S.; KARELIN, P.
ISOMORPHISM OF GRAPHS AND FINITE AUTOMATA.

KHOM 1976 -1 KHOMENKO, M. P.; VIVROT, T. M.
ON THE 0-1-MATRICES (UKRAINIAN).

KIKU 1975 -1 KIKUST, P. B.
EXISTENCE OF A HAMILTONIAN CYCLE IN A REGULAR GRAPH OF DEGREE 5 (RUSSIAN), THE.

KIRY 1975 -1 KIRYUKIN, V. V.; POLONSKIJ, D. V.
ON THE SEARCH FOR THE HAMILTONIAN CYCLES OF A GRAPH (RUSSIAN).

KITA 1975 -1 KITAMURA, S.
ON THE EDGE-CHROMATIC NUMBER OF A GRAPH.

KNOE 1971 -1 KNOEDEL, W.
VERFAHREN ZUR FESTSTELLUNG DER ISOMORPHIE VON ENDLICHEN ZUSAMMENHAENGENDEN GRAPHEN, EIN.

KOCH 1976 -1 KOCH, E.; PERLES, M. A.
COVERING EFFICIENCY OF TREES AND K-TREES.

KOPY 1977 -1 KOPYLOV, G. N.
MAXIMAL PATHS AND CYCLES IN A GRAPH (RUSSIAN).

KORS 1974 -1 KORSHUNOV, A. D.
ON THE NUMBER OF PAIRS OF HAMILTONIAN CYCLES IN A COMPLETE GRAPH HAVING A FIXED NUMBER OF COMMON ARCS (RUSSIAN).

KORS 1976 -1 KORSHUNOV, A. D.
SOLUTION OF THE PROBLEM OF ERDOES-RENYI ON THE HAMILTONIAN CYCLES IN UNDIRECTED GRAPHS (RUSSIAN).

KOU 1977 -1 KOU, L. T.
ALPHA-SOURCE AND ALPHA-SINK O A GRAPH.

KOZY 1972 -1 KOZYREV, V. P.
GRAPH THEORY (RUSSIAN).

KOZY 1976 -1 KOZYREV, V. P.
FINDING ALL THE HAMILTONIAN PATHS IN A GRAPH AND INVESTIGATION OF HAMILTONIAN CONNECTED GRAPHS (RUSSIAN).

KOZY 1978 -2 KOZYREV, V. P.
HAMILTONIAN CONNECTIVITY OF THE SQUARE OF A GRAPH (RUSSIAN).

KRARC 1972 -1 KRARUP, J.; HARARY, F.; SCHWENK, A.
GRAPHS SUPPRESSIBLE TO AN EDGE.

KUHN 1973 -1 KUHN, W. W.
INVERSE LINE GRAPHS AND HAMILTONIAN CIRCUITS.

KUND 1975 -1 KUNDU, S.
MINIMUM DIAMETER STEINER TREES IN A GRAPH, THE.

KUND 1974 -2 KUNDU, S.
DISJOINT REPRESENTATION OF TREE REALIZABLE SEQUENCES.

LESN 1977 -1 LESNIAK-FOSTER, L.
SOME RECENT RESULTS IN HAMILTONIAN GRAPHS.

LESNA 1975 -1 LESNIAK, L.
ON LONGEST PATHS IN CONNECTED GRAPHS.

LESNA 1975 -2 LESNIAK, L.
ON 1-EDGE HAMILTONIAN GRAPHS.

LEVI 1974 -1 LEVI, G.
GRAPH ISOMORPHISM: A HEURISTIC EDGE-PARTITIONING-ORIENTED ALGORITHM, A.

LEVI 1972 -1 LEVI, G.
NOTE ON THE DERIVATION OF MAXIMAL COMMON SUBGRAPHS OF TWO DIRECTED OR UNDIRECTED GRAPHS.

LIEB 1970 -1 LIEBLING, T.
GRAPHENTHEORIE IN PLANUNGS- UND TOURENPROBLEMEN.

LING 1973 -1 LING, R. F.
EXPECTED NUMBER OF COMPONENTS IN RANDOM LINEAR GRAPHS, THE.

LITTA 1976 -1 LITTLE, C. H. C.; HOLTON, D. A.
NEW CHARACTERIZATION OF PLANAR GRAPHS, A.

LITVC 1974 -1 LITVAK, E. I.; ZAVARIN, A. N.
ON THE DETERMINATION OF THE LENGTH OF THE SHORTEST PATH IN A GRAPH (RUSSIAN).

LORE 1972 -1 LOREA, M.
ENSEMBLES STABLES DANS LES HYPERGRAPHES.

LOVA 1974 -1 LOVASZ, L.
MINIMAX THEOREMS FOR HYPERGRAPHS.

LOVA 1968 -1 LOVASZ, L.
ON CHROMATIC NUMBER OF FINITE SET SYSTEMS.

LOVA 1975 -2 LOVASZ, L.
2-MATCHINGS AND 2-COVERS OF HYPERGRAPHS.

LOVA 1976 -2 LOVASZ, L.
ON TWO MINIMAX THEOREMS IN GRAPHS.

LOVA 1977 -2 LOVASZ, L.
GRAPH THEORY AND INTEGER PROGRAMMING.

MAC A 1976 -1 MAC AVANEY, K. L.
NUMBER AND STABILITY INDICES OF CN-TREES, THE.

MAC E 1974 -1 MAC DIARMID, C. J. H.
PATH-PARTITION STRUCTURES OF GRAPHS AND DIGRAPHS.

MARKC 1976 -1 MARKOSYAN, S. E.; KARAPETYAN, I. A.
ON THE PERFECT GRAPHS (RUSSIAN).

MARS 1971 -1 MARSHALL, C.
APPLIED GRAPH THEORY.

MATU 1968 -1 MATULA, D. W.
MIN-MAX THEOREM FOR GRAPHS WITH APPLICATION TO GRAPH COLORING, A.

MATU 1978 -1 MATULA, D. W.
SUBTREE ISOMORPHISM IN O(N POWER 5/2).

MAURB 1976 -1 MAURRAS, J. F.
POLYTOPES A SOMMETS DANS (0,1)N.

MAURC 1975 -1 MAURRAS, J.-F.
SOME RESULTS ON THE CONVEX HULL OF THE HAMILTONIAN CYCLES OF SYMMETRIC COMPLETE GRAPHS.

MEEU 1976 -1 MEEUSEN, W.; CUYVERS, L.
CLIQUE-DETECTION IN DIRECTED GRAPHS (DUTCH).

MEIR 1977 -1 MEIR, A.; MOON, J. W.
PACKING AND COVERING CONSTANTS FOR CERTAIN FAMILIES OF TREES, I.

MEIR 1973 -1 MEIR, A.; MOON, J. W.
EXPECTED NODE-INDEPENDENCE NUMBER OF RANDOM TREES, THE.

MEIR 1975 -2 MEIR, A.; MOON, J. W.
RELATIONS BETWEEN PACKING AND COVERING NUMBERS OF A TREE.

MENG 1974 -1 MENG, D.
MATCHINGS AND COVERINGS FOR GRAPHS.

MEYEB 1975 -1 MEYER, J.-C.
ON THE GOOD COLORINGS, II. CASE OF THE COMPLETE H-PARTITE HYPERGRAPHS.

MIKO 1976 -1 MIKOLA, M.
EXTREMAL CONNECTIVITY OF THE STRICTLY WEAK DIGRAPH, THE.

MILLB 1977 -1 MILLER, G. L.
GRAPH ISOMORPHISM, GENERAL REMARKS.

MILNA 1976 -1 MILNER, E. C.; WELSH, D. J. A.
ON THE COMPUTATIONAL COMPLEXITY OF GRAPH THEORETICAL PROPERTIES.

GRAPH THEORETIC RESULTS RELATED TO INTEGER PROGRAMMING

MINOA 1976 -1 MINOUX, M.
CARACTERISATION DES COUPLAGES DE POIDS MAXIMAL ET DE CARDINALITE FIXEE DANS LES GRAPHES PONDERES FINIS, UNE.

MORP 1971 -1 MORPURGO, R.
HEURISTIC METHOD FOR THE VERIFICATION OF THE ISOMORPHISM OF TWO SIMPLE AND NON ORIENTED GRAPHS, A.

MYER 1974 -1 MYERS, B. R.
ON SPANNING TREES, WEIGHTED COMPOSITIONS, FIBONACCI NUMBERS AND RESISTOR NETWORKS.

NASH 1975 -1 NASH-WILLIAMS, C. ST. J. A.
HAMILTONIAN CIRCUITS.

NASH 1967 -1 NASH-WILLIAMS, C. ST. J. A.
APPLICATION OF MATROIDS TO GRAPH THEORY, AN.

NESE 1976 -1 NESETRIL, J.; ROEDL, V.
RAMSEY PROPERTY FOR GRAPHS WITH FORBIDDEN COMPLETE SUBGRAPHS, THE.

NIEM 1975 -1 NIEMINEN, J.
CHARACTERIZATION OF VERTEX CUTS OF AN UNDIRECTED GRAPH, A.

NIEM 1974 -2 NIEMINEN, J.
TWO BOUNDS FOR THE DOMINATION NUMBER OF A GRAPH.

NIEM 1975 -3 NIEMINEN, J.
SOME OBSERVATIONS ON THE DETERMINATION OF AN UPPER BOUND FOR THE CLIQUE NUMBER OF A GRAPH.

NINC 1974 -1 NINCAK, J.
CHROMATIC NUMBERS OF CERTAIN GRAPHS (RUSSIAN), THE.

NOOR 1975 -1 NOORVASH, S.
COVERING THE VERTICES OF A GRAPH BY VERTEX-DISJOINT PATHS.

OLAR 1973 -1 OLARU, E.
UEBER PERFEKTE UND KRITISCH IMPERFEKTE GRAPHEN.

ORLI 1977 -1 ORLIN, J.
MINIMAL INTEGRAL SEPARATOR OF A THRESHOLD GRAPH, THE.

ORLOB 1975 -1 ORLOVA, G. I.; DORFMAN, G. YA.
OPTIMAL PARTITIONING OF THE SET OF VERTICES OF A GRAPH BY REDUCTION TO A PROBLEM OF MAXIMAL CUT (RUSSIAN).

OVER 1975 -1 OVERBECK-LARISCH, M.
HAMILTONSCHE LINIEN IN DIGRAPHEN.

OZAW 1973 -1 OZAWA, T.; HIRAO, S.; HATTORI, Y.
SOME PROPERTIES OF MULTICOLORED-BRANCH GRAPHS.

PACA 1974 -1 PACAULT, J. F.
COMPUTING THE WEAK COMPONENTS OF A DIRECTED GRAPH.

PALA 1971 -1 PALASTI, I.
ON THE THRESHOLD DISTRIBUTION FUNCTION OF CYCLES IN A DIRECTED RANDOM GRAPH.

PARR 1968 -1 PARRIS, R.
CODING PROBLEM FOR GRAPHS, THE.

PARRA 1967 -1 PARRIS, R.; READ, R. C.
GRAPH ISOMORPHISM AND THE CODING OF GRAPHS.

PARRA 1968 -1 PARRIS, R.; READ, R. C.
CODING PROCEDURE FOR GRAPHS, A.

PELE 1976 -1 PELED, U. N.
MATROIDAL GRAPHS.

PETRB 1976 -1 PETROV, S. V.
ON A PROBLEM IN GRAPH THEORY (RUSSIAN).

PIEH 1976 -1 PIEHLER, J.
UEBER SCHNITTVERSCHAERFUNGEN IN DER GEMISCHT-GANZZAHLIGEN LINEAREN OPTIMIERUNG.

PIPP 1976 -1 PIPPENGER, N.; VALIANT, L. G.
SHIFTING GRAPHS AND THEIR APPLICATIONS.

POPE 1976 -1 POPESCU, D.
CLASSES OF MAXIMAL NON HAMILTONIAN GRAPHS (RUMANIAN).

PROS 1974 -1 PROSKUROWSKI, A.
SEARCH FOR A UNIQUE INCIDENCE MATRIX OF A GRAPH.

PULL 1977 -1 PULLEYBLANK, W. R.
DUAL INTEGRALITY IN B-MATCHING PROBLEMS.

PULL 1977 -2 PULLEYBLANK, W. R.
TWO-BICRITICAL GRAPHS AND NODE COVERS.

READA 1977 -1 READ, R. C.; CORNEIL, D. G.
GRAPH ISOMORPHISM DISEASE, THE.

REYN 1977 -1 REYNER, W. S.
ANALYSIS OF A GOOD ALGORITHM FOR THE SUBTREE PROBLEM, AN.

RIVE 1976 -1 RIVEST, R. L.; VUILLEMIN, J.
ON RECOGNIZING GRAPH PROPERTIES FROM ADJACENCY MATRICES.

ROBEA 1976 -1 ROBERTS, J.
SUFFICIENT CONDITION FOR K-PATH HAMILTONIAN DIGRAPHS, A.

ROEDA 1977 -1 ROEDL, V.
ON THE CHROMATIC NUMBER OF SUBGRAPHS OF A GIVEN GRAPH.

ROSEA 1971 -1 ROSENFELD, M.
ON CERTAIN POLYTOPES ASSOCIATED WITH GRAPHS.

ROY 1966 -2 ROY, B.
CONTRIBUTION TO THE THEORY OF GRAPHS IN THE STUDY OF SCHEDULING PROBLEMS (FRENCH).

RUSK 1977 -1 RUSKEY, F.; HU, T. C.
GENERATING BINARY TREES LEXICOGRAPHICALLY.

SABI 1977 -1 SABIDUSSI, G.
TRACING GRAPHS WITHOUT BACKTRACKING.

SAMP 1976 -1 SAMPATHKUMAR, E.; BHAVE, V. N.
PARTITION GRAPHS AND COLORING NUMBERS OF A GRAPH.

SAUC 1971 -1 SAUCIER, G.
ALGORITHME EFFICACE RECHERCHANT L'ISOMORPHISME DE 2 GRAPHES, UN.

SCHA 1976 -1 SCHAAR, G.
ON THE MAXIMUM NUMBER OF EDGE-DISJOINT PATHS WITH ENDPOINTS BELONGING TO A GIVEN SET OF VERTICES IN A FINITE GRAPH.

SHAH 1974 -1 SHAH, Y. J.; DAVIDA, G. T.; MAC CARTHY, M. K.
OPTIMUM FEATURES AND GRAPH ISOMORPHISM.

SHOS 1974 -1 SHOSTAK, R. E.
REFUTATION GRAPHS AND RESOLUTION THEOREM PROVING.

SIRO 1971 -1 SIROVICH, F.
ISOMORPHISM BETWEEN GRAPHS: AN EFFICIENT ALGORITHM TO FIND ALL ISOMORPHISMS.

SKUP 1976 -1 SKUPIEN, Z.
HAMILTONIAN SHORTAGE, PATH PARTITIONS OF VERTICES, AND MATCHINGS IN A GRAPH.

****

SLATA 1976 -1 SLATER, P. J.
CENTRAL VERTICES IN A GRAPH.
SMIT 1974 -1 SMITH, D. H.
HAMILTONIAN CIRCUITS ON THE N-CUBE.
SPEN 1976 -1 SPENCER, J.
MAXIMAL ASYMMETRY OF GRAPHS.
STAC 1976 -1 STACEY, K. C.; MAC AVANEY, K. L.; SIMS, J.
STABILITY INDEX OF THE PRODUCT OF A PATH AND A TREE, THE.
STAD 1978 -1 STADEL, M.
REMARK ON THE TIME COMPLEXITY OF THE SUBTREE PROBLEM, A.
STAH 1976 -1 STAHL, S.
N-TUPLE COLORINGS AND ASSOCIATED GRAPHS.
STER 1974 -1 STERBOUL, F.
ON THE CHROMATIC NUMBER OF THE DIRECT PRODUCT OF HYPERGRAPHS.
STER 1974 -2 STERBOUL, F.
SUR UNE CONJECTURE DE V. CHVATAL.
STOCA 1968 -1 STOCKTON, F.
LINEARIZATION AND STANDARDIZATION OF GRAPHS.
SUMM 1976 -1 SUMMER, D. P.
MINIMAL LINE GRAPHS.
SUSS 1965 -1 SUSSENGUTH, E. H.
GRAPH THEORETIC ALGORITHM FOR MATCHING CHEMICAL STRUCTURES, A.
TAIN 1975 -1 TAINITER, M.
STATISTICAL THEORY OF CONNECTIVITY. I. BASIC DEFINITIONS AND PROPERTIES.
TEH 1976 -1 TEH, H. H.; SHEE, S. C.
ALGEBRAIC THEORY OF GRAPHS.
THOMC 1976 -1 THOMSON, P. F.
ENUMERATION OF NETWORKS.
TINH 1976 -1 TINHOFER, G.
METHODEN DER ANGEWANDTEN GRAPHENTHEORIE.
TOFT 1975 -1 TOFT, B.
ON COLOUR-CRITICAL HYPERGRAPHS.
TOME 1976 -1 TOMESCU, I.
NOMBRE MAXIMAL DE COLORATIONS D'UN GRAPHE HAMILTONIEN, LE.
TOME 1975 -3 TOMESCU, I.
PROBLEMES EXTREMAUX CONCERNANT LE NOMBRE DES COLORATIONS DES SOMMETS D'UN GRAPHE FINI.
TRAN 1976 -1 TRAN DINH AM; TSUKIYAMA, S.; SHIRAKAWA, I.; OZAKI, H.
ALGORITHM FOR GENERATING ALL THE DIRECTED PATHS AND ITS APPLICATION, AN.
TROTC 1976 -1 TROTTER, W. T. JR.; MOORE, J. I.; SUMNER, D. P.
DIMENSION OF A COMPARABILITY GRAPH, THE.
TROTD 1976 -1 TROTTER, W. T. JR.; MOORE, J. I. JR.
CHARACTERIZATION PROBLEMS FOR GRAPHS, PARTIALLY ORDERED SETS, LATTICES, AND FAMILIES OF SETS.
TRUE 1977 -2 TRUEMPER, K.
UNIMODULAR MATRICES OF FLOW PROBLEMS WITH ADDITIONAL CONSTRAINTS.
TURN 1968 -1 TURNER, J.
GENERALIZED MATRIX FUNCTIONS AND THE GRAPH ISOMORPHISM PROBLEM.
TUTT 1976 -1 TUTTE, W. T.
HAMILTONIAN CIRCUITS.
UNGE 1964 -1 UNGER, S. H.
HEURISTIC PROGRAM FOR TESTING PAIRS OF DIRECTED LINE GRAPHS FOR ISOMORPHISM, A.
URQU 1967 -1 URQUHART, R. J.
DEGREE CONSTRAINED SUBGRAPHES OF LINEAR GRAPHES.
VELIA 1975 -1 VELIKOVIC, L. L.
SYMMETRIZATION OF GRAPHS (RUSSIAN).
VENT 1975 -1 VENTRE, A. G.
EXTREMAL PROBLEM CONCERNING GRAPHS WITH A HAMILTONIAN CYCLE (ITALIAN), AN.
WALK 1976 -1 WALKER, R. C.
GRAPH-COLOURING THEOREM, A.
WALLA 1976 -1 WALLER, D. A.
GENERAL SOLUTION TO THE SPANNING TREE ENUMERATION PROBLEM IN ARBITRARY MULTIGRAPH JOINS.
WEINA 1975 -1 WEINBERG, L.
DUALITY IN NETWORKS: ROSES, BOUQUETS, AND CUT-SETS; TREES, FORESTS AND POLYGONS.
WEINA 1965 -1 WEINBERG, L.
PLANE REPRESENTATIONS AND CODES FOR PLANAR GRAPHS.
WEINA 1960 -1 WEINBERG, L.
SIMPLE AND EFFICIENT ALGORITHM FOR DETERMINING ISOMORPHISM OF PLANAR TRIPLE-CONNECTED GRAPHS.
WERR 1974 -1 WERRA, D. DE
PARTITIONS OF GRAPHS INTO COVERINGS AND HYPERGRAPHS INTO TRANSVERSALS.
WERR 1976 -1 WERRA, D. DE
COLOR-FEASIBLE SEQUENCES OF ALMOST BIPARTITE GRAPHS.
WERR 1975 -3 WERRA, D. DE
HOW TO COLOR A GRAPH.
WERR 1976 -3 WERRA, D. DE
MULTIGRAPHS WITH QUASIWEAK ODD CYCLES.
WERR 1976 -5 WERRA, D. DE
REMARKS ON THE COLOR-FEASIBLE SEQUENCES OF A MULTIGRAPH.
WILL 1974 -1 WILLE, D.
KOMBINATORISCHE UNTERSUCHUNGEN AN (0,1)-MATRIZEN, EINIGE.
WIN 1975 -1 WIN, S.
EXISTENZ VON GERUESTEN MIT VORGESCHRIEBENEM MAXIMALGRAD IN GRAPHEN.
WOOD 1976 -1 WOODALL, D. R.
MAXIMAL CIRCUITS OF GRAPHS, I.
WOOD 1975 -2 WOODALL, D. R.
MAXIMAL CIRCUITS OF GRAPHS II.
WRIG 1976 -1 WRIGHT, E. M.
PROPORTION OF UNLABELLED GRAPHS WHICH ARE HAMILTONIAN, THE.
ZAMF 1976 -1 ZAMFIRESCU, T.
ON LONGEST PATHS AND CIRCUITS IN GRAPHS.
ZAMF 1975 -1 ZAMFIRESCU, T.
HISTOIRE ET L'ETAT PRESENT DES BORNES CONNUES POUR P(KJ), C(KJ), P(J)- ET C(KJ)-, L'.
ZAMF 1973 -1 ZAMFIRESCU, T.
CHARACTERIZATION OF HAMILTONIAN GRAPHS, A.

GROUP THEORETIC APPROACH
****

BACH 1977 -1 BACHEM, A.
FUNDAMENTALPUNKTALGORITHMEN ZUR LOESUNG GANZZAHLIGER PROGRAMMIERUNGSAUFGABEN.

BACH 1976 -1 BACHEM, A.
ALGORITHMUS ZUR MODULO-OPTIMIERUNG STRUKTURIERTER MATRIZEN, EIN.

BACH 1976 -2 BACHEM, A.
APPROXIMATION GANZZAHLIGER POLYEDER DURCH CORNER POLYEDER.

BACH 1977 -2 BACHEM, A.
FUNDAMENTALPUNKTALGORITHMEN ZUR LOESUNG GANZZAHLIGER PROGRAMMIERUNGSAUFGABEN(EXTENDED ABSTRACT).

BACH 1977 -3 BACHEM, A.
REDUCTION AND DECOMPOSITION OF INTEGER PROGRAMS OVER CONES.

BACH 1976 -3 BACHEM, A.
BEITRAEGE ZUR THEORIE DER CORNER POLYEDER.

BACH 1976 -4 BACHEM, A.
REDUCTION AND DECOMPOSITION OF MODULO OPTIMIZATION PROBLEMS.

BELL 1977 -1 BELL, D. E.
SIMPLE ALGORITHM FOR INTEGER PROGRAMS USING GROUP CONSTRAINTS, A.

BURD 1977 -1 BURDET, C.-A.; JOHNSON, E. L.
SUBADDITIVE APPROACH TO SOLVE LINEAR INTEGER PROGRAMS, A.

CAMB 1976 -1 CAMBINI, A.
APPLICATION OF THE METHOD OF DYNAMIC PROGRAMMING TO AN OPTIMIZATION PROBLEM OVER A FINITE ABELIAN GROUP, AN

CHENB 1976 -1 CHEN, DER-S.; ZIONTS, S.
COMPARISON OF SOME ALGORITHMS FOR SOLVING THE GROUP THEORETIC INTEGER PROGRAMMING PROBLEM.

FRIEA 1976 -1 FRIEZE, A. M.
SHORTEST PATH ALGORITHMS FOR KNAPSACK TYPE PROBLEMS.

GALLC 1972 -1 GALLO, G.; MARTINO. E.; SIMEONE, B.
GROUP OPTIMIZATION ALGORITHMS AND SOME NUMERICAL RESULTS VIA A BRANCH AND BOUND APPROACH.

GIUL 1976 -1 GIULIANELLI, S.; LUCERTINI. M.
DECOMPOSITION TECHNIQUE IN INTEGER LINEAR PROGRAMMING, A.

GOND 1971 -1 GONDRAN, M.
PROGRAMMATION ENTIERE NON LINEAIRE: DUALITE DIOPHANTIENNE.

GOND 1977 -1 GONDRAN, M.
GROUP THEORETIC AND LAGRANGEAN METHODS.

HABE 1977 -1 HABENICHT, W.
ZUR AUSWAHL VON BASISLOESUNGEN BEI VERFAHREN DER GANZZAHLIGEN OPTIMIERUNG.

HOUC 1977 -1 HOUCK, D. J.; VEMUGANTI, R. R.
ALGORITHM FOR THE VERTEX PACKING PROBLEM, AN.

JOHNE 1977 -1 JOHNSON, E. L.
ON THE GROUP PROBLEM AND A SUBADDITIVE APPROACH TO INTEGER PROGRAMMING.

KORTB 1976 -2 KORTE, B.
INTEGER PROGRAMMING (SURVEY PAPER).

LIBU 1974 -1 LIBURA, M.
REDUCTION OF A LINEAR INTEGER PROGRAMMING PROBLEM TO GROUP-THEORETIC FORM (POLISH).

LITVA 1975 -1 LITVAK, B. G.; NAIVEL'T, A. V.
SOLUTION OF THE INTEGER LINEAR PROGRAMMING PROBLEM BY MEANS OF A GROUP-THEORETIC APPROACH (RUSSIAN).

MARCC 1972 -1 MARCUS, M.; UNDERWOOD, E. E.
NOTE ON THE MULTIPLICATIVE PROPERTY OF THE SMITH NORMAL FORM, A.

NEWM 1974 -1 NEWMAN, M.
SMITH NORMAL FORM OF A PARTITIONED MATRIX, THE.

NEWM 1972 -1 NEWMAN, M.
INTEGRAL MATRICES.

PIEH 1977 -1 PIEHLER, J.
GANZZAHLIGE OPTIMIERUNG MIT KONGRUENZNEBENBEDINGUNGEN.

RARDA 1976 -2 RARDIN, R. L.; UNGER, V. E.
SOME COMPUTATIONALLY RELEVANT GROUP THEORETIC STRUCTURES OF FIXED CHARGE PROBLEMS.

RATH 1977 -1 RATHMANN, D.
UMFORMULIERUNG BEIM ASYMPTOTISCHEN ANSATZ FUER GANZZAHLIGE PROGRAMME.

RATH 1975 -1 RATHMANN, D.
ON THE ORDER OF THE GROUP OCCURING IN THE ASYMPTOTIC INTEGER LINEAR PROGRAMMING APPROACH.

RATH 1977 -2 RATHMANN, D.
ZUR LOESUNG DISKRETER OPTIMIERUNGSPROBLEME MIT GRUPPENTHEORETISCHEN ANSAETZEN.

RATH 1975 -2 RATHMANN, D.
SUCHVERFAHREN ZUM ASYMPTOTISCHEN ANSATZ FUER GANZZAHLIGE PROGRAMME.

SALK 1975 -1 SALKIN, H. M.
INTEGER PROGRAMMING.

STEIB 1976 -1 STEINER, J.
ZUR KONSTRUKTION VON SCHNITTEN FUER GEMISCHT-GANZZAHLIGE LO-PROBLEME.

SWAN 1974 -1 SWANEK, J. J.
LINEAR INTEGER PROGRAMMING.

VOLP 1976 -1 VOLPENTESTA, A.
FINITE GROUPS AND INTEGER LINEAR PROGRAMS.

****

CALY 1974 -1 CALYKOV, V.
CERTAIN HEURISTIC ALGORITHM FOR THE SOLUTION OF THE TRAVELLING SALESMAN PROBLEM (RUSSIAN), A.

GERT 1977 -1 GERTSBACH, I.
PROBABILISTIC IDEAS IN HEURISTIC ALGORITHMS FOR SOLVING SOME SCHEDULING PROBLEMS.

GIMAA 1976 -1 GIMADI, E. K.; GLEBOV, N. I.; PEREPELITSA, V. A.
ALGORITHMS WITH ESTIMATIONS FOR THE PROBLEMS OF DISCRETE OPTIMIZATION (RUSSIAN).

GOLDB 1977 -2 GOLDEN, B. L.
STATISTICAL APPROACH TO THE TSP, A.

KARPA 1978 -1 KARP, R. M.
PATCHING ALGORITHM FOR THE NONSYMMETRIC TRAVELING-SALESMAN PROBLEM, A.

KARPA 1975 -1 KARP, R. M.
FAST APPROXIMATE SOLUTION OF HARD COMBINATORIAL PROBLEMS, THE.

KARPA 1976 -1 KARP, R. M.
PROBABILISTIC ANALYSIS OF SOME COMBINATORIAL SEARCH ALGORITHMS, THE.

KARPA 1977 -1 KARP, R. M.
PROBABILISTIC ANALYSIS OF PARTITIONING ALGORITHMS FOR THE TRAVELING-SALESMAN PROBLEM IN THE PLANE.

KUCE 1977 -1 KUCERA, L.
EXPECTED BEHAVIOR OF GRAPH COLORING ALGORITHMS.

MORO 1974 -1 MOROZENSKII, L. YU.
ASYMPTOTIC BEHAVIOUR OF THE LENGTH OF THE PATH OF A TRAVELING SALEMAN WHEN THE CITIES ARE RANDOMLY DISTRIBUTED (RUSSIAN), THE.

NAND 1974 -1 NANDA, R.; WEINGARTEN, A.
PRACTICABLE SOLUTION FOR THE QUADRATIC ASSIGNMENT PROBLEM, A.

RABI 1976 -1 RABIN, M. O.
PROBABILISTIC ALGORITHMS.

RUBID 1971 -1 RUBINSHTEIN, M. I.
ON THE SYMMETRIC TRAVELING SALESMAN PROBLEM (RUSSIAN).

SAVAA 1976 -1 SAVAGE, S. L.
PROBABILISTIC APPROACH TO HEURISTIC DESIGN, A.

SAVAA 1978 -1 SAVAGE, S. L.
PROBABILISTIC APPROACH TO THE DESIGN OF HEURISTICS FOR DECISION PROBLEMS, A.

SAVAA 1976 -2 SAVAGE, S. L.
SOME THEORETICAL IMPLICATIONS OF LOCAL OPTIMIZATION.

HEURISTICS (WITHOUT ANALYSIS)

ADRA 1976 -1 ADRABINSKI, A.
ALGORITHM 45: A HEURISTIC ALGORITHM FOR THE TRAVELING-SALESMAN PROBLEM.

AHREA 1976 -1 AHRENS, W.
NUMERICAL EXPERIENCE WITH HEURISTIC RULES IN SPECIFIC BINARY OPTIMIZATION ALGORITHMS.

ALCO 1975 -1 ALCOUFFE, A.; ENJALBERT, M.; MURATET, G.
METHODES DE RESOLUTION DU PROBLEME DE TRANSPORT ET DE PRODUCTION D'UNE ENTREPRISE A ETABLISSEMENTS MULTIPLES EN PRESENCE DE COUTS FIXES.

ANDO 1976 -1 ANDON, F. I.; VARTANYAN, G. M.
PROBLEM OF PACKAGING DEMANDS WITH TIMING MARKS IN INFORMATION SYSTEMS (RUSSIAN).

BARRB 1976 -1 BARRER, J.; CHANDRASEKARAN, R.
ADJACENT PAIRWISE INTERCHANGE (API) ALGORITHM: A CLARIFICATION OF THE LITERATURE, THE.

BARRD 1977 -1 BARRY, D. L.; ROBINSON, D. F.
HEURISTIC FOR SCHEDULING JOBS ON SEVERAL IDENTICAL MACHINES, A.

BEEN 1976 -1 BEENHAKKER, H. L.; NARAYANAN, V.
ALGORITHMS FOR SCHEDULING PROJECTS WITH LIMITED RESOURCES.

BENN 1976 -1 BENNETT, G. R.; BYRD, J. JR.
TRAINABLE HEURISTIC PROCEDURE FOR THE ASSEMBLY LINE BALANCING PROBLEM, A.

BERMB 1976 -1 BERMOND, J.-C.; KODRATOFF, Y.
HEURISTIQUE POUR LE CALCUL DE L'INDICE DE TRANSITIVITE D'UN TOURNOI, UNE.

BEYE 1970 -1 BEYER, K.
EINIGE VERFAHREN ZUR TERMINPLANUNG IN NETZPLAENEN MIT BESCHRAENKTER [illegible]VERFUEGBARKEIT.

BHAL 1970 -1 BHALLA, H. S.
MULTI-TIME PERIOD FACILITIES LOCATION: A HEURISTIC ALGORITHM WITH APPLICATIONS TO WASTE WATER TREATMENT SYSTEMS.

BION 1973 -1 BIONDI, E.
OPTIMUM DELIVERY PROBLEMS.

BOESC 1977 -1 BOESCH, F. T.; GIMPEL, J. F.
COVERING THE POINTS OF A DIGRAPH WITH POINT-DISJOINT PATHS AND ITS APPLICATION TO CODE OPTIMIZATION.

BRADA 1971 -1 BRADLEY, G. H.
HEURISTIC SOLUTION METHODS AND TRANSFORMED INTEGER LINEAR PROGRAMMING PROBLEMS.

BURKJ 1966 -1 BURKOV, V. N.; LOVETSKII, S. E.
HEURISTIC APPROACH TO THE SOLUTION OF DYNAMIC PROGRAMMING PROBLEMS IN THE ALLOCATION OF RESOURCES (RUSSIAN).

BURNB 1977 -1 BURNS, R. N.
STATIONARY POINTS OF HEURISTICS IN PERMUTATION SCHEDULING.

CALY 1974 -1 CALYKOV, V.
CERTAIN HEURISTIC ALGORITHM FOR THE SOLUTION OF THE TRAVELLING SALESMAN PROBLEM (RUSSIAN), A.

CAME 1973 -1 CAMERINI, P. M.; FRATTA, L.; MAFFIOLI, F.
HEURISTICALLY GUIDED ALGORITHM FOR THE TRAVELING SALESMAN PROBLEM, A.

CAMEA 1974 -1 CAMERINI, P. M.; MAFFIOLI, F.
BOUNDS FOR 3-MATROID INTERSECTION PROBLEMS.

CAMEB 1976 -1 CAMERON, A. V.; VOILS, L. R.
HEURISTIC FOR TRANSPORTATION ROUTING AND VEHICLE LOADING UNDER UNCERTAINTY, A.

CAMP 1970 -1 CAMPBELL, H. G.; DUDEK, R. A.; SMITH, M. L.
HEURISTIC ALGORITHM FOR THE N-JOB, M-MACHINE SEQUENCING PROBLEM.

CARRA 1965 -1 CARROLL, D. C.
HEURISTIC SEQUENCING OF SINGLE AND MULTIPLE COMPONENT JOBS.

CHAL 1976 -1 CHALMET, L.
SOLVING ASSIGNMENT PROBLEMS IN WAREHOUSES (DUTCH).

CHANJ 1976 -1 CHANG, L.; KORSH, J. F.
CANONICAL COIN CHANGING AND GREEDY SOLUTIONS.

CHRI 1976 -1 CHRISTOFIDES, N.
VEHICLE ROUTING PROBLEM, THE.

CLAUA 1975 -1 CLAUS, A.; KLEITMAN, D. J.
HEURISTIC METHODS FOR SOLVING LARGE SCALE NETWORK ROUTING PROBLEMS: THE TELPACKING PROBLEM.

COI 1975 -1 COI, S.; SERMAN, A. I.
APPROXIMATE METHOD FOR THE SOLUTION OF INTEGER PROBLEMS OF LINEAR PROGRAMMING (THE METHOD OF ADMISSIBLE COMPONENTS) (RUSSIAN), AN.

CONT 1976 -1 CONTI, A.; ROSSETTO, S.
LAMINATION OF SEAMLESS TUBES: A HEURISTIC METHOD FOR CHOOSING THE BARS (ITALIAN).

COOP 1975 -1 COOPER, L.
EFFICIENT HEURISTIC ALGORITHM FOR THE TRANSPORTATION-LOCATION PROBLEM, AN.

COOP 1975 -2 COOPER, L.
FIXED CHARGE PROBLEM, I. A NEW HEURISTIC METHOD, THE.

COVE 1976 -1 COVERDALE, I. L.; WHARTON, F.
IMPROVED HEURISTIC PROCEDURE FOR A NONLINEAR CUTTING STOCK PROBLEM, AN.

DANN 1977 -1 DANNENBRING, D. G.
EVALUATION OF FLOW SHOP SEQUENCING HEURISTICS, AN.

DANN 1977 -2 DANNENBRING, D. G.
PROCEDURES FOR ESTIMATING OPTIMAL SOLUTION VALUES FOR LARGE COMBINATORIAL PROBLEMS.

DAVID 1977 -1 DAVIS, K. R.; WALTERS, J. E.
ADDRESSING THE N/1 SCHEDULING PROBLEM - A HEURISTIC APPROACH.

DAVIE 1976 -1 DAVIS, R. P.; TERRELL, M. P.
APPROXIMATION TECHNIQUE FOR PSEUDO-BOOLEAN MAXIMIZATION PROBLEMS, AN.

DOYL 1971 -1 DOYLE, M. S.
MINIMUM PATH COVERING ALGORITHMS.

DUNS 1975 -1 DUNSTAN, F. D. J.
GREEDY ALGORITHMS FOR OPTIMISATION PROBLEMS.

DYKS 1977 -1 DYKSTRA, D. P.; RIGGS, J. L.
APPLICATION OF FACILITIES LOCATION THEORY TO THE DESIGN OF FOREST HARVESTING AREAS, AN.

ECKE 1977 -1 ECKER, K.
ORGANISATION VON PARALLELEN PROZESSEN. THEORIE DETERMINISTISCHER SCHEDULES.

ECON 1977 -1 ECONOMIDES, S. C.; DUESTERHOEFT, W. C.
CONSTRUCTION SCHEDULING SYSTEM WITH RESOURCE ALLOCATION FOR ELECTRIC UTILITIES, A.

EDENA 1976 -1 EDEN, B. N.
HEURISTISCHES VERFAHREN ZUR DISTRIKTERMITTLUNG IN BEWERTETEN GRAPHEN UND SEINE ANWENDUNG AUF DIE BERECHNUNG VON SCHULEINZUGSBEREICHEN, EIN.

EDENB 1978 -1 EDEN, B. N.; PAPE, U.
HEURISTISCHES VERFAHREN ZUR DISTRIKTERMITTLUNG IN BEWERTETEN GRAPHEN, EIN.

EILOA 1977 -1 EILON, S.; CHOWDHURY, I. G.
MINIMISING WAITING TIME VARIANCE IN THE SINGLE MACHINE PROBLEM.

ELSH 1977 -1 ELSHAFEI, A. N.
HOSPITAL LAYOUT AS A QUADRATIC ASSIGNMENT PROBLEM.

****

ERLE 1975 -2 ERLENKOTTER, D.
CAPACITY PLANNING FOR LARGE MULTILOCATION SYSTEMS: APPROXIMATE AND INCOMPLETE DYNAMIC PROGRAMMING APPROACHES.

ESCHB 1974 -1 ESCHER, H. K.
SEQUENTIELLE ZUORDNUNGSPROBLEME.

EVANB 1977 -1 EVANS, J. R.; JARVIS, J. J.
OUT-OF-KILTER BASED HEURISTIC FOR THE INTEGER MULTICOMMODITY TRANSPORTATION PROBLEM, AN.

EVANB 1975 -1 EVANS, J. R.; JARVIS, J. J.
IMPROVED FORMULATION AND A HEURISTIC ALGORITHM FOR THE INTEGER MULTICOMMODITY TRANSPORTATION PROBLEM, AN.

FEIN 1977 -1 FEINBERG, I.
HEURISTIC APPROACH TO WAREHOUSE LOCATION USING MULTIPLE CENTERS OF GRAVITY, A.

FINKA 1976 -1 FINKEL'SHTEIN, YU. YU.
APPROXIMATIVE METHODS AND APPLIED PROBLEMS IN DISCRETE PROGRAMMING (RUSSIAN).

FISKB 1977 -1 FISK, J.; MACKEOWN, P. G.
APPLICATION OF INTEGER PROGRAMMING TO THE ASSIGNMENT OF STUDENTS TO SCHOOLS, AN.

FONGA 1974 -1 FONG, C. O.
MULTI-REGION DYNAMIC CAPACITY EXPANSION PROBLEM: EXACT AND HEURISTIC APPROACHES, THE.

FRANF 1971 -1 FRANK, H.; CHOU, W.
HEURISTIC APPROACH TO ROUTING IN COMPUTER NETWORKS, A.

FREH 1975 -2 FREHEL, J.
REGULARISATION DANS LES PROBLEMES COMBINATOIRES, PROBLEMES DE TOURNEES, DE PARTITION ET DE RECOUVREMENT.

FRIEA 1977 -1 FRIEZE, A. M.
ALGORITHMS FOR GENERALISATIONS OF MATROIDS.

GABA 1975 -1 GABASOV, R.; KIRILLOVA, F. M.
CONSTRUCTION OF SUBOPTIMAL PLANS OF A TRANSPORTATION PROBLEM (RUSSIAN), THE.

GALLD 1976 -1 GALLUS, G.
HEURISTISCHE VERFAHREN ZUR LOESUNG ALLGEMEINER GANZZAHLIGER LINEARER OPTIMIERUNGSPROBLEME (EIN UEBERBLICK).

GALLD 1976 -2 GALLUS, G.
HEURISTISCHE VERFAHREN ZUR LOESUNG LINEARER OPTIMIERUNGSPROBLEME MIT GANZZAHLIGEN VARIABLEN.

GAREB 1976 -2 GAREY, M. R.; JOHNSON, D. S.
APPROXIMATION ALGORITHMS FOR COMBINATORIAL PROBLEMS: PROSPECTS AND LIMITATIONS.

GARF 1977 -1 GARFINKEL, R. S.
MINIMIZING WALLPAPER WASTE, PART 1: A CLASS OF TRAVELING SALESMAN PROBLEMS.

GARFB 1977 -1 GARFINKEL, R. S.; NEEBE, A. W.; RAO, M. R.
M-CENTER PROBLEM: MINIMAX FACILITY LOCATION, THE.

GAUT 1977 -1 GAUTHIER, J.-M.; RIBIERE, G.
EXPERIMENTS IN MIXED-INTEGER LINEAR PROGRAMMING USING PSEUDOCOSTS.

GILLF 1976 -1 GILLETT, B. E.; JOHNSON, J. G.
MULTI-TERMINAL VEHICLE-DISPATCH ALGORITHM.

GIMAB 1974 -1 GIMADI, E. K.; PEREPELITSA, V. A.
ASYMPTOTIC APPROACH TO THE SOLUTION OF THE TRAVELING SALESMAN PROBLEM (RUSSIAN).

GLOV 1977 -2 GLOVER, F.
HEURISTICS FOR INTEGER PROGRAMMING USING SURROGATE CONSTRAINTS.

GOLD 1975 -1 GOLDEN, B.
VEHICLE ROUTING PROBLEMS: FORMULATIONS AND HEURISTIC SOLUTION TECHNIQUES.

GOLDB 1977 -1 GOLDEN, B. L.
EVALUATING A SEQUENTIAL VEHICLE ROUTING ALGORITHM.

GOLDB 1976 -1 GOLDEN, B. L.
APPROACHES TO THE CUTTING STOCK PROBLEM.

GOLDB 1976 -2 GOLDEN, B. L.
RECENT DEVELOPMENTS IN VEHICLE ROUTING.

GOLDD 1977 -1 GOLDEN, B. L.; MAGNANTI, T. L.; NGUYEN, H. Q.
IMPLEMENTING VEHICLE ROUTING ALGORITHMS.

GONDA 1977 -1 GONDRAN, M.; DOSTATNI, M.
TRAITEMENT DES EXCLUSIVES DANS PLANNEC, LE.

GORDB 1974 -1 GORDON, J. H.
HEURISTIC METHODS IN RESOURCE ALLOCATION.

GRAN 1974 -1 GRANDA, T. M.
HEURISTIC APPROACH TO A LARGE PERSONNEL ASSIGNMENT PROBLEM, A.

GUPT 1976 -1 GUPTA, J. N. D.
HEURISTIC ALGORITHM FOR THE FLOWSHOP SCHEDULING PROBLEM, A.

GUTE 1976 -1 GUTENBAUM, J.; NICOLO, F.
OPTIMAL AND SUB-OPTIMAL ALLOCATION OF RESOURCES IN SOME HIERARCHICAL SYSTEMS (POLISH).

HADJ 1974 -1 HADJERES, S.
QELQUES METHODES RECENTES DE RESOLUTION DU PROBLEME DU VOYAGEUR DE COMMERCE.

HAEN 1973 -1 HAENSLER, E.
EXPERIMENTAL HEURISTIC PROCEDURE TO OPTIMIZE A TELECOMMUNICATION NETWORK UNDER NONLINEAR COST FUNCTIONS.

HALP 1976 -1 HALPERN, J.
SEQUENTIAL COVERING PROBLEM UNDER UNCERTAINTY, THE (REVISED).

HALP 1977 -2 HALPERN, J.
SEQUENTIAL COVERING PROBLEM UNDER UNCERTAINTY, THE.

HART 1968 -1 HART, P. E.; NILLSON, N. L.; RAPHAEL, B.
FORMAL BASIS FOR THE HEURISTIC DETERMINATION OF MINIMUM COST PATHS.

HENDA 1976 -1 HENDERSON, W. B.; BERRY, W. L.
HEURISTIC METHODS FOR TELEPHONE OPERATOR SHIFT SCHEDULING: AN EXPERIMENTAL ANALYSIS.

HITC 1976 -1 HITCHINGS, G. G.; COTTAM, M.
EFFICIENT HEURISTIC PROCEDURE FOR SOLVING THE LAYOUT DESIGN PROBLEM, AN.

HONG 1974 -1 HONG, M. L.
EXPERIMENTS WITH A JOB SHOP PRODUCTION MODEL COUPLING A LINEAR PROGRAM AND A HEURISTIC SCHEDULING PROCEDURE.

HU A 1976 -1 HU, T. C.; LENARD, M. L.
OPTIMALITY OF A HEURISTIC SOLUTION FOR A CLASS OF KNAPSACK PROBLEMS.

HULS 1977 -1 HULST, W. G. VAN; LIESHOUT, J. TH. VAN
INVESTMENT/FINANCIAL PLANNING WITH ENDOGENOUS LIFETIMES: A HEURISTIC APPROACH TO MIXED-INTEGER PROGRAMMING.

HUNG 1973 -1 HUNG, H. K.
HEURISTIC ALGORITHM FOR THE MULTI-PERIOD FACILITY LOCATION PROBLEM, A.

JACOB 1977 -1 JACOBSON, R. E.; SADOWSKI, R. F.
SCHEDULING ALGORITHM FOR THE UNBALANCED PRODUCTION LINE, A.

JAER 1973 -1 JAERVINEN, P.
ON A ROUTING PROBLEM IN A GRAPH WITH TWO TYPES OF COSTS ASSOCIATED WITH THE EDGES.

HEURISTICS (WITHOUT ANALYSIS)

****

JOHNI 1974 -1 JOHNSON, P. A.
HEURISTIC FOR INTEGRATING PLANT LOCATION AND PLANT TIMING QUESTIONS, A.

JULL 1976 -1 JULLIEN, B.
DIFFERENTES APPROCHES LOGICIELLES POUR LA RESOLUTION DES PROBLEMES COMBINATOIRES EN TEMPS REEL.

KALD 1975 -1 KALDYBAEV, S. U.
ALGORITHM FOR THE APPROXIMATION-COMBINATORIAL METHOD IN SOLVING AN ALLOCATION PROBLEM USING BOOLEAN VARIABLES (RUSSIAN), A.

KAZA 1975 -1 KAZARIN, L. S.; L'VOV, G. N.; TIMOFEEV, E. A.
HEURISTIC ALGORITHM FOR CONSTRUCTING A TIME-TABLE (RUSSIAN), A.

KERN 1973 -1 KERNIGHAN, B. W; LIN, S.
HEURISTIC SOLUTION OF A SIGNAL DESIGN OPTIMIZATION PROBLEM.

KHAC 1976 -1 KHACHATUROV, V. R.; ASTAKHOV, N. B.
DYNAMIC LOCATION PROBLEMS (MODELS AND SOLUTION METHODS) (RUSSIAN).

KOHLA 1975 -1 KOHLER, W.
COMPUTATIONAL EXPERIENCE WITH EFFICIENT EXACT AND APPROXIMATE ALGORITHMS FOR AN NP-COMPLETE SCHEDULING PROBLEM.

KRAR 1972 -1 KRARUP, J.
QUADRATIC ASSIGNMENT (IN DANISH).

KREU 1969 -1 KREUZBERGER, H.
VERFAHREN ZUR LOESUNG GANZZAHLIGER LINEARER OPTIMIERUNGSPROBLEME.

KRON 1971 -1 KRONE, M.
HEURISTIC PROGRAMMING APPLIED TO SCHEDULING MODELS.

LEBL 1977 -1 LEBLANC, L. J.
HEURISTIC APPROACH OF LARGE SCALE DISCRETE STOCHASTIC TRANSPORTATION-LOCATION PROBLEMS, A.

LEE 1976 -1 LEE, D. H.
LOW COST DRAINAGE NETWORKS.

LEE A 1976 -1 LEE, J.; BOSE, N.; HWANG, F.
USE OF STEINER'S PROBLEM IN SUBOPTIMAL ROUTING IN RECTILINEAR METRIC.

LENS 1973 -1 LENSTRA, H. W.
ACYCLIC SUBGRAPH PROBLEM, THE.

LEVI 1974 -1 LEVI, G.
GRAPH ISOMORPHISM: A HEURISTIC EDGE-PARTITIONING-ORIENTED ALGORITHM, A.

LIN D 1971 -1 LIN, S.; KERNIGHAN, B. W.
HEURISTIC TECHNIQUE FOR SOLVING A CLASS OF COMBINATORIAL OPTIMIZATION PROBLEMS, A.

LIU 1975 -1 LIU, C. L.
APPROXIMATION ALGORITHMS FOR DISCRETE OPTIMIZATION PROBLEMS (EXTENDED ABSTRACT).

LOS 1978 -1 LOS, M.
SIMULTANEOUS OPTIMIZATION OF LAND USE AND TRANSPORTATION: A SYNTHESIS OF THE QUADRATIC ASSIGNMENT PROBLEM AND THE OPTIMAL NETWORK PROBLEM.

LOS A 1977 -1 LOS, M.; NGUYEN, S.
SPATIAL ALLOCATION ON A NETWORK WITH CONGESTION.

LUED 1975 -1 LUEDTKE, H.
COMPARISON OF HEURISTIC GRAPH SCHEDULING ALGORITHMS, A.

MAFF 1975 -1 MAFFIOLI, F.
MATROIDS AND HEURISTICALLY GUIDED SEARCH IN PROBLEM SOLVING.

MAHM 1976 -1 MAHMOUD, S.; RIORDON, J. S.
OPTIMAL ALLOCATION OF RESOURCES IN DISTRIBUTED INFORMATION NETWORKS.

MARU 1976 -1 MARUYAMA, K.; TANG, D. T.; FRATTA, L.
FLOW ASSIGNMENT ALGORITHM FOR COMPUTER COMMUNICATION NETWORK DESIGN WITH DIFFERENT CLASSES OF PACKETS.

MASS 1976 -1 MASSA, L.
SEMIECONOMIC METHOD FOR THE DETERMINATION OF OPTIMAL LOT SIZES, A.

MICHA 1968 -1 MICHIE, D.; FLEMING, J. G.; OLDFIELD, J. V.
COMPARISON OF HEURISTIC, INTERACTIVE, AND UNAIDED METHODS OF SOLVING A SHORTEST-ROUTE PROBLEM, A.

MILI 1976 -1 MILIOTIS, P.
INTEGER PROGRAMMING APPROACHES TO THE TRAVELLING SALESMAN PROBLEM.

MINIB 1975 -1 MININA, T. R.; PEREKREST, V. T.
APPROXIMATION TO THE SOLUTION OF THE TRAVELING-SALESMAN PROBLEM BY C-CYCLES.

MINOA 1975 -2 MINOUX, M.
RESOLUTION DES PROBLEMES DE MULTIFLOTS EN NOMBRES ENTIERS DANS LES GRANDS RESEAUX.

MINOA 1976 -3 MINOUX, M.
MULTIFLOTS DE COUT MINIMAL AVEC FONCTIONS DE COUT CONCAVES.

MOLZ 1976 -1 MOLZBERGER, P.
HEURISTISCHER ALGORITHMUS ZUR WEGOPTIMIERUNG BEI RECHTWINKELIGEN, ACHSENPARALLELEN HINDERNISSEN, EIN.

MORP 1971 -1 MORPURGO, R.
HEURISTIC METHOD FOR THE VERIFICATION OF THE ISOMORPHISM OF TWO SIMPLE AND NON ORIENTED GRAPHS, A.

MORT 1977 -1 MORTON, T. E.; DHARAN, B. G.
ALGORISTICS FOR SEQUENCING WITH PRECEDENCE CONSTRAINTS.

MUEL 1976 -1 MUELLER-MERBACH, H.
ANSAETZE ZU ENTWURFSMETHODOLOGIEN FUER ALGORITHMEN DER KOMBINATORISCHEN OPTIMIERUNG.

MUEL 1975 -1 MUELLER-MERBACH, H.
MODELLING TECHNIQUES AND HEURISTICS FOR COMBINATORIAL PROBLEMS.

MUEL 1976 -2 MUELLER-MERBACH, H.
HEURISTIC PROCEDURES FOR SOLVING COMBINATORIAL OPTIMIZATION PROBLEMS IN TRANSPORTATION.

MUEL 1976 -3 MUELLER-MERBACH, H.
MORPHOLOGIE HEURISTISCHER VERFAHREN.

NABE 1972 -1 NABESHIMA, I.
ALGORITHMS FOR MULTIPROJECT SCHEDULING WITH RESOURCE CONSTRAINTS AND RELATED PARALLEL SCHEDULING.

NABE 1973 -1 NABESHIMA, I.
BACKTRACK PROGRAMMING ALGORITHM AND RELIABLE HEURISTIC PROGRAMS FOR GENERAL SCHEDULING PROBLEM, A.

NELS 1977 -1 NELSON, R. T.; HOLLOWAY, C. A.; WONG, R. M.-L.
CENTRALIZED SCHEDULING AND PRIORITY IMPLEMENTATION HEURISTICS FOR A DYNAMIC JOB SHOP MODEL.

NORB 1977 -1 NORBACK, J. P.; LOVE, R. F.
GEOMETRIC APPROACHES TO SOLVING THE TRAVELING SALESMAN PROBLEM.

NUNNA 1977 -1 NUNNIKHOVEN, T. S.; EMMONS, H.
SCHEDULING ON PARALLEL MACHINES TO MINIMIZE TWO CRITERIA RELATED TO JOB TARDINESS.

OFFI 1975 -1 OFFIR, J. D.
TRAVELLING SALESMAN WITH A KNAPSACK(A CALL-GIRL'S HEURISTIC), A.

****

ORLO 1976 -1 ORLOFF, C.; CAPRERA, D.
REDUCTION AND SOLUTION OF LARGE SCALE VEHICLE ROUTING PROBLEMS.

ORLOA 1976 -2 ORLOFF, C. S.
ROUTE CONSTRAINED FLEET SCHEDULING.

ORON 1976 -1 ORON, G.; KARMELI, D.; WALKER, W. R.
APPROXIMATION SOLUTION OF NON-LINEAR MIXED-INTEGER OPTIMIZATION, AN.

OSLEA 1977 -1 OSLEEB, J. P.; CROMLEY, R. G.
LOCATION-PRODUCTION-ALLOCATION PROBLEM WITH NONLINEAR PRODUCTION COSTS, THE.

PARK 1977 -1 PARKER, R. G.; DEANE, R. H.; HOLMES, R. A.
ON THE USE OF A VEHICLE ROUTING ALGORITHM FOR THE PARALLEL PROCESSOR PROBLEM WITH SEQUENCE DEPENDENT CHANGEOVER COSTS.

PAULA 1976 -1 PAULAUSKAS, M. A.; RASHKILIS, A. V.; SHYARKSHNIS, G. A.
PROBLEM CONCERNING THE OPTIMAL ALLOCATION OF DEFICIENT RESOURCES FOR A DISCRETE CONSUMPTION, 2. ALGORITHMS FOR THE APPROXIMATE SOLUTION OF PROBLEMS WITH A GREAT NUMBER OF UNKNOWNS (RUSSIAN).

PAULA 1976 -2 PAULAUSKAS, M. A.; RASHKILIS, A. V.; SHYARKSHNIS, G. A.
PROBLEM CONCERNING THE OPTIMAL ALLOCATION OF DEFICIENT RESOURCES FOR A DISCRETE CONSUMPTION, 3. HEURISTIC ALGORITHMS FOR THE SOLUTION OF THE PROBLEM (RUSSIAN).

PEAR 1974 -1 PEARMAN, A.
HEURISTIC APPROACHES TO ROAD NETWORK OPTIMIZATION.

PLOTA 1974 -1 PLOTNIKOV, V. N.; ZVEREV, V. YU.
METHODS OF FAST DISTRIBUTION OF ALGORITHMS IN COMPUTATIONAL SYSTEMS (RUSSIAN).

POHL 1969 -1 POHL, I.
BI-DIRECTIONAL AND HEURISTIC SEARCH IN PATH PROBLEMS.

POHL 1973 -1 POHL, I.
AVOIDANCE OF (RELATIVE) CATASTROPHE HEURISTIC COMPETENCE GENUINE DYNAMIC WEIGHTING AND COMPUTATIONAL ISSUES IN HEURISTIC PROBLEM SOLVING, THE.

RAO 1975 -1 RAO, M. R.
MULTI-REGION, MULTI-PERIOD CAPACITY EXPANSION MODELS: A HEURISTIC AND A SPECIAL CASE.

RAWA 1975 -1 RAWASWAMY, K. V.; DUDEK, R. A.
HEURISTIC ALGORITHM FOR SINGLE FACILITY SEQUENCING WITH PENALTY COST, A.

ROBB 1975 -1 ROBBINS, J. A.; SHAMBLIN, J. E.; TURNER, W. C.; BYRD, D. F.
DEVELOPMENT OF AND COMPUTATIONAL EXPERIENCE WITH A COMBINATION TOUR CONSTRUCTION - TOUR IMPROVEMENT ALGORITHM FOR VEHICLE ROUTING PROBLEMS.

ROBE 1976 -1 ROBERS, P.; COOPER, L.
STUDY OF THE FIXED CHARGE TRANSPORTATION PROBLEM, A.

ROBI 1972 -1 ROBINSON, D. F.
OPTIMIZATION OF FINITE SETS.

RODG 1977 -1 RODGERS, J. R.; MAC CARTHY, O.
SERVICE FACILITY LOCATION PROBLEM: THE LOCATION OF FARM ADVISORY OFFICERS IN N. Z., A.

ROOS 1975 -1 ROOSE, J. K.; RIEMERSMA, P.
HEURISTIC OPTIMISATION OF A LOCATION-ALLOCATION PROBLEM IN A TRANSPORT NETWORK WITH MODAL CHOICE.

RUBLA 1973 -1 RUBLINECKII, V. I.; TOMCUK, E. JA.
APPROXIMATE PROCEDURE FOR THE SOLUTION OF THE TRAVELING SALESMAN PROBLEM (RUSSIAN), AN.

RUSS 1975 -1 RUSSEL, R.
EFFECTIVE HEURISTIC FOR THE M-TOUR TRAVELING SALESMAN PROBLEM WITH SOME SIDE CONDITIONS, AN.

RUSSB 1977 -1 RUSSELL, R.; IGO, W.
ASSIGNMENT ROUTING PROBLEM, AN.

RUSSC 1977 -1 RUSSELL, R. A.
EFFECTIVE HEURISTIC FOR THE M-TOUR TRAVELING SALESMAN PROBLEM WITH SOME SIDE CONDITIONS, AN.

RUSSD 1965 -1 RUSSO, F. J.
HEURISTIC APPROACH TO ALTERNATE ROUTING IN A JOB SHOP.

SAND 1975 -1 SANDEWALL, E. J.
HEURISTISCHE SUCHE: KONZEPTE UND METHODEN.

SAVAA 1973 -1 SAVAGE, S. L.
SOLUTION OF DISCRETE LINEAR OPTIMIZATION PROBLEMS BY NEIGHBORHOOD SEARCH TECHNIQUES, THE.

SAVAA 1973 -2 SAVAGE, S. L.
STATISTICAL INDICATORS OF OPTIMALITY.

SEVA 1975 -1 SEVAST'YANOV, S. V.
ASYMPTOTIC APPROACH TO SOME PROBLEMS SEQUENCING THEORY (RUSSIAN).

SHAPD 1977 -1 SHAPIRO, S. D.
PERFORMANCE OF HEURISTIC BIN PACKING ALGORITHMS WITH SEGMENTS OF RANDOM LENGTH.

SHOW 1977 -1 SHOWALTER, M. J.; KRAJEWSKI, L. J.; RITZMAN, L. P.
MANPOWER ALLOCATION IN U.S. POSTAL FACILITIES: A HEURISTIC APPROACH.

SKAF 1966 -1 SKAFF, M. S.
APPROXIMATE SOLUTION OF A WEIGHTED SCHEDULING PROBLEM WITH MULTIPLE CONSTRAINTS.

SMITF 1975 -1 SMITH, M. L.; PANWALKAR, S. S.; DUDEK, R. A.; ULGEN, O.
FLOWSHOP SEQUENCING WITH NONPERMUTATION SCHEDULES - A HEURISTIC APPROACH.

SPAE 1977 -1 SPAETH, H.
NUMERISCHE ERFAHRUNGEN ZU HEURISTISCHEN LOESUNGSVERFAHREN BEIM VARIANZKRITERIUM IN DER CLUSTER-ANALYSE.

STAI 1977 -1 STAINTON, R. S.
CUTTING STOCK PROBLEM FOR THE STOCKHOLDER OF STEEL REINFORCEMENT BARS, THE.

STANB 1969 -1 STANKARD, M. F.; GUPTA, S. K.
NOTE ON BOMBERGER'S APPROACH TO LOT SIZE SCHEDULING: HEURISTIC PROPOSED.

STEIA 1977 -1 STEINBERG, D. I.
DESIGNING A HEURISTIC FOR THE FIXED-CHARGE TRANSPORTATION PROBLEM.

STEIA 1976 -1 STEINBERG, D. I.
ON SOLVING LARGE-SCALE FIXED CHARGE PROBLEMS.

STRAB 1976 -1 STRATMANN, K.-H.; BURKARD, R. E.
NUMERISCHE UNTERSUCHUNGEN UEBER QUADRATISCHE ZUORDNUNGSPROBLEME.

SUHL 1976 -2 SUHL, U.
HEURISTISCHE LOESUNGSVERFAHREN ZUR LOESUNG VON (0-1)-PROGRAMMEN MIT NICHT-NEGATIVEN KOEFFIZIENTEN.

SZWA 1977 -2 SZWARC, W.
SPECIAL CASES OF THE FLOW-SHOP PROBLEM.

SZWAA 1977 -1 SZWARC, W.; HUTCHINSON, G. K.
JOHNSON'S APPROXIMATE METHOD FOR THE 3XN JOB SHOP PROBLEM.

THES 1976 -1 THESEN, A.
HEURISTIC SCHEDULING OF ACTIVITIES UNDER RESOURCE AND PRECEDENCE RESTRICTIONS.

TILQ 1975 -1 TILQUIN, C.
ON SCHEDULING WITH EARLIEST STARTS AND DUE DATES ON A GROUP OF IDENTICAL MACHINES.

HEURISTICS (WITHOUT ANALYSIS)

****

TOME 1975 -1 TOMESCU, I.
ALGORITHME POUR L'OBTENTION D'UNE CHAINE HAMILTONIENNE EN PARTANT DE L'ARBRE MINIMAL D'UN GRAPHE, UN.

ULRI 1975 -1 ULRICH, L. Y.
HEURISTIC ALGORITHMS FOR SOLVING A LARGE SCALE MULTICOMMODITY FLOW PROBLEM ON A NETWORK WITH A STEP FUNCTION COST.

UNGE 1964 -1 UNGER, S. H.
HEURISTIC PROGRAM FOR TESTING PAIRS OF DIRECTED LINE GRAPHS FOR ISOMORPHISM, A.

VOLO 1976 -1 VOLODARSKIJ, Y. A. M.; GABOVICH, E. YA.; ZAKHARIN, A. YA.
SOLVABLE CASE IN DISCRETE PROGRAMMING (RUSSIAN), A.

WALKA 1976 -1 WALKER, W. E.
HEURISTIC ADJACENT EXTREME POINT ALGORITHM FOR THE FIXED CHARGE PROBLEM, A.

WANGB 1974 -1 WANG, J. C.; MICKLE, M. H.; HOELZEMAN, R. G.
SET COVERING APPLICATION IN MINIMUM ROUTE GENERATION, A.

WEINC 1975 -1 WEINER, P.
HEURISTICS.

WHITE 1966 -1 WHITE, M.
PROJECT SCHEDULING WITH LIMITED RESOURCES. EXPERIMENT IN HEURISTIC PROGRAMMING.

WONGA 1976 -1 WONG, C. K.; YAO, A. C.-C.
COMBINATORIAL OPTIMIZATION PROBLEM RELATED TO DATA SET ALLOCATION, A.

YAGI 1975 -1 YAGIZ, OE.
HEURISTICALLY AIDED SINGLE BRANCH SEARCH PROCEDURES FOR PLANT LOCATION PROBLEMS.

ZAIT 1975 -1 ZAITSEVA, Z. N.; SHTEIN, M. E.
PLOTTING OF OPTIMAL HAMILTONIAN CYCLES IN THE MULTIVERTEX GRAPHS (RUSSIAN).

ZANC 1974 -1 ZANCEVA, F. N.
APPROXIMATION-COMBINATORIAL METHOD FOR THE CASE OF SEVERAL APPROXIMATING PROBLEMS (RUSSIAN), AN.

ZHUR 1977 -1 ZHURAVLEV, JU. I.
CORRECT ALGEBRAS OVER THE SETS OF INCORRECT (HEURISTIC) ALGORITHMS, II (RUSSIAN).

****

BRUC 1976 -1 BRUCKER, P.
ANMERKUNGEN ZU HEURISTISCHEN VERFAHREN.
BRUNB 1974 -1 BRUNO, J.; COFFMAN, E. G. JR.; SETHI, R.
ALGORITHMS FOR MINIMIZING MEAN FLOW TIME.
BUTE 1974 -1 BUTEN, R. E.
ANALYSIS OF THE FLOW-SHOP SCHEDULING PROBLEM IN A SYSTEM WITH TWO CLASSES OF PROCESSORS.
CHAMA 1977 -1 CHAMPEAUX, D. DE; SINT, L.
IMPROVED BIDIRECTIONAL HEURISTIC SEARCH ALGORITHM, AN.
CHAND 1976 -1 CHANDRA, A. K.; HIRSCHBERG, D. S.; WONG, C. K.
APPROXIMATE ALGORITHMS FOR SOME GENERALIZED KNAPSACK PROBLEMS.
CHANJ 1976 -1 CHANG, L.; KORSH, J. F.
CANONICAL COIN CHANGING AND GREEDY SOLUTIONS.
CHRI 1976 -2 CHRISTOFIDES, N.
WORST-CASE ANALYSIS OF A NEW HEURISTIC FOR THE TRAVELLING SALESMAN PROBLEM.
CODY 1976 -1 CODY, R. A.; COFFMAN, E. G. JR.
RECORD ALLOCATION FOR MINIMIZING EXPECTED RETRIEVAL COSTS ON DRUM-LIKE STORAGE DEVICES.
COFFA 1976 -1 COFFMAN, E. G. JR.; SETHI, R.
ALGORITHMS MINIMIZING MEAN FLOW TIME SCHEDULE-LENGTH PROPERTIES.
CORNC 1977 -1 CORNUEJOLS, G.; FISHER, M.; NEMHAUSER, G. L.
ON THE UNCAPACITATED LOCATION PROBLEM.
CORND 1977 -1 CORNUEJOLS, G.; FISHER, M. L.; NEMHAUSER, G. L.
LOCATION OF BANK ACCOUNTS TO OPTIMIZE FLOAT: AN ANALYTIC STUDY OF EXACT AND APPROXIMATE ALGORITHMS.
CORND 1976 -1 CORNUEJOLS, G.; FISHER, M. L.; NEMHAUSER, G. L.
ANALYSIS OF HEURISTICS AND RELAXATIONS FOR THE UNCAPACITATED LOCATION PROBLEM, AN.
CORNE 1978 -1 CORNUEJOLS, G.; NEMHAUSER, G. L.
TIGHT BOUNDS FOR CHRISTOFIDES' TRAVELING SALESMAN HEURISTIC.
DUNS 1975 -1 DUNSTAN, F. D. J.
GREEDY ALGORITHMS FOR OPTIMISATION PROBLEMS.
FISHB 1977 -1 FISHER, M. L.; NEMHAUSER, G. L.; WOLSEY, L. A.
ANALYSIS OF APPROXIMATIONS FOR MAXIMIZING SUBMODULAR FUNCTIONS.
FISKB 1977 -1 FISK, J.; MACKEOWN, P. G.
APPLICATION OF INTEGER PROGRAMMING TO THE ASSIGNMENT OF STUDENTS TO SCHOOLS, AN.
FRED 1976 -1 FREDERICKSON, G. N.; HECHT, M. S.; KIM, C. E.
APPROXIMATION ALGORITHMS FOR SOME ROUTING PROBLEMS.
FRIEA 1977 -2 FRIEZE, A. M.
GENERALISATION OF THE GREEDY ALGORITHM FOR MATROIDS, A.
GAREA 1976 -1 GAREY, M. R.; GRAHAM, R. L.; JOHNSON, D. S.; YAO, A. C.-C.
RESOURCE CONSTRAINED SCHEDULING AS GENERALIZED BIN PACKING.
GARED 1976 -1 GAREY, M. R.; JOHNSON, D. S.
APPROXIMATION ALGORITHMS FOR COMBINATORIAL PROBLEMS: AN ANNOTATED BIBLIOGRAPHY.
GAREB 1976 -3 GAREY, M. R.; JOHNSON, D. S.
COMPLEXITY OF NEAR-OPTIMAL GRAPH COLORING, THE.
GIMAA 1976 -1 GIMADI, E. K.; GLEBOV, N. I.; PEREPELITSA, V. A.
ALGORITHMS WITH ESTIMATIONS FOR THE PROBLEMS OF DISCRETE OPTIMIZATION (RUSSIAN).
GIMAB 1974 -1 GIMADI, E. K.; PEREPELITSA, V. A.
ASYMPTOTIC APPROACH TO THE SOLUTION OF THE TRAVELING SALESMAN PROBLEM (RUSSIAN).
GOLDB 1977 -1 GOLDEN, B. L.
EVALUATING A SEQUENTIAL VEHICLE ROUTING ALGORITHM.
GRAH 1976 -1 GRAHAM, R. L.
BOUNDS ON THE PERFORMANCE OF SCHEDULING ALGORITHMS.
HAUSA 1977 -1 HAUSMANN, D.; JENKYNS, T. A.; KORTE, B.
WORST CASE ANALYSIS OF GREEDY TYPE ALGORITHMS FOR INDEPENDENCE SYSTEMS.
HAUSC 1978 -2 HAUSMANN, D.; KORTE, B.
WORST-CASE ANALYSIS FOR A CLASS OF COMBINATORIAL ALGORITHMS.
HAUSC 1977 -2 HAUSMANN, D.; KORTE, B.
WORST-CASE BEHAVIOUR OF POLYNOMIAL BOUNDED ALGORITHMS FOR INDEPENDENCE SYSTEMS.
HORO 1976 -1 HOROWITZ, E.; SAHNI, S.
EXACT AND APPROXIMATE ALGORITHMS FOR SCHEDULING NONIDENTICAL PROCESSORS.
HORV 1977 -1 HORVATH, E. C.; LAM, S.; SETHI, R.
LEVEL ALGORITHM FOR PREEMPTIVE SCHEDULING, A.
HU A 1976 -1 HU, T. C.; LENARD, M. L.
OPTIMALITY OF A HEURISTIC SOLUTION FOR A CLASS OF KNAPSACK PROBLEMS.
IBAR 1976 -1 IBARAKI, T.
COMPUTATIONAL EFFICIENCY OF APPROXIMATE BRANCH-AND-BOUND ALGORITHMS.
IBARB 1974 -1 IBARRA, O.; SAHNI, S.
P-COMPLETE PROBLEMS AND APPROXIMATE SOLUTIONS.
IBARC 1977 -1 IBARRA, O. H.; KIM, C. E.
HEURISTIC ALGORITHMS FOR SCHEDULING INDEPENDENT TASKS ON NONIDENTICAL PROCESSORS.
IBARC 1974 -1 IBARRA, O. H.; KIM, C. E.
FAST APPROXIMATION ALGORITHMS FOR THE KNAPSACK AND SUM OF SUBSET PROBLEMS.
JENK 1974 -1 JENKYNS, T. A.
MATCHOIDS: A GENERALIZATION OF MATCHINGS AND MATROIDS.
JENK 1976 -1 JENKYNS, T. A.
EFFICACY OF THE "GREEDY" ALGORITHM, THE.
JENK 1977 -1 JENKYNS, T. A.
GREEDY TRAVELLING SALESMAN PROBLEM, THE.
JOHNC 1974 -1 JOHNSON, D. S.
WORST CASE BEHAVIOR OF GRAPH COLORING ALGORITHMS.
JOHNK 1970 -1 JOHNSON, S. C.; KERNIGHAN, B. W.
MAKING CHANGE WITH A MINIMUM NUMBER OF COINS.
KAFU 1977 -1 KAFURA, D. G.; SHEN, V. Y.
TASK SCHEDULING ON A MULTIPROCESSOR SYSTEM WITH INDEPENDENT MEMORIES.
KARPA 1975 -1 KARP, R. M.
FAST APPROXIMATE SOLUTION OF HARD COMBINATORIAL PROBLEMS, THE.
KAUF 1974 -1 KAUFMAN, M. T.
ALMOST-OPTIMAL ALGORITHM FOR THE ASSEMBLY LINE SCHEDULING PROBLEM, AN.
KOHLA 1975 -1 KOHLER, W.
COMPUTATIONAL EXPERIENCE WITH EFFICIENT EXACT AND APPROXIMATE ALGORITHMS FOR AN NP-COMPLETE SCHEDULING PROBLEM.
KORTB 1977 -1 KORTE, B.
ANALYSE VON HEURISTISCHEN ALGORITHMEN ZUR KOMBINATORISCHEN OPTIMIERUNG.
KORTB 1976 -2 KORTE, B.
INTEGER PROGRAMMING (SURVEY PAPER).

HEURISTICS (WORST CASE ANALYSIS)
****

KORTB 1977 -2 KORTE, B.
APPROXIMATIVE ALGORITHMS FOR DISCRETE OPTIMIZATION PROBLEMS.

KORTC 1978 -1 KORTE, B.; HAUSMANN, D.
ANALYSIS OF THE GREEDY HEURISTIC FOR INDEPENDENCE SYSTEMS, AN.

KORTC 1977 -1 KORTE, B.; HAUSMANN, D.
ANALYSE DER GREEDY-HEURISTIK FUER UNABHAENGIGKEITSSYSTEME, EINE.

LAM 1977 -1 LAM, S.; SETHI, R.
WORST CASE ANALYSIS OF TWO SCHEDULING ALGORITHMS.

LAWL 1977 -3 LAWLER, E. L.
FAST APPROXIMATION ALGORITHMS FOR KNAPSACK PROBLEMS.

LIU 1975 -1 LIU, C. L.
APPROXIMATION ALGORITHMS FOR DISCRETE OPTIMIZATION PROBLEMS (EXTENDED ABSTRACT).

LOVA 1976 -1 LOVASZ, L.
COVERS, PACKINGS AND SOME HEURISTIC ALGORITHMS.

LOVA 1975 -1 LOVASZ, L.
ON THE RATIO OF OPTIMAL INTEGRAL AND FRACTIONAL COVERS.

MOROA 1976 -1 MOROZOV, S. A.
PROBLEM OF FINDING THE SHORTEST TREE OF CONNECTION WITH CONSTRAINTS (RUSSIAN), THE.

NEMH 1976 -1 NEMHAUSER, G. L.; WOLSEY, L. A.
BEST ALGORITHMS FOR APPROXIMATING THE MAXIMUM OF A SUBMODULAR SET FUNCTION.

NEMHA 1976 -1 NEMHAUSER, G. L.; WOLSEY, L. A.; FISHER, M. L.
ANALYSIS OF APPROXIMATIONS FOR MAXIMIZING SUBMODULAR SET FUNCTIONS, AN.

NIGM 1975 -1 NIGMATULLIN, R. G.
COMPLEXITY OF THE APPROXIMATE SOLUTION OF COMBINATORIAL PROBLEMS.

PAPAA 1977 -1 PAPADIMITRIOU, C. H.; STEIGLITZ, K.
ON THE COMPLEXITY OF LOCAL SEARCH FOR THE TRAVELING SALESMAN PROBLEM.

ROSEB 1977 -1 ROSENKRANTZ, D. J.; STEARNS, R. E.; LEWIS, P. M.
ANALYSIS OF SEVERAL HEURISTICS FOR THE TRAVELING SALESMAN PROBLEM, AN.

ROSEB 1974 -1 ROSENKRANTZ, D. J.; STEARNS, R. E.; LEWIS, P. M.
APPROXIMATE ALGORITHMS FOR THE TRAVELING SALESPERSON PROBLEM.

SAHN 1977 -1 SAHNI, S.
GENERAL TECHNIQUES FOR COMBINATORIAL APPROXIMATION.

SAHN 1975 -1 SAHNI, S.
APPROXIMATE ALGORITHMS FOR THE 0/1 KNAPSACK PROBLEM.

SAHNA 1974 -1 SAHNI, S.; GONZALES, T.
P-COMPLETE PROBLEMS AND APPROXIMATE SOLUTIONS.

SAHNB 1976 -1 SAHNI, S.; GONZALEZ, T.
P-COMPLETE APPROXIMATION PROBLEMS.

SAHNC 1976 -1 SAHNI, S. K.
ALGORITHMS FOR SCHEDULING INDEPENDENT TASKS.

SAVAB 1976 -1 SAVAGE, S. L.; WEINER, P.; BAGCHI, A.
NEIGHBORHOOD SEARCH ALGORITHMS FOR GUARANTEEING OPTIMAL TRAVELING SALESMAN TOURS MUST BE INEFFICIENT.

SEVA 1975 -1 SEVAST'YANOV, S. V.
ASYMPTOTIC APPROACH TO SOME PROBLEMS SEQUENCING THEORY (RUSSIAN).

TIEN 1977 -1 TIEN, B. N.; HU, T. C.
ERROR BOUNDS AND THE APPLICABILITY OF THE GREEDY SOLUTION TO THE COIN-CHANGING PROBLEM.

WONG 1976 -1 WONG, C. K.
SOME DISCRETE OPTIMIZATION PROBLEMS WITH SECONDARY STORAGE APPLICATIONS.

WONGA 1976 -1 WONG, C. K.; YAO, A. C.-C.
COMBINATORIAL OPTIMIZATION PROBLEM RELATED TO DATA SET ALLOCATION, A.

ZIMMA 1977 -3 ZIMMERMANN, U.
THRESHOLD METHODS FOR BOOLEAN OPTIMIZATION PROBLEMS WITH SEPARABLE OBJECTIVES.

ALTH 1976 -1 ALTHERR, W.
ECKPUNKTBESTIMMUNG KONVEXER POLYEDER.
ALTH 1975 -1 ALTHERR, W.
ALGORITHM FOR ENUMERATING ALL VERTICES OF A CONVEX POLYHEDRON, AN.
BACH 1976 -2 BACHEM, A.
APPROXIMATION GANZZAHLIGER POLYEDER DURCH CORNER POLYEDER.
BACH 1976 -3 BACHEM, A.
BEITRAEGE ZUR THEORIE DER CORNER POLYEDER.
BALAI 1969 -1 BALAS, E.; JEROSLOW, R.
ON THE STRUCTURE OF THE UNIT HYPERCUBE.
BOLK 1976 -1 BOLKER, E. D.
SIMPLICIAL GEOMETRY AND TRANSPORTATION POLYTOPES.
BRUAA 1976 -1 BRUALDI, R. A.; GIBSON, P. M.
ASSIGNMENT POLYTOPE, THE.
CUNNA 1975 -1 CUNNINGHAM, W. H.
UNBOUNDED MATROID INTERSECTION POLYHEDRON, AN.
DURA 1973 -1 DURAND, J. A. A.
POLYHEDRAL NEOPOLARITIES.
EDEL 1970 -1 EDELBERG, M.
INTEGRAL CONVEX POLYHEDRA AND AN APPROACH TO INTEGRALIZATION.
EDMO 1975 -1 EDMONDS, J.
SOME WELL-SOLVED PROBLEMS IN COMBINATORIAL OPTIMIZATION.
EMEL 1976 -1 EMELICHEV, V. A.; KRAVTSOV, M. K.
CERTAIN PROPERTIES OF TRANSPORTATION POLYHEDRA WITH A MAXIMAL NUMBER OF VERTICES (RUSSIAN).
EMEL 1976 -2 EMELICHEV, V. A.; KRAVTSOV, M. K.
ON ENUMERATIVE PROBLEMS ON TRANSPORT POLYHEDRA (RUSSIAN).
EMELA 1976 -1 EMELICHEV, V. A.; KRAVTSOV, M. K.; AVERBUH, N. D.
MAXIMAL NUMBER OF VERTICES OF A TRANSPORTATION POLYHEDRON (RUSSIAN), THE.
EMELB 1974 -1 EMELICHEVA, V. S.; KONONEKO, A. M.
MINIMAL NUMBER OF VERTICES AND EDGES OF THE POLYTOPE OF A PLANAR TRANSPORTATION PROBLEM (RUSSIAN), THE.
EMELC 1974 -1 EMELICHEVA, V. S.; KONONENKO, A. M.
MINIMAL NUMBER OF INTEGER VERTICES OF THE POLYTOPE OF A MULTIINDEX TRANSPORTATION PROBLEM (RUSSIAN), THE.
FEDE 1975 -1 FEDERICO, P. J.
NUMBER OF POLYHEDRA, THE.
GILEA 1977 -1 GILES, F. R.; PULLEYBLANK, W. R.
TOTAL DUAL INTEGRALITY AND INTEGER POLYHEDRA.
GLOV 1973 -1 GLOVER, F.
POLYHEDRAL ANNEXATION IN MIXED INTEGER PROGRAMMING.
GLOV 1975 -3 GLOVER, F.
POLYHEDRAL ANNEXATION IN MIXED INTEGER AND COMBINATORIAL PROGRAMMING.
GROE 1977 -3 GROETSCHEL, M.
MONOTONE 2-MATCHING POLYTOPE ON A COMPLETE GRAPH, THE.
GROE 1977 -6 GROETSCHEL, M.
POLYEDRISCHE CHARAKTERISIERUNGEN KOMBINATORISCHER OPTIMIERUNGSPROBLEME.
GROEA 1977 -2 GROETSCHEL, M.; PADBERG, M. W.
ON THE SYMMETRIC TRAVELLING SALESMAN PROBLEM 1: INEQUALITIES.
GROEA 1977 -3 GROETSCHEL, M.; PADBERG, M. W.
ON THE SYMMETRIC TRAVELLING SALESMAN PROBLEM 2: LIFTING THEOREMS AND FACETS.
HERI 1969 -1 HERING, F.
UNTERSUCHUNGEN UEBER DIE KOMBINATORISCHE STRUKTUR VON POLYEDERN.
HOFF 1977 -1 HOFFMAN, A. J.
ON LATTICE POLYHEDRA 3: BLOCKERS AND ANTIBLOCKERS OF LATTICE CLUTTERS.
HOFF 1976 -1 HOFFMAN, A. J.
ON LATTICE POLYHEDRA 2: CONSTRUCTION AND EXAMPLES.
HOFFA 1978 -1 HOFFMAN, A. J.; OPPENHEIM, R.
LOCAL UNIMODULARITY IN THE MATCHING POLYTOPE.
HOFFB 1976 -1 HOFFMAN, A. J.; SCHWARTZ, D. E.
ON LATTICE POLYHEDRA.
HUAN 1977 -1 HUANG, H.-C.; TROTTER, L. E. JR.
TECHNIQUE FOR DETERMINING BLOCKING AND ANTIBLOCKING POLYHEDRAL DESCRIPTIONS, A.
JEROA 1977 -1 JEROSLOW, R. G.
CUTTING-PLANE GAME AND ITS ALGORITHMS, A.
JOHNE 1975 -2 JOHNSON, E. L.
ON CUT-SET INTEGER POLYHEDRA.
KATC 1977 -1 KATCHALSKI, M.
INTERSECTING ALL EDGES OF CONVEX POLYTOPES BY PLANES.
KRAV 1976 -1 KRAVTSOV, M. K.
ENUMERATION PROBLEMS ON THE DEGENERATED TRANSPORTATION POLYHEDRA (RUSSIAN).
LIHA 1975 -1 LIHACEV, V. M.
CERTAIN ESTIMATE FOR THE NUMBER OF VERTICES OF A TRANSPORTATION POLYTOPE (RUSSIAN), A.
LIHAA 1974 -1 LIHACEV, V. M.; EMELICHEV, V. A.
ON AN UPPER BOUND OF THE NUMBER OF VERTICES OF A TRANSPORTATION POLYTOPE (RUSSIAN).
MAURB 1976 -1 MAURRAS, J. F.
POLYTOPES A SOMMETS DANS (0,1)N.
MAURB 1977 -1 MAURRAS, J. F.
EXAMPLE OF DUAL POLYTOPES IN THE UNIT HYPERCUBE, AN.
NAKAA 1975 -1 NAKANO, H.; NAKANISHI, Y.
STRUCTURE OF SOLUTION SPACE IN DISCRETE OPTIMIZATION PROBLEM AND ITS APPLICATION TO MINIMUM FEEDBACK EDGE SET PROBLEM.
ROSEA 1971 -1 ROSENFELD, M.
ON CERTAIN POLYTOPES ASSOCIATED WITH GRAPHS.
TROTA 1977 -1 TROTTER, L. E. JR.; MONMA, C. L.
ON A THEOREM OF FULKERSON.
WOLS 1975 -1 WOLSEY, L. A.
VALID INEQUALITIES AND SUPERADDITIVITY FOR 0-1 INTEGER PROGRAMS.
YOUNA 1976 -1 YOUNG, H. P.
ON PERMUTATIONS AND PERMUTATION POLYTOPES.

INTEGER PROGRAMMING (GENERAL)
****

AGRA 1976 -2 AGRAWAL, S. C.
ON INTEGER SOLUTIONS TO LINEAR FRACTIONAL FUNCTIONALS BY A BRANCH AND BOUND TECHNIQUE.

ALEK 1977 -1 ALEKSEYEV, O. G.
CLASS OF INTEGER PROGRAMMING PROBLEMS (RUSSIAN), A.

ALEKB 1976 -1 ALEKSEYEV, O. G.; VOLODOS, I. F.
COMBINED USE OF DYNAMIC PROGRAMMING AND BRANCH-AND-BOUND METHODS IN DISCRETE-PROGRAMMING PROBLEMS (RUSSIAN).

ALLEA 1976 -1 ALLENDE, S.
TYPE OF IMPORTANT FUNCTIONS IN INTEGER PROGRAMMING (SPANISH), A.

ARKH 1975 -1 ARKHIPOWA, T. T.; SEGIENKO, I. V.
METHOD OF POSSIBLE AND A METHOD OF DROP VECTOR IN THE INTEGER PROGRAMMING (RUSSIAN), A.

ARMSA 1975 -1 ARMSTRONG, R. D.; CHARNES, A.; PHILLIPS, F.
PAGE CUTS FOR INTEGER INTERVAL LINEAR PROGRAMMING.

AUST 1976 -1 AUST, R. J.
DYNAMIC PROGRAMMING BRANCH AND BOUND ALGORITHM FOR PURE INTEGER PROGRAMMING, A.

BAER 1976 -1 BAER, G.
UMFORMUNGSMOEGLICHKEITEN FUER DIE RESTRIKTIONEN EINES 0-1-OPTIMIERUNGSPROBLEMS.

BALI 1964 -1 BALINSKI, M. L.
ON FINDING INTEGER SOLUTIONS TO LINEAR PROGRAMS.

BARTB 1977 -1 BARTHOLDI, J. J. III; RATLIFF, H. D.
FIELD GUIDE TO IDENTIFYING NETWORK FLOW AND MATCHING PROBLEMS, A.

BARTB 1977 -2 BARTHOLDI, J. J. III; RATLIFF, H. D.
UNNETWORKS, WITH APPLICATIONS TO IDLE TIME SCHEDULING.

BELL 1973 -1 BELL, D. E.
RESOLUTION OF DUALITY GAPS IN DISCRETE OPTIMIZATION, THE.

BELLA 1977 -1 BELL, D. E.; SHAPIRO, J. F.
CONVERGENT DUALITY THEORY FOR INTEGER PROGRAMMING, A.

BELO 1977 -1 BELOUSOV, E. G.
INTRODUCTION TO CONVEX ANALYSIS AND INTEGER PROGRAMMING (RUSSIAN).

BILD 1970 -1 BILDE, O.
NONLINEAR AND DISCRETE PROGRAMMING.

BLAIA 1975 -1 BLAIR, C. E.
TOPICS IN INTEGER PROGRAMMING.

BRADA 1971 -1 BRADLEY, G. H.
HEURISTIC SOLUTION METHODS AND TRANSFORMED INTEGER LINEAR PROGRAMMING PROBLEMS.

BRADB 1969 -1 BRADLEY, G. H.; WAHI, P. N.
INTEGER PROGRAMMING TEST PROBLEMS.

BREG 1972 -1 BREGMAN, L. M.
CERTAIN APPROACH TO INTEGER PROGRAMMING PROBLEMS (RUSSIAN), A.

BROC 1976 -1 BROCKLEHURST, E. R.
GENERALIZED BRANCH AND BOUND: A METHOD FOR INTEGER LINEAR PROGRAMS.

BRUC 1975 -1 BRUCKER, P.
GANZZAHLIGE LINEARE PROGRAMMIERUNG MIT OEKONOMISCHEN ANWENDUNGEN.

BULA 1974 -1 BULATOV, V. P.
NEW METHOD FOR THE SOLUTION OF DISCRETE PROGRAMMING PROBLEMS (RUSSIAN), A.

BURD 1977 -1 BURDET, C.-A.; JOHNSON, E. L.
SUBADDITIVE APPROACH TO SOLVE LINEAR INTEGER PROGRAMS, A.

BURKJ 1968 -1 BURKOV, V. N.; LOVETSKII, S. E.
METHODS FOR THE SOLUTION OF EXTREMAL PROBLEMS OF COMBINATORIAL TYPE (REVIEW) (RUSSIAN).

BUZY 1976 -1 BUZYCKII, P. L.
CERTAIN CUT AND BRANCH FOR PRIMAL INTEGER PROGRAMMING ALGORITHMS (RUSSIAN), A.

BUZY 1976 -2 BUZYCKII, P. L.
DIRECT ALGORITHM OF INTEGER PROGRAMMING FOR A PROBLEM WITH TWO-SIDED CONSTRAINTS (RUSSIAN), A.

BUZYA 1976 -1 BUZYCKII, P. L.; VOTJAKOV, A. A.
DIRECT ALGORITHM FOR INTEGER PROGRAMMING (RUSSIAN), A.

CHERF 1973 -1 CHERVAK, YU. YU.
METHOD OF LEXICOGRAPHIC SEARCH FOR THE SOLUTION OF DISCRETE LINEAR PROGRAMMING PROBLEMS (RUSSIAN), THE.

CHVA 1975 -1 CHVATAL, V.
SOME LINEAR PROGRAMMING ASPECTS OF COMBINATORICS.

COI 1975 -1 COI, S.; SERMAN, A. I.
APPROXIMATE METHOD FOR THE SOLUTION OF INTEGER PROBLEMS OF LINEAR PROGRAMMING (THE METHOD OF ADMISSIBLE COMPONENTS) (RUSSIAN), AN.

COOPB 1974 -1 COOPER, L.; COOPER, M. W.
ALL INTEGER LINEAR PROGRAMMING - A NEW APPROACH VIA DYNAMIC PROGRAMMING.

DANI 1977 -1 DANILENKO, S. E.; KAGAN, B. M.; SAHUNJANC, T. G.
DISCRETE ALGORITHMS IN OPTIMIZATION PROBLEMS (RUSSIAN).

DANTB 1975 -1 DANTZIG, G. B.; EAVES, B. C.
FOURIER-MOTZKIN ELIMINATION AND ITS DUAL WITH APPLICATION TO INTEGER PROGRAMMING.

DIGU 1976 -1 DIGUGLIELMO, F.
ESTIMATES OF THE DUALITY GAP FOR DISCRETE OPTIMIZATION.

DUEC 1977 -1 DUECK, W.
EINFLUSSBEREICH DISKRETER OPTIMIERUNGSMETHODEN, DER.

DYMA 1975 -1 DYMARSKII, JA. S.; PRUDOVSKII, B. D.; STALBO, A. K.
SEPARABLE INTEGER-VALUED PROBLEM (RUSSIAN), A.

EDMO 1975 -1 EDMONDS, J.
SOME WELL-SOLVED PROBLEMS IN COMBINATORIAL OPTIMIZATION.

FORG 1974 -1 FORGO, F.
SHADOW PRICES AND DECOMPOSITION FOR INTEGER PROGRAMS.

FORG 1976 -2 FORGO, F.
SOME TRANSFORMATIONS OF INTEGER PROGRAMMING PROBLEMS.

FRID 1977 -1 FRIDMAN, A. A.
ON SOME MODERN ASPECTS IN DISCRETE OPTIMIZATION (RUSSIAN).

FRIDA 1973 -1 FRIDMAN, A. A.; LITVAK, B. G.; RAPPOPORT, A. M.
ABSOLUTELY INTEGER LINEAR PROGRAMMING PROBLEMS (RUSSIAN).

FRUM 1976 -1 FRUMKIN, M. A.
ALGORITHMS FOR THE SOLUTION IN INTEGERS OF SYSTEMS OF LINEAR EQUATIONS (RUSSIAN).

FRUM 1976 -2 FRUMKIN, M. A.
EXPERIMENTAL TIME ALGORITHM FOR THE TRIANGULARIZATION OF AN INTEGER MATRIX (RUSSIN), AN.

GALI 1975 -1 GALIL, Z.
ON ENUMERATION PROCEDURES FOR THEOREM PROVING AND FOR INTEGER PROGRAMMING.

GALLD 1976 -1 GALLUS, G.
HEURISTISCHE VERFAHREN ZUR LOESUNG ALLGEMEINER GANZZAHLIGER LINEARER OPTIMIERUNGSPROBLEME (EIN UEBERBLICK).

****

GEOF 1974 -1 GEOFFRION, A. M.
GUIDED TOUR OF RECENT PRACTICAL ADVANCES IN INTEGER LINEAR PROGRAMMING, A.

GERR 1976 -1 GERRITSEN, R.
EFFICIENT MANIPULATION OF BINARY ARRAYS.

GHIL 1973 -1 GHILARDOTTI, P.
BILP CODE: BOUNDED INTEGER LINEAR PROGRAMMING CODE USER'S INSTRUCTIONS WITH SOME NOTES AND FIRST COMPUTATIONAL RESULTS.

GIANA 1976 -1 GIANNESSI, F.; NICCOLUCCI, F.
CONNECTIONS BETWEEN NONLINEAR AND INTEGER PROGRAMMING PROBLEMS.

GILM 1977 -1 GILMORE, P. C.
CUTTING STOCK, LINEAR PROGRAMMING, KNAPSACKING, DYNAMIC PROGRAMMING AND INTEGER PROGRAMMING SOME INTERCONNECTIONS.

GIMA 1976 -1 GIMADI, E. H.; GLEBOV, N. I.; PEREPELICA, V. A.
ALGORITHMS WITH BOUNDS FOR DISCRETE OPTIMIZATION PROBLEMS (RUSSIAN).. THIRD ALL-UNION CONFERENCE ON PROBLEMS OF THEORETICAL CYBERNETICS, NOVOSIBIRSK 1974.

GLOV 1977 -1 GLOVER, F.
COMPACT LP BASES FOR A CLASS OF IP PROBLEMS.

GLOV 1976 -1 GLOVER, F.
REDUCING THE SIZE OF SOME IP FORMULATIONS BY SUBSTITUTION.

GLOVD 1977 -1 GLOVER, F.; KLINGMAN, D.; MACMILLAN, C.
NETFORM CONCEPT: A MORE EFFECTIVE MODEL FORM AND SOLUTION PROCEDURE FOR LARGE SCALE NONLINEAR PROBLEMS, THE.

GRABB 1972 -1 GRABOWSKI, W.
SOLVABILITY OF AN INTEGER LINEAR PROGRAMMING PROBLEM (POLISH), THE.

GRAV 1975 -2 GRAVER, J. E.
ON THE FOUNDATIONS OF LINEAR AND INTEGER LINEAR PROGRAMMING I.

GUBA 1975 -1 GUBA, W.
MAXIMALER LOKALER ALGORITHMUS FUER AUFGABEN DER GANZZAHLIGEN LINEAREN OPTIMIERUNG, EIN.

GUIGA 1976 -1 GUIGNARD, M.; SPIELBERG, K.
EXPERIMENTAL INTERACTIVE SYSTEM FOR INTEGER PROGRAMMING, AN.

GUIGA 1976 -2 GUIGNARD, M.; SPIELBERG, K.
INTEGER PROGRAMMING IN APL.

HARI 1976 -1 HARITON, JA. R.
CERTAIN METHOD FOR THE SOLUTION OF DISCRETE PROGRAMMING PROBLEMS (RUSSIAN), A.

HOLM 1978 -1 HOLM, S.; KLEIN, D.
DISCRETE RIGHT HAND SIDE PARAMETRIZATION FOR LINEAR INTEGER PROGRAMS.

IBAR 1972 -1 IBARAKI, T.
FINITE STATE REPRESENTATIONS OF CERTAIN DISCRETE OPTIMIZATION PROBLEMS.

IBAR 1972 -2 IBARAKI, T.
REPRESENTATION THEOREMS FOR EQUIVALENT OPTIMIZATION PROBLEMS.

IBAR 1976 -2 IBARAKI, T.
INTEGER PROGRAMMING FORMULATION OF COMBINATORIAL OPTIMIZATION PROBLEMS.

IVANC 1974 -1 IVANOV, N. N.
REDUCTION OF A CLASS OF INTEGER LINEAR PROGRAMMING PROBLEMS TO A COVERING PROBLEM (RUSSIAN).

JOHNE 1964 -1 JOHNSON, E. L.
NETWORK FLOWS, GRAPHS, AND INTEGER PROGRAMMING.

JOHNF 1978 -1 JOHNSON, E. L.; NEWMAN, D.; WINSTON, K.
INEQUALITY ON BINOMIAL COEFFICIENTS, AN.

KNUTA 1973 -1 KNUTS, L.-R.
COMBINING TWO INTEGER PROGRAMMING ALGORITHMS.

KORTB 1976 -1 KORTE, B.
GANZZAHLIGE PROGRAMMIERUNG (STICHWORTARTIKEL FUER: HANDBUCH DER MATHEMATISCHEN WIRTSCHAFTSWISSENSCHAFTEN, HRSG. M. J. BECKMANN).

KORTB 1976 -2 KORTE, B.
INTEGER PROGRAMMING (SURVEY PAPER).

KOVA 1976 -1 KOVACS, L. B.
COMPUTER PROGRAM SYSTEM FOR THE SOLUTION OF PURE LINEAR DISCRETE PROBLEMS, A.

KOVAB 1977 -1 KOVALEV, M. M.
DISCRETE OPTIMIZATION (INTEGER PROGRAMMING) (RUSSIAN).

KRAR 1973 -1 KRARUP, J.
SELECTED EXAMPLES OF COMBINATORIAL OPTIMIZATION PROBLEMS. (DANISH).

KREU 1969 -1 KREUZBERGER, H.
VERFAHREN ZUR LOESUNG GANZZAHLIGER LINEARER OPTIMIERUNGSPROBLEME.

KUEP 1976 -1 KUEPPER, H.-U.
VERGLEICHENDE ANALYSE GANZZAHLIGER PROGRAMMIERUNGSMODELLE ZUR ABLAUFPLANUNG INDUSTRIELLER UNTERNEHMUNGEN.

LANDA 1978 -1 LAND, A. H.; LAPORTE, G.: M. , P.
UNIFIED FORMULATION OF THE MACHINE SCHEDULING PROBLEM, A.

LAURA 1976 -1 LAURIERE, J. L.
LANGAGE ET UN PROGRAMME POUR ENONCER ET RESOUDRE DES PROBLEMES COMBINATOIRES, UN.

LEBE 1975 -1 LEBEDEV, S. S.
GENERALIZED LAGRANGE MULTIPLIERS AND ESTIMATION PROBLEMS IN INTEGER PROGRAMMING (RUSSIAN).

LEBE 1974 -1 LEBEDEV, S. S.
USE OF GENERALIZED RESOLVENT MULTIPLIERS FOR THE SOLUTION OF INTEGER LINEAR PROGRAMMING PROBLEMS BY THE METHOD OF ORDERED SORTING (RUSSIAN), THE.

LEE B 1976 -1 LEE, S. M.
INTERACTIVE INTEGER GOAL PROGRAMMING: METHODS AND APPLICATION.

LEITA 1974 -1 LEITEN, A. K.
CERTAIN MATHEMATICAL MODELS FOR DISCRETE PROGRAMMING (RUSSIAN).

LEVYB 1976 -1 LEVYI, V. D.; VOLODIN, A. A.
INTEGER PROGRAMMING PROBLEM WITH VARIABLE COEFFICIENTS (TRANSLATED FROM RUSSIAN), AN.

LIN 1977 -1 LIN, B. W. Y.; RARDIN, R. L.
DEVELOPMENT OF A PARAMETRIC GENERATING PROCEDURE FOR INTEGER PROGRAMMING TEST PROBLEMS.

LITV 1973 -1 LITVAK, B. G.
REDUCIBILITY OF ABSOLUTELY INTEGER PROBLEMS WITH A FIXED CONSTRAINT VECTOR (RUSSIAN), THE.

MAC D 1977 -1 MAC DANIEL, D.; DEVINE, M.
MODIFIED BENDERS' PARTITIONING ALGORITHM FOR MIXED INTEGER PROGRAMMING, A.

MARSB 1978 -1 MARSTEN, R. E.; MORIN, T. L.
HYBRID APPROACH TO DISCRETE MATHEMATICAL PROGRAMMING, A.

MINOA 1977 -1 MINOUX, M.
ALGORITHMES GLOUTONS ET ALGORITHMES GLOUTONS ACCELERES POUR LA RESOLUTION DES GRANDS PROBLEMES COMBINATOIRES.

MITE 1977 -1 MITEV, J. G.
APPLICATION OF THE BRANCH AND BOUND METHOD TO SOME DISCRETE LINEAR PROGRAMMING PROBLEMS (RUSSIAN).

INTEGER PROGRAMMING (GENERAL)

MUEL 1975 -2 MUELLER-MERBACH, H.
ROLE OF PUZZLES IN TEACHING COMBINATORIAL PROGRAMMING, THE.

NICOA 1976 -1 NICOLAS, J.-L.
ALGORITHMES D'OPTIMISATION EN NOMBRES ENTIERS.

OTTA 1976 -1 OTTAVIANI, M.
CUTS ON THE FEASIBLE REGION OF A LINEAR PROGRAMMING PROBLEM WITH INTEGER VARIABLES (ITALIAN).

POPEC 1976 -1 POPESCU, M.-F.
QUELQUES RESULTATS DE REDUCIBILITE DANS LA PROGRAMMATION EN ENTIERS.

RARD 1977 -1 RARDIN, R. L.; LIN, B.
WHAT MAKES INTEGER PROGRAMMING PROBLEMS HARD TO SOLVE?

RATH 1977 -2 RATHMANN, D.
ZUR LOESUNG DISKRETER OPTIMIERUNGSPROBLEME MIT GRUPPENTHEORETISCHEN ANSAETZEN.

REEV 1976 -1 REEVES, G. R.
GOAL PROGRAMMING FORMULATIONS OF INTEGER PROGRAMMING PROBLEMS.

ROED 1976 -1 ROEDDER, W.
LEXIKOGRAPHISCHER SUCHALGORITHMUS ZUR GANZZAHLIGEN PROGRAMMIERUNG: LEXS, EIN.

RYND 1977 -2 RYNDIN, B. N.
MULTIPLICATIVE VERSION OF GOMORY'S FIRST ALGORITHM AND COMBINATIONS OF IT WITH THE METHOD OF BRANCHING (RUSSIAN), A.

SAHN 1977 -1 SAHNI, S.
GENERAL TECHNIQUES FOR COMBINATORIAL APPROXIMATION.

SAHNB 1976 -1 SAHNI, S.; GONZALEZ, T.
P-COMPLETE APPROXIMATION PROBLEMS.

SALK 1975 -1 SALKIN, H. M.
INTEGER PROGRAMMING.

SANKA 1973 -1 SANKIBAEV, B. N.; SERMAN, A. I.
ALGORITHM FOR THE SOLUTION OF A PURELY INTEGER LINEAR PROGRAMMING PROBLEM (RUSSIAN), AN.

SAVAA 1976 -2 SAVAGE, S. L.
SOME THEORETICAL IMPLICATIONS OF LOCAL OPTIMIZATION.

SCHO 1976 -1 SCHOCH, M.
ERWEITERUNGSPRINZIP UND SEINE ANWENDUNG ZUR ENTWICKLUNG VON ALGORITHMEN FUER DIE LOESUNG KOMBINATORISCHER OPTIMIERUNGSAUFGABEN, DAS.

SCHUA 1977 -1 SCHURAWLJEW, J. I.; FINKEL'SHTEIN, YU. YU.
AUFGABENTYPEN UND LOESUNGSMETHODEN IN DER DISKRETEN OPTIMIERUNG.

SCOT 1971 -1 SCOTT, A. J.
COMBINATORIAL PROGRAMMING, SPATIAL ANALYSIS, AND PLANNING.

SERG 1975 -1 SERGIENKO, I. V.
APPLICATION OF THE METHOD OF THE VECTOR OF DESCENT TO THE SOLUTION OF INTEGER PROGRAMMING PROBLEMS (RUSSIAN).

SHAR 1975 -1 SHARIF, M. N.; AGARWAL, R. L.
ALGORITHM FOR INTEGER GOAL PROGRAMMING, AN.

SHEV 1975 -1 SHEVCHENKO, V. N.
SOLUTION OF AN ELEMENTARY PROBLEM IN LINEAR INTEGER PROGRAMMING (RUSSIAN).

SHEVA 1973 -1 SHEVCHENKO, V. N.; REMIZOVA, O. L.
CONSTRUCTION OF REGULAR CUT-OFFS IN LINEAR INTEGER PROGRAMMING (RUSSIAN), THE.

SHME 1975 -1 SHMELEV, V. V.
PENALTY FUNCTIONS IN LINEAR INTEGER PROGRAMMING (RUSSIAN).

SUHL 1975 -1 SUHL, U.
ENTWICKLUNG VON ALGORITHMEN FUER GANZZAHLIGE OPTIMIERUNGSMODELLE.

TERZ 1974 -1 TERZI, D. G.
ON A PARTICULARITY OF SOME ALGORITHMS IN LINEAR INTEGER PROGRAMMING (RUSSIAN).

TODD 1977 -1 TODD, M. J.
NUMBER OF NECESSARY CONSTRAINTS IN AN INTEGER PROGRAM: A NEW PROOF OF SCARF'S THEOREM, THE.

TROT 1973 -1 TROTTER, L. E. JR.
USER'S INSTRUCTIONS FOR THE INTEGER PROGRAMMING CODE ENUMER8.

VOLP 1977 -1 VOLPENTESTA, A.
UNIMODULAR SETS AND EXTREMUM PROBLEMS WITH INTEGER VARIABLES (ITALIAN).

VOTJ 1973 -1 VOTJAKOV, A. A.
EQUIVALENT TRANSFORMATIONS OF A LINEAR INTEGER PROBLEM (RUSSIAN).

WILLA 1977 -1 WILLIAMS, H. P.
LOGICAL PROBLEMS AND INTEGER PROGRAMMING.

WILLA 1975 -1 WILLIAMS, H. P.
FORMULATION OF MATHEMATICAL PROGRAMMING MODELS, THE.

WILLA 1976 -1 WILLIAMS, H. P.
FOURIER-MOTZKIN ELIMINATION EXTENSION TO INTEGER PROGRAMMING PROBLEMS.

WOOD 1977 -1 WOODALL, D. R.
NOTE ON RANK FUNCTIONS AND INTEGER PROGRAMMING, A.

ZANC 1974 -1 ZANCEVA, F. N.
APPROXIMATION-COMBINATORIAL METHOD FOR THE CASE OF SEVERAL APPROXIMATING PROBLEMS (RUSSIAN), AN.

ZHER 1976 -1 ZHERNAK, A. N.
GROUP OF ALGORITHMS FOR THE SOLUTION OF QUASILINEAR INTEGER PROGRAMMING PROBLEMS (RUSSIAN), A.

ZHER 1976 -2 ZHERNAK, A. N.
NEW METHOD FOR THE SOLUTION OF INTEGER PROGRAMMING PROBLEMS (RUSSIAN), A.

ZION 1977 -1 ZIONTS, S.
INTEGER LINEAR PROGRAMMING WITH MULTIPLE OBJECTIVES.

****

AHRE 1975 -1 AHRENS, J. H.; FINKE, G.
MERGING AND SORTING APPLIED TO THE ZERO-ONE KNAPSACK PROBLEM.

BALAG 1976 -1 BALAS, E.
FACETS OF ONE-DIMENSIONAL AND MULTI-DIMENSIONAL KNAPSACK POLYTOPES.

BARRA 1975 -1 BARR, R. S.; ROSS, G. T.
LINKED LIST DATA STRUCTURE FOR A BINARY KNAPSACK ALGORITHM, A.

BART 1975 -1 BARTHES, J. P.
BRANCHING METHODS IN COMBINATORIAL OPTIMIZATION.

BENI 1977 -1 BENICHOU, M.; GAUTHIER, J. M.; HENTGES, G.; RIBIERE, G.
EFFICIENT SOLUTION OF LARGE-SCALE LINEAR PROGRAMMING PROBLEMS - SOME ALGORITHMIC TECHNIQUES AND COMPUTATIONAL RESULTS, THE.

BRAN 1971 -1 BRANDMAN, E. I.
CERTAIN METHODS FOR THE SOLUTION OF PROBLEMS OF PROJECTED ALLOCATION OF CAPITAL INVESTMENTS WITHIN AN INDUSTRY (RUSSIAN).

BULFA 1977 -1 BULFIN, R. L.; SHETTY, C. M.; PARKER, R. G.
COMPUTATIONAL RESULTS WITH A BRANCH-AND-BOUND ALGORITHM FOR THE GENERAL KNAPSACK PROBLEM.

BURKJ 1968 -1 BURKOV, V. N.; LOVETSKII, S. E.
METHODS FOR THE SOLUTION OF EXTREMAL PROBLEMS OF COMBINATORIAL TYPE (REVIEW) (RUSSIAN).

CHAL 1976 -1 CHALMET, L.
SOLVING ASSIGNMENT PROBLEMS IN WAREHOUSES (DUTCH).

CHAND 1976 -1 CHANDRA, A. K.; HIRSCHBERG, D. S.; WONG, C. K.
APPROXIMATE ALGORITHMS FOR SOME GENERALIZED KNAPSACK PROBLEMS.

CHANJ 1976 -1 CHANG, L.; KORSH, J. F.
CANONICAL COIN CHANGING AND GREEDY SOLUTIONS.

CHERF 1972 -1 CHERVAK, YU. YU.
DETERMINATION OF THE OPTIMAL INTEGER CONVEX LINEAR COMBINATION OF POINTS. THE GENERAL DISCRETE CASE (RUSSIAN).

CHERF 1976 -1 CHERVAK, YU. YU.
KNAPSACK PROBLEMS THAT PERMIT A SIMPLE METHOD FOR FINDING THEIR SOLUTIONS (RUSSIAN).

DOBK 1976 -1 DOBKIN, D.
NONLINEAR LOWER BOUND ON LINEAR SEARCH TREE PROGRAMS FOR SOLVING KNAPSACK PROBLEMS, A.

DOBKB 1976 -1 DOBKIN, D.; LIPTON, R. J.
LOWER BOUND OF 1/(N POWER 2) ON LINEAR SEARCH PROGRAMS FOR THE KNAPSACK PROBLEM, A.

DUNS 1977 -1 DUNSTAN, F. D. J.
ALGORITHM FOR SOLVING A RESOURCE ALLOCATION PROBLEM, AN.

FAYA 1976 -1 FAYARD, D.; PLATEAU, G.
TECHNIQUE DE RESOLUTION DU PROBLEME DU KNAPSACK EN VARIABLES BIVALENTES.

FINKA 1977 -1 FINKEL'SHTEIN, YU. YU.
E-APPROACH TO THE MULTIDIMENSIONAL KNAPSACK PROBLEM: THE POLYNOMIAL GROWTH OF THE TREE OF BRANCHING (RUSSIAN), THE.

FISKC 1977 -1 FISK, J. C.
INITIAL BOUNDING PROCEDURE FOR USE WITH 0-1 SINGLE KNAPSACK ALGORITHMS, AN.

FORG 1974 -1 FORGO, F.
SHADOW PRICES AND DECOMPOSITION FOR INTEGER PROGRAMS.

FRIEA 1976 -1 FRIEZE, A. M.
SHORTEST PATH ALGORITHMS FOR KNAPSACK TYPE PROBLEMS.

GALLB 1976 -1 GALLO, G.; HAMMER, P. L.; SIMEONE, B.
QUADRATIC KNAPSACK PROBLEMS.

GILM 1977 -1 GILMORE, P. C.
CUTTING STOCK, LINEAR PROGRAMMING, KNAPSACKING, DYNAMIC PROGRAMMING AND INTEGER PROGRAMMING SOME INTERCONNECTIONS.

GVOZ 1976 -1 GVOZDEV, S. E.
ON A GENERALIZATION OF THE KNAPSACK PROBLEM (RUSSIAN).

GVOZ 1974 -1 GVOZDEV, S. E.
SOME MODELS AND SOLUTION METHODS FOR THE KNAPSACK PROBLEM WITH SUPPLEMENTARY CONSTRAINTS (RUSSIAN).

HAMMA 1975 -1 HAMMER, P. L.
BOOLEAN ELEMENTS IN COMBINATORIAL OPTIMIZATION: A SURVEY.

HAUSB 1977 -1 HAUSMANN, D.; KANNAN, R.; KORTE, B.
EXPONENTIAL LOWER BOUNDS ON A CLASS OF KNAPSACK ALGORITHMS.

HIRSA 1976 -1 HIRSCHBERG, D. S.; WONG, C. K.
POLYNOMIAL-TIME ALGORITHM FOR THE KNAPSACK PROBLEM WITH TWO VARIABLES, A.

HU A 1976 -1 HU, T. C.; LENARD, M. L.
OPTIMALITY OF A HEURISTIC SOLUTION FOR A CLASS OF KNAPSACK PROBLEMS.

HUNGA 1976 -1 HUNG, M. S.; FISK, J. C.
ALGORITHM FOR 0-1 MULTIPLE KNAPSACK PROBLEMS, AN.

IBARC 1974 -1 IBARRA, O. H.; KIM, C. E.
FAST APPROXIMATION ALGORITHMS FOR THE KNAPSACK AND SUM OF SUBSET PROBLEMS.

INGAA 1977 -1 INGARGIOLA, G. P.; KORSH, J. F.
GENERAL ALGORITHM FOR ONE-DIMENSIONAL KNAPSACK PROBLEMS, A.

ISHI 1977 -1 ISHII, H.; IBARAKI, T.; MINE, H.
FRACTIONAL KNAPSACK PROBLEMS.

JOHNK 1970 -1 JOHNSON, S. C.; KERNIGHAN, B. W.
MAKING CHANGE WITH A MINIMUM NUMBER OF COINS.

KAES 1976 -1 KAESTING, H.
LEXIKOGRAPHISCHES SUCHVERFAHREN ZUR LOESUNG DES BIVALENTEN KNAPSACK-PROBLEMS, EIN.

KORTB 1977 -2 KORTE, B.
APPROXIMATIVE ALGORITHMS FOR DISCRETE OPTIMIZATION PROBLEMS.

KUZN 1974 -1 KUZNETSOV, V. F.
ALLOCATION OF RESOURCES FOR A GIVEN NETWORK (RUSSIAN).

LAURB 1976 -1 LAURIERE, M.
METHODE DE RESOLUTION RAPIDE POUR LE PROBLEME DU KNAPSACK, UNE.

LAURB 1978 -1 LAURIERE, M.
ALGORITHM FOR THE 0/1 KNAPSACK PROBLEM, AN.

LAWL 1977 -3 LAWLER, E. L.
FAST APPROXIMATION ALGORITHMS FOR KNAPSACK PROBLEMS.

LITVA 1976 -1 LITVAK, B. G.; NAIVEL'T, A. V.
MUTUALLY INCOMPARABLE SOLUTIONS IN THE KNAPSACK PROBLEM WITH SUPPLEMENTARY BLOCK TYPE CONSTRAINTS (RUSSIAN).

LITVA 1973 -1 LITVAK, B. G.; NAIVEL'T, A. V.
SOLUTION OF THE MULTIDIMENSIONAL KNAPSACK PROBLEM WITH SUPPLEMENTARY CONSTRAINTS (RUSSIAN), THE.

MAGAA 1976 -1 MAGAZINE, M. J.; NUTTLE, H. L. W.
LONG DIVISION: AN EXAMPLE OF WHEN THE GREEDY ALGORITHM SOLVES A KNAPSACK PROBLEM.

KNAPSACK

****

MARTC 1976 -1 MARTELLO, S.; TOTH, P.
ENUMERATIVE ALGORITHM FOR THE 0-1 KNAPSACK PROBLEM (ITALIAN), AN.

MARTC 1977 -1 MARTELLO, S.; TOTH, P.
COMPUTATIONAL STUDY ON LARGE-SIZE UNIDIMENSIONAL KNAPSACK PROBLEMS, A.

MARTC 1977 -3 MARTELLO, S.; TOTH, P.
SOLUTION OF ZERO-ONE MULTIPLE KNAPSACK PROBLEM.

MARTC 1977 -4 MARTELLO, S.; TOTH, P.
UPPER BOUND FOR THE ZERO-ONE KNAPSACK PROBLEM AND A BRANCH AND BOUND ALGORITHM, AN.

MICH 1977 -1 MICHAELI, I.; POLLATSCHEK, M. A.
ON SOME NONLINEAR KNAPSACK PROBLEMS.

NAUS 1976 -1 NAUSS, R. M.
ALGORITHM FOR THE 0-1 INTEGER LINEAR PROGRAMMING PROBLEM, AN.

NAUS 1976 -2 NAUSS, R. M.
EFFICIENT ALGORITHM FOR THE 0-1 KNAPSACK PROBLEM, AN.

NEUM 1975 -1 NEUMANN, K.
OPERATIONS RESEARCH C. DYNAMISCHE OPTIMIERUNG.

OFFI 1975 -1 OFFIR, J. D.
TRAVELLING SALESMAN WITH A KNAPSACK(A CALL-GIRL'S HEURISTIC), A.

PADBA 1977 -1 PADBERG, M. W.
COVERING, PACKING AND KNAPSACK PROBLEMS.

PERL 1976 -1 PERL, J.
ZUR ALGORITHMISCHEN BEHANDLUNG SPEZIELLER NP-VOLLSTAENDIGER PROBLEME AUF GRAPHEN.

RINE 1977 -1 RINE, D. C.
OPTIMIZING TWO TYPES OF DISCRETE FUNCTIONS, SUBJECT TO LINEAR RESTRICTIONS.

SAHN 1975 -1 SAHNI, S.
APPROXIMATE ALGORITHMS FOR THE 0/1 KNAPSACK PROBLEM.

SALK 1975 -1 SALKIN, H. M.
INTEGER PROGRAMMING.

SALKA 1975 -1 SALKIN, H. M.; KLUYVER, C. A. DE
KNAPSACK PROBLEM: A SURVEY, THE.

THES 1975 -1 THESEN, A.
RECURSIVE BRANCH AND BOUND ALGORITHM FOR THE MULTIDIMENSIONAL KNAPSACK PROBLEM, A.

TIEN 1977 -1 TIEN, B. N.; HU, T. C.
ERROR BOUNDS AND THE APPLICABILITY OF THE GREEDY SOLUTION TO THE COIN-CHANGING PROBLEM.

TOTH 1977 -1 TOTH, P.
DYNAMIC PROGRAMMING ALGORITHM FOR THE ZERO-ONE KNAPSACK PROBLEM, A.

WALU 1975 -2 WALUKIEWICZ, S.
SIZE REDUCTION OF A BINARY KNAPSACK PROBLEM, THE.

YORM 1975 -1 YORMARK, J. S.
ACCELERATING GREENBERG'S METHOD FOR THE COMPUTATION OF KNAPSACK FUNCTIONS.

****

AGAP 1976 -1 AGAPOVA, L. V.
ALLOCATION PROBLEM FOR RESOURCES (RUSSIAN).

AGIN 1976 -1 AGIN, N. I.
SUMMARY OF A NOVEL APPROACH TO THE WAREHOUSE LOCATION PROBLEM.

AGRE 1977 -1 AGRESTI, W. W.
DYNAMIC PROGRAMMING FOR COMPUTER REGISTER ALLOCATION.

AKIN 1974 -1 AKINC, U.
BRANCH AND BOUND PROCEDURE FOR SOLVING WAREHOUSE LOCATION PROBLEMS WITH CAPACITY CONSTRAINTS, A.

AKINA 1977 -1 AKINC, U.; KHUMAWALA, B. M.
EFFICIENT BRANCH AND BOUND ALGORITHM FOR THE CAPACITATED WAREHOUSE LOCATION PROBLEM, AN.

ALBE 1977 -1 ALBERS, S.
MIXED INTEGER NONLINEAR PROGRAMMING PROCEDURE FOR SIMULTANEOUSLY LOCATING MULTIPLE PRODUCTS IN AN ATTRIBUTE SPACE, A.

ALCOA 1976 -1 ALCOUFFE, A.; MURATET, G.
OPTIMAL LOCATION OF PLANTS.

ALIB 1975 -1 ALIBEKOV, B. I.
ON AN ALLOCATION PROBLEM WITH CONSTRAINED CAPACITIES (RUSSIAN).

ALIB 1971 -1 ALIBEKOV, B. I.
APPROXIMATE METHOD FOR THE SOLUTION OF A MULTIEXTREMAL ALLOCATION PROBLEM (RUSSIAN), AN.

ARMSE 1977 -1 ARMSTRONG, R. D.; WILLIS, C. E.
SIMULTANEOUS INVESTMENT AND ALLOCATION DECISIONS APPLIED TO WATER PLANNING.

ARMSE 1975 -1 ARMSTRONG, R. D.; WILLIS, C. E.
SIMULTANEOUS INVESTMENT-ALLOCATION - AN APPLICATION OF GENERALIZED BENDERS DECOMPOSITION TO WATER PLANNING.

ASHT 1977 -1 ASHTON, W. B.; CLARK, G. M.
LOCATION AND SIZING OF URBAN FREIGHT CONSOLIDATION TERMINALS WITH MULTIPLE PLANNING PERIODS, THE.

AUST 1976 -2 AUST, R. J.
IMPROVEMENT ALGORITHM FOR SCHOOL TIMETABLING, AN.

AVDO 1975 -1 AVDONIN, L. N.
SOLUTION ALGORITHM OF ONE-PRODUCT ENTERPRISE ALLOCATION PROBLEM (RUSSIAN).

AVRI 1977 -1 AVRIEL, M.; GUROVICH, V.
CONDENSATION ALGORITHM FOR A CLASS OF ALGEBRAIC PROGRAMS, A.

BABUB 1972 -1 BABUSIAUX, D.
ESTABLISHING A NETWORK OF WAREHOUSES FOR FINISHED PRODUCTS, TAKING INTO ACCOUNT SEASONAL FLUCTUATIONS IN DEMAND.

BAKE 1976 -1 BAKER, K. R.
WORKFORCE ALLOCATION IN CYCLICAL SCHEDULING PROBLEMS: A SURVEY.

BAKEB 1977 -1 BAKER, K. R.; MAGAZINE, M.
WORKFORCE SCHEDULING WITH CYCLIC DEMANDS AND DAY-OFF CONSTRAINTS.

BALAA 1977 -1 BALACHANDRAN, V.; DITTMAN, D.
OVERHEAD ALLOCATION VIA MATHEMATICAL PROGRAMMING.

BALLB 1970 -1 BALLOU, R. H.
COMPUTATIONAL LIMITATIONS OF DYNAMIC PROGRAMMING FOR WAREHOUSE LOCATION: A COMMENT.

BARB 1970 -1 BARBER, G. M.
DYNAMIC LOCATIONAL ANALYSIS: THE FIRM AND ITS OPERATING POLICY.

BARB 1975 -1 BARBER, G. M.
MATHEMATICAL PROGRAMMING APPROACH TO A NETWORK DEVELOPMENT PROGRAM, A.

BATT 1977 -1 BATTAGLIA, F.
LOCATION OF CENTERS IN AN EUCLIDEAN SPACE (ITALIAN), THE.

BECKA 1974 -1 BECKER, L.; YEH, W. W.-G.
OPTIMAL TIMING, SEQUENCING, AND SIZING OF MULTIPLE RESERVOIR SURFACE WATER SUPPLY FACILITIES.

BECKA 1974 -2 BECKER, L.; YEH, W. W.-G.
TIMING AND SIZING OF COMPLEX WATER RESOURCE SYSTEMS.

BECKC 1976 -1 BECKMANN, M. J.; BURNS, L. D.
THEORY OF HOUSEHOLD AUTOMOBILE ALLOCATION DECISIONS, A.

BELLC 1965 -1 BELLMORE, M.
OPTIMAL ALLOCATION OF RESOURCES IN A COMMUNICATIONS NETWORK.

BERGC 1968 -1 BERGENDAHL, G.
CAPACITY PLANNING MODELS FOR SYSTEMS OF PLANT AND ROAD INVESTMENTS.

BERGC 1969 -1 BERGENDAHL, G.
COMBINED LINEAR AND DYNAMIC PROGRAMMING MODEL FOR INTERDEPENDENT ROAD INVESTMENT PLANNING, A.

BERGC 1966 -1 BERGENDAHL, G.
MODELS FOR CAPACITY PLANNING.

BERGD 1974 -1 BERGENDORFF, H.
MODEL TO EVALUATE AND PLAN PROJECTS IN FORESTRY AND FOREST INDUSTRIES, A.

BERL 1976 -1 BERLIN, G. N.; REVELLE, C. S.; ELZINGA, D. J.
DETERMINING AMBULANCE-HOSPITAL LOCATIONS FOR ON-SCENE AND HOSPITAL SERVICES.

BESW 1977 -1 BESWICK, C. A.
ALLOCATING SELLING EFFORT VIA DYNAMIC PROGRAMMING.

BHAL 1970 -1 BHALLA, H. S.
MULTI-TIME PERIOD FACILITIES LOCATION: A HEURISTIC ALGORITHM WITH APPLICATIONS TO WASTE WATER TREATMENT SYSTEMS.

BIDE 1975 -1 BIDER, I. G.; BOL'SHAKOV, I. A.
NON-MONOTONIC STRATEGIES FOR AN ALLOCATION OF LIMITED RESOURCES (RUSSIAN).

BILDA 1967 -1 BILDE, O.; KRARUP, J.
COMPUTATION OF THE OPTIMAL LOCATION OF PRODUCTION DATES (DANISH).

BILDA 1977 -1 BILDE, O.; KRARUP, J.
SHARP LOWER BOUNDS AND EFFICIENT ALGORITHMS FOR THE SIMPLE PLANT LOCATION PROBLEM.

BIND 1961 -1 BINDSCHEDLER, A. E.; MOORE, J. M.
OPTIMAL LOCATION OF NEW MACHINES IN EXISTING PLANT LAYOUTS.

BOCKA 1975 -1 BOCKO, G. D.; KASIN, V. I.; MAKAROV, L. I.
ALGORITHMS FOR SUCCESSIVE ALLOCATION AND SHORTEST TRACING (RUSSIAN).

BOET 1974 -1 BOETTCHER, H.; DEHNERT, G.; DOMSCHKE, W.
VERFAHREN ZUR BESTIMMUNG OPTIMALER STANDORTE FUER REGIONALE SYSTEME.

BONA 1976 -1 BONA, B.; INAUDI, D.; MAURO, V.
"FUZZY" APPROACH TO A PROBLEM IN RESIDENTIAL LOCATION (ITALIAN), A.

BRAN 1971 -1 BRANDMAN, E. I.
CERTAIN METHODS FOR THE SOLUTION OF PROBLEMS OF PROJECTED ALLOCATION OF CAPITAL INVESTMENTS WITHIN AN INDUSTRY (RUSSIAN).

BREGA 1974 -1 BREGMAN, T. T.
BRANCH AND BOUND METHOD FOR THE SOLUTION OF ALLOCATION PROBLEMS (RUSSIAN), A.

LOCATION AND ALLOCATION
****

BREI 1976 -1 BREITENBACH, R. B.
COMPUTER ALGORITHM FOR REFINING GRID LOCATIONAL ANALYSIS, A.

BROW 1976 -1 BROWN, D. B.
ALLOCATION OF FEDERAL HIGHWAY SAFETY FUNDS USING DYNAMIC PROGRAMMING, THE.

BROWC 1973 -1 BROWN, W.
GENERALIZED PLANT LOCATION PROBLEM, A.

BUR' 1976 -1 BUR'JAN, S. B.; SEROV, S. S.; UZDEMIR, A. P.
DYNAMIC INDUSTRIAL-PLANT LOCATION PROBLEM AND A NUMERICAL METHOD FOR SOLVING IT, II (RUSSIAN).

BUR' 1976 -2 BUR'JAN, S. B.; SEROV, S. S.; UZDEMIR, A. P.
DYNAMIC INDUSTRIAL-PLANT-LOCATION PROBLEM AND A NUMERICAL METHOD FOR SOLVING IT, III (RUSSIAN).

BURKJ 1968 -1 BURKOV, V. N.; LOVETSKII, S. E.
METHODS FOR THE SOLUTION OF EXTREMAL PROBLEMS OF COMBINATORIAL TYPE (REVIEW) (RUSSIAN).

BURKJ 1966 -1 BURKOV, V. N.; LOVETSKII, S. E.
HEURISTIC APPROACH TO THE SOLUTION OF DYNAMIC PROGRAMMING PROBLEMS IN THE ALLOCATION OF RESOURCES (RUSSIAN).

BURN 1976 -1 BURNESS, R. C.; WHITE, J. A.
TRAVELING SALESMAN LOCATION PROBLEM, THE.

BURTB 1976 -1 BURTON, J. S.; GLEASON, J. M.
ALGORITHM FOR THE DYNAMIC LOT SIZE PROBLEM WITH PRICE BREAKS, AN.

CARD 1971 -1 CARDENAS, M. A.
SYSTEMS APPROACH TO INVESTMENT PLANNING IN THE ELECTRICAL INDUSTRY, THE.

CARL 1977 -1 CARLTON, D. W.
BIRTHS OF SINGLE ESTABLISHMENT FIRMS AND REGIONAL VARIATION IN ECONOMIC COSTS.

CARL 1977 -2 CARLTON, D. W.
LOCATION DECISIONS OF MANUFACTURING FIRMS.

CARL 1977 -3 CARLTON, D. W.
MODELS OF NEW BUSINESS LOCATION.

CARL 1977 -4 CARLTON, D. W.
MODELS OF SINGLE ESTABLISHMENT BIRTHS, STANDARD INDUSTRIAL CLASSIFICATION 3079.

CARN 1976 -1 CARNALL, C. A.; WILD, R.
LOCATION OF VARIABLE WORK STATIONS AND THE PERFORMANCE OF PRODUCTION FLOW LINES, THE.

CARPA 1976 -1 CARPENITO, T. A.; WHITE, J. A.
ALLOCATION OF NON-IDENTICAL MACHINES AMONG NON-IDENTICAL SERVERS, THE.

CHAI 1975 -1 CHAIKEN, J. M.
PATROL ALLOCATION METHODOLOGY FOR POLICE DEPARTMENTS.

CHAIA 1976 -1 CHAIKEN, J. M.; DORMANT, P.
PATROL CAR ALLOCATION MODEL, A.

CHALA 1975 -1 CHALMET, L.; GELDERS, L.
ALGORITHM FOR GENERATING GOOD WAREHOUSE LAYOUTS, AN.

CHALB 1976 -1 CHALMET, L. G.; GELDERS, L. F.
LAGRANGEAN RELAXATIONS FOR SOLVING A WAREHOUSING MODEL.

CHAN 1976 -1 CHAN, A. W.; FRANCIS, R. L.
LEAST TOTAL DISTANCE FACILITY CONFIGURATION PROBLEM INVOLVING LATTICE POINTS, A.

CHAN 1976 -2 CHAN, A. W.; FRANCIS, R. L.
ROUND-TRIP LOCATION PROBLEM ON A TREE GRAPH, A.

CHER 1970 -1 CHERENIN, V. P.
ALLOCATION OF WAREHOUSES WITH REGARD TO DEMANDS (RUSSIAN).

CHERA 1977 -1 CHERENKOV, A. P.
PROBLEM OF DIFFERENT TYPES OF RESOURCES ALLOCATION (RUSSIAN), A.

CHIA 1975 -1 CHIANG, C. H.
MODEL FOR MULTI-PERIOD REGIONAL WASTEWATER PLANNING, A.

CHUR 1976 -1 CHURCH, R. L.; GARFINKEL, R. S.
ON LOCATING AN OBNOXIOUS FACILITY ON A NETWORK: THE MAXIMUM MEDIAN PROBLEM.

CHURA 1977 -1 CHURCH, R. L.; MEADOWS, M. E.
RESULTS OF A NEW APPROACH TO SOLVING THE P-MEDIAN PROBLEM WITH MAXIMUM DISTANCE CONSTRAINTS.

CHURB 1976 -1 CHURCH, R. L.; REVELLE, C. S.
THEORETICAL AND COMPUTATIONAL LINKS BETWEEN THE P-MEDIAN, LOCATION SET-COVERING, AND THE MAXIMAL COVERING LOCATION PROBLEM.

CODY 1976 -1 CODY, R. A.; COFFMAN, E. G. JR.
RECORD ALLOCATION FOR MINIMIZING EXPECTED RETRIEVAL COSTS ON DRUM-LIKE STORAGE DEVICES.

COHE 1973 -1 COHEN, J.
SURVEY ON THE WAREHOUSE LOCATION PROBLEM.

COME 1966 -1 COMES, G.; BELLON, J.-L.
ETUDE D'UNE POLITIQUE SEQUENTIELLE D'INVESTISSEMENTS POUR LA RESTRUCTURATION D'UN RESEAU D'USINES.

COOP 1963 -1 COOPER, L.
LOCATION-ALLOCATION PROBLEM.

COOP 1975 -1 COOPER, L.
EFFICIENT HEURISTIC ALGORITHM FOR THE TRANSPORTATION-LOCATION PROBLEM, AN.

CORND 1977 -1 CORNUEJOLS, G.; FISHER, M. L.; NEMHAUSER, G. L.
LOCATION OF BANK ACCOUNTS TO OPTIMIZE FLOAT: AN ANALYTIC STUDY OF EXACT AND APPROXIMATE ALGORITHMS.

CORND 1976 -1 CORNUEJOLS, G.; FISHER, M. L.; NEMHAUSER, G. L.
ANALYSIS OF HEURISTICS AND RELAXATIONS FOR THE UNCAPACITATED LOCATION PROBLEM, AN.

COTT 1974 -1 COTTLE, R.; KRARUP, J.
OPTIMIZATION METHODS FOR RESOURCE ALLOCATION.

CROU 1976 -1 CROUCHER, J. S.
OPTIMAL RESOURCE ALLOCATION AMONG AREAS OF EQUAL VALUE.

DAVIA 1975 -1 DAVIES, O.
SIX LOCATION STRATEGIES WITH INCREASING DEMAND LEVELS.

DAVIB 1976 -1 DAVIES, O.; THOMAS, E. A. C.
FACILITY LOCATION WITH CHANGING DEMAND LEVELS.

DEARA 1976 -1 DEARING, P. M.; FRANCIS, R. L.; LOWE, T. J.
CONVEX LOCATION PROBLEMS ON TREE NETWORKS.

DEARB 1975 -1 DEARING, P. M.; LANGFORD, G. J.
MULTIFACILITY TOTAL COST LOCATION PROBLEM ON A TREE NETWORK, THE.

DEME 1968 -1 DEMENCHUK, V. M.
FLOW INTERPRETATION AND ALGORITHM FOR SOLVING THE PROBLEM OF TIME-OPTIMAL ALLOCATION OF LIMITED RESOURCES.

DERM 1975 -1 DERMAN, C.; LIEBERMAN, G. J.; ROSS, S. M.
OPTIMAL ALLOCATION OF RESOURCES IN SYSTEMS.

DERM 1975 -2 DERMAN, C.; LIEBERMAN, G. J.; ROSS, S. M.
STOCHASTIC SEQUENTIAL ALLOCATION MODEL, A.

****

DIEH 1972 -1 DIEHR, G.
ALGORITHM FOR THE P-MEDIANS PROBLEM, AN.
DOMS 1975 -1 DOMSCHKE, W.
VERFAHREN ZUR BESTIMMUNG OPTIMALER BETRIEBS- UND LAGERSTANDORTE.
DOMS 1974 -1 DOMSCHKE, W.
GRAPHENTHEORETISCHE VERFAHREN UND IHRE ANWENDUNGEN ZUR LOESUNG VON ZUORDNUNGS- UND STANDORTPROBLEMEN.
DREY 1977 -1 DREYFUS, S. E.; LAW A. M.
ART AND THEORY OF DYNAMIC PROGRAMMING, THE.
DUNS 1977 -1 DUNSTAN, F. D. J.
ALGORITHM FOR SOLVING A RESOURCE ALLOCATION PROBLEM, AN.
DYKS 1977 -1 DYKSTRA, D. P.; RIGGS, J. L.
APPLICATION OF FACILITIES LOCATION THEORY TO THE DESIGN OF FOREST HARVESTING AREAS, AN.
EDENB 1978 -1 EDEN, B. N.; PAPE, U.
HEURISTISCHES VERFAHREN ZUR DISTRIKTERMITTLUNG IN BEWERTETEN GRAPHEN, EIN.
EINB 1977 -1 EINBU, J. M.
ON SHIH'S INCREMENTAL METHOD IN RESOURCE ALLOCATIONS.
ELSH 1975 -1 ELSHAFEI, A. N.
APPROACH TO LOCATIONAL ANALYSIS, AN.
ELSH 1977 -1 ELSHAFEI, A. N.
HOSPITAL LAYOUT AS A QUADRATIC ASSIGNMENT PROBLEM.
ELSHA 1974 -1 ELSHAFEI, A. N.; HALEY, K. B.
FACILITIES LOCATION: SOME FORMULATIONS, METHODS OF SOLUTION, APPLICATIONS, AND COMPUTATIONAL EXPERIENCE.
ELZI 1976 -1 ELZINGA, J.; HEARN, D.; RANDOLPH, W. D.
MINIMAX MULTIFACILITY LOCATION WITH EUCLIDEAN DISTANCES.
ELZIA 1976 -1 ELZINGA, J.; VEMUGANTI, R. R.
OPTIMAL SCHEDULING OF A SALES FORCE.
ERLE 1973 -1 ERLENKOTTER, D.
DYNAMIC MULTI-LOCATION MODELS FOR CAPACITY EXPANSION.
ERLE 1976 -1 ERLENKOTTER, D.
DUAL-BASED PROCEDURE FOR UNCAPACITATED FACILITY LOCATION, A.
ERLE 1967 -1 ERLENKOTTER, D.
TWO PRODUCING AREAS-DYNAMIC PROGRAMMING SOLUTIONS.
ERLE 1972 -1 ERLENKOTTER, D.
CAPACITY PLANNING FOR COMPETITIVE EXPANSIONS: A REDUCED DYNAMIC PROGRAMMING APPROACH.
ERLE 1975 -1 ERLENKOTTER, D.
BIBLIOGRAPHY ON DYNAMIC LOCATION MODELS.
ERLE 1974 -1 ERLENKOTTER, D.
DYNAMIC FACILITY LOCATION AND SIMPLE NETWORK MODELS.
ERLE 1973 -2 ERLENKOTTER, D.
SEQUENCING OF INTERDEPENDENT HYDROELECTRIC PROJECTS.
ERLE 1975 -2 ERLENKOTTER, D.
CAPACITY PLANNING FOR LARGE MULTILOCATION SYSTEMS: APPROXIMATE AND INCOMPLETE DYNAMIC PROGRAMMING APPROACHES.
ERLE 1974 -2 ERLENKOTTER, D.
DYNAMIC PROGRAMMING APPROACH TO CAPACITY EXPANSION WITH SPECIALIZATION, A.
ERLE 1976 -2 ERLENKOTTER, D.
FACILITY LOCATION WITH PRICE-SENSITIVE DEMANDS: PRIVATE, PUBLIC, AND QUASI-PUBLIC.
ERLE 1972 -2 ERLENKOTTER, D.
ECONOMIC INTEGRATION AND DYNAMIC LOCATION PLANNING.
ERLE 1975 -3 ERLENKOTTER, D.
COMMENT ON "OPTIMAL TIMING, SEQUENCING, AND SIZING OF MULTIPLE RESERVOIR SURFACE WATER SUPPLY FACILITIES" BY L. BECKER AND W. W.-G. YEH.
ESCH 1975 -1 ESCHENBACH, T.; CARLSON, R. C.
CAPACITATED MULTI-PERIOD LOCATION-ALLOCATION PROBLEM, THE.
ESCHA 1975 -1 ESCHENBACH, T. G.
MULTI-PERIOD LOCATION-ALLOCATION PROBLEM, THE.
EVAN 1974 -1 EVANS, J. P.
USE OF A LOCATION MODEL IN A REGIONALIZATION STUDY, THE.
EYST 1973 -1 EYSTER, J. W.; WHITE, J. A.; WIERVILLE, W. W.
ON SOLVING MULTIFACILITY LOCATION PROBLEMS USING A HYPERBOLOID APPROXIMATION PROCEDURE.
FEIN 1977 -1 FEINBERG, I.
HEURISTIC APPROACH TO WAREHOUSE LOCATION USING MULTIPLE CENTERS OF GRAVITY, A.
FISK 1977 -1 FISK, J.
SOLUTION PROCEDURE FOR A SPECIAL TYPE OF CAPACITATED WAREHOUSE LOCATION PROBLEM, A.
FISKB 1977 -1 FISK, J.; MACKEOWN, P. G.
APPLICATION OF INTEGER PROGRAMMING TO THE ASSIGNMENT OF STUDENTS TO SCHOOLS, AN.
FONGA 1974 -1 FONG, C. O.
MULTI-REGION DYNAMIC CAPACITY EXPANSION PROBLEM: EXACT AND HEURISTIC APPROACHES, THE.
FONGC 1974 -1 FONG, C. O.; SRINIVASAN, V.
MULTI-REGION DYNAMIC CAPACITY EXPANSION PROBLEM - I, THE.
FONGC 1974 -2 FONG, C. O.; SRINIVASAN, V.
MULTI-REGION DYNAMIC CAPACITY EXPANSION PROBLEM - II, THE.
FORC 1974 -1 FORCINA, J. L.
DYNAMIC LOCATION STRATEGIES FOR PUBLIC FACILITIES.
FOULA 1976 -1 FOULDS, L. R.; ROBINSON, D. F.
STRATEGY FOR SOLVING THE PLANT LAYOUT PROBLEM, A.
FRAN 1974 -1 FRANCIS, R. L.
FACILITY LAYOUT AND LOCATION.
FRAN 1963 -1 FRANCIS, R. L.
NOTE ON THE OPTIMUM LOCATION OF NEW MACHINES IN EXISTING PLANT LAYOUTS, A.
FRANA 1976 -1 FRANCIS, R. L.; LOWE, T. J.; RATLIFF, H. D.
DISTANCE CONSTRAINTS FOR TREE NETWORK MULTIFACILITY LOCATION PROBLEMS.
FRANB 1974 -1 FRANCIS, R. L.; WHITE, J. A.
FACILITY LAYOUT AND LOCATION: AN ANALYTIC APPROACH.
FREE 1977 -1 FREELAND, J. R.; MOORE, J. H.
IMPLICATIONS OF RESOURCE DIRECTIVE ALLOCATION MODELS FOR ORGANIZATIONAL DESIGN.
FREI 1973 -1 FREIMER, M.; SCHWARZ, L. B.; SRINIVASAN, V.
NOTE ON OPTIMAL SINGLE CYCLE POLICIES FOR THE ONE-WAREHOUSE N-RETAILER INVENTORY PROBLEM,A.
FUKU 1976 -1 FUKUDA, Y.
IMPLICIT ENUMERATION ALGORITHM FOR A STORAGE ALLOCATION PROBLEM, AN.
FUNK 1968 -1 FUNK, M. L.; TILLMAN, F. A.
OPTIMAL CONSTRUCTION STAGING BY DYNAMIC PROGRAMMING.

LOCATION AND ALLOCATION

GALLA 1975 -1 GALLAGHER, J. M.
NUCLEAR FUEL REPROCESSING EXPANSION STRATEGIES.

GARFB 1977 -1 GARFINKEL, R. S.; NEEBE, A. W.; RAO, M. R.
M-CENTER PROBLEM: MINIMAX FACILITY LOCATION, THE.

GARFB 1973 -1 GARFINKEL, R. S.; NEEBE, A. W.; RAO, M. R.
ALGORITHM FOR THE M-MEDIAN PLANT LOCATION PROBLEM, AN.

GATE 1971 -1 GATELY, D. T.
INVESTMENT PLANNING FOR THE ELECTRIC POWER INDUSTRY: A MIXED-INTEGER PROGRAMMING APPROACH, WITH APPLICATIONS TO SOUTHERN INDIA.

GEN 1974 -1 GEN, M.; OKUNO, H.
OPTIMIZATION OF REDUNDANT ALLOCATION AND SELECTION PROBLEMS OF SYSTEMS RELIABILITY BY MEANS OF 0-1 PROGRAMMING (JAPANESE).

GEN B 1975 -1 GEN, M.; SASAKI, M.
OPTIMAL ALLOCATION OF SYSTEM AVAILABILITY.

GEOF 1976 -1 GEOFFRION, A. M.
CUSTOMER AGGREGATION IN DISTRIBUTION MODELING.

GEOF 1975 -1 GEOFFRION, A. M.
GUIDE TO COMPUTER-ASSISTED METHODS FOR DISTRIBUTION SYSTEMS PLANNING, A.

GEOFB 1977 -1 GEOFFRION, A. M.; GRAVES, G. W.; LEE, S.-J.
STRATEGIC DISTRIBUTION SYSTEM PLANNING: A STATUS REPORT.

GEOFC 1977 -1 GEOFFRION, A. M.; MAC BRIDE, R.
LAGRANGEAN RELAXATION APPLIED TO FACILITY LOCATION PROBLEMS.

GEOR 1971 -1 GEORGE, J. A.
APPLICATION OF INTEGER PROGRAMMING TO PLANT LOCATION AND INDIVISIBILITIES, AN.

GEORA 1974 -1 GEORGOBIANI, D. A.; TABLIASHVILI, O. M.
ALLOCATION PROBLEM IN TERMS OF DYNAMIC PROGRAMMING. QUESTIONS ON MATHEMATICAL AND ENGINEERING CYBERNETICS (RUSSIAN), THE.

GERM 1971 -1 GERMEIER, JU. B.; SUHORUCENKO, E. M.
OPTIMAL ALLOCATION OF RESOURCES WHILE STRIVING TO REACH A PRESCRIBED EFFICIENCY LEVEL ON ALL THE SUBDIVISIONS (RUSSIAN).

GILLD 1968 -1 GILLESPIE, C. M. JR.
LOCATION ABSOLUTE 2-CENTERS OF UNDIRECTED GRAPHS.

GINZ 1977 -1 GINZBURG, M. YA.; BURTMAN, M. I.
OPERATIONAL ALLOCATION OF RESOURCES IN LARGE TRANSPORTATION NETWORKS WITH TWO ENDPOINTS AND WITHOUT CYCLES (RUSSIAN).

GLEA 1975 -1 GLEASON, J. M.
SET COVERING APPROACH TO BUS STOP LOCATION, A.

GLEN 1975 -1 GLEN, J. J.
DYNAMIC PROGRAMMING MODEL FOR WORK SCHEDULING IN A SHIPYARD, A.

GOLE 1974 -1 GOLENKO, D. I.; GELEVER, K. G.; KESLER, S. S.; GOFMAN, E. F.
PROBLEMS ON OPTIMAL SCHEDULING AND OPTIMAL RESOURCE ALLOCATION FOR A TECHNOLOGICAL OPERATING PROCEDURE (RUSSIAN).

GONZA 1975 -1 GONZALEZ-VALENZUELA, F.
SIMPLE AND CAPACITATED WAREHOUSE LOCATION PROBLEMS WITH STOCHASTIC DEMANDS.

GORDB 1974 -1 GORDON, J. H.
HEURISTIC METHODS IN RESOURCE ALLOCATION.

GRABB 1975 -1 GRABOWSKI, W.
SOLUTION OF THE PLANT LOCATION PROBLEM BY THE BRANCH AND BOUND METHOD (POLISH), A.

GRAP 1977 -1 GRAPA, E.; BELFORD, G. G.
SOME THEOREMS TO AID IN SOLVING THE FILE ALLOCATION PROBLEM.

GRAVB 1976 -1 GRAVES, R. J.; THOMAS, W. H.
ANALYTICAL APPROACH TO ACADEMIC DEPARTMENT LOCATION IN A MULTI-CAMPUS UNIVERSITY, AN.

GRAY 1975 -1 GRAY, P.; HEITZMAN, W. R.
DETECTIVE ALLOCATION MODEL, A.

GRIS 1977 -1 GRISHUKHIN, V. P.
ON INTEGER SOLUTIONS IN THE ALLOCATION PROBLEM (RUSSIAN).

GROS 1975 -1 GROS, J.
POWER PLANT SITING: A PARETIAN ENVIRONMENTAL APPROACH.

GRUP 1974 -1 GRUPE, H. J. C.
NEO-CLASSICAL ANALYSIS OF THE OPTIMAL LOCATION PROBLEM OF INDUSTRY (SPANISH).

GUIGA 1977 -1 GUIGNARD, M.; SPIELBERG, K.
ALGORITHMS FOR EXPLOITING THE STRUCTURE OF THE SIMPLE PLANT LOCATION PROBLEM.

GUIGA 1974 -1 GUIGNARD, M.; SPIELBERG, K.
TRIANGULAR STRUCTURE IN UNCAPACITATED PLANT LOCATION.

GUIGA 1976 -2 GUIGNARD, M.; SPIELBERG, K.
INTEGER PROGRAMMING IN APL.

GUIGA 1977 -2 GUIGNARD, M.; SPIELBERG, K.
DUAL METHOD FOR THE MIXED PLANT LOCATION PROBLEM, A.

GULD 1977 -1 GULDMANN, J.-M.; SHEFER, D.
OPTIMAL PLANT LOCATION AND AIR QUALITY MANAGEMENT UNDER INDIVISIBILITIES AND ECONOMIES OF SCALE.

GUTE 1976 -1 GUTENBAUM, J.; NICOLO, F.
OPTIMAL AND SUB-OPTIMAL ALLOCATION OF RESOURCES IN SOME HIERARCHICAL SYSTEMS (POLISH).

GUZI 1975 -1 GUZIK, B.
SIMULATION MODEL OF THE OPTIMAL LOCATION OF A DISTRIBUTION CENTRE (POLISH).

HACAA 1976 -1 HACATUROV, V. R.; ASTAHOV, N. D.
DYNAMICAL ALLOCATION PROBLEMS (MODELS AND METHODS OF SOLUTION) (RUSSIAN).

HAND 1977 -1 HANDLER, G. Y.
M-CENTER PROBLEM: A RELAXATION APPROACH, THE.

HANSC 1976 -1 HANSEN, P.; KAUFMAN, L.
PUBLIC FACILITIES LOCATION UNDER AN INVESTMENT CONSTRAINT.

HANSD 1972 -1 HANSEN, P.; KAUFMANN, L.
ALGORITHM FOR THE LOCATION OF CENTRAL FACILITIES, AN.

HANSE 1977 -1 HANSEN, P.; THISSE, J.-F.
MULTIPLIANT LOCATION FOR PROFIT MAXIMISATION.

HARM 1976 -1 HARMATUCK, D. J.
COMPARISON OF TWO APPROACHES TO STOCK LOCATION, A.

HEARA 1976 -1 HEARN, D.; LOWE, T. J.
SUBGRADIENT PROCEDURE FOR THE SOLUTION OF MINIMAL PROBLEMS, A.

HEIS 1977 -1 HEISING, C. D.
PLANT LOCATION AND TRANSPORTATION MODELS OF THE LIGHT-WATER REACTOR FUEL CYCLE.

HENR 1967 -1 HENRY, B. B.; JONES, C. H.
LINEAR PROGRAMMING FOR PRODUCTION ALLOCATION.

HEVE 1964 -1 HEVERAN, R. D.
RESOURCE ALLOCATION.

HODG 1974 -1 HODGSON, M. J.
STAGING INTERURBAN HIGHWAY CONSTRUCTION BY DYNAMIC PROGRAMMING.

HOWA 1970 -1 HOWARD, D. D.
DYNAMIC LOCATION-ALLOCATION ANALYSIS.

HUNG 1973 -1 HUNG, H. K.
HEURISTIC ALGORITHM FOR THE MULTI-PERIOD FACILITY LOCATION PROBLEM, A.

HUNGA 1977 -1 HUNG, M. S.; FISK, J. C.
ASSORTMENT PROBLEMS AND EQUIVALENT INTEGER PROGRAMMING FORMULATIONS.

HURT 1975 1 HURTER, A. P. JR.; SCHAEFER, M. K.; WENDELL, R. E.
SOLUTIONS OF CONSTRAINED LOCATION PROBLEMS.

HYUNA 1974 -1 HYUN, K. N.; MITSUO, G.; OKUNO, H.
OPTIMIZATION OF THE REDUNDANT ALLOCATION IN A SYSTEM WITH SEVERAL FAILURE MODES BY ZERO-ONE PROGRAMMING.

JACO 1973 -1 JACOB, H.
ZUR STANDORTWAHL DER UNTERNEHMUNGEN.

JACOA 1974 -1 JACOBSEN, S. K.
ALGORITHM FOR SMALL-SCALE DYNAMIC PLANT-LOCATION PROBLEMS, AN.

JAND 1969 -1 JANDY, G.
PROBLEMS OF LOCATION IN OPERATIONS RESEARCH MODELS.

JARA 1976 -1 JARANCEV, JU. A.
CERTAIN PROCEDURE FOR DYNAMIC MEMORY ALLOCATION IN COMPUTERS (RUSSIAN), A.

JASI 1976 -1 JASINSKA, E.; RYSZKOWSKA, E.
WAREHOUSE LOCATION IN A TRANSPORTATION PROBLEM OF TWO STAGES (POLISH).

JOHNG 1976 -1 JOHNSON, E. L.; SPIELBERG, K.
SOLUTION OF A MULTI-COMMODITY (NO-SPLITTING) PRODUCTION-DISTRIBUTION PROBLEM.

JOHNI 1974 -1 JOHNSON, P. A.
HEURISTIC FOR INTEGRATING PLANT LOCATION AND PLANT TIMING QUESTIONS, A.

JOHNJ 1958 -1 JOHNSON, R. E.
USE OF OPERATIONS RESEARCH TECHNIQUES IN PLANT LOCATION PROBLEMS, THE.

JUCK 1976 -1 JUCKER, J. V.; CARLSON, R. C.
SIMPLE PLANT-LOCATION PROBLEM UNDER UNCERTAINTY, THE.

KALD 1975 -1 KALDYBAEV, S. U.
ALGORITHM FOR THE APPROXIMATION-COMBINATORIAL METHOD IN SOLVING AN ALLOCATION PROBLEM USING BOOLEAN VARIABLES (RUSSIAN), A.

KALL 1976 -1 KALLINA, C.; LYNN, J.
OPTIMALITY OF THE CUBE-PER-ORDER-INDEX RULE FOR STOCK LOCATION IN A DISTRIBUTION WAREHOUSE.

KASPA 1974 -1 KASPSHITSKAJA, M. F.; SERGIENKO, I. V.; HIL'CENKO, V. I.
CERTAIN APPROACH TO THE SOLUTION OF ALLOCATION PROBLEMS (RUSSIAN), A.

KATSA 1976 -1 KATSOULAS, P. C.; SADOWSKI, R. P.
BRANCH AND BOUND ALGORITHM FOR USE IN RESOURCE ALLOCATION, A.

KATZ 1976 -1 KATZ, I.; COOPER, L.
OPTIMAL FACILITY LOCATION FOR NORMALLY AND EXPONENTIALLY DISTRIBUTED POINTS.

KEL' 1973 -1 KEL'MANS, A. K.
SELECTION OF THE OPTIMAL VERTEX IN A GRAPH (RUSSIAN), THE.

KENDA 1975 -1 KENDRICK, D.; STOUTJESDIJK, A.
PLANNING OF INDUSTRIAL INVESTMENT PROGRAMS: A METHODOLOGY, THE.

KHAC 1976 -1 KHACHATUROV, V. R.; ASTAKHOV, N. B.
DYNAMIC LOCATION PROBLEMS (MODELS AND SOLUTION METHODS) (RUSSIAN).

KHAN 1977 -1 KHAN, S. U.; BARI, A.
PROCEDURE FOR INTEGER SOLUTION TO SOME ALLOCATION PROBLEMS, A.

KILG 1977 -1 KILGORE, W. C.
WAREHOUSE FACILITIES ANALYSIS.

KRAR 1967 -1 KRARUP, J.
FIXED-COST AND OTHER NETWORK FLOW PROBLEMS AS RELATED TO PLANT LOCATION AND TO THE DESIGN OF TRANSPORTATION AND COMPUTER SYSTEMS.

KRARA 1967 -1 KRARUP, J.; BILDE, O.
DETERMINATION OF THE OPTIMAL LOCATION OF PRODUCTION PLANTS (DANISH).

KRARA 1977 -1 KRARUP, J.; BILDE, O.
PLANT LOCATION, SET COVERING AND ECONOMIC LOT SIZE: AN O(MN)-ALGORITHM FOR STRUCTURED PROBLEMS.

KRARD 1977 -1 KRARUP, J.; PRUZAN, P. M.
SELECTED FAMILIES OF DISCRETE LOCATION PROBLEMS PART 1: CENTER PROBLEMS PART 2: MEDIAN PROBLEMS.

KRARD 1977 -2 KRARUP, J.; PRUZAN, P. M.
SELECTED FAMILIES OF DISCRETE LOCATION PROBLEMS PART 3: THE PLANT LOCATION FAMILY.

KULI 1975 -1 KULIKOWSKI, R.
OPTIMUM ALLOCATION OF RESOURCES IN DECENTRALIZED MACROECONOMIC MODEL.

LAI 1975 -1 LAI, A. W.
HEURISTIC SOLUTION TO THE URBAN GOODS CONSOLIDATION TERMINAL SITE SELECTION DECISION, A.

LAMBB 1963 -1 LAMBOURN, S.
RESOURCE ALLOCATION AND MULTIPROJECT SCHEDULING (RAMPS). A NEW TOOL IN PLANNING AND CONTROL.

LAUR 1975 -1 LAURIA, D. T.
REGIONAL SEWERAGE PLANNING BY MIXED INTEGER PROGRAMMING.

LAWRA 1975 -1 LAWRENCE, S. M.; LAWRENCE, K. D.
LOCATING GROUPS OF INDUSTRIES IN A REGION USING AN INPUT-OUTPUT MODEL STRUCTURE AND MULTIPLE OBJECTIVE LINEAR PROGRAMMING.

LEA 1973 -1 LEA, A. C.
LOCATION-ALLOCATION MODELS: A REVIEW.

LEBEA 1976 -1 LEBEDEV, S. S.; KOVALEVSKAJA, M. I.
LAGRANGE MULTIPLIERS IN THE SIMPLEST ALLOCATION PROBLEM (RUSSIAN).

LEBL 1977 -1 LEBLANC, L. J.
HEURISTIC APPROACH OF LARGE SCALE DISCRETE STOCHASTIC TRANSPORTATION-LOCATION PROBLEMS, A.

LEIF 1967 -1 LEIFMAN, L. Y. A.
PROBLEM OF ALLOCATION OF RESOURCES IN NETWORK SCHEDULING (RUSSIAN).

LERN 1969 -1 LERNER, A. Y.; MAKAROV, I. M.
OPTIMAL ALLOCATION OF RESOURCES (RUSSIAN).

LEVY 1966 -1 LEVY, J.
OPTIMAL LOCATION OF NETWORKS.

LINDA 1977 -1 LIND, N. C.
CONTINUOUS ALLOCATION PROBLEMS.

LIPT 1976 -1 LIPTON, S. L.
SERVICE CENTER LOCATION USING MIXED INTEGER PROGRAMMING.

LODI 1970 -1 LODISH, L. M.
COMPUTATIONAL LIMITATIONS OF DYNAMIC PROGRAMMING FOR WAREHOUSE LOCATION.

****

LOS A 1977 -1 LOS, M.; NGUYEN, S.
SPATIAL ALLOCATION ON A NETWORK WITH CONGESTION.
LOVE 1976 -1 LOVE, R. F.
ONE-DIMENSIONAL FACILITY LOCATION-ALLOCATION USING DYNAMIC PROGRAMMING.
LOVEA 1976 -1 LOVE, R. F.; WONG, J. Y.
ON SOLVING A ONE-DIMENSIONAL SPACE ALLOCATION PROBLEM WITH INTEGER PROGRAMMING.
LUND 1973 -1 LUNDQVIST, L.
INTEGRATED LOCATION-TRANSPORTATION ANALYSIS: A DECOMPOSITION APPROACH.
LUND 1975 -1 LUNDQVIST, L.
TRANSPORTATION ANALYSIS AND ACTIVITY LOCATION IN LAND-USE PLANNING - WITH APPLICATIONS TO THE STOCKHOLM REGION.
LUSS 1975 -1 LUSS, H.
OPTIMIZATION OF A MULTIPERIOD RESOURCE ALLOCATION MODEL.
LYON 1977 -1 LYONS, J. P.; YOUNG, J. P.
DEVELOPMENT AND IMPLEMENTATION OF A STAFF ALLOCATION MODEL FOR MENTAL HEALTH FACILITIES.
MAC F 1977 -1 MAC GINNIS, L. F.
SURVEY OF RESULTS FOR A CLASS OF FACILITIES LOCATION PROBLEMS, A.
MAHM 1976 -1 MAHMOUD, S.; RIORDON, J. S.
OPTIMAL ALLOCATION OF RESOURCES IN DISTRIBUTED INFORMATION NETWORKS.
MAIR 1977 -1 MAIRS, T. G.; WAKEFIELD, G. W.; JOHNSON, E. L.; SPIELBERG, K.
ON A PRODUCTION ALLOCATION AND DISTRIBUTION PROBLEM.
MANN 1967 -1 MANNE, A. S.
INVESTMENTS FOR CAPACITY EXPANSION: SIZE, LOCATION, AND TIME-PHASING.
MARCA 1976 -1 MARCUK, G. I.
ENVIRONMENT, AND PROBLEMS OF OPTIMIZING THE LOCATION OF ENTERPRISES (RUSSIAN), THE.
MARCB 1967 -1 MARCUS, B. T.; UNGAR, A. J.
MULTI-PERIOD TWO-STAGE MODEL FOR LOCATING ALUMINA AND ALUMINUM REDUCTION CAPACITY, A.
MARKA 1975 -1 MARKLAND, R. E.; SMITH, L. D.; BECKER, J. D.; HAACK, P. J.
SITE SELECTION MODEL FOR LAND DISPOSAL OF WASTEWATER, A.
MARKF 1971 -1 MARKS. D. H.; JIRKA, G. H.
ENVIRONMENTAL SCREENING MODEL FOR THE LOCATION OF POWER GENERATING FACILITIES, AN.
MARSA 1972 -1 MARSTEN, R. E.
ALGORITHM FOR FINDING ALMOST ALL OF THE MEDIANS OF A NETWORK, AN.
MARTG 1976 -1 MARTINEC, I.
ON A GENERALIZED LOCATION PROBLEM.
MERE 1975 -1 MEREKIN, YU. V.
ON A PROBLEM OF OPTIMAL ALLOCATION OF THE PRODUCTION (RUSSIAN).
MERL 1971 -1 MERLINO, E. L. A.
MULTI-NATIONAL CAPACITY EXPANSION MODEL FOR PRIMARY ALUMINUM PLANTS.
MEYE 1973 -1 MEYER, C. F.
LONG-RANGE SELECTION AND TIMING ANALYSIS SYSTEM FOR FACILITY LOCATION: IMPLEMENTATION, A.
MILLA 1976 -1 MILLER, D. M.; FYFFE, D. E.
ALLOCATING BUILDING INSPECTION MANPOWER FOR FIRE PREVENTION.
MIRC 1974 -1 MIRCHANDANI, P. B.
LOCATION OF FACILITIES ON A STOCHASTIC NETWORK.
MJEL 1976 -1 MJELDE, K. M.
EVALUATION AND INCREMENTAL DETERMINATION OF ALMOST OPTIMAL ALLOCATIONS OF RESOURCES.
MORR 1977 -1 MORRILL, R. L.; SYMONS, J.
EFFICIENCY AND EQUITY ASPECTS OF OPTIMUM LOCATION.
MORRA 1972 -1 MORRIS, J. G.
FACILITIES LOCATION AND LOCATION-ALLOCATION USING RECTANGULAR DISTANCES.
MOSHA 1963 -1 MOSHMAN, J.; JOHNSON, J.; LARSEN, M.
RAMPS-TECHNIQUE FOR RESOURCE ALLOCATION AND MULTIPROJECT SCHEDULING.
MOSL 1976 -1 MOSLER, K. C.
OPTIMALE TRANSPORTNETZE. ZUR BESTIMMUNG IHRES KOSTENGUENSTIGSTEN STANDORTS BEI GEGEBENER NACHFRAGE.
NARUA 1977 -1 NARULA, S. C.; OGBU, U. I.; SAMUELSSON, H. M.
ALGORITHM FOR THE P-MEDIAN PROBLEM, AN.
NASHA 1977 -1 NASH, P.; GITTINS, J. C.
HAMILTONIAN APPROACH TO OPTIMAL STOCHASTIC RESOURCE ALLOCATION, A.
NAUS 1975 -1 NAUSS, R. M.
EFFICIENT ALGORITHM FOR THE CAPACITATED FACILITY LOCATION PROBLEM, AN.
NEEB 1974 -1 NEEBE, A. W.
M-CENTER AND M-MEDIAN FACILITY LOCATION PROBLEMS, THE.
NEEBA 1975 -1 NEEBE, A. W.; RAO, M. R.
SUBGRADIENT APPROACH TO THE M-MEDIAN PROBLEM, A.
NEMHA 1976 -1 NEMHAUSER, G. L.; WOLSEY, L. A.; FISHER, M. L.
ANALYSIS OF APPROXIMATIONS FOR MAXIMIZING SUBMODULAR SET FUNCTIONS, AN.
NOVI 1978 -1 NOVITZKY, N. A.
ANALYSIS OF FACTORS INFLUENCE ON INDUSTRY ALLOCATION EFFICIENCY (RUSSIAN).
NYKO 1964 -1 NYKOWSKI, I.; ZURKOWSKI, J.
PRODUCTION ALLOCATION PROBLEMS WHICH CAN BE REDUCED TO TRANSPORTATION PROBLEMS (POLISH).
ODON 1974 -1 ODONI, A. R.
LOCATION OF FACILITIES ON A NETWORK: A SURVEY OF RESULTS.
OLSO 1976 -1 OLSON, E. N.; WENDELL, R. E.
ANALYSIS OF WAREHOUSE LOCATION IN THE BEEF INDUSTRY, AN.
ONOK 1976 -1 ONOKERHORAYE, A. G.
CONCEPTUAL FRAMEWORK FOR THE LOCATION OF PUBLIC FACILITIES IN THE URBAN AREAS OF DEVELOPING COUNTRIES: THE NIGERIAN CASE, A.
ORLOA 1977 -1 ORLOFF, C. S.
THEORETICAL MODEL OF NET ACCESSIBILITY IN PUBLIC FACILITY LOCATION, A.
OSLE 1974 -1 OSLEEB, J. P.
LOCATION THEORY FOR THE UNIFORM-PRICE MANUFACTURER, A.
OSLEA 1977 -1 OSLEEB, J. P.; CROMLEY, R. G.
LOCATION-PRODUCTION-ALLOCATION PROBLEM WITH NONLINEAR PRODUCTION COSTS, THE.
PAPI 1975 -1 PAPINEAU, R. L.; FRANCIS, R. L.; BARTHOLDI, J.
MINIMAX FACILITY LAYOUT PROBLEM INVOLVING DISTANCES BETWEEN AND WITHIN FACILITIES, A.
PATE 1976 -1 PATEL, A. M.; WALD, C. G. DE; COTES, L. C.
PRE-PARTITIONING AS A MEANS OF ENHANCING PAIRWISE INTERCHANGE SOLUTIONS TO QUADRATIC ASSIGNMENT PROBLEMS.
PAULA 1976 -1 PAULAUSKAS, M. A.; RASHKILIS, A. V.; SHYARKSHNIS, G. A.
PROBLEM CONCERNING THE OPTIMAL ALLOCATION OF DEFICIENT RESOURCES FOR A DISCRETE CONSUMPTION, 2. ALGORITHMS FOR THE APPROXIMATE SOLUTION OF PROBLEMS WITH A GREAT NUMBER OF UNKNOWNS (RUSSIAN).

****

PAULA 1976 -2 PAULAUSKAS, M. A.; RASHKILIS, A. V.; SHYARKSHNIS, G. A.
PROBLEM CONCERNING THE OPTIMAL ALLOCATION OF DEFICIENT RESOURCES FOR A DISCRETE CONSUMPTION, 3. HEURISTIC ALGORITHMS FOR THE SOLUTION OF THE PROBLEM (RUSSIAN).
PAULB 1975 -1 PAULAUSKAS, M. A.; RASHKILIS, A. V.; STANIONIS, K.; SHYARKSHNIS, G. A.
PROBLEM OF THE OPTIMAL ALLOCATION OF DEFICIT RESOURCES FOR DISCRETE CONSUMPTION, 1. FORMULATION OF THE PROBLEM AND MATHEMATICAL METHODS OF SOLUTION (RUSSIAN), THE.
PICAD 1975 -1 PICARD, J.-C.; RATLIFF, H. D.
CUT APPROACH TO THE RECTILINEAR DISTANCE FACILITY LOCATION PROBLEM, A.
PINT 1975 -1 PINTO, J. V.
BACKGROUND AND DEVELOPMENT OF LOCATION THEORY.
PLAN 1975 -1 PLANE, D. R.; MONARCHI, D. E.; HENDRICK, T. E.
HIERARCHICAL APPROACH FOR SOME LOCATION PROBLEMS, A.
POCH 1976 -1 POCHTMAN, YU. M.; RAFES, R. YU.
MODIFICATION OF THE DYNAMIC PROGRAMMING ALGORITHM FOR PROBLEMS CONCERNING THE ALLOCATION OF RESOURCES (RUSSIAN).
POPO 1973 -1 POPOVICIU, N.
ALLOCATION PROBLEM, WITH BIVALENT ZERO-ONE VARIABLES DEFINED BY MEANS OF THE ELEMENTS OF A TRANSPORTATION PROBLEM (RUMANIAN), AN.
POPO 1972 -1 POPOVICIU, N.
DYNAMIC PROGRAMMING: ONE-SIZE ALLOCATION PROBLEM WITH A SUPPLEMENTARY CONSTRAINT.
POUL 1975 -1 POULTON, M. C.; KANAFANI, A.
APPLICATION OF LOCATION MODELS TO OFF-AIRPORT TERMINALS, THE.
PRIT 1970 -1 PRITSKER, A. A. B.; GHARE, P. M.
LOCATING NEW FACILITIES WITH RESPECT TO EXISTING FACILITIES.
RAMAA 1976 -1 RAMAKRISHNAN, K. G.
SOLUTION OF CAPACITATED PLANT LOCATION PROBLEMS USING BENDERS' DECOMPOSITION.
RAO 1973 -1 RAO, M. R.
ON THE DIRECT SEARCH APPROACH TO THE RECTILINEAR FACILITIES LOCATION PROBLEM.
RAO 1975 -1 RAO, M. R.
MULTI-REGION, MULTI-PERIOD CAPACITY EXPANSION MODELS: A HEURISTIC AND A SPECIAL CASE.
RAO A 1974 -1 RAO, R.; RUTENBERG, D. P.
EFFICIENT COMPUTATION OF CONTINUOUS-TIME PLANT LOCATION.
RAZH 1967 -1 RAZHUMIKHIN, B. S.
PROBLEMS OF THE OPTIMAL ALLOCATION OF RESOURCES (RUSSIAN).
RAZH 1965 -1 RAZHUMIKHIN, B. S.
PROBLEM OF OPTIMAL RESOURCE ALLOCATION.
RAZHA 1967 -1 RAZHUMIKHIN, B. S.; STAVROKOVA, N. E.
PROBLEM OF DISTRIBUTING RESOURCES TO MINIMIZE TIME OF EXECUTION OF JOBS COMPLEX.
RIDE 1976 -1 RIDER, K. L.
PARAMETRIC MODEL FOR THE ALLOCATION OF FIRE COMPANIES IN NEW YORK CITY, A.
RIDE 1975 -1 RIDER, K. L.
PARAMETRIC MODEL FOR THE ALLOCATION OF FIRE COMPANIES: EXECUTIVE SUMMARY, A.
ROBEB 1972 -1 ROBERTS, K. J.
DISCUSSION OF 'SELECTING SOLID WASTE DISPOSAL FACILITIES' BY HELMS AND CLARK.
RODG 1977 -1 RODGERS, J. R.; MAC CARTHY, O.
SERVICE FACILITY LOCATION PROBLEM: THE LOCATION OF FARM ADVISORY OFFICERS IN N. Z., A.
ROOD 1977 -1 ROODMAN, G. M.; SCHWARZ, L. B.
EXTENSIONS OF THE MULTI-PERIOD FACILITY PHASE-OUT MODEL: NEW PROCEDURES AND APPLICATION TO A PHASE-IN/PHASE-OUT PROBLEM.
ROOS 1975 -1 ROOSE, J. R.; RIEMERSMA, P.
HEURISTIC OPTIMISATION OF A LOCATION-ALLOCATION PROBLEM IN A TRANSPORT NETWORK WITH MODAL CHOICE.
ROSS 1977 -1 ROSS, G. T.; SOLAND, R. M.
MODELING FACILITY LOCATION PROBLEMS AS GENERALIZED ASSIGNMENT PROBLEMS.
RUSN 1976 -1 RUSNAK, Z.
ATTEMPT TO SOLVE THE CLASSICAL LOCATION PROBLEM IN THE PLANE (POLISH), AN.
SAKA 1976 -1 SAKAGUCHI, M.
SEQUENTIAL ALLOCATION PROBLEM FOR RANDOMLY APPEARING TARGETS, A.
SANKB 1975 -1 SANKOFF, D.; ROUSSEAU, P.
LOCATING THE VERTICES OF A STEINER TREE IN AN ARBITRARY SPACE.
SCHU 1968 -1 SCHULTZ, G. P.
MANAGERIAL DECISION-MAKING IN LOCAL GOVERNMENT: FACILITY PLANNING FOR SOLID WASTE COLLECTION.
SCHU 1969 -1 SCHULTZ, G. P.
FACILITY PLANNING FOR A PUBLIC SERVICE SYSTEM: DOMESTIC SOLID WASTE COLLECTION.
SCOT 1975 -1 SCOTT, A. J.
DISCRETE DYNAMIC LOCATIONAL SYSTEMS: NOTES ON A STRUCTURAL FRAMEWORK.
SCOT 1970 -1 SCOTT, A. J.
ELEMENTARY THEORETICAL DYNAMICS OF A COMPETITIVE LOCATIONAL SYSTEM, THE.
SCOT 1971 -1 SCOTT, A. J.
COMBINATORIAL PROGRAMMING, SPATIAL ANALYSIS, AND PLANNING.
SCOT 1971 -2 SCOTT, A. J.
DYNAMIC LOCATION-ALLOCATION SYSTEMS: SOME BASIC PLANNING STRATEGIES.
SCOT 1971 -3 SCOTT, A. J.
OPTIMAL DECISION PROCESSES FOR A CLASS OF DYNAMIC LOCATIONAL PROBLEMS.
SCRI 1975 -1 SCRIABIN, M.; VERGIN, R. C.
COMPARISON OF COMPUTER ALGORITHMS AND VISUAL BASED METHODS FOR PLANT LAYOUT.
SERC 1970 -1 SERCK-HANSSEN, J.
OPTIMAL PATTERNS OF LOCATION.
SERO 1976 -1 SEROV, S. S.; UZDEMIR, A. P.
DYNAMIC INDUSTRIAL-PLANT-LOCATION PROBLEM AND NUMERICAL METHOD FOR SOLVING IT, I (RUSSIAN).
SHARF 1972 -1 SHARPE, R.; BROTCHIE, J. F.
URBAN SYSTEMS STUDY, AN.
SHARG 1975 -1 SHARPE, R.; BROTCHIE, J. F.; AHERN, P. A.
EVALUATION OF ALTERNATIVE GROWTH PATTERNS FOR MELBOURNE.
SHEL 1975 -1 SHELNESS, N. H.; STEPHENS, P. D.; WHITFIELD, H.
EDINBURG MULTI-ACCESS SYSTEM SCHEDULING AND ALLOCATION PROCEDURES IN THE RESIDENT SUPERVISOR, THE.
SHEP 1974 -1 SHEPPARD, E. S.
CONCEPTUAL FRAMEWORK FOR DYNAMIC LOCATION-ALLOCATION ANALYSIS, A.
SHER 1977 -1 SHERALI, A. D.; SHETTY, C. M.
RECTILINEAR DISTANCE LOCATION-ALLOCATION PROBLEM, THE.
SHET 1975 -1 SHETTY, C. M.
BILINEAR PROGRAMMING WITH APPLICATIONS TO LOCATION-ALLOCATION PROBLEMS.

****

SHIE 1977 -1 SHIER, D. R.
MIN-MAX THEOREM FOR NETWORK LOCATION PROBLEMS ON A TREE, A.

SHIE 1977 -2 SHIER, D. R.
MIN-MAX THEOREM FOR P-CENTER PROBLEMS ON A TREE, A.

SHIH 1977 -1 SHIH, W.
BRANCH AND BOUND PROCEDURE FOR A CLASS OF DISCRETE RESOURCE ALLOCATION PROBLEMS WITH SEVERAL CONSTRAINTS, A.

SHOW 1977 -1 SHOWALTER, M. J.; KRAJEWSKI, L. J.; RITZMAN, L. P.
MANPOWER ALLOCATION IN U.S. POSTAL FACILITIES: A HEURISTIC APPROACH.

SING 1968 -1 SINGER, S.
MULTI-CENTERS AND MULTI-MEDIANS OF A GRAPH WITH A APPLICATION TO OPTIMAL WAREHOUSE LOCATION.

SLATA 1976 -1 SLATER, P. J.
CENTRAL VERTICES IN A GRAPH.

SLATA 1975 -2 SLATER, P. J.
MAXIMIN FACILITY LOCATION.

SLATA 1976 -2 SLATER, P. J.
R-DOMINATION IN GRAPHS.

SNIC 1978 -1 SNICKARS, F.
CONVEXITY AND DUALITY PROPERTIES OF A QUADRATIC INTRAREGIONAL LOCATION MODEL.

STAHA 1976 -1 STAHL, W.
SENSITIVITAETSANALYSE MEHRSTUFIGER STANDORTPROBLEME.

STIE 1976 -1 STIEBLIN, C.
ZUR STANDORTBESTIMMUNG VON KRANKENWAGENSYSTEMEN.

STOJ 1976 -1 STOJAN, YU. G.; GIL, N. I.
METHODS AND ALGORITHMS FOR THE PLACING OF PLANE GEOMETRIC OBJECTS (RUSSIAN).

SWAI 1971 -1 SWAIN, R. W.
DECOMPOSITION ALGORITHM FOR A CLASS OF FACILITY LOCATION PROBLEM, A.

SWAI 1976 -1 SWAIN, R. W.
CENTRAL FACILITIES LOCATION WITH MULTIPLE PLANNING PERIODS.

SWAI 1971 -2 SWAIN, R. W.
PARAMETRIC DECOMPOSITION APPROACH FOR THE SOLUTION OF UNCAPACITATED LOCATION PROBLEMS.

SWEEA 1976 -1 SWEENEY, D. J.; TATHAM, R. L.
IMPROVED LONG-RUN MODEL FOR MULTIPLE WAREHOUSE LOCATION, AN.

TAKA 1966 -1 TAKAHASHI, I.
TREE ALGORITHM FOR SOLVING RESOURCE ALLOCATION PROBLEMS.

TAPI 1971 -1 TAPIERO, C. S.
TRANSPORTATION-LOCATION-ALLOCATION PROBLEMS OVER TIME.

THIR 1966 -1 THIRIET, L.; GAUSSENS, J.
METHODE DE CHOIX DES AFFECTATIONS ET DES INVESTISSEMENTS.

THIS 1976 -1 THISSE, J.
MODELES MATHEMATIQUES D'OTIMISATION DE LA LOCALISATION DES ENTREPRISES.

TRIF 1973 -1 TRIFON, R.; LIVNAT, A.
SPATIAL ALLOCATION OF SCHOOLS OVER TIME IN CITIES, THE.

TRIV 1976 -1 TRIVEDI, V. M.; WARNER, D. M.
BRANCH AND BOUND ALGORITHM FOR OPTIMUM ALLOCATION OF FLOAT NURSES, A.

TRUB 1976 -1 TRUBIN, V. A.
EFFECTIVE ALGORITHM FOR THE SOLUTION OF THE ALLOCATION PROBLEM ON A NETWORK IN THE SHAPE OF A TREE (RUSSIAN), AN.

TRUS 1976 -1 TRUSCOTT, W. G.
NOTE ON A CONCEPTUAL FRAMEWORK FOR DYNAMIC LOCATION-ALLOCATION ANALYSIS, A.

TRUS 1976 -2 TRUSCOTT, W. G.
SOME DYNAMIC EXTENSIONS OF A DISCRETE LOCATION-ALLOCATION PROBLEM.

TSYN 1968 -1 TSYNKOVA, E. N.
PROBLEM OF OPTIMAL ALLOCATION OF FINITE RESOURCES.

TURNA 1975 -1 TURNER, W. C.; HOUGLAND, E. S.
OPTIMAL ROUTING OF SOLID WASTE COLLECTION VEHICLES, THE.

USHA 1978 -1 USHAKOV, E. P.
OPTIMIZATION OF PRODUCTION DEVELOPMENT AND ALLOCATION IN A REGION INCLUDING REPRODUCTION OF WATER RESOURCES (RUSSIAN).

VESE 1977 -1 VESELOVSKY, V. E.
ON SOME ALGORITHMS OF ALLOCATION LARGE SCALE PROBLEM SOLUTION (RUSSIAN).

VLAC 1977 -1 VLACH, M.
MULTICOMMODITY LOCATION PROBLEMS (CZECH).

VU 1976 -1 VU, V. T.; TURNER, W. C.
SYSTEMS DESIGN FOR RURAL REFUSE COLLECTION.

WEGLA 1976 -1 WEGLARZ, J.; SLOWINSKI, R.
COMPUTATIONAL ASPECTS OF A CERTAIN CLASS OF TIME-OPTIMAL RESOURCE ALLOCATION PROBLEMS IN A COMPLEX OF OPERATIONS.

WEND 1977 -1 WENDELL, R. E.; HURTER, A. P. JR.; LOWE, T. J.
EFFICIENT POINTS IN LOCATION PROBLEMS.

WESO 1974 -1 WESOLOWSKY, G. O.
REJOINDER.

WESO 1977 -1 WESOLOWSKY, G. O.
PROBABILISTIC WEIGHTS IN THE ONE-DIMENSIONAL FACILITY LOCATION PROBLEM.

WESOA 1971 -1 WESOLOWSKY, G. O.; LOVE, R. F.
OPTIMAL LOCATION OF NEW FACILITIES USING RECTANGULAR DISTANCES, THE.

WESOB 1975 -1 WESOLOWSKY, G. O.; TRUSCOTT, W. G.
MULTIPERIOD LOCATION-ALLOCATION PROBLEM WITH RELOCATION OF FACILITIES, THE.

WESOB 1976 -1 WESOLOWSKY, G. O.; TRUSCOTT, W. G.
DYNAMIC LOCATION OF MULTIPLE FACILITIES FOR A CLASS OF CONTINUOUS PROBLEMS.

WILSB 1974 -1 WILSON, D.; SMITH, R. L.
LOCATION OF STOCHASTIC SERVICES, THE.

WONG 1976 -1 WONG, C. K.
SOME DISCRETE OPTIMIZATION PROBLEMS WITH SECONDARY STORAGE APPLICATIONS.

YAGI 1975 -1 YAGIZ, OE.
HEURISTICALLY-AIDED SINGLE BRANCH SEARCH PROCEDURES FOR PLANT LOCATION PROBLEMS.

YALO 1976 -1 YALOVIJ, V. V.
ON SOME PROBLEMS IN THE ALLOCATION OF RESOURCES (UKRAINIAN).

YOUN 1970 -1 YOUNG, G. K.; MOSELY, J. C.; EVENSON, D. E.
TIME SEQUENCING OF ELEMENT CONSTRUCTION IN A MULTI-RESERVOIR SYSTEM.

ZUSU 1977 -1 ZUSUPBAEV, A.
SOLUTION OF THE SEASONAL PRODUCTION ALLOCATION PROBLEM BY THE METHOD OF SUCCESSIVE CALCULATIONS (RUSSIAN).

ZUSU 1974 -1 ZUSUPBAEV, A.
PRODUCTION ALLOCATION PROBLEM WITH A CONVEX SEPARABLE FUNCTIONAL (RUSSIAN), THE.

****

ALAV 1977 -1 ALAVI, Y.; BEHZAD, M.; LESNIAK-FOSTER, L. M.; NORDHAUS, E. A.
TOTAL MATCHINGS AND TOTAL COVERINGS OF GRAPHS.

ANGL 1977 -1 ANGLUIN, D.; VALIANT, L. G.
FAST PROBABILISTIC ALGORITHMS FOR HAMILTONIAN CIRCUITS AND MATCHINGS.

BALAJ 1975 -1 BALAS, E.; PADBERG, M. W.
SET PARTITIONING: A SURVEY.

BARTB 1977 -1 BARTHOLDI, J. J. III; RATLIFF, H. D.
FIELD GUIDE TO IDENTIFYING NETWORK FLOW AND MATCHING PROBLEMS, A.

BOLL 1977 -1 BOLLOBAS, B.
EXTREMAL PROBLEMS IN GRAPH THEORY.

BOLLA 1976 -1 BOLLOBAS, B.; DAYKIN, D. E.; ERDOES, P.
SETS OF INDEPENDENT EDGES OF A HYPERGRAPH.

BOLLB 1976 -1 BOLLOBAS, B.; ELDRIDGE, S. E.
MAXIMAL MATCHINGS IN GRAPHS WITH GIVEN MINIMAL AND MAXIMAL DEGREES.

BRAU 1974 -1 BRAUN, J.
EDMONDS' MATCHING ALGORITHMUS.

CHRI 1973 -1 CHRISTOFIDES, N.
OPTIMUM TRANSVERSAL OF A GRAPH, THE.

CHVAD 1976 -1 CHVATAL, V.; HANSON, D.
DEGREES AND MATCHINGS.

CUNNB 1978 -1 CUNNINGHAM, W. H.; MARSH, A. B. III.
PRIMAL ALGORITHM FOR OPTIMUM MATCHING, A.

DAYK 1976 -1 DAYKIN, D. E.
CORRIGENDUM AND ADDENDUM TO: "PSEUDO-MATCHINGS OF A BIPARTITE GRAPH" BY A. BRACE AND DAYKIN (PROCEEDINGS OF THE AMS 42, 1974, 28-32).

DERI 1978 -1 DERIGS, U.
ALGEBRAISCHE MATCHINGPROBLEME.

DERI 1977 -1 DERIGS, U.
DUALITY AND THE ALGEBRAIC MATCHING PROBLEM.

DERI 1978 -2 DERIGS, U.
ON SOLVING SYMMETRIC ASSIGNMENT AND PERFECT MATCHING PROBLEMS WITH ALGEBRAIC OBJECTIVES.

EDMO 1967 -1 EDMONDS, J.
INTRODUCTION TO MATCHING, AN.

EDMO 1975 -1 EDMONDS, J.
SOME WELL-SOLVED PROBLEMS IN COMBINATORIAL OPTIMIZATION.

EDMOB 1969 -1 EDMONDS, J.; JOHNSON F. L.; LOCKHART, S.
BLOSSOM I, A CODE FOR MATCHING.

ELDR 1976 -1 ELDRIDGE, S. E.; BOLLOBAS, B.
MAXIMAL MATCHINGS IN GRAPHS WITH GIVEN MAXIMAL AND MINIMAL DEGREES.

ERDOC 1977 -1 ERDOES, P.; MEIR, A.
ON TOTAL MATCHING NUMBERS AND TOTAL COVERING NUMBERS OF COMPLEMENTARY GRAPHS.

EVENB 1975 -1 EVEN, S.; KARIV, O.
O(N POWER 2.5) ALGORITHM FOR MAXIMUM MATCHING IN GENERAL GRAPHS, AN.

FEIC 1975 -1 FEICHT, E.; HECK, J.; PAPE, U.
ALGORITHMEN VON EDMONDS UND GABOW ZUR ERMITTLUNG DES MAXIMALEN MATCHINGS IN SCHLICHTEN UND ENDLICHEN GRAPHEN.

GABOB 1976 -1 GABOW, H. N.
EFFICIENT IMPLEMENTATION OF EDMONDS' ALGORITHM FOR MAXIMUM MATCHING ON GRAPHS, AN.

GILE 1977 -1 GILES, F. R.
OPTIMUM MATCHING FORESTS 1: SPECIAL WEIGHTS.

GILE 1977 -2 GILES, F. R.
OPTIMUM MATCHING FORESTS 2: GENERAL WEIGHTS.

GILE 1977 -3 GILES, F. R.
OPTIMUM MATCHING FORESTS 3: FACETS OF MATCHING FOREST POLYHEDRA.

GILLC 1975 -1 GILLES, F. R.
SUBMODULAR FUNCTIONS, GRAPHS AND INTEGER POLYHEDRA.

GOODC 1976 -1 GOODMAN, S.; HEDETNIEMI, S.; TARJAN, R. E.
B-MATCHINGS IN TREES.

GORDA 1971 -1 GORDON, B. E.
MAXIMUM MATCHING PROBLEM - A COMPARISON OF THE EDMONDS AND BALINSKI ALGORITHMS, THE.

GRAVA 1977 -1 GRAVER, J. E.; WATKINS, M. E.
COMBINATORICS WITH EMPHASIS ON THE THEORY OF GRAPHS.

GREEB 1974 -1 GREENE, C.; KLEITMAN, D. J.; MAGNANTI, T. L.
COMPLEMENTARY TREES AND INDEPENDENT MATCHINGS.

GROE 1977 -3 GROETSCHEL, M.
MONOTONE 2-MATCHING POLYTOPE ON A COMPLETE GRAPH, THE.

GROE 1977 -6 GROETSCHEL, M.
POLYEDRISCHE CHARAKTERISIERUNGEN KOMBINATORISCHER OPTIMIERUNGSPROBLEME.

GRUN 1974 -1 GRUNBAUM, B.
MATCHINGS IN POLYTOPAL GRAPHS.

HAUS 1978 -1 HAUSMANN, D.
ADJAZENZCHARAKTERISIERUNGEN FUER POLYEDERECKEN IN DER KOMBINATORISCHEN OPTIMIERUNG.

HAUSA 1977 -1 HAUSMANN, D.; JENKYNS, T. A.; KORTE, B.
WORST CASE ANALYSIS OF GREEDY TYPE ALGORITHMS FOR INDEPENDENCE SYSTEMS.

HAUSC 1976 -1 HAUSMANN, D.; KORTE, B.
COLOURING CRITERIA FOR ADJACENCY ON 0-1-POLYHEDRA.

HOFFA 1978 -1 HOFFMAN, A. J.; OPPENHEIM, R.
LOCAL UNIMODULARITY IN THE MATCHING POLYTOPE.

HOPCA 1971 -1 HOPCROFT, J. E.; KARP. R. M.
(N POWER 5/2) ALGORITHM FOR MAXIMUM MATCHINGS IN BIPARTITE GRAPHS, A.

ITAI 1977 -2 ITAI, A.; RODEH, M.
SOME MATCHING PROBLEMS.

JENK 1974 -1 JENKYNS, T. A.
MATCHOIDS: A GENERALIZATION OF MATCHINGS AND MATROIDS.

KARI 1976 -1 KARIV, O.
O(N POWER 2.5) ALGORITHM FOR FINDING A MAXIMUM MATCHING IN A GENERAL GRAPH, AN.

KNUT 1976 -1 KNUTH, D. E.
MARIAGES STABLES ET LEURS RELATIONS AVEC D'AUTRES PROBLEMES COMBINATOIRES.

KORTC 1978 -1 KORTE, B.; HAUSMANN, D.
ANALYSIS OF THE GREEDY HEURISTIC FOR INDEPENDENCE SYSTEMS, AN.

KORTC 1977 -1 KORTE, B.; HAUSMANN, D.
ANALYSE DER GREEDY-HEURISTIK FUER UNABHAENGIGKEITSSYSTEME, EINE.

LOVA 1975 -1 LOVASZ, L.
ON THE RATIO OF OPTIMAL INTEGRAL AND FRACTIONAL COVERS.

MATCHING

****

LOVA 1975 -2 LOVASZ, L.
2-MATCHINGS AND 2-COVERS OF HYPERGRAPHS.

MAC 1977 -1 MAC ALLISTER, J. D.; SAKAROVITCH, M.
CLASSIFICATION DE CERTAIN MATRICES 0-1.

MAURB 1976 -1 MAURRAS, J. F.
POLYTOPES A SOMMETS DANS (0,1)N.

MENG 1974 -1 MENG, D.
MATCHINGS AND COVERINGS FOR GRAPHS.

MINIA 1975 -1 MINIEKA, F.
MAXIMAL NETWORK FLOWS, MATROIDS AND MATCHINGS.

MINOA 1976 -1 MINOUX, M.
CARACTERISATION DES COUPLAGES DE POIDS MAXIMAL ET DE CARDINALITE FIXEE DANS LES GRAPHES PONDERES FINIS, UNE.

MITC 1975 -1 MITCHELL, S.; HEDETNIEMI, S.; GOODMAN, S.
SOME LINEAR ALGORITHMS ON TREES.

NIEM 1974 -1 NIEMINEN, J.
LINEAR PSEUDO-BOOLEAN VIEWPOINT ON MATCHING AND OTHER CENTRAL CONCEPTS IN GRAPH THEORY, A.

PADB 1971 -1 PADBERG, M.
SIMPLE ZERO-ONE PROBLEMS: SET COVERING, MATCHING AND COVERING IN GRAPHS.

PAYA 1977 -1 PAYAN, C.
SUR QUELQUES PROBLEMES DE COUVERTURE ET DE COUPLAGE EN COMBINATOIRE.

PIPP 1976 -1 PIPPENGER, N.; VALIANT, L. G.
SHIFTING GRAPHS AND THEIR APPLICATIONS.

PULL 1977 -1 PULLEYBLANK, W. R.
DUAL INTEGRALITY IN B-MATCHING PROBLEMS.

PULL 1973 -1 PULLEYBLANK, W. R.
MULTI OPTIMIZATION IN MATCHING PROBLEMS.

PULLA 1974 -1 PULLEYBLANK, W. R.; EDMONDS, J.
FACETS OF 1-MATCHING POLYHEDRA.

REYN 1977 -1 REYNER, W. S.
ANALYSIS OF A GOOD ALGORITHM FOR THE SUBTREE PROBLEM, AN.

SERD 1975 -1 SERDYUKOV, A. I.
ON A COVERING PROBLEM (RUSSIAN).

STEFA 1977 -1 STEFFENS, K.
MATCHINGS IN COUNTABLE GRAPHS.

STRI 1970 -1 STRICKER, R.
PUBLIC SECTOR VEHICLE ROUTING: THE CHINESE POSTMAN PROBLEM.

TOBI 1975 -1 TOBIN, R. L.
MINIMAL COMPLETE MATCHINGS AND NEGATIVE CYCLES.

TOLB 1975 -1 TOLBIN, R. L.
MINIMAL COMPLETE MATCHINGS AND NEGATIVE CYCLES.

UHRY 1975 -1 UHRY, J. P.
SUR LE PROBLEME DU COUPLAGE MAXIMAL.

URQU 1967 -1 URQUHART, R. J.
DEGREE CONSTRAINED SUBGRAPHES OF LINEAR GRAPHES.

VERGC 1975 -1 VERGNAS, M. LAS
NOTE ON MATCHINGS IN GRAPHS, A.

WEINB 1976 -1 WEINBERGER, D. B.
NETWORK FLOWS, MINIMUM COVERINGS, AND THE FOUR-COLOR CONJECTURE.

WEIND 1974 -1 WEINSTEIN, J.
LARGE MATCHINGS IN GRAPHS.

WERR 1974 -1 WERRA, D. DE
PARTITIONS OF GRAPHS INTO COVERINGS AND HYPERGRAPHS INTO TRANSVERSALS.

WHITB 1977 -1 WHITE, D. E.; WILLIAMSON, S. G.
RECURSIVE MATCHING ALGORITHMS AND LINEAR ORDERS ON THE SUBSET LATTICE.

WHITD 1967 -2 WHITE, L. J.
PARAMETRIC STUDY OF MATCHINGS AND COVERINGS IN WEIGHTED GRAPHS, A.

AIGN 1976 -1 AIGNER, M.
KOMBINATORIK II. MATROIDE UND TRANSVERSALTHEORIE.

BALLA 1977 -1 BALL, M. O.; NEMHAUSER, G. L.
MATROIDS AND A RELIABILITY ANALYSIS PROBLEM.

BENNB 1976 -1 BENNINGER, H.; LIEBLING, T. M.
OPTIMIERUNGSPROBLEME MIT POLYNOMIAL BEGRENZTEN ALGORITHMEN.

BIXB 1977 -1 BIXBY, R. E.
KURATOWSKI'S AND WAGNER'S THEOREMS FOR MATROIDS.

BIXB 1978 -1 BIXBY, R. E.
SOLUTION TO A MATROID PROBLEM OF KNUTH, THE.

BLAN 1977 -1 BLAND, R. G.
COMBINATORIAL ABSTRACTION OF LINEAR PROGRAMMING, A.

BLANA 1978 -1 BLAND, R. G.; VERGNAS, M. LAS
ORIENTABILITY OF MATROIDS.

BOWM 1974 -1 BOWMAN, V. J.; STARR, J. H.
TWO-COMPARABLE PRIME IMPLICANTS AND CANONICAL SWITCHING FUNCTIONS.

BRUA 1975 -1 BRUALDI, R. A.
MATROIDS INDUCED BY DIRECTED GRAPHS, A SURVEY.

BRUA 1975 -2 BRUALDI, R. A.
TRANSVERSAL THEORY AND GRAPHS.

BRUN 1968 -1 BRUNO, J.
MATROIDS, GRAPHS, AND RESISTANCE NETWORKS.

BRUND 1976 -1 BRUNO, J.; WEINBERG, L.
GENERALIZED NETWORKS: NETWORKS EMBEDDED ON A MATROID, PART 2.

BRUND 1976 -2 BRUNO, J.; WEINBERG, L.
GENERALIZED NETWORKS: NETWORKS EMDEDDED ON A MATROID, PART I.

BRUT 1970 -1 BRUTER, C. P.
MATROIDES. NOUVEL OUTIL MATHEMATIQUE, LES.

CAMEA 1974 -1 CAMERINI, P. M.; MAFFIOLI, F.
BOUNDS FOR 3-MATROID INTERSECTION PROBLEMS.

CHANF 1977 -1 CHANDRASEKARAN, R.
MINIMAL RATIO SPANNING TREES.

CHVA 1977 -1 CHVATAL, V.
DETERMINING THE STABILITY NUMBER OF A GRAPH.

CRAP 1970 -1 CRAPO, H. H.; ROTA, G. C.
ON THE FOUNDATIONS OF COMBINATORIAL THEORY: COMBINATORIAL GEOMETRIES.

CUNN 1975 -1 CUNNINGHAM, W.
CHORDS AND DISJOINT PATHS IN MATROIDS.

CUNNA 1975 -1 CUNNINGHAM, W. H.
UNBOUNDED MATROID INTERSECTION POLYHEDRON, AN.

DAVI 1975 -1 DAVIES, J.
ON BASES OF INDEPENDENCE STRUCTURES INTERSECTING IN A SET OF PRESCRIBED CARDINALITY.

DOWL 1977 -1 DOWLING, T. A.; DENIG, W. A.
GEOMETRIES ASSOCIATED WITH PARTIALLY ORDERED SETS.

EDMO 1977 -1 EDMONDS, J.
MATROID INTERSECTION.

EDMO 1975 -1 EDMONDS, J.
SOME WELL-SOLVED PROBLEMS IN COMBINATORIAL OPTIMIZATION.

EDMO 1968 -1 EDMONDS, J.
MATROID PARTITION.

EDMOA 1977 -1 EDMONDS, J.; GILES, R.
MIN-MAX RELATION FOR SUBMODULAR FUNCTIONS ON GRAPHS, A.

EULE 1978 -1 EULER, R.
GENERATING MAXIMUM MEMBERS OF INDEPENDENCE SYSTEMS.

EULE 1977 -1 EULER, R.
BESTIMMUNG MAXIMALER ELEMENTE IN MONOTONEN MENGENSYSTEMEN, DIE.

EULE 1978 -2 EULER, R.
ON RANK FUNCTIONS OF GENERAL INDEPENDENCE SYSTEMS.

FISHB 1977 -1 FISHER, M. L.; NEMHAUSER, G. L.; WOLSEY, L. A.
ANALYSIS OF APPROXIMATIONS FOR MAXIMIZING SUBMODULAR FUNCTIONS.

FOLD 1975 -1 FOLDES, S.; HAMMER, P. L.
ON A CLASS OF MATROID-PRODUCING GRAPHS.

FOUR 1971 -1 FOURNIER, J. C.
REPRESENTATION SUR UN CORPS DES MATROIDES D'ORDRE <8 OU =8.

FRIEA 1977 -1 FRIEZE, A. M.
ALGORITHMS FOR GENERALISATIONS OF MATROIDS.

FRIEA 1977 -2 FRIEZE, A. M.
GENERALISATION OF THE GREEDY ALGORITHM FOR MATROIDS, A.

GABOA 1976 -1 GABOW, H.
DECOMPOSING SYMMETRIC EXCHANGES IN MATROID BASES.

GILLC 1975 -1 GILLES, F. R.
SUBMODULAR FUNCTIONS, GRAPHS AND INTEGER POLYHEDRA.

GOLU 1977 -1 GOLUMBIC, M. C.
COMPARABILITY GRAPHS AND A NEW MATROID.

GOND 1975 -1 GONDRAN, M.
ALGEBRE DES CHEMINS ET ALGORITHMES.

GRAV 1975 -1 GRAVER, J. E.
BOOLEAN DESIGNS AND SELF-DUAL MATROIDS.

GRAVA 1977 -1 GRAVER, J. E.; WATKINS, M. E.
COMBINATORICS WITH EMPHASIS ON THE THEORY OF GRAPHS.

GREEC 1975 -1 GREENE, C.; MAGNANTI, T. L.
SOME ABSTRACT PIVOT ALGORITHMS.

GROE 1977 -1 GROETSCHEL, M.
FACETTEN VON MATROID-POLYTOPEN.

GROE 1977 -6 GROETSCHEL, M.
POLYEDRISCHE CHARAKTERISIERUNGEN KOMBINATORISCHER OPTIMIERUNGSPROBLEME.

HAMMB 1977 -1 HAMMER, P. L.; JOHNSON, E. L.; PELED, U. N.
ROLE OF MASTER POLYTOPES IN THE UNIT CUBE, THE.

HAUS 1978 -1 HAUSMANN, D.
ADJAZENZCHARAKTERISIERUNGEN FUER POLYEDERECKEN IN DER KOMBINATORISCHEN OPTIMIERUNG.

HAUS 1977 -2 HAUSMANN, D.
BENACHBARTE ECKEN AUF 0-1-POLYEDERN.

HAUSA 1977 -1 HAUSMANN, D.; JENKYNS, T. A.; KORTE, B.
WORST CASE ANALYSIS OF GREEDY TYPE ALGORITHMS FOR INDEPENDENCE SYSTEMS.

MATROIDS AND INDEPENDENCE SYSTEMS
****

HAUSC 1977 -1 HAUSMANN, D.; KORTE, B.
LOWER BOUNDS ON THE WORST-CASE COMPLEXITY OF SOME ORACLE ALGORITHMS.

HAUSC 1977 -2 HAUSMANN, D.; KORTE, B.
WORST-CASE BEHAVIOUR OF POLYNOMIAL BOUNDED ALGORITHMS FOR INDEPENDENCE SYSTEMS.

HAUSC 1978 -2 HAUSMANN, D.; KORTE, B.
WORST-CASE ANALYSIS FOR A CLASS OF COMBINATORIAL ALGORITHMS.

HERO 1973 -1 HERON, A. P.
PROPERTY OF THE HYPERPLANES OF A MATROID AND AN EXTENSION OF DILWORTH'S THEOREM, A.

HOLZA 1973 -1 HOLZMANN, C. A.; NORTON, P. G.; TOBEY, M. D.
GRAPHICAL REPRESENTATION OF MATROIDS, A.

INGL 1977 -1 INGLETON, A. W.
TRANSVERSAL MATROIDS AND RELATED STRUCTURES.

INGL 1976 -1 INGLETON, A. W.
NON-BASE-ORDERABLE MATROIDS.

IRI A 1975 -1 IRI, M.; TOMIZAWA, N.
UNIFYING APPROACH TO FUNDAMENTAL PROBLEMS IN NETWORK THEORY BY MEANS OF MATROIDS (JAPANESE), A.

ITAI 1977 -1 ITAI, A.; RODEH, M.
FINDING A MINIMUM CIRCUIT IN A GRAPH.

JENK 1974 -1 JENKYNS, T. A.
MATCHOIDS: A GENERALIZATION OF MATCHINGS AND MATROIDS.

JENK 1976 -1 JENKYNS, T. A.
EFFICACY OF THE "GREEDY" ALGORITHM, THE.

JENK 1977 -1 JENKYNS, T. A.
GREEDY TRAVELLING SALESMAN PROBLEM, THE.

JENK 1977 -2 JENKYNS, T. A.
P-SYSTEMS AS INTERSECTIONS OF MATROIDS.

KEL'A 1976 -1 KEL'MANS, A. K.; LOMONOSOV, M. V.; POLESSKIJ, V. P.
MINIMAL COVERINGS IN MATROIDS (RUSSIAN).

KNUT 1973 -1 KNUTH, D. E.
MATROID PARTITIONING.

KORTB 1977 -1 KORTE, B.
ANALYSE VON HEURISTISCHEN ALGORITHMEN ZUR KOMBINATORISCHEN OPTIMIERUNG.

KORTC 1978 -1 KORTE, B.; HAUSMANN, D.
ANALYSIS OF THE GREEDY HEURISTIC FOR INDEPENDENCE SYSTEMS, AN.

KORTC 1977 -1 KORTE, B.; HAUSMANN, D.
ANALYSE DER GREEDY-HEURISTIK FUER UNABHAENGIGKEITSSYSTEME, EINE.

KROG 1977 -1 KROGDAHL, S.
DEPENDENCE GRAPH FOR BASES IN MATROIDS, THE.

KROG 1976 -1 KROGDAHL, S.
COMBINATORIAL PROOF FOR LAWLER'S MATROID INTERSECTION ALGORITHM, A.

LAWL 1973 -1 LAWLER, E. L.
POLYNOMIAL-BOUNDED AND (APPARANTLY) NON-POLYNOMIAL-BOUNDED MATROID COMPUTATIONS.

LAWL 1971 -1 LAWLER, E. L.
MATROID INTERSECTION ALGORITHMS.

LAWL 1971 -2 LAWLER, E. L.
MATROIDS WITH PARITY CONDITIONS: A NEW CLASS OF COMBINATORIAL OPTIMIZATION PROBLEMS.

LAWL 1975 -4 LAWLER, E. L.
INTRODUCTION TO MATROID OPTIMIZATION, AN.

LIM 1977 -1 LIM, M.-H.
PRODUCT OF MATROIDS AND ITS AUTOMORPHISM GROUP, A.

LINS 1975 -1 LINS, S.
USE OF MATROID THEORY IN OPTIMAL MULTI-ASSIGNMENT OVER TOTALLY ORDERED SETS. THE.

LORE 1976 -1 LOREA, M.
STRUCTURE DE CERTAINS PROBLEMES D'AFFECTATION, LA.

LORE 1975 -1 LOREA, M.
HYPERGRAPHES ET MATROIDES.

LORE 1975 -2 LOREA, M.
MATROIDES ET LEURS APPLICATIONS A DES PROBLEMES DE RECHERCHE OPERATIONNELLE, LES.

MAC C 1975 -1 MAC CARTHY, P. J.
NUMBER OF INDEPENDENT SYSTEMS OF REPRESENTATIVES, THE.

MAC E 1974 -1 MAC DIARMID, C. J. H.
PATH-PARTITION STRUCTURES OF GRAPHS AND DIGRAPHS.

MAC E 1975 -1 MAC DIARMID, C. J. H.
RADO'S THEOREM FOR POLYMATROIDS.

MAFF 1976 -1 MAFFIOLI, F.
SUBGRADIENT OPTIMIZATION, MATROID PROBLEMS AND HEURISTIC EVALUATION.

MAFF 1975 -1 MAFFIOLI, F.
MATROIDS AND HEURISTICALLY GUIDED SEARCH IN PROBLEM SOLVING.

MASO 1977 -1 MASON, J. H.
MATROIDS AS THE STUDY OF GEOMETRICAL CONFIGURATIONS.

MATTA 1978 -1 MATTHEWS, L. R.
CLOSURE IN INDEPENDENCE SYSTEMS.

MATTA 1977 -1 MATTHEWS, L. R.
BICIRCU'AR MATROIDS.

MATTA 1978 -2 MATTHEWS, L. R.
MATROIDS ON THE EDGE SETS OF DIRECTED GRAPHS.

MAURA 1973 -1 MAURER, S. B.
MATROID BASIS GRAPHS, II.

MAURB 1976 -1 MAURRAS, J. F.
POLYTOPES A SOMMETS DANS (0,1)N.

MILN 1975 -1 MILNER, E. C.
TRANSVERSAL THEORY.

MINI 1976 -1 MINIEKA, E.
FINDING THE CIRCUITS OF A MATROID.

MINIA 1975 -1 MINIEKA, F.
MAXIMAL NETWORK FLOWS, MATROIDS AND MATCHINGS.

MINOA 1976 -5 MINOUX, M.
STRUCTURES ALGEBRIQUES GENERALISEES DES PROBLEMES DE CHEMINEMENT DANS LES GRAPHES.

MINT 1975 -1 MINTY, G. J.
ON THE AXIOMATIC FOUNDATIONS OF THE THEORIES OF DIRECTED LINEAR GRAPHS, ELECTRICAL NETWORKS AND NETWORK-PROGRAMMING.

MIRS 1971 -1 MIRSKY, L.
TRANSVERSAL THEORY.

****

MULLA 1978 -1 MULLIN, R. C.; STANTON, R. G.
CHARACTERIZATION OF PSEUDO-AFFINE DESIGNS AND THEIR RELATIONS TO A PROBLEM OF CORDES, A.
MURTC 1971 -1 MURTY, U. S. R.
ON THE NUMBER OF BASES OF A MATROID.
NASH 1967 -1 NASH-WILLIAMS, C. ST. J. A.
APPLICATION OF MATROIDS TO GRAPH THEORY, AN.
NEMH 1976 -1 NEMHAUSER, G. L.; WOLSEY, L. A.
BEST ALGORITHMS FOR APPROXIMATING THE MAXIMUM OF A SUBMODULAR SET FUNCTION.
NEMHA 1976 -1 NEMHAUSER, G. L.; WOLSEY, L. A.; FISHER, M. L.
ANALYSIS OF APPROXIMATIONS FOR MAXIMIZING SUBMODULAR SET FUNCTIONS, AN.
OUBI 1976 -1 OUBINA, L.
LOCALISATION D'UNE PROPRIETE DU GRAPHE DES BASES D'UNE MATROIDE BINAIRE.
PELE 1976 -1 PELED, U. N.
MATROIDAL GRAPHS.
RECS 1976 -1 RECSKI, A.
MATROIDS AND INDEPENDENT STATE VARIABLES.
RECS 1976 -2 RECSKI, A.
MATROIDS AND N-PORTS.
SCHR 1976 -1 SCHRIJVER, L.
LINKING SYSTEMS, MATROIDS AND BIPARTITE GRAPHS.
SEYM 1977 -1 SEYMOUR, P. D.
MATROIDS WITH THE MAX-FLOW MIN-CUT PROPERTY, THE.
SEYM 1976 -1 SEYMOUR, P. D.
MAX-FLOW MIN-CUT PROPERTY IN MATROIDS, THE.
SEYM 1975 -1 SEYMOUR, P. D.
MATROIDS, HYPERGRAPHS AND THE MAX-FLOW MIN-CUT THEOREM.
SHUO 1977 -1 SHUO-YEN, R. L.
EXTREMAL PROBLEM AMONG SUBSETS OF A SET, AN.
SIMO 1974 -1 SIMOES-PEREIRA, J. M. S.
MATROIDS, GRAPHS AND TOPOLOGY.
STEFA 1976 -1 STEFFENS, K.
ERGEBNISSE AUS DER TRANSVERSALENTHEORIE, I + II.
TODD 1976 -1 TODD, M. J.
COMBINATORIAL GENERALIZATION OF POLYTOPES, A.
TOMI 1973 -1 TOMIZAWA, N.
ON A SPECIALIZATION SEQUENCE FROM GENERAL MATROIDS TO LADDER GRAPHS WITH SPECIAL EMPHASIS ON THE CHARACTERIZATION OF LADDER MATROIDS.
TROT 1977 -1 TROTTER, L. E. JR.
ON THE GENERALITY OF MULTI-TERMINAL FLOW THEORY.
TROTD 1976 -1 TROTTER, W. T. JR.; MOORE, J. I. JR.
CHARACTERIZATION PROBLEMS FOR GRAPHS, PARTIALLY ORDERED SETS, LATTICES, AND FAMILIES OF SETS.
VERGC 1976 -1 VERGNAS, M. LAS
ON CERTAIN CONSTRUCTIONS FOR MATROIDS.
WEINA 1977 -1 WEINBERG, L.
MATROIDS, GENERALIZED NETWORKS, AND ELECTRIC NETWORK SYNTHESIS.
WEINB 1976 -2 WEINBERGER, D. B.
TRANSVERSAL MATROID INTERSECTIONS AND RELATED PACKINGS.
WELS 1976 -1 WELSH, D. J. A.
MATROID THEORY.
WELS 1970 -1 WELSH, D. J. A.
ON MATROID THEOREMS OF EDMONDS AND RADO.
WILSC 1973 -1 WILSON, R. J.
INTRODUCTION TO MATROID THEORY, AN.
WOOD 1974 -1 WOODALL, D. R.
EXCHANGE THEOREM FOR BASES OF MATROIDS, AN.
WOOD 1975 -1 WOODALL, D. R.
INDUCTION OF MATROIDS BY GRAPHS, THE.
WOOD 1977 -1 WOODALL, D. R.
NOTE ON RANK FUNCTIONS AND INTEGER PROGRAMMING, A.
ZIMMA 1977 -1 ZIMMERMANN, U.
DUALITY AND THE ALGEBRAIC MATROID INTERSECTION PROBLEM.
ZIMMA 1976 -1 ZIMMERMANN, U.
BOOLE'SCHE OPTIMIERUNGSPROBLEME MIT SEPARABLER ZIELFUNKTION UND MATROIDALEN RESTRIKTIONEN.
ZIMMA 1977 -2 ZIMMERMANN, U.
SOME PARTIAL ORDERS RELATED TO BOOLEAN OPTIMIZATION AND THE GREEDY-ALGORITHM.
ZIMMA 1977 -3 ZIMMERMANN, U.
THRESHOLD METHODS FOR BOOLEAN OPTIMIZATION PROBLEMS WITH SEPARABLE OBJECTIVES.

MIXED INTEGER PROGRAMMING
****

ABAD 1976 -1 ABADIE, J.; DAYAN, H.; AKOKA, J.
QUELQUES EXPERIENCES NUMERIQUES SUR LA PROGRAMMATION NON LINEAIRE EN NOMBRES ENTIERS.

ALBE 1977 -1 ALBERS, S.
MIXED INTEGER NONLINEAR PROGRAMMING PROCEDURE FOR SIMULTANEOUSLY LOCATING MULTIPLE PRODUCTS IN AN ATTRIBUTE SPACE, A.

ANDRA 1976 -1 ANDRONICO, A.; RANDI, P.; SCALAS, M. R.
INTERACTIVE CODE FOR MIXED INTEGER LINEAR PROGRAMS, AN.

ARMS 1978 -1 ARMSTRONG, R. D.
NONBASIC BRANCHING AND DUAL RELAXATION IN A BRANCH-AND-BOUND MIXED INTEGER PROGRAMMING ALGORITHM.

ARMSE 1975 -1 ARMSTRONG, R. D.; WILLIS, C. E.
SIMULTANEOUS INVESTMENT-ALLOCATION - AN APPLICATION OF GENERALIZED BENDERS DECOMPOSITION TO WATER PLANNING.

BACHB 1977 -1 BACHEM, A.; SCHRADER, R.
MINIMAL INEQUALITIES AND DUALITY FOR MIXED INTEGER PROGRAMS.

BECKC 1976 -1 BECKMANN, M. J.; BURNS, L. D.
THEORY OF HOUSEHOLD AUTOMOBILE ALLOCATION DECISIONS, A.

BENI 1977 -1 BENICHOU, M.; GAUTHIER, J. M.; HENTGES, G.; RIBIERE, G.
EFFICIENT SOLUTION OF LARGE-SCALE LINEAR PROGRAMMING PROBLEMS - SOME ALGORITHMIC TECHNIQUES AND COMPUTATIONAL RESULTS, THE.

BENNA 1972 -1 BENNETT, J. M.; EDWARDS, J.
EXPERIENCE WITH A METHOD FOR HANDLING LARGE STRUCTURAL LINEAR PROGRAMMING PROBLEMS WITH INTEGER RESTRICTIONS.

BESW 1977 -1 BESWICK, C. A.
ALLOCATING SELLING EFFORT VIA DYNAMIC PROGRAMMING.

BLAIB 1977 -1 BLAIR, C. E.; JEROSLOW, R. G.
VALUE FUNCTION OF A MIXED INTEGER PROGRAM: I, THE.

BRIC 1977 -1 BRICKER, D. L.
REFORMULATION OF SPECIAL ORDERED SETS FOR IMPLICIT ENUMERATION ALGORITHMS, WITH APPLICATIONS IN NONCONVEX SEPARABLE PROGRAMMING.

BROC 1976 -1 BROCKLEHURST, E. R.
GENERALIZED BRANCH AND BOUND: A METHOD FOR INTEGER LINEAR PROGRAMS.

BURK 1973 -1 BURKARD, R. E.
METHOD FOR MIXED-INTEGER CONVEX PROGRAMMING, A.

CHERE 1975 -1 CHERVAK, N. K.
FINITE ALGORITHM FOR THE TRUNCATION METHOD FOR INTEGER LINEAR PROGRAMMING PROBLEMS WITH NONINTEGER TARGET FUNCTION (RUSSIAN), A.

CHERF 1973 -1 CHERVAK, YU. YU.
METHOD OF LEXICOGRAPHIC SEARCH FOR THE SOLUTION OF DISCRETE LINEAR PROGRAMMING PROBLEMS (RUSSIAN), THE.

DEBA 1976 -1 DEBANNE, J. G.
GENERALIZED NETWORK AND INTEGER PROGRAMMING IN TECHNOLOGY ASSESSMENT AND PLANNING AN ENERGY SCENARIO STUDY.

EL-S 1975 -1 EL-SHAIEB, A. M.
MIXED-INTEGER PROGRAMMING MODEL FOR ECONOMIC RESIDENTIAL ENERGY USE, A.

ERLE 1976 -2 ERLENKOTTER, D.
FACILITY LOCATION WITH PRICE-SENSITIVE DEMANDS: PRIVATE, PUBLIC, AND QUASI-PUBLIC.

ESCU 1975 -1 ESCUDERO, L. F.
CONTINUOUS, INTEGER, AND MIXED LINEAR PROGRAMMING, VOL. 1 (SPANISH).

FOX B 1977 -1 FOX, B. L.; LENSTRA, J. K.; RINNOOY KAN, A. H. G.; SCHRAGE, L. E.
BRANCHING FROM THE LARGEST UPPER BOUND: FOLKLORE AND FACTS.

GARTA 1975 -1 GARTNER, N. H.; LITTLE, J. D. C.; GABBAY, H.
OPTIMIZATION OF TRAFFIC SIGNAL SETTINGS BY MIXED-INTEGER LINEAR PROGRAMMING. PART I: THE NETWORK COORDINATION PROBLEM, PART II: THE NETWORK SYNCHRONIZATION PROBLEM.

GATE 1971 -1 GATELY, D. T.
INVESTMENT PLANNING FOR THE ELECTRIC POWER INDUSTRY: A MIXED-INTEGER PROGRAMMING APPROACH, WITH APPLICATIONS TO SOUTHERN INDIA.

GAUT 1977 -1 GAUTHIER, J.-M.; RIBIERE, G.
EXPERIMENTS IN MIXED-INTEGER LINEAR PROGRAMMING USING PSEUDOCOSTS.

GEOFA 1973 -1 GEOFFRION, A. M.; GRAVES, G. W.
MULTICOMMODITY DISTRIBUTION SYSTEM DESIGN BY BENDERS DECOMPOSITION.

GLOV 1973 -1 GLOVER, F.
POLYHEDRAL ANNEXATION IN MIXED INTEGER PROGRAMMING.

GLOV 1975 -3 GLOVER, F.
POLYHEDRAL ANNEXATION IN MIXED INTEGER AND COMBINATORIAL PROGRAMMING.

GLOVI 1976 -1 GLOVER, F.; SOMMER, D.
INEQUALITIES FOR MIXED INTEGER PROGRAMS WITH STRUCTURE.

GODE 1976 -1 GODEFROID, P.
LOESUNG EINES GEMISCHT-GANZZAHLIGEN MODELLS ZUR OPTIMALEN BAUFINANZIERUNG, DIE.

GRANA 1977 -1 GRANOT, D.; GRANOT, F.
ON INTEGER AND MIXED INTEGER FRACTIONAL PROGRAMMING PROBLEMS.

GULD 1977 -1 GULDMANN, J.-M.; SHEFER, D.
OPTIMAL PLANT LOCATION AND AIR QUALITY MANAGEMENT UNDER INDIVISIBILITIES AND ECONOMIES OF SCALE.

HARTC 1975 -1 HARTMANN, K.
GEMISCHT GANZZAHLIGE LINEARE QUOTIENTENOPTIMIERUNG NACH DEM SCHNITTVERFAHREN VON GOMORY.

HEIS 1977 -1 HEISING, C. D.
PLANT LOCATION AND TRANSPORTATION MODELS OF THE LIGHT-WATER REACTOR FUEL CYCLE.

HROU 1975 -1 HROUDA, J.
MODIFIED BENDERS' ALGORITHM (CZECH).

HULS 1977 -1 HULST, W. G. VAN; LIESHOUT, J. TH. VAN
INVESTMENT/FINANCIAL PLANNING WITH ENDOGENOUS LIFETIMES: A HEURISTIC APPROACH TO MIXED-INTEGER PROGRAMMING.

IBM. 1972 -1 IBM.
INTRODUCTION TO MODELLING USING MIXED INTEGER PROGRAMMING, AN.

IVANA 1976 -1 IVANCHEV, M. D.; MITEV, I. G.; JANEV, N. I.
REALIZATION OF THE BRANCH AND BOUND METHOD FOR THE SOLUTION OF THE GENERAL MIXED INTEGER PROGRAMMING PROBLEM (RUSSIAN), A.

JOHNE 1976 -1 JOHNSON, E. L.
FACES OF POLYHEDRA OF MIXED INTEGER PROGRAMMING PROBLEMS.

KING 1976 -1 KINGSTON, P. L.; LIPTON, S. L.; STOJKA, J. P. JR.
MIXED INTEGER PROGRAMMING MODELS FOR GENERATOR MAINTENANCE SCHEDULING.

KNAP 1976 -1 KNAPEK, S.
EINSATZ HEURISTISCHER VERFAHREN IN DER GEMISCHT-GANZZAHLIGEN INVESTITIONS- UND FINANZIERUNGSPLANUNG, DER.

****

KOEHA 1977 -1 KOEHLER, J.
NUTZUNG DER DUALITAET ZUR LOESUNG EINES GEMISCHTEN LINEAREN 0-1-OPTIMALPROBLEMS.
KUEH 1977 -1 KUEHNER, J.; HARRINGTON, J. J.
WIRKSAME ANWENDUNG DES ALGORITHMUS FUER DIE GEMISCHT-GANZZAHLIGE PROGRAMMIERUNG MPSX-MIP VON IBM ZUR LOESUNG VON GROSSEN OEFFENTLICHEN INVESTITIONSPROBLEMEN MIT VIELEN BETEILIGTEN, TEIL II, DIE.
KUEH 1977 -2 KUEHNER, J.; HARRINGTON, J. J.
WIRKSAME ANWENDUNG DES ALGORITHMUS FUER DIE GEMISCHT-GANZZAHLIGE PROGRAMMIERUNG MPSX-MIP VON IBM ZUR LOESUNG VON GROSSEN OEFFENTLICHEN INVESTITIONSPROBLEMEN MIT VIELEN BETEILIGTEN, TEIL 1, DIE.
LAUR 1975 -1 LAURIA, D. T.
REGIONAL SEWERAGE PLANNING BY MIXED INTEGER PROGRAMMING.
LEBEB 1977 -1 LEBEDEV, V. YU.
METHOD FOR THE SOLUTION OF MIXED-INTEGER MATHEMATICAL PROGRAMMING PROBLEMS (RUSSIAN), THE.
LEBEC 1977 -1 LEBEDEVA, T. T.; ROSHCHIN, V. A.; SERGIENKO, I. V.
APPROACH TO SOLUTION OF MATHEMATICAL PROGRAMMING PROBLEMS WITH MIXED VARIABLES (RUSSIAN), AN.
LEBL 1975 -1 LEBLANC, L. J.
ALGORITHM FOR THE DISCRETE NETWORK DESIGN PROBLEM, AN.
LIND 1973 -1 LIND, C. B.; LORENTSEN, R.
APPLICATION OF A MIXED INTEGER LINEAR PROGRAMMING MODEL FOR THE DIMENSIONING OF TANK TRUCK FLEETS.
LIPT 1976 -1 LIPTON, S. L.
SERVICE CENTER LOCATION USING MIXED INTEGER PROGRAMMING.
LOVEA 1976 -2 LOVE, R. F.; WONG, J. Y.
SOLVING QUADRATIC ASSIGNMENT PROBLEMS WITH RECTANGULAR DISTANCES AND INTEGER PROGRAMMING.
LYON 1977 -1 LYONS, J. P.; YOUNG, J. P.
DEVELOPMENT AND IMPLEMENTATION OF A STAFF ALLOCATION MODEL FOR MENTAL HEALTH FACILITIES.
MAC B 1976 -1 MAC CALL, E. H.
ZERO-ONE MIXED INTEGER PROGRAMMING USING INTERACTIVE GRAPHICS.
MAC D 1977 -1 MAC DANIEL, D.; DEVINE, M.
MODIFIED BENDERS' PARTITIONING ALGORITHM FOR MIXED INTEGER PROGRAMMING, A.
MEYEC 1976 -1 MEYER, R. R.
MIXED INTEGER MINIMIZATION MODELS FOR PIECEWISE-LINEAR FUNCTIONS OF A SINGLE VARIABLE.
MUCK 1977 -1 MUCKSTADT, J. A.; KOENIG, S. A.
APPLICATION OF LAGRANGIAN RELAXATION TO SCHEDULING IN POWER-GENERATION SYSTEMS, AN.
NOON 1977 -1 NOONAN, F.; GIGLIO, R. J.
PLANNING ELECTRIC POWER GENERATION: A NONLINEAR MIXED INTEGER MODEL EMPLOYING BENDERS DECOMPOSITION.
ORON 1976 -1 ORON, G.; KARMELI, D.; WALKER, W. R.
APPROXIMATION SOLUTION OF NON-LINEAR MIXED-INTEGER OPTIMIZATION, AN.
PALM 1974 -1 PALM, K.-P.
BEMERKUNG ZUM GEMISCHT-GANZZAHLIGEN SPIELBERG-MODELL, EINE.
PHIL 1976 -1 PHILLIPS, L. W.; UNGER, P. S.
MATHEMATICAL PROGRAMMING SOLUTION OF A HOIST SCHEDULING PROGRAM.
RARDA 1976 -3 RARDIN, R. L.; UNGER, V. E.
SURROGATE CONSTRAINTS AND THE STRENGTH OF BOUNDS DERIVED FROM 0-1 BENDERS' PARTITIONING PROCEDURES.
RICH 1976 -1 RICHARDSON, R.
OPTIMIZATION APPROACH TO ROUTING AIRCRAFT, AN.
RICHB 1976 -1 RICHTER, D.
ALGORITHMUS ZUR LOESUNG GEMISCHT GANZZAHLIGER KONVEXER OPTIMIERUNGSPROBLEME, EIN.
RICHB 1976 -2 RICHTER, D.
CONTRIBUTION TO AN ALGORITHM FOR SOLVING MIXED INTEGER COVEX PROGRAMMING PROBLEMS, A.
ROCH 1976 -1 ROCHAT, J. C.
ON AN OPTIMIZATION PROBLEM IN QUEUEING.
ROSH 1977 -1 ROSHCHIN, V. A.; SERGIENKO, I. V.
ON A CONVERGENT ALGORITHM FOR SOLVING THE MATHEMATICAL PROGRAMMING PROBLEMS WITH MIXED VARIABLES (RUSSIAN).
SCAR 1977 -1 SCARBROUGH, J. E.
MIXED-INTEGER PROGRAMMING MODEL FOR THE SCHEDULING OF AIRCRAFT BY TIME OF DAY WITHIN A NETWORK OF AIRPORTS, A.
SCHI 1970 -1 SCHINZINGER, R.
ON THE REDUCTION OF THE CONTINUOUS VARIABLE COMPONENT IN MIXED INTEGER LINEAR PROGRAMMING.
SHAPB 1974 -1 SHAPIRO, J. F.
NEW DUALITY THEORY FOR MIXED INTEGER PROGRAMMING, A.
SHAPB 1976 -1 SHAPIRO, J. F.
MULTIPLE CRITERIA PUBLIC INVESTMENT DECISION MAKING BY MIXED INTEGER PROGRAMMING.
SHAPB 1976 -2 SHAPIRO, J. F.
NEW CONSTRUCTIVE DUALITY THEORY FOR MIXED INTEGER PROGRAMMING, A.
SIEL 1976 -1 SIELKEN, R. M. JR.
SEQUENCING WITH SETUP COSTS BY ZERO-ONE MIXED INTEGER LINEAR PROGRAMMING.
SLAT 1976 -1 SLATE, L.; SPIELBERG, K.
EXTENDED CONTROL LANGUAGE OF MPSX/370 (AND POSSIBLE APPLICATIONS), THE.
SMITA 1976 -1 SMITH, E. M.; OLSON, R. E.
MIXED INTEGER LINEAR PROGRAMMING MODEL APPLIED TO PETROLEUM PRODUCTS BLENDING.
SMITC 1976 -1 SMITH, K. R.; OVER, A. M. JR.; HANSEN, M. F.; GOLLADAY, F. L.; DAVENPORT, E. J.
ANALYTIC FRAMEWORK AND MEASUREMENT STRATEGY FOR INVESTIGATING OPTIMAL STAFFING IN MEDICAL PRACTICE.
STEIB 1976 -1 STEINER, J.
ZUR KONSTRUKTION VON SCHNITTEN FUER GEMISCHT-GANZZAHLIGE LO-PROBLEME.
STYC 1974 -1 STYCZEN, K.
TRANSPORTATION SYSTEM IN THE RAW MATERIAL PREPARATION DEPARTMENT OF A CEMENT PLANT: MATHEMATICAL MODEL AND CONTROL ALGORITHM (POLISH), A.
SWEEA 1976 -1 SWEENEY, D. J.; TATHAM, R. L.
IMPROVED LONG-RUN MODEL FOR MULTIPLE WAREHOUSE LOCATION, AN.
SZWA 1974 -1 SZWARC, W.
MIXED INTEGER LINEAR PROGRAMMING PROBLEM WHEN THE INTEGER VARIABLES ARE ZERO OR ONE, THE.
TONE 1976 -1 TONE, K.
PARAMETRIZATION OF BENDERS DECOMPOSITION, A.
TRUS 1976 -1 TRUSCOTT, W. G.
NOTE ON A CONCEPTUAL FRAMEWORK FOR DYNAMIC LOCATION-ALLOCATION ANALYSIS, A.
WESOB 1975 -1 WESOLOWSKY, G. O.; TRUSCOTT, W. G.
MULTIPERIOD LOCATION-ALLOCATION PROBLEM WITH RELOCATION OF FACILITIES, THE.

****

WEST 1977 -1 WESTERBERG, C.-H.; BJORKLUND, B.; HULTMAN, E.
APPLICATION OF MIXED INTEGER PROGRAMMING IN A SWEDISH STEEL MILL, AN.

****

ALES 1975 -1 ALESKEROV, S. A.; GRIGOROVIC, A. F.
MULTICOMMODITY FLOWS IN COMMUNICATION NETWORKS (RUSSIAN).

ASSA 1975 -1 ASSAD, A.
MULTICOMMODITY NETWORK FLOWS - A SURVEY.

ASSAA 1976 -1 ASSAD, A. A.
MULTICOMMODITY NETWORK FLOWS - COMPUTATIONAL EXPERIENCE.

CHERC 1976 -1 CHERKASSKII, B. V.
MULTIPORT TWO-PRODUCT PROBLEMS (RUSSIAN).

EVANA 1976 -1 EVANS, J. R.
COMBINATORIAL EQUIVALENCE BETWEEN A CLASS OF MULTICOMMODITY FLOW PROBLEMS AND THE CAPACITATED TRANSPORTATION PROBLEM, A.

EVANA 1977 -2 EVANS, J. R.
SURVEY OF INTEGRALITY IN MULTICOMMODITY NETWORKS, A.

EVANB 1975 -1 EVANS, J. R.; JARVIS, J. J.
IMPROVED FORMULATION AND A HEURISTIC ALGORITHM FOR THE INTEGER MULTICOMMODITY TRANSPORTATION PROBLEM, AN.

EVANB 1977 -1 EVANS, J. R.; JARVIS, J. J.
OUT-OF-KILTER BASED HEURISTIC FOR THE INTEGER MULTICOMMODITY TRANSPORTATION PROBLEM, AN.

EVANC 1976 -1 EVANS, J. R.; MARTIN, K.
FEASIBLE FLOWS IN MULTICOMMODITY NETWORKS WITH LOWER BOUNDS.

EVENA 1975 -1 EVEN, S.; ITAI, A.; SHAMIR, A.
ON THE COMPLEXITY OF TIMETABLE AND MULTI-COMMODITY FLOW PROBLEMS.

GHIU 1975 -1 GHIUS, V.
SUPPLY AND DEMAND THEOREM FOR THE FLOW OF TWO PRODUCTS (RUMANIAN), A.

GOLDB 1975 -1 GOLDEN, B. L.
MINIMUM-COST MULTICOMMODITY NETWORK FLOW PROBLEM CONCERNING IMPORTS AND EXPORTS, A.

GRAT 1970 -1 GRATZER, F.
COMPUTER SOLUTION OF LARGE MULTICOMMODITY FLOW PROBLEMS.

HELG 1977 -2 HELGASON, R. V.; KENNINGTON, J. L.
PRODUCT FORM REPRESENTATION OF THE INVERSE OF A MULTICOMMODITY CYCLE MATRIX, A.

IRI 1971 -1 IRI, M.
ON AN EXTENSION OF THE MAXIMUM-FLOW MINIMUM-CUT THEOREM TO MULTICOMMODITY FLOWS.

KANT 1973 -1 KANT, E.
MULTICOMMODITY FLOW ALGORITHM, A.

KAPL 1977 -1 KAPLAN, E. L.
ALGORITHMS FOR SOLUTION OF TRANSPORTATION PRODUCTION PROBLEMS (RUSSIAN).

KENN 1975 -1 KENNINGTON, J. L.
MULTICOMMODITY NETWORK FLOWS: A STATE-OF-THE-ART SURVEY OF LINEAR MODELS AND SOLUTION TECHNIQUES.

KENN 1977 -1 KENNINGTON, J. L.
SOLVING MULTICOMMODITY TRANSPORTATION PROBLEMS USING A PRIMAL PARTITIONING SIMPLEX TECHNIQUE.

KENN 1977 -2 KENNINGTON, J. L.
STATE-OF-THE-ART SURVEY OF LINEAR COST MULTICOMMODITY NETWORK FLOWS, A (REVISED VERSION).

KENNA 1977 -1 KENNINGTON, J. L.; SHALABY, M.
EFFECTIVE SUBGRADIENT PROCEDURE FOR MINIMAL COST MULTICOMMODITY FLOW PROBLEMS, AN.

MINOA 1975 -2 MINOUX, M.
RESOLUTION DES PROBLEMÉS DE MULTIFLOTS EN NOMBRES ENTIERS DANS LES GRANDS RESEAUX.

MINOA 1976 -3 MINOUX, M.
MULTIFLOTS DE COUT MINIMAL AVEC FONCTIONS DE COUT CONCAVES.

MURCB 1964 -1 MURCHLAND, J. D.
CIRCULATING FLOWS IN MULTI-COMMODITY NETWORKS.

PAPEA 1976 -1 PAPERNOV, B. A.
REALIZABILITY OF MULTIPRODUCT FLOWS (RUSSIAN).

ROTH 1968 -1 ROTHFARB, B.
COMBINATORIC METHODS FOR MULTICOMMODITY FLOWS.

ROTHA 1969 -1 ROTHFARB, R.; FRISCH, I. T.
ON THE THREE-COMMODITY FLOW PROBLEM.

SAHNB 1976 -1 SAHNI, S.; GONZALEZ, T.
P-COMPLETE APPROXIMATION PROBLEMS.

SAIG 1969 -1 SAIGAL, R.
MULTICOMMODITY FLOWS IN DIRECTED NETWORKS.

TCHE 1977 -1 TCHERKASSKY, B. V.
PROBLEM SOLUTION OF MULTIPRODUCT FLOWS IN A NETWORK (RUSSIAN).

TIFF 1974 -1 TIFFIN, J. E.; FOLIE, G. M.
RECENT DEVELOPMENTS IN TRANSPORTATION ALGORITHMS.

TRUE 1977 -2 TRUEMPER, K.
UNIMODULAR MATRICES OF FLOW PROBLEMS WITH ADDITIONAL CONSTRAINTS.

ULRI 1975 -1 ULRICH, L. Y.
HEURISTIC ALGORITHMS FOR SOLVING A LARGE SCALE MULTICOMMODITY FLOW PROBLEM ON A NETWORK WITH A STEP FUNCTION COST.

WEINE 1976 -1 WEINTRAUB, A.
OPTIMAL FLOWS IN NETWORKS AND GAMES: THE MULTICOMMODITY FLOW PROBLEM IN INTEGERS.

MULTI DIMENSIONAL TRANSPORTATION
****

ARET 1974 -1 ARET, L.
SOLUTION OF A CERTAIN MULTI-INDEX PROBLEM OF TRANSPORTATION TYPE (RUSSIAN), THE.

CORB 1972 -1 CORBAN, A.
THREE DIMENSIONAL STOCHASTIC TRANSPORTATION PROBLEM (RUMANIAN), THE.

CORB 1975 -1 CORBAN, A.
MULTIDIMENSIONAL TRANSPORTATION PROBLEM, II (RUMANIAN), A.

CORB 1973 -1 CORBAN, A.
TRANSPORTATION PROBLEMS WITH FIXED COSTS (RUMANIAN).

CORB 1975 -2 CORBAN, A.
TIME CRITERION IN TRANSPORTATION PROBLEMS (RUMANIAN), THE.

EMELC 1974 -1 EMELICHEVA, V. S.; KONONENKO, A. M.
MINIMAL NUMBER OF INTEGER VERTICES OF THE POLYTOPE OF A MULTIINDEX TRANSPORTATION PROBLEM (RUSSIAN), THE.

JUNG 1976 -1 JUNGINGER, W.
NUMERISCHE ERFAHRUNGEN BEI DER LOESUNG DES DREIDIMENSIONALEN TRANSPORTPROBLEMS.

KAPL 1977 -1 KAPLAN, E. L.
ALGORITHMS FOR SOLUTION OF TRANSPORTATION PRODUCTION PROBLEMS (RUSSIAN).

KONO 1976 -1 KONONENKO, A. M.; MIKUL'SKII, V. E.; TRUHANOVSKII, N. N.
SOLVABILITY CONDITIONS FOR THREE-INDEX PLANAR TRANSPORT PROBLEMS (RUSSIAN).

NEY 1972 -1 NEY, P.
RESEARCH ON THE TRANSPORTATION PROBLEM, I (RUMANIAN).

NEY 1973 -1 NEY, P.
RESEARCH ON THE TRANSPORTATION PROBLEM, II (RUMANIAN).

POPEA 1976 -1 POPESCU, D. C.
ALGORITHM FOR SOLVING THE TRANSPORTATION TRIDIMENSIONAL PROBLEM, AN.

SHARC 1978 -1 SHARMA, J. K.; SWARUP, K.
TIME MINIMIZATION IN TRANSPORTATION PROBLEMS.

STAN 1974 -1 STANCU-MINASIAN, I. M.
THREE-DIMENSIONAL TRANSPORTATION PROBLEM WITH A SPECIAL STRUCTURED OBJECTIVE FUNCTION, A.

SWANA 1976 -1 SWANSON, L. W.
N-DIMENSIONAL SCHEDULING.

AASH 1976 -1 AASHTIANI, H. A.; MAGNANTI, T. L.
IMPLEMENTING PRIMAL-DUAL NETWORK FLOW ALGORITHMS.
ADEL 1975 -1 ADEL'SON-VEL'SKIJ, G. M.; DINITS, E. A.; KARZANOV, A. V.
FLOW ALGORITHMS (RUSSIAN).
ADOL 1976 -1 ADOLPHSON, D. L.
NOTE ON MINIMUM COST CONVEX FLOWS, A.
AKAD 1976 -1 AKADEMIA NAUK SSSR, TSENTR. EHKON.-MAT. INST.
SOME QUESTIONS CONCERNING THE STUDY OF TRANSPORTATION FLOWS (RUSSIAN).
AKUL 1977 -1 AKULIC, F. E.
DECOMPOSITION METHOD FOR PROBLEMS OF MAXIMAL FLOW IN LARGE RANGED NETWORKS (RUSSIAN), A.
ALI 1977 -1 ALI, A. I.; HELGASON, R. V.; KENNINGTON, J. L.; LALL, H. S.
PRIMAL SIMPLEX NETWORK CODES: STATE-OF-THE-ART IMPLEMENTATION TECHNOLOGY (REVISED VERSION).
ALIA 1976 -1 ALIA, G.; MAESTRINI, P.
PROCEDURE TO DETERMINE OPTIMAL PARTITIONS OF WEIGHTED HYPERGRAPHS THROUGH A NETWORK-FLOW ANALOGY, A.
AMIN 1973 -1 AMIN, A. T.; HAKIMI, S. L.
MAXIMALLY EDGE-CONNECTED GRAPHS WITH LEAST NUMBER OF MINIMUM EDGE CUT-SETS.
ARIS 1977 -1 ARISAWA, S.; ELMAGHRABY, S. E.
"HUB" AND "WHEEL" SCHEDULING PROBLEMS, I. THE "HUB" SCHEDULING PROBLEM: THE MYOPIC CASE, THE.
ARIS 1977 -2 ARISAWA, S.; ELMAGHRABY, S. E.
"HUB" AND "WHEEL" SCHEDULING PROBLEMS, II. THE HUB OPERATION SCHEDULING PROBLEM (HOSP): MULTI-PERIOD AND INFINITE HORIZON, AND THE WHEEL OPERATION SCHEDULING PROBLEM (WOSP), THE.
BAGB 1976 -1 BAGBY, G.; THESEN, A.
NETWORK FLOW MODEL FOR ALLOCATION OF LAND USES TO SECTORS, A.
BAKO 1976 -2 BAKO, A.
SOLUTION OF THE MULTITERMINAL MINIMAL PATH PROBLEM IN A NETWORK WITH GAINS.
BARTA 1977 -1 BARTHOLDI, J. J. III; ORLIN, J. B.; RATLIFF, H. D.
CIRCULAR 1'S AND CYCLIC STAFFING.
BARTB 1977 -1 BARTHOLDI, J. J. III; RATLIFF, H. D.
FIELD GUIDE TO IDENTIFYING NETWORK FLOW AND MATCHING PROBLEMS, A.
BARTB 1977 -2 BARTHOLDI, J. J. III; RATLIFF, H. D.
UNNETWORKS, WITH APPLICATIONS TO IDLE TIME SCHEDULING.
BAZAC 1977 -1 BAZARAA, M. S.; JARVIS, J. J.
LINEAR PROGRAMMING AND NETWORK FLOWS.
BENNB 1976 -1 BENNINGER, H.; LIEBLING, T. M.
OPTIMIERUNGSPROBLEME MIT POLYNOMIAL BEGRENZTEN ALGORITHMEN.
BHAU 1974 -1 BHAUMIK, G.; JENSON, P.
COMPUTATIONALLY EFFICIENT ALGORITHM FOR THE NETWORK WITH GAINS PROBLEM, A.
BICH 1972 -1 BICHARA, J.
FLUESSE IN NETZWERKEN.
BOYC 1973 -1 BOYCE, D. E.; FARHI, A.; WEISCHEDEL, R.
OPTIMAL NETWORK PROBLEM: A BRANCH-AND-BOUND ALGORITHM.
BOYCB 1973 -1 BOYCE, W. M.; SEERY, J. B.
STEINER 72, AN IMPROVED VERSION OF COCKAYNE AND SCHILLER'S PROGRAM STEINER FOR THE MINIMAL NETWORK PROBLEM.
BREL 1977 -1 BRELAZ, D.
PROBLEME D'HORAIRE DE LABORATOIRE DE LANGUES, UN.
BRUND 1976 -1 BRUNO, J.; WEINBERG, L.
GENERALIZED NETWORKS: NETWORKS EMBEDDED ON A MATROID, PART 2.
BURK 1976 -1 BURKARD, R. E.
FLUESSE IN NETZWERKEN MIT ALLGEMEINEN KOSTEN.
BURKB 1976 -1 BURKARD, R. E.; HAMACHER, H.; ZIMMERMANN, U.
ALGEBRAIC NETWORK FLOW PROBLEM, THE.
BURKB 1977 -1 BURKARD, R. E.; HAMACHER, H.; ZIMMERMANN, U.
BEMERKUNGEN ZUM ALGEBRAISCHEN FLUSSPROBLEM, EINIGE.
BURKB 1977 -2 BURKARD, R. E.; HAMACHER, H.; ZIMMERMANN, U.
FLUSSPROBLEME MIT ALLGEMEINEN KOSTEN.
CARR 1971 -1 CARRE, B. A.
ELIMINATION METHOD FOR MINIMAL-COST NETWORK FLOW PROBLEMS, AN.
CAZA 1975 -1 CAZALET, J.; MEHR, P.
AMELIORATION DE L'UTILISATION D'UNE FLOTTE AERIENNE PARTAGEE ENTRE DEUX BASES PRINCIPALES (GENERATION DE CONVOYAGES).
CHANA 1976 -1 CHAN, Y.
TRANSPORTATION NETWORK INVESTMENT PROBLEM, THE; A SURVEY OF TREE SEARCH SOLUTION ALGORITHMS.
CHEB 1975 -1 CHEBOTAREV, A. S.
SEQUENTIAL OPTIMIZATION IN ONE SCHEDULING PROBLEM (RUSSIAN).
CHENF 1977 -1 CHEN, S.; SAIGAL, R.
PRIMAL ALGORITHM FOR SOLVING A CAPACITATED NETWORK FLOW PROBLEM WITH ADDITIONAL LINEAR CONSTRAINTS, A.
CHEU 1976 -1 CHEUNG, T.-Y.
NETWORK FLOW APPROACH TO STEINER PROBLEMS, A.
CRUM 1976 -1 CRUM, R.
CASH MANAGEMENT IN THE MULTINATIONAL FIRM: A CONSTRAINED GENERALIZED NETWORK APPROACH.
CRUMA 1977 -1 CRUM, R. L.; KLINGMAN, D.; TAVIS, L. A.
OPERATIONAL APPROACH TO INTEGRATED WORKING CAPITAL PLANNING, AN.
CRUS 1975 -1 CRUSE, A. B.
PROOF OF FULKERSON'S CHARACTERIZATION OF PERMUTATION MATRICES, A.
CUNNA 1976 -1 CUNNINGHAM, W. H.
NETWORK SIMPLEX METHOD, A.
DAFE 1976 -1 DAFERMOS, S. C.
INTEGRATED EQUILIBRIUM FLOW MODELS FOR TRANSPORTATION PLANNING.
DEBA 1976 -1 DEBANNE, J. G.
GENERALIZED NETWORK AND INTEGER PROGRAMMING IN TECHNOLOGY ASSESSMENT AND PLANNING AN ENERGY SCENARIO STUDY.
DEHN 1976 -1 DEHNERT, G.
BRANCH AND BOUND METHOD FOR SOLVING FIXED CHARGE NETWORK FLOW PROBLEMS, A.
DETR 1976 -1 DETROYE, J.
SYNTHESE OPTIMALE D'UN RESEAU PARCOURU PAR UN MULTIFLOT.
DINE 1972 -1 DINESCU, C.
DETERMINING THE MEAN FLOW IN A TRANSPORTATION NETWORK (RUMANIAN).
DINI 1973 -1 DINITS, E. A.
ALGORITHM FOR THE DIGITWISE DECREASE OF DISCREPANCIES, AND TRANSPORT PROBLEMS (RUSSIAN), AN.
DINI 1970 -1 DINITS, E. A.
ALGORITHM FOR SOLUTION OF A PROBLEM OF MAXIMUM FLOW IN A NETWORK WITH POWER ESTIMATOR.

NETWORK FLOWS

****

DOMS 1976 -1 DOMSCHKE, W.
NEW ALGORITHM FOR MINIMAL COST FLOW PROBLEMS AND THEIR APPLICATIONS.

DOULA 1969 -1 DOULLIEZ, P.; RAO, M. R.
MAXIMUM FLOW IN A MULTITERMINAL NETWORK WITH INCREASING REQUIREMENTS AND ARCS SUBJECT TO FAILURES.

DOULB 1976 -1 DOULLIEZ, P. J.; RAO, M. R.
NETWORK PROBLEM WITH PARAMETRIC DEMANDS AND ARC FAILURES - A LABELING ALGORITHM.

DRAG 1976 -1 DRAGAN, I.
NETWORK FLOW MODEL FOR SOLVING A COOPERATIVE GAME, A.

EDMOA 1977 -1 EDMONDS, J.; GILES, R.
MIN-MAX RELATION FOR SUBMODULAR FUNCTIONS ON GRAPHS, A.

ELAM 1977 -1 ELAM, J.; GLOVER, F.; KLINGMAN, D.
STRONGLY CONVERGENT PRIMAL ALGORITHM FOR GENERALIZED NETWORKS, A.

ERMO 1976 -1 ERMOL'EV, YU. M.; KRIVETS, T. A.; PETUKHOV, V. S.
PROBLEM OF PLANNING SEA SHIPPINGS OF EMPTY CONTAINERS (RUSSIAN), A.

EVANA 1977 -1 EVANS, J. R.
SOME NETWORK FLOW MODELS AND HEURISTICS FOR MULTIPRODUCT PRODUCTION AND INVENTORY PLANNING.

EVANA 1976 -3 EVANS, J. R.
MAXIMUM FLOW IN PROBABILISTIC GRAPHS - THE DISCRETE CASE.

EVANB 1977 -1 EVANS, J. R.; JARVIS, J. J.
OUT-OF-KILTER BASED HEURISTIC FOR THE INTEGER MULTICOMMODITY TRANSPORTATION PROBLEM, AN.

EVANC 1976 -1 EVANS, J. R.; MARTIN, K.
FEASIBLE FLOWS IN MULTICOMMODITY NETWORKS WITH LOWER BOUNDS.

EVEN 1974 -1 EVEN, S.
MAXIMUM FLOW IN A NETWORK AND THE CONNECTIVITY OF A GRAPH.

EVENA 1975 -1 EVEN, S.; ITAI, A.; SHAMIR, A.
ON THE COMPLEXITY OF TIMETABLE AND MULTI-COMMODITY FLOW PROBLEMS.

EVEND 1975 -2 EVEN, S.; TARJAN, R. E.
NETWORK FLOW AND TESTING GRAPH CONNECTIVITY.

FAUR 1976 -1 FAURE, R.; ROUCAIROL, C.; TOLLA, P.
CHEMINS ET FLOTS, ORDONNANCEMENTS.

FONG 1977 -1 FONG, A. C.; ULLMAN, J. D.
FINDING THE DEPTH OF A FLOW GRAPH.

FORG 1976 -2 FORGO, F.
SOME TRANSFORMATIONS OF INTEGER PROGRAMMING PROBLEMS.

FOX A 1975 -1 FOX, B. L.
PROBABILISTIC NETWORKS.

FRIDA 1973 -1 FRIDMAN, A. A.; LITVAK, B. G.; RAPPOPORT, A. M.
ABSOLUTELY INTEGER LINEAR PROGRAMMING PROBLEMS (RUSSIAN).

FRIS 1963 -2 FRISCH, I. T.
OPTIMUM ROUTES IN COMMUNICATION SYSTEMS WITH CHANNEL CAPACITIES AND CHANNEL RELIABILITIES.

FRON 1975 -1 FRONC, M. B.
ADDRESSED FLOWS IN SPECIAL CASES OF PLANAR NETWORKS (CZECH).

FUJIA 1963 -1 FUJISAWA, T.
MAXIMUM FLOWS IN A LOSSY NETWORK.

FULK 1975 -1 FULKERSON, D. R.
FLOW NETWORKS AND COMBINATORIAL OPERATIONS RESEARCH.

FULK 1975 -2 FULKERSON, D. R.
STUDIES IN GRAPH THEORY, I.

FULK 1975 -3 FULKERSON, D. R.
STUDIES IN GRAPH THEORY, II.

GAUL 1974 -1 GAUL, W.
R-SATISFYING MULTI-COMMODITY FLOWS AND CAPACITY-ALTERATIONS OF NETWORKS.

GAUL 1977 -1 GAUL, W.
PROBLEME BEI GLEICHZEITIGEN FLUESSEN IN GRAPHEN.

GAUL 1976 -2 GAUL, W.
UEBER FLUSSPROBLEME IN NETZWERKEN.

GAUL 1976 -3 GAUL, W.
ZUR BERECHNUNG GLEICHZEITIGER FLUESSE VERSCHIEDENER BESCHAFFENHEIT.

GINZ 1977 -1 GINZBURG, M. YA.; BURTMAN, M. I.
OPERATIONAL ALLOCATION OF RESOURCES IN LARGE TRANSPORTATION NETWORKS WITH TWO ENDPOINTS AND WITHOUT CYCLES (RUSSIAN).

GLAS 1978 -1 GLASSEY, C. R.
QUADRATIC NETWORK OPTIMIZATION MODEL FOR EQUILIBRIUM SINGLE COMMODITY TRADE FLOW, A.

GLOVA 1977 -1 GLOVER, F.; HULTZ, J.; KLINGMAN, D.
IMPROVED COMPUTER-BASED PLANNING TECHNIQUES.

GLOVB 1977 -1 GLOVER, F.; HULTZ, J.; KLINGMAN, D.; STUTZ, J.
NEW COMPUTER-BASED PLANNING TOOL, A.

GLOVC 1975 -1 GLOVER, F.; KLINGMAN, D.
CAPSULE VIEW OF FUTURE DEVELOPMENTS ON LARGE SCALE NETWORK AND NETWORK-RELATED PROBLEMS.

GLOVC 1976 -1 GLOVER, F.; KLINGMAN, D.
PRACTITIONER'S GUIDE TO THE STATE OF LARGE SCALE NETWORK AND NETWORK-RELATED PROBLEMS, A.

GLOVC 1976 -2 GLOVER, F.; KLINGMAN, D.
SOME RECENT PRACTICAL MISCONCEPTIONS ABOUT THE STATE OF THE ART OF NETWORK ALGORITHMS.

GLOVC 1975 -3 GLOVER, F.; KLINGMAN, D.
IMPROVED LABELING OF L. P. BASES IN NETWORKS.

GLOVC 1975 -4 GLOVER, F.; KLINGMAN, D.
NETWORK APPLICATION IN INDUSTRY AND GOVERNMENT.

GLOVC 1975 -5 GLOVER, F.; KLINGMAN, D.
NEW ADVANCES IN THE SOLUTION OF LARGE SCALE NETWORK AND NETWORK-RELATED PROBLEMS.

GLOVD 1977 -1 GLOVER, F.; KLINGMAN, D.; MACMILLAN, C.
NETFORM CONCEPT: A MORE EFFECTIVE MODEL FORM AND SOLUTION PROCEDURE FOR LARGE SCALE NONLINEAR PROBLEMS, THE.

GLOVE 1976 -1 GLOVER, F.; KLINGMAN, D.; NAPIER, A.
EQUIVALENCE OF GENERALIZED NETWORK AND GENERALIZED TRANSPORTATION PROBLEMS.

GLOVF 1973 -1 GLOVER, F.; KLINGMAN, D.; STUTZ, J.
IMPLEMENTATION AND COMPUTATIONAL STUDY OF A GENERALIZED NETWORK CODE.

GLOVG 1975 -1 GLOVER, F.; MULVEY, J.
EQUIVALENCE OF THE 0-1 INTEGER PROGRAMMING PROBLEM TO DISCRETE GENERALIZED AND DISCRETE PURE NETWORK MODELS.

GLOVG 1975 -2 GLOVER, F.; MULVEY, J.
EQUIVALENCE OF THE 0-1 INTEGER PROGRAMMING PROBLEM TO DISCRETE GENERALIZED AND PURE NETWORKS.

GOLDH 1971 -1 GOLDSTEIN, M. C.
CONSIDERATION IN THE DESIGN OF LARGE SCALE NETWORKS.

****

GOMO 1975 -1 GOMORY, R. E.; HU, T. C.
MULTI-TERMINAL FLOWS IN A NETWORK.
GRAV 1975 -2 GRAVER, J. E.
ON THE FOUNDATIONS OF LINEAR AND INTEGER LINEAR PROGRAMMING I.
GRAVA 1977 -1 GRAVER, J. E.; WATKINS, M. E.
COMBINATORICS WITH EMPHASIS ON THE THEORY OF GRAPHS.
HAEN 1973 -1 HAENSLER, E.
EXPERIMENTAL HEURISTIC PROCEDURE TO OPTIMIZE A TELECOMMUNICATION NETWORK UNDER NONLINEAR COST FUNCTIONS.
HALP 1977 -1 HALPERN, J.
GENERALIZED DYNAMIC FLOWS PROBLEM, A.
HASL 1975 -1 HASLINGER, A.
BALANCED GROWTH AND RELATIVE STABILITY IN DISCRETE, NONLINEAR PERSON-FLOW MODELS.
HATC 1975 -1 HATCH, R. S.
BENCH MARKS COMPARING TRANSPORTATION CODES BASED ON PRIMAL SIMPLEX AND PRIMAL-DUAL ALGORITHMS.
HELG 1977 -1 HELGASON, R. V.; KENNINGTON, J. L.
EFFICIENT PROCEDURE FOR IMPLEMENTING A DUAL SIMPLEX NETWORK FLOW ALGORITHM, AN.
HOLMA 1977 -1 HOLMES, J. H.; HAGGETT, P.
GRAPH THEORY INTERPRETATION OF FLOW MATRICES: A NOTE ON MAXIMIZATION PROCEDURES FOR IDENTIFYING SIGNIFICANT LINKS.
HOR 1976 -1 HOR, C. G.
MAXIMUM FLOWS IN A VERTEX WEIGHTED NETWORK.
HORS 1976 -1 HORST, R.
GRAPHENTHEORETISCHER ALGORITHMUS FUER EINIGE FLUSSPROBLEME IN NETZWERKEN MIT KANTENGEWINNEN, EIN.
HORS 1975 -1 HORST, R.
MAXIMALE NETZWERKFLUESSE MIT KANTENGEWINNEN.
HULT 1976 -1 HULTZ, J.
ALGORITHMS AND APPLICATIONS FOR GENERALIZED NETWORKS.
HULTA 1976 -1 HULTZ, J.; KLINGMAN, D.; RUSSELL, R.
ADVANCED DUAL BASIC FEASIBLE SOLUTION FOR A CLASS OF CAPACITATED GENERALIZED NETWORKS, AN.
IMOM 1976 -1 IMOMNAZAROV, B.
MAXIMAL FLOW AND MINIMAL CUT PROBLEM (RUSSIAN), THE.
IRI 1971 -1 IRI, M.
ON AN EXTENSION OF THE MAXIMUM-FLOW MINIMUM-CUT THEOREM TO MULTICOMMODITY FLOWS.
IRI 1960 -1 IRI, M.
NEW METHOD OF SOLVING TRANSPORTATION-NETWORK PROBLEMS, A.
IRI A 1975 -1 IRI, M.; TOMIZAWA, N.
UNIFYING APPROACH TO FUNDAMENTAL PROBLEMS IN NETWORK THEORY BY MEANS OF MATROIDS (JAPANESE), A.
JOHNE 1964 -1 JOHNSON, E. L.
NETWORK FLOWS, GRAPHS, AND INTEGER PROGRAMMING.
JOHNE 1975 -1 JOHNSON, E. L.
FLOWS IN NETWORKS.
KARN 1976 -1 KARNEY, D.; KLINGMAN, D.
IMPLEMENTATION AND COMPUTATIONAL STUDY ON AN IN-CORE, OUT-OF-CORE PRIMAL NETWORK CODE.
KARZ 1974 -1 KARZANOV, A.
DETERMINING THE MAX FLOW IN A NETWORK.
KAS 1975 -1 KAS, P.
ON QUADRATIC MINIMAL COST FLOW PROBLEMS (HUNGARIAN).
KINA 1977 -1 KINARIWALA, B.; RAO, A. G.
FLOW SWITCHING APPROACH TO THE MAXIMUM FLOW PROBLEM: I.
KOH A 1974 -1 KOH, K. M.; TEH, H. H.
ON DIRECTED POSTMAN PROBLEM.
KOMA 1975 -1 KOMARNICKI, J.
AUTOMATISCHE GENERIERUNG ANWENDUNGSSPEZIFISCHER NETZPLAN-TECHNIK-PROGRAMMSYSTEME.
KORN 1969 -1 KORNILOVA, L. E.
METHOD OF MINIMAL FLOW FOR THE EXACT SOLUTION OF THE PROBLEM OF MINIMIZATION OF THE MAXIMAL LEVEL OF RESOURCE UTILIZATION IN NETWORK PLANNING (RUSSIAN), THE.
KRAR 1967 -1 KRARUP, J.
FIXED-COST AND OTHER NETWORK FLOW PROBLEMS AS RELATED TO PLANT LOCATION AND TO THE DESIGN OF TRANSPORTATION AND COMPUTER SYSTEMS.
KRAR 1967 -2 KRARUP, J.
FIXED-COST NETWORK FLOW PROBLEMS.
KYAN 1975 -1 KYAN, S.; ONAGA, K.
MINIMUM-DELAY MAXIMUM-GAUGE ROUTES IN A TIME-VARYING DYNAMIC NETWORK.
LANGB 1973 -1 LANGLEY, R.
CONTINUOUS AND INTEGER GENERALIZED FLOW PROBLEMS.
LAWL 1977 -4 LAWLER, E. L.
SHORTEST PATH AND NETWORK FLOW ALGORITHMS.
LAWL 1975 -6 LAWLER, E. L.
OVERVIEW OF NETWORK FLOW THEORIE.
LAWL 1975 -7 LAWLER, E. L.
QUADRATIC ASSIGNMENT PROBLEM: A BRIEF REVIEW, THE.
LEBL 1976 -1 LEBLANC, L. J.
CONJUGATE GRADIENT TECHNIQUE FOR CERTAIN QUADRATIC NETWORK PROBLEMS, THE.
LIN B 1974 -1 LIN, P. M.; LEON, B. J.
IMPROVING THE EFFICIENCY OF LABELING ALGORITHMS FOR MAXIMUM FLOWS IN NETWORKS.
LITV 1973 -1 LITVAK, B. G.
REDUCIBILITY OF ABSOLUTELY INTEGER PROBLEMS WITH A FIXED CONSTRAINT VECTOR (RUSSIAN), THE.
LUCE 1975 -1 LUCE, B.
DUALITY METHODS FOR MAXIMAL FLOW AND SHORTEST PATH PROBLEMS.
MAC J 1976 -1 MAC KINNON, R. D.
OPTIMIZATION MODELS OF TRANSPORTATION NETWORK IMPROVEMENT: REVIEW AND FUTURE PROSPECTS.
MADS 1976 -1 MADSEN, K.; SCHJAER-JACOBSEN, H.
SINGULARITIES IN MINIMAX OPTIMIZATION OF NETWORKS.
MARQ 1977 -1 MARQUEZ, J.; MEDINA-MORA, O.
NETWORK SOLUTION TO A GENERAL VEHICLE SCHEDULING PROBLEM, A.
MARU 1976 -1 MARUYAMA, K.; TANG, D. T.; FRATTA, L.
FLOW ASSIGNMENT ALGORITHM FOR COMPUTER COMMUNICATION NETWORK DESIGN WITH DIFFERENT CLASSES OF PACKETS.
MAURB 1972 -1 MAURRAS, J. F.
FLOT OPTIMAL SUR UN GRAPHE AVEC MULTIPLICATEURS.

MEGI 1977 -1 MEGIDDO, N.
GOOD ALGORITHM FOR LEXICOGRAPHICALLY OPTIMAL FLOWS IN MULTI-TERMINAL NETWORKS, A.

MINIA 1975 -1 MINIEKA, F.
MAXIMAL NETWORK FLOWS, MATROIDS AND MATCHINGS.

MINOA 1976 -2 MINOUX, M.
FLOTS EQUILIBRES ET FLOTS AVEC SECURITE.

MINOA 1976 -3 MINOUX, M.
MULTIFLOTS DE COUT MINIMAL AVEC FONCTIONS DE COUT CONCAVES.

MOOR 1976 -1 MOORE, M. H.
ON THE FASTEST ROUTE FOR CONVOY-TYPE TRAFFIC IN FLOWRATE-CONSTRAINED NETWORKS.

MOSH 1963 -1 MOSHER, W. W. JR.
CAPACITY-RESTRAINT ALGORITHM FOR ASSIGNING FLOW TO A TRANSPORT NETWORK, A.

MULVA 1975 -2 MULVEY, J. M.
SPECIAL STRUCTURES IN NETWORK NODELS AND ASSOCIATED APPLICATIONS.

MULVA 1975 -3 MULVEY, J. M.
TESTING OF LARGE-SCALE NETWORK OPTIMIZATION PROGRAM.

MURTB 1976 -1 MURTY, K. G.
LINEAR AND COMBINATORIAL PROGRAMMING.

MYER 1967 -1 MYERS, B. R.
MAX-CUT MIN-FLOW IN A CLASS OF DIRECTED WEIGHTED NETS.

NAKA 1977 -1 NAKAMURA, T.; YAMASAKI, M.
MINIMAL CUT PROBLEMS ON AN INFINITE NETWORK.

OICE 1972 -1 OICER, V. I.
MAXIMAL FLOW IN A NETWORK WITH CAPACITIES (RUSSIAN).

ORLOB 1975 -1 ORLOVA, G. I.; DORFMAN, G. YA.
OPTIMAL PARTITIONING OF THE SET OF VERTICES OF A GRAPH BY REDUCTION TO A PROBLEM OF MAXIMAL CUT (RUSSIAN).

OSYC 1976 -1 OSYCZKA, A.
ALGORITHM OF OPTIMIZATION FOR A SPECIAL CLASS OF NETWORKS, AN.

PANA 1976 -1 PANAGIOTAKOPOULOS, D.
MULTI-COMMODITY MULTI-TRANSFORMED NETWORK FLOWS WITH AN APPLICATION TO RESIDUALS MANAGEMENT.

PAPA 1976 -2 PAPADIMITRIOU, C. H.
ON THE COMPLEXITY OF EDGE TRAVERSING.

PETE 1975 -1 PETER, K.
FLOWS IN NETWORKS WITH QUADRATIC ECONOMIC FUNCTIONS (HUNGARIAN).

PICAA 1975 -1 PICARD, J. C.; RATLIFF, H. D.
MINIMUM CUTS AND RELATED PROBLEMS.

PICAC 1977 -1 PICARD, J.-C.; QUEYRANNE, M.
NETWORKS, GRAPHS AND SOME NONLINEAR 0-1 PROGRAMMING PROBLEMS.

PICAD 1975 -1 PICARD, J.-C.; RATLIFF, H. D.
CUT APPROACH TO THE RECTILINEAR DISTANCE FACILITY LOCATION PROBLEM, A.

PINO 1976 -1 PINOCHET, M. R.
CENTRALIZATION AND DECENTRALIZATION OF DECISION-MAKING IN TRANSPORTATION NETWORKS.

PLESA 1976 -1 PLESTCHINSKY, A. S.
E-APPROACH TO A NETWORK MODEL OPTIMIZATION (RUSSIAN).

POTT 1972 -1 POTTS, R.; OLIVER, R.
FLOWS IN TRANSPORTATION NETWORKS.

RAMA 1977 -1 RAMAKRISHNAN, C. S.
NOTE ON THE TIME MINIMISING TRANSPORTATION PROBLEM, A.

RANDA 1977 -1 RANDOLPH, P.; RINGEISEN, R.
SHORTEST PATHS THROUGH MULTIPLE CRITERIA NETWORKS.

RAZH 1969 -1 RAZHUMIKHIN, B. S.
APPLICATION OF THE METHOD OF DYNAMIC PROGRAMMING TO CERTAIN PROBLEMS OF NETWORK OPTIMIZATION. I + II (RUSSIAN), AN.

RUBIC 1975 -1 RUBIN, I.
MULTITERMINAL PATH FLOWS FOR PACKET-SWITCHING COMMUNICATION NETWORKS.

SAHN 1974 -1 SAHNI, S.
SOME RELATED PROBLEMS FROM NETWORK FLOWS, GAME THEORY, SOLUTIONS.

SCHN 1978 -1 SCHNORR, C. P.
MULTITERMINAL NETWORK FLOW AND CONNECTIVITY IN UNSYMMETRICAL NETWORKS.

SCHW 1977 -1 SCHWARTZ, M.
COMPUTER-COMMUNICATION NETWORK. DESIGN AND ANALYSIS.

SCWA 1968 -1 SCWARTZ, J. G.
NETWORK FLOW AND SYNTHESIS MODELS FOR TRANSPORTATION PLANNING: A SURVEY.

SEYM 1975 -1 SEYMOUR, P. D.
MATROIDS, HYPERGRAPHS AND THE MAX-FLOW MIN-CUT THEOREM.

SHAPB 1977 -1 SHAPIRO, J. F.
NOTE ON THE PRIMAL-DUAL AND OUT-OF-KILTER ALGORITHMS FOR NETWORK OPTIMIZATION PROBLEMS, A.

SITR 1974 -1 SITRUK, G.
CHEMINS EXTREMAUX ET K-EXTREMAUX APPLICATION AUX PROBLEMES DE FLOTS.

STANA 1976 -1 STANFORD, R. E.
ANALYTICAL SOLUTION OF A DYNAMIC TRANSACTION FLOW PROBLEM.

STEF 1969 -1 STEFFEN, M. R.; DENT, W.
CONSTRUCTION OF REDUNDANT NETWORKS WITH MINIMUM DISTANCE.

SUMA 1976 -1 SUMAROKOV, S. V.
SOLUTION METHOD FOR A NETWORK PROBLEM WITH SEVERAL EXTREMA (RUSSIAN).

TARJ 1978 -1 TARJAN, R. E.
COMPLEXITY OF MONOTONE NETWORKS FOR COMPUTING CONJUNCTIONS.

TIND 1977 -2 TIND, J.
ON ANTIBLOCKING SETS AND POLYHEDRA.

TOBI 1975 -1 TOBIN, R. L.
MINIMAL COMPLETE MATCHINGS AND NEGATIVE CYCLES.

TROT 1977 -1 TROTTER, L. E. JR.
ON THE GENERALITY OF MULTI-TERMINAL FLOW THEORY.

TRUE 1977 -1 TRUEMPER, K.
ON MAX FLOWS WITH GAINS AND PURE MIN-COST FLOWS.

TRUE 1976 -3 TRUEMPER, K.
EFFICIENT SCALING PROCEDURE FOR GAIN NETWORKS, AN.

TUCK 1977 -1 TUCKER, A.
NOTE ON CONVERGENCE OF THE FORD-FULKERSON FLOW ALGORITHM, A.

WALSA 1976 -1 WALSH, T. S.
COMPLEXITE DES ALGORITHMES DE FLOT.

WEGL 1975 -1 WEGLARZ, J.
CERTAIN NETWORK PROGRAMMING PROBLEM (POLISH), A.

****

WEINB 1976 -1 WEINBERGER, D. B.
NETWORK FLOWS, MINIMUM COVERINGS, AND THE FOUR-COLOR CONJECTURE.
YAGE 1969 -1 YAGED, B. J.
ROUTING INFORMATION IN MINIMUM COST NETWORKS.

NONLINEAR INTEGER PROGRAMMING
****

ABAD 1976 -1 ABADIE, J.; DAYAN, H.; AKOKA, J.
QUELQUES EXPERIENCES NUMERIQUES SUR LA PROGRAMMATION NON LINEAIRE EN NOMBRES ENTIERS.

AGRA 1976 -1 AGRAWAL, S. C.
ALTERNATE METHOD ON INTEGER SOLUTIONS TO QUADRATIC PROGRAM BY A BRANCH AND BOUND TECHNIQUE, AN.

ALBE 1977 -1 ALBERS, S.
MIXED INTEGER NONLINEAR PROGRAMMING PROCEDURE FOR SIMULTANEOUSLY LOCATING MULTIPLE PRODUCTS IN AN ATTRIBUTE SPACE, A.

ANZA 1974 -1 ANZAI, Y.
ON INTEGER FRACTIONAL PROGRAMMING.

ARMSA 1975 -1 ARMSTRONG, R. D.; CHARNES, A.; PHILLIPS, F.
PAGE CUTS FOR INTEGER INTERVAL LINEAR PROGRAMMING.

ARMSC 1977 -1 ARMSTRONG, R. D.; COOK, W. D.; PALACIOS-GOMESZ, F. E.
APPLICATION OF QUASI-INTEGER PROGRAMMING TO A CAPITAL BUDGETING PROBLEM IN PAVEMENT MAINTENANCE, AN.

AROR 1977 -1 ARORA, S. R.
NOTE ON FRACTIONAL FIXED CHARGE PROBLEMS, A.

ARORA 1977 -1 ARORA, S. R.; PURI, M. C.
ENUMERATION TECHNIQUE FOR THE SET COVERING PROBLEM WITH A LINEAR FRACTIONAL FUNCTIONAL AS ITS OBJECTIVE FUNCTION.

ARORB 1977 -1 ARORA, S. R.; SWARUP, K.; PURI, M. C.
CUTTING PLANE TECHNIQUE FOR SET COVERING PROBLEM WITH LINEAR FRACTIONAL FUNCTIONAL.

ASHT 1977 -1 ASHTON, W. B.; CLARK, G. M.
LOCATION AND SIZING OF URBAN FREIGHT CONSOLIDATION TERMINALS WITH MULTIPLE PLANNING PERIODS, THE.

BACHA 1977 -1 BACHEM, A.; KORTE, B.
ALGORITHM FOR QUADRATIC OPTIMIZATION OVER TRANSPORTATION POLYTOPES, AN.

BACHA 1977 -2 BACHEM, A.; KORTE, B.
ESTIMATING INPUT-OUTPUT-MATRICES.

BACHA 1977 -3 BACHEM, A.; KORTE, B.
QUADRATIC OPTIMIZATION OVER TRANSPORTATION POLYTOPES.

BACHA 1977 -4 BACHEM, A.; KORTE, B.
QUADRATISCHE OPTIMIERUNG UEBER TRANSPORTPOLYEDER ZUR LOESUNG VON SCHAETZPROBLEMEN VON INPUT-OUTPUT MATRIZEN.

BACHC 1977 -1 BACHEM, A.; SIMON, H.
OPTIMAL PRODUCT POSITIONING IN AN ATTRIBUTE SPACE.

BAND 1975 -1 BANDLER, J. W.; CHEN, J. H. K.; DALSGAARD, P.; LIU, P. C.
TOLOPT - A PROGRAM FOR OPTIMAL, CONTINUOUS OR DISCRETE, DESIGN CENTERING AND TOLERANCING, PART 1 - USER'S GUIDE, PART 2 - FORTRAN LISTING.

BAR 1975 -1 BAR, G.
UEBER DEN EINSATZ LOGISCHER MITTEL ZUR LINEARISIERUNG DISKRETER OPTIMIERUNGSPROBLEME.

BAZA 1975 -1 BAZARAA, M. S.
COMPUTERIZED LAYOUT DESIGN: A BRANCH AND BOUND APPROACH.

BRIC 1977 -1 BRICKER, D. L.
REFORMULATION OF SPECIAL ORDERED SETS FOR IMPLICIT ENUMERATION ALGORITHMS, WITH APPLICATIONS IN NONCONVEX SEPARABLE PROGRAMMING.

BURK 1973 -1 BURKARD, R. E.
METHOD FOR MIXED-INTEGER CONVEX PROGRAMMING, A.

BURK 1977 -2 BURKARD, R. E.
ZUR BESTIMMUNG VON SCHREIBMASCHINENTASTATUREN DURCH QUADRATISCHE ZUORDNUNGSPROBLEME.

BURKC 1977 -1 BURKARD, R. E.; OFFERMANN, J.
ENTWURF VON SCHREIBMASCHINENTASTATUREN MITTELS QUADRATISCHER ZUORDNUNGSPROBLEME.

BURKI 1972 -1 BURKOV, V. N.; LAZEBNIK, A. I.; KHRANOVICH, I. L.
METHOD OF BRANCHES AND BOUNDS AS A REGULAR METHOD FOR THE SOLUTION OF IRREGULAR MATHEMATICAL PROGRAMMING PROBLEMS, I (RUSSIAN).

CHANC 1977 -1 CHAND, M.; AGRAWAL, S. C.
ON INTEGER SOLUTIONS TO COMPLEMENTARY PROGRAMMING PROBLEMS.

CHEND 1974 -1 CHEN, J. H. K.
DISOPT - A GENERAL PROGRAM FOR CONTINUOUS AND DISCRETE NONLINEAR PROGRAMMING PROBLEMS.

CHEND 1975 -1 CHEN, J. H. K.
DISOPT - A GENERAL PROGRAM FOR CONTINUOUS AND DISCRETE NONLINEAR PROGRAMMING PROBLEMS.

CHERE 1975 -1 CHERVAK, N. K.
FINITE ALGORITHM FOR THE TRUNCATION METHOD FOR INTEGER LINEAR PROGRAMMING PROBLEMS WITH NONINTEGER TARGET FUNCTION (RUSSIAN), A.

COOPB 1976 -1 COOPER, L.; COOPER, M. W.
EXACT ALGORITHM FOR THE CONCAVE TRANSPORTATION PROBLEM, AN.

COOPB 1975 -1 COOPER, L.; COOPER, M. W.
NON-LINEAR INTEGER PROGRAMMING.

COOPC 1977 -1 COOPER, L.; LEBLANC, L. J.
STOCHASTIC TRANSPORTATION PROBLEMS AND OTHER NETWORK RELATED CONVEX PROBLEMS.

COOPD 1977 -1 COOPER, M. W.
IMPROVED ALGORITHM FOR NON-LINEAR INTEGER PROGRAMMING, AN.

DAEN 1977 -1 DAENINCK, G.; SMEERS, Y.
USING SHORTEST PATHS IN SOME TRANSSHIPMENT PROBLEM WITH CONCAVE COSTS.

DAVY 1972 -1 DAVYDOV, E. G.; SIGAL, I. KH.
APPLICATION OF THE PENALTY FUNCTION METHOD IN INTEGER PROGRAMMING PROBLEMS (RUSSIAN).

DEARA 1976 -1 DEARING, P. M.; FRANCIS, R. L.; LOWE, T. J.
CONVEX LOCATION PROBLEMS ON TREE NETWORKS.

DOLO 1976 -1 DOLODARENKO, V. A.; FEDAN, E. I.
ASSIGNMENT PROBLEM WITH A NONLINEAR STOCHASTIC OBJECTIVE FUNCTION (RUSSIAN), THE.

ELSH 1977 -1 ELSHAFEI, A. N.
HOSPITAL LAYOUT AS A QUADRATIC ASSIGNMENT PROBLEM.

ERLA 1966 -1 ERLANDER, S.
NONLINEAR TRANSPORTATION AND INVENTORY PROBLEM.

ESOG 1977 -1 ESOGBUE, A. O.; MARKS, B. R.
DYNAMIC PROGRAMMING MODELS OF THE NONSERIAL CRITICAL PATH-COST PROBLEM.

EVANA 1976 -2 EVANS, J. R.
FUNDAMENTAL CIRCUITS IN COMPLETE GRAPHS AND THE TRAVELING SALESMAN PROBLEM.

FABI 1976 -1 FABIAN, CS.
SOME ALGORITHMS FOR NONLINEAR INTEGER PROGRAMMING.

FABIA 1977 -1 FABIAN, CS.; MOGA, S.
TWO SIMULATION PROCEDURES FOR SOLVING CONVEX INTEGER PROGRAMMING PROBLEMS.

FLORA 1970 -1 FLORIAN, M.; ROBILLARD, P.
HYPERBOLIC PROGRAMMING WITH BIVALENT VARIABLES.

****

FORG 1976 -2 FORGO, F.
SOME TRANSFORMATIONS OF INTEGER PROGRAMMING PROBLEMS.
FORG 1974 -2 FORGO, F.
SOME TRANSFORMATIONS OF INTEGER PROGRAMMING PROBLEMS (HUNGARIAN).
GAIK 1976 -2 GAIKOV, N. E.
MINIMIZATION OF A QUADRATIC FORM WITH BOOLEAN VARIABLES ON A SYMMETRIC GROUP (RUSSIAN), THE.
GALLB 1976 -1 GALLO, G.; HAMMER, P. L.; SIMEONE, B.
QUADRATIC KNAPSACK PROBLEMS.
GEOF 1977 -1 GEOFFRION, A. M.
HOW CAN SPECIALIZED DISCRETE AND CONVEX OPTIMIZATION METHODS BE MARRIED.
GLAS 1978 -1 GLASSEY, C. R.
QUADRATIC NETWORK OPTIMIZATION MODEL FOR EQUILIBRIUM SINGLE COMMODITY TRADE FLOW, A.
GLEB 1973 -1 GLEBOV, N. I.
ON A CLASS OF CONVEX INTEGER PROGRAMMING PROBLEMS (RUSSIAN).
GLOV 1975 -2 GLOVER, F.
IMPROVED LINEAR INTEGER PROGRAMMING FORMULATIONS OF NONLINEAR INTEGER PROBLEMS.
GOND 1971 -1 GONDRAN, M.
PROGRAMMATION ENTIERE NON LINEAIRE: DUALITE DIOPHANTIENNE.
GRANA 1977 -1 GRANOT, D.; GRANOT, F.
ON INTEGER AND MIXED INTEGER FRACTIONAL PROGRAMMING PROBLEMS.
GRANA 1976 -1 GRANOT, D.; GRANOT, F.
ON SOLVING FRACTIONAL (0,1) PROGRAMS BY IMPLICIT ENUMERATION.
HANSA 1977 -1 HANSEN, P.
METHODS OF NONLINEAR 0-1 PROGRAMMING.
HANSA 1975 -2 HANSEN, P.
FONCTIONS D'EVALUATION ET PENALITES POUR LES PROGRAMMES QUADRATIQUES EN VARIABLES 0-1.
HARTC 1975 -1 HARTMANN, K.
GEMISCHT GANZZAHLIGE LINEARE QUOTIENTENOPTIMIERUNG NACH DEM SCHNITTVERFAHREN VON GOMORY.
HARTC 1973 -1 HARTMANN, K.
ASPEKTE DER GANZZAHLIGEN LINEAREN QUOTIENTENOPTIMIERUNG, EINIGE.
HASL 1975 -1 HASLINGER, A.
BALANCED GROWTH AND RELATIVE STABILITY IN DISCRETE, NONLINEAR PERSON-FLOW MODELS.
HEIS 1977 -1 HEISING, C. D.
PLANT LOCATION AND TRANSPORTATION MODELS OF THE LIGHT-WATER REACTOR FUEL CYCLE.
HODGA 1977 -1 HODGSON, T. J.; KILPATRICK, K. E.; LONGINI, I. M.
SCHEDULING MULTISPECIALTY CLINICS USING INTEGER PROGRAMMING.
HYUN 1975 -1 HYUN, K. N.
RELIABILITY OPTIMIZATION BY 0-1 PROGRAMMING FOR A SYSTEM WITH SEVERAL FAILURE MODES.
ISHI 1977 -1 ISHII, H.; IBARAKI, T.; MINE, H.
FRACTIONAL KNAPSACK PROBLEMS.
ISHI 1976 -1 ISHII, H.; IBARAKI, T.; MINE, H.
PRIMAL CUTTING PLANE ALGORITHM FOR INTEGER FRACTIONAL PROGRAMMING PROBLEMS, A.
KAS 1975 -1 KAS, P.
ON QUADRATIC MINIMAL COST FLOW PROBLEMS (HUNGARIAN).
KEL' 1973 -1 KEL'MANS, A. K.
SELECTION OF THE OPTIMAL VERTEX IN A GRAPH (RUSSIAN), THE.
KIRC 1965 -1 KIRCHGAESSNER, K.
GRAPH THEORETICAL SOLUTION OF A NONLINEAR ASSIGNMENT PROBLEM (GERMAN).
KOVAB 1974 -1 KOVALEV, M. M.
CERTAIN INTEGER PROGRAMMING PROBLEM WITH A CONVEX SYMMETRIC TARGET FUNCTION (RUSSIAN), A.
KUZN 1974 -1 KUZNETSOV, V. F.
ALLOCATION OF RESOURCES FOR A GIVEN NETWORK (RUSSIAN).
LAWL 1975 -7 LAWLER, E. L.
QUADRATIC ASSIGNMENT PROBLEM: A BRIEF REVIEW, THE.
LECH 1976 -1 LECHEVALLIER, Y.
CLASSIFICATION AUTOMATIQUE OPTIMALE SOUS CONTRAINTE D'ORDRE TOTAL.
LOVEA 1976 -2 LOVE, R. F.; WONG, J. Y.
SOLVING QUADRATIC ASSIGNMENT PROBLEMS WITH RECTANGULAR DISTANCES AND INTEGER PROGRAMMING.
MAHM 1976 -1 MAHMOUD, S.; RIORDON, J. S.
OPTIMAL ALLOCATION OF RESOURCES IN DISTRIBUTED INFORMATION NETWORKS.
MEYEC 1977 -1 MEYER, R. R.
CLASS OF NONLINEAR INTEGER PROGRAMS SOLVABLE BY A SINGLE LINEAR PROGRAM, A.
MICH 1977 -1 MICHAELI, I.; POLLATSCHEK, M. A.
ON SOME NONLINEAR KNAPSACK PROBLEMS.
MINOA 1976 -3 MINOUX, M.
MULTIFLOTS DE COUT MINIMAL AVEC FONCTIONS DE COUT CONCAVES.
MITE 1977 -3 MITEV, J. G.
SOME ALGORITHMS FOR SOLVING INTEGER NON-LINEAR MATHEMATICAL PROGRAMMING PROBLEMS (RUSSIAN).
MJEL 1976 -1 MJELDE, K. M.
EVALUATION AND INCREMENTAL DETERMINATION OF ALMOST OPTIMAL ALLOCATIONS OF RESOURCES.
MOLN 1976 -1 MOLNAR, Z.; TRCKA, V.
SOLVING ZERO-ONE HYPERBOLIC PROGRAMMING PROBLEM BY MEANS OF BRANCH AND BOUND METHOD (CZECH).
NAND 1974 -1 NANDA, R.; WEINGARTEN, A.
PRACTICABLE SOLUTION FOR THE QUADRATIC ASSIGNMENT PROBLEM, A.
NJAR 1973 -1 NJARIPJA, H. K.
DIRECT ALGORITHM FOR THE SOLUTION OF INTEGER CONVEX PROGRAMMING PROBLEMS (RUSSIAN), A.
NOON 1977 -1 NOONAN, F.; GIGLIO, R. J.
PLANNING ELECTRIC POWER GENERATION: A NONLINEAR MIXED INTEGER MODEL EMPLOYING BENDERS DECOMPOSITION.
ORON 1976 -1 ORON, G.; KARMELI, D.; WALKER, W. R.
APPROXIMATION SOLUTION OF NON-LINEAR MIXED-INTEGER OPTIMIZATION, AN.
OVCH 1976 -1 OVCHINNIKOV, V. G.
ON A PROBLEM OF INTEGER LINEAR PROGRAMMING (RUSSIAN).
PANC 1976 -1 PANCHENKO, A. A.
SOME ALGORITHMS FOR SOLUTION OF QUOTIENTS OF LINEAR PROGRAMMING WITH BOOLEAN VARIABLES (RUSSIAN).
PATE 1976 -1 PATEL, A. M.; WALD, C. G. DE; COTES, L. C.
PRE-PARTITIONING AS A MEANS OF ENHANCING PAIRWISE INTERCHANGE SOLUTIONS TO QUADRATIC ASSIGNMENT PROBLEMS.
PENT 1976 -1 PENTICO, D. W.
ASSORTMENT PROBLEM WITH NONLINEAR COST FUNCTIONS, THE.
PERC 1975 -1 PERCOV, E. E.; BOGDANOVA, O. V.
ALGORITHM FOR THE MINIMIZATION OF A PSEUDO-BOOLEAN QUADRATIC FUNCTION (RUSSIAN), AN.
PHUO 1976 -1 PHUONG, T. H.
SOLUTION OF INTEGER PROGRAMS WITH A QUADRATIC OBJECTIVE FUNCTION.

****

PICAC 1977 -1 PICARD, J.-C.; QUEYRANNE, M.
NETWORKS, GRAPHS AND SOME NONLINEAR 0-1 PROGRAMMING PROBLEMS.

RICHB 1976 -1 RICHTER, D.
ALGORITHMUS ZUR LOESUNG GEMISCHT GANZZAHLIGER KONVEXER OPTIMIERUNGSPROBLEME, EIN.

RICHB 1976 -2 RICHTER, D.
CONTRIBUTION TO AN ALGORITHM FOR SOLVING MIXED INTEGER COVEX PROGRAMMING PROBLEMS, A.

SHARD 1977 -1 SHARMA, S. R.; SWARUP, K.
TRANSPORTATION FRACTIONAL PROGRAMMING WITH RESPECT TO TIME.

SHER 1977 -1 SHERALI, A. D.; SHETTY, C. M.
RECTILINEAR DISTANCE LOCATION-ALLOCATION PROBLEM, THE.

SHET 1975 -1 SHETTY, C. M.
BILINEAR PROGRAMMING WITH APPLICATIONS TO LOCATION-ALLOCATION PROBLEMS.

SNIC 1978 -1 SNICKARS, F.
CONVEXITY AND DUALITY PROPERTIES OF A QUADRATIC INTRAREGIONAL LOCATION MODEL.

SPAE 1977 -1 SPAETH, H.
NUMERISCHE ERFAHRUNGEN ZU HEURISTISCHEN LOESUNGSVERFAHREN BEIM VARIANZKRITERIUM IN DER CLUSTER-ANALYSE.

STRAA 1976 -1 STRATMANN, K.-H.
NUMERISCHE UNTERSUCHUNGEN UEBER QUADRATISCHE ZUORDNUNGSPROBLEME.

TRUB 1977 -1 TRUBIN, V. A.
UNIVERSALITY OF ONE CLASS OF QUADRATIC INTEGER PROBLEMS (RUSSIAN).

VIZV 1975 -1 VIZVARI, B.
ENUMERATIVE ALGORITHMS IN POLYNOMIAL 0-1 PROGRAMMING (HUNGARIAN).

VYAT 1975 -1 VYATKIN, A. K.; KRASIN, L. A.
PROBLEM OF OPTIMIZING A HIERARCHICAL ORGANIZATION STRUCTURE (RUSSIAN), A.

WOEH 1973 -1 WOEHNER, W.
GUETERZUGBILDUNG ALS KOMBINATORISCHE OPTIMIERUNGSAUFGABE.

ZANG 1966 -1 ZANGWILL, W.
SHORTEST ROUTE PROBLEM UNDER EITHER CONCAVE OR CONVEX SHIPPING COSTS, THE.

ZHER 1976 -1 ZHERNAK, A. N.
GROUP OF ALGORITHMS FOR THE SOLUTION OF QUASILINEAR INTEGER PROGRAMMING PROBLEMS (RUSSIAN), A.

****

ALBEA 1976 -1 ALBERTSON, M.; BOLLOBAS, B.; TUCKER, S.
INDEPENDENCE RATIO AND MAXIMUM DEGREE OF A GRAPH, THE.
ALBEB 1974 -1 ALBERTSON, M. O.
FINDING AN INDEPENDENT SET IN A PLANAR GRAPH.
ALBEB 1976 -1 ALBERTSON, M. O.
LOWER BOUND FOR THE INDEPENDENCE NUMBER OF A PLANAR GRAPH, A.
ALBED 1975 -1 ALBERTSON, M. O.; HUTCHINSON, J. P.
MAXIMUM SIZE OF AN INDEPENDENT SET IN A NONPLANAR GRAPH, THE.
ANIS 1976 -1 ANISIMOV, V. V.
ON OPTIMUM PACKING OF TREES (RUSSIAN).
BALAJ 1975 -1 BALAS, E.; PADBERG, M. W.
SET PARTITIONING: A SURVEY.
BALAJ 1976 -1 BALAS, E.; PADBERG, M. W.
SET PARTITIONING: A SURVEY.
BALAL 1977 -1 BALAS, E.; ZEMEL, E.
GRAPH SUBSTITUTION AND SET PACKING POLYTOPES.
BALAL 1976 -1 BALAS, E.; ZEMEL, E.
CRITICAL CUTSETS OF GRAPHS AND CANONICAL FACETS OF SET PACKING POLYTOPES.
BECK 1976 -1 BECK, L. L.; MATULA, D. W.
CLIQUE DETERMINATION IN LARGE DENSE GRAPHS.
BENIA 1976 -1 BENITO, F.; GROEFLIN, H.
OPTIMIERUNGSPROBLEME MIT NICHT POLYNOMIAL BEGRENZTEN ALGORITHMEN (NP-PROBLEME).
BOLLC 1976 -2 BOLLOBAS, B.; ERDOES, P.
CLIQUES IN RANDOM GRAPHS.
COCKA 1976 -1 COCKAYNE, E.; HEDETNIEMI, S.
LINEAR ALGORITHM FOR THE MAXIMUM WEIGHT OF AN INDEPENDENT SET IN A TREE, A.
COCKB 1976 -1 COCKAYNE, E. J.; HEDETNIEMI, S. T.
DISJOINT INDEPENDENT DOMINATING SETS IN GRAPHS.
DESL 1970 -1 DESLER, J. F.; HAKIMI, S. L.
ON FINDING A MAXIMUM INTERNALLY STABLE SET OF A GRAPH.
ECKE 1977 -1 ECKER, K.
ORGANISATION VON PARALLELEN PROZESSEN. THEORIE DETERMINISTISCHER SCHEDULES.
EDMOA 1977 -1 EDMONDS, J.; GILES, R.
MIN-MAX RELATION FOR SUBMODULAR FUNCTIONS ON GRAPHS, A.
ETCH 1976 -2 ETCHEBERRY, J.
SET REPRESENTATION PROBLEM: A NEW IMPLICIT ENUMERATION ALGORITHM, THE (REVISED VERSION).
FOUR 1975 -1 FOURNIER, J. C.
SUR LES STABLES D'UN GRAPHE.
FRAND 1976 -1 FRANK, A.
SOME POLYNOMIAL ALGORITHMS FOR CERTAIN GRAPHS AND HYPERGRAPHS.
FREDA 1975 -2 FREDMAN, M. L.
SOME ARITHMETICAL PACKING PROBLEMS.
GAREB 1976 -3 GAREY, M. R.; JOHNSON, D. S.
COMPLEXITY OF NEAR-OPTIMAL GRAPH COLORING, THE.
GOLU 1977 -2 GOLUMBIC, M. C.
COMPLEXITY OF COMPARABILITY GRAPH RECOGNITION AND COLORING, THE.
GUPTA 1969 -1 GUPTA, R. P.
INDEPENDENCE AND COVERING NUMBERS OF LINE GRAPHS AND TOTAL GRAPHS.
HANSA 1975 -1 HANSEN, P.
DEGRES ET NOMBRE DE STABILITE D'UN GRAPHE.
HANSA 1977 -2 HANSEN, P.
SUBDEGREES AND CHROMATIC NUMBERS OF HYPERGRAPHS.
HAUS 1978 -1 HAUSMANN, D.
ADJAZENZCHARAKTERISIERUNGEN FUER POLYEDERECKEN IN DER KOMBINATORISCHEN OPTIMIERUNG.
HAUSA 1977 -1 HAUSMANN, D.; JENKYNS, T. A.; KORTE, B.
WORST CASE ANALYSIS OF GREEDY TYPE ALGORITHMS FOR INDEPENDENCE SYSTEMS.
HAUSC 1976 -1 HAUSMANN, D.; KORTE, B.
COLOURING CRITERIA FOR ADJACENCY ON 0-1-POLYHEDRA.
HOUC 1977 -1 HOUCK, D. J.; VEMUGANTI, R. R.
ALGORITHM FOR THE VERTEX PACKING PROBLEM, AN.
JOHNE 1964 -1 JOHNSON, E. L.
NETWORK FLOWS, GRAPHS, AND INTEGER PROGRAMMING.
JOHNH 1976 -1 JOHNSON, L. F.
DETERMINING CLIQUES OF A GRAPH.
KACZ 1976 -1 KACZMARSKI, J.
NOMBRE DE STABILITE ET CERTAINES PROPRIETES DE GRAPHES CRITIQUES.
KARPA 1975 -1 KARP, R. M.
FAST APPROXIMATE SOLUTION OF HARD COMBINATORIAL PROBLEMS, THE.
KHAI 1976 -1 KHAIKIN, N. M.; ZAK, E. I.
CONSTRUCTION OF MINIMUM TREE PACKING IN THE PROBLEM OF OPTIMUM MEMORY DISTRIBUTION (RUSSIAN).
KORTC 1977 -1 KORTE, B.; HAUSMANN, D.
ANALYSE DER GREEDY-HEURISTIK FUER UNABHAENGIGKEITSSYSTEME, EINE.
KORTC 1978 -1 KORTE, B.; HAUSMANN, D.
ANALYSIS OF THE GREEDY HEURISTIC FOR INDEPENDENCE SYSTEMS, AN.
KOU A 1978 -1 KOU, L. T.; STOCKMEYER, L. J.; WONG, C. K.
COVERING EDGES BY CLIQUES WITH REGARD TO KEYWORD CONFLICTS AND INTERSECTION GRAPHS.
KUZJ 1977 -1 KUZJURIN, N. N.
MINIMAL COVERINGS AND MAXIMAL PACKINGS OF (K-1)-SUBSETS BY K-SUBSETS (RUSSIAN).
LAWR 1978 -1 LAWRENCE, L.
COVERING THE VERTEX SET OF A GRAPH WITH SUBGRAPHS OF SMALLER DEGREE.
LORE 1972 -1 LOREA, M.
ENSEMBLES STABLES DANS LES HYPERGRAPHES.
LOVA 1976 -1 LOVASZ, L.
COVERS, PACKINGS AND SOME HEURISTIC ALGORITHMS.
MAC A 1976 -1 MAC AVANEY, K. L.
NUMBER AND STABILITY INDICES OF CN-TREES, THE.
MAHU 1976 -1 MAHUAL, D. W.
LARGEST CLIQUE IN A RANDOM GRAPH, THE.
MEEU 1975 -1 MEEUSEN, W.; CUYVERS, L.
CLIQUE DETECTION IN DIRECTED GRAPHS: A NEW ALGORITHM.
MEIR 1977 -1 MEIR, A.; MOON, J. W.
PACKING AND COVERING CONSTANTS FOR CERTAIN FAMILIES OF TREES, I.
MEIR 1975 -1 MEIR, A.; MOON, J. W.
EXPECTED NODE-INDEPENDENCE NUMBER OF VARIOUS TYPES OF TREES, THE.

PACKING

****

MEIR 1973 -1 MEIR, A.; MOON, J. W.
EXPECTED NODE-INDEPENDENCE NUMBER OF RANDOM TREES, THE.

MEIR 1975 -2 MEIR, A.; MOON, J. W.
RELATIONS BETWEEN PACKING AND COVERING NUMBERS OF A TREE.

MEYEA 1972 -1 MEYER, J. C.
ENSEMBLES STABLES MAXIMAUX DANS LES HYPERGRAPHES.

MITC 1975 -1 MITCHELL, S.; HEDETNIEMI, S.; GOODMAN, S.
SOME LINEAR ALGORITHMS ON TREES.

MORE 1977 -1 MOREFIELD, C. L.
APPLICATION OF 0-1 INTEGER PROGRAMMING TO MULTITARGET TRACKING PROBLEMS.

NIEM 1975 -3 NIEMINEN, J.
SOME OBSERVATIONS ON THE DETERMINATION OF AN UPPER BOUND FOR THE CLIQUE NUMBER OF A GRAPH.

OLAR 1973 -1 OLARU, E.
UEBER PERFEKTE UND KRITISCH IMPERFEKTE GRAPHEN.

PADBA 1977 -1 PADBERG, M. W.
COVERING, PACKING AND KNAPSACK PROBLEMS.

PADBA 1976 -1 PADBERG, M. W.
ON THE COMPLEXITY OF SET PACKING POLYHEDRA (REVISED).

PADBA 1977 -2 PADBERG, M. W.
ON THE COMPLEXITY OF SET PACKING POLYHEDRA.

PELE 1976 -1 PELED, U. N.
MATROIDAL GRAPHS.

PERL 1976 -1 PERL, J.
ZUR ALGORITHMISCHEN BEHANDLUNG SPEZIELLER NP-VOLLSTAENDIGER PROBLEME AUF GRAPHEN.

PICAC 1977 -1 PICARD, J.-C.; QUEYRANNE, M.
NETWORKS, GRAPHS AND SOME NONLINEAR 0-1 PROGRAMMING PROBLEMS.

PICAC 1977 -2 PICARD, J.-C.; QUEYRANNE, M.
ON THE INTEGER-VALUED VARIABLES IN THE LINEAR VERTEX PACKING PROBLEM.

POLJ 1974 -1 POLJAK, S.
NOTE ON STABLE SETS AND COLORINGS OF GRAPHS, A.

PULL 1977 -2 PULLEYBLANK, W. R.
TWO-BICRITICAL GRAPHS AND NODE COVERS.

SAATA 1975 -1 SAATY, T. L.; ALEXANDER, J. M.
OPTIMIZATION AND THE GEOMETRY OF NUMBERS: PACKING AND COVERING.

SCHOD 1975 -1 SCHONHEIM, J.; BIALOSTOCKI, A.
PACKING AND COVERING OF THE COMPLETE GRAPH WITH 4-CYCLES.

SEDL 1975 -1 SEDLACEK, J.
PERFECT AND QUASIPERFECT GRAPHS (CZECH).

SHAK 1977 -1 SHAKHBASJAN, K. V.; TUSHKINA, T. A.
ON STATISTICAL EFFICIENCY OF SOLUTION ALGORITHMS FOR A PROBLEM OF VERTEX DIVISION (RUSSIAN).

SHAPD 1977 -1 SHAPIRO, S. D.
PERFORMANCE OF HEURISTIC BIN PACKING ALGORITHMS WITH SEGMENTS OF RANDOM LENGTH.

SONN 1974 -1 SONNEMANN, E.; KRAFFT, O.
INDEPENDENCE NUMBERS OF PRODUCT GRAPHS.

STAC 1976 -1 STACEY, K. C.; MAC AVANEY, K. L.; SIMS, J.
STABILITY INDEX OF THE PRODUCT OF A PATH AND A TREE, THE.

STAR 1975 -1 STAROBINEC, S. M.
LARGEST INTERNALLY STABLE SET OF A GRAPH (RUSSIAN), THE.

STAR 1972 -1 STAROBINEC, S. M.
ALGORITHM FOR FINDING THE GREATEST INTERNALLY STABLE SET OF A GRAPH (RUSSIAN), AN.

STER 1974 -2 STERBOUL, F.
SUR UNE CONJECTURE DE V. CHVATAL.

TARJ 1972 -1 TARJAN, R. E.
FINDING A MAXIMUM CLIQUE.

TARJA 1977 -1 TARJAN, R. E.; TROJANOWSKI, A. E.
FINDING A MAXIMUM INDEPENDENT SET.

TOME 1975 -2 TOMESCU, I.
INTRODUCTION TO COMBINATORICS (TRANSLATED FROM RUMANIAN).

TSUK 1977 -1 TSUKIYAMA, S.; IDE, M.; ARIYOSHI, H.; SHIRAKAWA, I.
NEW ALGORITHM FOR GENERATING ALL THE MAXIMAL INDEPENDENT SETS, A.

WALS 1974 -1 WALSH, A. M.
ON FINDING A K-CLIQUE IN A RANDOM GRAPH.

WEINB 1976 -2 WEINBERGER, D. B.
TRANSVERSAL MATROID INTERSECTIONS AND RELATED PACKINGS.

WOLS 1976 -2 WOLSEY, L. A.
FURTHER FACET GENERATING PROCEDURES FOR VERTEX PACKING POLYTOPES.

WONG 1976 -1 WONG, C. K.
SOME DISCRETE OPTIMIZATION PROBLEMS WITH SECONDARY STORAGE APPLICATIONS.

ZAHN 1973 -1 ZAHN, C. T.
ALTERNATING EULER PATHS FOR PACKINGS AND COVERS.

****

BALAD 1977 -1 BALACHANDRAN, V.; SRINIVASAN, V.; THOMPSON, G. L.
SURVEY OF THE APPLICATIONS OF THE OPERATOR THEORY OF PARAMETRIC PROGRAMMING FOR THE TRANSPORTATION AND GENERALIZED TRANSPORTATION PROBLEM, A.

BALAE 1975 -1 BALACHANDRAN, V.; THOMPSON, G. L.
OPERATOR THEORY OF PARAMETRIC PROGRAMMING FOR THE GENERALIZED TRANSPORTATION PROBLEM - 3 - WEIGHT OPERATORS, AN.

BALAE 1975 -2 BALACHANDRAN, V.; THOMPSON, G. L.
OPERATOR THEORY OF PARAMETRIC PROGRAMMING FOR THE GENERALIZED TRANSPORTATION PROBLEM - 4 - GLOBAL OPERATORS, AN.

BALAE 1975 -3 BALACHANDRAN, V.; THOMPSON, G. L.
OPERATOR THEORY OF PARAMETRIC PROGRAMMING FOR THE GENERALIZED TRANSPORTATION PROBLEM: 1. BASIC THEORY, AN.

BALAE 1975 -4 BALACHANDRAN, V.; THOMPSON, G. L.
OPERATOR THEORY OF PARAMETRIC PROGRAMMING FOR THE GENERALIZED TRANSPORTATION PROBLEM: 2. RIM, COST AND BOUND OPERATORS, AN.

BLAIB 1977 -1 BLAIR, C. E.; JEROSLOW, R. G.
VALUE FUNCTION OF A MIXED INTEGER PROGRAM: I, THE.

CARP 1977 -1 CARPANETO, G.; TOTH, P.
EFFICIENT ALGORITHM FOR THE ASYMMETRIC TRAVELLING SALESMAN PROBLEM, AN.

ESOG 1977 -1 ESOGBUE, A. O.; MARKS, B. R.
DYNAMIC PROGRAMMING MODELS OF THE NONSERIAL CRITICAL PATH-COST PROBLEM.

FONGB 1975 -1 FONG, C. O.; RAO, M. R.
PARAMETRIC STUDIES IN TRANSPORTATION-TYPE PROBLEMS.

GEOFD 1977 -1 GEOFFRION, A. M.; NAUSS, R.
PARAMETRIC AND POSTOPTIMALITY ANALYSIS IN INTEGER LINEAR PROGRAMMING.

HOLM 1978 -1 HOLM, S.; KLEIN, D.
DISCRETE RIGHT HAND SIDE PARAMETRIZATION FOR LINEAR INTEGER PROGRAMS.

HOLM 1976 -1 HOLM, S.; KLEIN, D.
PARAMETRIC ANALYSIS FOR INTEGER PROGRAMMING PROBLEMS.

HROU 1976 -1 HROUDA, J.
BENDERS METHOD AND PARAMETRIZATION OF THE RIGHT-HAND SIDES IN THE MIXED INTEGER LINEAR PROGRAMMING PROBLEM (CZECH), THE.

KASS 1977 -1 KASSAY, F.
APPROACH TO SENSITIVITY ANALYSIS OF THE TRANSPORTATION PROBLEM (CZECH), AN.

LAVI 1976 -1 LAVIT, C.
CLASSES DE SOLUTIONS PARAMETRIQUES D'UNE EQUATION BOOLEENNE.

MARSB 1977 -2 MARSTEN, R. E.; MORIN, T. L.
PARAMETRIC INTEGER PROGRAMMING: THE RIGHT-HAND-SIDE CASE.

MEYEC 1976 -1 MEYER, R. R.
MIXED INTEGER MINIMIZATION MODELS FOR PIECEWISE-LINEAR FUNCTIONS OF A SINGLE VARIABLE.

NOON 1977 -1 NOONAN, F.; GIGLIO, R. J.
PLANNING ELECTRIC POWER GENERATION: A NONLINEAR MIXED INTEGER MODEL EMPLOYING BENDERS DECOMPOSITION.

PIPE 1975 -1 PIPER, C.
COMPUTATIONAL STUDIES IN OPTIMIZING AND POSTOPTIMIZING LINEAR PROGRAMS IN ZERO-ONE VARIABLES.

PIPEA 1975 -1 PIPER, C. J.; ZOLTNERS, A. A.
IMPLICIT ENUMERATION BASED ALGORITHMS FOR POSTOPTIMIZING ZERO-ONE PROGRAMS.

PIPEA 1976 -1 PIPER, C. J.; ZOLTNERS, A. A.
SOME EASY POSTOPTIMALITY ANALYSIS FOR ZERO-ONE PROGRAMMING.

RIDE 1976 -1 RIDER, K. L.
PARAMETRIC MODEL FOR THE ALLOCATION OF FIRE COMPANIES IN NEW YORK CITY, A.

SHAPB 1977 -2 SHAPIRO, J. F.
SENSITIVITY ANALYSIS IN INTEGER PROGRAMMING.

SRINA 1977 -1 SRINIVASAN, V.; THOMPSON, G. L.
COST OPERATOR ALGORITHMS FOR THE TRANSPORTATION PROBLEM.

STAHA 1976 -1 STAHL, W.
SENSITIVITAETSANALYSE MEHRSTUFIGER STANDORTPROBLEME.

SWAI 1971 -1 SWAIN, R. W.
DECOMPOSITION ALGORITHM FOR A CLASS OF FACILITY LOCATION PROBLEM, A.

SWAI 1971 -2 SWAIN, R. W.
PARAMETRIC DECOMPOSITION APPROACH FOR THE SOLUTION OF UNCAPACITATED LOCATION PROBLEMS.

TONE 1976 -1 TONE, K.
PARAMETRIZATION OF BENDERS DECOMPOSITION, A.

WHITD 1967 -2 WHITE, L. J.
PARAMETRIC STUDY OF MATCHINGS AND COVERINGS IN WEIGHTED GRAPHS, A.

PARTITIONING

ALIA 1976 -1 ALIA, G.; MAESTRINI, P.
PROCEDURE TO DETERMINE OPTIMAL PARTITIONS OF WEIGHTED HYPERGRAPHS THROUGH A NETWORK-FLOW ANALOGY, A.

ANZA 1974 -1 ANZAI, Y.
ON INTEGER FRACTIONAL PROGRAMMING.

BALAG 1977 -3 BALAS, E.
SOME VALID INEQUALITIES FOR THE SET PARTITIONING PROBLEM.

BALAH 1974 -1 BALAS, E.; GERRITSEN, R.; PADBERG, M. W.
ALL-BINARY COLUMN GENERATING ALGORITHM FOR SET PARTITIONING, AN.

BALAJ 1976 -1 BALAS, E.; PADBERG, M. W.
SET PARTITIONING: A SURVEY.

BALAJ 1975 -1 BALAS, E.; PADBERG, M. W.
SET PARTITIONING: A SURVEY.

BALAL 1977 -1 BALAS, E.; ZEMEL, E.
GRAPH SUBSTITUTION AND SET PACKING POLYTOPES.

BALLA 1977 -1 BALL, M. O.; NEMHAUSER, G. L.
MATROIDS AND A RELIABILITY ANALYSIS PROBLEM.

BENIA 1976 -1 BENITO, F.; GROEFLIN, H.
OPTIMIERUNGSPROBLEME MIT NICHT POLYNOMIAL BEGRENZTEN ALGORITHMEN (NP-PROBLEME).

BRUC 1978 -1 BRUCKER, P.
ON THE COMPLEXITY OF CLUSTERING PROBLEMS.

COVE 1976 -1 COVERDALE, I. L.; WHARTON, F.
IMPROVED HEURISTIC PROCEDURE FOR A NONLINEAR CUTTING STOCK PROBLEM, AN.

ETCH 1976 -2 ETCHEBERRY, J.
SET REPRESENTATION PROBLEM: A NEW IMPLICIT ENUMERATION ALGORITHM, THE (REVISED VERSION).

FREH 1975 -1 FREHEL, J.
PROBLEME DE PARTITION SOUS CONTRAINTE, LE.

FREH 1975 -2 FREHEL, J.
REGULARISATION DANS LES PROBLEMES COMBINATOIRES, PROBLEMES DE TOURNEES, DE PARTITION ET DE RECOUVREMENT.

GOND 1976 -2 GONDRAN, M.
PROBLEMES DE PARTITION ET DE RECOUVREMENT. APPLICATIONS ET ALGORITHMES, LES.

HANSB 1976 -1 HANSEN, P.; DELATTRE, M.
CLUSTER ANALYSIS AND GRAPH COLORING.

HANSB 1976 -3 HANSEN, P.; DELATTRE, M.
COMPLETE-LINK CLUSTER ANALYSIS BY GRAPH COLORING.

HAUS 1978 -1 HAUSMANN, D.
ADJAZENZCHARAKTERISIERUNGEN FUER POLYEDERECKEN IN DER KOMBINATORISCHEN OPTIMIERUNG.

HAUSC 1976 -1 HAUSMANN, D.; KORTE, B.
COLOURING CRITERIA FOR ADJACENCY ON 0-1-POLYHEDRA.

HEUR 1975 -1 HEURGON, E.; DELORME, J.
PROBLEMES DE PARTITIONNEMENT: EXPLORATION ARBORESCENTE OU METHODE DE TRONCATURER?

JOHNE 1978 -1 JOHNSON, E. L.
ON THE EDGE-COLORING PROPERTY FOR THE CLOSURE OF THE COMPLETE HYPERGRAPHS.

KACPA 1975 -1 KACPRZYK, J.; STANCZAK, W.
APPLICATION OF THE METHOD OF MINIMAL GROUPS TO THE SOLUTION OF THE PROBLEM OF PARTITIONING A GROUP OF FIRMS INTO SUB-GROUPS (POLISH).

KARAA 1976 -1 KARAPETJAN, S. R.
(X-O,M)-MINIMAL CUT AND PARTITIONING OF A TREE (RUSSIAN).

KATS 1977 -1 KATSEV, S. B.
SOLUTION OF A GENERALIZED PROBLEM ON SET PARTITIONING (RUSSIAN).

KOVC 1976 -1 KOVCAS, L. B.; RUPP, M.
GRAPH PARTITIONING FOR MATRIX INVERSION BY INTEGER PROGRAMMING.

KUNDA 1977 -1 KUNDU, S.; MISRA, J.
LINEAR TREE PARTITIONING ALGORITHM, A.

LAMBA 1975 -1 LAMBERT, G.
INDICE DE C-PARTITION DES SOMMETS D'UN GRAPHE. ALGORITHMES DE RECOUVREMENT ET APPLICATION AUX FICHIERS.

LECH 1976 -1 LECHEVALLIER, Y.
CLASSIFICATION AUTOMATIQUE OPTIMALE SOUS CONTRAINTE D'ORDRE TOTAL.

LIN D 1971 -1 LIN, S.; KERNIGHAN, B. W.
HEURISTIC TECHNIQUE FOR SOLVING A CLASS OF COMBINATORIAL OPTIMIZATION PROBLEMS, A.

MAC E 1974 -1 MAC DIARMID, C. J. H.
PATH-PARTITION STRUCTURES OF GRAPHS AND DIGRAPHS.

MARSA 1975 -1 MARSTEN, R. E.
ALGORITHM FOR LARGE SET PARTITIONING PROBLEMS, AN.

MORE 1977 -1 MOREFIELD, C. L.
APPLICATION OF 0-1 INTEGER PROGRAMMING TO MULTITARGET TRACKING PROBLEMS.

MULV 1978 -1 MULVEY, J.
NETWORK RELAXATIONS FOR SET COVERING, SET PARTITIONING AND OTHER INTEGER PROGRAMMING PROBLEMS.

MULVA 1975 -1 MULVEY, J. M.
NETWORK RELAXATION APPROACH FOR THE SET PARTITIONING AND SET COVERING MODELS, A.

O'BR 1977 -1 O'BRIAN, R. C.
UPPER BOUND ON THE PATH NUMBER OF A DIGRAPH, AN.

ORLOB 1975 -1 ORLOVA, G. I.; DORFMAN, G. YA.
OPTIMAL PARTITIONING OF THE SET OF VERTICES OF A GRAPH BY REDUCTION TO A PROBLEM OF MAXIMAL CUT (RUSSIAN).

PARKA 1977 -1 PARKER, R. G.; GRAVES, R. J.; SHERALI, H.
APPLICATION OF A GRAPH-PARTITIONING ALGORITHM TO THE SCHEDULING OF CURRICULUM REQUIREMENTS, THE.

PETR 1976 -1 PETRICH, F.; WAGNER, K.
ALGORITHMUS FUER DAS LINEARE 0-1 OPTIMIERUNGSPROBLEM, EIN.

PRES 1975 -1 PRESENT, M.
PROBLEMES DE PARTITIONNEMENT COMMENT UTILISER DES METHODES DE REDUCTION?

PURC 1975 -1 PURCHECK, G. F. K.
LINEAR-PROGRAMMING METHOD FOR THE COMBINATORIAL GROUPING OF AN INCOMPLETE POWER SET, A.

RECS 1976 -3 RECSKI, A.
ON RANDOM PARTITIONS.

RIHA 1976 -1 RIHA, W.; JAMES, K. R.
ALGORITHM 29. EFFICIENT ALGORITHMS FOR DOUBLY AND MULTIPLY RESTRICTED PARTITIONS.

ROES 1976 -1 ROESSLER, M.
CLUSTER-ANALYSE.

SAHNA 1974 -1 SAHNI, S.; GONZALES, T.
P-COMPLETE PROBLEMS AND APPROXIMATE SOLUTIONS.

****

SALK 1975 -1 SALKIN, H. M.
INTEGER PROGRAMMING.
SHAK 1977 -1 SHAKHBASJAN, K. V.; TUSHKINA, T. A.
ON STATISTICAL EFFICIENCY OF SOLUTION ALGORITHMS FOR A PROBLEM OF VERTEX DIVISION (RUSSIAN).
SPAE 1977 -1 SPAETH, H.
NUMERISCHE ERFAHRUNGEN ZU HEURISTISCHEN LOESUNGSVERFAHREN BEIM VARIANZKRITERIUM IN DER CLUSTER-ANALYSE.
WERR 1974 -1 WERRA, D. DE
PARTITIONS OF GRAPHS INTO COVERINGS AND HYPERGRAPHS INTO TRANSVERSALS.
WHITD 1967 -1 WHITE, L. J.
MINIMUM PARTITIONS AND THE SHANNON SWITCHING GAME.

PROBABILISTIC ANALYSIS IN INTEGER PROGRAMMING
****

ALLE 1976 -1 ALLEN, B.; MUNRO, I.
SELF-ORGANIZING BINARY SEARCH TREES.

BOLLC 1976 -2 BOLLOBAS, B.; ERDOES, P.
CLIQUES IN RANDOM GRAPHS.

BONA 1976 -1 BONA, B.; INAUDI, D.; MAURO, V.
"FUZZY" APPROACH TO A PROBLEM IN RESIDENTIAL LOCATION (ITALIAN), A.

BURTA 1977 -1 BURTIN, YU. D.
PROBABILITY OF THE CONNECTIVITY OF A RANDOM SUBGRAPH OF AN M-DIMENSIONAL CUBE (RUSSIAN).

CERF 1974 -1 CERF, V. G.; COWAN, D. D.; MULLIN, R. C.; STANTON, R. G.
LOWER BOUND ON THE AVERAGE SHORTEST PATH LENGTH IN REGULAR GRAPHS, A.

CRAN 1976 -1 CRANDALL, K. C.
PROBABILISTIC TIME SCHEDULING.

DINE 1972 -1 DINESCU, C.
DETERMINING THE MEAN FLOW IN A TRANSPORTATION NETWORK (RUMANIAN).

ERDOD 1959 -1 ERDOES, P.; RENYI, A.
ON RANDOM GRAPHS, I.

ERDOE 1974 -1 ERDOES, P.; SPENCER, J.
PROBABILISTIC METHODS IN COMBINATORICS.

ERDOF 1977 -1 ERDOES, P.; WILSON, R. J.
ON THE CHROMATIC INDEX OF ALMOST ALL GRAPHS.

EVANA 1976 -3 EVANS, J. R.
MAXIMUM FLOW IN PROBABILISTIC GRAPHS - THE DISCRETE CASE.

FABR 1966 -1 FABRYCKY, W. J.; SHAMBLIN, J. E.
PROBABILITY BASED SEQUENCING ALGORITHM.

FOX A 1975 -1 FOX, B. L.
PROBABILISTIC NETWORKS.

FRANH 1977 -1 FRANK, O.
ESTIMATION OF GRAPH TOTALS.

GOLDB 1977 -2 GOLDEN, B. L.
STATISTICAL APPROACH TO THE TSP, A.

KARPA 1975 -1 KARP, R. M.
FAST APPROXIMATE SOLUTION OF HARD COMBINATORIAL PROBLEMS, THE.

KARPA 1977 -1 KARP, R. M.
PROBABILISTIC ANALYSIS OF PARTITIONING ALGORITHMS FOR THE TRAVELING-SALESMAN PROBLEM IN THE PLANE.

KORD 1976 -1 KORDECKI, W.
ELEMENTS OF THE THEORY OF RANDOM GRAPHS (POLISH).

KOVAA 1975 -1 KOVALENKO, I. N.
ON THE INVARIANCE THEOREMS FOR BOOLEAN RANDOM MATRICES (RUSSIAN).

KRAVA 1975 -1 KRAVTSOV, M. K.; SHERMAN, A. K.; AVERBUKH, N. D.
ALGORITHM FOR THE SOLUTION OF THE ASSIGNMENT PROBLEM (RUSSIAN), AN.

KUCE 1977 -1 KUCERA, L.
EXPECTED BEHAVIOR OF GRAPH COLORING ALGORITHMS.

LEBE 1975 -1 LEBEDEV, S. S.
GENERALIZED LAGRANGE MULTIPLIERS AND ESTIMATION PROBLEMS IN INTEGER PROGRAMMING (RUSSIAN).

LING 1975 -1 LING, R. F.
EXACT PROBABILITY DISTRIBUTION ON THE CONNECTIVITY OF RANDOM GRAPHS, AN.

LING 1973 -1 LING, R. F.
EXPECTED NUMBER OF COMPONENTS IN RANDOM LINEAR GRAPHS, THE.

LITVB 1975 -1 LITVAK, E. I.
ON THE PROBABILITY OF CONNECTEDNESS OF A GRAPH (RUSSIAN).

MAHU 1976 -1 MAHUAL, D. W.
LARGEST CLIQUE IN A RANDOM GRAPH, THE.

MEIR 1977 -1 MEIR, A.; MOON, J. W.
PACKING AND COVERING CONSTANTS FOR CERTAIN FAMILIES OF TREES, I.

MEIR 1973 -1 MEIR, A.; MOON, J. W.
EXPECTED NODE-INDEPENDENCE NUMBER OF RANDOM TREES, THE.

MEIR 1975 -1 MEIR, A.; MOON, J. W.
EXPECTED NODE-INDEPENDENCE NUMBER OF VARIOUS TYPES OF TREES, THE.

MEIR 1976 -1 MEIR, A.; MOON, J. W.
CLIMBING CERTAIN TYPES OF ROOTED TREES, I.

MIRC 1974 -1 MIRCHANDANI, P. B.
LOCATION OF FACILITIES ON A STOCHASTIC NETWORK.

MIRC 1976 -1 MIRCHANDANI, P. B.
SHORTEST DISTANCE AND RELIABILITY OF PROBABILISTIC NETWORKS.

MORO 1974 -1 MOROZENSKII, L. YU.
ASYMPTOTIC BEHAVIOUR OF THE LENGTH OF THE PATH OF A TRAVELING SALEMAN WHEN THE CITIES ARE RANDOMLY DISTRIBUTED (RUSSIAN), THE.

MURC 1973 -1 MURCHLAND, J.
CALCULATING THE PROBABILITY THAT A GRAPH IS DISCONNECTED.

MURCA 1973 -1 MURCHLAND, J.; SHIER, D.
CALCULATING THE PROBABILITY THAT AN UNDIRECTED GRAPH IS DISCONNECTED.

PALA 1971 -1 PALASTI, I.
ON THE THRESHOLD DISTRIBUTION FUNCTION OF CYCLES IN A DIRECTED RANDOM GRAPH.

PULL 1977 -2 PULLEYBLANK, W. R.
TWO-BICRITICAL GRAPHS AND NODE COVERS.

RECS 1976 -3 RECSKI, A.
ON RANDOM PARTITIONS.

SAVAA 1976 -1 SAVAGE, S. L.
PROBABILISTIC APPROACH TO HEURISTIC DESIGN, A.

SAVAA 1973 -2 SAVAGE, S. L.
STATISTICAL INDICATORS OF OPTIMALITY.

SCHN 1976 -1 SCHNORR, C. P.
ALGORITHM FOR TRANSITIVE CLOSURE WITH LINEAR EXPECTED TIME, AN.

TAIN 1975 -1 TAINITER, M.
STATISTICAL THEORY OF CONNECTIVITY. I. BASIC DEFINITIONS AND PROPERTIES.

TARJ 1977 -2 TARJAN, R. E.
FINDING OPTIMUM BRANCHINGS.

WRIG 1975 -1 WRIGHT, E. M.
PROBABILITY OF CONNECTEDNESS OF A LARGE UNLABELLED GRAPH, THE.

YAO B 1976 -1 YAO, A. C.; YAO, F. F.
COMPLEXITY OF SEARCHING AN ORDERED RANDOM TABLE, THE.

ZIMM 1975 -1 ZIMMERMANN, H.-J.; POLLATSCHEK, M. A.
PROBABILITY DISTRIBUTION FUNCTION OF THE OPTIMUM OF A 0-1 LINEAR PROGRAM WITH RANDOMLY DISTRIBUTED COEFFICIENTS OF THE OBJECTIVE FUNCTION AND THE RIGHT-HAND SIDE, THE.

****

ARMS 1978 -1 ARMSTRONG, R. D.
NONBASIC BRANCHING AND DUAL RELAXATION IN A BRANCH-AND-BOUND MIXED INTEGER PROGRAMMING ALGORITHM.
AZAR 1977 -1 AZARYAN, L. L.; LEBEDEV, S. S.; MESTETZKY, L. M.
INTEGER TRANSPORT TYPE PROBLEMS SOLVING BASED ON GENERALIZED LAGRANGE MULTIPLIERS (RUSSIAN).
BALAJ 1975 -1 BALAS, E.; PADBERG, M. W.
SET PARTITIONING: A SURVEY.
BAZAB 1977 -1 BAZARAA, M. S.; GOODE, J. J.
TRAVELING SALESMAN: A DUALITY APPROACH, THE.
BILDA 1977 -1 BILDE, O.; KRARUP, J.
SHARP LOWER BOUNDS AND EFFICIENT ALGORITHMS FOR THE SIMPLE PLANT LOCATION PROBLEM.
CAME 1974 -1 CAMERINI, P. M.; FRATTA, L.; MAFFIOLI, F.
RELAXATION METHODS IMPROVED BY MODIFIED GRADIENT TECHNIQUES.
CAMEA 1974 -1 CAMERINI, P. M.; MAFFIOLI, F.
BOUNDS FOR 3-MATROID INTERSECTION PROBLEMS.
CARP 1977 -1 CARPANETO, G.; TOTH, P.
EFFICIENT ALGORITHM FOR THE ASYMMETRIC TRAVELLING SALESMAN PROBLEM, AN.
CHALB 1976 -1 CHALMET, L. G.; GELDERS, L. F.
LAGRANGEAN RELAXATIONS FOR SOLVING A WAREHOUSING MODEL.
CORNC 1977 -1 CORNUEJOLS, G.; FISHER, M.; NEMHAUSER, G. L.
ON THE UNCAPACITATED LOCATION PROBLEM.
CORND 1977 -1 CORNUEJOLS, G.; FISHER, M. L.; NEMHAUSER, G. L.
LOCATION OF BANK ACCOUNTS TO OPTIMIZE FLOAT: AN ANALYTIC STUDY OF EXACT AND APPROXIMATE ALGORITHMS.
CORND 1976 -1 CORNUEJOLS, G.; FISHER, M. L.; NEMHAUSER, G. L.
ANALYSIS OF HEURISTICS AND RELAXATIONS FOR THE UNCAPACITATED LOCATION PROBLEM, AN.
CROW 1976 -1 CROWDER, H.
COMPUTATIONAL IMPROVEMENTS FOR SUBGRADIENT OPTIMIZATION.
DANN 1977 -2 DANNENBRING, D. G.
PROCEDURES FOR ESTIMATING OPTIMAL SOLUTION VALUES FOR LARGE COMBINATORIAL PROBLEMS.
ERLE 1976 -2 ERLENKOTTER, D.
FACILITY LOCATION WITH PRICE-SENSITIVE DEMANDS: PRIVATE, PUBLIC, AND QUASI-PUBLIC.
ETCH 1976 -1 ETCHEBERRY, J.
NEW BRANCHING AND BOUNDING STRATEGIES FOR 0-1 INTEGER LINEAR PROGRAMMING.
ETCH 1976 -2 ETCHEBERRY, J.
SET REPRESENTATION PROBLEM: A NEW IMPLICIT ENUMERATION ALGORITHM, THE (REVISED VERSION).
EVANB 1977 -1 EVANS, J. R.; JARVIS, J. J.
OUT-OF-KILTER BASED HEURISTIC FOR THE INTEGER MULTICOMMODITY TRANSPORTATION PROBLEM, AN.
FISHA 1976 -1 FISHER, M. L.
DUAL ALGORITHM FOR THE ONE-MACHINE SCHEDULING PROBLEM, A.
GEOFC 1977 -1 GEOFFRION, A. M.; MAC BRIDE, R.
LAGRANGEAN RELAXATION APPLIED TO FACILITY LOCATION PROBLEMS.
GLOVC 1975 -2 GLOVER, F.; KLINGMAN, D.
FINDING MINIMUM SPANNING TREES WITH A FIXED NUMBER OF LINKS AT A NODE.
GLOVH 1975 -1 GLOVER, F.; ROSS, G. T.
IMPROVED BOUNDS FOR A CLASS OF 0-1 INTEGER PROGRAMS USING COMBINED SURROGATE-LAGRANGEAN TECHNIQUES.
GOND 1977 -1 GONDRAN, M.
GROUP THEORETIC AND LAGRANGEAN METHODS.
GUNT 1977 -1 GUNTHER, T. M.
SUBGRADIENT APPROACH TO 0,1 INTEGER PROGRAMMING.
HABE 1977 -1 HABENICHT, W.
ZUR AUSWAHL VON BASISLOESUNGEN BEI VERFAHREN DER GANZZAHLIGEN OPTIMIERUNG.
HAND 1977 -1 HANDLER, G. Y.
M-CENTER PROBLEM: A RELAXATION APPROACH, THE.
HEARA 1976 -1 HEARN, D.; LOWE, T. J.
SUBGRADIENT PROCEDURE FOR THE SOLUTION OF MINIMAL PROBLEMS, A.
HONGA 1977 -1 HONG, S.
ASYMMETRIC TRAVELLING SALESMAN PROBLEM AND SPANNING ARBORESCENCES OF DIRECTED GRAPHS, THE.
HUNGA 1976 -1 HUNG, M. S.; FISK, J. C.
ALGORITHM FOR 0-1 MULTIPLE KNAPSACK PROBLEMS, AN.
KARW 1977 -1 KARWAN, M. H.; RARDIN, R. L.
SEARCHABILITY OF THE COMPOSITE AND MULTIPLE SURROGATE DUAL FUNCTIONS.
KARW 1976 -2 KARWAN, M. H.; RARDIN, R. L.
AURROGATE DUAL MULTIPLIER SEARCH PROCEDURES IN INTEGER PROGRAMMING.
KARW 1977 -2 KARWAN, M. H.; RARDIN, R. L.
SURROGATE DUALITY IN A BRANCH-AND-BOUND PROCEDURE.
KOEH 1977 -1 KOEHLER, G. J.; KUMAR, S.
LOOK AT FINDING POINTS OF A CONVEX POLYHEDRON USING RELAXATION AND SUBGRADIENT PROCEDURES, A.
LAWLA 1976 -1 LAWLER, E. L.; LABETOULLE, J.
SCHEDULING OF PARALLEL MACHINES WITH PREEMPTIONS.
LEBE 1975 -1 LEBEDEV, S. S.
GENERALIZED LAGRANGE MULTIPLIERS AND ESTIMATION PROBLEMS IN INTEGER PROGRAMMING (RUSSIAN).
LEBEA 1976 -1 LEBEDEV, S. S.; KOVALEVSKAJA, M. I.
LAGRANGE MULTIPLIERS IN THE SIMPLEST ALLOCATION PROBLEM (RUSSIAN).
LENSB 1975 -1 LENSTRA, J. K.; RINNOOY KAN, A. H. G.
ON THE JOB-SHOP SCHEDULING PROBLEM.
MAFF 1976 -1 MAFFIOLI, F.
SUBGRADIENT OPTIMIZATION, MATROID PROBLEMS AND HEURISTIC EVALUATION.
MUCK 1977 -1 MUCKSTADT, J. A.; KOENIG, S. A.
APPLICATION OF LAGRANGIAN RELAXATION TO SCHEDULING IN POWER-GENERATION SYSTEMS, AN.
MULVA 1975 -1 MULVEY, J. M.
NETWORK RELAXATION APPROACH FOR THE SET PARTITIONING AND SET COVERING MODELS, A.
NARUA 1977 -1 NARULA, S. C.; OGBU, U. I.; SAMUELSSON, H. M.
ALGORITHM FOR THE P-MEDIAN PROBLEM, AN.
NAUS 1976 -1 NAUSS, R. M.
ALGORITHM FOR THE 0-1 INTEGER LINEAR PROGRAMMING PROBLEM, AN.
NEEBA 1975 -1 NEEBE, A. W.; RAO, M. R.
SUBGRADIENT APPROACH TO THE M-MEDIAN PROBLEM, A.
PICAC 1976 -1 PICARD, J.-C.; QUEYRANNE, M.
TIME-DEPENDENT TRAVELING SALESMAN PROBLEM AND APPLICATION TO THE TARDINESS PROBLEM IN ONE-MACHINE SCHEDULING, THE.
SAYA 1976 -1 SAYAMA, H.; FAN, L. T.; FAN, L. S.
ON LAGRANGE MULTIPLIERS AND CONSTRAINTS, I. LAGRANGIAN APPROACH.

****

SAYA 1976 -2 SAYAMA, H.; FAN, L. T.; FAN, L. S.
ON LAGRANGE MULTIPLIERS AND CONSTRAINTS, II. AUGMENTED LANGRANGIAN APPROACH.
SCHO 1976 -1 SCHOCH, M.
ERWEITERUNGSPRINZIP UND SEINE ANWENDUNG ZUR ENTWICKLUNG VON ALGORITHMEN FUER DIE LOESUNG KOMBINATORISCHER OPTIMIERUNGSAUFGABEN, DAS.
SHAPB 1977 -3 SHAPIRO, J. F.
SURVEY OF LAGRANGEAN TECHNIQUES FOR DISCRETE OPTIMIZATION.
SMITH 1975 -1 SMITH, T. H. C.
LIFO IMPLICIT ENUMERATION ALGORITHM FOR THE ASYMMETRIC TRAVELING SALESMAN PROBLEM USING A ONE-ARBORESCENCE RELAXATION, A.
SWEE 1977 -1 SWEENEY, D. J.; MURPHY, R. A.
DECOMPOSITION PRINCIPLE FOR INTEGER PROGRAMS, A (REVISED VERSION).

****

AGNE 1976 -1 AGNEW, C. E.
ON QUADRATIC ADAPTIVE ROUTING ALGORITHMS.
ASAN 1976 -1 ASANO, T.; SHIBUI, M.; TAKANAMI, I.
GENERAL RESULTS ON TOUR LENGTHS IN MACHINES AND DIGRAPHS.
BABU 1977 -1 BABU, A. J. G.; HSIA, W. S.
MORE EFFICIENT ALGORITHM FOR AN OPTIMAL TOUR OF A HEALTH CARE CONSUMER WITH MULTIPLE HEALTH CARE FACILITIES IN EACH COUNTY, A.
BARL 1978 -1 BARLOW, J. F.
ALGORITHM FOR THE SOLUTION OF THE POSTMAN'S PROBLEM, AN.
BELL 1973 -1 BELL, D. E.
RESOLUTION OF DUALITY GAPS IN DISCRETE OPTIMIZATION, THE.
BERGE 1975 -1 BERGMANN, D. R.
INTEGER PROGRAMMING FORMULATION FOR DERIVING MINIMUM DISPATCH INTERVALS ON A GUIDEWAY ACCOMODATING THROUGH AND LOCAL PUBLIC TRANSPORTATION SERVICES.
BJOE 1977 -1 BJOERKLUND, B.; HOLMBERG, G.
VEHICLE ROUTING AND SCHEDULING FOR DELIVERING NEWSPAPERS IN THE STOCKHOLM METROPOLITAN AREA.
CANT 1974 -1 CANTOR, D. G.; GERLA, M.
OPTIMAL ROUTING IN A PACKET-SWITCHED COMPUTER NETWORK.
CHRI 1976 -1 CHRISTOFIDES, N.
VEHICLE ROUTING PROBLEM, THE.
CLAUA 1975 -1 CLAUS, A.; KLEITMAN, D. J.
HEURISTIC METHODS FOR SOLVING LARGE SCALE NETWORK ROUTING PROBLEMS: THE TELPACKING PROBLEM.
DINK 1976 -1 DINKEL, J. J.; KLEINDORFER, G. B.; KOCHENBERGER, G. A.; WONG, S.-N.
ENVIRONMENTAL INSPECTION ROUTES AND THE CONSTRAINED TRAVELLING SYSTEM SALESMAN PROBLEM.
DIXO 1976 -1 DIXON, E. T.; GOODMAN, S. E.
ALGORITHM FOR THE LONGEST CYCLE PROBLEM, AN.
FRANF 1971 -1 FRANK, H.; CHOU, W.
HEURISTIC APPROACH TO ROUTING IN COMPUTER NETWORKS, A.
FRAT 1975 -1 FRATTA, L.
BASIC ANALYTICAL TECHNIQUES AND ROUTING PROCEDURES.
FRED 1976 -1 FREDERICKSON, G. N.; HECHT, M. S.; KIM, C. E.
APPROXIMATION ALGORITHMS FOR SOME ROUTING PROBLEMS.
GABB 1974 -1 GABBAY, H.
OVERVIEW OF VEHICULAR SCHEDULING PROBLEMS, AN.
GALL 1977 -1 GALLAGER, R. G.
MINIMUM DELAY ROUTING ALGORITHM USING DISTRIBUTED COMPUTATION. SPECIAL ISSUE ON COMPUTER COMMUNICATIONS, A.
GAUL 1976 -1 GAUL, W.
ON CONSTRAINED SHORTEST-ROUTE PROBLEMS.
GIAN 1976 -1 GIANNESSI, F.
DECOMPOSITION METHOD FOR THE TRAVELLING SALESMAN PROBLEM AND ITS APPLICATIONS, A.
GOLD 1976 -1 GOLDEN, B.
LARGE-SCALE VEHICLE ROUTING AND RELATED COMBINATORIAL PROBLEMS.
GOLD 1975 -1 GOLDEN, B.
VEHICLE ROUTING PROBLEMS: FORMULATIONS AND HEURISTIC SOLUTION TECHNIQUES.
GOLDB 1977 -1 GOLDEN, B. L.
EVALUATING A SEQUENTIAL VEHICLE ROUTING ALGORITHM.
GOLDB 1976 -2 GOLDEN, B. L.
RECENT DEVELOPMENTS IN VEHICLE ROUTING.
GOLDD 1977 -1 GOLDEN, B. L.; MAGNANTI, T. L.; NGUYEN, H. Q.
IMPLEMENTING VEHICLE ROUTING ALGORITHMS.
GOND 1975 -2 GONDRAN, M.
PATH ALGEBRA AND ALGORITHMS.
GRUNA 1975 -1 GRUNDLER, G.
ALGORITHMEN ZUR LOESUNG VERSCHIEDENER VARIANTEN DES TOURENPROBLEMS.
GRUNB 1977 -1 GRUNDLER, G.; SCHOCH, M.
TOURENPROBLEM MIT ZENTRALEM FUHRPARK BEI UNTERSCHIEDLICHEN FAHRZEUGTYPEN, DAS.
HECK 1976 -1 HECK, J.
ALGORITHMUS ZUR LOESUNG DES CHINESE-POSTMAN PROBLEMS, EIN.
HEIN 1977 -2 HEIN, O.
LOESUNGSVORSCHLAG ZUM VERALLGEMEINERTEN RUNDFAHRTPROBLEM, EIN.
HERS 1965 -1 HERSHDORFER, A. M.
OPTIMAL ROUTING OF URBAN TRAFFIC.
HILL 1976 -1 HILL, A. V.; WHYBARK, D. C.
COMPARING EXACT SOLUTION PROCEDURES FOR THE MULTI-VEHICLE ROUTING PROBLEM.
ISHI 1975 -2 ISHII, H.; IBARAKI, T.; MINE, H.
TRAVELING SALESMAN PROBLEMS WITH A CAPACITY CONSTRAINT OF THE DELIVERY TRUCK.
JAER 1973 -1 JAERVINEN, P.
ON A ROUTING PROBLEM IN A GRAPH WITH TWO TYPES OF COSTS ASSOCIATED WITH THE EDGES.
JEFF 1965 -1 JEFFERIS, R. P.; FEGLEY, K. A.
APPLICATION OF DYNAMIC PROGRAMMING TO ROUTING PROBLEMS.
KAAZ 1973 -1 KAAZIK, YU. YA.; PREEM, M. K.
PROBLEM OF FINDING ROUTES (RUSSIAN), A.
KAPP 1975 -1 KAPPAUF, C. H.; KOEHLER, G. J.
POSTMAN PROBLEM, THE.
KERS 1976 -1 KERSHENBAUM, A.; HSIEH, W.; GOLDEN, B.
CONSTRAINED ROUTING IN LARGE SPARSE NETWORKS.
KOH A 1974 -1 KOH, K. M.; TEH, H. H.
ON DIRECTED POSTMAN PROBLEM.
KOZY 1972 -1 KOZYREV, V. P.
GRAPH THEORY (RUSSIAN).
KRAR 1975 -1 KRARUP, J.
PERIPATETIC SALESMAN AND SOME RELATED UNSOLVED PROBLEMS, THE.
LEE A 1976 -1 LEE, J.; BOSE, N.; HWANG, F.
USE OF STEINER'S PROBLEM IN SUBOPTIMAL ROUTING IN RECTILINEAR METRIC.
LEITA 1973 -1 LEITEN, A. K.
CERTAIN MODIFICATIONS OF THE TRAVELLING SALESMAN PROBLEM (RUSSIAN).
LEMA 1971 -1 L LEMAIRE, B.
PROBLEMES DE TOURNEES AVEC CONTRAINTES MULTIPLES.
LENSB 1976 -2 LENSTRA, J. K.; RINNOOY KAN, A. H. G.
ON GENERAL ROUTING PROBLEMS.
LENSB 1976 -3 LENSTRA, J. K.; RINNOOY KAN, A. H. G.
TRAVELLING ON A GRAPH (DUTCH).
MAC L 1977 -1 MAC NEELY, R. A.; MALSTROM, E. M.
COMPUTER GENERATED PROCESS ROUTINGS.

****

MALE 1977 -1 MALE, J. W.; LIEBMAN, J. C.; ORLOFF, C. S.
IMPROVEMENT OF ORLOFF'S GENERAL ROUTING PROBLEM, AN.

MICK 1975 -1 MICKLE, M. H.
METHOD OF OPTIMUM ROUTE REGENERATION, A.

MICKA 1969 -1 MICKLE, M. H.; VOGT, W. G.
DISCRETE CONTROL AS A TRANSPORTATION PROBLEM.

MICKA 1975 -1 MICKLE, M. H.; VOGT, W. G.
ON THE DYNAMICS OF A CLASS OF OPTIMUM ROUTING PROBLEMS.

MICKB 1977 -1 MICKLE, M. H.; WANG, J. C.; HOELZEMAN, R. G.
ON THE STORAGE AND REGENERATION OF OPTIMUM ROUTES.

MONT 1976 -1 MONTICONE, L. C.
SOLUTION TO AN OPTIMAL ROUTING PROBLEM IN A LARGE SCALE NETWORK VIA A MINIMUM PERMUTATION SET.

NGUY 1975 -1 NGUYEN, H.
MULTI-DEPOT VEHICLE ROUTING PROBLEMS.

ORLO 1976 -1 ORLOFF, C.; CAPRERA, D.
REDUCTION AND SOLUTION OF LARGE SCALE VEHICLE ROUTING PROBLEMS.

ORLOA 1976 -1 ORLOFF, C. S.
ON GENERAL ROUTING PROBLEMS: COMMENTS.

PAPA 1976 -2 PAPADIMITRIOU, C. H.
ON THE COMPLEXITY OF EDGE TRAVERSING.

PARK 1977 -1 PARKER, R. G.; DEANE, R. H.; HOLMES, R. A.
ON THE USE OF A VEHICLE ROUTING ALGORITHM FOR THE PARALLEL PROCESSOR PROBLEM WITH SEQUENCE DEPENDENT CHANGEOVER COSTS.

PERK 1965 -1 PERKO, A.
SOME COMPUTATIONAL NOTES ON THE SHORTEST ROUTE PROBLEM.

PETEB 1977 -1 PETERSEN, J. P.
MANUAL PROCEDURE FOR DESIGNING MAILMAN-ROUTES.

PIER 1975 -1 PIERCE, A. R.
BIBLIOGRAPHY ON ALGORITHMS FOR SHORTEST PATH, SHORTEST SPANNING TREE, AND RELATED CIRCUIT ROUTING PROBLEMS (1956-1974).

RICH 1976 -1 RICHARDSON, R.
OPTIMIZATION APPROACH TO ROUTING AIRCRAFT, AN.

RINA 1975 -1 RINALDI, S.
OPTIMAL CONSTRAINED CYCLES IN GRAPHS.

RINA 1975 -2 RINALDI, S.
OPTIMAL UNCONSTRAINED CYCLES IN GRAPHS.

ROBB 1975 -1 ROBBINS, J. A.; SHAMBLIN, J. E.; TURNER, W. C.; BYRD, D. F.
DEVELOPMENT OF AND COMPUTATIONAL EXPERIENCE WITH A COMBINATION TOUR CONSTRUCTION - TOUR IMPROVEMENT ALGORITHM FOR VEHICLE ROUTING PROBLEMS.

ROY 1975 -1 ROY, B.
CHEMINS ET CIRCUITS: ENUMERATION ET OPTIMISATION.

ROY 1975 -2 ROY, B.
CHEMINS ET CIRCUITS: ENUMERATION ET OPTIMISATION: A SURVEY.

RUSS 1975 -1 RUSSEL, R.
EFFECTIVE HEURISTIC FOR THE M-TOUR TRAVELING SALESMAN PROBLEM WITH SOME SIDE CONDITIONS, AN.

RUSSB 1977 -1 RUSSELL, R.; IGO, W.
ASSIGNMENT ROUTING PROBLEM, AN.

RUSSD 1965 -1 RUSSO, F. J.
HEURISTIC APPROACH TO ALTERNATE ROUTING IN A JOB SHOP.

SCHMB 1975 -1 SCHMITZ, W.
UEBER LAENGSTE WEGE UND KREISE IN GRAPHEN.

SCHO 1976 -1 SCHOCH, M.
ERWEITERUNGSPRINZIP UND SEINE ANWENDUNG ZUR ENTWICKLUNG VON ALGORITHMEN FUER DIE LOESUNG KOMBINATORISCHER OPTIMIERUNGSAUFGABEN, DAS.

SCHW 1977 -1 SCHWARTZ, M.
COMPUTER-COMMUNICATION NETWORK. DESIGN AND ANALYSIS.

SMITD 1976 -1 SMITH, L. D.
APPLICATION OF AN INTERACTIVE ALGORITHM TO DEVELOP CYCLICAL ROTATIONAL SCHEDULES FOR NURSING PERSONNEL, THE.

SPAT 1975 -1 SPATHIS-DIMITROPOULOU, V.
ETUDE ALGEBRIQUE D'UN ARBRE. UN PROBLEME DE CIRCULATION, UNE.

STEW 1976 -1 STEWART, W. R. JR.
DELIVERY TRUCK ROUTING PROBLEM WITH STOCHASTIC DEMANDS, THE.

STRE 1974 -1 STREET, A. P.
EULERIAN WASHING MACHINES.

SYSLA 1976 -1 SYSLO, M. M.
GENERALIZATIONS OF THE STANDARD TRAVELLING SALESMAN PROBLEM.

TURNB 1976 -1 TURNER, W. C.; TERRELL, M. P.
VEHICLE ROUTING ACTIVITY AT OKLAHOMA STATE UNIVERSITY.

YAGE 1969 -1 YAGED, B. J.
ROUTING INFORMATION IN MINIMUM COST NETWORKS.

YEN 1972 -1 YEN, J. Y.
FINDING THE OPTIMAL CYCLES IN MULTIPLY WEIGHTED NETWORKS.

****

ACKE 1977 -1 ACKERMAN, L. J.; LUSS, H.; BERKOWITZ, R. S.
APPLICATION OF SEQUENCING POLICIES TO TELEPHONE SWITCHING FACILITIES.

ACKE 1977 -2 ACKERMAN, L. J.; LUSS, H.; BERKOWITZ, R. S.
IMPLICIT ENUMERATION ALGORITHM FOR SEQUENCING POLICIES APPLIED TO TELEPHONE SWITCHING FACILITIES, AN.

ACKEA 1963 -1 ACKERMAN, S. S.
EVEN-FLOW. A SCHEDULING METHOD FOR REDUCING LATENESS IN JOB SHOPS.

ADAM 1977 -1 ADAM, N.; SURKIS, J.
COMPARISON OF CAPACITY PLANNING TECHNIQUES IN A JOB SHOP CONTROL SYSTEM, A.

ADOL 1977 -1 ADOLPHSON, D. L.
SINGLE MACHINE JOB SEQUENCING WITH PRECEDENCE CONSTRAINTS.

AKHP 1977 -1 AKHPATELOV, E. A.; CHERENIN, V. P.
OPTIMAL SEQUENCE FOR A SINGLE-DEVICE SERVICE (RUSSIAN), THE.

ALEKA 1977 -1 ALEKSEYEV, O. G.; MERZHANOV, V. YU.
ALGORITHM FOR JOB ALLOCATION ACCORDING TO A MINIMAX CRITERION (RUSSIAN), AN.

ALEKC 1964 -1 ALEKSEYEV, O. G.; YAKUSHEV, V. I.
ALGORITHM OF OPTIMAL MACHINE SCHEDULING (RUSSIAN).

ANDO 1976 -1 ANDON, F. I.; VARTANYAN, G. M.
PROBLEM OF PACKAGING DEMANDS WITH TIMING MARKS IN INFORMATION SYSTEMS (RUSSIAN).

ARIS 1977 -1 ARISAWA, S.; ELMAGHRABY, S. E.
"HUB" AND "WHEEL" SCHEDULING PROBLEMS, I. THE "HUB" SCHEDULING PROBLEM: THE MYOPIC CASE, THE.

ARIS 1977 -2 ARISAWA, S.; ELMAGHRABY, S. E.
"HUB" AND "WHEEL" SCHEDULING PROBLEMS, II. THE HUB OPERATION SCHEDULING PROBLEM (HOSP): MULTI-PERIOD AND INFINITE HORIZON, AND THE WHEEL OPERATION SCHEDULING PROBLEM (WOSP), THE.

ARMSB 1976 -1 ARMSTRONG, R. D.; COOK, W. D.
OPTIMAL LINKING OF TASKS AND SCHEDULE DEVELOPMENT FOR MILITARY AIR TRANSPORT.

BABUA 1977 -1 BABUSHKIN, A. I.; BASHTA, A. L.; BELOV, I. S.
CONSTRUCTION OF THE SCHEDULE FOR THE COUNTER ROUTES PROBLEM (RUSSIAN).

BABUA 1976 -1 BABUSHKIN, A. I.; BASHTA, A. L.; BELOV, I. S.
SCHEDULING A THREE-MACHINE JOB-SHOP (RUSSIAN).

BAGG 1973 -1 BAGGA, P. C.
TWO MACHINE SCHEDULING PROBLEM WHEN PASSING ALLOWED.

BAGGA 1967 -1 BAGGA, P. C.; CHAKRAVARTI, N. K.
OPTIMAL M-STAGE PRODUCTION SCHEDULE.

BAKE 1977 -1 BAKER, K. R.
COMPUTATIONAL EXPERIENCE WITH A SEQUENCING ALGORITHM ADAPTED TO THE TARDINESS PROBLEM.

BAKE 1976 -1 BAKER, K. R.
WORKFORCE ALLOCATION IN CYCLICAL SCHEDULING PROBLEMS: A SURVEY.

BAKEA 1977 -1 BAKER, K. R.; BURNS, R. N.; CARTER, M.
STAFF SCHEDULING WITH DAY-OFF AND WORKSTRETCH CONSTRAINTS.

BAKEB 1977 -1 BAKER, K. R.; MAGAZINE, M.
WORKFORCE SCHEDULING WITH CYCLIC DEMANDS AND DAY-OFF CONSTRAINTS.

BALAG 1968 -1 BALAS, E.
PROJECT SCHEDULING WITH RESOURCE CONSTRAINTS.

BALL 1977 -1 BALL, B. C.; WEBSTER, D. B.
OPTIMAL SCHEDULING FOR EVEN-NUMBERED TEAM ATHLETIC CONFERENCES.

BANE 1965 -1 BANERJEE, B. P.
SINGLE FACILITY SEQUENCING WITH RANDOM EXECUTION TIMES.

BANS 1976 -1 BANSAL, S. P.
GLOBAL LOWER BOUND IN THE FLOW-SHOP PROBLEM.

BANS 1977 -1 BANSAL, S. P.
MINIMIZING THE SUM OF COMPLETION TIMES OF N JOBS OVER M MACHINES IN A FLOWSHOP - A BRANCH AND BOUND APPROACH.

BARN 1977 -1 BARNES, J. W.; BRENNAN, J. J.
IMPROVED ALGORITHM FOR SCHEDULING JOBS ON IDENTICAL MACHINES, AN.

BARRB 1976 -1 BARRER, J.; CHANDRASEKARAN, R.
ADJACENT PAIRWISE INTERCHANGE (API) ALGORITHM: A CLARIFICATION OF THE LITERATURE, THE.

BARRD 1977 -1 BARRY, D. L.; ROBINSON, D. F.
HEURISTIC FOR SCHEDULING JOBS ON SEVERAL IDENTICAL MACHINES, A.

BARTA 1977 -1 BARTHOLDI, J. J. III; ORLIN, J. B.; RATLIFF, H. D.
CIRCULAR 1'S AND CYCLIC STAFFING.

BARTB 1977 -2 BARTHOLDI, J. J. III; RATLIFF, H. D.
UNNETWORKS, WITH APPLICATIONS TO IDLE TIME SCHEDULING.

BECKA 1974 -1 BECKER, L.; YEH, W. W.-G.
OPTIMAL TIMING, SEQUENCING, AND SIZING OF MULTIPLE RESERVOIR SURFACE WATER SUPPLY FACILITIES.

BEEN 1976 -1 BEENHAKKER, H. L.; NARAYANAN, V.
ALGORITHMS FOR SCHEDULING PROJECTS WITH LIMITED RESOURCES.

BELK 1965 -1 BELKINA, M. V.; GROMOVA, A. P.; KYUNNAPU, E. P.; OSKALKOV, I. O.
OPTIMIZATION OF OPERATIONAL SEQUENCES.

BELOA 1974 -1 BELOV, I. S.; STOLIN, JA. N.
ALGORITHM FOR THE SINGLE-ROUTE SCHEDULING PROBLEM (RUSSIAN), AN.

BENN 1976 -1 BENNETT, G. R.; BYRD, J. JR.
TRAINABLE HEURISTIC PROCEDURE FOR THE ASSEMBLY LINE BALANCING PROBLEM, A.

BERR 1977 -1 BERRY, W. L.
AUTHOR'S REPLY TO GRAVES' NOTE ON THE PAPER "CRITICAL RATIO SCHEDULING: AN EXPERIMENTAL ANALYSIS".

BERRA 1976 -1 BERRY, W. L.; FINLAY, R. A.
CRITICAL RATIO SCHEDULING WITH QUEUE WAITING TIME INFORMATION: AN EXPERIMENTAL ANALYSIS.

BERRB 1975 -1 BERRY, W. L.; RAO, V.
CRITICAL RATIO SCHEDULING: AN EXPERIMENTAL ANALYSIS.

BEST 1976 -1 BESTWICK, P. F.; HASTINGS, N. A. J.
NEW BOUND FOR MACHINE SCHEDULING, A.

BEYE 1970 -1 BEYER, K.
EINIGE VERFAHREN ZUR TERMINPLANUNG IN NETZPLAENEN MIT BESCHRAENKTER RESOURCENVERFUEGBARKEIT.

BJOE 1977 -1 BJOERKLUND, B.; HOLMBERG, G.
VEHICLE ROUTING AND SCHEDULING FOR DELIVERING NEWSPAPERS IN THE STOCKHOLM METROPOLITAN AREA.

BLAS 1967 -1 BLASYUK, B. A.
OPTIMAL ASSIGNMENT OF PROCESSING OF ITEMS ON THREE SEQUENTIAL MACHINES (RUSSIAN).

BLAZ 1976 -1 BLAZEWICZ, J.
DEADLINE SCHEDULING OF TASKS - A SURVEY.

BLAZA 1976 -1 BLAZEWICZ, J.; CELLARY, W.
SOME COMPUTATIONAL PROBLEMS OF SCHEDULING DEPENDENT AND PREEMPTABLE TASKS TO MINIMIZE SCHEDULE LENGTH.

****

BLAZB 1976 -1 BLAZEWICZ, J.; CELLARY, W.; SLOWINSKI, R.; WEGLARZ, J.
DETERMINISTIC PROBLEMS OF SCHEDULING TASKS ON PARALLEL PROCESSORS, I. SETS OF INDEPENDENT TASKS (POLISH).

BLAZB 1976 -2 BLAZEWICZ, J.; CELLARY, W.; SLOWINSKI, R.; WEGLARZ, J.
DETERMINISTIC PROBLEMS OF SCHEDULING TASKS ON PARALLEL PROCESSORS, II. SETS OF DEPENDENT TASKS (POLISH).

BODI 1976 -1 BODIN, L. D.; ROSENFIELD, D.
ESTIMATION OF THE OPERATING COST OF MASS TRANSIT SYSTEMS.

BOEB 1976 -1 BOEBION, J.; PUN, L.
SERIES-PARALLEL MULTIPLE-CRITERIA MODEL FOR A SCHEDULING PROBLEM IN THE DRESS-MAKING INDUSTRY, A.

BOER 1977 -1 BOERE, N, J,
AIR CANADA SAVES WITH AIRCRAFT MAINTENANCE SCHEDULING.

BOGA 1976 -1 BOGARDI, J. J.; DUCKSTEIN, L.; PLATE, E. J.
SCHEDULING AND SEQUENCING THE CONSTRUCTION OF FLOOD CONTROL RESERVOIRS.

BONNA 1976 -1 BONNEY, M. C.; GUNDRY, S. W.
SOLUTIONS TO THE CONSTRAINED FLOWSHOP SEQUENCING PROBLEM.

BOROD 1976 -1 BOROVITS, I.; EIN-DOR, P.
TWO ISSUES IN THE EVALUATION OF JOB-SHOP SCHEDULES.

BOURA 1970 -1 BOURKOV, V. N.
RESOLUTION DE QUELQUES PROBLEMES D'OPTIMISATION.

BRAT 1975 -1 BRATLEY, P.; FLORIAN, M.; ROBILLARD, P.
SCHEDULING WITH EARLIEST START AND DUE DATE CONSTRAINTS ON MULTIPLE MACHINES.

BRELA 1977 -1 BRELAZ, D.; NICOLIER, Y.; WERRA, D. DE
COMPACTNESS AND BALANCING IN SCHEDULING.

BROWD 1976 -1 BROWNELL, W. S.; LOWERRE, J. M.
SCHEDULING OF WORK FORCES REQUIRED IN CONTINUOUS OPERATIONS UNDER ALTERNATIVE LABOR POLICIES.

BRUC 1977 -1 BRUCKER, P.
PROZESSOR SCHEDULING IN VERBINDUNG MIT MULTIPROGRAMMING.

BRUC 1976 -2 BRUCKER, P.
KOMPLEXITAET VON SCHEDULING PROBLEMEN, DIE.

BRUCA 1977 -1 BRUCKER, P.; GAREY, M. R.; JOHNSON, D. S.
SCHEDULING EQUAL-LENGTH TASKS UNDER TREE-LIKE PRECEDENCE CONSTRAINTS TO MINIMIZE MAXIMUM LATENESS.

BRUCB 1975 -1 BRUCKER, P.; LENSTRA, J. K.; RINNOOY KAN, A. H. G.
COMPLEXITY OF MACHINE SCHEDULING PROBLEMS.

BRUNA 1976 -1 BRUNO, J.; COFFMAN, E. G. JR.; JOHNSON, D. B.
ON BATCH SCHEDULING OF JOBS WITH STOCHASTIC SERVICE TIMES AND COST STRUCTURES ON A SINGLE SERVER.

BRUNB 1974 -1 BRUNO, J.; COFFMAN, E. G. JR.; SETHI, R.
ALGORITHMS FOR MINIMIZING MEAN FLOW TIME.

BRUNC 1975 -1 BRUNO, J.; HOFRI, M.
ON SCHEDULING CHAINS OF JOBS ON ONE PROCESSOR WITH LIMITED PREEMPTION.

BRUNE 1976 -1 BRUNO, J. L.
SCHEDULING ALGORITHMS FOR MINIMISING THE MEAN WEIGHTED FLOW-TIME CRITERION.

BRUNE 1976 -2 BRUNO, J. L.
SEQUENCING JOBS WITH STOCHASTIC TASK STRUCTURES ON A SINGLE MACHINE.

BRUNF 1975 -1 BRUNS, M.
REIHENFOLGEPROBLEM DER NETZPLANTECHNIK, DAS.

BULF 1976 -1 BULFIN, R. L.; PARKER, R. G.
ON A TWO FACILITY SCHEDULING PROBLEM WITH SEQUENCE DEPENDENT PROCESSING TIME.

BULF 1976 -2 BULFIN, R. L.; PARKER, R. G.
ON THE ASSIGNMENT AND SEQUENCING OF OPERATIONS ON A CREW-SERVED PROJECT.

BUNT 1977 -1 BUNT, R. B.; HUME, J. N. P.
ADAPTIVE PROCESSOR SCHEDULING BASED ON APPROXIMATING DEMAND DISTRIBUTION.

BURKJ 1968 -1 BURKOV, V. N.; LOVETSKII, S. E.
METHODS FOR THE SOLUTION OF EXTREMAL PROBLEMS OF COMBINATORIAL TYPE (REVIEW) (RUSSIAN).

BURNA 1976 -1 BURNS, F.; ROOKER, J.
JOHNSON'S THREE-MACHINE FLOW-SHOP CONJECTURE.

BURNA 1975 -1 BURNS, F.; ROOKER, J.
SPECIAL CASE OF THE 3 X N FLOW SHOP PROBLEM, A.

BURNB 1976 -1 BURNS, R. N.
SCHEDULING TO MINIMIZE THE WEIGHTED SUM OF COMPLETION TIMES WITH SECONDARY CRITERIA.

BURNB 1977 -1 BURNS, R. N.
STATIONARY POINTS OF HEURISTICS IN PERMUTATION SCHEDULING.

BUSH 1976 -1 BUSHELL, G.; FERLAND, J.; GUERIN, G.; NASTANSKY, L.
ENGINE SCHEDULING PROBLEM IN A RAILWAY NETWORK, THE.

BUTE 1974 -1 BUTEN, R. E.
ANALYSIS OF THE FLOW-SHOP SCHEDULING PROBLEM IN A SYSTEM WITH TWO CLASSES OF PROCESSORS.

BUTL 1965 -1 BUTLAR, W. H.
SHORT INTERVAL SCHEDULING; TURNS LARGE JOBS INTO SMALL TASKS.

BUZA 1976 -1 BUZACOTT, J. A.
PRODUCTION CAPACITY OF JOB SHOPS WITH LIMITED STORAGE SPACE, THE.

CAMP 1970 -1 CAMPBELL, H. G.; DUDEK, R. A.; SMITH, M. L.
HEURISTIC ALGORITHM FOR THE N-JOB, M-MACHINE SEQUENCING PROBLEM.

CARRA 1965 -1 CARROLL, D. C.
HEURISTIC SEQUENCING OF SINGLE AND MULTIPLE COMPONENT JOBS.

CAVA 1976 -1 CAVAILLE, J. B.; DELMAS, J.
ORDONNANCEMENT OPTIMAL A TRES COURT TERME D'UN ATELIER DE DECOUPE.

CHANG 1975 -1 CHANDRASEKARAN, R.; EMMONS, H.
SEQUENCING JOBS FOR A SINGLE WORKER TO PROCESS AND DELIVER.

CHANH 1975 -1 CHANDRASEKARAN, R.; MITTAL, A. K.
OPTIMAL REARRANGEMENT OF OBJECT.

CHEB 1975 -1 CHEBOTAREV, A. S.
SEQUENTIAL OPTIMIZATION IN ONE SCHEDULING PROBLEM (RUSSIAN).

CHEBA 1975 -1 CHEBOTAREV, A. S.; SHIBANOV, S. E.
PROBABILISTIC PROBLEM OF SEQUENCING PROJECTS IN SEVERAL STEPS.

CHENA 1977 -1 CHEN, DER-S.
ALGORITHM FOR A MANPOWER SCHEDULING MODEL, AN.

CHENE 1976 -1 CHEN, R.; CROWDER, H.; JOHNSON, E. L.
INTEGER PROGRAMMING FORMULATION OF THE INSTALLATION SCHEDULING PROBLEM, AN.

CHENI 1971 -1 CHEN, Y. E.; SHEN, V. Y.
SCHEDULING PROBLEM IN A MULTIPROCESSING ENVIRONMENT WITH TWO CLASSES OF PROCESSORS.

****

CHEW 1975 -1 CHEW, K. L.; YAP, K. Y.
OPTIMAL PRODUCTION RUN LENGTHS FOR TWO-PRODUCT SCHEDULING.

CHEWA 1974 -1 CHEW, K. L.; YAP, K. Y.
NON-OPTIMAL FLOW-SHOP SCHEDULES.

CHOE 1976 -1 CHOE, S. R.
ON A NEW OPTIMAL DECISION METHOD FOR SCHEDULING PROBLEMS (KOREAN).

CIOB 1975 -1 CIOBANU, G.
GENERAL ALGORITHM FOR SOLVING FLOW SHOP SCHEDULING, A.

CIRI 1975 -1 CIRILLO, F.
NEW METHOD FOR THE SCHEDULING OF A PROJECT (ITALIAN), A.

COFF 1975 -1 COFFMAN, E. G. JR.
DETERMINISTIC SEQUENCING: COMPLEXITY AND OPTIMAL ALGORITHMS.

COFF 1976 -1 COFFMAN, E. G. JR.
COMPUTER AND JOB/SHOP SCHEDULING THEORY.

COFFA 1976 -1 COFFMAN, E. G. JR.; SETHI, R.
ALGORITHMS MINIMIZING MEAN FLOW TIME SCHEDULE-LENGTH PROPERTIES.

CONW 1966 -1 CONWAY, R. W.; MAXWELL, W. L.; OLDZIEY, J. W.
SEQUENCING AGAINST DUE DATES.

CRAN 1976 -1 CRANDALL, K. C.
PROBABILISTIC TIME SCHEDULING.

DANN 1977 -2 DANNENBRING, D. G.
PROCEDURES FOR ESTIMATING OPTIMAL SOLUTION VALUES FOR LARGE COMBINATORIAL PROBLEMS.

DANTC 1974 -1 DANTZIG, G. B.; FRIEL, J.; GOLIGHTLY, R.; HARVEY, R. P.; MAC KNIGHT, R. D.
SOLUTION OF A LARGE SCALE AIRFORCE ORDNANCE PLANNING PROBLEM BY MATHEMATICAL PROGRAMMING.

DAR- 1975 -1 DAR-EL, E. M.; COTHER, R. F.
ASSEMBLY LINE SEQUENCING FOR MODEL MIX.

DAR-A 1977 -1 DAR-EL, E. M.; TUR, Y.
MULTI-RESOURCE PROJECT SCHEDULING ALGORITHM, A.

DAVID 1977 -1 DAVIS, K. R.; WALTERS, J. E.
ADDRESSING THE N/1 SCHEDULING PROBLEM - A HEURISTIC APPROACH.

DAY 1975 -1 DAY, J. E.; HOTTENSTEIN, M. P.
IMPACT OF ADVANCING DUE DATES IN A PURE JOB SHOP, THE.

DEAR 1977 -1 DEAR, R. G.
EFFICIENT SEQUENCING AND SCHEDULING OF AIRCRAFT IN THE TERMINAL AREA, THE.

DEMIA 1976 -1 DEMIN, V. K.; CHEBOTAREV, A. S.
OPTIMAL SERVICE OF A DETERMINISTIC FLOW OF DEMANDS, I. LOWER BOUND (RUSSIAN).

DESC 1964 -1 DESCAMPS, R.; CHEVIGNON, P.
ALGORITHMS FOR OPTIMIZING A GENERAL CLASS OF SCHEDULING PROBLEMS WITH LIMITED RESOURCES (FRENCH).

DIKE 1964 -1 DIKE, S. H.
PROJECT SCHEDULING WITH RESOURCE CONSTRAINTS.

DINI 1976 -1 DINITS, E. A.
SOLUTION OF TWO SCHEDULING PROBLEMS (RUSSIAN), THE.

DOERA 1977 -1 DOERSCH, R. H.; PATTERSON, J. H.
SCHEDULING A PROJECT TO MAXIMIZE ITS PRESENT VALUE: A ZERO-ONE PROGRAMMING APPROACH.

DRIS 1976 -1 DRISCOLL, W. C.
BRANCH-BOUND ALGORITHM FOR SCHEDULING PRODUCTION ON ONE MACHINE WITH CHANGEOVER TIMES, A.

DRIS 1975 -1 DRISCOLL, W. C.
SCHEDULING PRODUCTION ON ONE MACHINE WITH CHANGEOVER TIMES.

DRUT 1975 -1 DRUTOV, V. E.
SOLUTION OF A SCHEDULING PROBLEM (RUSSIAN).

ECKE 1977 -1 ECKER, K.
ORGANISATION VON PARALLELEN PROZESSEN. THEORIE DETERMINISTISCHER SCHEDULES.

ECON 1977 -1 ECONOMIDES, S. C.; DUESTERHOEFT, W. C.
CONSTRUCTION SCHEDULING SYSTEM WITH RESOURCE ALLOCATION FOR ELECTRIC UTILITIES, A.

EILO 1975 -1 EILON S.; CHOWDHURY, I. G.; SERGHIOU, S. S.
EXPERIMENTS WITH THE SI RULE IN JOB-SHOP SCHEDULING.

EILOA 1975 -1 EILON, S.; CHOWDHURY, I. G.
STUDIES IN A SIMULATED JOB SHOP.

EILOA 1976 -1 EILON, S.; CHOWDHURY, I. G.
CONDITIONS FOR OPTIMAL SCHEDULING FOR THE THREE-MACHINE PROBLEM.

EILOA 1977 -1 EILON, S.; CHOWDHURY, I. G.
MINIMISING WAITING TIME VARIANCE IN THE SINGLE MACHINE PROBLEM.

EILOA 1976 -2 EILON, S.; CHOWDHURY, I. G.
DUE DATES IN JOB SHOP SCHEDULING.

EILOB 1967 -1 EILON, S.; HODGSON, M.
JOB SHOP SCHEDULING WITH DUE DATES.

EMDE 1975 -1 EMDE BOAS, P. V.
PRESERVING ORDER IN A FOREST IN LESS THAN LOGARITHMIC TIME.

EMMO 1968 -1 EMMONS, H.
ONE MACHINE SEQUENCING TO MINIMIZE TOTAL TARDINESS.

ERLE 1973 -2 ERLENKOTTER, D.
SEQUENCING OF INTERDEPENDENT HYDROELECTRIC PROJECTS.

ERLE 1975 -3 ERLENKOTTER, D.
COMMENT ON "OPTIMAL TIMING, SEQUENCING, AND SIZING OF MULTIPLE RESERVOIR SURFACE WATER SUPPLY FACILITIES" BY L. BECKER AND W. W.-G. YEH.

ERLE 1975 -4 ERLENKOTTER, D.
COORDINATING SCALE AND SEQUENCING DECISIONS FOR WATER RESOURCES PROJECTS.

ERLE 1975 -5 ERLENKOTTER, D.
OPTIMAL INVESTMENT SCHEDULING FOR PUBLIC SECTOR PROJECTS.

ERLEA 1975 -1 ERLENKOTTER, D.; ROGERS, J. S.
SEQUENCING COMPETITIVE EXPANSION PROJECTS.

ERLEB 1975 -1 ERLENKOTTER, D.; TRIPPI, R. R.
OPTIMAL INVESTMENT SCHEDULING WITH PRICE-SENSITIVE DYNAMIC DEMAND.

ERSC 1976 -1 ERSCHLER, J.
ANALYSE SOUS CONTRAINTES ET AIDE A LA DECISION POUR CERTAINS PROBLEMES D'ORDONNANCEMENT.

ESKE 1975 -1 ESKEW, J. D.; PARKER, R. G.
COMPUTATIONAL EXPERIENCE WITH A COST-BASED ALGORITHM FOR THE SHOP SCHEDULING PROBLEM.

FABR 1966 -1 FABRYCKY, W. J.; SHAMBLIN, J. E.
PROBABILITY BASED SEQUENCING ALGORITHM.

FAUR 1976 -1 FAURE, R.; ROUCAIROL, C.; TOLLA, P.
CHEMINS ET FLOTS, ORDONNANCEMENTS.

FEIG 1972 -1 FEIGIN, L. I.
GENERAL PROBLEM OF SCHEDULING THEORY WITH INCOMPLETE INFORMATION (RUSSIAN).

****

FEIG 1967 -1 FEIGIN, L. I.
EFFECT OF DISTURBANCES ON SCHEDULING.
FERN 1976 -1 FERNANDEZ, E. B.; LANG, T.
SCHEDULING AS A GRAPH TRANSFORMATION.
FERN 1975 -1 FERNANDEZ, E. B.; LANG, T.
COMPUTATION OF LOWER BOUNDS FOR MULTIPROCESSOR SCHEDULES.
FISHA 1976 -1 FISHER, M. L.
DUAL ALGORITHM FOR THE ONE-MACHINE SCHEDULING PROBLEM, A.
FOST 1976 -1 FOSTER, B. A.; RYAN, D. M.
INTEGER PROGRAMMING APPROACH TO THE VEHICLE SCHEDULING PROBLEM, AN.
FRIE 1975 -1 FRIEDMAN, M.
SCHEDULING A TRANSPORTATION NETWORK.
FROM 1976 -1 FROMOVITZ, S.; GASS, S. I.
SCHEDULING OF SUBMARINES FOR FLEET OPERATIONS SUPPORT, THE.
GABB 1974 -1 GABBAY, H.
OVERVIEW OF VEHICULAR SCHEDULING PROBLEMS, AN.
GAREA 1976 -1 GAREY, M. R.; GRAHAM, R. L.; JOHNSON, D. S.; YAO, A. C.-C.
RESOURCE CONSTRAINED SCHEDULING AS GENERALIZED BIN PACKING.
GAREB 1974 -1 GAREY, M. R.; JOHNSON, D. S.
COMPLEXITY RESULTS FOR MULTIPROCESSOR SCHEDULING UNDER RESOURCE CONSTRAINTS.
GAREB 1977 -2 GAREY, M. R.; JOHNSON, D. S.
TWO-PROCESSOR SCHEDULING WITH START TIMES AND DEADLINES.
GAREB 1976 -4 GAREY, M. R.; JOHNSON, D. S.
SCHEDULING TASKS WITH NONUNIFORM DEADLINES ON TWO PROCESSORS.
GAREC 1976 -1 GAREY, M. R.; JOHNSON, D. S.; SETHI, R.
COMPLEXITY OF FLOWSHOP AND JOBSHOP SCHEDULING, THE.
GELD 1976 -1 GELDERS, L.; WASSENHOVE, L. VAN
SHORT-TIME PLANNING OF THE "JOB-SHOP" TYPE FOR A PRODUCTION SITE (DUTCH).
GELDA 1976 -1 GELDERS, L. F.
SOLVING AN OVERALL PRODUCTION PLANNING PROBLEM IN A SIMPLE JOB-SHOP.
GEOFA 1976 -1 GEOFFRION, A. M.; GRAVES, G. W.
SCHEDULING PARALLEL PRODUCTION LINES WITH CHANGEOVER COSTS: PRACTICAL APPLICATION OF A QUADRATIC ASSIGNMENT/LP APPROACH.
GERO 1970 -1 GEROV, G. H.
SOLUTION OF A CERTAIN SCHEDULING PROBLEM IN A SPECIAL CASE (RUSSIAN).
GERT 1977 -1 GERTSBACH, I.
PROBABILISTIC IDEAS IN HEURISTIC ALGORITHMS FOR SOLVING SOME SCHEDULING PROBLEMS.
GERTA 1977 -1 GERTSBACH, I.; GUREVICH, YU.
CONSTRUCTING AN OPTIMAL FLEET FOR A TRANSPORTATION SCHEDULE.
GILLE 1975 -1 GILLETT, B. E.
COMPUTERIZED CREW SCHEDULING IN LEAD MINES.
GIMAA 1976 -1 GIMADI, E. K.; GLEBOV, N. I.; PEREPELITSA, V. A.
ALGORITHMS WITH ESTIMATIONS FOR THE PROBLEMS OF DISCRETE OPTIMIZATION (RUSSIAN).
GLAZ 1976 -1 GLAZEBROOK, K. D.
STOCHASTIC SCHEDULING WITH ORDER CONSTRAINTS.
GLAZA 1976 -1 GLAZEBROOK, K. D.; NASH, P.
ON MULTI-SERVER STOCHASTIC SCHEDULING.
GLEBA 1970 -1 GLEBOV, N. I.; PEREPELITSA, V. A.
LOWER AND UPPER ESTIMATES FOR A CERTAIN PROBLEM IN SCHEDULING THEORY (RUSSIAN), THE.
GLEBB 1963 -1 GLEBSKII, Y.; SHEVCHENKO, V. N.
CONSTRUCTION OF AN OPTIMAL WORK SCHEDULE (RUSSIAN).
GLEN 1975 -1 GLEN, J. J.
DYNAMIC PROGRAMMING MODEL FOR WORK SCHEDULING IN A SHIPYARD, A.
GLOVL 1965 -1 GLOVER, P. C.; NORMAN, J. T.
ASSEMBLY LINE BALANCING.
GOLE 1974 -1 GOLENKO, D. I.; GELEVER, K. G.; KESLER, S. S.; GOFMAN, E. F.
PROBLEMS ON OPTIMAL SCHEDULING AND OPTIMAL RESOURCE ALLOCATION FOR A TECHNOLOGICAL OPERATING PROCEDURE (RUSSIAN).
GOLL 1969 -1 GOLLER, S.
CONTRIBUTION TO THE THEORY OF PURPOSEFUL TIMETABLE MAKING (CZECH).
GOND 1976 -1 GONDRAN, M.
PROBLEME DU PLUS COURT CHEMIN DANS UN GRAPHE, LE.
GONDA 1977 -1 GONDRAN, M.; DOSTATNI, M.
TRAITEMENT DES EXCLUSIVES DANS PLANNEC, LE.
GONZB 1977 -1 GONZALEZ, M. J. JR.
DETERMINISTIC PROCESSOR SCHEDULING.
GONZC 1976 -1 GONZALEZ, M. J. JR.; SOH, J. W.
PERIODIC JOB SCHEDULING IN A DISTRIBUTED PROCESSOR SYSTEM.
GONZD 1977 -1 GONZALEZ, T.; IBARRA, O. H.; SAHNI, S.
BOUNDS FOR LPT SCHEDULES ON UNIFORM PROCESSORS.
GONZE 1978 -1 GONZALEZ, T.; SAHNI, S.
PREEMPTIVE SCHEDULING OF UNIFORM PROCESSOR SYSTEMS.
GONZE 1976 -1 GONZALEZ, T.; SAHNI, S.
OPEN SHOP SCHEDULING TO MINIMIZE FINISH TIME.
GORD 1976 -1 GORDON, A. JA.
ALGORITHM FOR THE SOLUTION OF THE MINIMAX SCHEDULING PROBLEM (RUSSIAN), AN.
GOYA 1976 -1 GOYAL, S. K.
GRAPHICAL AIDS FOR SCHEDULING A MULTI-PRODUCT SINGLE MACHINE SYSTEM.
GOYA 1977 -1 GOYAL, S. K.
FLOW-SHOP SEQUENCING PROBLEM WITH NO WAIT IN PROCESS AND NON-ZERO SET-UP TIMES.
GOYA 1975 -1 GOYAL, S. K.
SCHEDULING A MULTI-PRODUCT SINGLE MACHINE SYSTEM. A NEW APPROACH.
GOYA 1975 -2 GOYAL, S. K.
SCHEDULING A SINGLE MACHINE SYSTEM: A MULTI-PRODUCT MULTI-ITEM CASE.
GRAB 1976 -1 GRABOWSKI, J.
NEW FORMULATION AND SOLUTION OF THE SEQUENCING PROBLEM: ALGORITHM, A.
GRAB 1976 -2 GRABOWSKI, J.
NEW FORMULATION AND SOLUTION OF THE SEQUENCING PROBLEM: MATHEMATICAL MODEL, A.
GRABA 1975 -1 GRABOWSKI, J.; SYSLO, M. M.
NEW SOLVABLE CASES OF M-MACHINE AND N-ELEMENT SEQUENCING PROBLEMS.
GRAH 1976 -1 GRAHAM, R. L.
BOUNDS ON THE PERFORMANCE OF SCHEDULING ALGORITHMS.
GRAHA 1977 -1 GRAHAM, R. L.; LAWLER, E. L.; LENSTRA, J. K.; RINNOOY KAN, A. H. G.
OPTIMIZATION AND APPROXIMATION IN DETERMINISTIC SEQUENCING AND SCHEDULING: A SURVEY.

****

GRANC 1976 -1 GRANT, D. D.
HOW TO RUN A HOCKEY TOURNAMENT - A COMBINATORIAL SCHEDULING PROBLEM.
GRAVC 1977 -1 GRAVES, S. C.
NOTE ON "CRITICAL RATIO SCHEDULING: AN EXPERIMENTAL ANALYSIS", A.
GUPT 1976 -1 GUPTA, J. N. D.
HEURISTIC ALGORITHM FOR THE FLOWSHOP SCHEDULING PROBLEM, A.
GUPT 1975 -1 GUPTA, J. N. D.
OPTIMAL SCHEDULES FOR SPECIAL STRUCTURE FLOWSHOPS.
GUPT 1969 -1 GUPTA, J. N. D.
SEQUENCING N-JOBS ON M-PARALLEL PROCESSORS.
GUPT 1976 -2 GUPTA, J. N. D.
OPTIMAL FLOWSHOP SCHEDULES WITH NO INTERMEDIATE STORAGE SPACE.
HAES 1976 -1 HAESSLER, R. W.; HOGUE, S. L.
NOTE ON THE SINGLE-MACHINE MULTI-PRODUCT LOT SCHEDULING PROBLEM, A.
HEAR 1976 -1 HEARD, E. L.
DUE-DATES AND INSTANTANEOUS LOAD IN THE ONE-MACHINE SHOP.
HEIS 1977 -1 HEISING, C. D.
PLANT LOCATION AND TRANSPORTATION MODELS OF THE LIGHT WATER REACTOR FUEL CYCLE.
HELLA 1960 -1 HELLER, J.; LOGEMANN, G.
ALGORITHM FOR CONSTRUCTING FEASIBLE SCHEDULES AND COMPUTING THEIR SCHEDULE TIMES.
HEND 1976 -1 HENDERSON, P. B.; STEIGLITZ, K.
PROBLEM IN SINGLE-MACHINE SEQUENCING WITH NONLINEAR DELAY COSTS, A.
HEPP 1976 -1 HEPPES, A.; LUGOSI, G.
PRODUCTION SCHEDULING TO MINIMIZE THE SET-UP TIME OF PARALLEL MACHINES WITH ALTERNATIVE CHOICES (HUNGARIAN).
HERZ 1975 -1 HERZOG, U.
OPTIMAL SCHEDULING STRATEGIES FOR REAL-TIME COMPUTERS.
HINS 1976 -1 HINSON, J.; MULHERKAR, S.
IMPROVEMENTS TO THE CLARKE AND WRIGHT ALGORITHM AS APPLIED TO AN AIRLINE SCHEDULING PROBLEM.
HOLL 1977 -1 HOLLOWAY, C. A.; NELSON, R. T.
INTERMITTEN PRODUCTION SCHEDULING: PRACTICAL FORMULATIONS, RELAXATION, AND DECOMPOSITION PROCEDURES.
HOLMB 1976 -1 HOLMES, R. A.; PARKER, R. G.
VEHICLE SCHEDULING PROCEDURE BASED UPON SAVINGS AND A SOLUTION PERTURBATION SCHEME, A.
HOLZ 1975 -1 HOLZMAN, A. G.
MASTER TIME SCHEDULE GENERATOR - AN ALGORITHM APPROACH.
HONG 1974 -1 HONG, M. L.
EXPERIMENTS WITH A JOB SHOP PRODUCTION MODEL COUPLING A LINEAR PROGRAM AND A HEURISTIC SCHEDULING PROCEDURE.
HORE 1977 -1 HOREN, J. H.
EFFECT OF SCHEDULING NETWORK TELEVISION PROGRAMS ON AUDIENCE RATINGS, THE.
HORN 1972 -1 HORN, R. L. VAN
OPTIMIZING TREE SEARCH SIMULATOR FOR SCHEDULING MAINTENANCE ON AIRCRAFT, AN.
HORO 1976 -1 HOROWITZ, E.; SAHNI, S.
EXACT AND APPROXIMATE ALGORITHMS FOR SCHEDULING NONIDENTICAL PROCESSORS.
HORV 1977 -1 HORVATH, E. C.; LAM, S.; SETHI, R.
LEVEL ALGORITHM FOR PREEMPTIVE SCHEDULING, A.
HUME 1960 -1 HUME, J. N. P.
SCHEDULING PRODUCTION IN JOB SHOPS.
IAZE 1968 -1 IAZEOLLA, G.
TIME SHARING SYSTEMS: THEORETICAL TREATMENT OF A PRIORITY SCHEDULING ALGORITHM (ITALIAN).
IBARC 1977 -1 IBARRA, O. H.; KIM, C. E.
HEURISTIC ALGORITHMS FOR SCHEDULING INDEPENDENT TASKS ON NONIDENTICAL PROCESSORS.
IBARC 1975 -1 IBARRA, O. H.; KIM, C. E.
ON TWO-PROCESSOR SCHEDULING OF ONE- OR TWO-UNIT TIME TASKS WITH PRECEDENCE CONSTRAINTS.
IGUD 1975 -1 IGUDIN, R. V.
PROBLEMS OF SEQUENCING THEORY IN TRANSPORTATION AND ALGORITHMS FOR THEIR SOLUTION (RUSSIAN).
JACOB 1977 -1 JACOBSON, R. E.; SADOWSKI, R. F.
SCHEDULING ALGORITHM FOR THE UNBALANCED PRODUCTION LINE, A.
JAINA 1975 -1 JAIN, S. K.
SIMULATION-BASED SCHEDULING AND MANAGEMENT INFORMATION SYSTEM FOR A MACHINE SHOP, A.
JOHNE 1976 -3 JOHNSON, E. L.
NOTE ON MUTUALLY EXCLUSIVE SCHEDULING AND CRITICAL PATHS, A.
JULL 1976 -1 JULLIEN, B.
DIFFERENTES APPROCHES LOGICIELLES POUR LA RESOLUTION DES PROBLEMES COMBINATOIRES EN TEMPS REEL.
KAFU 1977 -1 KAFURA, D. G.; SHEN, V. Y.
TASK SCHEDULING ON A MULTIPROCESSOR SYSTEM WITH INDEPENDENT MEMORIES.
KALY 1976 -1 KALYUZHNYJ, B. I.; SOLDATOV, V. S.
MODEL OF SEQUENCING DISCRETE PRODUCTION PROCESSES (RUSSIAN), A.
KAMI 1975 -1 KAMINSKAS, V.; PUKAS, V.
OPTIMAL MEAN LOAD SCHEDULING IN A POWER PLANT (RUSSIAN).
KARA 1976 -1 KARAGEZOV, Z. A.
DECOMPOSITIONAL ALGORITHM FOR NON-CORRECT PROBLEM OF TIME SCHEDULE OPTIMIZATION (RUSSIAN).
KAUF 1974 -1 KAUFMAN, M. T.
ALMOST-OPTIMAL ALGORITHM FOR THE ASSEMBLY LINE SCHEDULING PROBLEM, AN.
KAZA 1975 -1 KAZARIN, L. S.; L'VOV, G. N.; TIMOFEEV, E. A.
HEURISTIC ALGORITHM FOR CONSTRUCTING A TIME-TABLE (RUSSIAN), A.
KHEN 1966 -1 KHENKIN, V. E.
PROBLEM IN SCHEDULING THEORY (2 TIMES M ORDERING PROBLEM).
KIBA 1976 -1 KIBALOV, E. B.; KHUTORETZKY, A. B.
VOLUME-TIME SCHEDULE PLANNING PROBLEM OF BUILDING (RUSSIAN), A.
KIM 1976 -1 KIM, K. V.; KRAYNOV, G. A.; SURMENEV, E. V.
ON A PROBLEM OF TIME SCHEDULE COMPILATION (RUSSIAN).
KLUY 1977 -1 KLUYVER, C. A. DE
EXTENSION OF ELZINGA AND VEMUGANTI'S DECOMPOSITION SCHEME TO MULTIPERIOD ADVERTISING MEDIA SCHEDULING, AN.
KOHLA 1975 -1 KOHLER, W.
COMPUTATIONAL EXPERIENCE WITH EFFICIENT EXACT AND APPROXIMATE ALGORITHMS FOR AN NP-COMPLETE SCHEDULING PROBLEM.
KOHLB 1976 -1 KOHLER, W. H.; STEIGLITZ, K.
ENUMERATIVE AND ITERATIVE COMPUTATIONAL APPROACHES.
KORN 1969 -1 KORNILOVA, L. E.
METHOD OF MINIMAL FLOW FOR THE EXACT SOLUTION OF THE PROBLEM OF MINIMIZATION OF THE MAXIMAL LEVEL OF RESOURCE UTILIZATION IN NETWORK PLANNING (RUSSIAN), THE.

****

KRON 1971 -1 KRONE, M.
HEURISTIC PROGRAMMING APPLIED TO SCHEDULING MODELS.

KUBA 1974 -1 KUBALE, M.
HEURISTIC ALGORITHM FOR ESTABLISHING TIME-TABLES IN A FACULTY (POLISH).

KURI 1973 -1 KURISU, T.
TWO-MACHINE SEQUENCING PROBLEM WITH EXPONENTIAL PROCESSING TIMES.

KURI 1976 -1 KURISU, T.
TWO-MACHINE SCHEDULING UNDER REQUIRED PRECEDENCE AMONG JOBS.

LAGE 1976 -1 LAGEWEG, B. J.; LENSTRA, J. K.; RINNOOY KAN, A. H .G.
GENERAL BOUNDING SCHEME FOR THE PERMUTATION FLOW-SHOP, A.

LAGEA 1976 -1 LAGEWEG, B. J.; LENSTRA, J. K.; RINNOOY KAN, A. H. G.
GENERAL BOUNDING SCHEME FOR THE PERMUTATION FLOW-SHOP, A.

LAGEA 1976 -2 LAGEWEG, B. J.; LENSTRA, J. K.; RINNOOY KAN, A. H. G.
JOB-SHOP SCHEDULING BY IMPLICIT ENUMERATION.

LAGEA 1976 -3 LAGEWEG, B. J.; LENSTRA, J. K.; RINNOOY KAN, A. H. G.
JOB-SHOP SCHEDULING BY IMPLICIT ENUMERATION.

LAGEA 1976 -4 LAGEWEG, B. J.; LENSTRA, J. K.; RINNOOY KAN, A. H. G.
MINIMIZING MAXIMUM LATENESS ON ONE MACHINE: COMPUTATIONAL EXPERIENCE AND SOME APPLICATIONS.

LAM 1977 -1 LAM, S.; SETHI, R.
WORST CASE ANALYSIS OF TWO SCHEDULING ALGORITHMS.

LAMBB 1963 -1 LAMBOURN, S.
RESOURCE ALLOCATION AND MULTIPROJECT SCHEDULING (RAMPS). A NEW TOOL IN PLANNING AND CONTROL.

LANDA 1978 -1 LAND, A. H.; LAPORTE, G.: M. , P.
UNIFIED FORMULATION OF THE MACHINE SCHEDULING PROBLEM, A.

LANG 1977 -1 LANG, T.; FERNANDEZ, E. B.
IMPROVING THE COMPUTATION OF LOWER BOUNDS FOR OPTIMAL SCHEDULES.

LANGA 1962 -1 LANGEFORS, B.
ACTIVITY NETWORK FOR PLANNING AND SCHEDULING.

LAPI 1976 -1 LAPIN, V. V.
ON A SCHEDULING ALGORITHM FOR SERVICING MANY DEMANDS (RUSSIAN).

LATE 1976 -1 LATECOERE, L.; BOEBION, J.; PUN, L.
MODEL ADAPTIVE SCHEDULING FOR SOLVING PERTURBATION PROBLEMS IN THE CLOTH INDUSTRY.

LAVE 1976 -1 LAVENBERG, S. S.; SHEDLER, G. S.
STOCHASTIC MODELING OF PROCESSOR SCHEDULING WITH APPLICATION TO DATA BASE MANAGEMENT SYSTEMS.

LAWL 1977 -1 LAWLER, E. L.
'PSEUDOPOLYNOMIAL' ALGORITHM FOR SEQUENCING JOBS TO MINIMIZE TOTAL TARDINESS, A.

LAWL 1978 -1 LAWLER, E. L.
SEQUENCING JOBS TO MINIMIZE TOTAL WEIGHTED COMPLETION TIME SUBJECT TO PRECEDENCE CONSTRAINTS.

LAWL 1975 -5 LAWLER, E. L.
OPTIMAL SEQUENCING OF JOBS SUBJECT TO SERIES PARALLEL PRECEDENCE CONSTRAINTS.

LAWLA 1976 -1 LAWLER, E. L.; LABETOULLE, J.
SCHEDULING OF PARALLEL MACHINES WITH PREEMPTIONS.

LAZA 1977 -1 LAZAK, D.; GERRITSEN, S. M.
TOPOLOGISCHE KANAELE BEI JOB-SHOP-SCHEDULING-VERFAHREN.

LEIF 1967 -1 LEIFMAN, L. Y. A.
PROBLEM OF ALLOCATION OF RESOURCES IN NETWORK SCHEDULING (RUSSIAN).

LENSA 1977 -1 LENSTRA, J. K.
SEQUENCING BY ENUMERATIVE METHODS.

LENSB 1975 -1 LENSTRA, J. K.; RINNOOY KAN, A. H. G.
ON THE JOB-SHOP SCHEDULING PROBLEM.

LENSB 1976 -1 LENSTRA, J. K.; RINNOOY KAN, A. H. G.
COMPLEXITY OF SCHEDULING UNDER PRECEDENCE CONSTRAINTS.

LENSC 1977 -1 LENSTRA, J. K.; RINNOOY KAN, A. H. G.; BRUCKER, P.
COMPLEXITY OF MACHINE SCHEDULING PROBLEMS.

LEVN 1973 -1 LEVNER, E. V.
NETWORK APPROACH TO PROBLEMS OF SCHEDULING THEORY (RUSSIAN), A.

LIET 1977 -1 LIETAER, B. A.
PLANNING MODEL FOR UNDERGROUND MINES. AN APPLICATION IN A DEVELOPING COUNTRY, A.

LINK 1976 -1 LINKE, K.-F.
BEMERKUNGEN ZUR MATHEMATISCH-OEKONOMISCHEN MODELLIERUNG UND NUMERISCHEN LOESUNG VON REIHENFOLGEAUFGABEN AUS DER INDUSTRIELLEN FERTIGUNG.

LINSA 1969 -1 LINSKIY, V. S.; KORNEV, M. D.
SCHEDULING TWO-STATE PROCESSES WITHOUT A BUFFER.

LITT 1976 -1 LITTGER, K. W.
NEW APPROACH FOR THE OPTIMUM SOLUTION OF THE M BY J PRODUCTION SCHEDULING PROBLEM, A.

LIU 1976 -1 LIU, C. L.
COMBINATORIAL STUDY OF SOME SCHEDULING ALGORITHMS, A.

LIU 1973 -1 LIU, C. L.
COMBINATORIAL STUDY OF SOME SCHEDULING ALGORITHMS (SUMMARY), A.

LONG 1977 -1 LONG, C. F.
STATE OF THE ART IN SCHEDULING PRACTICE, THE.

LUED 1975 -1 LUEDTKE, H.
COMPARISON OF HEURISTIC GRAPH SCHEDULING ALGORITHMS, A.

LUEDA 1976 -1 LUEDTKE, H.; SCHINDLER, S.
COMPLEXITY OF SCHEDULING ALGORITHMS FOR THREE-PROCESSOR SYSTEMS, THE.

LYTK 1974 -1 LYTKIN, I. P.
INVERSE SCHEDULING PROBLEM (RUSSIAN), THE.

LYTK 1974 -2 LYTKIN, I. P.
SCHEDULING OF MEANS OF TRANSPORTATION (RUSSIAN).

MAC L 1977 -1 MAC NEELY, R. A.; MALSTROM, E. M.
COMPUTER GENERATED PROCESS ROUTINGS.

MAGA 1977 -1 MAGAZINE, M. J.
SCHEDULING A PATIENT TRANSPORTATION SERVICE IN A HOSPITAL.

MAKI 1966 -1 MAKINO, T.
SCHEDULING PROBLEM.

MALI 1974 -1 MALICKA-WASOWSKA, J.
SEQUENCING MODEL WITH CONTINUOUS MACHINE OPERATION (POLISH), A.

MANS 1964 -1 MANSOOR, E. M.
ASSEMBLY LINE BALANCING. IMPROVEMENT ON RANKED POSITIONAL WEIGHT TECHNIQUE.

MARK 1977 -1 MARK, J. W.
GLOBAL SCHEDULING APPROACH TO CONFLICT-FREE MULTIACCESS: PART II-DELAY ANALYSIS FOR A BUS ARCHITECTURE.

****

MARKD 1974 -1 MARKOV, L. A.; PERVOZVANSKII, A. A.
RULE FOR "BOTTLENECKS" IN THE JOHNSON PROBLEM (RUSSIAN), A.
MARQ 1977 -1 MARQUEZ, J.; MEDINA-MORA, O.
NETWORK SOLUTION TO A GENERAL VEHICLE SCHEDULING PROBLEM, A.
MARTD 1976 -1 MARTIN-VEGA, L. A.; RATLIFF, H. D.
SCHEDULING RULES FOR PARALLEL PROCESSORS.
MARTH 1976 -1 MARTINEZ Y RODRIGUEZ, J. B.; RUSSELL, S. O.
SEQUENTIAL TECHNIQUES FOR PROJECT SCREENING.
MAUR 1976 -1 MAUREL, E.; ROUX, D.; DUPONT, D.
TECHNIQUES OPERATIONNELLES D'ORDONNANCEMENT FONDEES SUR LA METHODE PERT "POTENTIELS-TACHES".
MAWD 1974 -1 MAWDESLEY, M. J.; CULLINGFORD, G.
OPTIMAL RESOURCE SCHEDULING.
MAXW 1964 -1 MAXWELL, W. L.
SCHEDULING OF SINGLE MACHINE SYSTEMS. A REVIEW.
MAY A 1974 -1 MAY, J. G.; STONE, C. A.; WAGNER, N. S.
SCHEDULING MODEL FOR THE INBAND SIGNALING SHOP, A.
MEL'A 1976 -1 MEL'NIKOV, O. I.
PROPERTIES OF OPTIMAL SCHEDULINGS OF THE BELLMAN-JOHNSON PROBLEM FOR TWO MACHINES (RUSSIAN).
MEL'A 1975 -1 MEL'NIKOV, O. I.
ON THE BELLMAN-JOHNSON PROBLEM (RUSSIAN).
MEL'A 1976 -2 MEL'NIKOV, O. I.
STABILITY OF THE OPTIMAL SCHEDULING OF THE BELLMAN-JOHNSON PROBLEM (RUSSIAN).
MIHO 1965 -1 MIHOC, G.; BALAS, E.
PROBLEM OF OPTIMAL SCHEDULING (RUMANIAN).
MIKHA 1966 -1 MIKHALEVICH, V. S.; SHKURBA, V. V.
SEQUENTIAL SCHEMES OF OPTIMIZATION IN PROBLEMS OF SCHEDULING THE COMPLETION OF TASKS (RUSSIAN).
MILLC 1976 -1 MILLER, H. E.
PERSONNEL SCHEDULING AND WORK ASSIGNMENT PROBLEM, A.
MILLC 1976 -2 MILLER, H. E.
PERSONNEL SCHEDULING IN PUBLIC SYSTEMS: A SURVEY.
MILLD 1976 -1 MILLER, H. E.; PIERSKALLA, W. P.; RATH, G. J.
NURSE SCHEDULING USING MATHEMATICAL PROGRAMMING.
MING 1976 -1 MING-I, Y.
ON THE N JOB, M MACHINE SEQUENCING PROBLEM OF FLOW-SHOP.
MITT 1977 -1 MITTAL, B. S.; BAGGA, P. C.
PRIORITY PROBLEM IN SEQUENCING WITH STOCHASTIC SERVICE TIMES, A.
MITTA 1967 -1 MITTAL, M. L.
OPTIMAL SEQUENCING OF MULTISTAGE FLOW SHOP OPERATION.
MIYA 1977 -1 MIYAKAWA, M.; YUBA, T.; SUGITO, Y.; HOSHI, M.
OPTIMUM SEQUENCE TREES.
MOLE 1976 -1 MOLE, R. H.; JAMESON, S. R.
SEQUENTIAL ROUTE-BUILDING ALGORITHM EMPLOYING A GENERALISED SAVINGS CRITERION, A.
MONA 1976 -1 MONAHAN, J. P.
PRODUCTIONS SCHEDULING OF PERISHABLE PRODUCTS.
MONM 1977 -1 MONMA, C. L.
OPTIMAL MXN FLOW SHOP SEQUENCING WITH PRECEDENCE CONSTRAINTS AND LAG TIMES.
MORI 1975 -1 MORIN, T. , L.; MARSTEN, R. E.
HYBRID DYNAMIC PROGRAMMING/BRANCH-AND-BOUND APPROACH TO A CLASS OF SEQUENCING PROBLEMS, A.
MORT 1977 -1 MORTON, T. E.; DHARAN, B. G.
ALGORISTICS FOR SEQUENCING WITH PRECEDENCE CONSTRAINTS.
MUCK 1977 -1 MUCKSTADT, J. A.; KOENIG, S. A.
APPLICATION OF LAGRANGIAN RELAXATION TO SCHEDULING IN POWER-GENERATION SYSTEMS, AN.
MUNC 1977 -1 MUNCH-ANDERSEN, B.; ZAHLE, T. U.
SCHEDULING ACCORDING TO JOB PRIORITY WITH PREVENTION OF DEADLOCK AND PERMANENT BLOCKING.
MURA 1974 -1 MURAV'EV, V. I.; ROMANOVSKII, I. V.
USE OF THE VARIABLE BASIS METHOD FOR THE SOLUTION OF THE CONTINUOUS ANALOGUE OF JOHNSON'S PROBLEM (RUSSIAN), A.
MURP 1975 -1 MURPHY, F. H.; STOHR, E. A.
MATHEMATICAL PROGRAMMING APPROACH TO THE SCHEDULING OF CHECK SORTING OPERATIONS, A.
MUTH 1963 -1 MUTH, J. F.; THOMPSON, G. L.
INDUSTRIAL SCHEDULING.
NABE 1965 -1 NABESHIMA, I.
COMPUTATIONAL SOLUTION TO THE M-MACHINE SCHEDULING PROBLEM.
NABE 1972 -1 NABESHIMA, I.
ALGORITHMS FOR MULTIPROJECT SCHEDULING WITH RESOURCE CONSTRAINTS AND RELATED PARALLEL SCHEDULING.
NABE 1967 -1 NABESHIMA, I.
GENERAL SOLUTIONS OF SOME SEQUENCING PROBLEMS (JAPANESE).
NABE 1973 -1 NABESHIMA, I.
BACKTRACK PROGRAMMING ALGORITHM AND RELIABLE HEURISTIC PROGRAMS FOR GENERAL SCHEDULING PROBLEM, A.
NABEA 1975 -1 NABESHIMA, I.; MARUYAMA, S.
COMPUTATIONAL EXPERIENCE ON MULTIPROJECT SCHEDULING WITH RESOURCE CONSTRAINTS.
NARA 1977 -1 NARASIMHAN, S. L.
EFFECTIVENESS OF PRIORITY RULES FOR SCHEDULING MULTISTAGE CONTINUOUS FLOW SHOPS.
NEID 1967 -1 NEIDHART, F.
METHODS FOR SCHEDULING AND SEQUENCING PROBLEMS (GERMAN).
NELS 1977 -1 NELSON, R. T.; HOLLOWAY, C. A.; WONG, R. M.-L.
CENTRALIZED SCHEDULING AND PRIORITY IMPLEMENTATION HEURISTICS FOR A DYNAMIC JOB SHOP MODEL.
NEPO 1973 -1 NEPOMIASTCHY, P.
NON-COMBINATORIAL APPROACH TO THE SCHEDULING PROBLEM, A.
NICH 1968 -1 NICHOLSON, D. A. J.; PULLEN, R. D.
PERMUTATION PROCEDURE FOR JOB SHOP SCHEDULING.
NOLT 1976 -1 NOLTEMEIER, H.
REDUKTION VON PRAEZEDENZSTRUKTUREN.
NOMU 1974 -1 NOMURA, S.; UI, T.; KURIYAMA, S.
STUDY OF A SEQUENCING PROBLEM THAT TAKES INTO ACCOUNT THE SWITCH-OVER TIME (JAPANESE), A.
NUGE 1964 -1 NUGENT, C. E.
SAMPLING APPROACHES TO THE SOLUTION OF THE N-BY-M STATIC SEQUENCING PROBLEM.
NUNE 1978 -1 NUNEN, J. A. E. E. VAN; WESSELS, J.
MULTI-ITEM LOT SIZE DETERMINATION AND SCHEDULING UNDER CAPACITY CONSTRAINTS.
NUNNA 1977 -1 NUNNIKHOVEN, T. S.; EMMONS, H.
SCHEDULING ON PARALLEL MACHINES TO MINIMIZE TWO CRITERIA RELATED TO JOB TARDINESS.

****

O'BRA 1975 -1 O'BRIEN, J. J.
VPM SCHEDULING FOR HIGH-RISE BUILDINGS.

OCHI 1975 -1 OCHI, T.; MINE, H.
SCHEDULING ALGORITHM OF CONTAINER LOADING AND UNLOADING MACHINES (JAPANESE).

ORLOA 1976 -2 ORLOFF, C. S.
ROUTE CONSTRAINED FLEET SCHEDULING.

OSTE 1976 -1 OSTERMANN, R.
QUELQUES EXPERIENCES DE CONFECTION AUTOMATIQUE D'HORAIRES.

PANDA 1975 -1 PANDIT, S. N. N.; SUBRAHMANYAM, Y. V.
ENUMERATION OF ALL OPTIMAL JOB SEQUENCES.

PANW 1977 -1 PANWALKAR, S. S.; ISKANDER, W.
SURVEY OF SCHEDULING RULES, A.

PANWA 1975 -1 PANWALKAR, S. S.; KHAN, A. W.
IMPROVED BRANCH AND BOUND PROCEDURE FOR N X M FLOW SHOP PROBLEMS, AN.

PANWA 1977 -1 PANWALKAR, S. S.; KHAN, A. W.
CONVEX PROPERTY OF AN ORDERED FLOW SHOP SEQUENCING PROBLEM, A.

PANWA 1976 -1 PANWALKAR, S. S.; KHAN, A. W.
ORDERED FLOW SHOP PROBLEM WITH IDENTICAL DUE DATES AND TOTAL TARDINESS CRITERION, AN.

PANWA 1976 -2 PANWALKAR, S. S.; KHAN, A. W.
ORDERED FLOW-SHOP SEQUENCING PROBLEM WITH MEAN COMPLETION TIME CRITERION, AN.

PARK 1977 -1 PARKER, R. G.; DEANE, R. H.; HOLMES, R. A.
ON THE USE OF A VEHICLE ROUTING ALGORITHM FOR THE PARALLEL PROCESSOR PROBLEM WITH SEQUENCE DEPENDENT CHANGEOVER COSTS.

PARKA 1977 -1 PARKER, R. G.; GRAVES, R. J.; SHERALI, H.
APPLICATION OF A GRAPH-PARTITIONING ALGORITHM TO THE SCHEDULING OF CURRICULUM REQUIREMENTS, THE.

PARS 1970 -1 PARSONS, J. A.
SEQUENCING TASKS AND OPERATIONS.

PARS 1970 -2 PARSONS, J. A.
SEQUENCING TASKS WITH PROXIMITY CONSTRAINTS.

PATT 1976 -1 PATTERSON, J. H.
PROJECT SCHEDULING: THE EFFECTS OF PROBLEM STRUCTURE ON HEURISTIC PERFORMANCE.

PATTA 1976 -1 PATTERSON, J. H.; ROTH, G. W.
SCHEDULING A PROJECT UNDER MULTIPLE RESOURCE CONSTRAINTS: A ZERO-ONE PROGRAMMING APPROACH.

PERE 1966 -1 PEREPELITSA, V. A.
PROBLEM IN SCHEDULE THEORY.

PEREA 1976 -1 PEREPELITSA, V. A.; SEVAST'YANOV, S. V.
ON A SEQUENCING PROBLEM IN A NETWORK (RUSSIAN).

PHIL 1976 -1 PHILLIPS, L. W.; UNGER, P. S.
MATHEMATICAL PROGRAMMING SOLUTION OF A HOIST SCHEDULING PROGRAM.

PICAC 1976 -1 PICARD, J.-C.; QUEYRANNE, M.
TIME-DEPENDENT TRAVELING SALESMAN PROBLEM AND APPLICATION TO THE TARDINESS PROBLEM IN ONE-MACHINE SCHEDULING, THE.

PLAX 1976 -1 PLAXTON, A.
LINEAR PROGRAMMING MODEL TO OPTIMALLY SCHEDULE CHECK PROCESSING ACTIVITIES, A.

POLL 1977 -1 POLLACK, M.
SOME ELEMENTS OF THE AIRLINE FLEET PLANNING PROBLEM.

PORT 1976 -1 PORTUGAL, V. M.
ASYMPTOTIC BEHAVIOUR OF SOLUTIONS FOR PROBLEMS OF THE THEORY OF SCHEDULES BY DIFFERENT CRITERIA (RUSSIAN).

PORT 1976 -2 PORTUGAL, V. M.
ASYMPTOTIC BEHAVIOUR OF THE SCHEDULE THEORY PROBLEMS BY DIFFERENT CRITERIA (RUSSIAN).

POWE 1977 -1 POWELL, J.; GLEASON, J.; BURTON, J.
POWER GENERATOR SCHEDULING BY DYNAMIC PROGRAMMING.

PRIV 1974 -1 PRIVALOV, V. V.; TETERIN, G. P.
PROBLEM OF CHARGING A FORGE PRESS WITH 0-1-VARIABLES (RUSSIAN).

PUN 1975 -1 PUN, L.; BOEBION, J.; LATECOERE, L.; ALMAR, A.; PHILIPE, J.; DEROUAULT, Y.
ORDONNANCEMENT ADAPTIF DANS LES ENTREPRISES INDUSTRIELLES DE TAILLE MOYENNE.

RAJA 1977 -1 RAJARAMAN, M. K.
PARALLEL SEQUENCING ALGORITHM FOR MINIMIZING TOTAL COST, A.

RAMAB 1975 -1 RAMALINGAM, P.
SCHEDULING JOBS WITH DUE DATES AND COST FUNCTIONS ON MULTI-STAGE PROCESSORS.

RAMAC 1971 -1 RAMAMOORTHY, C. V.; CHANDY, K. M.; GONZALES, M. J.
TASK SCHEDULING ALGORITHMS IN A MULTIPROCESSOR ENVIRONMENT.

RAMAD 1976 -1 RAMAMOORTHY, C. V.; FOX, T. F.; LI, H. F.
SCHEDULING PARALLEL PROCESSABLE TASKS FOR A UNIPROCESSOR.

RAMAE 1975 -1 RAMASWAMY, K. H.
OPTIMAL ROUTE FOR PROCESSING A JOB LOT IN A JOB SHOP.

RAMAF 1976 -1 RAMASWAMY, K. V.
SCHEDULING OF N JOBS ON M DISTINCT PARALLEL PROCESSORS WITH PROCESS DEPENDENT TIMES AND COSTS.

RANDB 1976 -1 RANDOLPH, P. H.
JOB SHOP SCHEDULING: A CASE STUDY.

RAWA 1975 -1 RAWASWAMY, K. V.; DUDEK, R. A.
HEURISTIC ALGORITHM FOR SINGLE FACILITY SEQUENCING WITH PENALTY COST, A.

RAZHA 1967 -1 RAZHUMIKHIN, B. S.; STAVROKOVA, N. E.
PROBLEM OF DISTRIBUTING RESOURCES TO MINIMIZE TIME OF EXECUTION OF JOBS COMPLEX.

REDD 1976 -1 REDDI, S. S.
INTERDEPENDENCE OF INDUSTRIAL ENGINEERING AND COMPUTING: SELECTION OF SCHEDULING CRITERIA IN COMPUTER SYSTEMS, THE.

REDD 1976 -2 REDDI, S. S.
SEQUENCING WITH FINITE INTERMEDIATE STORAGE.

REKL 1977 -1 REKLAJTIS, V. K.; TAMULINAS, B. V.
AUTOMATIC CONSTRUCTION OF SEQUENCING RULES IN THE PROBLEM OF JOB SCHEDULING FOLLOWING A CALENDAR (RUSSIAN).

RINNA 1975 -1 RINNOOY KAN, A. H. G.; LAGEWEG, B. J.; LENSTRA, J. K.
MINIMIZING TOTAL COSTS IN ONE-MACHINE SCHEDULING.

ROTHB 1975 -1 ROTHKOPF, M. H.
NOTE ON ALLOCATING JOBS TO TWO MACHINES, A.

ROTHB 1966 -1 ROTHKOPF, M. H.
SCHEDULING WITH RANDOM SERVICE TIMES.

ROWE 1960 -1 ROWE, A. J.
TOWARD A THEORY OF SCHEDULING.

ROWI 1976 -1 ROWICKI, A.
ON OPTIMAL SCHEDULING FOR TWO-PROCESSOR SYSTEMS.

****

ROY 1966 -2 ROY, B.
CONTRIBUTION TO THE THEORY OF GRAPHS IN THE STUDY OF SCHEDULING PROBLEMS (FRENCH).
ROY A 1964 -1 ROY, B.; VIGGIONI, S.
SCHEDULING PROBLEMS (ITALIAN).
RUSC 1977 -1 RUSCHITZKA, M.; FABRY, R. S.
UNIFYING APPROACH TO SCHEDULING, A.
RUYG 1976 -1 RUYGROK, J. T.
FESTLEGUNG DER REIHENFOLGE DES LOESCHENS VON SEESCHIFFEN MIT HILFE DER GANZZAHLIGEN PROGRAMMIERUNG.
RYAN 1976 -1 RYAN, D. M.; FOSTER, B. A.; CHESHIRE, I. M.
COMPLEXITY OF REAL-WORLD SCHEDULING PROBLEMS, THE.
RYZH 1974 -1 RYZHKOV, A. P.
ORGANIZATION OF COMPUTER COMPUTATION OF INFORMATIONALLY CONNECTED PROGRAMS (RUSSIAN).
SAHB 1975 -1 SAHBAZJAN, K. V.; TUSKINA, T. A.
SURVEY OF METHODS FOR COMPILING SCHEDULES FOR MULTIPROCESSOR SYSTEMS. COMPUTATIONAL METHODS OF LINEAR ALGEBRA. PARALLEL COMPUTATIONS (RUSSIAN).
SAHBA 1974 -1 SAHBAZJAN, K. V.; TUSKINA, T. A.; LEBEDINSKAJA, N. B.
ALGORITHM FOR COMPILING AN OPTIMAL SCHEDULE FOR A PARALLEL-SEQUENTIAL NETWORK (RUSSIAN), AN.
SAHN 1977 -1 SAHNI, S.
GENERAL TECHNIQUES FOR COMBINATORIAL APPROXIMATION.
SAHNC 1976 -1 SAHNI, S. K.
ALGORITHMS FOR SCHEDULING INDEPENDENT TASKS.
SAKAC 1970 -1 SAKATA, M.; NOGUCHI, S.; OIZUMI, J.
OPTIMAL SCHEDULING IN SINGLE PROCESSOR SYSTEMS.
SANA 1968 -1 SANADA, H.; LINGST, S. T.; TEZUKA, Y.; KASAHARA, Y.
OPTIMAL SCHEDULING USING SLACK TIME ARRAY (JAPANESE).
SANAA 1967 -1 SANADA, H.; NAKAGAMI, K.; TEZUKA, Y.; KASAHARA, Y.
SCHEDULING PROBLEM OF OPTIMAL CYCLIC PRODUCTION (JAPANESE).
SANC 1976 -1 SANCHO, N. G. F.
OPTIMAL SOLUTION FOR SCHEDULING OF PRINTING AND BINDING, AN.
SCAR 1977 -1 SCARBROUGH, J. E.
MIXED-INTEGER PROGRAMMING MODEL FOR THE SCHEDULING OF AIRCRAFT BY TIME OF DAY WITHIN A NETWORK OF AIRPORTS, A.
SCHAB 1972 -1 SCHALLEHN, W.; SEBASTIAN, H.-J.
REIHENFOLGEPROBLEM IN DER NETZPLANTECHNIK, EIN.
SCHAC 1974 -1 SCHARGE, L.
OPTIMAL SCHEDULING RULES FOR QUADRATIC WAITING COSTS.
SCHUA 1977 -1 SCHURAWLJEW, J. I.; FINKEL'SHTEIN, YU. YU.
AUFGABENTYPEN UND LOESUNGSMETHODEN IN DER DISKRETEN OPTIMIERUNG.
SEME 1974 -1 SEMENOV, V. M.; HAIKIN, N. M.; MOGILA, V. G.
CERTAIN METHOD OF SOLVING A VOLUME SCHEDULING PROBLEM (RUSSIAN), A.
SETH 1976 -1 SETHI, R.
SCHEDULING GRAPHS ON TWO PROCESSORS.
SEVA 1975 -1 SEVAST'YANOV, S. V.
ASYMPTOTIC APPROACH TO SOME PROBLEMS SEQUENCING THEORY (RUSSIAN).
SHAP 1975 -1 SHAPIRO, A. D.
SOME ASYMPTOTIC PROPERTIES OF A SCHEDULE THEORY PROBLEM (RUSSIAN).
SHAPC 1976 -1 SHAPIRO, R. D.
SCHEDULING COUPLED TASKS.
SHARE 1975 -1 SHARP, G. P.; CHU, J.
COMBINATORIAL APPROACH TO SCHEDULING PUBLIC TRANSIT OPERATORS, A.
SHAW 1975 -1 SHAW, G. C.
QUALITY CONTROL MEASURES FOR CANADIAN AIRLINE SCHEDULES.
SHEL 1975 -1 SHELNESS, N. H.; STEPHENS, P. D.; WHITFIELD, H.
EDINBURG MULTI-ACCESS SYSTEM SCHEDULING AND ALLOCATION PROCEDURES IN THE RESIDENT SUPERVISOR, THE.
SHKU 1965 -1 SHKURBA, V. V.
COMPUTATIONAL SCHEMES OF SOLUTION OF PROBLEMES IN SCHEDULING THEORY (RUSSIAN).
SIDN 1977 -1 SIDNEY, J. B.
OPTIMAL SINGLE-MACHINE SCHEDULING WITH EARLINESS AND TARDINESS PENALTIES.
SIEL 1976 -1 SIELKEN, R. M. JR.
SEQUENCING WITH SETUP COSTS BY ZERO-ONE MIXED INTEGER LINEAR PROGRAMMING.
SKAF 1966 -1 SKAFF, M. S.
APPROXIMATE SOLUTION OF A WEIGHTED SCHEDULING PROBLEM WITH MULTIPLE CONSTRAINTS.
SMITD 1975 -1 SMITH, L. D.
SCHEDULING NURSING PERSONNEL: APPLICATION OF LIST PROCESSING AND PROBLEM ORIENTED DATA STRUCTURES.
SMITE 1977 -1 SMITH, L. D.; WIGGINS, A.
COMPUTER-BASED NURSE SCHEDULING SYSTEM, A.
SMITF 1975 -1 SMITH, M. L.; PANWALKAR, S. S.; DUDEK, R. A.; ULGEN, O.
FLOWSHOP SEQUENCING WITH NONPERMUTATION SCHEDULES - A HEURISTIC APPROACH.
SMITG 1975 -1 SMITH, R. S.; GIGLIO, R. J.
TWO PHASE EMPLOYEE SCHEDULING ALGORITHM FOR REDUCING LABOR COSTS IN FOOD SERVICE OPERATIONS, A.
SMOL 1966 -1 SMOLYAR, L. I.
MATHEMATICAL MODELS OF THE SCHEDULING OF SERIAL PRODUCTION (RUSSIAN).
SOVE 1976 -1 SOVETOV, B. YA.; TISHKIN, A. I.
DOMINANT DISTRIBUTIONS IN CERTAIN PROBLEMS OF SCHEDULING THEORY (RUSSIAN).
SOVE 1976 -2 SOVETOV, B. YA.; TISHKIN, A. I.
ON THE SOLUTION OF A CLASS OF PROBLEMS IN SEQUENCING THEORY (RUSSIAN).
SPIR 1976 -1 SPIRN, J. R.
MULTI-QUEUE SCHEDULING OF TWO TASKS.
SPRA 1967 -1 SPRADLIN, B. C.; PIERCE, D. A.
PRODUCTION SCHEDULING UNDER LEARNING EFFECT BY DYNAMIC PROGRAMMING.
STANB 1969 -1 STANKARD, M. F.; GUPTA, S. K.
NOTE ON BOMBERGER'S APPROACH TO LOT SIZE SCHEDULING: HEURISTIC PROPOSED.
STAU 1974 -1 STAUDER, E.
SIMULATION MODELLING FOR ANALYSIS OF OPTIMUM SEQUENCING IN THE SHOP-FLOOR.
STERA 1977 -1 STERN, H. I.; PODRIGUEZ, E. P.; UTTER, M. L.
CYCLICAL JOB SEQUENCING ON MULTIPLE SETS OF IDENTICAL MACHINES.
STOI 1977 -1 STOICA, M.
SPLITORIAL CONCEPTS USED IN SCHEDULING.
STOL 1977 -1 STOLZENBURG, I.
OPTIMIERUNGSVERFAHREN FUER REIHENFOLGEPROBLEME IN DER CONTAINER-SCHIFFAHRT.

****

STRA 1967 -1 STRADLIN, B. C.; PIERCE, D. A.
PRODUCTION SCHEDULING UNDER A LEARNING EFFECT BY DYNAMIC PROGRAMMING.
SWANA 1976 -1 SWANSON, L. W.
N-DIMENSIONAL SCHEDULING.
SZWA 1977 -1 SZWARC, W.
OPTIMAL TWO-MACHINE ORDERINGS IN THE 3 X N FLOW SHOP PROBLEM.
SZWA 1977 -2 SZWARC, W.
SPECIAL CASES OF THE FLOW-SHOP PROBLEM.
SZWAA 1977 -1 SZWARC, W.; HUTCHINSON, G. K.
JOHNSON'S APPROXIMATE METHOD FOR THE 3XN JOB SHOP PROBLEM.
TABO 1969 -1 TABOURIER, Y.
SCHEDULING WITH PURELY DISJUNCTIVE CONSTRAINTS (FRENCH).
TAKAA 1974 -1 TAKAHASHI, Y.
SEQUENCING MODEL WITH AN APPLICATION TO SPEED CLASS SEQUENCING IN AIR TRAFFIC CONTROL, A.
TALW 1967 -1 TALWAR, P. P.
SEQUENCING PROBLEM WITH UNCERTAIN JOB TIMES.
TANA 1975 -1 TANAEV, V. S.; SHKURBA, V. V.
INTRODUCTION INTO SEQUENCING THEORY (RUSSIAN).
THES 1976 -1 THESEN, A.
HEURISTIC SCHEDULING OF ACTIVITIES UNDER RESOURCE AND PRECEDENCE RESTRICTIONS.
TILQ 1975 -1 TILQUIN, C.
ON SCHEDULING WITH EARLIEST STARTS AND DUE DATES ON A GROUP OF IDENTICAL MACHINES.
TOWN 1977 -1 TOWNSEND, W.
BRANCH-AND-BOUND METHOD FOR SEQUENCING PROBLEMS WITH LINEAR AND EXPONENTIAL PENALTY FUNCTIONS, A.
TOWN 1977 -2 TOWNSEND, W.
MINIMISING THE MAXIMUM PENALTY IN THE TWO-MACHINE FLOW SHOP.
TOWS 1977 -1 TOWSNEND, W.
SEQUENCING N JOBS ON M MACHINES TO MINIMISE MAXIMUM TARDINESS: A BRANCH-AND-BOUND SOLUTION.
ULLM 1976 -2 ULLMAN, J. D.
COMPLEXITY OF SEQUENCING PROBLEMS.
ULLR 1976 -1 ULLRICH, M.
KONSTRUKTION OPTIMALER REIHENFOLGEN BEI SPEZIELLEN REIHENFOLGEPROBLEMEN.
ULLR 1976 -2 ULLRICH, M.
ZUSAMMENHANG ZWISCHEN DEM RUNDREISEPROBLEM UND EINEM REIHENFOLGEPROBLEM DER MASCHINENBELEGUNGSPLANUNG.
ULPH 1977 -1 ULPH, A.
EXTENSION OF A RESULT IN SCHOOL TIMETABLING, AN.
VAIR 1976 -1 VAIRAVAN, K.; MILLO, R. A. DE
ON THE COMPUTATIONAL COMPLEXITY OF A GENERALIZED SCHEDULING PROBLEM.
VELI 1974 -1 VELICKIN, V. Z.
CALCULATION OF THE OPTIMAL INTENSITY OF VARIOUS TYPES OF JOBS IN THE CASE OF A NONRHYTMIC FLOW OF WORK IN CONSTRUCTION (RUSSIAN).
VERG 1976 -1 VERGIN, R. C.
NOTE ON "LOT SIZE SCHEDULING ON A SINGLE MACHINE FOR STOCHASTIC DEMAND" BY GOYAL, A.
VERGA 1977 -1 VERGIN, R. C.; SCRIABIN, M.
MAINTENANCE SCHEDULING FOR MULTICOMPONENT EQUIPMENT.
VISH 1976 -1 VISHNEVSKIJ, L. D.
ON A SCHEDULING PROBLEM FOLLOWING A CALENDAR (RUSSIAN).
VLAS 1966 -1 VLASYUK, B. A.
OPTIMAL SCHEDULING FOR PARALLEL-SERIES MACHINES.
WALL 1977 -1 WALLACH, Y.
SCHEDULING ALGORITHMS FOR CONCURRENT EXECUTION.
WASO 1976 -1 WASOWSKI, J.
SELECTION METHOD FOR A CERTAIN PROBLEM OF ORDERING PRODUCTS BY 3 MACHINES (RUSSIAN), A.
WEEK 1976 -1 WEEKS, J. K.; FRYER, J. S.
SIMULATION STUDY OF OPERATING POLICIES IN A HYPOTHETICAL DUAL-CONSTRAINED JOB SHOP, A.
WELL 1976 -1 WELLS, D. W.
POWER SYSTEM SCHEDULING USING INTEGER PROGRAMMING.
WERR 1975 -1 WERRA, D. DE
FEW REMARKS ON CHROMATIC SCHEDULING, A.
WERR 1976 -2 WERRA, D. DE
EVERYTHING YOU ALWAYS WANTED TO KNOW ABOUT S.EX.
WERR 1975 -2 WERRA, D. DE
GRAPHS AND SCHEDULING.
WERR 1976 -4 WERRA, D. DE
ON A MULTI-PERIOD ASSIGNMENT PROBLEM.
WERR 1975 -4 WERRA, D. DE
ON A PARTICULAR CONFERENCE SCHEDULING PROBLEM.
WHITA 1977 -1 WHITE, C. H.; WILSON, R. C.
SEQUENCE DEPENDENT SET-UP TIMES AND JOB SEQUENCING.
WHITE 1966 -1 WHITE, M.
PROJECT SCHEDULING WITH LIMITED RESOURCES. EXPERIMENT IN HEURISTIC PROGRAMMING.
WILLE 1976 -1 WILLIS, R. J.; HASTINGS, N. A. J.
PROJECT SCHEDULING WITH RESOURCE CONSTRAINTS USING BRANCH AND BOUND METHODS.
YAKI 1966 -1 YAKIMOV, R. M.
FUNCTIONAL EQUATION DESCRIBING THE OPTIMAL SEQUENCING OF PROCESSING OF ITEMS (RUSSIAN).
YAMA 1977 -1 YAMAMOTO, M.
APPROXIMATE SOLUTION OF MACHINE SCHEDULING PROBLEMS BY DECOMPOSITION METHOD, AN.
YEH 1976 -1 YEH, R. T.
APPLIED COMPUTATION THEORY: ANALYSIS, DESIGN, MODELING.
YOUN 1970 -1 YOUNG, G. K.; MOSELY, J. C.; EVENSON, D. E.
TIME SEQUENCING OF ELEMENT CONSTRUCTION IN A MULTI-RESERVOIR SYSTEM.
YUAN 1975 -1 YUAN, J. S. C.; HOREN, J. H.; WAGNER, H. M.
OPTIMAL MULTI-PRODUCT PRODUCTION SCHEDULING AND EMPLOYMENT SMOOTHING WITH DETERMINISTIC DEMANDS.
ZALO 1975 -1 ZALOOM, V.; VATZ, D.
NOTE ON THE OPTIMAL SCHEDULING OF TWO PARALLEL PROCESSORS, A.
ZALOA 1977 -1 ZALOOM, V. A.; SMYER, W. N.
SCHEDULING TO MINIMIZE MAKESPAN REQUIRED TO PROCESS INDEPENDENT JOBS ON PARALLEL NON-EQUIVALENT SERVERS.
ZEHN 1965 -1 ZEHNDER, C. A.
COMPUTATION OF TIME TABLES FOR SCHOOLS AND TRANSPORTATION.
ZHURA 1968 -1 ZHURZENKO, S. L.
METHOD OF SOLVING CLASS OF PROBLEMS IN SCHEDULING THEORY.

****

ZIMMB 1967 -1 ZIMMERMANN, W.; PFAFFENZELLER, D.
SIMULTANE MENGEN- UND ABLAUFPLANUNG INDUSTRIELLER FERTIGUNGSPROGRAMME.

SHORTEST PATHS
****

AKER 1967 -1 AKERS, S. A. JR.
MODIFICATION OF LEE'S PATH CONNECTION ALGORITHM, A.

ALEKE 1971 -1 ALEKSEYEVA, S. M.; ALEKSEYEV, O. G.
ALGORITHM FOR FINDING THE SHORTEST PATH IN A NETWORK GRAPH (TRANSLATED FROM RUSSIAN), AN.

AM 1976 -1 AM, T. D.; TSUKIYAMA, S.; SHIRAKAWA, I.; OZAKI, H.
ALGORITHM FOR GENERATING ALL THE PATHS BETWEEN TWO VERTICES IN A DIGRAPH AND ITS APPLICATION, AN.

ANDR 1975 -1 ANDRAS, B.
SOLUTION ALGORITHM FOR A MINIMAL PATH PROBLEM IN A MULTITERMINAL GENERALIZED NETWORK (HUNGARIAN), A.

ASTA 1971 -1 ASTAFEVA, N. A.
ALGORITHM FOR DETERMINING THE SHORTEST ROUTES THROUGH A CONNECTING NETWORK (RUSSIAN).

BAKO 1976 -1 BAKO, A.
ALL PATHS IN AN ACTIVITY NETWORK.

BAKO 1975 -1 BAKO, A.
SOLUTION OF A MULTITERMINAL MINIMAL PATH PROBLEM IN A GENERALIZED NETWORK (HUNGARIAN).

BAKO 1976 -2 BAKO, A.
SOLUTION OF THE MULTITERMINAL MINIMAL PATH PROBLEM IN A NETWORK WITH GAINS.

BASC 1977 -1 BASCA, O. C.
ON ALGORITHMS OF MINIMUM PATHS IN GRAPHS.

BEAR 1959 -1 BEARDWOOD, J.; HALTON, J. H.; HAMMERSLEY, J. M.
SHORTEST PATH THROUGH MANY POINTS, THE.

BECKB 1956 -1 BECKMANN, M.; MAC GUIRE, C. B.; WINSTON, C. B.
STUDIES IN THE ECONOMICS OF TRANSPORTATION.

BEHA 1969 -1 BEHARA, D. N.
COMPUTATIONAL MATRIX ALGORITHM FOR SHORTEST ROUTE PROBLEM, A.

BERN 1958 -1 BERNERS-LEE, C. M.
INDUSTRIAL APPLICATION OF COMPUTERS, AN.

BERZ 1971 -1 BERZTISS, A. T.
DATA STRUCTURES, THEORY AND PRACTICE.

BLEW 1977 -1 BLEWETT, W. J.; HU, T. C.
TREE DECOMPOSITION ALGORITHM FOR LARGE NETWORKS.

BOCK 1957 -1 BOCK, F. C.
ALGORITHM FOR FINDING SHORTEST PATHS BETWEEN PAIRS OF POINTS IN A NETWORK.

BORIB 1966 -1 BORILLE, M.
REMARKS ON THE DETERMINATION OF THE OPTIMAL ALGORITHM TO FIND THE SHORTEST PATH IN A CIRCUITLESS GRAPH.

BORU 1926 -1 BORUVKA, O.
ON A MINIMAL PROBLEM.

BOUR 1969 -1 BOURGOIN, M. H.; HEURGON, E. M. J.
STUDY AND COMPARISON OF ALGORITHMS OF THE SHORTEST PATH THROUGH PLANNED EXPERIMENTS.

BRUC 1972 -1 BRUCKER, P.
ALL SHORTEST DISTANCES IN NETWORKS WITH A LARGE NUMBER OF STRONG COMPONENTS.

BULL 1970 -1 BULLINGHAM, J. M.; MATHEWS, P. A.
ELECTRONIC SIMULATOR FOR QUICKEST ROUTES IN A ROAD NETWORK.

BURT 1970 -1 BURT, J. M. JR.
ALL SHORTEST DISTANCES IN LARGE SERIAL NETWORKS.

BUTA 1968 -1 BUTAS, L. F.
DIRECTIONALLY ORIENTED SHORTEST PATH ALGORITHM, A.

BUTR 1964 -1 BUTRIMENKO, A. V.
ON SEARCHING FOR THE SHORTEST PATHS ALONG A GRAPH DURING VARIATIONS IN IT.

CARS 1977 -1 CARSON, J. S.; LAW, A. M.
NOTE ON SPIRA'S ALGORITHM FOR THE ALL-PAIRS SHORTEST-PATH PROBLEM, A.

CEDE 1974 -1 CEDERBAUM, I.
OPTIMAL BACKBOARD ORDERING THROUGH THE SHORTEST PATH ALGORITHM.

CERF 1974 -1 CERF, V. G.; COWAN, D. D.; MULLIN, R. C.; STANTON, R. G.
LOWER BOUND ON THE AVERAGE SHORTEST PATH LENGTH IN REGULAR GRAPHS, A.

CHAMA 1977 -1 CHAMPEAUX, D. DE; SINT, L.
IMPROVED BIDIRECTIONAL HEURISTIC SEARCH ALGORITHM, AN.

CHARD 1967 -1 CHARTRES, B. A.
ANMERKUNG ZU DEM AUFSATZ VON NICHOLSON (COMPUTER JOURNAL 9, 1966, 275).

CHAT 1975 -1 CHATY, GUY.
CHEMINEMENTS REMARQUABLES DANS LES GRAPHES: EXISTENCE, OBTENTION, CONSERVATION.

CHU 1965 -1 CHU, Y. J.; LIU T. H.
ON THE SHORTEST ARBORESCENCE OF DIRECTED GRAPHS.

DAEN 1977 -1 DAENINCK, G.; SMEERS, Y.
USING SHORTEST PATHS IN SOME TRANSSHIPMENT PROBLEM WITH CONCAVE COSTS.

DAMM 1977 -1 DAMM, E.
KUERZESTE WEG IN EINEM GITTERNETZ MIT HINDERNISSEN, DER.

DANT 1975 -1 DANTZIG, G. B.
ON THE SHORTEST ROUTE THROUGH A NETWORK.

DANT 1967 -1 DANTZIG, G. B.
ALL SHORTEST ROUTES IN A GRAPH.

DANTA 1966 -1 DANTZIG, G. B.; BLATTNER, W. O.; RAO, M. R.
ALL SHORTEST ROUTES FROM A FIXED ORIGIN IN A GRAPH.

DEJO 1977 -1 DEJON, B.
BESTIMMUNG VON R KUERZESTEN WEGEN IN NETZWERKEN UNTER NEBENBEDINGUNGEN: VERFAHREN VOM HOFFMAN-PAVLEY-TYP.

DEJOA 1976 -1 DEJON, B.; GATTINGER, M.
INNER AND OUTER ALGORITHMS FOR DETERMINING R SHORTEST PATHS UNDER SIDE CONDITIONS.

DEJOA 1977 -1 DEJON, B.; GATTINGER, M.
BESTIMMUNG VON R-EXTREMA IN HALBGEORDNETEN MENGEN.

DIALA 1977 -1 DIAL, R.; GLOVER, F.; KARNEY, D.; KLINGMAN, D.
COMPUTATIONAL ANALYSIS OF ALTERNATIVE ALGORITHMS AND LABELING TECHNIQUES FOR FINDING SHORTEST PATH TREES, A.

DIJK 1960 -1 DIJKSTRA, E. W.
SOME THEOREMS ON SPANNING SUBTREES OF A GRAPH.

DINE 1976 -1 DINESCU, C.
OPTIMAL ELEMENTARY PATHS IN NETWORKS WITH CIRCUITS.

DORA 1966 -1 DORAN, J. E.; MICHIE, D.
EXPERIMENTS WITH THE GRAPH TRAVERSER PROGRAM.

DREY 1977 -1 DREYFUS, S. E.; LAW A. M.
ART AND THEORY OF DYNAMIC PROGRAMMING, THE.

DUBO 1976 -1 DUBOV, YU. A.; EPIKHOVA, N. V.
ON AN ALGORITHM SEARCHING OPTIMAL SOLUTIONS IN THE PROBLEMS OF AUTOMATIC DESIGN (RUSSIAN).

EVST 1969 -1 EVSTIGNEEV, V. A.; JASAKOVA, V. P.
CERTAIN ALGORITHM ON GRAPHS (RUSSIAN), A.
FALK 1963 -1 FALKENHAUSEN, H. VON
VERFAHREN ZUR PROGNOSE DER VERKEHRSVERTEILUNG IN EINEM GEPLANTEN STRASSENNETZ, EIN.
FARB 1965 -1 FARBEY, B. A.; LAND, A. H.; MURCHLAND, J. D.
CASCADE ALGORITHM FOR FINDING THE MINIMUM DISTANCES ON A GRAPH, THE.
FAUR 1976 -1 FAURE, R.; ROUCAIROL, C.; TOLLA, P.
CHEMINS ET FLOTS, ORDONNANCEMENTS.
FEW 1955 -1 FEW, L.
SHORTEST PATH AND THE SHORTEST ROAD THROUGH N POINTS, THE.
FILLA 1976 -1 FILLER, M. F.
ALGORITHM FOR FINDING SHORTEST PATHS (RUSSIAN), AN.
FONGC 1977 -1 FONG, C. O.; SRINIVASAN, V.
DETERMINING ALL NONDEGENERATE SHADOW PRICES FOR THE TRANSPORTATION PROBLEM.
FOOT 1965 -1 FOOT, D. H. S.
SHORTEST ROUTE PROBLEM: ALGOL-PROGRAMS, AND DISCUSSION OF COMPUTATIONAL PROBLEMS IN LARGE TRANSPORT NETWORK APPLICATIONS, THE.
FOX 1975 -1 FOX, B.
MORE ON KTH SHORTEST PATHS.
FREDA 1975 -1 FREDMAN, M. L.
ON THE DECISION TREE COMPLEXITY OF THE SHORTEST PATH PROBLEMS.
FREDA 1976 -1 FREDMAN, M. L.
NEW BOUNDS ON THE COMPLEXITY OF THE SHORTEST PATH PROBLEM.
FRIEA 1976 -1 FRIEZE, A. M.
SHORTEST PATH ALGORITHMS FOR KNAPSACK TYPE PROBLEMS.
FRIEA 1977 -3 FRIEZE, A. M.
MINIMUM PATHS IN DIRECTED GRAPHS.
FRIS 1963 -2 FRISCH, I. T.
OPTIMUM ROUTES IN COMMUNICATION SYSTEMS WITH CHANNEL CAPACITIES AND CHANNEL RELIABILITIES.
FULK 1975 -2 FULKERSON, D. R.
STUDIES IN GRAPH THEORY, I.
GATT 1976 -1 GATTINGER, M.
INNERE UND AEUSSERE ALGORITHMEN ZUR BESTIMMUNG R KUERZESTER WEGE UNTER NEBENBEDINGUNGEN.
GAUL 1977 -2 GAUL, W.
UEBER WEGE MAXIMALER KAPAZITAET(EXTENDED ABSTRACT).
GAUL 1977 -3 GAUL, W.
ZUR LOESUNG VON WEGPROBLEMEN MIT ZEITABHAENGIGEN BESCHRAENKUNGEN.
GILLA 1969 -1 GILL, A.; TRAIGER, I. L.
COMPUTATION OF OPTIMAL PATHS IN FINITE GRAPHS.
GILS 1970 -1 GILSINN, J.; WITZGALL, C.
COMPUTER COMPARISATION OF LABELLING ALGORITHMS FOR CALCULATING SHORTEST PATH TREES.
GLOVK 1972 -1 GLOVER, F., ET AL.
NOTE ON FINDING ALL SHORTEST PATHS.
GOLD 1976 -2 GOLDEN, B.
SHORTEST-PATH ALGORITHMS: A COMPARISON.
GOLDA 1977 -1 GOLDEN, B.; BALL, M.
SHORTEST PATHS WITH EUCLIDEAN DISTANCES: AN EXPLANATORY MODEL.
GOLDF 1967 -1 GOLDMAN, A. J.; NEMHAUSER, G.
TRANSPORT IMPROVEMENT PROBLEM TRANSFORMABLE TO A BEST-PATH PROBLEM, A.
GOND 1974 -1 GONDRAN, M.
ALGEBRE DES CHEMINS ET ALGORITHMES.
GOND 1976 -1 GONDRAN, M.
PROBLEME DU PLUS COURT CHEMIN DANS UN GRAPHE, LE.
GOND 1975 -2 GONDRAN, M.
PATH ALGEBRA AND ALGORITHMS.
GORS 1970 -1 GORSHKOW, A. F.; KIRPICHNIKOV, V. M.
PRACTICAL ALGORITHM FOR FINDING PATHS OF A GRAPH (RUSSIAN), A.
GRIG 1974 -1 GRIGOR'EV, E. G.
CERTAIN MODIFICATION OF FORD'S ALGORITHM (RUSSIAN), A.
HADL 1977 -1 HADLOCK, F. O.
SHORTEST PATH ALGORITHM FOR GRID GRAPHS, A.
HAKIA 1972 -1 HAKIMI, S. L.
SHORTEST PATHS IN GRAPHS - A REVIEW.
HALP 1977 -3 HALPERN, J.
SHORTEST ROUTE WITH TIME DEPENDENT LENGTH OF EDGES AND LIMITED DELAY POSSIBILITIES IN NODES.
HAMM 1969 -1 HAMMER, G.
BEMERKUNG ZUM TRIPEL-ALGORITHMUS ZUR BESTIMMUNG DER KUERZESTEN WEGE IN EINEM GRAPHEN, EINE.
HAMM 1969 -2 HAMMER, G.
TRIPEL-ALGORITHMUS ZUR BESTIMMUNG KUERZESTER WEGE IN EINEM GRAPHEN.
HARD 1977 -1 HARDLOCK, F. O.
SHORTEST PATH ALGORITHM FOR GRID GRAPHS, A.
HART 1960 1 HART, P. F.; NILLSON, N. L.; RAPHAEL, B.
FORMAL BASIS FOR THE HEURISTIC DETERMINATION OF MINIMUM COST PATHS.
HEDEA 1963 -1 HEDETNIEMI, S.; LEVEN, B. M.
DETERMINING FASTEST ROUTES USING FIXED SCHEDULES.
HELD 1965 -1 HELD, M.; KARP, R. M.
CONSTRUCTION OF DISCRETE DYNAMIC PROGRAMMING ALGORITHMS, THE.
HITCA 1968 -1 HITCHNER, L.
COMPARATIVE INVESTIGATION OF THE COMPUTATIONAL EFFICIENCY OF SHORTEST PATH ALGORITHMS, A.
HOFFC 1958 -1 HOFFMANN, W.; PAVLEY, R.
APPLICATIONS OF DIGITAL COMPUTERS TO PROBLEMS IN THE STUDY OF VEHICULAR TRAFFIC.
HOUC 1976 -1 HOUCK, D. J.; VEMUGANTI, R. R.
TRAVELING SALESMAN PROBLEM AND SHORTEST N-PATHS, THE.
HU 1966 -1 HU, T. C.
DECOMPOSITION ALGORITHM FOR SHORTEST PATHS IN A NETWORK, A.
HU 1965 -1 HU, T. C.
MULTI-TERMINAL SHORTEST PATHS.
IBAR 1970 -1 IBARAKI, T.
SHORTEST PATH PROBLEMS VISITING SPECIFIED NODES (JAPANESE).
JOHNA 1973 -1 JOHNSON, D. B.
ALGORITHMS FOR SHORTEST PATHS.
JOHNA 1977 -1 JOHNSON, D. B.
EFFICIENT ALGORITHMS FOR SHORTEST PATHS IN SPARSE NETWORKS.
JOHND 1977 -1 JOHNSON, D. S.; LENSTRA, J. K.; RINNOOY KAN, A. H. G.
COMPLEXITY OF THE NETWORK DESIGN PROBLEM, THE.

****

KATA 1968 -1 KATAYAMA, H.; WATANABE, H.
ALGORITHM FOR THE SHORTEST PATH PROBLEM IN A NETWORK, A.
KAUFA 1977 -2 KAUFMANN, A.; PICHAT, E.
METHODES MATHEMATIQUES NON NUMERIQUES ET LEURS ALGORITHMES. TOME 2 ALGORITHMES DE RECHERCHEDE CHEMINS ET PROBLEMES ASSOCIES.
KERS 1976 -1 KERSHENBAUM, A.; HSIEH, W.; GOLDEN, B.
CONSTRAINED ROUTING IN LARGE SPARSE NETWORKS.
KIMB 1959 -1 KIMBALL, G. E.; HOWARD, R. A.
SEQUENTIAL DECISION PROCESSES.
KINS 1975 -1 KINSNER, W.
ALGORITHM AND A PROGRAM FOR NETWORK ROUTING PROBLEMS, AN.
KIRB 1966 -1 KIRBY, R. F.
MINIMUM PATH ALGORITHM FOR A ROAD NETWORK WITH TURN PENALTIES, A.
KIRBA 1968 -1 KIRBY, R. F.; POTTS, R. B.
MINIMUM ROUTE PROBLEM FOR NETWORKS WITH TURN PENALTIES AND PROHIBITIONS, THE.
KLEE 1977 -1 KLEE, V.; LARMAN, D.
FLOYD-WARSHALL REVISITED: SHORTEST RESTRICTED PATHS (PRELIMINARY VERSION).
KNUT 1976 -1 KNUTH, D. E.
MARIAGES STABLES ET LEURS RELATIONS AVEC D'AUTRES PROBLEMES COMBINATOIRES.
KOH 1976 -1 KOH, K. M.
ON THE SHORTEST-PATH PROBLEM AND DIJKSTRA'S ALGORITHM.
KOH A 1974 -1 KOH, K. M.; TEH, H. H.
UN DIRECTED POSTMAN PROBLEM.
KOLK 1963 -1 KOLKER, R.; LISS, D.; OKADA, S.
ALGEBRAIC TECHNIQUES OF PATH FINDING AND MINIMUM PATH FINDING IN GRAPHS.
KOPY 1977 -1 KOPYLOV, G. N.
MAXIMAL PATHS AND CYCLES IN A GRAPH (RUSSIAN).
KYAN 1975 -1 KYAN, S.; ONAGA, K.
MINIMUM-DELAY MAXIMUM-GAUGE ROUTES IN A TIME-VARYING DYNAMIC NETWORK.
LANDB 1966 -1 LAND, A. H.; MURCHLAND, J. D.
SUBSEQUENT ASSIGNMENT OF TRAFFIC TO SHORTEST PATHS, THE.
LAWL 1977 -2 LAWLER, E. L.
COMMENT ON COMPUTING THE K SHORTEST PATHS IN A GRAPH.
LAWL 1975 -3 LAWLER, E. L.
COMPUTING SHORTEST PATH IN NETWORKS.
LAWL 1977 -4 LAWLER, E. L.
SHORTEST PATH AND NETWORK FLOW ALGORITHMS.
LEE. 1962 -1 LEE. C. Y.
NOTE ON THE N-TH SHORTEST PATH PROBLEM, A.
LENI 1971 -1 LENIN, A. Y. U.
ALGORITHM FOR THE SHORTEST CONNECTION BETWEEN A GROUP OF GRAPH VERTICES (TRANSLATED FROM RUSSIAN), AN.
LENSB 1976 -3 LENSTRA, J. K.; RINNOOY KAN, A. H. G.
TRAVELLING ON A GRAPH (DUTCH).
LEON 1976 -1 LEONT'EV, V. K.
STABILITY IN COMBINATORIAL PROBLEMS OF SELECTION (RUSSIAN).
LESNA 1975 -1 LESNIAK, L.
ON LONGEST PATHS IN CONNECTED GRAPHS.
LEVIA 1963 -1 LEVIN, B. M.; HEDETNIEMI, S.
DETERMINING FASTEST ROUTES USING FIXED SCHEDULES.
LEVY 1966 -1 LEVY, J.
OPTIMAL LOCATION OF NETWORKS.
LIN A 1969 -1 LIN, P. M.; ALDERSON, G. E.
SYMBOLIC NETWORK FUNCTIONS BY A SINGLE PATH-FINDING ALGORITHM.
LITVC 1974 -1 LITVAK, E. I.; ZAVARIN, A. N.
ON THE DETERMINATION OF THE LENGTH OF THE SHORTEST PATH IN A GRAPH (RUSSIAN).
LUCE 1975 -1 LUCE, B.
DUALITY METHODS FOR MAXIMAL FLOW AND SHORTEST PATH PROBLEMS.
MAC K 1966 -1 MAC LAUGHLIN, W. A.
MULTI PATH SYSTEM TRAFFIC ASSIGNMENT ALGORITHM.
MAHR 1976 -1 MAHR, B.; PERL, J.
GRUENDE, TECHNIKEN UND KONSEQUENZEN DER VERALLGEMEINERUNG VON WEGEALGORITHMEN.
MAJO 1964 -1 MAJORANI, E.
SIMPLIFICATION OF LEE'S ALGORITHM FOR SPECIAL PROBLEMS.
MARTF 1963 -1 MARTIN, B. V.
MINIMUM PATH ALGORITHMS FOR TRANSPORTATION PLANNING.
MEYL 1972 -1 MEYL, H.
SHORTEST ROUTE PROBLEM WITH ADDITIONAL PROBLEMS.
MICHA 1968 -1 MICHIE, D.; FLEMING, J. G.; OLDFIELD, J. V.
COMPARISON OF HEURISTIC, INTERACTIVE, AND UNAIDED METHODS OF SOLVING A SHORTEST-ROUTE PROBLEM, A.
MICKA 1975 -1 MICKLE, M. H.; VOGT, W. G.
ON THE DYNAMICS OF A CLASS OF OPTIMUM ROUTING PROBLEMS.
MINOA 1975 -1 MINOUX, M.
PLUS COURT-CHEMIN AVEC CONTRAINTES: ALGORITHMES ET APPLICATIONS.
MINOA 1976 -4 MINOUX, M.
STRUCTURES ALGEBRIQUES GENERALISEES DES PROBLEMES DE CHEMINEMENT DANS LES GRAPHES FINIS, ALGORITHMES ET APPLICATIONS.
MIRC 1976 -1 MIRCHANDANI, P. B.
SHORTEST DISTANCE AND RELIABILITY OF PROBABILISTIC NETWORKS.
MOLZ 1976 -1 MOLZBERGER, P.
HEURISTISCHER ALGORITHMUS ZUR WEGOPTIMIERUNG BEI RECHTWINKELIGEN, ACHSENPARALLELEN HINDERNISSEN, EIN.
MOOR 1976 -1 MOORE, M. H.
ON THE FASTEST ROUTE FOR CONVOY-TYPE TRAFFIC IN FLOWRATE-CONSTRAINED NETWORKS.
MOSH 1963 -1 MOSHER, W. W. JR.
CAPACITY-RESTRAINT ALGORITHM FOR ASSIGNING FLOW TO A TRANSPORT NETWORK, A.
MURCB 1964 -1 MURCHLAND, J. D.
CIRCULATING FLOWS IN MULTI-COMMODITY NETWORKS.
MURCB 1971 -1 MURCHLAND, J. D.
UPPER BOUND FOR THE AVERAGE CANDIDATE CREATION FACTOR IN ARC TYPE SHORTEST DISTANCE CALCULATIONS, AN.
MURCB 1966 -1 MURCHLAND, J. D.
ALGOL PROCEDURE FOR ALL SHORTEST PATHS IN A SYMMETRIC NETWORK, AN.

****

MURCB 1967 -1 MURCHLAND, J. D.
EFFECT OF INCREASING OR DECREASING THE LENGTH OF A SINGLE ARC ON ALL SHORTEST DISTANCES IN A GRAPH, THE.

MURCB 1970 -1 MURCHLAND, J. D.
EXPLANATION OF DIAL'S SHORTEST DISTANCE ALGORITHM, AN.

MURCB 1969 -1 MURCHLAND, J. D.
FIXED MATRIX METHOD FOR ALL SHORTEST DISTANCES IN A DIRECTED GRAPH AND FOR THE INVERSE PROBLEM, A.

MURCB 1966 -2 MURCHLAND, J. D.
EXTENSION OF THE CASCADE ALGORITHM TO LARGE GRAPHS, THE.

MURCB 1968 -2 MURCHLAND, J. D.
ALGORITHMS FOR DETERMINING ALL CONNECTIVITIES IN DIRECTED AND UNDIRECTED GRAPHS.

MURCB 1967 -2 MURCHLAND, J. D.
FALSE PROPOSITION CONCERNING THE SECOND SHORTEST ELEMENTARY PATH IN A GRAPH, A.

MURCB 1966 -3 MURCHLAND, J. D.
INDUCTIVE MATRIX METHOD FOR FINDING ALL SHORTEST PATHS IN A DIRECTED GRAPH, AN.

MURCB 1967 -3 MURCHLAND, J. D.
NUMBER OF OPERATIONS IN FIXED MATRIX METHODS FOR FINDING ALL SHORTEST DISTANCES IN A GRAPH, THE.

MURCB 1968 -3 MURCHLAND, J. D.
EASY CASE WHEN TESTING A DIRECTED GRAPH FOR THE PRESENCE OF A CIRCUIT OF NEGATIVE LENGHT, AN.

MURCB 1967 -4 MURCHLAND, J. D.
TWO REMARKS ON CONGESTED ASSIGNMENT.

NIEM 1976 -1 NIEMINEN, J.
REALIZING THE DISTANCE MATRIX OF A GRAPH.

NOLT 1975 -1 NOLTEMEIER, H.
ALGORITHM FOR THE DETERMINATION OF LONGEST DISTANCES IN A GRAPH, AN.

NORD 1969 -1 NORDBECK, S.; RYSTEDT, B.
COMPUTER CARTOGRAPHY SHORTEST ROUTE PROBLEMS.

NOUR 1968 -1 NOUR, M. E.
NEW LINEAR PROGRAMMING TECHNIQUES FOR THE SHORTEST ROUTE PROBLEM.

NOURA 1970 -1 NOUR, M. E.; RATHBONE, D. E.; GRANSBORG, B. S. M.
SHORTEST ROUTE MATRIX TECHNIQUE.

OREL 1977 -1 OREL, E. N.
FINDING A SHORTEST PATH IN A GRAPH USING THE MINORANT OF THE BELLMAN FUNCTION (RUSSIAN).

OSYC 1976 -1 OSYCZKA, A.
ALGORITHM OF OPTIMIZATION FOR A SPECIAL CLASS OF NETWORKS, AN.

OUDH 1977 -1 OUDHEUSDEN, D. L. VAN
SHORTEST DISTANCES IN A MODIFIED TRANSPORTATION NETWORK.

PAND 1962 -1 PANDIT, S. N. N.
MINADDITION AND AN ALGORITHM TO FIND THE MOST RELIABLE PATHS IN A NETWORK.

PAPE 1969 -1 PAPE, U.
KUERZESTE WEGE IN ASYMMETRISCHEN NETZWERKEN ZWISCHEN EINEM FESTEN KNOTEN UND BELIEBIGEN KNOTEN.

PAPE 1977 -1 PAPE, U.
BIBLIOGRAPHIE ZU "KUERZESTE WEGLAENGEN UND WEGE IN GRAPHEN", EINE (STAND: ENDE 1976).

PAPE 1969 -2 PAPE, U.
VERAENDERUNGEN IN GERICHTETEN NETZWERKEN UND KORREKTUR KUERZESTER WEGLAENGEN ZWISCHEN EINEM FESTEN KNOTEN UND ANDEREN KNOTEN.

PAPE 1976 -4 PAPE, U.
ZUR WIRTSCHAFTLICHKEIT VON KORREKTURVERFAHREN BEI NETZWERKVERAENDERUNGEN.

PICAC 1976 -1 PICARD, J.-C.; QUEYRANNE, M.
TIME-DEPENDENT TRAVELING SALESMAN PROBLEM AND APPLICATION TO THE TARDINESS PROBLEM IN ONE-MACHINE SCHEDULING, THE.

PIER 1975 -1 PIERCE, A. R.
BIBLIOGRAPHY ON ALGORITHMS FOR SHORTEST PATH, SHORTEST SPANNING TREE, AND RELATED CIRCUIT ROUTING PROBLEMS (1956-1974).

POHL 1969 -1 POHL, I.
BI-DIRECTIONAL AND HEURISTIC SEARCH IN PATH PROBLEMS.

POHL 1970 -1 POHL, I.
HEURISTIC SEARCH VIEWED AS PATH FINDING IN A GRAPH.

RAND 1976 -1 RANDOLPH, P.
SHORTEST PATHS THROUGH MULTIPLE CRITERIA NETWORKS.

RANDA 1977 -1 RANDOLPH, P.; RINGEISEN, R.
SHORTEST PATHS THROUGH MULTIPLE CRITERIA NETWORKS.

RAPA 1959 -1 RAPAPORT, H.; ABRAMSON, P.
ANALOG COMPUTER FOR FINDING AN OPTIMUM ROUTE THROUGH A COMMUNICATIONS NETWORK, AN.

RAYB 1966 -1 RAYBOV, G. G.
ALGORITHM FOR THE SOLUTION OF A LABYRINTH ON A DISCRETE FIELD AND ITS APPLICATIONS (TRANSLATED FROM RUSSIAN), AN.

REIN 1977 -1 REINGOLD, E. M.; NIEVERGELT, J.; DEO, N.
COMBINATORIAL ALGORITHMS: THEORY AND PRACTICE.

RIBB 1972 -1 RIBBECK, K. F.
ROUTENSUCHEN IN STADTSTRASSENNETZEN.

RIBE 1971 -1 RIBEILL, G.; TABOURIER, Y.
CHEMINS EXTREMAUX DANS UN GRAPHE ENTRE TOUS COUPLES DE SOMMETS: QUELQUES ALGORITHMES ET LEUR MISE EN OEUVRE.COMPARAISONS SUR LES PERFORMANCES.

RIDL 1965 -1 RIDLEY, T. M.
INVESTMENT IN A NETWORK TO REDUCE LENGTH OF A SHORTEST ROUTE.

RILE 1970 -1 RILEY, D.
LONGEST PATH IN A NETWORK, THE.

ROSEC 1975 -1 ROSENTHAL, A.
BIT-PUSHING SHORTEST DISTANCE ALGORITHM, A.

ROY 1975 -1 ROY, B.
CHEMINS ET CIRCUITS: ENUMERATION ET OPTIMISATION.

ROY 1967 -1 ROY, B.
NOMBRE CHROMATIQUE ET PLUS LONGS CHEMINS D'UN GRAPHE.

ROY 1966 -1 ROY, B.
CHEMINS DE LONGUEUR EXTREMALES.

ROY 1975 -2 ROY, B.
CHEMINS ET CIRCUITS: ENUMERATION ET OPTIMISATION: A SURVEY.

RUIT 1968 -1 RUITER, E. R.
TRAFFIC ASSIGNMENT RESEARCH, FINAL REPORT.

****

SAKAB 1966 -1 SAKAROVITCH, M.
K SHORTEST ROUTES AND THE K SHORTEST CHAINS IN A GRAPH, THE.

SAKAB 1968 -1 SAKAROVITCH, M.
K SHORTEST CHAINS IN A GRAPH, THE.

SCHAA 1971 -1 SCHAERF, S.; SHAHAM, J.
PUBLIC TRANSPORTATION SIMULATION ALGORITHM, A.

SCHMB 1975 -1 SCHMITZ, W.
UEBER LAENGSTE WEGE UND KREISE IN GRAPHEN.

SCLE 1977 -1 SCLEE, W.
MATRIXALGORITHMEN BEI STOCHASTISCHER BEWERTUNG.

SHAM 1972 -1 SHAMMA, M. M.; WILLIAMS, T. A.
CONSTRUCTING SAMPLE NETWORKS FOR ANALYZING AND COMPARING SHORTEST ROUTE ALGORITHMS.

SHIE 1976 -1 SHIER, D. R.
ALGORITHMS FOR FINDING THE K SHORTEST PATHS IN A NETWORK.

SHIE 1976 -3 SHIER, D. R.
ITERATIVE METHODS FOR DETERMINING THE K SHORTEST PATHS IN A NETWORK.

SHIM 1953 -1 SHIMBEL, A.
STRUCTURAL PARAMETERS OF COMMUNICATION NETWORKS.

SHIM 1951 -1 SHIMBEL, A.
APPLICATION OF MATRIX ALGEBRA TO COMMUNICATION NETS.

SITR 1974 -1 SITRUK, G.
CHEMINS EXTREMAUX ET K-EXTREMAUX APPLICATION AUX PROBLEMES DE FLOTS.

SWEN [illegible] -1 SWEN, G.
DEVELOPMENT OF THE SPANNING TREE CONCEPT IN GRAPH THEORY WITH APPLICATIONS TO THE SHORTEST ROUTE PROBLEM.

SYSL 1972 -1 SYSLO, M.
ALGORITHMS 17-18: ALGORITHMS FOR SHORTEST PATH PROBLEMS.

SYSLA 1975 -1 SYSLO, M. M.
APPLIED GRAPH THEORY II. EXTREMAL PATH PROBLEMS IN GRAPHS AND NETS (POLISH).

TABO 1973 -1 TABOURIER, Y.
DIVERS ASPECTS DU PROBLEME DES PLUS COURTS CHEMINS PRINCIPAUX ALGORITHMES.

TERN 1975 -1 TERNO, J.; GIERTH, G.; WAGNER, H.
BESTIMMUNG DES KUERZESTEN WEGES MITTELS DYNAMISCHER OPTIMIERUNG, DIE.

THOMA 1976 -1 THOMAS, R. E.
ON OPTIMUM PATH PROBLEMS.

THOR 1966 -1 THORELLI, L. E.
ALGORITHM FOR COMPUTING ALL PATHS IN A GRAPH, AN.

THUE 1976 -1 THUEMMLER, S.
KUERZESTE WEGE IN GEMISCHTEN NETZEN.

THUM 1975 -1 THUMER, R.
VERFAHREN ZUR AUSWAHL VON TEILNETZEN BEI DER BESTIMMUNG KUERZESTER WEGE IN VERKEHRSNETZEN, EIN.

TOME 1966 -1 TOMESCU, I.
METHOD FOR DETERMINING THE SHORTEST PATH BETWEEN TWO NODES OF A FINITE GRAPH (RUMANIAN), A.

TOME 1975 -2 TOMESCU, I.
INTRODUCTION TO COMBINATORICS (TRANSLATED FROM RUMANIAN).

TRYG 1976 -1 TRYGAR, T. A.; WRIGHT, G.
APPLICATION OF DECOMPOSITION TECHNIQUES TO THE CALCULATION OF A MINIMUM DISTANCE MATRIX FOR A LARGE SCALE STREET NETWORK MODEL, THE.

TYUR 1963 -1 TYURENKOV, V. A.
ALGORITHM FOR FINDING THE SHORTEST PATH (RUSSIAN), AN.

UNIT 1964 -1 UNITED STATES DEPT. OF COMMERCE BUREAU OF PUBLIC ROADS.
TRAFFIC ASSIGNMENT MANUAL.

VERGB 1951 -1 VERGLUNSKY, S.
ON THE SHORTEST PATH THROUGH A NUMBER OF POINTS.

VIDY 1975 -1 VIDYASANKAR, K.; YOUNGER, D.
MINIMAX EQUALITY RELATED TO THE LONGEST DIRECTED PATH IN AN ACYCLIC GRAPH, A.

WAGN 1976 -1 WAGNER, R. A.
SHORTEST PATH ALGORITHM FOR EDGE-SPARSE GRAPHS, A.

WANGB 1974 -1 WANG, J. C.; MICKLE, M. H.; HOELZEMAN, R. G.
SET COVERING APPLICATION IN MINIMUM ROUTE GENERATION, A.

WATT 1963 -1 WATTLEWORTH, J. A.; SHULDINER, P. W.
ANALYTICAL METHODS IN TRANSPORTATION; LEFT-TURN PENALTIES IN TRAFFIC ASSIGNMENT MODELS.

WEGE 1974 -1 WEGENER, M.
ROUTENSUCHALGORITHMEN IN POLIS.

WEIG 1973 -1 WEIGAND, M. M.
UEBER EIN VERFAHREN ZUR BESTIMMUNG VON K-KUERZESTEN WEGEN IN EINEM GRAPHEN.

WEIG 1976 -1 WEIGAND, M. M.
ALGORITHMUS 27. EIN NEUER ALGORITHMUS ZUR BESTIMMUNG VON K-KUERZESTEN WEGEN IN EINEM GRAPHEN.

WEINF 1976 -1 WEINTRAUB, A.
OPTIMAL FLOWS IN NETWORKS AND GAMES: THE MULTICOMMODITY FLOW PROBLEM IN INTEGERS.

WHIT 1977 -1 WHITAKER, R. A.
THREE ALGORITHMS FOR CALCULATING SOME OR ALL OF THE SHORTEST PATHS IN A SPARSE NETWORK.

WILLB 1975 -1 WILLIAMS, T.
IMPROVED SHORTEST PATH ALGORITHM FOR NETWORKS IN WHICH THE MAXIMUM OUTWARD DEGREE IS LESS THAN N/2, AN.

WILLC 1972 -1 WILLIAMS, T. A.; WHITE, G. P.
COMPARING SHORTEST ROUTE ALGORITHMS FOR DIRECTED GRAPHS.

WILLC 1972 -2 WILLIAMS, T. A.; WHITE, G. P.
NOTE ON YEN'S ALGORITHM FOR FINDING THE LENGTH OF ALL COMPARING SHORTEST ROUTE ALGORITHMS FOR DIRECTED GRAPHS, A.

WINGA 1965 -1 WING, O.; GOLDMAN, A. J.
MOST PROFITABLE ROUTINE BEFORE MAINTENANCE.

WONGB 1971 -1 WONG, C. K.; YUE, P. C.
UPPER-BOUNDS FOR VARIOUS TYPES OF PATH LENGTHS OF BINARY TREES.

WONGD 1976 -1 WONGSEELASHOTE, A.
ALGEBRA FOR DETERMINING ALL PATH-VALUES IN A NETWORK WITH APPLICATION TO K-SHORTEST-PATHS PROBLEMS, AN.

YAO A 1977 -1 YAO, A. C.; AVIS, D. M.; RIVEST, R. L.
OMEGA (N POWER 2 LOG N) LOWER BOUND TO THE SHORTEST PATHS PROBLEM, AN.

YEN 1975 -1 YEN, J. Y.
SHORTEST PATH NETWORK PROBLEMS.

YEN 1969 -1 YEN, J. Y.
SOME ALGORITHMS FOR FINDING THE SHORTEST ROUTES THROUGH THE GENERAL NETWORKS.

****

YEN 1971 -1 YEN, J. Y.
ON THE EFFICIENCIES OF ALGORITHMS FOR DETECTING NEGATIVE LOOPS IN NETWORKS.

YUVA 1976 -1 YUVAL, G.
ALGORITHM FOR FINDING ALL SHORTEST PATHS USING (N POWER 2.81) INFINITE PRECISION MULTIPLICATIONS, AN.

ZABL 1976 -1 ZABLUDOWSKI, A.; CHMARA, K.
METHOD FOR FINDING A PATH IN A GRAPH (POLISH), A.

ZAMF 1976 -1 ZAMFIRESCU, T.
ON LONGEST PATHS AND CIRCUITS IN GRAPHS.

ZANG 1966 -1 ZANGWILL, W.
SHORTEST ROUTE PROBLEM UNDER EITHER CONCAVE OR CONVEX SHIPPING COSTS, THE.

SPANNING TREE

****

ALI 1977 -1 ALI, A. I.; HELGASON, R. V.; KENNINGTON, J. L.; LALL, H. S.
PRIMAL SIMPLEX NETWORK CODES: STATE-OF-THE-ART IMPLEMENTATION TECHNOLOGY (REVISED VERSION).

ANIS 1976 -1 ANISIMOV, V. V.
ON OPTIMUM PACKING OF TREES (RUSSIAN).

BIRD 1976 -1 BIRD, C. G.
ON COST ALLOCATION FOR A SPANNING TREE: A GAME THEORETIC APPROACH.

BLAZC 1976 -1 BLAZHKEVICH, B. I.; SEMOTYUK, V. N.
TREE SEARCH ALGORITHM IN A GRAPH (RUSSIAN).

BURNC 1976 -1 BURNS, R. N.; HAFF, C. E.
COMBINATORIAL RANKING PROBLEM, A.

CAHI 1975 -1 CAHIT, I.; CAHIT, R.
MINIMUM SPANNING TREE.

CHANF 1977 -1 CHANDRASEKARAN, R.
MINIMAL RATIO SPANNING TREES.

CHERB 1976 -1 CHERITON, D.; TARJAN, R. E.
FINDING MINIMUM SPANNING TREES.

CHU 1965 -1 CHU, Y. J.; LIU T. H.
ON THE SHORTEST ARBORESCENCE OF DIRECTED GRAPHS.

CHVAA 1978 -1 CHVATAL, V.; ERDOES P.
BIASED POSITIONAL GAMES.

DALA 1976 -1 DALAL, Y. K.
DISTRIBUTED ALGORITHM FOR CONSTRUCTING MINIMAL SPANNING TREES IN COMPUTER-COMMUNICATION NETWORKS, A.

DIJK 1976 -1 DIJKSTRA, E. W.
DISCIPLINE OF PROGRAMMING, A.

DIJK 1960 -1 DIJKSTRA, E. W.
SOME THEOREMS ON SPANNING SUBTREES OF A GRAPH.

DINIA 1977 -1 DINITS, E. A.; ZAJTSEV, M. A.
ALGORITHMS WHICH GENERATE NON-ISOMORPHIC TREES (RUSSIAN).

EDMO 1975 -1 EDMONDS, J.
SOME WELL-SOLVED PROBLEMS IN COMBINATORIAL OPTIMIZATION.

EVST 1969 -1 EVSTIGNEEV, V. A.; JASAKOVA, V. P.
CERTAIN ALGORITHM ON GRAPHS (RUSSIAN), A.

FONG 1977 -1 FONG, A. C.; ULLMAN, J. D.
FINDING THE DEPTH OF A FLOW GRAPH.

GABOB 1977 -1 GABOW, H. N.
TWO ALGORITHMS FOR GENERATING WEIGHTED SPANNING TREES IN ORDER.

GELDB 1975 -1 GELDMACHER, R. C.; LIU, C. S.
PRODUCT OF INDEPENDENT SETS OF THE VERTEX-INCIDENCE VECTORS OF A GRAPH, THE.

GLOVC 1975 -2 GLOVER, F.; KLINGMAN, D.
FINDING MINIMUM SPANNING TREES WITH A FIXED NUMBER OF LINKS AT A NODE.

GRIM 1976 -1 GRIMMETT, G. R.
UPPER BOUND FOR THE NUMBER OF SPANNING TREES OF A GRAPH, AN.

GROE 1977 -7 GROETSCHEL, M.
STRONG BLOCKS AND THE OPTIMUM BRANCHING PROBLEM.

HADL 1974 -1 HADLOCK, F. O.
MINIMUM SPANNING FORESTS OF BOUNDED TREES.

HAFF 1973 -1 HAFF, C. E.; BURNS, R. N.
RANKING PROBLEM IN GRAPHS, A.

HARTD 1976 -1 HARTNELL, B. L.
SOME PROBLEMS ON SPANNING TREES.

HAUS 1978 -1 HAUSMANN, D.
ADJAZENZCHARAKTERISIERUNGEN FUER POLYEDERECKEN IN DER KOMBINATORISCHEN OPTIMIERUNG.

HONGA 1977 -1 HONG, S.
ASYMMETRIC TRAVELLING SALESMAN PROBLEM AND SPANNING ARBORESCENCES OF DIRECTED GRAPHS, THE.

HU 1973 -1 HU, T. C.
OPTIMUM COMMUNICATION SPANNING TREES.

KAME 1976 -1 KAMEDA, T.
ON MAXIMALLY DISTANT SPANNING TREES OF A GRAPH.

KEL' 1976 -1 KEL'MANS, A. K.
NUMBER OF TREES IN A GRAPH WHICH CONTAINS A GIVEN FOREST (RUSSIAN).

KHAI 1976 -1 KHAIKIN, N. M.; ZAK, E. I.
CONSTRUCTION OF MINIMUM TREE PACKING IN THE PROBLEM OF OPTIMUM MEMORY DISTRIBUTION (RUSSIAN).

KRAR 1975 -1 KRARUP, J.
PERIPATETIC SALESMAN AND SOME RELATED UNSOLVED PROBLEMS, THE.

KUND 1974 -1 KUNDU, S.
BOUNDS ON THE NUMBER OF DISJOINT SPANNING TREES.

MANA 1973 -1 MANABE, R.; KOTANI, S.
SEARCH METHODS OF THE OPTIMUM MAXIMAL TREE IN THE ORIENTED GRAPH.

MOROA 1976 -1 MOROZOV, S. A.
PROBLEM OF FINDING THE SHORTEST TREE OF CONNECTION WITH CONSTRAINTS (RUSSIAN), THE.

MURCB 1965 -1 MURCHLAND, J. D.
PRIM'S RULE FOR CONSTRUCTING A MINIMAL SPANNING TREE.

MYER 1974 -1 MYERS, B. R.
ON SPANNING TREES, WEIGHTED COMPOSITIONS, FIBONACCI NUMBERS AND RESISTOR NETWORKS.

NASH 1961 -1 NASH-WILLIAMS, C. ST. J. A.
EDGE-DISJOINT SPANNING TREES OF FINITE GRAPHS.

NASH 1961 -2 NASH-WILLIAMS, C. ST. J. A.
EDGE-DISJOINT SPANNING TREES OF FINITE GRAPHS.

OLAH 1975 -1 OLAH, GY.
ENUMERATION OF SPANNING TREES CONTAINING GIVEN SUBTREES.

PICA 1975 -1 PICARD, C.-F.
ARBRE MINIMAL D'UNE SOMME DE GRAPHES.

PIER 1975 -1 PIERCE, A. R.
BIBLIOGRAPHY ON ALGORITHMS FOR SHORTEST PATH, SHORTEST SPANNING TREE, AND RELATED CIRCUIT ROUTING PROBLEMS (1956-1974).

READB 1975 -1 READ, R. C.; TARJAN, R. E.
BOUNDS ON BACKTRACK ALGORITHMS FOR LISTING CYCLES, PATHS, AND SPANNING TREES.

REIN 1977 -1 REINGOLD, E. M.; NIEVERGELT, J.; DEO, N.
COMBINATORIAL ALGORITHMS: THEORY AND PRACTICE.

ROGE 1971 -1 ROGER, J.
UPDATING A MINIMUM SPANNING TREE.

ROSHA 1977 -1 ROSHIN, G. B.
AN ALGORITHM FOR FINDING A MINIMAL SPANNING TREE IN A GROWING GRAPH (RUSSIAN).

****

ROUX 1975 -1 ROUX, M.
NOTES SUR L'ARBRE DE LONGUEUR MINIMA.

SEDL 1975 -1 SEDLACEK, J.
PERFECT AND QUASIPERFECT GRAPHS (CZECH).

SHIE 1974 -1 SHIER, D. R.
MAXIMIZING THE NUMBER OF SPANNING TREES IN A GRAPH WITH N NODES AND M EDGES.

SWEN 1967 -1 SWEN, G.
DEVELOPMENT OF THE SPANNING TREE CONCEPT IN GRAPH THEORY WITH APPLICATIONS TO THE SHORTEST ROUTE PROBLEM.

TARJ 1975 1 TARJAN, R. E.
FINDING EDGE-DISJOINT SPANNING TREES.

TARJ 1975 -2 TARJAN, R. E.
FINDING MINIMUM SPANNING TREES.

TOME 1975 -1 TOMESCU, I.
ALGORITHME POUR L'OBTENTION D'UNE CHAINE HAMILTONIENNE EN PARTANT DE L'ARBRE MINIMAL D'UN GRAPHE, UN.

WALLA 1976 -1 WALLER, D. A.
GENERAL SOLUTION TO THE SPANNING TREE ENUMERATION PROBLEM IN ARBITRARY MULTIGRAPH JOINS.

WASS 1975 -1 WASSON, W. D.; MAC ISAAC, R.
NEW SPANNING TREE ALGORITHM, A.

WIN 1975 -1 WIN, S.
EXISTENZ VON GERUESTEN MIT VORGESCHRIEBENEM MAXIMALGRAD IN GRAPHEN.

YAO 1975 -1 YAO, A. C.
O (|E|LOG LOG|V|) ALGORITHM FOR FINDING MINIMUM SPANNING TREES, AN.

STOCHASTIC INTEGER PROGRAMMING
****

BRUNA 1976 -1 BRUNO, J.; COFFMAN, E. G. JR.; JOHNSON, D. B.
ON BATCH SCHEDULING OF JOBS WITH STOCHASTIC SERVICE TIMES AND COST STRUCTURES ON A SINGLE SERVER.

BRUNE 1976 -2 BRUNO, J. L.
SEQUENCING JOBS WITH STOCHASTIC TASK STRUCTURES ON A SINGLE MACHINE.

CHEBA 1975 -1 CHEBOTAREV, A. S.; SHIBANOV, S. E.
PROBABILISTIC PROBLEM OF SEQUENCING PROJECTS IN SEVERAL STEPS.

CHVA 1977 -1 CHVATAL, V.
DETERMINING THE STABILITY NUMBER OF A GRAPH.

COOPC 1977 -1 COOPER, L.; LEBLANC, L. J.
STOCHASTIC TRANSPORTATION PROBLEMS AND OTHER NETWORK RELATED CONVEX PROBLEMS.

CORB 1972 -1 CORBAN, A.
THREE DIMENSIONAL STOCHASTIC TRANSPORTATION PROBLEM (RUMANIAN), THE.

CRAN 1976 -1 CRANDALL, K. C.
PROBABILISTIC TIME SCHEDULING.

DAGA 1977 -1 DAGANZO, C. F.; SHEFFI, Y.
ON STOCHASTIC MODELS OF TRAFFIC ASSIGNMENT.

DIAL 1970 -1 DIAL, R.
PROBABILISTIC ASSIGNMENT: A MULTIPATH TRAFFIC ASSIGNMENT MODEL WHICH OBVIATES PATH ENUMERATION.

DOLO 1976 -1 DOLODARENKO, V. A.; FEDAN, E. I.
ASSIGNMENT PROBLEM WITH A NONLINEAR STOCHASTIC OBJECTIVE FUNCTION (RUSSIAN), THE.

FISB 1971 -1 FISBEIN, S. M.
ON THE SOLUTION OF A CERTAIN COMBINATORIAL PROBLEM (RUSSIAN).

GLAZ 1976 -1 GLAZEBROOK, K. D.
STOCHASTIC SCHEDULING WITH ORDER CONSTRAINTS.

HALP 1976 -1 HALPERN, J.
SEQUENTIAL COVERING PROBLEM UNDER UNCERTAINTY, THE (REVISED).

HALP 1977 -2 HALPERN, J.
SEQUENTIAL COVERING PROBLEM UNDER UNCERTAINTY, THE.

JUCK 1976 -1 JUCKER, J. V.; CARLSON, R. C.
SIMPLE PLANT-LOCATION PROBLEM UNDER UNCERTAINTY, THE.

LAVE 1976 -1 LAVENBERG, S. S.; SHEDLER, G. S.
STOCHASTIC MODELING OF PROCESSOR SCHEDULING WITH APPLICATION TO DATA BASE MANAGEMENT SYSTEMS.

LEBL 1977 -1 LEBLANC, L. J.
HEURISTIC APPROACH OF LARGE SCALE DISCRETE STOCHASTIC TRANSPORTATION-LOCATION PROBLEMS, A.

MASS 1976 -1 MASSA, L.
SEMIECONOMIC METHOD FOR THE DETERMINATION OF OPTIMAL LOT SIZES, A.

MIRC 1976 -1 MIRCHANDANI, P. B.
SHORTEST DISTANCE AND RELIABILITY OF PROBABILISTIC NETWORKS.

NASHA 1977 -1 NASH, P.; GITTINS, J. C.
HAMILTONIAN APPROACH TO OPTIMAL STOCHASTIC RESOURCE ALLOCATION, A.

SCLE 1977 -1 SCLEE, W.
MATRIXALGORITHMEN BEI STOCHASTISCHER BEWERTUNG.

STEW 1976 -1 STEWART, W. R. JR.
DELIVERY TRUCK ROUTING PROBLEM WITH STOCHASTIC DEMANDS, THE.

SUST 1975 -1 SUSTAKASVILI, I. I.
ALGORITHMIZATION OF THE SELECTION OF OPTIMAL VARIANTS IN STOCHASTIC NETWORK GRAPHS THAT INCLUDE CONNECTING AND DISCONNECTING PATHS (RUSSIAN).

TOBI 1977 -1 TOBIN, R. L.
EXTENSION OF DIAL'S ALGORITHM UTILIZING A MODEL OF TRIPMAKERS' PRECEPTIONS, AN.

VERG 1976 -1 VERGIN, R. C.
NOTE ON "LOT SIZE SCHEDULING ON A SINGLE MACHINE FOR STOCHASTIC DEMAND" BY GOYAL, A.

WESO 1977 -1 WESOLOWSKY, G. O.
PROBABILISTIC WEIGHTS IN THE ONE-DIMENSIONAL FACILITY LOCATION PROBLEM.

WILSA 1975 -1 WILSON, D.
MEAN COST APPROXIMATION FOR TRANSPORTATION PROBLEMS WITH STOCHASTIC DEMAND, A.

WITT 1978 -1 WITTEN, P.; ZIMMERMANN, H.-G.
STOCHASTISCHE TRANSPORTPROBLEME.

YUDIA 1974 -1 YUDIN, D. B.; TSOY, E. V.
INTEGER-VALUED STOCHASTIC PROGRAMMING (RUSSIAN).

ALI 1977 -1 ALI, A. I.; HELGASON, R. V.; KENNINGTON, J. L.; LALL, H. S.
PRIMAL SIMPLEX NETWORK CODES: STATE-OF-THE-ART IMPLEMENTATION TECHNOLOGY (REVISED VERSION).

ASSA 1975 -1 ASSAD, A.
MULTICOMMODITY NETWORK FLOWS - A SURVEY.

BAKE 1976 -1 BAKER, K. R.
WORKFORCE ALLOCATION IN CYCLICAL SCHEDULING PROBLEMS: A SURVEY.

BALAD 1977 -1 BALACHANDRAN, V.; SRINIVASAN, V.; THOMPSON, G. L.
SURVEY OF THE APPLICATIONS OF THE OPERATOR THEORY OF PARAMETRIC PROGRAMMING FOR THE TRANSPORTATION AND GENERALIZED TRANSPORTATION PROBLEM, A.

BALAG 1977 -1 BALAS, E.
DISJUNCTIVE PROGRAMMING.

BALAJ 1975 -1 BALAS, E.; PADBERG, M. W.
SET PARTITIONING: A SURVEY.

BART 1975 -1 BARTHES, J. P.
BRANCHING METHODS IN COMBINATORIAL OPTIMIZATION.

BEAL 1977 -1 BEALE, E. M. L.
BRANCH AND BOUND METHODS.

BEAL 1977 -2 BEALE, E. M. L.
INTEGER PROGRAMMING.

BEEN 1976 -1 BEENHAKKER, H. L.; NARAYANAN, V.
ALGORITHMS FOR SCHEDULING PROJECTS WITH LIMITED RESOURCES.

BENI 1977 -1 BENICHOU, M.; GAUTHIER, J. M.; HENTGES, G.; RIBIERE, G.
EFFICIENT SOLUTION OF LARGE-SCALE LINEAR PROGRAMMING PROBLEMS - SOME ALGORITHMIC TECHNIQUES AND COMPUTATIONAL RESULTS, THE.

BERG 1977 -1 BERGE, C.
THEORIE FRACTIONNAIRE DES GRAPHES.

BLAZ 1976 -1 BLAZEWICZ, J.
DEADLINE SCHEDULING OF TASKS - A SURVEY.

BOOKA 1977 -1 BOOKBINDER, J. H.; SETHI, S. P.
DYNAMIC TRANSPORTATION PROBLEM: A SURVEY, THE.

BOUR 1969 -1 BOURGOIN, M. H.; HEURGON, E. M. J.
STUDY AND COMPARISON OF ALGORITHMS OF THE SHORTEST PATH THROUGH PLANNED EXPERIMENTS.

BRAE 1975 -1 BRAEMERT, D.
GRAPH THEORY: A SELECTED BIBLIOGRAPHY.

BROC 1976 -1 BROCKLEHURST, E. R.
GENERALIZED BRANCH AND BOUND: A METHOD FOR INTEGER LINEAR PROGRAMS.

BRUA 1975 -1 BRUALDI, R. A.
MATROIDS INDUCED BY DIRECTED GRAPHS, A SURVEY.

BRUA 1975 -2 BRUALDI, R. A.
TRANSVERSAL THEORY AND GRAPHS.

BURK 1977 -1 BURKARD, R. E.
TRAVELLING SALESMAN AND ASSIGNMENT PROBLEMS: A SURVEY.

BURKJ 1968 -1 BURKOV, V. N.; LOVETSKII, S. E.
METHODS FOR THE SOLUTION OF EXTREMAL PROBLEMS OF COMBINATORIAL TYPE (REVIEW) (RUSSIAN).

CHARC 1973 -1 CHARTRAND, G.; KAPOOR, S. F.; KRONK, H. V.
MANY FACETS OF HAMILTONIAN GRAPHS, THE.

CHENB 1976 -1 CHEN, DER-S.; ZIONTS, S.
COMPARISON OF SOME ALGORITHMS FOR SOLVING THE GROUP THEORETIC INTEGER PROGRAMMING PROBLEM.

CHERB 1976 -1 CHERITON, D.; TARJAN, R. E.
FINDING MINIMUM SPANNING TREES.

CHRI 1976 -1 CHRISTOFIDES, N.
VEHICLE ROUTING PROBLEM, THE.

CHRI 1975 -1 CHRISTOFIDES, N.
HAMILTONIAN CIRCUITS AND THE TRAVELLING SALESMAN PROBLEM: A SURVEY.

CHVA 1975 -1 CHVATAL, V.
SOME LINEAR PROGRAMMING ASPECTS OF COMBINATORICS.

COCKB 1977 -1 COCKAYNE, E. J.; HEDETNIEMI, S. T.
TOWARDS A THEORY OF DOMINATION IN GRAPHS.

COHE 1973 -1 COHEN, J.
SURVEY ON THE WAREHOUSE LOCATION PROBLEM.

CORN 1974 -1 CORNEIL, D. G.
ANALYSIS OF GRAPH THEORETICAL ALGORITHMS, THE.

DEO 1969 -1 DEO, N.
EXTENSIVE ENGLISH LANGUAGE BIBLIOGRAPHY ON GRAPH THEORY AND ITS APPLICATIONS, AN.

DIALA 1977 -1 DIAL, R.; GLOVER, F.; KARNEY, D.; KLINGMAN, D.
COMPUTATIONAL ANALYSIS OF ALTERNATIVE ALGORITHMS AND LABELING TECHNIQUES FOR FINDING SHORTEST PATH TREES, A.

DUEC 1977 -1 DUECK, W.
EINFLUSSBEREICH DISKRETER OPTIMIERUNGSMETHODEN, DER.

EDMO 1977 -1 EDMONDS, J.
MATROID INTERSECTION.

ERLE 1975 -1 ERLENKOTTER, D.
BIBLIOGRAPHY ON DYNAMIC LOCATION MODELS.

EVANA 1977 -2 EVANS, J. R.
SURVEY OF INTEGRALITY IN MULTICOMMODITY NETWORKS, A.

EVANA 1976 -4 EVANS, J. R.
TRANSPORTATION PROBLEM AND ITS EXTENSIONS: A SURVEY, THE.

FAYA 1976 -1 FAYARD, D.; PLATEAU, G.
TECHNIQUE DE RESOLUTION DU PROBLEME DU KNAPSACK EN VARIABLES BIVALENTES.

FINKA 1976 -1 FINKEL'SHTEIN, YU. YU.
APPROXIMATIVE METHODS AND APPLIED PROBLEMS IN DISCRETE PROGRAMMING (RUSSIAN).

FIOR 1974 -1 FIORINI, S.; WILSON, R. J.
ON THE CHROMATIC INDEX OF A GRAPH, II.

FIOR 1973 -1 FIORINI, S.; WILSON, R. J.
ON THE CHROMATIC INDEX OF A GRAPH, I.

FRANG 1970 -1 FRANK, H.; FRISCH, I.
NETWORK ANALYSIS.

GABB 1974 -1 GABBAY, H.
OVERVIEW OF VEHICULAR SCHEDULING PROBLEMS, AN.

GALLD 1976 -1 GALLUS, G.
HEURISTISCHE VERFAHREN ZUR LOESUNG ALLGEMEINER GANZZAHLIGER LINEARER OPTIMIERUNGSPROBLEME (EIN UEBERBLICK).

GAREB 1976 -1 GAREY, M. R.; JOHNSON, D. S.
APPROXIMATION ALGORITHMS FOR COMBINATORIAL PROBLEMS: AN ANNOTATED BIBLIOGRAPHY.

SURVEYS, BIBLIOGRAPHIES

****

GAREB 1976 -2 GAREY, M. R.; JOHNSON, D. S.
APPROXIMATION ALGORITHMS FOR COMBINATORIAL PROBLEMS: PROSPECTS AND LIMITATIONS.

GEOF 1977 -1 GEOFFRION, A. M.
HOW CAN SPECIALIZED DISCRETE AND CONVEX OPTIMIZATION METHODS BE MARRIED.

GEOF 1974 -1 GEOFFRION, A. M.
GUIDED TOUR OF RECENT PRACTICAL ADVANCES IN INTEGER LINEAR PROGRAMMING, A.

GILM 1977 -1 GILMORE, P. C.
CUTTING STOCK, LINEAR PROGRAMMING, KNAPSACKING, DYNAMIC PROGRAMMING AND INTEGER PROGRAMMING SOME INTERCONNECTIONS.

GLOVC 1975 -5 GLOVER, F.; KLINGMAN, D.
NEW ADVANCES IN THE SOLUTION OF LARGE SCALE NETWORK AND NETWORK-RELATED PROBLEMS.

GOLD 1975 -1 GOLDEN, B.
VEHICLE ROUTING PROBLEMS: FORMULATIONS AND HEURISTIC SOLUTION TECHNIQUES.

GOLD 1976 -2 GOLDEN, B.
SHORTEST-PATH ALGORITHMS: A COMPARISON.

GOLDB 1976 -1 GOLDEN, B. L.
APPROACHES TO THE CUTTING STOCK PROBLEM.

GOLDB 1976 -2 GOLDEN, B. L.
RECENT DEVELOPMENTS IN VEHICLE ROUTING.

GOLDC 1977 -1 GOLDEN, B. L.; MAGNANTI, T. L.
DETERMINISTIC NETWORK OPTIMIZATION: A BIBLIOGRAPHY

GOND 1977 -1 GONDRAN, M.
GROUP THEORETIC AND LAGRANGEAN METHODS.

GOND 1976 -1 GONDRAN, M.
PROBLEME DU PLUS COURT CHEMIN DANS UN GRAPHE, LE.

GOND 1976 -2 GONDRAN, M.
PROBLEMES DE PARTITION ET DE RECOUVREMENT. APPLICATIONS ET ALGORITHMES, LES.

GOODB 1974 -1 GOODMAN, S.; HEDETNIEMI, S.; SLATER, P.
SOME RESULTS ON THE OPTIMAL AND NON-OPTIMAL HAMILTONIAN COMPLETION PROBLEM.

GRAHA 1977 -1 GRAHAM, R. L.; LAWLER, E. L.; LENSTRA, J. K.; RINNOOY KAN, A. H. G.
OPTIMIZATION AND APPROXIMATION IN DETERMINISTIC SEQUENCING AND SCHEDULING: A SURVEY.

HADJ 1974 -1 HADJERES, S.
QELQUES METHODES RECENTES DE RESOLUTION DU PROBLEME DU VOYAGEUR DE COMMERCE.

HAKIA 1972 -1 HAKIMI, S. L.
SHORTEST PATHS IN GRAPHS - A REVIEW.

HAMMA 1975 -1 HAMMER, P. L.
BOOLEAN ELEMENTS IN COMBINATORIAL OPTIMIZATION: A SURVEY.

HAMMA 1977 -1 HAMMER, P. L.
BOOLEAN ELEMENTS IN COMBINATORIAL OPTIMIZATION. A SURVEY.

HANSA 1977 -1 HANSEN, P.
METHODS OF NONLINEAR 0-1 PROGRAMMING.

HANSA 1975 -3 HANSEN, P.
PROCEDURES D'EXPLORATION ET D'OPTIMISATION PAR SEPARATION ET EVALUATION, 1: CLASSIFICATION ET FORME DES TESTS, LES.

HANSA 1975 -4 HANSEN, P.
PROCEDURES D'EXPLORATION ET D'OPTIMISATION PAR SEPARATION ET EVALUATION: A SURVEY, LES.

HANSB 1976 -1 HANSEN, P.; DELATTRE, M.
CLUSTER ANALYSIS AND GRAPH COLORING.

HITCA 1968 -1 HITCHNER, L.
COMPARATIVE INVESTIGATION OF THE COMPUTATIONAL EFFICIENCY OF SHORTEST PATH ALGORITHMS, A.

HUBA 1975 -1 HUBAUT, X. L.
STRONGLY REGULAR GRAPHS.

IBAR 1977 -1 IBARAKI, T.
BRANCH-AND-BOUND ENUMERATION.

INGL 1977 -1 INGLETON, A. W.
TRANSVERSAL MATROIDS AND RELATED STRUCTURES.

JERO 1977 -2 JEROSLOW, R.
INTRODUCTION TO THE THEORY OF CUTTING-PLANES, AN.

JEROA 1977 -2 JEROSLOW, R. G.
CUTTING-PLANE THEORY: DISJUNCTIVE METHODS.

JOHNE 1977 -1 JOHNSON, E. L.
ON THE GROUP PROBLEM AND A SUBADDITIVE APPROACH TO INTEGER PROGRAMMING.

KARPA 1975 -1 KARP, R. M.
FAST APPROXIMATE SOLUTION OF HARD COMBINATORIAL PROBLEMS, THE.

KAST 1976 -1 KASTNING, C.
INTEGER PROGRAMMING AND RELATED AREAS. A CLASSIFIED BIBLIOGRAPHY.

KENN 1975 -1 KENNINGTON, J. L.
MULTICOMMODITY NETWORK FLOWS: A STATE-OF-THE-ART SURVEY OF LINEAR MODELS AND SOLUTION TECHNIQUES.

KENN 1977 -2 KENNINGTON, J. L.
STATE-OF-THE-ART SURVEY OF LINEAR COST MULTICOMMODITY NETWORK FLOWS, A (REVISED VERSION).

KORTB 1976 -1 KORTE, B.
GANZZAHLIGE PROGRAMMIERUNG (STICHWORTARTIKEL FUER: HANDBUCH DER MATHEMATISCHEN WIRTSCHAFTSWISSENSCHAFTEN, HRSG. M. J. BECKMANN).

KORTB 1977 -2 KORTE, B.
APPROXIMATIVE ALGORITHMS FOR DISCRETE OPTIMIZATION PROBLEMS.

KORTB 1976 -2 KORTE, B.
INTEGER PROGRAMMING (SURVEY PAPER).

KOZY 1972 -1 KOZYREV, V. P.
GRAPH THEORY (RUSSIAN).

KRAR 1972 -1 KRARUP, J.
QUADRATIC ASSIGNMENT (IN DANISH).

KRAR 1973 -1 KRARUP, J.
SELECTED EXAMPLES OF COMBINATORIAL OPTIMIZATION PROBLEMS. (DANISH).

KRARD 1977 -1 KRARUP, J.; PRUZAN, P. M.
SELECTED FAMILIES OF DISCRETE LOCATION PROBLEMS PART 1: CENTER PROBLEMS PART 2: MEDIAN PROBLEMS.

KRARD 1977 -2 KRARUP, J.; PRUZAN, P. M.
SELECTED FAMILIES OF DISCRETE LOCATION PROBLEMS PART 3: THE PLANT LOCATION FAMILY.

LAGEA 1976 -2 LAGEWEG, B. J.; LENSTRA, J. K.; RINNOOY KAN, A. H. G.
JOB-SHOP SCHEDULING BY IMPLICIT ENUMERATION.

LAGEA 1976 -3 LAGEWEG, B. J.; LENSTRA, J. K.; RINNOOY KAN, A. H. G.
JOB-SHOP SCHEDULING BY IMPLICIT ENUMERATION.

LAND 1977 -1 LAND, A.; POWELL, S.
COMPUTER CODES FOR PROBLEMS OF INTEGER PROGRAMMING.

LAWL 1975 -1 LAWLER, E. L.
ALGORITHMS, GRAPHS, AND COMPLEXITY.

LAWL 1976 -1 LAWLER, E. L.
GRAPHICAL ALGORITHMS AND THEIR COMPLEXITY.

LAWL 1975 -2 LAWLER, E. L.
COMPLEXITY OF COMBINATORIAL COMPUTATIONS.

LAWL 1975 -3 LAWLER, E. L.
COMPUTING SHORTEST PATH IN NETWORKS.

LAWL 1975 -4 LAWLER, E. L.
INTRODUCTION TO MATROID OPTIMIZATION, AN.

LAWL 1977 -4 LAWLER, E. L.
SHORTEST PATH AND NETWORK FLOW ALGORITHMS.

LAWL 1975 -6 LAWLER, E. L.
OVERVIEW OF NETWORK FLOW THEORIE.

LAWL 1975 -8 LAWLER, E. L.
SOME ASPECTS OF DUALITY IN COMBINATORIAL OPTIMIZATION.

LEA 1973 -1 LEA, A. C.
LOCATION-ALLOCATION MODELS: A REVIEW.

LENS 1973 -1 LENSTRA, H. W.
ACYCLIC SUBGRAPH PROBLEM, THE.

LENSB 1975 -1 LENSTRA, J. K.; RINNOOY KAN, A. H. G.
ON THE JOB-SHOP SCHEDULING PROBLEM.

LENSB 1977 -1 LENSTRA, J. K.; RINNOOY KAN, A. H. G.
COMPUTATIONAL COMPLEXITY OF DISCRETE OPTIMIZATION PROBLEMS.

LENSC 1977 -1 LENSTRA, J. K.; RINNOOY KAN, A. H. G.; BRUCKER, P.
COMPLEXITY OF MACHINE SCHEDULING PROBLEMS.

LESN 1977 -1 LESNIAK-FOSTER, L.
SOME RECENT RESULTS IN HAMILTONIAN GRAPHS.

LIU 1976 -1 LIU, C. L.
COMBINATORIAL STUDY OF SOME SCHEDULING ALGORITHMS, A.

LONG 1977 -1 LONG, C. F.
STATE OF THE ART IN SCHEDULING PRACTICE, THE.

LOVA 1977 -2 LOVASZ, L.
GRAPH THEORY AND INTEGER PROGRAMMING.

MAC F 1977 -1 MAC GINNIS, L. F.
SURVEY OF RESULTS FOR A CLASS OF FACILITIES LOCATION PROBLEMS, A.

MAFF 1975 -2 MAFFIOLI, F.
TRAVELING SALESMAN PROBLEM AND ITS IMPLICATIONS, THE.

MARS 1971 -1 MARSHALL, C.
APPLIED GRAPH THEORY.

MARTC 1977 -1 MARTELLO, S.; TOTH, P.
COMPUTATIONAL STUDY ON LARGE-SIZE UNIDIMENSIONAL KNAPSACK PROBLEMS, A.

MATE 1976 -1 MATETI, P.; DEO, N.
ON ALGORITHMS FOR ENUMERATING ALL CIRCUITS OF A GRAPH.

MATUA 1972 -1 MATULA, D. W.; MARBLE, G.; ISAACSON, J. D.
GRAPH COLORING ALGORITHMS.

MICHA 1968 -1 MICHIE, D.; FLEMING, J. G.; OLDFIELD, J. V.
COMPARISON OF HEURISTIC, INTERACTIVE, AND UNAIDED METHODS OF SOLVING A SHORTEST-ROUTE PROBLEM, A.

MILN 1975 -1 MILNER, E. C.
TRANSVERSAL THEORY.

MITC 1975 -1 MITCHELL, S.; HEDETNIEMI, S.; GOODMAN, S.
SOME LINEAR ALGORITHMS ON TREES.

MITCA 1976 -1 MITCHEN, J.
ON VARIOUS ALGORITHMS FOR ESTIMATING THE CHROMATIC NUMBER OF A GRAPH.

MUEL 1976 -1 MUELLER-MERBACH, H.
ANSAETZE ZU ENTWURFSMETHODOLOGIEN FUER ALGORITHMEN DER KOMBINATORISCHEN OPTIMIERUNG.

MUEL 1976 -3 MUELLER-MERBACH, H.
MORPHOLOGIE HEURISTISCHER VERFAHREN.

MULV 1978 -1 MULVEY, J.
NETWORK RELAXATIONS FOR SET COVERING, SET PARTITIONING AND OTHER INTEGER PROGRAMMING PROBLEMS.

NICOA 1976 -1 NICOLAS, J.-L.
ALGORITHMES D'OPTIMISATION EN NOMBRES ENTIERS.

NIEM 1974 -1 NIEMINEN, J.
LINEAR PSEUDO-BOOLEAN VIEWPOINT ON MATCHING AND OTHER CENTRAL CONCEPTS IN GRAPH THEORY, A.

NORT 1976 -1 NORTHWESTERN UNIVERSITY; TRANSPORTATION CENTER LIBRARY
DOCTORAL DISSERTATION ON TRANSPORTATION (6. ISSUE, 1974-1975).

ODON 1974 -1 ODONI, A. R.
LOCATION OF FACILITIES ON A NETWORK: A SURVEY OF RESULTS.

PADBA 1977 -1 PADBERG, M. W.
COVERING, PACKING AND KNAPSACK PROBLEMS.

PANW 1977 -1 PANWALKAR, S. S.; ISKANDER, W.
SURVEY OF SCHEDULING RULES, A.

PAPE 1977 -1 PAPE, U.
BIBLIOGRAPHIE ZU "KUERZESTE WEGLAENGEN UND WEGE IN GRAPHEN", EINE (STAND: ENDE 1976).

PIER 1975 -1 PIERCE, A. R.
BIBLIOGRAPHY ON ALGORITHMS FOR SHORTEST PATH, SHORTEST SPANNING TREE, AND RELATED CIRCUIT ROUTING PROBLEMS (1956-1974).

POLL 1977 -1 POLLACK, M.
SOME ELEMENTS OF THE AIRLINE FLEET PLANNING PROBLEM.

RABI 1977 -1 RABIN, M. O.
COMPLEXITY OF COMPUTATIONS.

RATH 1977 -2 RATHMANN, D.
ZUR LOESUNG DISKRETER OPTIMIERUNGSPROBLEME MIT GRUPPENTHEORETISCHEN ANSAETZEN.

READA 1977 -1 READ, R. C.; CORNEIL, D. G.
GRAPH ISOMORPHISM DISEASE, THE.

RIBE 1971 -1 RIBEILL, G.; TABOURIER, Y.
CHEMINS EXTREMAUX DANS UN GRAPHE ENTRE TOUS COUPLES DE SOMMETS: QUELQUES ALGORITHMES ET LEUR MISE EN OEUVRE.COMPARAISONS SUR LES PERFORMANCES.

ROY 1975 -2 ROY, B.
CHEMINS ET CIRCUITS: ENUMERATION ET OPTIMISATION: A SURVEY.

SAHB 1975 -1 SAHBAZJAN, K. V.; TUSKINA, T. A.
SURVEY OF METHODS FOR COMPILING SCHEDULES FOR MULTIPROCESSOR SYSTEMS. COMPUTATIONAL METHODS OF LINEAR ALGEBRA. PARALLEL COMPUTATIONS (RUSSIAN).

****

SALKA 1975 -1 SALKIN, H. M.; KLUYVER, C. A. DE
KNAPSACK PROBLEM: A SURVEY, THE.

SCHUA 1977 -1 SCHURAWLJEW, J. I.; FINKEL'SHTEIN, YU. YU.
AUFGABENTYPEN UND LOESUNGSMETHODEN IN DER DISKRETEN OPTIMIERUNG.

SCWA 1968 -1 SCWARTZ, J. G.
NETWORK FLOW AND SYNTHESIS MODELS FOR TRANSPORTATION PLANNING: A SURVEY.

SHAPB 1977 -3 SHAPIRO, J. F.
SURVEY OF LAGRANGEAN TECHNIQUES FOR DISCRETE OPTIMIZATION.

SHIE 1976 -1 SHIER, D. R.
ALGORITHMS FOR FINDING THE K SHORTEST PATHS IN A NETWORK.

SPIE 1977 -1 SPIELBERG, K.
ENUMERATIVE METHODS IN INTEGER PROGRAMMING.

STER 1976 -1 STERBOUL, F.
SOME EXTREMAL COLORING PROBLEMS: A SURVEY.

THIS 1976 -1 THISSE, J.
MODELES MATHEMATIQUES D'OTIMISATION DE LA LOCALISATION DES ENTREPRISES.

THOM 1976 -1 THOMAS, M. E.
SURVEY OF THE STATE OF THE ART IN DYNAMIC PROGRAMMING, A.

TIFF 1974 -1 TIFFIN, J. E.; FOLIE, G. M.
RECENT DEVELOPMENTS IN TRANSPORTATION ALGORITHMS.

TIND 1977 -1 TIND, J.
BLOCKING AND ANTIBLOCKING POLYHEDRA.

WATE 1976 -1 WATERS, W. G. II; TSE, A.
BIBLIOGRAPHY OF THE MAJOR TRANSPORTATION JOURNALS 1965-1974, A.

WEINC 1975 -1 WEINER, P.
HEURISTICS.

WERR 1975 -2 WERRA, D. DE
GRAPHS AND SCHEDULING.

WERR 1975 -3 WERRA, D. DE
HOW TO COLOR A GRAPH.

WHITG 1973 -1 WHITEHEAD, E. G. JR.
COMBINATORIAL ALGORITHMS.

WONGC 1976 -1 WONG, R.
SURVEY OF SOME NETWORK DESIGN PROBLEMS, A.

YEH 1976 -1 YEH, R. T.
APPLIED COMPUTATION THEORY: ANALYSIS, DESIGN, MODELING.

YEN 1971 -1 YEN, J. Y.
ON THE EFFICIENCIES OF ALGORITHMS FOR DETECTING NEGATIVE LOOPS IN NETWORKS.

****

ADEL 1975 -1 ADEL'SON-VEL'SKIJ, G. M.; DINITS, E. A.; KARZANOV, A. V.
FLOW ALGORITHMS (RUSSIAN).

AHO B 1974 -1 AHO, A. V.; HOPCROFT, J. E.; ULLMAN, J. D.
DESIGN AND ANALYSIS OF COMPUTER ALGORITHMS, THE.

AIGN 1976 -1 AIGNER, M.
KOMBINATORIK II. MATROIDE UND TRANSVERSALTHEORIE.

ANDE 1970 -1 ANDERSON, S.
GRAPH THEORY AND FINITE COMBINATORICS.

BACH 1976 -3 BACHEM, A.
BEITRAEGE ZUR THEORIE DER CORNER POLYEDER.

BAZAC 1977 -1 BAZARAA, M. S.; JARVIS, J. J.
LINEAR PROGRAMMING AND NETWORK FLOWS.

BELO 1977 -1 BELOUSOV, E. G.
INTRODUCTION TO CONVEX ANALYSIS AND INTEGER PROGRAMMING (RUSSIAN).

BERG 1973 -1 BERGE, C.
INTRODUCTION A LA THEORIE DES HYPERGRAPHES.

BEUE 1975 -1 BEUERMANN-WEHMEYER, A.
PRIMAL-DUAL-BEZIEHUNGEN IN GRAPHENALGORITHMEN.

BIES 1976 -1 BIESS, G.
GRAPHENTHEORIE. MATHEMATIK FUER INGENIEURE, NATURWISSENSCHAFTLER, OEKONOMEN UND LANDWIRTE 21/2.

BIRDA 1976 -1 BIRD, R.
PROGRAMS AND MACHINES. AN INTRODUCTION TO THE THEORY OF COMPUTING.

BOHE 1977 -1 BOHEME, G.
ANWENDUNGSORIENTIERTE MATHEMATIK. BD. 4: AKTUELLE ANWENDUNGEN DER MATHEMATIK.

BRADD 1977 -1 BRADLEY, S. P.; HAX, A. C.; MAGNANTI, T. L.
APPLIED MATHEMATICAL PROGRAMMING.

BRUC 1975 -1 BRUCKER, P.
GANZZAHLIGE LINEARE PROGRAMMIERUNG MIT OEKONOMISCHEN ANWENDUNGEN.

BRUT 1970 -1 BRUTER, C. P.
MATROIDES. NOUVEL OUTIL MATHEMATIQUE, LES.

CABO 1977 -1 CABOT, A. V.; HARNETT, D. L.
INTRODUCTION TO MANAGEMENT SCIENCE, AN.

CHARB 1977 -1 CHARTRAND, G.
GRAPHS AS MATHEMATICAL MODELS.

CHENH 1976 -1 CHEN, W.-K.
APPLIED GRAPH THEORY. GRAPHS AND ELECTRICAL NETWORKS (2ND REVISED ED.).

CHEV 1977 -1 CHEVALIER, A.
PROGRAMMATION DYNAMIQUE. DE LA MAITRISE DE LA CONJONCTURE A LA PLANIFICATION, LA.

COFF 1976 -1 COFFMAN, E. G. JR.
COMPUTER AND JOB/SHOP SCHEDULING THEORY.

COOPA 1977 -1 COOPER, L.; BHAT, U. N.; LEBLANC, L. J.
INTRODUCTION TO OPERATIONS RESEARCH MODELS.

CRAP 1970 -1 CRAPO, H. H.; ROTA, G. C.
ON THE FOUNDATIONS OF COMBINATORIAL THEORY: COMBINATORIAL GEOMETRIES.

DIJK 1976 -1 DIJKSTRA, E. W.
DISCIPLINE OF PROGRAMMING, A.

DREY 1977 -1 DREYFUS, S. E.; LAW A. M.
ART AND THEORY OF DYNAMIC PROGRAMMING, THE.

ECKE 1977 -1 ECKER, K.
ORGANISATION VON PARALLELEN PROZESSEN. THEORIE DETERMINISTISCHER SCHEDULES.

ERDOE 1974 -1 ERDOES, P.; SPENCER, J.
PROBABILISTIC METHODS IN COMBINATORICS.

ESCU 1975 -1 ESCUDERO, L. F.
CONTINUOUS, INTEGER, AND MIXED LINEAR PROGRAMMING, VOL. 1 (SPANISH).

FAUR 1976 -1 FAURE, R.; ROUCAIROL, C.; TOLLA, P.
CHEMINS ET FLOTS, ORDONNANCEMENTS.

FILL 1977 -1 FILL, J. A.; REINGOLD, E. M.
SOLUTIONS MANUAL. COMBINATORIAL ALGORITHMS: THEORY AND PRACTICE (BY E.M. REINGOLD, J. NIEVERGELT, N. DEO).

FRAN 1974 -1 FRANCIS, R. L.
FACILITY LAYOUT AND LOCATION.

FRANB 1974 -1 FRANCIS, R. L.; WHITE, J. A.
FACILITY LAYOUT AND LOCATION: AN ANALYTIC APPROACH.

GRAVA 1977 -1 GRAVER, J. E.; WATKINS, M. E.
COMBINATORICS WITH EMPHASIS ON THE THEORY OF GRAPHS.

GROE 1977 -6 GROETSCHEL, M.
POLYEDRISCHE CHARAKTERISIERUNGEN KOMBINATORISCHER OPTIMIERUNGSPROBLEME.

HARTA 1976 -1 HARTLEY, R. V.
OPERATIONS RESEARCH: A MANAGERIAL EMPHASIS.

HEIN 1977 -1 HEIN, O.
GRAPHENTHEORIE FUER ANWENDER.

JAKS 1974 -1 JAKSCH, H. J.
THEORIE LINEARER WIRTSCHAFTSMODELLE 1. MATHEMATISCHE GRUNDLAGEN.

KAUFA 1977 -1 KAUFMANN, A.; PICHAT, E.
METHODES MATHEMATIQUES NON NUMERIQUES ET LEURS ALGORITHMES. TOME 1 ALGORITHMES DE RECHERCHE DES ELEMENTS MAXIMAUX.

KAUFA 1977 -2 KAUFMANN, A.; PICHAT, E.
METHODES MATHEMATIQUES NON NUMERIQUES ET LEURS ALGORITHMES. TOME 2 ALGORITHMES DE RECHERCHEDE CHEMINS ET PROBLEMES ASSOCIES.

KNUT 1976 -1 KNUTH, D. E.
MARIAGES STABLES ET LEURS RELATIONS AVEC D'AUTRES PROBLEMES COMBINATOIRES.

KOVAB 1977 -1 KOVALEV, M. M.
DISCRETE OPTIMIZATION (INTEGER PROGRAMMING) (RUSSIAN).

LENSA 1977 -1 LENSTRA, J. K.
SEQUENCING BY ENUMERATIVE METHODS.

LIEB 1970 -1 LIEBLING, T.
GRAPHENTHEORIE IN PLANUNGS- UND TOURENPROBLEMEN.

MANN 1967 -1 MANNE, A. S.
INVESTMENTS FOR CAPACITY EXPANSION: SIZE, LOCATION, AND TIME-PHASING.

MIRS 1971 -1 MIRSKY, L.
TRANSVERSAL THEORY.

MURTB 1976 -1 MURTY, K. G.
LINEAR AND COMBINATORIAL PROGRAMMING.

****

MUTH 1963 -1 MUTH, J. F.; THOMPSON, G. L.
INDUSTRIAL SCHEDULING.
NEUM 1975 -1 NEUMANN, K.
OPERATIONS RESEARCH C. DYNAMISCHE OPTIMIERUNG.
NEWM 1972 -1 NEWMAN, M.
INTEGRAL MATRICES.
NIJE 1975 -1 NIJENHUIS, A.; WILF, H.
COMBINATORIAL ALGORITHMS.
PFAF 1976 -1 PFAFFENBERGER, R. C.; WALKER, D. A.
MATHEMATICAL PROGRAMMING FOR ECONOMICS AND BUSINESS.
POTT 1972 -1 POTTS, R.; OLIVER, R.
FLOWS IN TRANSPORTATION NETWORKS.
REIN 1977 -1 REINGOLD, E. M.; NIEVERGELT, J.; DEO, N.
COMBINATORIAL ALGORITHMS: THEORY AND PRACTICE.
SALK 1975 -1 SALKIN, H. M.
INTEGER PROGRAMMING.
SAVA 1976 -1 SAVAGE, J. E.
COMPLEXITY OF COMPUTING, THE.
SCHW 1977 -1 SCHWARTZ, M.
COMPUTER-COMMUNICATION NETWORK. DESIGN AND ANALYSIS.
SCOT 1971 -1 SCOTT, A. J.
COMBINATORIAL PROGRAMMING, SPATIAL ANALYSIS, AND PLANNING.
SERC 1970 -1 SERCK-HANSSEN, J.
OPTIMAL PATTERNS OF LOCATION.
STAHA 1976 -1 STAHL, W.
SENSITIVITAETSANALYSE MEHRSTUFIGER STANDORTPROBLEME.
STUH 1966 -1 STUHR, R.
ANWENDUNGEN UND LOESUNGSVERFAHREN DER GANZZAHLIGEN LINEAREN PROGRAMMIERUNG.
TAHA 1975 -1 TAHA, H. A.
INTEGER PROGRAMMING.
TANA 1975 -1 TANAEV, V. S.; SHKURBA, V. V.
INTRODUCTION INTO SEQUENCING THEORY (RUSSIAN).
TEH 1976 -1 TEH, H. H.; SHEE, S. C.
ALGEBRAIC THEORY OF GRAPHS.
TINH 1976 -1 TINHOFER, G.
METHODEN DER ANGEWANDTEN GRAPHENTHEORIE.
TOME 1975 -2 TOMESCU, I.
INTRODUCTION TO COMBINATORICS (TRANSLATED FROM RUMANIAN).
WELLA 1971 -1 WELLS, M.
ELEMENTS OF COMBINATORIAL COMPUTING.
WELS 1976 -1 WELSH, D. J. A.
MATROID THEORY.
WHITG 1973 -1 WHITEHEAD, E. G. JR.
COMBINATORIAL ALGORITHMS.
YEN 1975 -1 YEN, J. Y.
SHORTEST PATH NETWORK PROBLEMS.

****

ADAMA 1976 -1 ADAMSKA-TCHON, J.; JACAK, W.; TCHON, K.
FORMAL DESCRIPTION AND ALGORITHMIZATION OF A CERTAIN TRANSPORT PROBLEM (POLISH), A.
ADOLA 1977 -1 ADOLPHSON, D. L.; THOMAS, G. N.
LINEAR TIME ALGORITHM FOR A 2 X N TRANSPORTATION PROBLEM, A.
AGINA 1975 -1 AGIN, N. I.; CULLEN, D. E.
ALGORITHM FOR TRANSPORTATION ROUTING AND VEHICLE LOADING, AN.
AHREA 1976 -1 AHRENS, W.
NUMERICAL EXPERIENCE WITH HEURISTIC RULES IN SPECIFIC BINARY OPTIMIZATION ALGORITHMS.
AKAD 1976 -1 AKADEMIA NAUK SSSR, TSENTR. EHKON.-MAT. INST.
SOME QUESTIONS CONCERNING THE STUDY OF TRANSPORTATION FLOWS (RUSSIAN).
ALCO 1975 -1 ALCOUFFE, A.; ENJALBERT, M.; MURATET, G.
METHODES DE RESOLUTION DU PROBLEME DE TRANSPORT ET DE PRODUCTION D'UNE ENTREPRISE A ETABLISSEMENTS MULTIPLES EN PRESENCE DE COUTS FIXES.
ALEM 1972 -1 ALEMANY, R.; DIEGO, A.
SEMITENSIONS IN THE SOLVING OF THE "POTENTIAL'S RESTRICTED PROBLEM".
ALI 1977 -1 ALI, A. I.; HELGASON, R. V.; KENNINGTON, J. L.; LALL, H. S.
PRIMAL SIMPLEX NETWORK CODES: STATE-OF-THE-ART IMPLEMENTATION TECHNOLOGY (REVISED VERSION).
ASTAA 1976 -1 ASTAHOV, N. D.
GRADIENT ALGORITHMS FOR SOLVING SOME TRANSPORTATION PROBLEMS (RUSSIAN).
AZAR 1977 -1 AZARYAN, L. L.; LEBEDEV, S. S.; MESTETZKY, L. M.
INTEGER TRANSPORT TYPE PROBLEMS SOLVING BASED ON GENERALIZED LAGRANGE MULTIPLIERS (RUSSIAN).
BACHA 1977 -1 BACHEM, A.; KORTE, B.
ALGORITHM FOR QUADRATIC OPTIMIZATION OVER TRANSPORTATION POLYTOPES, AN.
BACHA 1977 -2 BACHEM, A.; KORTE, B.
ESTIMATING INPUT-OUTPUT-MATRICES.
BACHA 1977 -3 BACHEM, A.; KORTE, B.
QUADRATIC OPTIMIZATION OVER TRANSPORTATION POLYTOPES.
BACHA 1977 -4 BACHEM, A.; KORTE, B.
QUADRATISCHE OPTIMIERUNG UEBER TRANSPORTPOLYEDER ZUR LOESUNG VON SCHAETZPROBLEMEN VON INPUT-OUTPUT MATRIZEN.
BALA 1976 -1 BALACHANDRAN, V.
INTEGER GENERALIZED TRANSPORTATION MODEL FOR OPTIMAL JOB ASSIGNMENT IN COMPUTER NETWORKS, AN.
BALAA 1977 -1 BALACHANDRAN, V.; DITTMAN, D.
OVERHEAD ALLOCATION VIA MATHEMATICAL PROGRAMMING.
BALAC 1976 -1 BALACHANDRAN, V.; PERRY, A.
TRANSPORTATION TYPE PROBLEMS WITH QUANTITY DISCOUNTS.
BALAD 1977 -1 BALACHANDRAN, V.; SRINIVASAN, V.; THOMPSON, G. L.
SURVEY OF THE APPLICATIONS OF THE OPERATOR THEORY OF PARAMETRIC PROGRAMMING FOR THE TRANSPORTATION AND GENERALIZED TRANSPORTATION PROBLEM, A.
BALAE 1975 -1 BALACHANDRAN, V.; THOMPSON, G. L.
OPERATOR THEORY OF PARAMETRIC PROGRAMMING FOR THE GENERALIZED TRANSPORTATION PROBLEM - 3 - WEIGHT OPERATORS, AN.
BALAE 1975 -2 BALACHANDRAN, V.; THOMPSON, G. L.
OPERATOR THEORY OF PARAMETRIC PROGRAMMING FOR THE GENERALIZED TRANSPORTATION PROBLEM - 4 - GLOBAL OPERATORS, AN.
BALAE 1975 -3 BALACHANDRAN, V.; THOMPSON, G. L.
OPERATOR THEORY OF PARAMETRIC PROGRAMMING FOR THE GENERALIZED TRANSPORTATION PROBLEM: 1. BASIC THEORY, AN.
BALAE 1975 -4 BALACHANDRAN, V.; THOMPSON, G. L.
OPERATOR THEORY OF PARAMETRIC PROGRAMMING FOR THE GENERALIZED TRANSPORTATION PROBLEM: 2. RIM, COST AND BOUND OPERATORS, AN.
BARA 1972 -1 BARANKIEWICZ, M. N.
CERTAIN GENERALIZATIONS OF THE MULTISTAGE TRANSPORTATION PROBLEM (POLISH).
BARR 1977 -2 BARR, R. S.; GLOVER, F.; KLINGMAN, D.
ALTERNATING BASIS ALGORITHM FOR ASSIGNMENT PROBLEMS, THE.
BARR 1977 -3 BARR, R. S.; GLOVER, F.; KLINGMAN, D.
GENERALIZED ALTERNATING PATH ALGORITHM FOR TRANSPORTATION PROBLEMS, THE.
BENNB 1976 -1 BENNINGER, H.; LIEBLING, T. M.
OPTIMIERUNGSPROBLEME MIT POLYNOMIAL BEGRENZTEN ALGORITHMEN.
BERGE 1975 -1 BERGMANN, D. R.
INTEGER PROGRAMMING FORMULATION FOR DERIVING MINIMUM DISPATCH INTERVALS ON A GUIDEWAY ACCOMODATING THROUGH AND LOCAL PUBLIC TRANSPORTATION SERVICES.
BHAT 1976 -1 BHATIA, H. L.; SWARUP, K.; PURI, M. C.
TIME-COST TRADE-OFF IN A TRANSPORTATION PROBLEM.
BIONA 1969 -1 BIONDI, E.; DIVIETI, L.; ROVEDA, C.; SCHMID, R.
OPTIMAL SOLUTION FOR A CLASS OF ASSIGNMENT AND TRANSPORTATION PROBLEMS.
BODI 1976 -1 BODIN, L. D.; ROSENFIELD, D.
ESTIMATION OF THE OPERATING COST OF MASS TRANSIT SYSTEMS.
BOLK 1976 -1 BOLKER, E. D.
SIMPLICIAL GEOMETRY AND TRANSPORTATION POLYTOPES.
BOOKA 1977 -1 BOOKBINDER, J. H.; SETHI, S. P.
DYNAMIC TRANSPORTATION PROBLEM: A SURVEY, THE.
BRAD 1976 -1 BRADLEY, G.; BROWN, G.; GRAVES, G.
DESIGN AND IMPLEMENTATION OF LARGE SCALE PRIMAL TRANSSHIPMENT ALGORITHMS.
BREGA 1974 -1 BREGMAN, T. T.
BRANCH AND BOUND METHOD FOR THE SOLUTION OF ALLOCATION PROBLEMS (RUSSIAN), A.
BURK 1976 -2 BURKARD, R. E.
GENERAL HUNGARIAN METHOD FOR THE ALGEBRAIC TRANSPORTATION PROBLEM, A.
BURKE 1978 -1 BURKARD, R. E.; ZIMMERMANN, U.
SOLUTION OF ALGEBRAIC ASSIGNMENT AND TRANSPORTATION PROBLEMS, THE.
CAMEB 1976 -1 CAMERON, A. V.; VOILS, L. R.
HEURISTIC FOR TRANSPORTATION ROUTING AND VEHICLE LOADING UNDER UNCERTAINTY, A.
CHANA 1976 -1 CHAN, Y.
TRANSPORTATION NETWORK INVESTMENT PROBLEM, THE; A SURVEY OF TREE SEARCH SOLUTION ALGORITHMS.
CHANI 1977 -1 CHANDRASEKARAN, R.; RAO, S. S.
SPECIAL CASE OF THE TRANSPORTATION PROBLEM, A.
CHAR 1975 -1 CHARNES, A.; GLOVER, F.; KLINGMAN, D.
PAST, PRESENT AND FUTURE OF LARGE SCALE TRANSSHIPMENT COMPUTER CODES AND APPLICATIONS.
COOP 1975 -1 COOPER, L.
EFFICIENT HEURISTIC ALGORITHM FOR THE TRANSPORTATION-LOCATION PROBLEM, AN.
COOPB 1976 -1 COOPER, L.; COOPER, M. W.
EXACT ALGORITHM FOR THE CONCAVE TRANSPORTATION PROBLEM, AN.
COOPC 1977 -1 COOPER, L.; LEBLANC, L. J.
STOCHASTIC TRANSPORTATION PROBLEMS AND OTHER NETWORK RELATED CONVEX PROBLEMS.

TRANSPORTATION
****

CORB 1973 -1 CORBAN, A.
TRANSPORTATION PROBLEMS WITH FIXED COSTS (RUMANIAN).

CRUS 1975 -1 CRUSE, A. B.
PROOF OF FULKERSON'S CHARACTERIZATION OF PERMUTATION MATRICES, A.

CUNNA 1976 -1 CUNNINGHAM, W. H.
NETWORK SIMPLEX METHOD, A.

DAEN 1977 -1 DAENINCK, G.; SMEERS, Y.
USING SHORTEST PATHS IN SOME TRANSSHIPMENT PROBLEM WITH CONCAVE COSTS.

DINE 1972 -1 DINESCU, C.
DETERMINING THE MEAN FLOW IN A TRANSPORTATION NETWORK (RUMANIAN).

DINI 1973 -1 DINITS, E. A.
ALGORITHM FOR THE DIGITWISE DECREASE OF DISCREPANCIES, AND TRANSPORT PROBLEMS (RUSSIAN), AN.

DOUL 1970 -1 DOULLIEZ, P.
POLITIQUE OPTIMALE D'INVESTISSEMENT DANS UN RESEAU DE TRANSPORT.

DWYE 1975 -1 DWYER, P. S.
TRANSPORTATION PROBLEMS WITH SOME X(IJ) NEGATIVE AND TRANSSHIPMENT PROBLEMS.

EMEL 1976 -1 EMELICHEV, V. A.; KRAVTSOV, M. K.
CERTAIN PROPERTIES OF TRANSPORTATION POLYHEDRA WITH A MAXIMAL NUMBER OF VERTICES (RUSSIAN).

EMEL 1976 -2 EMELICHEV, V. A.; KRAVTSOV, M. K.
ON ENUMERATIVE PROBLEMS ON TRANSPORT POLYHEDRA (RUSSIAN).

EMELA 1976 -1 EMELICHEV, V. A.; KRAVTSOV, M. K.; AVERBUH, N. D.
MAXIMAL NUMBER OF VERTICES OF A TRANSPORTATION POLYHEDRON (RUSSIAN), THE.

EMELB 1974 -1 EMELICHEVA, V. S.; KONONEKO, A. M.
MINIMAL NUMBER OF VERTICES AND EDGES OF THE POLYTOPE OF A PLANAR TRANSPORTATION PROBLEM (RUSSIAN), THE.

ERLA 1966 -1 ERLANDER, S.
NONLINEAR TRANSPORTATION AND INVENTORY PROBLEM.

EVANA 1976 -1 EVANS, J. R.
COMBINATORIAL EQUIVALENCE BETWEEN A CLASS OF MULTICOMMODITY FLOW PROBLEMS AND THE CAPACITATED TRANSPORTATION PROBLEM, A.

EVANA 1976 -4 EVANS, J. R.
TRANSPORTATION PROBLEM AND ITS EXTENSIONS: A SURVEY, THE.

FAN 1964 -1 FAN, L. T.; WANG, G. S.
APPLICATION OF DISCRETE MAXIMUM PRINCIPLE TO TRANSPORTATION PROBLEM.

FAUR 1976 -1 FAURE, R.; ROUCAIROL, C.; TOLLA, P.
CHEMINS ET FLOTS, ORDONNANCEMENTS.

FISK 1976 -1 FISK, J.
ALGORITHM FOR ASSIGNING USES TO SOURCES IN A TRANSPORTATION TABLEAU, AN.

FISK 1976 -2 FISK, J.
SOLUTION PROCEDURE FOR A SPECIAL CLASS OF TRANSPORTATION PROBLEM CONTAINING MULTIPLE CHOICE CONSTRAINTS, A.

FISKA 1977 -1 FISK, J.; MACKEOWN, P.
DETERMINING THE MAXIMUM DEGENERATE SOLUTION FOR TRANSPORTATION PROBLEMS.

FONGB 1975 -1 FONG, C. O.; RAO, M. R.
PARAMETRIC STUDIES IN TRANSPORTATION-TYPE PROBLEMS.

FONGC 1977 -1 FONG, C. O.; SRINIVASAN, V.
DETERMINING ALL NONDEGENERATE SHADOW PRICES FOR THE TRANSPORTATION PROBLEM.

FRIE 1975 -1 FRIEDMAN, M.
SCHEDULING A TRANSPORTATION NETWORK.

FRIEA 1975 -1 FRIEZE, A. M.
BOTTLENECK LINEAR PROGRAMMING.

GABA 1975 -1 GABASOV, R.; KIRILLOVA, F. M.
CONSTRUCTION OF SUBOPTIMAL PLANS OF A TRANSPORTATION PROBLEM (RUSSIAN), THE.

GAVI 1977 -1 GAVISH, B.; SCHWEITZER, P.; SHLIFER, E.
ZERO PIVOT PHENOMENON IN TRANSPORTATION AND ASSIGNMENT PROBLEMS AND ITS COMPUTATIONAL IMPLICATIONS, THE.

GAZI 1975 -1 GAZIS, D. C.
TRANSPORTATION NETWORKS.

GEOF 1976 -1 GEOFFRION, A. M.
CUSTOMER AGGREGATION IN DISTRIBUTION MODELING.

GIBS 1976 -1 GIBSON, P. M.
FACETS OF FACES OF TRANSPORTATION POLYTOPES.

GILL 1960 -1 GILL, A.
CLASS OF TRANSPORTATION PROBLEMS.

GIMAA 1976 -1 GIMADI, E. K.; GLEBOV, N. I.; PEREPELITSA, V. A.
ALGORITHMS WITH ESTIMATIONS FOR THE PROBLEMS OF DISCRETE OPTIMIZATION (RUSSIAN).

GLIC 1977 -1 GLICKMAN, T. S.; BERGER, P. D.
COST/COMPLETION-DATE TRADEOFFS IN THE TRANSPORTATION PROBLEM.

GLOVB 1977 -1 GLOVER, F.; HULTZ, J.; KLINGMAN, D.; STUTZ, J.
NEW COMPUTER-BASED PLANNING TOOL, A.

GLOVC 1974 -1 GLOVER, F.; KLINGMAN, D.
IMPORTANT PRACTICAL MISCONCEPTIONS OF OPTIMIZING LARGE SCALE ASSIGNMENT AND TRANSPORTATION PROBLEMS.

GLOVC 1976 -1 GLOVER, F.; KLINGMAN, D.
PRACTITIONER'S GUIDE TO THE STATE OF LARGE SCALE NETWORK AND NETWORK-RELATED PROBLEMS, A.

GLOVE 1976 -1 GLOVER, F.; KLINGMAN, D.; NAPIER, A.
EQUIVALENCE OF GENERALIZED NETWORK AND GENERALIZED TRANSPORTATION PROBLEMS.

GOLDB 1976 -2 GOLDEN, B. L.
RECENT DEVELOPMENTS IN VEHICLE ROUTING.

GOLDE 1977 -1 GOLDFARB, D.
STEEPEST EDGE SIMPLEX ALGORITHMS FOR NETWORK FLOW PROBLEMS.

GRABB 1976 -1 GRABOWSKI, W.
LEXICOGRAPHIC AND TIME MINIMIZATION IN THE TRANSPORTATION PROBLEM.

GRIGA 1974 -1 GRIGOR'EVA, L. I.
MINIMIZATION OF CARGO LOSSES IN TRANSFERS WITH ALLOWANCE FOR THE DELIVERY DATES (RUSSIAN).

HARR 1976 -1 HARRIS, B.
CODE FOR THE TRANSPORTATION PROBLEM OF LINEAR PROGRAMMING, A.

HATC 1975 -1 HATCH, R. S.
BENCH MARKS COMPARING TRANSPORTATION CODES BASED ON PRIMAL SIMPLEX AND PRIMAL-DUAL ALGORITHMS.

HIRC 1970 -1 HIRCHE, J.
UEBER ZWEI MODIFIKATIONEN DES DEKOMPOSITIONSALGORITHMUS VON WILLIAMS.

HIRCA 1975 -1 HIRCHE, J.; KOEHLER, J.; STIEBITZ, V.
EFFEKTIVES LOESUNGSVERFAHREN FUER DAS HYPERBOLISCHE TRANSPORTPROBLEM, EIN.

****

HULTA 1976 -1 HULTZ, J.; KLINGMAN, D.; RUSSELL, R.
ADVANCED DUAL BASIC FEASIBLE SOLUTION FOR A CLASS OF CAPACITATED GENERALIZED NETWORKS, AN.
INTR 1976 -1 INTRATOR, J.; LEV, B.
METHODS FOR IDENTIFICATION OF VANISHING VARIABLES IN TRANSPORTATION PROBLEMS AND ITS POSSIBLE APPLICATIONS.
INTRA 1977 -1 INTRATOR, J.; PAROUSH, J.
SENSITIVITY ANALYSIS OF THE CLASSICAL TRANSPORTATION PROBLEM. A COMBINATORIAL APPROACH.
IRI 1960 -1 IRI, M.
NEW METHOD OF SOLVING TRANSPORTATION-NETWORK PROBLEMS, A.
IVANB 1974 -1 IVANISCEV, V. V.
DECOMPOSITION METHOD FOR THE SOLUTION OF THE PROBLEM OF THE OPTIMAL USE OF MEANS OF TRANSPORTATION ON A NETWORK (RUSSIAN), A.
JACOA 1978 -1 JACOBSEN, S. K.
ON THE USE OF TREE-INDEXING METHODS IN TRANSPORTATION ALGORITHMS.
JAKO 1974 -1 JAKOVLEVA, M. A.
TRANSPORTATION PROBLEM WITH BORDERING (RUSSIAN), THE.
JAKS 1974 -1 JAKSCH, H. J.
THEORIE LINEARER WIRTSCHAFTSMODELLE 1. MATHEMATISCHE GRUNDLAGEN.
JAKS 1975 -1 JAKSCH, H. J.
THEORIE LINEARER WIRTSCHAFTSMODELLE 2. TRANSPORT- UND STANDORTMODELLE.
KACZ 1976 -2 KACZMARSKI, J.
STABILITY OF THE SOLUTIONS OF THE TRANSPORT PROBLEM (POLISH), THE.
KACZ 1976 -3 KACZMARSKI, J.
TRANSPORT PROBLEM IN CONNECTION WITH THE CONSTRUCTION OF A NETWORK (POLISH), THE.
KAPL 1977 -1 KAPLAN, E. L.
ALGORITHMS FOR SOLUTION OF TRANSPORTATION PRODUCTION PROBLEMS (RUSSIAN).
KASS 1977 -1 KASSAY, F.
APPROACH TO SENSITIVITY ANALYSIS OF THE TRANSPORTATION PROBLEM (CZECH), AN.
KENN 1976 -1 KENNINGTON, J. L.
FIXED-CHARGE TRANSPORTATION PROBLEM: A COMPUTATIONAL STUDY WITH A BRANCH-AND-BOUND CODE, THE.
KENN 1977 -1 KENNINGTON, J. L.
SOLVING MULTICOMMODITY TRANSPORTATION PROBLEMS USING A PRIMAL PARTITIONING SIMPLEX TECHNIQUE.
KRAV 1976 -1 KRAVTSOV, M. K.
ENUMERATION PROBLEMS ON THE DEGENERATED TRANSPORTATION POLYHEDRA (RUSSIAN).
KURC 1974 -1 KURCEVIC, K. A.; KRAVTSOV, M. K.
METHOD FOR LOWERING THE DIMENSION OF THE TRANSPORTATION PROBLEM OF LINEAR PROGRAMMING (RUSSIAN), A.
LAMB 1975 -1 LAMBE, T. A.
INITIAL BASIS FOR THE TRANSPORTATION PROBLEM.
LEBL 1977 -1 LEBLANC, L. J.
HEURISTIC APPROACH OF LARGE SCALE DISCRETE STOCHASTIC TRANSPORTATION-LOCATION PROBLEMS, A.
LIHA 1975 -1 LIHACEV, V. M.
CERTAIN ESTIMATE FOR THE NUMBER OF VERTICES OF A TRANSPORTATION POLYTOPE (RUSSIAN), A.
LIHAA 1974 -1 LIHACEV, V. M.; EMELICHEV, V. A.
ON AN UPPER BOUND OF THE NUMBER OF VERTICES OF A TRANSPORTATION POLYTOPE (RUSSIAN).
LOS 1978 -1 LOS, M.
SIMULTANEOUS OPTIMIZATION OF LAND USE AND TRANSPORTATION: A SYNTHESIS OF THE QUADRATIC ASSIGNMENT PROBLEM AND THE OPTIMAL NETWORK PROBLEM.
LUND 1975 -1 LUNDQVIST, L.
TRANSPORTATION ANALYSIS AND ACTIVITY LOCATION IN LAND-USE PLANNING - WITH APPLICATIONS TO THE STOCKHOLM REGION.
LUND 1973 -1 LUNDQVIST, L.
INTEGRATED LOCATION-TRANSPORTATION ANALYSIS: A DECOMPOSITION APPROACH.
LYTK 1974 -2 LYTKIN, I. P.
SCHEDULING OF MEANS OF TRANSPORTATION (RUSSIAN).
MAC I 1975 -1 MAC KEOWN, P. G.; RUBIN, D. S.
ADJACENT VERTICES ON TRANSPORTATION POLYTOPES.
MAC J 1976 -1 MAC KINNON, R. D.
OPTIMIZATION MODELS OF TRANSPORTATION NETWORK IMPROVEMENT: REVIEW AND FUTURE PROSPECTS.
MAIL 1977 -1 MAILLARD, P.
CONTRIBUTION A L'ETUDE DES PROBLEMES DE TRANSPORT.
MEDN 1975 -1 MEDNITSKII, V. G.
THEOREM ON TRANSPORTATION ZONES IN THE PRODUCTION-TRANSPORTATION PROBLEM (RUSSIAN), THE.
MICK 1975 -1 MICKLE, M. H.
METHOD OF OPTIMUM ROUTE REGENERATION, A.
MICKA 1969 -1 MICKLE, M. H.; VOGT, W. G.
DISCRETE CONTROL AS A TRANSPORTATION PROBLEM.
MOORA 1973 -1 MOORE, S. A.; PEARSON, J. M.; LINDSAY, M. A.
TRANSPORTATION PROBLEMS.
MOSL 1976 -1 MOSLER, K. C.
OPTIMALE TRANSPORTNETZE. ZUR BESTIMMUNG IHRES KOSTENGUENSTIGSTEN STANDORTS BEI GEGEBENER NACHFRAGE.
MURT 1976 -1 MURTHY, M. S.
BULK TRANSPORTATION PROBLEM, A.
MURTB 1976 -1 MURTY, K. G.
LINEAR AND COMBINATORIAL PROGRAMMING.
NAUS 1976 -3 NAUSS, R. M.
FIXED CHARGE TRANSPORTATION PROBLEM WITH ADDITIONAL CONSTRAINTS, THE.
NEY 1973 -1 NEY, P.
RESEARCH ON THE TRANSPORTATION PROBLEM, II (RUMANIAN).
NEY 1972 -1 NEY, P.
RESEARCH ON THE TRANSPORTATION PROBLEM, I (RUMANIAN).
NYKO 1964 -1 NYKOWSKI, I.; ZURKOWSKI, J.
PRODUCTION ALLOCATION PROBLEMS WHICH CAN BE REDUCED TO TRANSPORTATION PROBLEMS (POLISH).
PINO 1976 -1 PINOCHET, M. R.
CENTRALIZATION AND DECENTRALIZATION OF DECISION-MAKING IN TRANSPORTATION NETWORKS.
PIRO 1975 -1 PIROSKA, T.
APPLICATION OF THE DECOMPOSITION PRINCIPLE OF DANTZIG AND WOLFE TO THE SOLUTION OF A GENERALIZED TRANSPORTATION PROBLEM (HUNGARIAN), AN.
POPO 1973 -1 POPOVICIU, N.
ALLOCATION PROBLEM, WITH BIVALENT ZERO-ONE VARIABLES DEFINED BY MEANS OF THE ELEMENTS OF A TRANSPORTATION PROBLEM (RUMANIAN), AN.

POTT 1972 -1 POTTS, R.; OLIVER, R.
FLOWS IN TRANSPORTATION NETWORKS.

RAJAA 1977 -1 RAJAT, K. D.
ON GENERATION OF TEST DATA AND MINIMAL COVER OF DIRECTED GRAPHS.

RAMA 1977 -1 RAMAKRISHNAN, C. S.
NOTE ON THE TIME MINIMISING TRANSPORTATION PROBLEM, A.

RAMAA 1976 -1 RAMAKRISHNAN, K. G.
SOLUTION OF CAPACITATED PLANT LOCATION PROBLEMS USING BENDERS' DECOMPOSITION.

RAPP 1976 -1 RAPPOPORT, A. M.
TRANSPORT PROBLEMS WITH SUPPLEMENTARY CONSTRAINTS (RUSSIAN).

ROBE 1976 -1 ROBERS, P.; COOPER, L.
STUDY OF THE FIXED CHARGE TRANSPORTATION PROBLEM, A.

ROSSB 1974 -1 ROSS, G. T. JR.
ALGORITHMIC AND COMPUTATIONAL RESULTS FOR CONSTRAINED TRANSPORTATION, NETWORK AND ASSIGNMENT MODELS.

SAFA 1970 -1 SAFAEVA, K.
CERTAIN DESIGN PROBLEM (RUSSIAN), A.

SALE 1976 -1 SALERNO, G.
"BRANCH AND BACK-TRACK" ALGORITHM FOR THE DESIGN OF AN OPTIMAL TRANSPORTATION NETWORK (ITALIAN), A.

SALI 1970 -1 SALIHOV, Z. M.; PAIZIEV, E.; BURKOV, V. N.; KANCEL'SON, M. B.
AGGREGATED ALGORITHM FOR THE SOLUTION OF THE TRANSPORT PROBLEM (RUSSIAN), AN.

SCHO 1976 -1 SCHOCH, M.
ERWEITERUNGSPRINZIP UND SEINE ANWENDUNG ZUR ENTWICKLUNG VON ALGORITHMEN FUER DIE LOESUNG KOMBINATORISCHER OPTIMIERUNGSAUFGABEN, DAS.

SCIO 1964 -1 SCIONS, H. I.
COMPACT REPRESENTATION OF A ROOTED TREE AND THE TRANSPORTATION PROBLEM, THE.

SHARB 1976 -1 SHARMA, J. K.; AGRAWAL, S. C.
FIXED-CHARGE TRANSPORTATION PROBLEM, A.

SHARB 1976 -2 SHARMA, J. K.; AGRAWAL, S. C.
TIME-COST TRADE-OFF IN A TRANSPORTATION PROBLEM.

SHARC 1978 -1 SHARMA, J. K.; SWARUP, K.
TIME MINIMIZATION IN TRANSPORTATION PROBLEMS.

SHARD 1977 -1 SHARMA, S. R.; SWARUP, K.
TRANSPORTATION FRACTIONAL PROGRAMMING WITH RESPECT TO TIME.

SRINA 1977 -1 SRINIVASAN, V.; THOMPSON, G. L.
COST OPERATOR ALGORITHMS FOR THE TRANSPORTATION PROBLEM.

SRINA 1976 -1 SRINIVASAN, V.; THOMPSON, G. L.
ALGORITHMS FOR MINIMIZING TOTAL COST, BOTTLENECK TIME AND BOTTLENECK SHIPMENT IN TRANSPORTATION PROBLEMS.

STEIA 1977 -1 STEINBERG, D. I.
DESIGNING A HEURISTIC FOR THE FIXED-CHARGE TRANSPORTATION PROBLEM.

STRO 1977 -1 STROBEL, H.; GENSER, R.; ETSCHMAIER, M.
OPTIMIZATION APPLIED TO TRANSPORTATION SYSTEMS.

STYC 1974 -1 STYCZEN, K.
TRANSPORTATION SYSTEM IN THE RAW MATERIAL PREPARATION DEPARTMENT OF A CEMENT PLANT: MATHEMATICAL MODEL AND CONTROL ALGORITHM (POLISH), A.

SVAR 1975 -1 SVARTIN, S. M.
CERTAIN ALGORITHM FOR THE PLANNING OF MINIMAL TIME TRANSFERS (RUSSIAN), A.

TIFF 1974 -1 TIFFIN, J. E.; FOLIE, G. M.
RECENT DEVELOPMENTS IN TRANSPORTATION ALGORITHMS.

TRUE 1977 -2 TRUEMPER, K.
UNIMODULAR MATRICES OF FLOW PROBLEMS WITH ADDITIONAL CONSTRAINTS.

TURCA 1975 -1 TURCHANYI, P.
DECOMPOSITION METHOD FOR THE SOLUTION OF A TRANSPORTATION PROBLEM WITH PIECEWISE LINEAR CONVEX COST FUNCTIONS (HUNGARIAN), A.

VLAC 1969 -1 VLACH, M.
ON SOME TRANSPORTATION PROBLEMS.

VLACA 1971 -1 VLACH, M.; ZEGELING, H.
NOTE ON A PAPER BY A. BRANDT AND Y. INTRATOR, A.

VODO 1976 -1 VODOLAZOV, V. S.; MUCNIK, M. I.
CONDITIONS FOR THE SOLVABILITY OF TRANSPORT TYPE PROBLEMS WITH LIMITED CUT CAPACITIES (RUSSIAN).

WATE 1976 -1 WATERS, W. G. II; TSE, A.
BIBLIOGRAPHY OF THE MAJOR TRANSPORTATION JOURNALS 1965-1974, A.

WENDA 1976 -1 WENDLER, K.
ERWEITERTES TRANSPORTPROBLEM MIT PARAMETERN ALS KAPAZITAETSBESCHRAENKUNGEN, EIN.

WILSA 1975 -1 WILSON, D.
MEAN COST APPROXIMATION FOR TRANSPORTATION PROBLEMS WITH STOCHASTIC DEMAND, A.

WITT 1978 -1 WITTEN, P.; ZIMMERMANN, H.-G.
STOCHASTISCHE TRANSPORTPROBLEME.

ZAYC 1975 -1 ZAYCHENKO, YU. P.; PECHURIN, M. K.; TOMASHEVSKIY, V. M.
GENERALIZED ASSIGNMENT PROBLEM AND ITS SOLUTION ALGORITHM (UKRAINIAN), THE.

ZEHN 1965 -1 ZEHNDER, C. A.
COMPUTATION OF TIME TABLES FOR SCHOOLS AND TRANSPORTATION.

ZVJA 1974 -1 ZVJAGINA, R. A.
SOLUTION OF A DYNAMICAL TRANSPORTATION PROBLEM WITH INCREASED PRODUCTION (RUSSIAN), THE.

****

ADRA 1976 -1 ADRABINSKI, A.
ALGORITHM 45: A HEURISTIC ALGORITHM FOR THE TRAVELING-SALESMAN PROBLEM.

ALEKD 1974 -1 ALEKSEYEVA, S. M.
METHODS OF SCHEDULE THEORY IN TECHNOLOGICAL CHECKING PROBLEMS (RUSSIAN).

ANGL 1977 -1 ANGLUIN, D.; VALIANT, L. G.
FAST PROBABILISTIC ALGORITHMS FOR HAMILTONIAN CIRCUITS AND MATCHINGS.

ANN 1977 -1 ANN, T.-K.
TRAITEMENT DU PROBLEME DE VOYAGEUR DE COMMERCE PAR UNE METHODE DE PROGRAMMATION LINEAIRE.

BAZAA 1977 -1 BAZARAA, M. S.; ELSHAFEI, A. N.
ON THE USE OF FICTITIOUS BOUNDS IN TREE SEARCH ALGORITHMS.

BAZAB 1977 -1 BAZARAA, M. S.; GOODE, J. J.
TRAVELING SALESMAN: A DUALITY APPROACH, THE.

BELL 1973 -1 BELL, D. E.
RESOLUTION OF DUALITY GAPS IN DISCRETE OPTIMIZATION, THE.

BENIA 1976 -1 BENITO, F.; GROEFLIN, H.
OPTIMIERUNGSPROBLEME MIT NICHT POLYNOMIAL BEGRENZTEN ALGORITHMEN (NP-PROBLEME).

BERM 1976 -1 BERMOND, J. C.
ON HAMILTONIAN WALKS.

BHAUA 1974 -1 BHAUMIK, G.; STANTON, L. W.
APPLICATION OF STOPPING RULES THROUGH MONTE CARLO SIMULATION TO THE TRAVELING SALESMAN PROBLEM, AN.

BOURA 1970 -1 BOURKOV, V. N.
RESOLUTION DE QUELQUES PROBLEMES D'OPTIMISATION.

BURDA 1976 -1 BURDYUK, V. YA.; TROFIMOV, V. N.
GENERALIZATION OF RESULTS BY GILMORE-GOMORY CONCERNING THE TRAVELLING SALESMAN PROBLEM (RUSSIAN).

BURK 1977 -1 BURKARD, R. E.
TRAVELLING SALESMAN AND ASSIGNMENT PROBLEMS: A SURVEY.

BURKJ 1968 -1 BURKOV, V. N.; LOVETSKII, S. E.
METHODS FOR THE SOLUTION OF EXTREMAL PROBLEMS OF COMBINATORIAL TYPE (REVIEW) (RUSSIAN).

BURN 1976 -1 BURNESS, R. C.; WHITE, J. A.
TRAVELING SALESMAN LOCATION PROBLEM, THE.

CALY 1974 -1 CALYKOV, V.
CERTAIN HEURISTIC ALGORITHM FOR THE SOLUTION OF THE TRAVELLING SALESMAN PROBLEM (RUSSIAN), A.

CAME 1973 -1 CAMERINI, P. M.; FRATTA, L.; MAFFIOLI, F.
HEURISTICALLY GUIDED ALGORITHM FOR THE TRAVELING SALESMAN PROBLEM, A.

CAMEA 1974 -1 CAMERINI, P. M.; MAFFIOLI, F.
BOUNDS FOR 3-MATROID INTERSECTION PROBLEMS.

CARP 1977 -1 CARPANETO, G.; TOTH, P.
EFFICIENT ALGORITHM FOR THE ASYMMETRIC TRAVELLING SALESMAN PROBLEM, AN.

CHARC 1973 -1 CHARTRAND, G.; KAPOOR, S. F.; KRONK, H. V.
MANY FACETS OF HAMILTONIAN GRAPHS, THE.

CHRI 1974 -1 CHRISTOFIDES, N.
HAMILTONIAN CIRCUITS AND THE TRAVELLING SALESMAN PROBLEM.

CHRI 1975 -1 CHRISTOFIDES, N.
HAMILTONIAN CIRCUITS AND THE TRAVELLING SALESMAN PROBLEM: A SURVEY.

CHRI 1976 -2 CHRISTOFIDES, N.
WORST-CASE ANALYSIS OF A NEW HEURISTIC FOR THE TRAVELLING SALESMAN PROBLEM.

CHRIB 1976 -1 CHRISTOS, H.; PAPADIMITRIOU, H.; STEIGLITZ, K.
SOME COMPLEXITY RESULTS FOR THE TRAVELING SALESMAN PROBLEM.

CHVAA 1978 -1 CHVATAL, V.; ERDOES P.
BIASED POSITIONAL GAMES.

CIOBA 1977 -1 CIOBOTARU, S.
LINGUISTIC METHOD FOR DETERMINING THE HAMILTONIAN PATHS (RUMANIAN), A.

CORNE 1978 -1 CORNUEJOLS, G.; NEMHAUSER, G. L.
TIGHT BOUNDS FOR CHRISTOFIDES' TRAVELING SALESMAN HEURISTIC.

DEMI 1976 -1 DEMIDENKO, V. M.
SPECIAL CASE OF THE TRAVELLING SALESMAN PROBLEM (RUSSIAN), A.

DINK 1976 -1 DINKEL, J. J.; KLEINDORFER, G. B.; KOCHENBERGER, G. A.; WONG, S.-N.
ENVIRONMENTAL INSPECTION ROUTES AND THE CONSTRAINED TRAVELLING SYSTEM SALESMAN PROBLEM.

DREY 1977 -1 DREYFUS, S. E.; LAW A. M.
ART AND THEORY OF DYNAMIC PROGRAMMING, THE.

EVANA 1976 -2 EVANS, J. R.
FUNDAMENTAL CIRCUITS IN COMPLETE GRAPHS AND THE TRAVELING SALESMAN PROBLEM.

FEW 1955 -1 FEW, L.
SHORTEST PATH AND THE SHORTEST ROAD THROUGH N POINTS, THE.

FRATA 1972 -1 FRATTA, L.; MAFFIOLI, F.
ON THE SHORTEST HAMILTONIAN CHAIN OF A NETWORK.

FREH 1975 -2 FREHEL, J.
REGULARISATION DANS LES PROBLEMES COMBINATOIRES, PROBLEMES DE TOURNEES, DE PARTITION ET DE RECOUVREMENT.

GABO 1976 -1 GABOVICH, E. JA.
CONSTANT PROBLEMS OF DISCRETE PROGRAMMING ON SUBSTITUTION SETS (RUSSIAN).

GARED 1976 -1 GAREY, M. R.; JOHNSON, D. S.; TARJAN, R. E.
PLANAR HAMILTONIAN CIRCUIT PROBLEM IS NP-COMPLETE, THE.

GARF 1977 -1 GARFINKEL, R. S.
MINIMIZING WALLPAPER WASTE, PART 1: A CLASS OF TRAVELING SALESMAN PROBLEMS.

GARFA 1976 -1 GARFINKEL, R. S.; GILBERT, K. C.
BOTTLENECK TRAVELING SALESMAN PROBLEM, THE.

GIAN 1976 -1 GIANNESSI, F.
DECOMPOSITION METHOD FOR THE TRAVELLING SALESMAN PROBLEM AND ITS APPLICATIONS, A.

GIMAA 1976 -1 GIMADI, E. K.; GLEBOV, N. I.; PEREPELITSA, V. A.
ALGORITHMS WITH ESTIMATIONS FOR THE PROBLEMS OF DISCRETE OPTIMIZATION (RUSSIAN).

GIMAB 1974 -1 GIMADI, E. K.; PEREPELITSA, V. A.
ASYMPTOTIC APPROACH TO THE SOLUTION OF THE TRAVELING SALESMAN PROBLEM (RUSSIAN).

GLOVC 1975 -2 GLOVER, F.; KLINGMAN, D.
FINDING MINIMUM SPANNING TREES WITH A FIXED NUMBER OF LINKS AT A NODE.

GOLDB 1977 -2 GOLDEN, B. L.
STATISTICAL APPROACH TO THE TSP, A.

GOLDD 1977 -1 GOLDEN, B. L.; MAGNANTI, T. L.; NGUYEN, H. Q.
IMPLEMENTING VEHICLE ROUTING ALGORITHMS.

GONZ 1962 -1 GONZALES, R. H.
SOLUTION OF THE TRAVELING SALESMAN PROBLEM BY DYNAMIC PROGRAMMING ON THE HYPERCUBE.

****

GOODD 1975 -1 GOODMAN, S. E.; HEDETNIEMI, S. T.; SLATER, P. J.
ADVANCES ON THE HAMILTONIAN COMPLETION PROBLEM.

GOYA 1977 -1 GOYAL, S. K.
FLOW-SHOP SEQUENCING PROBLEM WITH NO WAIT IN PROCESS AND NON-ZERO SET-UP TIMES.

GRAB 1976 -3 GRABOWSKI, J.
SOLVING THE TRAVELLING SALESMAN PROBLEM BY USING DISJUNCTIVE GRAPHS.

GROE 1977 -4 GROETSCHEL, M.
ON THE MONOTONE SYMMETRIC TRAVELLING SALESMAN PROBLEM: HYPOHAMILTONIAN/HYPOTRACEABLE GRAPHS AND FACETS.

GROE 1977 -5 GROETSCHEL, M.
ON THE SYMMETRIC TRAVELLING SALESMAN PROBLEM: SOLUTION OF A 120-CITY PROBLEM.

GROE 1977 -6 GROETSCHEL, M.
POLYEDRISCHE CHARAKTERISIERUNGEN KOMBINATORISCHER OPTIMIERUNGSPROBLEME.

GROE 1977 -7 GROETSCHEL, M.
STRONG BLOCKS AND THE OPTIMUM BRANCHING PROBLEM.

GROEA 1977 -1 GROETSCHEL, M.; PADBERG, M. W.
LINEARE CHARAKTERISIERUNGEN VON TRAVELLING SALESMAN PROBLEMEN.

GROEA 1978 -1 GROETSCHEL, M.; PADBERG, M. W.
ON THE SYMMETRIC TRAVELLING SALESMAN PROBLEM: THEORY AND COMPUTATION.

GROEA 1977 -2 GROETSCHEL, M.; PADBERG, M. W.
ON THE SYMMETRIC TRAVELLING SALESMAN PROBLEM 1: INEQUALITIES.

GROEA 1977 -3 GROETSCHEL, M.; PADBERG, M. W.
ON THE SYMMETRIC TRAVELLING SALESMAN PROBLEM 2: LIFTING THEOREMS AND FACETS.

GRUNA 1975 -1 GRUNDLER, G.
ALGORITHMEN ZUR LOESUNG VERSCHIEDENER VARIANTEN DES TOURENPROBLEMS.

GUHA 1977 -1 GUHA, D. K.
BRANCH AND BOUND ALGORITHM FOR SOLVING TRAVELING SALESMAN PROBLEM, A.

GUPT 1976 -2 GUPTA, J. N. D.
OPTIMAL FLOWSHOP SCHEDULES WITH NO INTERMEDIATE STORAGE SPACE.

HADJ 1974 -1 HADJERES, S.
QELQUES METHODES RECENTES DE RESOLUTION DU PROBLEME DU VOYAGEUR DE COMMERCE.

HANS 1971 -1 HANSEN, K. H.
SEQUENCING PROBLEMS WITH EMPHASIS ON A MODIFICATION OF HELD AND KARP'S ALGORITHM. OR: AN ALGORITHM FOR THE OPTIMAL SOLUTION OF THE SYMMETRIC TRAVELING SALESMAN PROBLEM. PART 2. (DANISH).

HANS 1971 -2 HANSEN, K. H.
SEQUENCING PROBLEMS WITH EMPHASIS ON A MODIFICATION OF HELD AND KARP'S ALGORITHM. OR: AN ALGORITHM FOR THE OPTIMAL SOLUTION OF THE SYMMETRIC TRAVELING SALESMAN PROBLEM. PART 1. (DANISH).

HARAB 1974 -1 HARARY, F.; MALLION, R. B.
ELEMENTARY NECESSARY CONDITION FOR HAMILTONIAN GRAPHS INVOLVING CONNECTIVITY, AN.

HAUS 1978 -1 HAUSMANN, D.
ADJAZENZCHARAKTERISIERUNGEN FUER POLYEDERECKEN IN DER KOMBINATORISCHEN OPTIMIERUNG.

HAUSA 1977 -1 HAUSMANN, D.; JENKYNS, T. A.; KORTE, B.
WORST CASE ANALYSIS OF GREEDY TYPE ALGORITHMS FOR INDEPENDENCE SYSTEMS.

HECK 1976 -1 HECK, J.
ALGORITHMUS ZUR LOESUNG DES CHINESE-POSTMAN PROBLEMS, EIN.

HEIN 1976 -1 HEIN, O.
ALGORITHMUS ZUR BERECHNUNG DER MINIMALEN KANTENBASIS IN GERICHTETEN GRAPHEN, EIN.

HEIN 1977 -2 HEIN, O.
LOESUNGSVORSCHLAG ZUM VERALLGEMEINERTEN RUNDFAHRTPROBLEM, EIN.

HELL 1955 -1 HELLER, I.
ON THE TRAVELING SALESMAN'S PROBLEM.

HONGA 1977 -1 HONG, S.
ASYMMETRIC TRAVELLING SALESMAN PROBLEM AND SPANNING ARBORESCENCES OF DIRECTED GRAPHS, THE.

HONGB 1977 -1 HONG, S.; PADBERG, M. W.
NOTE ON THE SYMMETRIC MULTIPLE TRAVELING SALESMAN PROBLEM WITH FIXED CHARGES, A.

HOUC 1976 -1 HOUCK, D. J.; VEMUGANTI, R. R.
TRAVELING SALESMAN PROBLEM AND SHORTEST N-PATHS, THE.

ISHI 1975 -1 ISHII, H.; IBARAKI, T.; MINE, H.
CERTAIN TRAVELING SALESMAN PROBLEM THAT HAS AN OBJECTIVE FUNCTION OF MINIMAX TYPE (JAPANESE), A.

ISHI 1975 -2 ISHII, H.; IBARAKI, T.; MINE, H.
TRAVELING SALESMAN PROBLEMS WITH A CAPACITY CONSTRAINT OF THE DELIVERY TRUCK.

JENK 1977 -1 JENKYNS, T. A.
GREEDY TRAVELLING SALESMAN PROBLEM, THE.

KALM 1975 -1 KALMANSON, K.
EDGECONVEX CIRCUITS AND THE TRAVELING SALESMAN PROBLEM.

KALM 1970 -1 KALMANSON, K.
CLASSES OF COMBINATORIAL EXTREMA IN CERTAIN METRIC SPACES.

KALM 1972 -1 KALMANSON, K.
ANALYSIS OF EXTREME HAMILTONIAN CIRCUITS, AN.

KARPA 1975 -1 KARP, R. M.
FAST APPROXIMATE SOLUTION OF HARD COMBINATORIAL PROBLEMS, THE.

KARPA 1977 -1 KARP, R. M.
PROBABILISTIC ANALYSIS OF PARTITIONING ALGORITHMS FOR THE TRAVELING-SALESMAN PROBLEM IN THE PLANE.

KARPA 1978 -1 KARP, R. M.
PATCHING ALGORITHM FOR THE NONSYMMETRIC TRAVELING-SALESMAN PROBLEM, A.

KASA 1975 -1 KASAI, T.; KURODA, K.
NUMBER OF HAMILTONIAN CIRCUITS CONTAINING SPECIFIED EDGES IN AN INCOMPLETE GRAPH AND ITS APPLICATIONS TO MULTIGRAPHS.

KIRY 1975 -1 KIRYUKIN, V. V.; POLONSKIJ, D. V.
ON THE SEARCH FOR THE HAMILTONIAN CYCLES OF A GRAPH (RUSSIAN).

KIRZ 1973 -1 KIRZNER, V. M.; RUBLINECKII, V. I.
PROCEDURE "PROCEED TO THE NEXT" IN THE TRAVELING SALESMAN PROBLEM (RUSSIAN), THE.

KIRZ 1974 -1 KIRZNER, V. M.; RUBLINECKII, V. I.
UPPER BOUND OF THE LENGTH OF THE MINIMAL ROUTE IN THE TRAVELING SALESMAN PROBLEM (RUSSIAN), THE.

KLIT 1976 -1 KLITZ, J. K.
TRAVELING-SALESMAN PROBLEM FORMULATION AND SOLUTION FOR A MASS STORAGE SYSTEM, A.

KLJA 1976 -1 KLJAUS, P. S.
STRUCTURE OF THE OPTIMAL SOLUTIONS OF SOME CLASSES OF TRAVELING SALESMAN PROBLEMS (RUSSIAN).

KORTC 1978 -1 KORTE, B.; HAUSMANN, D.
ANALYSIS OF THE GREEDY HEURISTIC FOR INDEPENDENCE SYSTEMS, AN.

KORTC 1977 -1 KORTE, B.; HAUSMANN, D.
ANALYSE DER GREEDY-HEURISTIK FUER UNABHAENGIGKEITSSYSTEME, EINE.

KRAR 1969 -1 KRARUP, J.
BRANCH-BOUND ALGORITHM FOR A WIRING PROBLEM, A.

KRAR 1965 -1 KRARUP, J.
TRAVELING SALESMAN I, THE; TRAVELING SALESMAN II, THE.

KUBO 1976 -1 KUBO, H.; OKINO, N.
SOLUTION FOR LARGE-SCALE TRAVELING SALESMAN PROBLEMS (JAPANESE), A.

LEBED 1974 -1 LEBENSOLD, K. W.
EXTREMA OF A CLASS OF FUNCTIONS ON A FINITE SET.

LEITA 1973 -1 LEITEN, A. K.
CERTAIN MODIFICATIONS OF THE TRAVELLING SALESMAN PROBLEM (RUSSIAN).

LEITA 1974 -1 LEITEN, A. K.
CERTAIN MATHEMATICAL MODELS FOR DISCRETE PROGRAMMING (RUSSIAN).

LEMA 1975 -1 LEMAIRE, B.
DENOMBREMENT DES CYCLES HAMILTONIENS DE KN ET KN,N EMPRUNTANT OU EVITANT DES ARETES DONNEES.

LEMA 1976 -1 LEMAIRE, B.
FONDEMENTS, GENERALISATION ET CRITIQUE DE LA NOTION D'AFFINITE (PROBLEME DU VOYAGEUR DE COMMERCE).

LENSA 1977 -1 LENSTRA, J. K.
SEQUENCING BY ENUMERATIVE METHODS.

LENSB 1976 -3 LENSTRA, J. K.; RINNOOY KAN, A. H. G.
TRAVELLING ON A GRAPH (DUTCH).

LEON 1975 -1 LEONT'EV, V. K.
STABILITY OF THE TRAVELING SALESMAN PROBLEM (RUSSIAN).

LEON 1976 -1 LEONT'EV, V. K.
STABILITY IN COMBINATORIAL PROBLEMS OF SELECTION (RUSSIAN).

LESN 1977 -1 LESNIAK-FOSTER, L.
SOME RECENT RESULTS IN HAMILTONIAN GRAPHS.

LIEB 1970 -1 LIEBLING, T.
GRAPHENTHEORIE IN PLANUNGS- UND TOURENPROBLEMEN.

LIN C 1975 -1 LIN, S.
COMPUTER SOLUTIONS OF THE TSP.

LIN D 1971 -1 LIN, S.; KERNIGHAN, B. W.
HEURISTIC TECHNIQUE FOR SOLVING A CLASS OF COMBINATORIAL OPTIMIZATION PROBLEMS, A.

MAFF 1976 -1 MAFFIOLI, F.
SUBGRADIENT OPTIMIZATION, MATROID PROBLEMS AND HEURISTIC EVALUATION.

MAFF 1975 -2 MAFFIOLI, F.
TRAVELING SALESMAN PROBLEM AND ITS IMPLICATIONS, THE.

MAURB 1976 -1 MAURRAS, J. F.
POLYTOPES A SOMMETS DANS (0,1)N.

MAURC 1975 -1 MAURRAS, J.-F.
SOME RESULTS ON THE CONVEX HULL OF THE HAMILTONIAN CYCLES OF SYMMETRIC COMPLETE GRAPHS.

MILI 1976 -1 MILIOTIS, P.
INTEGER PROGRAMMING APPROACHES TO THE TRAVELLING SALESMAN PROBLEM.

MINIB 1975 -1 MININA, T. R.; PEREKREST, V. T.
APPROXIMATION TO THE SOLUTION OF THE TRAVELING-SALESMAN PROBLEM BY C-CYCLES.

MORO 1974 -1 MOROZENSKII, L. YU.
ASYMPTOTIC BEHAVIOUR OF THE LENGTH OF THE PATH OF A TRAVELING SALEMAN WHEN THE CITIES ARE RANDOMLY DISTRIBUTED (RUSSIAN), THE.

NORB 1977 -1 NORBACK, J. P.; LOVE, R. F.
GEOMETRIC APPROACHES TO SOLVING THE TRAVELING SALESMAN PROBLEM.

OFFI 1975 -1 OFFIR, J. D.
TRAVELLING SALESMAN WITH A KNAPSACK(A CALL-GIRL'S HEURISTIC), A.

OVER 1976 -1 OVERBECK-LARISCH, M.
HAMILTONIAN PATHS IN ORIENTED GRAPHS.

PADBB 1977 -1 PADBERG, M. W.; HONG, S.
ON THE SYMMETRIC TRAVELLING SALESMAN PROBLEM: A COMPUTATIONAL STUDY.

PAPA 1977 -1 PAPADIMITRIOU, C. H.
EUCLIDEAN TRAVELING SALESMAN PROBLEM IS NP-COMPLETE, THE.

PAPA 1975 -1 PAPADIMITRIOU, C. H.
ON THE COMPLEXITY OF THE LOCAL STRUCTURE OF CERTAIN CONVEX POLYTOPES.

PAPAA 1977 -1 PAPADIMITRIOU, C. H.; STEIGLITZ, K.
ON THE COMPLEXITY OF LOCAL SEARCH FOR THE TRAVELING SALESMAN PROBLEM.

PAPAA 1976 -1 PAPADIMITRIOU, C. H.; STEIGLITZ, K.
SOME COMPLEXITY RESULTS FOR THE TRAVELING SALESMAN PROBLEM.

PARS 1969 -1 PARSONS, J. A.
BRANCH AND BOUND ALGORITHMS: THE TRAVELING SALESMAN PROBLEM.

PEVE 1969 -1 PEVEPELICHE, V. A.; GIMADI, E. X.
ON THE PROBLEM OF FINDING THE MINIMAL HAMILTONIAN CICUIT ON A GRAPH WITH WEIGHTED ARCS (RUSSIAN).

PICAC 1976 -1 PICARD, J.-C.; QUEYRANNE, M.
TIME DEPENDENT TRAVELING SALESMAN PROBLEM AND APPLICATION TO THE TARDINESS PROBLEM IN ONE-MACHINE SCHEDULING, THE.

PLOT 1974 -1 PLOTNIKOV, V. N.; GALKIN, A. V.
METHOD FOR THE SOLUTION OF CERTAIN EXTREMAL PROBLEMS OF COMBINATORIAL TYPE (RUSSIAN), A.

POPE 1976 -1 POPESCU, D.
CLASSES OF MAXIMAL NON HAMILTONIAN GRAPHS (RUMANIAN).

RAO 1976 -1 RAO, M. R.
ADJACENCY OF THE TRAVELING SALESMAN TOURS AND 0-1 VERTICES.

REIN 1977 -1 REINGOLD, E. M.; NIEVERGELT, J.; DEO, N.
COMBINATORIAL ALGORITHMS: THEORY AND PRACTICE.

RICHA 1976 -1 RICHMOND, L. B.
ON HAMILTONIAN POLYGONS.

ROBEA 1976 -1 ROBERTS, J.
SUFFICIENT CONDITION FOR K-PATH HAMILTONIAN DIGRAPHS, A.

ROSEB 1974 -1 ROSENKRANTZ, D. J.; STEARNS, R. E.; LEWIS, P. M.
APPROXIMATE ALGORITHMS FOR THE TRAVELING SALESPERSON PROBLEM.

ROSEB 1977 -1 ROSENKRANTZ, D. J.; STEARNS, R. E.; LEWIS, P. M.
ANALYSIS OF SEVERAL HEURISTICS FOR THE TRAVELING SALESMAN PROBLEM, AN.

RUBID 1971 -1 RUBINSHTEIN, M. I.
ON THE SYMMETRIC TRAVELING SALESMAN PROBLEM (RUSSIAN).

RUBL 1973 -1 RUBLINECKII, V. I.
ESTIMATES OF THE ACCURACY OF PROCEDURES IN THE TRAVELING SALESMAN PROBLEM (RUSSIAN).

****

RUBLA 1973 -1 RUBLINECKII, V. I.; TOMCUK, E. JA.
APPROXIMATE PROCEDURE FOR THE SOLUTION OF THE TRAVELING SALESMAN PROBLEM (RUSSIAN), AN.

RUSS 1975 -1 RUSSEL, R.
EFFECTIVE HEURISTIC FOR THE M-TOUR TRAVELING SALESMAN PROBLEM WITH SOME SIDE CONDITIONS, AN.

RUSSC 1977 -1 RUSSELL, R. A.
EFFECTIVE HEURISTIC FOR THE M-TOUR TRAVELING SALESMAN PROBLEM WITH SOME SIDE CONDITIONS, AN.

SAHNA 1974 -1 SAHNI, S.; GONZALES, T.
P-COMPLETE PROBLEMS AND APPROXIMATE SOLUTIONS.

SAHNB 1976 -1 SAHNI, S.; GONZALEZ, T.
P-COMPLETE APPROXIMATION PROBLEMS.

SAVAA 1978 -1 SAVAGE, S. L.
PROBABILISTIC APPROACH TO THE DESIGN OF HEURISTICS FOR DECISION PROBLEMS, A.

SAVAA 1973 -2 SAVAGE, S. L.
STATISTICAL INDICATORS OF OPTIMALITY.

SAVAB 1976 -1 SAVAGE, S. L.; WEINER, P.; BAGCHI, A.
NEIGHBORHOOD SEARCH ALGORITHMS FOR GUARANTEEING OPTIMAL TRAVELING SALESMAN TOURS MUST BE INEFFICIENT.

SCHO 1976 -1 SCHOCH, M.
ERWEITERUNGSPRINZIP UND SEINE ANWENDUNG ZUR ENTWICKLUNG VON ALGORITHMEN FUER DIE LOESUNG KOMBINATORISCHER OPTIMIERUNGSAUFGABEN, DAS.

SCHOE 1976 -1 SCHOENLEIN, A.
BEZUEGLICH KERNSPEICHER- UND RECHENZEITBEDARF EFFIZIENTES BRANCH-UND-BOUND-VERFAHREN ZUR LOESUNG DES TRAVELLING-SALESMAN-PROBLEMS, EIN.

SEIF 1974 -1 SEIFERAS, J. I.
NOTE ON REDUCTIONS TO DIRECTED HC, A.

SHAPA 1966 -1 SHAPIRO, D. M.
ALGORITHMS FOR THE SOLUTION OF THE OPTIMAL COST AND BOTTLENECK TRAVELING SALESMAN PROBLEMS.

SINGA 1975 -1 SINGMASTER D.
ENUMERATING UNLABELLED HAMILTONIAN CIRCUITS.

SIRI 1968 -1 SIRIK, A. F.
ON THE ALGORITHM OF M. HELD AND R. M. KARP FOR THE SOLUTION OF THE TRAVELLING SALESMAN PROBLEM (RUSSIAN).

SMITB 1977 -1 SMITH, J. F.; QUINN, M. J.
HAMILTONIAN CIRCUITS: A HIERARCHY OF EXAMPLES.

SMITI 1977 -1 SMITH, T. H. C.; SRINIVASAN, V.; THOMPSON, G. L.
COMPUTATIONAL PERFORMANCE OF THREE SUBTOUR ELIMINATION ALGORITHMS FOR SOLVING ASYMMETRIC TRAVELING SALESMAN PROBLEMS.

SMITJ 1977 -1 SMITH, T. H. C.; THOMPSON, G. L.
LIFO IMPLICIT ENUMERATION SEARCH ALGORITHM FOR THE SYMMETRIC TRAVELING SALESMAN PROBLEM USING HELD AND KARP'S 1-TREE RELAXATION, A.

SUPR 1975 -1 SUPRUNENKO, D. A.
ON THE TRAVELING SALESMAN PROBLEM (RUSSIAN).

SYSLA 1976 -1 SYSLO, M. M.
GENERALIZATIONS OF THE STANDARD TRAVELLING SALESMAN PROBLEM.

THOMB 1976 -1 THOMPSON, G. L.
SOLUTIONS OF LARGE STRUCTURED TRAVELLING SALESMAN PROBLEMS (COPIES OF TRANSPARANCIES).

TIMO 1975 -1 TIMOFEEV, E. A.
SOME ALGORITHMS FOR THE SOLUTION OF THE TRAVELLING SALESMAN PROBLEM ON THE PLANE WITH A LARGE NUMBER OF CITIES (RUSSIAN).

TOME 1975 -1 TOMESCU, I.
ALGORITHME POUR L'OBTENTION D'UNE CHAINE HAMILTONIENNE EN PARTANT DE L'ARBRE MINIMAL D'UN GRAPHE, UN.

TORR 1976 -1 TORRIERO, A.
DECOMPOSITION METHODS FOR THE TRAVELLING SALESMAN PROBLEM IN LARGE DIMENSIONS (ITALIAN).

TURNA 1975 -1 TURNER, W. C.; HOUGLAND, E. S.
OPTIMAL ROUTING OF SOLID WASTE COLLECTION VEHICLES, THE.

ULLR 1976 -2 ULLRICH, M.
ZUSAMMENHANG ZWISCHEN DEM RUNDREISEPROBLEM UND EINEM REIHENFOLGEPROBLEM DER MASCHINENBELEGUNGSPLANUNG.

VARS 1974 -1 VARSAVSKII, V. I.; EPIFANOV, G. V.; MININA, T. R.
USING EQUIVALENT TRANSFORMATIONS OF THE DISTANCE MATRIX TO SIMPLIFY THE SOLUTION OF THE TRAVELLING SALESMAN PROBLEM (RUSSIAN).

VISI 1977 -1 VISING, V. G.
ON COMPLEXITY OF THE PROBLEM OF A TRAVELLING SALESMAN (RUSSIAN).

VOLOA 1976 -1 VOLOVIK, L. N.; SKACHEK, B. YA.
ON MODIFICATION OF ALGORITHM FOR TRAVELLING SALESMAN PROBLEM (RUSSIAN).

WHITF 1969 -1 WHITE, W. W.
TRAVELING SALESMAN ALGORITHM.

ZAIT 1975 -1 ZAITSEVA, Z. N.; SHTEIN, M. E.
PLOTTING OF OPTIMAL HAMILTONIAN CYCLES IN THE MULTIVERTEX GRAPHS (RUSSIAN).

ZAIT 1976 -1 ZAITSEVA, Z. N.; SHTEIN, M. E.
HAMILTONIAN CYCLES IN SOLVING LAYING OUT PROBLEM (RUSSIAN).

ZAK 1976 -1 ZAK, YU. A.
GENERALIZED TRAVELLING-SALESMAN PROBLEM AND ITS PRACTICAL ASPECTS (RUSSIAN), THE.

ZAMF 1973 -1 ZAMFIRESCU, T.
CHARACTERIZATION OF HAMILTONIAN GRAPHS, A.

****

BARTA 1977 -1 BARTHOLDI, J. J. III; ORLIN, J. B.; RATLIFF, H. D.
CIRCULAR 1'S AND CYCLIC STAFFING.
BAUM 1977 -1 BAUM, S.; TROTTER, L. E. JR.
INTEGER ROUNDING AND POLYHEDRAL DECOMPOSITION FOR TOTALLY UNIMODULAR SYSTEMS.
BERG 1973 -1 BERGE, C.
INTRODUCTION A LA THEORIE DES HYPERGRAPHES.
BROWA 1977 -1 BROWN, D. P.
COMPOUND AND UNIMODULAR MATRICES.
BROWA 1976 -1 BROWN, D. P.
CIRCUIT AND UNIMODULAR MATRICES.
DEY 1976 -1 DEY, A.; MIDHA, C. K.
GENERALISED BALANCED MATRICES AND THEIR APPLICATIONS.
FEDO 1975 -1 FEDORENKO, R. P.
ON M-MATRICES REDUCING TO SIMPLEST UNIMODULATED ONE (RUSSIAN).
FRAND 1977 -1 FRANK, A.
ON A CLASS OF BALANCED HYPERGRAPHS.
GILLC 1975 -1 GILLES, F. R.
SUBMODULAR FUNCTIONS, GRAPHS AND INTEGER POLYHEDRA.
GRAVA 1977 -1 GRAVER, J. E.; WATKINS, M. E.
COMBINATORICS WITH EMPHASIS ON THE THEORY OF GRAPHS.
HOFF 1976 -2 HOFFMAN, A. J.
TOTAL UNIMODULARITY AND COMBINATORIAL THEOREMS.
HOFF 1977 -2 HOFFMAN, A. J.
ROLE OF UNIMODULARITY IN APPLYING LINEAR INEQUALITIES TO COMBINATORIAL THEOREMS, THE.
HOFFA 1978 -1 HOFFMAN, A. J.; OPPENHEIM, R.
LOCAL UNIMODULARITY IN THE MATCHING POLYTOPE.
MAC 1977 -1 MAC ALLISTER, J. D.; SAKAROVITCH, M.
CLASSIFICATION DE CERTAIN MATRICES 0-1.
MEYEC 1977 -1 MEYER, R. R.
CLASS OF NONLINEAR INTEGER PROGRAMS SOLVABLE BY A SINGLE LINEAR PROGRAM, A.
OLAR 1973 -1 OLARU, E.
UEBER PERFEKTE UND KRITISCH IMPERFEKTE GRAPHEN.
PADBA 1975 -1 PADBERG, M. W.
CHARACTERIZATIONS OF TOTALLY UNIMODULAR, BALANCED AND PERFECT MATRICES.
PALM 1974 -1 PALM, K.-P.
BEMERKUNG ZUM GEMISCHT-GANZZAHLIGEN SPIELBERG-MODELL, EINE.
RAGH 1976 -1 RAGHAVACHARI, M.
CONSTRUCTIVE METHOD TO RECOGNIZE THE TOTAL UNIMODULARITY OF A MATRIX, A.
SAKAB 1976 -1 SAKAROVITCH, M.
QUASI BALANCED MATRICES - AN ADDENDUM.
SEDL 1975 -1 SEDLACEK, J.
PERFECT AND QUASIPERFECT GRAPHS (CZECH).
TAMI 1976 -1 TAMIR, A.
ON TOTALLY UNIMODULAR MATRICES.
TRUE 1976 -1 TRUEMPER, K.
ALGEBRAIC CHARACTERIZATIONS OF TOTALLY UNIMODULAR MATRICES: A FRAMEWORK.
TRUE 1977 -2 TRUEMPER, K.
UNIMODULAR MATRICES OF FLOW PROBLEMS WITH ADDITIONAL CONSTRAINTS.
TRUE 1976 -2 TRUEMPER, K.
ALGEBRAIC CHARACTERIZATIONS OF UNIMODULAR MATRICES.
TRUEA 1977 -1 TRUEMPER, K.; CHANDRASEKARAN, R.
LOCAL UNIMODULARITY OF MATRIX-VECTOR PAIRS(EXTENDED ABSTRACT).
TRUEB 1977 -1 TRUEMPER, L.; CHANDRASEKARAN, R.
LOCAL UNIMODULARITY OF MATRIX-VECTOR PAIRS (REVISED VERSION).

ZERO-ONE PROGRAMMING
****

AHREA 1976 -1 AHRENS, W.
NUMERICAL EXPERIENCE WITH HEURISTIC RULES IN SPECIFIC BINARY OPTIMIZATION ALGORITHMS.
AHREA 1972 -1 AHRENS, W.
BINAERE OPTIMIERUNG ALS HILFSMITTEL BEI DER PLANUNG REGIONALER ABWASSERBESEITIGUNGSSYSTEME. II, DIE.
AUST 1976 -1 AUST, R. J.
DYNAMIC PROGRAMMING BRANCH AND BOUND ALGORITHM FOR PURE INTEGER PROGRAMMING, A.
BABAA 1976 -1 BABAYEV, D. A.
METHOD OF HYPERSPHERES FOR BOOLEAN PROGRAMMING PROBLEMS (RUSSIAN).
BABAA 1975 -1 BABAYEV, D. A.
METHOD OF HYPERSPHERES FOR THE SOLUTION OF DISCRETE PROGRAMMING PROBLEMS (RUSSIAN), THE.
BAER 1976 -2 BAER, G.
ZUR LINEARISIERUNG BELIEBIGER 0-1-OPTIMIERUNGSPROBLEME.
BALAG 1975 -1 BALAS, E.
BIVALENT PROGRAMMING BY IMPLICIT ENUMERATION.
BALAI 1969 -1 BALAS, E.; JEROSLOW, R.
ON THE STRUCTURE OF THE UNIT HYPERCUBE.
BALAL 1975 -1 BALAS, E.; ZEMEL, E.
ALL THE FACETS OF ZERO-ONE PROGRAMMING POLYTOPES WITH POSITIVE COEFFICIENTS.
BELLA 1975 -1 BELL, D. E.; SHAPIRO, J. F.
FINITELY CONVERGENT DUALITY THEORY FOR ZERO-ONE INTEGER PROGRAMMING, A.
BITR 1977 -1 BITRAN, G. R.
LINEAR MULTIPLE OBJECTIVE PROGRAMS WITH ZERO-ONE VARIABLES.
BRAU 1974 -1 BRAUN, J.
EDMONDS' MATCHING ALGORITHMUS.
DAVIE 1976 -1 DAVIS, R. P.; TERRELL, M. P.
APPROXIMATION TECHNIQUE FOR PSEUDO-BOOLEAN MAXIMIZATION PROBLEMS, AN.
DOERA 1977 -1 DOERSCH, R. H.; PATTERSON, J. H.
SCHEDULING A PROJECT TO MAXIMIZE ITS PRESENT VALUE: A ZERO-ONE PROGRAMMING APPROACH.
ELLE 1973 -1 ELLERVIK, P.
ANOTHER METHOD FOR THE SOLUTION OF COMBINATORIAL 0-1 LINEAR PROBLEMS (DANISH).
ELLW 1972 -1 ELLWEIN, L. B.
FLEXIBLE ENUMERATION SCHEME FOR ZERO-ONE PROGRAMMING, A.
ETCH 1976 -1 ETCHEBERRY, J.
NEW BRANCHING AND BOUNDING STRATEGIES FOR 0-1 INTEGER LINEAR PROGRAMMING.
FLORA 1970 -1 FLORIAN, M.; ROBILLARD, P.
HYPERBOLIC PROGRAMMING WITH BIVALENT VARIABLES.
FORG 1974 -2 FORGO, F.
SOME TRANSFORMATIONS OF INTEGER PROGRAMMING PROBLEMS (HUNGARIAN).
GAIK 1976 -2 GAIKOV, N. E.
MINIMIZATION OF A QUADRATIC FORM WITH BOOLEAN VARIABLES ON A SYMMETRIC GROUP (RUSSIAN), THE.
GEN 1974 -1 GEN, M.; OKUNO, H.
OPTIMIZATION OF REDUNDANT ALLOCATION AND SELECTION PROBLEMS OF SYSTEMS RELIABILITY BY MEANS OF 0-1 PROGRAMMING (JAPANESE).
GEN B 1975 -1 GEN, M.; SASAKI, M.
OPTIMAL ALLOCATION OF SYSTEM AVAILABILITY.
GLOVD 1977 -1 GLOVER, F.; KLINGMAN, D.; MACMILLAN, C.
NETFORM CONCEPT: A MORE EFFECTIVE MODEL FORM AND SOLUTION PROCEDURE FOR LARGE SCALE NONLINEAR PROBLEMS, THE.
GLOVG 1975 -1 GLOVER, F.; MULVEY, J.
EQUIVALENCE OF THE 0-1 INTEGER PROGRAMMING PROBLEM TO DISCRETE GENERALIZED AND DISCRETE PURE NETWORK MODELS.
GLOVG 1975 -2 GLOVER, F.; MULVEY, J.
EQUIVALENCE OF THE 0-1 INTEGER PROGRAMMING PROBLEM TO DISCRETE GENERALIZED AND PURE NETWORKS.
GRANB 1970 -1 GRANOT, F.; HAMMER, P.
ON THE USE OF BOOLEAN FUNCTIONS IN 0-1 PROGRAMMING.
GRIS 1976 -1 GRISHUKHIN, V. P.
BRANCH AND BOUNDARY ALGORITHMS IN PROBLEMS WITH BOOLEAN VARIABLES: ESTIMATION OF THEIR EFFECTIVENESS (RUSSIAN).
GRIS 1972 -1 GRISHUKHIN, V. P.
ESTIMATION OF THE COMPLEXITY OF THE BALAS ALGORITHM (RUSSIAN).
GRIS 1976 -2 GRISHUKHIN, V. P.
EFFECTIVENESS OF THE BRANCH AND BOUND METHOD IN PROBLEMS WITH BOOLEAN VARIABLES (RUSSIAN).
GUIG 1974 -1 GUIGNARD, M.
PREFERRED SHADOW PRICES IN 0-1 PROGRAMMING.
GUIGA 1976 -1 GUIGNARD, M.; SPIELBERG, K.
EXPERIMENTAL INTERACTIVE SYSTEM FOR INTEGER PROGRAMMING, AN.
GUIGA 1977 -3 GUIGNARD, M.; SPIELBERG, K.
PROPAGATION, PENALTY IMPROVEMENT AND USE OF LOGICAL INEQUALITIES IN INTERACTIVE (0,1) PROGRAMMING.
GUNT 1977 -1 GUNTHER, T. M.
SUBGRADIENT APPROACH TO 0,1 INTEGER PROGRAMMING.
HAMMA 1975 -1 HAMMER, P. L.
BOOLEAN ELEMENTS IN COMBINATORIAL OPTIMIZATION: A SURVEY.
HANSA 1976 -1 HANSEN, P.
CASCADE ALGORITHM FOR THE LOGICAL CLOSURE OF A SET OF BINARY RELATIONS, A.
HANSA 1975 -2 HANSEN, P.
FONCTIONS D'EVALUATION ET PENALITES POUR LES PROGRAMMES QUADRATIQUES EN VARIABLES 0-1.
HAUSC 1976 -1 HAUSMANN, D.; KORTE, B.
COLOURING CRITERIA FOR ADJACENCY ON 0-1-POLYHEDRA.
HYUNA 1974 -1 HYUN, K. N.; MITSUO, G.; OKUNO, H.
OPTIMIZATION OF THE REDUNDANT ALLOCATION IN A SYSTEM WITH SEVERAL FAILURE MODES BY ZERO-ONE PROGRAMMING.
KALD 1975 -1 KALDYBAEV, S. U.
ALGORITHM FOR THE APPROXIMATION-COMBINATORIAL METHOD IN SOLVING AN ALLOCATION PROBLEM USING BOOLEAN VARIABLES (RUSSIAN), A.
KAPLA 1976 -1 KAPLINSKII, A. I.; OHERNYSOVA, G. D.
METHOD OF ADAPTIVE ALGORITHMS FOR SOLVING OPTIMIZATION PROBLEMS WITH BOOLEAN VARIABLES (RUSSIAN).
KIAN 1976 -1 KIANFAR, F.
STRONGER INEQUALITIES FOR 0-1 INTEGER PROGRAMMING: COMPUTATIONAL REFINEMENTS.
KOCHA 1977 -1 KOCHURA, S. N.
ON AN OPTIMIZATION ALGORITHM OF CAPITAL INVESTMENTS DISTRIBUTION USING THE BOOLEAN VARIABLES (RUSSIAN).

****

KOEHA 1977 -1 KOEHLER, J.
NUTZUNG DER DUALITAET ZUR LOESUNG EINES GEMISCHTEN LINEAREN 0-1-OPTIMALPROBLEMS.

LOVEA 1976 -1 LOVE, R. F.; WONG, J. Y.
ON SOLVING A ONE-DIMENSIONAL SPACE ALLOCATION PROBLEM WITH INTEGER PROGRAMMING.

MAC B 1976 -1 MAC CALL, E. H.
ZERO-ONE MIXED INTEGER PROGRAMMING USING INTERACTIVE GRAPHICS.

MARSB 1977 -1 MARSTEN, R. E.; MORIN, T. L.
OPTIMAL SOLUTIONS FOUND FOR SENJU AND TOYODA'S 0/1 INTEGER PROGRAMMING PROBLEMS.

MOORB 1976 -1 MOORE, T. G.
SOLVING LARGE 0-1 PROBLEMS.

MULVA 1975 -2 MULVEY, J. M.
SPECIAL STRUCTURES IN NETWORK MODELS AND ASSOCIATED APPLICATIONS.

NAST 1975 -1 NASTANSKY, L.; GROSSE, H. W.
COMPUTATIONAL EXPERIMENTS WITH LINEAR PROGRAMMING ORIENTED IMPLICIT ENUMERATION ALGORITHMS FOR THE SOLUTION OF LINEAR (0,1) - PROGRAMS.

NAUS 1976 -1 NAUSS, R. M.
ALGORITHM FOR THE 0-1 INTEGER LINEAR PROGRAMMING PROBLEM, AN.

ORTHA 1972 -1 ORTH, H.
BINAERE OPTIMIERUNG ALS HILFSMITTEL BEI DER PLANUNG REGIONALER ABWASSERBESEITIGUNSSYSTEME. I, DIE.

PANC 1976 -1 PANCHENKO, A. A.
SOME ALGORITHMS FOR SOLUTION OF QUOTIENTS OF LINEAR PROGRAMMING WITH BOOLEAN VARIABLES (RUSSIAN).

PELE 1977 -1 PELED, U. N.
PROPERTIES OF FACETS OF BINARY POLYTOPES.

PIPE 1975 -1 PIPER, C.
COMPUTATIONAL STUDIES IN OPTIMIZING AND POSTOPTIMIZING LINEAR PROGRAMS IN ZERO-ONE VARIABLES.

PIPEA 1975 -1 PIPER, C. J.; ZOLTNERS, A. A.
IMPLICIT ENUMERATION BASED ALGORITHMS FOR POSTOPTIMIZING ZERO-ONE PROGRAMS.

PIPEA 1976 -1 PIPER, C. J.; ZOLTNERS, A. A.
SOME EASY POSTOPTIMALITY ANALYSIS FOR ZERO-ONE PROGRAMMING.

PLUK 1976 -1 PLUKENE, D. B.
APPLICATION OF SEMI-BOOLEAN METHODS TO THE SOLUTION OF 0-1 INTEGER LINEAR PROGRAMMING PROBLEMS (RUSSIAN).

RUBIA 1977 -1 RUBIN, D. S.
VERTEX GENERATION METHODS FOR PROBLEMS WITH LOGICAL CONSTRAINTS.

SAAT 1977 -1 SAATCIOGLU, O.
MATHEMATICAL PROGRAMMING MODELS FOR AIRPORT SITE SELECTION.

SALKB 1972 -1 SALKIN, H. M.; SPIELBERG, K.
ADAPTIVE ALGORITHM FOR BINARY PROGRAMMING, AN.

SANDA 1975 -1 SANDI, C.
SOLUTION OF THE MACHINE LOADING PROBLEM WITH BINARY VARIABLES.

SCHO 1976 -1 SCHOCH, M.
ERWEITERUNGSPRINZIP UND SEINE ANWENDUNG ZUR ENTWICKLUNG VON ALGORITHMEN FUER DIE LOESUNG KOMBINATORISCHER OPTIMIERUNGSAUFGABEN, DAS.

SIEL 1976 -1 SIELKEN, R. M. JR.
SEQUENCING WITH SETUP COSTS BY ZERO-ONE MIXED INTEGER LINEAR PROGRAMMING.

SPIE 1972 -1 SPIELBERG, K.
MINIMAL PREFERRED VARIABLE METHODS IN ZERO-ONE PROGRAMMING.

SPIE 1973 -1 SPIELBERG, K.
MINIMAL PREFERRED INEQUALITIES, PENALTIES AND STRUCTURE IN 0-1 PROGRAMMING.

SPIE 1974 -1 SPIELBERG, K.
MINIMAL PREFERRED VARIABLE METHODS IN ZERO-ONE PROGRAMMING.

STARA 1975 -1 STARR, J. H.
ZERO-ONE PROGRAMMING: A PARTIAL ORDERING AND ITS USE IN THE OPTIMIZATION OF NUMERICAL AND NON-NUMERICAL FUNCTIONS.

SUHL 1975 -1 SUHL, U.
ENTWICKLUNG VON ALGORITHMEN FUER GANZZAHLIGE OPTIMIERUNGSMODELLE.

SUHL 1976 -1 SUHL, U.
ALGORITHMUS ZUR LOESUNG EINER SPEZIELLEN KLASSE GANZZAHLIGER OPTIMIERUNGSMODELLE, EIN.

SZWA 1974 -1 SZWARC, W.
MIXED INTEGER LINEAR PROGRAMMING PROBLEM WHEN THE INTEGER VARIABLES ARE ZERO OR ONE, THE.

TAHA 1975 -2 TAHA, H. A.
NOTE ON THE RELATIONSHIP BETWEEN SOME SPECIALIZED BINARY ALGORITHMS AND THE ALGORITHM OF LAND AND DOIG, A.

TRYP 1977 -1 TRYPIA, M.
NON-BINARY TREE SEARCH METHOD FOR THE 0-1 LINEAR PROGRAMMING PROBLEM, A.

VIZV 1975 -1 VIZVARI, B.
ENUMERATIVE ALGORITHMS IN POLYNOMIAL 0-1 PROGRAMMING (HUNGARIAN).

WOEH 1973 -1 WOEHNER, W.
GUETERZUGBILDUNG ALS KOMBINATORISCHE OPTIMIERUNGSAUFGABE.

WOLS 1975 -1 WOLSEY, L. A.
VALID INEQUALITIES AND SUPERADDITIVITY FOR 0 1 INTEGER PROGRAMS.

ZAK 1977 -1 ZAK, YU. A.
METHOD OF DIRECTED SCANNING IN PROBLEMS OF INTEGRAL LINEAR PROGRAMMING WITH BOOLEAN VARIABLES (RUSSIAN).

ZIMMA 1976 -1 ZIMMERMANN, U.
BOOLE'SCHE OPTIMIERUNGSPROBLEME MIT SEPARABLER ZIELFUNKTION UND MATROIDALEN RESTRIKTIONEN.

****

```
PPPPPPPPPPP    AAAAAAAAAA    RRRRRRRRRRR    TTTTTTTTTTTT                    3333333333
PPPPPPPPPPPP  AAAAAAAAAAAA   RRRRRRRRRRRR   TTTTTTTTTTTT                   333333333333
PP        PP  AA        AA   RR        RR        TT                        33        33
PP        PP  AA        AA   RR        RR        TT                                  33
PP        PP  AA        AA   RR        RR        TT                                  33
PPPPPPPPPPPP  AAAAAAAAAAAA   RRRRRRRRRRRR        TT                               3333
PPPPPPPPPPP   AAAAAAAAAAAA   RRRRRRRRRRR         TT                               3333
PP            AA        AA   RR     RR           TT                                  33
PP            AA        AA   RR      RR          TT                                  33
PP            AA        AA   RR       RR         TT                        33        33
PP            AA        AA   RR        RR        TT                        333333333333
PP            AA        AA   RR         RR       TT                         3333333333
```

LIST OF AUTHORS

****
****

AASHTIANI, H. A.
AASH 1976 -1
ABADIE, J.
ABAD 1976 -1
ABELLA, R. J.
ABEL 1977 -1
ABRAMSON, P.
RAPA 1959 -1
ACKERMAN, L. J.
ACKE 1977 -1; ACKE 1977 -2
ACKERMAN, S. S.
ACKEA 1963 -1
ADAM, N.
ADAM 1977 -1
ADAMSKA-TCHON, J.
ADAMA 1976 -1
ADEL'SON-VEL'SKIJ, G. M.
ADEL 1975 -1
ADLEMAN, L.
MAND 1976 -1
ADLER, H. J.
ADLE 1975 -1
ADOLPHSON, D. L.
ADOL 1976 -1; ADOL 1977 -1; ADOLA 1977 -1
ADRABINSKI, A.
ADRA 1976 -1
AGAPOVA, L. V.
AGAP 1976 -1
AGARWAL, A. K.
AGAR 1976 -1
AGARWAL, R. L.
SHAR 1975 -1
AGIN, N. I.
AGIN 1976 -1; AGINA 1975 -1
AGNEW, C. E.
AGNE 1976 -1
AGRAWAL, S. C.
AGRA 1976 -1; AGRA 1976 -2; CHANC 1977 -1
SHARB 1976 -1; SHARB 1976 -2
AGRESTI, W. W.
AGRE 1977 -1
AHERN, P. A.
SHARG 1975 -1
AHO, A. V.
AHO 1977 -1; AHO A 1974 -1; AHO B 1974 -1
AHO C 1976 -1
AHRENS, J. H.
AHRE 1975 -1
AHRENS, W.
AHREA 1972 -1; AHREA 1974 -1; AHREA 1974 -2
AHREA 1975 -1; AHREA 1976 -1
AIGNER, M.
AIGN 1976 -1
AKADEMIA NAUK SSSR, TSENTR. EHKON.-MAT. INST.
AKAD 1976 -1
AKERS, S. A. JR.
AKER 1967 -1
AKHPATELOV, E. A.
AKHP 1977 -1
AKINC, U.
AKIN 1974 -1; AKINA 1977 -1
AKOKA, J.
ABAD 1976 -1
AKOPJAN, G. C.
AKOP 1974 -1
AKULIC, F. E.
AKUL 1977 -1
ALAVI, Y.
ALAV 1977 -1
ALBERS, S.
ALBE 1977 -1
ALBERTSON, M.
ALBEA 1976 -1
ALBERTSON, M. O.
ALBEB 1974 -1; ALBEB 1976 -1; ALBEC 1976 -1
ALBED 1975 -1
ALCOUFFE, A.
ALCO 1975 -1; ALCOA 1976 -1
ALDERSON, G. E.
LIN A 1969 -1
ALEKSEYEV, O. G.
ALEK 1974 -1; ALEK 1977 -1; ALEKA 1977 -1
ALEKB 1976 -1; ALEKC 1964 -1; ALEKE 1971 -1
ALEKSEYEVA, S. M.
ALEKD 1974 -1; ALEKE 1971 -1
ALEMANY, R.
ALEM 1972 -1; ALEM 1973 -1
ALESKEROV, S. A.
ALES 1975 -1
ALEXANDER, J. M.
SAATA 1975 -1
ALI, A. I.
ALI 1977 -1
ALIA, G.
ALIA 1976 -1
ALIBEKOV, B. I.
ALIB 1971 -1; ALIB 1975 -1
ALLEN, B.
ALLE 1976 -1
ALLENDE, S.
ALLEA 1976 -1
ALMAR, A.
PUN 1975 -1
ALSPACH, B.
ALSP 1976 -1
ALTHERR, W.
ALTH 1975 -1; ALTH 1976 -1; ALTHA 1976 -1
AM, T. D.
AM 1976 -1
AMIN, A. T.
AMIN 1973 -1
ANDERSON, S.
ANDE 1970 -1
ANDERSSEN, R. S.
ANDEA 1976 -1
ANDON, F. I.
ANDO 1976 -1
ANDRAS, B.
ANDR 1975 -1
ANDRONICO, A.
ANDRA 1976 -1
ANEJA, Y. P.
ANEJ 1977 -1
ANGLUIN, D.
ANGL 1977 -1
ANISIMOV, V. V.
ANIS 1976 -1
ANN, T.-K.
ANN 1977 -1
ANZAI, Y.
ANZA 1974 -1
AOSHIMA, K.
AOSH 1977 -1
APPEL, K.
APPE 1976 -1
ARDITTI, J. C.
ARDI 1977 -1
ARDITTI, J.-C.
ARDIA 1975 -1; ARDIA 1976 -1
ARET, L.
ARET 1974 -1
ARISAWA, S.
ARIS 1977 -1; ARIS 1977 -2
ARIYOSHI, H.
TSUK 1977 -1
ARJOMANDI, E.
ARJO 1975 -1
ARKHIPOWA, T. T.
ARKH 1975 -1
ARLAZAROV, V. L.
ARLA 1973 -1
ARMSTRONG, R. D.
ARMS 1978 -1; ARMSA 1975 -1; ARMSB 1976 -1
ARMSB 1977 -1; ARMSB 1977 -2; ARMSC 1977 -1
ARMSD 1976 -1; ARMSE 1975 -1; ARMSE 1977 -1
ARNOLD, L. R.
ARNO 1974 -1; ARNO 1974 -2
ARORA, S. R.
AGAR 1976 -1; AROR 1977 -1; ARORA 1977 -1
ARORB 1976 -1; ARORB 1977 -1
ARUIN, L. I.
ARUI 1971 -1
ASANO, T.
ASAN 1976 -1
ASHTON, W. B.
ASHT 1977 -1
ASSAD, A.
ASSA 1975 -1
ASSAD, A. A.
ASSAA 1976 -1
ASTAFEVA, N. A.
ASTA 1971 -1
ASTAHOV, N. D.
ASTAA 1976 -1; HACAA 1976 -1
ASTAKHOV, N. B.
KHAC 1976 -1
ATTMANN, I.
ATTM 1973 -1
AUSIELLO, G.
AUSI 1977 -1; AUSIA 1977 -1; AUSIB 1977 -1
AUST, R. J.
AUST 1976 -1; AUST 1976 -2
AVDONIN, L. N.
AVDO 1975 -1
AVERBUH, N. D.
EMELA 1976 -1
AVERBUKH, N. D.
KRAVA 1975 -1
AVIS, D. M.
YAO A 1977 -1

AVRIEL, M.
AVRI 1977 -1
AZARYAN, L. L.
AZAR 1977 -1
BABAT, L. G.
BABA 1977 -1
BABAYEV, D. A.
BABAA 1975 -1; BABAA 1976 -1
BABU, A. J. G.
BABU 1977 -1
BABUSHKIN, A. I.
BABUA 1976 -1; BABUA 1977 -1
BABUSIAUX, D.
BABUB 1972 -1
BACHEM, A.
BACH 1976 -1; BACH 1976 -2; BACH 1976 -3
BACH 1976 -4; BACH 1977 -1; BACH 1977 -2
BACH 1977 -3; BACH 1978 -1; BACHA 1977 -1
BACHA 1977 -2; BACHA 1977 -3; BACHA 1977 -4
BACHB 1977 -1; BACHC 1977 -1
BAER, G.
BAER 1976 -1; BAER 1976 -2
BAGBY, G.
BAGB 1976 -1
BAGCHI, A.
SAVAB 1976 -1
BAGGA, P. C.
BAGG 1973 -1; BAGGA 1967 -1; MITT 1977 -1
BAKER, K. R.
BAKE 1976 -1; BAKE 1977 -1; BAKEA 1977 -1
BAKEB 1977 -1
BAKER, T.
BAKEC 1975 -1
BAKO, A.
BAKO 1975 -1; BAKO 1976 -1; BAKO 1976 -2
BALACHANDRAN, V.
BALA 1976 -1; BALAA 1977 -1; BALAB 1977 -1
BALAC 1976 -1; BALAD 1977 -1; BALAE 1975 -1
BALAE 1975 -2; BALAE 1975 -3; BALAE 1975 -4
BALAGURU, P. N.
BALAF 1977 -1
BALAS, E.
BALAG 1968 -1; BALAG 1973 -1; BALAG 1974 -1
BALAG 1974 -2; BALAG 1975 -1; BALAG 1975 -2
BALAG 1975 -3; BALAG 1976 -1; BALAG 1976 -2
BALAG 1977 -1; BALAG 1977 -2; BALAG 1977 -3
BALAH 1974 -1; BALAI 1969 -1; BALAJ 1975 -1
BALAJ 1976 -1; BALAK 1977 -1; BALAL 1975 -1
BALAL 1976 -1; BALAL 1977 -1; BALAM 1975 -1
MIHO 1965 -1
BALERIAUX, H.
BALE 1970 -1
BALINSKI, M. L.
BALI 1964 -1; BALIA 1977 -1
BALITSKAYA, E. E.
MIKH 1975 -1
BALL, B. C.
BALL 1977 -1
BALL, M.
GOLDA 1977 -1
BALL, M. O.
BALLA 1977 -1
BALLOU, R. H.
BALLB 1970 -1
BANDLER, J. W.
BAND 1975 -1; BANDA 1975 -1
BANERJEE, B. P.
BANE 1965 -1
BANSAL, S. P.
BANS 1976 -1; BANS 1977 -1
BAR, G.
BAR 1975 -1
BARANKIEWICZ, M. N.
BARA 1972 -1
BARBER, G. M.
BARB 1970 -1; BARB 1975 -1
BARI, A.
KHAN 1977 -1
BARLOW, J. F.
BARL 1978 -1
BARNES, J. W.
BARN 1977 -1
BARR, R. S.
BARR 1977 -1; BARR 1977 -2; BARR 1977 -3
BARR 1977 -4; BARRA 1975 -1
BARRER, J.
BARRB 1976 -1
BARRY, D.
BARRC 1977 -1
BARRY, D. L.
BARRD 1977 -1
BARSKIY, A. B.
BARS 1974 -1
BARTHES, J. P.
BART 1975 -1
BARTHOLDI, J.
PAPI 1975 -1
BARTHOLDI, J. J. III
BARTA 1977 -1; BARTB 1977 -1; BARTB 1977 -2
BASCA, O. C.
BASC 1977 -1
BASHTA, A. L.
BABUA 1976 -1; BABUA 1977 -1
BATTAGLIA, F.
BATT 1977 -1
BAUERSFELD, G.
BAUE 1973 -1
BAUM, S.
BAUM 1977 -1
BAUMGARTNER, J. E.
BAUMA 1975 -1
BAUTZ, M.
BAUT 1973 -1
BAZARAA, M. S.
BAZA 1975 -1; BAZAA 1977 -1; BAZAB 1977 -1
BAZAC 1977 -1
BEALE, E. M. L.
BEAL 1977 -1; BEAL 1977 -2
BEAN, D. R.
BEAN 1976 -1
BEAN, D. W. T.
BEANA 1970 -1
BEARDWOOD, J.
BEAR 1959 -1
BECK, L. L.
BECK 1976 -1
BECKER, J. D.
MARKA 1975 -1
BECKER, L.
BECKA 1974 -1; BECKA 1974 -2
BECKMANN, M.
BECKB 1956 -1
BECKMANN, M. J.
BECKC 1976 -1
BEENHAKKER, H. L.
BEEN 1976 -1
BEHARA, D. N.
BEHA 1969 -1
BEHZAD, M.
ALAV 1977 -1
BELETSKIJ, S. A.
BELE 1977 -1
BELFORD, G. G.
GRAP 1977 -1
BELKINA, M. V.
BELK 1965 -1
BELL, D. E.
BELL 1973 -1; BELL 1977 -1; BELLA 1975 -1
BELLA 1977 -1
BELL, E. J.
BELLB 1965 -1
BELLMORE, M.
ARNO 1974 -1; ARNO 1974 -2; BELLC 1965 -1
BELLD 1970 -1
BELLON, J.-L.
COME 1966 -1
BELOUSOV, E. G.
BELO 1977 -1
BELOV, I. S.
BABUA 1976 -1; BABUA 1977 -1; BELOA 1974 -1
BENICHOU, M.
BENI 1977 -1
BENITO, F.
BENIA 1976 -1
BENNETT, G. R.
BENN 1976 -1
BENNETT, J. M.
BENNA 1972 -1
BENNINGER, H.
BENNB 1976 -1
BENZI, W. M. L.
SMYT 1974 -1
BERESNEV, V. L.
BERE 1974 -1
BERGE, C.
BERG 1973 -1; BERG 1973 -2; BERG 1975 -1
BERG 1975 -2; BERG 1977 -1; BERG 1978 -1
BERGA 1976 -1; BERGB 1977 -1
BERGENDAHL, G.
BERGC 1966 -1; BERGC 1968 -1; BERGC 1969 -1
BERGENDORFF, H.
BERGD 1974 -1
BERGER, P. D.
GLIC 1977 -1
BERGMANN, D. R.
BERGE 1975 -1
BERKOWITZ, R. S.
ACKE 1977 -1; ACKE 1977 -2
BERLIN, G. N.
BERL 1976 -1

BERMAN. L.
HARTB 1976 -1
BERMAN, D. M.
ALBEC 1976 -1
BERMOND, J. C.
BERM 1975 -1; BERM 1976 -1
BERMOND, J.-C.
BERMA 1974 -1; BERMB 1976 -1
BERNERS-LEE, C. M.
BERN 1958 -1
BERRY, W. L.
BERR 1977 -1; BERRA 1976 -1; BERRB 1975 -1
HENDA 1976 -1
BERSHTEYN, L. S.
KELI 1968 -1; MELI 1972 -1
BERSTEL, J.
BERS 1976 -1
BERZTISS, A. T.
BERZ 1971 -1; BERZ 1973 -1
BESTWICK, P. F.
BEST 1976 -1
BESWICK, C. A.
BESW 1977 -1
BEUERMANN-WEHMEYER, A.
BEUE 1975 -1
BEYER, K.
BEYE 1970 -1
BHALLA, H. S.
BHAL 1970 -1
BHAT, U. N.
COOPA 1977 -1
BHATIA, H. L.
BHAT 1976 -1
BHAUMIK, G.
BHAU 1974 -1; BHAUA 1974 -1; SCHOA 1970 -1
BHAVE, V. N.
SAMP 1976 -1
BIALOSTOCKI, A.
SCHOD 1975 -1
BICHARA, J.
BICH 1972 -1
BIDER, I. G.
BIDE 1975 -1
BIESS, G.
BAUT 1973 -1; BIES 1976 -1; BIESA 1977 -1
BILDE, O.
BILD 1970 -1; BILDA 1967 -1; BILDA 1977 -1
KRARA 1967 -1; KRARA 1977 -1
BINDSCHEDLER, A. E.
BIND 1961 -1
BIONDI, E.
BION 1973 -1; BIONA 1969 -1
BIRD, C. G.
BIRD 1976 -1
BIRD, R.
BIRDA 1976 -1
BITRAN, G. R.
BITR 1977 -1
BIXBY, R. E.
BIXB 1975 -1; BIXB 1977 -1; BIXB 1978 -1
BJOERKLUND, B.
BJOE 1977 -1
BJORKLUND, B.
WEST 1977 -1
BLAIR, C.
BLAI 1976 -1
BLAIR, C. E.
BLAIA 1975 -1; BLAIB 1976 -1; BLAIB 1977 -1
BLAND, R. G.
BLAN 1977 -1; BLANA 1978 -1
BLASYUK, B. A.
BLAS 1967 -1
BLATTNER, W. O.
DANTA 1966 -1
BLAZEWICZ, J.
BLAZ 1976 -1; BLAZA 1976 -1; BLAZB 1976 -1
BLAZB 1976 -2
BLAZHKEVICH, B. I.
BLAZC 1976 -1
BLEWETT, W. J.
BLEW 1977 -1
BLIN, J. M.
BLIN 1973 -1; BLIN 1976 -1
BLIN, J.-M.
BLINA 1976 -1
BLUM, M.
BLUM 1967 -1
BOCK, F. C.
BOCK 1957 -1
BOCKO, G. D.
BOCKA 1975 -1
BODIN, L. D.
BODI 1976 -1
BOEBION, J.
BOEB 1976 -1; LATE 1976 -1; PUN 1975 -1
BOERE, N, J,
BOER 1977 -1
BOESCH, F.
BOES 1976 -1; BOESA 1976 -1
BOESCH, F. T.
BOESB 1974 -1; BOESC 1977 -1
BOETTCHER, H.
BOET 1974 -1
BOGARDI, J. J.
BOGA 1976 -1
BOGART, K. P.
TROTB 1976 -1; TROTB 1976 -2
BOGDANOVA, O. V.
PERC 1975 -1
BOHEME, G.
BOHE 1977 -1
BOL'SHAKOV, I. A.
BIDE 1975 -1
BOLKER, E. D.
BOLK 1976 -1
BOLLOBAS, B.
ALBEA 1976 -1; BOLL 1977 -1; BOLLA 1976 -1
BOLLB 1976 -1; BOLLC 1975 -1; BOLLC 1976 -1
BOLLC 1976 -2; BOLLD 1974 -1; ELDR 1976 -1
BONA, B.
BONA 1976 -1
BONDY, J. A.
BOND 1978 -1; BONDA 1976 -1
BONNET, R.
BONN 1974 -1
BONNEY, M. C.
BONNA 1976 -1
BOOK, R. V.
BOOK 1974 -1
BOOKBINDER, J. H.
BOOKA 1977 -1
BOOTH, K. S.
BOOT 1976 -1; BOOTA 1975 -1; BOOTA 1976 -1
BORIES, F.
BORI 1975 -1; BORI 1975 -2; BORIA 1976 -1
BORILLE, M.
BORIB 1966 -1
BORODIN, A. B.
BORO 1973 -1; BORO 1973 -2; BOROA 1972 -1
BORODIN, O. V.
BOROB 1977 -1; BOROC 1974 -1
BOROVITS, I.
BOROD 1976 -1
BOROWIECKI, M.
BOROE 1972 -1
BORUVKA, O.
BORU 1926 -1
BOSAK, J.
BOSA 1976 -1
BOSE, N.
LEE A 1976 -1
BOSE, R. C.
BOSE 1973 -1
BOTT, H.
BOTT 1974 -1
BOURGOIN, M. H.
BOUR 1969 -1
BOURKOV, V. N.
BOURA 1970 -1
BOVA, T. E.
ABEL 1977 -1
BOWMAN, V. J.
BOWM 1974 -1
BOWMAN, V. J. JR.
BOWMA 1977 -1
BOX, R. M.
BOX 1976 -1
BOYCE, D. E.
BOYC 1973 -1
BOYCE, W. M.
BOYCA 1975 -1; BOYCB 1973 -1
BRACE, A.
BRAC 1975 -1
BRADLEY, G.
BRAD 1976 -1
BRADLEY, G. H.
BRADA 1971 -1; BRADB 1969 -1
BRADLEY, S. P.
BRADC 1965 -1; BRADD 1977 -1
BRAEMERT, D.
BRAE 1975 -1
BRANDMAN, E. I.
BRAN 1971 -1
BRATLEY, P.
BRAT 1975 -1
BRAUN, J.
BRAU 1974 -1
BREGMAN, L. M.
BREG 1972 -1

BREGMAN, T. T.
BREGA 1974 -1
BREITENBACH, R. B.
BREI 1976 -1
BRELAZ, D.
BREL 1977 -1; BRELA 1977 -1
BRENNAN, J. J.
BARN 1977 -1
BRENT, R. P.
ANDEA 1976 -1
BRESLAW, J. A.
BRES 1976 -1
BRICKER, D. L.
BRIC 1977 -1
BRITAVSKIJ, G. M.
BRIT 1976 -1
BROCKLEHURST, E. R.
BROC 1976 -1
BROTCHIE, J. F.
SHARF 1972 -1; SHARG 1975 -1
BROWN, D. B.
BROW 1976 -1
BROWN, D. P.
BROWA 1976 -1; BROWA 1977 -1
BROWN, G.
BRAD 1976 -1
BROWN, J. R.
BROWB 1976 -1
BROWN, W.
BROWC 1973 -1
BROWNELL, W. S.
BROWD 1976 -1
BRUALDI, R. A.
BRUA 1975 -1; BRUA 1975 -2; BRUAA 1976 -1
BRUAA 1977 -1; BRUAA 1977 -2; BRUAA 1977 -3
BRUAB 1978 -1
BRUCKER, P.
BRUC 1972 -1; BRUC 1975 -1; BRUC 1976 -1
BRUC 1976 -2; BRUC 1976 -3; BRUC 1977 -1
BRUC 1978 -1; BRUCA 1977 -1; BRUCB 1975 -1
LENSC 1977 -1
BRUNO, J.
BRUN 1968 -1; BRUNA 1976 -1; BRUNB 1974 -1
BRUNC 1975 -1; BRUND 1976 -1; BRUND 1976 -2
BRUNO, J. L.
BRUNE 1976 -1; BRUNE 1976 -2
BRUNS, M.
BRUNF 1975 -1
BRUTER, C. P.
BRUT 1970 -1
BULATOV, V. P.
BULA 1974 -1
BULFIN, R. L.
BULF 1976 -1; BULF 1976 -2; BULFA 1977 -1
BULLINGHAM, J. M.
BULL 1970 -1
BUNT, R. B.
BUNT 1977 -1
BUR'JAN, S. B.
BUR' 1976 -1; BUR' 1976 -2
BURDET, C.-A.
BURD 1977 -1
BURDYUK, V. YA.
BURDA 1976 -1
BURKARD, R. E.
BURK 1973 -1; BURK 1976 -1; BURK 1976 -2
BURK 1976 -3; BURK 1977 -1; BURK 1977 -2
BURKA 1976 -1; BURKA 1977 -1; BURKB 1976 -1
BURKB 1977 -1; BURKB 1977 -2; BURKC 1977 -1
BURKD 1976 -1; BURKE 1977 -1; BURKE 1978 -1
STRAB 1976 -1
BURKHARD, W. A.
BURKF 1973 -1; BURKF 1975 -1; BURKG 1971 -1
BURKOV, V. N.
BURKH 1975 -1; BURKI 1972 -1; BURKJ 1966 -1
BURKJ 1968 -1; SALI 1970 -1
BURNESS, R. C.
BURN 1976 -1
BURNS, F.
BURNA 1975 -1; BURNA 1976 -1
BURNS, L. D.
BECKC 1976 -1
BURNS, R. N.
BAKEA 1977 -1; BURNB 1976 -1; BURNB 1977 -1
BURNC 1976 -1; HAFF 1973 -1
BURR, S. A.
BURR 1974 -1
BURRELL, J. E.
BURRA 1976 -1
BURSHTEJN, M. I.
BURS 1976 -1; BURS 1976 -2
BURT, J. M. JR.
BURT 1970 -1
BURTIN, YU. D.
BURTA 1977 -1
BURTMAN, M. I.
GINZ 1977 -1
BURTON, J.
POWE 1977 -1
BURTON, J. S.
BURTB 1976 -1
BUSHELL, G.
BUSH 1976 -1
BUTAS, L. F.
BUTA 1968 -1
BUTEN, R. E.
BUTE 1974 -1
BUTLAR, W. H.
BUTL 1965 -1
BUTLER, D. E.
BOES 1976 -1
BUTRIMENKO, A. V.
BUTR 1964 -1
BUZACOTT, J. A.
BUZA 1976 -1
BUZYCKII, P. L.
BUZY 1976 -1; BUZY 1976 -2; BUZYA 1976 -1
BYRD, D. F.
ROBB 1975 -1
BYRD, J. JR.
BENN 1976 -1
CABOT, A. V.
CABO 1977 -1
CABOT, V. A.
CABOA 1976 -1
CACCETTA, L.
CACC 1976 -1
CAHIT, I.
CAHI 1975 -1
CAHIT, R.
CAHI 1975 -1
CALYKOV, V.
CALY 1974 -1
CAMBINI, A.
CAMB 1976 -1
CAMERINI, P. M.
CAME 1973 -1; CAME 1974 -1; CAMEA 1974 -1
CAMERON, A. V.
CAMEB 1976 -1
CAMERON, P. J.
CAMEC 1975 -1
CAMPBELL, H. G.
CAMP 1970 -1
CANTOR, D. G.
CANT 1974 -1
CAPRERA, D.
ORLO 1976 -1
CARDENAS, M. A.
CARD 1971 -1
CARLSON, R. C.
ESCH 1975 -1; JUCK 1976 -1
CARLTON, D. W.
CARL 1977 -1; CARL 1977 -2; CARL 1977 -3
CARL 1977 -4
CARNALL, C. A.
CARN 1976 -1
CARPANETO, G.
CARP 1977 -1
CARPENITO, T. A.
CARPA 1976 -1
CARRE, B. A.
CARR 1971 -1
CARROLL, D. C.
CARRA 1965 -1
CARSON, J. S.
CARS 1977 -1
CARTER, M.
BAKEA 1977 -1
CARUSO, F. R.
CARU 1965 -1
CARVALLO, M.
CARV 1977 -1
CAVAILLE, J. B.
CAVA 1976 -1
CAZALET, J.
CAZA 1975 -1
CEDERBAUM, I.
CEDE 1974 -1; CEDE 1976 -1
CELLARY, W.
BLAZA 1976 -1; BLAZB 1976 -1; BLAZB 1976 -2
CERF, V. G.
CERF 1974 -1
CHACHRA, V.
SHARA 1975 -1
CHAIKEN, J. M.
CHAI 1975 -1; CHAIA 1976 -1
CHAKRAVARTI, N. K.
BAGGA 1967 -1
CHALMET, L.
CHAL 1976 -1; CHALA 1975 -1

CHALMET, L. G.
CHALB 1976 -1
CHAMBERS, M. L.
CHAM 1976 -1
CHAMPEAUX, D. DE
CHAMA 1977 -1
CHAN, A. W.
CHAN 1976 -1; CHAN 1976 -2
CHAN, Y.
CHANA 1976 -1
CHAN, Y. P.
CHANB 1969 -1
CHAND, M.
CHANC 1977 -1
CHANDRA, A. K.
CHAND 1976 -1; CHANE 1976 -1
CHANDRASEKARAN, R.
BARRB 1976 -1; CHANF 1977 -1; CHANG 1975 -1
CHANH 1975 -1; CHANI 1977 -1; TRUEA 1977 -1
TRUEB 1977 -1
CHANDY, K. M.
RAMAC 1971 -1; SICK 1977 -1
CHANG, L.
CHANJ 1976 -1
CHANG, R. L. P.
CHANK 1977 -1
CHARNES, A.
ARMSA 1975 -1; CHAR 1975 -1; CHARA 1975 -1
CHARA 1977 -1
CHARTRAND, G.
CHARB 1977 -1; CHARC 1973 -1
CHARTRES, B. A.
CHARD 1967 -1
CHATY, GUY.
CHAT 1975 -1
CHEBOTAREV, A. S.
CHEB 1975 -1; CHEBA 1975 -1; DEMIA 1976 -1
CHEIN, M.
CHEI 1975 -1
CHEN, C. C.
CHEN 1975 -1
CHEN, DER-S.
CHENA 1977 -1; CHENB 1976 -1
CHEN, I. N.
CHENC 1975 -1
CHEN, J. H. K.
BAND 1975 -1; BANDA 1975 -1; CHEND 1974 -1
CHEND 1975 -1
CHEN, R.
CHENE 1976 -1
CHEN, S.
BOESB 1974 -1; CHENF 1977 -1
CHEN, W. K.
CHENG 1969 -1
CHEN, W.-K.
CHENH 1976 -1
CHEN, Y. E.
CHENI 1971 -1
CHERENIN, V. P.
AKHP 1977 -1; CHER 1970 -1
CHERENKOV, A. P.
CHERA 1977 -1
CHERITON, D.
CHERB 1976 -1
CHERKASSKII, B. V.
CHERC 1976 -1
CHERNYSHEV, YU. YU.
CHERD 1976 -1
CHERVAK, N. K.
CHERE 1975 -1
CHERVAK, YU. YU.
CHERF 1972 -1; CHERF 1973 -1; CHERF 1974 -1
CHERF 1976 -1; CHERF 1977 -1
CHESHIRE, I. M.
RYAN 1976 -1
CHEUNG, L. K.
OHTSA 1976 -1
CHEUNG, T.-Y.
CHEU 1976 -1; CHEU 1976 -2
CHEVALIER, A.
CHEV 1977 -1
CHEVIGNON, P.
DESC 1964 -1
CHEW, K. L.
CHEW 1975 -1; CHEWA 1974 -1
CHIANG, C. H.
CHIA 1975 -1
CHIAPPA, R. A.
CHIAA 1975 -1
CHISMAN, J. A.
CHIS 1977 -1
CHMARA, K.
ZABL 1976 -1
CHOE, S. R.
CHOE 1976 -1
CHOU, W.
CHOU 1975 -1; CHOUA 1973 -1; FRANF 1971 -1
CHOUDUM, S. A.
PART 1976 -1
CHOW, T. S.
CHOW 1976 -1
CHOWDHURY, I. G.
EILO 1975 -1; EILOA 1975 -1; EILOA 1976 -1
EILOA 1976 -2; EILOA 1977 -1
CHOY, D. M.
CHOY 1977 -1
CHRISTOFIDES, N.
CHRI 1973 -1; CHRI 1974 -1; CHRI 1975 -1
CHRI 1976 -1; CHRI 1976 -2; CHRIA 1977 -1
CHRISTOS, H.
CHRIB 1976 -1
CHU, J.
SHARF 1975 -1
CHU, Y. J.
CHU 1965 -1
CHUANG, Y. A.
CHUA 1965 -1
CHUNG, F. R. K.
CHUN 1978 -1
CHURCH, R. L.
CHUR 1976 -1; CHURA 1977 -1; CHURB 1976 -1
CHVATAL, V.
BONDA 1976 -1; CHVA 1975 -1; CHVA 1976 -1
CHVA 1977 -1; CHVAA 1978 -1; CHVAB 1978 -1
CHVAC 1977 -1; CHVAD 1976 -1
CIOBANU, G.
CIOB 1975 -1
CIOBOTARU, S.
CIOBA 1977 -1
CIRILLO, F.
CIRI 1975 -1
CLARK, C. R.
CLAR 1976 -1
CLARK, G. M.
ASHT 1977 -1
CLAUDER, G.
CLAU 1977 -1
CLAUS, A.
CLAUA 1975 -1
COCK, J. DE
COCK 1975 -1
COCKAYNE, E.
COCKA 1976 -1
COCKAYNE, E. J.
COCKB 1975 -1; COCKB 1976 -1; COCKB 1977 -1
COCKC 1977 -1
CODY, R. A.
CODY 1976 -1
COFFMAN, E. G. JR.
BRUNA 1976 -1; BRUNB 1974 -1; CODY 1976 -1
COFF 1975 -1; COFF 1976 -1; COFFA 1976 -1
COGIS, O.
COGI 1976 -1
COHEN, J.
COHE 1973 -1
COI, S.
COI 1975 -1
COLLATZ, L.
COLL 1975 -1
COMES, G.
COME 1966 -1
CONTI, A.
CONT 1976 -1
CONWAY, R. W.
CONW 1966 -1
COOK, R. J.
COOK 1974 -1; COOK 1975 -1
COOK, S. A.
COOKA 1974 -1
COOK, W. D.
ARMSB 1976 -1; ARMSB 1977 -1; ARMSB 1977 -2
ARMSC 1977 -1
COOPER, L.
COOP 1963 -1; COOP 1975 -1; COOP 1975 -2
COOPA 1977 -1; COOPB 1974 -1; COOPB 1975 -1
COOPB 1976 -1; COOPC 1977 -1; KATZ 1976 -1
ROBE 1976 -1
COOPER, M. W.
COOPB 1974 -1; COOPB 1975 -1; COOPB 1976 -1
COOPD 1977 -1
CORBAN, A.
CORB 1972 -1; CORB 1973 -1; CORB 1975 -1
CORB 1975 -2
CORNEIL, D. G.
ARJO 1975 -1; CORN 1968 -1; CORN 1970 -1
CORN 1972 -1; CORN 1974 -1; CORNA 1970 -1
CORNB 1978 -1; READA 1977 -1
CORNUEJOLS, G.
CORNC 1977 -1; CORND 1976 -1; CORND 1977 -1
CORNE 1978 -1

****

COT, N.
COT 1974 -1
COTES, L. C.
PATE 1976 -1
COTHER, R. F.
DAR- 1975 -1
COTTAM, M.
HITC 1976 -1
COTTLE, R.
COTT 1974 -1
COVERDALE, I. L.
COVE 1976 -1
COWAN, D. D.
CERF 1974 -1
CRANDALL, K. C.
CRAN 1976 -1
CRAPO, H. H.
CRAP 1970 -1
CROMLEY, R. G.
OSLEA 1977 -1
CROUCHER, J. S.
CROU 1976 -1
CROWDER, H.
CHENE 1976 -1; CROW 1976 -1
CRUM, R.
CRUM 1976 -1
CRUM, R. L.
CRUMA 1977 -1
CRUSE, A. B.
CRUS 1975 -1
CUBE, H. VON
IRWI 1971 -1
CULLEN, D. E.
AGINA 1975 -1
CULLINGFORD, G.
MAWD 1974 -1
CUMMINS, R. L.
GOOD 1965 -1
CUNNINGHAM, W.
CUNN 1975 -1
CUNNINGHAM, W. H.
CUNNA 1974 -1; CUNNA 1975 -1; CUNNA 1976 -1
CUNNB 1976 -1
CUYVERS, L.
MEEU 1975 -1; MEEU 1976 -1
CVETKOVIC, D. M.
CVET 1976 -1
CZERNIAKIEWICZ, A.
CZER 1977 -1
D'ATRI, A.
AUSIA 1977 -1; AUSIB 1977 -1
DAENINCK, G.
DAEN 1977 -1
DAFERMOS, S. C.
DAFE 1976 -1
DAGANZO, C. F.
DAGA 1977 -1
DALAL, Y. K.
DALA 1976 -1
DALSGAARD, P.
BAND 1975 -1
DAMM, E.
DAMM 1977 -1
DANARAJ, G.
DANA 1977 -1
DANILENKO, S. E.
DANI 1977 -1
DANNENBRING, D. G.
DANN 1977 -1; DANN 1977 -2
DANTZIG, G. B.
DANT 1967 -1; DANT 1975 -1; DANTA 1966 -1
DANTB 1975 -1; DANTC 1974 -1
DAR-EL, E. M.
DAR- 1975 -1; DAR-A 1977 -1
DAVENPORT, E. J.
SMITC 1976 -1
DAVIDA, G. T.
SHAH 1974 -1
DAVIES, J.
DAVI 1975 -1
DAVIES, O.
DAVIA 1975 -1; DAVIB 1976 -1
DAVIS, J. M.
DEO A 1977 -1
DAVIS, K. R.
DAVIC 1977 -1; DAVID 1977 -1
DAVIS, R. P.
DAVIE 1976 -1
DAVYDOV, E. G.
DAVY 1972 -1
DAY, J. E.
DAY 1975 -1
DAY, W. H. E.
DAY A 1977 -1
DAYAN, H.
ABAD 1976 -1
DAYKIN, D. E.
BAUMA 1975 -1; BOLLA 1976 -1; BRAC 1975 -1
CHEN 1975 -1; DAYK 1976 -1; DAYK 1976 -2
DEANE, R. H.
PARK 1977 -1
DEAR, R. G.
DEAR 1977 -1
DEARING, P. M.
DEARA 1976 -1; DEARB 1975 -1
DEBANNE, J. G.
DEBA 1976 -1
DEHNERT, G.
BOET 1974 -1; DEHN 1976 -1
DEININGER, R. A.
DEIN 1969 -1; DEINA 1971 -1
DEJON, B.
DEJO 1977 -1; DEJOA 1976 -1; DEJOA 1977 -1
DELATTRE, M.
HANSB 1976 -1; HANSB 1976 -2; HANSB 1976 -3
DELMAS, J.
CAVA 1976 -1
DELORME, J.
HEUR 1975 -1
DEMBO, R.
DEMB 1976 -1
DEMENCHUK, V. M.
DEME 1968 -1
DEMIDENKO, V. M.
DEMI 1976 -1
DEMIN, V. K.
DEMIA 1976 -1
DENIG, W. A.
DOWL 1977 -1
DENT, W.
STEF 1969 -1
DEO, N.
DEO 1969 -1; DEO A 1977 -1; MATE 1976 -1
REIN 1977 -1
DERIGS, U.
DERI 1975 -1; DERI 1977 -1; DERI 1977 -2
DERI 1977 -3; DERI 1978 -1; DERI 1978 -2
DERIA 1978 -1
DERMAN, C.
DERM 1975 -1; DERM 1975 -2; DERM 1976 -1
DEROUAULT, Y.
PUN 1975 -1
DESCAMPS, R.
DESC 1964 -1
DESLER, J. F.
DESL 1970 -1
DETROYE, J.
DETR 1976 -1
DEVINE, M.
MAC D 1977 -1
DEWESS, G.
DEWE 1976 -1
DEY, A.
DEY 1976 -1
DHARAN, B. G.
MORT 1977 -1
DIAL, R.
DIAL 1970 -1; DIAL 1971 -1; DIALA 1977 -1
DICKSON, L. E.
DICK 1971 -1
DIEGO, A.
ALEM 1972 -1; ALEM 1973 -1
DIEHR, G.
DIEH 1972 -1
DIGUGLIELMO, F.
DIGU 1976 -1; DIGU 1976 -2
DIJKSTRA, E. W.
DIJK 1960 -1; DIJK 1976 -1
DIKE, S. H.
DIKE 1964 -1
DINESCU, C.
DINE 1972 -1; DINE 1976 -1
DINITS, E. A.
ADEL 1975 -1; DINI 1970 -1; DINI 1973 -1
DINI 1976 -1; DINIA 1977 -1
DINKEL, J. J.
DINK 1976 -1
DITTMAN, D.
BALAA 1977 -1
DIVIETI, L.
BIONA 1969 -1
DIXON, E. T.
DIXO 1976 -1
DOBKIN, D.
DOBK 1976 -1; DOBKA 1975 -1; DOBKB 1976 -1
DODD, N.
IRWI 1971 -1
DOERFLER, W.
DOER 1978 -1

****

DOERSCH, R. H.
DOERA 1977 -1
DOLODARENKO, V. A.
DOLO 1976 -1
DOMSCHKE, W.
BOET 1974 -1; DOMS 1974 -1; DOMS 1975 -1
DOMS 1976 -1
DONETS, G. A.
DONE 1976 -1
DORAN, J. E.
DORA 1966 -1
DORFMAN, G. YA.
ORLOB 1975 -1
DORFMAN, YA. G.
ORLOC 1975 -1
DORMANT, P.
CHAIA 1976 -1
DOSTATNI, M.
GONDA 1977 -1
DOULLIEZ, P.
BALE 1970 -1; DOUL 1970 -1; DOULA 1969 -1
DOULLIEZ, P. J.
DOULB 1976 -1
DOWLING, T. A.
DOWL 1977 -1
DOYLE, M. S.
DOYL 1971 -1
DRAGAN, I.
DRAG 1976 -1
DREYFUS, S. E.
DREY 1977 -1
DRISCOLL, W. C.
DRIS 1975 -1; DRIS 1976 -1
DROESKE, F.
DROE 1976 -1
DRUFFEL, L. E.
DRUF 1975 -1; SCHMA 1976 -1
DRUTOV, V. E.
DRUT 1975 -1
DUBOV, YU. A.
DUBO 1976 -1
DUCKSTEIN, L.
BOGA 1976 -1
DUDEK, R. A.
CAMP 1970 -1; RAWA 1975 -1; SMITF 1975 -1
DUECK, W.
DUEC 1977 -1
DUERRE, K.
DUER 1976 -1; DUER 1976 -2
DUESTERHOEFT, W. C.
ECON 1977 -1
DULMAGE, A. L.
JOHNB 1962 -1
DUNSTAN, F. D. J.
DUNS 1975 -1; DUNS 1976 -1; DUNS 1977 -1
DUPONT, D.
MAUR 1976 -1
DURAND, J. A. A.
DURA 1973 -1
DWYER, P. S.
DWYE 1975 -1
DYKSTRA, D. P.
DYKS 1977 -1
DYMARSKII, JA. S.
DYMA 1975 -1
DYSON, R. G.
CHAM 1976 -1
DZIKIEWICS, J.
DZIK 1974 -1; DZIKA 1974 -1
EAVES, B. C.
DANTB 1975 -1
ECKER, K.
ECKE 1977 -1
ECONOMIDES, S. C.
ECON 1977 -1
EDELBERG, M.
EDEL 1970 -1
EDEN, B.
EDEN 1977 -1
EDEN, B. N.
EDENA 1976 -1; EDENB 1978 -1
EDMONDS, J.
EDMO 1967 -1; EDMO 1968 -1; EDMO 1975 -1
EDMO 1977 -1; EDMOA 1977 -1; EDMOB 1969 -1
PULLA 1974 -1
EDWARDS, J.
BENNA 1972 -1
EGIAZARJAN, E. V.
EGIA 1976 -1
EGOROVA, N. V.
EGOR 1977 -1
EILON S.
EILO 1975 -1
EILON, S.
EILOA 1975 -1; EILOA 1976 -1; EILOA 1976 -2
EILOA 1977 -1; EILOB 1967 -1
EIN-DOR, P.
BOROD 1976 -1
EINBU, J. M.
EINB 1977 -1
EL-SHAIEB, A. M.
EL-S 1975 -1
ELAM, J.
ELAM 1977 -1
ELDRIDGE, S. E.
BOLLB 1976 -1; ELDR 1976 -1
ELLERVIK, P.
ELLE 1973 -1
ELLWEIN, L. B.
ELLW 1972 -1
ELMAGHRABY, S. E.
ARIS 1977 -1; ARIS 1977 -2
ELSHAFEI, A. N.
BAZAA 1977 -1; ELSH 1975 -1; ELSH 1977 -1
ELSHA 1974 -1
ELZINGA, D. J.
BERL 1976 -1
ELZINGA, J.
ELZI 1976 -1; ELZIA 1976 -1
EMDE BOAS, P. V.
EMDE 1975 -1
EMDE BOAS, P. VAN
LEEU 1978 -1
EMELICHEV, V. A.
EMEL 1976 -1; EMEL 1976 -2; EMELA 1976 -1
LIHAA 1974 -1
EMELICHEVA, V. S.
EMELB 1974 -1; EMELC 1974 -1
EMMONS, H.
CHANG 1975 -1; EMMO 1968 -1; NUNNA 1977 -1
ENJALBERT, M.
ALCO 1975 -1
ENSCORE, E. E.
PEKI 1975 -1
EPIFANOV, G. V.
VARS 1974 -1
EPIHIN, V. V.
EPIH 1975 -1
EPIKHOVA, N. V.
DUBO 1976 -1
EPLETT, W. J. R.
EPLE 1977 -1
EPLEY, D. L.
EPLEA 1970 -1
ERDOES P.
CHVAA 1978 -1
ERDOES, P.
BOLLA 1976 -1; BOLLC 1975 -1; BOLLC 1976 -1
BOLLC 1976 -2; BOLLD 1974 -1; BONN 1974 -1
ERDO 1976 -1; ERDOA 1976 -1; ERDOB 1966 -1
ERDOC 1977 -1; ERDOD 1959 -1; ERDOE 1974 -1
ERDOF 1977 -1
ERLANDER, S.
ERLA 1966 -1
ERLENKOTTER, D.
ERLE 1967 -1; ERLE 1972 -1; ERLE 1972 -2
ERLE 1973 -1; ERLE 1973 -2; ERLE 1974 -1
ERLE 1974 -2; ERLE 1975 -1; ERLE 1975 -2
ERLE 1975 -3; ERLE 1975 -4; ERLE 1975 -5
ERLE 1976 -1; ERLE 1976 -2; ERLEA 1975 -1
ERLEB 1975 -1
ERMOL'EV, YU. M.
ERMO 1976 -1
ERSCHLER, J.
ERSC 1976 -1
ESCHENBACH, T.
ESCH 1975 -1
ESCHENBACH, T. G.
ESCHA 1975 -1
ESCHER, H. K.
ESCHB 1974 -1
ESCUDERO, L. F.
ESCU 1975 -1; ESCUA 1976 -1
ESKEW, J. D.
ESKE 1975 -1
ESOGBUE, A. O.
ESOG 1977 -1
ESSMANN, CH.
BAUE 1973 -1
ESWARAN, K. P.
ESWA 1976 -1
ETCHEBERRY, J.
ETCH 1976 -1; ETCH 1976 -2; ETCH 1977 -1
ETSCHMAIER, M.
STRO 1977 -1
EULER, R.
EULE 1977 -1; EULE 1978 -1; EULE 1978 -2

EVANS, J. P.
EVAN 1974 -1
EVANS, J. R.
EVANA 1976 -1; EVANA 1976 -2; EVANA 1976 -3
EVANA 1976 -4; EVANA 1977 -1; EVANA 1977 -2
EVANB 1975 -1; EVANB 1977 -1; EVANC 1976 -1
EVEN, S.
EVEN 1974 -1; EVENA 1975 -1; EVENB 1975 -1
EVENC 1970 -1; EVEND 1975 -1; EVEND 1975 -2
EVENSON, D. E.
YOUN 1970 -1
EVSTIGNEEV, V. A.
EVST 1969 -1
EYSTER, J. W.
EYST 1973 -1
FABIAN, CS.
FABI 1976 -1; FABIA 1977 -1
FABRY, R. S.
RUSC 1977 -1
FABRYCKY, W. J.
FABR 1966 -1
FALKENHAUSEN, H. VON
FALK [illegible] 1
FAN, L. S.
SAYA 1976 -1; SAYA 1976 -2
FAN, L. T.
FAN 1964 -1; HORI 1973 -1; SAYA 1976 -1
SAYA 1976 -2
FARADZEV, I. A.
ARLA 1973 -1
FARBEY, B. A.
FARB 1965 -1
FARHI, A.
BOYC 1973 -1
FAUDREE, R. J.
ERDOA 1976 -1; FAUD 1974 -1
FAURE, R.
FAUR 1976 -1
FAYARD, D.
FAYA 1976 -1
FEDAN, E. I.
DOLO 1976 -1
FEDERICO, P. J.
FEDE 1975 -1
FEDORENKO, R. P.
FEDO 1975 -1
FEGLEY, K. A.
JEFF 1965 -1
FEICHT, E.
FEIC 1975 -1
FEIGIN, L. I.
FEIG 1967 -1; FEIG 1972 -1
FEINBERG, I.
FEIN 1977 -1
FENSKE, R. W.
FENS 1967 -1
FERLAND, J.
BUSH 1976 -1
FERNANDEZ, E. B.
FERN 1975 -1; FERN 1976 -1; LANG 1977 -1
FERSTER, V. B.
FERS 1976 -1
FEW, L.
FEW 1955 -1
FIDUCCIA, C. M.
FIDU 1977 -1
FILL, J. A.
FILL 1977 -1
FILLER, M. F.
FILLA 1976 -1
FINKE, G.
AHRE 1975 -1; FINK 1975 -1
FINKEL'SHTEIN, YU. YU.
FINKA 1976 -1; FINKA 1977 -1; SCHUA 1977 -1
FINLAY, R. A.
BERRA 1976 -1
FIORINI, S.
FIOR 1973 -1; FIOR 1974 -1; FIOR 1976 -1
FISBEIN, S. M.
FISB 1971 -1
FISCHER, M. J.
FISC 1971 -1
FISHBURN, P. C.
FISH 1976 -1
FISHER, M.
CORNC 1977 -1
FISHER, M. L.
CORND 1976 -1; CORND 1977 -1; FISHA 1976 -1
FISHB 1977 -1; FISHC 1973 -1; NEMHA 1976 -1
FISK, J.
FISK 1976 -1; FISK 1976 -2; FISK 1977 -1
FISKA 1977 -1; FISKB 1977 -1
FISK, J. C.
FISKC 1977 -1; HUNGA 1976 -1; HUNGA 1977 -1
FISK, S.
FISKD 1977 -1
FITZGERALD, C. H.
FITZ 1974 -1
FLEISCHNER, H.
FLEI 1976 -1; FLEI 1976 -2
FLEISHER, J. M.
MEYED 1977 -1
FLEMING, J. G.
MICHA 1968 -1
FLORIAN, M.
BRAT 1975 -1; FLOR 1976 -1; FLORA 1970 -1
FOLDES, S.
FOLD 1975 -1; FOLD 1976 -1; FOLD 1976 -2
FOLD 1977 -1; FOLD 1978 -1
FOLIE, G. M.
TIFF 1974 -1
FONG, A. C.
FONG 1977 -1
FONG, C. O.
FONGA 1974 -1; FONGB 1975 -1; FONGC 1974 -1
FONGC 1974 [illegible]; [illegible] 1977 -1
FONTET, M.
FONT 1976 -1
FOOT, D. H. S.
FOOT 1965 -1
FORCINA, J. L.
FORC 1974 -1
FORGO, F.
FORG 1974 -1; FORG 1974 -2; FORG 1976 -1
FORG 1976 -2
FOSTER, B. A.
FOST 1976 -1; RYAN 1976 -1
FOULDS, L. R.
FOUL 1976 -1; FOULA 1976 -1
FOURNIER, J. C.
FOUR 1971 -1; FOUR 1973 -1; FOUR 1975 -1
FOUR 1976 -1
FOX, B.
FOX 1975 -1
FOX, B. L.
FOX A 1975 -1; FOX A 1976 -1; FOX B 1977 -1
FOX, T. F.
RAMAD 1976 -1
FRANCIS, R. L.
CHAN 1976 -1; CHAN 1976 -2; DEARA 1976 -1
FRAN 1963 -1; FRAN 1974 -1; FRANA 1976 -1
FRANB 1974 -1; PAPI 1975 -1
FRANCON, J.
FRANC 1976 -1
FRANK, A.
FRAND 1976 -1; FRAND 1977 -1
FRANK, H.
FRANE 1969 -1; FRANF 1971 -1; FRANG 1970 -1
FRANK, O.
FRANH 1977 -1
FRATTA, L.
CAME 1973 -1; CAME 1974 -1; FRAT 1975 -1
FRATA 1972 -1; FRATB 1975 -1; MARU 1976 -1
FRAZER, W. D.
FRAZ 1972 -1
FREDERICKSON, G. N.
FRED 1976 -1
FREDMAN, M. L.
FREDA 1975 -1; FREDA 1975 -2; FREDA 1976 -1
FREELAND, J. R.
FREE 1977 -1
FREHEL, J.
FREH 1975 -1; FREH 1975 -2
FREIMER, M.
FREI 1973 -1
FRENNC, A. DE
DROE 1976 -1
FRIDMAN, A. A.
FRID 1977 -1; FRIDA 1973 -1; FRIDB 1976 -1
FRIEDMAN, M.
FRIE 1975 -1; FRIE 1976 -1
FRIEL, J.
DANTC 1974 -1
FRIEZE, A. M.
FRIEA 1975 -1; FRIEA 1976 -1; FRIEA 1977 -1
FRIEA 1977 -2; FRIEA 1977 -3
FRISCH, I.
FRANG 1970 -1; PARI 1963 -1
FRISCH, I. T.
FRIS 1963 -1; FRIS 1963 -2; ROTHA 1969 -1
FROME, E. L.
ARMSD 1976 -1
FROMOVITZ, S.
FROM 1976 -1
FRONC, M. B.
FRON 1975 -1
FRUMKIN, M. A.
FRUM 1976 -1; FRUM 1976 -2

FRYER, J. S.
WEEK 1976 -1
FUJII, A.
FUJI 1966 -1
FUJISAWA, T.
FUJIA 1963 -1; FUJIB 1974 -1; OHTSA 1976 -1
FUJISHIGE, S.
FUJIC 1977 -1; FUJIC 1977 -2
FUKUDA, Y.
FUKU 1976 -1
FULKERSON, D. R.
FULK 1975 -1; FULK 1975 -2; FULK 1975 -3
FULKA 1976 -1
FULLER, R. V.
FULL 1976 -1
FUNK, M.
ROBEC 1964 -1
FUNK, M. L.
FUNK 1968 -1
FYFFE, D. E.
MILLA 1976 -1
GABASOV, R.
GABA 1975 -1
GABBAY, H.
GABB 1974 -1; GARTA 1975 -1
GABOVICH, E. JA.
GABO 1976 -1
GABOVICH, E. YA.
VOLO 1976 -1
GABOW, H.
GABOA 1976 -1
GABOW, H. N.
GABOB 1976 -1; GABOB 1976 -2; GABOB 1977 -1
GAGNON, G.
GAGN 1967 -1
GAIHA, P.
GAIH 1977 -1
GAIKOV, N. E.
GAIK 1976 -1; GAIK 1976 -2
GALIL, Z.
GALI 1975 -1; GALI 1976 -1
GALKIN, A. V.
PLOT 1974 -1
GALLAGER, R. G.
GALL 1977 -1
GALLAGHER, J. M.
GALLA 1975 -1
GALLO, G.
GALLB 1976 -1; GALLC 1972 -1
GALLUS, G.
GALLD 1976 -1; GALLD 1976 -2
GARBAI, L.
GARB 1976 -1
GAREY, M. R.
AHO 1977 -1; BRUCA 1977 -1; CHVAB 1978 -1
GARE 1976 -1; GARE 1977 -1; GAREA 1976 -1
GAREB 1974 -1; GAREB 1976 -1; GAREB 1976 -2
GAREB 1976 -3; GAREB 1976 -4; GAREB 1976 -5
GAREB 1976 -6; GAREB 1977 -1; GAREB 1977 -2
GAREC 1976 -1; GARED 1976 -1
GARFINKEL, R. S.
CHUR 1976 -1; GARF 1977 -1; GARFA 1976 -1
GARFB 1973 -1; GARFB 1977 -1
GARSIA, A. M.
GARS 1977 -1
GARTNER, N.
GART 1976 -1
GARTNER, N. H.
GARTA 1975 -1
GASS, S. I.
FROM 1976 -1
GATELY, D. T.
GATE 1971 -1
GATTINGER, M.
DEJOA 1976 -1; DEJOA 1977 -1; GATT 1976 -1
GAUDIANO, M.
AUSIA 1977 -1
GAUL, W.
GAUL 1974 -1; GAUL 1976 -1; GAUL 1976 -2
GAUL 1976 -3; GAUL 1977 -1; GAUL 1977 -2
GAUL 1977 -3
GAUSSENS, J.
THIR 1966 -1
GAUTHIER, J. M.
BENI 1977 -1
GAUTHIER, J.-M.
GAUT 1977 -1
GAVISH, B.
GAVI 1977 -1
GAVRIL, F.
GAVR 1977 -1
GAZIS, D. C.
GAZI 1975 -1
GELDERS, L.
CHALA 1975 -1; GELD 1976 -1
GELDERS, L. F.
CHALB 1976 -1; GELDA 1976 -1
GELDMACHER, R. C.
GELDB 1975 -1
GELEVER, K. G.
GOLE 1974 -1
GELLER, D.
GELL 1974 -1
GELLER, D. P.
GELLA 1976 -1; GELLB 1974 -1
GEN, M.
GEN 1974 -1; GEN A 1976 -1; GEN B 1975 -1
GENSER, R.
STRO 1977 -1
GEOFFRION, A. M.
GEOF 1974 -1; GEOF 1975 -1; GEOF 1976 -1
GEOF 1977 -1; GEOFA 1973 -1; GEOFA 1976 -1
GEOFB 1977 -1; GEOFC 1977 -1; GEOFD 1977 -1
GEOLECJAN, G. G.
GEOL 1975 -1
GEORGE, J. A.
GEOR 1971 -1
GEORGOBIANI, D. A.
GEORA 1974 -1
GERHARDT, C.
TILA 1976 -1
GERLA, M.
CANT 1974 -1
GERMEIER, JU. B.
GERM 1971 -1
GEROV, G. H.
GERO 1970 -1
GERRITSEN, R.
BALAH 1974 -1; GERR 1976 -1
GERRITSEN, S. M.
LAZA 1977 -1
GERTSBACH, I.
GERT 1977 -1; GERTA 1977 -1
GHARE, P. M.
PRIT 1970 -1
GHIGGI, C.
GHIG 1976 -1
GHILARDOTTI, P.
GHIL 1973 -1
GHIUS, V.
GHIU 1975 -1
GHOSH, A.
GHOS 1970 -1
GIANNESSI, F.
GIAN 1976 -1; GIANA 1976 -1
GIBBONS, P.
GIBB 1976 -1
GIBERT, A.
GIBE 1968 -1
GIBSON, P. M.
BRUAA 1976 -1; BRUAA 1977 -1; BRUAA 1977 -2
BRUAA 1977 -3; GIBS 1976 -1
GIERTH, G.
TERN 1975 -1
GIGLIO, R. J.
NOON 1977 -1; SMITG 1975 -1
GIL, N. I.
STOJ 1976 -1
GILBERT, E. N.
GILB 1968 -1
GILBERT, K. C.
GARFA 1976 -1
GILES, F. R.
GILE 1977 -1; GILE 1977 -2; GILE 1977 -3
GILEA 1977 -1
GILES, R.
EDMOA 1977 -1; GILEB 1976 -1
GILL, A.
GILL 1960 -1; GILLA 1969 -1
GILL, J.
BAKEC 1975 -1; GILLB 1977 -1
GILLES, F. R.
GILLC 1975 -1
GILLESPIE, C. M. JR.
GILLD 1968 -1
GILLETT, B. E.
GILLE 1975 -1; GILLF 1976 -1
GILMORE, P. C.
GILM 1977 -1
GILSINN, J.
GILS 1970 -1
GIMADI, E. H.
GIMA 1976 -1
GIMADI, E. K.
GIMAA 1976 -1; GIMAB 1974 -1
GIMADI, E. X.
PEVE 1969 -1
GIMPEL, J. F.
BOESC 1977 -1

GINOVKER, V. D.
GINO 1966 -1
GINZBURG, M. YA.
GINZ 1977 -1
GIRAUD, G.
GIRA 1976 -1
GITTINS, J. C.
NASHA 1977 -1
GIULIANELLI, S.
GIUL 1976 -1
GLASSEY, C. R.
GLAS 1978 -1
GLAZEBROOK, K. D.
GLAZ 1976 -1; GLAZA 1976 -1
GLEASON, J.
POWE 1977 -1
GLEASON, J. M.
BURTB 1976 -1; GLEA 1975 -1
GLEBOV, N. I.
GIMA 1976 -1; GIMAA 1976 -1; GLEB 1973 -1
GLEBA 1970 -1
GLEBSKII, Y.
GLEBB 1969 -1
GLEN, J. J.
GLEN 1975 -1
GLICKMAN, T. S.
GLIC 1977 -1
GLOVER, F.
BARR 1977 -1; BARR 1977 -2; BARR 1977 -3
BARR 1977 -4; CHAR 1975 -1; DIALA 1977 -1
ELAM 1977 -1; GLOV 1971 -1; GLOV 1973 -1
GLOV 1973 -2; GLOV 1975 -1; GLOV 1975 -2
GLOV 1975 -3; GLOV 1976 -1; GLOV 1977 -1
GLOV 1977 -2; GLOVA 1977 -1; GLOVB 1977 -1
GLOVC 1974 -1; GLOVC 1975 -1; GLOVC 1975 -2
GLOVC 1975 -3; GLOVC 1975 -4; GLOVC 1975 -5
GLOVC 1976 -1; GLOVC 1976 -2; GLOVD 1977 -1
GLOVE 1976 -1; GLOVF 1973 -1; GLOVG 1975 -1
GLOVG 1975 -2; GLOVH 1975 -1; GLOVI 1976 -1
GLOVJ 1976 -1
GLOVER, F., ET AL.
GLOVK 1972 -1
GLOVER, P. C.
GLOVL 1965 -1
GODEFROID, P.
GODE 1976 -1
GOFMAN, E. F.
GOLE 1974 -1
GOL'DBERG, M. K.
GOL' 1974 -1; GOL'A 1975 -1
GOL'DGABER, E. M.
GOL'B 1977 -1
GOLDEN, B.
GART 1976 -1; GOLD 1975 -1; GOLD 1976 -1
GOLD 1976 -2; GOLDA 1977 -1; KERS 1976 -1
GOLDEN, B. L.
GOLDB 1975 -1; GOLDB 1976 -1; GOLDB 1976 -2
GOLDB 1977 -1; GOLDB 1977 -2; GOLDC 1977 -1
GOLDD 1977 -1
GOLDFARB, D.
GOLDE 1977 -1
GOLDMAN, A. J.
GOLDF 1967 -1; WINGA 1965 -1
GOLDNER, A.
ROSED 1977 -1
GOLDNER, A. M.
GOLDG 1975 -1
GOLDSTEIN, M. C.
GOLDH 1971 -1
GOLENKO, D. I.
GOLE 1974 -1
GOLIGHTLY, R.
DANIC 1974 -1
GOLLADAY, F. L.
SMITC 1976 -1
GOLLER, S.
GOLL 1969 -1
GOLUMBIC, M. C.
GOLU 1977 -1; GOLU 1977 -2
GOMORY, R. E.
GOMO 1975 -1
GONDRAN, M.
GOND 1971 -1; GOND 1974 -1; GOND 1975 -1
GOND 1975 -2; GOND 1976 -1; GOND 1976 -2
GOND 1977 -1; GONDA 1977 -1; GONDB 1977 -1
GONZALES, M. J.
RAMAC 1971 -1
GONZALES, R. H.
GONZ 1962 -1
GONZALES, T.
SAHNA 1974 -1
GONZALEZ-VALENZUELA, F.
GONZA 1975 -1
GONZALEZ, M. J. JR.
GONZB 1977 -1; GONZC 1976 -1
GONZALEZ, T.
GONZD 1977 -1; GONZE 1976 -1; GONZE 1978 -1
SAHNB 1976 -1
GOODE, J. J.
BAZAB 1977 -1
GOODMAN, R. G.
GOOD 1965 -1
GOODMAN, S.
GOODA 1973 -1; GOODA 1974 -1; GOODB 1974 -1
GOODC 1976 -1; MITC 1975 -1; MORRB 1976 -1
GOODMAN, S. E.
DIXO 1976 -1; GOODD 1975 -1; SLATB 1976 -1
GORDON, A. JA.
GORD 1976 -1
GORDON, B. E.
GORDA 1971 -1
GORDON, J. H.
GORDB 1974 -1
GORSHKOW, A. F.
GORS 1970 -1
GOTLIEB, C. C.
CORNA 1970 -1
GOYAL, E. K.
GOYA 1975 -1; GOYA 1975 -2; GOYA 1976 -1
GOYA 1977 -1
GRABOWSKI, J.
GRAB 1976 -1; GRAB 1976 -2; GRAB 1976 -3
GRABA 1975 -1
GRABOWSKI, W.
GRABB 1972 -1; GRABB 1975 -1; GRABB 1976 -1
GRAHAM, R. L.
CHUN 1978 -1; GARE 1976 -1; GARE 1977 -1
GAREA 1976 -1; GRAH 1967 -1; GRAH 1976 -1
GRAHA 1977 -1; GRAHB 1974 -1
GRANDA, T. M.
GRAN 1974 -1
GRANOT, D.
CHARA 1975 -1; CHARA 1977 -1; GRANA 1976 -1
GRANA 1977 -1
GRANOT, F.
CHARA 1975 -1; CHARA 1977 -1; GRANA 1976 -1
GRANA 1977 -1; GRANB 1970 -1
GRANSBORG, B. S. M.
NOURA 1970 -1
GRANT, D. D.
GRANC 1976 -1
GRAPA, E.
GRAP 1977 -1
GRATZER, F.
GRAT 1970 -1
GRAVER, J. E.
GRAV 1975 -1; GRAV 1975 -2; GRAV 1975 -3
GRAVA 1977 -1
GRAVES, G.
BRAD 1976 -1
GRAVES, G. W.
GEOFA 1973 -1; GEOFA 1976 -1; GEOFB 1977 -1
GRAVES, R. J.
GRAVB 1976 -1; PARKA 1977 -1
GRAVES, S. C.
GRAVC 1977 -1
GRAY, P.
GRAY 1975 -1
GREEN, C. D.
GREE 1976 -1
GREENE, C.
GREEA 1977 -1; GREEB 1974 -1; GREEC 1975 -1
GRIER, B.
GRIE 1976 -1
GRIESSHABER, R.
GRIEA 1976 -1
GRIGOR'EV, E. G.
GRIG 1974 -1
GRIGOR'EVA, L. I.
GRIGA 1974 -1
GRIGOROVIC, A. F.
ALES 1975 -1
GRIMMETT, G. R.
GRIM 1976 -1; GRIMA 1975 -1
GRISHUKHIN, V. P.
GRIS 1972 -1; GRIS 1973 -1; GRIS 1976 -1
GRIS 1976 -2; GRIS 1977 -1
GROEFLIN, H.
BENIA 1976 -1
GROETSCHEL, M.
GROE 1976 -1; GROE 1977 -1; GROE 1977 -2
GROE 1977 -3; GROE 1977 -4; GROE 1977 -5
GROE 1977 -6; GROE 1977 -7; GROEA 1977 -1
GROEA 1977 -2; GROEA 1977 -3; GROEA 1978 -1
GROMOVA, A. P.
BELK 1965 -1
GROPPEN, V. O.
BURKH 1975 -1
GROS, J.
GROS 1975 -1

GROSS, O.
GROSA 1960 -1
GROSSE, H. W.
NAST 1975 -1
GRUNBAUM, B.
GRUN 1974 -1
GRUNDLER, G.
GRUNA 1975 -1; GRUNB 1977 -1
GRUPE, H. J. C.
GRUP 1974 -1
GUBA, W.
GUBA 1975 -1
GUERIN, G.
BUSH 1976 -1
GUGLIELMO, F. D.
GUGL 1976 -1
GUHA, D. K.
GUHA 1977 -1
GUIGNARD, M.
GUIG 1974 -1; GUIGA 1974 -1; GUIGA 1976 -1
GUIGA 1976 -2; GUIGA 1977 -1; GUIGA 1977 -2
GUIGA 1977 -3; GUIGA 1977 -4
GULDMANN, J.-M.
GULD 1977 -1
GUNDRY, S. W.
BONNA 1976 -1
GUNTHER, T. M.
GUNT 1977 -1
GUPTA, J. N. D.
GUPT 1969 -1; GUPT 1975 -1; GUPT 1976 -1
GUPT 1976 -2
GUPTA, R. P.
GUPTA 1969 -1
GUPTA, S. K.
GAIH 1977 -1; STANB 1969 -1
GUREVICH, YU.
GERTA 1977 -1
GUROVICH, V.
AVRI 1977 -1
GUTENBAUM, J.
GUTE 1976 -1
GUTMAN, I.
BAUMA 1975 -1
GUZIK, B.
GUZI 1975 -1
GVOZDEV, S. E.
GVOZ 1974 -1; GVOZ 1976 -1
GYULUMYAN, S. M.
GYUL 1975 -1
HAACK, P. J.
MARKA 1975 -1
HABENICHT, W.
HABE 1976 -1; HABE 1977 -1
HABIB, M.
HABI 1975 -1
HACATUROV, V. R.
HACA 1974 -1; HACAA 1976 -1
HADJERES, S.
HADJ 1974 -1
HADLOCK, F. O.
HADL 1974 -1; HADL 1977 -1
HAENSLER, E.
HAEN 1973 -1
HAESSLER, R. W.
HAES 1976 -1
HAFF, C. E.
BURNC 1976 -1; HAFF 1973 -1
HAGGETT, P.
HOLMA 1977 -1
HAHN, W.
BURKA 1976 -1; BURKA 1977 -1
HAIKIN, N. M.
SEME 1974 -1
HAJNAL, A.
ERDOB 1966 -1
HAKEN, W.
APPE 1976 -1
HAKIMI, S.
HAKI 1969 -1
HAKIMI, S. L.
AMIN 1973 -1; DESL 1970 -1; HAKIA 1972 -1
PATR 1974 -1
HALEY, K. B.
ELSHA 1974 -1
HALIN, R.
HALI 1975 -1
HALPERN, J.
HALP 1976 -1; HALP 1977 -1; HALP 1977 -2
HALP 1977 -3
HALTON, J. H.
BEAR 1959 -1
HAMACHER, H.
BURKB 1976 -1; BURKB 1977 -1; BURKB 1977 -2
HAMMER, G.
HAMM 1969 -1; HAMM 1969 -2; HAMM 1976 -1
HAMM 1977 -1
HAMMER, P.
GRANB 1970 -1
HAMMER, P. L.
CHVAC 1977 -1; FOLD 1975 -1; FOLD 1976 -1
FOLD 1976 -2; FOLD 1977 -1; FOLD 1978 -1
GALLB 1976 -1; HAMMA 1975 -1; HAMMA 1976 -1
HAMMA 1977 -1; HAMMB 1977 -1
HAMMERSLEY, J. M.
BEAR 1959 -1
HANDLER, G. Y.
HAND 1977 -1
HANSEN, K. H.
HANS 1971 -1; HANS 1971 -2
HANSEN, M. F.
SMITC 1976 -1
HANSEN, P.
HANSA 1975 -1; HANSA 1975 -2; HANSA 1975 -3
HANSA 1975 -4; HANSA 1976 -1; HANSA 1977 -1
HANSA 1977 -2; HANSB 1976 -1; HANSB 1976 -2
HANSB 1976 -3; HANSC 1976 -1; HANSD 1972 -1
HANSE 1977 -1
HANSON, D.
CHVAD 1976 -1
HANSON, H. H.
HANSF 1977 -1
HARARY, F.
BOESA 1976 -1; GOLDG 1975 -1; HARA 1975 -1
HARAA 1974 -1; HARAB 1974 -1; HARAB 1975 -1
KRARB 1973 -1; KRARC 1972 -1
HARDING, G. C.
FULKA 1976 -1
HARDLOCK, F. O.
HARD 1977 -1
HARDOUIN, D. J.
HARDA 1975 -1
HARITON, JA. R.
HARI 1976 -1
HARKEVIC, A. D.
EPIH 1975 -1
HARMATUCK, D. J.
HARM 1976 -1
HARNETT, D. L.
CABO 1977 -1
HARRINGTON, J. J.
KUEH 1977 -1; KUEH 1977 -2
HARRIS, B.
HARR 1976 -1
HART, P. E.
HART 1968 -1
HARTLEY, R. V.
HARTA 1976 -1
HARTMANIS, J.
HARTB 1976 -1
HARTMANN, K.
HARTC 1973 -1; HARTC 1975 -1
HARTNELL, B. L.
HARTD 1976 -1
HARVEY, R. P.
DANTC 1974 -1
HASEGAWA, T.
SHIR 1976 -1
HASLINGER, A.
HASL 1975 -1
HASTINGS, N. A. J.
BEST 1976 -1; WILLE 1976 -1
HATCH, R. S.
HATC 1975 -1
HATTORI, Y.
OZAW 1973 -1; SAKAA 1976 -1
HAUSMANN, D.
HAUS 1977 -1; HAUS 1977 -2; HAUS 1977 -3
HAUS 1978 -1; HAUSA 1977 -1; HAUSB 1977 -1
HAUSC 1976 -1; HAUSC 1977 -1; HAUSC 1977 -2
HAUSC 1978 -1; HAUSC 1978 -2; KORTC 1977 -1
KORTC 1978 -1
HAX, A. C.
BRADD 1977 -1
HAY, L.
HAY 1978 -1
HEARD, E. L.
HEAR 1976 -1
HEARN, D.
ELZI 1976 -1; HEARA 1976 -1
HECHLER, S. H.
HECH 1977 -1
HECHT, M. S.
FRED 1976 -1
HECK, J.
FEIC 1975 -1; HECK 1976 -1

HEDETNIEMI, S.
COCKA 1976 -1; GOODA 1973 -1; GOODA 1974 -1
GOODB 1974 -1; GOODC 1976 -1; HEDE 1971 -1
HEDEA 1963 -1; LEVIA 1963 -1; MITC 1975 -1
HEDETNIEMI, S. T.
COCKB 1975 -1; COCKB 1976 -1; COCKB 1977 -1
GOODD 1975 -1; SLATB 1976 -1
HEIN, O.
HEIN 1976 -1; HEIN 1977 -1; HEIN 1977 -2
HEINRICH, K.
HEINA 1976 -1
HEISING, C. D.
HEIS 1977 -1
HEITZMAN, W. R.
GRAY 1975 -1
HELD, M.
HELD 1965 -1
HELGASON, R. V.
ALI 1977 -1; HELG 1977 -1; HELG 1977 -2
HELLER, I.
HELL 1955 -1; HELL 1956 -1
HELLER, J.
HELLA 1960 -1
HENDERSON, P. B.
HEND 1976 -1
HENDERSON, W. B.
HENDA 1976 -1
HENDRICK, T. E.
PLAN 1975 -1
HENRY, B. B.
HENR 1967 -1
HENTGES, G.
BENI 1977 -1
HEPPES, A.
HEPP 1976 -1
HERING, F.
HERI 1969 -1
HERON, A. P.
HERO 1973 -1
HERSHDORFER, A. M.
HERS 1965 -1
HERZOG, U.
HERZ 1975 -1
HESSELBACH, J.
HESS 1977 -1
HEURGON, E.
HEUR 1975 -1
HEURGON, E. M. J.
BOUR 1969 -1
HEVERAN, R. D.
HEVE 1964 -1
HIL'CENKO, V. I.
KASPA 1974 -1
HILL, A. V.
HILL 1976 -1
HILL, E. JR.
HILLA 1978 -1
HILTON, A. J. W.
GELLB 1974 -1; HILT 1975 -1; HILT 1975 -2
HILT 1976 -1; HILT 1977 -1; HILTA 1973 -1
HINSON, J.
HINS 1976 -1
HINTEREGGER, J.
HINT 1977 -1
HIRAO, S.
OZAW 1973 -1
HIRCHE, J.
HIRC 1970 -1; HIRCA 1975 -1
HIRSCHBERG, D. S.
AHO A 1974 -1; CHAND 1976 -1; HIRS 1977 -1
HIRSA 1976 -1; HIRSA 1976 -2
HITCHINGS, G. G.
HITC 1976 -1
HITCHNER, L.
HITCA 1968 -1
HODGSON, M.
EILOB 1967 -1
HODGSON, M. J.
HODG 1974 -1
HODGSON, T. J.
HODGA 1977 -1
HOELZEMAN, R. G.
MICKB 1977 -1; WANGB 1974 -1
HOEY, D.
SHAMA 1975 -1
HOFFMAN, A. J.
BERGA 1976 -1; HOFF 1976 -1; HOFF 1976 -2
HOFF 1977 -1; HOFF 1977 -2; HOFFA 1978 -1
HOFFB 1976 -1
HOFFMANN, W.
HOFFC 1958 -1
HOFRI, M.
BRUNC 1975 -1
HOGUE, S. L.
HAES 1976 -1
HOLLOWAY, C. A.
HOLL 1977 -1; NELS 1977 -1
HOLM, S.
HOLM 1976 -1; HOLM 1978 -1
HOLMBERG, G.
BJOE 1977 -1
HOLMES, J. H.
HOLMA 1977 -1
HOLMES, R. A.
HOLMB 1976 -1; PARK 1977 -1
HOLTON, D. A.
LITTA 1976 -1
HOLZMAN, A. G.
HOLZ 1975 -1
HOLZMANN, C. A.
HOLZA 1973 -1
HONG, M. L.
HONG 1974 -1
HONG, S.
HONGA 1977 -1; HONGB 1977 -1; PADBB 1977 -1
HONKANEN, P. A.
HSU 1973 -1
HOPCROFT, J. E.
AHO B 1974 -1; HOPC 1971 -1; HOPCA 1971 -1
HOPCB 1971 -1; HOPCB 1972 -1; HOPCB 1973 -1
HOPCC 1974 -1
HOR, C. G.
HOR 1976 -1
HOREN, J. H.
HORE 1977 -1; YUAN 1975 -1
HORIE, Y.
HORI 1973 -1
HORN, R. L. VAN
HORN 1972 -1
HOROWITZ, E.
HORO 1976 -1
HORST, R.
HORS 1975 -1; HORS 1976 -1
HORVATH, E. C.
HORV 1977 -1
HOSHI, M.
MIYA 1977 -1
HOTTENSTEIN, M. P.
DAY 1975 -1
HOUCK, D. J.
HOUC 1976 -1; HOUC 1977 -1
HOUGLAND, E. S.
TURNA 1975 -1
HOWARD, D. D.
HOWA 1970 -1
HOWARD, R. A.
KIMB 1959 -1
HROUDA, J.
HROU 1975 -1; HROU 1976 -1; HROU 1977 -1
HSIA, W. S.
BABU 1977 -1
HSIEH, W.
KERS 1976 -1
HSU, H. T.
HSU 1973 -1
HU, T. C.
BLEW 1977 -1; GOMO 1975 -1; HU 1965 -1
HU 1966 -1; HU 1973 -1; HU A 1976 -1
RUSK 1977 -1; TIEN 1977 -1
HUANG, H.-C.
HUAN 1977 -1
HUBAUT, X. L.
HUBA 1975 -1
HUBICKA, E.
HUBI 1975 -1
HULME, B. L.
HULM 1975 -1
HULST, W. G. VAN
HULS 1977 -1
HULTMAN, E.
WEST 1977 -1
HULTZ, J.
GLOVA 1977 -1; GLOVB 1977 -1; HULT 1976 -1
HULTA 1976 -1
HUME, J. N. P.
BUNT 1977 -1; HUME 1960 -1
HUNG, H. K.
HUNG 1973 -1
HUNG, M. S.
HUNGA 1976 -1; HUNGA 1977 -1
HURTER, A. P. JR.
HURT 1975 -1; WEND 1977 -1
HUTCHINSON, G. K.
SZWAA 1977 -1
HUTCHINSON, J. P.
ALBED 1975 -1
HWANG, F.
LEE A 1976 -1
HWANG, F. K.
AHO 1977 -1; HWAN 1976 -1; HWANA 1969 -1

HYAFIL, L.
HYAF 1976 -1
HYUN, K. N.
HYUN 1975 -1; HYUNA 1974 -1
IAZEOLLA, G.
IAZE 1968 -1
IBARAKI, T.
IBAR 1970 -1; IBAR 1972 -1; IBAR 1972 -2
IBAR 1973 -1; IBAR 1976 -1; IBAR 1976 -2
IBAR 1976 -3; IBAR 1977 -1; IBAR 1977 -2
IBARA 1976 -1; ISHI 1975 -1; ISHI 1975 -2
ISHI 1976 -1; ISHI 1977 -1; SHIR 1976 -1
IBARRA, O.
IBARB 1974 -1
IBARRA, O. H.
GONZD 1977 -1; IBARC 1974 -1; IBARC 1975 -1
IBARC 1977 -1
IBM.
IBM. 1972 -1
IDE, M.
TSUK 1977 -1
IGO, W.
RUSSB 1977 -1
IGUDIN, R. V.
IGUD 1975 -1
IMOMNAZAROV, B.
IMOM 1976 -1
INAUDI, D.
BONA 1976 -1
INGARGIOLA, G.
INGA 1975 -1
INGARGIOLA, G. P.
INGAA 1977 -1
INGLETON, A. W.
INGL 1976 -1; INGL 1977 -1
INTRATOR, J.
INTR 1976 -1; INTRA 1977 -1
IORDANSKII, M. A.
IORD 1974 -1; IORD 1976 -1
IRI, M.
AOSH 1977 -1; IRI 1960 -1; IRI 1971 -1
IRI A 1975 -1; IRI A 1976 -1
IRWIN, N. A.
IRWI 1971 -1
ISAACSON, J. D.
MATUA 1972 -1
ISHII, H.
IBARA 1976 -1; ISHI 1975 -1; ISHI 1975 -2
ISHI 1976 -1; ISHI 1977 -1
ISKANDER, W.
PANW 1977 -1
ITAI, A.
EVENA 1975 -1; ITAI 1977 -1; ITAI 1977 -2
IVANCHEV, M. D.
IVAN 1976 -1; IVANA 1976 -1
IVANISCEV, V. V.
IVANB 1974 -1
IVANOV, N. N.
IVANC 1974 -1; IVANC 1975 -1
JACAK, W.
ADAMA 1976 -1
JACOB, H.
JACO 1973 -1
JACOBSEN, S. K.
JACOA 1974 -1; JACOA 1978 -1
JACOBSON, R. E.
JACOB 1977 -1
JAENICKE, W.
CLAU 1977 -1
JAERVINEN, P.
JAER 1973 -1
JAIN, R. C.
JAIN 1975 -1
JAIN, S.
BALAB 1977 -1
JAIN, S. K.
JAINA 1975 -1
JAKOVLEVA, M. A.
JAKO 1974 -1
JAKSCH, H. J.
JAKS 1974 -1; JAKS 1975 -1
JAMES, K. R.
JAME 1976 -1; RIHA 1976 -1
JAMESON, S. R.
MOLE 1976 -1
JAMOULLE, E.
BALE 1970 -1
JANDY, G.
JAND 1969 -1
JANEV, N. I.
IVANA 1976 -1
JANICKE, W.
BIESA 1977 -1; JANI 1976 -1
JARANCEV, JU. A.
JARA 1976 -1
JARVIS, J. J.
BAZAC 1977 -1; EVANB 1975 -1; EVANB 1977 -1
FULL 1976 -1
JASAKOVA, V. P.
EVST 1969 -1
JASINSKA, E.
JASI 1976 -1
JEFFERIS, R. P.
JEFF 1965 -1
JENKYNS, T. A.
HAUSA 1977 -1; JENK 1974 -1; JENK 1976 -1
JENK 1977 -1; JENK 1977 -2
JENSCH, P.
JENS 1975 -1
JENSEN, P.
BELLD 1970 -1
JENSON, P.
BHAU 1974 -1
JEROSLOW, R.
BALAI 1969 -1; JERO 1977 -1; JERO 1977 -2
JEROSLOW, R. G.
BLAIB 1976 -1; BLAIB 1977 -1; JEROA 1977 -1
JEROA 1977 -2; JEROA 1977 -3
JEWELL, W.
JEWE 1967 -1
JIRKA, G. H.
MARKF 1971 -1
JOHNSON F. L.
EDMOB 1969 -1
JOHNSON, C. R.
JOHN 1976 -1
JOHNSON, D. B.
BRUNA 1976 -1; JOHNA 1973 -1; JOHNA 1975 -1
JOHNA 1977 -1
JOHNSON, D. M.
JOHNB 1962 -1
JOHNSON, D. S.
BRUCA 1977 -1; CHVAB 1978 -1; GARE 1976 -1
GARE 1977 -1; GAREA 1976 -1; GAREB 1974 -1
GAREB 1976 -1; GAREB 1976 -2; GAREB 1976 -3
GAREB 1976 -4; GAREB 1976 -5; GAREB 1976 -6
GAREB 1977 -1; GAREB 1977 -2; GAREC 1976 -1
GARED 1976 -1; JOHNC 1974 -1; JOHNC 1978 -1
JOHND 1977 -1
JOHNSON, E. L.
BERGB 1977 -1; BURD 1977 -1; CHENE 1976 -1
HAMMB 1977 -1; JOHNE 1964 -1; JOHNE 1975 -1
JOHNE 1975 -2; JOHNE 1976 -1; JOHNE 1976 -2
JOHNE 1976 -3; JOHNE 1977 -1; JOHNE 1978 -1
JOHNF 1978 -1; JOHNG 1976 -1; MAIR 1977 -1
JOHNSON, J.
MOSHA 1963 -1
JOHNSON, J. G.
GILLF 1976 -1
JOHNSON, L. F.
JOHNH 1976 -1
JOHNSON, P. A.
JOHNI 1974 -1
JOHNSON, R. E.
JOHNJ 1958 -1
JOHNSON, S. C.
JOHNK 1970 -1
JOLIVET, J.-L.
BORIA 1976 -1; JOLI 1975 -1
JONES, C. H.
HENR 1967 -1
JONES, N. D.
JONE 1973 -1
JUCKER, J. V.
JUCK 1976 -1
JUCOVIC, E.
JUCO 1974 -1
JUDIN, D. B.
JUDI 1976 -1
JUENICKE, W.
JUEN 1976 -1
JUKHIMENKO, B. I.
BRIT 1976 -1
JULLIEN, B.
JULL 1976 -1
JUNGINGER, W.
JUNG 1976 -1
JUSKA, F.
JUSK 1975 -1
KAAZIK, YU. YA.
KAAZ 1973 -1
KACPRZYK, J.
KACP 1975 -1; KACPA 1975 -1
KACZMARSKI, J.
KACZ 1976 -1; KACZ 1976 -2; KACZ 1976 -3
KAESTING, H.
KAES 1976 -1
KAFURA, D. G.
KAFU 1977 -1

KAGAN, B. M.
DANI 1977 -1
KALDYBAEV, S. U.
KALD 1975 -1
KALLINA, C.
KALL 1976 -1
KALMANSON, K.
KALM 1970 -1; KALM 1972 -1; KALM 1975 -1
KALYUZHNYJ, B. I.
KALY 1976 -1
KAMEDA, T.
KAME 1976 -1
KAMINSKAS, V.
KAMI 1975 -1
KANAFANI, A.
POUL 1975 -1
KANCEL'SON, M. B.
SALI 1970 -1
KANNAN, R.
HAUSB 1977 -1; KANN 1978 -1
KANT, E.
KANT 1973 -1
KAPLAN, E. L.
KAPL 1977 -1
KAPLINSKII, A. I.
KAPLA 1976 -1
KAPOOR, S. F.
CHARC 1973 -1
KAPPAUF, C. H.
KAPP 1975 -1
KARAGEZOV, Z. A.
KARA 1976 -1
KARAPETJAN, S. R.
KARAA 1976 -1
KARAPETYAN, I. A.
MARKC 1976 -1
KARELIN, P.
KELI 1968 -1
KARIV, O.
KARI 1976 -1
KARIV, Q.
EVENB 1975 -1
KARMELI, D.
ORON 1976 -1
KARNAUKH, V. G.
GOL'B 1977 -1
KARNEY, D.
DIALA 1977 -1; KARN 1976 -1
KARP
KARP 1975 -1
KARP. R. M.
HOPCA 1971 -1
KARP, R. M.
HELD 1965 -1; KARPA 1975 -1; KARPA 1975 -2
KARPA 1976 -1; KARPA 1977 -1; KARPA 1978 -1
KARWAN, M. H.
KARW 1976 -1; KARW 1976 -2; KARW 1977 -1
KARW 1977 -2
KARZANOV, A.
KARZ 1974 -1
KARZANOV, A. V.
ADEL 1975 -1
KAS, P.
KAS 1975 -1
KAS'JANOV, V. N.
KAS' 1975 -1
KASAHARA, Y.
SANA 1968 -1; SANAA 1967 -1
KASAI, T.
KASA 1975 -1
KASIN, V. I.
BOCKA 1975 -1
KASPSHITSKAJA, M. F.
KASP 1977 -1; KASPA 1974 -1
KASSAY, F.
KASS 1977 -1
KASTNING, C.
KAST 1976 -1
KASZONYI, L.
KASZ 1973 -1
KATAYAMA, H.
KATA 1968 -1
KATCHALSKI, M.
KATC 1977 -1
KATSEV, S. B.
KATS 1977 -1
KATSOULAS, P. C.
KATSA 1976 -1
KATZ, I.
KATZ 1976 -1
KAUFMAN, L.
HANSC 1976 -1
KAUFMAN, M. T.
KAUF 1974 -1
KAUFMANN, A.
KAUFA 1977 -1; KAUFA 1977 -2
KAUFMANN, L.
HANSD 1972 -1
KAZARIN, L. S.
KAZA 1975 -1
KEL'MANS, A. K.
KEL' 1973 -1; KEL' 1976 -1; KEL'A 1976 -1
KELECOM, J. VAN
BALE 1970 -1
KELIKHOV, A. N.
KELI 1968 -1
KENDALL, K. E.
KEND 1977 -1
KENDRICK, D.
KENDA 1975 -1
KENNINGTON, J. L.
ALI 1977 -1; HELG 1977 -1; HELG 1977 -2
KENN 1975 -1; KENN 1976 -1; KENN 1977 -1
KENN 1977 -2; KENNA 1977 -1
KERNIGHAN, B. W
KERN 1973 -1
KERNIGHAN, B. W.
JOHNK 1970 -1; LIN D 1971 -1
KERSHENBAUM, A.
CHOUA 1973 -1; KERS 1976 -1
KESLER, S. S.
GOLE 1974 -1
KEYS, C. D.
KEYS 1975 -1
KHACHATUROV, V. R.
KHAC 1976 -1
KHAIKIN, N. M.
KHAI 1976 -1
KHAN, A. W.
PANWA 1975 -1; PANWA 1976 -1; PANWA 1976 -2
PANWA 1977 -1
KHAN, S. U.
KHAN 1977 -1
KHENKIN, V. E.
KHEN 1966 -1
KHOMENKO, M. P.
KHOM 1976 -1
KHRANOVICH, I. L.
BURKI 1972 -1
KHUMAWALA, B. M.
AKINA 1977 -1
KHUTORETZKY, A. B.
KIBA 1976 -1
KIANFAR, F.
KIAN 1976 -1
KIBALOV, E. B.
KIBA 1976 -1
KICHA, L. I.
KICH 1975 -1
KIKUST, P. B.
KIKU 1975 -1
KILGORE, W. C.
KILG 1977 -1
KILPATRICK, K. E.
HODGA 1977 -1
KIM, C. E.
FRED 1976 -1; IBARC 1974 -1; IBARC 1975 -1
IBARC 1977 -1
KIM, K. V.
KIM 1976 -1
KIMBALL, G. E.
KIMB 1959 -1
KINARIWALA, B.
KINA 1977 -1
KINGSTON, P. L.
KINC 1976 -1
KINSNER, W.
KINS 1975 -1
KIRBY, R. F.
KIRB 1966 -1; KIRBA 1968 -1
KIRCHGAESSNER, K.
KIRC 1965 -1
KIRILLOVA, F. M.
GABA 1975 -1
KIRPICHNIKOV, V. M.
GORS 1970 -1
KIRYUKIN, V. V.
KIRY 1975 -1
KIRZNER, V. M.
KIRZ 1973 -1; KIRZ 1974 -1
KITAMURA, S.
KITA 1975 -1
KIVISTIK, L.
KIVI 1976 -1
KLEE, V.
DANA 1977 -1; KLEE 1977 -1; KLEEA 1976 -1
KLEIN, D.
HOLM 1976 -1; HOLM 1978 -1

KLEINDORFER, G. B.
DINK 1976 -1
KLEITMAN, D. J.
CLAUA 1975 -1; GREEB 1974 -1; KLEI 1969 -1
KLEI 1970 -1; STEI 1969 -1; WANGA 1973 -1
KLEITMAN, D. J. ET AL.
BAUMA 1975 -1
KLINGMAN, D.
BARR 1977 -1; BARR 1977 -2; BARR 1977 -3
BARR 1977 -4; CHAR 1975 -1; CRUMA 1977 -1
DIALA 1977 -1; ELAM 1977 -1; GLOVA 1977 -1
GLOVB 1977 -1; GLOVC 1974 -1; GLOVC 1975 -1
GLOVC 1975 -2; GLOVC 1975 -3; GLOVC 1975 -4
GLOVC 1975 -5; GLOVC 1976 -1; GLOVC 1976 -2
GLOVD 1977 -1; GLOVE 1976 -1; GLOVF 1973 -1
HULTA 1976 -1; KARN 1976 -1
KLIPKER, I. A.
GOL'A 1975 -1
KLITZ, J. K.
KLIT 1976 -1
KLJAUS, P. S.
KLJA 1976 -1
KLUYVER, C. A. DE
KLUY 1977 -1; SALKA 1975 -1
KNAPEK, S.
KNAP 1976 -1
KNOEDEL, W.
KNOE 1971 -1
KNUTH, D. E.
KNUT 1973 -1; KNUT 1976 -1
KNUTS, L.-R.
KNUTA 1973 -1
KOCH, E.
KOCH 1976 -1
KOCHENBERGER, G. A.
DINK 1976 -1
KOCHURA, S. N.
KOCHA 1977 -1
KODRATOFF, Y.
BERMB 1976 -1
KOEHLER, G. J.
KAPP 1975 -1; KOEH 1977 -1
KOEHLER, J.
HIRCA 1975 -1; KOEHA 1977 -1
KOENIG, S. A.
MUCK 1977 -1
KOH, K. M.
KOH 1976 -1; KOH A 1974 -1
KOHLER, C.
KOHL 1975 -1
KOHLER, W.
KOHLA 1975 -1
KOHLER, W. H.
KOHLB 1976 -1
KOHOV, V. A.
KOHO 1974 -1
KOKOTOV, V. M.
SADY 1976 -1
KOLKER, R.
KOLK 1963 -1
KOMARNICKI, J.
KOMA 1975 -1
KOMAROMI, E.
KOMAA 1974 -1
KONONEKO, A. M.
EMELB 1974 -1
KONONENKO, A. M.
EMELC 1974 -1; KONO 1976 -1
KOPYLOV, G. N.
KOPY 1977 -1
KORDECKI, W.
KORD 1976 -1
KORNEV, M. D.
LINSA 1969 -1
KORNILOVA, L. E.
KORN 1969 -1
KOROLEV, A. G.
KURE 1977 -1
KORSH, J. F.
CHANJ 1976 -1; INGA 1975 -1; INGAA 1977 -1
KORSHUNOV, A. D.
KORS 1974 -1; KORS 1976 -1; KORSA 1976 -1
KORTANEK, K. O.
KORT 1966 -1
KORTE B.
KORTA 1977 -1
KORTE, B.
BACHA 1977 -1; BACHA 1977 -2; BACHA 1977 -3
BACHA 1977 -4; HAUSA 1977 -1; HAUSB 1977 -1
HAUSC 1976 -1; HAUSC 1977 -1; HAUSC 1977 -2
HAUSC 1978 -1; HAUSC 1978 -2; KORTB 1976 -1
KORTB 1976 -2; KORTB 1977 -1; KORTB 1977 -2
KORTC 1977 -1; KORTC 1978 -1
KOSTOCHKA, A. V.
BOROB 1977 -1
KOSTOCKA, A. V.
KOST 1977 -1
KOTANI, S.
MANA 1973 -1
KOU, L. T.
KOU 1977 -1; KOU 1977 -2; KOU A 1978 -1
KOVACS, L. B.
KOVA 1976 -1
KOVALENKO, I. N.
KOVAA 1975 -1
KOVALEV, M. M.
KOVAB 1974 -1; KOVAB 1977 -1
KOVALEVSKAJA, M. I.
LEBEA 1976 -1
KOVCAS, L. B.
KOVC 1976 -1
KOZYREV, V. P.
KOZY 1972 -1; KOZY 1976 -1; KOZY 1976 -2
KRAFFT, O.
SONN 1974 -1
KRAJEWSKI, L. J.
SHOW 1977 -1
KRARUP, J.
BILDA 1967 -1; BILDA 1977 -1; COTT 1974 -1
KRAR 1965 -1; KRAR 1967 -1; KRAR 1967 -2
KRAR 1969 -1; KRAR 1969 -2; KRAR 1972 -1
KRAR 1973 -1; KRAR 1975 -1; KRARA 1967 -1
KRARA 1977 -1; KRARB 1973 -1; KRARC 1972 -1
KRARD 1968 -1; KRARD 1977 -1; KRARD 1977 -2
KRASIN, L. A.
VOLY 1974 -1; VYAT 1975 -1
KRATKO, M. I.
KRAT 1977 -1
KRAVTSOV, M. K.
EMEL 1976 -1; EMEL 1976 -2; EMELA 1976 -1
KRAV 1976 -1; KRAVA 1975 -1; KURC 1974 -1
KRAYNOV, G. A.
KIM 1976 -1
KREUZBERGER, H.
KREU 1969 -1
KRIVETS, T. A.
ERMO 1976 -1
KROGDAHL, S.
KROG 1976 -1; KROG 1977 -1
KRONE, M.
KRON 1971 -1
KRONK, H.
GELL 1974 -1
KRONK, H. V.
CHARC 1973 -1
KROON, F. W.
BURKG 1971 -1
KRUSHEVSKII, A. V.
KRUS 1964 -1
KUBALE, M.
KUBA 1974 -1
KUBO, H.
KUBO 1976 -1
KUCERA, L.
KUCE 1977 -1
KUEHNER, J.
KUEH 1977 -1; KUEH 1977 -2
KUEPPER, H.-U.
KUEP 1976 -1
KUHN, W. W.
KUHN 1973 -1
KULIKOWSKI, R.
KULI 1975 -1
KUMAR, S.
KOEH 1977 -1
KUNDU, S.
KUND 1974 -1; KUND 1974 -2; KUND 1975 -1
KUND 1976 -1; KUNDA 1977 -1
KUNG, H. T.
KUNG 1974 -1
KURCEVIC, K. A.
KURC 1974 -1
KUREJCHIK, V. M.
KURE 1977 -1
KUREYCHIK, V. M.
MELI 1972 -1
KURISU, T.
KURI 1973 -1; KURI 1976 -1
KURIYAMA, S.
NOMU 1974 -1
KURODA, K.
KASA 1975 -1
KUSJURIN, M. N.
KUSJ 1975 -1
KUZJURIN, N. N.
KUZJ 1977 -1
KUZNETSOV, V. F.
KUZN 1974 -1
KYAN, S.
KYAN 1975 -1

****

KYUNNAPU, E. P.
BELK 1965 -1
L'VOV, G. N.
KAZA 1975 -1
LABETOULLE, J.
LAWLA 1976 -1
LACKOVIC, I. B.
CVET 1976 -1
LADNER, R. E.
LADN 1975 -1
LAGEWEG, B. J.
LAGE 1976 -1; LAGEA 1976 -1; LAGEA 1976 -2
LAGEA 1976 -3; LAGEA 1976 -4; RINNA 1975 -1
LAI, A. W.
LAI 1975 -1
LALL, H. S.
ALI 1977 -1
LAM, S.
HORV 1977 -1; LAM 1977 -1
LAMBE, T. A.
LAMB 1975 -1
LAMBERT, G.
[illegible] -1; LAMBA 1975 -1
LAMBOURN, S.
LAMBB 1963 -1
LAND, A.
LAND 1977 -1
LAND, A. H.
FARB 1965 -1; LANDA 1978 -1; LANDB 1966 -1
LANG, T.
FERN 1975 -1; FERN 1976 -1; LANG 1977 -1
LANGEFORS, B.
LANGA 1962 -1
LANGFORD, G. J.
DEARB 1975 -1
LANGLEY, R.
LANGB 1973 -1
LAPIN, V. V.
LAPI 1976 -1
LAPORTE, G.: M. , P.
LANDA 1978 -1
LARMAN, D.
KLEE 1977 -1
LARSEN, M.
MOSHA 1963 -1
LATECOERE, L.
LATE 1976 -1; PUN 1975 -1
LAUER, P. E.
SZWAB 1976 -1
LAURIA, D. T.
LAUR 1975 -1
LAURIERE, J. L.
LAURA 1974 -1; LAURA 1976 -1
LAURIERE, M.
LAURB 1976 -1; LAURB 1978 -1
LAVENBERG, S. S.
LAVE 1976 -1
LAVIT, C.
LAVI 1976 -1
LAW A. M.
DREY 1977 -1
LAW, A. M.
CARS 1977 -1
LAWLER, E. L.
GRAHA 1977 -1; LAWL 1971 -1; LAWL 1971 -2
LAWL 1973 -1; LAWL 1975 -1; LAWL 1975 -2
LAWL 1975 -3; LAWL 1975 -4; LAWL 1975 -5
LAWL 1975 -6; LAWL 1975 -7; LAWL 1975 -8
LAWL 1976 -1; LAWL 1976 -2; LAWL 1977 -1
LAWL 1977 -2; LAWL 1977 -3; LAWL 1977 -4
LAWL 1978 -1; LAWLA 1976 -1
LAWLER, E. W.
LAWLB 1976 -1
LAWRENCE, K. D.
LAWRA 1975 -1
LAWRENCE, L.
LAWR 1978 -1
LAWRENCE, S. M.
LAWRA 1975 -1
LAZAK, D.
LAZA 1977 -1
LAZEBNIK, A. I.
BURKI 1972 -1
LEA, A. C.
LEA 1973 -1
LEBEDEV, S. S.
AZAR 1977 -1; LEBE 1974 -1; LEBE 1975 -1
LEBEA 1976 -1
LEBEDEV, V. YU.
LEBEB 1977 -1
LEBEDEVA, T. T.
LEBEC 1977 -1
LEBEDINSKAJA, N. B.
SAHBA 1974 -1
LEBENSOLD, K. W.
LEBED 1974 -1
LEBLANC, L. J.
COOPA 1977 -1; COOPC 1977 -1; LEBL 1975 -1
LEBL 1976 -1; LEBL 1977 -1; LEBLA 1974 -1
LEBLB 1974 -1
LECHEVALLIER, Y.
LECH 1976 -1
LECHNER, R. J.
LECHA 1970 -1
LEE. C. Y.
LEE. 1962 -1
LEE, D. H.
LEE 1976 -1
LEE, G. M.
BRUAB 1978 -1
LEE, J.
LEE A 1976 -1
LEE, S. M.
LEE B 1976 -1
LEE, S.-J.
[illegible]
LEEUWEN, J. VAN
LEEU 1978 -1
LEGENDRE, J.-P.
LEGE 1977 -1
LEIFMAN, L. Y. A.
LEIF 1967 -1
LEIGHTON, F. T.
JOHN 1976 -1
LEITEN, A.
LEIT 1972 -1
LEITEN, A. K.
LEITA 1973 -1; LEITA 1974 -1
LEMAIRE, B.
LEMA 1971 -1; LEMA 1975 -1; LEMA 1976 -1
LEMPEL, A.
EVENC 1970 -1; HARAA 1974 -1
LENARD, M. L.
HU A 1976 -1
LENIN, A. Y. U.
LENI 1971 -1
LENSTRA, H. W.
LENS 1973 -1
LENSTRA, J. K.
BRUCB 1975 -1; FOX B 1977 -1; GRAHA 1977 -1
JOHND 1977 -1; LAGE 1976 -1; LAGEA 1976 -1
LAGEA 1976 -2; LAGEA 1976 -3; LAGEA 1976 -4
LENSA 1977 -1; LENSB 1975 -1; LENSB 1976 -1
LENSB 1976 -2; LENSB 1976 -3; LENSB 1977 -1
LENSC 1977 -1; RINNA 1975 -1
LEON, B. J.
LIN B 1974 -1
LEONT'EV, V. K.
LEON 1975 -1; LEON 1976 -1
LEPESHINSKIJ, N. A.
LEPE 1976 -1
LERDA, F.
LERD 1975 -1
LERNER, A. Y.
LERN 1969 -1
LEROY, P.
LERO 1975 -1
LESNIAK-FOSTER, L.
LESN 1977 -1
LESNIAK-FOSTER, L. M.
ALAV 1977 -1
LESNIAK, L.
LESNA 1975 -1; LESNA 1975 -2
LETICHEVSKIY, A. A.
LETI 1976 -1
LEV, B.
INTR 1976 -1
LEV, O. E.
LEV 1976 -1
LEVEN, B. M.
HEDEA 1963 -1
LEVI, G.
LEVI 1972 -1; LEVI 1974 -1
LEVIN, B. M.
LEVIA 1963 -1
LEVNER, E. V.
LEVN 1973 -1
LEVY, J.
LEVY 1966 -1
LEVY, L. S.
LEVYA 1973 -1
LEVYI, V. D.
LEVYB 1976 -1
LEWIS, P. M.
ROSEB 1974 -1; ROSEB 1977 -1
LI, H. F.
RAMAD 1976 -1
LIBURA, M.
LIBU 1974 -1

LIEBERMAN, G. J.
DERM 1975 -1; DERM 1975 -2; DERM 1976 -1
LIEBLING, T.
LIEB 1970 -1
LIEBLING, T. M.
BENNB 1976 -1
LIEBMAN, J. C.
LIEBA 1964 -1; MALE 1977 -1
LIEN, Y. E.
LIEN 1974 -1
LIESHOUT, J. TH. VAN
HULS 1977 -1
LIETAER, B. A.
LIET 1977 -1
LIHACEV, V. M.
LIHA 1975 -1; LIHAA 1974 -1
LIM, M.-H.
LIM 1977 -1
LIN, B.
RARD 1977 -1
LIN, B. W. Y.
LIN 1977 -1
LIN, P. M.
LIN A 1969 -1; LIN B 1974 -1
LIN, S.
HWANA 1969 -1; KERN 1973 -1; LIN C 1975 -1
LIN D 1971 -1
LIND, C. B.
LIND 1973 -1
LIND, N. C.
LINDA 1977 -1
LINDSAY, M. A.
MOORA 1973 -1
LING, R. F.
LING 1973 -1; LING 1975 -1
LINGST, S. T.
SANA 1968 -1
LINKE, K.-F.
LINK 1976 -1
LINS, S.
LINS 1975 -1
LINSKIY, V. S.
LINSA 1969 -1
LIPTON, R.
DOBKA 1975 -1
LIPTON, R. J.
DOBKB 1976 -1
LIPTON, S. L.
KING 1976 -1; LIPT 1976 -1
LISKOVEC, V. A.
LISK 1977 -1
LISS, D.
KOLK 1963 -1
LISYAK, V. V.
MELI 1972 -1
LITTGER, K. W.
LITT 1976 -1
LITTLE, C. H. C.
LITTA 1976 -1
LITTLE, J. D. C.
GARTA 1975 -1
LITVAK, B. G.
FRIDA 1973 -1; LITV 1973 -1; LITVA 1973 -1
LITVA 1975 -1; LITVA 1976 -1
LITVAK, E. I.
KICH 1975 -1; LITVB 1975 -1; LITVC 1974 -1
LIU T. H.
CHU 1965 -1
LIU, C. L.
LIU 1973 -1; LIU 1975 -1; LIU 1976 -1
LIU, C. S.
GELDB 1975 -1
LIU, J.
LIU A 1976 -1
LIU, P. C.
BAND 1975 -1; BANDA 1975 -1
LIVNAT, A.
TRIF 1973 -1
LIVOVSCHI, L.
LIVO 1976 -1
LIVSIK, E. M.
LIVS 1972 -1; LIVS 1972 -2
LOCKHART, S.
EDMOB 1969 -1
LODISH, L. M.
LODI 1970 -1
LOEHLE, H.
BAUE 1973 -1
LOGEMANN, G.
HELLA 1960 -1
LOHSE, D.
LOHS 1976 -1
LOMONOSOV, M. V.
KEL'A 1976 -1
LONG, C. F.
LONG 1977 -1
LONGINI, I. M.
HODGA 1977 -1
LORD, R. E.
DEO A 1977 -1
LOREA, M.
LORE 1972 -1; LORE 1975 -1; LORE 1975 -2
LORE 1976 -1
LORENTSEN, R.
LIND 1973 -1
LOS, M.
LOS 1978 -1; LOS A 1977 -1
LOVASZ, L.
LOVA 1968 -1; LOVA 1974 -1; LOVA 1975 -1
LOVA 1975 -2; LOVA 1976 -1; LOVA 1976 -2
LOVA 1977 -1; LOVA 1977 -2; LOVA 1977 -3
LOVE, R. F.
LOVE 1976 -1; LOVEA 1976 -1; LOVEA 1976 -2
NORB 1977 -1; WESOA 1971 -1
LOVETSKII, S. E.
BURKJ 1966 -1; BURKJ 1968 -1
LOWE, T. J.
DEARA 1976 -1; FRANA 1976 -1; HEARA 1976 -1
WEND 1977 -1
LOWERRE, J. M.
BROWD 1976 -1
LUBELL, D.
LUBE 1975 -1
LUCE, B.
LUCE 1975 -1
LUCERTINI. M.
GIUL 1976 -1
LUDOT, J. P.
RUBIE 1976 -1
LUDOT, J.-P.
LUDO 1971 -1
LUEDTKE, H.
LUED 1975 -1; LUEDA 1976 -1
LUEKER, G. S.
BOOTA 1975 -1; BOOTA 1976 -1; ROSE 1976 -1
LUGOSI, G.
HEPP 1976 -1
LUNDQVIST, L.
LUND 1973 -1; LUND 1975 -1
LUSS, H.
ACKE 1977 -1; ACKE 1977 -2; LUSS 1975 -1
LYNN, J.
KALL 1976 -1
LYONS, J. P.
LYON 1977 -1
LYTKIN, I. P.
LYTK 1974 -1; LYTK 1974 -2
MAC ALLISTER, J. D.
MAC 1977 -1
MAC AVANEY, K. L.
MAC A 1976 -1; STAC 1976 -1
MAC BRIDE, R.
GEOFC 1977 -1
MAC CALL, E. H.
MAC B 1976 -1
MAC CARTHY, M. K.
SHAH 1974 -1
MAC CARTHY, O.
RODG 1977 -1
MAC CARTHY, P. J.
MAC C 1975 -1
MAC DANIEL, D.
MAC D 1977 -1
MAC DIARMID, C. J. H.
GRIMA 1975 -1; MAC E 1974 -1; MAC E 1975 -1
MAC GINNIS, L. F.
MAC F 1977 -1
MAC GUIRE, C. B.
BECKB 1956 -1
MAC HUGH, J. A. M.
BOESB 1974 -1
MAC ISAAC, R.
WASS 1975 -1
MAC KEOWN, P.
MAC G 1975 -1
MAC KEOWN, P. G.
MAC H 1975 -1; MAC I 1975 -1
MAC KINNON, R. D.
MAC J 1976 -1
MAC KNIGHT, R. D.
DANTC 1974 -1
MAC LAUGHLIN, W. A.
MAC K 1966 -1
MAC NEELY, R. A.
MAC L 1977 -1
MACHOL, R. E.
MACH 1976 -1
MACKEOWN, P.
FISKA 1977 -1

MACKEOWN, P. G.
FISKB 1977 -1
MACMILLAN, C.
GLOVD 1977 -1
MADSEN, K.
MADS 1976 -1
MAESTRINI, P.
ALIA 1976 -1
MAFFIOLI, F.
CAME 1973 -1; CAME 1974 -1; CAMEA 1974 -1
FRATA 1972 -1; MAFF 1975 -1; MAFF 1975 -2
MAFF 1976 -1
MAGAZINE, M.
BAKEB 1977 -1
MAGAZINE, M. J.
MAGA 1977 -1; MAGAA 1976 -1
MAGNANTI, T. L.
AASH 1976 -1; BRADD 1977 -1; GOLDC 1977 -1
GOLDD 1977 -1; GREEB 1974 -1; GREEC 1975 -1
MAHMOUD, S.
MAHM 1976 -1
MAHR, B.
MAHR 1976 -1
MAHUAL, D. W.
MAHU 1976 -1
MAIER, S. F.
MAIE 1976 -1
MAILLARD, P.
MAIL 1977 -1
MAIMON, A.
MAIM 1975 -1
MAIRS, T. G.
MAIR 1977 -1
MAJORANI, E.
LERD 1975 -1; MAJO 1964 -1
MAKAROV, I. M.
LERN 1969 -1
MAKAROV, L. I.
BOCKA 1975 -1
MAKINO, T.
MAKI 1966 -1
MALE, J. W.
MALE 1977 -1
MALICKA-WASOWSKA, J.
MALI 1974 -1
MALLION, R. B.
HARAB 1974 -1; HARAB 1975 -1
MALSTROM, E. M.
MAC L 1977 -1
MALYSHKO, V. V.
LEPE 1976 -1; MALY 1976 -1
MANABE, R.
MANA 1973 -1
MANDERS, K.
MAND 1976 -1
MANHEIM, M.
MARTE 1965 -1
MANNE, A. S.
MANN 1967 -1
MANSOOR, E. M.
MANS 1964 -1
MANTEL, F.
MANT 1977 -1
MARBLE, G.
MATUA 1972 -1
MARCU, D.
MARC 1976 -1
MARCUK, G. I.
MARCA 1976 -1
MARCUS, B. T.
MARCB 1967 -1
MARCUS, M.
MARÇC 1972 -1
MARK, J. W.
MARK 1977 -1
MARKLAND, R. E.
MARKA 1975 -1
MARKOSYAN, A. G.
MARKB 1976 -1
MARKOSYAN, S. E.
MARKB 1976 -1; MARKC 1976 -1
MARKOV, L. A.
MARKD 1974 -1
MARKOWSKY, G.
CHANE 1976 -1; MARKE 1976 -1
MARKS. D. H.
MARKF 1971 -1
MARKS, B. R.
ESOG 1977 -1
MARQUEZ, J.
MARQ 1977 -1
MARSH, A. B. III.
CUNNB 1976 -1
MARSHALL, C.
MARS 1971 -1
MARSTEN, R. E.
MARSA 1972 -1; MARSA 1975 -1; MARSB 1977 -1
MARSB 1977 -2; MARSB 1978 -1; MORI 1975 -1
MORIA 1976 -1
MARTELLI, A.
MART 1976 -1; MARTA 1975 -1
MARTELLO, S.
MARTB 1977 -1; MARTC 1976 -1; MARTC 1977 -1
MARTC 1977 -2; MARTC 1977 -3; MARTC 1977 -4
MARTIN-VEGA, L. A.
MARTD 1976 -1
MARTIN, B.
MARTE 1965 -1
MARTIN, B. V.
MARTF 1963 -1
MARTIN, K.
EVANC 1976 -1
MARTINEC, I.
MARTG 1976 -1
MARTINEZ Y RODRIGUEZ, J. B.
MARTH 1976 -1
MARTINO, F.
GALLC 1972 -1
MARTINOVA, M. K.
MARTI 1974 -1
MARUYAMA, K.
MARU 1976 -1
MARUYAMA, S.
NABEA 1975 -1
MASON, D. W.
ALSP 1976 -1
MASON, J. H.
MASO 1977 -1
MASSA, L.
MASS 1976 -1
MATETI, P.
MATE 1976 -1
MATHEWS, P. A.
BULL 1970 -1
MATHON, R. A.
CORNB 1978 -1
MATIJASEVIC, JU. V.
MATI 1974 -1
MATSUI, H.
MATS 1971 -1
MATTHAEUS, F. W.
MATT 1976 -1
MATTHEISS, T. H.
SCHM 1976 -1
MATTHEWS, L. R.
MATTA 1977 -1; MATTA 1978 -1; MATTA 1978 -2
MATULA, D. W.
BECK 1976 -1; MATU 1968 -1; MATU 1975 -1
MATU 1977 -1; MATU 1978 -1; MATUA 1972 -1
MAUREL, E.
MAUR 1976 -1
MAURER, S. B.
MAURA 1973 -1
MAURO, V.
BONA 1976 -1
MAURRAS, J. F.
MAURB 1972 -1; MAURB 1976 -1; MAURB 1977 -1
MAURRAS, J.-F.
MAURC 1975 -1
MAWDESLEY, M. J.
MAWD 1974 -1
MAXWELL, W. L.
CONW 1966 -1; MAXW 1964 -1
MAY, A. D.
MAY 1976 -1
MAY, J. G.
MAY A 1974 -1
MAYER, J.
ALTHA 1976 -1; MAYE 1976 -1
MEADOWS, M. E.
CHURA 1977 -1
MEDINA-MORA, O.
MARQ 1977 -1
MEDNITSKII, V. G.
MEDN 1975 -1
MEEUSEN, W.
MEEU 1975 -1; MEEU 1976 -1
MEGIDDO, N.
MEGI 1977 -1
MEHLHORN, K.
MEHL 1975 -1
MEHR, P.
CAZA 1975 -1
MEIR, A.
ERDOC 1977 -1; MEIR 1973 -1; MEIR 1975 -1
MEIR 1975 -2; MEIR 1976 -1; MEIR 1977 -1
MEISSLER, H. G.
BIESA 1977 -1
MEL'NIKOV, L. S.
BOROC 1974 -1; MEL' 1977 -1

****

MEL'NIKOV, O. I.
MEL'A 1975 -1; MEL'A 1976 -1; MEL'A 1976 -2
MELIKHOV, A. N.
MELI 1972 -1
MELZAK, Z. A.
MELZ 1961 -1
MENDELSOHN, N. S.
JOHNB 1962 -1
MENG, D.
MENG 1974 -1
MENSCH, G.
MENS 1975 -1
MERCHANT, D. K.
MERC 1974 -1; MERCA 1976 -1
MEREKIN, YU. V.
MERE 1975 -1
MERLINO, E. L. A.
MERL 1971 -1
MERZHANOV, V. YU.
ALEKA 1977 -1
MESTETZKY, L. M.
AZAR 1977 -1
MEVERT, P.
MEVE 1977 -1
MEYER, A. R.
FISC 1971 -1
MEYER, C. F.
MEYE 1973 -1
MEYER, J. C.
MEYEA 1972 -1; MEYEA 1978 -1
MEYER, J.-C.
MEYEB 1975 -1; MEYEB 1976 -1
MEYER, R. R.
MEYEC 1976 -1; MEYEC 1977 -1; MEYED 1977 -1
MEYER, W.
MEYEE 1974 -1
MEYL, H.
MEYL 1972 -1
MICHAELI, I.
MICH 1977 -1
MICHIE, D.
DORA 1966 -1; MICHA 1968 -1
MICKLE, M. H.
MICK 1975 -1; MICKA 1969 -1; MICKA 1975 -1
MICKB 1977 -1; WANGB 1974 -1
MIDHA, C. K.
DEY 1976 -1
MIHALYFFY, L.
MIHA 1976 -1
MIHOC, G.
MIHO 1965 -1
MIKHAJLOVA, E. D.
MIKH 1975 -1
MIKHALEVICH, V. S.
MIKHA 1966 -1
MIKOLA, M.
MIKO 1976 -1
MIKUL'SKII, V. E.
KONO 1976 -1
MILIOTIS, P.
MILI 1976 -1
MILLER, D. M.
MILL 1976 -1; MILLA 1976 -1
MILLER, G. L.
MILLB 1977 -1
MILLER, H. E.
MILLC 1976 -1; MILLC 1976 -2; MILLD 1976 -1
MILLO, R. A. DE
VAIR 1976 -1
MILLS. W. H.
MILLE 1973 -1
MILNER, E. C.
MILN 1975 -1; MILNA 1976 -1
MINE, H.
IBARA 1976 -1; ISHI 1975 -1; ISHI 1975 -2
ISHI 1976 -1; ISHI 1977 -1; OCHI 1975 -1
MING-I, Y.
MING 1976 -1
MINIEKA, E.
MINI 1976 -1
MINIEKA, F.
MINIA 1975 -1
MININA, T. R.
MINIB 1975 -1; VARS 1974 -1
MINOLI, D.
MINO 1975 -1
MINOUX, M.
GONDB 1977 -1; LEGE 1977 -1; MINOA 1975 -1
MINOA 1975 -2; MINOA 1976 -1; MINOA 1976 -2
MINOA 1976 -3; MINOA 1976 -4; MINOA 1976 -5
MINOA 1977 -1
MINTY, G. J.
MINT 1975 -1
MIRCHANDANI, P. B.
MIRC 1974 -1; MIRC 1976 -1

MIRSKY, L.
MIRS 1971 -1
MISRA, J.
KUNDA 1977 -1
MITCHELL, S.
MITC 1975 -1
MITCHEN, J.
MITCA 1976 -1
MITEV, I. G.
IVANA 1976 -1
MITEV, J. G.
MITE 1977 -1; MITE 1977 -2; MITE 1977 -3
MITSUO, G.
HYUNA 1974 -1
MITTAL, A. K.
CHANH 1975 -1
MITTAL, B. S.
MITT 1977 -1
MITTAL, M. L.
MITTA 1967 -1
MITTEN, L. G.
MITTB 1973 -1
MIYAKAWA, M.
MIYA 1977 -1
MJELDE, K. M.
MJEL 1976 -1
MOGA, S.
FABIA 1977 -1
MOGILA, V. G.
SEME 1974 -1
MOJENA, R.
MOJE 1976 -1
MOLE, R. H.
MOLE 1976 -1
MOLNAR, L.
GARB 1976 -1
MOLNAR, Z.
MOLN 1976 -1
MOLZBERGER, P.
MOLZ 1976 -1
MONAHAN, J. P.
MONA 1976 -1
MONARCHI, D. E.
PLAN 1975 -1
MONMA, C. L.
KANN 1978 -1; KORTA 1977 -1; MONM 1977 -1
TROTA 1977 -1
MONTANARI, U.
FRATB 1975 -1; MARTA 1975 -1
MONTICONE, L. C.
MONT 1976 -1
MOON, J. W.
MEIR 1973 -1; MEIR 1975 -1; MEIR 1975 -2
MEIR 1976 -1; MEIR 1977 -1
MOORE, J. H.
FREE 1977 -1
MOORE, J. I.
TROTC 1976 -1
MOORE, J. I. JR.
TROTD 1976 -1
MOORE, J. M.
BIND 1961 -1
MOORE, M. H.
MOOR 1976 -1
MOORE, S. A.
MOORA 1973 -1
MOORE, T. G.
MOORB 1976 -1
MOREFIELD, C. L.
MORE 1977 -1
MORIN, T. , L.
MORI 1975 -1
MORIN, T. L.
MARSB 1977 -1; MARSD 1977 -2; MAREB 1978 -1
MORIA 1976 -1
MORLOK, L. J.
LEBLA 1974 -1; LEBLB 1974 -1
MOROZENSKII, L. YU.
MORO 1974 -1
MOROZOV, S. A.
MOROA 1976 -1
MORPURGO, R.
MORP 1971 -1
MORRILL, R. L.
MORR 1977 -1
MORRIS, J. G.
MORRA 1972 -1
MORROW, C.
MORRB 1976 -1
MORTON, T. E.
MORT 1977 -1
MOSELY, J. C.
YOUN 1970 -1
MOSHER, W. W. JR.
MOSH 1963 -1

MOSHMAN, J.
MOSHA 1963 -1
MOSLER, K. C.
MOSL 1976 -1
MUCHNICK, S. S.
MUCH 1976 -1
MUCKSTADT, J. A.
MUCK 1977 -1
MUCNIK, M. I.
VODO 1976 -1
MUELLER-MERBACH, H.
MUEL 1975 -1; MUEL 1975 -2; MUEL 1976 -1
MUEL 1976 -2; MUEL 1976 -3
MUELLER, H. CH.
MUELA 1973 -1
MULHERKAR, S.
HINS 1976 -1
MULLABAEV, A. A.
MULL 1974 -1
MULLIN, R. C.
CERF 1974 -1; MULLA 1978 -1
MULVEY, J.
DEMB 1976 -1; GLOVE 1975 -1; GLOVE 1975 -2
MULV 1978 -1
MULVEY, J. M.
MULVA 1975 -1; MULVA 1975 -2; MULVA 1975 -3
MUNCH-ANDERSEN, B.
MUNC 1977 -1
MUNRO, I.
ALLE 1976 -1; BOROA 1972 -1
MUNRO, J. I.
MUNR 1971 -1
MURATET, G.
ALCO 1975 -1; ALCOA 1976 -1
MURAV'EV, V. I.
MURA 1974 -1
MURCHLAND, J.
MURC 1973 -1; MURCA 1973 -1
MURCHLAND, J. D.
FARB 1965 -1; LANDB 1966 -1; MURCB 1964 -1
MURCB 1965 -1; MURCB 1966 -1; MURCB 1966 -2
MURCB 1966 -3; MURCB 1967 -1; MURCB 1967 -2
MURCB 1967 -3; MURCB 1967 -4; MURCB 1968 -1
MURCB 1968 -2; MURCB 1968 -3; MURCB 1969 -1
MURCB 1970 -1; MURCB 1971 -1
MURDASOV, A. B.
MURD 1976 -1
MURPHY, F. H.
MURP 1975 -1; MURP 1976 -1
MURPHY, R. A.
SWEE 1977 -1
MURTHY, M. S.
MURT 1976 -1
MURTY, K.
MURTA 1967 -1
MURTY, K. G.
MURTB 1976 -1
MURTY, U. S. R.
MURTC 1971 -1
MUTH, J. F.
MUTH 1963 -1
MYERS, B. R.
MYER 1967 -1; MYER 1974 -1
NAAMAN, A. E.
BALAF 1977 -1
NABESHIMA, I.
NABE 1965 -1; NABE 1967 -1; NABE 1972 -1
NABE 1973 -1; NABEA 1975 -1
NAIVEL'T, A. V.
LITVA 1973 -1; LITVA 1975 -1; LITVA 1976 -1
NAKAGAMI, K.
SANAA 1967 -1
NAKAMURA, T.
NAKA 1977 -1
NAKANISHI, Y.
NAKAA 1975 -1
NAKANO, H.
NAKAA 1975 -1
NAKASHIMA, K.
SAKAA 1976 -1
NANDA, R.
NAND 1974 -1
NAPIER, A.
GLOVE 1976 -1
NARASIMHAN, S. L.
NARA 1977 -1
NARAYANAN, V.
BEEN 1976 -1
NARULA, S. C.
NARU 1976 -1; NARUA 1977 -1
NASEKIN, V. YA.
CHERD 1976 -1
NASH-WILLIAMS, C. ST. J. A.
NASH 1961 -1; NASH 1961 -2; NASH 1964 -1
NASH 1967 -1; NASH 1975 -1
NASH, P.
GLAZA 1976 -1; NASHA 1977 -1
NASTANSKY, L.
BUSH 1976 -1; NAST 1975 -1
NATALI, L.
NICOB 1976 -1
NAUSS, R.
GEOFD 1977 -1
NAUSS, R. M.
NAUS 1975 -1; NAUS 1976 -1; NAUS 1976 -2
NAUS 1976 -3
NEEBE, A. W.
GARFB 1973 -1; GARFB 1977 -1; NEEB 1974 -1
NEEBA 1975 -1
NEIDHART, F.
NEID 1967 -1
NELSON, R. T.
HOLL 1977 -1; NELS 1977 -1
NEMHAUSER, G.
GOLDF 1967 -1
NEMHAUSER, G. L.
BALLA 1977 -1; CORNC 1977 -1; CORND 1976 -1
CORND 1977 -1; CORNE 1978 -1; FISHB 1977 -1
MERC 1974 -1; NEMH 1976 -1; NEMHA 1976 -1
NEMIROVSKIJ, A. S.
YUDI 1976 -1
NEMIROVSKY, A. S.
JUDI 1976 -1
NEPOMIASTCHY, P.
NEPO 1973 -1
NESETRIL, J.
BOSA 1976 -1; NESE 1976 -1
NEUMANN, K.
NEUM 1975 -1
NEWMAN, D.
JOHNF 1978 -1
NEWMAN, M.
NEWM 1972 -1; NEWM 1974 -1
NEY, P.
NEY 1972 -1; NEY 1973 -1
NGUYEN, H.
NGUY 1975 -1
NGUYEN, H. Q.
GOLDD 1977 -1
NGUYEN, S.
FLOR 1976 -1; LOS A 1977 -1
NICCOLUCCI, F.
GIANA 1976 -1
NICHOLSON, D. A. J.
NICH 1968 -1
NICOGOSYAN, Z. G.
NICO 1975 -1
NICOLAS, J.-L.
NICOA 1976 -1
NICOLETTI, B.
NICOB 1976 -1
NICOLIER, Y.
BRELA 1977 -1
NICOLO, F.
GUTE 1976 -1
NIEMINEN, J.
NIEM 1974 -1; NIEM 1974 -2; NIEM 1975 -1
NIEM 1975 -2; NIEM 1975 -3; NIEM 1976 -1
NIEVERGELT, J.
REIN 1977 -1
NIGMATULLIN, R. G.
NIGM 1975 -1
NIJENHUIS, A.
NIJE 1975 -1
NILLSON, N. L.
HART 1968 -1
NINCAK, J.
NINC 1974 -1
NINCAK, YA.
KORSA 1976 -1
NJARIPJA, H. K.
NJAR 1973 -1
NOGUCHI, S.
SAKAC 1970 -1
NOLTEMEIER, H.
NOLT 1975 -1; NOLT 1976 -1
NOMURA, S.
NOMU 1974 -1
NOONAN, F.
NOON 1977 -1
NOORVASH, S.
NOOR 1975 -1
NORBACK, J. P.
NORB 1977 -1
NORDBECK, S.
NORD 1969 -1
NORDHAUS, E. A.
ALAV 1977 -1
NORMAN, J. M.
NORM 1975 -1

NORMAN, J. T.
GLOVL 1965 -1
NORTHWESTERN UNIVERSITY
NORT 1976 -1
NORTON, P. G.
HOLZA 1973 -1
NOUR, M. E.
NOUR 1968 -1; NOURA 1970 -1
NOVANSKY, R.
NOVA 1974 -1
NOVITZKY, N. A.
NOVI 1978 -1
NUGENT, C. E.
NUGE 1964 -1
NUNEN, J. A. E. E. VAN
NUNE 1978 -1
NUNNER, O.
NUNN 1976 -1
NUNNIKHOVEN, T. S.
NUNNA 1977 -1
NUTTLE, H. L. W.
MAGAA 1976 -1
NYKOWSKI, I.
NYKO 1964 -1
O'BRIAN, R. C.
O'BR 1977 -1
O'BRIEN, J. J.
O'BRA 1975 -1
O'NEIL, E. J.
O'NE 1973 -1
O'NEIL, P. E.
O'NE 1973 -1
OCHI, T.
OCHI 1975 -1
ODONI, A. R.
ODON 1974 -1
OFFERMANN, J.
BURKC 1977 -1
OFFIR, J. D.
OFFI 1975 -1
OGBU, U. I.
NARU 1976 -1; NARUA 1977 -1
OHERNYSOVA, G. D.
KAPLA 1976 -1
OHTSUKI, T.
OHTS 1976 -1; OHTSA 1976 -1
OICER, V. I.
OICE 1972 -1
OIZUMI, J.
SAKAC 1970 -1
OKADA, S.
KOLK 1963 -1
OKAMOTO, Y.
MOJE 1976 -1
OKINO, N.
KUBO 1976 -1
OKUNO, H.
GEN 1974 -1; GEN A 1976 -1; HYUNA 1974 -1
OLAH, GY.
OLAH 1975 -1
OLARU, E.
OLAR 1973 -1
OLDFIELD, J. V.
MICHA 1968 -1
OLDZIEY, J. W.
CONW 1966 -1
OLEJNIK, F.
JUCO 1974 -1
OLIVER, R.
POTT 1972 -1
OLSON, E. N.
OLSO 1976 -1
OLSON, R. E.
SMITA 1976 -1
ONAGA, K.
KYAN 1975 -1
ONOKERHORAYE, A. G.
ONOK 1976 -1
OPPENHEIM, R.
HOFFA 1978 -1
OREL, E. N.
OREL 1977 -1
ORINO, H.
FUJIB 1974 -1
ORLIN, J.
ORLI 1977 -1
ORLIN, J. B.
BARTA 1977 -1
ORLOFF, C.
ORLO 1976 -1
ORLOFF, C. S.
MALE 1977 -1; ORLOA 1976 -1; ORLOA 1976 -2
ORLOA 1977 -1
ORLOVA, G. I.
ORLOB 1975 -1; ORLOC 1975 -1
ORON, G.
ORON 1976 -1
ORTH, E.
ORTH 1973 -1
ORTH, H.
ORTHA 1972 -1
OSKALKOV, I. O.
BELK 1965 -1
OSLEEB, J. P.
OSLE 1974 -1; OSLEA 1977 -1
OSTERMANN, R.
OSTE 1976 -1
OSYCZKA, A.
OSYC 1976 -1
OTTAVIANI, M.
OTTA 1976 -1
OUBINA, L.
OUBI 1976 -1
OUDHEUSDEN, D. L. VAN
OUDH 1977 -1
OVCHINNIKOV, V. G.
OVCH 1976 -1
OVER, A. M. JR.
SMITC 1976 -1
OVERBECK-LARISCH, M.
OVER 1975 -1; OVER 1976 -1
OZAKI, H.
AM 1976 -1; TRAN 1976 -1; TSUKA 1975 -1
OZAWA, T.
OZAW 1973 -1
PACAULT, J. F.
PACA 1974 -1
PADBERG, M.
PADB 1971 -1
PADBERG, M. W.
BALAH 1974 -1; BALAJ 1975 -1; BALAJ 1976 -1
GROEA 1977 -1; GROEA 1977 -2; GROEA 1977 -3
GROEA 1978 -1; HONGB 1977 -1; PADBA 1973 -1
PADBA 1975 -1; PADBA 1976 -1; PADBA 1977 -1
PADBA 1977 -2; PADBB 1977 -1
PAIZIEV, E.
SALI 1970 -1
PALACIOS-GOMESZ, F. E.
ARMSC 1977 -1
PALASTI, I.
PALA 1971 -1
PALM, K.-P.
PALM 1974 -1
PANAGIOTAKOPOULOS, D.
PANA 1976 -1; PANA 1977 -1
PANCHENKO, A. A.
PANC 1976 -1
PANDIT, S. N. N.
PAND 1962 -1; PANDA 1975 -1
PANWALKAR, S. S.
PANW 1977 -1; PANWA 1975 -1; PANWA 1976 -1
PANWA 1976 -2; PANWA 1977 -1; SMITF 1975 -1
PAPADIMITRIOU, C. H.
PAPA 1975 -1; PAPA 1976 -1; PAPA 1976 -2
PAPA 1977 -1; PAPAA 1976 -1; PAPAA 1977 -1
PAPADIMITRIOU, H.
CHRIB 1976 -1
PAPE, U.
EDEN 1977 -1; EDENB 1978 -1; FEIC 1975 -1
PAPE 1969 -1; PAPE 1969 -2; PAPE 1976 -1
PAPE 1976 -2; PAPE 1976 -3; PAPE 1976 -4
PAPE 1977 -1
PAPERNOV, B. A.
PAPEA 1976 -1
PAPINEAU, R. L.
PAPI 1975 -1
PARIKH, S.
PARI 1963 -1
PARKER, R. G.
BULF 1976 -1; BULF 1976 -2; BULFA 1977 -1
ESKE 1975 -1; HOLMB 1976 -1; PARK 1977 -1
PARKA 1977 -1
PAROUSH, J.
INTRA 1977 -1
PARRIS, R.
PARR 1968 -1; PARRA 1967 -1; PARRA 1968 -1
PARSONS, J. A.
PARS 1969 -1; PARS 1970 -1; PARS 1970 -2
PARTHASARATHY, K. R.
PART 1976 -1
PASSY, U.
PASS 1975 -1
PATEL, A. M.
PATE 1976 -1
PATKIN, B. P.
EGOR 1977 -1
PATRINOS, A. N.
PATR 1974 -1
PATTERSON, J. H.
DOERA 1977 -1; PATT 1976 -1; PATTA 1976 -1

PAUCK, R.
PAUC 1976 -1
PAUL, W. J.
PAUL 1977 -1
PAULAUSKAS, M. A.
PAULA 1976 -1; PAULA 1976 -2; PAULB 1975 -1
PAVLEY, R.
HOFFC 1958 -1
PAVLIDIS, T.
CHANK 1977 -1
PAYAN, C.
PAYA 1977 -1
PEARMAN, A.
PEAR 1974 -1
PEARSON, J. M.
MOORA 1973 -1
PECHURIN, M. K.
ZAYC 1975 -1
PEKICH, T. J.
PEKI 1975 -1
PELED, U. N.
HAMMB 1977 -1; PELE 1976 -1; PELE 1977 -1
PENAUD, U. G.
PENA 1976 -1
PENTICO, D. W.
PENT 1976 -1
PERCOV, E. E.
PERC 1975 -1
PEREKREST, V. T.
MINIB 1975 -1
PEREPELICA, V. A.
GIMA 1976 -1
PEREPELITSA, V. A.
GIMAA 1976 -1; GIMAB 1974 -1; GLEBA 1970 -1
PERE 1966 -1; PEREA 1976 -1
PERKO, A.
PERK 1965 -1
PERL, J.
MAHR 1976 -1; PERL 1976 -1
PERLES, M. A.
KOCH 1976 -1
PERLMAN, B. L.
PERLA 1975 -1
PERRY, A.
BALAC 1976 -1
PERVOZVANSKII, A. A.
MARKD 1974 -1
PETER, K.
PETE 1975 -1
PETERSEN, E. R.
PETEA 1975 -1
PETERSEN, J. P.
PETEB 1977 -1
PETERSON, R. M.
MAIM 1975 -1
PETRICH, F.
PETR 1976 -1
PETROV, I. P.
PETRA 1975 -1
PETROV, S. V.
PETRB 1976 -1
PETUKHOV, V. S.
ERMO 1976 -1
PEVEPELICHE, V. A.
PEVE 1969 -1
PFAFFENBERGER, R. C.
PFAF 1976 -1
PFAFFENZELLER, D.
ZIMMB 1967 -1
PHILIPE, J.
PUN 1975 -1
PHILLIPS, F.
ARMSA 1975 -1
PHILLIPS, I. W.
PHIL 1976 -1
PHUONG, T. H.
PHUO 1976 -1
PICARD, C.-F.
PICA 1975 -1
PICARD, J. C.
PICAA 1975 -1
PICARD, J.-C.
PICAB 1976 -1; PICAC 1976 -1; PICAC 1977 -1
PICAC 1977 -2; PICAD 1975 -1
PICHAT, E.
KAUFA 1977 -1; KAUFA 1977 -2
PICHOCKA, M.
PICH 1976 -1
PIEHLER, J.
PIEH 1975 -1; PIEH 1976 -1; PIEH 1977 -1
PIERCE, A. R.
PIER 1975 -1
PIERCE, D. A.
SPRA 1967 -1; STRA 1967 -1
PIERSKALLA, W.
LEBLB 1974 -1
PIERSKALLA, W. P.
MILLD 1976 -1
PILIPOSJAN, T. E.
PILI 1976 -1
PINOCHET, M. R.
PINO 1976 -1
PINTO, J. V.
PINT 1975 -1
PIPER, C.
PIPE 1975 -1
PIPER, C. J.
PIPEA 1975 -1; PIPEA 1976 -1
PIPPENGER, N.
PIPP 1976 -1
PIROSKA, T.
PIRO 1975 -1
PLANE, D. R.
PLAN 1975 -1
PLATE, E. J.
HOCA 1976 -1
PLATEAU, G.
FAYA 1976 -1
PLATEK, M.
NOVA 1974 -1
PLAXTON, A.
PLAX 1976 -1
PLESNIK, JAN.
PLES 1976 -1
PLESTCHINSKY, A. S.
PLESA 1976 -1
PLOTNIKOV, V. N.
PLOT 1974 -1; PLOTA 1974 -1
PLUKENE, D. B.
PLUK 1976 -1
PNUELI, A.
EVENC 1970 -1
POCHTMAN, YU. M.
POCH 1976 -1
PODRIGUEZ, E. P.
STERA 1977 -1
POHL, I.
POHL 1969 -1; POHL 1970 -1; POHL 1973 -1
POLESSKIJ, V. P.
KEL'A 1976 -1; POLE 1976 -1
POLJAK, S.
POLJ 1974 -1
POLJAKOV, A. S.
POLJA 1974 -1; POLJA 1975 -1
POLLACK, M.
POLL 1977 -1
POLLAK H. O.
GILB 1968 -1
POLLATSCHECK, M. A.
POLLA 1976 -1
POLLATSCHEK, M. A.
MICH 1977 -1; ZIMM 1975 -1
POLONSKIJ, D. V.
KIRY 1975 -1
POLYMERIS, A.
POLY 1976 -1
POPESCU, D.
POPE 1976 -1
POPESCU, D. C.
POPEA 1976 -1
POPESCU, E.
POPEB 1975 -1
POPESCU, M.-F.
POPEC 1976 -1
POPOVICIU, N.
POPO 1972 -1; POPO 1973 -1
PORTUGAL, V. M.
PORT 1976 -1; PORT 1976 -2
POTTS, R.
POTT 1972 -1
POTTS, R. B.
KIRBA 1968 -1
POULTON, M. C.
POUL 1975 -1
POWELL, J.
POWE 1977 -1
POWELL, S.
LAND 1977 -1
PRATT, V. R.
PRAT 1973 -1
PREEM, M. K.
KAAZ 1973 -1
PRESENT, M.
PRES 1975 -1
PRESIA, H.-J.
PRESA 1975 -1; PRESA 1975 -2
PRIMAK, M. E.
KRUS 1964 -1

PRITSKER, A. A. B.
PRIT 1970 -1
PRIVALOV, V. V.
PRIV 1974 -1
PROSKUROWSKI, A.
PROS 1974 -1
PROTASI, M.
AUSIA 1977 -1; AUSIB 1977 -1
PRUDOVŠKII, B. D.
DYMA 1975 -1
PRUZAN, P. M.
KRARD 1968 -1; KRARD 1977 -1; KRARD 1977 -2
PUHOV, G. E.
PUHO 1976 -1
PUKAS, V.
KAMI 1975 -1
PULIAFITO, P. P.
GHIG 1976 -1
PULLEN, R. D.
NICH 1968 -1
PULLEYBLANK, W. R.
CILEA 1977 -1; PULL 1973 -1; PULL 1977 -1
PULL 1977 -2; PULLA 1974 -1
PULLMAN, N. J.
ALSP 1976 -1
PUN, L.
BOEB 1976 -1; LATE 1976 -1; PUN 1975 -1
PURCHECK, G. F. K.
PURC 1975 -1
PURI, M. C.
ARORA 1977 -1; ARORB 1976 -1; ARORB 1977 -1
BHAT 1976 -1; PURI 1975 -1
QUAIFE, H.
KLEEA 1976 -1
QUEYRANNE, M.
PICAC 1976 -1; PICAC 1977 -1; PICAC 1977 -2
QUINN, M. J.
SMITB 1977 -1
RABIN, M. O.
RABI 1959 -1; RABI 1960 -1; RABI 1976 -1
RABI 1977 -1
RABINOWITZ, P.
MANT 1977 -1
RADENSKI, A. A.
RADE 1975 -1
RADO. R.
HILTA 1973 -1
RAFES, R. YU.
POCH 1976 -1
RAGHAVACHARI, M.
RAGH 1976 -1
RAJARAMAN, M. K.
RAJA 1977 -1
RAJAT, K. D.
RAJAA 1977 -1
RAMAKRISHNAN, C. S.
RAMA 1977 -1
RAMAKRISHNAN, K. G.
RAMAA 1976 -1
RAMAKRISHNAN, V.
BALAF 1977 -1
RAMALINGAM, P.
RAMAB 1975 -1
RAMAMOORTHY, C. V.
RAMAC 1971 -1; RAMAD 1976 -1
RAMASWAMI, V.
RAVI 1977 -1
RAMASWAMY, K. H.
RAMAE 1975 -1
RAMASWAMY, K. V.
RAMAF 1976 -1
RANDI, P.
ANDRA 1976 -1
RANDOLPH, P.
RAND 1976 -1; RANDA 1977 -1
RANDOLPH, P. H.
RANDB 1976 -1
RANDOLPH, W. D.
ELZI 1976 -1
RAO, A. G.
KINA 1977 -1
RAO, M. R.
DANTA 1966 -1; DOULA 1969 -1; DOULB 1976 -1
FONGB 1975 -1; GARFB 1973 -1; GARFB 1977 -1
MERCA 1976 -1; NEEBA 1975 -1; RAO 1973 -1
RAO 1975 -1; RAO 1976 -1
RAO, R.
RAO A 1974 -1
RAO, S. S.
CHANI 1977 -1
RAO, V.
BERRB 1975 -1
RAPAPORT, H.
RAPA 1959 -1
RAPHAEL, B.
HART 1968 -1
RAPPOPORT, A. M.
FRIDA 1973 -1; RAPP 1976 -1
RARDIN, R. L.
KARW 1976 -1; KARW 1976 -2; KARW 1977 -1
KARW 1977 -2; LIN 1977 -1; RARD 1977 -1
RARDA 1976 -1; RARDA 1976 -2; RARDA 1976 -3
RASHKILIS, A. V.
PAULA 1976 -1; PAULA 1976 -2; PAULB 1975 -1
RATH, G. J.
MILLD 1976 -1
RATHBONE, D. E.
NOURA 1970 -1
RATHMANN, D.
RATH 1975 -1; RATH 1975 -2; RATH 1977 -1
RATH 1977 -2
RATLIFF, H. D.
BARTA 1977 -1; BARTB 1977 -1; BARTB 1977 -2
FRANA 1976 -1; MARTD 1976 -1; PICAA 1975 -1
PICAD 1975 -1
RAVINDRAN, A.
RAVI 1977 -1
RAWASWAMY, K. V.
RAWA 1975 -1
RAYBOV, G. G.
RAYB 1966 -1
RAZHUMIKHIN, B. S.
RAZH 1965 -1; RAZH 1967 -1; RAZH 1969 -1
RAZHA 1967 -1
READ, R. C.
PARRA 1967 -1; PARRA 1968 -1; READ 1978 -1
READA 1977 -1; READB 1975 -1
RECKHOW, R. A.
COOKA 1974 -1
RECSKI, A.
RECS 1976 -1; RECS 1976 -2; RECS 1976 -3
REDDI, S. S.
REDD 1976 -1; REDD 1976 -2
REEVES, G. R.
REEV 1976 -1
REINGOLD, E. M.
FILL 1977 -1; REIN 1977 -1
REKLAJTIS, V. K.
REKL 1977 -1
REMIZOVA, O. L.
SHEVA 1973 -1
RENYI, A.
ERDOD 1959 -1
REVELLE, C. S.
BERL 1976 -1; CHURB 1976 -1
REVIN, O. M.
KRAT 1977 -1
REYNER, W. S.
REYN 1977 -1
RIBBECK, K. F.
RIBB 1972 -1
RIBEILL, G.
RIBE 1971 -1
RIBIERE, G.
BENI 1977 -1; GAUT 1977 -1
RICHARDSON, R.
RICH 1976 -1
RICHMOND, L. B.
RICHA 1976 -1
RICHTER, D.
RICHB 1976 -1; RICHB 1976 -2; RICHB 1976 -3
RICHTER, K.
RICHC 1976 -1; RICHC 1977 -1
RIDER, K. L.
RIDE 1975 -1; RIDE 1976 -1
RIDLEY, T. M.
RIDL 1965 -1
RIEMERSMA, P.
ROOS 1975 -1
RIGGS, J. L.
DYKS 1977 -1
RIHA, W.
JAME 1976 -1; RIHA 1976 -1
RILEY, D.
RILE 1970 -1
RINALDI, S.
RINA 1975 -1; RINA 1975 -2
RINE, D. C.
RINE 1977 -1
RINGEISEN, R.
RANDA 1977 -1
RINNOOY KAN, A. H .G.
LAGE 1976 -1

RINNOOY KAN, A. H. G.
BRUCB 1975 -1; FOX B 1977 -1; GRAHA 1977 -1
JOHND 1977 -1; LAGEA 1976 -1; LAGEA 1976 -2
LAGEA 1976 -3; LAGEA 1976 -4; LENSB 1975 -1
LENSB 1976 -1; LENSB 1976 -2; LENSB 1976 -3
LENSB 1977 -1; LENSC 1977 -1; RINN 1976 -1
RINNA 1975 -1
RIORDON, J. S.
MAHM 1976 -1
RITZMAN, L. P.
SHOW 1977 -1
RIVEST, R. L.
HYAF 1976 -1; RIVE 1976 -1; YAO A 1977 -1
RIVIERE, M.
RIVI 1975 -1
ROBBINS, J. A.
ROBB 1975 -1
ROBERS, P.
ROBE 1976 -1
ROBERTS, F. D. K.
COCKC 1977 -1
ROBERTS, J.
ROBEA 1976 -1
ROBERTS, K. J.
ROBEB 1972 -1
ROBERTS, R.
ROBEC 1964 -1
ROBILLARD, P.
BRAT 1975 -1; FLORA 1970 -1
ROBINSON, D. F.
BARRC 1977 -1; BARRD 1977 -1; FOULA 1976 -1
ROBI 1972 -1
ROBINSON, D. R.
ROBIA 1977 -1
ROCHAT, J. C.
ROCH 1976 -1
RODEH, M.
ITAI 1977 -1; ITAI 1977 -2
RODGERS, J. R.
RODG 1977 -1
ROEDDER, W.
ROED 1976 -1
ROEDL, V.
NESE 1976 -1; ROEDA 1977 -1
ROESSLER, M.
ROES 1976 -1
ROGER, J.
ROGE 1971 -1
ROGERS, J. S.
ERLEA 1975 -1
ROMANOVSKII, I. V.
MURA 1974 -1
ROODMAN, G. M.
ROOD 1977 -1
ROOKER, J.
BURNA 1975 -1; BURNA 1976 -1
ROOSE, J. R.
ROOS 1975 -1
ROSE, D. J.
ROSE 1976 -1
ROSENFELD, M.
ROSEA 1971 -1
ROSENFIELD, D.
BODI 1976 -1
ROSENKRANTZ, D. J.
ROSEB 1974 -1; ROSEB 1977 -1
ROSENTHAL, A.
ROSEC 1975 -1; ROSEC 1977 -1; ROSED 1977 -1
ROSHCHIN, V. A.
LEBEC 1977 -1; ROSH 1977 -1
ROSHIN, G. B.
ROSHA 1977 -1
ROSS, G. T.
BARRA 1975 -1; CLOVI 1975 -1; ROSS 1977 -1
ROSSA 1976 -1
ROSS, G. T. JR.
ROSSB 1974 -1
ROSS, S. M.
DERM 1975 -1; DERM 1975 -2; DERM 1976 -1
ROSSETTO, S.
CONT 1976 -1
ROSSMAN, L. A.
ROSSC 1975 -1
ROTA, G. C.
CRAP 1970 -1
ROTH, G. W.
PATTA 1976 -1
ROTHFARB, B.
ROTH 1968 -1
ROTHFARB, R.
ROTHA 1969 -1
ROTHKOPF, M. H.
ROTHB 1966 -1; ROTHB 1975 -1
ROTHSCHILD, B. L.
GRAHB 1974 -1
ROUCAIROL, C.
FAUR 1976 -1
ROUSSEAU, C. C.
ERDOA 1976 -1
ROUSSEAU, P.
SANKB 1975 -1
ROUX, D.
MAUR 1976 -1
ROUX, M.
ROUX 1975 -1
ROVEDA, C.
BIONA 1969 -1
ROWE, A. J.
ROWE 1960 -1
ROWICKI, A.
ROWI 1976 -1
ROY, B.
ROY 1966 -1; ROY 1966 -2; ROY 1967 -1
ROY 1975 -1; ROY 1975 -2; ROY A 1964 -1
RUBIN, D. P.
RUBI 1972 -1
RUBIN, D. J.
MAC I 1975 -1; RUBIA 1977 -1
RUBIN, F.
RUBIB 1975 -1
RUBIN, I.
RUBIC 1975 -1
RUBINSHTEIN, M. I.
RUBID 1971 -1
RUBINSTEIN, M. C.
RUBIE 1976 -1
RUBINSTEIN, M.-C.
LUDO 1971 -1; RUBIF 1974 -1
RUBLINECKII, V. I.
KIRZ 1973 -1; KIRZ 1974 -1; LIVS 1972 -1
LIVS 1972 -2; RUBL 1973 -1; RUBLA 1973 -1
RUITER, E. R.
RUIT 1968 -1
RUPP, M.
KOVC 1976 -1
RUSCHITZKA, M.
RUSC 1977 -1
RUSCITTI, A.
RUSCA 1976 -1
RUSKEY, F.
RUSK 1977 -1
RUSNAK, Z.
RUSN 1976 -1
RUSSEL, R.
RUSS 1975 -1
RUSSELL, A. H.
RUSSA 1976 -1
RUSSELL, R.
HULTA 1976 -1; RUSSB 1977 -1
RUSSELL, R. A.
RUSSC 1977 -1
RUSSELL, S. O.
MARTH 1976 -1
RUSSO, F. J.
RUSSD 1965 -1
RUTENBERG, D. P.
RAO A 1974 -1
RUYGROK, J. T.
RUYG 1976 -1
RYAN, D. M.
FOST 1976 -1; RYAN 1976 -1
RYNDIN, B. N.
RYND 1977 -1; RYND 1977 -2
RYSTEDT, B.
NORD 1969 -1
RYSZKOWSKA, E.
JASI 1976 -1
RYZHKOV, A. P.
RYZH 1974 -1; RYZH 1975 -1; RYZH 1975 -2
SAARI, D. G.
SAAR 1977 -1
SAATCIOGLU, O.
SAAT 1977 -1
SAATY, T. L.
SAATA 1975 -1
SABIDUSSI, G.
SABI 1977 -1
SADOWSKI, R. F.
JACOB 1977 -1
SADOWSKI, R. P.
KATSA 1976 -1
SADYKOV, I. H.
SADY 1976 -1
SAFAEVA, K.
SAFA 1970 -1
SAGUEZ, C.
SAGU 1976 -1
SAHBAZJAN, K. V.
SAHB 1975 -1; SAHBA 1974 -1

SAHNI, S.
GONZD 1977 -1; GONZE 1976 -1; GONZE 1978 -1
HORO 1976 -1; IBARB 1974 -1; SAHN 1974 -1
SAHN 1975 -1; SAHN 1977 -1; SAHNA 1974 -1
SAHNB 1976 -1
SAHNI, S. K.
SAHNC 1976 -1
SAHUNJANC, T. G.
DANI 1977 -1
SAIGAL, R.
CHENF 1977 -1; SAIG 1969 -1
SAIPE, A. L.
SAIP 1975 -1
SAKAGUCHI, M.
SAKA 1976 -1
SAKAKI, T.
SAKAA 1976 -1
SAKAROVITCH, M.
MAC 1977 -1; SAKAB 1966 -1; SAKAB 1968 -1
SAKAB 1976 -1
SAKATA, M.
SAKAC 1970 -1
SALERNO, G.
SALE 1976 -1
SALIHOV, Z. M.
SALI 1970 -1
SALKIN, H. M.
SALK 1975 -1; SALKA 1975 -1; SALKB 1972 -1
SAMPATHKUMAR, E.
SAMP 1976 -1
SAMUELSSON, H.
BALAK 1977 -1
SAMUELSSON, H. M.
NARUA 1977 -1
SANADA, H.
SANA 1968 -1; SANAA 1967 -1
SANCHO, N. G. F.
SANC 1976 -1
SANDEWALL, E. J.
SAND 1975 -1
SANDI, C.
SANDA 1975 -1
SANKIBAEV, B. N.
SANK 1975 -1; SANKA 1973 -1
SANKOFF, D.
SANKB 1975 -1
SARASIDZE, G. K.
PUHO 1976 -1
SARKAR, H.
GHOS 1970 -1
SARVANOV, V. I.
SARV 1976 -1
SARVANOV, V.I.
SARVA 1976 -1
SASAKI, M.
GEN B 1975 -1
SAUCIER, G.
SAUC 1971 -1
SAVAGE, J. E.
SAVA 1976 -1
SAVAGE, S. L.
SAVAA 1973 -1; SAVAA 1973 -2; SAVAA 1976 -1
SAVAA 1976 -2; SAVAA 1978 -1; SAVAB 1976 -1
SAVITCH, W. J.
SAVI 1974 -1
SAYAMA, H.
SAYA 1976 -1; SAYA 1976 -2
SCALAS, M. R.
ANDRA 1976 -1; NICOB 1976 -1
SCARBROUGH, J. E.
SCAR 1977 -1
SCHAAR, G.
SCHA 1976 -1
SCHAEFER, M. K.
HURT 1975 -1
SCHAERF, S.
SCHAA 1971 -1
SCHALLEHN, W.
SCHAB 1972 -1
SCHARGE, L.
SCHAC 1974 -1
SCHELP, R. H.
ERDOA 1976 -1; FAUD 1974 -1
SCHINDLER, S.
LUEDA 1976 -1
SCHINZINGER, R.
SCHI 1970 -1
SCHJAER-JACOBSEN, H.
MADS 1976 -1
SCHMID, R.
BIONA 1969 -1
SCHMIDT, B. K.
SCHM 1976 -1
SCHMIDT, D. C.
SCHMA 1976 -1
SCHMITZ, W.
SCHMB 1975 -1
SCHNORR, C. P.
SCHN 1976 -1; SCHN 1976 -2; SCHN 1978 -1
SCHN 1978 -2
SCHOCH, M.
GRUNB 1977 -1; SCHO 1976 -1
SCHOEMAN, M. E. F.
SCHOA 1970 -1
SCHOENE, W.
SCHOB 1975 -1
SCHOENLEIN, A.
SCHOC 1976 -1; SCHOC 1977 -1
SCHONHEIM, J.
SCHOD 1975 -1
SCHRADER, R.
BACHB 1977 -1
SCHRAGE, L. E.
FOX B 1977 -1
SCHRIJVER, L.
SCHR 1976 -1
SCHULTZ, G. P.
SCHU 1968 -1; SCHU 1969 -1
SCHURAWLJEW, J. I.
SCHUA 1977 -1
SCHWARTZ, D. E.
HOFFB 1976 -1
SCHWARTZ, M.
SCHW 1977 -1
SCHWARZ, L. B.
FREI 1973 -1; ROOD 1977 -1
SCHWEITZER, P.
GAVI 1977 -1
SCHWENK, A.
KRARC 1972 -1
SCIONS, H. I.
SCIO 1964 -1
SCLEE, W.
SCLE 1977 -1
SCOTT, A. J.
SCOT 1970 -1; SCOT 1971 -1; SCOT 1971 -2
SCOT 1971 -3; SCOT 1975 -1
SCOTT, S. H.
HILTA 1973 -1; SCOTA 1975 -1
SCOTT, T. B.
SCOTB 1976 -1
SCRIABIN, M.
SCRI 1975 -1; VERGA 1977 -1
SCWARTZ, J. G.
SCWA 1968 -1
SEBASTIAN, H.-J.
SCHAB 1972 -1
SEDLACEK, J.
SEDL 1975 -1
SEELAENDER, J.
SEEL 1974 -1; SEEL 1976 -1; SEEL 1977 -1
SEERY, J. B.
BOYCB 1973 -1
SEGIENKO, I. V.
ARKH 1975 -1
SEIDVASSER, M. A.
SEID 1974 -1; SEID 1974 -2
SEIFERAS, J. I.
SEIF 1974 -1
SEMENOV, V. M.
SEME 1974 -1
SEMOTYUK, V. N.
BLAZC 1976 -1
SERCK-HANSSEN, J.
SERC 1970 -1
SERDYUKOV, A. I.
SERD 1975 -1; SERD 1976 -1
SERGHIOU, S. S.
EILO 1975 -1
SERGIENKO, I. V.
KASP 1977 -1; KASPA 1974 -1; LEBEC 1977 -1
ROSH 1977 -1; SERG 1975 -1
SERMAN, A. I.
COI 1975 -1; SANKA 1973 -1
SEROV, S. S.
BUR' 1976 -1; BUR' 1976 -2; SERO 1976 -1
SETHI, R.
BRUNB 1974 -1; COFFA 1976 -1; GAREC 1976 -1
HORV 1977 -1; LAM 1977 -1; SETH 1976 -1
SETHI, S. P.
BOOKA 1977 -1
SEVAST'YANOV, S. V.
PEREA 1976 -1; SEVA 1975 -1
SEYMOUR, P. D.
SEYM 1975 -1; SEYM 1976 -1; SEYM 1977 -1
SHAH, Y. J.
SHAH 1974 -1
SHAHAM, J.
SCHAA 1971 -1

****

SHAKHBASJAN, K. V.
SHAK 1977 -1
SHALABY, M.
KENNA 1977 -1
SHAMBLIN, J. E.
FABR 1966 -1; ROBB 1975 -1
SHAMIR, A.
EVENA 1975 -1
SHAMMA, M. M.
SHAM 1972 -1
SHAMOS, M. I.
SHAMA 1975 -1
SHAPIRO, A. D.
SHAP 1975 -1
SHAPIRO, D. M.
SHAPA 1966 -1
SHAPIRO, J. F.
BELLA 1975 -1; BELLA 1977 -1; FISHC 1973 -1
SHAPB 1974 -1; SHAPB 1976 -1; SHAPB 1976 -2
SHAPB 1977 -1; SHAPB 1977 -2; SHAPB 1977 -3
SHAPIRO, R. D.
SHAPC 1976 -1
SHAPIRO, S. D.
SHAPD 1977 -1
SHARIF, M. N.
SHAR 1975 -1
SHARIQ, S. Z.
SHARA 1975 -1
SHARMA, J. K.
SHARB 1976 -1; SHARB 1976 -2; SHARC 1978 -1
SHARMA, S. R.
SHARD 1977 -1
SHARP, G. P.
SHARE 1975 -1
SHARPE, R.
SHARF 1972 -1; SHARG 1975 -1
SHAW, G. C.
SHAW 1975 -1
SHAW, J. F. B.
SHAWA 1974 -1
SHEDLER, G. S.
LAVE 1976 -1
SHEE, S. C.
TEH 1976 -1
SHEEHAN, J.
SHEE 1975 -1
SHEFER, D.
GULD 1977 -1
SHEFFI, Y.
DAGA 1977 -1
SHELNESS, N. H.
SHEL 1975 -1
SHEN, V. Y.
CHENI 1971 -1; KAFU 1977 -1
SHEPPARD, E. S.
SHEP 1974 -1
SHEPSLE, K. A.
SHEPA 1975 -1; SHEPA 1975 -2
SHERALI, A. D.
SHER 1977 -1
SHERALI, H.
PARKA 1977 -1
SHERMAN, A. K.
KRAVA 1975 -1
SHERMAN, E. B.
VASE 1975 -1
SHETTY, C. M.
BULFA 1977 -1; SHER 1977 -1; SHET 1975 -1
SHEVCHENKO, V. N.
GLEBB 1963 -1; SHEV 1975 -1; SHEV 1976 -1
SHEVA 1973 -1
SHIBANOV, S. E.
CHEBA 1975 -1
SHIBUI, M.
ASAN 1976 -1
SHIER, D.
MURCA 1973 -1
SHIER, D. R.
SHIE 1974 -1; SHIE 1976 -1; SHIE 1976 -2
SHIE 1976 -3; SHIE 1977 -1; SHIE 1977 -2
SHIH, W.
SHIH 1977 -1
SHIMBEL, A.
SHIM 1951 -1; SHIM 1953 -1
SHINOFUJI, S.
GEN A 1976 -1
SHIRAI, Y.
SHIR 1976 -1
SHIRAKAWA, I.
AM 1976 -1; TRAN 1976 -1; TSUK 1977 -1
TSUKA 1975 -1
SHIREY, R. W.
SHIRA 1969 -1
SHKURBA, V. V.
MIKHA 1966 -1; SHKU 1965 -1; TANA 1975 -1
SHLIFER, E.
GAVI 1977 -1
SHMELEV, V. V.
SHME 1975 -1
SHOSTAK, R. E.
SHOS 1974 -1
SHOWALTER, M. J.
SHOW 1977 -1
SHTEIN, M. E.
ZAIT 1975 -1; ZAIT 1976 -1
SHULDINER, P. W.
WATT 1963 -1
SHUO-YEN, R. L.
SHUO 1977 -1
SHYARKSHNIS, G. A.
PAULA 1976 -1; PAULA 1976 -2; PAULB 1975 -1
SIBIRSKIJ, V. K.
SIBI 1976 -1
SICKLE, L. VAN
SICK 1977 -1
SIDNEY, J. B.
SIDN 1977 -1
SIELKEN, R. M. JR.
SIEL 1976 -1
SIGAL, I. KH.
DAVY 1972 -1
SILMAN, L.
PASS 1975 -1
SIMEONE, B.
GALLB 1976 -1; GALLC 1972 -1
SIMIC, S. K.
CVET 1976 -1
SIMMONS, L. F.
DAVIC 1977 -1
SIMOES-PEREIRA, J. M. S.
SIMO 1974 -1
SIMON, H.
BACHC 1977 -1
SIMS, J.
STAC 1976 -1
SINGER, S.
SING 1968 -1
SINGMASTER D.
SINGA 1975 -1
SINT, L.
CHAMA 1977 -1
SIRIK, A. F.
SIRI 1968 -1
SIROVICH, F.
SIRO 1971 -1
SITRUK, G.
SITR 1974 -1
SKACHEK, B. YA.
VOLOA 1976 -1
SKAFF, M. S.
SKAF 1966 -1
SKOROBOGATOV, V. A.
SKOR 1975 -1
SKUPIEN, Z.
SKUP 1976 -1
SLATE, L.
SLAT 1976 -1
SLATER, P.
GOODB 1974 -1
SLATER, P. J.
GOODD 1975 -1; SLATA 1975 -1; SLATA 1975 -2
SLATA 1976 -1; SLATA 1976 -2; SLATB 1976 -1
SLOBODTCHIKOVA, R. I.
EGOR 1977 -1
SLOMINSKI, L.
SLOM 1975 -1
SLOWINSKI, R.
BLAZB 1976 -1; BLAZB 1976 -2; WEGLA 1976 -1
SMEERS, Y.
DAEN 1977 -1
SMELEV, V. V.
SMEL 1975 -1
SMITH, D. H.
SMIT 1974 -1
SMITH, E. M.
SMITA 1976 -1
SMITH, J. F.
SMITB 1977 -1
SMITH, K. R.
SMITC 1976 -1
SMITH, L. D.
MARKA 1975 -1; SMITD 1975 -1; SMITD 1976 -1
SMITE 1977 -1
SMITH, M. L.
CAMP 1970 -1; SMITF 1975 -1
SMITH, R. L.
WILSB 1974 -1
SMITH, R. S.
SMITG 1975 -1

SMITH, T. H. C.
SMITH 1975 -1; SMITI 1977 -1; SMITJ 1977 -1
SMITH, W. G.
CHUA 1965 -1
SMOLYAR, L. I.
SMOL 1966 -1
SMYER, W. N.
ZALOA 1977 -1
SMYTH, W. F.
SMYT 1974 -1
SNICKARS, F.
SNIC 1978 -1
SODARDO, D.
KORT 1966 -1
SOH, J. W.
GONZC 1976 -1
SOLAND, R. M.
ROSS 1977 -1; ROSSA 1976 -1
SOLDATOV, V. S.
KALY 1976 -1
SOLOVAY, R.
BAKEC 1975 -1
SOLOVAY, R. M.
SOLO 1976 -1
SOMMER, D.
GLOVI 1976 -1
SOMMER, D. C.
SOMM 1972 -1
SOMNEA, D.
LIVO 1976 -1
SONNEMANN, E.
SONN 1974 -1
SOUKUP, J.
SOUK 1975 -1
SOVETOV, B. YA.
SOVE 1976 -1; SOVE 1976 -2
SPAETH, H.
SPAE 1977 -1
SPANEL, J.
SPAN 1976 -1
SPATHIS-DIMITROPOULOU, V.
SPAT 1975 -1
SPENCER, J.
ERDOE 1974 -1; SPEN 1976 -1
SPERANZA, F.
SPER 1975 -1
SPIELBERG, K.
GUIGA 1974 -1; GUIGA 1976 -1; GUIGA 1976 -2
GUIGA 1977 -1; GUIGA 1977 -2; GUIGA 1977 -3
GUIGA 1977 -4; JOHNG 1976 -1; MAIR 1977 -1
SALKB 1972 -1; SLAT 1976 -1; SPIE 1972 -1
SPIE 1973 -1; SPIE 1973 -2; SPIE 1974 -1
SPIE 1977 -1
SPIRN, J. R.
SPIR 1976 -1
SPRADLIN, B. C.
SPRA 1967 -1
SRINIVASAN, V.
BALAD 1977 -1; FONGC 1974 -1; FONGC 1974 -2
FONGC 1977 -1; FREI 1973 -1; SMITI 1977 -1
SRIN 1976 -1; SRINA 1976 -1; SRINA 1977 -1
STACEY, K. C.
STAC 1976 -1
STADEL, M.
STAD 1978 -1
STAHL, S.
STAH 1976 -1
STAHL, W.
STAHA 1976 -1
STAINTON, R. S.
STAI 1977 -1
STALBO, A. K.
DYMA 1975 -1
STANCU-MINASIAN, I. M.
STAN 1974 -1
STANCZAK, W.
KACPA 1975 -1
STANFORD, R. E.
STANA 1976 -1
STANIONIS, K.
PAULB 1975 -1
STANKARD, M. F.
STANB 1969 -1
STANTON, L. W.
BHAUA 1974 -1
STANTON, R. G.
CERF 1974 -1; MULLA 1978 -1
STAROBINEC, S. M.
STAR 1972 -1; STAR 1975 -1
STARR, J. H.
BOWM 1974 -1; BOWMA 1977 -1; STARA 1975 -1
STAUDER, E.
STAU 1974 -1
STAVROKOVA, N. E.
RAZHA 1967 -1
STEARNS, R. E.
ROSEB 1974 -1; ROSEB 1977 -1
STEFFEN, M. R.
STEF 1969 -1
STEFFENS, K.
STEFA 1976 -1; STEFA 1977 -1
STEIGLITZ, K.
CHRIB 1976 -1; HEND 1976 -1; KOHLB 1976 -1
PAPAA 1976 -1; PAPAA 1977 -1; STEI 1969 -1
STEINBERG, D. I.
STEIA 1976 -1; STEIA 1977 -1
STEINBERG, R.
MEL' 1977 -1
STEINER, J.
STEIB 1976 -1
STEPHENS, P. D.
SHEL 1975 -1
STERBOUL, F.
STER 1974 -1; STER 1974 -2; STER 1975 -1
STER 1976 -1; STER 1977 -1
STERN, H. I.
STERA 1977 -1
STEWART, W. R. JR.
STEW 1976 -1
STIEBITZ, V.
HIRCA 1975 -1
STIERLIN, C.
STIE 1976 -1
STOCKMEYER, L.
STOC 1973 -1
STOCKMEYER, L. J.
KOU A 1978 -1
STOCKTON, F.
STOCA 1968 -1
STOHR, E. A.
MURP 1975 -1; MURP 1976 -1
STOICA, M.
STOI 1977 -1
STOJAN, YU. G.
STOJ 1976 -1
STOJKA, J. P. JR.
KING 1976 -1
STOLIN, JA. N.
BELOA 1974 -1
STOLZENBURG, I.
STOL 1977 -1
STONE, C. A.
MAY A 1974 -1
STOUTJESDIJK, A.
KENDA 1975 -1
STRADLIN, B. C.
STRA 1967 -1
STRATMANN, K.-H.
BURKD 1976 -1; STRAA 1976 -1; STRAB 1976 -1
STRAUS, E. G.
BOLLD 1974 -1
STREET, A. P.
STRE 1974 -1; STRE 1976 -1
STRICKER, R.
STRI 1970 -1
STROBEL, H.
STRO 1977 -1
STROEKER, R. J.
STROA 1976 -1
STUHR, R.
STUH 1966 -1
STUTZ, J.
GLOVB 1977 -1; GLOVF 1973 -1
STYCZEN, K.
STYC 1974 -1
SU, S. Y.
DEINA 1971 -1
SUBRAHMANYAM, Y. V.
PANDA 1975 -1
SUGITO, Y.
MIYA 1977 -1
SUHL, U.
MEVE 1977 -1; SUHL 1975 -1; SUHL 1976 -1
SUHL 1976 -2
SUHORUCENKO, E. M.
GERM 1971 -1
SUL'GA, YU. N.
SUL' 1974 -1
SUMAROKOV, S. V.
SUMA 1976 -1
SUMMER, D. P.
SUMM 1976 -1
SUMNER, D. P.
TROTC 1976 -1
SUPRUNENKO, D. A.
SUPR 1975 -1
SURKIS, J.
ADAM 1977 -1
SURMENEV, E. V.
KIM 1976 -1

SUSSENGUTH, E. H.
SUSS 1965 -1
SUSTAKASVILI, I. I.
SUST 1975 -1
SVARTIN, S. M.
SVAR 1975 -1
SWAIN, R. W.
SWAI 1971 -1; SWAI 1971 -2; SWAI 1976 -1
SWANEK, J. J.
SWAN 1974 -1
SWANSON, L. W.
SWANA 1976 -1
SWARUP, K.
ARORB 1976 -1; ARORB 1977 -1; BHAT 1976 -1
PURI 1975 -1; SHARC 1978 -1; SHARD 1977 -1
SWEENEY, D. J.
SWEE 1977 -1; SWEEA 1976 -1
SWEN, G.
SWEN 1967 -1
SYMONS, J.
MORR 1977 -1
SYSLO, M.
SYSL 1972 -1; SYSL 1973 -1; SYSL 1974 -1
SYSLO, M. M.
DZIKA 1974 -1; GRABA 1975 -1; HUBI 1975 -1
SYSLA 1975 -1; SYSLA 1976 -1
SZWARC, W.
SZWA 1974 -1; SZWA 1977 -1; SZWA 1977 -2
SZWAA 1977 -1
SZWARCFITER, J. L.
SZWAB 1976 -1
TABLIASHVILI, O. M.
GEORA 1974 -1
TABOURIER, Y.
RIBE 1971 -1; TABO 1969 -1; TABO 1973 -1
TAHA, H. A.
TAHA 1975 -1; TAHA 1975 -2
TAINITER, M.
TAIN 1975 -1
TAKAHASHI, I.
TAKA 1966 -1
TAKAHASHI, Y.
TAKAA 1974 -1
TAKANAMI, I.
ASAN 1976 -1
TAKAZAWA, T.
FUJI 1966 -1
TALWAR, P. P.
TALW 1967 -1
TAMIR, A.
TAMI 1976 -1
TAMULINAS, B. V.
REKL 1977 -1
TANAEV, V. S.
TANA 1975 -1
TANG, D. T.
MARU 1976 -1
TANGEDAHL, L.
GLOVJ 1976 -1
TAPIERO, C. S.
TAPI 1971 -1
TARJAN, R. E.
CHERB 1976 -1; ESWA 1976 -1; EVEND 1975 -1
EVEND 1975 -2; GARED 1976 -1; GOODC 1976 -1
HOPCB 1971 -1; HOPCB 1972 -1; HOPCB 1973 -1
READB 1975 -1; ROSE 1976 -1; TARJ 1971 -1
TARJ 1972 -1; TARJ 1973 -1; TARJ 1973 -2
TARJ 1974 -1; TARJ 1975 -1; TARJ 1975 -2
TARJ 1976 -1; TARJ 1976 -2; TARJ 1977 -1
TARJ 1977 -2; TARJ 1978 -1; TARJA 1977 -1
TARTAR, J.
TART 1976 -1
TATHAM, R. L.
SWEEA 1976 -1
TAVIS, L. A.
CRUMA 1977 -1
TAYLOR, P. E.
TAYL 1975 -1
TCHERKASSKY, B. V.
TCHE 1977 -1
TCHON, K.
ADAMA 1976 -1
TEH, H. H.
KOH A 1974 -1; TEH 1976 -1
TEHRANI, A.
TEHR 1975 -1
TEPLICKII, M. G.
TEPL 1975 -1
TERNO, J.
TERN 1975 -1
TERRELL, M. P.
DAVIE 1976 -1; TURNB 1976 -1
TERZI, D. G.
TERZ 1974 -1
TETERIN, G. P.
PRIV 1974 -1
TEZUKA, Y.
SANA 1968 -1; SANAA 1967 -1
THESEN, A.
BAGB 1976 -1; THES 1975 -1; THES 1976 -1
THIRIET, L.
THIR 1966 -1
THISSE, J.
THIS 1976 -1
THISSE, J.-F.
HANSE 1977 -1
THOMAS, E. A. C.
DAVIB 1976 -1
THOMAS, G. N.
ADOLA 1977 -1
THOMAS, M. E.
THOM 1976 -1
THOMAS, R. E.
THOMA 1976 -1
THOMAS, W. H.
GRAVB 1976 -1
THOMPSON, G. L.
BALAD 1977 -1; BALAE 1975 -1; BALAE 1975 -2
BALAE 1975 -3; BALAE 1975 -4; MUTH 1963 -1
SMITI 1977 -1; SMITJ 1977 -1; SRINA 1976 -1
SRINA 1977 -1; THOMB 1976 -1
THOMSON, P. F.
THOMC 1976 -1
THORELLI, L. E.
THOR 1966 -1
THUEMMLER, S.
THUE 1976 -1
THUMER, R.
THUM 1975 -1
TIEN, B. N.
TIEN 1977 -1
TIERNAN, J. C.
TIER 1974 -1
TIFFIN, J. E.
TIFF 1974 -1
TILANUS, C. B.
TILA 1976 -1
TILLMAN, F. A.
FUNK 1968 -1
TILQUIN, C.
TILQ 1975 -1
TIMOFEEV, A. V.
PETRA 1975 -1
TIMOFEEV, E. A.
KAZA 1975 -1; TIMO 1975 -1; TIMO 1977 -1
TIND, J.
TIND 1977 -1; TIND 1977 -2
TINHOFER, G.
HINT 1977 -1; TINH 1976 -1; TINH 1976 -2
TISHKIN, A. I.
SOVE 1976 -1; SOVE 1976 -2
TOBEY, M. D.
HOLZA 1973 -1
TOBIN, R. L.
TOBI 1975 -1; TOBI 1977 -1
TODD, M. J.
TODD 1976 -1; TODD 1977 -1
TOFT, B.
TOFT 1975 -1
TOLBIN, R. L.
TOLB 1975 -1
TOLCHAN, A. YA.
TOLC 1976 -1
TOLLA, P.
FAUR 1976 -1
TOMASHEVSKIY, V. M.
ZAYC 1975 -1
TOMCUK, E. JA.
RUBLA 1973 -1
TOMESCU, I.
TOME 1966 -1; TOME 1968 -1; TOME 1975 -1
TOME 1975 -2; TOME 1975 -3; TOME 1976 -1
TOMIZAWA, N.
IRI A 1975 -1; IRI A 1976 -1; TOMI 1973 -1
TOMLIN, J. A.
TOML 1966 -1
TONE, K.
TONE 1976 -1
TORRIERO, A.
TORR 1976 -1
TOTH, P.
CARP 1977 -1; MARTC 1976 -1; MARTC 1977 -1
MARTC 1977 -2; MARTC 1977 -3; MARTC 1977 -4
TOTH 1977 -1
TOWNSEND, W.
TOWN 1977 -1; TOWN 1977 -2
TOWSNEND, W.
TOWS 1977 -1

TRAIGER, I. L.
GILLA 1969 -1
TRAN DINH AM
TRAN 1976 -1
TRANSPORTATION CENTER LIBRARY
NORT 1976 -1
TRCKA, V.
MOLN 1976 -1
TRIFON, R.
TRIF 1973 -1
TRIPPI, R. R.
ERLEB 1975 -1
TRIVEDI, V. M.
TRIV 1976 -1
TROFIMOV, V. N.
BURDA 1976 -1
TROJANOWSKI, A. E.
TARJA 1977 -1
TROTTER, L. E. JR.
BAUM 1977 -1; HUAN 1977 -1; TROT 1973 -1
TROT 1977 -1; TROTA 1977 -1
TROTTER, W. T. JR.
TROTB 1976 -1; TROTB 1976 -2; TROTC 1976 -1
TROTD 1976 -1
TRUBIN, V. A.
TRUB 1976 -1; TRUB 1977 -1
TRUEMPER, K.
TRUE 1976 -1; TRUE 1976 -2; TRUE 1976 -3
TRUE 1977 -1; TRUE 1977 -2; TRUEA 1977 -1
TRUEMPER, L.
TRUEB 1977 -1
TRUHANOVSKII, N. N.
KONO 1976 -1
TRUSCOTT, W. G.
TRUS 1976 -1; TRUS 1976 -2; WESOB 1975 -1
WESOB 1976 -1
TRYGAR, T. A.
TRYG 1976 -1
TRYPIA, M.
TRYP 1977 -1
TSE, A.
WATE 1976 -1
TSOY, E. V.
YUDIA 1974 -1
TSUKIYAMA, S.
AM 1976 -1; TRAN 1976 -1; TSUK 1977 -1
TSUKA 1975 -1
TSYNKOVA, E. N.
TSYN 1968 -1
TUCKER, A.
TUCK 1977 -1
TUCKER, S.
ALBEA 1976 -1
TUERK, P.
TUER 1977 -1
TUR, Y.
DAR-A 1977 -1
TURCAT, C.
TURC 1974 -1
TURCHANYI, P.
TURCA 1975 -1
TURK, P.
TURK 1977 -1
TURNER, J.
TURN 1968 -1
TURNER, W. C.
ROBB 1975 -1; TURNA 1975 -1; TURNB 1976 -1
VU 1976 -1
TUSHKINA, T. A.
SHAK 1977 -1
TUSKINA, T. A.
SAHB 1975 -1; SAHBA 1974 -1
TUTTE, W. T.
TUTT 1961 -1; TUTT 1976 -1
TYURENKOV, V. A.
TYUR 1963 -1
UEDA, T.
UEDA 1977 -1
UELKUECUE, A.
UELK 1975 -1
UHRY, J. P.
UHRY 1975 -1
UI, T.
NOMU 1974 -1
ULGEN, O.
SMITF 1975 -1
ULLMAN, J. D.
AHO A 1974 -1; AHO B 1974 -1; AHO C 1976 -1
FONG 1977 -1; ULLM 1976 -1; ULLM 1976 -2
ULLMANN, J. R.
ULLMA 1976 -1
ULLRICH, M.
ULLR 1976 -1; ULLR 1976 -2
ULPH, A.
ULPH 1977 -1
ULRICH, L. Y.
ULRI 1975 -1
UNDERWOOD, E. E.
MARCC 1972 -1
UNGAR, A. J.
MARCB 1967 -1
UNGER, P. S.
PHIL 1976 -1
UNGER, S. H.
UNGE 1964 -1
UNGER, V. E.
RARDA 1976 -1; RARDA 1976 -2; RARDA 1976 -3
UNITED STATES DEPT. OF COMMERCE BUREAU OF PUBLIC
UNIT 1964 -1
URQUHART, R. J.
URQU 1967 -1
USHAKOV, E. P.
USHA 1978 -1
USKOV, A. V.
ARLA 1973 -1
USMANOV, A. A.
USMA 1970 -1
UTTER, M. L.
STERA 1977 -1
UZDEMIR, A. P.
BUR' 1976 -1; BUR' 1976 -2; SERO 1976 -1
VAIRAVAN, K.
VAIR 1976 -1
VALIANT, L. G.
ANGL 1977 -1; PIPP 1976 -1; VALI 1976 -1
VANEMDEBOAS, P.
VANL 1976 -1
VANEMDEN, M. H.
VANE 1975 -1
VANLEEUWEN, J.
VANL 1976 -1
VARSAVSKII, V. I.
VARS 1974 -1
VARTANYAN, G. M.
ANDO 1976 -1
VASERSHTEJN, L. N.
VASE 1975 -1
VATZ, D.
ZALO 1975 -1
VAZQUUEZ MUNIZ, A.
ESCUA 1976 -1
VELICKIN, V. Z.
VELI 1974 -1
VELIKOVIC, L. L.
VELIA 1975 -1
VEMUGANTI, R. R.
ANEJ 1977 -1; ELZIA 1976 -1; HOUC 1976 -1
HOUC 1977 -1
VENTRE, A. G.
VENT 1975 -1
VERGIN, R. C.
SCRI 1975 -1; VERG 1976 -1; VERGA 1977 -1
VERGLUNSKY, S.
VERGB 1951 -1
VERGNAS, M. LAS
BLANA 1978 -1; VERGC 1975 -1; VERGC 1976 -1
VESELOVSKY, V. E.
VESE 1977 -1
VIDYASANKAR, K.
VIDY 1975 -1
VIGGIONI, S.
ROY A 1964 -1
VINCENTELLI, S. A.
VINC 1975 -1
VINOGRADSKAJA, S. V.
VINO 1975 -1
VISHNEVSKIJ, L. D.
VISH 1976 -1
VISING, V. G.
VISI 1977 -1
VIVROT, T. M.
KHOM 1976 -1
VIZVARI, B.
VIZV 1975 -1
VJATKIN, A. K.
VOLY 1974 -1
VLACH, M.
VLAC 1969 -1; VLAC 1977 -1; VLACA 1971 -1
VLASYUK, B. A.
VLAS 1966 -1
VODOLAZOV, V. S.
VODO 1976 -1
VOGT, W. G.
MICKA 1969 -1; MICKA 1975 -1
VOILS, L. R.
CAMEB 1976 -1
VOLLMANN, T. E.
MOJE 1976 -1
VOLODARSKIJ, Y. A. M.
VOLO 1976 -1

****

VOLODIN, A. A.
LEVYB 1976 -1
VOLODOS, I. F.
ALEKB 1976 -1
VOLOVIK, L. N.
VOLOA 1976 -1
VOLPENTESTA, A.
VOLP 1976 -1; VOLP 1977 -1
VOLYNSKII, E. I.
VOLY 1974 -1
VOTJAKOV, A. A.
BUZYA 1976 -1; FRIDB 1976 -1; VOTJ 1973 -1
VU, V. T.
VU 1976 -1
VUILLEMIN, J.
RIVE 1976 -1
VYATKIN, A. K.
VYAT 1975 -1
WACHS, M. L.
GARS 1977 -1
WAGNER, H.
TEON 1975 -1
WAGNER, H. M.
YUAN 1975 -1
WAGNER, K.
PETR 1976 -1
WAGNER, N. S.
MAY A 1974 -1
WAGNER, R. A.
WAGN 1976 -1
WAHI, P. N.
BRADB 1969 -1
WAKEFIELD, G. W.
MAIR 1977 -1
WALD, C. G. DE
PATE 1976 -1
WALDVOGEL, J.
WALD 1976 -1
WALKER, D. A.
PFAF 1976 -1
WALKER, R. C.
WALK 1976 -1
WALKER, W. E.
WALKA 1976 -1
WALKER, W. R.
ORON 1976 -1
WALLACH, Y.
WALL 1977 -1
WALLER, D. A.
WALLA 1976 -1
WALSH, A. M.
WALS 1974 -1
WALSH, T. S.
WALSA 1976 -1
WALTERS, J. E.
DAVID 1977 -1
WALTERS, J. R.
WALT 1976 -1
WALUKIEWICZ, S.
WALU 1975 -1; WALU 1975 -2
WANG, D. L.
WANG 1976 -1
WANG, D.-L.
WANGA 1973 -1
WANG, G. S.
FAN 1964 -1
WANG, J. C.
MICKB 1977 -1; WANGB 1974 -1
WARBURTON, A. R.
MITTB 1973 -1
WARNER, D. M.
TRIV 1976 -1
WARREN, H. S. JR.
WARR 1977 -1
WASOWSKI, J.
WASO 1976 -1
WASSENHOVE, L. VAN
GELD 1976 -1
WASSON, W. D.
WASS 1975 -1
WATANABE, H.
KATA 1968 -1
WATERS, W. G. II
WATE 1976 -1
WATKINS, M. E.
GRAVA 1977 -1
WATTLEWORTH, J. A.
WATT 1963 -1
WDOWIAK, J.
WDOW 1973 -1
WEBSTER, D. B.
BALL 1977 -1
WEEKS, J. K.
WEEK 1976 -1
WEGENER, M.
WEGE 1974 -1
WEGLARZ, J.
BLAZB 1976 -1; BLAZB 1976 -2; WEGL 1975 -1
WEGL 1976 -1; WEGLA 1976 -1
WEIGAND, M. M.
WEIG 1973 -1; WEIG 1976 -1
WEINBERG, F.
WEIN 1977 -1
WEINBERG, L.
BRUND 1976 -1; BRUND 1976 -2; WEINA 1960 -1
WEINA 1965 -1; WEINA 1975 -1; WEINA 1977 -1
WEINBERGER, D. B.
WEINB 1976 -1; WEINB 1976 -2
WEINER, P.
SAVAB 1976 -1; STEI 1969 -1; WEINC 1975 -1
WEINGARTEN, A.
NAND 1974 -1
WEINSTEIN, J.
WEIND 1974 -1
WEINSTEIN, J. M.
WEINE 1975 -1
WEINTRAUB, A.
WEINF 1976 -1
WEISCHEDEL, R.
BOYC 1973 -1
WELLS, D. W.
WELL 1976 -1
WELLS, M.
WELLA 1971 -1
WELSH, D. J. A.
MILNA 1976 -1; WELS 1970 -1; WELS 1976 -1
WENDELL, R. E.
HURT 1975 -1; OLSO 1976 -1; WEND 1977 -1
WENDLER, K.
WENDA 1976 -1
WERRA, D. DE
BRELA 1977 -1; WERR 1974 -1; WERR 1975 -1
WERR 1975 -2; WERR 1975 -3; WERR 1975 -4
WERR 1975 -5; WERR 1976 -1; WERR 1976 -2
WERR 1976 -3; WERR 1976 -4; WERR 1976 -5
WERR 1976 -6; WERR 1977 -1; WERR 1977 -2
WESOLOWSKY, G. O.
WESO 1974 -1; WESO 1977 -1; WESOA 1971 -1
WESOB 1975 -1; WESOB 1976 -1
WESSELS, H.
WESS 1972 -1
WESSELS, J.
NUNE 1978 -1
WESTERBERG, C.-H.
WEST 1977 -1
WHARTON, F.
COVE 1976 -1
WHITAKER, R. A.
WHIT 1977 -1
WHITE, C. H.
WHITA 1977 -1
WHITE, D. E.
WHITB 1977 -1
WHITE, D. J.
WHITC 1975 -1
WHITE, G. P.
WILLC 1972 -1; WILLC 1972 -2
WHITE, J. A.
BURN 1976 -1; CARPA 1976 -1; EYST 1973 -1
FRANB 1974 -1
WHITE, L. J.
WHITD 1967 -1; WHITD 1967 -2
WHITE, M.
WHITE 1966 -1
WHITE, W. W.
WHITF 1969 -1
WHITEHEAD, E. G. JR.
WHITG 1973 -1
WHITFIELD, H.
SHEL 1975 -1
WHITLOCK, C.
CHRIA 1977 -1
WHYBARK, D. C.
HILL 1976 -1
WIEN, M.
MACH 1976 -1
WIERVILLE, W. W.
EYST 1973 -1
WIGGINS, A.
SMITE 1977 -1
WIJNANDS, J.
HESS 1977 -1
WILD, R.
CARN 1976 -1
WILF, H.
NIJE 1975 -1
WILF, H. S.
WILF 1978 -1

WILLE, D.
WILL 1974 -1
WILLIAMS, H. P.
WILLA 1975 -1; WILLA 1976 -1; WILLA 1977 -1
WILLIAMS, T.
WILLB 1975 -1
WILLIAMS, T. A.
SHAM 1972 -1; WILLC 1972 -1; WILLC 1972 -2
WILLIAMSON, S. G.
WHITB 1977 -1; WILLD 1976 -1
WILLIS, C. E.
ARMSE 1975 -1; ARMSE 1977 -1
WILLIS, R. J.
WILLE 1976 -1
WILSON, A. G.
WILS 1973 -1
WILSON, D.
WILSA 1975 -1; WILSB 1974 -1
WILSON, R. C.
WHITA 1977 -1
WILSON, R. J.
ERDOF 1977 -1; FIOR 1973 -1; FIOR 1974 -1
FIOR 1976 -1; WILSC 1973 -1
WIN, S.
WIN 1975 -1
WING, O.
WING 1961 -1; WINGA 1965 -1
WINSTON, C. B.
BECKB 1956 -1
WINSTON, K.
JOHNF 1978 -1
WITTEN, P.
WITT 1978 -1
WITZGALL, C.
GILS 1970 -1
WOEHNER, W.
WOEH 1973 -1
WOLSEY, L. A.
FISHB 1977 -1; NEMH 1976 -1; NEMHA 1976 -1
WOLS 1975 -1; WOLS 1976 -1; WOLS 1976 -2
WOLS 1976 -3
WONG, C. K.
CHAND 1976 -1; CHOY 1977 -1; HIRSA 1976 -1
HIRSA 1976 -2; KOU A 1978 -1; WONG 1976 -1
WONGA 1976 -1; WONGB 1971 -1
WONG, J. K.
HOPCC 1974 -1
WONG, J. Y.
LOVEA 1976 -1; LOVEA 1976 -2
WONG, R.
GART 1976 -1; WONGC 1976 -1
WONG, R. M.-L.
NELS 1977 -1
WONG, S.-N.
DINK 1976 -1
WONGSEELASHOTE, A.
WONGD 1976 -1
WOODALL, D. R.
WOOD 1974 -1; WOOD 1975 -1; WOOD 1975 -2
WOOD 1976 -1; WOOD 1977 -1
WRIGHT, E. M.
WRIG 1975 -1; WRIG 1976 -1
WRIGHT, G.
TRYG 1976 -1
YAGED, B. J.
YAGE 1969 -1
YAGIZ, OE.
YAGI 1975 -1
YAKIMOV, R. M.
YAKI 1966 -1
YAKUSHEV, V. I.
ALEKC 1964 -1
YALOVIJ, V. V.
YALO 1976 -1
YAMAMOTO, M.
YAMA 1977 -1
YAMASAKI, M.
NAKA 1977 -1
YAO, A. C.
YAO 1975 -1; YAO A 1977 -1; YAO B 1976 -1
YAO, A. C.-C.
GAREA 1976 -1; WONGA 1976 -1
YAO, F. F.
YAO B 1976 -1
YAP. K. Y.
CHEW 1975 -1
YAP, K. Y.
CHEWA 1974 -1
YEH, R. T.
YEH 1976 -1
YEH, W. W.-G.
BECKA 1974 -1; BECKA 1974 -2
YELOWITZ, L.
YELO 1976 -1
YEN, J. Y.
YEN 1969 -1; YEN 1971 -1; YEN 1972 -1
YEN 1973 -1; YEN 1975 -1
YORMARK, J. S.
YORM 1975 -1
YOUNG, G. K.
YOUN 1970 -1
YOUNG, H. P.
BALIA 1977 -1; YOUNA 1976 -1
YOUNG, J. P.
LYUN 1977 -1
YOUNG, P.
YOUNB 1977 -1
YOUNG, R. D.
YOUNC 1975 -1
YOUNGER, D.
VIDY 1975 -1
YUAN, J. S. C.
YUAN 1975 -1
YUBA, T.
MIYA 1977 -1
YUDIN, D. B.
YUDI 1976 -1; YUDIA 1974 -1
YUE, P. C.
WONGB 1971 -1
YUVAL, G.
YUVA 1976 -1; YUVA 1976 -2
ZABLUDOWSKI, A.
ZABL 1976 -1
ZAHLE, T. U.
MUNC 1977 -1
ZAHN, C. T.
ZAHN 1973 -1
ZAITSEVA, Z. N.
ZAIT 1975 -1; ZAIT 1976 -1
ZAJTSEV, M. A.
DINIA 1977 -1
ZAK, E. I.
KHAI 1976 -1
ZAK, YU. A.
ZAK 1976 -1; ZAK 1977 -1
ZAKHARIN, A. YA.
VOLO 1976 -1
ZALCSTEIN, Y.
FIDU 1977 -1
ZALOOM, V.
ZALO 1975 -1
ZALOOM, V. A.
ZALOA 1977 -1
ZAMBICKII, D. K.
ZAMB 1974 -1
ZAMFIRESCU, T.
ZAMF 1973 -1; ZAMF 1975 -1; ZAMF 1976 -1
ZANCEVA, F. N.
ZANC 1974 -1
ZANGWILL, W.
ZANG 1966 -1
ZAVARIN, A. N.
LITVC 1974 -1
ZAVILEVICH, M. L.
ZAVI 1975 -1
ZAYCHENKO, YU. P.
ZAYC 1975 -1
ZEGELING, H.
VLACA 1971 -1
ZEHNDER, C. A.
ZEHN 1965 -1
ZEMEL, E.
BALAL 1975 -1; BALAL 1976 -1; BALAL 1977 -1
ZEVIN, A. A.
ZEVI 1976 -1
ZHERNAK, A. N.
ZHER 1976 -1; ZHER 1976 -2
ZHURAVLEV, JU. I.
ZHUR 1977 -1
ZHURZENKO, S. L.
ZHURA 1968 -1
ZIMMERMANN, H.-G.
WITT 1978 -1
ZIMMERMANN, H.-J.
ZIMM 1975 -1
ZIMMERMANN, U.
BURKA 1976 -1; BURKA 1977 -1; BURKB 1976 -1
BURKB 1977 -1; BURKB 1977 -2; BURKE 1977 -1
BURKE 1978 -1; DERIA 1978 -1; ZIMMA 1976 -1
ZIMMA 1977 -1; ZIMMA 1977 -2; ZIMMA 1977 -3
ZIMMERMANN, W.
ZIMMB 1967 -1
ZIONTS, S.
CHENB 1976 -1; KEND 1977 -1; ZION 1977 -1
ZOLTNERS, A.
BALAM 1975 -1
ZOLTNERS, A. A.
PIPEA 1975 -1; PIPEA 1976 -1; ROSSA 1976 -1

ZOPPOLI, R.
GHIG 1976 -1
ZURKOWSKI, J.
NYKO 1964 -1
ZUSUPBAEV, A.
ZUSU 1974 -1; ZUSU 1977 -1
ZVEREV, V. YU.
PLOTA 1974 -1
ZVJAGINA, R. A.
ZVJA 1974 -1

Vol. 59: J. A. Hanson, Growth in Open Economies. V, 128 pages. 1971.

Vol. 60: H. Hauptmann, Schätz- und Kontrolltheorie in stetigen dynamischen Wirtschaftsmodellen. V, 104 Seiten. 1971.

Vol. 61: K. H. F. Meyer, Wartesysteme mit variabler Bearbeitungsrate. VII, 314 Seiten. 1971.

Vol. 62: W. Krelle u. G. Gabisch unter Mitarbeit von J. Burgermeister, Wachstumstheorie. VII, 223 Seiten. 1972.

Vol. 63: J. Kohlas, Monte Carlo Simulation im Operations Research. VI, 162 Seiten. 1972.

Vol. 64: P. Gessner u. K. Spremann, Optimierung in Funktionenräumen. IV, 120 Seiten. 1972.

Vol. 65: W. Everling, Exercises in Computer Systems Analysis. VIII, 184 pages. 1972.

Vol. 66: F. Bauer, P. Garabedian and D. Korn, Supercritical Wing Sections. V, 211 pages. 1972.

Vol. 67: I. V. Girsanov, Lectures on Mathematical Theory of Extremum Problems. V, 136 pages. 1972.

Vol. 68: J. Loeckx, Computability and Decidability. An Introduction for Students of Computer Science. VI, 76 pages. 1972.

Vol. 69: S. Ashour, Sequencing Theory. V, 133 pages. 1972.

Vol. 70: J. P. Brown, The Economic Effects of Floods. Investigations of a Stochastic Model of Rational Investment. Behavior in the Face of Floods. V, 87 pages. 1972.

Vol. 71: R. Henn und O. Opitz, Konsum- und Produktionstheorie II. V, 134 Seiten. 1972.

Vol. 72: T. P. Bagchi and J. G. C. Templeton, Numerical Methods in Markov Chains and Bulk Queues. XI, 89 pages. 1972.

Vol. 73: H. Kiendl, Suboptimale Regler mit abschnittweise linearer Struktur. VI, 146 Seiten. 1972.

Vol. 74: F. Pokropp, Aggregation von Produktionsfunktionen. VI, 107 Seiten. 1972.

Vol. 75: GI-Gesellschaft für Informatik e.V. Bericht Nr. 3. 1. Fachtagung über Programmiersprachen · München, 9.–11. März 1971. Herausgegeben im Auftrag der Gesellschaft für Informatik von H. Langmaack und M. Paul. VII, 280 Seiten. 1972.

Vol. 76: G. Fandel, Optimale Entscheidung bei mehrfacher Zielsetzung. II, 121 Seiten. 1972.

Vol. 77: A. Auslender, Problèmes de Minimax via l'Analyse Convexe et les Inégalités Variationelles: Théorie et Algorithmes. VII, 132 pages. 1972.

Vol. 78: GI-Gesellschaft für Informatik e.V. 2. Jahrestagung, Karlsruhe, 2.–4. Oktober 1972. Herausgegeben im Auftrag der Gesellschaft für Informatik von P. Deussen. XI, 576 Seiten. 1973.

Vol. 79: A. Berman, Cones, Matrices and Mathematical Programming. V, 96 pages. 1973.

Vol. 80: International Seminar on Trends in Mathematical Modelling, Venice, 13–18 December 1971. Edited by N. Hawkes. VI, 288 pages. 1973.

Vol. 81: Advanced Course on Software Engineering. Edited by F. L. Bauer. XII, 545 pages. 1973.

Vol. 82: R. Saeks, Resolution Space, Operators and Systems. X, 267 pages. 1973.

Vol. 83: NTG/GI-Gesellschaft für Informatik, Nachrichtentechnische Gesellschaft. Fachtagung „Cognitive Verfahren und Systeme", Hamburg, 11.–13. April 1973. Herausgegeben im Auftrag der NTG/GI von Th. Einsele, W. Giloi und H.-H. Nagel. VIII, 373 Seiten. 1973.

Vol. 84: A. V. Balakrishnan, Stochastic Differential Systems I. Filtering and Control. A Function Space Approach. V, 252 pages. 1973.

Vol. 85: T. Page, Economics of Involuntary Transfers: A Unified Approach to Pollution and Congestion Externalities. XI, 159 pages. 1973.

Vol. 86: Symposium on the Theory of Scheduling and its Applications. Edited by S. E. Elmaghraby. VIII, 437 pages. 1973.

Vol. 87: G. F. Newell, Approximate Stochastic Behavior of n-Server Service Systems with Large n. VII, 118 pages. 1973.

Vol. 88: H. Steckhan, Güterströme in Netzen. VII, 134 Seiten. 1973.

Vol. 89: J. P. Wallace and A. Sherret, Estimation of Product. Attributes and Their Importances. V, 94 pages. 1973.

Vol. 90: J.-F. Richard, Posterior and Predictive Densities for Simultaneous Equation Models. VI, 226 pages. 1973.

Vol. 91: Th. Marschak and R. Selten, General Equilibrium with Price-Making Firms. XI, 246 pages. 1974.

Vol. 92: E. Dierker, Topological Methods in Walrasian Economics. IV, 130 pages. 1974.

Vol. 93: 4th IFAC/IFIP International Conference on Digital Computer Applications to Process Control, Part I. Zürich/Switzerland, March 19–22, 1974. Edited by M. Mansour and W. Schaufelberger. XVIII, 544 pages. 1974.

Vol. 94: 4th IFAC/IFIP International Conference on Digital Computer Applications to Process Control, Part II. Zürich/Switzerland, March 19–22, 1974. Edited by M. Mansour and W. Schaufelberger. XVIII, 546 pages. 1974.

Vol. 95: M. Zeleny, Linear Multiobjective Programming. X, 220 pages. 1974.

Vol. 96: O. Moeschlin, Zur Theorie von Neumannscher Wachstumsmodelle. XI, 115 Seiten. 1974.

Vol. 97: G. Schmidt, Über die Stabilität des einfachen Bedienungskanals. VII, 147 Seiten. 1974.

Vol. 98: Mathematical Methods in Queueing Theory. Proceedings 1973. Edited by A. B. Clarke. VII, 374 pages. 1974.

Vol. 99: Production Theory. Edited by W. Eichhorn, R. Henn, O. Opitz, and R. W. Shephard. VIII, 386 pages. 1974.

Vol. 100: B. S. Duran and P. L. Odell, Cluster Analysis. A Survey. VI, 137 pages. 1974.

Vol. 101: W. M. Wonham, Linear Multivariable Control. A Geometric Approach. X, 344 pages. 1974.

Vol. 102: Analyse Convexe et Ses Applications. Comptes Rendus, Janvier 1974. Edited by J.-P. Aubin. IV, 244 pages. 1974.

Vol. 103: D. E. Boyce, A. Farhi, R. Weischedel, Optimal Subset Selection. Multiple Regression, Interdependence and Optimal Network Algorithms. XIII, 187 pages. 1974.

Vol. 104: S. Fujino, A Neo-Keynesian Theory of Inflation and Economic Growth. V, 96 pages. 1974.

Vol. 105: Optimal Control Theory and its Applications. Part I. Proceedings 1973. Edited by B. J. Kirby. VI, 425 pages. 1974.

Vol. 106: Optimal Control Theory and its Applications. Part II. Proceedings 1973. Edited by B. J. Kirby. VI, 403 pages. 1974.

Vol. 107: Control Theory, Numerical Methods and Computer Systems Modeling. International Symposium, Rocquencourt, June 17–21, 1974. Edited by A. Bensoussan and J. L. Lions. VIII, 757 pages. 1975.

Vol. 108: F. Bauer et al., Supercritical Wing Sections II. A Handbook. V, 296 pages. 1975.

Vol. 109: R. von Randow, Introduction to the Theory of Matroids. IX, 102 pages. 1975.

Vol. 110: C. Striebel, Optimal Control of Discrete Time Stochastic Systems. III. 208 pages. 1975.

Vol. 111: Variable Structure Systems with Application to Economics and Biology. Proceedings 1974. Edited by A. Ruberti and R. R. Mohler. VI, 321 pages. 1975.

Vol. 112: J. Wilhelm, Objectives and Multi-Objective Decision Making Under Uncertainty. IV, 111 pages. 1975.

Vol. 113: G. A. Aschinger, Stabilitätsaussagen über Klassen von Matrizen mit verschwindenden Zeilensummen. V, 102 Seiten. 1975.

Vol. 114: G. Uebe, Produktionstheorie. XVII, 301 Seiten. 1976.

Vol. 115: Anderson et al., Foundations of System Theory: Finitary and Infinitary Conditions. VII, 93 pages. 1976

Vol. 116: K. Miyazawa, Input-Output Analysis and the Structure of Income Distribution. IX, 135 pages. 1976.

Vol. 117: Optimization and Operations Research. Proceedings 1975. Edited by W. Oettli and K. Ritter. IV, 316 pages. 1976.

Vol. 118: Traffic Equilibrium Methods, Proceedings 1974. Edited by M. A. Florian. XXIII, 432 pages. 1976.

Vol. 119: Inflation in Small Countries. Proceedings 1974. Edited by H. Frisch. VI, 356 pages. 1976.

Vol. 120: G. Hasenkamp, Specification and Estimation of Multiple-Output Production Functions. VII, 151 pages. 1976.

Vol. 121: J. W. Cohen, On Regenerative Processes in Queueing Theory. IX, 93 pages. 1976.

Vol. 122: M. S. Bazaraa, and C. M. Shetty, Foundations of Optimization VI. 193 pages. 1976

Vol. 123: Multiple Criteria Decision Making. Kyoto 1975. Edited by M. Zeleny. XXVII, 345 pages. 1976.

Vol. 124: M. J. Todd. The Computation of Fixed Points and Applications. VII, 129 pages. 1976.

Vol. 125: Karl C. Mosler. Optimale Transportnetze. Zur Bestimmung ihres kostengünstigsten Standorts bei gegebener Nachfrage. VI, 142 Seiten. 1976.

Vol. 126: Energy, Regional Science and Public Policy. Energy and Environment I. Proceedings 1975. Edited by M. Chatterji and P. Van Rompuy. VIII, 316 pages. 1976.

Vol. 127: Environment, Regional Science and Interregional Modeling. Energy and Environment II. Proceedings 1975. Edited by M. Chatterji and P. Van Rompuy. IX, 211 pages. 1976.

Vol. 128: Integer Programming and Related Areas. A Classified Bibliography. Edited by C. Kastning. XII, 495 pages. 1976.

Vol. 129: H.-J. Lüthi, Komplementaritäts- und Fixpunktalgorithmen in der mathematischen Programmierung. Spieltheorie und Ökonomie. VII, 145 Seiten. 1976.

Vol. 130: Multiple Criteria Decision Making, Jouy-en-Josas, France. Proceedings 1975. Edited by H. Thiriez and S. Zionts. VI, 409 pages. 1976.

Vol. 131: Mathematical Systems Theory. Proceedings 1975. Edited by G. Marchesini and S. K. Mitter. X, 408 pages. 1976.

Vol. 132: U. H. Funke, Mathematical Models in Marketing. A Collection of Abstracts. XX, 514 pages. 1976.

Vol. 133: Warsaw Fall Seminars in Mathematical Economics 1975. Edited by M. W. Loś, J. Loś, and A. Wieczorek. V. 159 pages. 1976.

Vol. 134: Computing Methods in Applied Sciences and Engineering. Proceedings 1975. VIII, 390 pages. 1976.

Vol. 135: H. Haga, A Disequilibrium – Equilibrium Model with Money and Bonds. A Keynesian – Walrasian Synthesis. VI, 119 pages. 1976.

Vol. 136: E. Kofler und G. Menges, Entscheidungen bei unvollständiger Information. XII, 357 Seiten. 1976.

Vol. 137: R. Wets, Grundlagen Konvexer Optimierung. VI, 146 Seiten. 1976.

Vol. 138: K. Okuguchi, Expectations and Stability in Oligopoly Models. VI, 103 pages. 1976.

Vol. 139: Production Theory and Its Applications. Proceedings. Edited by H. Albach and G. Bergendahl. VIII, 193 pages. 1977.

Vol. 140: W. Eichhorn and J. Voeller, Theory of the Price Index. Fisher's Test Approach and Generalizations. VII, 95 pages. 1976.

Vol. 141: Mathematical Economics and Game Theory. Essays in Honor of Oskar Morgenstern. Edited by R. Henn and O. Moeschlin. XIV, 703 pages. 1977.

Vol. 142: J. S. Lane, On Optimal Population Paths. V, 123 pages. 1977.

Vol. 143: B. Näslund, An Analysis of Economic Size Distributions. XV, 100 pages. 1977.

Vol. 144: Convex Analysis and Its Applications. Proceedings 1976. Edited by A. Auslender. VI, 219 pages. 1977.

Vol. 145: J. Rosenmüller, Extreme Games and Their Solutions. IV, 126 pages. 1977.

Vol. 146: In Search of Economic Indicators. Edited by W. H. Strigel. XVI, 198 pages. 1977.

Vol. 147: Resource Allocation and Division of Space. Proceedings. Edited by T. Fujii and R. Sato. VIII, 184 pages. 1977.

Vol. 148: C. E. Mandl, Simulationstechnik und Simulationsmodelle in den Sozial- und Wirtschaftswissenschaften. IX, 173 Seiten. 1977.

Vol. 149: Stationäre und schrumpfende Bevölkerungen: Demographisches Null- und Negativwachstum in Österreich. Herausgegeben von G. Feichtinger. VI, 262 Seiten. 1977.

Vol. 150: Bauer et al., Supercritical Wing Sections III. VI, 179 pages. 1977.

Vol. 151: C. A. Schneeweiß, Inventory-Production Theory. VI, 116 pages. 1977.

Vol. 152: Kirsch et al., Notwendige Optimalitätsbedingungen und ihre Anwendung. VI, 157 Seiten. 1978.

Vol. 153: Kombinatorische Entscheidungsprobleme: Methoden und Anwendungen. Herausgegeben von T. M. Liebling und M. Rössler. VIII, 206 Seiten. 1978.

Vol. 154: Problems and Instruments of Business Cycle Analysis. Proceedings 1977. Edited by W. H. Strigel. VI, 442 pages. 1978.

Vol. 155: Multiple Criteria Problem Solving. Proceedings 1977. Edited by S. Zionts. VIII, 567 pages. 1978.

Vol. 156: B. Näslund and B. Sellstedt, Neo-Ricardian Theory. With Applications to Some Current Economic Problems. VI, 165 pages. 1978.

Vol. 157: Optimization and Operations Research. Proceedings 1977. Edited by R. Henn, B. Korte, and W. Oettli. VI, 270 pages. 1978.

Vol. 158: L. J. Cherene, Set Valued Dynamical Systems and Economic Flow. VIII, 83 pages. 1978.

Vol. 159: Some Aspects of the Foundations of General Equilibrium Theory: The Posthumous Papers of Peter J. Kalman. Edited by J. Green. VI, 167 pages. 1978.

Vol. 160: Integer Programming and Related Areas. A Classified Bibliography. Edited by D. Hausmann. XIV, 314 pages. 1978.